micromechanics: overall properties of heterogeneous materials

New Topics Include:
Dynamic Failure in Compression
Piezoelectricity with New Bounds on Coupled Moduli
Physically-Based Plasticity Models
Homogenization and Multi-Scale Asymptotics
Uniform Field Theory
New Bounds for Finite Composites
Duality Principles in Elasticity

micromechanics: overall properties of heterogeneous materials

Sia Nemat-Nasser
Department of Applied Mechanics and Engineering Sciences
University of California, San Diego
La Jolla, CA 6 92093-0416, USA

Muneo Hori
Earthquake Research Institute
University of Tokyo
Tokyo, Japan

Second Revised Edition

1999
ELSEVIER
Amsterdam - Lausanne - New York - Oxford - Shannon - Singapore - Tokyo

ELSEVIER SCIENCE B.V.
Sara Burgerhartstraat 25
P.O. Box 211, 1000 AE Amsterdam, The Netherlands

First edition 1993 ISBN: 0 444 89881 6
Second revised edition 1999

Library of Congress Cataloging in Publication Data
A catalog record from the Library of Congress has been applied for.

ISBN: 0 444 50084 7

∞ The paper used in this publication meets the requirements of ANSI/NISO Z39.48-1992 (Permanence of Paper).
Printed in The Netherlands.

PREFACE

In this second edition of *Micromechanics: Overall Properties of Heterogeneous Materials*, several new topics of technological interest are added. The added topics include: coupled mechanical and nonmechanical overall properties of heterogeneous piezoelectric materials (Subsection 2.8), new upper and lower bounds for these coupled properties (Subsection 9.8), a systematic comparison between the *average-field theory* and the results obtained using *multi-scale perturbation theory* (Appendix B), an account of the uniform-field theory (Appendix C), improveable bounds on overall moduli of heterogeneous materials, which remain finite even when isolated cavities and rigid inclusions are present (Appendix D), and a brief account of a fundamental duality principle in anisotropic elasticity (Subsection 21.6). In addition, better explanations of a number of topics are given, more recent references are added, the Subject Index is expanded, and printing and typographical errors are corrected.

The structure of the first edition remains intact. As before, the book consists of two parts. The content of each part is briefly reviewed in the corresponding précis. The Précis to Part 1 starts on page 3, and the Précis to Part 2, on page 617. Part 1 consists of four chapters which are organized into fourteen sections and four appendixes. It deals with materials with microdefects such as cavities, cracks, and inclusions, as well as with elastic composites. Part 2 consists of two chapters which are divided into seven sections, Sections fifteen to twenty-one. It provides an introduction to the theory of linear elasticity, added to make the book self-contained, since linear elasticity serves as the basis of the development of small-deformation micromechanics.

The material for the first part of this book grew out of lecture notes of the first author in a course on micromechanics, which was initiated at the University of California, San Diego (UCSD), in response to the existing need for a fundamental understanding of the micromechanics of the overall response and failure modes of advanced materials, such as ceramics and ceramic and other composites. These advanced materials have become the focus of systematic and extensive research at the Center of Excellence for Advanced Materials (CEAM), which was established at UCSD in 1986, to include the U.S. Government's University Research Initiative on Dynamic Performance of Materials. The course, given in the Spring of 1987, was intended to furnish a basic background for rigorous micromechanical modeling of the mechanical behavior and failure regimes of a broad class of brittle materials.

Class notes taken by some of the graduate students were totally rewritten by the second author in early 1988. Then this new version was completely reworked by both authors, and used when the micromechanics course was given for the second time in the Fall of 1988. A new version of the notes was then completed and a major part of it was used again in the micromechanics course in the Winter Quarter of 1990,[1] and with more extensive revisions and additions, again in the Winter Quarter of 1991. The manuscript was read by many graduate students who helped to correct misprints and related errors. The material contained in this book as Part 1, is a thoroughly reexamined, modified, expanded, and amended version of these class instruction notes, as well as the results of the published research of many graduate students and scholars who have provided fundamental contributions in micromechanics of heterogeneous materials. Included in Part 1 are new results on many basic issues in micromechanics, which we hope will be helpful to graduate students and researchers dealing with rigorous physically-based modeling of overall properties of heterogeneous materials; see Précis of Part 1, for comments.

Part 2 of this book, except for Sections 15, 19, and 21, is essentially part of the lecture notes on elasticity which the first author wrote in the late 1960's, while teaching at UCSD. Many sections were done at that time, in collaboration with the late Professor William Prager, and it was intended then to publish a monograph on elasticity in a series of books in solid mechanics, by Blaisdell Publishing Company. Unfortunately, before the monograph was completed, that series was discontinued, and the authors were released from contractual obligations. The notes were completed and used many times at UCSD and at Northwestern University, for class instruction by the first author. The section on variational methods included some new results of such generality that even today, much of it is not known to many researchers. These were the results obtained in collaboration with Professor Prager. The variational principles in the current version, given in Section 19, are even more general than those contained in the original class notes. These principles include some new elements which should prove useful for application to advanced modeling, as well as solutions of composites and related heterogeneous bodies. Section 15 is a brief modern version of elements in vector and tensor algebra, and is particularly tailored to provide background for the rest of this book. Even for students of mechanics and mechanics of materials familiar with the basic elements of linear elasticity, a quick examination of Subsection 15.5 should prove helpful for a better understanding of the notation and some of the details of the manipulations involved in Part 1.

The material in Part 2 is mostly standard, given for background information. In preparing Part 1, the authors have benefited from the contribution of many modern researchers and numerous publications which have appeared over the past few decades. Since the book is intended for use as a graduate text, only a few selected references are cited within the text. Important references may have been left out inadvertently, for which the authors apologize.

While the authors take full responsibility for any errors that remain, they wish to express their appreciation to Dr. L. Ni and graduate students, B. Balendran, Hang Deng, Mark Rashid, G. Subhash, John Wehrs, Niann-i Yu, Vinod Sharma, Abbas Azhdari, Yeou-Fong Li, Anil Thakur, Tomoo Okinaka, and others who read with care various parts of various versions of the manuscript of the

---

[1] Nemat-Nasser, S. and Hori, M. (1989), *Micromechanics: Overall properties of heterogeneous elastic solids*, Lecture Notes Initiated at UCSD, 439 pgs.

first edition, and sought to remove errors of minor or major importance; Dr. Ni also read Section 21 of this second edition and pointed out several misprints. Thanks are also due John Willis whose seminars and comments provided inspiration and insight, and led to considerable improvement of several sections of the first edition, as well as due Éva who proofread the entire manuscript of each edition, Shiba who helped with word processing of the first edition, and Dr. Masoud Beizaie who assisted in preparing the new figures, references, and helped with several other tasks. The authors also gratefully acknowledge the Army Research Office Contract DAAL-03-86-K-0169 which provided the core support for the Center of Excellence for Dynamic Performance of Materials at UCSD, and partially supported the two authors during the preparation of both editions of this book. The book has been formatted by the authors, using *ditroff*. The figures and the graphs have been constructed by *pic* and *grap*.

Sia Nemat-Nasser, La Jolla, California
Muneo Hori, Tokyo, Japan

August, 1998

TABLE OF CONTENTS

PART 1

OVERALL PROPERTIES OF HETEROGENEOUS MATERIALS

PART 2

INTRODUCTION TO BASIC ELEMENTS OF ELASTICITY THEORY

PART 1

OVERALL PROPERTIES OF HETEROGENEOUS MATERIALS

PRÉCIS: PART 1

Part 1 of this book contains a systematic development of the overall response parameters of materials with microheterogeneities and defects such as cracks, cavities, and inclusions. The work deals with small deformations, particularly relevant to advanced materials such as ceramics and ceramic composites, as well as metals and polymeric composites, in a deformation range where overall geometrical dimensions and shapes are not altered substantially by material deformation. While, for the most part, a linearly elastic matrix containing linearly elastic inclusions or cavities and cracks, is considered, overall material nonlinearity caused, for example, by the formation and growth of microcracks, is included. In addition, as discussed in Subsection 12.8 and in Appendix A of Part 1, the basic results apply directly to composites and heterogeneous solids consisting of elastoplastic or rate-dependent elastoviscoplastic materials, as long as the small-deformation theory applies, or small incremental steps with updated geometry and properties are used.

Within the above-mentioned general framework, the subject matter of micromechanics is treated in a deliberate and systematic manner, at each stage beginning with the fundamentals which are then treated in depth with considerable care, leading to illustrative examples to bring out in a concrete fashion the involved basic steps, and then providing a number of major results with broad applicability.

More specifically, in Sections 1 and 2, Chapter I, the basic idea of a heterogeneous *representative volume element* (RVE) is discussed. The associated boundary-value problem is formulated, both in terms of the rate of change of the field variables, as well as in terms of the total quantities. Averaging methods are examined in Section 2, and a series of important basic identities are presented. Essentially all (unless otherwise explicitly stated) the results in Sections 1 and 2 are valid for small deformations of solids consisting of any (elastic or inelastic) constituents with any material properties. These comments apply to results presented in Subsections 2.1 to 2.5. In Subsection 2.5, detailed discussions are given for solids consisting of material constituents which admit (at the local level) stress and/or strain potentials; the response, however, need not be linear. It is then shown how the overall macropotentials relate directly to the average of the corresponding micropotentials. A set of *exact* relations is obtained in this manner, and based on this, the notion of a representative volume element is examined. The question of the effects of the boundary data considered for an RVE, on the resulting overall energy density (elastic but materially nonlinear), and on the overall effective moduli for a linearly elastic RVE, is given a thorough examination, leading to a set of universal inequalities and two *universal theorems* which provide exact ordering relations when uniform boundary tractions, linear boundary displacements, or general boundary data are considered for an RVE.[1] In particular, general results developed in Subsections 2.5.6 and 2.5.7, relating to bounds on macropotentials and, therefore, on the overall strain energy and complementary energy functionals, prove useful as guiding elements for specific results outlined in later sections, using the *assumption (postulate)* of *statistical homogeneity,* and employing simple *models* to calculate the local average quantities. These universal theorems are used to obtain rigorous, computable bounds of considerable generality in Section 9, as well as improvable bounds which remain finite even if isolated cavities and isolated rigid inclusions are present.[2]

In Subsection 2.6, questions of statistical homogeneity and representative volume elements are again examined to clarify the nature of the boundary data which may be assigned to an RVE, and their influence on the resulting average stresses, strains, and their potentials. Conditions under which the average of the product of the stress and strain (or their rates) equals the product of the corresponding averages, are discussed.

Included in Sections 1 and 2 also is a brief discussion of nonmechanical properties of inhomogeneous media. General results developed for mechanical properties are specialized and applied to electrostatic, magnetostatic, thermal, and diffusional properties of a heterogeneous RVE. In this second edition, a discussion of the coupled mechanical and nonmechanical properties are added as Subsection 2.8, and the corresponding bounds are obtained in Subsection 9.7.

Chapter II is devoted to estimating, in a systematic manner, the overall elasticity and compliance tensors of a linearly elastic matrix containing microcavities and microcracks, using two simple models for the averaging procedure. These are: the *dilute distribution model* which assumes that the inhomogeneities are small and far apart, so that their interaction may be neglected; and the *self-consistent model* which takes into account the corresponding interaction, in a certain, overall, approximate manner. In particular, in Section 3 the stress-strain relations of linear elasticity are reviewed and the necessary background is provided for subsequent sections. In Section 4, a systematic discussion of the overall stress and strain in a porous RVE is given for two limiting alternative boundary data, namely, uniform tractions, and linear displacements. It is shown, directly and in a simple manner, how the corresponding overall compliance and elasticity tensors can be estimated, using the reciprocal theorem and simple estimates of the cavity boundary displacements. These results are generally valid for cavities of any shape or distribution, and, except for the assumption of linear elasticity, no approximations are involved. In Section 5 the general results of Section 4 are applied to porous, linearly elastic solids. The dilute-distribution and the self-consistent models are used. The relation between these models is

---

[1] These *theorems* are called *universal* because they apply to any heterogeneous solid of any shape and size, consisting of inhomogeneities of any arbitrary shape, size, and distribution, and subjected to any general (consistent) boundary data. The only (minimal) requirement is that the constituents admit convex potentials.

[2] See Appendix D and the following references: Balendran, B. and Nemat-Nasser, S. (1995), "Bounds on Elastic Moduli of Composites," *J. Mech. Phys. Solids*, Vol. 43, 1825-1853; and Nemat-Nasser, S. and Hori, M. (1995), "Universal Bounds for Overall Properties of Linear and Nonlinear Heterogeneous Solids," ASME's *J. Engrg Mat. Tech.* Vol. 117, 412-432.

discussed in terms of specific problems. Section 6 deals with elastic solids with microcracks. Here again, the same two *models* are used in a number of illustrative examples of scientific and technological importance, and the corresponding results are compared and discussed. The effective moduli of solids with various microcrack distributions are calculated explicitly in Subsections 6.1 to 6.8, emphasizing the phenomenon of microcrack-induced anisotropy and the role of friction (Subsections 6.3 and 6.4). Both slit microcracks (two-dimensional problems) and penny-shaped ones (three-dimensional problems), with various distributions (various alignments, random), are examined in considerable detail, providing explicit results. In addition, a brief overview of recent advances in theoretical and experimental evaluation of brittle failure in compression is presented in Subsection 6.9 for the static loadings, and in Subsection 6.10 for the dynamic cases.

Chapter III is devoted to linearly elastic solids with elastic micro-inclusions, as well as the elastic response of polycrystals. First, in Section 7, for micro-inclusions of any geometry and elasticity, exact general expressions for the overall elastic modulus and compliance tensors are obtained, for overall uniform boundary tractions and overall linear boundary displacements, respectively. Then in Subsection 7.3, the concepts of *eigenstrain* and *eigenstress* required to *homogenize* the heterogeneous RVE are introduced and examined in some detail. In particular, Eshelby's tensor for an ellipsoidal inclusion embedded in a uniform, infinitely extended, linearly elastic solid, is presented, together with its dual tensor (the first associated with an eigenstrain, and the second associated with the corresponding eigenstress), their properties examined, together with their dual relations, and they are used to obtain *consistency conditions* associated with the homogenization. These results are then related to the **H**- and **J**-tensors, introduced in Section 4, to homogenize an elastic RVE containing microcavities and microcracks. The results are then used to formulate overall modulus and compliance tensors of an elastic RVE with elastic inclusions, on the basis of the dilute-distribution and the self-consistent models; Subsections 7.4 and 7.5. The formulation of the overall elasticity and compliance tensors, in terms of the overall elastic energy of the RVE, is discussed in Subsection 7.6, focusing particularly on the required symmetry for the overall elasticity and compliance tensors. Section 8 contains specific illustrative examples for elastic solids with micro-inclusions. A number of problems are worked out in detail, and numerical illustrations are presented.

Section 8 focuses on elastic solids which contain elastic micro-inclusions, i.e., elastic composites. Explicit results are presented for random distribution of spherical inclusions (Subsection 8.1), aligned fiber reinforcements (Subsection 8.2), and the three-dimensional analysis of the plane-strain and plane-stress states, often assumed for fiber-reinforced composites (Subsection 8.3).

Upper and lower bounds for the overall elastic moduli are presented in Section 9. First, the Hashin-Shtrikman variational principle (Subsection 9.1), as generalized by Willis, is presented, when either the eigenstrains or the eigenstresses are used to homogenize the corresponding heterogeneous RVE, leading to two functionals: one, when the overall uniform boundary tractions are prescribed, where the eigenstrains are used for homogenization, and the other, when the overall linear displacement boundary data are assigned, in which case the eigenstresses are used to homogenize the RVE. This leads to an elegant dual principle, with the corresponding Euler equations defining the associated consistency conditions.

The upper and lower bounds for the energy functionals are presented in Subsection 9.2, and their generalization is given in Subsection 9.3. Direct estimates of the overall moduli, using approximate correlation tensors, are presented in Subsection 9.4. In Subsection 9.5, the Hashin-Shtrikman variational principle is generalized for boundary data other than uniform tractions and linear displacements, and the corresponding generalized bounds are obtained. With the aid of the universal theorems of Subsection 2.5.6, these bounds are then related to the bounds for the uniform traction and linear displacement boundary data. It is proved that two out of four possible approximate expressions that result are indeed rigorous bounds. Explicit, computable, exact upper and lower bounds for the overall moduli are then given, when the composite is statistically homogeneous and isotropic. Finally, it is shown in Subsection 9.6 that these new observations lead to *universal* bounds on two overall moduli of multi-phase composites, valid for any shape or distribution of phases.[3] Furthermore, it is established that the bounds are valid for any *finite* elastic solid of ellipsoidal shape, consisting of any distribution of inhomogeneities of any shape and elasticity. (In Section 13, it is proved that the same bounds emerge for multi-phase composites with periodic, but otherwise completely arbitrary, microstructure.)

For historical reasons, the bounds on the overall properties in Section 9 are based on the Hashin-Shtrikman variational principle. An alternative formulation of *exact computable bounds* is to use the universal theorems of Subsection 2.5.6, together with proper choices of the reference elasticity or compliance tensors. This is presented in Subsection 9.5.6. It is also used in Subsection 9.7.2 to formulate bounds on parameters which define nonmechanical properties (e.g., conductivity and resistivity tensors) of composites, and in Subsection 9.8 to obtain exact upper and lower bounds for the coupled mechanical and nonmechanical moduli, using piezoelectricity as a specific case.

A number of averaging methods (*models*) are studied in a systematic manner in Section 10. This includes the dilute-distribution (Subsection 10.1) and the self-consistent methods (Subsection 10.2), as well as the differential scheme (Subsection 10.3), and the two- and three-phase models. The double-inclusion method is discussed in Subsection 10.4, together with the Mori-Tanaka result, which leads to a number of interesting conclusions; for example, the self-consistent estimate is shown to be a special case of the double-inclusion model, the results are related to the Hashin-Shtrikman bounds, and using the universal theorems of Section 2.5.6, it is shown that the double-inclusion model does in fact provide exact bounds. The double-inclusion model is then generalized to

[3] These *bounds* are called *universal* because they are valid for any periodic structure with the unit cell consisting of inhomogeneities of any arbitrary shape, size, and distribution. They are also valid for a finite-sized composite of ellipsoidal shape with any arbitrary shape, size, and distribution of inhomogeneities.

multi-inclusion models, where, again, all the average field quantities are estimated analytically. For a set of nested ellipsoidal regions of arbitrary aspect ratios, relative locations, and orientation, which is embedded in an infinitely extended homogeneous elastic solid of arbitrary elasticity, and which undergoes transformations with uniform but distinct transformation strains within each annulus, the resulting average strain and stress in each annulus are computed *exactly* and in closed form; the transformation strains in the innermost region need not be uniform. Explicit results are presented for an embedded double inclusion, as well as a *nested* set of n inclusions. As examples of the application of the multi-inclusion model, a composite containing multi-layer inclusions and a composite consisting of several distinct materials are considered, and their overall moduli are analytically estimated. Then, relations among various *approximate* techniques are studied (Subsection 10.5), and comments on other averaging schemes are made (Subsection 10.6).

The development of Eshelby's tensor in terms of the infinite-space Green function, is contained in Section 11, where the properties of this tensor and its dual are studied (Subsections 11.1 to 11.3). Given in this section is the Mori-Tanaka result, and its generalization to the case when arbitrary nonuniform eigenstrains (or transformation strains) are distributed in a region of arbitrary geometry which is contained in an ellipsoidal domain which, in turn, is embedded in an infinite homogeneous domain. This result provides a powerful tool for the study of, for example, fiber-reinforced composites with coated and/or partially debonded fibers. Then, relations among various average quantities are examined (Subsection 11.4), and the energy associated with heterogeneity and, hence, with the homogenizing eigenstrains or eigenstresses, is quantified explicitly.

Chapter IV covers the fundamentals of heterogeneous elastic solids with periodically distributed inhomogeneities, such as inclusions, fibers, cavities, *and cracks*. This chapter includes a number of new results, while at the same time presenting the theory in considerable depth, starting at an elementary level. Section 12 provides background information and gives a number of illustrative examples, including periodically distributed interacting cracks. For a periodic structure, the concept of a unit cell is introduced (Subsections 12.1 and 12.2) and, using Fourier series, the general solution is obtained (Subsection 12.3). The concept of homogenization using (variable, periodic) eigenstrains or eigenstresses, is introduced, and the corresponding consistency conditions are obtained (Subsection 12.4). Specific classes of problems are then solved as illustration, and the overall average elastic parameters are obtained in terms of the geometry and properties of the representative unit cell: two-phase microstructure (Subsection 12.5); inclusions and cavities (Subsection 12.6); and microcracks (Subsection 12.7). Application of the general results to a unit cell with rate-dependent or rate-independent constituents is examined in Subsection 12.8.

Section 13 focuses on the overall response of solids with periodic microstructure: e.g., linearly elastic uniform micro-inclusions periodically embedded in a linearly elastic uniform matrix. First, an equivalent homogeneous solid is defined by introducing suitable periodic eigenstrain or eigenstress fields (Subsection 13.1). Then, the Hashin-Shtrikman variational principle is applied to solids with periodic microstructure, and bounds on the overall moduli are obtained by defining energy functionals for the eigenstrain or eigenstress fields in the equivalent homogeneous solid (Subsection 13.2). The bounds for the periodic microstructure are *exact* and can be computed to any desired degree of accuracy, using the Fourier series representation (Subsection 13.3), as exemplified in Subsections 13.4 to 13.6. It is shown in Subsection 13.5, that there are always two overall elastic parameters whose bounds, obtained on the basis of the periodic and random (RVE) microstructures, are identical, and hence exact. These bounds are valid for inclusions of any shape or elasticity. In addition, the minimum potential and complementary potential energies are used to obtain *directly* bounds on the overall parameters by a suitable choice of a *reference elasticity or compliance tensor.*

In Section 14 the concept of mirror images of points and vectors is introduced and then used to decompose tensor-valued functions defined on the unit cell, to their symmetric and antisymmetric parts. The decomposition is applied to Fourier series representations of tensor-valued field quantities such as strain, stress, and elastic moduli, resulting in considerable economy in numerical computation and clarity in restrictions which must be imposed on the boundary data.

In Appendix A of Part 1, application of the basic results to nonlinear rate-dependent and rate-independent inelastic heterogeneous solids is briefly examined. Illustrative constitutive relations for phenomenological and slip-induced plasticity models are briefly presented, and their implementation in terms of the general theories of the preceding sections is pointed out. First, certain rate-independent phenomenological plasticity theories are outlined, with a brief examination of slip-induced crystal plasticity. Then their interpretation in terms of rate-dependent processes is mentioned.

Appendix B of Part 1, addresses the general question of *homogenization*, using the average-field theory and the multi-scale perturbation theory. The relation between the corresponding results is discussed, and based on the universal theorems of Subsection 2.5.6, a general theory is developed, which is capable of rigorously predicting the effective moduli even when the strain gradients are so large that the average-field theory no longer applies.

In Appendix C of Part 1, the uniform-field theory is discussed. Unlike the average-field theory or the homogenization theory, the uniform-field theory does not seek to determine the overall material properties. Instead, it provides conditions that must be satisfied by the overall moduli which are predicted by applying some other theoretical methods or are actually measured experimentally. Thus, it provides a mean of assessing whether or not the results of a given model or a set of experiments are reasonable.

Appendix D of Part 1 is devoted to an account of recent bounds obtained for composites which consist of a matrix material that completely surrounds the other inhomogeneities, e.g., isolated cracks, isolated cavities, or isolated inclusions, or collections of these. Based on the universal theorems of Subsection 2.5.6, improvable upper and lower bounds are developed and the results are illustrated. These bounds are valid for composites of any geometry (finite, periodic, or an RVE). They also apply to nonlinear composites whose constituents admit convex potentials.

CHAPTER I

AGGREGATE PROPERTIES AND AVERAGING METHODS

In this chapter the concept of representative volume element (RVE) is introduced and some averaging techniques for obtaining aggregate properties in terms of microstructure are presented. While attention is confined to small-deformation theories, no additional restrictions are imposed on the constitutive properties of the micro-elements which comprise an RVE. The general relations obtained in this chapter are used throughout the remainder of this book. Familiarity with the basic concepts and field equations of small-deformation continuum mechanics, and hence with associated tensor fields and tensorial operations is assumed. A brief account of tensors and tensor fields, and basic field equations of small-deformation continuum mechanics, particularly linear elasticity, is presented in Part 2 of this book.

The basic idea of a heterogeneous representative volume element (RVE) is introduced and illustrated in Subsection 1.1. The scope of the book is defined in Subsection 1.2, the basic field equations and the boundary conditions for an RVE are listed in Subsection 1.3.

The theorems which give the average stress and its rate (Subsection 2.1), the average strain and its rate (Subsection 2.2), and the average rate of stress-work (Subsection 2.3), in terms of the associated boundary data, are presented in Section 2, together with conditions at surfaces of discontinuities (Subsection 2.4). All these results are valid for small deformations of solids consisting of any (elastic or inelastic) constituents with any material properties. A comprehensive account of the relations between the micro- and macropotentials is presented in Subsection 2.5. Of particular importance are two universal theorems, given in Subsection 2.5.6, which provide exact ordering relations when uniform boundary tractions, linear boundary displacements, or general boundary data are considered for an RVE. These theorems are used in

Section 9 and Appendix D to establish rigorous bounds on the overall properties of heterogeneous solids. The notion of statistical homogeneity is examined in Subsection 2.6, and in Subsections 2.7 and 2.8 the nonmechanical and the coupled nonmechanical and mechanical problems are discussed.

SECTION 1 AGGREGATE PROPERTIES

The relation between the continuum properties of a material neighborhood and its microstructure and microconstituents is discussed in general terms. The physical basis of the transition from the microscale to the macroscale is examined and illustrated, arriving at the notion of a representative volume element (RVE). Then the associated boundary-value problems are formulated, in terms of the total field quantities and their rates. Included also is a brief mention of nonmechanical properties such as overall thermal, electrical, magnetic, and diffusional measures for microscopically heterogeneous media.

1.1. REPRESENTATIVE VOLUME ELEMENT

Continuum mechanics deals with idealized materials consisting of material points and material neighborhoods. It assumes that the material distribution, the stresses, and the strains within an infinitesimal material neighborhood of a typical particle (or a material element) can be regarded as essentially uniform. On the microscale, however, the infinitesimal material neighborhood, in general, is not uniform, consisting of various constituents with differing properties and shapes, i.e., an infinitesimal material element has its own complex and, in general, evolving microstructure. Hence, the stress and strain fields within the material element likewise are not uniform at the microscale level. One of the main objectives of micromechanics is to express in a systematic and rigorous manner the *continuum quantities* associated with an infinitesimal material neighborhood in terms of the parameters that characterize the *microstructure and properties of the microconstituents* of the material neighborhood.

To this end, the concept of a *representative volume element* (RVE) is introduced; Hill (1963), Hashin (1964, 1983), Kröner (1977), Willis (1981), and Nemat-Nasser (1986). An RVE for a material point of a continuum mass is a material volume which is statistically representative of the infinitesimal material neighborhood of that material point. The continuum material point is called a *macro-element*. The corresponding microconstituents of the RVE are called the *micro-elements*. An RVE must include a very large number of micro-elements, and be statistically representative of the local continuum properties.

Figure 1.1.1a shows a continuum, and identifies a typical material point P surrounded by an infinitesimal material element. When the macro-element is magnified, as sketched in Figure 1.1.1b, it may have its own complex microstructure. It may consist of grains separated by grain boundaries, voids, inclusions, cracks, and other similar defects. To be representative, this RVE must include a very large number of such microheterogeneities.

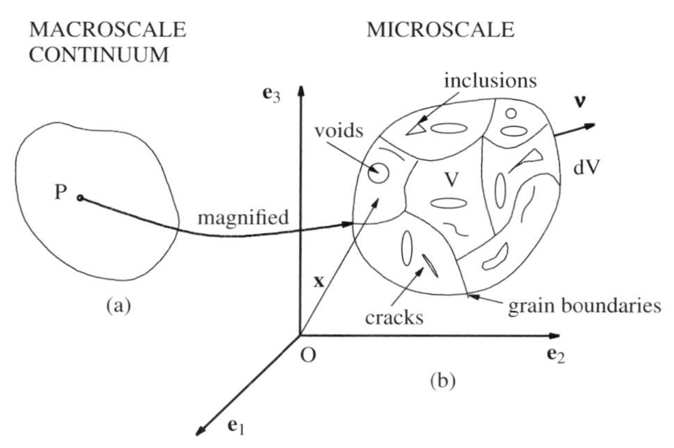

Figure 1.1.1

(a) P is a material point or material element surrounded by a material neighborhood, i.e., a macro-element; (b) Possible microstructure of an RVE for the material neighborhood of P

Figure 1.1.2 shows the microstructure in magnesia-partially stabilized zirconia (Mg-PSZ). Figures 1.1.2a,b are optical micrographs of a zirconia sample which has been subjected to a single compressive pulse (uniaxial stress) in the direction of the arrows, producing phase transformation (Figure 1.1.2a) from a meta-stable tetragonal to a stable monoclinic crystalline structure in PSZ, as well as creating microcracks (Figure 1.1.2b) essentially parallel to the direction of compression; Rogers and Nemat-Nasser (1989) and Subhash and Nemat-Nasser (1993).[1] If these cracks are regarded as approximately flat, their normals then fall on a plane normal to the direction of compression, having an essentially uniform distribution. Figure 1.1.2c is a micrograph showing the intersection of these cracks with a plane normal to the direction of compression. While these cracks are not "flat", they are randomly oriented.

After the first loading discussed above, the sample, which has a cubical geometry, is subjected to another single compression pulse (uniaxial stress) in a direction normal to the direction of the first loading. Figure 1.1.2d is the corresponding micrograph showing new microcracks which have been formed in the direction of the second loading, essentially normal to the first set of cracks.

Phase transformation from tetragonal to monoclinic occurs in platelet precipitates. This transformation involves both shear deformation and volumetric expansion. The constraint imposed by the surrounding matrix, forces the precipitates to accommodate the transformation shear strain through twinning, as is

[1] For qualitative experimental and quantitative theoretical modeling of compression-induced cracks in brittle solids, see Nemat-Nasser and Horii (1982), Horii and Nemat-Nasser (1985, 1986), and Ashby and Hallam (1986), where references to other works, especially in rock mechanics, are also given; for a review, see Nemat-Nasser (1989), and Subsection 6.9.

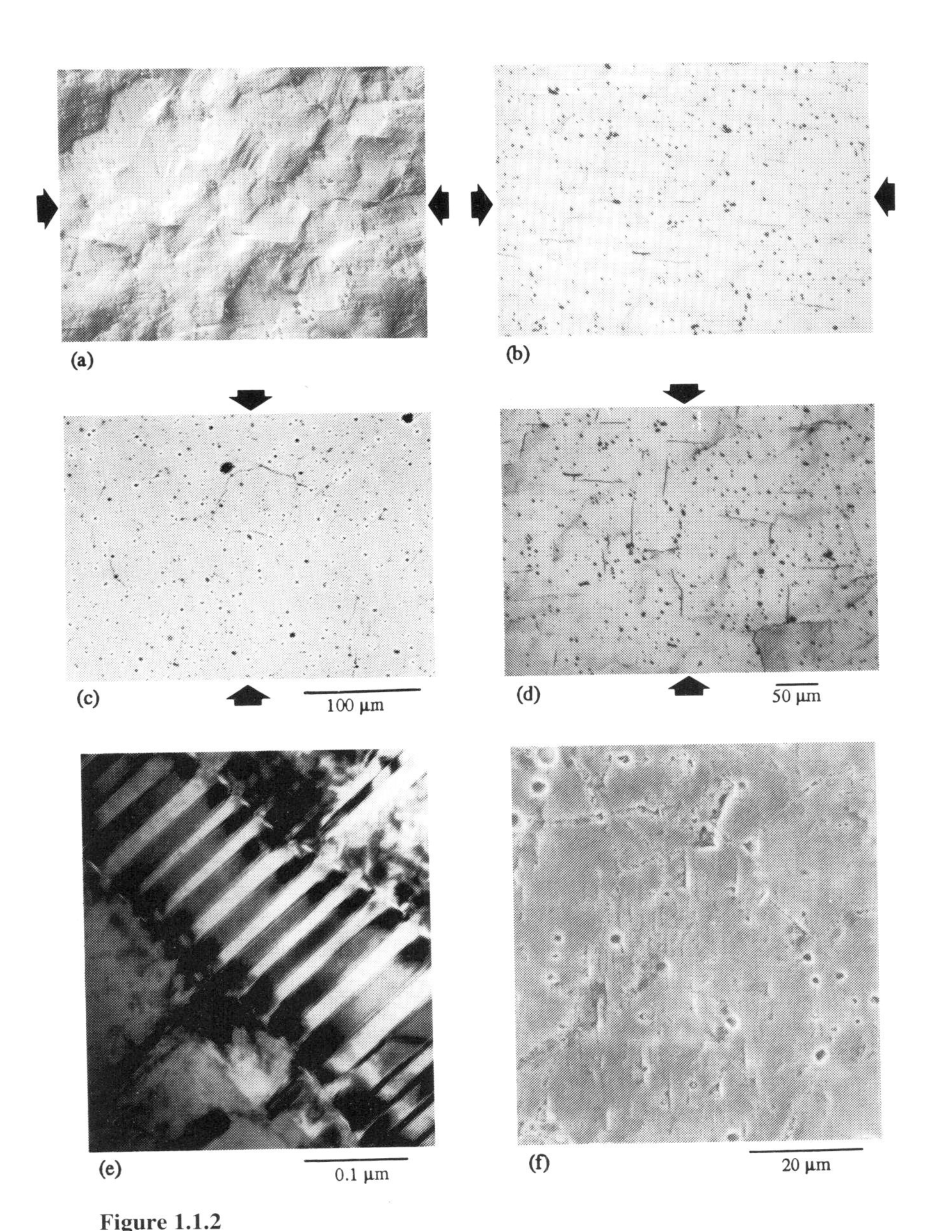

Figure 1.1.2

(a) Optical micrograph of surface rumples due to phase transformation in Mg-PSZ; (b) Microcracks in the direction of applied compression pulse; (c) Essentially randomly oriented microcracks normal to applied compression; (d) Additional microcracks in the direction of second loading; (e) Transmission electron micrograph showing twinning of a transformed precipitate and microcracks at interface with matrix; (f) Microcavities, grain boundaries, microcracks, etc. in Mg-PSZ (from Subhash, 1991)

shown in the transmission electron micrograph of Figure 1.1.2e. Twinning introduces additional minute cracks at the interfaces between the PSZ precipitates and the elastic matrix, as is evident in this last figure.

Therefore, in a tested sample of Mg-PSZ, in addition to pre-existing microcavities and grain boundaries (Figure 1.1.2f), there are numerous microcracks with a rather special distribution, depending on the loading history. The phase transformation and twinning strains within small precipitates which are distributed in a cubic manner within each crystal of this polycrystalline ceramic, produce additional minute microcracks at the interfaces between the transformed precipitates and the matrix. Tension cracks are also observed to form normal to the applied compression, upon unloading.

Figures 1.1.3a,b,c are optical micrographs of a metal-matrix composite (MMC) consisting of an aluminum matrix with alumina inclusions of a special arrangement caused by the processing technique, which involves a final uniaxial extrusion of the composite. The alumina particles are more or less aligned in the direction of extrusion. Figures 1.1.3a,b,c show the cross sections of a typical thin plate of this material, taken, respectively, through the plate thickness, normal to the direction of the extrusion; parallel to the direction of extrusion; and parallel to both the extrusion-direction and the upper and lower surfaces of the extruded plate.

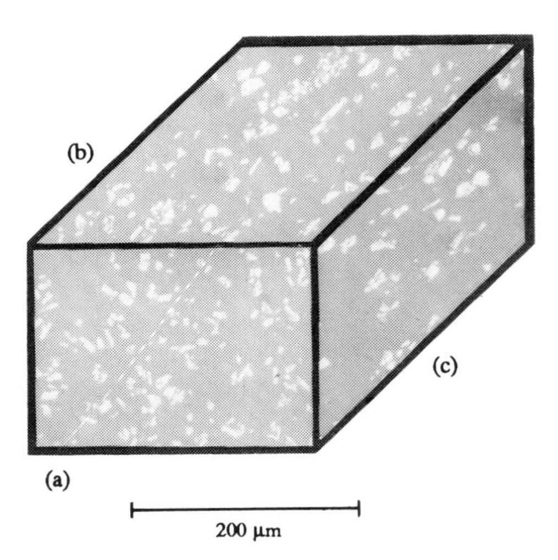

Figure 1.1.3

Optical micrographs of an aluminum-alumina metal-matrix composite: (a) Normal to the extrusion-direction; (b) Parallel to the extrusion-direction and through the plate thickness; (c) Parallel to the extrusion-direction and plate faces (from Altman *et al.*, 1992)

In these examples, a sample of a typical dimension of several millimeters may be used as an RVE. As may be inferred from these illustrations, to quantify the concept of an RVE, two length-scales are necessary: one is the continuum- or *macro-length-scale*, by which the infinitesimal material neighborhood is measured; the second is the *micro-length-scale* which corresponds to the smallest microconstituent whose properties and shape are judged to have direct, first-order effects on the overall response and properties of the continuum infinitesimal material neighborhood or macro-element. In general, the typical dimension of the macro-element, D, must be orders of magnitude larger than the typical dimension of the micro-element, d; i.e., $D/d >> 1$. For example, if the continuum is a polycrystalline solid which is viewed as a homogenized continuum, and one is interested in describing the aggregate or polycrystal properties (the polycrystal being the macro-element) in terms of single-crystal properties (each crystal being a micro-element), then the dimension, D, of the RVE should be much larger than the typical size, d, of the individual crystals. As a second example, if one is interested in estimating the elastic moduli of a whisker-reinforced composite in terms of the matrix (assumed uniform and homogeneous) and the whisker parameters, then the size of the RVE must be such that it includes a large number of whiskers. In either example, whether or not the micro-elements have a random, periodic, or other distribution does not affect the requirement of $D/d >> 1$, although, of course, the corresponding overall properties of the RVE are directly affected by this distribution. In the illustrations of Figures 1.1.2a~f and Figures 1.1.3a,b,c, the macroscopic samples used in experimentally obtaining the overall mechanical properties are clearly good candidates for the corresponding RVE, since their macro-dimensions (of the order of several millimeters) are orders of magnitude greater than the dimension of the cavities, microcracks, precipitates, individual crystals, and inclusions, which are no greater than tens of microns.

Note that the absolute dimensions of the microconstituents may be very large or very small, depending on the size of the continuum mass and the objectives of the analysis. It is only the relative dimensions that are of concern. For example, in characterizing the overall properties of a mass of compacted fine powder in powder-metallurgy, with grains of submicron size, a neighborhood of a dimension of 100 microns would be sufficient as an RVE, whereas in characterizing an earth dam as a continuum, with aggregates of many centimeters in size, the absolute dimension of an RVE would be of the order of tens of meters.

Another important question is what constitutes an underlying *essential* microconstituent. This also is a relative concept, depending on the particular problem and the particular objective. It must be addressed through systematic microstructural observation at the level of interest, and must be guided by experimental results. Perhaps *one of the most vital decisions that the analyst makes is the definition of the* RVE. An optimum choice would be one that includes the most dominant features that have first-order influence on the overall properties of interest and, at the same time, yields the simplest model. This can only be done through a coordinated sequence of microscopic (small-scale) and macroscopic (continuum-scale) observation, experimentation, and analysis. In many problems in the mechanics of materials, suitable choices often emerge naturally in the course of the examination of the corresponding physical attributes and the experimental results.

1.2. SCOPE OF THE BOOK

The extraction of macroscopic properties of microscopically heterogeneous media, on the basis of systematic modeling, has taken varied paths in the literature. The path chosen in this book rests heavily on the basic rigorous approach of applied mechanics, in the spirit pioneered by Hill (1952, 1963, 1964a,b, 1965a,b), Kröner (1953, 1958, 1977, 1978), Hershey (1954a,b), Hashin (1964, 1965a,b, 1968, 1970, 1983), Hashin and Shtrikman (1962a,b, 1963), Budiansky (1965), Walpole (1966a,b), and Willis (1977); see also, Christensen (1979), Nabarro (1979), Walpole (1981), Willis (1981, 1982), Bilby *et al.* (1985), Mura (1987), Weng *et al.* (1990), and references cited therein. Roughly speaking, the approach begins with a simple *model*, exploits fundamental principles of continuum mechanics, especially linear elasticity and the associated extremum principles, and, estimating local quantities in an RVE in terms of global boundary data, seeks to compute the overall properties and the associated *bounds*.

The book is organized in two parts. In Part 1, a fundamental and general framework for quantitative, rigorous analysis of the overall response and failure modes of microstructurally heterogeneous solids is systematically developed. Based on the theory of elasticity, particularly basic variational principles, and general averaging techniques, exact expressions are obtained for parameters which describe the overall mechanical and nonmechanical properties of heterogeneous solids and composites, in terms of the corresponding microstructure. These expressions apply to broad classes of materials with inhomogeneities and defects. The inhomogeneities may be precipitates, inclusions, whiskers, and reinforcing fibers, or they may be voids, microcracks, or plastically-induced slips, twins, and transformed materials. While, for the most part, the general framework is set within linear elasticity, the results directly translate to heterogeneous solids with rate-dependent or rate-independent inelastic constituents. This application is specifically pointed out at various suitable places within the book.

The general exact relations obtained between the overall properties and the microstructure, are then used together with simple *models*, to develop techniques for direct quantitative evaluation of the overall response which is generally described in terms of instantaneous overall moduli or compliances. These techniques include the dilute-distribution, the self-consistent, the differential, the double- and multi-inclusion, and the periodic models. The relations among the corresponding results for a variety of problems are examined in great detail, illustrated by specific, technologically significant, problems, and discussed in relation to rigorous computable bounds. Examples include solids with microcavities, microcracks, micro-inclusions, and fibers. The bounds, as well as the specific results, include new observations and original developments, as well as a careful account of the state of the art.

More specifically, in Section 1, the basic concept of representative volume element (RVE) is introduced, and in Section 2, general averaging theorems are presented. For heterogeneous solids whose constituents admit stress or strain potentials, exact relations are obtained between the macro- and microquantities.

In addition, two universal theorems are given, which provide clear ordering for the strain energy and the complementary strain energy of any heterogeneous elastic solid (not necessarily linear) subjected to various boundary data (different, but consistent) which produce either the same overall strains or the same overall stresses. These universal bounds are then used in a novel manner later, in Section 9, to develop exact computable bounds on the overall energies and moduli or compliances of a broad class of composites. In Sections 3-8, simple illustrative examples are worked out in great detail, to show the application of the fundamental relations. Section 9 deals with the general concept of variational principles and bounds on the overall parameters. In Section 10, the results of various models are reexamined with care and compared, and in Section 11, certain necessary mathematical background information, particularly on Green's functions, is given. Sections 12, 13, and 14 are dedicated to basic results and illustrative examples of heterogeneous solids with periodic microstructure, including inclusions, voids, and cracks. Exact, computable bounds are given for periodic microstructures with unit cells consisting of any number, shape, or distribution of phases. In particular, universal bounds on two overall parameters of the composite are developed, and it is shown that the same exact bounds remain valid for any volume element (not necessarily with periodic microstructure) of any heterogeneous elastic solid. In Appendix A of Part 1, application of the basic results to nonlinear rate-dependent and rate-independent inelastic heterogeneous solids is briefly examined.

To render the book self-contained, fundamentals of continuum mechanics, particularly linear elasticity, essential for micromechanics, are briefly presented in Sections 15-20 of Part 2. Section 21 reviews the mathematical tools for the solution of two-dimensional elasticity problems with singularities, including the Hilbert problem formulation in terms of singular integral equations, both Cauchy singular and Hadamard's finite-part integral, for general anisotropic materials.

There are other, equally rigorous and useful, approaches which provide at least complementary information on the overall behavior of microscopically heterogeneous solids. One such approach is the *explicitly statistical* formulation, where an RVE is viewed as a member of an ensemble of RVE's, from which ensemble averages are sought, estimated, and used to represent the corresponding macroscopic constitutive parameters, as well as the material response; for discussion and references, see Beran (1968, 1971), Kröner (1971), Batchelor (1974), and McCoy (1981). The statistical approach usually seeks to define the required overall properties of a microstructurally randomly heterogeneous material in terms of the so-called correlation tensors. The n-point correlation tensor is the probability of finding certain material phases at n different points within an RVE. The simplest case is a two-point correlation tensor which provides the probability of finding, say, the αth phase at two points within an RVE, $\mathbf{x}^1$ and $\mathbf{x}^2$. In a similar way, cross correlation of two and several phases are defined. Early work in this area is by Brown (1955), Miller (1969a,b), and Hori and Yonezawa (1974, 1975); for a discussion and references, see Torquato (1991). Such statistical information has been used to develop improved bounds on effective properties of microstructurally randomly heterogeneous materials.

This book does not deal with the statistical approach. The above comments and references, therefore, are given as an entry to the vast literature on the subject, focused on statistical estimations of continuum properties. Since the assumption of *ergodicity* allows replacing ensemble averages with sample averages, it may be viewed as a bridge between the explicitly statistical and the approach chosen in this book. At this early point in the discourse, suffice it to say that an RVE may be regarded as a representative part of a very large heterogeneous solid (infinitely extended), any of whose suitably large subregions may be used to obtain essentially the same overall macroscopic material properties and local continuum field variables. A large solid of this kind is called *statistically homogeneous*. The assumption (hypothesis) of ergodicity then allows extracting ensemble statistics from averages obtained over such a statistically homogeneous, very large, but microscopically heterogeneous continuum. In this context, one may consider three length-scales, namely, a *microscale* defining the heterogeneity within an RVE, a *miniscale* defining the size of an RVE, and a *macroscale* associated with the laboratory (or the continuum) sample; for discussion and references, see Hashin (1983). Since, within the infinitely extended, statistically homogeneous solid, translation and rotation of an RVE (if isotropic) are assumed to leave the corresponding averages essentially unchanged, these are also referred to as *moving averages*. Certain mathematical aspects of this concept are examined in Subsections 2.5 and 2.6.

In addition to the above alternatives, there is considerable literature on the engineering approach to estimating material stiffness and strength, mainly focused on engineering composites. Much of the material covered in this book can and does serve as a fundamental framework for other, more application-oriented techniques. There are journals and proceedings of national and international conferences on composite materials, which cover a broad spectrum of approaches of this kind; see, as illustration, Vinson and Sierakowski (1986), Talreja (1987), Wilde and Blain (1990), the *Delaware Composite Design Encyclopedia* volume 1-6 edited by Carlsson and Gillespie (1989-90), Nayfeh (1995), and Herakovich (1998). As an entry to the vast literature of the mechanics-related materials aspect of micromechanics, particularly relating to the properties of ceramics and ceramic composites, the following general references are mentioned: Khachaturyan (1983), Pask and Evans (1987), Rühle *et al.* (1990), Mazdiyasni (1990), and Suresh (1991).

An approach of recent origin, akin to phenomenological plasticity, is *damage mechanics*, with an already rather extensive series of contributions. For general reference, see, e.g., Talreja (1985), a comprehensive review by Krajcinovic (1989), and a symposium proceedings edited by Ju (1992). A recent book by Krajcinovic (1996) provides a broad overview as well as a detailed presentation of major topics in damage mechanics. For basic application of thermodynamics with internal variables, modeling the nonlinear response of composites, see Schapery (1987, 1990, 1995); see also Valanis (1966), Coleman and Gurtin (1967), and Rice (1971) for the fundamentals.

1.3. DESCRIPTION OF RVE

In micromechanics the concept of an RVE is used to estimate the continuum properties at a continuum material point, in terms of the microstructure and microconstituents that comprise that material point and its infinitesimal material neighborhood, i.e., *to obtain the continuum constitutive properties in terms of the properties and structure of the microconstituents*. These constitutive properties, often expressed as *constitutive relations*, are then used in the balance equations to calculate the overall response of the continuum mass to applied loads and prescribed boundary data. The balance equations include the equations of the conservation of mass, linear and angular momenta, and energy. These equations contain the body forces representing the effect of the materials not in contact with the considered continuum and the inertia forces due to the motion of the continuum itself, as well as the associated force and displacement boundary data which represent the effect of the other continua in contact with the considered continuum. Therefore, in formulating boundary-value problems associated with an RVE, it is not necessary to include the body forces. Nor is it necessary to include the inertia terms for a broad range of problems[2]. The basic requirement is to obtain the overall average properties of the RVE, when subjected to the boundary data corresponding to the uniform fields in the continuum infinitesimal material neighborhood which the RVE is aimed to represent. In other words, an RVE may be viewed as a *heterogeneous* medium under prescribed boundary data which correspond to the uniform local continuum fields. The aim then is to calculate its overall response parameters, and use these to describe the local properties of the continuum material element.

Since the microstructure of the material, in general, changes in the course of deformation, the overall properties of its RVE also, in general, change. Hence, an incremental formulation is often necessary. For certain problems in elasticity, however, this may not be necessary, and a formulation in terms of the total stresses and strains may suffice.

Consider an RVE with volume V bounded by a regular surface ∂V. A typical point in V is identified by its position vector, $\mathbf{x}$, with components, x_i ($i = 1, 2, 3$), relative to a fixed rectangular Cartesian coordinate system. The unit base vectors of this coordinate system are denoted by $\mathbf{e}_i$ ($i = 1, 2, 3$), and the position vector $\mathbf{x}$ is given by

$$\mathbf{x} = x_i\,\mathbf{e}_i, \tag{1.3.1}$$

where repeated subscripts are summed. For the purpose of micromechanical calculations, the RVE is regarded as a *heterogeneous continuum with spatially variable, but known, constitutive properties*. In many cases, the objective then is to estimate the overall (average), say, strain increment, as a function of the corresponding prescribed incremental surface forces or, conversely, the average stress increment, as a function of the prescribed incremental surface

[2] An example in which the inertia forces are of prime importance is the description of the gas laws in terms of the corresponding molecular motion. Another example is the description of the heat capacity of solids in terms of atomic vibrations.

displacements. For uniform macrofields, the prescribed incremental surface tractions may be taken as spatially uniform, or, in the converse case, the prescribed incremental surface displacements may be assumed as spatially linear.

Under the prescribed surface data, the RVE must be in equilibrium and its overall deformation compatible. In constitutive modeling, body forces and inertia terms are absent. The prescribed surface tractions must hence be self-equilibrating. In the same manner, the prescribed surface displacements must be self-compatible so that they do not include rigid-body translations or rotations. Moreover, if the prescribed surface displacements are associated with a strain field, this field must be compatible. *These conditions are assumed to hold throughout this chapter and elsewhere in this book, whenever we deal with an* RVE *with prescribed boundary data.*

Whether boundary displacements or boundary tractions are regarded as prescribed, a viable micromechanical approach should produce *equivalent overall constitutive parameters* for the corresponding macro-element. For example, if the instantaneous overall moduli and compliances are being calculated, then the resulting instantaneous modulus tensor obtained for the prescribed incremental surface displacements should be the inverse of the instantaneous compliance tensor obtained for the prescribed incremental surface tractions on the RVE.

The displacement, $\mathbf{u} = \mathbf{u}(\mathbf{x})$, strain, $\boldsymbol{\varepsilon} = \boldsymbol{\varepsilon}(\mathbf{x})$, and stress, $\boldsymbol{\sigma} = \boldsymbol{\sigma}(\mathbf{x})$, fields within volume V of the RVE, vary from point to point, even if the boundary tractions are uniform or the boundary displacements are linear. The governing field equations at a typical point $\mathbf{x}$ in V, include the balance of linear and angular momenta,[3]

$$\boldsymbol{\nabla} \cdot \boldsymbol{\sigma} = \mathbf{0}, \qquad \boldsymbol{\sigma} = \boldsymbol{\sigma}^{\mathrm{T}} \quad \text{in V,} \tag{1.3.2a,b}$$

and the strain-displacement relation,

$$\boldsymbol{\varepsilon} = \frac{1}{2}\{\boldsymbol{\nabla} \otimes \mathbf{u} + (\boldsymbol{\nabla} \otimes \mathbf{u})^{\mathrm{T}}\} \quad \text{in V,} \tag{1.3.3a}$$

where $\boldsymbol{\nabla}$ is the del operator defined by

$$\boldsymbol{\nabla} \equiv \partial_i \, \mathbf{e}_i \equiv \frac{\partial}{\partial x_i} \, \mathbf{e}_i, \tag{1.3.4}$$

and superscript T denotes transpose; see Part 2, especially Section 15 of this book, for additional discussion and comments. In rectangular Cartesian component form, (1.3.2) and (1.3.3) become

$$\sigma_{ji,j} = 0, \qquad \sigma_{ij} = \sigma_{ji} \quad \text{in V,} \tag{1.3.2c,d}$$

and

$$\varepsilon_{ij} = \frac{1}{2}(u_{i,j} + u_{j,i}) \quad \text{in V,} \tag{1.3.3b}$$

---

[3] Here the stress and deformation fields are assumed to be continuous. Interfaces and discontinuities are considered in Subsection 2.4.

where i, j = 1, 2, 3, and a comma followed by an index denotes partial differentiation with respect to the corresponding coordinate variable.

When the self-equilibrating tractions (not necessarily uniform), $\mathbf{t}^o$, are assumed prescribed on the boundary ∂V of the RVE, as shown in Figure 1.3.1a, then

$$\boldsymbol{\nu} \cdot \boldsymbol{\sigma} = \mathbf{t}^o \quad \text{on } \partial V, \tag{1.3.5a}$$

or

$$\sigma_{ji}\, \nu_j = t_i^o \quad \text{on } \partial V, \tag{1.3.5b}$$

where $\boldsymbol{\nu}$ is the outer unit normal vector of ∂V. On the other hand, when the displacements (not necessarily linear), $\mathbf{u}^o$, are assumed prescribed on the boundary of the RVE, as shown in Figure 1.3.1b, it follows that

$$\mathbf{u} = \mathbf{u}^o \quad \text{on } \partial V, \tag{1.3.6a}$$

or

$$u_i = u_i^o \quad \text{on } \partial V. \tag{1.3.6b}$$

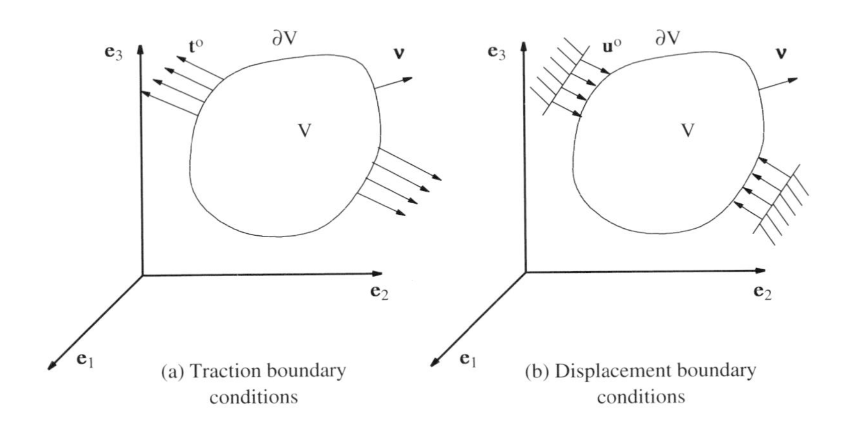

(a) Traction boundary conditions

(b) Displacement boundary conditions

Figure 1.3.1

For the incremental formulation it is necessary to consider a rate problem, where traction rates $\dot{\mathbf{t}}^o$, or velocity $\dot{\mathbf{u}}^o$, but not both, as discussed above, may be regarded as prescribed on the boundary of the RVE. Here the rates may be measured in terms of a monotone increasing parameter, since no inertia effects are included. For a rate-dependent material response, however, the actual time must be used. The basic field equations are obtained from (1.3.2~6) by substituting the corresponding rate quantities, e.g., $\dot{\boldsymbol{\sigma}}$ for $\boldsymbol{\sigma}$, $\dot{\boldsymbol{\varepsilon}}$ for $\boldsymbol{\varepsilon}$, and $\dot{\mathbf{u}}$ for $\mathbf{u}$, arriving at

$$\nabla \cdot \dot{\boldsymbol{\sigma}} = \mathbf{0}, \qquad \dot{\boldsymbol{\sigma}} = \dot{\boldsymbol{\sigma}}^{\mathrm{T}} \qquad \text{in V,} \tag{1.3.7a,b}$$

and

$$\dot{\boldsymbol{\varepsilon}} = \frac{1}{2}\{\nabla \otimes \dot{\mathbf{u}} + (\nabla \otimes \dot{\mathbf{u}})^{\mathrm{T}}\} \qquad \text{in V.} \tag{1.3.8}$$

When the self-equilibrating boundary traction rates, $\dot{\mathbf{t}}^{\mathrm{o}}$, are prescribed,

$$\boldsymbol{\nu} \cdot \dot{\boldsymbol{\sigma}} = \dot{\mathbf{t}}^{\mathrm{o}} \qquad \text{on } \partial\mathrm{V}, \tag{1.3.9}$$

and when the self-compatible boundary velocities, $\dot{\mathbf{u}}^{\mathrm{o}}$, are prescribed,

$$\dot{\mathbf{u}} = \dot{\mathbf{u}}^{\mathrm{o}} \qquad \text{on } \partial\mathrm{V}. \tag{1.3.10}$$

For the most part, this book focuses on the mechanical properties of heterogeneous media. However, essentially all of the results can be reduced and directly applied to the nonmechanical properties of microscopically heterogeneous materials. This will be pointed out at appropriate places throughout Part 1, providing guidance for this kind of application.

As an illustration, consider thermal conduction problems, and let $\mathrm{u} = \mathrm{u}(\mathbf{x})$ be the temperature. With $\mathbf{q} = \mathbf{q}(\mathbf{x})$ defining the corresponding heat *flux*, in the absence of any heat sources the steady-state regime corresponds to

$$\nabla \cdot \mathbf{q} = 0 \qquad \text{in V.} \tag{1.3.11a}$$

The boundary conditions may be expressed, either in terms of the normal component of the flux,

$$\boldsymbol{\nu} \cdot \mathbf{q} = \mathrm{q}^{\mathrm{o}} \qquad \text{on } \partial\mathrm{V}, \tag{1.3.12a}$$

or in terms of the temperature field,

$$\mathrm{u} = \mathrm{u}^{\mathrm{o}} \qquad \text{on } \partial\mathrm{V}. \tag{1.3.13}$$

In component form, (1.3.11a) and (1.3.12a), respectively, become

$$\mathrm{q}_{\mathrm{i,i}} = 0 \qquad \text{in V,} \tag{1.3.11b}$$

$$\nu_{\mathrm{i}}\, \mathrm{q}_{\mathrm{i}} = \mathrm{q}^{\mathrm{o}} \qquad \text{on } \partial\mathrm{V}. \tag{1.3.12b}$$

Note from (1.3.11a) that the boundary flux q^{o} must be *self-balanced* in the sense that

$$\int_{\mathrm{V}} \nabla \cdot \mathbf{q} \, \mathrm{dV} = \int_{\partial\mathrm{V}} \mathrm{q}^{\mathrm{o}} \, \mathrm{dS} = 0. \tag{1.3.12c}$$

For future use, define $\mathbf{p} = \mathbf{p}(\mathbf{x})$ by

$$\mathbf{p} = -\, \nabla \mathrm{u} \qquad \text{in V,} \tag{1.3.14a}$$

or

$$\mathrm{p}_{\mathrm{i}} = -\, \mathrm{u}_{,\mathrm{i}} \qquad \text{in V.} \tag{1.3.14b}$$

The negative of the temperature gradient, $\mathbf{p}$, may be viewed as the *force* which drives the heat flux, $\mathbf{q}$. In the terminology of irreversible thermodynamics, at least for convenience, $\mathbf{p}$ may be called the *force* conjugate to the *flux* $\mathbf{q}$. Finally, for an incremental formulation, the rate quantities, namely, $\dot{u}$, $\dot{\mathbf{q}}$, and $\dot{\mathbf{p}}$, are used.

In the context of estimating the overall material parameters of an RVE, the steady-state thermal, diffusional, electrical, and magnetic field equations are quite similar. For example, u may be identified with the electric potential (usually denoted by ϕ), $\mathbf{p}$ with the electric field (usually denoted by $\mathbf{E}$), and $\mathbf{q}$ with the electric displacement (usually denoted by $\mathbf{D}$). The relation to mass diffusion is obvious. For magnetostatics, $\mathbf{q}$ is identified with the magnetic induction (usually denoted by $\mathbf{B}$) and $\mathbf{p}$ with the magnetic field intensity (usually denoted by $\mathbf{H}$).

1.4. REFERENCES

Altman, B. S., Nemat-Nasser, S., Vecchio, K. S., and Isaacs, J. B., (1992), Homogeneous deformation of a particulate reinforced metal-matrix composite, *Shock Compression of Condensed Matter 1991* (Schmidt S. C., Dick, R. D., Forbes, J. W., and Tasker, D. G., eds.), Elsevier, Amsterdam, 543-546.

Ashby, M. F. and Hallam, D. (1986), The failure of brittle solids containing small cracks under compressive stress states, *Acta. Metall.*, Vol. 34, 497-510.

Batchelor, G. K. (1974), Transport properties of two-phase materials with random structure, *Annu. Rev. Fluid Mech.*, Vol. 6, 227-255.

Beran, M. J. (1968), *Statistical continuum theories*, Wiley-Interscience, New York.

Beran, M. J. (1971), Application of statistical theories of heterogeneous materials, *Phys. Status Solidi A*, Vol. 6, 365-384.

Bilby, B. A., Miller, K. J., and Willis, J. R., eds. (1985), *Fundamentals of deformation and fracture - Eshelby Memorial Symposium*, Cambridge University Press, Cambridge.

Brown, W. F. (1955), Solid mixture permittivities, *J. Chem. Phys.*, Vol. 23, 1514-1517.

Budiansky, B. (1965), On the elastic moduli of some heterogeneous materials, *J. Mech. Phys. Solids*, Vol. 13, 223-227.

Carlsson, L. A. and Gillespie, J. W., eds. (1989-1990), *Mechanical behavior and properties of composite materials*, Vol. 1-6, Tecnomic Publishing Co., Lancaster.

Christensen, R. M. (1979), *Mechanics of composite materials*, Wiley-Interscience, New York.

Coleman, B. and Gurtin, M. (1967), Thermodynamics with internal state variables, *J. Chem. Phys.*, Vol. 47, 597-613.

Hashin, Z. (1964), Theory of mechanical behaviour of heterogeneous media, *Appl. Mech. Rev.*, Vol. 17, 1-9.

Hashin, Z. (1965a), Elasticity of random media, *Trans. Soc. Rheol.*, Vol. 9, 381-406.

Hashin, Z. (1965b), On elastic behaviour of fiber reinforced materials of arbitrary transverse phase geometry, *J. Mech. Phys. Solids*, Vol. 13, 119-134.

Hashin, Z. (1968), Assessment of the self-consistent scheme approximation, *J. Compos. Mater.*, Vol. 2, 284-300.

Hashin, Z. (1970), Theory of composite materials, in *Mechanics of composite materials* (F. W. Went, H. Liebowitz, and N. Perrone, eds.), Pergamon Press, Oxford, 201-242.

Hashin, Z. (1983), Analysis of composite materials - A survey, *J. Appl. Mech.*, Vol. 50, 481-505.

Hashin, Z. and Shtrikman, S. (1962a), On some variational principles in anisotropic and nonhomogeneous elasticity, *J. Mech. Phys. Solids*, Vol. 10, 335-342.

Hashin, Z. and Shtrikman, S. (1962b), A variational approach to the theory of the elastic behaviour of polycrystals, *J. Mech. Phys. Solids*, Vol. 10, 343-352.

Hashin, Z. and Shtrikman, S. (1963), A variational approach to the theory of the elastic behaviour of multiphase materials, *J. Mech. Phys. Solids*, Vol. 11, 127-140.

Herakovich, C. T. (1998), *Mechanics of fibrous composites*, Wiley, New York.

Hershey, A. V. (1954a), The elasticity of an isotropic aggregate of anisotropic cubic crystals, *J. Appl. Mech.*, Vol. 21, 236-240.

Hershey, A. V. (1954b), The plasticity of an isotropic aggregate of anisotropic face-centered cubic crystals, *J. Appl. Mech.*, Vol. 21, 241-249.

Hill, R. (1952), The elastic behaviour of a crystalline aggregate, *Proc. Phys. Soc., London, Sect. A*, Vol. 65, 349-354.

Hill, R. (1963), Elastic properties of reinforced solids: Some theoretical principles, *J. Mech. Phys. Solids*, Vol. 11, 357-372.

Hill, R. (1964a), Theory of mechanical properties of fibre-strengthened materials - I: Elastic behaviour, *J. Mech. Phys. Solids*, Vol. 12, 199-212.

Hill, R. (1964b), Theory of mechanical properties of fibre-strengthened materials - II: Inelastic behaviour, *J. Mech. Phys. Solids*, Vol. 12, 213-218.

Hill, R. (1965a), Theory of mechanical properties of fibre-strengthened materials - III: Self-consistent model, *J. Mech. Phys. Solids*, Vol. 13, 189-198.

Hill, R. (1965b), A self-consistent mechanics of composite materials, *J. Mech. Phys. Solids*, Vol. 13, 213-222.

Hori, M. and Yonezawa, F. (1974), Statistical theory of effective electrical, thermal, and magnetic properties of random heterogeneous materials. III. Perturbation treatment of the effective permittivity in completely random heterogeneous materials, *J. Math. Phys.*, Vol. 15, 2177-2185.

Hori, M. and Yonezawa, F. (1975), Statistical theory of effective electrical, thermal, and magnetic properties of random heterogeneous materials. IV. Effective-medium theory and cumulant expansion method, *J. Math. Phys.*, Vol. 16, 352-364.

Horii, H. and Nemat-Nasser, S. (1985), Compression-induced microcrack growth in brittle solids: Axial splitting and shear failure, *J. Geophys. Res.*, Vol. 90, B4, 3105-3125.

Horii, H. and Nemat-Nasser, S. (1986), Brittle failure in compression: Splitting, faulting and brittle-ductile transition, *Phil. Trans. R. Soc. Lond.*, Vol. A319, 337-374.

Ju, J. W., ed. (1992), *Recent advances in damage mechanics and plasticity*, AMD-Vol. 132, MD-Vol. 30, ASME, New York.

Khachaturyan, A. G. (1983), *Theory of structural transformations in solids*, John Wiley & Sons, New York.

Krajcinovic, D. (1989), Damage Mechanics, *Mech. Mat.*, Vol. 8, 117-197.

Krajcinovic, D. (1996), *Damage Mechanics*, North-Holland, New York.

Kröner, E. (1953), Das Fundamentalintegral der anisotropen elastischen Differentialgleichungen, *Z. Phys.*, Vol. 136, 402-410.

Kröner, E. (1958), Berechnung der elastischen Konstanten des Vielkristalls aus den Konstanten des Einkristalls, *Z. Phys.*, Vol. 151, 504-518.

Kröner, E. (1971), *Statistical continuum mechanics*, Springer-Verlag, Udine.

Kröner, E. (1977), Bounds for effective elastic moduli of disordered materials, *J. Mech. Phys. Solids*, Vol. 25, 137-155.

Kröner, E. (1978), Self-consistent scheme and graded disorder in polycrystal elasticity, *J. Phys. F*, Vol. 8, 2261-2267.

Mazdiyasni, K. S. (1990), *Fiber reinforced ceramic composites*, Noyes, New Jersey.

McCoy, J. J. (1981), Macroscopic response of continua with random microstructure, in *Mechanics Today* (Nemat-Nasser, S., ed.), Pergamon, Oxford, Vol. 6, 1-40.

Miller, M. N. (1969a), Bounds for effective electrical, thermal, and magnetic properties of heterogeneous materials, *J. Math. Phys.*, Vol. 10, 1988-2004.

Miller, M. N. (1969b), Bounds for effective bulk modulus of heterogeneous materials, *J. Math. Phys.*, Vol. 10, 2005-2013.

Mura, T. (1987), *Micromechanics of defects in solids (2nd Edition)*, Martinus Nijhoff Publishers, Dordrecht.

Nabarro, F. R. N. (1979), *Dislocations in Solids*, Volume 1: The Elasticity Theory, North-Holland, Amsterdam.

Nayfeh, A. H. (1995), *Wave Propagation in Layered Anisotropic Media with Application to Composites*, Elsevier, Amsterdam.

Nemat-Nasser, S. (1986), Overall stresses and strains in solids with microstructure, in *Modeling small deformations of polycrystals* (J. Gittus and J. Zarka, eds.), Elsevier Publishers, Netherlands, 41-64.

Nemat-Nasser, S. (1989), Compression-induced ductile flow of brittle materials and brittle fracturing of ductile materials, in *Advances in fracture research*, Proceedings of ICF7, Vol. 1, 423-445.

Nemat-Nasser, S. and Horii, H. (1982), Compression-induced nonplanar crack extension with application to splitting, exfoliation, and rockburst, *J. Geophys. Res.*, Vol. 87, 6805-6821.

Pask, J. A. and Evans, A. G. (1987), eds. *Ceramic microstructures '86, Role of interfaces*, Plenum, New York.

Rice, J. R. (1971), Inelastic constitutive relations for solids: an internal-variable theory and its application to metal plasticity, *J. Mech. Phys. Solids*, Vol. 19, 433-455.

Rogers, W. P. and Nemat-Nasser, S. (1989), Transformation plasticity at high strain rate in magnesia-partially-stabilized zirconia, *J. Am. Cer. Soc.*, Vol. 73, 136-139.

Rühle, M., Evans, A. G., Ashby, M. F., and Hirth, J. P., eds. (1990), *Metal-ceramic interfaces, Proceedings of an international workshop*, Pergamon Press, Oxford.

Schapery, R. A. (1987), Deformation and fracture characterization of inelastic

composite materials using potentials, *Polymer Engrg. Sci.*, Vol. 27, 63-76.
Schapery, R. A. (1990), A theory of material behaviour of elastic media with growing damage and other changes in structure, *J. Mech. Phys. Solids*, Vol. J. Mech. Phys. Solids, Vol. 38, No. 2, 215-253.
Schapery, R. A. (1995), Prediction of compressive strength and kink bands in composites using a work potential, *Int. J. Solids Struct.*, Vol. 32, 739-765.
Subhash, G. (1991), Dynamic behavior of zirconia ceramics in uniaxial compression, Ph.D. Thesis, University of California, San Diego.
Subhash, G. and Nemat-Nasser, S. (1993), Dynamic stress-induced transformation and texture formation in uniaxial compression of zirconia ceramics, *J. Am. Cer. Soc.*, Vol. 76, 153-165.
Suresh, S. (1991), *Fatigue of materials*, Cambridge University Press, New York.
Talreja, R. (1985), A continuum mechanics characterization of damage in composite materials, *Proc. R. Soc. London*, Series A, Vol. 399, 195-216.
Talreja, R. (1987), *Fatigue of Composite Materials*, Technomic Pub. Co., Lancaster.
Torquato, S. (1991), Random heterogeneous media: microstructure and improved bounds on effective properties, *Appl. Mech. Rev.*, Vol. 42, 37-76.
Valanis, K. C. (1966), Thermodynamics of large viscoelastic deformations, *J. of Mathematics and Physics*, Vol. 45, No. 2, 197-212.
Vinson, J. R. and Sierakowski, R. L. (1986), *The behavior of structures composed of composite materials*, Martinus Nijhoff Publisher, Dordrecht.
Walpole, L. J. (1966a), On bounds for the overall elastic moduli of inhomogeneous systems - I, *J. Mech. Phys. Solids*, Vol. 14, 151-162.
Walpole, L. J. (1966b), On bounds for the overall elastic moduli of inhomogeneous systems - II, *J. Mech. Phys. Solids*, Vol. 14, 289-301.
Walpole, L. J. (1981), Elastic behavior of composite materials: Theoretical foundations, *Advances in Applied Mechanics*, Vol. 21, 169-242.
Weng, G. J., Taya, M., and Abé, H., eds. (1990), *Micromechanics and inhomogeneity - The T. Mura 65th anniversary volume*, Springer-Verlag, New York.
Wilde, W. P. and Blain, W. R., eds. (1990), *Composite materials design and analysis, Proceedings of the second international conference on computer aided design in composite material technology*, Springer-Verlag, Berlin.
Willis, J. R. (1977), Bounds and self-consistent estimates for the overall properties of anisotropic composites, *J. Mech. Phys. Solids*, Vol. 25, 185-202.
Willis, J. R. (1981), Variational and related methods for the overall properties of composites, *Advances in Applied Mechanics*, Vol. 21, 1-78.
Willis, J. R. (1982), Elasticity Theory of Composites, in *Mechanics of Solids - The Rodney Hill 60th Anniversary Volume* (H. G. Hopkins and M. J. Sewell eds.), Wheaton, London, 653-686.

SECTION 2 AVERAGING METHODS

Fundamental averaging theorems necessary to extract the overall quantities are presented in this section. Many of the results apply to heterogeneous solids with constituents of arbitrary material properties, linear or nonlinear, rate-dependent or rate-independent. Then attention is focused on heterogeneous solids whose constituents admit stress and/or strain potentials. Relations between macropotentials and corresponding micropotentials are examined in some detail for various boundary conditions. A number of bounding theorems are developed, which provide ordering for the overall stress and strain potentials when uniform tractions, linear displacements, or general (mixed, but consistent) boundary data for an RVE are considered. In light of these basic results, the notions of statistical homogeneity and representative volume element are reexamined and precise conditions implied by, and implying, statistical homogeneity are studied in detail. This section, therefore, lays the theoretical foundation for many of the results developed in subsequent sections.

2.1. AVERAGE STRESS AND STRESS RATE

Whether the prescribed self-equilibrating boundary tractions on ∂V are spatially uniform or not, the *unweighted volume average* of the *variable stress field* $\boldsymbol{\sigma}(\mathbf{x})$, taken over the volume V of the RVE, *is completely defined in terms of the prescribed boundary tractions.* To show this, denote the volume average of a typical, spatially variable, integrable quantity, $\mathbf{T}(\mathbf{x})$, by

$$< \mathbf{T} > \equiv \frac{1}{V}\int_V \mathbf{T}(\mathbf{x})\, dV. \tag{2.1.1}$$

Then the unweighted volume average stress, denoted by $\bar{\boldsymbol{\sigma}}$, is

$$\bar{\boldsymbol{\sigma}} \equiv < \boldsymbol{\sigma} >. \tag{2.1.2}$$

The gradient of $\mathbf{x}$ satisfies

$$(\boldsymbol{\nabla} \otimes \mathbf{x})^T = \partial_j x_i\, \mathbf{e}_i \otimes \mathbf{e}_j = x_{i,j}\, \mathbf{e}_i \otimes \mathbf{e}_j = \delta_{ij}\, \mathbf{e}_i \otimes \mathbf{e}_j = \mathbf{1}^{(2)}, \tag{2.1.3}$$

where δ_{ij} is the Kronecker delta ($\delta_{ij} = 1$ when $i = j$, and $= 0$ otherwise), and $\mathbf{1}^{(2)}$ is the second-order unit tensor. From equations of equilibrium (1.3.2), and since the stress tensor is divergence-free,

$$\boldsymbol{\sigma} = \mathbf{1}^{(2)} \cdot \boldsymbol{\sigma} = (\boldsymbol{\nabla} \otimes \mathbf{x})^T \cdot \boldsymbol{\sigma} = \{\boldsymbol{\nabla} \cdot (\boldsymbol{\sigma} \otimes \mathbf{x})\}^T. \tag{2.1.4}$$

By means of the Gauss theorem, the average stress $\bar{\boldsymbol{\sigma}}$ is expressed as

$$< \boldsymbol{\sigma} > = \frac{1}{V}\int_V \{\boldsymbol{\nabla}\cdot(\boldsymbol{\sigma}\otimes\mathbf{x})\}^T \, dV = \frac{1}{V}\int_{\partial V} \{\mathbf{v}\cdot(\boldsymbol{\sigma}\otimes\mathbf{x})\}^T \, dS, \tag{2.1.5}$$

and in view of (1.3.5a),

$$\bar{\boldsymbol{\sigma}} = \frac{1}{V}\int_{\partial V} \mathbf{x}\otimes\mathbf{t}^o \, dS, \tag{2.1.6a}$$

or

$$\bar{\sigma}_{ij} = \frac{1}{V}\int_{\partial V} x_i \, t_j^o \, dS. \tag{2.1.6b}$$

It should be noted that since the prescribed surface tractions, $\mathbf{t}^o$, are self-equilibrating, their resultant total force and total moment about a fixed point vanish, i.e.,

$$\int_{\partial V} \mathbf{t}^o \, dS = \mathbf{0}, \qquad \int_{\partial V} \mathbf{x}\times\mathbf{t}^o \, dS = \mathbf{0}, \tag{2.1.7a,b}$$

or

$$\int_{\partial V} t_i^o \, dS = 0, \qquad \int_{\partial V} e_{ijk} \, x_j \, t_k^o \, dS = 0, \tag{2.1.7c,d}$$

where e_{ijk} is the permutation symbol of the third order; $e_{ijk} = (+1, -1, 0)$ when i, j, k form (even, odd, no) permutation of 1, 2, 3. Hence, the average stress $\bar{\boldsymbol{\sigma}}$ defined by (2.1.6) is symmetric and independent of the origin of the coordinate system. Indeed, from (2.1.7c),

$$\int_{\partial V} \mathbf{x}\otimes\mathbf{t}^o \, dS = \int_{\partial V} \mathbf{t}^o\otimes\mathbf{x} \, dS, \tag{2.1.7e}$$

and, hence, $\bar{\boldsymbol{\sigma}}^T = \bar{\boldsymbol{\sigma}}$. Also, for any constant vector $\mathbf{x}^o$,

$$\int_{\partial V} (\mathbf{x}-\mathbf{x}^o)\otimes\mathbf{t}^o \, dS = \int_{\partial V} \mathbf{x}\otimes\mathbf{t}^o \, dS. \tag{2.1.7f}$$

Therefore, the average stress defined by (2.1.6) is meaningful only if the prescribed surface tractions are self-equilibrating.

For the rate problem, the traction rates $\dot{\mathbf{t}}^o$ are prescribed, (1.3.9), producing a stress rate $\dot{\boldsymbol{\sigma}} = \dot{\boldsymbol{\sigma}}(\mathbf{x})$ in accord with equilibrium conditions (1.3.7a,b). The traction rates $\dot{\mathbf{t}}^o$ must be self-equilibrating so that (2.1.7a~d) written for $\dot{\mathbf{t}}^o$, are satisfied. The average stress rate is then given in terms of the prescribed boundary traction rates by

$$\bar{\dot{\boldsymbol{\sigma}}} \equiv < \dot{\boldsymbol{\sigma}} > = \frac{1}{V}\int_{\partial V} \mathbf{x}\otimes\dot{\mathbf{t}}^o \, dS, \tag{2.1.8a}$$

or

$$\bar{\dot{\sigma}}_{ij} \equiv < \dot{\sigma}_{ij} > = \frac{1}{V}\int_{\partial V} x_i \, \dot{t}_j^o \, dS. \tag{2.1.8b}$$

Hence,

$$\bar{\dot{\boldsymbol{\sigma}}} \equiv < \dot{\boldsymbol{\sigma}} > = \frac{d}{dt} < \boldsymbol{\sigma} > \equiv \dot{\bar{\boldsymbol{\sigma}}}. \tag{2.1.9}$$

It is noted in passing that only for small-deformation theories does the average of the Cauchy stress rate equal the rate of the average Cauchy stress. For finite deformations, in general, this is not valid; see Hill (1972), Havner (1982, 1992), Nemat-Nasser (1983), and Iwakuma and Nemat-Nasser (1984).

2.2. AVERAGE STRAIN AND STRAIN RATE

Whether the prescribed boundary displacements on ∂V are spatially linear or not, the *unweighted volume average of the variable displacement gradient* $\nabla \otimes \mathbf{u}$, *taken over volume* V *of the* RVE, *is completely defined in terms of the prescribed boundary displacements.* From the Gauss theorem, and in view of the boundary conditions (1.3.6a),

$$\int_V \nabla \otimes \mathbf{u} \, dV = \int_{\partial V} \boldsymbol{\nu} \otimes \mathbf{u} \, dS = \int_{\partial V} \boldsymbol{\nu} \otimes \mathbf{u}^o \, dS. \tag{2.2.1}$$

Thus, the average displacement gradient for the RVE is

$$\overline{\nabla \otimes \mathbf{u}} \equiv < \nabla \otimes \mathbf{u} > = \frac{1}{V} \int_{\partial V} \boldsymbol{\nu} \otimes \mathbf{u}^o \, dS, \tag{2.2.2a}$$

or

$$\overline{u_{j,i}} \equiv < u_{j,i} > = \frac{1}{V} \int_{\partial V} \nu_i \, u_j^o \, dS. \tag{2.2.2b}$$

Since the strain $\boldsymbol{\varepsilon}$ is the symmetric part of the displacement gradient, (1.3.3a), and the infinitesimal rotation $\boldsymbol{\omega}$ is the corresponding antisymmetric part,

$$\boldsymbol{\omega} = \frac{1}{2}\{\nabla \otimes \mathbf{u} - (\nabla \otimes \mathbf{u})^T\}, \tag{2.2.3a}$$

or

$$\omega_{ij} = \frac{1}{2}(u_{j,i} - u_{i,j}), \tag{2.2.3b}$$

the average strain, denoted by $\bar{\boldsymbol{\varepsilon}}$, and the average rotation, denoted by $\bar{\boldsymbol{\omega}}$, are, respectively, given in terms of the boundary displacements by

$$\bar{\boldsymbol{\varepsilon}} \equiv < \boldsymbol{\varepsilon} > = \frac{1}{V} \int_{\partial V} \frac{1}{2}(\boldsymbol{\nu} \otimes \mathbf{u}^o + \mathbf{u}^o \otimes \boldsymbol{\nu}) \, dS, \tag{2.2.4a}$$

or

$$\bar{\varepsilon}_{ij} \equiv < \varepsilon_{ij} > = \frac{1}{V} \int_{\partial V} \frac{1}{2}(\nu_i \, u_j^o + u_i^o \, \nu_j) \, dS, \tag{2.2.4b}$$

and

$$\bar{\boldsymbol{\omega}} \equiv < \boldsymbol{\omega} > = \frac{1}{V} \int_{\partial V} \frac{1}{2}(\boldsymbol{\nu} \otimes \mathbf{u}^o - \mathbf{u}^o \otimes \boldsymbol{\nu}) \, dS, \tag{2.2.5a}$$

or

$$\bar{\omega}_{ij} \equiv < \omega_{ij} > = \frac{1}{V} \int_{\partial V} \frac{1}{2}(\nu_i \, u_j^o - u_i^o \, \nu_j) \, dS. \tag{2.2.5b}$$

As mentioned before, the prescribed surface displacements are assumed to be self-compatible in the sense that they do not include a rigid-body translation or rotation of the RVE. Note, however, that the average strain $\bar{\boldsymbol{\varepsilon}}$ defined by (2.2.4), is unchanged even if a rigid-body translation or rotation is added to the surface data. At a typical point $\mathbf{x}$ in the RVE, a rigid translation $\mathbf{u}^r$ and a rigid-body rotation associated with an antisymmetric, constant, infinitesimal rotation tensor $\boldsymbol{\omega}^r$, produce an additional displacement given by $\mathbf{u}^r + \mathbf{x} \cdot \boldsymbol{\omega}^r$. The corresponding additional average displacement gradient then is

$$< \boldsymbol{\nabla} \otimes (\mathbf{u}^r + \mathbf{x} \cdot \boldsymbol{\omega}^r) > = \{\frac{1}{V} \int_{\partial V} \boldsymbol{\nu} \, dS\} \otimes \mathbf{u}^r + \{\frac{1}{V} \int_{\partial V} \boldsymbol{\nu} \otimes \mathbf{x} \, dS\} \cdot \boldsymbol{\omega}^r. \tag{2.2.6a}$$

Making use of the Gauss theorem, it follows that

$$\frac{1}{V} \int_{\partial V} \boldsymbol{\nu} \, dS = \frac{1}{V} \int_{\partial V} \boldsymbol{\nu} \cdot \mathbf{1}^{(2)} \, dS = \frac{1}{V} \int_{V} \boldsymbol{\nabla} \cdot \mathbf{1}^{(2)} \, dV = \mathbf{0},$$

$$\frac{1}{V} \int_{\partial V} \boldsymbol{\nu} \otimes \mathbf{x} \, dS = \frac{1}{V} \int_{V} \boldsymbol{\nabla} \otimes \mathbf{x} \, dV = \frac{1}{V} \int_{V} \mathbf{1}^{(2)} \, dV = \mathbf{1}^{(2)}. \tag{2.2.6b,c}$$

Hence,

$$< \boldsymbol{\nabla} \otimes (\mathbf{u}^r + \mathbf{x} \cdot \boldsymbol{\omega}^r) > = \boldsymbol{\omega}^r \tag{2.2.6d}$$

which does not affect $\boldsymbol{\varepsilon}$. Therefore, whether or not the prescribed surface displacements $\mathbf{u}^o$ include rigid-body translation or rotation, is of no significance in estimating the relations between the average stresses and strains or their increments. For simplicity, however, it will be assumed that the prescribed boundary displacements are self-compatible.

In general, the average displacement, denoted by $\bar{\mathbf{u}}$, cannot be expressed in terms of the surface data. For example, in view of the identity $\mathbf{1}^{(2)} = \boldsymbol{\nabla} \otimes \mathbf{x}$, the displacement field may be written as

$$\mathbf{u} = \mathbf{u} \cdot (\boldsymbol{\nabla} \otimes \mathbf{x}) = \boldsymbol{\nabla} \cdot (\mathbf{u} \otimes \mathbf{x}) - (\boldsymbol{\nabla} \cdot \mathbf{u}) \mathbf{x}. \tag{2.2.7}$$

Therefore, the volume average of the displacement field, $\bar{\mathbf{u}}$, is given by

$$\bar{\mathbf{u}} \equiv < \mathbf{u} > = \frac{1}{V} \int_{\partial V} \boldsymbol{\nu} \cdot (\mathbf{u}^o \otimes \mathbf{x}) \, dS - \frac{1}{V} \int_{V} (\boldsymbol{\nabla} \cdot \mathbf{u}) \mathbf{x} \, dV, \tag{2.2.8a}$$

or

$$\bar{u}_i = < u_i > = \frac{1}{V} \int_{\partial V} \nu_j \, u_j^o \, x_i \, dS - \frac{1}{V} \int_{V} u_{j,j} \, x_i \, dV, \tag{2.2.8b}$$

which includes the volumetric strain. For *incompressible materials*, however, the displacement field is divergence-free,

$$\boldsymbol{\nabla} \cdot \mathbf{u} \equiv 0 \qquad \text{in V}, \tag{2.2.9}$$

and the average displacement, $\bar{\mathbf{u}}$, can be expressed in terms of the prescribed surface displacements, $\mathbf{u}^o$, by

$$\bar{\mathbf{u}} = \frac{1}{V} \int_{\partial V} \boldsymbol{\nu} \cdot (\mathbf{u}^o \otimes \mathbf{x}) \, dS, \tag{2.2.10a}$$

or

$$\bar{u}_i = \frac{1}{V}\int_{\partial V} \nu_j\, u_j^o\, x_i\, dS. \tag{2.2.10b}$$

For the incremental formulation, the velocity field on the boundary ∂V of the RVE is prescribed, (1.3.10). All the above relations hold for the rate fields, if $\mathbf{u}$ is replaced by $\dot{\mathbf{u}}$, $\boldsymbol{\varepsilon}$ by $\dot{\boldsymbol{\varepsilon}}$, and $\boldsymbol{\omega}$ by $\dot{\boldsymbol{\omega}}$. In particular, the average velocity gradient becomes

$$< \boldsymbol{\nabla}\otimes\dot{\mathbf{u}} > = \frac{1}{V}\int_{\partial V} \boldsymbol{\nu}\otimes\dot{\mathbf{u}}^o\, dS, \tag{2.2.11}$$

from which the average strain rate and the average rotation rate are obtained, as follows:

$$\bar{\dot{\boldsymbol{\varepsilon}}} \equiv < \dot{\boldsymbol{\varepsilon}} > = \frac{1}{V}\int_{\partial V} \frac{1}{2}(\boldsymbol{\nu}\otimes\dot{\mathbf{u}}^o + \dot{\mathbf{u}}^o\otimes\boldsymbol{\nu})\, dS,$$

$$\bar{\dot{\boldsymbol{\omega}}} \equiv < \dot{\boldsymbol{\omega}} > = \frac{1}{V}\int_{\partial V} \frac{1}{2}(\boldsymbol{\nu}\otimes\dot{\mathbf{u}}^o - \dot{\mathbf{u}}^o\otimes\boldsymbol{\nu})\, dS, \tag{2.2.12a,b}$$

or

$$\bar{\dot{\varepsilon}}_{ij} \equiv < \dot{\varepsilon}_{ij} > = \frac{1}{V}\int_{\partial V} \frac{1}{2}(\nu_i\, \dot{u}_j^o + \dot{u}_i^o\, \nu_j)\, dS,$$

$$\bar{\dot{\omega}}_{ij} \equiv < \dot{\omega}_{ij} > = \frac{1}{V}\int_{\partial V} \frac{1}{2}(\nu_i\, \dot{u}_j^o - \dot{u}_i^o\, \nu_j)\, dS. \tag{2.2.12c,d}$$

It is seen that, *for the small deformations considered here*, the average strain rate equals the rate of change of the average strain,

$$\bar{\dot{\boldsymbol{\varepsilon}}} \equiv < \dot{\boldsymbol{\varepsilon}} > = \frac{d}{dt} < \boldsymbol{\varepsilon} > \equiv \dot{\bar{\boldsymbol{\varepsilon}}}, \tag{2.2.13a}$$

and, similarly, the average rotation rate equals the rate of change of the average rotation,

$$\bar{\dot{\boldsymbol{\omega}}} \equiv < \dot{\boldsymbol{\omega}} > = \frac{d}{dt} < \boldsymbol{\omega} > \equiv \dot{\bar{\boldsymbol{\omega}}}. \tag{2.2.13b}$$

From (2.2.13a,b), or by direct use of (2.2.2) and (2.2.11), it follows that

$$\overline{\boldsymbol{\nabla}\otimes\dot{\mathbf{u}}} \equiv < \boldsymbol{\nabla}\otimes\dot{\mathbf{u}} > = \frac{d}{dt} < \boldsymbol{\nabla}\otimes\mathbf{u} > \equiv \dot{\overline{\boldsymbol{\nabla}\otimes\mathbf{u}}}. \tag{2.2.13c}$$

2.3. AVERAGE RATE OF STRESS-WORK

Whether or not the deformation is small, or whether or not the effects of inertia and body forces are included, the balance of energy leads to the following local equation for the rate of change of the internal energy density $\dot{e}$:

$$\rho\,\dot{e} + \boldsymbol{\nabla}\boldsymbol{\cdot}\mathbf{q} = \boldsymbol{\sigma} : \dot{\boldsymbol{\varepsilon}} + \rho\, h \qquad \text{in V}, \tag{2.3.1}$$

where ρ is the (current) mass-density, $\mathbf{q}$ is the heat flux vector, and h is the heat supplied through radiation or other energy source fields. In (2.3.1), $\boldsymbol{\sigma} : \dot{\boldsymbol{\varepsilon}}$ is the rate of *stress-work* per unit volume.

For the moment, consider elastic materials, and let $\phi = \phi(\boldsymbol{\varepsilon}, \theta)$ be the Helmholtz free energy per unit volume, where θ is the temperature. The stress, $\boldsymbol{\sigma}$, and entropy, η, are given by[4]

$$\boldsymbol{\sigma} = \frac{\partial \phi}{\partial \boldsymbol{\varepsilon}}, \qquad \eta = -\frac{\partial \phi}{\partial \theta}. \tag{2.3.2a,b}$$

At constant temperature (isothermal change),

$$\dot{\phi} = \boldsymbol{\sigma} : \dot{\boldsymbol{\varepsilon}}, \tag{2.3.2c}$$

so that the rate of stress-work equals the rate of change of the Helmholtz free energy at constant temperature. It is often convenient to introduce the complementary energy function $\psi = \psi(\boldsymbol{\sigma}, \theta)$ such that

$$\phi(\boldsymbol{\varepsilon}, \theta) + \psi(\boldsymbol{\sigma}, \theta) = \boldsymbol{\sigma} : \boldsymbol{\varepsilon}, \tag{2.3.3a}$$

and obtain

$$\boldsymbol{\varepsilon} = \frac{\partial \psi}{\partial \boldsymbol{\sigma}}, \qquad \eta = \frac{\partial \psi}{\partial \theta}. \tag{2.3.3b,c}$$

In general, the material constituents of an RVE need not be elastic. For example, rate-independent and rate-dependent elastic-plastic models may have to be used to describe certain classes of materials. Whatever the specific constitutive properties of the material within an RVE may be, it is of interest to calculate the average rate of stress-work and to explore conditions under which $<\boldsymbol{\sigma} : \dot{\boldsymbol{\varepsilon}}>$ equals $<\boldsymbol{\sigma}> : <\dot{\boldsymbol{\varepsilon}}>$.

To this end it is observed that, in view of (1.3.2a,b),

$$\boldsymbol{\sigma} : \dot{\boldsymbol{\varepsilon}} = \boldsymbol{\sigma} : (\dot{\boldsymbol{\varepsilon}} + \dot{\boldsymbol{\omega}}) = \boldsymbol{\sigma} : (\boldsymbol{\nabla} \otimes \dot{\mathbf{u}}) = \boldsymbol{\nabla} \cdot (\boldsymbol{\sigma} \cdot \dot{\mathbf{u}}) - (\boldsymbol{\nabla} \cdot \boldsymbol{\sigma}) \cdot \dot{\mathbf{u}} = \boldsymbol{\nabla} \cdot (\boldsymbol{\sigma} \cdot \dot{\mathbf{u}}). \tag{2.3.4a}$$

Hence,

$$<\boldsymbol{\sigma} : \dot{\boldsymbol{\varepsilon}}> = \frac{1}{V} \int_{\partial V} \mathbf{t} \cdot \dot{\mathbf{u}} \, dS, \tag{2.3.4b}$$

where $\mathbf{t}$ $(= \boldsymbol{\nu} \cdot \boldsymbol{\sigma})$ are the surface tractions on ∂V. Since $\dot{\boldsymbol{\varepsilon}}$ is unchanged by the addition of rigid-body motions, such motions do not affect $<\boldsymbol{\sigma} : \dot{\boldsymbol{\varepsilon}}>$.

Now the difference between $<\boldsymbol{\sigma} : \dot{\boldsymbol{\varepsilon}}>$ and $<\boldsymbol{\sigma}> : <\dot{\boldsymbol{\varepsilon}}>$ is expressed in terms of the boundary data by (Hill, 1963, 1967; and Mandel, 1980)

$$<\boldsymbol{\sigma} : \dot{\boldsymbol{\varepsilon}}> - <\boldsymbol{\sigma}> : <\dot{\boldsymbol{\varepsilon}}>$$

$$= \frac{1}{V} \int_{\partial V} \{\dot{\mathbf{u}} - \mathbf{x} \cdot <\boldsymbol{\nabla} \otimes \dot{\mathbf{u}}>\} \cdot \{\boldsymbol{\nu} \cdot (\boldsymbol{\sigma} - <\boldsymbol{\sigma}>)\} \, dS. \tag{2.3.5}$$

---

[4] In the present context, $\phi(\boldsymbol{\varepsilon}, \theta)$ is interpreted as $\phi(\varepsilon_{ij}, \theta) \equiv \phi(\varepsilon_{11}, \varepsilon_{22}, \ldots, \theta)$ and $\frac{\partial \phi}{\partial \boldsymbol{\varepsilon}}$ is interpreted as $\frac{\partial \phi}{\partial \varepsilon_{ij}} \mathbf{e}_i \otimes \mathbf{e}_j$. Similar notation is used throughout this book.

The proof is straightforward: set $\mathbf{t} = \boldsymbol{\nu} \cdot \boldsymbol{\sigma}$ on ∂V and compute the integrand in the right-hand side of (2.3.5) to obtain

$$\{\dot{\mathbf{u}} - \mathbf{x} \cdot < \boldsymbol{\nabla} \otimes \dot{\mathbf{u}} >\} \cdot \{\boldsymbol{\nu} \cdot (\boldsymbol{\sigma} - < \boldsymbol{\sigma} >)\}$$

$$= \dot{\mathbf{u}} \cdot \mathbf{t} - \dot{\mathbf{u}} \cdot (\boldsymbol{\nu} \cdot < \boldsymbol{\sigma} >) - (\mathbf{x} \cdot < \boldsymbol{\nabla} \otimes \dot{\mathbf{u}} >) \cdot \mathbf{t} + (\mathbf{x} \cdot < \boldsymbol{\nabla} \otimes \dot{\mathbf{u}} >) \cdot (\boldsymbol{\nu} \cdot < \boldsymbol{\sigma} >)$$

$$= \dot{\mathbf{u}} \cdot \mathbf{t} - (\boldsymbol{\nu} \otimes \dot{\mathbf{u}}) : < \boldsymbol{\sigma} > - (\mathbf{x} \otimes \mathbf{t}) : < \boldsymbol{\nabla} \otimes \dot{\mathbf{u}} >$$

$$+ (\mathbf{x} \otimes \boldsymbol{\nu}) : (< \boldsymbol{\nabla} \otimes \dot{\mathbf{u}} > \cdot < \boldsymbol{\sigma} >^{T}), \tag{2.3.6a}$$

or in component form,

$$\{\dot{u}_i - x_j < \dot{u}_{i,j} >\}\{\nu_k (\sigma_{ki} - < \sigma_{ki} >)\} = \dot{u}_i t_i - \nu_i \dot{u}_j < \sigma_{ij} > - x_i t_j < \dot{u}_{j,i} >$$

$$+ x_i \nu_j < \dot{u}_{k,i} > < \sigma_{jk} >. \tag{2.3.6b}$$

Integrate (2.3.6a) over ∂V to obtain

$$\frac{1}{V}\int_{\partial V} \dot{\mathbf{u}} \cdot \mathbf{t} \, dS - \{\frac{1}{V}\int_{\partial V} \boldsymbol{\nu} \otimes \dot{\mathbf{u}} \, dS\} : < \boldsymbol{\sigma} > - \{\frac{1}{V}\int_{\partial V} \mathbf{x} \otimes \mathbf{t} \, dS\} : < \boldsymbol{\nabla} \otimes \dot{\mathbf{u}} >$$

$$+ \{\frac{1}{V}\int_{\partial V} \mathbf{x} \otimes \boldsymbol{\nu} \, dS\} : (< \boldsymbol{\nabla} \otimes \dot{\mathbf{u}} > \cdot < \boldsymbol{\sigma} >^{T})$$

$$= < \boldsymbol{\sigma} : \dot{\boldsymbol{\varepsilon}} > - < \dot{\boldsymbol{\varepsilon}} > : < \boldsymbol{\sigma} > - < \boldsymbol{\sigma} > : < \dot{\boldsymbol{\varepsilon}} > + \mathbf{1}^{(2)} : (< \boldsymbol{\nabla} \otimes \dot{\mathbf{u}} > \cdot < \boldsymbol{\sigma} >^{T})$$

$$= < \boldsymbol{\sigma} : \dot{\boldsymbol{\varepsilon}} > - < \boldsymbol{\sigma} > : < \dot{\boldsymbol{\varepsilon}} >, \tag{2.3.6c}$$

where the following is used:

$$\mathbf{1}^{(2)} : (< \boldsymbol{\nabla} \otimes \dot{\mathbf{u}} > \cdot < \boldsymbol{\sigma} >^{T}) = < \boldsymbol{\nabla} \otimes \dot{\mathbf{u}} > : < \boldsymbol{\sigma} > = < \dot{\boldsymbol{\varepsilon}} > : < \boldsymbol{\sigma} >. \tag{2.3.6d}$$

Whatever the material properties, and whether or not the prescribed boundary data of the RVE are uniform, identity (2.3.5) is valid in the context of the small-deformation theory. For special boundary data, the right-hand side of (2.3.5) may vanish, and then the average rate of stress-work equals the rate of work of the average stress. Two such boundary data of particular importance in micromechanics are considered below.

2.3.1. Uniform Boundary Tractions

When the prescribed boundary tractions for an RVE are uniform, they can be expressed in terms of a *constant* symmetric second-order tensor, $\boldsymbol{\sigma}^o$, as

$$\mathbf{t}^o = \boldsymbol{\nu} \cdot \boldsymbol{\sigma}^o, \qquad \text{or} \qquad t_i^o = \nu_j \sigma_{ji}^o. \tag{2.3.7a,b}$$

In view of identity (2.2.6c), the average stress from (2.1.6) becomes

$$\bar{\boldsymbol{\sigma}} = < \boldsymbol{\sigma} > = \{\frac{1}{V}\int_{\partial V} \mathbf{x} \otimes \boldsymbol{\nu} \, dS\} \cdot \boldsymbol{\sigma}^o = \boldsymbol{\sigma}^o. \tag{2.3.8}$$

Then $\boldsymbol{\nu}\cdot(\boldsymbol{\sigma} - <\boldsymbol{\sigma}>)$ is zero on ∂V. Substitution into (2.3.5) yields

$$<\boldsymbol{\sigma} : \dot{\boldsymbol{\varepsilon}}> = <\boldsymbol{\sigma}> : <\dot{\boldsymbol{\varepsilon}}> = \bar{\boldsymbol{\sigma}} : \dot{\bar{\boldsymbol{\varepsilon}}} = \boldsymbol{\sigma}^o : \dot{\bar{\boldsymbol{\varepsilon}}}. \tag{2.3.9}$$

2.3.2. Linear Boundary Velocities

When the prescribed velocities on the boundary of an RVE are spatially linear, they can be represented in terms of a *constant* second-order tensor which may be split into a symmetric part, denoted by $\dot{\boldsymbol{\varepsilon}}^o$, and an antisymmetric part, denoted by $\boldsymbol{\omega}^o$. The velocity of a typical point $\mathbf{x}$ on ∂V is given by

$$\dot{\mathbf{u}}^o = \mathbf{x}\cdot(\dot{\boldsymbol{\varepsilon}}^o + \dot{\boldsymbol{\omega}}^o). \tag{2.3.10}$$

In view of identity (2.2.6c), the average velocity gradient becomes

$$<\boldsymbol{\nabla}\otimes\dot{\mathbf{u}}> = \{\frac{1}{V}\int_{\partial V} \boldsymbol{\nu}\otimes\mathbf{x}\, dS\}\cdot(\dot{\boldsymbol{\varepsilon}}^o + \dot{\boldsymbol{\omega}}^o) = \dot{\boldsymbol{\varepsilon}}^o + \dot{\boldsymbol{\omega}}^o. \tag{2.3.11a}$$

Then $\dot{\mathbf{u}} - \mathbf{x}\cdot<\boldsymbol{\nabla}\otimes\dot{\mathbf{u}}>$ is zero on ∂V. Hence,

$$<\dot{\boldsymbol{\varepsilon}}> = \dot{\boldsymbol{\varepsilon}}^o, \qquad <\dot{\boldsymbol{\omega}}> = \dot{\boldsymbol{\omega}}^o, \tag{2.3.11b,c}$$

and substitution into (2.3.5) yields

$$<\boldsymbol{\sigma} : \dot{\boldsymbol{\varepsilon}}> = <\boldsymbol{\sigma}> : <\dot{\boldsymbol{\varepsilon}}> = \bar{\boldsymbol{\sigma}} : \dot{\bar{\boldsymbol{\varepsilon}}} = \bar{\boldsymbol{\sigma}} : \dot{\boldsymbol{\varepsilon}}^o. \tag{2.3.12}$$

2.3.3. Other Useful Identities

The identity (2.3.5) remains valid if $\boldsymbol{\sigma}$ is replaced by $\dot{\boldsymbol{\sigma}}$, and $\dot{\boldsymbol{\varepsilon}}$ is replaced by $\boldsymbol{\varepsilon}$, provided that the variable stress rate, $\dot{\boldsymbol{\sigma}}$, is self-equilibrating; $\boldsymbol{\nabla}\cdot\dot{\boldsymbol{\sigma}} = \mathbf{0}$ and $\dot{\boldsymbol{\sigma}}^T = \dot{\boldsymbol{\sigma}}$ in V. Then

$$<\dot{\boldsymbol{\sigma}} : \boldsymbol{\varepsilon}> - <\dot{\boldsymbol{\sigma}}> : <\boldsymbol{\varepsilon}>$$

$$= \frac{1}{V}\int_{\partial V} \{\mathbf{u} - \mathbf{x}\cdot<\boldsymbol{\nabla}\otimes\mathbf{u}>\}\cdot\{\boldsymbol{\nu}\cdot(\dot{\boldsymbol{\sigma}} - <\dot{\boldsymbol{\sigma}}>)\}\, dS, \tag{2.3.13}$$

as can be verified by direct computation, in line with (2.3.6). Therefore, when the boundary conditions on ∂V are given by the *uniform* traction rates, i.e., $\dot{\mathbf{t}}^o = \boldsymbol{\nu}\cdot\dot{\boldsymbol{\sigma}}^o$ (with constant $\dot{\boldsymbol{\sigma}}^o$), or by the *linear* displacements, i.e., $\mathbf{u} = \mathbf{x}\cdot\boldsymbol{\varepsilon}^o$ (with constant $\boldsymbol{\varepsilon}^o$), then

$$<\dot{\boldsymbol{\sigma}} : \boldsymbol{\varepsilon}> = <\dot{\boldsymbol{\sigma}}> : <\boldsymbol{\varepsilon}>. \tag{2.3.14}$$

Similarly, in terms of $\boldsymbol{\sigma}$ and $\boldsymbol{\varepsilon}$,

$$<\boldsymbol{\sigma} : \boldsymbol{\varepsilon}> - <\boldsymbol{\sigma}> : <\boldsymbol{\varepsilon}>$$

$$= \frac{1}{V}\int_{\partial V} \{\mathbf{u} - \mathbf{x}\cdot<\boldsymbol{\nabla}\otimes\mathbf{u}>\}\cdot\{\boldsymbol{\nu}\cdot(\boldsymbol{\sigma} - <\boldsymbol{\sigma}>)\}\, dS. \tag{2.3.15}$$

Hence, for either *uniform* boundary tractions or *linear* boundary displacements,

$$< \boldsymbol{\sigma} : \boldsymbol{\varepsilon} > = < \boldsymbol{\sigma} > : < \boldsymbol{\varepsilon} >. \tag{2.3.16}$$

It is important to note that identity (2.3.5) is valid whether or not the self-equilibrating stress field $\boldsymbol{\sigma}$ is related to the self-compatible strain-rate field $\dot{\boldsymbol{\varepsilon}}$. Similar comments apply to $\dot{\boldsymbol{\sigma}}$ and $\boldsymbol{\varepsilon}$, as well as to $\boldsymbol{\sigma}$ and $\boldsymbol{\varepsilon}$; (2.3.13) and (2.3.15). It is also important to note that all results in Subsections 2.1, 2.2, and 2.3, except for expressions (2.3.2) and (2.3.3), are valid for materials of any constitutive properties, since only equilibrium and compatibility are required. With proper interpretation, most of these results also apply to finite deformations; Hill (1972), Havner (1982), and Nemat-Nasser (1983).

2.3.4. Virtual Work Principle

A stress field, $\boldsymbol{\sigma} = \boldsymbol{\sigma}(\mathbf{x})$, which satisfies equilibrium conditions (1.3.2a,b) in V, and the stress-boundary conditions (1.3.5) on any part of ∂V where the tractions are prescribed, is called *statically admissible*. A displacement field $\mathbf{u} = \mathbf{u}(\mathbf{x})$ which is suitably smooth, so that it yields a suitable strain field through (1.3.3), and satisfies all prescribed displacement boundary conditions, is called *kinematically admissible*. If $\mathbf{u} = \mathbf{u}(\mathbf{x})$ is kinematically admissible, then any variation $\delta\mathbf{u} = \delta\mathbf{u}(\mathbf{x})$ in this field, which produces a smooth kinematically admissible displacement field $\mathbf{u} + \delta\mathbf{u}$ must be such that $\delta\mathbf{u} = \mathbf{0}$ on any part of ∂V where $\mathbf{u}$ is prescribed. Setting $\delta\boldsymbol{\varepsilon} = \{\boldsymbol{\nabla} \otimes \delta\mathbf{u} + (\boldsymbol{\nabla} \otimes \delta\mathbf{u})^{\mathrm{T}}\}/2$, and using an analysis similar to (2.3.4), it follows that

$$< \boldsymbol{\sigma} : \delta\boldsymbol{\varepsilon} > - \frac{1}{V}\int_{\partial V} \mathbf{t}^{\mathrm{o}} \boldsymbol{\cdot} \delta\mathbf{u}\, dS = 0, \tag{2.3.17}$$

where $\mathbf{t}^{\mathrm{o}} = \boldsymbol{\nu} \boldsymbol{\cdot} \boldsymbol{\sigma}$ denotes the boundary tractions. This is the statement of the virtual work principle, valid for any statically admissible stress field $\boldsymbol{\sigma}$ and any unrelated or related virtual or real kinematically admissible variation $\delta\mathbf{u}$ of the displacement field; see Subsection 19.2 of Part 2.

2.4. INTERFACES AND DISCONTINUITIES

In general, the overall properties of an RVE are strongly affected by the structure, chemical composition, strength, and other relevant attributes of the interfaces among its individual microconstituents. For example, both the strength and toughness of fiber-reinforced ceramics are directly related to the nature of the interface bonding between the fiber and the matrix. Extensive debonding often deflects cracks, consumes mechanical energy, and leads to greater toughness. Such debonding can be modeled as displacement discontinuities whose effects must be included in the overall response of the RVE. Similarly, intergranular and transgranular microcracks can be treated as displacement discontinuities, and must be included in estimating the overall deformation and its increments. In this section the effects of the discontinuities within an RVE on the overall quantities are examined, without reference to the specific physical nature of such discontinuities.

Let S be the collection of all surfaces within an RVE, across which certain field quantities may suffer jump discontinuities; see Figure 2.4.1. S includes three types of surfaces: (1) a closed surface, totally within the volume V of the RVE, which separates materials of the RVE into those inside of S and those outside of S; (2) an isolated surface bounded by a curve ∂S, totally within the volume V; and (3) the discontinuity surface S which intersects the boundary ∂V of the RVE. In all three cases, S is regarded piecewise continuous, with piecewise continuously turning tangent planes. Examples are: (1) a debonded inclusion within V, across which tangential displacements may be discontinuous; (2) an interior penny-shaped crack, across which the tangential component of the displacement and also the normal component (only for the opening mode) may be discontinuous; and (3) cracks intersecting the boundary ∂V. A crack may be viewed as a cavity with one dimension which is infinitesimally small. For such a cavity, one can consider an inside and an outside part. In this manner, the volume V is divided by the collection of all discontinuity surfaces, S, into two parts, V^- and V^+. Let the unit normal $\mathbf{n}$ point from V^- toward V^+, and for simplicity, set

$$\partial V^- = S^- \equiv S, \qquad \partial V^+ = \partial V + S^+, \tag{2.4.1}$$

where $\mathbf{n}$ now is the exterior unit normal on S^-, the exterior unit normal on S^+ being $-\mathbf{n}$. Whether V^+ and V^- are simply or multiply connected regions, the discontinuity surface S can be defined in the above manner. Denote the jump in a typical field quantity, $\mathbf{T}$, across S by $\Delta\mathbf{T}$. The stress, displacement, and strain jumps at a typical point $\boldsymbol{\xi}$ on S are defined, respectively, by

$$\Delta\boldsymbol{\sigma}(\boldsymbol{\xi}) \equiv \lim_{\mathbf{x}^+\to\boldsymbol{\xi}} \boldsymbol{\sigma}(\mathbf{x}^+) - \lim_{\mathbf{x}^-\to\boldsymbol{\xi}} \boldsymbol{\sigma}(\mathbf{x}^-),$$

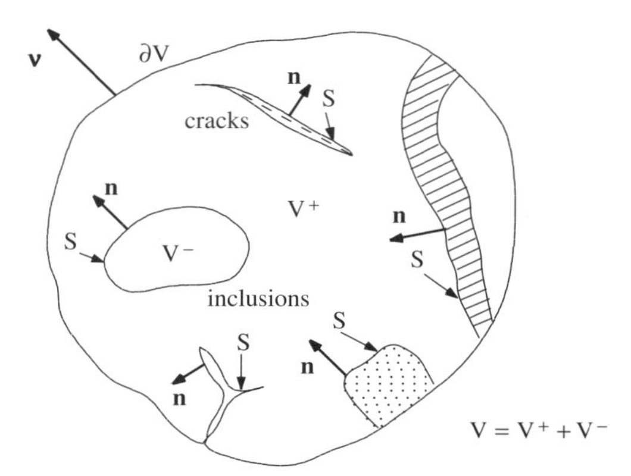

Figure 2.4.1

Discontinuity surfaces

$$\Delta \mathbf{u}(\boldsymbol{\xi}) \equiv \lim_{\mathbf{x}^+ \to \boldsymbol{\xi}} \mathbf{u}(\mathbf{x}^+) - \lim_{\mathbf{x}^- \to \boldsymbol{\xi}} \mathbf{u}(\mathbf{x}^-),$$

$$\Delta \boldsymbol{\varepsilon}(\boldsymbol{\xi}) \equiv \lim_{\mathbf{x}^+ \to \boldsymbol{\xi}} \boldsymbol{\varepsilon}(\mathbf{x}^+) - \lim_{\mathbf{x}^- \to \boldsymbol{\xi}} \boldsymbol{\varepsilon}(\mathbf{x}^-), \tag{2.4.2a~c}$$

where $\mathbf{x}^+$ and $\mathbf{x}^-$ are points in V^+ and V^-, respectively.

In general, if the material properties change abruptly across S, jumps $\Delta\boldsymbol{\sigma}$ and $\Delta\boldsymbol{\varepsilon}$ in the stress and strain fields occur to compensate for the material mismatch across the interface. However, even when $\Delta\boldsymbol{\sigma}$ is nonzero, the jump in the tractions, defined by

$$\Delta \mathbf{t}(\boldsymbol{\xi}) \equiv \mathbf{n} \cdot \Delta \boldsymbol{\sigma}(\boldsymbol{\xi}), \tag{2.4.2d}$$

must vanish to ensure equilibrium. On the other hand, the displacement jump $\Delta\mathbf{u}$ may not vanish when the bonding across the interface is imperfect. The jump $\Delta\mathbf{u}$ can be decomposed into an *opening gap*, $\Delta\mathbf{u}_n$, and a *sliding gap*, $\Delta\mathbf{u}_s$, as follows:

$$\Delta \mathbf{u}_n \equiv (\mathbf{n} \cdot \Delta \mathbf{u})\, \mathbf{n}, \qquad \Delta \mathbf{u}_s \equiv \Delta \mathbf{u} - \Delta \mathbf{u}_n, \tag{2.4.3a,b}$$

where only nonnegative $\mathbf{n} \cdot \Delta\mathbf{u}$ is admitted, since the interpenetration of microconstituents must be excluded on physical grounds.

The stress field $\boldsymbol{\sigma}$ is regarded continuous and smooth in V^+ and V^-. The average stress, $\bar{\boldsymbol{\sigma}}$, may be calculated by applying the Gauss theorem to V^+ and V^- separately, arriving at

$$\begin{aligned}
\bar{\boldsymbol{\sigma}} &= \frac{1}{V}\{\int_{V^+} \boldsymbol{\sigma}\, dV + \int_{V^-} \boldsymbol{\sigma}\, dV\} \\
&= \frac{1}{V}\{\int_{\partial V^+} \mathbf{x}^+ \otimes \mathbf{t}\, dS + \int_{\partial V^-} \mathbf{x}^- \otimes \mathbf{t}\, dS\} \\
&= \frac{1}{V}\{\int_{\partial V} \mathbf{x} \otimes \mathbf{t}\, dS - \int_S \boldsymbol{\xi} \otimes \Delta \mathbf{t}\, dS\},
\end{aligned} \tag{2.4.4}$$

where (2.4.1) is used; the integral on S, in general, vanishes since the tractions are continuous there. In a similar manner, the average strain, $\bar{\boldsymbol{\varepsilon}}$, can be calculated as

$$\begin{aligned}
\bar{\boldsymbol{\varepsilon}} &= \frac{1}{V}\{\int_{V^+} \boldsymbol{\varepsilon}\, dV + \int_{V^-} \boldsymbol{\varepsilon}\, dV\} \\
&= \frac{1}{V}\{\int_{\partial V^+} \frac{1}{2}(\mathbf{v} \otimes \mathbf{u} + \mathbf{u} \otimes \mathbf{v})\, dS + \int_{\partial V^-} \frac{1}{2}(\mathbf{v} \otimes \mathbf{u} + \mathbf{u} \otimes \mathbf{v})\, dS\} \\
&= \frac{1}{V}\{\int_{\partial V} \frac{1}{2}(\mathbf{v} \otimes \mathbf{u} + \mathbf{u} \otimes \mathbf{v})\, dS - \int_S \frac{1}{2}(\mathbf{n} \otimes \Delta \mathbf{u} + \Delta \mathbf{u} \otimes \mathbf{n})\, dS\}.
\end{aligned} \tag{2.4.5}$$

It should be noted that if a microconstituent translates or rotates as a rigid body, the corresponding displacement jump can be expressed in terms of constant $\mathbf{u}^r$ and constant $\boldsymbol{\omega}^r$, as

$$\Delta \mathbf{u} = \mathbf{u}^r + \boldsymbol{\xi} \cdot \boldsymbol{\omega}^r \qquad \text{on S.} \tag{2.4.6}$$

In this case, the integral on S vanishes, and (2.4.5) reduces to (2.2.4).

In a similar manner, the effect of the discontinuity on the average rate of stress-work is examined. For continuous tractions across the discontinuity surface S, it follows that

$$<\boldsymbol{\sigma}:\dot{\boldsymbol{\varepsilon}}> = \frac{1}{V}\{\int_{V^+} \boldsymbol{\sigma}:\dot{\boldsymbol{\varepsilon}}\, dV + \int_{V^-} \boldsymbol{\sigma}:\dot{\boldsymbol{\varepsilon}}\, dV\}$$

$$= \frac{1}{V}\{\int_{\partial V^+} \mathbf{t}\cdot\dot{\mathbf{u}}\, dS + \int_{\partial V^-} \mathbf{t}\cdot\dot{\mathbf{u}}\, dS\}$$

$$= \frac{1}{V}\{\int_{\partial V} \mathbf{t}\cdot\dot{\mathbf{u}}\, dS - \int_{S} \mathbf{t}\cdot\Delta\dot{\mathbf{u}}\, dS\}, \tag{2.4.7}$$

where $\Delta\dot{\mathbf{u}}$ is the velocity jump across S, defined by

$$\Delta\dot{\mathbf{u}}(\boldsymbol{\xi}) \equiv \lim_{\mathbf{x}^+\to\boldsymbol{\xi}} \dot{\mathbf{u}}(\mathbf{x}^+) - \lim_{\mathbf{x}^-\to\boldsymbol{\xi}} \dot{\mathbf{u}}(\mathbf{x}^-). \tag{2.4.2e}$$

The integral on S in (2.4.7) is the rate of work of the interface tractions due to the relative interfacial motion. In particular, if the velocity components are continuous across S, (2.4.7) reduces to (2.3.4b). Again, if the velocity jump $\dot{\mathbf{u}}$ is given by

$$\Delta\dot{\mathbf{u}} = \dot{\mathbf{u}}^r + \boldsymbol{\xi}\cdot\dot{\boldsymbol{\omega}}^r \quad \text{on S}, \tag{2.4.8}$$

with constant $\dot{\mathbf{u}}^r$ and $\dot{\boldsymbol{\omega}}^r$, then the integral of $\mathbf{t}\cdot\Delta\dot{\mathbf{u}}$ on S vanishes. It is seen that the discontinuity in the field quantities does not influence the average stress, strain, strain rate, and the rate of stress-work, if the tractions, the displacements, and the velocity fields are continuous in V.

2.5. POTENTIAL FUNCTION FOR MACRO-ELEMENTS

As pointed out before, an RVE represents the microstructure of a macro-element in a continuum mass. The stress and strain fields and their rates are, in general, functions of the position of the macro-elements within the continuum. Denote the position of a typical continuum macro-element by $\mathbf{X}$, and the stress and strain fields of the continuum by $\boldsymbol{\Sigma}$ and $\mathbf{E}$, respectively. These fields, in general, are functions of $\mathbf{X}$ and time t, $\boldsymbol{\Sigma} = \boldsymbol{\Sigma}(\mathbf{X}, t)$ and $\mathbf{E} = \mathbf{E}(\mathbf{X}, t)$. To distinguish these fields from the stress and strain fields within an RVE which represents the microstructure of a typical continuum material neighborhood, the continuum stress and strain fields are referred to as *macrostress* and *macrostrain* fields, and those of an RVE as *microstress* and *microstrain* fields, respectively. That is, instead of, for example, "the stress field within the continuum" or "the stress field within the RVE which corresponds to the material neighborhood of particle $\mathbf{X}$", the expressions *macrostress* field and *microstress* field are used. In a similar manner, the continuum displacement, mass-density, temperature, and other physical quantities are identified by an appropriate use of the prefix "macro", and those of an RVE by the prefix "micro".

The macrofields must satisfy the continuum balance equations summarized in Part 2 of this book. In particular, the equations of motion are

$$\boldsymbol{\nabla} \cdot \boldsymbol{\Sigma} + \mathbf{F} = \mathrm{R}\,\ddot{\mathbf{U}}, \tag{2.5.1}$$

where $\boldsymbol{\nabla} \equiv \mathbf{e}_i\,\partial/\partial X_i$, and $\mathbf{F}$, R, and $\mathbf{U} = \mathbf{U}(\mathbf{X})$ are the macroscopic body force, mass density, and macrodisplacement fields, respectively, and superposed dot denotes time differentiation. Moreover, the macrostrain-macrodisplacement relation is

$$\mathbf{E} = \frac{1}{2}\{\boldsymbol{\nabla} \otimes \mathbf{U} + (\boldsymbol{\nabla} \otimes \mathbf{U})^{\mathrm{T}}\}. \tag{2.5.2}$$

In general, at a typical point $\mathbf{X}$ in the continuum, at a fixed instant t, the values of the macrostress and macrostrain tensors, $\boldsymbol{\Sigma}$ and $\mathbf{E}$, can be determined by the average microstress and microstrain, $\bar{\boldsymbol{\sigma}}$ and $\bar{\boldsymbol{\varepsilon}}$, over the RVE which represents the corresponding macro-element. In micromechanics it is assumed that $\boldsymbol{\Sigma}$ and $\mathbf{E}$ are *equal* to $\bar{\boldsymbol{\sigma}}$ and $\bar{\boldsymbol{\varepsilon}}$,

$$\boldsymbol{\Sigma} = \bar{\boldsymbol{\sigma}}, \qquad \mathbf{E} = \bar{\boldsymbol{\varepsilon}}. \tag{2.5.3a,b}$$

Conversely, the macrostress and macrostrain tensors, $\boldsymbol{\Sigma}$ and $\mathbf{E}$, provide the uniform traction or linear displacement boundary data for the RVE. Hence, when the traction boundary data for the RVE are prescribed,

$$\mathbf{t}^{\mathrm{o}} = \boldsymbol{\nu} \cdot \boldsymbol{\Sigma} \qquad \text{on } \partial V, \tag{2.5.4a}$$

and when the displacements are assumed to be prescribed on ∂V of the RVE,

$$\mathbf{u}^{\mathrm{o}} = \mathbf{x} \cdot \mathbf{E} \qquad \text{on } \partial V. \tag{2.5.4b}$$

Furthermore, when thermal effects are also of interest, the value of the *macrotemperature*, Θ, at the considered macro-element must equal the average *microtemperature* $\bar{\theta}$ over the RVE,

$$\Theta = \bar{\theta} \equiv \langle \theta \rangle. \tag{2.5.5}$$

In general, the response of the macro-element characterized by, for example, relations among macrostress $\boldsymbol{\Sigma}$, macrostrain $\mathbf{E}$, and macrotemperature Θ, will be inelastic and history-dependent, even if the microconstituents of the corresponding RVE are elastic. This is because, in the course of deformation, flaws, microcracks, cavities, and other microdefects develop within the RVE, and the microstructure of the RVE changes with changes of the overall applied loads. Therefore, the stress-strain relations for the macro-elements must, in general, include additional parameters which describe the current microstructure of the corresponding RVE. This section focuses on a broad class of materials whose microconstituents are elastic (linear or nonlinear) and, therefore, the inelastic response of their macro-elements stems from the generation and evolution of defects and hence from microstructural changes.

For a typical macro-element, denote the current state of its microstructure, collectively, by $\mathbf{S}$, which may stand for a set of parameters, scalar or possibly tensorial, that completely defines the microstructure. For example, if the microdefects are penny-shaped cracks, $\mathbf{S}$ will stand for the sizes, orientations, and distribution of these cracks. The matrix material is elastic, and the inelasticity is produced by the growth of the cracks; see, e.g., Schapery (1995) and Krajcinovic (1996). *If there is no change in the microstructure, e.g., no crack growth, the response of the macro-element will be elastic.* Hence, a Helmholtz free energy,

$$\Phi = \Phi(\mathbf{E}, \Theta; \mathbf{S}), \tag{2.5.6a}$$

exists, which at *constant* **S**, yields

$$\boldsymbol{\Sigma} = \frac{\partial \Phi}{\partial \mathbf{E}}, \qquad \mathrm{H} = -\frac{\partial \Phi}{\partial \Theta}, \tag{2.5.6b,c}$$

where H is the *macro-entropy*. Then a *macrostrain potential,*

$$\Psi = \Psi(\boldsymbol{\Sigma}, \Theta; \mathbf{S}), \tag{2.5.7a}$$

is introduced through the Legendre transformation

$$\Phi + \Psi = \boldsymbol{\Sigma} : \mathbf{E}, \tag{2.5.7b}$$

with the result that, at *constant* **S**,

$$\mathbf{E} = \frac{\partial \Psi}{\partial \boldsymbol{\Sigma}}, \qquad \mathrm{H} = \frac{\partial \Psi}{\partial \Theta}. \tag{2.5.7c,d}$$

The aim is to express the macropotential functions Φ and Ψ in terms of the volume averages of the microstress and microstrain potentials of the microconstituents.

Since the material within the RVE is assumed to be elastic, it admits a stress potential, $\phi = \phi(\mathbf{x}, \boldsymbol{\varepsilon}, \theta)$, and a strain potential, $\psi = \psi(\mathbf{x}, \boldsymbol{\sigma}, \theta)$, such that (2.3.2) and (2.3.3) hold. Consider the cases of the prescribed boundary tractions and the prescribed boundary displacements for the RVE separately, as follows, assuming a uniform constant temperature and a fixed microstructure for the RVE; hence, the dependence on θ and **S** will *not* be displayed explicitly.

2.5.1. Stress Potential

For prescribed *constant* macrostrain **E**, the *variable* microstrain and microstress fields in the RVE are

$$\boldsymbol{\varepsilon} = \boldsymbol{\varepsilon}(\mathbf{x}; \mathbf{E}), \qquad \boldsymbol{\sigma} = \boldsymbol{\sigma}(\mathbf{x}; \mathbf{E}), \tag{2.5.8a,b}$$

where the argument **E** emphasizes that the *displacement* boundary data are prescribed through the macrostrain **E**. Hence $\mathbf{E} = < \boldsymbol{\varepsilon}(\mathbf{x}; \mathbf{E}) >$. The corresponding microstress potential then becomes

$$\phi = \phi(\mathbf{x}, \boldsymbol{\varepsilon}(\mathbf{x}; \mathbf{E})) = \phi^{\mathrm{E}}(\mathbf{x}; \mathbf{E}); \tag{2.5.9}$$

the superscript E on ϕ emphasizes the fact that the microstress potential is associated with the prescribed macrostrain **E**.

Consider now an infinitesimally small variation $\delta\mathbf{E}$ in the macrostrain, which produces a variation in the microstrain field given by

$$\delta \boldsymbol{\varepsilon}(\mathbf{x};\, \mathbf{E}) = \delta \mathrm{E}_{ij}\, \frac{\partial \boldsymbol{\varepsilon}}{\partial \mathrm{E}_{ij}}(\mathbf{x};\, \mathbf{E}). \tag{2.5.10a}$$

Then,

$$\begin{aligned} <\boldsymbol{\sigma} : \delta\boldsymbol{\varepsilon}> &= <(\frac{\partial \phi}{\partial \boldsymbol{\varepsilon}}(\mathbf{x},\, \boldsymbol{\varepsilon})) : (\delta \mathrm{E}_{ij}\, \frac{\partial \boldsymbol{\varepsilon}}{\partial \mathrm{E}_{ij}}(\mathbf{x};\, \mathbf{E}))> \\ &= <\delta\mathbf{E} : (\frac{\partial \phi^{\mathrm{E}}}{\partial \mathbf{E}}(\mathbf{x};\, \mathbf{E}))> = (\frac{\partial}{\partial \mathbf{E}} <\phi^{\mathrm{E}}>) : \delta\mathbf{E}. \end{aligned} \tag{2.5.10b}$$

It now follows that

$$<\boldsymbol{\sigma}(\mathbf{x};\, \mathbf{E})> = \frac{\partial}{\partial \mathbf{E}} <\phi^{\mathrm{E}}>. \tag{2.5.11}$$

Therefore *define* the macrostress potential by

$$\boldsymbol{\Phi}^{\mathrm{E}} = \boldsymbol{\Phi}^{\mathrm{E}}(\mathbf{E}) \equiv <\phi^{\mathrm{E}}> \equiv \frac{1}{\mathrm{V}} \int_{\mathrm{V}} \phi^{\mathrm{E}}(\mathbf{x};\, \mathbf{E})\, \mathrm{dV}, \tag{2.5.12a}$$

the corresponding macrostress (as before) by

$$\boldsymbol{\Sigma}^{\mathrm{E}} \equiv <\boldsymbol{\sigma}(\mathbf{x};\, \mathbf{E})>, \tag{2.5.12b}$$

and conclude that (Hutchinson, 1987)

$$\boldsymbol{\Sigma}^{\mathrm{E}} = \frac{\partial \boldsymbol{\Phi}^{\mathrm{E}}}{\partial \mathbf{E}}, \tag{2.5.12c}$$

where the superscript E on $\boldsymbol{\Sigma}$ emphasizes that $\boldsymbol{\Sigma}^{\mathrm{E}}$ is *the average stress produced by the constant macrostrain* $\mathbf{E}$. Note that the integral in (2.5.12a) depends on the current microstructure and hence on $\mathbf{S}$. For example, $\phi^{\mathrm{E}} = 0$ in cavities and cracks. As cavities and cracks grow, local strains (microstrains) change. Hence, $\boldsymbol{\Phi}^{\mathrm{E}} = <\phi^{\mathrm{E}}>$ changes. This is expressed by writing

$$\boldsymbol{\Phi}^{\mathrm{E}} = \boldsymbol{\Phi}^{\mathrm{E}}(\mathbf{E},\, \Theta;\, \mathbf{S}) \tag{2.5.12d}$$

which also includes the macrotemperature.

2.5.2. Strain Potential

With macrotemperature Θ and microstructure $\mathbf{S}$ fixed, let the RVE be subjected to uniform boundary tractions defined through a constant macrostress $\boldsymbol{\Sigma}$. The microstrain and microstress fields may be expressed as

$$\boldsymbol{\varepsilon} = \boldsymbol{\varepsilon}(\mathbf{x};\, \boldsymbol{\Sigma}), \qquad \boldsymbol{\sigma} = \boldsymbol{\sigma}(\mathbf{x};\, \boldsymbol{\Sigma}), \tag{2.5.13a,b}$$

where the argument $\boldsymbol{\Sigma}$ emphasizes the fact that a traction boundary-value problem with constant macrostress $\boldsymbol{\Sigma}$ is being considered. The microstrain potential then becomes

$$\psi = \psi(\mathbf{x},\, \boldsymbol{\sigma}(\mathbf{x};\, \boldsymbol{\Sigma})) = \psi^{\Sigma}(\mathbf{x};\, \boldsymbol{\Sigma}). \tag{2.5.14}$$

For an arbitrary change $\delta\boldsymbol{\Sigma}$ in the macrostress,

$$\delta\boldsymbol{\sigma}(\mathbf{x};\,\boldsymbol{\Sigma}) = \delta\Sigma_{ij}\,\frac{\partial\boldsymbol{\sigma}}{\partial\Sigma_{ij}}(\mathbf{x};\,\boldsymbol{\Sigma}), \tag{2.5.15a}$$

and, hence,

$$\begin{aligned} <\delta\boldsymbol{\sigma} : \boldsymbol{\varepsilon}> &= <(\delta\Sigma_{ij}\,\frac{\partial\boldsymbol{\sigma}}{\partial\Sigma_{ij}}(\mathbf{x};\,\boldsymbol{\Sigma})) : (\frac{\partial\psi}{\partial\boldsymbol{\sigma}}(\mathbf{x},\,\boldsymbol{\sigma}))> \\ &= <\delta\boldsymbol{\Sigma} : (\frac{\partial\psi^{\Sigma}}{\partial\boldsymbol{\Sigma}}(\mathbf{x};\,\boldsymbol{\Sigma}))> \\ &= (\frac{\partial}{\partial\boldsymbol{\Sigma}} <\psi^{\Sigma}>) : \delta\boldsymbol{\Sigma}. \end{aligned} \tag{2.5.15b}$$

Thus, it follows that

$$<\boldsymbol{\varepsilon}(\mathbf{x};\,\boldsymbol{\Sigma})> = \frac{\partial}{\partial\boldsymbol{\Sigma}} <\psi^{\Sigma}>. \tag{2.5.16}$$

Therefore *define* the macrostrain potential by

$$\Psi^{\Sigma} = \Psi^{\Sigma}(\boldsymbol{\Sigma}) \equiv <\psi^{\Sigma}> \equiv \frac{1}{V}\int_V \psi^{\Sigma}(\mathbf{x};\,\boldsymbol{\Sigma})\,dV, \tag{2.5.17a}$$

the corresponding macrostrain by

$$\mathbf{E}^{\Sigma} \equiv <\boldsymbol{\varepsilon}(\mathbf{x};\,\boldsymbol{\Sigma})>, \tag{2.5.17b}$$

and obtain (Nemat-Nasser and Hori, 1990)

$$\mathbf{E}^{\Sigma} = \frac{\partial\Psi^{\Sigma}}{\partial\boldsymbol{\Sigma}}, \tag{2.5.17c}$$

where the superscript Σ on $\mathbf{E}$ emphasizes that $\mathbf{E}^{\Sigma}$ is *the average strain produced by the prescribed macrostress* $\boldsymbol{\Sigma}$. Like the macrostress potential $\Phi^E = \Phi^E(\mathbf{E},\,\Theta;\,\mathbf{S})$, the macrostrain potential Ψ^{Σ} is also a function of the current macrotemperature Θ and microstructure $\mathbf{S}$. This is expressed by

$$\Psi^{\Sigma} = \Psi^{\Sigma}(\boldsymbol{\Sigma},\,\Theta;\,\mathbf{S}). \tag{2.5.17d}$$

2.5.3. Relation between Macropotentials

In the preceding subsection, the macrostrain potential is defined as the volume average of the microstrain potential $\psi^{\Sigma}(\mathbf{x};\,\boldsymbol{\Sigma})$, when the macrostress $\boldsymbol{\Sigma}$ is prescribed. Then, with the corresponding macrostrain defined by

$$\mathbf{E}^{\Sigma} = \mathbf{E}^{\Sigma}(\boldsymbol{\Sigma}) \equiv <\boldsymbol{\varepsilon}(\mathbf{x};\,\boldsymbol{\Sigma})>, \tag{2.5.18a}$$

it is concluded that

$$\Psi^{\Sigma} = \Psi^{\Sigma}(\boldsymbol{\Sigma}) \equiv <\psi^{\Sigma}(\mathbf{x};\,\boldsymbol{\Sigma})>, \qquad \mathbf{E}^{\Sigma} = \frac{\partial\Psi^{\Sigma}}{\partial\boldsymbol{\Sigma}}(\boldsymbol{\Sigma}). \tag{2.5.18b,c}$$

The notation in (2.5.18a) shows that $\mathbf{E}^{\Sigma}$ is the macrostrain produced by the *prescribed* macrostress $\boldsymbol{\Sigma}$. Define now a *new* macrostress potential function

$$\mathbf{\Phi}^{\Sigma} = \mathbf{\Phi}^{\Sigma}(\mathbf{E}^{\Sigma}) \equiv \mathbf{\Sigma} : \mathbf{E}^{\Sigma} - \Psi^{\Sigma}(\mathbf{\Sigma}), \tag{2.5.19a}$$

where, as usual, $\mathbf{\Sigma}$ is regarded as a function of $\mathbf{E}^{\Sigma}$ through (2.5.18a).

At the local level, on the other hand,

$$\phi^{\Sigma} = \phi(\mathbf{x}, \boldsymbol{\varepsilon}(\mathbf{x}; \mathbf{\Sigma})) = \phi^{\Sigma}(\mathbf{x}; \mathbf{\Sigma}),$$

$$\psi^{\Sigma} = \psi(\mathbf{x}, \boldsymbol{\sigma}(\mathbf{x}; \mathbf{\Sigma})) = \psi^{\Sigma}(\mathbf{x}; \mathbf{\Sigma}), \tag{2.5.20a,b}$$

and hence

$$\phi^{\Sigma} + \psi^{\Sigma} = \boldsymbol{\sigma}(\mathbf{x}; \mathbf{\Sigma}) : \boldsymbol{\varepsilon}(\mathbf{x}; \mathbf{\Sigma}). \tag{2.5.20c}$$

The volume average over V yields

$$< \phi^{\Sigma} > + < \psi^{\Sigma} > = \mathbf{\Sigma} : \mathbf{E}^{\Sigma}, \tag{2.5.21a}$$

and comparison with (2.5.19a) shows that

$$\Phi^{\Sigma} = < \phi^{\Sigma} >, \qquad \mathrm{E}^{\Sigma} = \frac{\partial \Psi^{\Sigma}}{\partial \mathbf{\Sigma}}(\mathbf{\Sigma}) \Longleftrightarrow \mathbf{\Sigma} = \frac{\partial \Phi^{\Sigma}}{\partial \mathbf{E}^{\Sigma}}(\mathbf{E}^{\Sigma}). \tag{2.5.21b,c}$$

In all these expressions, the superscript Σ shows that the corresponding quantity is obtained for the prescribed macrostress $\mathbf{\Sigma}$.

In a similar manner, when the macrostrain $\mathbf{E}$ is prescribed through linear boundary displacements, $\mathbf{u} = \mathbf{x} \cdot \mathbf{E}$ on ∂V,

$$\mathbf{\Sigma}^{\mathrm{E}} = < \boldsymbol{\sigma}(\mathbf{x}; \mathbf{E}) >,$$

$$\mathbf{\Phi}^{\mathrm{E}} = \mathbf{\Phi}^{\mathrm{E}}(\mathbf{E}) \equiv < \phi^{\mathrm{E}}(\mathbf{x}; \mathbf{E}) >, \quad \mathbf{\Sigma}^{\mathrm{E}} = \frac{\partial \mathbf{\Phi}^{\mathrm{E}}}{\partial \mathbf{E}}(\mathbf{E}). \tag{2.5.22a~c}$$

Hence *define* a new macrostrain potential Ψ^{E} by

$$\Psi^{\mathrm{E}} = \Psi^{\mathrm{E}}(\mathbf{\Sigma}^{\mathrm{E}}) \equiv \mathbf{\Sigma}^{\mathrm{E}} : \mathbf{E} - \mathbf{\Phi}^{\mathrm{E}}(\mathbf{E}), \tag{2.5.23}$$

where, again, $\mathbf{E}$ is viewed as a function of $\mathbf{\Sigma}^{\mathrm{E}}$. For the microquantities, furthermore, set

$$\phi^{\mathrm{E}} = \phi(\mathbf{x}, \boldsymbol{\varepsilon}(\mathbf{x}; \mathbf{E})) = \phi^{\mathrm{E}}(\mathbf{x}; \mathbf{E}),$$

$$\psi^{\mathrm{E}} = \psi(\mathbf{x}, \boldsymbol{\sigma}(\mathbf{x}; \mathbf{E})) = \psi^{\mathrm{E}}(\mathbf{x}; \mathbf{E}), \tag{2.5.24a,b}$$

and

$$\phi^{\mathrm{E}} + \psi^{\mathrm{E}} = \boldsymbol{\sigma}(\mathbf{x}; \mathbf{E}) : \boldsymbol{\varepsilon}(\mathbf{x}; \mathbf{E}). \tag{2.5.24c}$$

The volume average over V yields

$$< \phi^{\mathrm{E}} > + < \psi^{\mathrm{E}} > = \mathbf{\Sigma}^{\mathrm{E}} : \mathbf{E}, \tag{2.5.25a}$$

and comparison with (2.5.23) shows that

$$\Psi^{\mathrm{E}} = < \psi^{\mathrm{E}} >, \qquad \mathbf{\Sigma}^{\mathrm{E}} = \frac{\partial \mathbf{\Phi}^{\mathrm{E}}}{\partial \mathbf{E}}(\mathbf{E}) \Longleftrightarrow \mathbf{E} = \frac{\partial \Psi^{\mathrm{E}}}{\partial \mathbf{\Sigma}^{\mathrm{E}}}(\mathbf{\Sigma}^{\mathrm{E}}). \tag{2.5.25b,c}$$

2.5.4. On Definition of RVE

When the boundary tractions are given by

$$\mathbf{t}^{\mathrm{o}} = \boldsymbol{\nu} \cdot \boldsymbol{\Sigma} \quad \text{on } \partial V, \tag{2.5.26a}$$

the microstress and microstrain fields are

$$\boldsymbol{\sigma} = \boldsymbol{\sigma}(\mathbf{x}; \boldsymbol{\Sigma}), \qquad \boldsymbol{\varepsilon} = \boldsymbol{\varepsilon}(\mathbf{x}; \boldsymbol{\Sigma}), \tag{2.5.26b,c}$$

and $\mathbf{E}^{\Sigma} = < \boldsymbol{\varepsilon}(\mathbf{x}; \boldsymbol{\Sigma}) >$ is the overall macrostrain. Suppose that the boundary displacements are defined for this macrostrain by

$$\mathbf{u}^{\mathrm{o}} = \mathbf{x} \cdot \mathbf{E}^{\Sigma} \quad \text{on } \partial V, \tag{2.5.27a}$$

resulting in the microstress and microstrain fields,

$$\boldsymbol{\sigma} = \boldsymbol{\sigma}(\mathbf{x}; \mathbf{E}^{\Sigma}), \qquad \boldsymbol{\varepsilon} = \boldsymbol{\varepsilon}(\mathbf{x}; \mathbf{E}^{\Sigma}). \tag{2.5.27b,c}$$

In general, these fields are *not* identical with (2.5.26b,c). Furthermore, while $\mathbf{E}^{\Sigma} = < \boldsymbol{\varepsilon}(\mathbf{x}; \mathbf{E}^{\Sigma}) >$, there is no *a priori* reason that $< \boldsymbol{\sigma}(\mathbf{x}; \mathbf{E}^{\Sigma}) >$ should be equal to $\boldsymbol{\Sigma}$ for an arbitrary heterogeneous elastic solid.

The RVE is regarded as *statistically representative* of the macroresponse of the continuum material neighborhood, if and only if any arbitrary constant macrostress $\boldsymbol{\Sigma}$ produces through (2.5.26a) a macrostrain $\mathbf{E}^{\Sigma} = < \boldsymbol{\varepsilon}(\mathbf{x}; \boldsymbol{\Sigma}) >$ such that when the displacement boundary conditions (2.5.27a) are imposed instead, then the macrostress, $< \boldsymbol{\sigma}(\mathbf{x}; \mathbf{E}^{\Sigma}) > \approx \boldsymbol{\Sigma}$, is obtained, where the equality is to hold to a given degree of accuracy. Conversely, when the macrostrain $\mathbf{E}$ produces microstress and microstrain fields, $\boldsymbol{\sigma} = \boldsymbol{\sigma}(\mathbf{x}; \mathbf{E})$ and $\boldsymbol{\varepsilon} = \boldsymbol{\varepsilon}(\mathbf{x}; \mathbf{E})$, then the RVE is regarded statistically representative if and only if the prescribed macrostress, $\boldsymbol{\Sigma}^{\mathrm{E}} = < \boldsymbol{\sigma}(\mathbf{x}; \mathbf{E}) >$, leads to a microstrain field $\boldsymbol{\varepsilon}(\mathbf{x}; \boldsymbol{\Sigma}^{\mathrm{E}})$ such that $< \boldsymbol{\varepsilon}(\mathbf{x}; \boldsymbol{\Sigma}^{\mathrm{E}}) > \approx \mathbf{E}$. The relation between this definition of the RVE and an energy-based definition involving stress and strain potentials is discussed in Subsection 2.5.6, where several interesting inequalities are also developed.

Based on the above definitions for an RVE, the macrostrain potential, $\Psi^{\Sigma}(\boldsymbol{\Sigma})$, given by (2.5.18b), and the macrostress potential, $\Phi^{\mathrm{E}}(\mathbf{E})$, given by (2.5.22b), correspond to each other in the sense that

$$\frac{\partial \Psi^{\Sigma}}{\partial \boldsymbol{\Sigma}}(\boldsymbol{\Sigma}) \approx \mathbf{E} \Longleftrightarrow \frac{\partial \Phi^{\mathrm{E}}}{\partial \mathbf{E}}(\mathbf{E}) \approx \boldsymbol{\Sigma}, \tag{2.5.28a}$$

and in accordance with the Legendre transformation,

$$\Psi^{\Sigma}(\boldsymbol{\Sigma}) + \Phi^{\mathrm{E}}(\mathbf{E}) \approx \boldsymbol{\Sigma} : \mathbf{E}. \tag{2.5.28b}$$

It should be noted that $\boldsymbol{\sigma}(\mathbf{x}; \boldsymbol{\Sigma}) \neq \boldsymbol{\sigma}(\mathbf{x}; \mathbf{E})$ and $\boldsymbol{\varepsilon}(\mathbf{x}; \boldsymbol{\Sigma}) \neq \boldsymbol{\varepsilon}(\mathbf{x}; \mathbf{E})$, even for $\boldsymbol{\Sigma}$ and $\mathbf{E}$ which satisfy (2.5.28a). Moreover, in general,

$$\psi^{\Sigma}(\mathbf{x}; \boldsymbol{\Sigma}) + \phi^{\mathrm{E}}(\mathbf{x}; \mathbf{E}) \neq \boldsymbol{\sigma}(\mathbf{x}; \boldsymbol{\Sigma}) : \boldsymbol{\varepsilon}(\mathbf{x}; \mathbf{E}). \tag{2.5.28c}$$

Similarly, the complementary macropotentials, $\Phi^{\Sigma}(\mathbf{E}^{\Sigma})$ and $\Psi^{\mathrm{E}}(\boldsymbol{\Sigma}^{\mathrm{E}})$, are related through

$$\frac{\partial \Phi^{\Sigma}}{\partial \mathbf{E}^{\Sigma}}(\mathbf{E}^{\Sigma}) \approx \boldsymbol{\Sigma}^{\mathrm{E}} \Longleftrightarrow \frac{\partial \Psi^{\mathrm{E}}}{\partial \boldsymbol{\Sigma}^{\mathrm{E}}}(\boldsymbol{\Sigma}^{\mathrm{E}}) \approx \mathbf{E}^{\Sigma}, \tag{2.5.29a}$$

$$\Phi^{\Sigma}(\mathbf{E}^{\Sigma}) + \Psi^{E}(\boldsymbol{\Sigma}^{E}) \approx \mathbf{E}^{\Sigma} : \boldsymbol{\Sigma}^{E}, \tag{2.5.29b}$$

whereas the corresponding micropotentials do not satisfy a similar relation, i.e., in general,

$$\phi^{\Sigma}(\mathbf{x};\, \mathbf{E}^{\Sigma}) + \psi^{E}(\mathbf{x};\, \boldsymbol{\Sigma}^{E}) \neq \boldsymbol{\sigma}(\mathbf{x};\, \mathbf{E}) : \boldsymbol{\varepsilon}(\mathbf{x};\, \boldsymbol{\Sigma}). \tag{2.5.29c}$$

Table 2.5.1 provides a summary of the results presented in this subsection. Subsection 2.5.6 gives additional results on relations between the potentials. In Subsection 2.6 the notion of statistical homogeneity is discussed, and several important results on equivalence of the displacement and traction boundary conditions are obtained.

2.5.5. Linear Versus Nonlinear Response

When the microstructure is fixed and the material of the RVE is linearly elastic, then the corresponding overall response will also be linearly elastic. In this case, for a prescribed macrostrain $\mathbf{E}$, the macrostress $\boldsymbol{\Sigma}^{E}$ will be proportional to $\mathbf{E}$,

$$\boldsymbol{\Sigma}^{E} = <\boldsymbol{\sigma}(\mathbf{x};\, \mathbf{E})> = \bar{\mathbf{C}} : \mathbf{E}, \tag{2.5.30a}$$

where $\bar{\mathbf{C}}$ is the overall elasticity tensor.

Similarly, for a prescribed macrostress $\boldsymbol{\Sigma}$,

$$\mathbf{E}^{\Sigma} = <\boldsymbol{\varepsilon}(\mathbf{x};\, \boldsymbol{\Sigma})> = \bar{\mathbf{D}} : \boldsymbol{\Sigma}, \tag{2.5.30b}$$

where $\bar{\mathbf{D}}$ is the overall compliance tensor.

Now, if the RVE is statistically representative, then $\bar{\mathbf{C}} = \bar{\mathbf{D}}^{-1}$. A *consistent* averaging technique is expected to satisfy this inverse relation.

When the material of an RVE is nonlinearly elastic, then, for a fixed microstructure, the overall response will be nonlinearly elastic. In this case, the first gradient of the overall macrostress potential with respect to the overall macrostrain $\mathbf{E}$, and that of the macrostrain potential with respect to the overall macrostress $\boldsymbol{\Sigma}$, satisfy the relation (2.5.28a), when the RVE is statistically representative and a consistent averaging technique is employed. However, there is no *a priori* reason to believe that a similar correspondence between higher-order gradients of these potentials should continue to hold, even if a consistent averaging technique is employed. The relations between macropotentials are further discussed in the following section, considering an RVE with possibly nonlinearly elastic materials. In the remaining part of this book, however, attention is confined to RVE's with linearly elastic constituents, unless otherwise stated.

2.5.6. General Relations Between Macropotentials

The stress potential $\phi = \phi(\mathbf{x};\, \boldsymbol{\varepsilon})$ is said to be *convex* with respect to the argument $\boldsymbol{\varepsilon}$, if for every pair of admissible but nonidentical $\boldsymbol{\varepsilon}^{(1)}$ and $\boldsymbol{\varepsilon}^{(2)}$,

Table 2.5.1

Relation between macro- and micro-potentials for prescribed macrostress and macrostrain

	$\boldsymbol{\Sigma}$	$\mathbf{E}$
Microstress	$\boldsymbol{\sigma}(\mathbf{x}; \boldsymbol{\Sigma})$	$\boldsymbol{\sigma}(\mathbf{x}; \mathbf{E})$
Microstrain	$\boldsymbol{\varepsilon}(\mathbf{x}; \boldsymbol{\Sigma})$	$\boldsymbol{\varepsilon}(\mathbf{x}; \mathbf{E})$
Macrostress	$\boldsymbol{\Sigma} = < \boldsymbol{\sigma}(\mathbf{x}; \boldsymbol{\Sigma}) >$	$\boldsymbol{\Sigma}^{\mathrm{E}} = < \boldsymbol{\sigma}(\mathbf{x}; \mathbf{E}) >$
Macrostrain	$\mathbf{E}^{\Sigma} = < \boldsymbol{\varepsilon}(\mathbf{x}; \boldsymbol{\Sigma}) >$	$\mathbf{E} = < \boldsymbol{\varepsilon}(\mathbf{x}; \mathbf{E}) >$
Microstress potential	$\phi^{\Sigma}(\mathbf{x}; \boldsymbol{\Sigma}) = \phi(\mathbf{x}, \boldsymbol{\varepsilon}(\mathbf{x}; \boldsymbol{\Sigma}))$ $\boldsymbol{\sigma}(\mathbf{x}; \boldsymbol{\Sigma}) = \frac{\partial \phi}{\partial \boldsymbol{\varepsilon}}(\mathbf{x}, \boldsymbol{\varepsilon}(\mathbf{x}; \boldsymbol{\Sigma}))$	$\phi^{\mathrm{E}}(\mathbf{x}; \mathbf{E}) = \phi(\mathbf{x}, \boldsymbol{\varepsilon}(\mathbf{x}; \mathbf{E}))$ $\boldsymbol{\sigma}(\mathbf{x}; \mathbf{E}) = \frac{\partial \phi}{\partial \boldsymbol{\varepsilon}}(\mathbf{x}, \boldsymbol{\varepsilon}(\mathbf{x}; \mathbf{E}))$
Microstrain potential	$\psi^{\Sigma}(\mathbf{x}; \boldsymbol{\Sigma}) = \psi(\mathbf{x}, \boldsymbol{\sigma}(\mathbf{x}; \boldsymbol{\Sigma}))$ $\boldsymbol{\varepsilon}(\mathbf{x}; \boldsymbol{\Sigma}) = \frac{\partial \psi}{\partial \boldsymbol{\sigma}}(\mathbf{x}, \boldsymbol{\sigma}(\mathbf{x}; \boldsymbol{\Sigma}))$	$\psi^{\mathrm{E}}(\mathbf{x}; \mathbf{E}) = \psi(\mathbf{x}, \boldsymbol{\sigma}(\mathbf{x}; \mathbf{E}))$ $\boldsymbol{\varepsilon}(\mathbf{x}; \mathbf{E}) = \frac{\partial \psi}{\partial \boldsymbol{\sigma}}(\mathbf{x}, \boldsymbol{\sigma}(\mathbf{x}; \mathbf{E}))$
Macrostress potential	$\Phi^{\Sigma} = \Phi^{\Sigma}(\mathbf{E}^{\Sigma}) = < \phi^{\Sigma} >$ $\boldsymbol{\Sigma} = \frac{\partial \Phi^{\Sigma}}{\partial \mathbf{E}^{\Sigma}}(\mathbf{E}^{\Sigma})$	$\Phi^{\mathrm{E}} = \Phi^{\mathrm{E}}(\mathbf{E}) = < \phi^{\mathrm{E}} >$ $\boldsymbol{\Sigma}^{\mathrm{E}} = \frac{\partial \Phi^{\mathrm{E}}}{\partial \mathbf{E}}(\mathbf{E})$
Macrostrain potential	$\Psi^{\Sigma} = \Psi^{\Sigma}(\boldsymbol{\Sigma}) = < \psi^{\Sigma} >$ $\mathbf{E} = \frac{\partial \Psi^{\Sigma}}{\partial \boldsymbol{\Sigma}}(\boldsymbol{\Sigma})$	$\Psi^{\mathrm{E}} = \Psi^{\mathrm{E}}(\boldsymbol{\Sigma}^{\mathrm{E}}) = < \psi^{\mathrm{E}} >$ $\mathbf{E} = \frac{\partial \Psi^{\mathrm{E}}}{\partial \boldsymbol{\Sigma}^{\mathrm{E}}}(\boldsymbol{\Sigma}^{\mathrm{E}})$
MicroLegendre transformation	$\phi^{\Sigma} + \psi^{\Sigma} = \boldsymbol{\sigma}(\mathbf{x}; \boldsymbol{\Sigma}) : \boldsymbol{\varepsilon}(\mathbf{x}; \boldsymbol{\Sigma})$	$\phi^{\mathrm{E}} + \psi^{\mathrm{E}} = \boldsymbol{\sigma}(\mathbf{x}; \mathbf{E}) : \boldsymbol{\varepsilon}(\mathbf{x}; \mathbf{E})$
MacroLegendre transformation	$\Phi^{\Sigma} + \Psi^{\Sigma} = \boldsymbol{\Sigma} : \mathbf{E}^{\Sigma}$	$\Phi^{\mathrm{E}} + \Psi^{\mathrm{E}} = \boldsymbol{\Sigma}^{\mathrm{E}} : \mathbf{E}$
Approximated macroLegendre transformation	$\Phi^{\mathrm{E}}(\mathbf{E}) + \Psi^{\Sigma}(\boldsymbol{\Sigma}) \approx \boldsymbol{\Sigma} : \mathbf{E}$	$\Phi^{\Sigma}(\mathbf{E}^{\Sigma}) + \Psi^{\mathrm{E}}(\boldsymbol{\Sigma}^{\mathrm{E}}) \approx \boldsymbol{\Sigma}^{\mathrm{E}} : \mathbf{E}^{\Sigma}$
Corresponding microLegendre transformation	$\phi^{\mathrm{E}} + \psi^{\Sigma} \neq \boldsymbol{\sigma}(\mathbf{x}; \boldsymbol{\Sigma}) : \boldsymbol{\varepsilon}(\mathbf{x}; \mathbf{E})$	$\phi^{\Sigma} + \psi^{\mathrm{E}} \neq \boldsymbol{\sigma}(\mathbf{x}; \mathbf{E}) : \boldsymbol{\varepsilon}(\mathbf{x}; \boldsymbol{\Sigma})$

$$\phi(\mathbf{x}; \boldsymbol{\varepsilon}^{(1)}) - \phi(\mathbf{x}; \boldsymbol{\varepsilon}^{(2)}) > (\boldsymbol{\varepsilon}^{(1)} - \boldsymbol{\varepsilon}^{(2)}) : \frac{\partial \phi}{\partial \boldsymbol{\varepsilon}}(\mathbf{x}; \boldsymbol{\varepsilon}^{(2)}). \tag{2.5.31}$$

Consider RVE's consisting of convex elastic materials.

Examine now *two different* boundary conditions for the same RVE which consists of convex elastic constituents: (I) uniform tractions $\mathbf{t} = \boldsymbol{\nu} \cdot \boldsymbol{\Sigma}$ prescribed on boundary ∂V of V; and (II) any mixed uniform or nonuniform *consistent* displacements and tractions prescribed on ∂V. Denote the strain and stress fields for the first boundary data (i.e., uniform boundary tractions) by $\boldsymbol{\varepsilon}^\Sigma = \boldsymbol{\varepsilon}^\Sigma(\mathbf{x})$ and $\boldsymbol{\sigma}^\Sigma = \boldsymbol{\sigma}^\Sigma(\mathbf{x})$, and those for the second (i.e., general, possibly mixed) by $\boldsymbol{\varepsilon}^G = \boldsymbol{\varepsilon}^G(\mathbf{x})$ and $\boldsymbol{\sigma}^G = \boldsymbol{\sigma}^G(\mathbf{x})$, respectively. The corresponding average quantities are denoted as follows:

$$\mathbf{E}^\Sigma = < \boldsymbol{\varepsilon}^\Sigma >, \qquad \boldsymbol{\Sigma}^\Sigma = < \boldsymbol{\sigma}^\Sigma > = \boldsymbol{\Sigma}, \tag{2.5.32a,b}$$

and

$$\mathbf{E}^G = < \boldsymbol{\varepsilon}^G >, \qquad \boldsymbol{\Sigma}^G = < \boldsymbol{\sigma}^G >, \tag{2.5.33a,b}$$

for case (I) and case (II), respectively. Consider the overall macrostress potentials when the boundary data in cases (I) and (II) are adjusted such that $\mathbf{E}^\Sigma = \mathbf{E}^G$, i.e., *they both produce the same overall macrostrains.*

Theorem 1: *The macrostress potential* $\Phi^\Sigma(\mathbf{E}^\Sigma)$ *associated with the uniform traction boundary data of case* (I) *cannot exceed the macrostress potential* $\Phi^G(\mathbf{E}^G)$ *associated with the second (general, possibly mixed) boundary data of case* (II) *for* $\mathbf{E}^\Sigma = \mathbf{E}^G$, *when the corresponding* RVE *consists of convex elastic constituents.*[5]

Proof: Calculate the difference, $\Phi^G(\mathbf{E}^G) - \Phi^\Sigma(\mathbf{E}^\Sigma)$, as follows:

$$\Phi^G(\mathbf{E}^G) - \Phi^\Sigma(\mathbf{E}^\Sigma) = < \phi(\mathbf{x};\, \boldsymbol{\varepsilon}^G) - \phi(\mathbf{x};\, \boldsymbol{\varepsilon}^\Sigma) > \; > \; < (\boldsymbol{\varepsilon}^G - \boldsymbol{\varepsilon}^\Sigma) : \frac{\partial \phi}{\partial \boldsymbol{\varepsilon}}(\mathbf{x};\, \boldsymbol{\varepsilon}^\Sigma) >$$

$$= \{< \boldsymbol{\varepsilon}^G > - < \boldsymbol{\varepsilon}^\Sigma >\} : < \boldsymbol{\sigma}^\Sigma > = (\mathbf{E}^G - \mathbf{E}^\Sigma) : \boldsymbol{\Sigma}. \tag{2.5.34a}$$

The last expression in (2.5.34a) vanishes if the boundary data are adjusted such that $\mathbf{E}^G = \mathbf{E}^\Sigma$, in which case,

$$\Phi^G(\mathbf{E}^G) > \Phi^\Sigma(\mathbf{E}^\Sigma) \qquad \text{for } \mathbf{E}^G = \mathbf{E}^\Sigma, \tag{2.5.34b}$$

which completes the proof (Willis, 1989).

Clearly, this result remains valid for the case when the RVE consists of linearly elastic constituents. Moreover, it also shows that the macrostress potential corresponding to linear displacement boundary data, $\mathbf{u} = \mathbf{x} \cdot \mathbf{E}$, i.e., $\Phi^E(\mathbf{E})$, cannot be less than the macrostress potential $\Phi^\Sigma(\mathbf{E}^\Sigma)$ which corresponds to uniform traction boundary data, when $\mathbf{E} = \mathbf{E}^\Sigma$, i.e.,

$$\Phi^E(\mathbf{E}) \geq \Phi^\Sigma(\mathbf{E}). \tag{2.5.34c}$$

Therefore, an RVE is considered to be statistically representative if the difference between the macrostress potential for uniform traction and linear displacement boundary data which produce the same overall average macrostrain, is less than a prescribed small value. Note that Φ^Σ *can only approach* Φ^E *from below.*

Consider now a *third* loading case involving linear displacement boundary data for the *same* RVE, as follows: (III) on the boundary ∂V of the RVE, linear displacements $\mathbf{u} = \mathbf{x} \cdot \mathbf{E}$ are prescribed such that they produce the same overall

[5] The superscript G on Φ denotes the fact that the corresponding boundary data are *general* and unrestricted.

macrostress corresponding to case (II) of the general boundary data. Denote the microstrain and microstress fields for case (III) by $\boldsymbol{\varepsilon}^{\mathrm{E}} = \boldsymbol{\varepsilon}^{\mathrm{E}}(\mathbf{x})$ and $\boldsymbol{\sigma}^{\mathrm{E}} = \boldsymbol{\sigma}^{\mathrm{E}}(\mathbf{x})$, and set

$$\mathbf{E} = < \boldsymbol{\varepsilon}^{\mathrm{E}} >, \qquad \boldsymbol{\Sigma}^{\mathrm{E}} = < \boldsymbol{\sigma}^{\mathrm{E}} >. \tag{2.5.35a,b}$$

Theorem 2: *The macrostrain potential* $\Psi^{\mathrm{E}}(\boldsymbol{\Sigma}^{\mathrm{E}})$ *associated with the linear displacement boundary data of case* (III) *cannot exceed the macrostrain potential* $\Psi^{\mathrm{G}}(\boldsymbol{\Sigma}^{\mathrm{G}})$ *associated with the second (general, possibly mixed) boundary data of case* (II) *for* $\boldsymbol{\Sigma}^{\mathrm{E}} = \boldsymbol{\Sigma}^{\mathrm{G}}$, *when the corresponding RVE consists of convex elastic constituents.*

Proof: Calculate the difference, $\Psi^{\mathrm{G}}(\boldsymbol{\Sigma}^{\mathrm{G}}) - \Psi^{\mathrm{E}}(\boldsymbol{\Sigma}^{\mathrm{E}})$, as follows:

$$\Psi^{\mathrm{G}}(\boldsymbol{\Sigma}^{\mathrm{G}}) - \Psi^{\mathrm{E}}(\boldsymbol{\Sigma}^{\mathrm{E}}) = < \psi(\mathbf{x};\, \boldsymbol{\sigma}^{\mathrm{G}}) - \psi(\mathbf{x};\, \boldsymbol{\sigma}^{\mathrm{E}}) > \; > \; < (\boldsymbol{\sigma}^{\mathrm{G}} - \boldsymbol{\sigma}^{\mathrm{E}}) : \frac{\partial \psi}{\partial \boldsymbol{\sigma}}(\mathbf{x};\, \boldsymbol{\sigma}^{\mathrm{E}}) >$$

$$= \{< \boldsymbol{\sigma}^{\mathrm{G}} > - < \boldsymbol{\sigma}^{\mathrm{E}} >\} : < \boldsymbol{\varepsilon}^{\mathrm{E}} > = (\boldsymbol{\Sigma}^{\mathrm{G}} - \boldsymbol{\Sigma}^{\mathrm{E}}) : \mathbf{E}. \tag{2.5.36a}$$

The last expression vanishes when $\boldsymbol{\Sigma}^{\mathrm{G}} = \boldsymbol{\Sigma}^{\mathrm{E}}$, leading to

$$\Psi^{\mathrm{G}}(\boldsymbol{\Sigma}^{\mathrm{G}}) > \Psi^{\mathrm{E}}(\boldsymbol{\Sigma}^{\mathrm{E}}), \qquad \text{for } \boldsymbol{\Sigma}^{\mathrm{G}} = \boldsymbol{\Sigma}^{\mathrm{E}}, \tag{2.5.36b}$$

which completes the proof (Willis, 1989). In particular, the macrostrain potential associated with uniform traction boundary data always exceeds that associated with the linear displacement boundary data when both boundary conditions produce the same overall macrostress, i.e.,

$$\Psi^{\mathrm{E}}(\boldsymbol{\Sigma}) \leq \Psi^{\Sigma}(\boldsymbol{\Sigma}). \tag{2.5.36c}$$

Again, a statistically representative RVE can be defined by requiring that the difference in the macrostrain potential for uniform traction and linear displacement boundary conditions which produce the same overall macrostress, be less than a prescribed small value. Note that Ψ^{E} *can only approach* Ψ^{Σ} *from below.*

Theorems 1 and 2 may be stated in the form of the following minimum principles:

Theorem I: *For an elastic* RVE, *among all consistent boundary data which produce the same overall macrostrain, the uniform boundary tractions render the total stress potential* Φ *an absolute minimum.*

Theorem II: *For an elastic* RVE, *among all consistent boundary data which produce the same overall macrostress, the linear boundary displacements render the total strain potential* Ψ *an absolute minimum.*

Consider now a linearly elastic RVE, and set

$$\Phi^{\mathrm{G}}(\mathbf{E}^{\mathrm{G}}) = \frac{1}{2}\mathbf{E}^{\mathrm{G}} : \bar{\mathbf{C}}^{\mathrm{G}} : \mathbf{E}^{\mathrm{G}}, \qquad \Psi^{\mathrm{G}}(\boldsymbol{\Sigma}^{\mathrm{G}}) = \frac{1}{2}\boldsymbol{\Sigma}^{\mathrm{G}} : \bar{\mathbf{D}}^{\mathrm{G}} : \boldsymbol{\Sigma}^{\mathrm{G}}, \tag{2.5.37a,b}$$

corresponding to general consistent boundary data which produce an overall macrostrain $\mathbf{E}^{\mathrm{G}}$ and an overall macrostress $\boldsymbol{\Sigma}^{\mathrm{G}}$, satisfying

$$\frac{1}{V}\int_{\partial V} \{\mathbf{u}^G - \mathbf{x}\cdot\mathbf{E}^G\}\cdot\{\boldsymbol{\nu}\cdot(\boldsymbol{\sigma} - \boldsymbol{\Sigma}^G)\}\, dS = 0, \tag{2.5.38a}$$

so that

$$< \boldsymbol{\varepsilon}^G : \boldsymbol{\sigma}^G > = < \boldsymbol{\varepsilon}^G > : < \boldsymbol{\sigma}^G > = \mathbf{E}^G : \boldsymbol{\Sigma}^G; \tag{2.5.38b}$$

see (2.3.5). Here $\overline{\mathbf{C}}^G$ and $\overline{\mathbf{D}}^G$ are the overall elasticity and compliance tensors defined through

$$\boldsymbol{\Sigma}^G = \overline{\mathbf{C}}^G : \mathbf{E}^G, \qquad \mathbf{E}^G = \overline{\mathbf{D}}^G : \boldsymbol{\Sigma}^G, \tag{2.5.37c,d}$$

and hence *are each other's inverse*.

Each of the boundary conditions of case (I) (uniform tractions) and case (III) (linear displacements) satisfies (2.5.38a). Define $\overline{\mathbf{C}}^\Sigma$ and $\overline{\mathbf{D}}^E$, respectively, by

$$\Phi^\Sigma(\mathbf{E}^\Sigma) = \frac{1}{2}\mathbf{E}^\Sigma : \overline{\mathbf{C}}^\Sigma : \mathbf{E}^\Sigma, \tag{2.5.39a}$$

and

$$\Psi^E(\boldsymbol{\Sigma}^E) = \frac{1}{2}\boldsymbol{\Sigma}^E : \overline{\mathbf{D}}^E : \boldsymbol{\Sigma}^E. \tag{2.5.39b}$$

Since $\overline{\mathbf{C}}^\Sigma$ in (2.5.39a) and $\overline{\mathbf{D}}^E$ in (2.5.39b) correspond to *different* boundary conditions, they are not necessarily each other's inverse. Their inverses are denoted by $\overline{\mathbf{D}}^\Sigma = (\overline{\mathbf{C}}^\Sigma)^{-1}$ and $\overline{\mathbf{C}}^E = (\overline{\mathbf{D}}^E)^{-1}$, respectively.

Now, according to Theorems I and II, for any $\mathbf{E}$,

$$\mathbf{E} : (\overline{\mathbf{C}}^G - \overline{\mathbf{C}}^\Sigma) : \mathbf{E} \geq 0, \tag{2.5.40a}$$

and, for any $\boldsymbol{\Sigma}$,

$$\boldsymbol{\Sigma} : (\overline{\mathbf{D}}^G - \overline{\mathbf{D}}^E) : \boldsymbol{\Sigma} \geq 0. \tag{2.5.40b}$$

Hence, uniform traction and linear displacement boundary data can be used to obtain lower and upper bounds of the elastic moduli associated with any other consistent boundary data.

As will be discussed in Subsection 3.1, a second-order symmetric tensor can be expressed by a six by one column vector, and a fourth-order symmetric tensor can be expressed by a six by six matrix; see (3.1.4) and (3.1.7). Then, in this matrix form, (2.5.40a,b) are written as

$$[\Gamma_a]^T([\overline{C}^G_{ab}] - [\overline{C}^\Sigma_{ab}])[\Gamma_b] \geq 0, \qquad [T_a]^T([\overline{D}^G_{ab}] - [\overline{D}^E_{ab}])[T_b] \geq 0, \tag{2.5.41a,b}$$

where $[\Gamma_a]$ and $[T_a]$ correspond to $\mathbf{E}$ and $\boldsymbol{\Sigma}$, and $[\overline{C}^G_{ab}]$, $[\overline{C}^\Sigma_{ab}]$, $[\overline{D}^G_{ab}]$, and $[\overline{D}^E_{ab}]$ correspond to $\overline{\mathbf{C}}^G$, $\overline{\mathbf{C}}^\Sigma$, $\overline{\mathbf{D}}^G$, and $\overline{\mathbf{D}}^E$, respectively. Since the six by six matrices, $[\overline{C}^A_{ab}]$ and $[\overline{D}^A_{ab}]$ for A = G, Σ, E, are symmetric and positive-definite, they can be written as

$$[\overline{C}^A_{ab}] = [Q^A_{ap}][\lambda^A_p\,\delta_{pq}][Q^A_{bq}]^T, \qquad [\overline{D}^A_{ab}] = [Q^A_{ap}][\lambda^A_p\,\delta_{pq}]^{-1}[Q^A_{bq}]^T, \tag{2.5.42a,b}$$

where $[\lambda^A_a\,\delta_{ab}]$ is the diagonal matrix of the characteristic values, λ^A_a, of $[\overline{C}^A_{ab}]$, and $[Q^{\mathrm{A}}_{\mathrm{ab}}]$ is the rotation matrix satisfying $[Q^{\mathrm{A}}_{\mathrm{ab}}][Q^{\mathrm{A}}_{\mathrm{ab}}]^{\mathrm{T}} = [\delta_{\mathrm{ab}}]$.

When the three rotation matrices, $[Q^{\mathrm{G}}_{\mathrm{ab}}]$, $[Q^{\Sigma}_{\mathrm{ab}}]$, and $[Q^{\mathrm{E}}_{\mathrm{ab}}]$, are identical, i.e.,

$$[Q^{\mathrm{G}}_{\mathrm{ab}}] = [Q^{\Sigma}_{\mathrm{ab}}] = [Q^{\mathrm{E}}_{\mathrm{ab}}], \tag{2.5.43a}$$

inequalities (2.5.41a,b) imply that

$$\lambda^{\mathrm{G}}_{\mathrm{a}} - \lambda^{\Sigma}_{\mathrm{a}} \geq 0, \qquad \frac{1}{\lambda^{\mathrm{G}}_{\mathrm{a}}} - \frac{1}{\lambda^{\mathrm{E}}_{\mathrm{a}}} \geq 0 \qquad \text{for a} = 1, 2, ..., 6. \tag{2.5.43b,c}$$

Therefore, the characteristic values $\lambda^{\mathrm{G}}_{\mathrm{a}}$ of the overall elasticity tensor $\bar{\mathbf{C}}^{\mathrm{G}}$, corresponding to general consistent boundary data, are bounded by

$$\lambda^{\Sigma}_{\mathrm{a}} \leq \lambda^{\mathrm{G}}_{\mathrm{a}} \leq \lambda^{\mathrm{E}}_{\mathrm{a}}. \tag{2.5.44a}$$

This implies that the corresponding quadratic forms of the macropotentials satisfy similar relations, i.e.,

$$\mathbf{E} : \bar{\mathbf{C}}^{\Sigma} : \mathbf{E} \leq \mathbf{E} : \bar{\mathbf{C}}^{\mathrm{G}} : \mathbf{E} \leq \mathbf{E} : \bar{\mathbf{C}}^{\mathrm{E}} : \mathbf{E} \Longleftrightarrow \Phi^{\Sigma}(\mathbf{E}) \leq \Phi^{\mathrm{G}}(\mathbf{E}) \leq \Phi^{\mathrm{E}}(\mathbf{E}), \tag{2.5.44b}$$

or

$$\boldsymbol{\Sigma} : \bar{\mathbf{D}}^{\Sigma} : \boldsymbol{\Sigma} \geq \boldsymbol{\Sigma} : \bar{\mathbf{D}}^{\mathrm{G}} : \boldsymbol{\Sigma} \geq \boldsymbol{\Sigma} : \bar{\mathbf{D}}^{\mathrm{E}} : \boldsymbol{\Sigma} \Longleftrightarrow \Psi^{\Sigma}(\boldsymbol{\Sigma}) \geq \Psi^{\mathrm{G}}(\boldsymbol{\Sigma}) \geq \Psi^{\mathrm{E}}(\boldsymbol{\Sigma}). \tag{2.5.44c}$$

The rotation matrix $[Q^{\mathrm{G}}_{\mathrm{ab}}]$ depends on the structure and properties of the RVE, as well as on the loading conditions. However, if the overall response of the RVE is expected to be isotropic, then, $[Q^{\mathrm{G}}_{\mathrm{ab}}]$'s are the same under all loading conditions. Hence, for all possible general boundary data, equality (2.5.43a) is satisfied, and the bounds (2.5.44a~c) hold in the isotropic case. Note that the *energy* bounds (2.5.44b,c) are valid whether or not the general loading satisfies (2.5.38a), as is evident from (2.5.34) to (2.5.37a,b). However, for the *general loading* (but *not* for the uniform tractions or linear displacements), the overall moduli defined in terms of the *total energy* by (2.5.37a,b) will not equal those defined by relating the average stress, $\boldsymbol{\Sigma}^{\mathrm{G}}$, and the average strain, $\mathbf{E}^{\mathrm{G}}$, as in (2.5.37c,d), unless this loading satisfies (2.5.38a). The validity of the inverse relation between the overall elasticity and compliance tensors also depends on whether or not (2.5.38a) holds.

2.5.7. Bounds on Macropotential Functions

Although, due to the heterogeneity of the RVE, it is extremely difficult to compute the exact macrostress potential $\boldsymbol{\Phi}$ or macrostrain potential $\boldsymbol{\Psi}$, strict upper and lower bounds for these quantities can be obtained, using variational principles.[6] In this subsection, the macropotential functions, defined as *functions* of the macrofield quantities, are related to *functionals* of the displacement or stress fields, when linear displacements or uniform tractions are prescribed on

---

[6] See Section 19 for detailed discussions of variational principles in linear elasticity.

the boundary of the RVE.

First, consider a bound for the macrostress potential, Φ^E, when macrostrain $\mathbf{E} = \boldsymbol{\varepsilon}^o$ is prescribed. For a displacement field which satisfies the linear displacement boundary condition, $\mathbf{u} = \mathbf{x} \cdot \boldsymbol{\varepsilon}^o$ on ∂V, define a functional I^E, by

$$I^E(\mathbf{u}; \boldsymbol{\varepsilon}^o) \equiv < \phi(\mathbf{x}, \boldsymbol{\varepsilon}(\mathbf{x}; \boldsymbol{\varepsilon}^o) >, \tag{2.5.45}$$

where $\boldsymbol{\varepsilon}$ is given by the symmetric part of the gradient of $\mathbf{u}$, $sym\,(\boldsymbol{\nabla} \otimes \mathbf{u})$, with sym standing for the symmetric part of its second-order tensor argument. Since $\delta\phi = \boldsymbol{\sigma}(\boldsymbol{\varepsilon}) : \delta\boldsymbol{\varepsilon}$, the first variation of I^E is

$$\delta I^E = < \boldsymbol{\sigma}(\boldsymbol{\varepsilon}) : \delta\boldsymbol{\varepsilon}(\mathbf{x}) >$$

$$= \frac{1}{V} \int_{\partial V} (\mathbf{v} \cdot \boldsymbol{\sigma}) \cdot \delta\mathbf{u} \, dS - < \{\boldsymbol{\nabla} \cdot \boldsymbol{\sigma}(\boldsymbol{\varepsilon})\} \cdot \delta\mathbf{u} >. \tag{2.5.46a}$$

Since the surface integral vanishes when the displacements are prescribed on ∂V, the solution of $\delta I^E = 0$, denoted by $\mathbf{u}^{ex}$, yields a stress field, $\boldsymbol{\sigma}^{ex} = \partial\phi/\partial\boldsymbol{\varepsilon}(\boldsymbol{\varepsilon}^{ex})$ for the strain field $\boldsymbol{\varepsilon}^{ex} = sym\,(\boldsymbol{\nabla} \otimes \mathbf{u}^{ex})$, which satisfies the equations of equilibrium. Here, $\mathbf{u}^{ex}$ is the actual displacement field produced by macrostrain $\mathbf{E} = \boldsymbol{\varepsilon}^o$, and the corresponding macrostress potential is given by

$$I^E(\mathbf{u}^{ex}; \boldsymbol{\varepsilon}^o) = \Phi^E(\boldsymbol{\varepsilon}^o). \tag{2.5.46b}$$

If the microstress potential is positive-definite, i.e., if $\delta^2\phi > 0$, then, among all suitably smooth displacement fields which satisfy the displacement boundary conditions,[7] the exact displacement field renders functional I^E an absolute minimum. Hence, for any kinematically admissible displacement field, $\mathbf{u}$, the following inequality holds:

$$I^E(\mathbf{u}^{ex}; \boldsymbol{\varepsilon}^o) \le I^E(\mathbf{u}; \boldsymbol{\varepsilon}^o), \tag{2.5.46c}$$

where equality holds if and only if $\mathbf{u} = \mathbf{u}^{ex}$. Since the linear displacement field, $\mathbf{u}(\mathbf{x}) = \mathbf{x} \cdot \boldsymbol{\varepsilon}^o$ for $\mathbf{x}$ in V, is kinematically admissible and produces the constant strain field $\boldsymbol{\varepsilon}(\mathbf{x}) = \boldsymbol{\varepsilon}^o$, (2.5.46c) yields

$$I^E(\mathbf{u}^{ex}; \boldsymbol{\varepsilon}^o) < I^E(\mathbf{x} \cdot \boldsymbol{\varepsilon}^o; \boldsymbol{\varepsilon}^o) = < \phi(\mathbf{x}; \boldsymbol{\varepsilon}^o) >, \tag{2.5.47a}$$

where $\phi(\mathbf{x}; \boldsymbol{\varepsilon}^o)$ is the microstress potential at $\mathbf{x}$, evaluated at strain $\boldsymbol{\varepsilon}^o$. From (2.5.46b) and (2.5.47a), an upper bound is obtained for the macrostress potential,

$$\Phi^E(\boldsymbol{\varepsilon}^o) < < \phi(\mathbf{x}; \boldsymbol{\varepsilon}^o) >. \tag{2.5.47b}$$

Next, consider a bound for the macrostrain potential, Ψ^Σ, when macrostress $\boldsymbol{\Sigma} = \boldsymbol{\sigma}^o$ is prescribed. For a suitably smooth symmetric stress field which satisfies the equations of equilibrium, $\boldsymbol{\nabla} \cdot \boldsymbol{\sigma} = \mathbf{0}$ in V, and traction boundary conditions,[8] $\mathbf{v} \cdot \boldsymbol{\sigma} = \mathbf{v} \cdot \boldsymbol{\sigma}^o$ on ∂V, define functional I^Σ, by

---

[7] Such a displacement field is called kinematically admissible; see Section 19, Part 2.

[8] Such a stress field is called statically admissible; see Section 19, Part 2.

$$I^{\Sigma}(\boldsymbol{\sigma};\ \boldsymbol{\sigma}^{o}) = <\ \psi(\mathbf{x};\ \boldsymbol{\varepsilon}(\mathbf{x};\ \boldsymbol{\sigma}^{o})) + \boldsymbol{\mu}(\mathbf{x}) \cdot (\boldsymbol{\nabla} \cdot \boldsymbol{\sigma}(\mathbf{x}))\ >, \tag{2.5.48}$$

where $\boldsymbol{\mu}$ is a Lagrange multiplier[9] which, in a weak sense, enforces the equations of equilibrium and, hence, the static admissibility of $\boldsymbol{\sigma}$. Since $\delta\psi = \boldsymbol{\varepsilon}(\boldsymbol{\sigma}) : \delta\boldsymbol{\sigma}$, the first variation of I^{Σ} is

$$\delta I^{\Sigma} = <\ \{\boldsymbol{\varepsilon}(\boldsymbol{\sigma}) : \delta\boldsymbol{\sigma} + \delta(\boldsymbol{\mu} \cdot (\boldsymbol{\nabla} \cdot \boldsymbol{\sigma}))\}\ >$$

$$= \frac{1}{V} \int_{\partial V} (\boldsymbol{\nu} \cdot \delta\boldsymbol{\sigma}) \cdot \boldsymbol{\mu}\, dS + <\ (\boldsymbol{\varepsilon} - sym\, \boldsymbol{\nabla} \otimes \boldsymbol{\mu}) : \delta\boldsymbol{\sigma} + \delta\boldsymbol{\mu} \cdot (\boldsymbol{\nabla} \cdot \boldsymbol{\sigma})\ >. \tag{2.5.49a}$$

Since the surface integral vanishes for the considered traction boundary conditions, the solution of $\delta I^{\Sigma} = 0$, denoted by $\boldsymbol{\sigma}^{ex}$, produces a compatible strain field, $\boldsymbol{\varepsilon}^{ex} = \partial\psi/\partial\boldsymbol{\sigma}(\boldsymbol{\sigma}^{ex})$; see Subsection 19.5. For $\boldsymbol{\sigma}^{ex}$ produced by $\boldsymbol{\Sigma} = \boldsymbol{\sigma}^{o}$, the macrostrain potential is given by

$$I^{\Sigma}(\boldsymbol{\sigma}^{ex};\ \boldsymbol{\sigma}^{o}) = \Psi^{\Sigma}(\boldsymbol{\sigma}^{o}). \tag{2.5.49b}$$

Furthermore, if the microstrain potential is positive-definite, then, for any statically admissible stress field, $\boldsymbol{\sigma}(\mathbf{x})$,

$$I^{\Sigma}(\boldsymbol{\sigma}^{ex};\ \boldsymbol{\sigma}^{o}) \leq I^{\Sigma}(\boldsymbol{\sigma};\ \boldsymbol{\sigma}^{o}), \tag{2.5.49c}$$

where equality holds if and only if $\boldsymbol{\sigma} = \boldsymbol{\sigma}^{ex}$. The uniform stress field, $\boldsymbol{\sigma}(\mathbf{x}) = \boldsymbol{\sigma}^{o}$ for $\mathbf{x}$ in V, satisfies the equations of equilibrium, as well as the prescribed traction boundary conditions. Hence, (2.5.49c) yields

$$I^{\Sigma}(\boldsymbol{\sigma}^{ex};\ \boldsymbol{\sigma}^{o}) < I^{\Sigma}(\boldsymbol{\sigma}^{o};\ \boldsymbol{\sigma}^{o}) = <\ \psi(\mathbf{x};\ \boldsymbol{\sigma}^{o})\ >, \tag{2.5.50a}$$

where $\psi(\mathbf{x};\ \boldsymbol{\sigma}^{o})$ is the microstrain potential at $\mathbf{x}$, evaluated at uniform stress $\boldsymbol{\sigma}^{o}$. The upper bound for the macrostrain potential is now given by

$$\Psi^{\Sigma}(\boldsymbol{\sigma}^{o}) < <\ \psi(\mathbf{x};\ \boldsymbol{\sigma}^{o})\ >. \tag{2.5.50b}$$

The above results are direct consequences of the minimum energy principles in elasticity. They show that the uniform strain field, $\boldsymbol{\varepsilon} = \boldsymbol{\varepsilon}^{o}$ (= $\mathbf{E}$), leads to an upper bound for macrostress potential $\Phi^{E}(\mathbf{E})$, while the uniform stress field, $\boldsymbol{\sigma} = \boldsymbol{\sigma}^{o}$ (= $\boldsymbol{\Sigma}$), provides an upper bound for macrostrain potential $\Psi^{\Sigma}(\boldsymbol{\Sigma})$. In linear elasticity, the macrostress and macrostrain potentials correspond, respectively, to the average strain energy and the average complementary strain energy. The overall elastic parameters may be defined in terms of either potential. Hence, upper bounds can be obtained for the overall moduli from the uniform strain fields, and lower bounds from the uniform stress fields. The estimates of the overall moduli and the overall compliances of a linearly elastic composite, using uniform strains and uniform stresses, are due to Voigt (1889) and Reuss (1929), respectively; see Subsections 7.2 and 7.1, and (7.2.9) and (7.1.14). The fact that these are actually bounds has been shown by Hill (1952, 1963), and Paul (1960).

[9] $\boldsymbol{\mu}$ is a vector field in V.

2.6. STATISTICAL HOMOGENEITY, AVERAGE QUANTITIES, AND OVERALL PROPERTIES

To obtain further insight into the relation between the microstructure and the overall properties, imagine a very large volume of the heterogeneous solid with the property that any suitably large subvolume can be used as an RVE to obtain the overall macroscopic parameters. This large body is then called *statistically homogeneous*. The term "large volume" here refers to a material neighborhood which is several orders of magnitude larger than the corresponding RVE. On the other hand, the RVE must be several orders of magnitude larger than the size of its microconstituents. For simplicity, consider only linearly elastic materials in this subsection.

To be specific, let B be the large solid and V be a typical representative volume element within B. Denote the length scales of B, V, and the microconstituents, respectively, by L, D, and d. It is assumed that these length scales satisfy

$$\frac{\mathrm{d}}{\mathrm{D}} << 1, \qquad \frac{\mathrm{D}}{L} << 1. \tag{2.6.1a,b}$$

Note that the length scales D and d are the same macro-length-scale and micro-length-scale as mentioned in Subsection 1.1. The statistical homogeneity of B may be described in terms of average fields over V.

First, define average fields in the following manner. Consider only interior regions within B that are at least the distance D away from the boundary ∂B of B. Denote the collection of all such interior regions by B′; see Figure 2.6.1. Let V^{o} be a suitably large region whose centroid is located at the origin, and denote points in V^{o} by $\mathbf{z}$. By rigid-body translation, say, $\mathbf{y}$, points in V^{o} form a region whose shape and orientation are the same as V^{o}, i.e., a region given by the collection of all $\mathbf{x} = \mathbf{y} + \mathbf{z}$ with $\mathbf{y}$ fixed and $\mathbf{z}$ varying in V^{o}; Figure 2.6.1. Regard such a region as an RVE, and denote it by V. By definition, the centroid of V is $\mathbf{y}$. With $\mathbf{y}$ fixed, the strain and stress within V are expressed as

$$\boldsymbol{\varepsilon} = \boldsymbol{\varepsilon}(\mathbf{x}) = \boldsymbol{\varepsilon}(\mathbf{y}+\mathbf{z}), \qquad \boldsymbol{\sigma} = \boldsymbol{\sigma}(\mathbf{x}) = \boldsymbol{\sigma}(\mathbf{y}+\mathbf{z}) \qquad \mathbf{z} \text{ in } \mathrm{V}^{\mathrm{o}}. \tag{2.6.2a,b}$$

The corresponding average strain and stress over V are given by

$$\bar{\boldsymbol{\varepsilon}}(\mathbf{y}) \equiv \frac{1}{\mathrm{V}} \int_{\mathrm{V}} \boldsymbol{\varepsilon}(\mathbf{x})\, \mathrm{dV}_{\mathbf{x}} \equiv \frac{1}{\mathrm{V}^{\mathrm{o}}} \int_{\mathrm{V}^{\mathrm{o}}} \boldsymbol{\varepsilon}(\mathbf{y}+\mathbf{z})\, \mathrm{dV}_{\mathbf{z}},$$

$$\bar{\boldsymbol{\sigma}}(\mathbf{y}) \equiv \frac{1}{\mathrm{V}} \int_{\mathrm{V}} \boldsymbol{\sigma}(\mathbf{x})\, \mathrm{dV}_{\mathbf{x}} \equiv \frac{1}{\mathrm{V}^{\mathrm{o}}} \int_{\mathrm{V}^{\mathrm{o}}} \boldsymbol{\sigma}(\mathbf{y}+\mathbf{z})\, \mathrm{dV}_{\mathbf{z}}. \tag{2.6.3a,b}$$

In general, $\bar{\boldsymbol{\varepsilon}}(\mathbf{y})$ and $\bar{\boldsymbol{\sigma}}(\mathbf{y})$ depend on $\mathbf{y}$ and on the shape and size of V or V^{o}. With V^{o} fixed, for various values of $\mathbf{y}$, (2.6.3) defines *moving averages* of the strain and stress fields.

In terms of the moving average strain and stress fields, the statistical homogeneity is described as follows: Let suitable uniform farfield boundary data associated with constant strains and stresses, $\boldsymbol{\varepsilon}^{\mathrm{o}}$ and $\boldsymbol{\sigma}^{\mathrm{o}}$, be prescribed for the large body B. The large volume B is statistically homogeneous if the moving average strains and stresses taken over *any sufficiently large volume* V *are*

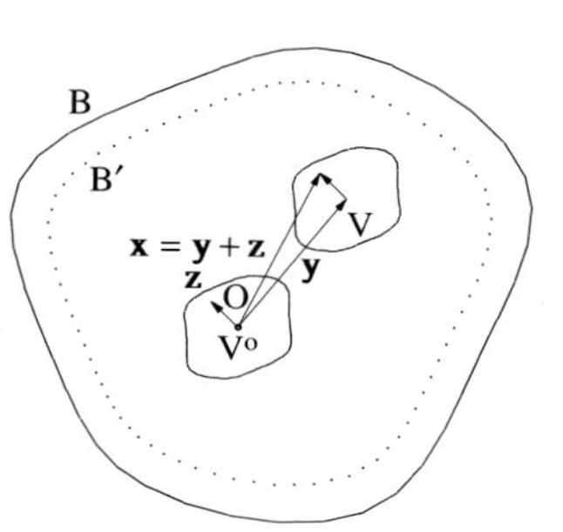

Figure 2.6.1

Large body B, its interior region B′, and RVE, V, obtained by translation of V^o

independent of the location of the centroid **y** *of* V *within* B′ *and the shape and size of* V. Thus, averages taken over any RVE in B′ are essentially the same as those taken over B when the RVE is suitably large. For statistically homogeneous B, it therefore follows that

$$\bar{\boldsymbol{\varepsilon}}(\mathbf{y}) \approx \boldsymbol{\varepsilon}^o, \qquad \bar{\boldsymbol{\sigma}}(\mathbf{y}) \approx \boldsymbol{\sigma}^o, \tag{2.6.4a,b}$$

where $\boldsymbol{\varepsilon}^o$ and $\boldsymbol{\sigma}^o$ are the prescribed uniform farfield strains and stresses.

It is of interest to examine whether or not the statistical homogeneity of B implies that the volume average of the product of the strain and stress tensors taken over any RVE equals the product of their respective volume averages, i.e., whether or not

$$\overline{(\boldsymbol{\sigma} : \boldsymbol{\varepsilon})}(\mathbf{y}) \approx \bar{\boldsymbol{\sigma}}(\mathbf{y}) : \bar{\boldsymbol{\varepsilon}}(\mathbf{y}) \approx \boldsymbol{\sigma}^o : \boldsymbol{\varepsilon}^o, \tag{2.6.5a}$$

where

$$\overline{(\boldsymbol{\sigma} : \boldsymbol{\varepsilon})}(\mathbf{y}) \equiv \frac{1}{V^o} \int_{V^o} \boldsymbol{\sigma}(\mathbf{y}+\mathbf{z}) : \boldsymbol{\varepsilon}(\mathbf{y}+\mathbf{z})\, dV_z. \tag{2.6.5b}$$

The validity of (2.6.5a) ensures that the overall moduli defined through the average strain energy or complementary strain energy are essentially the same as those defined through the average stress-strain relations. Before discussing this issue, consider an alternative volume averaging method which involves a *smooth weighting function.*[10]

[10] The following brief account is closely related to the work of Murat and Tartar (1985) and Francfort and Murat (1986), although it is cast in a less mathematical language; comments by Willis (1992) are gratefully acknowledged.

2.6.1. Local Average Fields

Volume averaging over any volume V may be performed in terms of smooth weighting functions. In the following, volume V is always viewed as the collection of points obtained by rigid translation of points within V^o. Thus, as $\mathbf{z}$ varies within V^o, $\mathbf{x} = \mathbf{y} + \mathbf{z}$ defines V for given $\mathbf{y}$ and V^o; Figure 2.6.1. In this manner, the domain of variation of $\mathbf{z} = \mathbf{x} - \mathbf{y}$ is always V^o. The *unweighted* volume average of field $(...) = (...)(\mathbf{x})$ taken over V is written as

$$\overline{(...)}(\mathbf{y}) \equiv \frac{1}{V^o} \int_B H(\mathbf{x}-\mathbf{y};\, V^o)\,(...)(\mathbf{x})\, dV_x, \tag{2.6.6a}$$

where $H(\mathbf{z};\, V^o)$ is the Heaviside step function,[11] having the value 1 for $\mathbf{z}$ in V^o, and zero otherwise. Since H is discontinuous, the computation of the derivatives of $\overline{(...)}(\mathbf{y};\, V^o)$ with respect to $\mathbf{y}$ is not straightforward. Hence, instead of H, introduce a suitably *smooth* function $\phi = \phi(\mathbf{z};\, V^o)$ with the following properties:

$$\phi(\mathbf{z};\, V^o) \begin{cases} \geq 0 & \text{for } \mathbf{z} \text{ in } V^o \\ = 0 & \text{otherwise,} \end{cases}$$

$$\frac{1}{V^o} \int_{V^o} \phi(\mathbf{z};\, V^o)\, dV_z = \frac{1}{V^o} \int_B \phi(\mathbf{x};\, V^o)\, dV_x = 1. \tag{2.6.7a,b}$$

The function ϕ and its derivatives may be required to vary as smoothly as needed, within and close to the boundary ∂V^o of V^o. Except for this smoothness property, $\phi(\mathbf{z};\, V^o)$ plays a role similar to the Heaviside step function $H(\mathbf{z};\, V^o)$.

The averaging operation (2.6.6a) may be replaced by the operation which uses the weighting function ϕ to define a *local average field value at* $\mathbf{y}$,

$$(...)^a(\mathbf{y};\, \phi) \equiv \frac{1}{V^o} \int_B \phi(\mathbf{x}-\mathbf{y};\, V^o)\,(...)(\mathbf{x})\, dV_x. \tag{2.6.6b}$$

The differentiation of the local average field $(...)^a(\mathbf{y};\, \phi)$ with respect to $\mathbf{y}$ is given by

$$\begin{aligned} \frac{\partial}{\partial y_i}(...)^a(\mathbf{y};\, \phi) &= \frac{1}{V^o} \int_B \left[\frac{\partial}{\partial y_i} \phi(\mathbf{x}-\mathbf{y};\, V^o) \right] (...)(\mathbf{x})\, dV_x \\ &= \frac{1}{V^o} \int_B \phi(\mathbf{x}-\mathbf{y};\, V^o) \left[\frac{\partial}{\partial x_i}(...)(\mathbf{x}) \right] dV_x \\ &= \left[\frac{\partial}{\partial x_i}(...) \right]^a(\mathbf{y};\, \phi), \end{aligned} \tag{2.6.8}$$

since $\phi(\mathbf{x}-\mathbf{y};\, V^o) = \phi(\mathbf{z};\, V^o)$ is smooth and vanishes smoothly at points close to ∂V^o. Hence, *the operation of differentiation commutes with the operation of local averaging with weighting function* ϕ. This weighting function, therefore, serves as the *smoothing function*.

The moving average of the strain and stress fields, (2.6.3a,b), is now replaced by the local average in terms of the smoothing function ϕ, i.e.,

---

[11] Function $H(\mathbf{z};\, V^o)$ is also called the characteristic function of V^o.

$$\boldsymbol{\varepsilon}^a(\mathbf{y};\,\phi) \equiv \frac{1}{V^o}\int_B \phi(\mathbf{x}-\mathbf{y};\,V^o)\,\boldsymbol{\varepsilon}(\mathbf{x})\,dV_\mathbf{x},$$

$$\boldsymbol{\sigma}^a(\mathbf{y};\,\phi) \equiv \frac{1}{V^o}\int_B \phi(\mathbf{x}-\mathbf{y};\,V^o)\,\boldsymbol{\sigma}(\mathbf{x})\,dV_\mathbf{x}. \tag{2.6.9a,b}$$

Since local averaging commutes with differentiation, $\boldsymbol{\varepsilon}^a$ is regarded compatible, and $\boldsymbol{\sigma}^a$ divergence-free, i.e.,

$$\boldsymbol{\varepsilon}^a(\mathbf{y};\,\phi) = sym\,\{\boldsymbol{\nabla}_\mathbf{y}\otimes\mathbf{u}^a(\mathbf{y};\,\phi)\}, \qquad \boldsymbol{\nabla}_\mathbf{y}\boldsymbol{\cdot}\boldsymbol{\sigma}(\mathbf{y};\,\phi) = \mathbf{0}, \tag{2.6.10a,b}$$

where $\mathbf{u}^a$ is the displacement field associated with $\boldsymbol{\varepsilon}^a$, and *sym* stands for the symmetric part of the second-order tensor.

Now, suppose that the local average strain and stress fields satisfy

$$\boldsymbol{\varepsilon}^a(\mathbf{y};\,\phi) \approx \boldsymbol{\varepsilon}^o, \qquad \boldsymbol{\sigma}^a(\mathbf{y};\,\phi) \approx \boldsymbol{\sigma}^o, \tag{2.6.11a,b}$$

for some suitably smooth weighting function ϕ which satisfies (2.6.7a,b), and for any sufficiently large V^o. Since $\phi(\mathbf{x};\,V^o)$ can be chosen to be close to $H(\mathbf{x};\,V^o)$, condition (2.6.11a,b) ensures that the statistical homogeneity, (2.6.4a), remains valid. Hence, if $(\boldsymbol{\sigma}:\boldsymbol{\varepsilon})^a$ which is defined by

$$(\boldsymbol{\sigma}:\boldsymbol{\varepsilon})^a(\mathbf{y};\,\phi) \equiv \frac{1}{V^o}\int_B \phi(\mathbf{x}-\mathbf{y};\,V^o)\,\boldsymbol{\sigma}(\mathbf{x}):\boldsymbol{\varepsilon}(\mathbf{x})\,dV_\mathbf{x}, \tag{2.6.9c}$$

satisfies

$$(\boldsymbol{\sigma}:\boldsymbol{\varepsilon})^a(\mathbf{y};\,\phi) \approx \boldsymbol{\sigma}^a(\mathbf{y};\,\phi):\boldsymbol{\varepsilon}^a(\mathbf{y};\,\phi) \approx \boldsymbol{\sigma}^o:\boldsymbol{\varepsilon}^o, \tag{2.6.11c}$$

for some suitable ϕ, then (2.6.5a) holds. Therefore, instead of seeking to obtain the conditions for the validity of (2.6.5a), consider the condition for the validity of (2.6.11c).

Let $\boldsymbol{\varepsilon}^0(\mathbf{x}) = \boldsymbol{\varepsilon}^o$ and $\boldsymbol{\sigma}^0(\mathbf{x}) = \boldsymbol{\sigma}^o$ be constant fields in B, and let $\mathbf{u}^0(\mathbf{x}) = \mathbf{x}\boldsymbol{\cdot}\boldsymbol{\varepsilon}^o$ be the linear displacement field associated with $\boldsymbol{\varepsilon}^0$. Based on the definition of $\boldsymbol{\varepsilon}^a$ and $\boldsymbol{\sigma}^a$, (2.6.11c) is rewritten as

$$\frac{1}{V^o}\int_B \phi(\mathbf{x}-\mathbf{y};\,V^o)\,\{\boldsymbol{\sigma}(\mathbf{x}):\boldsymbol{\varepsilon}(\mathbf{x})-\boldsymbol{\sigma}^0(\mathbf{x}):\boldsymbol{\varepsilon}^0(\mathbf{x})\}\,dV_\mathbf{x} \approx 0, \tag{2.6.12}$$

since ϕ satisfies (2.6.7b). The integrand in the left side of (2.6.12) is rearranged as

$$\boldsymbol{\sigma}:\boldsymbol{\varepsilon}-\boldsymbol{\sigma}^0:\boldsymbol{\varepsilon}^0 = (\boldsymbol{\sigma}-\boldsymbol{\sigma}^0):(\boldsymbol{\varepsilon}-\boldsymbol{\varepsilon}^0)+\boldsymbol{\sigma}^0:(\boldsymbol{\varepsilon}-\boldsymbol{\varepsilon}^0)+(\boldsymbol{\sigma}-\boldsymbol{\sigma}^0):\boldsymbol{\varepsilon}^0. \tag{2.6.13a}$$

Since $\boldsymbol{\sigma}^0$ and $\boldsymbol{\varepsilon}^0$ are constant fields, the local average of the last two terms is approximately zero; in view of (2.6.11a,b),

$$\boldsymbol{\sigma}^0:\left[\frac{1}{V^o}\int_B \phi(\mathbf{x}-\mathbf{y};\,V^o)\,\{\boldsymbol{\varepsilon}(\mathbf{x})-\boldsymbol{\varepsilon}^0(\mathbf{x})\}\,dV_\mathbf{x}\right] \approx 0,$$

$$\boldsymbol{\varepsilon}^0:\left[\frac{1}{V^o}\int_B \phi(\mathbf{x}-\mathbf{y};\,V^o)\,\{\boldsymbol{\sigma}(\mathbf{x})-\boldsymbol{\sigma}^0(\mathbf{x})\}\,dV_\mathbf{x}\right] \approx 0. \tag{2.6.14a,b}$$

Hence, the local average of (2.6.13a) becomes

$$\frac{1}{V^o}\int_B \phi(\mathbf{x}-\mathbf{y};\, V^o)\,\{\boldsymbol{\sigma}(\mathbf{x}) : \boldsymbol{\varepsilon}(\mathbf{x}) - \boldsymbol{\sigma}^0(\mathbf{x}) : \boldsymbol{\varepsilon}^0(\mathbf{x})\}\, dV_x$$

$$\approx \frac{1}{V^o}\int_B \phi(\mathbf{x}-\mathbf{y};\, V^o)\,\{\boldsymbol{\sigma}(\mathbf{x}) - \boldsymbol{\sigma}^0(\mathbf{x})\} : \{\boldsymbol{\varepsilon}(\mathbf{x}) - \boldsymbol{\varepsilon}^0(\mathbf{x})\}\, dV_x. \tag{2.6.13b}$$

Since $\boldsymbol{\varepsilon}(\mathbf{x})$ and $\boldsymbol{\varepsilon}^0(\mathbf{x})$ are the symmetric parts of the gradient of $\mathbf{u}(\mathbf{x})$ and $\mathbf{u}^0(\mathbf{x}) = \mathbf{x} \cdot \boldsymbol{\varepsilon}^o$, respectively, using the divergence theorem, rewrite the right side of (2.6.13b) as

$$\frac{1}{V^o}\int_B \phi(\mathbf{x}-\mathbf{y};\, V^o)\,\{\boldsymbol{\sigma}(\mathbf{x}) - \boldsymbol{\sigma}^0(\mathbf{x})\} : \{sym\,(\boldsymbol{\nabla}\otimes\mathbf{u})(\mathbf{x}) - sym\,(\boldsymbol{\nabla}\otimes\mathbf{u}^0)(\mathbf{x})\}\, dV_x$$

$$= \frac{1}{V^o}\int_{\partial B} \phi(\mathbf{x}-\mathbf{y};\, V^o)\,\{\boldsymbol{\sigma}(\mathbf{x}) - \boldsymbol{\sigma}^0(\mathbf{x})\} : \{\boldsymbol{\nu}(\mathbf{x})\otimes(\mathbf{u}(\mathbf{x}) - \mathbf{u}^0(\mathbf{x}))\}\, dV_x$$

$$- \frac{1}{V^o}\int_B \phi(\mathbf{x}-\mathbf{y};\, V^o)\,\{\boldsymbol{\nabla}\cdot(\boldsymbol{\sigma}(\mathbf{x}) - \boldsymbol{\sigma}^0(\mathbf{x}))\}\cdot\{\mathbf{u}(\mathbf{x}) - \mathbf{u}^0(\mathbf{x})\}\, dV_x$$

$$- \frac{1}{V^o}\int_B \{\boldsymbol{\sigma}(\mathbf{x}) - \boldsymbol{\sigma}^0(\mathbf{x})\} : \left[\,\{\boldsymbol{\nabla}\otimes\phi(\mathbf{x}-\mathbf{y})\}\otimes\{\mathbf{u}(\mathbf{x}) - \mathbf{u}^0(\mathbf{x})\}\right] dV_x. \tag{2.6.13c}$$

The first integral vanishes, since ϕ is zero outside of V^o, and the second integral vanishes, since both $\boldsymbol{\sigma}$ and $\boldsymbol{\sigma}^0$ are divergence-free. With the aid of the Schwartz inequality, the absolute value of the last integral is evaluated to be

$$\frac{1}{V^o}\left|\int_B \{\boldsymbol{\sigma}(\mathbf{x}) - \boldsymbol{\sigma}^0(\mathbf{x})\} : \left[\,\{\boldsymbol{\nabla}\otimes\phi(\mathbf{x}-\mathbf{y})\}\otimes\{\mathbf{u}(\mathbf{x}) - \mathbf{u}^0(\mathbf{x})\}\right] dV_x\right|^2$$

$$= \frac{1}{V^o}\left|\int_V \{\boldsymbol{\sigma}(\mathbf{x}) - \boldsymbol{\sigma}^0(\mathbf{x})\} : \left[\,\{\boldsymbol{\nabla}\otimes\phi(\mathbf{x}-\mathbf{y})\}\otimes\{\mathbf{u}(\mathbf{x}) - \mathbf{u}^0(\mathbf{x})\}\right] dV_x\right|^2$$

$$< \left[\frac{1}{V^o}\int_{V^o} |\boldsymbol{\nabla}\otimes\phi(\mathbf{z})|^2\, dV_z\right]\left[\frac{1}{V}\int_V |\boldsymbol{\sigma}(\mathbf{x}) - \boldsymbol{\sigma}^0(\mathbf{x})|^2\, dV_x\right]$$

$$\left[\frac{1}{V}\int_V |\mathbf{u}(\mathbf{x}) - \mathbf{u}^0(\mathbf{x})|^2\, dV_x\right]. \tag{2.6.15}$$

The first two integrals are bounded, the first because the smoothing function $\phi(\mathbf{x};\, V^o)$ can be chosen to satisfy this condition, and the second, since the average of the squared deviation of the stress field $\boldsymbol{\sigma}(\mathbf{x})$ from $\boldsymbol{\sigma}^0(\mathbf{x}) = \boldsymbol{\sigma}^o$ taken over V should remain bounded when the total complementary energy density is bounded.

Therefore, if it is assumed that

$$\frac{1}{V}\int_V |\mathbf{u}(\mathbf{x}) - \mathbf{u}^0(\mathbf{x})|^2\, dV_x \approx 0, \tag{2.6.16}$$

then the validity of (2.6.11c), and hence the validity of (2.6.5a), are ensured. While the strain field $\boldsymbol{\varepsilon}$ may have wild variations about $\boldsymbol{\varepsilon}^0$, the displacement field $\mathbf{u}$ remains continuous. It is intuitively clear that the magnitude of the fluctuation of $\mathbf{u}$ about $\mathbf{u}^0$ should decrease as the size of the microconstituents decreases. Hence, it is expected that (2.6.6) should hold if d is suitably small, as discussed below.

2.6.2. Limiting Process and Limit Fields

Condition (2.6.1a) is satisfied if the size of the RVE, D, is fixed and the size of the micro-inhomogeneities, d, becomes infinitesimally small. Physically, this corresponds to the case where the "gauge length" (in this case D) is fixed and the size of the micro-inhomogeneities becomes vanishingly small, i.e., the material behaves essentially as if *microstructurally homogeneous*. It is clear that, in this case, the difference between the actual displacement field, **u**, and the smooth field $\mathbf{u}^0$ (or equivalently, the local averaged displacement field $\mathbf{u}^a$) becomes vanishingly small, although the first represents the actual particle displacement at the microscale, and the second relates to the locally averaged deformation field. At the limit as $d \to 0$, with D fixed, the integral in the left side of (2.6.16) is identically zero, and hence the energy density averaged over an RVE, $\langle \boldsymbol{\sigma} : \boldsymbol{\varepsilon} \rangle / 2$, equals the energy density associated with the corresponding average fields, $\bar{\boldsymbol{\sigma}} : \bar{\boldsymbol{\varepsilon}} / 2$.

From the physical point of view, on the other hand, it is generally necessary to deal with finite-sized microheterogeneities. That is, the physics of the problem at hand dictates the required minimum microscale d, and it is the gauge length D which must be chosen suitably large to accommodate the required reproducible macroscopic measurements, i.e., the averaging is now satisfied by keeping d fixed (at its minimum value) and choosing D to be suitably large. In this case, the amplitude of the fluctuation of the actual displacement field **u** measured at the microscale relative to the smooth local average field $\mathbf{u}^0$, does not necessarily become infinitesimally small, as the size of the RVE, namely D, is increased. Indeed, counterexamples can be constructed, e.g., in terms of the periodic or quasi-periodic microstructures, which show that the density of $|\mathbf{u} - \mathbf{u}^0|^2$ may remain essentially unchanged,[12] with d fixed and $D \to \infty$. For problems of this kind, therefore, relations based on $d/D \to 0$ with vanishingly small micro-inhomogeneities, may not be relevant.

In practical applications, it may, however, often happen that the inhomogeneity size d is indeed small enough to only introduce tolerable errors, so that, while the left side of (2.6.16) remains finite, it may be regarded as essentially zero. Notwithstanding this, the distinction between $d/D \to 0$ with D fixed and $d \to 0$ on the one hand, and with d fixed and $D \to \infty$ on the other hand, should be kept in mind when interpreting the corresponding results.

---

[12] Although the absolute value of $|\mathbf{u} - \mathbf{u}^0|^2$ may remain essentially unchanged if d is fixed, the relative value of $|\mathbf{u} - \mathbf{u}^0|^2$ with respect to D decreases as D increases. This means that even if the size of the microconstituents is kept the same, the fluctuation of the displacement relative to the size of a sufficiently large gauge length, becomes vanishingly small.

2.7. NONMECHANICAL PROPERTIES

The results of the preceding subsections can easily be specialized for application to electrostatic, magnetostatic, thermal, and diffusional properties of a heterogeneous RVE. Except for different physical interpretations of the field quantities and the associated material parameters, the basic steady state field equations necessary for the present application, are essentially the same. Steady state thermal conduction, mass diffusion, and electrostatics can be considered simultaneously, as commented on at the end of Subsection 1.3.

With reference to equations (1.3.11) to (1.3.14), let u stand for {temperature, or pore pressure, or electric potential} when {thermal conduction, or mass diffusion, or electrostatics} is considered. Then $\mathbf{p} = -\nabla u$ may be viewed as the corresponding (thermodynamic) *force*, i.e., {– pressure gradient, or – temperature gradient, or electric field}, with $\mathbf{q}$ identifying the associated flux, i.e., {mass flux, or heat flux, or electric displacement}. In this subsection, general results of the preceding subsections are reduced and applied to this class of problems.

2.7.1. Averaging Theorems

Consider the volume average of the force field $\mathbf{p} = \mathbf{p}(\mathbf{x})$ and the flux field $\mathbf{q} = \mathbf{q}(\mathbf{x})$ over an RVE. Whatever the nature of the boundary conditions, these averages, denoted by $\bar{\mathbf{p}}$ and $\bar{\mathbf{q}}$, are completely defined in terms of the corresponding boundary data.

Indeed, from definition (1.3.14a) and the application of the Gauss theorem, it immediately follows that

$$\bar{\mathbf{p}} \equiv < \mathbf{p} > = -\frac{1}{V}\int_{\partial V} \boldsymbol{\nu}\, u^o\, dS, \tag{2.7.1}$$

when the boundary temperature[13] (not necessarily linear), u^o, is prescribed. Similarly, from (1.3.11a) and (1.3.12a) it is deduced that

$$\bar{\mathbf{q}} \equiv < \mathbf{q} > = \frac{1}{V}\int_{\partial V} q^o\, \mathbf{x}\, dS, \tag{2.7.2}$$

for any self-balanced surface flux $\boldsymbol{\nu} \cdot \mathbf{q}^o = q^o$ (not necessarily uniform). Moreover, from identity

$$< \mathbf{q} \cdot \mathbf{p} > - < \mathbf{q} > \cdot < \mathbf{p} > = \frac{1}{V}\int_{\partial V} (-u + \mathbf{x} \cdot < \mathbf{p} >)\, \{\boldsymbol{\nu} \cdot (\mathbf{q} - < \mathbf{q} >)\}\, dS, \tag{2.7.3}$$

it follows that

$$< \mathbf{q} \cdot \mathbf{p} > = < \mathbf{q} > \cdot < \mathbf{p} >, \tag{2.7.4}$$

whenever either *uniform* boundary flux $\mathbf{q}^o$,

---

[13] For the sake of simplicity in referencing, the results will be illustrated in terms of the thermal conduction problems.

$$\boldsymbol{\nu}\cdot\mathbf{q} = \boldsymbol{\nu}\cdot\mathbf{q}^{o} \quad \text{on } \partial V, \tag{2.7.5a}$$

or *linear* boundary potential,

$$u = u^{o} = -\,\mathbf{x}\cdot\mathbf{p}^{o} \quad \text{on } \partial V, \tag{2.7.6a}$$

is prescribed. In the first case,

$$\bar{\mathbf{q}} \equiv \langle q \rangle = \mathbf{q}^{o}, \tag{2.7.5b}$$

and in the second case,

$$\bar{\mathbf{p}} \equiv \langle p \rangle = \mathbf{p}^{o}, \tag{2.7.6b}$$

exactly.

2.7.2. Macropotentials

Essentially all the results presented in Subsections 2.5 and 2.6 directly translate and apply to this class of problems. All that is needed is to identify $\mathbf{p} = -\,\boldsymbol{\nabla}u$ with the strain tensor, and the corresponding flux, $\mathbf{q}$, with the stress tensor; an important difference, however, being that $\mathbf{p}$ and $\mathbf{q}$ are now *vector* fields.

Suppose the material of the RVE admits potentials, $\psi = \psi(\mathbf{x}, \mathbf{p})$ and $\phi = \phi(\mathbf{x}, \mathbf{q})$, such that

$$\mathbf{q} = \frac{\partial \psi}{\partial \mathbf{p}}(\mathbf{x}, \mathbf{p}), \qquad \mathbf{p} = \frac{\partial \phi}{\partial \mathbf{q}}(\mathbf{x}, \mathbf{q}). \tag{2.7.8a,b}$$

The two potentials are, therefore, related through the Legendre transformation,

$$\psi + \phi = \mathbf{q}\cdot\mathbf{p}. \tag{2.7.8c}$$

Let $\mathbf{P}$ be a *constant* vector, and consider the linear boundary data,

$$u = -\,\mathbf{x}\cdot\mathbf{P} \quad \text{on } \partial V, \tag{2.7.9a}$$

for this heterogeneous RVE. Let the resulting force and flux fields be given by

$$\mathbf{p} = \mathbf{p}(\mathbf{x};\, \mathbf{P}), \qquad \mathbf{q} = \mathbf{q}(\mathbf{x};\, \mathbf{P}). \tag{2.7.9b,c}$$

Define the macropotential $\Psi^{P}(\mathbf{P})$ by

$$\Psi^{P}(\mathbf{P}) \equiv \langle \psi(\mathbf{x}, \mathbf{p}(\mathbf{x};\, \mathbf{P})) \rangle = \langle \psi^{P}(\mathbf{x};\, \mathbf{P}) \rangle, \tag{2.7.10a}$$

and obtain (see Subsection 2.5.1),

$$\mathbf{Q}^{P} \equiv \langle \mathbf{q}(\mathbf{x};\, \mathbf{P}) \rangle = \frac{\partial \Psi^{P}}{\partial \mathbf{P}}. \tag{2.7.10b}$$

In a similar manner, for uniform boundary data,

$$\boldsymbol{\nu}\cdot\mathbf{q} = \boldsymbol{\nu}\cdot\mathbf{Q} \quad \text{on } \partial V, \tag{2.7.11a}$$

where $\mathbf{Q}$ is a *constant* vector, write the resulting force and flux fields as

$$\mathbf{p} = \mathbf{p}(\mathbf{x};\, \mathbf{Q}), \qquad \mathbf{q} = \mathbf{q}(\mathbf{x};\, \mathbf{Q}). \tag{2.7.11b,c}$$

Then, defining the macropotential $\Phi^Q(\mathbf{Q})$ by

$$\Phi^Q(\mathbf{Q}) \equiv < \phi(\mathbf{x},\, \mathbf{q}(\mathbf{x};\, \mathbf{Q})) > = < \phi^Q(\mathbf{x};\, \mathbf{Q}) >, \tag{2.7.12a}$$

arrive at

$$\mathbf{P}^Q \equiv < \mathbf{p}(\mathbf{x};\, \mathbf{Q}) > = \frac{\partial \Phi^Q}{\partial \mathbf{Q}}. \tag{2.7.12b}$$

Finally, relations between the macropotentials and the corresponding definition of an RVE follow directly from Subsections 2.5.3 and 2.5.4; see also Table 2.5.1.

2.7.3. Basic Inequalities

When the potentials $\psi(\mathbf{x}, \mathbf{p})$ and $\phi(\mathbf{x}, \mathbf{q})$ are convex in the sense of[14] (2.5.31), then an analysis similar to that of Subsection 2.5.6 immediately leads to the following two basic theorems:

Theorem I: *For an RVE whose microconstituents admit convex potentials, among all consistent boundary data which produce the same overall macroforce,* $\mathbf{P}$, *the uniform flux boundary data render the total macroflux potential* $\Psi(\mathbf{P})$ *an absolute minimum.*

Theorem II: *For an RVE whose microconstituents admit convex potentials, among all consistent boundary data which produce the same overall macroflux,* $\mathbf{Q}$, *the linear boundary data associated with a uniform force, render the total macroforce potential* $\Phi(\mathbf{Q})$ *an absolute minimum.*

Thus, if the macroforce associated with the general boundary data is denoted by $\mathbf{P}^G$, and the corresponding macroflux potential by $\Psi^G(\mathbf{P}^G)$, then it follows that

$$\Psi^G(\mathbf{P}^G) > \Psi^Q(\mathbf{P}^Q) \qquad \text{for } \mathbf{P}^G = \mathbf{P}^Q. \tag{2.7.13a}$$

In particular, when the boundary data are defined by the linear relation (2.7.9a), then

$$\Psi^P(\mathbf{P}) \geq \Psi^Q(\mathbf{P}). \tag{2.7.13b}$$

In a similar manner, it follows from Theorem II that, for any general but consistent boundary data which produce the macroflux $\mathbf{Q}^G$,

$$\Phi^G(\mathbf{Q}^G) > \Phi^P(\mathbf{Q}^P) \qquad \text{for } \mathbf{Q}^G = \mathbf{Q}^P. \tag{2.7.14a}$$

Moreover, when uniform boundary fluxes are prescribed,

---

[14] Replace, e.g., $\boldsymbol{\varepsilon}$ in (2.5.31) by $\mathbf{p}$.

$$\Phi^Q(\mathbf{Q}) \geq \Phi^P(\mathbf{Q}). \tag{2.7.14b}$$

Consider now linear RVE's for which the local potentials are given by

$$\psi(\mathbf{x}, \mathbf{p}) = \frac{1}{2}\mathbf{p} \cdot \mathbf{K} \cdot \mathbf{p} = \frac{1}{2}K_{ij}\, p_i\, p_j, \tag{2.7.15a,b}$$

$$\phi(\mathbf{x}, \mathbf{q}) = \frac{1}{2}\mathbf{q} \cdot \mathbf{R} \cdot \mathbf{q} = \frac{1}{2}R_{ij}\, q_i\, q_j, \tag{2.7.16a,b}$$

where for thermal conduction, for example, $\mathbf{K}$ and $\mathbf{R}$ are conductivity and "resistivity" tensors[15] respectively, with $\mathbf{K} = \mathbf{K}^T = \mathbf{R}^{-1}$, $\mathbf{R} = \mathbf{R}^T = \mathbf{K}^{-1}$. Let

$$\Psi^G(\mathbf{P}^G) = \frac{1}{2}\mathbf{P}^G \cdot \bar{\mathbf{K}}^G \cdot \mathbf{P}^G, \qquad \Phi^G(\mathbf{Q}^G) = \frac{1}{2}\mathbf{Q}^G \cdot \bar{\mathbf{R}}^G \cdot \mathbf{Q}^G, \tag{2.7.17a,b}$$

correspond to any general consistent boundary data which produce an overall macroforce $\mathbf{P}^G$ and an overall macroflux $\mathbf{Q}^G$, satisfying

$$\frac{1}{V}\int_{\partial V} \{u^G + \mathbf{x} \cdot \mathbf{P}^G\}\{\boldsymbol{\nu} \cdot (\mathbf{q}^G - \mathbf{Q}^G)\}\, dS = 0, \tag{2.7.18a}$$

so that

$$< \mathbf{q}^G \cdot \mathbf{p}^G > = < \mathbf{q}^G > \cdot < \mathbf{p}^G > = \mathbf{Q}^G \cdot \mathbf{P}^G. \tag{2.7.18b}$$

In (2.7.17a,b), $\bar{\mathbf{K}}^G$ and $\bar{\mathbf{R}}^G$ are the overall, e.g., conductivity and resistivity tensors defined through

$$\mathbf{Q}^G = \bar{\mathbf{K}}^G \cdot \mathbf{P}^G, \qquad \mathbf{P}^G = \bar{\mathbf{R}}^G \cdot \mathbf{Q}^G, \tag{2.7.18c,d}$$

and hence *are each other's inverse*.

Each of the boundary conditions of uniform flux and linear, e.g., temperature, satisfies (2.7.18a). Define $\bar{\mathbf{K}}^Q$ and $\bar{\mathbf{R}}^P$, respectively, by

$$\Psi^Q(\mathbf{P}^Q) = \frac{1}{2}\mathbf{P}^Q \cdot \bar{\mathbf{K}}^Q \cdot \mathbf{P}^Q,$$

$$\Phi^P(\mathbf{Q}^P) = \frac{1}{2}\mathbf{Q}^P \cdot \bar{\mathbf{R}}^P \cdot \mathbf{Q}^P. \tag{2.7.19a,b}$$

The conductivity $\bar{\mathbf{K}}^Q$ is *not* the inverse of the resistivity $\bar{\mathbf{R}}^P$, since these tensors are defined for *different boundary data*.

According to Theorems I and II, it now follows that for any constant $\mathbf{P}$,

$$\mathbf{P} \cdot (\bar{\mathbf{K}}^G - \bar{\mathbf{K}}^Q) \cdot \mathbf{P} \geq 0, \tag{2.7.20a}$$

and for any constant $\mathbf{Q}$,

$$\mathbf{Q} \cdot (\bar{\mathbf{R}}^G - \bar{\mathbf{R}}^P) \cdot \mathbf{Q} \geq 0. \tag{2.7.20b}$$

Thus, uniform flux and linear, say, temperature boundary data provide lower and upper bounds for the conductivity associated with any other boundary data. In particular, when the second-order symmetric (and positive-definite) tensors

---

[15] In electrostatics, $\mathbf{K}$ is the dielectric tensor.

$\bar{\mathbf{K}}^G$, $\bar{\mathbf{K}}^Q$, $\bar{\mathbf{R}}^G$, and $\bar{\mathbf{R}}^P = (\bar{\mathbf{K}}^P)^{-1}$ are coaxial,[16] then denoting the corresponding principal values (all positive) by λ_a^G, λ_a^Q, and λ_a^P (a = 1, 2, 3), from coaxiality and Theorems I and II, it follows that

$$\lambda_a^G - \lambda_a^Q \geq 0, \qquad \frac{1}{\lambda_a^G} - \frac{1}{\lambda_a^P} \geq 0, \qquad \text{for } a = 1, 2, 3. \tag{2.7.21a,b}$$

In the special case when $\bar{\mathbf{K}}^G$, $\bar{\mathbf{K}}^Q$, and $\bar{\mathbf{K}}^P$ are isotropic, then each has only one distinct (positive) principal value, say, $\bar{K}^A$ (A = G, Q , P), and (2.7.21a,b) reduce to

$$\bar{K}^Q \leq \bar{K}^G \leq \bar{K}^P. \tag{2.7.21c}$$

All other comments regarding bounds in Subsections 2.5.6 and 2.5.7 also apply here. In particular, the energy bounds of Subsection 2.5.7 follow directly, producing lower and upper bounds for the effective conductivity; the derivation of these results is left as an exercise for the reader.

2.8. COUPLED MECHANICAL AND NONMECHANICAL PROPERTIES

The general results presented in the preceding subsections for uncoupled mechanical and nonmechanical properties of a heterogeneous RVE, directly carry over to the cases where the mechanical and nonmechanical properties are coupled. The case of piezoelectricity is considered in this subsection.[17] The results also apply to other cases, e.g., coupled thermoelasticity.[18] The field variables in piezoelectricity are the displacement, strain, and stress for the mechanical behavior, and the electric potential, electric field, and electric displacement for the nonmechanical response, i.e., $(\mathbf{u}, \boldsymbol{\varepsilon}, \boldsymbol{\sigma})$ and $(u, \mathbf{p}, \mathbf{q})$.

2.8.1. Field Equations

The spatial variation of the field variables, $(\mathbf{u}, \boldsymbol{\varepsilon}, \boldsymbol{\sigma})$ and $(u, \mathbf{p}, \mathbf{q})$, are governed by three sets of field equations. The first set ensures mechanical equilibrium and the absence of free charge. In the absence of body forces and free electric charges, $\boldsymbol{\sigma}$ and $\mathbf{q}$ are divergence-free,

$$\nabla \cdot \boldsymbol{\sigma} = \mathbf{0}, \qquad \nabla \cdot \mathbf{q} = 0, \tag{2.8.1a,b}$$

[16] I.e., they have the same principal directions.

[17] For an overview of the fundamentals of piezoelectricity, see Ikeda (1990), for linear vibration of piezoelectric plates, see Tiersten (1969), and for an account of nonlinear electromechanical effects, see Maugin (1985).

[18] For fundamentals of thermoelasticity with applications, see Boley and Weiner (1960), Nowacki (1962), Sneddon (1974), and Nowinski (1978).

or

$$\sigma_{ij,i} = 0, \qquad q_{i,i} = 0. \tag{2.8.1c,d}$$

The second set defines the strain tensor, $\boldsymbol{\varepsilon}$, and the electric field, $\mathbf{p}$, in terms of the displacement vector, $\mathbf{u}$, and the electric potential, u, as follows:

$$\boldsymbol{\varepsilon} = \frac{1}{2}((\boldsymbol{\nabla} \otimes \mathbf{u}) + (\boldsymbol{\nabla} \otimes \mathbf{u})^T), \qquad \mathbf{p} = -\boldsymbol{\nabla} u, \tag{2.8.2a,b}$$

or

$$\varepsilon_{ij} = \frac{1}{2}(u_{i,j} + u_{j,i}), \qquad p_i = -u_{,i}. \tag{2.8.2c,d}$$

The third set is constitutive, relating the strain and electric field, $\boldsymbol{\varepsilon}$ and $\mathbf{p}$, to the stress and electric displacement, $\boldsymbol{\sigma}$ and $\mathbf{q}$. For coupled mechanical and electric problems, consider[19] the linear relations which relate $\boldsymbol{\varepsilon}$ and $\mathbf{q}$ to $\boldsymbol{\sigma}$ and $\mathbf{p}$,

$$\boldsymbol{\sigma} = \mathbf{C} : \boldsymbol{\varepsilon} + \mathbf{h} \cdot \mathbf{q}, \qquad \mathbf{p} = \mathbf{h}^T : \boldsymbol{\varepsilon} + \mathbf{R} \cdot \mathbf{q}, \tag{2.8.3a,b}$$

or

$$\sigma_{ij} = C_{ijkl}\, \varepsilon_{kl} + h_{ijk}\, q_k, \qquad p_i = h_{kli}\, \varepsilon_{kl} + R_{ik}\, q_k, \tag{2.8.3c,d}$$

where $C_{ijkl} = C_{jikl} = C_{ijlk}$ and $h_{ijk} = h_{jik}$. Here, $\mathbf{h}$ is the coupling[20] modulus.

Instead of constitutive relations (2.8.3) which relate ($\boldsymbol{\varepsilon}$, $\mathbf{q}$) and ($\boldsymbol{\sigma}$, $\mathbf{p}$), alternative relations between ($\boldsymbol{\varepsilon}$, $\mathbf{p}$) and ($\boldsymbol{\sigma}$, $\mathbf{q}$) can be obtained from (2.8.3). These are

$$\boldsymbol{\sigma} = \mathbf{C}^P : \boldsymbol{\varepsilon} + \mathbf{i} \cdot \mathbf{p}, \qquad \mathbf{q} = -\mathbf{i}^T : \boldsymbol{\varepsilon} + \mathbf{K} \cdot \mathbf{p}, \tag{2.8.4a,b}$$

or

$$\sigma_{ij} = C^P_{ijkl}\, \varepsilon_{kl} + i_{ijk}\, p_k, \qquad q_i = -i_{kli}\, \varepsilon_{kl} + K_{ik}\, p_k, \tag{2.8.4c,d}$$

where $\mathbf{K} = \mathbf{R}^{-1}$, and $\mathbf{C}^P$ and $\mathbf{i}$ are given by

$$\mathbf{C}^P = \mathbf{C} - \mathbf{h} \cdot \mathbf{K} \cdot \mathbf{h}^T, \qquad \mathbf{i} = \mathbf{h} \cdot \mathbf{K}, \tag{2.8.5a,b}$$

or

$$C^P_{ijkl} = C_{ijkl} - h_{ijm}\, K_{mn}\, h_{kln}, \qquad i_{ijk} = h_{ijm}\, K_{mk}. \tag{2.8.5c,d}$$

Substituting (2.8.2) and (2.8.4) into (2.8.1), obtain the coupled governing equations for the mechanical displacement, $\mathbf{u}$, and electric potential, u, as follows:

---

[19] Constitutive relations (2.8.3) are proposed by Lothe and Barnett (1977); see also Lothe and Barnett (1976). In this form, the coupling moduli retain a certain symmetry, i.e., $h_{ijk}\, q_k$ and $h_{ijk}\, \varepsilon_{ij}$ occur in the expressions for σ_{ij} and p_k, respectively. This is in contrast to the representation (2.8.4) which uses anti-symmetric coupling moduli, i.e., $i_{ijk}\, p_k$ and $-i_{ijk}\, \varepsilon_{ij}$ occur in the expressions for σ_{ij} and q_k.

[20] The superscript T on a third-order tensor stands for *transpose* in a symbolic sense so that $\mathbf{h}^T : \boldsymbol{\varepsilon}$ corresponds to $h_{kli}\, \varepsilon_{kl}$.

$$\boldsymbol{\nabla}\boldsymbol{.}(\mathbf{C}^{\mathrm{P}} : (\boldsymbol{\nabla}\otimes\mathbf{u})) - \boldsymbol{\nabla}\boldsymbol{.}(\mathbf{i}\boldsymbol{.}(\boldsymbol{\nabla}\mathrm{u})) = \mathbf{0}, \qquad -\boldsymbol{\nabla}\boldsymbol{.}(\mathbf{i}^{\mathrm{T}} : (\boldsymbol{\nabla}\otimes\mathbf{u})) - \boldsymbol{\nabla}\boldsymbol{.}(\mathbf{K}\boldsymbol{.}(\boldsymbol{\nabla}\mathrm{u})) = 0, \tag{2.8.6a,b}$$

or

$$(\mathrm{C}^{\mathrm{P}}_{ijkl}\, \mathrm{u}_{k,l})_{,i} - (\mathrm{i}_{ijk}\, \mathrm{u}_{,k})_{,i} = 0, \qquad -(\mathrm{i}_{kli}\, \mathrm{u}_{k,l})_{,i} - (\mathrm{K}_{ik}\, \mathrm{u}_{,k})_{,i} = 0. \tag{2.8.6c,d}$$

2.8.2. Averaging Theorems

The averaging theorems presented in Subsections 2.1 to 2.4 and 2.7, for uncoupled mechanical and nonmechanical properties of heterogeneous materials, are independent of the associated constitutive properties. Thus, they hold for coupled systems.

Consider an RVE with appropriate (consistent) boundary data, as discussed in Sections 2.1 to 2.4, and 2.7. Denote the surface data (whether prescribed or not) by superscript o, e.g., $\mathbf{u}^{\mathrm{o}}$, $\mathbf{t}^{\mathrm{o}}$, u^{o}, and q^{o}, respectively, standing for the values of surface displacement vector, surface traction vector, surface electric potential, and normal surface flux. Then, the volume average of the associated field variables, taken over the RVE, is given in terms of the surface data by ($\boldsymbol{\nu}$ is the exterior unit normal of $\partial\mathrm{V}$)

$$<\boldsymbol{\sigma}> = \frac{1}{\mathrm{V}}\int_{\partial\mathrm{V}} \mathbf{x}\otimes\mathbf{t}^{\mathrm{o}}\, \mathrm{dS}, \qquad <\boldsymbol{\varepsilon}> = \frac{1}{\mathrm{V}}\int_{\partial\mathrm{V}} \frac{1}{2}((\boldsymbol{\nu}\otimes\mathbf{u}^{\mathrm{o}}) + (\boldsymbol{\nu}\otimes\mathbf{u}^{\mathrm{o}})^{\mathrm{T}})\, \mathrm{dS},$$

$$<\mathbf{q}> = \frac{1}{\mathrm{V}}\int_{\partial\mathrm{V}} \mathbf{x}\, \mathrm{q}^{\mathrm{o}}\, \mathrm{dS}, \qquad <\mathbf{p}> = -\frac{1}{\mathrm{V}}\int_{\partial\mathrm{V}} \boldsymbol{\nu}\, \mathrm{u}^{\mathrm{o}}\, \mathrm{dS}, \tag{2.8.7a~d}$$

or

$$<\sigma_{ij}> = \frac{1}{\mathrm{V}}\int_{\partial\mathrm{V}} \mathrm{x}_i\, \mathrm{t}^{\mathrm{o}}_j\, \mathrm{dS}, \qquad <\varepsilon_{ij}> = \frac{1}{\mathrm{V}}\int_{\partial\mathrm{V}} \frac{1}{2}(\nu_i\, \mathrm{u}^{\mathrm{o}}_j + \nu_j\, \mathrm{u}^{\mathrm{o}}_i)\, \mathrm{dS},$$

$$<\mathrm{q}_i> = \frac{1}{\mathrm{V}}\int_{\partial\mathrm{V}} \mathrm{x}_i\, \mathrm{q}^{\mathrm{o}}\, \mathrm{dS}, \qquad <\mathrm{p}_i> = -\frac{1}{\mathrm{V}}\int_{\partial\mathrm{V}} \nu_i\, \mathrm{u}^{\mathrm{o}}\, \mathrm{dS}. \tag{2.8.7e~h}$$

The averaging theorems for the product of the strain and stress and the product of the electric field and electric displacement also hold, i.e.,

$$<\boldsymbol{\sigma}:\boldsymbol{\varepsilon}> = \frac{1}{\mathrm{V}}\int_{\partial\mathrm{V}} \mathbf{t}^{\mathrm{o}}\boldsymbol{.}\mathbf{u}^{\mathrm{o}}\, \mathrm{dS}, \qquad <\mathbf{q}\boldsymbol{.}\mathbf{p}> = -\frac{1}{\mathrm{V}}\int_{\partial\mathrm{V}} \mathrm{q}^{\mathrm{o}}\, \mathrm{u}^{\mathrm{o}}\, \mathrm{dS}, \tag{2.8.8a,b}$$

or

$$<\sigma_{ij}\, \varepsilon_{ij}> = \frac{1}{\mathrm{V}}\int_{\partial\mathrm{V}} \mathrm{t}^{\mathrm{o}}_i\, \mathrm{u}^{\mathrm{o}}_i\, \mathrm{dS}, \qquad <\mathrm{q}_i\, \mathrm{p}_i> = -\frac{1}{\mathrm{V}}\int_{\partial\mathrm{V}} \mathrm{q}^{\mathrm{o}}\, \mathrm{u}^{\mathrm{o}}\, \mathrm{dS}. \tag{2.8.8c,d}$$

Combining (2.8.7) and (2.8.8), compute the difference of the average of the products and the product of the averages, as follows:

$$<\boldsymbol{\sigma}:\boldsymbol{\varepsilon}> - <\boldsymbol{\sigma}>:<\boldsymbol{\varepsilon}> = \frac{1}{\mathrm{V}}\int_{\partial\mathrm{V}} (\mathbf{u}^{\mathrm{o}} - \mathbf{x}\boldsymbol{.}<\boldsymbol{\varepsilon}>)\boldsymbol{.}(\mathbf{t}^{\mathrm{o}} - \boldsymbol{\nu}\boldsymbol{.}<\boldsymbol{\sigma}>)\, \mathrm{dS},$$

$$< \mathbf{q} \cdot \mathbf{p} > - < \mathbf{q} > \cdot < \mathbf{p} > = \frac{1}{V} \int_{\partial V} (-u^o + \mathbf{x} \cdot < \mathbf{p} >) \cdot (\mathbf{q}^o - \boldsymbol{\nu} \cdot < \mathbf{q} >) \, dS,$$

(2.8.9a,b)

or

$$< \sigma_{ij} \, \varepsilon_{ij} > - < \sigma_{ij} > < \varepsilon_{ij} > = \frac{1}{V} \int_{\partial V} (u^o{}_i - x_j < \varepsilon_{ji} >)(t^o{}_i - \nu_k < \sigma_{ki} >) \, dS,$$

$$< q_i \, p_i > - < q_i > < p_i > = \frac{1}{V} \int_{\partial V} (-u^o + x_i < p_i >)(q^o - \nu_j < q_j >) \, dS.$$

(2.8.9c,d)

As in the preceding subsections, these identities are all exact and valid independently of the constitutive properties.

2.8.3. Stress/Electric-field Potential

Let the material of a heterogeneous RVE admit a potential function, ϕ, depending on the strain tensor, $\boldsymbol{\varepsilon}$, and the electric displacement, $\mathbf{q}$, such that the stress, $\boldsymbol{\sigma}$, and the electric field, $\mathbf{p}$, are given by

$$\boldsymbol{\sigma} = \frac{\partial \phi}{\partial \boldsymbol{\varepsilon}}(\mathbf{x}; \boldsymbol{\varepsilon}, \mathbf{q}), \qquad \mathbf{p} = \frac{\partial \phi}{\partial \mathbf{q}}(\mathbf{x}; \boldsymbol{\varepsilon}, \mathbf{q}). \tag{2.8.10a,b}$$

Note that $\phi = \phi(\mathbf{x}; \boldsymbol{\varepsilon}, \mathbf{q})$ is defined in terms of the strain $\boldsymbol{\varepsilon}$ and the electric displacement $\mathbf{q}$, instead of the electric field $\mathbf{p}$ which is used in Subsection 2.7. For the present purposes, it turns out that (2.8.10) is the appropriate representation of the coupling effects.[21] Then, when the material is linear this potential is given in terms of $(\mathbf{C}, \mathbf{h}, \mathbf{R})$ by

$$\phi = \frac{1}{2} \boldsymbol{\varepsilon} : \mathbf{C} : \boldsymbol{\varepsilon} + \boldsymbol{\varepsilon} : \mathbf{h} \cdot \mathbf{q} + \frac{1}{2} \mathbf{q} \cdot \mathbf{R} \cdot \mathbf{q}. \tag{2.8.11}$$

In this form, ϕ is positive-definite, as shown by Lothe and Barnett (1977), and convex. *For the bounding theorems to hold in the coupled cases, it is essential that correct pairs of the field variables are used.*

Suppose that the RVE is subjected to suitable boundary conditions which satisfy

$$\int_{\partial V} \left\{ (\mathbf{u} - \mathbf{x} \cdot < \boldsymbol{\varepsilon} >) \cdot (\mathbf{t} - \boldsymbol{\nu} \cdot < \boldsymbol{\sigma} >) + (-u + \mathbf{x} \cdot < p >)(q - \boldsymbol{\nu} \cdot < q >) \right\} dS = 0.$$

(2.8.12)

[21] It is not possible to express a suitable potential in terms of the strain $\boldsymbol{\varepsilon}$ and the electric field $\mathbf{p}$, using the constitutive relations (2.8.4); the product of the stress and strain and the product of the electric field and displacement are given by $\boldsymbol{\varepsilon} : (\mathbf{C}^P : \boldsymbol{\varepsilon} + \mathbf{i} : \mathbf{p})$ and $\mathbf{p} \cdot (-\mathbf{i}^T : \boldsymbol{\varepsilon} + \mathbf{K} \cdot \mathbf{p})$, respectively, and the coupling terms in these expressions, $\boldsymbol{\varepsilon} : \mathbf{i} \cdot \mathbf{p}$ and $-\mathbf{p} \cdot \mathbf{i}^T : \boldsymbol{\varepsilon}$, vanish when these two products are summed.

In view of (2.8.9), it then follows that

$$<\boldsymbol{\sigma}:\boldsymbol{\varepsilon}>+<\mathbf{q}\cdot\mathbf{p}>=<\boldsymbol{\sigma}>:<\boldsymbol{\varepsilon}>+<\mathrm{q}>\cdot<\mathrm{p}>. \tag{2.8.13}$$

The following uniform boundary conditions

$$(\mathbf{u}^{\mathrm{o}},\ \mathrm{q}^{\mathrm{o}}) = (\mathbf{x}\cdot\mathbf{E},\ \boldsymbol{\nu}\cdot\mathbf{Q}) \qquad \text{on } \partial\mathrm{V}, \tag{2.8.14}$$

necessarily satisfy (2.8.12), where $\mathbf{E}$ is a constant strain tensor, and $\mathbf{Q}$ is a constant electric displacement vector. The resulting fields then are functions of $\mathbf{E}$ and $\mathbf{Q}$, which define linear displacement and uniform electric displacement boundary data. Denote the corresponding macropotential by Φ^{EQ}. Here, superscript EQ emphasizes that the macropotential is defined for the case when linear displacements, $\mathbf{u}^{\mathrm{o}} = \mathbf{x}\cdot\mathbf{E}$, and uniform electric displacements, $\mathrm{q}^{\mathrm{o}} = \boldsymbol{\nu}\cdot\mathbf{Q}$, are prescribed on $\partial\mathrm{V}$. The macropotential is hence given by

$$\Phi^{\mathrm{EQ}}(\mathbf{E},\ \mathbf{Q}) = <\phi(\mathbf{x};\ \boldsymbol{\varepsilon}(\mathbf{x};\ \mathrm{E},\ \mathbf{Q}),\ \mathbf{q}(\mathbf{x};\ \mathrm{E},\ \mathbf{Q}))>, \tag{2.8.15}$$

where the argument $(\mathbf{E},\ \mathbf{Q})$ in $\boldsymbol{\varepsilon}$ and $\mathbf{q}$ indicates that these fields depend on the prescribed constant $\mathbf{E}$ and $\mathbf{Q}$. By definition, this macropotential yields the corresponding macrostress and macro-electric fields (both constant), as follows:

$$\boldsymbol{\Sigma}^{\mathrm{EQ}} = \frac{\partial\Phi^{\mathrm{EQ}}}{\partial\mathbf{E}}, \qquad \mathbf{P}^{\mathrm{EQ}} = \frac{\partial\Phi^{\mathrm{EQ}}}{\partial\mathbf{Q}}, \tag{2.8.16a,b}$$

where $\boldsymbol{\Sigma}^{\mathrm{EQ}} = <\boldsymbol{\sigma}(\mathbf{x};\ \mathbf{E},\ \mathbf{Q})>$ and $\mathbf{P}^{\mathrm{EQ}} = <\mathrm{p}(\mathbf{x};\ \mathbf{E},\ \mathbf{Q})>$. These exact results can be proved following the steps outlined in Subsections 2.5.1 and 2.5.2.

2.8.4. Strain/Electric-displacement Potential

In a similar manner, consider another micropotential, ψ, a function of the stress and the electric field, $\psi = \psi(\mathbf{x};\ \boldsymbol{\sigma},\ \mathbf{p})$, such that, at each point $\mathbf{x}$ in the corresponding RVE, the strain, $\boldsymbol{\varepsilon}$, and the electric displacement, $\mathbf{q}$, are given by[22]

$$\boldsymbol{\varepsilon} = \frac{\partial\psi}{\partial\boldsymbol{\sigma}}(\mathbf{x};\ \boldsymbol{\sigma},\ \mathbf{p}), \qquad \mathbf{q} = \frac{\partial\psi}{\partial\mathbf{p}}(\mathbf{x};\ \boldsymbol{\sigma},\ \mathbf{p}). \tag{2.8.17a,b}$$

The potential function ψ is related to ϕ through the following Legendre transform:

$$\phi + \psi = \boldsymbol{\sigma}:\boldsymbol{\varepsilon} + \mathbf{q}\cdot\mathbf{p}. \tag{2.8.18}$$

For the linear constitutive relations, this ψ is also positive-definite and convex.

Suppose that the RVE is subjected to the following uniform boundary conditions:

$$(\mathbf{t}^{\mathrm{o}},\ \mathrm{u}^{\mathrm{o}}) = (\boldsymbol{\nu}\cdot\boldsymbol{\Sigma},\ -\mathbf{x}\cdot\mathbf{P}) \qquad \text{on } \partial\mathrm{V}, \tag{2.8.19}$$

where $\boldsymbol{\Sigma}$ and $\mathbf{P}$ are the constant stress and the constant electric field. These boundary conditions satisfy (2.8.12). A macropotential, $\Psi^{\Sigma\mathrm{P}}$, defined by

[22] Note that, here again, a correct pair of field variables must be used, i.e., $\boldsymbol{\sigma}$ and $\mathbf{p}$, instead of $\boldsymbol{\varepsilon}$ and $\mathbf{q}$. A suitable potential cannot be defined if $\boldsymbol{\sigma}$ and $\mathbf{q}$ are used, as in the case of $\boldsymbol{\varepsilon}$ and $\mathbf{p}$.

$$\Psi^{\Sigma P}(\mathbf{\Sigma}, \mathbf{P}) = < \psi(\mathbf{x}, \boldsymbol{\sigma}(\mathbf{x}; \mathbf{\Sigma}, \mathbf{P}), \mathbf{p}(\mathbf{x}; \mathbf{\Sigma}, \mathbf{P}) >, \tag{2.8.20}$$

exists such that

$$\mathbf{E}^{\Sigma P} = \frac{\partial \Phi^{\Sigma P}}{\partial \mathbf{\Sigma}}, \qquad \mathbf{Q}^{\Sigma P} = \frac{\partial \Phi^{\Sigma P}}{\partial \mathbf{P}}, \tag{2.8.21a,b}$$

where $\mathbf{E}^{\Sigma P} = < \boldsymbol{\varepsilon}(\mathbf{x}; \mathbf{\Sigma}, \mathbf{P}) >$ and $\mathbf{Q}^{\Sigma P} = < \mathbf{q}(\mathbf{x}; \mathbf{\Sigma}, \mathbf{P}) >$. Here, again, superscript ΣP emphasizes the fact that the RVE is subjected to the uniform tractions $\mathbf{t}^o = \boldsymbol{\nu} \cdot \mathbf{\Sigma}$, and the linear electric field, $u^o = -\mathbf{x} \cdot \mathbf{P}$. These exact results can be confirmed using the method of Subsections 2.5.1 and 2.5.2.

2.8.5. Basic Inequalities

Suppose that the micropotentials ϕ and ψ are convex in the sense of (2.5.31), i.e., for any two sets of the strain and the electric displacement, $(\boldsymbol{\varepsilon}^{(\alpha)}, \mathbf{q}^{(\alpha)})$, and the stress and the electric field, $(\boldsymbol{\sigma}^{(\alpha)}, \mathbf{p}^{(\alpha)})$ $(\alpha = 1,2)$, the following conditions hold:

$$\phi(\mathbf{x}; \boldsymbol{\varepsilon}^{(1)}, \mathbf{q}^{(1)}) - \phi(\mathbf{x}; \boldsymbol{\varepsilon}^{(2)}, \mathbf{q}^{(2)})$$

$$> (\boldsymbol{\varepsilon}^{(1)} - \boldsymbol{\varepsilon}^{(2)}) : \frac{\partial \phi}{\partial \boldsymbol{\varepsilon}}(\mathbf{x}; \boldsymbol{\varepsilon}^{(2)}; \mathbf{q}^{(2)}) + (\mathbf{q}^{(1)} - \mathbf{q}^{(2)}) : \frac{\partial \phi}{\partial \mathbf{q}}(\mathbf{x}; \boldsymbol{\varepsilon}^{(2)}; \mathbf{q}^{(2)}),$$

$$\psi(\mathbf{x}; \boldsymbol{\sigma}^{(1)}, \mathbf{p}^{(1)}) - \psi(\mathbf{x}; \boldsymbol{\sigma}^{(2)}, \mathbf{p}^{(2)})$$

$$> (\boldsymbol{\sigma}^{(1)} - \boldsymbol{\sigma}^{(2)}) : \frac{\partial \psi}{\partial \boldsymbol{\sigma}}(\mathbf{x}; \boldsymbol{\sigma}^{(2)}; \mathbf{p}^{(2)}) + (\mathbf{p}^{(1)} - \mathbf{p}^{(2)}) : \frac{\partial \psi}{\partial \mathbf{p}}(\mathbf{x}; \boldsymbol{\sigma}^{(2)}; \mathbf{p}^{(2)}).$$

(2.8.22a,b)

Taking advantage of the averaging theorems, the following two theorems are then proved:

Theorem I: *For an RVE whose microconstituents admit convex potentials,* $\phi(\mathbf{x}; \boldsymbol{\varepsilon}, \mathbf{q})$, *among all consistent boundary data which produce the same overall macrostrain and macro-electric displacement,* $\mathbf{E}$ *and* $\mathbf{Q}$, *the boundary data of (2.8.19) associated with the uniform stress and the linear electric potential render the total macropotential* $\Phi(\mathbf{E}, \mathbf{Q})$ *an absolute minimum.*

Theorem II: *For an RVE whose microconstituents admit convex potentials,* $\psi(\mathbf{x}; \boldsymbol{\sigma}, \mathbf{p})$, *among all consistent boundary data which produce the same overall macrostress and macro-electric fields,* $\mathbf{\Sigma}$ *and* $\mathbf{P}$, *the boundary data of (2.8.14) associated with the linear displacement and the uniform electric displacement render the total macropotential* $\Psi(\mathbf{\Sigma}, \mathbf{P})$ *an absolute minimum.*

The proof of these theorems is essentially the same as that for the uncoupled mechanical and nonmechanical problems, although some attention must be paid to choosing the correct pair of field variables, $(\boldsymbol{\varepsilon}, \mathbf{q})$ and $(\boldsymbol{\sigma}, \mathbf{p})$.

Consider now any general boundary data (mixed or otherwise, but self-consistent) which produce a strain field, $\boldsymbol{\varepsilon} = \boldsymbol{\varepsilon}(\mathbf{x};\, \mathbf{E}^{G},\, \mathbf{Q}^{G})$, whose volume average is $\mathbf{E}^{G}$, and an electric displacement field, $\mathbf{q} = \mathbf{q}(\mathbf{x};\, \mathbf{E}^{G},\, \mathbf{Q}^{G})$, whose volume average is $\mathbf{Q}^{G}$. Denote the volume average of the corresponding potential, $\phi(\mathbf{x};\, \boldsymbol{\varepsilon}(\mathbf{x};\, \mathbf{E}^{G},\, \mathbf{Q}^{G}),\, \mathbf{q}(\mathbf{x};\, \mathbf{E}^{G},\, \mathbf{Q}^{G}))$, by $\Phi^{G} = \Phi^{G}(\mathbf{E}^{G},\, \mathbf{Q}^{G})$. From Theorem I, it follows that

$$\Phi^{G}(\mathbf{E}^{G},\, \mathbf{Q}^{G}) > \Phi^{\Sigma P}(\mathbf{E}^{\Sigma P},\, \mathbf{Q}^{\Sigma P}) \qquad \text{for } (\mathbf{E}^{G},\, \mathbf{Q}^{G}) = (\mathbf{E}^{\Sigma P},\, \mathbf{Q}^{\Sigma P}). \tag{2.8.23a}$$

In particular,

$$\Phi^{EQ}(\mathbf{E},\, \mathbf{Q}) > \Phi^{\Sigma P}(\mathbf{E},\, \mathbf{Q}), \tag{2.8.23b}$$

for any constant strain tensor $\mathbf{E}$ and electric displacement vector $\mathbf{Q}$.

Similarly, it follows from Theorem II, that

$$\Psi^{G}(\boldsymbol{\Sigma}^{G},\, \mathbf{P}^{G}) > \Psi^{EQ}(\boldsymbol{\Sigma}^{EQ},\, \mathbf{P}^{EQ}) \qquad \text{for } (\boldsymbol{\Sigma}^{G},\, \mathbf{P}^{G}) = (\boldsymbol{\Sigma}^{EQ},\, \mathbf{P}^{EQ}), \tag{2.8.24a}$$

where $\Psi^{G}(\boldsymbol{\Sigma}^{G},\, \mathbf{P}^{G})$ is the volume average of $\psi(\mathbf{x};\, \boldsymbol{\sigma}(\mathbf{x};\, \boldsymbol{\Sigma}^{G},\, \mathbf{P}^{G}),\, \mathbf{p}(\mathbf{x};\, \boldsymbol{\Sigma}^{G},\, \mathbf{p}^{G}))$, corresponding to some general self-consistent boundary data which produce the average stress $\boldsymbol{\Sigma}^{G}$ and the average electric field $\mathbf{P}^{G}$. In particular,

$$\Psi^{\Sigma P}(\boldsymbol{\Sigma},\, \mathbf{P}) > \Psi^{EQ}(\boldsymbol{\Sigma},\, \mathbf{P}), \tag{2.8.24b}$$

for any constant stress tensor and electric field vector, $\boldsymbol{\Sigma}$ and $\mathbf{P}$.

Expressions (2.8.23) and (2.8.24) are valid for any linear or *nonlinear* material, as long as the potentials ϕ and ψ exist and are convex. Consider now linear RVE's. The micropotentials are then given by

$$\phi = \frac{1}{2}\boldsymbol{\varepsilon} : \mathbf{C} : \boldsymbol{\varepsilon} + \boldsymbol{\varepsilon} : \mathbf{h}\cdot\mathbf{q} + \frac{1}{2}\mathbf{q}\cdot\mathbf{R}\cdot\mathbf{q},$$

$$\psi = \frac{1}{2}\boldsymbol{\sigma} : \mathbf{D}^{P} : \boldsymbol{\sigma} + \boldsymbol{\sigma} : \mathbf{i}^{P}\cdot\mathbf{p} + \frac{1}{2}\mathbf{p}\cdot\mathbf{K}^{P}\cdot\mathbf{p}, \tag{2.8.25a,b}$$

or

$$\phi = \frac{1}{2}\,\varepsilon_{ij}\,C_{ijkl}\,\varepsilon_{kl} + \varepsilon_{ij}\,h_{ijk}\,q_k + \frac{1}{2}\,q_i R_{ij}\,q_j,$$

$$\psi = \frac{1}{2}\,\sigma_{ij}\,D^{P}_{ijkl}\,\sigma_{kl} + \sigma_{ij}\,i^{P}_{ijk}\,p_k + \frac{1}{2}\,p_i\,K^{P}_{ij}\,p_j. \tag{2.8.25c,d}$$

Here, $\mathbf{D}^{P}$, $\mathbf{i}^{P}$, and $\mathbf{K}^{P}$ are defined[23] by

$$\mathbf{D}^{P} = (\mathbf{C} - \mathbf{h}\cdot\mathbf{R}^{-1}\cdot\mathbf{h}^{T})^{-1}, \qquad \mathbf{K}^{P} = (\mathbf{R} - \mathbf{h}^{T} : \mathbf{C}^{-1} : \mathbf{h})^{-1},$$

$$\mathbf{i}^{P} = -\mathbf{C}^{-1} : \mathbf{h}\cdot\mathbf{K}^{P}, \tag{2.8.26a~c}$$

or

$$D^{P}_{ijkl} = (C_{ijkl} - h_{ijp}\,R^{-1}_{pq}\,h_{klq})^{-1}, \qquad K^{P}_{ij} = (R_{ij} - h_{pqi}\,C^{-1}_{pqrs}\,h_{rsj})^{-1},$$

$$i^{P}_{ijk} = -C^{-1}_{ijpq}\,h_{pqr}\,K^{P}_{rk}. \tag{2.8.26d~f}$$

Let the macropotentials for any general consistent boundary data be

$$\Phi^G = \frac{1}{2}\mathbf{E}^G : \bar{\mathbf{C}}^G : \mathbf{E}^G + \mathbf{E}^G : \bar{\mathbf{h}}^G \boldsymbol{.} \mathbf{Q}^G + \frac{1}{2}\mathbf{Q}^G \boldsymbol{.} \bar{\mathbf{R}}^G \boldsymbol{.} \mathbf{Q}^G,$$

$$\Psi^G = \frac{1}{2}\boldsymbol{\Sigma}^G : (\bar{\mathbf{D}}^P)^G : \boldsymbol{\Sigma}^G + \boldsymbol{\Sigma}^G : (\bar{\mathbf{i}}^P)^G \boldsymbol{.} \mathbf{P}^G + \frac{1}{2}\mathbf{P}^G \boldsymbol{.} (\bar{\mathbf{K}}^P)^G \boldsymbol{.} \mathbf{P}^G. \qquad (2.8.27a,b)$$

These macropotentials are defined as follows: Considering Φ^G, assume that some general consistent, but otherwise arbitrary, boundary data are prescribed on ∂V such that they produce the average potential,

$$\Phi^G = \frac{1}{2} < \boldsymbol{\varepsilon} : \mathbf{C} : \boldsymbol{\varepsilon} > + < \boldsymbol{\varepsilon} : \mathbf{h} : \mathbf{q} > + \frac{1}{2} < \mathbf{q} : \mathbf{R} : \mathbf{q} >,$$

$$\Psi^G = \frac{1}{2} < \boldsymbol{\sigma} : \mathbf{D}^P : \boldsymbol{\sigma} > + < \boldsymbol{\sigma} : \mathbf{i}^P : \mathbf{p} > + \frac{1}{2} < \mathbf{p} : \mathbf{K}^P : \mathbf{p} >. \qquad (2.8.28a,b)$$

Let the corresponding average strain and stress be given by $\mathbf{E}^G$ and $\boldsymbol{\Sigma}^G$, respectively, and the average electric field and electric displacement by $\mathbf{P}^G$ and $\mathbf{Q}^G$. Then define $(\bar{\mathbf{C}}^G, \bar{\mathbf{h}}^G, \bar{\mathbf{R}}^G)$ and $((\bar{\mathbf{D}}^P)^G, (\bar{\mathbf{i}}^P)^G, (\bar{\mathbf{K}}^P)^G)$ such that (2.8.27a) and (2.8.27b) hold, respectively. Note that, in general, $\boldsymbol{\Sigma}^G$ and $\mathbf{P}^G$ will *not* be equal to the gradients of Φ^G with respect to $\mathbf{E}^G$ and $\mathbf{Q}^G$. Similarly, $\mathbf{E}^G$ and $\mathbf{Q}^G$ will *not* be the gradients of Ψ^G with respect to $\boldsymbol{\Sigma}^G$ and $\mathbf{P}^G$. That is, in general,

$$(\boldsymbol{\Sigma}^G, \mathbf{P}^G) \neq (\partial\Phi^G/\partial\mathbf{E}^G, \partial\Phi^G/\partial\mathbf{Q}^G), \qquad (\mathbf{E}^G, \mathbf{Q}^G) \neq (\partial\Psi^G/\partial\boldsymbol{\Sigma}^G, \partial\Psi^G/\partial\mathbf{P}^G),$$

unless the boundary data are such that (2.8.12) is satisfied.

Suppose that the boundary data are such that (2.8.12) is satisfied. Then, in view of the identity $< \boldsymbol{\sigma} : \boldsymbol{\varepsilon} + \mathbf{q} \boldsymbol{.} \mathbf{p} > = < \boldsymbol{\sigma} > : < \boldsymbol{\varepsilon} > + < \mathbf{q} > \boldsymbol{.} < \mathbf{p} >$, it follows that

$$\boldsymbol{\Sigma}^G = \bar{\mathbf{C}}^G : \mathbf{E}^G + \bar{\mathbf{h}}^G \boldsymbol{.} \mathbf{Q}^G, \qquad \mathbf{P}^G = (\bar{\mathbf{h}}^G)^T : \mathbf{E}^G + \bar{\mathbf{R}}^G \boldsymbol{.} \mathbf{Q}^G. \qquad (2.8.29a,b)$$

Similarly, when (2.8.12) is satisfied for a set of boundary data with average $\boldsymbol{\Sigma}^G$ and $\mathbf{P}^G$, it follows that

$$\mathbf{E}^G = (\bar{\mathbf{D}}^P)^G : \boldsymbol{\Sigma}^G + (\bar{\mathbf{i}}^P)^G \boldsymbol{.} \mathbf{P}^G, \qquad \mathbf{Q}^G = ((\bar{\mathbf{i}}^P)^G)^T : \boldsymbol{\Sigma}^G + (\bar{\mathbf{K}}^P)^G \boldsymbol{.} \mathbf{P}^G. \quad (2.8.29c,d)$$

Therefore, $((\bar{\mathbf{D}}^P)^G, (\bar{\mathbf{i}}^P)^G, (\bar{\mathbf{K}}^P)^G)$ in (2.8.29c,d) is[24] related to $(\bar{\mathbf{C}}^G, \bar{\mathbf{h}}^G, \bar{\mathbf{R}}^G)$ in (2.8.29a,b) in the same manner as $(\mathbf{D}^P, \mathbf{i}^P, \mathbf{K}^P)$ is related to $(\mathbf{C}, \mathbf{h}, \mathbf{R})$ in (2.8.26a~c).

Finally, Theorems I and II yield the following inequalities:

$$\mathbf{E} : (\bar{\mathbf{C}}^G - \bar{\mathbf{C}}^{\Sigma P}) : \mathbf{E} + 2\mathbf{E} : (\bar{\mathbf{h}}^G - \bar{\mathbf{h}}^{\Sigma P}) \boldsymbol{.} \mathbf{Q} + \mathbf{Q} \boldsymbol{.} (\bar{\mathbf{R}}^G - \bar{\mathbf{R}}^{\Sigma P}) \boldsymbol{.} \mathbf{Q} \geq 0, \qquad (2.8.29a)$$

for any pair of constant strain tensor $\mathbf{E}$ and electric displacement vector $\mathbf{Q}$, and

$$\boldsymbol{\Sigma} : ((\bar{\mathbf{D}}^P)^G - (\bar{\mathbf{D}}^P)^{EQ}) : \boldsymbol{\Sigma} + 2\boldsymbol{\Sigma} : ((\bar{\mathbf{i}}^P)^G - (\bar{\mathbf{i}}^P)^{EQ}) \boldsymbol{.} \mathbf{P} + \mathbf{P} \boldsymbol{.} ((\bar{\mathbf{K}}^P)^G - (\bar{\mathbf{K}}^P)^{EQ}) \boldsymbol{.} \mathbf{P} \geq 0,$$

---

[23] Tensor $\mathbf{D}^P$ is the inverse of $\mathbf{C}^P$ which is defined by (2.8.5).

[24] In this sense, $(\bar{\mathbf{C}}^G, \bar{\mathbf{h}}^G, \bar{\mathbf{R}}^G)$ and $((\bar{\mathbf{D}}^P)^G, (\bar{\mathbf{i}}^P)^G, (\bar{\mathbf{K}}^P)^G)$ are regarded as corresponding to each other's inverse.

(2.8.29b)

for any pair of constant stress tensor $\mathbf{\Sigma}$ and electric field $\mathbf{P}$.

2.9. REFERENCES

Boley, B. A. and Weiner, J. (1960), *Theory of thermal stresses*, Wiley, New York.

Francfort, G. A. and Murat, F. (1986), Homogenization and optimal bounds in linear elasticity, *Archive Rat. Mech. and Analysis*, Vol. 94, 307-334.

Havner, K. S. (1982), The theory of finite plastic deformation of crystalline solids, in *Mechanics of solids (The Rodney Hill 60th anniversary volume)*, Pergamon Press, Oxford, 265-302.

Havner, K.S. (1992), *Finite plastic deformation of crystalline solids*, Cambridge University Press.

Hill, R. (1952), The elastic behaviour of a crystalline aggregate, *Proc. Phys. Soc., London, Sect. A*, Vol. 65, 349-354.

Hill, R. (1963), Elastic properties of reinforced solids: Some theoretical principles, *J. Mech. Phys. Solids*, Vol. 11, 357-372.

Hill, R. (1967), The essential structure of constitutive laws for metal composites and polycrystals, *J. Mech. Phys. Solids*, Vol. 15, 79-95.

Hill, R. (1972), On constitutive macro-variables for heterogeneous solids at finite strain, *Proc. Roy. Soc. Lond.*, Vol. A326, 131.

Hutchinson, J. W. (1987), Micro-mechanics of damage in deformation and fracture, Technical University of Denmark, Denmark.

Ikeda, M. (1990), *Fundamentals of piezoelectricity*, Oxford science publications, Oxford University Press.

Iwakuma, T. and Nemat-Nasser, S. (1984), Finite elastic-plastic deformation of polycrystalline metals, *Proc. Royal Soc. Lond. A*, Vol. 394, 87-119.

Krajcinovic, D. (1996), *Damage Mechanics*, North-Holland, New York.

Lothe, and Barnett, D. (1976), Integral formalism for surface waves in piezoelectric crystals. Existence considerations, *J. Appl. Phys.*, Vol. 47, 1799-1807.

Lothe, and Barnett, D. (1977), Further development of he theory for surface waves in piezoelectric crystals, *Physica Norvegica*, Vol. 4, 239-254.

Mandel, J. (1980), Généralization dans R9 de la régle du potential plastique pour un élément polycrystallin, *Compt. Rend. Acad. Sci. Paris*, Vol. 290, 481-484.

Maugin, G. A. (1985), *Nonlinear electromechanical effects and applications*, World Scientific Pub., Singapore, Distributed by Taylor & Francis, Philadelphia.

Murat, F. and Tartar, L. (1985), Calcul des variations et homogenization, in *Les methodes d'homogenization: theorie et applications en physique. Coll. de la Dir. des Etudes et Recherches D'Electricite de France*, Eyrolles, Paris, 319.

Nemat-Nasser, S. (1983), On finite plastic flow of crystalline solids and geomaterials, *J. Appl. Mech. (50th Anniversary Issue)*, Vol. 50, 1114-1126.

Nemat-Nasser, S. and Hori, M. (1990), Elastic solids with microdefects, in *Micromechanics and inhomogeneity - The Toshio Mura anniversary volume*, Springer-Verlag, New York, 297-320.

Nowacki, W. (1962), *Thermoelasticity*, Addison-Wesley, Reading, Mass.

Nowinski, J. L. (1978), *Theory of thermoelasticity with applications*, Sijthoff & Noordhoff International Publishers.

Paul, B. (1960), Prediction of the elastic constants of multiphase materials, *Trans. Am. Inst. Min. Metall. Pet. Eng.*, Vol. 218, 36-41.

Reuss, A. (1929), Berechnung der Fliessgrenze von Mischkristallen auf Grund der Plastizitätsbedingung für Einkristalle, *Z. Angew. Math. Mech.*, Vol. 9, 49-58.

Schapery, R. A. (1995), Prediction of compressive strength and kink bands in composites using a work potential, *Int. J. Solids Struct.*, Vol. 32, 739-765.

Sneddon, I. N. (1974), *The linear theory of thermoelasticity*, Courses and lectures; International Centre for Mechanical Sciences ; no. 119; Springer-Verlag, New York.

Tiersten, H. F. (1969), *Linear piezoelectric plate vibrations*, Plenum Press, New York.

Voigt, W. (1889), Über die Beziehung zwischen den beiden Elastizitätskonstanten isotroper Körper, *Wied. Ann. Physik*, Vol. 38, 573-587.

Willis, J. R. (1989), *Private communication*.

Willis, J. R. (1992), *Private communication*.

CHAPTER II

ELASTIC SOLIDS WITH MICROCAVITIES AND MICROCRACKS

In this chapter the overall elastic modulus and compliance tensors are established for a macro-element represented by an RVE which consists of linearly elastic constituents containing microcavities and/or microcracks. For a fixed microstructure, the increment of the average stress, $\delta\bar{\boldsymbol{\sigma}}$, *relates linearly to the corresponding increment of the average strain,* $\delta\bar{\boldsymbol{\varepsilon}}$, *by the relation* $\delta\bar{\boldsymbol{\sigma}} = \bar{\mathbf{C}} : \delta\bar{\boldsymbol{\varepsilon}}$, *and the objective of the analysis is to calculate the overall modulus tensor* $\bar{\mathbf{C}}$, *in terms of the corresponding moduli of the constituents and the microstructure of the RVE. The modulus tensor* $\bar{\mathbf{C}}$, *in general, depends on and changes with the microstructure. In this chapter, first some fundamental results in linear elasticity are briefly reviewed, and then the results are used in a systematic manner to estimate the overall properties of the macro-element, in terms of the properties and geometry of its microconstituents. Throughout the chapter, attention is focused on the stress-strain relations, ignoring the temperature and the associated thermal effects. The Helmholtz free energy* ϕ *then reduces to the strain energy density function which is denoted by* $\mathrm{w} = \mathrm{w}(\boldsymbol{\varepsilon})$, *and the corresponding complementary energy function* ψ *reduces to the complementary strain energy density function which is denoted by* $\mathrm{w}^{\mathrm{c}} = \mathrm{w}^{\mathrm{c}}(\boldsymbol{\sigma})$. *When the fact that, for a heterogeneous* RVE, w *and* w^{c} *also depend on the position* $\mathbf{x}$ *of the material in the RVE needs to be emphasized, then* $\mathrm{w} = \mathrm{w}(\mathbf{x}, \boldsymbol{\varepsilon})$ *and* $\mathrm{w}^{\mathrm{c}} = \mathrm{w}^{\mathrm{c}}(\mathbf{x}, \boldsymbol{\sigma})$; *note that even for a homogeneous material,* w *and* w^{c} *are implicit functions of* $\mathbf{x}$ *through* $\boldsymbol{\varepsilon} = \boldsymbol{\varepsilon}(\mathbf{x})$ *and* $\boldsymbol{\sigma} = \boldsymbol{\sigma}(\mathbf{x})$, *respectively.*

More specifically, the overall elasticity and compliance tensors of a linearly elastic matrix containing microcavities and microcracks, are estimated using two simple models for the averaging procedure: the dilute distribution model which assumes that the inhomogeneities are small and far apart, so that their interaction may be neglected; and the self-consistent model which takes into account the corresponding interaction, in a certain, overall, approximate manner. In particular, in

Section 3 the stress-strain relations of linear elasticity are reviewed and the necessary background is provided for subsequent sections. In Section 4, a systematic discussion of the overall stress and strain in a porous RVE is given for two limiting alternative boundary data, namely, uniform tractions, and linear displacements. It is shown, directly and in a simple manner, how the corresponding overall compliance and elasticity tensors can be estimated, using the reciprocal theorem and simple estimates of the cavity boundary displacements. These results are generally valid for cavities of any shape or distribution, and, except for the assumption of linear elasticity, no approximations are involved. In Section 5 the general results of Section 4 are applied to porous, linearly elastic solids. The dilute-distribution and the self-consistent models are used. The relation between these models is discussed in terms of specific problems. Section 6 deals with elastic solids with microcracks. Here again, the same two models are used in a number of illustrative examples of scientific and technological importance, and the corresponding results are compared and discussed. The effective moduli of solids with various microcrack distributions are calculated explicitly in Subsections 6.1 to 6.8, emphasizing the phenomenon of microcrack-induced anisotropy and the role of friction (Subsections 6.3 and 6.4). Both slit microcracks (two-dimensional problems) and penny-shaped ones (three-dimensional problems), with various distributions (various alignments, random), are examined in considerable detail, providing explicit results. In addition, a brief overview of recent advances in theoretical and experimental evaluation of brittle failure in compression is presented in Subsection 6.9 for the static loadings, and in Subsection 6.10 for the dynamic cases.

SECTION 3 LINEARLY ELASTIC SOLIDS

This section presents stress-strain relations in linear elasticity and summarizes elasticity and compliance tensors for materials with several commonly considered symmetries. General three-dimensional, as well as plane-strain and plane-stress conditions are briefly examined. Then the reciprocal theorem of linear elasticity is introduced, followed by the principle of superposition and a brief discussion of Green's function. The material in this section will be used throughout the remaining sections in Part 1.

3.1. HOOKE'S LAW AND MATERIAL SYMMETRY

Constitutive relations in linear elasticity are given by the generalized Hooke law which linearly relates the stress and strain tensors through the elasticity and/or compliance tensors. The coefficients of these linear relations are the elastic and/or compliance moduli, as is detailed in this subsection.[1]

3.1.1. Elastic Moduli

Consider a typical homogeneous and linearly elastic constituent of an RVE, and denote its stress-strain relation by

$$\boldsymbol{\sigma} = \mathbf{C} : \boldsymbol{\varepsilon}, \qquad \text{or} \qquad \sigma_{ij} = C_{ijkl}\,\varepsilon_{kl}. \tag{3.1.1a,b}$$

Since both the stress and strain tensors are symmetric, $\boldsymbol{\sigma} = \boldsymbol{\sigma}^T$ and $\boldsymbol{\varepsilon} = \boldsymbol{\varepsilon}^T$, the elastic modulus tensor $\mathbf{C}$ must also possess a similar symmetry,

$$C_{ijkl} = C_{jikl} = C_{ijlk} = C_{jilk}. \tag{3.1.1c}$$

Therefore, out of the eighty-one components of $\mathbf{C}$, only thirty-six are independent. The number of independent components reduces to twenty-one for the most general anisotropic linearly elastic case, if the material admits a strain energy density function $w = w(\boldsymbol{\varepsilon})$, such that

$$\boldsymbol{\sigma} = \frac{\partial w}{\partial \boldsymbol{\varepsilon}}, \qquad \text{or} \qquad \sigma_{ij} = \frac{\partial w}{\partial \varepsilon_{ij}}. \tag{3.1.2a,b}$$

Because of the linearity of stress-strain relation (3.1.1), to within an additive constant, the strain energy density function is given by the following quadratic

---

[1] See Love (1944), Sokolnikoff (1956), Hearmon (1961), Lekhnitskii (1963), and Jones (1975).

form:

$$w = \frac{1}{2}\boldsymbol{\varepsilon} : \mathbf{C} : \boldsymbol{\varepsilon} = \frac{1}{2} C_{ijkl}\, \varepsilon_{ij}\, \varepsilon_{kl}. \tag{3.1.2c}$$

The elasticity tensor **C** is therefore symmetric with respect to the first and second pairs of its indices,

$$C_{ijkl} = C_{klij}, \tag{3.1.1d}$$

which, together with (3.1.1c), leaves only twenty-one independent components for **C**.

It is often convenient to express the stress-strain relation (3.1.1a) in terms of a six-dimensional matrix.[2] To this end, the stress and strain tensors are represented by six by one column vectors, and the elasticity tensor **C** by a six by six matrix, as follows. First define the six by one column vectors $[\tau_a]$ and $[\gamma_a]$ for the stress and strain tensors, respectively:

$$\tau_1 = \sigma_{11}, \qquad \tau_2 = \sigma_{22}, \qquad \tau_3 = \sigma_{33},$$

$$\tau_4 = \sigma_{23} = \sigma_{32}, \qquad \tau_5 = \sigma_{31} = \sigma_{13}, \qquad \tau_6 = \sigma_{12} = \sigma_{21}, \tag{3.1.3a}$$

and

$$\gamma_1 = \varepsilon_{11}, \qquad \gamma_2 = \varepsilon_{22}, \qquad \gamma_3 = \varepsilon_{33},$$

$$\gamma_4 = 2\varepsilon_{23} = 2\varepsilon_{32}, \qquad \gamma_5 = 2\varepsilon_{31} = 2\varepsilon_{13}, \qquad \gamma_6 = 2\varepsilon_{12} = 2\varepsilon_{21}. \tag{3.1.3b}$$

Then denote the first two subscripts ij by, say, a, and the second two subscripts kl by, say, b, to obtain the matrix of the elastic moduli, $[C_{ab}]$, as

$$[C_{ab}] = \begin{bmatrix} C_{1111} & C_{1122} & C_{1133} & C_{1123} & C_{1131} & C_{1112} \\ C_{2211} & C_{2222} & C_{2233} & C_{2223} & C_{2231} & C_{2212} \\ C_{3311} & C_{3322} & C_{3333} & C_{3323} & C_{3331} & C_{3312} \\ C_{2311} & C_{2322} & C_{2333} & C_{2323} & C_{2331} & C_{2312} \\ C_{3111} & C_{3122} & C_{3133} & C_{3123} & C_{3131} & C_{3112} \\ C_{1211} & C_{1222} & C_{1233} & C_{1223} & C_{1231} & C_{1212} \end{bmatrix}, \tag{3.1.3c}$$

where a, b = 1, 2, ..., 6. The stress-strain relation (3.1.1) now becomes

$$[\tau_a] = [C_{ab}][\gamma_b], \qquad \text{or} \qquad \tau_a = C_{ab}\,\gamma_b, \tag{3.1.4a,b}$$

where repeated indices are summed, and a and b take on values from 1 to 6. The strain energy density function w may be expressed as

$$w = \frac{1}{2}[\gamma_a]^T[C_{ab}][\gamma_b] = \frac{1}{2} C_{ab}\,\gamma_a\,\gamma_b, \tag{3.1.5a}$$

so that

[2] This notation has been introduced by Voigt; see Hearmon (1961).

$$[\tau_a] = \frac{\partial w}{\partial [\gamma_a]}, \qquad \text{or} \qquad \tau_a = \frac{\partial w}{\partial \gamma_a}. \tag{3.1.5b,c}$$

3.1.2. Elastic Compliances

The six by six matrix of the elastic moduli $[C_{ab}]$ is positive-definite and symmetric, admitting a unique inverse matrix $[D_{ab}]$, i.e.,

$$[D_{ab}] = [C_{ab}]^{-1} \qquad \text{or} \qquad [C_{ab}] = [D_{ab}]^{-1}. \tag{3.1.6a,b}$$

Observe that the relation between the components of the *matrix* $[C_{ab}]$ and the components of the *tensor* **C** differs from the relation between the components of the *matrix* $[D_{ab}]$ and the components of the *tensor* **D**. Defining a diagonal matrix $[W_{ab}]$, by

$$[W_{ab}] = \begin{bmatrix} 1 & 0 & 0 & 0 & 0 & 0 \\ 0 & 1 & 0 & 0 & 0 & 0 \\ 0 & 0 & 1 & 0 & 0 & 0 \\ 0 & 0 & 0 & 2 & 0 & 0 \\ 0 & 0 & 0 & 0 & 2 & 0 \\ 0 & 0 & 0 & 0 & 0 & 2 \end{bmatrix}, \tag{3.1.6c}$$

and associating the components of tensor **D** with matrix $[D_{ab}]$ in a manner similar to (3.1.3c), it follows that

$$[D_{ab}] = [W_{ap}][\mathrm{D}_{pq}][W_{qb}]$$

$$= \begin{bmatrix} \mathrm{D}_{1111} & \mathrm{D}_{1122} & \mathrm{D}_{1133} & 2\mathrm{D}_{1123} & 2\mathrm{D}_{1131} & 2\mathrm{D}_{1112} \\ \mathrm{D}_{2211} & \mathrm{D}_{2222} & \mathrm{D}_{2233} & 2\mathrm{D}_{2223} & 2\mathrm{D}_{2231} & 2\mathrm{D}_{2212} \\ \mathrm{D}_{3311} & \mathrm{D}_{3322} & \mathrm{D}_{3333} & 2\mathrm{D}_{3323} & 2\mathrm{D}_{3331} & 2\mathrm{D}_{3312} \\ 2\mathrm{D}_{2311} & 2\mathrm{D}_{2322} & 2\mathrm{D}_{2333} & 4\mathrm{D}_{2323} & 4\mathrm{D}_{2331} & 4\mathrm{D}_{2312} \\ 2\mathrm{D}_{3111} & 2\mathrm{D}_{3122} & 2\mathrm{D}_{3133} & 4\mathrm{D}_{3123} & 4\mathrm{D}_{3131} & 4\mathrm{D}_{3112} \\ 2\mathrm{D}_{1211} & 2\mathrm{D}_{1222} & 2\mathrm{D}_{1233} & 4\mathrm{D}_{1223} & 4\mathrm{D}_{1231} & 4\mathrm{D}_{1212} \end{bmatrix}, \tag{3.1.6d}$$

where *matrix* $[\mathrm{D}_{ab}]$ in expression $[W_{ap}][\mathrm{D}_{pq}][W_{qb}]$ is obtained by replacing in the right-hand side of (3.1.3c), the letter C by the letter D. In Part 2 of this book, Section 15 gives a detailed discussion of the relation between the tensor operation and the corresponding six-dimensional matrix operation.[3]

The matrix $[D_{ab}]$ and the compliance tensor **D** are both symmetric, and, in general, each has at most twenty-one independent components. The strain-stress relation can be written as

$$[\gamma_a] = [D_{ab}][\tau_b], \qquad \text{or} \qquad \gamma_a = D_{ab}\,\tau_b. \tag{3.1.7a,b}$$

In tensor representation this becomes

[3] An alternative formulation is by Kelvin (1856), which has been extended by Mehrabadi and Cowin (1990) and Cowin and Mehrabadi (1992). The approach outlined in Section 15 appeared in the 1988 version of the notes (Nemat-Nasser and Hori, 1988) which evolved into the present book.

$$\boldsymbol{\varepsilon} = \mathbf{D} : \boldsymbol{\sigma}, \qquad \text{or} \qquad \varepsilon_{ij} = D_{ijkl}\,\sigma_{kl}. \tag{3.1.8a,b}$$

The complementary strain energy density function,

$$w^c = \frac{1}{2}[\tau_a]^T[D_{ab}][\tau_b] = \frac{1}{2}D_{ab}\,\tau_a\,\tau_b, \tag{3.1.9a}$$

is such that

$$[\gamma_a] = \frac{\partial w^c}{\partial[\tau_a]}, \qquad \text{or} \qquad \gamma_a = \frac{\partial w^c}{\partial \tau_a}. \tag{3.1.9b,c}$$

The tensor representation of (3.1.9a~c) becomes

$$w^c = \frac{1}{2}\boldsymbol{\sigma} : \mathbf{D} : \boldsymbol{\sigma} = \frac{1}{2}D_{ijkl}\,\sigma_{ij}\,\sigma_{kl}, \tag{3.1.10a}$$

and

$$\boldsymbol{\varepsilon} = \frac{\partial w^c}{\partial \boldsymbol{\sigma}}, \qquad \text{or} \qquad \varepsilon_{ij} = \frac{\partial w^c}{\partial \sigma_{ij}}. \tag{3.1.10b,c}$$

3.1.3. Elastic Symmetry

The number of independent elastic parameters (elastic moduli or elastic compliances) further reduces from the maximum of twenty-one, if the material possesses elastic symmetries. The greatest symmetry exists in *isotropic* materials for which any plane is a plane of symmetry. Recall that a plane normal to an orientation constitutes a plane of elastic symmetry, if reflection about this plane leaves the elastic parameters unchanged.

For the isotropic case, there are only two elastic parameters, and the elasticity tensor becomes

$$\mathbf{C} = \lambda \mathbf{1}^{(2)} \otimes \mathbf{1}^{(2)} + 2\mu \mathbf{1}^{(4s)}, \tag{3.1.11a}$$

or

$$C_{ijkl} = \lambda\,\delta_{ij}\,\delta_{kl} + \mu\,(\delta_{ik}\,\delta_{jl} + \delta_{il}\,\delta_{jk}), \tag{3.1.11b}$$

where $\mathbf{1}^{(2)}$ is the second-order unit tensor and $\mathbf{1}^{(4s)}$ is the *symmetric* fourth-order unit tensor; the fourth-order unit tensor $\mathbf{1}^{(4)}$ is given by

$$1^{(4)}_{ijkl} \equiv \delta_{ik}\,\delta_{jl} = \frac{1}{2}(\delta_{ik}\,\delta_{jl} + \delta_{il}\,\delta_{jk}) + \frac{1}{2}(\delta_{ik}\,\delta_{jl} - \delta_{il}\,\delta_{jk}) \equiv 1^{(4s)}_{ijkl} + 1^{(4a)}_{ijkl}. \tag{3.1.11c}$$

With second-order contraction, $\mathbf{1}^{(4s)}$ ($\mathbf{1}^{(4a)}$) maps a symmetric (antisymmetric) second-order tensor to itself, but maps an antisymmetric (symmetric) second-order tensor to $\mathbf{0}$, while $\mathbf{1}^{(4)}$ maps any second-order tensor to itself. The parameters λ and μ are called the Lamé constants (for heterogeneous elastic materials they, of course, are not constant); μ is the shear modulus. If Young's modulus is denoted by E, Poisson's ratio by ν, and the bulk modulus by K, then these elastic parameters are all related, such that they can be expressed in terms of any two chosen from them. Table 3.1.1 gives the relations among these parameters.

Table 3.1.1

Relations among isotropic elastic moduli

Moduli	Relations
λ	$\dfrac{2\mu\nu}{1-2\nu} = \dfrac{\mu(E-2\mu)}{3\mu-E} = K - \dfrac{2}{3}\mu = \dfrac{E\nu}{(1+\nu)(1-2\nu)} = \dfrac{3K\nu}{1+\nu} = \dfrac{3K(3K-E)}{9K-E}$
μ	$\dfrac{\lambda(1-2\nu)}{2\nu} = \dfrac{3}{2}(K-\lambda) = \dfrac{E}{2(1+\nu)} = \dfrac{3K(1-2\nu)}{2(1+\nu)} = \dfrac{3KE}{9K-E}$
ν	$\dfrac{\lambda}{2(\lambda+\mu)} = \dfrac{\lambda}{3K-\lambda} = \dfrac{E}{2\mu} - 1 = \dfrac{3K-2\mu}{2(3K+\mu)} = \dfrac{3K-E}{6K}$
E	$\dfrac{\mu(3\lambda+2\mu)}{\lambda+\mu} = \dfrac{\lambda(1+\nu)(1-2\nu)}{\nu} = \dfrac{9K(K-\lambda)}{3K-\lambda} = 2\mu(1+\nu) = \dfrac{9K\mu}{3K+\mu} = 3K(1-2\nu)$
K	$\lambda + \dfrac{2}{3}\mu = \dfrac{\lambda(1+\nu)}{3\nu} = \dfrac{2\mu(1+\nu)}{3(1-2\nu)} = \dfrac{\mu E}{3(3\mu-E)} = \dfrac{E}{3(1-2\nu)}$

For the isotropic case, all the components of the matrix $[C_{ab}]$ can be expressed in terms of two components, say, C_{11} and C_{12} which, together with C_{44}, are defined in terms of Young's modulus E and Poisson's ratio ν, as follows:

$$C_{11} = \frac{E}{1+\nu}\frac{1-\nu}{1-2\nu}, \qquad C_{12} = \frac{E}{1+\nu}\frac{\nu}{1-2\nu},$$

$$C_{44} = \frac{E}{2(1+\nu)} (= \mu). \tag{3.1.12a~c}$$

It then follows that

$$[C_{ab}] = \begin{bmatrix} C_{11} & C_{12} & C_{12} & 0 & 0 & 0 \\ C_{12} & C_{11} & C_{12} & 0 & 0 & 0 \\ C_{12} & C_{12} & C_{11} & 0 & 0 & 0 \\ 0 & 0 & 0 & (C_{11}-C_{12})/2 & 0 & 0 \\ 0 & 0 & 0 & 0 & (C_{11}-C_{12})/2 & 0 \\ 0 & 0 & 0 & 0 & 0 & (C_{11}-C_{12})/2 \end{bmatrix}. \tag{3.1.12d}$$

In a similar manner, the compliance matrix $[D_{ab}]$ is expressed in terms of, say, D_{11} and D_{12}, arriving at an expression similar to (3.1.12d):

$$[D_{ab}] = \begin{bmatrix} D_{11} & D_{12} & D_{12} & 0 & 0 & 0 \\ D_{12} & D_{11} & D_{12} & 0 & 0 & 0 \\ D_{12} & D_{12} & D_{11} & 0 & 0 & 0 \\ 0 & 0 & 0 & 2(D_{11}-D_{12}) & 0 & 0 \\ 0 & 0 & 0 & 0 & 2(D_{11}-D_{12}) & 0 \\ 0 & 0 & 0 & 0 & 0 & 2(D_{11}-D_{12}) \end{bmatrix}. \tag{3.1.12e}$$

In terms of E and ν, we have

$$D_{11} = \frac{1}{\mathrm{E}}, \qquad D_{12} = -\frac{\nu}{\mathrm{E}}, \qquad D_{44} = \frac{2(1+\nu)}{\mathrm{E}} (= \frac{1}{\mu}). \tag{3.1.12f~h}$$

Next, we consider *transversely isotropic materials* which have five independent elastic parameters. In this case, there exists a plane of isotropy. Choose this plane to coincide with the x_1,x_2-plane. The compliance matrix $[D_{ab}]$ then becomes

$$[D_{ab}] = \begin{bmatrix} D_{11} & D_{12} & D_{13} & 0 & 0 & 0 \\ D_{12} & D_{11} & D_{13} & 0 & 0 & 0 \\ D_{13} & D_{13} & D_{33} & 0 & 0 & 0 \\ 0 & 0 & 0 & D_{44} & 0 & 0 \\ 0 & 0 & 0 & 0 & D_{44} & 0 \\ 0 & 0 & 0 & 0 & 0 & 2(D_{11}-D_{12}) \end{bmatrix}. \tag{3.1.13a}$$

In this case, the Young modulus associated with any direction in the x_1,x_2-plane is the same, say, $\mathrm{E}_1 = \mathrm{E}_2 = \mathrm{E}$, and the Poisson ratio associated with any two orthogonal directions in this plane is also the same, say, $\nu_{12} = \nu_{21} = \nu$. The corresponding shear modulus $\mu_{12} = \mu_{21} = \mu$ is given by $\mu = \mathrm{E}/2(1+\nu)$. If Young's modulus in the x_3-direction is denoted by E_3, Poisson's ratio associated with the x_3-direction and a direction in the x_1,x_2-plane by, say, $\nu_{13} = \nu_{23} \equiv \nu_3$, with the corresponding shear modulus $\mu_{13} = \mu_{23} \equiv \mu_3$, then

$$D_{11} = \frac{1}{\mathrm{E}}, \qquad D_{12} = -\frac{\nu}{\mathrm{E}}, \qquad D_{13} = -\frac{\nu_3}{\mathrm{E}_3},$$

$$D_{33} = \frac{1}{\mathrm{E}_3}, \qquad D_{44} = \frac{1}{\mu_3}. \tag{3.1.13b~f}$$

A similar expression holds for the elastic modulus matrix $[C_{ab}]$:

$$[C_{ab}] = \begin{bmatrix} C_{11} & C_{12} & C_{13} & 0 & 0 & 0 \\ C_{12} & C_{11} & C_{13} & 0 & 0 & 0 \\ C_{13} & C_{13} & C_{33} & 0 & 0 & 0 \\ 0 & 0 & 0 & C_{44} & 0 & 0 \\ 0 & 0 & 0 & 0 & C_{44} & 0 \\ 0 & 0 & 0 & 0 & 0 & (C_{11}-C_{12})/2 \end{bmatrix}, \tag{3.1.13g}$$

where

$$C_{11} = \mathrm{D}\{\frac{1}{\mathrm{E}\,\mathrm{E}_3} - \frac{\nu_3^2}{\mathrm{E}_3^2}\}, \qquad C_{12} = \mathrm{D}\{\frac{\nu}{\mathrm{E}\,\mathrm{E}_3} + \frac{\nu_3^2}{\mathrm{E}_3^2}\},$$

$$C_{13} = \mathrm{D}\,\frac{(1+\nu)\nu_3}{\mathrm{E}\,\mathrm{E}_3}, \qquad C_{33} = \mathrm{D}\frac{1-\nu^2}{\mathrm{E}^2},$$

$$C_{44} = \mu_3, \qquad \mathrm{D} \equiv \frac{\mathrm{E}^2\,\mathrm{E}_3^2}{(1+\nu)\{(1-\nu)\mathrm{E}_3 - 2\nu_3^2\mathrm{E}\}}. \tag{3.1.13h~l}$$

The material is called *orthotropic*, if it possesses two mutually orthogonal planes of symmetry. In this case the material will also be symmetric with respect to a plane perpendicular to the two planes of symmetry. The number of independent elastic parameters reduces to nine, and the elasticity matrix, for example, can be expressed in a coordinate system coincident with the material symmetry directions, as

$$[C_{\mathrm{ab}}] = \begin{bmatrix} C_{11} & C_{12} & C_{13} & 0 & 0 & 0 \\ C_{12} & C_{22} & C_{23} & 0 & 0 & 0 \\ C_{13} & C_{23} & C_{33} & 0 & 0 & 0 \\ 0 & 0 & 0 & C_{44} & 0 & 0 \\ 0 & 0 & 0 & 0 & C_{55} & 0 \\ 0 & 0 & 0 & 0 & 0 & C_{66} \end{bmatrix}, \tag{3.1.14a}$$

with a similar expression holding for $[D_{\mathrm{ab}}]$. With E, ν, and μ standing for Young's modulus, Poisson's ratio, and the shear modulus, respectively, and with subscripts representing quantities with respect to the planes of symmetry, the compliance matrix in this case becomes

$$[D_{\mathrm{ab}}] = \begin{bmatrix} 1/\mathrm{E}_1 & -\nu_{21}/\mathrm{E}_2 & -\nu_{31}/\mathrm{E}_3 & 0 & 0 & 0 \\ -\nu_{12}/\mathrm{E}_1 & 1/\mathrm{E}_2 & -\nu_{32}/\mathrm{E}_3 & 0 & 0 & 0 \\ -\nu_{13}/\mathrm{E}_1 & -\nu_{23}/\mathrm{E}_2 & 1/\mathrm{E}_3 & 0 & 0 & 0 \\ 0 & 0 & 0 & 1/\mu_{23} & 0 & 0 \\ 0 & 0 & 0 & 0 & 1/\mu_{13} & 0 \\ 0 & 0 & 0 & 0 & 0 & 1/\mu_{12} \end{bmatrix}. \tag{3.1.14b}$$

Note that, since $D_{\mathrm{ab}} = D_{\mathrm{ba}}$ for a, b = 1, 2, 3,

$$\frac{\nu_{21}}{\mathrm{E}_2} = \frac{\nu_{12}}{\mathrm{E}_1}, \qquad \frac{\nu_{32}}{\mathrm{E}_3} = \frac{\nu_{23}}{\mathrm{E}_2},$$

$$\frac{\nu_{13}}{\mathrm{E}_1} = \frac{\nu_{31}}{\mathrm{E}_3}. \tag{3.1.14c}$$

When there is only one plane of symmetry, the total number of independent elastic parameters is thirteen, and the material is called *monoclinic*. Taking the x_3-direction normal to the plane of symmetry, the corresponding elasticity matrix $[C_{\mathrm{ab}}]$ becomes

$$[C_{ab}] = \begin{bmatrix} C_{11} & C_{12} & C_{13} & 0 & 0 & C_{16} \\ C_{12} & C_{22} & C_{23} & 0 & 0 & C_{26} \\ C_{13} & C_{23} & C_{33} & 0 & 0 & C_{36} \\ 0 & 0 & 0 & C_{44} & C_{45} & 0 \\ 0 & 0 & 0 & C_{45} & C_{55} & 0 \\ C_{16} & C_{26} & C_{36} & 0 & 0 & C_{66} \end{bmatrix}, \tag{3.1.15}$$

with a similar expression for the compliance matrix $[D_{ab}]$.

There are other symmetry conditions which can be considered. Many of these are related to crystal structure.[4] For example, in a crystal with cubic symmetry, there are three independent elastic parameters. These may be chosen to be

$$C_{12} = C_{23} = C_{31} = \lambda, \qquad C_{44} = C_{55} = C_{66} = \mu,$$

$$C_{11} = C_{22} = C_{33} = \lambda + 2\mu + \mu'. \tag{4.1.16a~c}$$

Here λ and μ are the usual Lamé constants, and the quantity

$$\mu' = 2\{\frac{C_{11} - C_{12}}{2} - C_{66}\} \tag{3.1.16d}$$

measures the degree of cubic anisotropy.

3.1.4. Plane Strain/Plane Stress

Consider a case when either the strain tensor or the stress tensor has vanishing components in a certain direction, say, the x_3-direction. If the strain satisfies

$$\mathbf{e}_3 \cdot \boldsymbol{\varepsilon} = \mathbf{0}, \qquad \text{or} \qquad \varepsilon_{33} = \varepsilon_{32}\ (= \varepsilon_{23}) = \varepsilon_{31}\ (= \varepsilon_{13}) = 0, \tag{3.1.17a,b}$$

then, it is called a *plane strain* state of deformation. On the other hand, if the stress satisfies

$$\mathbf{e}_3 \cdot \boldsymbol{\sigma} = \mathbf{0}, \qquad \text{or} \qquad \sigma_{33} = \sigma_{32}\ (= \sigma_{23}) = \sigma_{31}\ (= \sigma_{13}) = 0, \tag{3.1.18a,b}$$

then, it is called a *plane stress* state of deformation. For both plane strain and plane stress, strain components, ε_{11}, ε_{22}, and $\varepsilon_{12} = \varepsilon_{21}$, and stress components, σ_{11}, σ_{22}, and $\sigma_{12} = \sigma_{21}$, are called the inplane strain and stress components, while strain components, ε_{33}, $\varepsilon_{23} = \varepsilon_{32}$, $\varepsilon_{31} = \varepsilon_{13}$, and stress components, σ_{33}, $\sigma_{23} = \sigma_{32}$, $\sigma_{31} = \sigma_{13}$, are called the out-of-plane strain and stress components. It should be noted that although the word "plane" appears, the stress and/or strain fields are three-dimensional. That is, the plane strain or plane stress state corresponds to only $\mathbf{e}_3 \cdot \boldsymbol{\varepsilon} = \mathbf{0}$ or $\mathbf{e}_3 \cdot \boldsymbol{\sigma} = \mathbf{0}$, respectively, but not to both.

Since the plane strain and plane stress correspond to special deformation and stress states, the constitutive relations discussed in the previous subsections

---

[4] See Nye (1957).

do not change, and relation (3.1.1) or (3.1.8) holds between strain and stress tensors, i.e., $\boldsymbol{\sigma} = \mathbf{C} : \boldsymbol{\varepsilon}$ or $\boldsymbol{\varepsilon} = \mathbf{D} : \boldsymbol{\sigma}$. In matrix notation, (3.1.4a) for plane strain becomes

$$\begin{bmatrix} \tau_1 \\ \tau_2 \\ \tau_3 \\ \tau_4 \\ \tau_5 \\ \tau_6 \end{bmatrix} = \begin{bmatrix} C_{11} & C_{12} & C_{13} & C_{14} & C_{15} & C_{16} \\ C_{12} & C_{22} & C_{23} & C_{24} & C_{25} & C_{26} \\ C_{13} & C_{23} & C_{33} & C_{34} & C_{35} & C_{36} \\ C_{14} & C_{24} & C_{34} & C_{44} & C_{45} & C_{46} \\ C_{15} & C_{25} & C_{35} & C_{45} & C_{55} & C_{56} \\ C_{16} & C_{26} & C_{36} & C_{46} & C_{56} & C_{66} \end{bmatrix} \begin{bmatrix} \gamma_1 \\ \gamma_2 \\ 0 \\ 0 \\ 0 \\ \gamma_6 \end{bmatrix}, \qquad (3.1.19a)$$

or, when decomposed into the inplane stress components, $[\tau_1, \tau_2, \tau_6]$, and the out-of-plane stress components, $[\tau_3, \tau_4, \tau_5]$,

$$\begin{bmatrix} \tau_1 \\ \tau_2 \\ \tau_6 \end{bmatrix} = \begin{bmatrix} C_{11} & C_{12} & C_{16} \\ C_{12} & C_{22} & C_{26} \\ C_{16} & C_{26} & C_{66} \end{bmatrix} \begin{bmatrix} \gamma_1 \\ \gamma_2 \\ \gamma_6 \end{bmatrix},$$

$$\begin{bmatrix} \tau_3 \\ \tau_4 \\ \tau_5 \end{bmatrix} = \begin{bmatrix} C_{31} & C_{32} & C_{36} \\ C_{41} & C_{42} & C_{46} \\ C_{51} & C_{52} & C_{56} \end{bmatrix} \begin{bmatrix} \gamma_1 \\ \gamma_2 \\ \gamma_6 \end{bmatrix}. \qquad (3.1.19b,c)$$

Similarly, (3.1.7a) for plane stress becomes

$$\begin{bmatrix} \gamma_1 \\ \gamma_2 \\ \gamma_3 \\ \gamma_4 \\ \gamma_5 \\ \gamma_6 \end{bmatrix} = \begin{bmatrix} D_{11} & D_{12} & D_{13} & D_{14} & D_{15} & D_{16} \\ D_{12} & D_{22} & D_{23} & D_{24} & D_{25} & D_{26} \\ D_{13} & D_{23} & D_{33} & D_{34} & D_{35} & D_{36} \\ D_{14} & D_{24} & D_{34} & D_{44} & D_{45} & D_{46} \\ D_{15} & D_{25} & D_{35} & D_{45} & D_{55} & D_{56} \\ D_{16} & D_{26} & D_{36} & D_{46} & D_{56} & D_{66} \end{bmatrix} \begin{bmatrix} \tau_1 \\ \tau_2 \\ 0 \\ 0 \\ 0 \\ \tau_6 \end{bmatrix}, \qquad (3.1.20a)$$

or, when decomposed into the inplane strain components, $[\gamma_1, \gamma_2, \gamma_6]$, and the out-of-plane strain components, $[\gamma_3, \gamma_4, \gamma_5]$,

$$\begin{bmatrix} \gamma_1 \\ \gamma_2 \\ \gamma_6 \end{bmatrix} = \begin{bmatrix} D_{11} & D_{12} & D_{16} \\ D_{12} & D_{22} & D_{26} \\ D_{16} & D_{26} & D_{66} \end{bmatrix} \begin{bmatrix} \tau_1 \\ \tau_2 \\ \tau_6 \end{bmatrix},$$

$$\begin{bmatrix} \gamma_3 \\ \gamma_4 \\ \gamma_5 \end{bmatrix} = \begin{bmatrix} D_{31} & D_{32} & D_{36} \\ D_{41} & D_{42} & D_{46} \\ D_{51} & D_{52} & D_{56} \end{bmatrix} \begin{bmatrix} \tau_1 \\ \tau_2 \\ \tau_6 \end{bmatrix}. \qquad (3.1.20b,c)$$

These are *apparent constitutive relations*, expressing all stress components in terms of the inplane strain components for the plane strain case, and for the plane stress case, expressing all strain components in terms of the inplane stress components, i.e., (3.1.19b,c) and (3.1.20b,c), respectively. Since the out-of-plane components of stress (in plane strain) and the components of strain (in plane stress) are defined, once the corresponding inplane quantities are obtained, in what follows only the inplane relations (3.1.19b) and (3.1.20b) will be considered. Because

$$\begin{bmatrix} C_{11} & C_{12} & C_{16} \\ C_{12} & C_{22} & C_{26} \\ C_{16} & C_{26} & C_{66} \end{bmatrix} \neq \begin{bmatrix} D_{11} & D_{12} & D_{16} \\ D_{12} & D_{22} & D_{26} \\ D_{16} & D_{26} & D_{66} \end{bmatrix}^{-1} \tag{3.1.21a}$$

and

$$\begin{bmatrix} D_{11} & D_{12} & D_{16} \\ D_{12} & D_{22} & D_{26} \\ D_{16} & D_{26} & D_{66} \end{bmatrix} \neq \begin{bmatrix} C_{11} & C_{12} & C_{16} \\ C_{12} & C_{22} & C_{26} \\ C_{16} & C_{26} & C_{66} \end{bmatrix}^{-1} \tag{3.1.21b}$$

for $\mathbf{D} = \mathbf{C}^{-1}$ or $\mathbf{C} = \mathbf{D}^{-1}$, these *apparent* constitutive relations do *not* enjoy the usual reciprocal relation displayed by (3.1.1) and (3.1.8).

Now examine the apparent constitutive relations for the inplane strain and stress components, when the material is isotropic or transversely isotropic, with the plane of symmetry being the x_1,x_2-plane. Since isotropy is a special case of transverse isotropy, the latter case is considered. From (3.1.13), the three by three matrix for the apparent constitutive relations is as follows: for plane strain,

$$[C_{ab}^{(\varepsilon)}] = \frac{E^2 E_3}{(1+\nu)\{(1-\nu)E_3 - 2\nu_3^2 E\}} \times \begin{bmatrix} 1/E - \nu_3^2/E_3 & \nu/E + \nu_3^2/E_3 & 0 \\ \nu/E + \nu_3^2/E_3 & 1/E - \nu_3^2/E_3 & 0 \\ 0 & 0 & \{(1-\nu)E_3 - 2\nu_3^2 E\}/2EE_3 \end{bmatrix},$$

$$[D_{ab}^{(\varepsilon)}] = \begin{bmatrix} 1/E - \nu_3^2/E_3 & -\nu/E - \nu_3^2/E_3 & 0 \\ -\nu/E - \nu_3^2/E_3 & 1/E - \nu_3^2/E_3 & 0 \\ 0 & 0 & 2(1+\nu)/E \end{bmatrix}, \tag{3.1.22a,b}$$

and for plane stress,

$$[D_{ab}^{(\sigma)}] = \frac{1}{E}\begin{bmatrix} 1 & -\nu & 0 \\ -\nu & 1 & 0 \\ 0 & 0 & 2(1+\nu) \end{bmatrix},$$

$$[C_{ab}^{(\sigma)}] = \frac{E}{1-\nu^2}\begin{bmatrix} 1 & \nu & 0 \\ \nu & 1 & 0 \\ 0 & 0 & (1-\nu)/2 \end{bmatrix}, \tag{3.1.23a,b}$$

where $[D_{ab}^{(\varepsilon)}] \equiv [C_{ab}^{(\varepsilon)}]^{-1}$ and $[C_{ab}^{(\varepsilon)}] \equiv [D_{ab}^{(\varepsilon)}]^{-1}$, and the superscript (ε) or (σ) emphasizes whether the three by three matrix of the apparent constitutive relations is for plane strain or plane stress.

Since in (3.1.22) and (3.1.23), $\gamma_6 = \tau_6/\mu$ or $\tau_6 = \mu\gamma_6$($\varepsilon_{12} = \sigma_{12}/2\mu$ or $\sigma_{12} = 2\mu\varepsilon_{12}$), where μ is the shear modulus given by $E/2(1+\nu)$, expressions (3.1.22b) and (3.1.23a) become

$$[D_{ab}^{(\varepsilon)}] = \frac{1}{\mu}\begin{bmatrix} 1/2 - \mu(\nu_3^2/E_3 + \nu/E) & -\mu(\nu_3^2/E_3 + \nu/E) & 0 \\ -\mu(\nu_3^2/E_3 + \nu/E) & 1/2 - \mu(\nu_3^2/E_3 + \nu/E) & 0 \\ 0 & 0 & 1 \end{bmatrix},$$

$$[D_{ab}^{(\sigma)}] = \frac{1}{\mu}\begin{bmatrix} 1/2 - \nu/2(1+\nu) & -\nu/2(1+\nu) & 0 \\ -\nu/2(1+\nu) & 1/2 - \nu/2(1+\nu) & 0 \\ 0 & 0 & 1 \end{bmatrix}. \tag{3.1.24a,b}$$

As shown in later sections, these are two-dimensional isotropic matrices. Define κ by

$$\kappa \equiv \frac{3D_{11}^{(\alpha)} + D_{12}^{(\alpha)}}{D_{11}^{(\alpha)} - D_{12}^{(\alpha)}} = \frac{3D_{1111}^{(\alpha)} + D_{1122}^{(\alpha)}}{D_{1111}^{(\alpha)} - D_{1122}^{(\alpha)}} \qquad (\alpha = \varepsilon,\, \sigma), \tag{3.1.25a}$$

where $D_{ijkl}^{(\alpha)}$ is a *two-dimensional* fourth-order isotropic tensor corresponding to the three by three matrix $[D_{ab}^{(\alpha)}]$. Then, (3.1.24a) and (3.1.24b) yield

$$\kappa = \begin{cases} 3 - 8\mu\left(\dfrac{\nu_3^2}{E_3} + \dfrac{\nu}{E}\right) & \text{for plane strain} \\ \dfrac{3-\nu}{1+\nu} & \text{for plane stress.} \end{cases} \tag{3.1.25b}$$

Hence, the apparent constitutive relations (3.1.24a,b) and their inverses are given by

$$[D_{ab}^{(\alpha)}] = \frac{1}{\mu}\begin{bmatrix} (\kappa+1)/8 & (\kappa-3)/8 & 0 \\ (\kappa-3)/8 & (\kappa+1)/8 & 0 \\ 0 & 0 & 1 \end{bmatrix},$$

$$[C_{ab}^{(\alpha)}] = \mu\begin{bmatrix} (\kappa+1)/(\kappa-1) & -(\kappa-3)/(\kappa-1) & 0 \\ -(\kappa-3)/(\kappa-1) & (\kappa+1)/(\kappa-1) & 0 \\ 0 & 0 & 1 \end{bmatrix}, \tag{3.1.26a,b}$$

where $\alpha = \varepsilon$ for plane strain, and $\alpha = \sigma$ for plane stress. While μ relates the inplane shear stress and shear strain, the relation between the inplane hydrostatic stress, $(\sigma_{11} + \sigma_{22})/2$, and the inplane volumetric strain, $\varepsilon_{11} + \varepsilon_{22}$, is given by κ, as

$$\frac{\sigma_{11} + \sigma_{22}}{2} = \frac{2\mu}{\kappa - 1}(\varepsilon_{11} + \varepsilon_{22}). \tag{3.1.27}$$

Note that when the material is isotropic, $E_3 = E$ and $\nu_3 = \nu$, and κ for plane strain reduces to $3 - 4\nu$.

The apparent constitutive relations for the out-of-plane stress and strain components are easily determined. Since the constitutive relations for an isotropic or a transversely isotropic material with the x_3-axis as the axis of symmetry, are given by

$$\tau_3 = C_{31}(\gamma_1 + \gamma_2) + C_{33}\gamma_3 \qquad \text{or} \qquad \gamma_3 = D_{31}(\tau_1 + \tau_2) + D_{33}\tau_3,$$

$$\tau_4 = C_{44}\gamma_4, \qquad \tau_5 = C_{44}\gamma_5, \tag{3.1.28a~d}$$

it immediately follows that the out-of-plane normal stress and strain components satisfy

$$\tau_3 = C_{31}(\gamma_1 + \gamma_2) \qquad \text{or} \qquad \sigma_{33} = C_{3311}(\varepsilon_{11} + \varepsilon_{22}), \tag{3.1.22c,d}$$

for plane strain, and

$$\gamma_3 = D_{31}(\tau_1 + \tau_2) \qquad \text{or} \qquad \varepsilon_{33} = D_{3311}(\sigma_{11} + \sigma_{22}), \tag{3.1.23c,d}$$

for plane stress, where

$$C_{31} = C_{3311} = \frac{\nu_3 E E_3}{(1-\nu)E_3 - 2\nu_3^2 E}, \qquad D_{31} = D_{3311} = -\frac{\nu_3}{E_3}. \tag{3.1.29a,b}$$

In particular, for the isotropic case, $C_{31} = C_{3311} = (1-\nu)E/(1+\nu)(1-2\nu)$ and $D_{31} = D_{3311} = -\nu/E$, and the out-of-plane shear strain and stress components must vanish for either plane strain or plane stress. For other symmetries, however, the out-of-plane shear stress or shear strain may not be zero in the plane strain or plane stress case.

3.2. RECIPROCAL THEOREM, SUPERPOSITION, AND GREEN'S FUNCTION

When the microstructure of an RVE which consists of linearly elastic constituents, is fixed (i.e., existing cavities and cracks do not grow, and there is no frictional sliding of microcracks), the response of the RVE will be linear, and the reciprocal theorem applies.[5]

Consider two *separate* loadings of an RVE, with two different sets of self-equilibrating tractions, each applied separately on the surface ∂V of the RVE with fixed microstructure; note that these are *different* loadings of the same RVE. Denote the first set of tractions by $\mathbf{t}^{(1)}$, and the second set by $\mathbf{t}^{(2)}$, and refer to both collectively by $\mathbf{t}^{(\alpha)}$ ($\alpha = 1,\ 2$); see Figure 3.2.1. The displacement, the strain, and the stress produced by $\mathbf{t}^{(\alpha)}$ are designated as follows:

$$\{\mathbf{u}, \boldsymbol{\varepsilon}, \boldsymbol{\sigma}\} = \{\mathbf{u}^{(\alpha)}, \boldsymbol{\varepsilon}^{(\alpha)}, \boldsymbol{\sigma}^{(\alpha)}\} \qquad (\alpha = 1, 2). \tag{3.2.1a}$$

These fields satisfy the equilibrium equations, (1.3.2), the strain-displacement relations, (1.3.3a), the traction boundary conditions, (1.3.5a), and the linear stress-strain relations, (3.1.1a). In particular, the stress boundary conditions are

$$\mathbf{t}^{(\alpha)} = \boldsymbol{\nu} \cdot \boldsymbol{\sigma}^{(\alpha)} \qquad \text{on } \partial V, \tag{3.2.1b}$$

and the stress-strain relations become

$$\boldsymbol{\sigma}^{(\alpha)} = \mathbf{C} : \boldsymbol{\varepsilon}^{(\alpha)}, \tag{3.2.1c}$$

where the elasticity tensor, in general, is a function of position $\mathbf{x}$ in V, since the RVE is, in general, heterogeneous, i.e., $\mathbf{C} = \mathbf{C}(\mathbf{x})$.

---

[5] According to Love (1944, p.173) the theorem is due to E. Betti (*Il nuve Cimento* (Ser. 2), tt. 7 and 8 (1872)). More general theorems are given by Rayleigh (1873), and by Lamb (1889) who includes inertia.

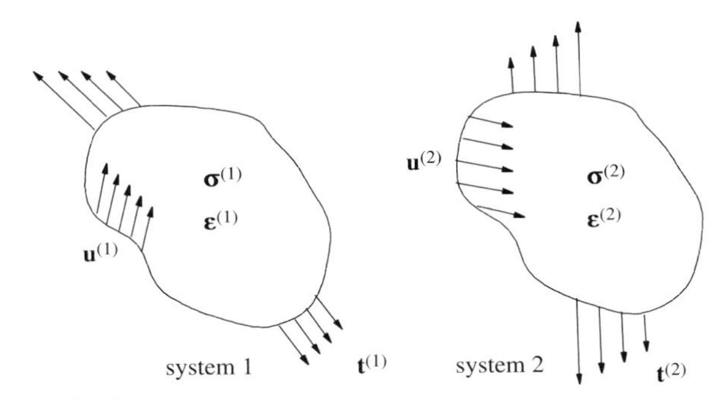

Figure 3.2.1

An RVE subjected to two sets of surface tractions, $\mathbf{t}^{(1)}$ and $\mathbf{t}^{(2)}$

3.2.1. Reciprocal Theorem

The *reciprocal theorem* states that the work done by the self-equilibrating surface tractions $\mathbf{t}^{(1)}$ going through the displacements $\mathbf{u}^{(2)}$ which are produced by the self-equilibrating surface tractions $\mathbf{t}^{(2)}$, equals the work done by the tractions $\mathbf{t}^{(2)}$ going through the displacements $\mathbf{u}^{(1)}$ which are produced by the tractions $\mathbf{t}^{(1)}$, i.e.,

$$\int_{\partial V} \mathbf{t}^{(1)} \cdot \mathbf{u}^{(2)} \, dS = \int_{\partial V} \mathbf{t}^{(2)} \cdot \mathbf{u}^{(1)} \, dS. \tag{3.2.2}$$

The proof follows from the symmetry of the elasticity tensor, which yields

$$\boldsymbol{\sigma}^{(\alpha)} : \boldsymbol{\varepsilon}^{(\beta)} = (\mathbf{C} : \boldsymbol{\varepsilon}^{(\alpha)}) : \boldsymbol{\varepsilon}^{(\beta)} = (\mathbf{C} : \boldsymbol{\varepsilon}^{(\beta)}) : \boldsymbol{\varepsilon}^{(\alpha)} = \boldsymbol{\sigma}^{(\beta)} : \boldsymbol{\varepsilon}^{(\alpha)}, \tag{3.2.3a}$$

and the fact that the two stress fields are symmetric and divergence-free, so that

$$\boldsymbol{\sigma}^{(\alpha)} : \boldsymbol{\varepsilon}^{(\beta)} = \boldsymbol{\sigma}^{(\alpha)} : (\boldsymbol{\nabla} \otimes \mathbf{u}^{(\beta)}) = \boldsymbol{\nabla} \cdot (\boldsymbol{\sigma}^{(\alpha)} \cdot \mathbf{u}^{(\beta)}). \tag{3.2.3b}$$

Now, substitution of (3.2.1b) into (3.2.2), and the use of the Gauss theorem and (3.2.3b) yield

$$\int_{\partial V} \mathbf{t}^{(\alpha)} \cdot \mathbf{u}^{(\beta)} \, dS = \int_V \boldsymbol{\sigma}^{(\alpha)} : \boldsymbol{\varepsilon}^{(\beta)} \, dV = \int_V \boldsymbol{\sigma}^{(\beta)} : \boldsymbol{\varepsilon}^{(\alpha)} \, dV = \int_{\partial V} \mathbf{t}^{(\beta)} \cdot \mathbf{u}^{(\alpha)} \, dS. \tag{3.2.3c}$$

The reciprocal theorem also holds when the self-compatible surface displacements instead of the surface tractions are prescribed on ∂V. In this case, the surface tractions are determined by the resulting stress field.

3.2.2. Superposition

The linearity of the RVE for a fixed microstructure permits construction of various solutions by means of superposition. This means that if the solutions for two different boundary data are known, then the solution when both boundary data are applied, is obtained by the addition of the corresponding field quantities. For example, for any two constants $a^{(1)}$ and $a^{(2)}$, the solution for the self-equilibrating surface tractions

$$\mathbf{t} = a^{(1)}\mathbf{t}^{(1)} + a^{(2)}\mathbf{t}^{(2)} = a^{(\alpha)}\mathbf{t}^{(\alpha)} \qquad \text{on } \partial V \ (\alpha \text{ summed}), \tag{3.2.4a}$$

is given by

$$\{\mathbf{u}, \boldsymbol{\varepsilon}, \boldsymbol{\sigma}\} = \{a^{(\alpha)}\mathbf{u}^{(\alpha)}, a^{(\alpha)}\boldsymbol{\varepsilon}^{(\alpha)}, a^{(\alpha)}\boldsymbol{\sigma}^{(\alpha)}\} \qquad (\alpha \text{ summed}). \tag{3.2.4b}$$

Note that the superposition follows if, instead of surface tractions, surface displacements are prescribed, or if a suitable combination of surface tractions and surface displacements are given.

3.2.3. Green's Function

The Dirac delta function, $\delta(\mathbf{x})$, has the property that, for any suitably smooth scalar- or tensor-valued function, $f(\mathbf{x})$, defined on ∂V,

$$\int_{\partial V} f(\mathbf{y})\, \delta(\mathbf{y} - \mathbf{x})\, dS = f(\mathbf{x}), \tag{3.2.5}$$

where the integration is with respect to $\mathbf{y}$ on ∂V. The Dirac delta function, interpreted in the distributional sense,[6] can be used to represent concentrated forces. For example, if $\mathbf{T}$ is a concentrated force applied at a *fixed* point $\mathbf{x}$ on ∂V, the corresponding tractions on ∂V can be defined as

$$\mathbf{t}(\mathbf{y}) = \mathbf{T}\, \delta(\mathbf{y} - \mathbf{x}) \qquad \text{for } \mathbf{y} \text{ on } \partial V. \tag{3.2.6}$$

The concept of Green's function can be effectively used to obtain general results for *linear problems*, where superposition applies. The actual calculation of Green's function will generally be unnecessary. It is only the concept that is applied. Green's function here is introduced in terms of the concentrated boundary forces, but the basic idea also applies when concentrated body forces are involved.

Green's function $\mathbf{G} = \mathbf{G}(\mathbf{x}, \mathbf{y})$ is a second-order tensor with components $G_{ij}(\mathbf{x}, \mathbf{y})$, representing the displacement component $u_i(\mathbf{x})$ at a point $\mathbf{x}$, due to the unit concentrated force applied at a point $\mathbf{y}$ in the $\mathbf{e}_j$-direction;[7] see Figure 3.2.2. $\mathbf{G}$ is *a two-point tensor field.* According to the reciprocal theorem, $\mathbf{G}$ has the property that

$$G_{ij}(\mathbf{x}, \mathbf{y}) = G_{ji}(\mathbf{y}, \mathbf{x}). \tag{3.2.7}$$

Suppose the self-equilibrating surface tractions $\mathbf{t} = \mathbf{t}(\mathbf{y})$ are prescribed on the boundary ∂V of an RVE. With the use of Green's function and

---

[6] See, for example, Stakgold (1967).

[7] When V is finite, conceptually, suitable body forces may be distributed within the body in order to equilibrate the applied concentrated force.

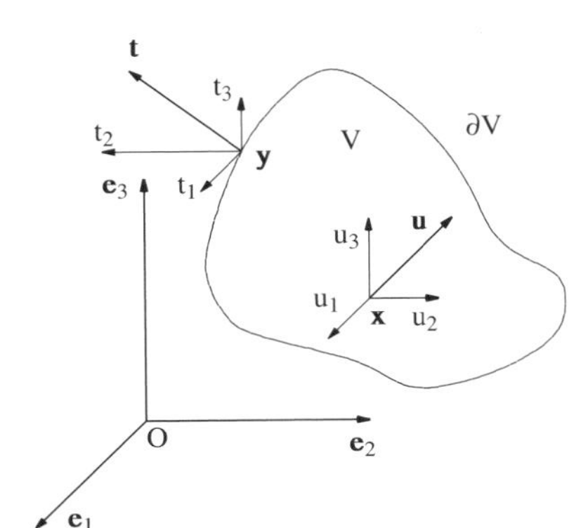

Figure 3.2.2

Green's function

superposition, the resulting displacement field is given by

$$\mathbf{u}(\mathbf{x}) = \int_{\partial V} \mathbf{G}(\mathbf{x},\, \mathbf{y}) \cdot \mathbf{t}(\mathbf{y})\, dS, \tag{3.2.8a}$$

where the integration is with respect to $\mathbf{y}$. The corresponding strain and stress fields are now obtained by direct calculation, arriving at

$$\boldsymbol{\varepsilon}(\mathbf{x}) = \int_{\partial V} \frac{1}{2} \{[\boldsymbol{\nabla} \otimes \mathbf{G}(\mathbf{x},\, \mathbf{y})] \cdot \mathbf{t}(\mathbf{y}) + \{[\boldsymbol{\nabla} \otimes \mathbf{G}(\mathbf{x},\, \mathbf{y})] \cdot \mathbf{t}(\mathbf{y})\}^T\}\, dS,$$

$$\boldsymbol{\sigma}(\mathbf{x}) = \int_{\partial V} \mathbf{C} : [\boldsymbol{\nabla} \otimes \mathbf{G}(\mathbf{x},\, \mathbf{y})] \cdot \mathbf{t}(\mathbf{y})\, dS, \tag{3.2.8b,c}$$

where $\boldsymbol{\nabla}$ is with respect to $\mathbf{x}$. In component form, (3.2.8a~c) become

$$u_i(\mathbf{x}) = \int_{\partial V} G_{ij}(\mathbf{x},\, \mathbf{y})\, t_j(\mathbf{y})\, dS,$$

$$\varepsilon_{ij}(\mathbf{x}) = \int_{\partial V} \frac{1}{2} [G_{ik,j}(\mathbf{x},\, \mathbf{y}) + G_{jk,i}(\mathbf{x},\, \mathbf{y})]\, t_k(\mathbf{y})\, dS,$$

$$\sigma_{ij}(\mathbf{x}) = \int_{\partial V} C_{ijkl}\, G_{km,l}(\mathbf{x},\, \mathbf{y})\, t_m(\mathbf{y})\, dS, \tag{3.2.9a~c}$$

where, for example, $G_{ik,j}(\mathbf{x},\, \mathbf{y}) \equiv \partial G_{ik}(\mathbf{x},\, \mathbf{y})/\partial x_j$.[8]

In the next section the response of an RVE with microcavities to both prescribed macrostresses and prescribed macrostrains is considered. For the case when the macrostrains are prescribed, it is convenient to consider the inverse of (3.2.8a). To this end, let $\mathbf{x}$ and $\mathbf{y}$ be two arbitrary points, *both* on the boundary ∂V of the RVE; note that in (3.2.8a~c), $\mathbf{x}$ is a typical point in V or on ∂V. When tractions $\mathbf{t}(\mathbf{y})$ are prescribed on ∂V, then the *resulting* surface displacements are given by (3.2.8a) for $\mathbf{x}$ on ∂V. On the other hand, and in view of the uniqueness of the solution to linearly elastic problems, if the self-compatible

[8] See, e.g., Roach (1982) for a detailed discussion of Green's function.

displacements $\mathbf{u}(\mathbf{x})$ are prescribed over ∂V, then the *corresponding* self-equilibrating tractions which must be applied in order to attain such displacements are given by the solution $\mathbf{t}(\mathbf{y})$ of the integral equation (3.2.8a). Thus, (3.2.8a) admits an inverse. This is expressed by

$$\mathbf{t}(\mathbf{y}) = \int_{\partial V} \mathbf{G}^{-1}(\mathbf{y}, \mathbf{x}) \cdot \mathbf{u}(\mathbf{x}) \, dS, \tag{3.2.10a}$$

where the integration is with respect to $\mathbf{x}$ over ∂V. The second-order tensor-valued function $\mathbf{G}^{-1}(\mathbf{y}, \mathbf{x})$, is also a Green function. It is the "inverse" of $\mathbf{G}(\mathbf{x}, \mathbf{y})$, in the sense that

$$\int_{\partial V} \mathbf{G}(\mathbf{x}, \mathbf{z}) \cdot \mathbf{G}^{-1}(\mathbf{z}, \mathbf{y}) \, dS = \mathbf{1}^{(2)} \, \delta(\mathbf{y} - \mathbf{x}), \tag{3.2.10b}$$

where $\mathbf{1}^{(2)} = \delta_{ij} \, \mathbf{e}_i \otimes \mathbf{e}_j$ is the second-order identity tensor, $\mathbf{x}$ and $\mathbf{y}$ are typical points on ∂V, and the integration is with respect to point $\mathbf{z}$; see Figure 3.2.3. Then, upon substitution of (3.2.10a) into (3.2.8a), it follows that

$$\begin{aligned} \mathbf{u}(\mathbf{x}) &= \int_{\partial V} \{ \int_{\partial V} \mathbf{G}(\mathbf{x}, \mathbf{z}) \cdot \mathbf{G}^{-1}(\mathbf{z}, \mathbf{y}) \, dS \} \cdot \mathbf{u}(\mathbf{y}) \, dS \\ &= \int_{\partial V} \delta(\mathbf{y} - \mathbf{x}) \, \mathbf{u}(\mathbf{y}) \, dS = \mathbf{u}(\mathbf{x}). \end{aligned} \tag{3.2.10c}$$

Similarly,

$$\int_{\partial V} \mathbf{G}^{-1}(\mathbf{x}, \mathbf{z}) \cdot \mathbf{G}(\mathbf{z}, \mathbf{y}) \, dS = \mathbf{1}^{(2)} \, \delta(\mathbf{y} - \mathbf{x}). \tag{3.2.10d}$$

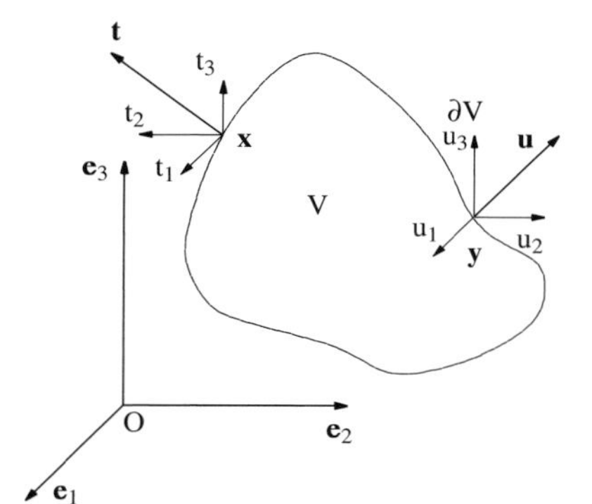

Figure 3.2.3

Inverse of Green's function

It is emphasized that (3.2.10b,c) are valid if and only if points $\mathbf{x}$ and $\mathbf{y}$ are on the same boundary ∂V where the tractions $\mathbf{t}(\mathbf{z})$ *correspond* to the prescribed displacements $\mathbf{u}(\mathbf{x})$ and, conversely, the displacements $\mathbf{u}(\mathbf{x})$ *correspond* to the prescribed tractions $\mathbf{t}(\mathbf{z})$.

3.3. REFERENCES

Cowin, S. C. and Mehrabadi, M. M. (1992), The structure of the linear anisotropic elastic symmetries, *J. Mech. Phys. Solids*, Vol. 40, 1459-1471.

Hearmon, R. F. S. (1961), *An introduction to applied anisotropic elasticity*, Oxford University Press, London.

Jones, R. M. (1975), *Mechanics of composite materials*, McGraw-Hill, New York.

Kelvin, Lord (1856), *Phil. Trans. R. Soc.*, Vol. 166, 481.

Lamb, H. (1889), *London Math. Soc. Proc.,* Vol. 19, 144.

Lekhnitskii, A. G. (1963), *Theory of elasticity of an anisotropic elastic body*, Holder Day, San Francisco.

Love, A. E. H. (1944), *A Treatise on the mathematical theory of elasticity*, Dover Publications, New York.

Mehrabadi, M. M. and Cowin, S. C. (1990), Eigentensors of linear anisotropic elastic materials, *Q J. Mech. Appl. Math.*, Vol. 43, Pt. 1.

Nemat-Nasser, S. and Hori, M. (1988), *Micromechanics: Overall properties of heterogeneous elastic solids*, Class Notes, UCSD.

Nye, J. F. (1957), *Physical properties of crystals*, Clarendon Press, Oxford.

Rayleigh, Lord (1873), *London Math. Soc. Proc.,* Vol. 4 or *Scientific Papers*, Vol. 1, 179.

Roach, G. F. (1982), *Green's functions* (second edition), Cambridge University Press.

Sokolnikoff, I. S. (1956), *Mathematical theory of elasticity*, McGraw-Hill, New York.

Stakgold, I. (1967), *Boundary value problems of mathematical physics*, Vol. I, Macmillan, New York.

SECTION 4 ELASTIC SOLIDS WITH TRACTION-FREE DEFECTS

In this section, an RVE consisting of a linearly elastic material which contains stress-free cavities, is considered. The overall stress-strain/strain-stress relations are developed. The results are then illustrated by a number of simple examples. This section is intended as a simple but concrete illustration of how macroquantities are related to microquantities and microstructure. Subsequent sections will then treat more general cases, including open and closed microcracks, micro-inclusions, and related problems, using various averaging techniques, and comparing results by means of illustrative examples. Bounds on overall moduli are given in Sections 9 and 13.[1]

4.1. STATEMENT OF PROBLEM AND NOTATION

Consider an RVE with total volume V, bounded *externally* by surface ∂V. On this surface, either uniform tractions,

$$\mathbf{t}^{\mathrm{o}} = \boldsymbol{\nu} \cdot \boldsymbol{\sigma}^{\mathrm{o}} \qquad \text{on } \partial V, \tag{4.1.1a}$$

or linear displacements,

$$\mathbf{u}^{\mathrm{o}} = \mathbf{x} \cdot \boldsymbol{\varepsilon}^{\mathrm{o}} \qquad \text{on } \partial V, \tag{4.1.1b}$$

are assumed to be prescribed, where $\boldsymbol{\sigma}^{\mathrm{o}}$ and $\boldsymbol{\varepsilon}^{\mathrm{o}}$ are second-order symmetric *constant* stress and strain tensors for the macro-element. It is emphasized that either (4.1.1a) or (4.1.1b), but not both, can be prescribed. In other words, if the traction boundary data (4.1.1a) corresponding to the constant macrostress $\boldsymbol{\Sigma} = \boldsymbol{\sigma}^{\mathrm{o}}$, are prescribed, then the surface displacements on ∂V, corresponding to these tractions, in general, are *not* spatially linear, being affected by the microstructure of the RVE. Similarly, if the linear displacement boundary data (4.1.1b) corresponding to the constant macrostrain $\mathbf{E} = \boldsymbol{\varepsilon}^{\mathrm{o}}$, are prescribed, then the surface tractions on ∂V, produced by these displacements, are *not*, in general, spatially uniform. In the sequel, therefore, the two cases are treated separately and independently, and then the relation between the results is discussed.[2]

---

[1] For application of the results of this section and references to related issues, see Sections 5 and 6.

[2] See also general results presented in Section 2, especially Subsections 2.5 and 2.6.

Assume that the material of the RVE is linearly elastic and *homogeneous* (but not necessarily isotropic). The inhomogeneity, therefore, stems solely from the presence of cavities. Denote a typical cavity by Ω_α, with the boundary $\partial\Omega_\alpha$ ($\alpha = 1, 2, ..., n$), so that there are a total of n individual cavities in V. The union of these cavities is denoted by Ω, having the boundary $\partial\Omega$ which is the union of all $\partial\Omega_\alpha$, i.e.,

$$\Omega \equiv \bigcup_{\alpha=1}^{n} \Omega_\alpha, \qquad \partial\Omega \equiv \bigcup_{\alpha=1}^{n} \partial\Omega_\alpha. \tag{4.1.2a,b}$$

The remainder of the RVE (i.e, when Ω is excluded) is called the *matrix*. The matrix is denoted by[3] M. The boundary of M is the sum of ∂V and $\partial\Omega$, Figure 4.1.1,

$$M \equiv V - \Omega, \qquad \partial M \equiv \partial V + \partial\Omega. \tag{4.1.3a,b}$$

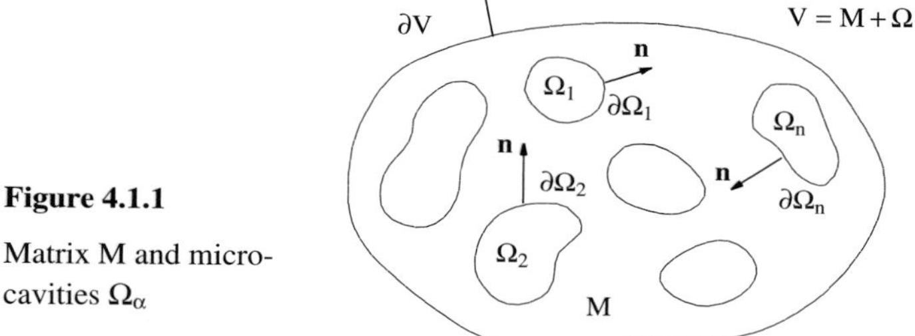

Figure 4.1.1

Matrix M and micro-cavities Ω_α

The total boundary surface of the RVE can include some portion of $\partial\Omega$. For simplicity, however, exclude this possibility. Thus, all cavities are within the RVE, each being fully surrounded by the matrix material. For a typical cavity, Ω_α, two faces of its surface boundary, $\partial\Omega_\alpha$, may be distinguished, as follows: (1) the *exterior* face of the cavity, denoted by $\partial\Omega_\alpha^c$, which is the face toward the matrix material, defined by the direction of the exterior unit normal **n** of the cavity; and (2) the *exterior* face of the *surrounding matrix*, denoted by $\partial\Omega_\alpha^M$, which is the face toward the interior of the cavity, defined by the direction of the exterior unit normal $(-\mathbf{n})$ of the matrix (i.e., the interior unit normal of the cavity). $\partial\Omega_\alpha$ coincides with $\partial\Omega_\alpha^c$ for the cavity Ω_α, while ∂M at the cavity Ω_α coincides with $\partial\Omega_\alpha^M$; see Figure 4.1.2. In view of this convention, the integral of a surface quantity taken over ∂M can always be decomposed as

$$\int_{\partial M} (.)\, dS = \int_{\partial V} (.)\, dS + \sum_{\alpha=1}^{n} \int_{\partial\Omega_\alpha^M} (.)\, dS = \int_{\partial V} (.)\, dS - \sum_{\alpha=1}^{n} \int_{\partial\Omega_\alpha^c} (.)\, dS$$

[3] In general, V, Ω, and M are open sets, but sometimes they may be treated as closed, if it is felt that this does not cause any confusion.

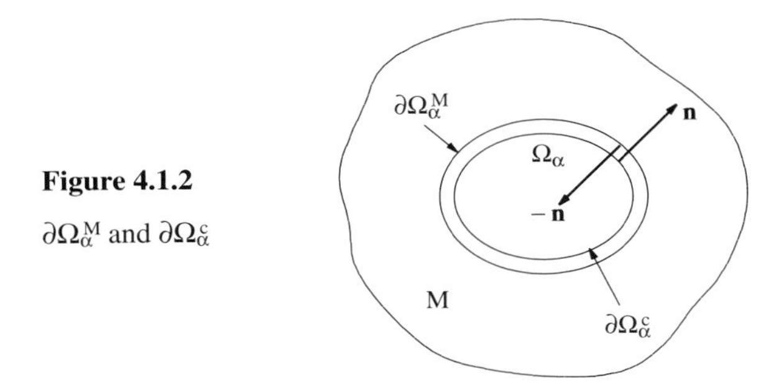

Figure 4.1.2

$\partial\Omega_\alpha^M$ and $\partial\Omega_\alpha^c$

$$= \int_{\partial V} (.)\, dS - \int_{\partial\Omega} (.)\, dS. \tag{4.1.4}$$

Thus $\partial\Omega$ *always stands for the union of* $\partial\Omega_\alpha^c$ ($\alpha = 1, 2, ..., n$).

To distinguish the boundary of M at the cavities from that at the exterior of the RVE, which is ∂V, the exterior unit normal on ∂V is systematically denoted by $\mathbf{v}$ (as before), and the *exterior* unit normal on the surface $\partial\Omega_\alpha$ for a typical cavity Ω_α, by $\mathbf{n}$, *pointing from the inside of the cavity toward the matrix* M.

The matrix material is linearly elastic and homogeneous. Denote the corresponding constant elasticity tensor by **C** and the compliance tensor by **D**.

4.2. AVERAGE STRAIN FOR PRESCRIBED MACROSTRESS

Suppose that uniform tractions $\mathbf{t}^o = \mathbf{v} \cdot \boldsymbol{\sigma}^o$ are prescribed on ∂V, associated with the constant symmetric macrostress $\boldsymbol{\Sigma} = \boldsymbol{\sigma}^o$. If the RVE is homogeneous, having *no* cavities, then the corresponding average strain associated with the average stress $\boldsymbol{\sigma}^o$ would be

$$\boldsymbol{\varepsilon}^o \equiv \mathbf{D} : \boldsymbol{\sigma}^o, \tag{4.2.1}$$

and hence, in conjunction with $\bar{\boldsymbol{\sigma}} = \boldsymbol{\sigma}^o$, the average strain would be $\boldsymbol{\varepsilon}^o$. The presence of cavities disturbs the uniform stress and strain fields, producing the variable stress field $\boldsymbol{\sigma} = \boldsymbol{\sigma}(\mathbf{x})$ and strain field $\boldsymbol{\varepsilon} = \boldsymbol{\varepsilon}(\mathbf{x})$, in M, with $\boldsymbol{\sigma} = \mathbf{0}$ in Ω. Nevertheless, from the results of Section 2,

$$\bar{\boldsymbol{\sigma}} = < \boldsymbol{\sigma} > = \frac{1}{V} \int_V \boldsymbol{\sigma}\, dV = \frac{1}{V} \int_M \boldsymbol{\sigma}\, dV = \boldsymbol{\sigma}^o. \tag{4.2.2}$$

On the other hand, the average strain is *not*, in general, equal to $\boldsymbol{\varepsilon}^o$. Instead,

$$\bar{\boldsymbol{\varepsilon}} = < \boldsymbol{\varepsilon} > = \boldsymbol{\varepsilon}^o + \bar{\boldsymbol{\varepsilon}}^c, \tag{4.2.3}$$

where $\boldsymbol{\varepsilon}^o$ is *defined* by (4.2.1), and $\bar{\boldsymbol{\varepsilon}}^c$ is the additional strain due to the presence of cavities.

To calculate the additional strain $\bar{\boldsymbol{\varepsilon}}^c$ due to cavities, one may apply the reciprocal theorem, as follows. Consider two sets of loads, one defined by

$$\mathbf{t}^{(1)} = \begin{cases} \boldsymbol{\nu} \cdot \delta\boldsymbol{\sigma}^o & \text{on } \partial V \\ -\mathbf{n} \cdot \delta\boldsymbol{\sigma}^o & \text{on } \partial\Omega \end{cases} \tag{4.2.4a}$$

which corresponds to uniform *virtual* stress $\delta\boldsymbol{\sigma}^o$ and strain $\delta\boldsymbol{\varepsilon}^o = \mathbf{D} : \delta\boldsymbol{\sigma}^o$ within the entire RVE (as illustrated in Figure 4.1.2, $-\mathbf{n}$ is the *interior* unit normal on the *cavity* surface $\partial\Omega$, or the *exterior* unit normal to the boundary of the *matrix*), and the other defined by

$$\mathbf{t}^{(2)} = \begin{cases} \boldsymbol{\nu} \cdot \boldsymbol{\sigma}^o & \text{on } \partial V \\ \mathbf{0} & \text{on } \partial\Omega \end{cases} \tag{4.2.4b}$$

which is the actual loading considered for the RVE.

Denote the displacement, strain, and stress fields associated with the first loading (4.2.4a) by

$$\{\mathbf{u}^{(1)},\ \boldsymbol{\varepsilon}^{(1)},\ \boldsymbol{\sigma}^{(1)}\} = \{(\mathbf{x} \cdot \delta\boldsymbol{\varepsilon}^o),\ \delta\boldsymbol{\varepsilon}^o,\ \delta\boldsymbol{\sigma}^o\} \tag{4.2.5a}$$

which follows from the fact that, for loading (4.2.4a), the strain and stress fields are both uniform throughout the matrix M. And denote the fields associated with the second (i.e., the actual) loading (4.2.4b) by

$$\{\mathbf{u}^{(2)},\ \boldsymbol{\varepsilon}^{(2)},\ \boldsymbol{\sigma}^{(2)}\} = \{\mathbf{u},\ \boldsymbol{\varepsilon},\ \boldsymbol{\sigma}\}. \tag{4.2.5b}$$

From the reciprocal theorem, (3.2.2), it follows that

$$\int_{\partial V} (\boldsymbol{\nu} \cdot \boldsymbol{\sigma}^o) \cdot (\mathbf{x} \cdot \delta\boldsymbol{\varepsilon}^o)\, dS = \int_{\partial V} (\boldsymbol{\nu} \cdot \delta\boldsymbol{\sigma}^o) \cdot \mathbf{u}\, dS - \int_{\partial\Omega} (\mathbf{n} \cdot \delta\boldsymbol{\sigma}^o) \cdot \mathbf{u}\, dS \tag{4.2.6a}$$

which can be written as

$$\delta\boldsymbol{\sigma}^o : \{\int_{\partial V} \mathbf{D} : \{(\mathbf{x} \otimes \boldsymbol{\nu}) \cdot \boldsymbol{\sigma}^o\}\, dS - \int_{\partial V} \boldsymbol{\nu} \otimes \mathbf{u}\, dS + \int_{\partial\Omega} \mathbf{n} \otimes \mathbf{u}\, dS\} = 0. \tag{4.2.6b}$$

Since $\delta\boldsymbol{\sigma}^o$ is an arbitrary symmetric tensor, the symmetric part of the quantity within the braces must vanish identically. Noting that the first integral within the braces yields

$$\frac{1}{V} \int_{\partial V} \mathbf{D} : \{(\mathbf{x} \otimes \boldsymbol{\nu}) \cdot \boldsymbol{\sigma}^o\}\, dS = \mathbf{D} : \{\mathbf{1}^{(2)} \cdot \boldsymbol{\sigma}^o\} = \boldsymbol{\varepsilon}^o, \tag{4.2.7a}$$

and using the averaging scheme presented in Section 2, it follows that

$$\begin{aligned} \bar{\boldsymbol{\varepsilon}} &= \frac{1}{V} \int_V \frac{1}{2} \{\boldsymbol{\nabla} \otimes \mathbf{u} + (\boldsymbol{\nabla} \otimes \mathbf{u})^T\}\, dV \\ &= \boldsymbol{\varepsilon}^o + \frac{1}{V} \int_{\partial\Omega} \frac{1}{2} (\mathbf{n} \otimes \mathbf{u} + \mathbf{u} \otimes \mathbf{n})\, dS. \end{aligned} \tag{4.2.7b}$$

Comparison with (4.2.3) shows that the additional strain $\bar{\boldsymbol{\varepsilon}}^c$ due to cavities, is given by[4]

$$\bar{\boldsymbol{\varepsilon}}^{\mathrm{c}} = \frac{1}{\mathrm{V}} \int_{\partial\Omega} \frac{1}{2} (\mathbf{n} \otimes \mathbf{u} + \mathbf{u} \otimes \mathbf{n}) \, \mathrm{dS}, \tag{4.2.8a}$$

or, in component form,

$$\bar{\varepsilon}^{\mathrm{c}}_{\mathrm{ij}} = \frac{1}{\mathrm{V}} \int_{\partial\Omega} \frac{1}{2} (\mathrm{n_i\, u_j} + \mathrm{n_j\, u_i}) \, \mathrm{dS}. \tag{4.2.8b}$$

4.3. OVERALL COMPLIANCE TENSOR FOR POROUS ELASTIC SOLIDS

Define *the overall compliance* $\bar{\mathbf{D}}$ of the porous RVE with a linearly elastic homogeneous matrix, through

$$\bar{\boldsymbol{\varepsilon}} = \bar{\mathbf{D}} : \bar{\boldsymbol{\sigma}} = \bar{\mathbf{D}} : \boldsymbol{\sigma}^{\mathrm{o}}, \tag{4.3.1}$$

where the macrostress, $\boldsymbol{\Sigma} = \boldsymbol{\sigma}^{\mathrm{o}}$, is regarded prescribed, and the average strain is given by (4.2.3). To obtain the overall compliance in an explicit form, the strain $\bar{\boldsymbol{\varepsilon}}^{\mathrm{c}}$ due to cavities will now be expressed in terms of the applied stress $\boldsymbol{\sigma}^{\mathrm{o}}$. Since the matrix of the RVE is linearly elastic, for a given microstructure the displacement $\mathbf{u}(\mathbf{x})$ at a point $\mathbf{x}$ on $\partial\Omega$ is linearly dependent on the uniform overall stress $\boldsymbol{\sigma}^{\mathrm{o}}$. This is easily seen if the applied tractions (4.1.1a) are substituted into (3.2.8a), to arrive at

$$\mathbf{u}(\mathbf{x}) = \int_{\partial\mathrm{V}} \mathbf{G}(\mathbf{x}, \mathbf{y}) \cdot \{\boldsymbol{\nu}(\mathbf{y}) \cdot \boldsymbol{\sigma}^{\mathrm{o}}\} \, \mathrm{dS}, \tag{4.3.2a}$$

where the integration is taken with respect to $\mathbf{y}$ over the boundary $\partial\mathrm{V}$ of the RVE. Since $\boldsymbol{\sigma}^{\mathrm{o}}$ is a symmetric constant tensor, (4.3.2a) can be expressed as

$$\mathrm{u_i}(\mathbf{x}) = \mathrm{K_{ijk}}(\mathbf{x})\, \sigma^{\mathrm{o}}_{\mathrm{jk}}, \tag{4.3.2b}$$

where the third-order tensor,

$$\mathrm{K_{ijk}}(\mathbf{x}) = \mathrm{K_{ikj}}(\mathbf{x}) = \int_{\partial\mathrm{V}} \frac{1}{2} \{\mathrm{G_{ij}}(\mathbf{x}, \mathbf{y})\, \nu_{\mathrm{k}}(\mathbf{y}) + \mathrm{G_{ik}}(\mathbf{x}, \mathbf{y})\, \nu_{\mathrm{j}}(\mathbf{y})\} \, \mathrm{dS}, \tag{4.3.2c}$$

depends on the geometry and the elastic properties of the matrix of the RVE.

To obtain the additional overall strain, $\bar{\boldsymbol{\varepsilon}}^{\mathrm{c}}$, due to the presence of cavities in terms of the prescribed overall stress, $\boldsymbol{\sigma}^{\mathrm{o}}$, substitute from (4.3.2c) into (4.2.8), to arrive at

$$\bar{\varepsilon}^{\mathrm{c}}_{\mathrm{ij}} = \mathrm{H_{ijkl}}\, \sigma^{\mathrm{o}}_{\mathrm{kl}}, \tag{4.3.3a}$$

where the *constant* fourth-order tensor, $\mathbf{H}$, is given by

$$\mathrm{H_{ijkl}} \equiv \mathrm{H_{jikl}} \equiv \mathrm{H_{ijlk}} \equiv \frac{1}{\mathrm{V}} \int_{\partial\Omega} \frac{1}{2} \{\mathrm{n_i}(\mathbf{x})\, \mathrm{K_{jkl}}(\mathbf{x}) + \mathrm{n_j}(\mathbf{x})\, \mathrm{K_{ikl}}(\mathbf{x})\} \, \mathrm{dS}. \tag{4.3.3b}$$

---

[4] For a direct evaluation of cavity strain, (4.2.8) is due to Horii and Nemat-Nasser (1983), where application to frictional cracks is formulated, as discussed in Subsection 6.4.

Hence, for an RVE with a linearly elastic matrix containing cavities of *arbitrary shapes and sizes*, the following general result is obtained, when the overall macrostress is regarded prescribed (Horii and Nemat-Nasser, 1983)[5]:

$$\bar{\boldsymbol{\varepsilon}}^c = \mathbf{H} : \boldsymbol{\sigma}^o, \qquad \text{or} \qquad \bar{\varepsilon}^c_{ij} = H_{ijkl}\, \sigma^o_{kl}. \tag{4.3.4a,b}$$

It should be noted that this exact result is valid whether or not the linearly elastic constituent of the RVE is homogeneous. The requirements are: (1) the matrix of the RVE is linearly elastic, and (2) the microstructure of the RVE remains unchanged under the applied macrostress $\boldsymbol{\Sigma} = \boldsymbol{\sigma}^o$.

It may be instructive to re-examine (4.3.2b) by first introducing for a pair of points, $\mathbf{x}$ on $\partial\Omega$ and $\mathbf{y}$ on ∂V, a fourth-order tensor, $\mathbf{h}(\mathbf{x}, \mathbf{y})$, as

$$h_{ijkl}(\mathbf{x}, \mathbf{y}) = \frac{1}{4} \{n_i(\mathbf{x})\, G_{jk}(\mathbf{x}, \mathbf{y})\, \nu_l(\mathbf{y}) + n_i(\mathbf{x})\, G_{jl}(\mathbf{x}, \mathbf{y})\, \nu_k(\mathbf{y})$$

$$+ n_j(\mathbf{x})\, G_{ik}(\mathbf{x}, \mathbf{y})\, \nu_l(\mathbf{y}) + n_j(\mathbf{x})\, G_{il}(\mathbf{x}, \mathbf{y})\, \nu_k(\mathbf{y})\}, \tag{4.3.5a}$$

and then obtain

$$\mathbf{H} = \frac{1}{V} \int_{\partial\Omega} \int_{\partial V} \mathbf{h}(\mathbf{x}, \mathbf{y})\, dS\, dS, \tag{4.3.5b}$$

where the $\mathbf{y}$-integral is over ∂V, and the $\mathbf{x}$-integral is over $\partial\Omega$.

To obtain the overall elastic compliance tensor $\bar{\mathbf{D}}$, in terms of the constant compliance of the matrix, $\mathbf{D}$, and the *constant* tensor $\mathbf{H}$, substitute (4.2.1), (4.3.1), and (4.3.4) into (4.2.3), and noting that the resulting equation must hold for any macrostress $\boldsymbol{\sigma}^o$, arrive at

$$\bar{\mathbf{D}} = \mathbf{D} + \mathbf{H}, \tag{4.3.6a}$$

or

$$\bar{D}_{ijkl} = D_{ijkl} + H_{ijkl}. \tag{4.3.6b}$$

In many situations, the tensor $\mathbf{H}$ can be computed *directly*, using (4.2.8). Several examples of this are given later on, in Sections 5 and 6.

4.4. AVERAGE STRESS FOR PRESCRIBED MACROSTRAIN

Suppose that, instead of the uniform tractions, the linear displacements $\mathbf{u}^o = \mathbf{x} \cdot \boldsymbol{\varepsilon}^o$ (associated with the constant symmetric macrostrain $\mathbf{E} = \boldsymbol{\varepsilon}^o$) are prescribed on ∂V. The matrix of the RVE is assumed to be homogeneous, as in Subsection 4.2. *In the absence of cavities*, the corresponding average stress

[5] This procedure has been followed by Wang *et al.* (1986) to estimate the overall properties of composites with special distributions of microcracks. It has also been reproduced by Talreja (1989) who incorrectly attributes the method to Wang *et al.* (1986). As pointed out by Nemat-Nasser (1987), the method is particularly suited for estimating the overall properties of solids with cavities and cracks of arbitrary shapes, and with frictional cracks.

associated with the prescribed macrostrain, $\boldsymbol{\varepsilon}^o$, would be

$$\boldsymbol{\sigma}^o = \mathbf{C} : \boldsymbol{\varepsilon}^o. \tag{4.4.1}$$

Due to the presence of cavities, the actual field quantities are nonuniform. From the results of Section 2,

$$\bar{\boldsymbol{\varepsilon}} = <\boldsymbol{\varepsilon}> = \frac{1}{V}\int_V \boldsymbol{\varepsilon}\, dV = \frac{1}{V}\int_{\partial V} \frac{1}{2}(\mathbf{v}\otimes\mathbf{u} + \mathbf{u}\otimes\mathbf{v})\, dS = \boldsymbol{\varepsilon}^o \tag{4.4.2}$$

which is valid for any RVE of any material and microstructure. Note that the surface integral in (4.4.2) extends over the exterior boundary, ∂V, of the RVE only. It does *not* include the cavity boundaries $\partial\Omega$. Equation (4.4.2) is the direct consequence of the fact that the average strain for an RVE is given in terms of its boundary displacements which are prescribed here to be $\mathbf{u}^o = \mathbf{x}\cdot\boldsymbol{\varepsilon}^o$. The *macrostrain* $\mathbf{E} = \boldsymbol{\varepsilon}^o$ is *prescribed* here and should not be confused with the quantity $\boldsymbol{\varepsilon}^o$ *defined* in (4.2.1), in terms of the *prescribed macrostress* $\boldsymbol{\Sigma} = \boldsymbol{\sigma}^o$. Similarly, the quantity $\boldsymbol{\sigma}^o$ is *defined* in the present subsection by (4.4.1), in terms of the *prescribed* $\boldsymbol{\varepsilon}^o$, and should not be confused with the macrostress $\boldsymbol{\Sigma} = \boldsymbol{\sigma}^o$ which is the *prescribed* quantity in (4.2.1) of Subsection 4.2.1.[6] Indeed, for a prescribed macrostrain, the average stress is not, in general, equal to $\boldsymbol{\sigma}^o$. Instead,

$$\bar{\boldsymbol{\sigma}} = <\boldsymbol{\sigma}> = \boldsymbol{\sigma}^o + \bar{\boldsymbol{\sigma}}^c, \tag{4.4.3}$$

where $\boldsymbol{\sigma}^o$ is *defined* by (4.4.1), and $\bar{\boldsymbol{\sigma}}^c$ is the decrement in the overall stress due to the presence of cavities.

As in Subsection 4.2, the reciprocal theorem will be applied to calculate the average stress $\bar{\boldsymbol{\sigma}}$ in (4.4.3). To this end, a third set of boundary data is defined by

$$\begin{aligned} \mathbf{u}^{(3)} &= \mathbf{x}\cdot\boldsymbol{\varepsilon}^o && \text{on } \partial V, \\ \mathbf{t}^{(3)} &= \mathbf{0} && \text{on } \partial\Omega. \end{aligned} \tag{4.2.4c}$$

The displacement, strain, and stress fields associated with these boundary conditions are denoted by

$$\{\mathbf{u}^{(3)},\ \boldsymbol{\varepsilon}^{(3)},\ \boldsymbol{\sigma}^{(3)}\} = \{\mathbf{u},\ \boldsymbol{\varepsilon},\ \boldsymbol{\sigma}\}, \tag{4.2.5c}$$

which are actual fields, in general, different from those given by (4.2.5b) for the boundary conditions (4.2.4b). The actual tractions on the boundary of the RVE now are

$$\mathbf{t}(\mathbf{x}) = \mathbf{v}(\mathbf{x})\cdot\boldsymbol{\sigma}(\mathbf{x}), \tag{4.4.4}$$

where $\mathbf{x}$ is on ∂V. These tractions are required in order to impose the boundary displacements prescribed by (4.2.4c).

Applying the reciprocal theorem, (3.2.2), to the two sets of loads, (4.2.4a) and (4.2.4c), it follows that

---

[6] Note, however, that (4.2.1) and (4.4.1) represent *the same* set of equations for the homogeneous linearly elastic RVE *without* any cavities, whether $\boldsymbol{\sigma}^o$ or $\boldsymbol{\varepsilon}^o$ is regarded prescribed.

$$\int_{\partial V} \mathbf{t} \cdot (\mathbf{x} \cdot \delta\boldsymbol{\varepsilon}^o) \, dS = \int_{\partial V} (\mathbf{v} \cdot \delta\boldsymbol{\sigma}^o) \cdot (\mathbf{x} \cdot \boldsymbol{\varepsilon}^o) \, dS - \int_{\partial\Omega} (\mathbf{n} \cdot \delta\boldsymbol{\sigma}^o) \cdot \mathbf{u} \, dS \tag{4.4.5a}$$

which can be written as

$$\delta\boldsymbol{\varepsilon}^o : \{\int_{\partial V} \mathbf{t} \otimes \mathbf{x} \, dS - \int_{\partial V} \mathbf{C} : \{(\mathbf{x} \otimes \mathbf{v}) \cdot \boldsymbol{\varepsilon}^o\} \, dS + \int_{\partial\Omega} \mathbf{C} : (\mathbf{n} \otimes \mathbf{u}) \, dS\} = 0, \tag{4.4.5b}$$

where, in using loading (4.2.5a), the quantity $\delta\boldsymbol{\varepsilon}^o$ is regarded as a virtual spatially constant strain field with the corresponding stress field, $\delta\boldsymbol{\sigma}^o = \mathbf{C} : \delta\boldsymbol{\varepsilon}^o$. Since $\delta\boldsymbol{\varepsilon}^o$ is an arbitrary symmetric tensor, the symmetric part of the quantity within the braces in (4.4.5b) must vanish identically. Noting that the second integral within the parentheses can be expressed as

$$\frac{1}{V} \int_{\partial V} \mathbf{C} : \{(\mathbf{x} \otimes \mathbf{v}) \cdot \boldsymbol{\varepsilon}^o\} \, dS = \mathbf{C} : \{1^{(2)} \cdot \boldsymbol{\varepsilon}^o\} = \boldsymbol{\sigma}^o, \tag{4.4.6a}$$

and using the averaging procedure of Section 2, it now follows that

$$\bar{\boldsymbol{\sigma}} = \frac{1}{V} \int_{\partial V} \mathbf{t} \otimes \mathbf{x} \, dS = \boldsymbol{\sigma}^o - \mathbf{C} : \{\frac{1}{V} \int_{\partial\Omega} \frac{1}{2} (\mathbf{n} \otimes \mathbf{u} + \mathbf{u} \otimes \mathbf{n}) \, dS\}. \tag{4.4.6b}$$

Comparison with (4.4.3) shows that the decremental stress $\bar{\boldsymbol{\sigma}}^c$ due to the presence of cavities, is given by

$$\bar{\boldsymbol{\sigma}}^c = -\mathbf{C} : \bar{\boldsymbol{\varepsilon}}^c, \tag{4.4.7}$$

where $\bar{\boldsymbol{\varepsilon}}^c$ is the strain due to the presence of cavities given by (4.2.8a,b), which now must be computed for the prescribed boundary displacements $\mathbf{u}^o = \mathbf{x} \cdot \boldsymbol{\varepsilon}^o$.

4.5. OVERALL ELASTICITY TENSOR FOR POROUS ELASTIC SOLIDS

When the overall macrostrain is regarded prescribed, $\mathbf{E} = \boldsymbol{\varepsilon}^o$, designate the overall elasticity tensor of the porous RVE with a linearly elastic and homogeneous matrix, by $\bar{\mathbf{C}}$, and define it through

$$\bar{\boldsymbol{\sigma}} = \bar{\mathbf{C}} : \boldsymbol{\varepsilon}^o. \tag{4.5.1}$$

Substitution of (4.4.1), (4.4.7), and (4.5.1) into (4.4.3) then yields

$$(\bar{\mathbf{C}} - \mathbf{C}) : \boldsymbol{\varepsilon}^o + \mathbf{C} : \bar{\boldsymbol{\varepsilon}}^c = \mathbf{0}. \tag{4.5.2}$$

For a given microstructure (i.e., for existing cavities with fixed shapes, sizes, and distribution), the response of the RVE is linear. Hence, the displacement field anywhere within the linearly elastic matrix of the RVE is a linear and homogeneous function of the prescribed overall constant strain $\boldsymbol{\varepsilon}^o$. Therefore, in line with result (4.3.2b,c) for the case when the macrostresses were considered to be prescribed, at a typical point $\mathbf{x}$ on the boundary of the cavities, $\partial\Omega$,

$$u_i(\mathbf{x}) = L_{ijk}(\mathbf{x}) \, \varepsilon^o_{jk}, \tag{4.5.3a}$$

where $\mathbf{L}(\mathbf{x})$ is a third-order tensor-valued function with the symmetry property, $L_{ijk} = L_{ikj}$. Now, from the definition of $\bar{\boldsymbol{\varepsilon}}^c$, (4.2.8a,b),

$$\bar{\varepsilon}^c_{ij} = J_{ijkl}\, \varepsilon^o_{kl}, \tag{4.5.3b}$$

where the *constant* fourth-order tensor, $\mathbf{J}$, is given by

$$J_{ijkl} \equiv J_{jikl} \equiv J_{ijlk} \equiv \frac{1}{V}\int_{\partial\Omega} \frac{1}{2}\{n_i(\mathbf{x})\, L_{jkl}(\mathbf{x}) + n_j(\mathbf{x})\, L_{ikl}(\mathbf{x})\}\, dS. \tag{4.5.3c}$$

Hence, for an RVE with a linearly elastic matrix (whether homogeneous or not) containing cavities of arbitrary shapes and sizes, the following general result is obtained, when the overall macrostrains are regarded prescribed:

$$\bar{\boldsymbol{\varepsilon}}^c = \mathbf{J} : \boldsymbol{\varepsilon}^o, \qquad \text{or} \qquad \bar{\varepsilon}^c_{ij} = J_{ijkl}\, \varepsilon^o_{kl}. \tag{4.5.4a,b}$$

To obtain an expression for the overall elastic moduli of the porous RVE, substitute (4.5.4) into (4.5.2) and, noting that the resulting expression must be valid for any constant symmetric macrostrain $\boldsymbol{\varepsilon}^o$, arrive at

$$\bar{\mathbf{C}} = \mathbf{C} - \mathbf{C} : \mathbf{J}, \qquad \bar{C}_{ijkl} = C_{ijkl} - C_{ijmn}\, J_{mnkl}. \tag{4.5.5a,b}$$

It should be noted that in many practical problems the tensor $\mathbf{J}$, similarly to the tensor $\mathbf{H}$, can be calculated *directly* from (4.2.8), and therefore, the overall elastic moduli can be estimated from (4.5.5). It may, however, be instructive to seek to construct the tensor $\mathbf{J}$ in terms of the Green functions $\mathbf{G}(\mathbf{x}, \mathbf{y})$ and $\mathbf{G}^{-1}(\mathbf{y}, \mathbf{x})$, which are discussed in Subsection 3.2.3.

To this end, for the linear displacements, $\mathbf{u}^o = \mathbf{z} \cdot \boldsymbol{\varepsilon}^o$, *prescribed* on the outer boundary ∂V of the RVE, express the *resulting* tractions, $\mathbf{t}(\mathbf{y})$, using (3.2.10a), as

$$\mathbf{t}(\mathbf{y}) = \int_{\partial V} \mathbf{G}^{-1}(\mathbf{y}, \mathbf{z}) \cdot (\mathbf{z} \cdot \boldsymbol{\varepsilon}^o)\, dS, \tag{4.5.6a}$$

where the integration is taken with respect to $\mathbf{z}$ over the outer boundary ∂V (excluding the traction-free cavity boundaries) of the RVE. Substituting (4.5.6a) into (3.2.8a), the displacement field for points on $\partial\Omega$ is obtained in terms of the prescribed macrostrain $\boldsymbol{\varepsilon}^o$, as

$$\mathbf{u}(\mathbf{x}) = \int_{\partial V} \mathbf{G}(\mathbf{x}, \mathbf{y}) \cdot \{\int_{\partial V} \mathbf{G}^{-1}(\mathbf{y}, \mathbf{z}) \cdot (\mathbf{z} \cdot \boldsymbol{\varepsilon}^o)\, dS\}\, dS, \tag{4.5.6b}$$

where both the $\mathbf{y}$- and $\mathbf{z}$-integral are taken over ∂V. Noting that $\boldsymbol{\varepsilon}^o$ is a symmetric tensor, tensor $\mathbf{L}$ in (4.5.3a) may now be written in terms of $\mathbf{G}$ and $\mathbf{G}^{-1}$, as

$$L_{ijk}(\mathbf{x}) = \int_{\partial V} G_{im}(\mathbf{x}, \mathbf{y}) \left\{ \int_{\partial V} \frac{1}{2}\{G^{-1}_{mj}(\mathbf{y}, \mathbf{z})\, z_k + G^{-1}_{mk}(\mathbf{y}, \mathbf{z})\, z_j\}\, dS \right\} dS. \tag{4.5.6c}$$

Therefore, from comparison of (4.5.3b) with (4.5.6c), a fourth-order tensor, $\mathbf{j}(\mathbf{x}, \mathbf{y})$, can be introduced as

$$j_{ijkl}(\mathbf{x}, \mathbf{y})$$
$$= \int_{\partial V} \frac{1}{4}\Big\{ n_i(\mathbf{x})\, G_{jm}(\mathbf{x}, \mathbf{y})\, G^{-1}_{mk}(\mathbf{y}, \mathbf{z})\, z_l + n_i(\mathbf{x})\, G_{jm}(\mathbf{x}, \mathbf{y})\, G^{-1}_{ml}(\mathbf{y}, \mathbf{z})\, z_k$$

$$+ n_j(\mathbf{x})\, G_{im}(\mathbf{x}, \mathbf{y})\, G^{-1}_{mk}(\mathbf{y}, \mathbf{z})\, z_l + n_j(\mathbf{x})\, G_{im}(\mathbf{x}, \mathbf{y})\, G^{-1}_{ml}(\mathbf{y}, \mathbf{z})\, z_k \Big\}\, dS, \tag{4.5.6d}$$

where the integral is taken with respect to **z** over ∂V. The constant tensor **J** in (4.5.3b) now becomes

$$\mathbf{J} = \frac{1}{V} \int_{\partial \Omega} \int_{\partial V} \mathbf{j}(\mathbf{x}, \mathbf{y})\, dS\, dS, \tag{4.5.6e}$$

where the **y**-integration is over ∂V, and the **x**-integration is over $\partial \Omega$.

4.6. REFERENCES

Horii, H. and Nemat-Nasser, S. (1983), Overall moduli of solids with microcracks: Load-induced anisotropy, *J. Mech. Phys. Solids*, Vol. 31, 155-177.

Nemat-Nasser, S. (1987), Discussion of: Mechanics of fatigue damage and degradation in random short-fiber composites, Part II - Analysis of anisotropic property degradation, by S.S. Wang, E.S.-M. Chim, and H. Suemasu, published in *J. Appl. Mech.*, Vol. 53 (1986), 347-353; *ibid.*, Vol. 54, 479-480.

Talreja, R. (1989), Damage development in composite: mechanics and modeling, *J. Strain Analysis*, Vol. 24, 215-222.

Wang, S. S., Chim, E.S.-M., and Suemasu, H. (1986), Mechanics of fatigue damage and degradation in random short-fiber composites, Part II - Analysis of anisotropic property degradation, *J. Appl. Mech.*, Vol. 53 (1986), 347-353.

SECTION 5 ELASTIC SOLIDS WITH MICROCAVITIES

By means of simple examples involving elementary solutions in linear elasticity, it is *illustrated* in this section how the overall elasticity and compliance tensors of a porous RVE may be *estimated* for small porosities in a straightforward manner. The objective is to show: (1) how elasticity solutions to simple problems can be used in conjunction with the general results reported in this chapter, to calculate the effective elastic moduli and compliances of elastic solids containing cavities; and (2) how the tensors **H** and **J** in (4.2.8) or (4.3.5b) and (4.5.6e) can be estimated *directly*, without invoking the corresponding Green function.

Two extreme cases are considered: (1) when the elastic solid contains a *dilute* distribution of cavities, so that typical cavities are so far apart that their interaction may be neglected; and (2) when the cavities are randomly distributed. In this latter case, the idea of the *self-consistent method* is introduced; this will be discussed in some detail later on. The range of validity of these *approximating* methods is limited to relatively small volume fractions of inhomogeneities.[1]

5.1. EFFECTIVE MODULI OF AN ELASTIC PLATE CONTAINING CIRCULAR HOLES

In this subsection, the problem of estimating the effective moduli of a linearly elastic homogeneous solid containing circular cylindrical cavities, is worked out in some detail. Assume either plane stress which then corresponds to a thin plate containing circular holes, or plane strain which then corresponds to a long cylindrical body containing cylindrical holes with circular cross sections and a common generator.[2] Both cases deal with a two-dimensional problem; the first case, with generalized plane stress and the second case, with plane strain. A rectangular Cartesian coordinate system is chosen such that

$$\sigma_{3i} = \sigma_{i3} = 0 \qquad \text{for plane stress,}$$

[1] Several averaging techniques and their limitations are discussed in Section 10, and elastic solids with periodically distributed cavities and inclusions are examined in Section 12.

[2] In this case, the RVE will be transversely isotropic, when the defects are randomly distributed and the matrix is isotropic.

$$\varepsilon_{3i} = \varepsilon_{i3} = 0 \qquad \text{for plane strain,} \tag{5.1.1a,b}$$

for i = 1, 2, 3. All field quantities, hence, are functions of two space variables, x_1 and x_2, or when polar coordinates are used, r and θ.

For simplicity, the matrix of the RVE is assumed to be isotropic, linearly elastic, and homogeneous. Then, the corresponding two-dimensional stress-strain and the strain-stress relations become

$$\sigma_{ij} = \mu \{ -\frac{\kappa - 3}{\kappa - 1} \delta_{ij} \delta_{kl} + (\delta_{ik} \delta_{jl} + \delta_{il} \delta_{jk}) \} \varepsilon_{kl},$$

$$\varepsilon_{ij} = \frac{1}{\mu} \{ \frac{\kappa - 3}{8} \delta_{ij} \delta_{kl} + \frac{1}{4} (\delta_{ik} \delta_{jl} + \delta_{il} \delta_{jk}) \} \sigma_{kl}, \tag{5.1.2a,b}$$

or in matrix notation,

$$\begin{bmatrix} \sigma_{11} \\ \sigma_{22} \\ \sigma_{12} \end{bmatrix} = \mu \begin{bmatrix} (\kappa+1)/(\kappa-1) & -(\kappa-3)/(\kappa-1) & 0 \\ -(\kappa-3)/(\kappa-1) & (\kappa+1)/(\kappa-1) & 0 \\ 0 & 0 & 1 \end{bmatrix} \begin{bmatrix} \varepsilon_{11} \\ \varepsilon_{22} \\ 2\varepsilon_{12} \end{bmatrix},$$

$$\begin{bmatrix} \varepsilon_{11} \\ \varepsilon_{22} \\ 2\varepsilon_{12} \end{bmatrix} = \frac{1}{\mu} \begin{bmatrix} (\kappa+1)/8 & (\kappa-3)/8 & 0 \\ (\kappa-3)/8 & (\kappa+1)/8 & 0 \\ 0 & 0 & 1 \end{bmatrix} \begin{bmatrix} \sigma_{11} \\ \sigma_{22} \\ \sigma_{12} \end{bmatrix}, \tag{5.1.2c,d}$$

where μ is the shear modulus, ν is the Poisson ratio, and

$$\kappa = \begin{cases} 3 - 4\nu & \text{for plane strain} \\ (3 - \nu)/(1 + \nu) & \text{for plane stress.} \end{cases} \tag{5.1.2e}$$

The relations involving σ_{3i} or ε_{3i} are not considered in this subsection, but they are readily written down with the aid of the results of Subsection 3.1.4.

5.1.1. Estimates of Three-Dimensional Moduli from Two-Dimensional Results

In general, when an elastic RVE contains cylindrical cavities, cracks, or elastic fibers, all aligned in, say, the x_3-direction, i.e., when the generator of the cylindrical inhomogeneities is parallel to the x_3-direction, and when the distribution of these inhomogeneities is otherwise random, then the overall response of the elastic RVE will be transversely isotropic. In this case, the *inplane* effective shear modulus, $\bar{\mu}$, and the Poisson ratio, $\bar{\nu}$, may be estimated, using either a plane-stress or plane-strain formulation, with two-dimensional inplane stress-strain or strain-stress relations, defined by (5.1.2). In other words, the two-dimensional formulation of a transversely isotropic RVE, in terms of the inplane effective shear modulus $\bar{\mu}$ and the inplane effective parameter $\bar{\kappa}$, circumvents the immediate consideration of the effect of the third dimension. However, when one wishes to obtain the effective inplane Young modulus, $\bar{E}$, and the Poisson ratio, $\bar{\nu}$, the influence of the third dimension must be taken into account. In this subsection the relation between plane-stress and plane-strain solutions for this class of problems is examined, in a rather general setting. The results apply to problems of an elastic RVE with cylindrical cavities, slit cracks, and elastic fibers, all aligned, say, in the x_3-direction. Examples are considered in this section and in Sections 6 and 8.

With x_3 as the axis of symmetry, the overall compliance matrix, $[\bar{D}_{ab}]$, for a transversely isotropic RVE, takes on the form

$$[\bar{D}_{ab}] = \begin{bmatrix} 1/\bar{E} & -\bar{\nu}/\bar{E} & -\bar{\nu}_3/\bar{E}_3 & 0 & 0 & 0 \\ -\bar{\nu}/\bar{E} & 1/\bar{E} & -\bar{\nu}_3/\bar{E}_3 & 0 & 0 & 0 \\ -\bar{\nu}_3/\bar{E}_3 & -\bar{\nu}_3/\bar{E}_3 & 1/\bar{E}_3 & 0 & 0 & 0 \\ 0 & 0 & 0 & 1/\bar{\mu}_3 & 0 & 0 \\ 0 & 0 & 0 & 0 & 1/\bar{\mu}_3 & 0 \\ 0 & 0 & 0 & 0 & 0 & 1/\bar{\mu} \end{bmatrix}, \qquad (5.1.3)$$

where $\bar{E}$, $\bar{\nu}$, and $\bar{\mu} \equiv \bar{E}/2(1+\bar{\nu})$ are the inplane effective Young modulus, Poisson ratio, and shear modulus, respectively, and $\bar{E}_3$, $\bar{\nu}_3$, and $\bar{\mu}_3$ are the effective Young modulus in the x_3-direction, Poisson ratio, and shear modulus common to the x_1,x_3- and x_2,x_3-directions, respectively.

For simplicity, consider only normal stresses and normal strains, and partition the six by six matrix, $[\bar{D}_{ab}]$, into two three by three matrices, $[\bar{D}^{(1)}_{ab}]$ and $[\bar{D}^{(2)}_{ab}]$,

$$[\bar{D}^{(1)}_{ab}] = \begin{bmatrix} 1/\bar{E} & -\bar{\nu}/\bar{E} & -\bar{\nu}_3/\bar{E}_3 \\ -\bar{\nu}/\bar{E} & 1/\bar{E} & -\bar{\nu}_3/\bar{E}_3 \\ -\bar{\nu}_3/\bar{E}_3 & -\bar{\nu}_3/\bar{E}_3 & 1/\bar{E}_3 \end{bmatrix}, \qquad [\bar{D}^{(2)}_{ab}] = \begin{bmatrix} 1/\bar{\mu}_3 & 0 & 0 \\ 0 & 1/\bar{\mu}_3 & 0 \\ 0 & 0 & 1/\bar{\mu} \end{bmatrix}. \qquad (5.1.4a,b)$$

Moreover, define two three by two matrices, $[P_{ab}]$ and $[P'_{ab}]$, by

$$[P_{ab}] = \begin{bmatrix} 1 & 0 \\ 0 & 1 \\ 0 & 0 \end{bmatrix}, \qquad [P'_{ab}] = \begin{bmatrix} 1 & 0 \\ 0 & 1 \\ \bar{\nu}_3 & \bar{\nu}_3 \end{bmatrix}. \qquad (5.1.5a,b)$$

Then the *nominal* compliance tensors, relating the normal inplane stress and strain components, for plane-strain and plane-stress states, may be obtained by multiplying $[P_{ab}]^T[\bar{D}^{(1)}_{ab}]$ from the right by $[P'_{ab}]$ for plane strain and by $[P_{ab}]$ for plane stress, respectively, i.e., for plane strain,

$$[P_{ab}]^T[\bar{D}^{(1)}_{bc}][P'_{cd}] = \begin{bmatrix} 1/\bar{E} - \bar{\nu}_3^2/\bar{E}_3 & -\bar{\nu}/\bar{E} - \bar{\nu}_3^2/\bar{E}_3 \\ -\bar{\nu}/\bar{E} - \bar{\nu}_3^2/\bar{E}_3 & 1/\bar{E} - \bar{\nu}_3^2/\bar{E}_3 \end{bmatrix}, \qquad (5.1.6a)$$

and for plane stress,

$$[P_{ab}]^T[\bar{D}^{(1)}_{bc}][P_{cd}] = \begin{bmatrix} 1/\bar{E} & -\bar{\nu}/\bar{E} \\ -\bar{\nu}/\bar{E} & 1/\bar{E} \end{bmatrix}; \qquad (5.1.6b)$$

see Subsection 3.1. The parameter $\bar{\kappa}$ for these nominal compliance tensors is then given by

$$\bar{\kappa} = \begin{cases} 3 - 8\bar{\mu}\{\dfrac{\bar{\nu}}{\bar{E}} + \dfrac{\bar{\nu}_3^2}{\bar{E}_3}\} & \text{plane strain} \\ \dfrac{3-\bar{\nu}}{1+\bar{\nu}} & \text{plane stress.} \end{cases} \tag{5.1.7}$$

For the isotropic case, $\bar{\nu} = \bar{\nu}_3 \, (= \nu)$ and $\bar{E} = \bar{E}_3 \, (= E)$, and $\bar{\kappa}$ reduces to κ, given by $3 - 4\nu$ in plane strain, and $(3 - \nu)/(1 + \nu)$ in plane stress.

5.1.2. Effective Moduli: Dilute Distribution of Cavities

Since plane problems are considered, an RVE of *unit thickness* in the x_3-direction is used. Furthermore, because all the field quantities are assumed to be uniform throughout this thickness, integration over the thickness is not displayed, but simply implied.

Figure 5.1.1 is a typical portion of an RVE in the x_1,x_2-plane. The circular holes for the dilute distribution of cavities are far apart. Let there be n holes in the RVE. Denote the volume of a typical hole by Ω_α, being bounded by the surface area $\partial\Omega_\alpha$. From (4.2.8a), the overall strain due to the cavities becomes

$$\bar{\boldsymbol{\varepsilon}}^c = \sum_{\alpha=1}^{n} \frac{\Omega_\alpha}{V} \, \bar{\boldsymbol{\varepsilon}}^\alpha, \tag{5.1.8a}$$

where $\bar{\boldsymbol{\varepsilon}}^\alpha$ is given by

$$\bar{\boldsymbol{\varepsilon}}^\alpha \equiv \frac{1}{\Omega_\alpha} \int_{\partial\Omega_\alpha} \frac{1}{2}(\mathbf{n} \otimes \mathbf{u} + \mathbf{u} \otimes \mathbf{n}) \, dS. \tag{5.1.8b}$$

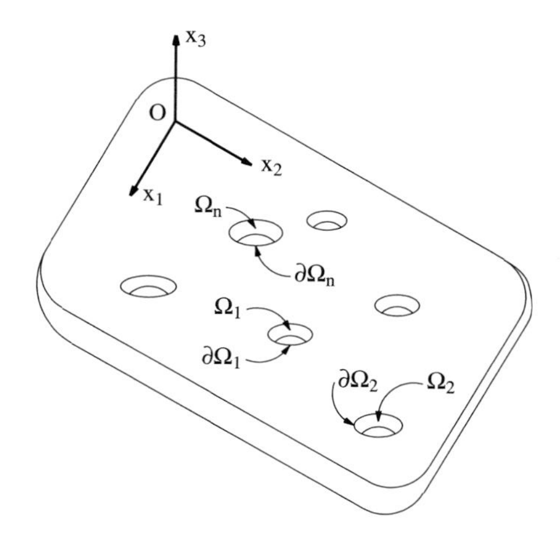

Figure 5.1.1

A typical portion of an RVE containing microcavities

The right-hand side of (5.1.8b) is now estimated, using the assumption that the cavities are not interacting with each other, since they are far apart.

Consider three loading conditions: (1) only σ^o_{11} is nonzero; (2) only σ^o_{22} is nonzero; and (3) only $\sigma^o_{12} = \sigma^o_{21}$ is nonzero. The general loading case is then obtained by a suitable combination of these.

Figure 5.1.2 shows a typical cavity of radius a, subjected to farfield stress σ^o_{11}, with all other stress components at infinity being zero. The displacement field for this problem is[3]

$$u_r = u_r(r,\,\theta) = \frac{\sigma^o_{11}}{8\mu r}\,\{(\kappa-1)r^2 + 2a^2 + 2[a^2(\kappa+1) + r^2 - \frac{a^4}{r^2}\,]\cos 2\theta\},$$

$$u_\theta = u_\theta(r,\,\theta) = -\,\frac{\sigma^o_{11}}{4\mu r}\,\{r^2 + a^2(\kappa-1) + \frac{a^4}{r^2}\}\,\sin 2\theta, \qquad (5.1.9a,b)$$

where u_r and u_θ are the displacement components with respect to the polar coordinates r, θ. The rectangular Cartesian components of the displacements are then given by

$$u_1 = u_r\cos\theta - u_\theta\sin\theta, \qquad u_2 = u_r\sin\theta + u_\theta\cos\theta. \qquad (5.1.9c,d)$$

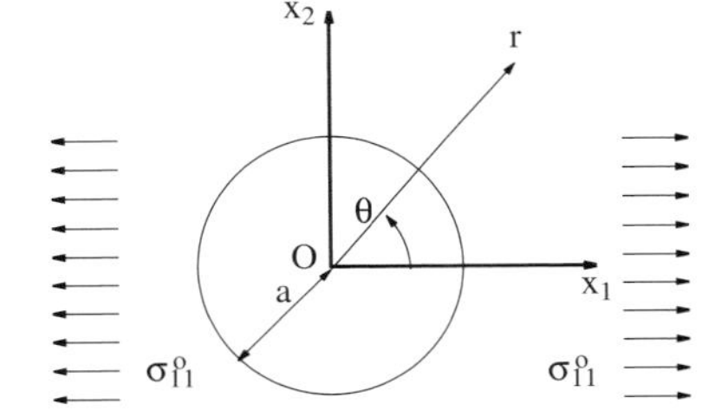

Figure 5.1.2

A typical cavity of radius a, subjected to a farfield stress σ^o_{11}

The components of the unit vector **n**, normal to the boundary of the circular cavity, are

$$n_1 = \cos\theta, \qquad n_2 = \sin\theta. \qquad (5.1.10a,b)$$

Substitution of (5.1.9a~d) and (5.1.10a,b) into (5.1.8b) with $r = a = a_\alpha$, and simple integration yield

$$\bar{\boldsymbol{\varepsilon}}^\alpha = \frac{\kappa+1}{8\mu}\,(3\mathbf{e}_1\otimes\mathbf{e}_1 - \mathbf{e}_2\otimes\mathbf{e}_2)\,\sigma^o_{11}. \qquad (5.1.11a)$$

Similarly, when only σ^o_{22} is nonzero,

$$\bar{\boldsymbol{\varepsilon}}^\alpha = \frac{\kappa+1}{8\mu}\,(-\,\mathbf{e}_1\otimes\mathbf{e}_1 + 3\mathbf{e}_2\otimes\mathbf{e}_2)\,\sigma^o_{22}. \qquad (5.1.11b)$$

The results for the case when only $\sigma^o_{12} = \sigma^o_{21}$ is nonzero, can be obtained from (5.1.11a,b) by a 45-degree rotation of the coordinate system and superposition;

[3] See Michell (1899), Love (1944), and Timoshenko and Goodier (1951).

see Figure 5.1.3.

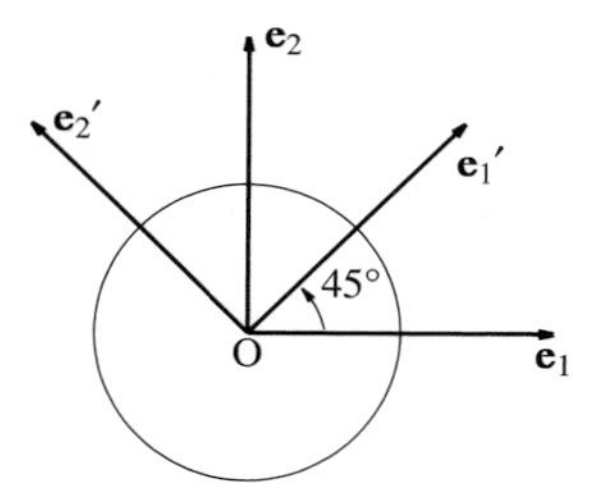

Figure 5.1.3
Coordinate system rotated by 45 degrees

To this end, let

$$\mathbf{e}_1' \equiv \frac{1}{\sqrt{2}}\,(\mathbf{e}_1 + \mathbf{e}_2), \qquad \mathbf{e}_2' \equiv \frac{1}{\sqrt{2}}\,(-\,\mathbf{e}_1 + \mathbf{e}_2) \tag{5.1.12a,b}$$

define the unit base vectors of the rotated coordinate system. Then,

$$\begin{aligned}\bar{\boldsymbol{\varepsilon}}^{\alpha} &= \frac{\kappa+1}{8\mu}\,\{(3\mathbf{e}_1'\otimes\mathbf{e}_1' - \mathbf{e}_2'\otimes\mathbf{e}_2') - (-\,\mathbf{e}_1'\otimes\mathbf{e}_1' + 3\mathbf{e}_2'\otimes\mathbf{e}_2')\}\,\sigma^{o}_{12} \\ &= \frac{\kappa+1}{8\mu}\,(4\mathbf{e}_1\otimes\mathbf{e}_2 + 4\mathbf{e}_2\otimes\mathbf{e}_1)\,\sigma^{o}_{12} \\ &= \frac{\kappa+1}{8\mu}\,(2\mathbf{e}_1\otimes\mathbf{e}_2 + 2\mathbf{e}_2\otimes\mathbf{e}_1)\,(\sigma^{o}_{12} + \sigma^{o}_{21}).\end{aligned} \tag{5.1.12c}$$

Note that the results in (5.1.12a~c) are independent of the void size. From (5.1.12a~c), the average strain due to the presence of cavities is given by

$$\begin{aligned}\bar{\boldsymbol{\varepsilon}}^{c} = \mathrm{f}\,\frac{\kappa+1}{8\mu}\,\{&(3\mathbf{e}_1\otimes\mathbf{e}_1 - \mathbf{e}_2\otimes\mathbf{e}_2)\,\sigma^{o}_{11} + (-\,\mathbf{e}_1\otimes\mathbf{e}_1 + 3\mathbf{e}_2\otimes\mathbf{e}_2)\,\sigma^{o}_{22} \\ &+ 2(\mathbf{e}_1\otimes\mathbf{e}_2 + \mathbf{e}_2\otimes\mathbf{e}_1)\,(\sigma^{o}_{12} + \sigma^{o}_{21})\},\end{aligned} \tag{5.1.13a}$$

where f is *the void volume fraction* of the RVE, defined by

$$\mathrm{f} \equiv \sum_{\alpha=1}^{n} \frac{\Omega_\alpha}{V}. \tag{5.1.13b}$$

Therefore, the tensor **H** in (4.3.6a) takes on the form

$$\begin{aligned}\mathbf{H} = \mathrm{f}\,\frac{\kappa+1}{8\mu}\,\{&(3\mathbf{e}_1\otimes\mathbf{e}_1 - \mathbf{e}_2\otimes\mathbf{e}_2)\otimes(\mathbf{e}_1\otimes\mathbf{e}_1) + (-\,\mathbf{e}_1\otimes\mathbf{e}_1 + 3\mathbf{e}_2\otimes\mathbf{e}_2)\otimes(\mathbf{e}_2\otimes\mathbf{e}_2) \\ &+ 2(\mathbf{e}_1\otimes\mathbf{e}_2 + \mathbf{e}_2\otimes\mathbf{e}_1)\otimes(\mathbf{e}_1\otimes\mathbf{e}_2 + \mathbf{e}_2\otimes\mathbf{e}_1)\}.\end{aligned} \tag{5.1.14a}$$

In matrix notation corresponding to (5.1.2d), this becomes

$$[H_{\mathrm{ab}}] = \mathrm{f}\,\frac{\kappa+1}{\mu}\begin{bmatrix} 3/8 & -\,1/8 & 0 \\ -\,1/8 & 3/8 & 0 \\ 0 & 0 & 1 \end{bmatrix}. \tag{5.1.14b}$$

Since $H_{33} = 2(H_{11} - H_{12})$, the tensor **H** is isotropic in the x_1,x_2-plane. Substitution of (5.1.14b) into (4.3.6) yields the overall compliance tensor $\bar{\mathbf{D}}$, where, for the two-dimensional case, the compliance matrix $[D_{ab}]$ can be read off (5.1.2b). In particular, the effective shear modulus, $\bar{\mu}$, and the effective Poisson ratio, $\bar{\nu}$, are obtained from

$$\frac{\bar{\mu}}{\mu} = \{1 + f(\kappa + 1)\}^{-1} = 1 - f(\kappa + 1) + O(f^2),$$

$$\frac{\bar{\kappa}}{\kappa} = \{1 + f\frac{2(\kappa + 1)}{\kappa}\}\{1 + f(\kappa + 1)\}^{-1} = 1 - f\frac{(\kappa + 1)(\kappa - 2)}{\kappa} + O(f^2), \tag{5.1.15a,b}$$

where $\bar{\kappa} = 3 - 8\mu(\bar{\nu}/\bar{E} + \bar{\nu}_3^2/\bar{E}_3)$ for plane strain, and $\bar{\kappa} = (3 - \bar{\nu})/(1 + \bar{\nu})$ for plane stress. For plane stress, the effective Young modulus, $\bar{E}$, and Poisson ratio, $\bar{\nu}$, are

$$\frac{\bar{E}}{E} = (1 + 3f)^{-1} = 1 - 3f + O(f^2),$$

$$\frac{\bar{\nu}}{\nu} = (1 + f\frac{1}{\nu})(1 + 3f)^{-1} = 1 - (3 - \frac{1}{\nu})f + O(f^2). \tag{5.1.15c,d}$$

Figure 5.1.4 shows the variation of the shear and Young moduli with respect to the void volume fraction f, for plane stress with $\nu = 1/3$; note that, for the present case, this value of ν leads to $\bar{\nu}/\nu = \bar{\kappa}/\kappa = 1$, and $\bar{\mu}/\mu = \bar{E}/E = (1 + 3f)^{-1}$. Since a dilute distribution of cavities is assumed, the applicability of these results is limited to small values of the void volume fraction f. Figure 5.1.4 also includes the curve for the case when the macrostrain is prescribed, as discussed below.

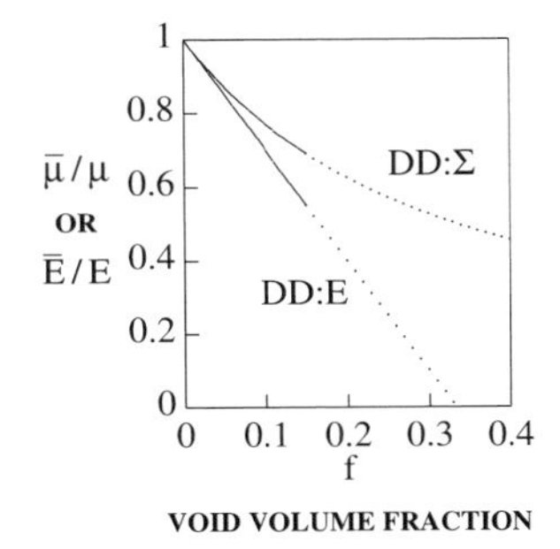

Figure 5.1.4

Normalized overall shear, $\bar{\mu}/\mu$, and Young, $\bar{E}/E$, moduli for $\nu = 1/3$
DD:Σ ≡ dilute distribution with macrostress prescribed
DD:E ≡ dilute distribution with macrostrain prescribed

In the same manner as the overall compliance tensor, $\bar{\mathbf{D}}$, is obtained for a prescribed macrostress, $\boldsymbol{\Sigma} = \boldsymbol{\sigma}^o$, the overall elasticity tensor, $\bar{\mathbf{C}}$, is obtained for a prescribed macrostrain, $\mathbf{E} = \boldsymbol{\varepsilon}^o$, by computing the decremental stress $\bar{\boldsymbol{\sigma}}^c$ $(= -\mathbf{C} : \bar{\boldsymbol{\varepsilon}}^c)$ from the displacements of $\partial\Omega_\alpha$. In the model of an *infinite* body containing a single cavity, shown in Figure 5.1.2, farfield stress $\boldsymbol{\sigma}^o$ may be replaced by the farfield strain $\boldsymbol{\varepsilon}^o = \mathbf{D} : \boldsymbol{\sigma}^o$, and (5.1.13a) written in terms of the prescribed $\boldsymbol{\varepsilon}^o$, as

$$\bar{\boldsymbol{\varepsilon}}^{c} = \mathbf{H} : (\mathbf{C} : \boldsymbol{\varepsilon}^{o}), \tag{5.1.16}$$

where $\mathbf{C}$ is the elasticity tensor of the matrix material corresponding to (5.1.2a).[4] Then, the tensor $\mathbf{J}$ in (4.5.3b) is given by

$$\mathbf{J} = \mathbf{H} : \mathbf{C} = f\,\frac{\kappa+1}{2}\,\{(\frac{\kappa}{\kappa-1}\,\mathbf{e}_1\otimes\mathbf{e}_1 - \frac{\kappa-2}{\kappa-1}\,\mathbf{e}_2\otimes\mathbf{e}_2)\otimes(\mathbf{e}_1\otimes\mathbf{e}_1)$$

$$+(-\frac{\kappa-2}{\kappa-1}\mathbf{e}_1\otimes\mathbf{e}_1 + \frac{\kappa}{\kappa-1}\,\mathbf{e}_2\otimes\mathbf{e}_2)\otimes(\mathbf{e}_2\otimes\mathbf{e}_2)$$

$$+\frac{1}{2}(\mathbf{e}_1\otimes\mathbf{e}_2+\mathbf{e}_2\otimes\mathbf{e}_1)\otimes(\mathbf{e}_1\otimes\mathbf{e}_2+\mathbf{e}_2\otimes\mathbf{e}_1)\}, \tag{5.1.17a}$$

or, in matrix notation corresponding to (5.1.2a),[5]

$$[J_{ab}] = f\,\frac{\kappa+1}{2}\begin{bmatrix} \kappa/(\kappa-1) & -(\kappa-2)/(\kappa-1) & 0 \\ -(\kappa-2)/(\kappa-1) & \kappa/(\kappa-1) & 0 \\ 0 & 0 & 2 \end{bmatrix}. \tag{5.1.17b}$$

Since $J_{33} = J_{11} - J_{12}$, the tensor $\mathbf{J}$ is isotropic in the x_1,x_2-plane. In the same manner as (5.1.15a,b) are obtained, the effective shear modulus, $\bar{\mu}$, and Poisson ratio, $\bar{\nu}$, in the present case are calculated from

$$\frac{\bar{\mu}}{\mu} = 1 - f(\kappa+1),$$

$$\frac{\bar{\kappa}}{\kappa} = \{1 - f\,\frac{(\kappa+1)(\kappa^2-2\kappa+2)}{\kappa(\kappa-1)}\}(1 - f\,\frac{\kappa+1}{\kappa-1})^{-1}$$

$$= 1 - f\,\frac{(\kappa+1)(\kappa-2)}{\kappa} + O(f^2). \tag{5.1.18a,b}$$

For plane stress, the effective Young modulus, $\bar{E}$, and Poisson ratio, $\bar{\nu}$, are

$$\frac{\bar{E}}{E} = (1 - f\,\frac{4}{1+\nu})\,(1 - f\,\frac{2}{1-\nu})\,(1 - f\,\frac{3-2\nu+3\nu^2}{1-\nu^2})^{-1} = 1 - 3f + O(f^2),$$

$$\frac{\bar{\nu}}{\nu} = \{1 + f\,\frac{1-6\nu+\nu^2}{\nu(1-\nu^2)}\}\,(1 - f\,\frac{3-2\nu+3\nu^2}{1-\nu^2})^{-1} = 1 - (3 - \frac{1}{\nu})\,f + O(f^2). \tag{5.1.18c,d}$$

For $\nu = 1/3$, it follows that $\bar{\nu}/\nu = \bar{\kappa}/\kappa = 1$, and $\bar{\mu}/\mu = \bar{E}/E = (1-3f)$. These results are displayed in Figure 5.1.4. It turns out that the self-consistent method also yields the same estimates for $\bar{E}/E$ and $\bar{\mu}/\mu$, as discussed in the next section.

The product of the overall compliance tensor $\bar{\mathbf{D}}$ and the overall elasticity tensor $\bar{\mathbf{C}}$ is

$$\bar{\mathbf{D}} : \bar{\mathbf{C}} = \mathbf{1}^{(4s)} - \mathbf{H} : \mathbf{C} : \mathbf{H} : \mathbf{C} = \mathbf{1}^{(4s)} + O(f^2), \tag{5.1.19a}$$

---

[4] It should be noted that only for this *model*, $\mathbf{J} = \mathbf{H} : \mathbf{C}$. In general, $\mathbf{J}$ obtained from (4.5.6e) may not relate to $\mathbf{H}$ obtained in (4.3.5b) by this simple expression.

[5] The matrix $[J_{ab}]$ is defined by $[\gamma_a^c] = [J_{ab}][\gamma_b^o]$, and hence $J_{33} = 2J_{1212}$; compare (3.1.3c) and (3.1.6d).

or

$$\bar{\mathbf{C}} : \bar{\mathbf{D}} = \mathbf{1}^{(4s)} - \mathbf{C} : \mathbf{H} : \mathbf{C} : \mathbf{H} = \mathbf{1}^{(4s)} + \mathrm{O}(\mathrm{f}^2), \tag{5.1.19b}$$

where $\mathbf{1}^{(4s)}$ is the fourth-order symmetric identity tensor. If $\bar{\mathbf{D}}$ and $\bar{\mathbf{C}}$, obtained respectively from $\mathbf{H}$ and $\mathbf{J}$ of (5.1.14a) and (5.1.17a), are regarded as estimates of the overall compliance and overall modulus tensors, then they are each other's inverse only to the first order in the void volume fraction f. This is reasonable, since the corresponding equations are valid for dilute distributions of voids, and hence, small f only; see Section 10 for additional comments and illustrations.

It is seen from Figure 5.1.4 that the relation between the overall elastic moduli for the uniform overall stress and uniform overall strain, obtained by the dilute-distribution model, is exactly the reverse of that required by the energy Theorem I of Subsection 2.5, namely, by inequality (2.5.44). The contradiction is indeed a consequence of the *modeling* procedure used in the dilute-distribution approach, to estimate the *concentration tensors,* or equivalently, the **H**- and **J**-tensors. This is discussed in some detail in Subsection 10.1.1.

In this connection, it may be instructive to note that the estimate of the three-dimensional overall inplane Young modulus $\bar{E}$ and Poisson ratio $\bar{\nu}$, given by (5.1.18c,d), does *not* coincide with the corresponding results obtained using the plane-*strain* conditions. This contradiction also stems from the errors inherent in the dilute-distribution modeling procedure. In view of these observations, the results of this approximate averaging technique should be used for very small values of f, where these discrepancies are negligibly small.

5.1.3. Effective Moduli: Self-Consistent Estimates

The "self-consistent" estimate of the overall properties of an RVE with microstructure, refers to a very special averaging procedure. In this approach, for the present problem, a single typical cavity is embedded in a homogeneous linearly elastic solid *which has the yet-unknown overall moduli of the* RVE, and then the necessary local quantities are estimated and used to obtain the overall moduli; Kröner (1958), Budiansky (1965), Hill (1965), and Hashin (1968). For example, in estimating the tensor **H** or the tensor **J**, for the elastic matrix, the unknown average shear modulus $\bar{\mu}$ and Poisson ratio $\bar{\nu}$ are used. Substitution of the result into (5.1.14) or (5.1.17), for example, then gives a system of two equations for $\bar{\mu}$ and $\bar{\nu}$.

In view of these comments, for the case when the macrostress $\boldsymbol{\Sigma} = \boldsymbol{\sigma}^{\mathrm{o}}$ is prescribed, the additional overall strain, $\bar{\boldsymbol{\varepsilon}}^{\mathrm{c}}$, estimated by the self-consistent method, is obtained from (5.1.13a), by replacing μ by $\bar{\mu}$, and in the expression for κ, by replacing ν by $\bar{\nu}$; κ is replaced by $\bar{\kappa}$, with $\bar{\kappa} = 3 - 8\bar{\mu}(\bar{\nu}/\bar{E} + \bar{\nu}_3^2/\bar{E}_3)$ for plane strain and $\bar{\kappa} = (3-\bar{\nu})/(1+\bar{\nu})$ for plane stress. Equation (5.1.14a) now takes on the form

$$\bar{\mathbf{H}} = \mathrm{f}\,\frac{\bar{\kappa}+1}{8\bar{\mu}}\,\{(3\mathbf{e}_1\otimes\mathbf{e}_1 - \mathbf{e}_2\otimes\mathbf{e}_2)\otimes(\mathbf{e}_1\otimes\mathbf{e}_1) + (-\,\mathbf{e}_1\otimes\mathbf{e}_1 + 3\mathbf{e}_2\otimes\mathbf{e}_2)\otimes(\mathbf{e}_2\otimes\mathbf{e}_2)$$

$$+ (2\mathbf{e}_1 \otimes \mathbf{e}_2 + 2\mathbf{e}_2 \otimes \mathbf{e}_1) \otimes (\mathbf{e}_1 \otimes \mathbf{e}_2 + \mathbf{e}_2 \otimes \mathbf{e}_1)\}, \tag{5.1.20a}$$

with similar modification for (5.1.14b). Using (5.1.20a) and (4.3.6),

$$\overline{\mathbf{D}} = \mathbf{D} + \overline{\mathbf{H}}, \tag{5.1.20b}$$

or in matrix form,

$$\frac{1}{\bar{\mu}} \begin{bmatrix} (\bar{\kappa}+1)/8 & (\bar{\kappa}-3)/8 & 0 \\ (\bar{\kappa}-3)/8 & (\bar{\kappa}+1)/8 & 0 \\ 0 & 0 & 1 \end{bmatrix} = \frac{1}{\mu} \begin{bmatrix} (\kappa+1)/8 & (\kappa-3)/8 & 0 \\ (\kappa-3)/8 & (\kappa+1)/8 & 0 \\ 0 & 0 & 1 \end{bmatrix}$$

$$+ f \frac{\bar{\kappa}+1}{\bar{\mu}} \begin{bmatrix} 3/8 & -1/8 & 0 \\ -1/8 & 3/8 & 0 \\ 0 & 0 & 1 \end{bmatrix}. \tag{5.1.20c}$$

Since $\mathbf{D}$, $\overline{\mathbf{D}}$, and $\overline{\mathbf{H}}$ are isotropic tensors, there are only two linearly independent relations among the three equations in (5.1.20c). From these, a system of two equations is obtained for the two unknowns $\bar{\mu}$ and $\bar{\kappa}$,

$$\frac{\bar{\kappa}+1}{8\bar{\mu}} = \frac{\kappa+1}{8\mu} + f\,\frac{3(\bar{\kappa}+1)}{8\bar{\mu}}, \qquad \frac{1}{\bar{\mu}} = \frac{1}{\mu} + f\,\frac{\bar{\kappa}+1}{\bar{\mu}}, \tag{5.1.20d,e}$$

from which $\bar{\mu}$ and $\bar{\kappa}$ are determined as

$$\frac{\bar{\mu}}{\mu} = (1-3f)\,\{1+f(\kappa-2)\}^{-1} = 1 - f(\kappa+1) + O(f^2),$$

$$\frac{\bar{\kappa}}{\kappa} = (1 - f\,\frac{\kappa-2}{\kappa})\,\{1+f(\kappa-2)\}^{-1} = 1 - f\,\frac{(\kappa+1)(\kappa-2)}{\kappa} + O(f^2). \tag{5.1.21a,b}$$

In particular, for plane stress, the effective Young modulus, $\bar{E}$, and Poisson ratio, $\bar{\nu}$, are given by

$$\frac{\bar{E}}{E} = 1 - 3f,$$

$$\frac{\bar{\nu}}{\nu} = 1 - (3 - \frac{1}{\nu})\,f. \tag{5.1.21c,d}$$

Again, for $\nu = 1/3$, it follows that $\bar{\nu}/\nu = \bar{\kappa}/\kappa = 1$, and $\bar{E}/E$ and $\bar{\mu}/\mu$ are exactly the same as those estimated for the dilute distribution model with macrostrain prescribed. Figure 5.1.5 shows the graph of the normalized overall shear modulus over a range of the void volume fraction f, for $\nu = 1/4$. For comparison, the corresponding estimates obtained for a dilute distribution of cavities are also shown (dotted lines). As is seen, the self-consistent method yields an estimate very close to that of the dilute distribution with the overall strain prescribed. Indeed, for $\nu = 1/4$, it follows that $\bar{\mu}/\mu = (1+16f/5)^{-1}$, $(1-3f)/(1+f/5)$, and $(1-16f/5)$, for a dilute distribution with Σ prescribed, the self-consistent method, and the dilute distribution with E prescribed, respectively. It is emphasized that these results are valid for only relatively small void volume fractions.

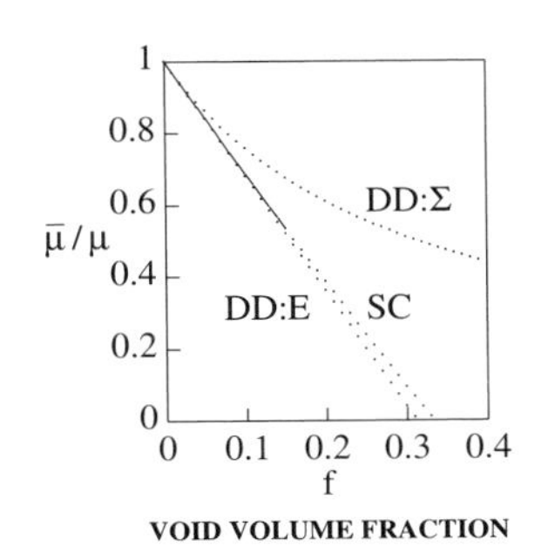

Figure 5.1.5

Normalized overall shear modulus $\bar{\mu}/\mu$ for $\nu = 1/4$
SC ≡ self-consistent
DD:Σ ≡ dilute distribution with macrostress prescribed
DD:E ≡ dilute distribution with macrostrain prescribed

In a similar manner, one obtains the decremental overall stress, $\bar{\boldsymbol{\sigma}}^c$, estimated by the self-consistent method for the case when the macrostrain $\mathbf{E} = \boldsymbol{\varepsilon}^o$ is prescribed. The overall elasticity tensor $\bar{\mathbf{C}}$ then becomes

$$\bar{\mathbf{C}} = \mathbf{C} - \mathbf{C} : \bar{\mathbf{J}}, \tag{5.1.22a}$$

where $\bar{\mathbf{J}}$ is obtained from (5.1.17a), by replacing μ and κ by $\bar{\mu}$ and $\bar{\kappa}$, respectively. For the present model (but not in general), $\mathbf{H}$ and $\mathbf{J}$ are related by

$$\bar{\mathbf{H}} = \bar{\mathbf{J}} : \bar{\mathbf{C}}^{-1} \quad \text{or} \quad \bar{\mathbf{J}} = \bar{\mathbf{H}} : \bar{\mathbf{C}}. \tag{5.1.22b,c}$$

Substitution of (5.1.22c) into (5.1.22a) gives

$$\bar{\mathbf{C}} = \mathbf{C} : (\mathbf{1}^{(4s)} - \bar{\mathbf{H}} : \bar{\mathbf{C}}). \tag{5.1.22d}$$

This yields two equations for $\bar{\mu}$ and $\bar{\kappa}$, which are identical with those obtained from (5.1.20b). Indeed, multiplying (5.1.22d) by $\mathbf{C}^{-1}$ from the left-hand side and by $\bar{\mathbf{C}}^{-1}$ from the right-hand side, it follows that

$$\mathbf{C}^{-1} = \bar{\mathbf{C}}^{-1} - \bar{\mathbf{H}} \tag{5.1.22e}$$

which is identical to (5.1.20b) if $\bar{\mathbf{C}}^{-1}$ is identified with $\bar{\mathbf{D}}$. Hence the overall elasticity tensor $\bar{\mathbf{C}}$ for prescribed macrostrains, is given by the inverse of the overall compliance tensor $\bar{\mathbf{D}}$ for prescribed macrostresses. Therefore, the *self-consistent method yields a unique overall compliance tensor (or elasticity tensor) whether the macrostress or the macrostrain is regarded prescribed.*[6] In particular, a unique set of overall $\bar{\mu}$ and $\bar{\kappa}$ is obtained from (5.1.21a,b), for a porous RVE.

5.1.4. Effective Moduli in x_3-Direction

Finally, consider the effective Young modulus in the x_3-direction, and the Poisson ratio in the x_1,x_3- and x_2,x_3-directions, $\bar{E}_3$ and $\bar{\nu}_3$, for a solid containing

[6] The term "self-consistent" is used in the literature to emphasize the existence of this inverse property; the method dates back to Bruggeman (1935).

circular cylindrical cavities, all parallel to the x_3-direction.

Assume that the stress field in the RVE is given by

$$\sigma_{33} = \begin{cases} \sigma_{33}^{M} & \text{in M} \\ 0 & \text{in } \Omega_\alpha \ (\alpha = 1, 2, ..., n), \end{cases} \tag{5.1.23}$$

where σ_{33}^{M} is constant, and all other stress components are zero. This stress field satisfies traction-free conditions at cavities on any plane normal to the x_3-direction. Since the stress field in M is uniform, the strain field there is uniform, given by

$$\varepsilon_{11} = \varepsilon_{22} = -\frac{\nu}{E}\,\sigma_{33}^{M}, \qquad \varepsilon_{33} = \frac{1}{E}\,\sigma_{33}^{M} \qquad \text{in M.} \tag{5.1.24a,b}$$

The boundary displacement at the cavities is *compatible* with that of the matrix, if the strain field in Ω_α is defined by

$$\varepsilon_{11} = \varepsilon_{22} = -\frac{\nu}{E}\,\sigma_{33}^{M}, \qquad \varepsilon_{33} = \frac{1}{E}\,\sigma_{33}^{M} \qquad \text{in } \Omega_\alpha \ (\alpha = 1, 2, ..., n). \tag{5.1.24c,d}$$

Therefore, denoting the volume average over $V = M + \sum_{\alpha=1}^{n} \Omega_\alpha$ by $<\ >_V$, the average stress and strain over V become

$$<\sigma_{33}>_V = (1 - f)\,\sigma_{33}^{M} \tag{5.1.25}$$

and

$$<\varepsilon_{11}>_V = <\varepsilon_{22}>_V = -\frac{\nu}{E}\,\sigma_{33}^{M}, \qquad <\varepsilon_{33}>_V = \frac{1}{E}\,\sigma_{33}^{M}, \tag{5.1.26a,b}$$

with other components of $<\boldsymbol{\sigma}>_V$ and $<\boldsymbol{\varepsilon}>_V$ being zero. These results are exact.

Since $<\sigma_{33}>_V$ is the only nonzero component of $<\boldsymbol{\sigma}>_V$, the ratio of $<\sigma_{33}>_V$ to $<\varepsilon_{33}>_V$ determines $\bar{E}_3$, and the ratio of $<\varepsilon_{11}>_V = <\varepsilon_{22}>_V$ to $<\varepsilon_{33}>_V$ determines $\bar{\nu}_3$. That is,

$$\bar{E}_3 \equiv \frac{<\sigma_{33}>_V}{<\varepsilon_{33}>_V} = (1 - f)\,E,$$

$$\bar{\nu}_3 \equiv -\frac{<\varepsilon_{11}>_V}{<\varepsilon_{33}>_V} = -\frac{<\varepsilon_{22}>_V}{<\varepsilon_{33}>_V} = \nu. \tag{5.1.27a,b}$$

Hence, $\bar{E}_3$ decreases in proportion to the volume fraction of the cavities, while $\bar{\nu}_3$ remains the same as that of the matrix. In particular, when cavities reduce to slit cracks parallel to the x_3-direction, it follows from (5.1.27a,b) that $\bar{E}_3 = E$, and $\bar{\nu}_3 = \nu$, that is, in linear elasticity, slit cracks parallel to the x_3-direction do not affect the x_3-stiffness, and, hence the corresponding stress-strain relation; see Isida and Nemat-Nasser (1987a,b).

5.2. EFFECTIVE BULK MODULUS OF AN ELASTIC BODY CONTAINING SPHERICAL CAVITIES

In this subsection, the effective bulk modulus of a linearly elastic homogeneous solid containing microcavities is estimated. For simplicity, assume an isotropic matrix containing spherical microcavities, Ω_α, of radius a_α ($\alpha = 1, 2, ..., n$); see Figure 5.2.1. The bulk modulus of an isotropic elastic material is defined in terms of the Lamé constants, λ and μ, by

$$K = \lambda + \frac{2}{3}\mu. \tag{5.2.1a}$$

The mean stress, $\sigma_{ii}/3 \equiv \sigma$, and the volumetric strain, $\varepsilon_{ii} \equiv \varepsilon$, are then related by

$$\sigma = K\,\varepsilon. \tag{5.2.1b}$$

Consider the response of the RVE, subjected to the prescribed macrostress $\boldsymbol{\Sigma} = \sigma^o \mathbf{1}^{(2)}$, or to the prescribed macrostrain $\mathbf{E} = \varepsilon^o \mathbf{1}^{(2)}$. First the overall bulk modulus is estimated, assuming a dilute distribution of microcavities, i.e., neglecting the interaction among them. Then consider the self-consistent estimate of this modulus.

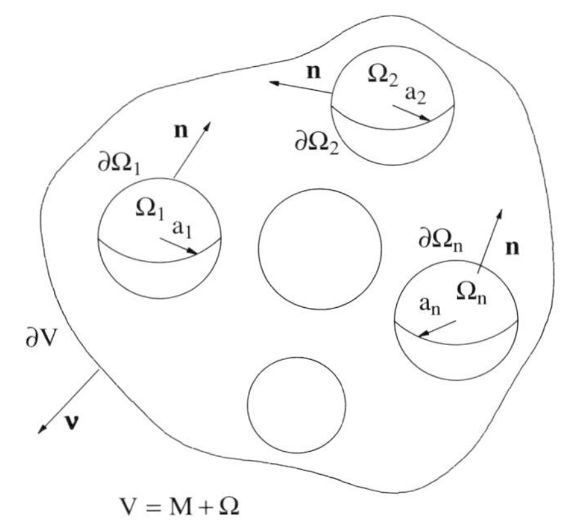

Figure 5.2.1

An RVE containing spherical microcavities

For a typical cavity Ω_α of radius a_α, the field variables in the neighborhood of Ω_α are assumed to be spherically symmetric; see Figure 5.2.2. The additional strain or the decremental stress is computed, using the spherical coordinates (r, θ, ψ) with the origin at the center of the cavity. Under the farfield stress $\sigma^o \mathbf{1}^{(2)}$, the displacement components are

$$u_r = u_r(r) = \frac{\sigma^o}{3\lambda + 2\mu} r + \frac{\sigma^o}{4\mu}\frac{a^3}{r^2}, \qquad u_\theta = 0, \qquad u_\psi = 0, \tag{5.2.2a~c}$$

where a is the radius of the cavity. Since the unit normal $\mathbf{n}$ on $\partial\Omega_\alpha$ coincides with the radial base vector $\mathbf{e}_r$, the average strain for Ω_α, (5.1.8b), becomes

Figure 5.2.2

A spherical cavity and spherical coordinates

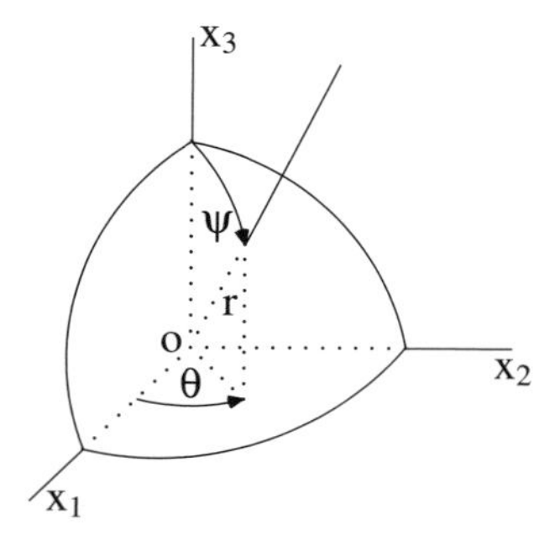

$$\bar{\varepsilon}^{\alpha}_{ii} = \bar{\varepsilon}^{\alpha} = \frac{1}{\Omega_{\alpha}} \int_{\partial\Omega_{\alpha}} u_r(a)\, dS = \frac{1}{\Omega_{\alpha}} \{\partial\Omega_{\alpha}\, u_r(a)\}, \tag{5.2.3}$$

where $\Omega_{\alpha} = 4\pi a^3/3$, and $\partial\Omega_{\alpha} = 4\pi a^2$. Therefore, the additional volumetric strain due to the presence of cavities is given by

$$\bar{\varepsilon}^{c}_{ii} = \bar{\varepsilon}^{c} = \frac{1}{K}\, f\,(1 + \frac{3K}{4\mu})\,\sigma^{o}, \tag{5.2.4a}$$

where f is the void volume fraction. From (5.2.4a), the "dilute estimate" of the effective bulk modulus, $\bar{K}$, is obtained when the macrostress is prescribed,

$$\frac{\bar{K}}{K} = \{1 + f\,\frac{3(1-\nu)}{2(1-2\nu)}\}^{-1} = 1 - f\,\frac{3(1-\nu)}{2(1-2\nu)} + O(f^2). \tag{5.2.4b}$$

If the macrostrain $\mathbf{E} = \varepsilon^{o}\mathbf{1}^{(2)}$ is prescribed, an infinite body subjected to the farfield stress given by $K\,\varepsilon^{o}\mathbf{1}^{(2)}$ is considered. Then, by replacing σ^{o} with $K\,\varepsilon^{o}$ in (5.2.4a), the corresponding additional volumetric strain due to the cavities becomes

$$\bar{\varepsilon}^{c} = f\,(1 + \frac{3K}{4\mu})\,\varepsilon^{o}. \tag{5.2.5a}$$

From (5.2.5a) the effective bulk modulus, $\bar{K}$, is estimated for the prescribed macrostrain, as

$$\frac{\bar{K}}{K} = 1 - f\,(1 + \frac{3K}{4\mu}) = 1 - f\,\frac{3(1-\nu)}{2(1-2\nu)}. \tag{5.2.5b}$$

Comparing (5.2.4b) and (5.2.5b), it is observed that the two expressions agree with each other to within the first order in the void volume fraction f. For $\nu = 1/3$, $\bar{\kappa}/\kappa = (1-3f)^{-1}$ for the case when the macrostress is prescribed, and $\bar{\kappa}/\kappa = (1-3f)$ when the macrostrain is prescribed. These are identical with the corresponding $\bar{\mu}/\mu$ and $\bar{E}/E$ of (5.1.15) and (5.1.18); see Figure 5.1.4. As is seen, here again, the result of Theorem I of Subsection 2.5 is contradicted, revealing the limitation of the dilute-distribution *model*.

The estimates (5.2.4b) and (5.2.5b) do not include any interaction among the cavities. To include this interaction for a random distribution of cavities, the self-consistent method may be used. Then, for the case when the macrostress is regarded prescribed, (5.2.4a) is replaced by

$$\bar{\varepsilon}^c = \frac{1}{K} f (1 + \frac{3\bar{K}}{4\bar{\mu}}) \sigma^o, \tag{5.2.6a}$$

and instead of (5.2.4b), it now follows that

$$\frac{\bar{K}}{K} = 1 - f \frac{3(1-\bar{\nu})}{2(1-2\bar{\nu})} \tag{5.2.6b}$$

which also requires an estimate of the overall Poisson ratio $\bar{\nu}$. Similarly, a self-consistent estimate of the overall bulk modulus can be obtained when the macrostrain is regarded prescribed. As pointed out in connection with (5.1.22a~e), the result is identical with that for the prescribed macrostress, i.e., with (5.2.6b).

5.3. ENERGY CONSIDERATION AND SYMMETRY PROPERTIES OF TENSOR H

The overall compliance tensor $\bar{\mathbf{D}}$ and elasticity tensor $\bar{\mathbf{C}}$ are defined, respectively, by (4.3.1) and (4.5.1). They are given for a linearly elastic RVE with microcavities, by (4.3.6a) and (4.5.5a).

The overall quantities $\bar{\mathbf{D}}$ and $\bar{\mathbf{C}}$ may also be defined in terms of the total elastic energy stored in the RVE, in the sense that if the RVE is replaced by an equivalent linearly elastic and *homogeneous* solid, it must store the same amount of elastic energy as the actual RVE for the same macrostress, $\mathbf{\Sigma} = \boldsymbol{\sigma}^o$, when the overall stress is prescribed, or for the same macrostrain, $\mathbf{E} = \boldsymbol{\varepsilon}^o$, when the overall strain is prescribed. The two cases of prescribed macrostress and prescribed macrostrain are treated separately, starting with the former.

Denote the macro-complementary strain energy function by $W^c = W^c(\mathbf{\Sigma}) = W^c(\boldsymbol{\sigma}^o)$, when the macrostress is given by $\mathbf{\Sigma} = \boldsymbol{\sigma}^o$. From (2.5.17a),

$$W^c(\mathbf{\Sigma}) = < w^c > = \frac{1}{V} \int_V w^c(\boldsymbol{\sigma}(\mathbf{x}; \mathbf{\Sigma}))\, dV, \tag{5.3.1}$$

where $w^c(\boldsymbol{\sigma})$ is the complementary energy density function of the matrix material at point $\mathbf{x}$. Since the RVE is linearly elastic and subjected to uniform tractions $\mathbf{t}^o = \boldsymbol{\nu} \cdot \boldsymbol{\sigma}^o$ on ∂V,

$$2W^c(\boldsymbol{\sigma}^o) = < \boldsymbol{\sigma} : \boldsymbol{\varepsilon} > = \boldsymbol{\sigma}^o : < \boldsymbol{\varepsilon} > = \boldsymbol{\sigma}^o : \bar{\boldsymbol{\varepsilon}}$$

$$= \boldsymbol{\sigma}^o : (\boldsymbol{\varepsilon}^o + \bar{\boldsymbol{\varepsilon}}^c) = \boldsymbol{\sigma}^o : (\mathbf{D} + \mathbf{H}) : \boldsymbol{\sigma}^o, \tag{5.3.2}$$

where definitions (4.2.1) and (4.3.3a), and expression (4.2.3) are also used. Hence, whatever the structure of the microcavities, only the symmetric part of $\mathbf{H}$ contributes to the stored elastic energy; note that the microstructure is fixed and no frictional effects are included. Therefore the definition of $\mathbf{H}$ given by (4.3.5b), is replaced by

$$H_{ijkl} = \frac{1}{V}\int_{\partial\Omega}\int_{\partial V}\frac{1}{2}(h_{ijkl}+h_{klij})\,dS\,dS. \tag{5.3.3}$$

The effective compliance of the RVE may now be defined as the constant symmetric tensor $\bar{\mathbf{D}}$ with the property that, for any macrostress $\mathbf{\Sigma} = \boldsymbol{\sigma}^o$, the overall complementary energy density is

$$W^c(\boldsymbol{\sigma}^o) = \frac{1}{2}\boldsymbol{\sigma}^o : \bar{\mathbf{D}} : \boldsymbol{\sigma}^o. \tag{5.3.4a}$$

Comparison with (5.3.2) shows that $\bar{\mathbf{D}}$ is defined by

$$\bar{\mathbf{D}} = \mathbf{D} + \mathbf{H}. \tag{5.3.4b}$$

In a similar manner, when the macrostrain is prescribed to be $\mathbf{E} = \boldsymbol{\varepsilon}^o$, from (2.5.12a), the overall elastic energy density of the RVE becomes

$$W(\mathbf{E}) = < w > = \frac{1}{V}\int_V w(\boldsymbol{\varepsilon}(\mathbf{x};\,\mathbf{E}))\,dV. \tag{5.3.5}$$

Moreover,

$$2W(\boldsymbol{\varepsilon}^o) = < \boldsymbol{\sigma} : \boldsymbol{\varepsilon} > = < \boldsymbol{\sigma} > : \boldsymbol{\varepsilon}^o = \bar{\boldsymbol{\sigma}} : \boldsymbol{\varepsilon}^o$$

$$= (\boldsymbol{\sigma}^o + \bar{\boldsymbol{\sigma}}^c) : \boldsymbol{\varepsilon}^o = \boldsymbol{\varepsilon}^o : (\mathbf{C} - \mathbf{C} : \mathbf{J}) : \boldsymbol{\varepsilon}^o, \tag{5.3.6}$$

where (4.4.7) and (4.5.3b) are used. Defining the overall effective elasticity tensor, $\bar{\mathbf{C}}$, such that

$$W(\boldsymbol{\varepsilon}^o) = \frac{1}{2}\boldsymbol{\varepsilon}^o : \bar{\mathbf{C}} : \boldsymbol{\varepsilon}^o \tag{5.3.7}$$

for any prescribed constant strain $\boldsymbol{\varepsilon}^o$, it is concluded from (5.3.6) and (5.3.7) that

$$\bar{\mathbf{C}} = \mathbf{C} - \mathbf{C} : \mathbf{J}, \tag{5.3.8a}$$

where $\mathbf{C} : \mathbf{J}$ is required to have the following symmetry property:

$$C_{ijrs}\,J_{rskl} = C_{klrs}\,J_{rsij}. \tag{5.3.8b}$$

5.4. CAVITY STRAIN

When the macrostress is prescribed, $\mathbf{\Sigma} = \boldsymbol{\sigma}^o$, the expression (4.2.1) *defines* $\boldsymbol{\varepsilon}^o = \mathbf{D} : \boldsymbol{\sigma}^o$ which is the uniform strain in a homogeneous matrix with compliance $\mathbf{D}$ subjected to constant stress $\boldsymbol{\sigma}^o$. The additional strain $\bar{\boldsymbol{\varepsilon}}^c$ due to the presence of cavities, is then given by (4.2.8), namely

$$\bar{\boldsymbol{\varepsilon}}^c = \frac{1}{V}\int_{\partial\Omega}\frac{1}{2}(\mathbf{n}\otimes\mathbf{u} + \mathbf{u}\otimes\mathbf{n})\,dS; \tag{5.4.1a}$$

note that $\mathbf{n}$ is the *exterior* unit normal of the *cavity*. The cavity may be regarded as an elastic continuum with zero elastic resistance. Then $\bar{\boldsymbol{\varepsilon}}^c$ becomes the weighted sum of the average strain over each individual cavity, i.e.,

$$\bar{\boldsymbol{\varepsilon}}^c = \sum_{\alpha=1}^{n} \frac{\Omega_\alpha}{V} \{\frac{1}{\Omega_\alpha} \int_{\partial\Omega_\alpha} \frac{1}{2}(\mathbf{n}\otimes\mathbf{u}+\mathbf{u}\otimes\mathbf{n})\,dS\}$$

$$= \sum_{\alpha=1}^{n} f_\alpha \{\frac{1}{\Omega_\alpha} \int_{\Omega_\alpha} \frac{1}{2}\{\boldsymbol{\nabla}\otimes\mathbf{u}+(\boldsymbol{\nabla}\otimes\mathbf{u})^T\}\,dV\}$$

$$= \sum_{\alpha=1}^{n} f_\alpha \bar{\boldsymbol{\varepsilon}}^\alpha, \tag{5.4.1b}$$

where

$$f_\alpha \equiv \frac{\Omega_\alpha}{V} \tag{5.4.1c}$$

is the volume fraction of the αth cavity Ω_α, there are n cavities in V, and the Gauss theorem is used to obtain

$$\bar{\boldsymbol{\varepsilon}}^\alpha = \frac{1}{\Omega_\alpha} \int_{\Omega_\alpha} \boldsymbol{\varepsilon}(\mathbf{x})\,dV. \tag{5.4.2}$$

It is shown later on, how the average cavity strain $\bar{\boldsymbol{\varepsilon}}^\alpha$ relates to *the transformation strain* or *the eigenstrain* introduced by Eshelby (1957) for ellipsoidal inclusions. Note in this connection that *the result* $\bar{\boldsymbol{\varepsilon}}^c = \mathbf{H} : \boldsymbol{\sigma}^o$ *is valid for cavities of any shape in an elastic solid of any (finite or infinite) dimension, with* $\mathbf{H}$ *a constant fourth-order tensor which depends only on the geometry and elastic properties of the matrix.* Eshelby's results, on the other hand, are for an ellipsoidal inclusion in an infinitely extended elastic solid.

5.5. REFERENCES

Bruggeman, D. A. G. (1935), Berechnung verschiedener physikalischer Konstanten von heterogenen substanzen, *Annalen der Physik*, Vol. 24, 636-679.

Budiansky, B. (1965), On the elastic moduli of some heterogeneous materials, *J. Mech. Phys. Solids*, Vol. 13, 223-227.

Eshelby, J. D. (1957), The determination of the elastic field of an ellipsoidal inclusion, and related problems, *Proc. Roy. Soc.*, Vol. A241, 376-396.

Hashin, Z. (1968), Assessment of the self-consistent scheme approximation, *J. Compos. Mater.*, Vol. 2, 284-300.

Hill, R. (1965), A self-consistent mechanics of composite materials, *J. Mech. Phys. Solids*, Vol. 13, 213-222.

Isida, M. and Nemat-Nasser, S. (1987a), A unified analysis of various problems relating to circular holes with edge cracks, *Eng. Fract. Mech.*, Vol. 27, 571-591.

Isida, M. and Nemat-Nasser, S. (1987b), On mechanics of crack growth and its effects on the overall response of brittle porous solids, *Acta. Metall.*, Vol 12, 2887-2898.

Kröner, E. (1958), Berechnung der elastischen Konstanten des Vielkristalls aus

den Konstanten des Einkristalls, *Z. Phys.*, Vol. 151, 504-518.
Love, A. E. H. (1944), *A treatise on the mathematical theory of elasticity*, Dover, New York.
Michell, J. H. (1899), On the direct determination of stress in an elastic solid, with application to the theory of plates, *Proc. London Math. Soc.* Vol. 31, 100-124.
Timoshenko, S. P. and Goodier, J. N. (1951), *Theory of elasticity*, McGraw-Hill, New York.

SECTION 6 ELASTIC SOLIDS WITH MICROCRACKS

In this section, the general results obtained in Section 4 are specialized for application to linearly elastic solids which contain microcracks. Problems of this kind arise when one seeks to understand the overall response and failure modes of brittle materials such as ceramics, ceramic composites, rocks, and cement-like and related materials. A variety of microcrack arrangements are examined and the overall properties of the solid are estimated, using various averaging techniques.

6.1. OVERALL STRAIN DUE TO MICROCRACKS

In Subsection 4.2, the additional strain, $\bar{\boldsymbol{\varepsilon}}^c$, due to the presence of cavities is calculated; see (4.2.8). A crack is a cavity, one of whose dimensions is very small relative to the other two dimensions. For example, an elliptical crack can be regarded as an ellipsoidal cavity, the length of one of whose principal axes becomes very small in comparison with the length of the other two principal axes. Similarly, a penny-shaped crack can be regarded as a limiting case of a cavity in the shape of an ellipsoidal of revolution. In general, a crack is identified by two identical surfaces which are separated by *the crack-opening-displacement*, representing the relative displacement of the corresponding points on the two identical surfaces. These surfaces are often called "crack faces". Figure 6.1.1 illustrates this for an elliptical crack. Here the "upper" surface or

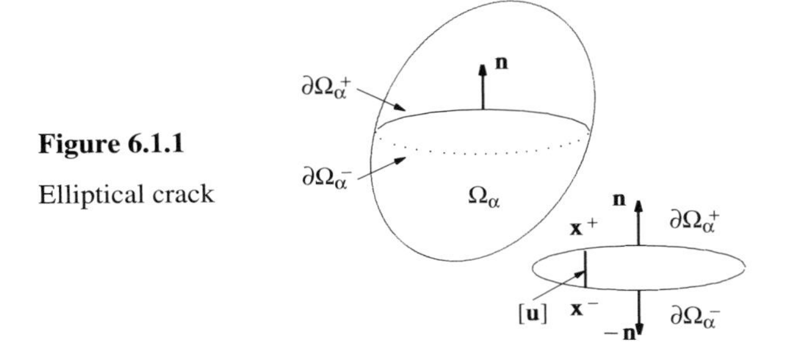

Figure 6.1.1

Elliptical crack

the "upper" face of the crack is denoted by $\partial\Omega^+$, and the "lower" one by $\partial\Omega^-$. The *exterior* unit normal of the crack on $\partial\Omega^+$ is denoted by $\mathbf{n}$. Hence, the exterior unit normal of the crack on $\partial\Omega^-$ is $-\mathbf{n}$.

In general, the boundary $\partial\Omega_\alpha$ of a typical αth crack in an RVE is divided into $\partial\Omega_\alpha^+$ and $\partial\Omega_\alpha^-$; $\partial\Omega_\alpha = \partial\Omega_\alpha^+ + \partial\Omega_\alpha^-$. For a function $\mathrm{f}(\mathbf{x})$ defined on $\partial\Omega_\alpha$, the surface integral over the entire $\partial\Omega_\alpha$ can be reduced to the integral over $\partial\Omega_\alpha^+$ only, as follows:

$$\int_{\partial\Omega_\alpha} \mathrm{f}(\mathbf{x})\, \mathrm{dS} = \int_{\partial\Omega_\alpha^+} \mathrm{f}(\mathbf{x}^+)\, \mathrm{dS} + \int_{\partial\Omega_\alpha^-} \mathrm{f}(\mathbf{x}^-)\, \mathrm{dS}$$

$$= \int_{\partial\Omega_\alpha^+} \{\mathrm{f}(\mathbf{x}^+) - \mathrm{f}(\mathbf{x}^-)\}\, \mathrm{dS}, \tag{6.1.1}$$

where $\mathrm{f}(\mathbf{x}^+)$ and $\mathrm{f}(\mathbf{x}^-)$ are calculated at the corresponding points, $\mathbf{x}^+$ and $\mathbf{x}^-$, on $\partial\Omega_\alpha^+$ and $\partial\Omega_\alpha^-$, respectively.

Let $\partial\Omega$, $\partial\Omega^+$, and $\partial\Omega^-$ respectively denote the union of all crack surfaces, their "plus" or "upper" surfaces, and their "minus" or "lower" surfaces. Hence, $\partial\Omega = \partial\Omega^+ + \partial\Omega^-$. The expression for the additional strain due to cracks becomes

$$\bar{\boldsymbol{\varepsilon}}^{\mathrm{c}} = \frac{1}{\mathrm{V}} \sum_{\alpha=1}^{\mathrm{n}} \int_{\partial\Omega_\alpha} \frac{1}{2}\{\mathbf{n}(\mathbf{x})\otimes\mathbf{u}(\mathbf{x}) + \mathbf{u}(\mathbf{x})\otimes\mathbf{n}(\mathbf{x})\}\, \mathrm{dS}$$

$$= \frac{1}{\mathrm{V}} \sum_{\alpha=1}^{\mathrm{n}} \int_{\partial\Omega_\alpha^+} \left\{ \frac{1}{2}\{\mathbf{n}(\mathbf{x}^+)\otimes\mathbf{u}(\mathbf{x}^+) + \mathbf{u}(\mathbf{x}^+)\otimes\mathbf{n}(\mathbf{x}^+)\} \right.$$

$$\left. - \frac{1}{2}\{\mathbf{n}(\mathbf{x}^+)\otimes\mathbf{u}(\mathbf{x}^-) + \mathbf{u}(\mathbf{x}^-)\otimes\mathbf{n}(\mathbf{x}^+)\} \right\} \mathrm{dS}$$

$$= \frac{1}{\mathrm{V}} \sum_{\alpha=1}^{\mathrm{n}} \int_{\partial\Omega_\alpha^+} \frac{1}{2}\{\mathbf{n}(\mathbf{x}^+)\otimes[\mathbf{u}](\mathbf{x}^+) + [\mathbf{u}](\mathbf{x}^+)\otimes\mathbf{n}(\mathbf{x}^+)\}\, \mathrm{dS}$$

$$= \frac{1}{\mathrm{V}} \int_{\partial\Omega^+} \frac{1}{2}\{\mathbf{n}\otimes[\mathbf{u}] + [\mathbf{u}]\otimes\mathbf{n}\}\, \mathrm{dS}, \tag{6.1.2a}$$

where $\mathbf{n}$ is the exterior unit normal of $\partial\Omega^+$, and $[\mathbf{u}] \equiv [\mathbf{u}](\mathbf{x}^+)$ is *the crack-opening-displacement*, COD, defined by

$$[\mathbf{u}] \equiv [\mathbf{u}](\mathbf{x}^+) = \mathbf{u}(\mathbf{x}^+) - \mathbf{u}(\mathbf{x}^-), \tag{6.1.2b}$$

where $\mathbf{x}^+$ and $\mathbf{x}^-$ are the corresponding points on the "plus" and "minus" faces of the crack.

6.2. OVERALL COMPLIANCE AND MODULUS TENSORS OF HOMOGENEOUS LINEARLY ELASTIC SOLIDS WITH MICROCRACKS

When the macrostress for an RVE is prescribed, $\mathbf{\Sigma} = \boldsymbol{\sigma}^{o}$, an analysis similar to that in Section 4, immediately shows that the overall effective compliance tensor $\bar{\mathbf{D}}$ of the cracked RVE with a homogeneous linearly elastic matrix of compliance tensor $\mathbf{D}$, is given by (4.3.6), where the constant tensor $\mathbf{H}$ is now defined through

$$\bar{\boldsymbol{\varepsilon}}^{c} = \frac{1}{V} \int_{\partial\Omega^{+}} \frac{1}{2} \{\mathbf{n} \otimes [\mathbf{u}] + [\mathbf{u}] \otimes \mathbf{n}\} \, dS = \mathbf{H} : \boldsymbol{\sigma}^{o}, \tag{6.2.1}$$

where $\boldsymbol{\sigma}^{o}$ is an arbitrary constant prescribed macrostress. This expression is valid when all microcracks are *open*. It is also valid when some microcracks are *closed* and undergo *frictional sliding*; Horii and Nemat-Nasser (1983). Indeed, it remains valid even if the microcracks exhibit resistance to the relative displacement of their surfaces, as long as the COD is a linear and homogeneous function of the overall prescribed macrostress $\boldsymbol{\sigma}^{o}$, i.e., as long as

$$[\mathbf{u}] = \mathbf{K}^{c}(\mathbf{x}) : \boldsymbol{\sigma}^{o}, \tag{6.2.2}$$

for some linear operator $\mathbf{K}^{c}(\mathbf{x})$ which is a third-order tensor; here $\mathbf{x}$ is a point on the "upper" or "plus" surface of the crack at which the COD is being measured.

Similarly, when the macrostrain is prescribed, $\mathbf{E} = \boldsymbol{\varepsilon}^{o}$, under the conditions stated above,

$$\bar{\boldsymbol{\varepsilon}}^{c} = \frac{1}{V} \int_{\partial\Omega^{+}} \frac{1}{2} \{\mathbf{n} \otimes [\mathbf{u}] + [\mathbf{u}] \otimes \mathbf{n}\} \, dS = \mathbf{J} : \boldsymbol{\varepsilon}^{o}. \tag{6.2.3}$$

The overall elasticity tensor then is given by (4.5.5). In the remaining parts of this section a number of illustrative examples of some practical importance are given. As for the case of an elastic solid with microcavities, the overall moduli are obtained with: (1) the assumption that the crack distribution is dilute so that crack interaction may be neglected, and (2) the self-consistent approach which approximately accounts for the interaction effects.[1]

The effect of a dilute concentration of randomly oriented ribbon-shaped and penny-shaped cracks on the overall moduli of a solid is examined by Bristow (1960), using the elastic energy associated with a single crack in an unbounded uniform elastic matrix. Walsh (1969) considers a similar model with circular cracks, and also examines the influence of fluid filled cracks. The dilute distribution model has also been employed by Salganik (1973) to study the effect of elliptical cracks on the elasticity of cracked solids, by Griggs *et al.* (1975) who use circular cracks to model source regions associated with anomalous pressure/shear wave velocities encountered in earthquake events, and by Garbin and Knopoff (1973, 1975a,b) who address the effects of randomly distributed circular cracks (both with and without fluids) on the overall elastic properties of the solid, using the zero-frequency limit of scattering elastic

---

[1]See also Section 10 for the differential scheme and the double- and multi-inclusion methods, and Section 12 for a periodic distribution of microcracks.

waves. The self-consistent model (Kröner, 1958; Budiansky, 1965; Hill, 1965) is used by Budiansky and O'Connell (1976) who consider elliptical cracks and identify parameters which adapt the elliptic crack-results to other convex crack shapes; see O'Connell and Budiansky (1974, 1977). Related studies are by Hoenig (1979) who considers non-random distribution of cracks; Horii and Nemat-Nasser (1983) who examine the anisotropy induced by preferential crack opening and closing in the presence of friction; and Nemat-Nasser and Horii (1982) and Horii and Nemat-Nasser (1985a, 1986) who study failure in compression, including axial splitting, faulting, and brittle to ductile transition under increasing confining pressures (see also Nemat-Nasser and Obata, 1988; and Nemat-Nasser, 1989); the effects of crack geometry and distribution on the overall elastic moduli is studied by Laws *et al.* (1983), Laws and Brockenbrough (1987), and Laws and Dvorak (1987). A different method, called the differential scheme, has been used by Hashin (1988), and by Nemat-Nasser and Hori (1990), where comparison of various techniques is provided.

6.3. EFFECTIVE MODULI OF AN ELASTIC SOLID CONTAINING ALIGNED SLIT MICROCRACKS

In this subsection two-dimensional problems are considered, i.e., plane strain, plane stress, and antiplane shear. The overall elastic moduli of an RVE containing microcracks are then estimated. All cracks are assumed to be planar and parallel to the x_3-direction. Their intersections with the x_1,x_2-plane, therefore, are lines identifying the crack faces; see Figure 6.3.1.

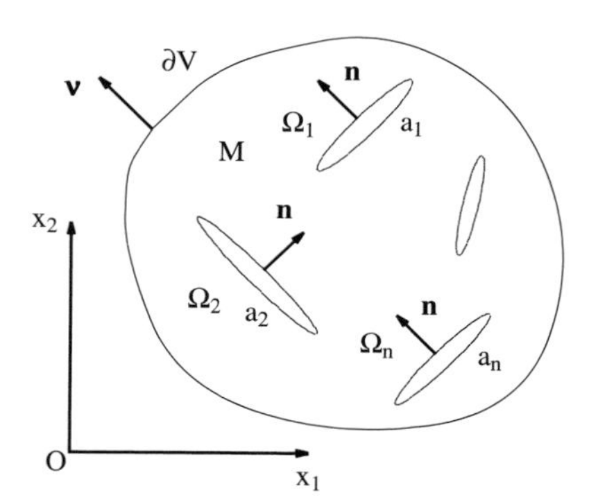

Figure 6.3.1

A typical portion of a two-dimensional RVE containing microcracks

6.3.1. Crack Opening Displacements

Consider a typical crack lying on the x_1,x_3-plane, from $x_1 = -a$ to $x_1 = a$, with $|x_3| < \infty$, in a homogeneous linearly elastic isotropic solid subjected to farfield uniform stresses σ_{22}^{∞} and $\sigma_{12}^{\infty} = \sigma_{21}^{\infty}$. The crack opening displacements (COD) are

$$[u_i] = \sqrt{a^2 - x_1^2}\,\frac{4}{E'}\,\sigma_{i2}^{\infty}, \qquad |x_1| \leq a \; (i = 1, 2), \tag{6.3.1a}$$

where

$$\frac{1}{E'} = \frac{\kappa+1}{8\mu} = \begin{cases} \dfrac{1-\nu^2}{E} & \text{for plane strain} \\ \dfrac{1}{E} & \text{for plane stress,} \end{cases} \tag{6.3.1b}$$

with $\kappa = 3 - 4\nu$ for plane strain, and $\kappa = (3 - \nu)/(1 + \nu)$ for plane stress; see Figure 6.3.2. Modulus E' may be regarded as a *nominal* two-dimensional Young modulus.

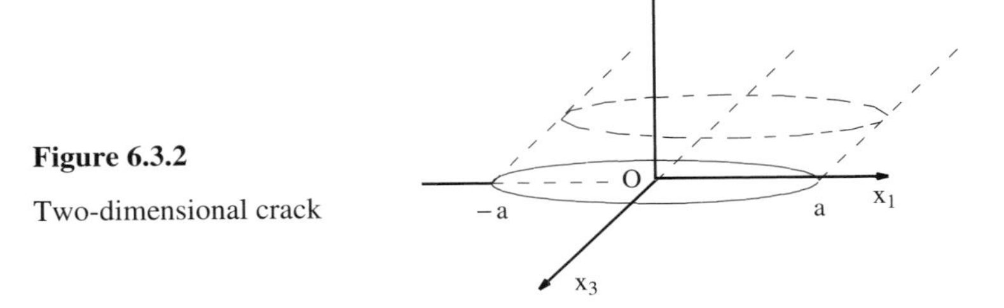

Figure 6.3.2

Two-dimensional crack

Similarly, for antiplane shear, with the farfield uniform stress $\sigma_{23}^{\infty} = \sigma_{32}^{\infty}$ applied to an infinitely extended solid containing a planar crack on $|x_1| \leq a$, $|x_3| < \infty$, the crack opening displacement is

$$[u_3] = \sqrt{a^2 - x_1^2}\,\frac{2}{\mu}\,\sigma_{23}^{\infty}, \qquad |x_1| \leq a. \tag{6.3.1c}$$

6.3.2. Effective Moduli: Dilute Distribution of Aligned Microcracks

Consider a dilute distribution of microcracks all aligned and parallel to the x_1-axis; see Figure 6.3.3. Throughout this section microcracks are parallel to the x_3-direction as discussed before, but for conciseness the x_3-configuration is not mentioned. The crack sizes and the location of their centers in the x_1,x_2-plane are assumed to be random. Initially the cracks are all closed. Under an overall compressive uniaxial macrostress $\Sigma_{22} = -p^o$ $(p^o > 0)$, applied in the x_2-direction, the cracks will have no effect, since they remain closed, transmitting the uniform normal compression. The overall Young modulus will be the same as that of the matrix material, i.e., E. On the other hand, under tensile uniaxial macrostress, $\Sigma_{22} = p^o$ ($p^o > 0$), applied in the x_2-direction, the cracks will open and hence make a contribution to the overall macrostrain in the x_2-direction. The solid then is more compliant in tension than in compression when loaded in the x_2-direction.

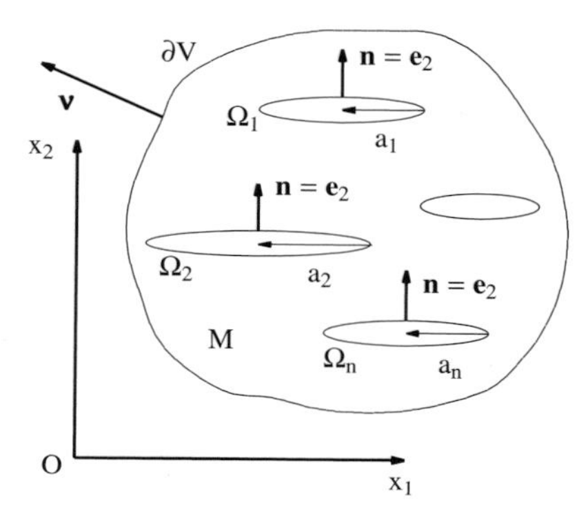

Figure 6.3.3

Microcracks aligned parallel to the x_3-axis

The change in Young's modulus is now estimated for uniaxial *tension* in the x_2-direction by considering a single crack in an infinitely extended solid subjected at infinity to uniform *tension* $\Sigma_{22} = \sigma^o_{22}$; see Figure 6.3.4.

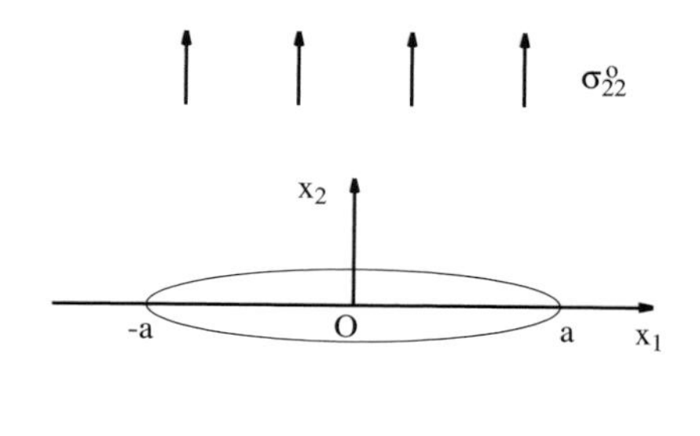

Figure 6.3.4

A single crack in an infinitely extended solid subjected to uniform tension at infinity

For a typical crack Ω_α of length $2a_\alpha$, it follows from (6.3.1a) that

$$\bar{\varepsilon}^{\alpha}_{22} \equiv \frac{1}{a^2_\alpha} \int_{-a_\alpha}^{a_\alpha} [u_2]\, dx_1 = \frac{2\pi}{E'}\, \sigma^o_{22}. \tag{6.3.2a}$$

Define f_α by

$$f_\alpha = N_\alpha\, a^2_\alpha \qquad (\alpha \text{ not summed}), \tag{6.3.3a}$$

where N_α is the number of cracks of length $2a_\alpha$ per unit area in the x_1,x_2-plane. The parameter f_α measures the density of the cracks of length $2a_\alpha$; Budiansky and O'Connell (1976). When there are n sets of cracks, Ω_α, in an RVE, each set having its own length $2a_\alpha$ ($\alpha = 1, 2, ..., n$), the overall *crack density parameter* f is defined by

$$f = \sum_{\alpha=1}^{n} f_\alpha = \sum_{\alpha=1}^{n} N_\alpha\, a_\alpha^2, \tag{6.3.3b}$$

with the constraint

$$N = \sum_{\alpha=1}^{n} N_\alpha, \tag{6.3.3c}$$

where N is the total number of cracks per unit area. When all cracks have a common length 2a, then $f = N\,a^2$.

The total contribution to the overall strain by the opening of the microcracks which are all aligned in the x_1-direction, now becomes

$$\bar{\varepsilon}_{22}^{c} = \sum_{\alpha=1}^{n} N_\alpha \int_{-a_\alpha}^{a_\alpha} [u_2]\, dx_1 = \sum_{\alpha=1}^{n} (N_\alpha\, a_\alpha^2) \frac{1}{a_\alpha^2} \int_{-a_\alpha}^{a_\alpha} [u_2]\, dx_1 = f\, \frac{2\pi}{E'}\, \sigma_{22}^{o}. \tag{6.3.2b}$$

From (4.3.6), the nominal Young modulus[2] $\bar{E}_2'$, is obtained,

$$\frac{\bar{E}_2'}{E'} = \{1 + 2\pi f\, H(\sigma_{22}^{o})\}^{-1}, \tag{6.3.4a}$$

or

$$\frac{\bar{E}_2'}{E} = \frac{4}{(\kappa+1)(1+\nu)}\, \{1 + 2\pi f\, H(\sigma_{22}^{o})\}^{-1}, \tag{6.3.4b}$$

where H(x) is the Heaviside step function, being zero if its argument x is nonpositive, and one if x is positive, and $\bar{E}_2'$ is the overall nominal Young modulus in the x_2-direction. Since the presence of the cracks all aligned in the x_1-direction does not affect the nominal Young modulus in the x_1-direction, $\bar{E}_1 = E'$ which is the nominal matrix Young modulus. Therefore, with the microcracks aligned in the x_1-direction, the response of the RVE is anisotropic. From (6.3.4a,b) and the definition of $\bar{E}_2'$, therefore, it follows that

$$\frac{\bar{E}_2'}{E} = \{1 + 2\pi(1 - \eta\nu^2)\, f\, H(\sigma_{22}^{o})\}^{-1}, \tag{6.3.4c}$$

where $\eta = 0, 1$ for plane stress and plane strain, respectively. Thus, the results for plane strain are obtained from those for plane stress if f is replaced by $(1-\nu^2)\,f$.

Assuming that the cracks are *frictionless*, the overall shear moduli $\bar{\mu}_{12}$ and $\bar{\mu}_{23}$, are estimated as follows. To estimate $\bar{\mu}_{12}$ for a dilute distribution of microcracks aligned in the x_1-direction, consider the applied macrostress

---

[2] Because of anisotropic overall response, $\bar{E}_2'$ is defined by $1/\bar{E}_2' = 1/\bar{E}_2 - \eta\nu^2/E$, where $\eta = 0$ for plane stress and $\eta = 1$ for plane strain.

$\Sigma_{12} = \Sigma_{21} = \sigma^o_{12}$, and obtain, using (6.3.1a),

$$\bar{\varepsilon}^c_{12} = \sum_{\alpha=1}^{n} N_\alpha \int_{-a_\alpha}^{a_\alpha} \frac{1}{2}[u_1]\,dx_1 = \sum_{\alpha=1}^{n} (N_\alpha\, a_\alpha^2)\, \frac{1}{a_\alpha^2} \int_{-a_\alpha}^{a_\alpha} \frac{1}{2}[u_1]\,dx_1 = f\,\frac{\pi}{E'}\,\sigma^o_{12}. \tag{6.3.5a}$$

Direct application of (4.2.3) now gives

$$\bar{\varepsilon}_{12} = \varepsilon^o_{12} + \bar{\varepsilon}^c_{12} = \frac{1}{2\mu}\,\sigma^o_{12} + f\,\frac{\pi(\kappa+1)}{8\mu}\,\sigma^o_{12}. \tag{6.3.5b}$$

Then from $\bar{\varepsilon}_{12} = \sigma^o_{12}/2\bar{\mu}_{12}$, it follows that

$$\frac{\bar{\mu}_{12}}{\mu} = \{1 + f\,\frac{\pi(\kappa+1)}{4}\}^{-1} = 1 - f\,\frac{\pi(\kappa+1)}{4} + O(f^2). \tag{6.3.6}$$

In a similar manner, $\bar{\mu}_{23}$ is estimated from $\bar{\varepsilon}^c_{23} = f\,\pi\sigma^o_{23}/2\mu$, arriving at

$$\frac{\bar{\mu}_{23}}{\mu} = \{1 + f\,\pi\}^{-1} = 1 - f\,\pi + O(f^2), \tag{6.3.7}$$

where (6.3.1c) is used. Note that the overall shear modulus in the x_1,x_3-plane is not affected by the presence of cracks aligned in the x_1-direction; hence, $\bar{\mu}_{13} = \mu$.

Summarizing the above results, observe that the matrix $[H_{ab}]$ for the *plane* problem is given by

$$[H_{ab}] = f\,\frac{\pi(\kappa+1)}{4\mu} \begin{bmatrix} 0 & 0 & 0 \\ 0 & H(\sigma^o_{22}) & 0 \\ 0 & 0 & 1 \end{bmatrix}, \tag{6.3.8a}$$

when the overall stress is prescribed. Moreover, for the *antiplane shear* problem,

$$[H_{ab}] = f\,\frac{\pi}{\mu} \begin{bmatrix} 0 & 0 \\ 0 & 1 \end{bmatrix}, \tag{6.3.9a}$$

again when the macrostress is prescribed. On the other hand, when the overall macrostrains are regarded given, an analysis similar to that presented in Subsection 5.1.2, readily shows that, for the *plane* problem,

$$[J_{ab}] = f\,\frac{\pi(\kappa+1)}{4(\kappa-1)} \begin{bmatrix} 0 & 0 & 0 \\ -(\kappa-3)\,H(\sigma^o_{22}) & (\kappa+1)\,H(\sigma^o_{22}) & 0 \\ 0 & 0 & \kappa-1 \end{bmatrix}, \tag{6.3.8b}$$

where $\sigma^o_{22} = \{\mu(\kappa+1)\,\varepsilon^o_{22} - \mu(\kappa-3)\,\varepsilon^o_{11}\}/(\kappa-1)$. For the *antiplane shear* problem,

$$[J_{ab}] = f\,\pi \begin{bmatrix} 0 & 0 \\ 0 & 1 \end{bmatrix}. \tag{6.3.9b}$$

Although **J** for the plane problem is not symmetric, i.e., $J_{ijkl} \neq J_{klij}$, the product $\mathbf{C} : \mathbf{J}$ is symmetric:

$$[C_{ab}][J_{bc}] = f\,\frac{\pi(\kappa+1)\mu}{4(\kappa-1)^2}$$

$$\times\begin{bmatrix} (\kappa-3)^2\,H(\sigma^o_{22}) & -(\kappa-3)(\kappa+1)\,H(\sigma^o_{22}) & 0 \\ -(\kappa-3)(\kappa+1)\,H(\sigma^o_{22}) & (\kappa+1)^2\,H(\sigma^o_{22}) & 0 \\ 0 & 0 & (\kappa-1)^2 \end{bmatrix}. \tag{6.3.8c}$$

Indeed, if **H** and **C** are symmetric, then $(\mathbf{C}:\mathbf{J})^T = (\mathbf{C}:\mathbf{H}:\mathbf{C})^T = \mathbf{C}:\mathbf{H}:\mathbf{C} = \mathbf{C}:\mathbf{J}$. From (6.3.8) and (6.3.9), various components of $[\bar{C}_{ab}]$ for plane and anti-plane problems are obtained. For example, the overall shear modulus of plane problems, $\bar{\mu}_{12}$, is

$$\frac{\bar{\mu}_{12}}{\mu} = 1 - f\,\frac{\pi(\kappa+1)}{4}; \tag{6.3.10}$$

the overall shear modulus of antiplane problems, $\bar{\mu}_{23}$, is

$$\frac{\bar{\mu}_{23}}{\mu} = 1 - f\,\pi; \tag{6.3.11}$$

and $\bar{\mu}_{13} = \mu$. As in the case of microcavities, Subsection 5.1, $\bar{\mu}_{12}$ given by (6.3.6) and (6.3.10), and $\bar{\mu}_{23}$ given by (6.3.7) and (6.3.11) agree only to the first order in the crack density parameter f.

6.3.3. Effective Moduli: Dilute Distribution of Aligned Frictional Microcracks

The results of the preceding subsection show that the presence of frictionless microcracks aligned in the x_1-direction renders the elastic response of the RVE both anisotropic and history-dependent. The history dependence in this case refers to the response to uniaxial loading in the x_2-direction, where, in tension, the value of the Young modulus is reduced due to the presence of cracks, compared with the Young modulus in pure compression. Furthermore, if uniaxial tension in the x_2-direction is applied first, then the corresponding Young modulus for the superimposed incremental tension or compression in the x_2-direction will have a different value than in the case when uniaxial compression in the x_2-direction is applied first, and then incremental loading or unloading is superimposed in the same direction. Because of the assumption of *frictionless* microcracks, the response to shearing is independent of the history, but of course, dependent on the direction of shearing.

The history dependence of the response of an RVE containing aligned microcracks becomes more pronounced when the cracks are *frictional*. To illustrate this, let the coefficient of friction be denoted by η, and consider a dilute distribution of microcracks, all aligned in the x_1-direction. Consider loadings of such an RVE, following different paths, all of which lead to the final overall macrostresses $\Sigma_{11} = \Sigma_{22} = -p^o$ and $\Sigma_{12} = \Sigma_{21} = \tau^o$ (p^o, $\tau^o > 0$). Furthermore, assume that

$$\tau^o < \eta \, p^o. \tag{6.3.12}$$

Load Path I: As the first load path, first apply a small hydrostatic tension, $\Sigma_{11} = \Sigma_{22} = \sigma > 0$, so that all the cracks are open. Then apply the shear stress, τ^o, which produces the shear strain, $\bar{\varepsilon}_{12}$, given by

$$\bar{\varepsilon}_{12} = \bar{\varepsilon}_{21} = \frac{1}{2\bar{\mu}_{12}} \tau^o = \frac{1}{2\mu} \{1 + f\pi(1-\nu)\} \tau^o, \tag{6.3.13a}$$

where (6.3.5b) is used, and plane strain is assumed; for plane stress, a similar result is obtained when the corresponding value of κ is substituted into (6.3.6), and the result is inserted in $\bar{\varepsilon}_{12} = \tau^o/2\bar{\mu}_{12}$.

Apply a new uniform macropressure, $-(p^o + \sigma)$, so that the final macrostress state is given by $\Sigma_{11} = \Sigma_{22} = -p^o$ and $\Sigma_{12} = \Sigma_{21} = \tau^o$. The corresponding shear stress-shear strain relation is shown in Figure 6.3.5, and consists of the straight line OA_I.

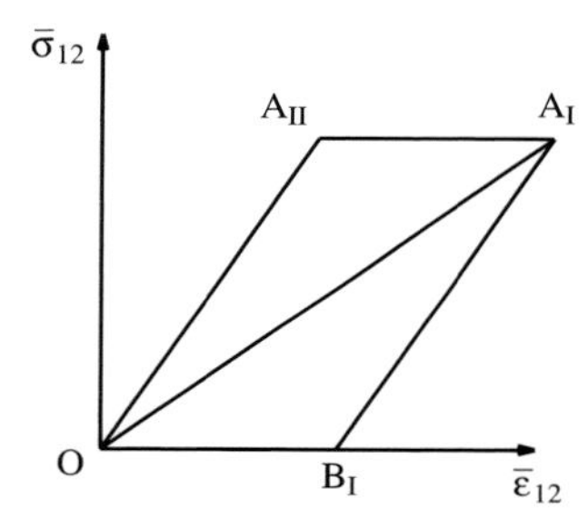

Figure 6.3.5

Shear stress-shear strain relation

Now consider unloading by removing the macroshear stress τ^o. All the cracks remain locked under the action of macrocompression $\Sigma_{22} = -p^o$. The stress-strain path then follows the straight line A_IB_I whose slope is given by 2μ, rather than $2\bar{\mu}_{12}$. At point B_I, a residual strain of

$$\bar{\varepsilon}^c_{12} = f\,\frac{\pi(1-\nu)}{2\mu}\,\tau^o \tag{6.3.13b}$$

is locked in the RVE.

If the applied macropressure is removed, then the RVE will undergo an overall macroshear strain of the magnitude given by (6.3.13b). The area within the triangle OA_IB_I is half of the total elastic energy per unit volume, which is lost in this cycle by frictional sliding.

Load Path II: Consider an alternative load path which starts from the origin in the stress-strain coordinates of Figure 6.3.5, but first, apply the uniform macrocompression of magnitude p^o, so that all cracks are locked in their initial state of zero COD. Then, the overall macroshear stress of magnitude τ^o is applied. Since τ^o is restricted in magnitude by (6.3.12), all microcracks remain locked. The overall response of the RVE to this shear stress of magnitude τ^o would be an elastic strain,

$$\bar{\varepsilon}_{12} = \frac{1}{2\mu}\,\tau^o. \tag{6.3.13c}$$

This corresponds to the straight line OA_{II} in Figure 6.3.5. If the uniform macropressure is removed while the same uniform macroshear stress is maintained, the overall macroshear strain will increase to the value given by (6.3.13b), and the representative point in the stress-strain diagram will move from point A_{II} to point A_I. If the macroshear stress is then removed, the path A_IO will be followed.

The above example has been used by Nemat-Nasser (1987) to illustrate the complexity of the response of elastic materials containing *frictional* microcracks. This complexity increases manyfold, when the microcracks in addition to opening and closing, may also grow preferentially in response to the applied macrostresses; see Nemat-Nasser and Horii (1982), Horii and Nemat-Nasser (1985a, 1986), and Ashby and Hallam (1986) for several illustrations, including model experiments.

6.4. EFFECTIVE MODULI OF AN ELASTIC SOLID CONTAINING RANDOMLY DISTRIBUTED SLIT MICROCRACKS

6.4.1. Effective Moduli: Random Dilute Distribution of Open Microcracks

Consider two-dimensional problems, where the unit normals of the microcracks lie in the x_1,x_2-plane. Furthermore, assume that the distribution of these cracks is such that their interaction can be ignored in the estimate of the overall elastic moduli. For simplicity, consider the additional assumption that all the cracks are open and remain so for the considered class of loading. As was illustrated in Subsection 6.3.3, the effect of crack closure due to the applied loading can be quite complex, as illustrated in Subsection 6.4.5; see Horii and Nemat-Nasser (1983).

To obtain the overall moduli for a random dilute distribution of microcracks, consider a typical microcrack, Ω_α, lying in the x_1^α-direction which makes the angle θ_α with the fixed coordinate x_1-axis. The unit normal of this crack is in the positive direction of the local x_2^α-axis. Hence, the x_1^α,x_2^α-axes attached to the αth crack are obtained by rotating the x_1,x_2-coordinate system by the angle θ_α about the center of the crack, O_α, in a counterclockwise direction; see Figure 6.4.1. Suppose that the macrostress $\boldsymbol{\Sigma} = \boldsymbol{\sigma}^o$ is prescribed. According to linearity (see Subsection 3.2), $\mathbf{H}^\alpha$ can be defined such that the average strain due to the presence of a typical crack Ω_α is given by $\mathbf{H}^\alpha : \boldsymbol{\sigma}^o$. Denoting the components of $\mathbf{H}^\alpha$ in the x_1^α,x_2^α-coordinate system (the α-coordinates) by $\hat{H}^\alpha_{ijkl}$, it follows that

$$\hat{\varepsilon}^\alpha_{ij} \equiv \frac{1}{a_\alpha^2}\int_{-a_\alpha}^{a_\alpha} \frac{1}{2}(\hat{n}_i[\hat{u}_j] + \hat{n}_j[\hat{u}_i])\,dx_1^\alpha = \hat{H}^\alpha_{ijkl}\,\hat{\sigma}^o_{kl}, \tag{6.4.1a}$$

where all quantities are expressed in terms of components in the α-coordinates;

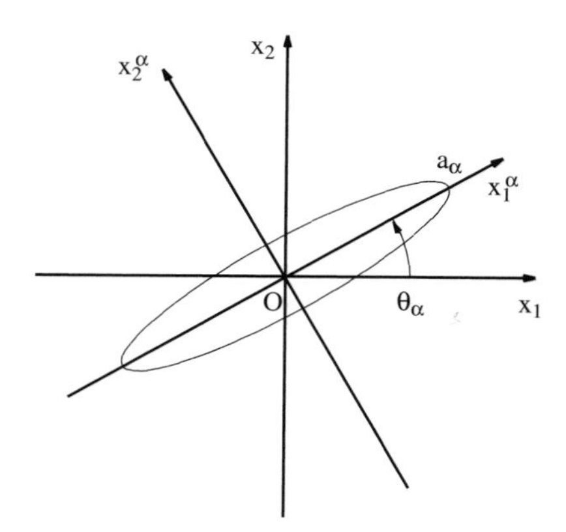

Figure 6.4.1

αth crack and local α-coordinate system

these components are designated by superimposed carets, e.g., $\hat{\sigma}^o_{ij}$ are the components of the applied uniform macrostress $\mathbf{\Sigma} = \boldsymbol{\sigma}^o$ in the α-coordinates. From (6.3.1a) it follows that

$$\hat{H}^\alpha_{2222} = \frac{2\pi}{E'}, \qquad \hat{H}^\alpha_{1212} = \hat{H}^\alpha_{2121} = \hat{H}^\alpha_{1221} = \hat{H}^\alpha_{2112} = \frac{\pi}{2E'}, \tag{6.4.1b,c}$$

with all other components of $\mathbf{H}^\alpha$ being zero, where E′ is defined by (6.3.1b).

To obtain the tensor **H** for the dilute *random* distribution of microcracks of various sizes, transform the components in the α-coordinates into the x_1,x_2-coordinates, and integrate the results over all values of the crack orientation angle θ_α, assuming a *uniform* distribution. Let $\mathbf{Q}^\alpha$ be the corresponding orthonormal tensor, transferring components in the α-coordinates into those in the x_1,x_2-coordinates. Denoting the unit vector in the x_i^α-direction by $\mathbf{e}_i^\alpha$, define $\mathbf{Q}^\alpha$ by

$$\mathbf{Q}^\alpha \equiv \mathbf{e}_i^\alpha \otimes \mathbf{e}_i, \tag{6.4.2a}$$

or

$$\mathbf{e}_i^\alpha \equiv \mathbf{Q}^\alpha \cdot \mathbf{e}_i \equiv Q^\alpha_{ji}\,\mathbf{e}_j, \qquad (i = 1, 2), \tag{6.4.2b}$$

where the components of $\mathbf{Q}^\alpha$ in the x_1,x_2-coordinates, denoted by Q^α_{ij}, are,

$$Q^\alpha_{11} = \cos\theta_\alpha, \qquad Q^\alpha_{21} = \sin\theta_\alpha,$$
$$Q^\alpha_{12} = -\sin\theta_\alpha, \qquad Q^\alpha_{22} = \cos\theta_\alpha, \tag{6.4.2c~f}$$

and where $Q^\alpha_{ij} = \mathbf{Q}^\alpha : (\mathbf{e}_i \otimes \mathbf{e}_j) = \mathbf{e}_i \cdot \mathbf{e}_j^\alpha$. From (6.4.2a), the following identity is obtained:

$$Q^\alpha_{ij}\, Q^\alpha_{kj} = Q^\alpha_{ji}\, Q^\alpha_{jk} = \delta_{ik} \qquad (\alpha \text{ not summed}). \tag{6.4.2g}$$

In terms of Q^α_{ij} and the overall crack density parameter f defined by (6.3.3b), the **H**-tensor is finally expressed as

$$\mathbf{H} = H_{ijkl}\, \mathbf{e}_i \otimes \mathbf{e}_j \otimes \mathbf{e}_k \otimes \mathbf{e}_l$$

$$= \frac{f}{2\pi} \int_0^{2\pi} \hat{H}^{\alpha}_{pqrs}\, \mathbf{e}^{\alpha}_p \otimes \mathbf{e}^{\alpha}_q \otimes \mathbf{e}^{\alpha}_r \otimes \mathbf{e}^{\alpha}_s\, d\theta_\alpha$$

$$= \left\{ \frac{f}{2\pi} \int_0^{2\pi} Q^{\alpha}_{ip}\, Q^{\alpha}_{jq}\, Q^{\alpha}_{kr}\, Q^{\alpha}_{ls}\, \hat{H}^{\alpha}_{pqrs}\, d\theta_\alpha \right\} \mathbf{e}_i \otimes \mathbf{e}_j \otimes \mathbf{e}_k \otimes \mathbf{e}_l, \tag{6.4.3a}$$

where the components of $\mathbf{Q}^\alpha$ depend on θ_α. Note that, although an individual microcrack Ω_α has its own $\mathbf{H}^\alpha$-tensor, the components of all $\mathbf{H}^\alpha$'s in their own α-coordinates are given by (6.4.1b,c).

Since the crack distribution is random, $\mathbf{H}$ is an isotropic tensor of the form (in two-dimensional space)

$$H_{ijkl} = h_1\, \delta_{ij}\, \delta_{kl} + h_2\, \frac{1}{2}(\delta_{ik}\, \delta_{jl} + \delta_{il}\, \delta_{jk}), \tag{6.4.3b}$$

where h_1 and h_2 are obtained from (6.4.3a) with the aid of (6.4.2g), as follows:

$$H_{iijj} = 4h_1 + 2h_2 = f\, \hat{H}^{\alpha}_{ppqq},$$

$$H_{ijij} = 2h_1 + 3h_2 = f\, \hat{H}^{\alpha}_{pqpq}. \tag{6.4.4a,b}$$

Substitute (6.4.1b,c) into (6.4.4a,b), to arrive at

$$4h_1 + 2h_2 = f\, \hat{H}^{\alpha}_{2222} = f\, \frac{2\pi}{E'}$$

$$2h_1 + 3h_2 = f\, (\hat{H}^{\alpha}_{2222} + 2\hat{H}^{\alpha}_{1212}) = f\, \frac{3\pi}{E'}. \tag{6.4.4c,d}$$

Therefore, h_1 and h_2 are

$$h_1 = 0, \qquad h_2 = f\, \frac{\pi}{E'}. \tag{6.4.4e,f}$$

In matrix form, $\mathbf{H}$ is given by

$$[H_{ab}] = f\, \frac{\pi}{E'} \begin{bmatrix} 1 & 0 & 0 \\ 0 & 1 & 0 \\ 0 & 0 & 2 \end{bmatrix}. \tag{6.4.4g}$$

Since the overall compliance tensor is

$$\bar{\mathbf{D}} = \mathbf{D} + \mathbf{H}, \tag{6.4.5}$$

the overall shear modulus, $\bar{\mu}$, becomes

$$\frac{\bar{\mu}}{\mu} = \{1 + f\, \frac{\pi(\kappa+1)}{4}\}^{-1} = 1 - f\, \frac{\pi(\kappa+1)}{4} + O(f^2), \tag{6.4.6a}$$

and the overall Poisson ratio, $\bar{\nu}$, for plane strain and plane stress can be calculated from

$$\frac{\bar{\kappa}}{\kappa} = \{1 + f\, \frac{3\pi(\kappa+1)}{4\kappa}\}\{1 + f\, \frac{\pi(\kappa+1)}{4}\}^{-1}$$

$$= 1 - f\,\frac{\pi(\kappa+1)(\kappa-3)}{4\kappa} + O(f^2), \tag{6.4.6b}$$

where $\bar{\kappa} = 3 - 8\bar{\mu}(\bar{\nu}/\bar{E} + \nu^2/E)$ for plane strain and $\bar{\kappa} = (3-\bar{\nu})/(1+\bar{\nu})$ for plane stress. For plane stress, the overall Young modulus, $\bar{E}$, and Poisson ratio, $\bar{\nu}$, are

$$\frac{\bar{E}}{E} = (1 + f\,\pi)^{-1} = 1 - f\,\pi + O(f^2),$$

$$\frac{\bar{\nu}}{\nu} = (1 + f\,\pi)^{-1} = 1 - f\,\pi + O(f^2). \tag{6.4.6c,d}$$

Note that the Young modulus in the x_3-direction coincides with that of the matrix, since all microcracks are parallel to the x_3-axis. Note also that the overall inplane Young modulus for plane strain is given by

$$\frac{\bar{E}}{E} = \{1 + f\,(1-\nu^2)\pi\}^{-1}. \tag{6.4.6e}$$

Consider now the case when the macrostrain $\mathbf{E} = \boldsymbol{\varepsilon}^o$ is prescribed. The average strain $\bar{\boldsymbol{\varepsilon}}^\alpha$ contributed by the αth microcrack Ω_α then is

$$\bar{\boldsymbol{\varepsilon}}^\alpha = \mathbf{J}^\alpha : \boldsymbol{\varepsilon}^o, \tag{6.4.7a}$$

where, as shown in Subsection 5.1.2, the tensor $\mathbf{J}^\alpha$ is

$$\mathbf{J}^\alpha = \mathbf{H}^\alpha : \mathbf{C}. \tag{6.4.7b}$$

Therefore, if the dilute distribution of microcracks is *uniform*, from $\mathbf{H}$ given by (6.4.4), $\mathbf{J}$ is obtained,

$$J_{ijkl} = H_{ijmn}\,C_{mnkl}$$

$$= f\,\frac{\pi(\kappa+1)}{8}\,\{-\,\frac{\kappa-3}{\kappa-1}\,\delta_{ij}\,\delta_{kl} + (\delta_{ik}\,\delta_{jl} + \delta_{il}\,\delta_{jk})\}, \tag{6.4.8a}$$

or in matrix form,

$$[J_{ab}] = f\,\frac{\pi(\kappa+1)}{8(\kappa-1)} \begin{bmatrix} \kappa+1 & -(\kappa-3) & 0 \\ -(\kappa-3) & \kappa+1 & 0 \\ 0 & 0 & 2(\kappa-1) \end{bmatrix}. \tag{6.4.8b}$$

Then, the overall elasticity tensor becomes

$$\bar{\mathbf{C}} = \mathbf{C} - \mathbf{C} : \mathbf{J}. \tag{6.4.9}$$

In particular, the overall shear modulus, $\bar{\mu}$, and non-dimensionalized parameter, $\bar{\kappa}$, are obtained from

$$\frac{\bar{\mu}}{\mu} = 1 - f\,\frac{\pi(\kappa+1)}{4},$$

$$\frac{\bar{\kappa}}{\kappa} = \{1 - f\,\frac{\pi(\kappa+1)(\kappa^2-2\kappa+3)}{4\kappa(\kappa-1)}\}\{1 - f\,\frac{\pi(\kappa+1)}{2(\kappa-1)}\}^{-1}$$

$$= 1 - f\,\frac{\pi(\kappa+1)(\kappa-3)}{4\kappa} + O(f^2), \tag{6.4.10a,b}$$

where $\bar{\kappa} = 3 - 8\bar{\mu}(\bar{\nu}/\bar{E} + \nu^2/E)$ for plane strain and $\bar{\kappa} = (3-\bar{\nu})/(1+\bar{\nu})$ for plane stress. For plane stress, the overall Young modulus, $\bar{E}$, and Poisson ratio, $\bar{\nu}$, are

$$\frac{\bar{E}}{E} = (1 - f\frac{\pi}{1+\nu})\,(1 - f\frac{\pi}{1-\nu})\,\{1 - f\frac{\pi(1+\nu^2)}{1-\nu^2}\}^{-1} = 1 - f\pi + O(f^2)$$

$$\frac{\bar{\nu}}{\nu} = (1 - f\frac{2\pi}{1-\nu^2})\,\{1 - f\frac{\pi(1+\nu^2)}{1-\nu^2}\}^{-1} = 1 - f\pi + O(f^2). \qquad (6.4.10c,d)$$

As explained in Subsection 5.1.2, $\bar{\mathbf{D}}$ and $\bar{\mathbf{C}}$, given by (6.4.5) and (6.4.9), are each other's inverse only to the first order in the overall crack density parameter f, i.e.,

$$\bar{\mathbf{C}} : \bar{\mathbf{D}} = \mathbf{1}^{(4s)} + O(f^2), \quad \text{and} \quad \bar{\mathbf{D}} : \bar{\mathbf{C}} = \mathbf{1}^{(4s)} + O(f^2). \qquad (6.4.11a,b)$$

6.4.2. Effective Moduli: Self-Consistent Estimate

Suppose the distribution of microcracks in the x_1,x_2-plane is random, and it is desirable to include their interaction to a certain extent. As in Subsection 5.1.3, the self-consistent scheme may be applied to estimate the overall compliance $\bar{\mathbf{D}}$ (or the elasticity tensor $\bar{\mathbf{C}}$) of the RVE. The resulting overall compliance tensor then is isotropic, due to the random distribution of microcracks. In terms of the unknown overall shear modulus, $\bar{\mu}$, and non-dimensionalized parameter, $\bar{\kappa}$, $\bar{\mathbf{H}}$ is expressed as

$$\bar{H}_{ijkl} = f\,\frac{\pi}{\bar{E}'}\,\frac{1}{2}(\delta_{ik}\,\delta_{jl} + \delta_{il}\,\delta_{jk}) = f\,\frac{\pi(\bar{\kappa}+1)}{16\bar{\mu}}\,(\delta_{ik}\,\delta_{jl} + \delta_{il}\,\delta_{jk}), \qquad (6.4.12)$$

where $\bar{\kappa} = 3 - 8\bar{\mu}(\bar{\nu}/\bar{E} + \nu^2/E)$ for plane strain and $\bar{\kappa} = (3-\bar{\nu})/(1+\bar{\nu})$ for plane stress. Then, the overall compliance $\bar{\mathbf{D}}$ satisfies

$$\bar{\mathbf{D}} = \mathbf{D} + \bar{\mathbf{H}}, \qquad (6.4.13a)$$

or in matrix form

$$\frac{1}{8\bar{\mu}}\begin{bmatrix} \bar{\kappa}+1 & \bar{\kappa}-3 & 0 \\ \bar{\kappa}-3 & \bar{\kappa}+1 & 0 \\ 0 & 0 & 8 \end{bmatrix} = \frac{1}{8\mu}\begin{bmatrix} \kappa+1 & \kappa-3 & 0 \\ \kappa-3 & \kappa+1 & 0 \\ 0 & 0 & 8 \end{bmatrix} + f\,\frac{\pi(\bar{\kappa}+1)}{8\bar{\mu}}\begin{bmatrix} 1 & 0 & 0 \\ 0 & 1 & 0 \\ 0 & 0 & 2 \end{bmatrix}. \qquad (6.4.13b)$$

Since $\mathbf{D}$, $\bar{\mathbf{D}}$, and $\bar{\mathbf{H}}$ are isotropic tensors, there are only two linearly independent relations among the three equations in (6.4.13b). From these, a system of two equations is obtained for the two unknowns $\bar{\mu}$ and $\bar{\kappa}$,

$$\frac{\bar{\kappa}+1}{8\bar{\mu}} = \frac{\kappa+1}{8\mu} + f\,\frac{\pi(\bar{\kappa}+1)}{8\bar{\mu}}, \qquad \frac{1}{\bar{\mu}} = \frac{1}{\mu} + f\,\frac{\pi(\bar{\kappa}+1)}{4\bar{\mu}}. \qquad (6.4.13c,d)$$

Then, $\bar{\mu}$ and $\bar{\kappa}$ are given by,

$$\frac{\bar{\mu}}{\mu} = (1 - f\pi)\,\{1 + f\,\frac{\pi(\kappa-3)}{4}\}^{-1} = 1 - f\,\frac{\pi(\kappa+1)}{4} + O(f^2),$$

$$\frac{\bar{\kappa}}{\kappa} = \{1 - f\,\frac{\pi(\kappa-3)}{4\kappa}\}\,\{1 + f\,\frac{\pi(\kappa-3)}{4}\}^{-1}$$

$$= 1 - f\,\frac{\pi(\kappa+1)(\kappa-3)}{4\kappa} + O(f^2). \tag{6.4.14a,b}$$

In particular, for plane stress, the effective shear modulus, $\bar{\mu}$, Young modulus, $\bar{E}$, and Poisson ratio, $\bar{\nu}$, become,

$$\frac{\bar{\mu}}{\mu} = (1 - f\pi)\,(1 - f\,\frac{\pi\nu}{1+\nu})^{-1} = 1 - f\,\frac{\pi}{1-\nu} + O(f^2),$$

$$\frac{\bar{E}}{E} = 1 - f\pi,$$

$$\frac{\bar{\nu}}{\nu} = 1 - f\pi. \tag{6.4.14c~e}$$

Figure 6.4.2 shows the variation of the overall shear and Young moduli with respect to the crack density parameter f, for dilute distributions (6.4.6,10) and self-consistent estimates (6.4.14). The Poisson ratio ν is set equal to 1/3. It should be noted that in the self-consistent estimate, the overall elasticity tensor, $\bar{\mathbf{C}}$, which satisfies

$$\bar{\mathbf{C}} = \mathbf{C} - \mathbf{C} : \bar{\mathbf{J}} = \mathbf{C} - \mathbf{C} : \bar{\mathbf{H}} : \bar{\mathbf{C}}, \tag{6.4.15a}$$

is in fact the inverse of $\bar{\mathbf{D}}$ given by (6.4.13a), i.e.,

$$\bar{\mathbf{C}} = \bar{\mathbf{D}}^{-1}. \tag{6.4.15b}$$

See Subsection 5.1.3 for a detailed derivation.

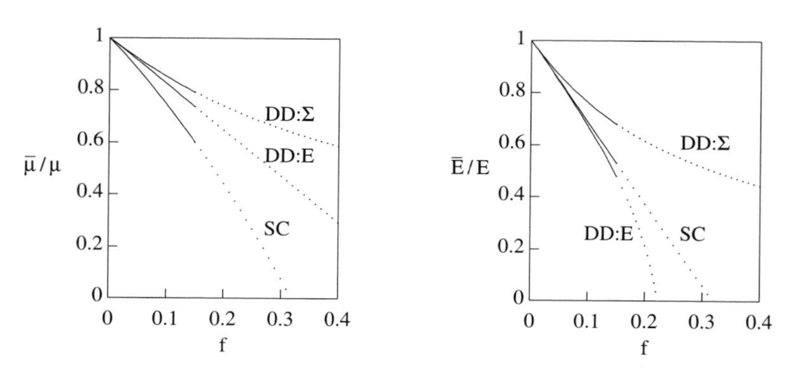

Figure 6.4.2

Normalized overall shear, $\bar{\mu}/\mu$, and Young's, $\bar{E}/E$, moduli for dilute random distribution of slit microcracks, all parallel to the x_3-axis; $\nu = 1/3$
SC ≡ self-consistent
DD:Σ ≡ dilute distribution with macrostress prescribed
DD:E ≡ dilute distribution with macrostrain prescribed

6.4.3. Effective Moduli in Antiplane Shear: Random Dilute Distribution of Frictionless Microcracks

Next, consider the antiplane shearing of the RVE, where, again, the unit normals of the microcracks lie in the x_1,x_2-plane. The lengths and orientations of the microcracks in the x_1,x_2-plane are random such that the overall shear moduli for the x_1,x_3-plane and the x_2,x_3-plane are the same, i.e., $\bar{\mu}_{13} = \bar{\mu}_{23} \equiv \bar{\mu}_3$. The procedure outlined in Subsections 6.4.1 and 6.4.2 for two-dimensional inplane problems will be used to estimate the overall shear modulus $\bar{\mu}_3$ for a dilute distribution of frictionless microcracks, and then the self-consistent estimate will be worked out.

Consider a typical microcrack Ω_α and the corresponding x_1^α,x_2^α-coordinate system (the α-coordinates) as defined in Subsection 6.4.1; see Figure 6.4.1. The x_3^α-axis coincides with the x_3-axis. Suppose that the macrostress $\mathbf{\Sigma} = \boldsymbol{\sigma}^o$ is prescribed such that its nonzero components in the x_1,x_2-coordinates are σ_{13}^o $(= \sigma_{31}^o)$ and σ_{23}^o $(= \sigma_{32}^o)$. Then, the contribution to the average strain by a typical crack Ω_α is given by (6.4.1a), i.e.,

$$\bar{\boldsymbol{\varepsilon}}^\alpha = \frac{1}{a_\alpha^2} \int_{-a_\alpha}^{a_\alpha} \frac{1}{2} \{\mathbf{n} \otimes [\mathbf{u}] + [\mathbf{u}] \otimes \mathbf{n}\} \, dx_1^\alpha = \mathbf{H}^\alpha : \boldsymbol{\sigma}^o. \tag{6.4.16a}$$

From (6.3.1c), the components of $\mathbf{H}^\alpha$ in the α-coordinates, $\hat{H}_{ijkl}^\alpha$, are easily determined,

$$\hat{H}_{2323}^\alpha = \hat{H}_{3223}^\alpha = \hat{H}_{2332}^\alpha = \hat{H}_{3232}^\alpha = \frac{\pi}{4\mu}, \tag{6.4.16b}$$

with all other components being zero. Recall that the tensor $\mathbf{H}^\alpha$ depends on the orientation of the microcrack Ω_α, but its components in the corresponding α-coordinates are constant and independent of the crack length; they are given by (6.4.16b). Then, in transforming from the α-coordinates to the x_1,x_2-coordinates through (6.4.2), the $\mathbf{H}$-tensor for a uniform (random) distribution of cracks is given by (6.4.3a), i.e.,

$$\mathbf{H} = H_{ijkl} \, \mathbf{e}_i \otimes \mathbf{e}_j \otimes \mathbf{e}_k \otimes \mathbf{e}_l$$

$$= \{ \frac{f}{2\pi} \int_0^{2\pi} Q_{ip}^\alpha \, Q_{jq}^\alpha \, Q_{kr}^\alpha \, Q_{ls}^\alpha \, \hat{H}_{pqrs}^\alpha \, d\theta_\alpha \} \, \mathbf{e}_i \otimes \mathbf{e}_j \otimes \mathbf{e}_k \otimes \mathbf{e}_l, \tag{6.4.16c}$$

where θ_α is the angle of the x_1^α-axis with respect to the x_1-axis.

Since the crack distribution is random, $\mathbf{H}$ is isotropic. For the antiplane shearing considered here, the nonzero components of $\mathbf{H}$ are

$$H_{1313} = H_{3113} = H_{1331} = H_{3131} = H_{2323} = H_{3223} = H_{2332} = H_{3232}. \tag{6.4.17a}$$

Taking advantage of (6.4.3b) and using (6.4.16b), now obtain

$$H_{ijij} = 4H_{1313} = f \, \hat{H}_{pqpq}^\alpha = 2f \, \hat{H}_{2323}^\alpha = f \, \frac{\pi}{2\mu} \tag{6.4.17b}$$

and hence

$$H_{1313} = H_{2323} = \ldots = f \, \frac{\pi}{8\mu}. \tag{6.4.17c}$$

In matrix form, the **H**-tensor for this antiplane shearing can be expressed by

$$[H_{ab}] = f\,\frac{\pi}{2\mu}\begin{bmatrix}1 & 0\\ 0 & 1\end{bmatrix}. \tag{6.4.17d}$$

Then, the corresponding overall compliance matrix takes on the following form:

$$\frac{1}{\bar{\mu}_3}\begin{bmatrix}1 & 0\\ 0 & 1\end{bmatrix} = \frac{1}{\mu}\begin{bmatrix}1 & 0\\ 0 & 1\end{bmatrix} + f\,\frac{\pi}{2\mu}\begin{bmatrix}1 & 0\\ 0 & 1\end{bmatrix}. \tag{6.4.17e}$$

Therefore, the overall shear modulus, $\bar{\mu}_3 = \bar{\mu}_{13} = \bar{\mu}_{23}$, is obtained,

$$\frac{\bar{\mu}_3}{\mu} = (1 + f\,\frac{\pi}{2})^{-1} = 1 - f\,\frac{\pi}{2} + O(f^2). \tag{6.4.18}$$

Next, suppose that the macrostrain $\mathbf{E} = \boldsymbol{\varepsilon}^o$ is prescribed such that its nonzero components in the x_1,x_2-coordinates are ε^o_{13} $(= \varepsilon^o_{31})$ and ε^o_{23} $(= \varepsilon^o_{32})$. The average strain due to the microcrack Ω_α is given by (6.4.7a), i.e.,

$$\bar{\boldsymbol{\varepsilon}}^\alpha = \mathbf{J}^\alpha : \boldsymbol{\varepsilon}^o, \tag{6.4.19a}$$

where, in terms of $\mathbf{H}^\alpha$ defined by (6.4.1), the $\mathbf{J}^\alpha$-tensor for the antiplane shear becomes

$$\mathbf{J}^\alpha = \mathbf{H}^\alpha : \mathbf{C} = 2\mu\mathbf{H}^\alpha. \tag{6.4.19b}$$

Therefore, since the dilute distribution of microcracks is random, in terms of **H** given by (6.4.17c), the **J**-tensor is equal to $2\mu\mathbf{H}$, and its nonzero components in the x_1,x_2-coordinates are

$$J_{1313} = J_{2323} = \ldots = f\,\frac{\pi}{4}, \tag{6.4.20a}$$

or, in matrix form,

$$[J_{ab}] = f\,\frac{\pi}{2}\begin{bmatrix}1 & 0\\ 0 & 1\end{bmatrix}. \tag{6.4.20b}$$

The relevant components of the overall elasticity tensor, $\bar{\mathbf{C}}$, are now given by

$$\bar{\mu}_3\begin{bmatrix}1 & 0\\ 0 & 1\end{bmatrix} = \mu\begin{bmatrix}1 & 0\\ 0 & 1\end{bmatrix} - f\,\frac{\pi\mu}{2}\begin{bmatrix}1 & 0\\ 0 & 1\end{bmatrix}. \tag{6.4.20c}$$

Therefore, the overall shear modulus, $\bar{\mu}_3$ $(= \bar{\mu}_{13} = \bar{\mu}_{23})$, becomes

$$\frac{\bar{\mu}_3}{\mu} = 1 - f\,\frac{\pi}{2}. \tag{6.4.21}$$

The overall shear modulus $\bar{\mu}_3$, given by (6.4.18), and that given by (6.4.21) agree with each other only up to the first order in the overall crack density parameter f.

Finally, consider the case when the distribution of microcracks in the x_1,x_2-plane is random (and still dilute), and their interaction effects are to be included through the self-consistent approach. The resulting overall compliance tensor $\bar{\mathbf{D}}$ is isotropic in the x_1,x_2-plane. The overall shear modulus $\bar{\mu}_3$ ($= \bar{\mu}_{13} = \bar{\mu}_{23}$) will now be estimated. In matrix form, the $\bar{\mathbf{H}}$-tensor is expressed in terms of $\bar{\mu}_3$ by

$$[\bar{H}_{ab}] = f\,\frac{\pi}{2\bar{\mu}_3}\begin{bmatrix} 1 & 0 \\ 0 & 1 \end{bmatrix}, \tag{6.4.22a}$$

and the relevant components of the overall compliance tensor, $\bar{\mathbf{D}}$, become

$$\frac{1}{\bar{\mu}_3}\begin{bmatrix} 1 & 0 \\ 0 & 1 \end{bmatrix} = \frac{1}{\mu}\begin{bmatrix} 1 & 0 \\ 0 & 1 \end{bmatrix} - f\,\frac{\pi}{2\bar{\mu}_3}\begin{bmatrix} 1 & 0 \\ 0 & 1 \end{bmatrix}. \tag{6.4.22b}$$

Therefore, the overall shear modulus, $\bar{\mu}_3$, is

$$\frac{\bar{\mu}_3}{\mu} = 1 - f\,\frac{\pi}{2}. \tag{6.4.23}$$

As is seen from (6.4.21) and (6.4.23), the self-consistent estimate of the overall shear modulus coincides with the estimate obtained from a dilute distribution of microcracks when the macrostrain is regarded prescribed; see Figure 6.4.3. It is emphasized that, like the dilute distribution assumption, the self-consistent method is valid only for small values of the crack density f. However, the dilute distribution assumption leads to results in violation of Theorem I of Subsection 2.5, whereas the self-consistent method does not. Indeed, unlike the dilute distribution assumption, the self-consistent approach yields the same overall moduli, whether the macrostress or the macrostrain is regarded prescribed. Since the relevant components of the $\bar{\mathbf{J}}$-tensor are

$$[\bar{J}_{ac}] = [\bar{H}_{ab}][\bar{C}_{bc}] = f\,\frac{\pi}{2}\begin{bmatrix} 1 & 0 \\ 0 & 1 \end{bmatrix} = [J_{ac}], \tag{6.4.24}$$

the corresponding components of the overall elasticity tensor, $\bar{\mathbf{C}}$ ($= \mathbf{C} - \mathbf{C} : \mathbf{J}$), coincide with those given by (6.4.20c).

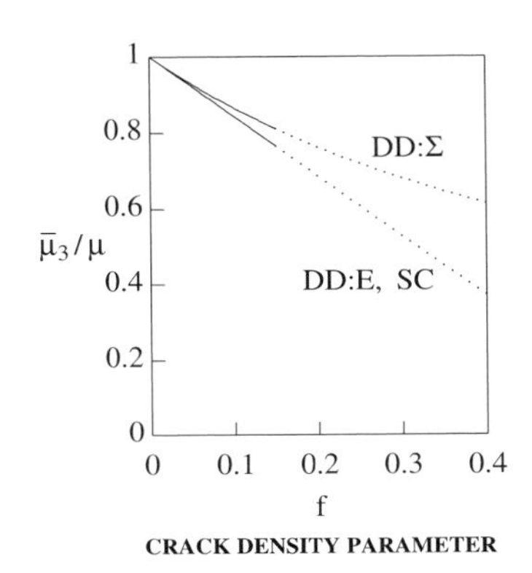

Figure 6.4.3

Normalized overall shear modulus, $\bar{\mu}_3/\mu$, for dilute distribution of slit cracks, all parallel to the x_3-axis
SC ≡ self-consistent
DD:Σ ≡ dilute distribution with macrostress prescribed
DD:E ≡ dilute distribution with macrostrain prescribed

6.4.4. Plane Stress, Plane Strain, and Three-Dimensional Overall Moduli

Since the slit cracks are parallel to the x_3-direction, having an otherwise random distribution, the overall response of the RVE is transversely isotropic, with the x_1,x_2-plane being the plane of isotropy. The formulation in terms of $\bar{\mu}$ and $\bar{\kappa}$, in Subsections 6.4.1 and 6.4.2, permits direct evaluation of the overall moduli without specific reference to whether plane stress or plane strain conditions are assumed; see also, Subsections 3.1.4, 5.1.1, and 8.3. For plane stress, $\bar{\kappa} = (3 - \bar{\nu})/(1 + \bar{\nu})$, and the results illustrated in Figure 6.4.2 and relations (6.4.14) are obtained. It may be instructive to compare these results with the corresponding estimates of the overall moduli obtained under plane *strain* conditions. This is done below for the self-consistent model.

Since, for plane strain, $\kappa = 3 - 4\nu$ and $\bar{\kappa} = 3 - 8\bar{\mu}(\frac{\bar{\nu}}{\bar{E}} + \frac{\nu^2}{E})$, from (6.4.13) it follows that

$$\frac{\bar{\mu}}{\mu} = (1 - f\pi)(1 - f\pi\nu)^{-1},$$

$$\frac{\bar{E}}{E} = (1 - f\pi)(1 - f\pi\nu^2)^{-1},$$

$$\frac{\bar{\nu}}{\nu} = (1 - f\pi)(1 - f\pi\nu^2)^{-1}. \qquad (6.4.25a\sim c)$$

The first two equations in (6.4.25) have been obtained by Laws and Brockenbrough (1987) using a different procedure. Results (6.4.13a~d) unify the necessary analyses, and lead to the interesting conclusion that[3] $\bar{E}/E = \bar{\nu}/\nu$ for both plane stress and plane strain.

It is reasonable to expect that the overall moduli in the plane of isotropy, i.e., in the x_1,x_2-plane, should not depend on whether plane strain or plane stress conditions are employed in order to calculate these overall parameters; see also Subsection 8.3.1. The difference between the corresponding expression in (6.4.14c~e) for plane stress, and in (6.4.25a~c) for plane strain, therefore, is a manifestation of the modeling approximation. This difference, however, is small enough to be neglected. Indeed, essentially the same estimates are obtained even if the effect of anisotropy associated with the x_3-direction is altogether neglected. In this case, $\bar{\kappa} \approx 3 - 4\bar{\nu}$ and it follows that (Horii and Nemat-Nasser, 1983)

$$\frac{\bar{E}}{E} = \frac{(1 - f\pi)(1 + \nu - f2\pi\nu)}{(1 + \nu)(1 - f\pi\nu)^2},$$

$$\frac{\bar{\nu}}{\nu} = (1 - f\pi)(1 - f\pi\nu)^{-1}. \qquad (6.4.26a,b)$$

In Figure 6.4.4, $\bar{E}/E$ and $\bar{\nu}/\nu$, given by (6.4.25b,c) and (6.4.26a,b), are plotted

---

[3]This equality holds for the self-consistent model and the dilute distribution model with overall *stresses* prescribed. It does *not* hold for the dilute distribution model with overall *strains* prescribed.

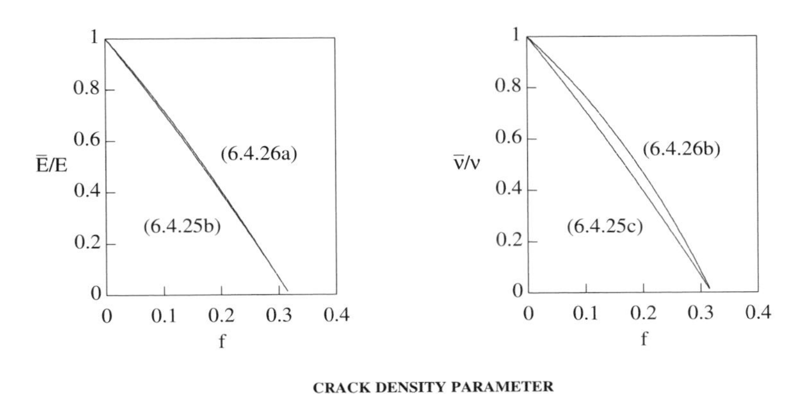

Figure 6.4.4

Comparison of normalized overall Young's modulus $\bar{E}/E$ and Poisson's ratio $\bar{\nu}/\nu$, for slit microcracks (parallel to the x_3-axis), obtained for plane strain, assuming transverse isotropy, (6.4.25b) and (6.4.25c), and isotropy, (6.4.26a) and (6.4.26b); $\nu = 1/3$

for plane strain. As is seen, the differences are insignificant, especially for the overall Young modulus.[4]

6.4.5. Effect of Friction and Load-Induced Anisotropy

When pre-existing cracks are closed and undergo frictional sliding under overall compressive loads, the overall response may become anisotropic even when the matrix material is isotropic and the distribution of pre-existing cracks is random. This anisotropy is load-dependent, often affected by the load history. Hence, in general, the corresponding overall shear and bulk moduli may depend on the history of the applied overall hydrostatic pressure and overall shear stresses.

Unlike an open crack, the deformation of a frictional crack involves a complex set of conditions. To simplify the analysis and yet illustrate the basic phenomenon, assume that: 1) the normal and shear stresses transmitted across the crack surfaces are constant; and 2) the crack sliding is governed by a simple friction law. Let the crack shown in Figure 6.4.1 be closed, and consider the following governing conditions:

---

[4]The corresponding results for $\bar{\mu}/\mu$ are identical, given by (6.4.25a).

$$
\begin{aligned}
&\text{if } |\overline{\sigma}_{21}| < -\eta\,\overline{\sigma}_{22}, && \text{then} \quad [u_1'] = [u_2'] = 0, \quad \hat{\sigma}^c_{22} = \overline{\sigma}_{22} \\
& && \qquad\qquad \text{and} \quad \hat{\sigma}^c_{21} = \overline{\sigma}_{21}, \\
&\text{if } |\overline{\sigma}_{21}| > -\eta\,\overline{\sigma}_{22}, && \text{then} \quad \hat{\sigma}^c_{21} = -\eta\, sgn(\overline{\sigma}_{21})\, \hat{\sigma}^c_{22} \\
& && \qquad\qquad \text{and} \quad [u_2'] = 0,
\end{aligned}
\tag{6.4.27}
$$

where $\hat{\sigma}^c_{22}$ and $\hat{\sigma}^c_{21}$ are the normal and shear stresses transmitted across the crack surfaces, η is the coefficient of sliding friction (regarded to be a positive constant), and $sgn(x)$ is 1, 0, or -1, depending on whether x is positive, zero, or negative.

Let the applied loads be as in Subsection 6.3.3, i.e., uniform pressure, $\sigma^o_{11} = \sigma^o_{22} = -p^o$, accompanied by pure shear, $\sigma^o_{12} = \sigma^o_{21} = \tau^o$ (p^o, $\tau^o > 0$). Consider the sliding displacement of closed cracks. Since the stress field is symmetric with respect to the lines $x_1 = \pm x_2$, consider the range $-\pi/4 < \theta_\alpha < 3\pi/4$. In the αth coordinate system, the macrostresses are

$$
\hat{\sigma}^o_{11} = -p^o + \tau^o \sin 2\theta_\alpha, \qquad \hat{\sigma}^o_{22} = -p^o - \tau^o \sin 2\theta_\alpha,
$$

$$
\hat{\sigma}^o_{12} = \tau^o \cos 2\theta_\alpha. \tag{6.4.28}
$$

Hence, slip conditions (6.4.27) determine the behavior of the αth crack, as follows: the αth crack is

$$
\left\{ \begin{array}{l} \text{open} \\ \text{closed with slip} \\ \text{closed without slip} \end{array} \right\} \quad \text{for} \quad \left\{ \begin{array}{l} \pi/2 > |\theta_\alpha - \pi/4| > \pi/4 - \theta_c \\ \pi/4 - \theta_c > |\theta_\alpha - \pi/4| > \pi/4 - \theta_s \\ \pi/4 - \theta_s > |\theta_\alpha - \pi/4| > 0 \end{array} \right\},
$$

where θ_c and θ_s are defined by

$$
\begin{aligned}
\sin 2\theta_c &= -\frac{p^o}{\tau^o}, && (|\theta_c| < \pi/4) \\
\cos 2\theta_s &= \eta\,(\frac{p^o}{\tau^o} + \sin 2\theta_s), && (|\theta_s| < \pi/4).
\end{aligned}
\tag{6.4.29a,b}
$$

Hence, all cracks are closed if $p^o/\tau^o > 1$.

In view of (6.3.1), the crack opening displacements for the closed crack with slip are given by

$$
[\hat{u}_1] = \left\{ \begin{array}{lll} 4\sqrt{a^2 - \hat{x}_1^2}\,(\hat{\sigma}^o_{12} + \eta\hat{\sigma}^o_{22})/E' & \text{for } \hat{\sigma}^o_{12} > 0 & (\theta_\alpha < \pi/4) \\ 4\sqrt{a^2 - \hat{x}_1^2}\,(\hat{\sigma}^o_{12} - \eta\hat{\sigma}^o_{22})/E' & \text{for } \hat{\sigma}^o_{12} < 0 & (\theta_\alpha > \pi/4), \end{array} \right.
$$

$$
[\hat{u}_2] = 0, \tag{6.4.30a,b}
$$

where $\hat{\sigma}^c_{22} = \hat{\sigma}^o_{22}$. For this closed crack with slip, the $\mathbf{H}^\alpha$-tensor defined by (6.4.1) is given by

$$
\hat{H}^\alpha_{1212} = \frac{\pi}{2E'},
$$

$$\hat{H}^{\alpha}_{1222} = \begin{cases} \frac{\pi}{E'}\,\eta & \text{for } \theta_\alpha < \pi/4 \\ -\frac{\pi}{E'}\,\eta & \text{for } \theta_\alpha > \pi/4. \end{cases} \tag{6.4.31a,b}$$

Therefore, the **H**-tensor defined by (6.4.3) is

$$H_{1111} = H_{2222} = f\,\frac{1}{E'}\,\{\,2[\theta]_{-\pi/4}^{\theta_c} + \frac{1}{8}\,[4\theta - \sin 4\theta - \eta\cos 4\theta]_{\theta_c}^{\theta_s}\,\},$$

$$H_{1212} = f\,\frac{1}{4E'}\,\{\,4[\theta]_{-\pi/4}^{\theta_c} + \frac{1}{2}\,[4\theta + \sin 4\theta + \eta\cos 4\theta]_{\theta_c}^{\theta_s}\,\},$$

$$H_{1122} = H_{2211} = f\,\frac{1}{E'}\,\{\,-\frac{1}{8}\,[4\theta - \sin 4\theta - \eta\cos 4\theta]_{\theta_c}^{\theta_s}\,\},$$

$$H_{1112} = H_{2212} = f\,\frac{\pi}{2E'}\,\{\,[\cos 2\theta]_{-\pi/4}^{\theta_c}\,\},$$

$$H_{1211} = H_{1222} = f\,\frac{1}{2E'}\,\{\,[\cos 2\theta]_{-\pi/4}^{\theta_c} + [\eta\sin 2\theta]_{\theta_c}^{\theta_s}\,\}, \tag{6.4.32a~e}$$

where, again, $f = a^2N$.

When the macrostrains are regarded as given, an analysis similar to that presented in the preceding subsections shows that the **J**-tensor is given by **H** : **C**. Unlike in the previous cases, however, neither **C** : **J** nor **J** is symmetric. Indeed, **C** : **J** is given by

$$(\mathbf{C}:\mathbf{J})_{1111} = (\mathbf{C}:\mathbf{J})_{2222}$$

$$= f\,\frac{(\kappa+1)^2 E'}{64(\kappa-1)^2}\,\{4(4+(\kappa-1)^2)\,[\theta]_{-\pi/4}^{\theta_c} + \frac{(\kappa-1)^2}{2}\,[4\theta - \sin 4\theta - \eta\cos 4\theta]_{\theta_c}^{\theta_s}\},$$

$$(\mathbf{C}:\mathbf{J})_{1212} = f\,\frac{(\kappa+1)^2 E'}{64(\kappa-1)}\,\{4[\theta]_{-\pi/4}^{\theta_c} + \frac{1}{2}\,[4\theta + \sin 4\theta + \eta\cos 4\theta]_{\theta_c}^{\theta_s}\},$$

$$(\mathbf{C}:\mathbf{J})_{1122} = (\mathbf{C}:\mathbf{J})_{2211} = f\,\frac{(\kappa+1)^2 E'}{64(\kappa-1)^2}\,\{4(4-(\kappa-1)^2)\,[\theta]_{-\pi/4}^{\theta_c} - \frac{(\kappa-1)^2}{2}\,[4\theta - \sin 4\theta - \eta\cos 4\theta]_{\theta_c}^{\theta_s}\},$$

$$(\mathbf{C}:\mathbf{J})_{1112} = (\mathbf{C}:\mathbf{J})_{2212} = f\,\frac{(\kappa+1)^2 E'}{16(\kappa-1)}\,[\cos 2\theta]_{-\pi/4}^{\theta_c},$$

$$(\mathbf{C}:\mathbf{J})_{1211} = (\mathbf{C}:\mathbf{J})_{1222} = f\,\frac{(\kappa+1)^2 E'}{16(\kappa-1)}\,\{[\cos 2\theta]_{-\pi/4}^{\theta_c} + [\eta \sin 2\theta]_{\theta_c}^{\theta_s}\}. \tag{6.4.33a~e}$$

To simplify the discussion, define the overall shear and bulk moduli by

$$\bar{\varepsilon}_{12} = \frac{1}{2\bar{\mu}}\,\sigma^o_{12}, \qquad \bar{\varepsilon}_{11} + \bar{\varepsilon}_{22} = \frac{1}{2\bar{K}}\,(\sigma^o_{11} + \sigma^o_{22}). \tag{6.4.34a,b}$$

Figure 6.4.5 shows the variation of $\bar{\mu}$ and $\bar{K}$ with respect to the load parameter, p^o/τ^o, for indicated values of the crack density, f. All cracks are closed if $p^o/\tau^o = 1$, and are open if $p^o/\tau^o << -1$. As p^o/τ^o decreases, some cracks open up, and hence $\bar{\mu}$ decreases with decreasing p^o/τ^o. The influence of the applied shear stress on the overall volumetric stiffness is revealed by considering the principal stresses, $-p^o \pm \tau^o$. As τ^o increases, one of the principal stresses becomes tensile, causing a large number of suitably oriented cracks to remain open.

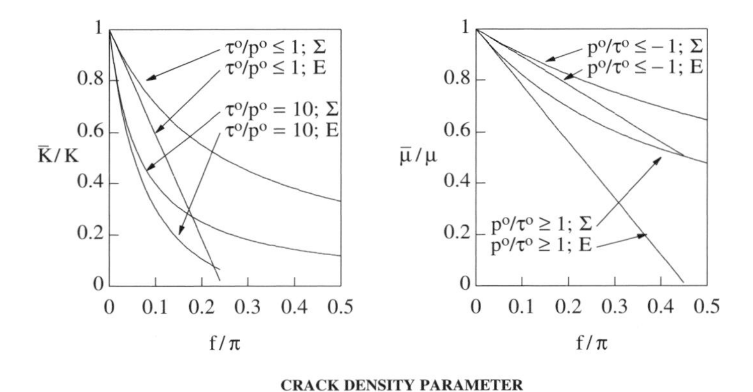

Figure 6.4.5

Normalized overall bulk and shear moduli for dilute distribution of closed cracks; Σ ≡ macrostress prescribed and E ≡ macrostrain prescribed

Next, consider the self-consistent method which accounts for crack interaction in a certain manner. Let the αth crack be embedded in a two-dimensionally anisotropic solid, with the yet-unknown overall compliance tensor, $\bar{\mathbf{D}}$. For an open crack under the farfield stress $\boldsymbol{\sigma}^o$, the COD's are

$$[\hat{u}_1] = 2\sqrt{a^2 - \hat{x}_1^2}\,\hat{\bar{D}}_{1111}\,\{(\alpha_1\beta_2 + \alpha_2\beta_1)\,\hat{\sigma}^o_{22} + (\beta_1 + \beta_2)\,\hat{\sigma}^o_{12}\},$$

$$[\hat{u}_2] = 2\sqrt{a^2 - \hat{x}_1^2}\,\hat{\bar{D}}_{2222}\,\frac{1}{(\alpha_1^2 + \beta_1^2)(\alpha_2^2 + \beta_2^2)}$$

$$\times \Big[\{\beta_1(\alpha_2^2+\beta_2^2)+\beta_2(\alpha_1^2+\beta_1^2)\}\,\hat{\sigma}_{22}^{o}+(\alpha_1\beta_2+\alpha_2\beta_1)\,\hat{\sigma}_{12}^{o}\Big], \qquad (6.4.35a,b)$$

where α_1, α_2, β_1, and β_2 (β_1, $\beta_2 > 0$) are given by $\lambda_1 = \alpha_1 \pm \iota\beta_1$ and $\lambda_2 = \alpha_2 \pm \iota\beta_2$, which are the roots of the characteristic equation,

$$\hat{\hat{D}}_{1111}\,\lambda^4 - 4\,\hat{\hat{D}}_{1112}\,\lambda^3 + 2\,(\hat{\hat{D}}_{1122}+2\,\hat{\hat{D}}_{1212})\,\lambda^2 - 4\,\hat{\hat{D}}_{2212}\,\lambda + \hat{\hat{D}}_{2222} = 0; \quad (6.4.36)$$

see Sih, Paris, and Irwin (1965), and Section 21.

Assume that the sliding condition (6.4.27) holds for this anisotropic case. If the crack is closed and undergoes sliding, the transmitted normal stress, $\hat{\sigma}_{22}^{c}$, is computed from (6.4.35), as

$$\hat{\sigma}_{22}^{c} = \frac{\{\beta_1(\alpha_2^2+\beta_2^2)+\beta_2(\alpha_1^2+\beta_1^2)\}\,\hat{\sigma}_{22}^{o}+(\alpha_1\beta_2+\alpha_2\beta_1)\,\hat{\sigma}_{12}^{o}}{\beta_1(\alpha_2^2+\beta_2^2)+\beta_2(\alpha_1^2+\beta_1^2)-\eta\,sgn\,(\hat{\sigma}_{12}^{o})(\alpha_1\beta_2+\alpha_2\beta_1)} < 0, \qquad (6.4.37)$$

and the resulting slip is

$$[\hat{u}_1] = 2\sqrt{a^2-\hat{x}_1^2}\,\hat{\hat{D}}_{1111}\,\{(\alpha_1\beta_2+\alpha_2\beta_1)\,(\hat{\sigma}_{22}^{o}-\hat{\sigma}_{22}^{c})$$

$$+(\beta_1+\beta_2)\,(\hat{\sigma}_{12}^{o}+\eta\,sgn\,(\hat{\sigma}_{12}^{o})\,\hat{\sigma}_{22}^{c})\}. \qquad (6.4.38)$$

As is seen, the symmetry with respect to the $x_1 = \pm\, x_2$-lines does not hold, due to material anisotropy.

With the aid of (6.4.36) and (6.4.38), the $\hat{\mathbf{H}}^{\alpha}$-tensor can be computed. The results are summarized as follows:

(a) for an open crack,

$$\hat{H}_{2222}^{\alpha} = \pi\left(\frac{\beta_1}{\alpha_1^2+\beta_1^2}+\frac{\beta_2}{\alpha_2^2+\beta_2^2}\right)\hat{\hat{D}}_{2222},$$

$$\hat{H}_{1222}^{\alpha} = \hat{H}_{2212}^{\alpha} = \frac{1}{2}\,\pi\,(\alpha_1\beta_2+\alpha_2\beta_1)\,\hat{\hat{D}}_{1111},$$

$$\hat{H}_{1212}^{\alpha} = \frac{1}{4}\,\pi\,(\beta_1+\beta_2)\,\hat{\hat{D}}_{1111}, \qquad (6.4.39a\text{~}c)$$

with the other $\hat{H}_{ijkl}^{\alpha}$ vanishing;

(b) for a closed crack without slip, i.e., when $|\hat{\sigma}_{12}^{o}| < -\eta\,\hat{\sigma}_{22}^{o}$,

$$\hat{H}_{ijkl}^{\alpha} = 0; \qquad (6.4.40)$$

(c) and for a closed crack with slip, i.e., when $|\hat{\sigma}_{12}^{o}| > -\eta\hat{\sigma}_{22}^{o}$ and $\hat{\sigma}_{22}^{c} < 0$,

$$\hat{H}_{1222}^{\alpha} = \frac{1}{2}\,\pi\Big\{(\alpha_1\beta_2+\alpha_2\beta_1)$$

$$-\frac{\{(\alpha_1\beta_2+\alpha_2\beta_1)-\eta sgn\,(\hat{\sigma}_{12}^{o})(\beta_1+\beta_2)\}\{\beta_1(\alpha_2^2+\beta_2^2)+\beta_2(\alpha_1^2+\beta_1^2)\}}{\{\beta_1(\alpha_2^2+\beta_2^2)+\beta_2(\alpha_1^2+\beta_1^2)\}-\eta sgn\,(\hat{\sigma}_{12}^{o})(\alpha_1\beta_2+\alpha_2\beta_1)}\Big\}\,\hat{\hat{D}}_{1111},$$

$$\hat{H}^{\alpha}_{2222} = \frac{1}{4}\,\pi\Big\{(\beta_1+\beta_2)$$

$$-\frac{\{(\alpha_1\beta_2+\alpha_2\beta_1)-\eta\,sgn\,(\hat{\sigma}^o_{12})(\beta_1+\beta_2)\}(\alpha_1\beta_2+\alpha_2\beta_1)}{\{\beta_1(\alpha_2^2+\beta_2^2)+\beta_2(\alpha_1^2+\beta_1^2)\}-\eta\,sgn\,(\hat{\sigma}^o_{12})(\alpha_1\beta_2+\alpha_2\beta_1)}\Big\}\,\hat{\bar{D}}_{1111}, \tag{6.4.41}$$

with the other $\hat{H}^{\alpha}_{ijkl}$ being zero. Once $\hat{H}^{\alpha}_{ijkl}$ is determined, the overall compliance tensor $\bar{\mathbf{D}}$ is computed from, say, (6.4.13). Note that the overall shear and bulk moduli defined by (6.4.34) are expressed in terms of this anisotropic overall compliance $\bar{\mathbf{D}}$, as

$$\frac{1}{\bar{\mu}} = -2\,(\bar{D}_{1211}+\bar{D}_{1222})\,\frac{p^o}{\tau^o} + 4\,\bar{D}_{1212},$$

$$\frac{1}{\bar{K}} = \bar{D}_{1111}+\bar{D}_{2222}+2\bar{D}_{1122}-2\,(\bar{D}_{1122}+\bar{D}_{2212})\,\frac{p^o}{\tau^o}. \tag{6.4.42a,b}$$

As shown in Subsection 6.4.2, the overall elasticity tensor $\bar{\mathbf{C}}$ obtained under the assumption of macrostrains prescribed, coincides with the inverse of the compliance tensor $\bar{\mathbf{D}}$.

It is seen that $\mathbf{H}^{\alpha}$ and hence $\bar{\mathbf{D}}$ are not symmetric when sliding occurs on closed frictional cracks, unless the coefficient of friction, η, is zero. For illustration, consider two extreme cases, one with $\eta = 0$ (sliding with no frictional loss) and the other with $\eta >> 1$ (no sliding). Figure 6.4.6 shows $\bar{\mu}$ and $\bar{K}$ as functions of f for several extreme loading conditions. Here, also, the effect of the loading, p^o/τ^o or τ^o/p^o, is clearly seen.[5]

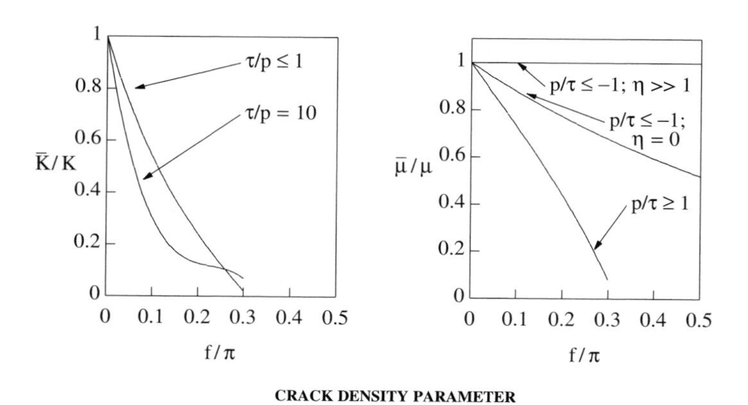

Figure 6.4.6

Normalized overall bulk and shear moduli for dilute-distribution of closed cracks; self-consistent method

[5] See Mao and Sunder (1992a,b) who examine polycrystalline ice, and Ju (1991) who presents damage models for microcracked solids.

6.5. EFFECTIVE MODULI OF AN ELASTIC BODY CONTAINING ALIGNED PENNY-SHAPED MICROCRACKS

In this subsection the overall elastic moduli of an RVE consisting of a homogeneous linearly elastic isotropic matrix material which contains penny-shaped microcracks, are estimated. Similarly to the case of the two-dimensional problems examined in Subsections 6.3 and 6.4, the overall response of the RVE may be isotropic or anisotropic, depending on the distribution of microcracks. This and related issues are brought into focus through several illustrative examples.

6.5.1. Crack Opening Displacements

Consider a penny-shaped crack of radius a, lying in the x_1,x_2-plane with its center at the origin of the coordinate system. The unit normal **n** of the positive crack face, therefore, coincides with the unit base vector $\mathbf{e}_3$. Under the action of farfield stresses, $\sigma_{13}^{\infty} = \sigma_{31}^{\infty}$, $\sigma_{23}^{\infty} = \sigma_{32}^{\infty}$, and $\sigma_{33}^{\infty} > 0$, the COD's are

$$[u_i] = \sqrt{a^2 - r^2}\,\frac{16(1-\nu^2)}{\pi E(2-\nu)}\,\sigma_{i3}^{\infty}, \qquad r \leq a \;(i = 1, 2),$$

$$[u_3] = \sqrt{a^2 - r^2}\,\frac{8(1-\nu^2)}{\pi E}\,\sigma_{33}^{\infty}, \qquad r \leq a, \tag{6.5.1a,b}$$

where $r^2 = x_1^2 + x_2^2$; see Figure 6.5.1.

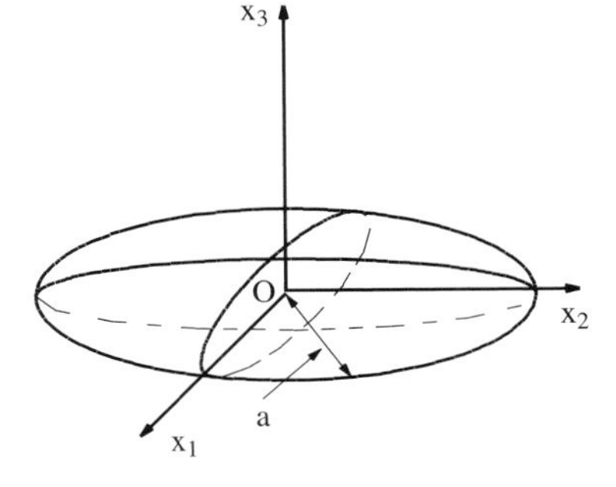

Figure 6.5.1

A penny-shaped crack

6.5.2. Effective Moduli: Dilute Distribution of Aligned Microcracks

Consider a dilute distribution of penny-shaped microcracks, all parallel to the x_1,x_2-plane, as shown in Figure 6.5.2. The crack sizes and the locations of their centers are assumed to be random. This means that the overall response of the RVE is transversely isotropic, with the x_1,x_2-plane defining the plane of isotropy. Hence, there are a total of no more than five independent overall elastic moduli. Furthermore, when the cracks are closed, the Young modulus in the

x_3-direction, i.e., $\bar{E}_3$, will be the same as the Young modulus of the isotropic matrix, E. Under uniaxial tension in the x_3-direction, however, $\bar{E}_3 < E$, since, due to the crack opening displacement, the RVE is more compliant for such uniaxial tensile loading. If it is further assumed that the cracks are *frictionless*, then the overall moduli will have the form given by (3.1.13), with the exception that

$$\bar{E}_3 \begin{cases} = E & \text{in compression} \\ < E & \text{in tension.} \end{cases} \tag{6.5.2}$$

When the microcracks are frictional, the response will be history-dependent, similar to the case discussed in Subsections 6.3.3 and 6.4.5.

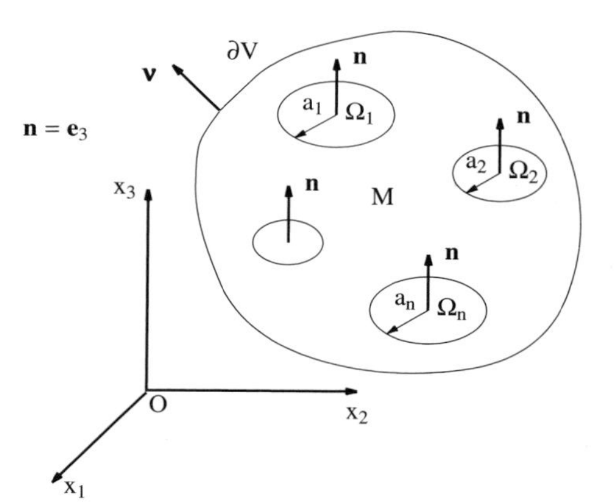

Figure 6.5.2

A dilute distribution of penny-shaped microcracks, parallel to the x_1,x_2-plane

To estimate the overall moduli for the dilute distribution of penny-shaped microcracks aligned normal to the x_3-axis, let N_α be the number of cracks of radius a_α per unit volume of the RVE. Then the additional overall strain due to the presence of microcracks, $\bar{\boldsymbol{\varepsilon}}^c$, can be expressed as

$$\bar{\boldsymbol{\varepsilon}}^c = \sum_{\alpha=1}^{n} f_\alpha \, \bar{\boldsymbol{\varepsilon}}^\alpha, \qquad f_\alpha \equiv N_\alpha \, a_\alpha^3,$$

$$\bar{\boldsymbol{\varepsilon}}^\alpha \equiv \frac{1}{a_\alpha^3} \int_{\partial\Omega_\alpha^+} \frac{1}{2} \{\mathbf{n} \otimes [\mathbf{u}] + [\mathbf{u}] \otimes \mathbf{n}\} \, dS, \tag{6.5.3a~c}$$

where f_α measures the density of cracks with radius a_α, and the total crack density f and the total number of cracks per unit volume, N, respectively are

$$f = \sum_{\alpha=1}^{n} f_\alpha, \qquad N = \sum_{\alpha=1}^{n} N_\alpha. \tag{6.5.3d,e}$$

Direct calculation, based on (6.5.1) and (6.5.3c), readily yields

$$\bar{\varepsilon}^{\alpha}_{i3} = \frac{16(1-\nu^2)}{3E(2-\nu)}\,\sigma^{\infty}_{i3} \qquad (i = 1, 2),$$

$$\bar{\varepsilon}^{\alpha}_{33} = \frac{16(1-\nu^2)}{3E}\,\sigma^{\infty}_{33}. \tag{6.5.4a,b}$$

Hence, $\bar{\boldsymbol{\varepsilon}}^{\alpha}$ is independent of the crack radius a_{α}. The matrix $[H_{ab}]$ is obtained from (6.5.4a,b),

$$[H_{ab}] = f\,\frac{16(1-\nu^2)}{3E}\begin{bmatrix} 0 & 0 & 0 & 0 & 0 & 0 \\ 0 & 0 & 0 & 0 & 0 & 0 \\ 0 & 0 & 1 & 0 & 0 & 0 \\ 0 & 0 & 0 & 2/(2-\nu) & 0 & 0 \\ 0 & 0 & 0 & 0 & 2/(2-\nu) & 0 \\ 0 & 0 & 0 & 0 & 0 & 0 \end{bmatrix}, \tag{6.5.5a}$$

or, when $[H_{ab}]$ is divided into two non-trivial parts,

$$[H_{ab}] = \begin{bmatrix} [H^{(1)}_{ab}] & [0] \\ [0] & [H^{(2)}_{ab}] \end{bmatrix}, \tag{6.5.5b}$$

it follows that

$$[H^{(1)}_{ab}] = f\,\frac{16(1-\nu^2)}{3E}\begin{bmatrix} 0 & 0 & 0 \\ 0 & 0 & 0 \\ 0 & 0 & 1 \end{bmatrix}, \tag{6.5.5c}$$

$$[H^{(2)}_{ab}] = f\,\frac{32(1-\nu^2)}{3E(2-\nu)}\begin{bmatrix} 1 & 0 & 0 \\ 0 & 1 & 0 \\ 0 & 0 & 0 \end{bmatrix}. \tag{6.5.5d}$$

From this and (4.3.6), it is seen that the overall compliance tensor $\bar{\mathbf{D}}$ is transversely isotropic. Then the overall elastic moduli are

$$\frac{\bar{E}}{E} = 1, \qquad \frac{\bar{\nu}}{\nu} = 1, \qquad \frac{\bar{\mu}}{\mu} = 1, \tag{6.5.6a~c}$$

where the notation $\bar{E} = \bar{E}_1 = \bar{E}_2$, $\bar{\nu} = \bar{\nu}_{12}$, and $\bar{\mu} = \bar{\mu}_{12}$ is used for the overall elastic moduli in the x_1,x_2-plane; and

$$\frac{\bar{E}_3}{E} = \{1 + f\,\frac{16(1-\nu^2)}{3}\}^{-1} = 1 - f\,\frac{16(1-\nu^2)}{3} + O(f^2),$$

$$\frac{\bar{\nu}_3}{\nu} = \{1 + f\,\frac{16(1-2\nu)(\nu^2-1)}{3\nu(2-\nu)}\}\{1 + f\,\frac{16(1-\nu^2)}{3}\}^{-1}$$

$$= 1 - f\,\frac{16(1-\nu^2)(1+\nu-\nu^2)}{3\nu(2-\nu)} + O(f^2),$$

$$\frac{\bar{\mu}_3}{\mu} = \{1 + f\,\frac{16(1-\nu)}{3(2-\nu)}\}^{-1} = 1 - f\,\frac{16(1-\nu)}{3(2-\nu)} + O(f^2), \tag{6.5.6d~f}$$

where the notation $\bar{\nu}_3 = \bar{\nu}_{13} = \bar{\nu}_{23}$ and $\bar{\mu}_3 = \bar{\mu}_{13} = \bar{\mu}_{23}$ is used; see Subsection 3.1 and (3.1.13).

When the macrostrains are regarded prescribed, the overall elasticity tensor is given by (4.5.5). The tensor $\mathbf{J}$ in (4.5.5) relates to the tensor $\mathbf{H}$ by

$\mathbf{J} = \mathbf{H} : \mathbf{C}$, for the model of a dilute distribution of microcracks. To calculate $\mathbf{J}$, in view of[6] (6.5.5), set

$$[J_{ab}] = \begin{bmatrix} [J^{(1)}_{ab}] & [0] \\ [0] & [J^{(2)}_{ab}] \end{bmatrix}, \qquad [C_{ab}] = \begin{bmatrix} [C^{(1)}_{ab}] & [0] \\ [0] & [C^{(2)}_{ab}] \end{bmatrix}. \tag{6.5.7a,b}$$

Then,

$$[J^{(1)}_{ac}] = [H^{(1)}_{ab}][C^{(1)}_{bc}], \qquad [J^{(2)}_{ac}] = [H^{(2)}_{ab}][C^{(2)}_{bc}]. \tag{6.5.7c,d}$$

Since $[C_{ab}]$ is isotropic, it follows that

$$[C^{(1)}_{ab}] = \frac{E}{(1+\nu)(1-2\nu)} \begin{bmatrix} 1-\nu & \nu & \nu \\ \nu & 1-\nu & \nu \\ \nu & \nu & 1-\nu \end{bmatrix},$$

$$[C^{(2)}_{ab}] = \frac{E}{2(1+\nu)} \begin{bmatrix} 1 & 0 & 0 \\ 0 & 1 & 0 \\ 0 & 0 & 1 \end{bmatrix}. \tag{6.5.8a,b}$$

Thus, in view of (6.5.5),

$$[J^{(1)}_{ab}] = f\,\frac{16(1-\nu)}{3(1-2\nu)} \begin{bmatrix} 0 & 0 & 0 \\ 0 & 0 & 0 \\ \nu & \nu & 1-\nu \end{bmatrix}, \qquad [J^{(2)}_{ab}] = f\,\frac{16(1-\nu)}{3(2-\nu)} \begin{bmatrix} 1 & 0 & 0 \\ 0 & 1 & 0 \\ 0 & 0 & 0 \end{bmatrix}. \tag{6.5.9a,b}$$

Although $[J^{(1)}_{ab}]$ is not symmetric, the product$[C^{(1)}_{ab}][J^{(1)}_{bc}]$ is symmetric,

$$[C^{(1)}_{ab}][J^{(1)}_{bc}] = f\,\frac{16(1-\nu)E}{3(1+\nu)(1-2\nu)^2} \begin{bmatrix} \nu^2 & \nu^2 & \nu(1-\nu) \\ \nu^2 & \nu^2 & \nu(1-\nu) \\ \nu(1-\nu) & \nu(1-\nu) & (1-\nu)^2 \end{bmatrix},$$

$$[C^{(2)}_{ab}][J^{(2)}_{ab}] = f\,\frac{8(1-\nu)E}{3(1+\nu)(2-\nu)} \begin{bmatrix} 1 & 0 & 0 \\ 0 & 1 & 0 \\ 0 & 0 & 0 \end{bmatrix}. \tag{6.5.9c,d}$$

Note that, since

$$(\mathbf{C} : \mathbf{J})_{1212} = \frac{1}{2}\{(\mathbf{C} : \mathbf{J})_{1111} - (\mathbf{C} : \mathbf{J})_{1122}\} = 0, \tag{6.5.9e}$$

$$([C^{(2)}_{ab}][J^{(2)}_{bc}])_{33} = \frac{1}{2}\{([C^{(1)}_{ab}][J^{(1)}_{bc}])_{11} - ([C^{(1)}_{ab}][J^{(1)}_{bc}])_{12}\} = 0, \tag{6.5.9f}$$

the tensor $\mathbf{C} : \mathbf{J}$ is transversely isotropic. From (6.5.9c,d) and (4.5.5), various components of $[\bar{C}_{ab}]$ are obtained. For example, the overall shear moduli, $\bar{\mu} = \bar{\mu}_{12}$ and $\bar{\mu}_3 = \bar{\mu}_{13} = \bar{\mu}_{23}$, are

$$\frac{\bar{\mu}}{\mu} = 1, \qquad \frac{\bar{\mu}_3}{\mu} = 1 - f\,\frac{16(1-\nu)}{3(2-\nu)}. \tag{6.5.10a,b}$$

---

[6] Note that the matrix representing a tensor is denoted by the corresponding italic letter; see Sections 3 and 15 for details.

These shear moduli are equivalent to those given by (6.5.6c) and (6.5.6e), up to the first order in the crack density parameter f.

For the self-consistent model, the **H**-tensor is to be computed for a penny-shaped crack in an unbounded transversely isotropic elastic solid, with the crack normal to the axis of elastic isotropy. This problem is not examined here; see Section 21. The self-consistent results, however, are given by Hoenig (1979) using a different method; see also Laws and Brockenbrough (1987).

6.6. EFFECTIVE MODULI OF AN ELASTIC BODY CONTAINING RANDOMLY DISTRIBUTED PENNY-SHAPED MICROCRACKS

6.6.1. Dilute Open Microcracks with Prescribed Distribution

A typical penny-shaped microcrack is defined by its radius, a_α, and its orientation given by the unit normal, $\mathbf{n}_\alpha = \mathbf{n}$. The components of $\mathbf{n}$ in fixed Cartesian coordinates may be expressed by

$$n_1 = \sin\psi \cos\theta, \qquad n_2 = \sin\psi \sin\theta, \qquad n_3 = \cos\psi; \tag{6.6.1}$$

see Figure 6.6.1. When there are a very large number of microcracks with radii ranging from a_m to a_M, and with unit normals ranging over all orientations, a *density function*, $w = w(a, \theta, \psi)$, may be introduced such that the number of cracks per unit volume with radii in the range of a to a + da, and orientations in the range of (θ, ψ) to $(\theta + d\theta, \psi + d\psi)$, is given by $w(a, \theta, \psi) \sin\psi \, da \, d\theta \, d\psi$. Then the total number of cracks per unit volume, N, is

$$N = \frac{1}{4\pi} \int_{a_m}^{a_M} \int_0^{2\pi} \int_0^{\pi} w(a, \theta, \psi) \sin\psi \, da \, d\theta \, d\psi. \tag{6.6.2a}$$

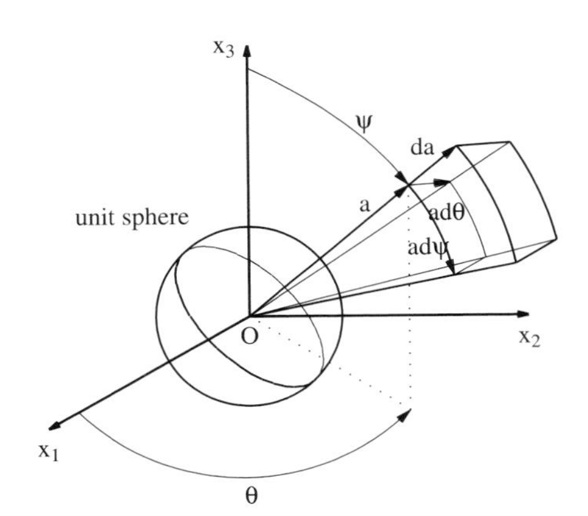

Figure 6.6.1

Distribution of radii and unit normals of penny-shaped cracks

Note that $\sin\psi\, d\theta\, d\psi$ defines the elementary solid angle with orientation (θ, ψ). With θ ranging from 0 to 2π and ψ ranging from 0 to π, the corresponding unit vector traces a unit sphere.

When the crack size distribution is independent of the crack orientation, the density function may be expressed as

$$w(a, \theta, \psi) = w_r(a)\, w_o(\theta, \psi), \tag{6.6.2b}$$

and it follows that

$$N = \int_{a_m}^{a_M} w_r(a)\, da, \qquad 1 = \frac{1}{4\pi} \int_0^{2\pi} \int_0^{\pi} w_o(\theta, \psi) \sin\psi\, d\theta\, d\psi. \tag{6.6.2c,d}$$

To estimate the elastic moduli of an RVE with a prescribed *dilute* distribution of penny-shaped microcracks, first consider a typical microcrack, Ω_α, of radius a_α and orientation $(\theta_\alpha, \psi_\alpha)$, and calculate the corresponding $\mathbf{H}^\alpha$-tensor. Then integrate the result over all possible radii and orientations, using the corresponding weighting function $w(a, \theta, \psi)$. Let $(x_1^\alpha, x_2^\alpha, x_3^\alpha)$ be the local rectangular Cartesian coordinate system (the α-coordinates) for the microcrack Ω_α, where the unit base vectors in the α-coordinates are $\mathbf{e}_i^\alpha$ (i = 1, 2, 3), and the origin O_α is at the center of Ω_α. The crack Ω_α lies in the x_1^α, x_2^α-plane. Its unit normal $\mathbf{n}$ $(= \mathbf{e}_3^\alpha)$ is in the x_3^α-direction. Since the crack is penny-shaped, the x_1^α-direction in the x_1,x_2-plane may be chosen arbitrarily. For simplicity, choose the x_1^α-axis parallel to the x_1,x_2-plane, i.e., on the intersection of the x_1,x_2- and the crack-plane. This uniquely determines the α-coordinates; see Figure 6.6.2. The angle between the x_2- and the x_1^α-axis is θ_α, and the angle between the x_3- and the x_3^α-axis is ψ_α. Therefore, the orthogonal tensor $\mathbf{Q}^\alpha$, defined by (6.4.2a), has the following components in the x_i-coordinates:

$$Q_{11}^\alpha = -\sin\theta_\alpha, \qquad Q_{21}^\alpha = \cos\theta_\alpha, \qquad Q_{31}^\alpha = 0,$$

$$Q_{12}^\alpha = -\cos\psi_\alpha \cos\theta_\alpha, \qquad Q_{22}^\alpha = -\cos\psi_\alpha \sin\theta_\alpha, \qquad Q_{32}^\alpha = \sin\psi_\alpha,$$

$$Q_{13}^\alpha = \sin\psi_\alpha \cos\theta_\alpha, \qquad Q_{23}^\alpha = \sin\psi_\alpha \sin\theta_\alpha, \qquad Q_{33}^\alpha = \cos\psi_\alpha. \tag{6.6.3a~i}$$

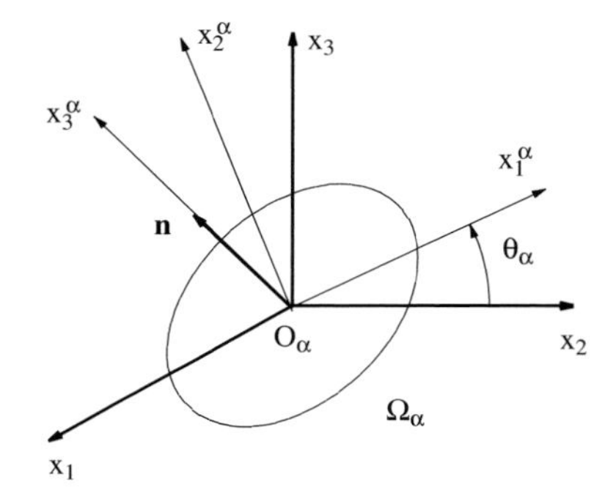

Figure 6.6.2

A typical microcrack in α-coordinates

Suppose that the macrostress $\mathbf{\Sigma} = \boldsymbol{\sigma}^o$ is regarded prescribed. From the contribution to the overall strain by the single crack Ω_α, the corresponding $\mathbf{H}^\alpha$-tensor is defined by

$$\bar{\boldsymbol{\varepsilon}}^\alpha = \frac{1}{a_\alpha^3} \int_{\partial\Omega_\alpha^+} \frac{1}{2}\{\mathbf{n}\otimes[\mathbf{u}] + [\mathbf{u}]\otimes\mathbf{n}\}\, dS = \mathbf{H}^\alpha : \boldsymbol{\sigma}^o. \tag{6.6.4a}$$

The components of $\mathbf{H}^\alpha$ in the *local* α-coordinates can be read off (6.5.4a,b). Denoting these components by $\hat{H}^\alpha_{ijkl}$, obtain

$$\hat{H}^\alpha_{3333} = \frac{16(1-\nu^2)}{3E}, \qquad \hat{H}^\alpha_{i3i3} = \hat{H}^\alpha_{3ii3} = \hat{H}^\alpha_{i33i} = \hat{H}^\alpha_{3i3i} = \frac{8(1-\nu^2)}{3E(2-\nu)}$$

$$(i = 1,\, 2;\ i \text{ not summed}), \tag{6.6.4b,c}$$

with other components being zero. It should be noted that: (1) the $\mathbf{H}^\alpha$-tensor depends on the crack orientation $(\theta_\alpha, \psi_\alpha)$, but is independent of the crack size a_α; and (2) the components of $\mathbf{H}^\alpha$ in the corresponding α-coordinates are constant, given by (6.6.4b,c). The components of $\mathbf{H}^\alpha$ in the x_i-coordinates, therefore, are functions of the orientation angles θ_α and ψ_α.

Since the distribution of microcracks is prescribed (i.e., a crack density function $w(a, \theta, \psi)$ is given), to obtain the $\mathbf{H}$-tensor, integrate the $\mathbf{H}^\alpha$-tensor over all possible radii and orientations, using the corresponding crack density function:

$$\mathbf{H} = \frac{1}{4\pi} \int_{a_m}^{a_M} \int_0^{2\pi} \int_0^{\pi} a^3\, \mathbf{H}^\alpha(\psi, \theta)\, w(a, \theta, \psi) \sin\psi\, da\, d\theta\, d\psi. \tag{6.6.5a}$$

In particular, when the crack size distribution is independent of the crack orientation, from (6.6.2b) and (6.6.4c), $\mathbf{H}$ becomes

$$\mathbf{H} = \{\int_{a_m}^{a_M} a^3\, w_r(a)\, da\}\, \hat{H}^\alpha_{ijkl}$$

$$\times \{\frac{1}{4\pi} \int_0^{2\pi} \int_0^{\pi} \mathbf{e}_i^\alpha \otimes \mathbf{e}_j^\alpha \otimes \mathbf{e}_k^\alpha \otimes \mathbf{e}_l^\alpha\, w_o(\theta, \psi) \sin\psi\, d\theta\, d\psi\}, \tag{6.6.5b}$$

where the base vectors $\mathbf{e}_i^\alpha$ are functions of θ and ψ, defined by (6.6.3a~i).

Next, consider the case when the macrostrain $\mathbf{E} = \boldsymbol{\varepsilon}^o$ is prescribed. For a typical microcrack Ω_α of radius a_α and orientation $(\theta_\alpha, \psi_\alpha)$, define a tensor $\mathbf{J}^\alpha$ by (6.4.7a), which is related to the tensor $\mathbf{H}^\alpha$ through

$$\mathbf{J}^\alpha = \mathbf{H}^\alpha : \mathbf{C}, \tag{6.6.6a}$$

where the components of $\mathbf{J}^\alpha$ in the x_i-coordinates are functions of θ_α and ψ_α. In the corresponding α-coordinates, however, these components (denoted by $\hat{J}^\alpha_{ijkl}$) are,

$$\hat{J}^\alpha_{3333} = \frac{16(1-\nu)^2}{3(1-2\nu)}, \qquad \hat{J}^\alpha_{33ii} = \frac{16\nu(1-\nu)}{3(1-2\nu)},$$

$$\hat{J}^\alpha_{i3i3} = \hat{J}^\alpha_{3ii3} = \hat{J}^\alpha_{i33i} = \hat{J}^\alpha_{3i3i} = \frac{8(1+\nu)}{3(2-\nu)} \quad (i = 1,\, 2; i \text{ not summed}), \tag{6.6.6b~d}$$

with other components being zero; see (6.5.9a) and (6.5.9b). Similarly to $\hat{H}^{\alpha}_{ijkl}$, the components $\hat{J}^{\alpha}_{ijkl}$ are independent of crack size. In a manner similar to (6.6.5a), the tensor **J** is given by

$$\mathbf{J} = \frac{1}{4\pi} \int_{a_m}^{a_M} \int_0^{2\pi} \int_0^{\pi} a^3\, \mathbf{J}^{\alpha}(\psi, \theta)\, w(a, \theta, \psi) \sin\psi \, da\, d\theta\, d\psi = \mathbf{H} : \mathbf{C}. \tag{6.6.7a}$$

In particular, when the crack size distribution is independent of the crack orientation, from (6.6.2b) and (6.6.6b~d), **J** becomes

$$\mathbf{J} = \{\int_{a_m}^{a_M} a^3\, w_r(a)\, da\} \, \hat{J}^{\alpha}_{ijkl}$$

$$\times \{\frac{1}{4\pi} \int_0^{2\pi} \int_0^{\pi} \mathbf{e}_i^{\alpha} \otimes \mathbf{e}_j^{\alpha} \otimes \mathbf{e}_k^{\alpha} \otimes \mathbf{e}_l^{\alpha}\, w_o(\theta, \psi) \sin\psi \, d\theta\, d\psi\}, \tag{6.6.7b}$$

where, again, $\mathbf{e}_i^{\alpha}$ and w_o are functions of θ and ψ. It should be noted that, unlike $\mathbf{H}^{\alpha}$ and $\mathbf{H}$, $\mathbf{J}^{\alpha}$ and $\mathbf{J}$ may not be symmetric with respect to the first and last pair of their indices, i.e., $J^{\alpha}_{ijkl} \neq J^{\alpha}_{klij}$. However, the tensors $\mathbf{C} : \mathbf{J}^{\alpha}$ and $\mathbf{C} : \mathbf{J}$, which determine the overall elasticity tensor $\bar{\mathbf{C}}$, have this symmetry; see Subsections 6.3.2 and 6.5.2.

6.6.2. Effective Moduli: Random Dilute Distribution of Microcracks

Consider a simple case where: (1) the distribution of microcracks is dilute; (2) the crack orientation distribution is random; and (3) the crack size distribution is independent of the crack orientation; Figure 6.6.3. Then, the crack orientation distribution function, $w_o(\theta, \psi)$, given by (6.6.2b), becomes

$$w_o(\theta, \psi) = \text{constant} = 1. \tag{6.6.8}$$

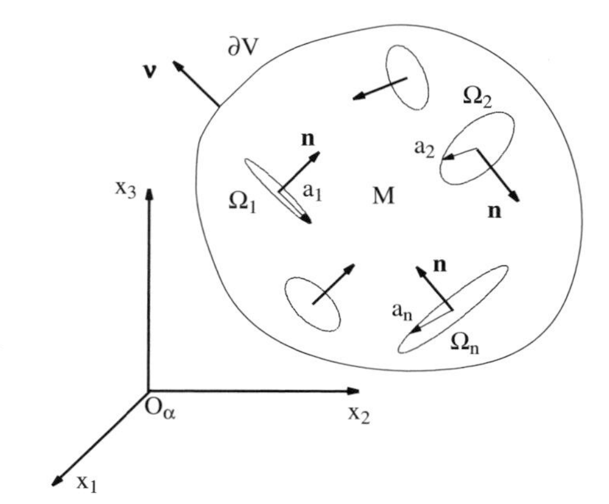

Figure 6.6.3

A random dilute distribution of microcracks

Furthermore, the **H**- and **J**-tensors, respectively defined by (6.6.5b) and (6.6.7b), become

$$\mathbf{H} = \hat{H}^{\alpha}_{ijkl}\,\{\frac{f}{4\pi}\int_0^{2\pi}\int_0^{\pi}\mathbf{e}_i^{\alpha}(\theta,\,\psi)\otimes\mathbf{e}_j^{\alpha}(\theta,\,\psi)\otimes\mathbf{e}_k^{\alpha}(\theta,\,\psi)\otimes\mathbf{e}_l^{\alpha}(\theta,\,\psi)\,\sin\psi\,d\theta\,d\psi\},$$

$$\mathbf{J} = \hat{J}^{\alpha}_{ijkl}\,\{\frac{f}{4\pi}\int_0^{2\pi}\int_0^{\pi}\mathbf{e}_i^{\alpha}(\theta,\,\psi)\otimes\mathbf{e}_j^{\alpha}(\theta,\,\psi)\otimes\mathbf{e}_k^{\alpha}(\theta,\,\psi)\otimes\mathbf{e}_l^{\alpha}(\theta,\,\psi)\,\sin\psi\,d\theta\,d\psi\}, \tag{6.6.9a,b}$$

where f is the crack density parameter, defined by

$$f \equiv \int_{a_m}^{a_M} a^3\,w_r(a)\,da, \tag{6.6.10a}$$

and the dependence of the base vectors $\mathbf{e}_i^{\alpha}$ on θ and ψ is explicitly indicated. It should be noted that due to the dilute distribution of microcracks, $f << 1$. In particular, when the crack size distribution is uniform, from (6.6.2c), $w_r(a)$ becomes

$$w_r(a) = \text{constant} = \frac{N}{a_M - a_m}. \tag{6.6.10b}$$

The crack density parameter f then is

$$f = \frac{N}{a_M - a_m}\int_{a_m}^{a_M} a^3\,da = \frac{N}{4}\,(a_M^3 + a_M^2\,a_m + a_M\,a_m^2 + a_m^3). \tag{6.6.10c}$$

When all microcracks in the RVE have the same radius a, f becomes

$$f = N\,a^3. \tag{6.6.10d}$$

Due to the assumption of a random distribution of cracks, the overall response of the RVE is isotropic. Hence, **H** and **J** may be expressed as

$$\mathbf{H} = h_1\,\mathbf{1}^{(2)}\otimes\mathbf{1}^{(2)} + h_2\,\mathbf{1}^{(4s)}, \qquad \mathbf{J} = j_1\,\mathbf{1}^{(2)}\otimes\mathbf{1}^{(2)} + j_2\,\mathbf{1}^{(4s)}, \tag{6.6.11a,b}$$

where $(h_1,\,h_2)$ and $(j_1,\,j_2)$ are unknown functions of the crack density parameter f. Therefore, the resulting overall elasticity and compliance tensors, $\bar{\mathbf{C}}$ and $\bar{\mathbf{D}}$, are isotropic and can be expressed as

$$\bar{\mathbf{C}} = \frac{\bar{\nu}\,\bar{E}}{(1+\bar{\nu})(1-2\bar{\nu})}\,\mathbf{1}^{(2)}\otimes\mathbf{1}^{(2)} + \frac{\bar{E}}{1+\bar{\nu}}\,\mathbf{1}^{(4s)},$$

$$\bar{\mathbf{D}} = -\frac{\bar{\nu}}{\bar{E}}\,\mathbf{1}^{(2)}\otimes\mathbf{1}^{(2)} + \frac{1+\bar{\nu}}{\bar{E}}\,\mathbf{1}^{(4s)}, \tag{6.6.11c,d}$$

where the overall Young modulus, $\bar{E}$, and Poisson ratio, $\bar{\nu}$, depend on the crack density parameter f, as well as on the moduli of the isotropic matrix material.

First, **H** is obtained and the overall elastic moduli of the RVE are estimated, when the macrostress $\boldsymbol{\Sigma}$ is prescribed. Using the rotation tensor $\mathbf{Q}^{\alpha}$ defined by (6.4.2a) or (6.6.3a~i), express the components H_{ijkl} of **H** in the x_i-coordinates in terms of $\hat{H}^{\alpha}_{ijkl}$. Then use suitable summations of H_{ijkl} to obtain (6.6.11a), given below; see also Subsection 6.4.1. Since

$$(\mathbf{1}^{(2)} \otimes \mathbf{1}^{(2)})::(\mathbf{e}_i^\alpha \otimes \mathbf{e}_j^\alpha \otimes \mathbf{e}_k^\alpha \otimes \mathbf{e}_l^\alpha) = \delta_{ij}\,\delta_{kl},$$

$$\mathbf{1}^{(4s)}::(\mathbf{e}_i^\alpha \otimes \mathbf{e}_j^\alpha \otimes \mathbf{e}_k^\alpha \otimes \mathbf{e}_l^\alpha) = \frac{1}{2}(\delta_{ik}\,\delta_{jl} + \delta_{il}\,\delta_{jk}), \tag{6.6.12a,b}$$

and

$$(\mathbf{1}^{(2)} \otimes \mathbf{1}^{(2)})::(\mathbf{1}^{(2)} \otimes \mathbf{1}^{(2)}) = \delta_{ii}\,\delta_{jj} = 9,$$

$$(\mathbf{1}^{(2)} \otimes \mathbf{1}^{(2)})::\mathbf{1}^{(4s)} = \mathbf{1}^{(4s)}::(\mathbf{1}^{(2)} \otimes \mathbf{1}^{(2)}) = \delta_{ij}\,\delta_{ij} = 3,$$

$$\mathbf{1}^{(4s)}::\mathbf{1}^{(4s)} = \frac{1}{2}(\delta_{ij}\,\delta_{ij} + \delta_{ii}\,\delta_{jj}) = 6, \tag{6.6.12c~e}$$

where :: denotes fourth-order contraction, it follows that fourth-order contractions of $\mathbf{H}$, (6.6.9a) and (6.6.11a), yield

$$\mathbf{H}::(\mathbf{1}^{(2)} \otimes \mathbf{1}^{(2)}) = f\,\hat{H}^\alpha_{iijj} = f\,\hat{H}^\alpha_{3333} = 9h_1 + 3h_2,$$

$$\mathbf{H}::\mathbf{1}^{(4s)} = f\,\frac{1}{2}(\hat{H}^\alpha_{ijij} + \hat{H}^\alpha_{ijji}) = f\,(\hat{H}^\alpha_{3333} + 2\hat{H}^\alpha_{1313} + 2\hat{H}^\alpha_{2323})$$

$$= 3h_1 + 6h_2. \tag{6.6.13a,b}$$

Substitution from (6.6.4b,c) for components $\hat{H}^\alpha_{ijkl}$, now leads to the following set of equations for unknowns h_1 and h_2:

$$9h_1 + 3h_2 = f\,\frac{16(1-\nu^2)}{3E},$$

$$3h_1 + 6h_2 = f\,\{\frac{16(1-\nu^2)}{3E} + \frac{32(1-\nu^2)}{3E(2-\nu)}\}. \tag{6.6.13c,d}$$

Hence,

$$h_1 = -f\,\frac{16\nu(1-\nu^2)}{45(2-\nu)}\,\frac{1}{E}, \qquad h_2 = f\,\frac{32(1-\nu^2)(5-\nu)}{45(2-\nu)}\,\frac{1}{E}. \tag{6.6.13e,f}$$

In matrix form, $\mathbf{H}$ is

$$[H_{ab}] = f\,\frac{16(1-\nu^2)}{45(2-\nu)E}\begin{bmatrix} 10-3\nu & -\nu & -\nu & 0 & 0 & 0 \\ -\nu & 10-3\nu & -\nu & 0 & 0 & 0 \\ -\nu & -\nu & 10-3\nu & 0 & 0 & 0 \\ 0 & 0 & 0 & 4(5-\nu) & 0 & 0 \\ 0 & 0 & 0 & 0 & 4(5-\nu) & 0 \\ 0 & 0 & 0 & 0 & 0 & 4(5-\nu) \end{bmatrix}. \tag{6.6.13g}$$

From (4.3.6) and (6.6.11d), the overall Young modulus, $\bar{E}$, and Poisson ratio, $\bar{\nu}$, are

$$\frac{\bar{E}}{E} = \{1 + f\,\frac{16(1-\nu^2)(10-3\nu)}{45(2-\nu)}\}^{-1} = 1 - f\,\frac{16(1-\nu^2)(10-3\nu)}{45(2-\nu)} + O(f^2),$$

$$\frac{\bar{\nu}}{\nu} = \{1+f\,\frac{16(1-\nu^2)}{45(2-\nu)}\}\,\{1+f\,\frac{16(1-\nu^2)(10-3\nu)}{45(2-\nu)}\}^{-1}$$

$$= 1-f\,\frac{16(1-\nu^2)(3-\nu)}{15(2-\nu)} + O(f^2), \qquad (6.6.14a,b)$$

and the overall shear modulus, $\bar{\mu} = \bar{E}/2(1+\bar{\nu})$, becomes

$$\frac{\bar{\mu}}{\mu} = \{1+f\,\frac{32(1-\nu)(5-\nu)}{45(2-\nu)}\}^{-1} = 1-f\,\frac{32(1-\nu)(5-\nu)}{45(2-\nu)} + O(f^2). \quad (6.6.14c)$$

Next, $\mathbf{J}$ is obtained and the overall elastic moduli of the RVE are estimated when the macrostrain $\mathbf{E}$ is prescribed. In view of

$$(\mathbf{1}^{(2)}\otimes\mathbf{1}^{(2)}):(\mathbf{1}^{(2)}\otimes\mathbf{1}^{(2)}) = 3\mathbf{1}^{(2)}\otimes\mathbf{1}^{(2)},$$

$$(\mathbf{1}^{(2)}\otimes\mathbf{1}^{(2)}):\mathbf{1}^{(4s)} = \mathbf{1}^{(4s)}:(\mathbf{1}^{(2)}\otimes\mathbf{1}^{(2)}) = \mathbf{1}^{(2)}\otimes\mathbf{1}^{(2)},$$

$$\mathbf{1}^{(4s)}:\mathbf{1}^{(4s)} = \mathbf{1}^{(4s)}, \qquad (6.6.15a\sim c)$$

obtain

$$\mathbf{H}:\mathbf{C} = \left\{ f\,\frac{16(1-\nu^2)}{45(2-\nu)E}\,\{-\nu\mathbf{1}^{(2)}\otimes\mathbf{1}^{(2)} + 2(5-\nu)\mathbf{1}^{(4s)}\}\right\}$$

$$:\left\{\frac{E}{1+\nu}\,\{\frac{\nu}{1-2\nu}\,\mathbf{1}^{(2)}\otimes\mathbf{1}^{(2)} + \mathbf{1}^{(4s)}\}\right\}$$

$$= f\,\frac{16(1-\nu)}{45(2-\nu)}\,\{\frac{3\nu(3-\nu)}{1-2\nu}\,\mathbf{1}^{(2)}\otimes\mathbf{1}^{(2)} + 2(5-\nu)\mathbf{1}^{(4s)}\}, \qquad (6.6.15d)$$

or in matrix form

$$[J_{ab}] = f\,\frac{16(1-\nu)}{45(2-\nu)}\times$$

$$\begin{bmatrix} \frac{10-13\nu+\nu^2}{1-2\nu} & \frac{3\nu(3-\nu)}{1-2\nu} & \frac{3\nu(3-\nu)}{1-2\nu} & 0 & 0 & 0 \\ \frac{3\nu(3-\nu)}{1-2\nu} & \frac{10-13\nu+\nu^2}{1-2\nu} & \frac{3\nu(3-\nu)}{1-2\nu} & 0 & 0 & 0 \\ \frac{3\nu(3-\nu)}{1-2\nu} & \frac{3\nu(3-\nu)}{1-2\nu} & \frac{10-13\nu+\nu^2}{1-2\nu} & 0 & 0 & 0 \\ 0 & 0 & 0 & 2(5-\nu) & 0 & 0 \\ 0 & 0 & 0 & 0 & 2(5-\nu) & 0 \\ 0 & 0 & 0 & 0 & 0 & 2(5-\nu) \end{bmatrix}. \qquad (6.6.15e)$$

Then, it follows that

$$\mathbf{C}:\mathbf{J} = f\,\frac{16(1-\nu)E}{45(1+\nu)(2-\nu)}\,\{\frac{\nu(19-16\nu+\nu^2)}{(1-2\nu)^2}\,\mathbf{1}^{(2)}\otimes\mathbf{1}^{(2)} + 2(5-\nu)\mathbf{1}^{(4s)}\}. \qquad (6.6.15f)$$

Therefore, the overall elastic parameters, $\bar{E}$, $\bar{\nu}$, and $\bar{\mu}$, are,

$$\frac{\bar{E}}{E} = \{1 - f\,\frac{32(1-\nu)(5-\nu)}{45(2-\nu)}\}\,\{1 - f\,\frac{16(1-\nu)}{45(2-\nu)}\,\frac{5(2+3\nu-\nu^3)}{(1+\nu)(1-2\nu)}\}$$

$$\times\ \{1 - f\,\frac{32(1-\nu)}{45(2-\nu)}\,\frac{5-2\nu+8\nu^2-3\nu^3}{1-2\nu}\}^{-1}$$

$$= 1 - f\,\frac{16(1-\nu^2)(10-3\nu)}{45(2-\nu)} + O(f^2),$$

$$\frac{\bar{\nu}}{\nu} = \{1 - f\,\frac{16(1-\nu)(19-16\nu+\nu^2)}{45(2-\nu)(1-2\nu)}\}$$

$$\times\ \{1 - f\,\frac{32(1-\nu)}{45(2-\nu)}\,\frac{5-2\nu+8\nu^2-3\nu^3}{1-2\nu}\}^{-1}$$

$$= 1 - f\,\frac{16(1-\nu^2)(3-\nu)}{15(2-\nu)} + O(f^2),$$

$$\frac{\bar{\mu}}{\mu} = 1 - f\,\frac{32(1-\nu)(5-\nu)}{45(2-\nu)}. \qquad (6.6.16a\sim c)$$

These overall elastic moduli, $\bar{E}$, $\bar{\nu}$, and $\bar{\mu}$, agree with those given by (6.6.14a~c), up to the first order in the crack density parameter f.

6.6.3. Effective Moduli: Self-Consistent Estimates

Suppose that the interaction effects are to be included to a certain extent. Assume that the crack orientation and size distribution are random, with size distribution being independent of orientation. Then, the RVE is isotropic. The self-consistent scheme may be applied to estimate the overall elastic moduli, as follows.

As discussed in Subsections 5.1.3 and 6.4.2, in the self-consistent method, a typical microcrack Ω_α is embedded in a homogeneous *isotropic* elastic solid which has the *yet-unknown* overall moduli, say, Young's modulus, $\bar{E}$, and Poisson's ratio, $\bar{\nu}$. Then, the $\mathbf{H}^\alpha$- and $\mathbf{J}^\alpha$-tensors defined by (6.6.4) and (6.6.6), are replaced by $\bar{\mathbf{H}}^\alpha$ and $\bar{\mathbf{J}}^\alpha$, by substituting $\bar{E}$ and $\bar{\nu}$ for E and ν. $\bar{\mathbf{H}}^\alpha$ or $\bar{\mathbf{J}}^\alpha$ is then integrated over all possible crack sizes and orientations, to arrive at

$$\bar{\mathbf{H}} = \frac{1}{4\pi}\int_{a_m}^{a_M}\int_0^{2\pi}\int_0^{\pi} a^3\,\bar{\mathbf{H}}^\alpha(\psi,\,\theta)\,w(a,\,\theta,\,\psi)\,\sin\psi\,da\,d\theta\,d\psi,$$

$$\bar{\mathbf{J}} = \frac{1}{4\pi}\int_{a_m}^{a_M}\int_0^{2\pi}\int_0^{\pi} a^3\,\bar{\mathbf{J}}^\alpha(\psi,\,\theta)\,w(a,\,\theta,\,\psi)\,\sin\psi\,da\,d\theta\,d\psi. \qquad (6.6.17a,b)$$

According to the self-consistent method, $\bar{\mathbf{J}}^\alpha = \bar{\mathbf{H}}^\alpha : \bar{\mathbf{C}}$, and hence $\bar{\mathbf{J}} = \bar{\mathbf{H}} : \bar{\mathbf{C}}$. This relation between $\bar{\mathbf{H}}$ and $\bar{\mathbf{J}}$ leads to the equivalence of the overall elasticity and compliance tensors, $\bar{\mathbf{C}}$ and $\bar{\mathbf{D}}$; Subsections 5.1.3 and 6.4.2. Hence, it suffices to consider the overall compliance tensor $\bar{\mathbf{D}}$, in order to obtain the overall elastic moduli, $\bar{E}$ and $\bar{\nu}$. Since the crack orientation distribution is random and the crack size distribution is independent of the orientation, $\bar{\mathbf{H}}$ is given by (see (6.6.9a) and (6.6.11a))

$$\bar{\mathbf{H}} = \bar{\hat{H}}^{\alpha}_{ijkl} \{ \frac{f}{4\pi} \int_0^{2\pi} \int_0^{\pi} \mathbf{e}_i^{\alpha}(\theta, \psi) \otimes \mathbf{e}_j^{\alpha}(\theta, \psi) \otimes \mathbf{e}_k^{\alpha}(\theta, \psi) \otimes \mathbf{e}_l^{\alpha}(\theta, \psi) \sin\psi \, d\theta \, d\psi \},$$

$$= \bar{h}_1 \mathbf{1}^{(2)} \otimes \mathbf{1}^{(2)} + \bar{h}_2 \mathbf{1}^{(4s)}, \tag{6.6.17c}$$

where, from (6.6.13e,f), $\bar{h}_1$ and $\bar{h}_2$ are determined by replacing (E, ν) by ($\bar{E}$, $\bar{\nu}$). Then, $\bar{\mathbf{H}}$ becomes

$$\bar{\mathbf{H}} = -f \frac{16\bar{\nu}(1-\bar{\nu}^2)}{45(2-\bar{\nu})} \frac{1}{\bar{E}} \mathbf{1}^{(2)} \otimes \mathbf{1}^{(2)} + f \frac{32(1-\bar{\nu}^2)(5-\bar{\nu})}{45(2-\bar{\nu})} \frac{1}{\bar{E}} \mathbf{1}^{(4s)}. \tag{6.6.18a}$$

Since the overall compliance tensor $\bar{\mathbf{D}}$ is equal to $\mathbf{D} + \bar{\mathbf{H}}$, from the coefficients of $\mathbf{1}^{(2)} \otimes \mathbf{1}^{(2)}$ and $\mathbf{1}^{(4s)}$, obtain the following set of equations for $\bar{E}$ and $\bar{\nu}$:

$$\frac{\bar{\nu}}{\bar{E}} = \frac{\nu}{E} + f \frac{16\bar{\nu}(1-\bar{\nu}^2)}{45(2-\bar{\nu})\bar{E}},$$

$$\frac{1+\bar{\nu}}{\bar{E}} = \frac{1+\nu}{E} + f \frac{32(1-\bar{\nu}^2)(5-\bar{\nu})}{45(2-\bar{\nu})\bar{E}}. \tag{6.6.18b,c}$$

From (6.6.18b,c), a non-linear equation for $\bar{\nu}$ results

$$\frac{\bar{\nu}}{\nu} = \{1 - f \frac{16(1-\bar{\nu}^2)(10-3\bar{\nu})}{45(2-\bar{\nu})}\} \{1 - f \frac{16(1-\bar{\nu}^2)}{45(2-\bar{\nu})}\}^{-1}, \tag{6.6.18d}$$

or

$$f = \frac{45(\nu-\bar{\nu})(2-\bar{\nu})}{16(1-\bar{\nu}^2)[10\nu - \bar{\nu}(1+3\nu)]}. \tag{6.6.18e}$$

Then, in terms of $\bar{\nu}$, the overall Young modulus, $\bar{E}$, and the overall shear modulus, $\bar{\mu}$, are given by

$$\frac{\bar{E}}{E} = 1 - f \frac{16(1-\bar{\nu}^2)(10-3\bar{\nu})}{45(2-\bar{\nu})},$$

$$\frac{\bar{\mu}}{\mu} = \{1 - f \frac{16(1-\bar{\nu}^2)(10-3\bar{\nu})}{45(2-\bar{\nu})}\} \frac{1+\nu}{1+\bar{\nu}},$$

$$= 1 - f \frac{32(1-\bar{\nu})(5-\bar{\nu})}{45(2-\bar{\nu})}. \tag{6.6.18f,g}$$

These results agree with those first obtained by Budiansky and O'Connell (1976) using a different method. Since the leading term in the expansion of $\bar{\nu}/\nu$ with respect to f is 1, it follows that

$$\frac{\bar{E}}{E} = 1 - f \frac{16(1-\nu^2)(10-3\nu)}{45(2-\nu)} + O(f^2),$$

$$\frac{\bar{\nu}}{\nu} = 1 - f \frac{16(1-\nu^2)(3-\nu)}{15(2-\nu)} + O(f^2),$$

$$\frac{\bar{\mu}}{\mu} = 1 - f\,\frac{32(1-\nu)(5-\nu)}{45(2-\nu)} + O(f^2). \tag{6.6.18h~j}$$

Figure 6.6.4 shows the graphs of the overall moduli, $\bar{E}$ and $\bar{\nu}$, with respect to the crack density parameter f; the self-consistent estimates, (6.6.18), and the estimates obtained from a dilute distribution, (6.6.14) and (6.6.16), are displayed. Here, for completeness, the overall modulus, $\bar{K}$, is also reported,

$$\frac{\bar{K}}{K} = 1 - f\,\frac{16(1-\bar{\nu}^2)}{9(1-2\bar{\nu})}. \tag{6.6.18k}$$

This follows from the definition of the bulk modulus and (6.6.18f,g).

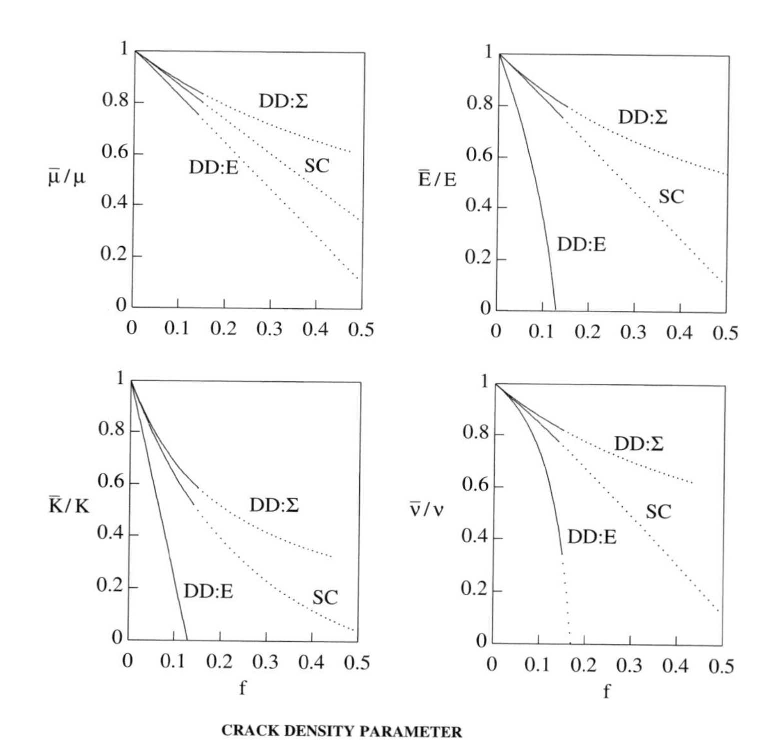

Figure 6.6.4

Normalized overall shear, $\bar{\mu}/\mu$, Young's, $\bar{E}/E$, and bulk, $\bar{K}/K$, moduli, and Poisson's ratio, $\bar{\nu}/\nu$, for random dilute distribution of penny-shaped microcracks; $\nu = 1/3$

SC ≡ self-consistent

DD:Σ ≡ dilute distribution with macrostress prescribed

DD:E ≡ dilute distribution with macrostrain prescribed

The procedure outlined in this subsection may be used to estimate the overall moduli of an elastic solid containing randomly distributed open *elliptical* microcracks. Problems of this kind have been considered by Budiansky and O'Connell (1976), following a different procedure. These authors show that, for elliptical microcracks of a common aspect ratio, b/a = constant, the self-consistent model leads to the following overall moduli:

$$\frac{\bar{\mu}}{\mu} = 1 - f\,\frac{32}{45}(1-\nu)\{1 + \frac{3}{4}T(b/a,\,\bar{\nu})\},$$

$$\frac{\bar{E}}{E} = 1 - f\,\frac{16}{45}(1-\bar{\nu}^2)\{3 + T(b/a,\,\bar{\nu})\}, \qquad (6.6.19a,b)$$

where $\bar{\nu}$ is the solution of the nonlinear equation,

$$f = \frac{45(\nu-\bar{\nu})}{16(1-\bar{\nu}^2)\{2(1+3\nu)-(1-2\nu)T(b/a,\,\bar{\nu})\}}, \qquad (6.6.19c)$$

and T is defined by

$$T(b/a,\,\bar{\nu}) \equiv k^2F(k)\,\{R(k,\,\bar{\nu}) + Q(k,\,\bar{\nu})\}, \qquad k = (1+b^2/a^2)^{1/2},$$

$$R(k,\,\bar{\nu}) \equiv \{(k^2-\bar{\nu})\,F(k) + \bar{\nu}(b/a)^2E(k)\}^{-1},$$

$$Q(k,\,\bar{\nu}) \equiv \{(k^2-\bar{\nu}(b/a)^2)\,F(k) - \bar{\nu}(b/a)^2E(k)\}^{-1}, \qquad (6.6.19d\sim g)$$

where F(k) and E(k) denote the complete elliptic integrals of the first and second kind. The crack density parameter, f, for this problem is defined by

$$f = \frac{2N}{\pi} < \frac{A^2}{P} >, \qquad (6.6.19h)$$

where $A = \pi ab$ is the area and $P = 4\pi F(k)$ is the perimeter of the elliptical crack. Budiansky and O'Connell conclude that the variation of the effective moduli with the crack density parameter f defined by (6.6.19h), is insensitive to the values of the aspect ratio b/a, and indeed, with this definition of f, these moduli may be represented to within a few percent by Equations (6.6.18f,g,k). In view of this conclusion, one may consider equivalent penny-shaped cracks of an effective radius $\bar{a}$, obtained by equating the expression for f given by (6.6.19h) to the corresponding expression for penny-shaped cracks of a common radius $\bar{a}$. This leads to

$$\bar{a} = \{\frac{\pi ab^2}{2E(k)}\}^{1/3}. \qquad (6.6.19i)$$

Thus, the dependence of the results on the aspect ratio b/a for all cracks may be relaxed, as long as the elliptic crack orientations remain random and uncorrelated with their size and aspect ratios. This issue has been further examined by Laws and Brockenbrough (1987) who show that knowledge of the aspect ratio b/a is essential for defining correct families of equivalent penny-shaped cracks, by illustrating that the radius of the equivalent penny-shaped crack is sensitive to the aspect ratio, as is also evident from the Budiansky and O'Connell result (6.6.19i).

6.7. EFFECTIVE MODULI OF AN ELASTIC BODY CONTAINING PENNY-SHAPED MICROCRACKS PARALLEL TO AN AXIS

In the preceding subsections, cases were considered where the overall response of the RVE satisfies some symmetry conditions due to a particular distribution of microcracks. In Subsection 6.5.2, microcracks parallel to the x_1,x_2-plane render the overall compliance and elasticity tensors transversely isotropic, while in Subsections 6.6.2 and 6.6.3, the random distribution of microcracks makes the overall response isotropic. In this subsection, another particular distribution of microcracks is considered, which renders the overall response of the RVE transversely isotropic. Suppose that all microcracks are parallel to the x_3-axis with their unit normals which lie in the x_1,x_2-plane, having a random distribution; see Figure 6.7.1. Then, due to the symmetry in the x_1,x_2-plane, the overall elastic response of the RVE is transversely isotropic, with the x_3-axis being the axis of symmetry. The transverse isotropy induced by randomly distributed microcracks *parallel* to the x_3-axis, and the transverse isotropy induced by microcracks parallel to the x_1,x_2-plane (or *perpendicular* to the x_3-axis), are compared later in this subsection.

As in Subsection 6.5.2, the distribution of the microcracks is assumed to be dilute, and the distribution of the crack sizes is also assumed to be independent of that of their orientations. Then, since the unit normals of the microcracks are uniformly distributed in the x_1,x_2-plane, the crack orientation distribution function $w_o(\theta, \psi)$ is independent of θ. It is given by

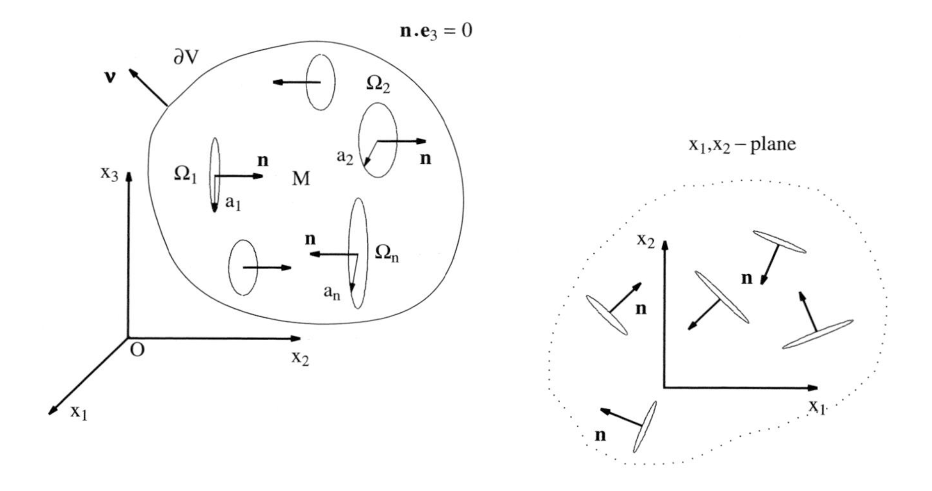

Figure 6.7.1

Microcracks parallel to the x_3-axis

$$w_o(\theta, \psi) = 2\,\delta(\psi - \frac{\pi}{2}), \tag{6.7.1}$$

where δ is the Dirac delta function. From (6.7.1) and (6.6.9a,b), the **H**- and **J**-tensors are obtained,

$$\mathbf{H} = \hat{H}^{\alpha}_{ijkl}\,\{\frac{f}{2\pi}\int_0^{2\pi} \mathbf{e}^{\alpha}_i(\theta, \frac{\pi}{2})\otimes\mathbf{e}^{\alpha}_j(\theta, \frac{\pi}{2})\otimes\mathbf{e}^{\alpha}_k(\theta, \frac{\pi}{2})\otimes\mathbf{e}^{\alpha}_l(\theta, \frac{\pi}{2})\,d\theta\,\},$$

$$\mathbf{J} = \hat{J}^{\alpha}_{ijkl}\,\{\frac{f}{2\pi}\int_0^{2\pi} \mathbf{e}^{\alpha}_i(\theta, \frac{\pi}{2})\otimes\mathbf{e}^{\alpha}_j(\theta, \frac{\pi}{2})\otimes\mathbf{e}^{\alpha}_k(\theta, \frac{\pi}{2})\otimes\mathbf{e}^{\alpha}_l(\theta, \frac{\pi}{2})\,d\theta\,\}, \tag{6.7.2a,b}$$

where f is the crack density parameter defined by (6.6.10a).

From (6.6.3a~i), the components of the coordinate transformation tensor, $\mathbf{Q}^{\alpha}(\theta, \pi/2)$, are

$$Q^{\alpha}_{11}(\theta, \frac{\pi}{2}) = -\sin\theta, \qquad Q^{\alpha}_{21}(\theta, \frac{\pi}{2}) = \cos\theta, \qquad Q^{\alpha}_{31}(\theta, \frac{\pi}{2}) = 0,$$

$$Q^{\alpha}_{12}(\theta, \frac{\pi}{2}) = 0, \qquad Q^{\alpha}_{22}(\theta, \frac{\pi}{2}) = 0, \qquad Q^{\alpha}_{32}(\theta, \frac{\pi}{2}) = 1, \tag{6.7.3a~i}$$

$$Q^{\alpha}_{13}(\theta, \frac{\pi}{2}) = \cos\theta, \qquad Q^{\alpha}_{23}(\theta, \frac{\pi}{2}) = \sin\theta, \qquad Q^{\alpha}_{33}(\theta, \frac{\pi}{2}) = 0.$$

Since $\mathbf{e}^{\alpha}_j = Q^{\alpha}_{ij}\,\mathbf{e}_i$, the integrands in (6.7.2a,b) are expressed in terms of θ, and the components of **H** and **J** in the x_i-coordinates are obtained.

On the other hand, if advantage is taken of transverse isotropy, **H** can be obtained without having to integrate (6.7.2a). Since the x_3-axis is the axis of symmetry, the components of **H** in the x_i-coordinates (denoted by H_{ijkl}) can be expressed in terms of five unknown parameters, say, h_i (i = 1, 2, ..., 5), which depend on the crack density parameter f and the matrix elastic moduli E and ν, as

$$H_{1111} = H_{2222} = h_1, \qquad H_{1122} = h_2, \qquad H_{1133} = H_{2233} = h_3,$$

$$H_{3333} = h_4, \qquad H_{1313} = H_{2323} = h_5, \tag{6.7.4a~e}$$

and also

$$H_{1212} = \frac{1}{2}(H_{1111} - H_{1122}) = \frac{1}{2}(h_1 - h_2) \tag{6.7.4f}$$

and $H_{ijkl} = H_{jikl} = H_{ijlk} = H_{jilk}$, with other components being zero; see (3.1.13).

From (6.6.12a,b), it follows that

$$H_{iijj} = f\,\hat{H}^{\alpha}_{iijj}, \qquad H_{ijij} = f\,\hat{H}^{\alpha}_{ijij}. \tag{6.7.5a,b}$$

Since $\mathbf{e}^{\alpha}_i(\theta, \pi/2)\cdot\mathbf{e}_3 = 1$ for i = 2 and 0 for i = 1 or 3,

$$\mathbf{H}::(\mathbf{e}_i\otimes\mathbf{e}_i\otimes\mathbf{e}_3\otimes\mathbf{e}_3) = H_{ii33} = f\,\hat{H}^{\alpha}_{ii22},$$

$$\mathbf{H}::(\mathbf{e}_i\otimes\mathbf{e}_3\otimes\mathbf{e}_i\otimes\mathbf{e}_3) = H_{i3i3} = f\,\hat{H}^{\alpha}_{i2i2},$$

$$\mathbf{H}::(\mathbf{e}_3 \otimes \mathbf{e}_3 \otimes \mathbf{e}_3 \otimes \mathbf{e}_3) = H_{3333} = f\,\hat{H}^{\alpha}_{2222}. \tag{6.7.5c~e}$$

Now, substitute (6.7.4a~f) and (6.6.4b,c) into H_{ijkl} and $\hat{H}^{\alpha}_{ijkl}$, respectively, to obtain a set of equations for the unknowns h_i ($i = 1, 2, ..., 5$), as follows:

$$2h_1 + 2h_2 + 4h_3 + h_4 = f\,\frac{16(1-\nu^2)}{3E},$$

$$3h_1 - h_2 + h_4 + 4h_5 = f\,\frac{16(1-\nu^2)(4-\nu)}{3E(2-\nu)},$$

$$h_4 + 2h_5 = f\,\frac{8(1-\nu^2)}{3E(2-\nu)},$$

$$2h_3 + h_4 = 0, \qquad h_4 = 0. \tag{6.7.5f~j}$$

Hence,

$$h_1 = f\,\frac{2(1-\nu^2)(8-3\nu)}{3(2-\nu)}\,\frac{1}{E}, \qquad h_2 = -f\,\frac{2\nu(1-\nu^2)}{3(2-\nu)}\,\frac{1}{E},$$

$$h_3 = 0, \qquad h_4 = 0, \qquad h_5 = f\,\frac{4(1-\nu^2)}{3(2-\nu)}\,\frac{1}{E}. \tag{6.7.5k~o}$$

When the RVE is subjected to uniaxial tractions in the x_3-direction, i.e., when only the macrostress component Σ_{33} is nonzero, the open microcracks parallel to the x_3-axis do not contribute to the overall strains. This physical observation implies that $H_{ii33} = 0$ for $i = 1, 2, 3$, and hence both h_3 and h_4 are zero; see (6.7.5m,n). From (4.3.6), the overall compliance tensor $\bar{\mathbf{D}}$ is $\bar{\mathbf{D}} = \mathbf{D} + \mathbf{H}$ which leads to the following overall transversely isotropic elastic moduli for the solid with microcracks:

$$\frac{\bar{E}}{E} = \{1 + f\,\frac{2(1-\nu^2)(8-3\nu)}{3(2-\nu)}\}^{-1} = 1 - f\,\frac{2(1-\nu^2)(8-3\nu)}{3(2-\nu)} + O(f^2),$$

$$\frac{\bar{\nu}}{\nu} = \{1 + f\,\frac{2(1-\nu^2)}{3(2-\nu)}\}\,\{1 + f\,\frac{2(1-\nu^2)(8-3\nu)}{3(2-\nu)}\}^{-1}$$

$$= 1 - f\,\frac{2(1-\nu^2)(7-3\nu)}{3(2-\nu)} + O(f^2),$$

$$\frac{\bar{\mu}}{\mu} = \{1 + f\,\frac{4(1-\nu)(4-\nu)}{3(2-\nu)}\}^{-1} = 1 - f\,\frac{4(1-\nu)(4-\nu)}{3(2-\nu)} + O(f^2), \tag{6.7.6a~c}$$

where $\bar{E}$, $\bar{\nu}$, and $\bar{\mu}$ are the overall Young modulus, Poisson ratio, and shear modulus in the x_1,x_2-plane; and

$$\frac{\bar{E}_3}{E} = 1, \qquad \frac{\bar{\nu}_3}{\nu} = 1,$$

$$\frac{\bar{\mu}_3}{\mu} = \{1 + f\,\frac{8(1-\nu)}{3(2-\nu)}\}^{-1} = 1 - f\,\frac{8(1-\nu)}{3(2-\nu)} + O(f^2), \tag{6.7.6d~f}$$

where $\bar{\nu}_3 = \bar{\nu}_{13} = \bar{\nu}_{23}$ and $\bar{\mu}_3 = \bar{\mu}_{13} = \bar{\mu}_{23}$; see (3.1.13).

Next, consider the **J**-tensor. As shown in Subsections 6.3.2 and 6.5.2, even though both **H** and **C** are transversely isotropic, **J**, given by **H** : **C**, may not be transversely isotropic. However, the tensor **C** : **J** = **C** : **H** : **C** is transversely isotropic, and the components of **C** : **J** satisfy relations similar to (6.7.4a~f). Hence calculation of **C** : **J** follows the steps (6.7.5a~o) outlined for **H**.

In this subsection, however, **J** is computed by direct matrix calculation. Using the same notation as (6.5.5b) and (6.5.7a), denote the matrices of **H** and **J** by

$$[H_{ab}] = \begin{bmatrix} [H^{(1)}_{ab}] & [0] \\ [0] & [H^{(2)}_{ab}] \end{bmatrix}, \qquad [J_{ab}] = \begin{bmatrix} [J^{(1)}_{ab}] & [0] \\ [0] & [J^{(2)}_{ab}] \end{bmatrix}, \tag{6.7.7a,b}$$

where $[H^{(1)}_{ab}]$ and $[H^{(2)}_{ab}]$ are

$$[H^{(1)}_{ab}] = f\,\frac{2(1-\nu^2)}{3(2-\nu)E} \begin{bmatrix} 8-3\nu & -\nu & 0 \\ -\nu & 8-3\nu & 0 \\ 0 & 0 & 0 \end{bmatrix},$$

$$[H^{(2)}_{ab}] = f\,\frac{8(1-\nu^2)}{3(2-\nu)E} \begin{bmatrix} 2 & 0 & 0 \\ 0 & 2 & 0 \\ 0 & 0 & 4-\nu \end{bmatrix}. \tag{6.7.7c,d}$$

From **J** = **H** : **C**,

$$[J^{(1)}_{ac}] = [H^{(1)}_{ab}][C^{(1)}_{bc}]$$

$$= f\,\frac{2(1-\nu)}{3(2-\nu)(1-2\nu)} \begin{bmatrix} 2\nu^2-11\nu+8 & -2\nu^2+7\nu & -4\nu^2+8\nu \\ -2\nu^2+7\nu & 2\nu^2-11\nu+8 & -4\nu^2+8\nu \\ 0 & 0 & 0 \end{bmatrix},$$

$$[J^{(2)}_{ac}] = [H^{(2)}_{ab}][C^{(2)}_{bc}] = f\,\frac{4(1-\nu)}{3(2-\nu)} \begin{bmatrix} 2 & 0 & 0 \\ 0 & 2 & 0 \\ 0 & 0 & 4-\nu \end{bmatrix}. \tag{6.7.8a,b}$$

Although $[J^{(1)}_{ab}]$ is not symmetric, $[C^{(1)}_{ab}][J^{(1)}_{bc}]$ is symmetric; see (6.3.8c) or (6.5.9c). Then,

$$[C^{(1)}_{ab}][J^{(1)}_{bc}] = f\,\frac{2(1-\nu)E}{3(1+\nu)(2-\nu)(1-2\nu)^2}$$

$$\times \begin{bmatrix} -4\nu^3+20\nu^2-19\nu+8 & 4\nu^3-20\nu^2+15\nu & -4\nu^2+8\nu \\ 4\nu^3-20\nu^2+15\nu & -4\nu^3+20\nu^2-19\nu+8 & -4\nu^2+8\nu \\ -4\nu^2+8\nu & -4\nu^2+8\nu & -8\nu^3+16\nu^2 \end{bmatrix},$$

$$[C^{(2)}_{ab}][J^{(2)}_{ab}] = f\,\frac{2(1-\nu)E}{3(1+\nu)(2-\nu)} \begin{bmatrix} 2 & 0 & 0 \\ 0 & 2 & 0 \\ 0 & 0 & 4-\nu \end{bmatrix}. \tag{6.7.8c,d}$$

Since

$$(\mathbf{C}:\mathbf{J})_{1212} = \frac{1}{2}\{(\mathbf{C}:\mathbf{J})_{1111} - (\mathbf{C}:\mathbf{J})_{1122}\} = f\,\frac{2(1-\nu)(4-\nu)E}{3(1+\nu)(2-\nu)}, \tag{6.7.8e}$$

or

$$([C^{(2)}_{ab}][J^{(2)}_{bc}])_{33} = \frac{1}{2}\{([C^{(1)}_{ab}][J^{(1)}_{bc}])_{11} - ([C^{(1)}_{ab}][J^{(1)}_{bc}])_{12}\}$$

$$= f\,\frac{2(1-\nu)(4-\nu)E}{3(1+\nu)(2-\nu)}; \tag{6.7.8f}$$

tensor $\mathbf{C}:\mathbf{J}$ is transversely isotropic. From (6.7.8c,d), various components of the overall elasticity tensor $\bar{\mathbf{C}}$ are obtained. For example, the overall shear moduli, $\bar{\mu} = \bar{\mu}_{12}$ and $\bar{\mu}_3 = \bar{\mu}_{13} = \bar{\mu}_{23}$, are

$$\frac{\bar{\mu}}{\mu} = 1 - f\,\frac{4(1-\nu)(4-\nu)}{3(2-\nu)}, \qquad \frac{\bar{\mu}_3}{\mu} = 1 - f\,\frac{8(1-\nu)}{3(2-\nu)}. \tag{6.7.9a,b}$$

The overall shear moduli $\bar{\mu}$ given by (6.7.6c) and (6.7.9a), and $\bar{\mu}_3$ given by (6.7.6f) and (6.7.9b), agree up to the first order in the crack density parameter f; see Figure 6.7.2.

Now, consider the transverse isotropies associated with two different random distributions of microcracks, one with all cracks perpendicular to the x_3-axis and the other with cracks parallel to this axis. The overall transversely isotropic elastic moduli for the first case are given by (6.7.6a~f) and (6.7.9a,b), and those for the microcracks parallel to the x_3-axis, are given by (6.5.6a~f) and (6.5.10a,b). From comparison of the moduli, $\bar{E}_3$, $\bar{E}$, $\bar{\mu}$, and $\bar{\mu}_3$, for the two cases, it is concluded that: (1) under uniaxial tension in the x_3-axis, the RVE containing microcracks parallel to the x_3-axis is stiffer than the one containing microcracks perpendicular to this axis (this is obvious on physical grounds); (2) under biaxial loading in the x_1,x_2-plane, the RVE containing microcracks parallel to the x_3-axis is less stiff than the one containing microcracks perpendicular to this

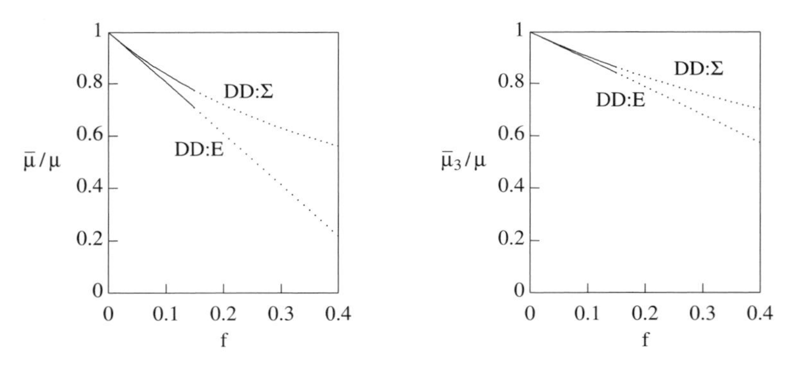

Figure 6.7.2

Comparison of normalized overall shear moduli for penny-shaped microcracks parallel to the x_3-axis; $\nu = 1/3$
DD:Σ ≡ dilute distribution with macrostress prescribed
DD:E ≡ dilute distribution with macrostrain prescribed

axis (this is also obvious on physical grounds); and (3) under shear loading in the x_1,x_3-plane or in the x_2,x_3-plane, the RVE containing microcracks parallel to the x_3-axis is stiffer, whereas under shear loading in the x_1,x_2-plane, the RVE with microcracks perpendicular to the x_3-axis is stiffer. The results for microcracks parallel to the x_3-axis have been used by Rogers and Nemat-Nasser (1990) to model damage evolution in magnesia-partially-stablized zirconia (Mg-PSZ) ceramic samples subjected to uniaxial compressive stress pulses; similar modeling has been used by Subhash and Nemat-Nasser (1993).

6.8. INTERACTION EFFECTS

When microcracks are closely spaced, their interaction may require a more effective modeling than that provided by the self-consistent method. The periodic model presented in Sections 12, 13, and 14 may then be an attractive alternative. As an intermediate step, one may consider an elastic solid containing a certain distribution of *rows of collinear* cracks, as recently proposed by Deng and Nemat-Nasser (1992a) in a two-dimensional setting. The slit crack arrays in this model are all assumed to be parallel to, say, the x_3-axis. In addition, they may be either all parallel or randomly oriented, resulting in an overall (two-dimensional, i.e., inplane) orthotropic or isotropic material response, respectively; three-dimensionally, the response then is orthotropic or transversely isotropic, respectively. The method of dilute distribution, the self-consistent method, and the differential scheme (see Section 10) are used by these authors to estimate the overall instantaneous moduli, focusing attention on two-dimensional problems, either plane strain or plane stress. This is accomplished by considering a crack array on the x_1-axis, Figure 6.8.1, and using the corresponding solution for the COD to calculate the required **H**- and **J**-tensors.

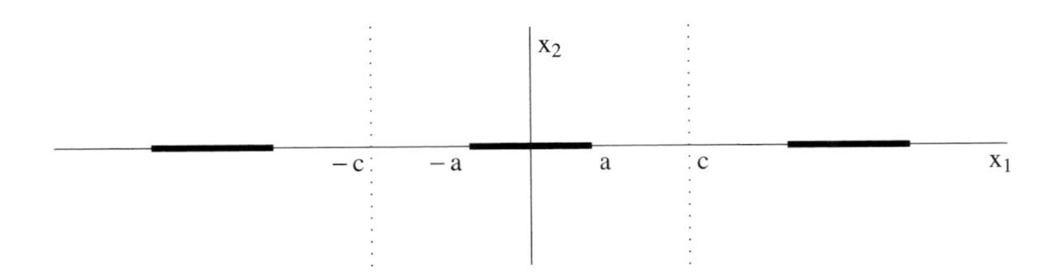

Figure 6.8.1

A row of equally-spaced equal collinear cracks on the x_1-axis

6.8.1. Crack Opening Displacements and Associated Strains

The elasticity problem for a collinear crack array has been addressed by Irwin (1958), Koiter (1959), England and Green (1963), Sneddon and Srivastav (1965), and Sneddon and Lowengrub (1969) for isotropic media. Deng and Nemat-Nasser (1992a) use the results of Nemat-Nasser and Hori (1987), and report closed-form expressions for the crack opening displacements, for a crack array in a transversely isotropic elastic solid, and also in an orthotropic elastic solid; see Section 21. They then use these results to estimate the corresponding overall moduli in plane strain and plane stress cases.

The COD's for a row of cracks (Figure 6.8.1) in a transversely isotropic elastic matrix, under uniform farfield stresses $\boldsymbol{\sigma}^\infty$, are given by

$$\{[u_1],\,[u_2],\,[u_3]\} = T(x_1)\frac{1}{\sqrt{2}}\{\frac{(\kappa+1)}{\mu}\,\sigma_{21}^\infty,\ \frac{(\kappa+1)}{\mu}\,\sigma_{22}^\infty,\ \frac{4}{\mu_{23}}\,\sigma_{23}^\infty\}, \tag{6.8.1a}$$

where μ and μ_{23} are the inplane and out-of-plane shear moduli, associated with the x_1,x_2- and x_2,x_3-coordinates, respectively, and

$$T(x_1) = \cos(\frac{\pi x_1}{2c})\int_{x_1}^{a} \tan(\frac{\pi\xi}{2c})\,\{\cos(\frac{\pi x_1}{c}) - \cos(\frac{\pi\xi}{c})\}^{-1/2}\,d\xi. \tag{6.8.1b}$$

The strain components due to a *single* crack in this array of collinear cracks now are

$$\{\varepsilon_{22}^\alpha,\ \varepsilon_{21}^\alpha,\ \varepsilon_{23}^\alpha\} = S\{\frac{2(\kappa+1)}{\mu}\,\sigma_{22}^\infty,\ \frac{(\kappa+1)}{\mu}\,\sigma_{21}^\infty,\ \frac{4}{\mu_{23}}\,\sigma_{23}^\infty\}, \tag{6.8.2a}$$

where

$$S = -\,\frac{c^2}{\pi a^2}\ln\cos(\frac{\pi a}{2c}). \tag{6.8.2b}$$

Hence, (6.8.2a) becomes

$$\{\varepsilon_{22}^\alpha,\,\varepsilon_{21}^\alpha,\,\varepsilon_{23}^\alpha\} = -\,\frac{c^2}{\pi a^2}\ln\cos(\frac{\pi a}{2c})\,\{\frac{2(\kappa+1)}{\mu}\,\sigma_{22}^\infty,\ \frac{(\kappa+1)}{\mu}\,\sigma_{21}^\infty,\ \frac{4}{\mu_{23}}\,\sigma_{23}^\infty\} \tag{6.8.2c}$$

which yields the following limiting expression:

$$\lim_{a/c\to 0}\{\varepsilon_{22}^\alpha,\ \varepsilon_{21}^\alpha,\ \varepsilon_{23}^\alpha\} = \frac{\pi}{8}\{\frac{2(\kappa+1)}{\mu}\,\sigma_{22}^\infty,\ \frac{(\kappa+1)}{\mu}\,\sigma_{21}^\infty,\ \frac{4}{\mu_{23}}\,\sigma_{23}^\infty\}, \tag{6.8.2d}$$

in agreement with the results for a single crack in an infinitely extended solid, given in Subsection 6.3.2.

As is shown in Section 21, the crack opening displacements $[u_\alpha]$ ($\alpha = 1, 2$) and the associated strains for a general two-dimensional anisotropic case (plane strain or plane stress), can be related to the corresponding results for the transversely isotropic case, as follows (Nemat-Nasser and Hori, 1987; Deng and Nemat-Nasser, 1992a):

$$[u_\alpha] = P_\alpha\,[u_\alpha^{iso}] \qquad (\alpha \text{ not summed}), \tag{6.8.3a}$$

where

$$P_1 = \left\{ \frac{1}{2D_{1111}} \{2D_{1212} + D_{1122} + (D_{1111}D_{2222})^{1/2}\} \right\}^{1/2},$$

$$P_2 = \frac{1}{2D_{1111}} \left\{ 2D_{2222} \{2D_{1212} + D_{1122} + (D_{1111}D_{2222})^{1/2}\} \right\}^{1/2}; \qquad (6.8.3b,c)$$

here, the components of the compliance tensor are interpreted for *plane* problems, as

$$D_{1111} = \frac{1}{E_1} - \eta \frac{\nu_{31}^2}{E_3}, \qquad D_{2222} = \frac{1}{E_2} - \eta \frac{\nu_{32}^2}{E_3},$$

$$D_{1122} = D_{2211} = -\frac{\nu_{21}}{E_2} - \eta \frac{\nu_{31}\nu_{32}}{E_3}, \qquad (6.8.4a\text{~}c)$$

where $\eta = 0$ in plane stress and $\eta = 1$ in plane strain; see Section 3 for notation. Therefore, for the two-dimensionally anisotropic matrix, the strains associated with *a single crack in a crack array* are,

$$\{\varepsilon_{22}^{\alpha}, \varepsilon_{21}^{\alpha}, \varepsilon_{23}^{\alpha}\} = -\frac{8c^2}{\pi a^2} \ln \cos(\frac{\pi a}{2c})$$

$$\times \{2P_2 D_{1111} \sigma_{22}^{\infty}, P_1 D_{1111} \sigma_{12}^{\infty}, 2D_{2323} \sigma_{23}^{\infty}\}, \qquad (6.8.5)$$

and, in view of these, the fourth-order tensor $\mathbf{H}^{\alpha}$,

$$\boldsymbol{\varepsilon}^{\alpha} = \mathbf{H}^{\alpha} : \boldsymbol{\sigma}^{\infty} \quad \text{or} \quad \varepsilon_{ij}^{\alpha} = H_{ijkl}^{\alpha} \sigma_{kl}^{\infty}, \quad (i, j, k, l = 1, 2, 3), \qquad (6.8.6a,b)$$

has the following nonzero components:

$$H_{2222}^{\alpha} = -\frac{16c^2}{\pi a^2} P_2 D_{1111} \ln \cos(\frac{\pi a}{2c}),$$

$$H_{2121}^{\alpha} = H_{2112}^{\alpha} = H_{1221}^{\alpha} = H_{1212}^{\alpha} = \frac{1}{4} H_{2222}^{\alpha} \{D_{1111} / D_{2222}\}^{1/2},$$

$$H_{2323}^{\alpha} = H_{2332}^{\alpha} = H_{3223}^{\alpha} = H_{3232}^{\alpha} = -\frac{8c^2}{\pi a^2} D_{2323} \ln \cos(\frac{\pi a}{2c}). \qquad (6.8.7a\text{~}c)$$

Since the strain components due to a crack in a row of cracks, depend only on the ratio, a/c, of the common crack length and spacing, the inelastic strains due to crack arrays of different crack lengths but identical a/c-ratio, are the same. For a parallel distribution of crack arrays, the overall strain due to a large number of crack arrays with the same a/c-ratio, therefore, depends only on the number of cracks measured per unit area, normal to the x_3-direction. Thus, for an estimate of the overall inelastic strains, an average crack length, 2a, and an average crack spacing, 2c, may be used in order to obtain an effective {crack length} / {crack spacing} = a/c, and then (6.8.2a) may be used, for an isotropic matrix, or (6.8.5) for an orthotropic matrix. For randomly oriented crack arrays, the overall strain is averaged over all orientations.

6.8.2. Dilute Distribution of Parallel Crack Arrays

For a dilute distribution of crack arrays, the interaction between any two *arrays* may be neglected. For an elastic matrix which contains collinear parallel rows of crack arrays *with the same* a/c-*ratio*, the corresponding overall inelastic strains due to cracks become,

$$\bar{\boldsymbol{\varepsilon}}^c = \sum_{\alpha=1}^{n} N_\alpha \int_{-a_\alpha}^{a_\alpha} \frac{1}{2}\{[\mathbf{u}]\otimes\mathbf{n}+\mathbf{n}\otimes[\mathbf{u}]\}\,dx = \sum_{\alpha=1}^{n} N_\alpha\, a_\alpha^2\, \boldsymbol{\varepsilon}^\alpha = f\,\boldsymbol{\varepsilon}^\alpha, \tag{6.8.8}$$

where N_α is the number of cracks of length $2a_\alpha$ per unit area. Then, for a three-dimensionally isotropic matrix, it follows that,

$$\bar{\varepsilon}_{22}^c = -f\,\frac{16D_{1111}c^2\,\sigma_{22}^\infty}{\pi a^2}\ln\cos(\frac{\pi a}{2c}), \qquad \bar{\varepsilon}_{12}^c = -f\,\frac{8D_{1111}c^2\,\sigma_{12}^\infty}{\pi a^2}\ln\cos(\frac{\pi a}{2c}),$$

$$\bar{\varepsilon}_{23}^c = -f\,\frac{16D_{2323}c^2\,\sigma_{23}^\infty}{\pi a^2}\ln\cos(\frac{\pi a}{2c}). \tag{6.8.9a~c}$$

With $\bar{\varepsilon}_{ij}^c = H_{ijkl}\,\sigma_{kl}^\infty$, for plane and anti-plane problems, arrive at

$$H_{2222} = -f\,\frac{16D_{1111}c^2}{\pi a^2}\ln\cos(\frac{\pi a}{2c}),$$

$$H_{1221} = H_{2112} = H_{2121} = H_{1212} = -f\,\frac{2D_{1212}c^2(\kappa+1)}{\pi a^2}\ln\cos(\frac{\pi a}{2c}),$$

$$H_{2332} = H_{3223} = H_{3232} = H_{2323} = -f\,\frac{8D_{2323}c^2}{\pi a^2}\ln\cos(\frac{\pi a}{2c}), \tag{6.8.10a~c}$$

where, for the isotropic matrix, $\kappa = 3-4\nu$ for plane strain and $\kappa = (3-\nu)/(1+\nu)$ for plane stress, and all other components of **H** are zero. The overall response is three-dimensionally orthotropic, with $\bar{E}_1 = \bar{E}_3 = E$, $\bar{\nu}_{31} = \bar{\nu}_{13} = \nu$, and $\bar{\mu}_{13} = \bar{\mu}_{31} = \mu$; see (3.1.14b).

For the case when the macrostress, $\boldsymbol{\Sigma} = \boldsymbol{\sigma}^o$, is prescribed, instead of (6.3.4), (6.3.6), and (6.3.7), which are obtained from a single-crack solution, now obtain, for crack arrays,

$$\frac{\bar{E}'_2}{E'} = \{1 - f\,\frac{16c^2}{\pi a^2}\ln\cos(\frac{\pi a}{2c})\}^{-1},$$

$$\frac{\bar{\mu}_{12}}{\mu} = \{1 - f\,\frac{2c^2(\kappa+1)}{\pi a^2}\ln\cos(\frac{\pi a}{2c})\}^{-1},$$

$$\frac{\bar{\mu}_{23}}{\mu} = \{1 - f\,\frac{8c^2}{\pi a^2}\ln\cos(\frac{\pi a}{2c})\}^{-1}. \tag{6.8.11a~c}$$

Indeed, for the special case of $a/c \to 0$, these results reduce to the corresponding equations of Subsection 6.3. If the matrix material is anisotropic, for example, it is orthotropic, and the crack arrays are aligned with one of the principal material directions, then, instead of (6.8.2a), expression (6.8.5) must be used to obtain the strains due to the cracks. Note that (6.8.11a) yields

$$\frac{\bar{E}_2}{E} = \{1 - f(1 - \eta\nu^2)\frac{16c^2}{\pi a^2} \ln\cos(\frac{\pi a}{2c})\}^{-1}, \tag{6.8.11d}$$

for both plane stress ($\eta = 0$) and plane strain ($\eta = 1$). This is identical with (6.3.4c) at the limit as $a/c \to 0$.

When the macrostrain, $\mathbf{E} = \boldsymbol{\varepsilon}^o$, is regarded prescribed, the dilute distribution model with parallel *rows* of cracks, yields

$$\frac{\bar{\mu}_{12}}{\mu} = 1 - 4\frac{C_{1212}\,H_{1212}\,C_{1212}}{\mu} = 1 + f\,\frac{2c^2(\kappa+1)}{\pi a^2} \ln\cos(\frac{\pi a}{2c}),$$

$$\frac{\bar{\mu}_{23}}{\mu} = 1 - 4\frac{C_{2323}\,H_{2323}\,C_{2323}}{\mu} = 1 + f\,\frac{8c^2}{\pi a^2} \ln\cos(\frac{\pi a}{2c}), \tag{6.8.12a,b}$$

which also reduce to the results of Subsection 6.3, as $a/c \to 0$. Note again here that the overall compliance and elasticity tensors are each other's inverse only to the first order in f.

Consider now the self-consistent averaging method. For *crack arrays parallel to the* x_1*-axis*, the $\bar{\mathbf{H}}^\alpha$-tensor for the self-consistent method is defined by (6.8.7), except that the relevant components of the compliance tensor in these equations must be replaced by the corresponding average overall quantities, i.e., all relevant D_{ijkl} are to be replaced by $\bar{D}_{ijkl}$. This results in the following nonzero components for the $\bar{\mathbf{H}}$-tensor:

$$\bar{H}_{2222} = f\,\bar{H}^\alpha_{2222} = -f\,\frac{16c^2}{\pi a^2}\bar{P}_2\,\bar{D}_{1111} \ln\cos(\frac{\pi a}{2c}),$$

$$\bar{H}_{1212} = f\,\bar{H}^\alpha_{1212} = f\,\frac{1}{4}\bar{H}_{2222}\,\{\bar{D}_{1111}/\bar{D}_{2222}\}^{1/2},$$

$$\bar{H}_{2323} = f\,\bar{H}^\alpha_{2323} = -f\,\frac{8c^2\bar{D}_{2323}}{\pi a^2} \ln\cos(\frac{\pi a}{2c}). \tag{6.8.13a~c}$$

Because of the special crack arrangement, however, it follows that

$$\bar{D}_{1111} = D_{1111}, \qquad \bar{D}_{1122} = D_{1122}. \tag{6.8.14a,b}$$

For an isotropic matrix, $D_{1111} = D_{2222} = \frac{(\kappa+1)}{8\mu}$ and $D_{1122} = \frac{(\kappa-3)}{8\mu}$, leading to

$$\frac{E'}{\bar{E}'_2} = 1 - f\,\frac{8c^2}{\pi a^2} \ln\cos(\frac{\pi a}{2c})$$
$$\times \left\{ \frac{E'}{\bar{E}'_2}\left[\frac{8\mu}{(\kappa+1)\bar{\mu}_{12}} + \frac{2(\kappa-3)}{\kappa+1} + 2\sqrt{E'/\bar{E}'_2}\right] \right\}^{1/2},$$

$$\frac{\mu}{\bar{\mu}_{12}} = 1 - f\,\frac{(\kappa+1)c^2}{\pi a^2} \ln\cos(\frac{\pi a}{2c})$$
$$\times \left\{ \frac{8\mu}{(\kappa+1)\bar{\mu}_{12}} + \frac{2(\kappa-3)}{\kappa+1} + 2\sqrt{E'/\bar{E}'_2}\right\}^{1/2},$$

$$\frac{\bar{\mu}_{23}}{\mu} = 1 + f \frac{8c^2}{\pi a^2} \ln\cos(\frac{\pi a}{2c}). \qquad (6.8.15a\sim c)$$

The first two equations in (6.8.15) are coupled, and hence the corresponding overall elastic moduli are calculated by iteration. As an illustration, Figure 6.8.2 gives the self-consistent estimate of $\bar{E}_2/E$ and $\bar{\mu}_{12}/\mu$ in plane stress, for indicated values of a/c. For comparison, the limiting results for a/c = 0 are also given.

6.8.3. Randomly Oriented Open Slit Crack Arrays Parallel to an Axis

Consider an isotropic elastic matrix containing collinear open slit crack arrays with random orientations, parallel to the x_3-axis, resulting in a transversely isotropic overall response. Based on the procedure outlined in Subsection 6.4, and in view of the exact correspondence between the $\mathbf{H}^{\alpha}$-tensor for a single crack, given in Subsection 6.4, and that given in the present section for an array of collinear cracks, it follows that the overall moduli can be written down by simple inspection.

Consider first the dilute distribution model with macrostresses regarded prescribed. With parameter S defined in (6.8.2b), the components of the compliance tensor are readily obtained to be

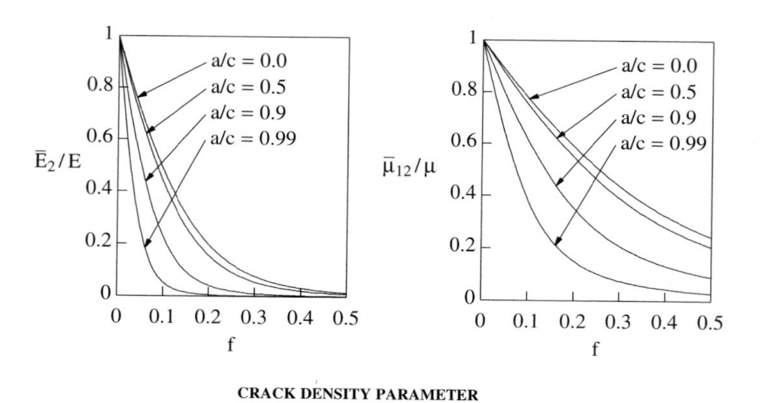

Figure 6.8.2

The self-consistent estimate of the normalized x_2-direction Young modulus, $\bar{E}_2/E$, and the inplane shear modulus, $\bar{\mu}_{12}/\mu$, for random distribution of collinear frictionless crack arrays, parallel to the x_1-axis; $\nu = 1/3$, and a/c as indicated

$$\frac{\bar{E}'}{E'} = (1 + 8S\,f)^{-1},$$

$$\frac{\bar{\mu}}{\mu} = \{1 + 2(\kappa + 1)S\,f\}^{-1},$$

$$\frac{\bar{\mu}_{23}}{\mu} = (1 + 4S\,f)^{-1}, \tag{6.8.16a~c}$$

where $\bar{E}'$ and $\bar{\mu}$ are the nominal inplane Young and the inplane shear moduli. The inplane Young modulus then is

$$\frac{\bar{E}}{E} = \{1 + 8(1 - \eta\nu^2)S\,f\}^{-1}, \tag{6.8.16d}$$

where $\eta = 0$ for plane stress and $\eta = 1$ for plane strain.

Next consider the self-consistent method. The components of the **H**-tensor then are,

$$\bar{H}_{1111} = \bar{H}_{2222} = \frac{2(\bar{\kappa} + 1)}{\bar{\mu}} S\,f,$$

$$\bar{H}_{1212} = \bar{H}_{2121} = \bar{H}_{1221} = \bar{H}_{2112} = \frac{1}{4}\,\bar{H}_{2222},$$

$$\bar{H}_{1313} = \bar{H}_{3131} = \bar{H}_{3113} = \bar{H}_{1331} = \bar{H}_{2323} = \bar{H}_{3232} = \bar{H}_{3223} = \bar{H}_{2332} = \frac{2S}{\bar{\mu}_{23}}\,f, \tag{6.8.17a~c}$$

where $\bar{\kappa} = (3 - \bar{\nu})/(1 + \bar{\nu})$ for plane stress and $\bar{\kappa} = 3 - 8\bar{\mu}(\bar{\nu}/\bar{E} + \nu^2/E)$ for plane strain; here, $\bar{\nu}$ is the inplane overall Poisson ratio. Thus, the relevant components of the overall transversely isotropic compliance tensor are estimated by the self-consistent method to be,

$$\frac{\bar{E}'}{E'} = 1 - 8S\,f,$$

$$\frac{\bar{\mu}}{\mu} = \xi\,(\xi - 8\nu S\,f)\,(1 - 8S\,f)^{-1},$$

$$\frac{\bar{\mu}_{23}}{\mu} = 1 - 4S\,f, \tag{6.8.18a~c}$$

where $\xi = 1$ for plane strain and $\xi = 1 + \nu$ for plane stress. From the first relation, the effective inplane Young modulus is given by

$$\frac{\bar{E}}{E} = (1 - 8S\,f)(1 - 8\eta\nu^2 S\,f)^{-1}, \tag{6.8.18d}$$

where $\eta = 0$ for plane stress and $\eta = 1$ for plane strain, respectively. For plane stress, $\bar{E}/E$ and $\bar{\mu}/\mu$ are plotted in Figure 6.8.3, showing the influence of a/c on these moduli.

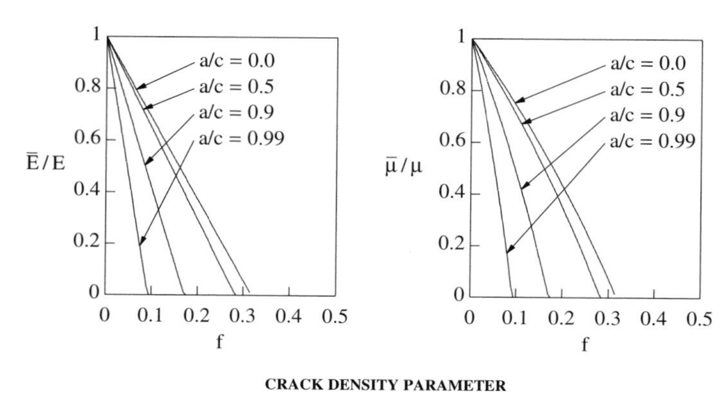

Figure 6.8.3

The self-consistent estimate of the normalized inplane Young modulus, $\bar{E}/E$, and the shear modulus, $\bar{\mu}/\mu$, for collinear frictionless crack arrays with randomly distributed orientation of the arrays; $\nu = 1/3$, and a/c as indicated

6.9. BRITTLE FAILURE IN COMPRESSION

The material developed in this section can be used to study failure of brittle solids with microdefects. Tensile cracking is a common mode of failure of many brittle materials. Even under all-around compressive loads, brittle materials tend to fail by the formation of tensile microcracks at microdefects such as cavities, grain boundaries, inclusions, and other inhomogeneities; see Figure 1.1.1 of Section 1, which represents vivid examples of compression-induced axial tensile cracks, and a recent review of the micromechanics of rock failure by Myer *et al.* (1992). As the overall confining pressure is increased, plastic flow may accompany microcracking, and eventually may become the dominant mechanism of the overall deformation. Hence, under great confining pressure, brittle materials such as rocks and ceramics may undergo plastic flow before rupture. In this subsection, some aspects of brittle failure in compression are briefly discussed; tensile failure of microflawed solids is discussed by Karihaloo and Huang (1991).

6.9.1. Introductory Comments

Failure of materials by formation and growth of tension cracks under *tensile* loading is extensively studied and to a large extent understood. Failure under overall compression, on the other hand, has received considerably less attention. Historically, experimental investigation of compressive failure of materials such as rocks has led to paradoxes. Bridgman (1931) demonstrates several failure modes peculiar to high pressures, leading to paradoxical results which impel him to express skepticism on whether there is such a thing as a genuine rupture criterion. The common feature of these paradoxes is that failure always occurs by the formation of *tension* cracks in specimens subjected to *pure compression*. Efforts to observe through electron microscopy the fracture pattern in failed specimens have raised further questions, since microcracks have been seen to have emanated from a variety of defects in various directions, although predominantly in the direction of maximum compression. These and related difficulties have led several authors to criticize micromechanical models that have been suggested for explaining brittle failure under compressive loads.

Over the past decade, several developments have helped to bring the issue of *brittle failure in compression* to a somewhat satisfactory level of basic understanding. The unexplained Bridgman paradoxes have been resolved (Scholz *et al.*, 1986), models which satisfactorily and quantitatively explain axial splitting, faulting, and transition from brittle to ductile modes of failure have been developed, and, most importantly, the mechanisms of fracturing in *loading* and *unloading* have been captured experimentally and by means of laboratory models. These have given credence to the simple but effective micromechanical modeling of brittle failure on the basis of preexisting flaws with frictional and cohesive resistance. Such a model, though an idealization of a rather complex process, seems to capture the observed phenomenon of axial splitting in the absence of confinement, as well as the related phenomena of exfoliation or sheet fracture, and rockburst; Holzhausen (1978), Nemat-Nasser and Horii (1982), and Ashby and Hallam (1986). In the presence of moderate confining pressures, furthermore, faulting by the interaction of preexisting microflaws has also been modeled, by considering the interactive growth of tension cracks from an echelon of suitably oriented microflaws; Horii and Nemat-Nasser (1985a). Failure by faulting through intensive cracking in the presence of confining pressure has been observed experimentally by, e.g., Hallbauer *et al.* (1973), Olsson (1974), and Kranz (1983); see also Myer *et al.* (1992). Furthermore, by including, in addition to tension cracks, possible zones of plastically deformed materials at high shear-stress regions around preexisting flaws, the transition from brittle-type failure to ductile flow under very high confining pressures has been modeled; Horii and Nemat-Nasser (1986). A series of accompanying experimental model studies lends qualitative support to these analytical results. In particular, the influence of confining pressure on the mode of failure of brittle materials seems to have been understood and modeled under quasi-static loads.

During *unloading*, however, microcracks may grow essentially *normal* to the direction of the applied compression. Indeed, even extremely ductile crystalline solids such as single-crystal copper (an fcc metal), and mild steel and pure iron (bcc metals) can undergo *tensile cracking normal to the direction of compression*, possibly during the unloading phase, under suitable conditions; see Nemat-Nasser and Chang (1990). In these experiments, the sample experiences *only compressive loading and unloading. Nevertheless, tension cracks are developed, basically normal to the applied compression*; see Nemat-Nasser and Hori (1987) for a model prediction of this phenomenon.

6.9.2. Bridgman Paradoxes

Bridgman performed a number of experiments on failure in compression which led to several paradoxes. In each case, a sample of essentially brittle material is subjected to high fluid pressures; see Figures 6.9.1a~e. Two of Bridgman's paradoxes, one called the *pinching-off effect* and the other the *ring paradox*, have been shown to be basically due to hydraulic fracturing; see Jaeger and Cook (1963) and Scholz *et al.* (1986).

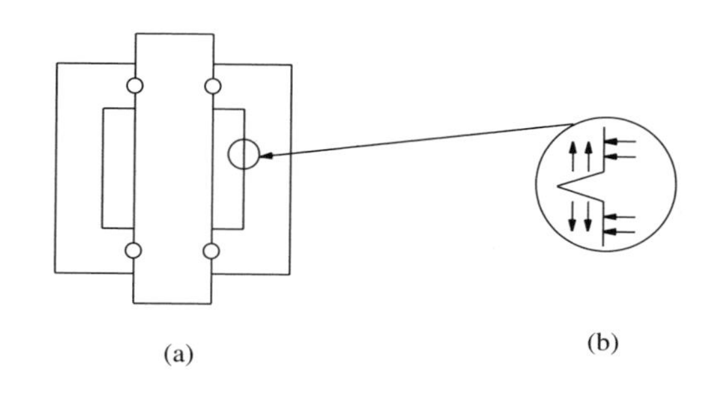

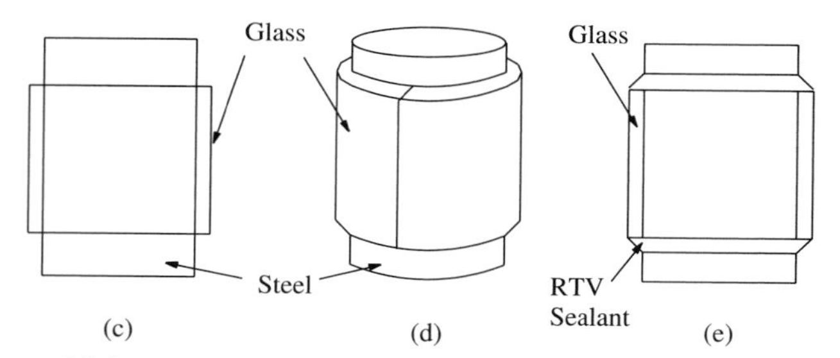

Figure 6.9.1

(a) Bridgman's pinching-off experiment; (b) Failure caused by hydraulic fracturing; (c) Bridgman's ring paradox; (d) Axial crack due to hydraulic fracturing; (e) Bridgman's second ring paradox

In the *pinching-off experiment*, a long cylindrical sample of circular cross section is placed in a chamber with the two ends of the rod extending out of the chamber, as shown in Figure 6.9.1a. The chamber contains pressurized fluid. At a certain pressure on the order of, but greater than the tensile strength of the sample, the sample fractures with a *crack normal to its axis*, somewhere close to its mid-length, and is explosively discharged from the chamber. Jaeger and Cook (1963) repeat the test and obtain similar results. They reason that the pressurized fluid penetrating into preexisting flaws can drive a crack in a direction normal to the axis of the cylinder, as sketched in Figure 6.9.1b. Since the stress intensity factor at the tip of such a crack increases essentially as the square root of the crack length, once the crack begins to grow, the growth accelerates with increasing crack length, leading to explosive dynamic failure.

The *ring paradox* which also is repeated and solved by Jaeger and Cook (1963), again involves hydraulic fracturing. The experiment consists of a thin cylindrical tube of a brittle material (Bridgman used hard rubber), tightly fitted over a solid steel cylinder (Figure 6.9.1c), and totally immersed in a fluid bath and pressurized; note that the entire package, i.e., the tube and the steel cylinder, is under hydrostatic fluid pressure. The tube fractures by a *single axial crack* which apparently starts from its interior surface and grows radially toward its exterior surface; see Figure 6.9.1d. Here again, hydraulic fracturing occurs from an axial flaw at the interior surface of the tube, since the hoop stress σ_θ is the smallest compressive principal stress in the tube. Failure occurs at a pressure on the order of the tensile strength of the tube. Bridgman repeats the same experiment, except that the ends of the tube are sealed, as sketched in Figure 6.9.1e, before submerging it in the fluid which is then pressurized. Again, an axial crack develops, this time presumably from the exterior surface inward.

Bridgman repeats the pinching-off experiment of Figure 6.9.1a, but this time encloses the cylindrical sample in a rubber jacket, as sketched in Figure 6.9.2a. The failure mode is quite different from the unjacketed sample, occurring at pressures close to the *compressive* (rather than the tensile) strength of the sample. Failure occurs by *stable growth of a number of tensile cracks normal to the axis of the cylinder*, breaking the sample into several disks. Jaeger and Cook (1963) repeat the test, arriving at similar results.

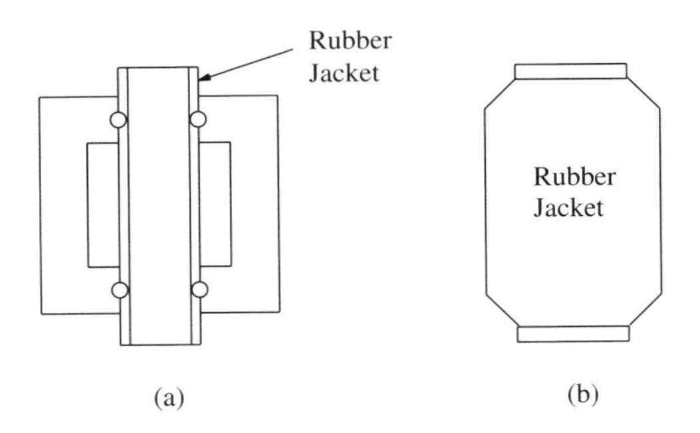

Figure 6.9.2

(a) Bridgman's pinching-off experiment with sealed sample (failure by axial splitting); (b) Bridgman's third ring experiment

Related to this paradox is another version of the *ring paradox*. Both of these paradoxes have been more elusive and indeed, touch on some rather subtle aspects of *brittle failure of brittle materials under all-around compressive loads*. In the third ring experiment, Bridgman jackets the sealed tube/steel construction of Figure 6.9.1e (as shown in Figure 6.9.2b), before submerging it in a fluid bath which is then pressurized. It is observed that axial tension cracks develop from the interior surface of the tube in the radial direction, growing axially, apparently in a stable manner, and never reaching the exterior surface of the tube.

All three *ring* experiments have been repeated by Scholz *et al.* (1986), using pyrex glass tubes which fit a steel rod with a tolerance better than 3μm, with a tube thickness exceeding mm size, and length and radius on the order of cm. By direct measurement, through strain gauges placed on the glass tube and by simple calculation, it is established that *all three principal stresses everywhere within the glass tube are compressive*. Nevertheless, 2 to 6 *axial tension cracks* are seen to form from the interior surface, growing radially and axially, without reaching the exterior surface of the glass tube. This paradox has been (quantitatively) explained by Scholz *et al.* (1986) in terms of model calculations of Nemat-Nasser and Horii (1982); see Figure 6.9.3. A similar explanation applies to the disking phenomenon of the jacketed cylindrical rod of Figure 6.9.2a.

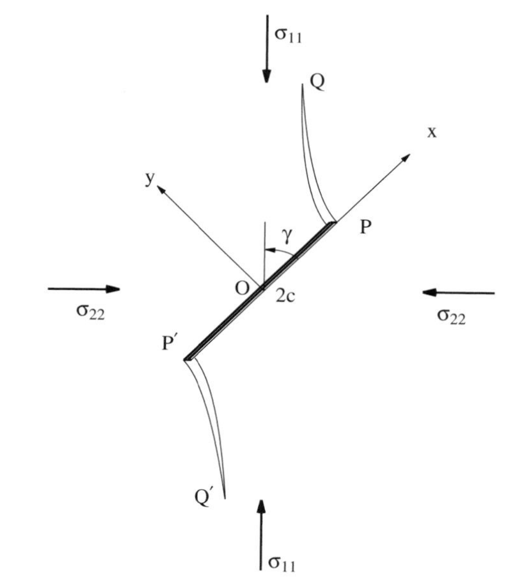

Figure 6.9.3

The sliding crack model of Brace and Bombolakis (1963) as analyzed by Nemat- Nasser and Horii (1982): preexisting flaw PP′ and curved tension cracks PQ and P′Q′ under biaxial compression

Scholz *et al.* (1986) consider a preexisting flaw with suitable inclination and estimate the required flaw size which, in the case of the jacketed ring experiment of Figure 6.9.2b, can produce tension cracks under the prevailing compressive stress state, using the *sliding crack model* (Figure 6.9.3), initially proposed by Brace and Bombolakis (1963), and later quantified analytically, as well as confirmed experimentally, by Nemat-Nasser and Horii (1982); see also Steif (1984), Ashby and Hallam (1986), Myer *et al.* (1992), and Yoshida and Horii (1992).

Calculations based on the Nemat-Nasser and Horii (1982) theory show that a preexisting flaw size of 10μm is sufficient to produce such axial tension cracks. SEM observations show that axial cracks emanate from preexisting flaws of about 20μm, and that the axial cracks consist of several individual cracks which seem to have been initiated from different preexisting flaws.

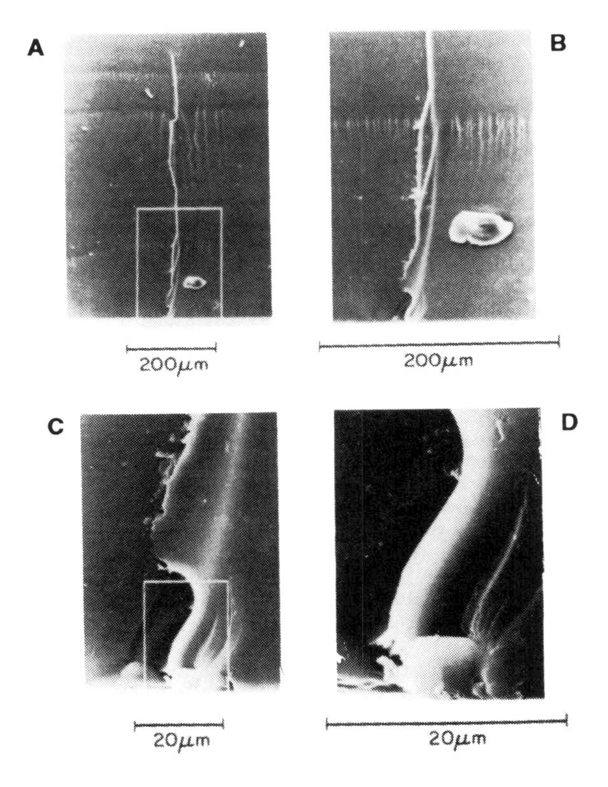

Figure 6.9.4

Scanning electron photomicrographs of axial crack observed in the jacketed ring (from Scholz *et al.*, 1986)

Figure 6.9.4 shows the scanning electron photomicrographs of an axial crack observed in the jacketed construction of Figure 6.9.2b, at successively greater magnification, A, B, C, and D. The sample has been subjected to several loading cycles. The corresponding crack front position can be seen in these photographs. The pictures are taken perpendicular to the fracture plane, with the inside surface of the pyrex ring appearing at the bottom of each photograph. The crack appears to have been initiated from a flaw of rather complex geometry through an abrupt kink, and then extended axially. The light vertical lines in photographs A and B mark the arrest lines in successive pressurizations.

This is perhaps one of the most conclusive laboratory experiments which not only resolves the Bridgman paradox, but also shows the role of preexisting flaws in generating tensile cracks under all-around compression in brittle solids. The fact that flaws in the pyrex glass in this experiment are few and far apart, precludes their interaction, leading to axial cracking. In a rock, ceramic, or similar specimen, there are numerous preexisting microflaws such as pores, grain boundaries, preexisting cracks, and inclusions, each of which can be and often is a source of producing local tensile stresses, even though the applied loads may all be compressive. Note that although the sliding-crack *model* seems to *represent* and *capture* the involved rather complex process of failure, it has not often been experimentally observed to be the major micromechanism of generating tensile cracks under all-around compressive loads; see Myer *et al.* (1992).

6.9.3. A New Look at Microcracking in Compression

The fact that *axial splitting* under uniaxial compression is caused essentially by nucleation at various flaws of tension cracks which grow essentially in the direction of compression, has been demonstrated in a recent series of experiments by Zheng *et al.* (1988)[7]. In these experiments the microstructure of the compressed sample is preserved by impregnating the specimen with molten Wood's metal which solidifies prior to the removal of the compressive loads. The sample is then sectioned and studied. Figure 6.9.5a is a photomicrograph of an axial section of a uniaxially compressed sample of Indiana limestone, showing a preponderance of nearly axial cracks, filled with Wood's metal (white). Figure 6.9.5b is a photomicrograph of a similar sample, with limestone grains removed by etching. Planar extension cracks are clearly seen in this three-dimensional photo. The presence of solidified Wood's metal which penetrates tube-like pores of diameters exceeding 0.15μm, and planar cavities with apertures exceeding 0.05μm, precludes further fracturing during unloading. Hence, microcracking produced solely during the application of compression can be studied. The authors conclude that a variety of microscopic mechanisms (bending, point loading, and sliding) produces tensile cracking parallel to the direction of maximum compression.

The model experiments by Brace and Bombolakis (1963), Hoek and Bieniawski (1965), Bieniawski (1967), and Nemat-Nasser and Horii (1982), involving preexisting, inclined, slit flaws, seem to capture the essence of this failure process. In addition, Horii and Nemat-Nasser (1985a, 1986) have provided illustration of transition from the axial splitting mode of failure to faulting, when axial compression is applied in the presence of lateral confinement, and transition from a brittle to a ductile mode of failure, when the confining pressure is suitably large.

---

[7] Another method of preserving the microstructure is to subject the sample to a single compressive pulse of limited total energy; see Nemat-Nasser *et al.* (1991). The results of Figure 1.1.1 of Section 1 are obtained in this manner; Subhash (1991), and Subhash and Nemat-Nasser (1993).

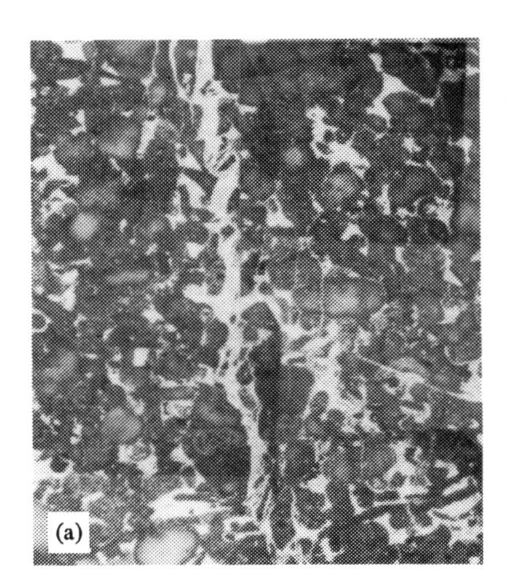

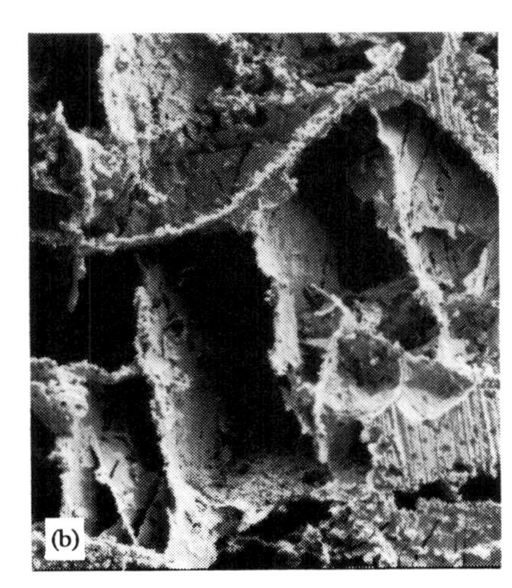

Figure 6.9.5

(a) Scanning electron photomicrographs of axial cracks observed in uniaxially compressed Indiana limestone (white is Wood's metal); (b) Sample with limestone grains removed by etching (from Myer *et al.*, 1992)

The experiments involve thin plates of relatively brittle material (e.g., Columbia resin CR39) containing thin slits (flaws) fitted with thin brass sheets, and subjected to inplane compression. Under inplane axial compression, tension cracks are observed to nucleate from the flaws, to curve toward the direction of maximum inplane compression, and to grow with increasing compression, eventually becoming parallel to this loading direction; see Figure 6.9.6. Of particular interest in these experiments is the fact that *the presence of slight inplane lateral tension can render a crack growth regime of this kind unstable*: once a critical crack extension length is attained, the crack would grow spontaneously, leading to axial splitting of the specimen. Nemat-Nasser and Horii (1982) seek to explain the phenomena of axial splitting, exfoliation or sheet fracture (Holzhausen, 1978), and rockburst, using this observation.

It therefore appears that, in the absence of lateral confinement, axial splitting may well be the result of the formation of axially oriented tension cracks at the most compliant inhomogeneities. These cracks then grow axially and lead to axial splitting. Once such a process is initiated, the specimen no longer remains homogeneous in a continuum sense. At this stage, the strength drops dramatically.

When lateral confinement accompanies axial compression, a profound change in the overall response of rocks, concrete, ceramics, and other brittle materials is often observed. Microscopic observation shows that, in this case also, microcracks are nucleated at various micro-inhomogeneities, and these

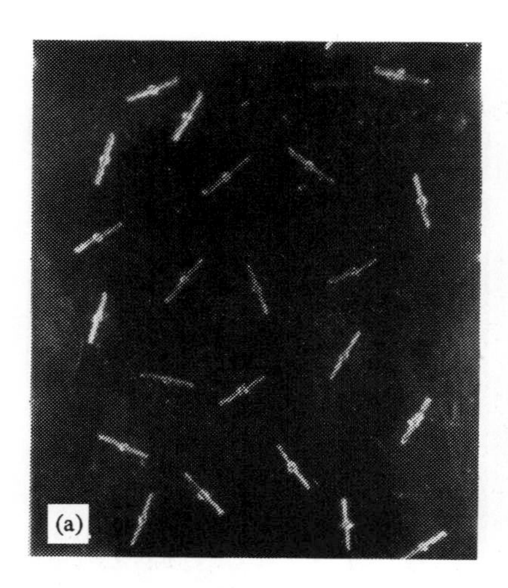

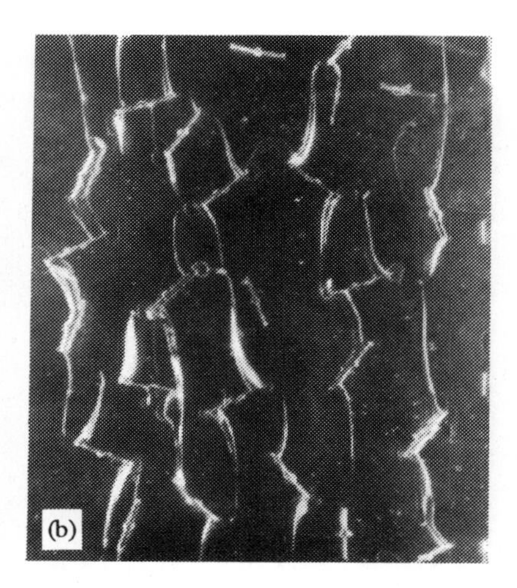

Figure 6.9.6

(a) Specimen with a number of randomly oriented cracks; (b) Failure pattern under overall axial compression (from Nemat-Nasser and Horii, 1982)

cracks grow essentially in the direction of maximum compression. However, the presence of confinement seems to arrest further growth of cracks of this kind. Indeed, electron microscopy, as well as optical microscopy, seem to suggest a more or less uniform distribution of microcracks within the sample, up to axial loads rather close to the peak stress; see, e.g., Hallbauer *et al.* (1973), Olsson and Peng (1976), Wong (1982), and Myer *et al.* (1992). Close to the peak stress a region of high-density microcracks begins to emerge, which eventually becomes the final failure plane. The sample fails by faulting at an angle somewhere between 10 and 30° with respect to the axial compression.

Horii and Nemat-Nasser (1985a, 1986) have suggested that such *faulting may be the result of the interactive unstable growth of tension cracks at suitable sets of interacting microflaws.* To verify this, a series of model experiments is performed on plates which contain sets of small flaws and a number of large flaws; a flaw here is a thin slit (0.4 mm thick) containing two thin brass sheets (0.2 mm each). Two identical specimens are tested, one without confining pressure, the other with some confinement; see Figure 6.9.7.

In the absence of confinement, cracks emanate from the tips of the longer flaws, grow in the direction of axial compression, and lead to axial splitting, while many of the smaller flaws have not even nucleated any cracks; Figure 6.9.7b. On the other hand, when some confinement accompanies axial compression, cracks emanating from the larger flaws are soon arrested. Then, at a certain stage of loading, suddenly, cracks emanating from many small flaws grow in an unstable manner, leading to eventual faulting; Figure 6.9.7c. The faulting is observed to initiate at some small preexisting flaws, and then run through the sample at a finite speed; Horii and Nemat-Nasser (1985a).

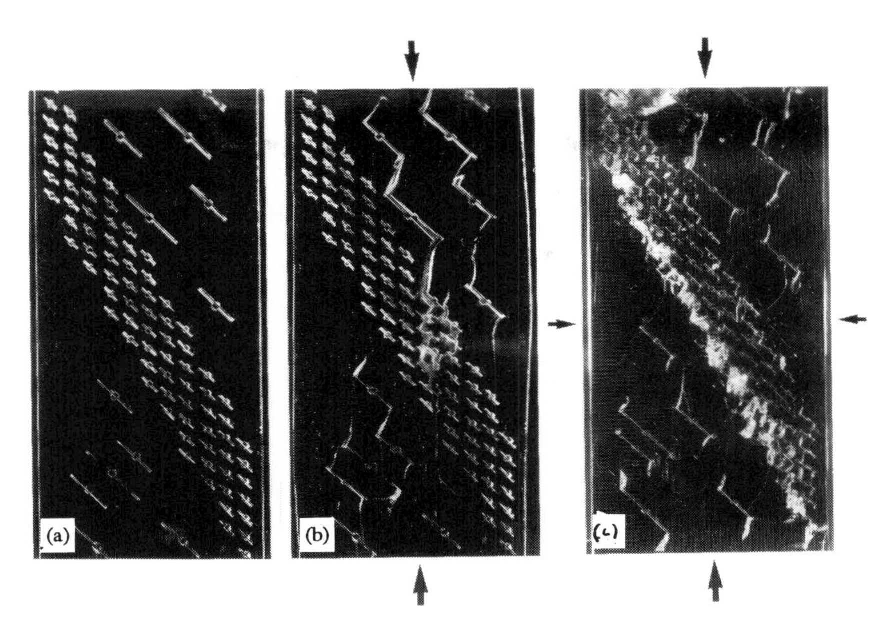

Figure 6.9.7

(a) Specimen containing rows of small flaws, and several larger flaws; (b) Axial splitting under overall axial compression *without* lateral confinement; (c) Shear failure (faulting) under axial compression *with* lateral confinement (from Horii and Nemat-Nasser, 1985a)

When the confining pressure is quite large, e.g., exceeding 25-30% of the peak stress, then a transition from brittle failure by faulting to a ductile response by overall plastic flow takes place. Microscopic observation shows a rather general distribution of microcracks accompanying extensive plastic deformation. The sample may fail by either localized plastic shearing or by barreling. Horii and Nemat-Nasser (1986) suggest a model which seems to illustrate the involved mechanism. Figure 6.9.8 shows a sample containing two collinear flaws. When inplane axial compression is applied in the presence of relatively large confining pressures, both tension cracks and plastically deformed zones develop close to the tips of the preexisting flaws. That is, under axial compression, cracks can emanate from the tips of the flaws, while at the same time plastic zones exist there. The crack length and the size of the plastic zone depend on the confining pressure. For moderate confinement, the tension crack length increases at a great rate, while the plastic zone size remains limited. On the other hand, once suitable confinement exists, the tension crack soon ceases to grow in response to the increasing axial compression, while the plastic zone size continues to increase. Indeed, if the confinement is large enough, the growth of the plastic zone may, at a certain stage, actually relax the stress field around the tension crack, resulting in partial closure of the tension crack. Moreover, plastic zones seem to form first and, in fact, often *shield* cracking which, once initiated, may then suddenly *snap* to a finite length, in an unstable growth mode. Based on this model, Horii and Nemat-Nasser (1986) estimate the brittle-ductile transition pressure, and obtain results in reasonable agreement with experimentally observed values. These experiments are performed at suitably low temperatures, where creep effects can be regarded insignificant. The phenomenon of creep in rock is rather complex and outside the scope of the present brief review; see, e.g., Kranz (1979, 1980), and Yoshida and Horii (1992).

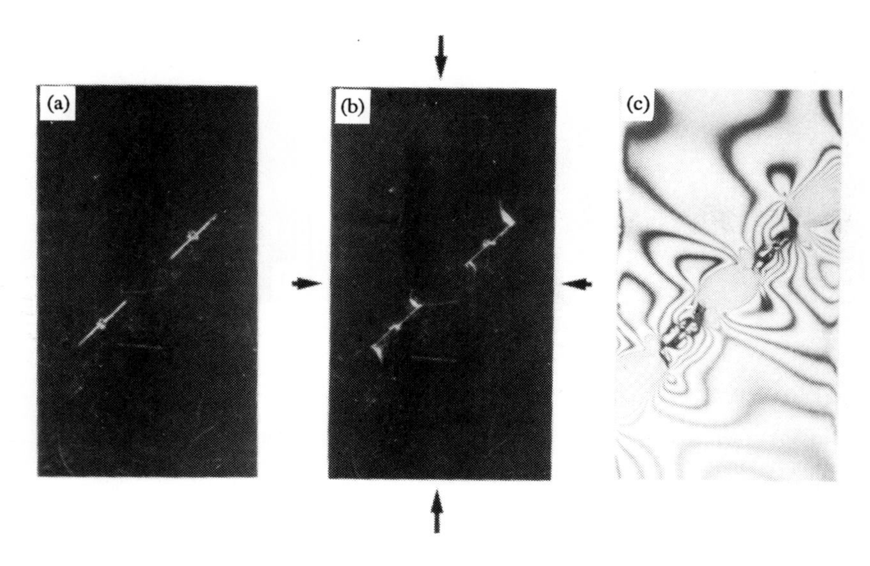

Figure 6.9.8

(a) Specimen containing two collinear flaws; (b) Arrested tension cracks emanating from the flaws under axial and lateral compressive stresses of constant ratio $\sigma_{22}/\sigma_{11} = 0.05$; (c) Photoelastic picture of unloaded specimen showing the residual strain distribution (from Horii and Nemat-Nasser, 1986)

6.9.4. Model Calculations: Axial Splitting

The two-dimensional elasticity boundary-value problem associated with the model shown in Figure 6.9.3 has been formulated in terms of singular integral equations and solved numerically; Nemat-Nasser and Horii (1982) and Horii and Nemat-Nasser (1983, 1985a,b). The boundary conditions on the flaw PP′ are

$$u_y^+ = u_y^-, \qquad \tau_{xy}^+ = \tau_{xy}^- = -\tau_c + \eta\,\sigma_y, \tag{6.9.1a,b}$$

and on the curved cracks PQ and P′Q′, it is required that

$$\sigma_\theta = \tau_{r\theta} = 0, \tag{6.9.1c}$$

where τ_c is the cohesive (or yield) stress, η is the frictional coefficient, u_y is the displacement in the y-direction, σ_y is the normal stress and τ_{xy} is the shear stress on PP′, and σ_θ is the hoop stress and $\tau_{r\theta}$ is the shear stress on PQ. Superscripts + and - denote the values of the considered quantities above and below the x-axis. Figure 6.9.9 shows some typical results.

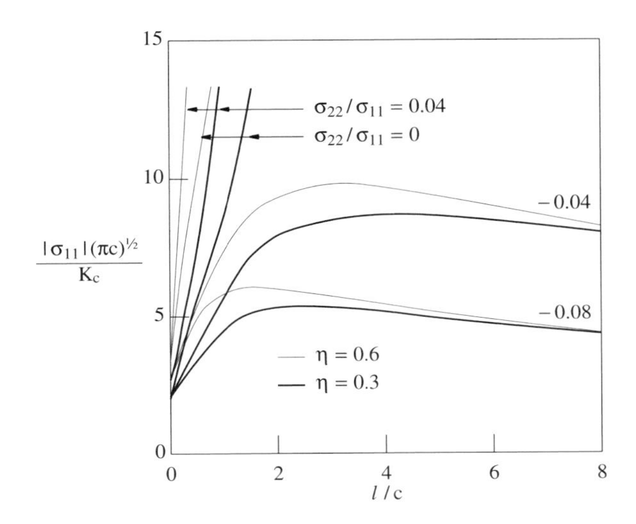

Figure 6.9.9

Normalized axial compression required to attain the associated crack extension length (from Horii and Nemat-Nasser, 1983)

As is seen, in the presence of small lateral tension, crack growth becomes unstable after a certain crack extension length is attained. This unstable crack growth is considered to be the fundamental mechanism of axial splitting of a uniaxially compressed rock specimen. Peng and Johnson (1972) report the presence of lateral tension in the uniaxially compressed specimen because of the end-boundary conditions. Different end inserts affect the ultimate strength. They report a radial tensile stress of 4-8% of the applied compression. These experimental data seem to support the analytical results.

Nemat-Nasser and Horii (1982) have made a series of model experiments and have shown that the unstable growth of tension cracks discussed above, may indeed be the basic micromechanism of axial splitting; see their Figs. 13-20.

The numerical calculations of the singular integral equation which corresponds to the elasticity model of Figure 6.9.3 are rather laborious. Furthermore, they preclude further modeling which often requires simple closed-form analytic expressions. Efforts have been made to develop such expressions for the model of Figure 6.9.3, by substituting for the curved cracks, equivalent straight cracks; see, e.g., Ashby and Hallam (1986), Steif (1984), and Horii and Nemat-Nasser (1986). Simple expressions which seem to yield accurate results over the entire range of crack lengths and orientations are given by Horii and Nemat-Nasser (1986). In a more recent article, Nemat-Nasser and Obata (1988) have used these analytical expressions to model the dilatancy and the hysteretic cycle observed in rocks. A brief outline of the approximate analytical solution of Horii and Nemat-Nasser (1986) is given below.

Figure 6.9.10a shows the flaw with straight cracks, and Figure 6.9.10b sh ows a crack of length $2l$ subjected to a pair of forces of common magnitude F, which represent the effect of the flaw on cracks PQ and P′Q′. These cracks are additionally subjected to farfield stresses σ_{11} and σ_{22}. The force F is estimated from the driving shear stress τ^*, on the preexisting flaw,

$$\tau^* = -\frac{1}{2}(\sigma_{11} - \sigma_{22}) \sin 2\text{gamma} - \tau_c + \frac{1}{2}\eta\,\{\,\sigma_{11} + \sigma_{22} - (\sigma_{11} - \sigma_{22}) \cos 2\gamma\,\},$$

$$F = 2c\,\tau^*. \tag{6.9.2a,b}$$

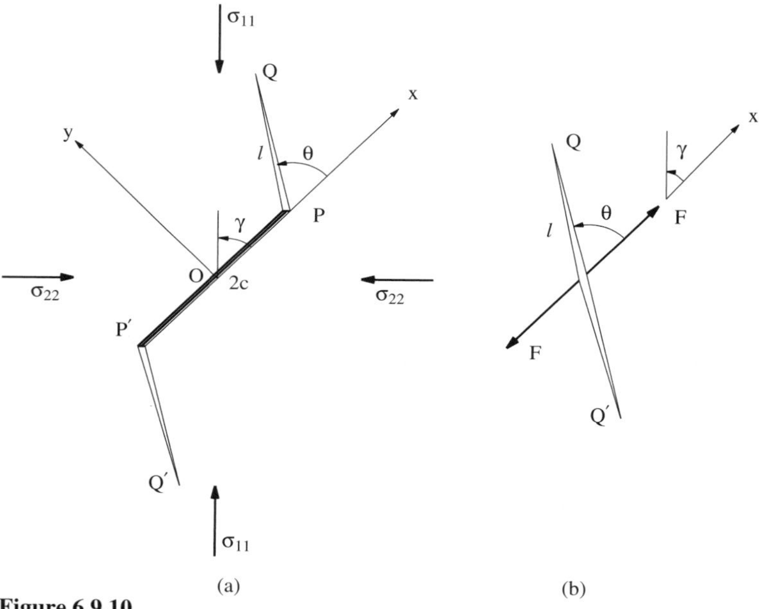

Figure 6.9.10

(a) Preexisting flaw PP′ and straight cracks PQ and P′Q′; (b) A representative tension crack QQ′ with splitting forces F

Then, under the action of the concentrated coaxial forces of magnitude F and the farfield stresses, the stress intensity factors at Q and Q′ are given by

$$K_I = \frac{2c\,\tau^* \sin\theta}{\{\,\pi(l + l^*)\}^{1/2}} + \tfrac{1}{2}(\pi l)^{1/2}\{\sigma_{11} + \sigma_{22} - (\sigma_{11} - \sigma_{22}) \cos 2(\theta - \gamma)\,\},$$

$$K_{II} = \frac{-2c\,\tau^* \cos\theta}{\{\,\pi(l + l^*)\}^{1/2}} - \tfrac{1}{2}(\pi l)^{1/2}(\sigma_{11} - \sigma_{22}) \sin 2(\theta - \gamma). \tag{6.9.3a,b}$$

In this equation, $l^*/c = 0.27$ is introduced so that when the crack length l is vanishingly small, the corresponding stress intensity factors are still accurately given by (6.9.3a,b). Note that, when l is large, the presence of l^* is of little consequence. Thus, (6.9.3a,b) are good estimates over the entire range of crack lengths. Alternative expressions are given by Ashby and Hallam (1986)[8].

6.9.5. Model Calculations: Faulting

In the presence of confining pressure, an axially compressed sample of rock fails by faulting or (macroscopic) shear failure. To explain the mechanics of such faulting, some authors have emphasized the role of Euler-type buckling associated with columnar regions formed in the sample because of axial cracking; see, for example, Fairhurst and Cook (1966), Janach (1977), and Holzhausen and Johnson (1979).

A different model has been suggested by Horii and Nemat-Nasser (1983, 1985b). This model considers a row of suitably oriented microflaws and seeks to estimate the axial compression at which out-of-plane cracks that nucleate from the tips of these flaws can suddenly grow in an unstable manner, leading to the formation of a fault; see Figure 6.9.11.

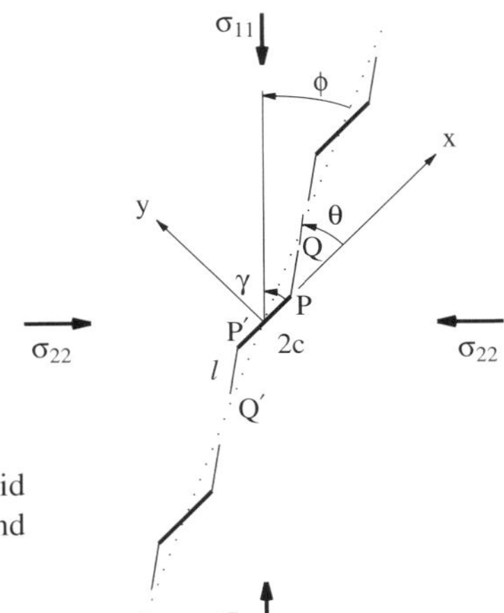

Figure 6.9.11

An unbounded two-dimensional solid with a row of preexisting flaws PP′ and tension cracks PQ and P′Q′

[8] In addition, other microcrack-induced failure models have been proposed; see, e.g., Kachanov (1982), Costin (1985), Ortiz (1985), Kemeny and Cook (1986), Talreja (1987), Krajcinovic (1989, 1996), Ju (1990), Karihaloo and Fu (1990), a special issue of Applied Mechanics Reviews, edited by Li (1992), and Nemat-Nasser et al. (1993).

The solution of the elasticity problem associated with a solid containing a row of periodically distributed flaws with out-of-plane microcracks, is given by Horii and Nemat-Nasser (1983, 1985b). Typical results are shown in Figure 6.9.12. For small values of ϕ, the axial compression first increases with increasing crack extension length, attains a peak value, decreases, and then begins to rise again. This suggests an unstable crack growth at a critical value of the axial stress, which may lead to the formation of a fault zone. It is seen from Figure 6.9.11 that the peak values of the axial stress for the values of ϕ from 29° to 36° fall in a very narrow range, i.e., $|\Delta\sigma_1|\sqrt{\pi c}/K_c \approx 0.3$. This implies that *the overall failure angle is sensitive to imperfection and other effects*. Indeed, the orientations of the fracture plane observed in experiments often scatter over a certain range. The range of the overall failure angle, however, may be limited, since the peak value of the axial stress increases sharply as ϕ decreases. The possible range of the overall orientation angle ϕ can be specified by prescribing the *stress barrier*, $|\Delta\sigma_1|\sqrt{\pi c}/K_c$, which can be overcome. Note that the value of γ in Figure 6.9.11 is chosen such that the required axial compression for instability is minimized; see Horii and Nemat-Nasser (1985a, 1986) for further comments and examples.

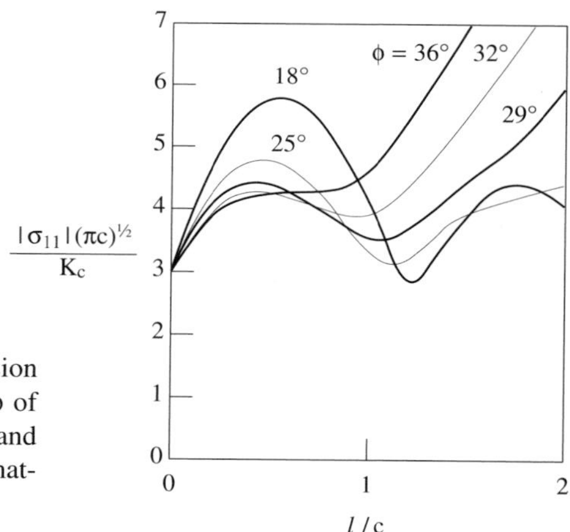

Figure 6.9.12

Axial stress versus crack extension length for indicated orientation ϕ of crack row, $\gamma = 0.24\pi$, $\tau_c = 0$, and $\eta = 0.4$ (from Horii and Nemat-Nasser, 1985a)

6.9.6. Model Calculations: Brittle-Ductile Transition

Brittle failure by faulting is suppressed by sufficiently high confining pressures that promote distributed inelastic deformation at various flaws, throughout the sample. Microscopically, the deformation remains highly heterogeneous, in view of the microstructure of the material. Depending on the material and the temperature, the inelastic deformation may stem from grain-size microcracking, plastic glide, or a combination of the two. For example, in marble and limestone, as well as in pyroxenes, microcracking and the associated cataclysmic flow can be inhibited at room temperature by large enough confining pressures, whereas for other materials, such as quartz and feldspar, this requires higher temperatures; Donath *et al.* (1971), Tobin and Donath (1971), Olsson and Peng (1976), Tullis and Yund (1977), Kirby and Kronenberg (1984), and Myer *et al.* (1992). This difference in response most likely stems from the microstructural differences, such as grain size, shape, and composition, among these materials. To gain insight, it is instructive to examine the influence of increasing lateral pressure on the interactive, unstable crack growth associated with a row of preexisting flaws, shown in Figure 6.9.11. For $\phi = 29°$, $\gamma = 43°$, and d/c = 4, the results are shown in Figure 6.9.13. It is seen that *increasing the lateral pressure suppresses the unstable growth of tension cracks emanating from the tips of the interacting flaws, and therefore suppresses the associated faulting.*

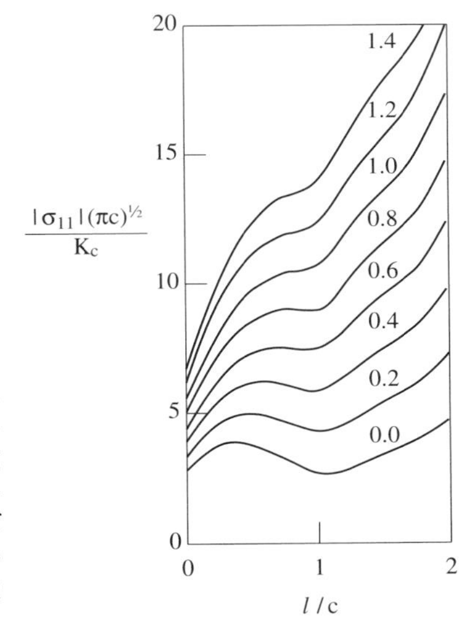

Figure 6.9.13

Compressive force required to attain the associated length of cracks emanating from a row of preexisting flaws, under the indicated normalized lateral stresses (contours of constant $|\sigma_{22}|(\pi c)^{1/2}/K_c$), with d/c = 4, $\gamma = 43°$, and $\phi = 29°$ (from Horii and Nemat-Nasser, 1986)

To estimate the brittle-ductile transition analytically , Horii and Nemat-Nasser consider the model shown in Figure 6.9.14. It consists of the frictional and cohesive flaw PP′ which, in addition to plastic zones PR and P′R′ of common length l_p, has produced at its tips, out-of-plane tension cracks PQ and P′Q′ of common length l_t. The boundary conditions on the preexisting flaw and the tension cracks are given, respectively, by (6.9.1a,b) and (6.9.1c). The conditions on the slip lines PR and P′R′ are

$$u_y^+ = u_y^-, \qquad \tau_{xy} = -\tau_Y, \tag{6.9.4a,b}$$

where τ_Y is the yield stress in shear. The principal stresses at infinity are prescribed to be σ_{11} and σ_{22}. In this model the tension cracks are assumed to be straight. The plastic zones are modeled by dislocation lines collinear with the preexisting flaw, as motivated by the model experiments, although it is not difficult to consider a non-collinear dislocation line or several such lines, depending on the circumstances. The use of collinear dislocation lines is reasonable as a starting point for modeling, and seems to yield adequate results.

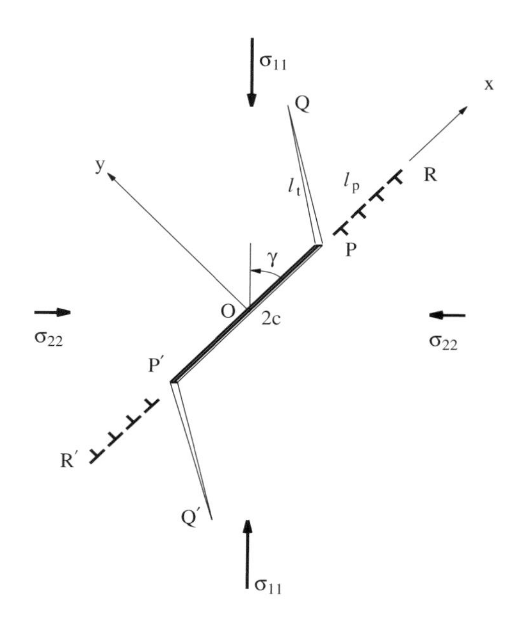

Figure 6.9.14

Preexisting flaw PP′, tension cracks PQ and P′Q′, and plastic zones PR and P′R′

The stresses at the ends of the plastic zones must be bounded. Consider a solution that renders the Mode II stress intensity factor at R and R′ zero, i.e., require that

$$K_{II}^{R} = 0, \quad \text{at R and R}'. \tag{6.9.4c}$$

Horii and Nemat-Nasser (1986) present an exact formulation and solutions for the problem sketched in Figure 6.9.14, in terms of singular integral equations. They also give approximate closed-form solutions which may prove effective for further modeling. The approximate analytical solution is based on the assumption that the *ductility* defined by

$$\Delta = \frac{K_c}{\tau_y(\pi c)^{1/2}} \tag{6.9.5}$$

is small, and the size of the tension crack is large relative to the size of the plastic zone. Hence, the interaction between the plastic zone and the crack is neglected. The stress intensity factor K_I at Q and Q′ is estimated using (6.9.3a). Then the Dugdale (Dugdale, 1960) model is used in Mode II to estimate the size of the plastic zone such that $K_{II}^{R} = 0$ at points R and R′ in Figure 6.9.14. This yields the following expression for σ_{11}/τ_Y :

$$\frac{\sigma_{11}}{\tau_Y} = \frac{-\left\{2 + \frac{4}{\pi}(\tau_c/\tau_Y - 1)\arcsin\{1 + l_p/c\}^{-1}\right\}}{(1 - \frac{\sigma_{22}}{\sigma_{11}})\sin 2\gamma - \eta\left\{1 + \frac{\sigma_{22}}{\sigma_{11}} - (1 - \frac{\sigma_{22}}{\sigma_{11}})\cos 2\gamma\right\}\frac{2}{\pi}\arcsin\{1 + l_p/c\}^{-1}}, \tag{6.9.6a}$$

and in view of (6.9.3a,b),

$$\frac{K_I}{\tau_Y(\pi c)^{1/2}} = \frac{-\sin\theta}{\pi\{\frac{l_t}{c} + \frac{l_t^*}{c}\}^{1/2}}$$

$$\times\left\{\frac{\sigma_{11}}{\tau_Y}\left\{(1 - \frac{\sigma_{22}}{\sigma_{11}})\sin 2\gamma - \eta\{1 + \frac{\sigma_{22}}{\sigma_{11}} - (1 - \frac{\sigma_{22}}{\sigma_{11}})\cos 2\gamma\}\right\} + \frac{\tau_c}{\tau_Y}\right\}$$

$$+ \tfrac{1}{2}(l_t/c)^{1/2}\frac{\sigma_{11}}{\tau_Y}\left\{1 + \frac{\sigma_{22}}{\sigma_{11}} - \{1 - \frac{\sigma_{22}}{\sigma_{11}}\}\cos 2(\theta - \gamma)\right\}. \tag{6.9.6b}$$

Horii and Nemat-Nasser (1986) examine the accuracy of (6.9.6a,b) by comparing the corresponding results with the numerical ones for the exact formulation. For ductility, Δ, less than about 0.1, the approximate results are quite good. One shortcoming of the approximate results (6.9.6a,b) is that they do not yield a maximum value for the size of the tension cracks, whereas the *exact* calculation does. Figure 6.9.15 shows the relation between l_t/c and l_p/c for $\Delta = 0.04$ and $\Delta = 0.08$, obtained by the numerical solution of the singular integral equations for the exact formulation of the boundary-value problem.

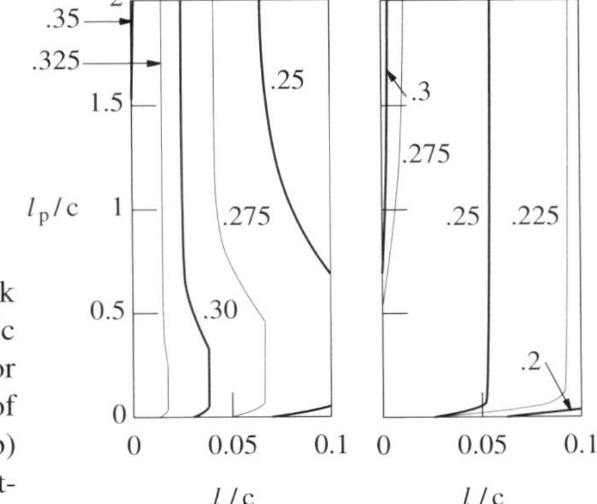

Figure 6.9.15

Relation between the tension crack length and the size of the plastic zone under proportional loading for indicated stress ratios (contours of σ_{22}/σ_{11}), for: (a) $\Delta = 0.04$; and (b) $\Delta = 0.08$ (from Horii and Nemat-Nasser, 1986)

From the results presented in Figure 6.9.15, it is seen that for small lateral compression, l_p/c remains very small as l_t/c increases rapidly, dominating the failure regime; the response of the solid is *brittle* in this case. With suitably large values of σ_{22}/σ_{11}, on the other hand, l_t/c ceases to increase after it attains a certain (negligibly small) value, while l_p/c continues to increase with increasing axial compression; the response of the solid in this case is *ductile*. Indeed, for large enough σ_{22}/σ_{11} (e.g., $\sigma_{22}/\sigma_{11} = 0.2$), the tension crack actually begins to relax and close, as the plastic zone extends. The model also suggests another possible failure mode, where, while a plastic zone develops first, once the tension crack is initiated, it grows to a finite length in an unstable manner, as the plastic zone relaxes; this is referred to as the *transitional mode*.

By examining the maximum size of the tension cracks and whether they grow in a stable or unstable manner, Horii and Nemat-Nasser produce from this two-dimensional model, the *brittle-ductile diagram*[9] of Figure 6.9.16.

For σ_{22}/σ_{11} exceeding 0.2 to 0.25, this figure shows a transition to the ductile response. This seems to be in accord with experimental observations summarized by Mogi (1966). Note from Figure 6.9.15 that, for σ_{22}/σ_{11} greater than certain values, the size of the plastic zone continues to grow with increasing compression (for proportional loading), once l_t/c attains certain maximum values. A deformation process of this kind characterizes a *ductile mode*. The change from the brittle to the ductile mode is illustrated in Figure 6.9.15a,b for $\Delta = 0.04$ and 0.08, respectively. It is seen that this change occurs when the stress ratio increases from 0.325 to 0.35 for $\Delta = 0.04$, and from 0.25 to 0.275 for $\Delta = 0.08$.

It thus appears that whether the failure is brittle, being dominated by the growth of tension cracks, or ductile, being dominated by the growth of plastic zones, depends on the magnitude of the stress ratio, σ_{22}/σ_{11}, and the overall ductility, Δ. The influence of temperature enters implicitly through the associated values of fracture toughness, K_c, and yield stress, τ_Y. Since the former increases

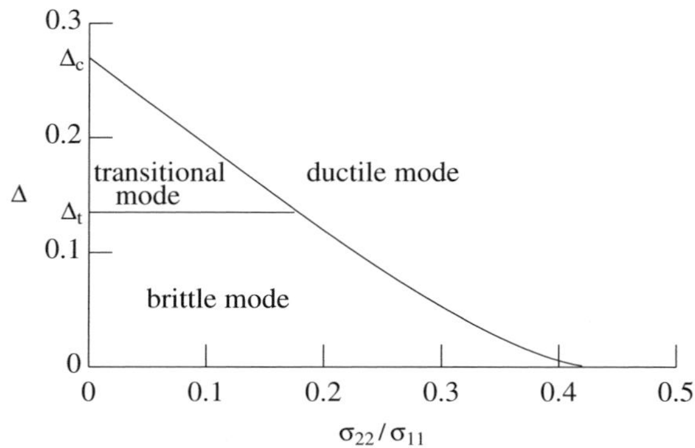

Figure 6.9.16

Brittle-ductile diagram (from Horii and Nemat-Nasser, 1986)

[9] The model does not include the effects of strain hardening, strain rates, and stress three-dimensionality, which affect the quantitative (but not the qualitative) nature of the results.

and the latter decreases with increasing temperature, Δ increases with increasing temperature. Also, the influence of grain size is implicitly included through the dependency of Δ on the flaw size c: the larger the c, the smaller the Δ. It is shown by Horii and Nemat-Nasser (1986) that when Δ is suitably large, the growth of tension cracks can be essentially suppressed by suitable confinement. For a Δ of the order of a few percent, however, both tension cracks and plastic deformation can occur. The material for a small Δ is inherently *brittle*. However, suitably large confining pressures suppress unstable growth of microcracks and promote plastic flow instead. Hence, it seems that compression may induce plastic flow of crystalline solids which, otherwise, are commonly classified as *brittle*.

An important aspect of brittle failure, not considered in this brief review, is the mechanism of microcracking ahead of an advancing tensile macrocrack.[10] It has been reasoned that the generation of such microcracks may result in increased toughness,[11] on account of the additional energy required to create microcracks. Furthermore, toughening by crack bridging, both over a small region in the neighborhood of the crack tip (Budiansky, 1986; Rose, 1987; and Budiansky *et al.*, 1988), as well as partial or full bridging by fiber reinforcement (Nemat-Nasser and Hori, 1987) are timely topics which require examination in their own right; see Subsection 21.5.4 for additional comments.

6.10. DYNAMIC BRITTLE FAILURE IN COMPRESSION

Experimental observations on dynamic failure of brittle solids under compression, indicate that compressive failure stress and the resulting fragment sizes depend on the strain rate and the stress state. The models discussed in Subsection 6.9 for quasi-static compressive failure can be modified and used to examine the rate effects in dynamic compression failure, as has been shown by Deng and Nemat-Nasser (1992b, 1994a,b) and Nemat-Nasser and Deng (1994) who examine the phenomena of damage initiation and evolution in brittle solids, using this approach. The analysis is particularly simplified by noting that the near-field stress field of a dynamically growing slit crack (a two-dimensional problem) in a linearly elastic homogeneous solid, depends on the crack-tip dynamics only through the instantaneous crack-tip velocity, say, v(t), which

[10] For mathematical analyses of interacting microcracks near the tip of a macrocrack, see, e.g., Chudnovsky and Kachanov (1983), Horii and Nemat-Nasser (1983, 1985b), Hori and Nemat-Nasser (1987), and Wu and Chudnovsky (1990).

[11] Evans (1990) provides an overview of possible methods for improving the toughness of ceramics; see also, Evans and McMeeking (1986), and Evans and Fu (1985). The mechanics of matrix cracking in brittle composites has been examined by Marshall *et al.* (1985), McCartney (1987), Marshall and Cox (1987), He and Hutchinson (1989), Laws and Lee (1989), McMeeking and Evans (1990), Hutchinson and Jensen (1990), Kim and Pagano (1991), and Barsoum *et al.* (1992). For issues pertaining to fracture resistance of reinforced ceramics, see, e.g., Kotil *et al.* (1990), Sonuparlak (1990), Llorca and Elices (1990), and Curtin (1991). Thermally induced cracks in ceramics have been modeled by Tvergaard and Hutchinson (1988), where other related references can be found.

need not be uniform (possibly nonsteady crack growth); see Freund and Clifton (1974), Nilsson (1974), Achenbach and Bazant (1975), and, for a comprehensive account, see Freund (1993).[12] For example, the Mode I stress field of a semi-infinite slit crack, located on the x_1-axis (see Figure 6.10.1), and growing with possibly nonuniform velocity v(t), has the following asymptotic expansion form:

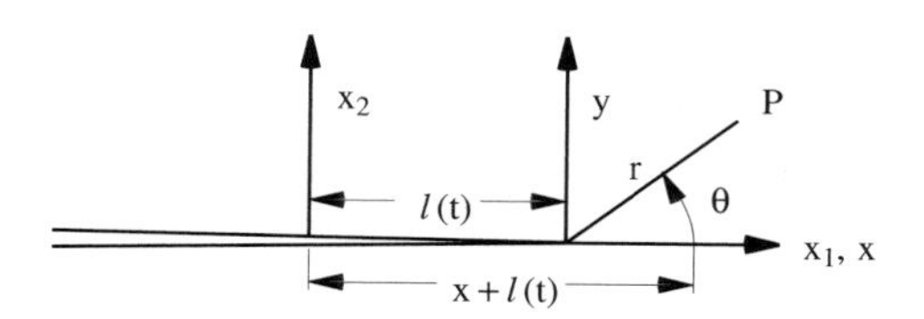

Figure 6.10.1

The coordinate systems for dynamically growing crack: x_1, x_2 are the stationary, and x, y are the moving coordinates

$$\sigma_{ij} = \frac{K_I(t)}{(2\pi r)^{1/2}} \Sigma^I_{ij}(\theta, v) + \ldots, \qquad \text{as } r \to 0, \tag{6.10.1a}$$

where r and θ are the polar coordinates with the origin at the crack tip, dots stand for terms of order unity, and the angular variation of the singular terms of the stress field is given by $\Sigma^I_{ij}(\theta, v)$ which is of order unity; see Freund (1993, page 163). The useful result is the fact that the *dynamic* stress intensity factor, $K_I(t)$, can be represented by

$$K_I(t) \approx k(v)\, K_{Is}, \tag{6.10.1b}$$

where K_{Is} is the associated *static* stress intensity factor, and k(v) is a function of the crack velocity v. This function can be evaluated using the solution of the corresponding steady crack-growing problem. For example, for concentrated loads at the center of a crack,

$$k(v) \approx \frac{C_R - v}{C_R - 0.75v}, \tag{6.10.1c}$$

and for the uniform farfield stress,

$$k(v) \approx \frac{C_R - v}{C_R - 0.5v}, \tag{6.10.1d}$$

[12] For earlier contributions, considering steady crack-growth problems, see Cotterell (1964), Rice (1968), and Sih (1970).

are acceptable approximations, where C_R is the Rayleigh wave speed.[13]

6.10.1. Strain-rate Effect on Brittle Failure in Compression

The model of Figure 6.9.11 with $\phi = 0$, can be used to study the strain-rate effect on brittle failure in compression. Figure 6.10.2a shows this model. Based on the sliding crack model of Figure 6.9.10a,b, the model of Figure 6.10.2a is simplified, as shown in Figure 6.10.2b. Since the collinear cracks in Figure 6.10.2b are under identical loads, their common dynamic stress intensity factor can be calculated by the superposition of the solution for a crack array under pairs of concentrated forces applied at the crack centers, and the solution for a crack array under uniform farfield stresses. The Mode I dynamic stress intensity factor at crack tips in a crack array under both concentrated and uniform loads is then given by

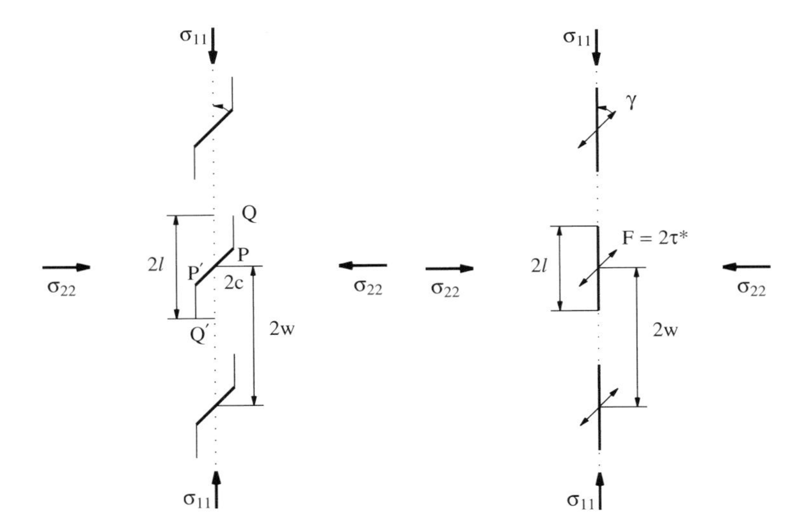

Figure 6.10.2

A model for dynamic compression failure: (a) uniformly spaced pre-existing flaws (PP′) along the maximum compression axis, with compression-induced, dynamically growing tension cracks PQ and P′Q′ (an array of dynamically growing wing cracks); and (b) an equivalent array of dynamically growing collinear cracks loaded by concentrated forces, F, and farfield stresses, σ_{11} and σ_{22} (from Nemat-Nasser and Deng, 1994)

---

[13] Nemat-Nasser and Deng (1994) show that (6.10.1c,d) are good approximations by comparing with the exact solutions of the corresponding self-similar crack growth problem, given by Willis (1973), Cherepanov (1979), and Freund (1993); see Figure 2 of Nemat-Nasser and Deng (1994).

$$K_{Id}^{array} = k_{Is_1}(\dot{l})\, K_{Is_1}^{array} + k_{Is_2}(\dot{l})\, K_{Is_2}^{array}, \tag{6.10.2a}$$

in which,

$$K_{Is_1}^{array} = F \sin\gamma \left[w \sin \frac{\pi(l + l_0)}{w} \right]^{-1/2},$$

$$K_{Is_2}^{array} = \sigma_{22} \left[2w \tan \frac{\pi l}{2w} \right]^{1/2}. \tag{6.10.2b,c}$$

The functions $k_{Is_1}(\dot{l})$ and $k_{Is_2}(\dot{l})$ are defined by (6.10.1c,d), and F is given by (6.9.2a,b) with $\theta = \gamma$. The small length l_0 is introduced to make the model applicable at $l = 0$. Horii and Nemat-Nasser (1985a) calculate l_0 such that, at crack nucleation, the same stress intensity factor results at points Q and Q′ whether a wing crack or a single crack under the corresponding concentrated forces is used; see Figure 6.9.10. These authors obtain $l_0 = 0.27c$ for the optimal crack nucleation angle of $\theta_n = 0.392\pi$. Since the tension cracks turn toward the maximum compression direction soon after their inception (Nemat-Nasser and Horii, 1982), assume this direction for crack growth and choose l_0 to be an arbitrary small number to avoid possible numerical difficulties; e.g., use $l_0 = 0.0001c$ before tension crack nucleation and $l_0 = 0$ for $l > 0$.

For the fracture criterion, assume that the dynamic stress intensity factor cannot exceed a constant fracture toughness, K_{Ic}, i.e.,

$$K_{Id}^{array} \leq K_{Ic}. \tag{6.10.3}$$

The common growth speed of the compression-induced tension cracks is now obtained from (6.10.2a) and (6.10.3) as

$$\dot{l} = C_R \frac{1.5K_{Is_1}^{array} + 1.75K_{Is_2}^{array} - 1.25rK_{Ic} - X^{1/2}}{K_{Is_1}^{array} + 1.5K_{Is_2}^{array} - 0.75K_{Ic}},$$

$$X = \{1.5K_{Is_1}^{array} + 1.75\, K_{Is_2}^{array} - 1.25\, K_{Ic}\}^2 - 4(K_{Is_1}^{array} + K_{Is_2}^{array} - K_{Ic})$$
$$\times (0.5K_{Is_1}^{array} + 0.75K_{Is_2}^{array} - 0.375K_{Ic}). \tag{6.10.4a,b}$$

The common tension crack length is calculated by integrating the crack tip speed until either failure or complete unloading is attained. To obtain the failure stress for a given stress pulse and the material microstructure, estimate the time t_f, at which compression-induced tension cracks coalescence. This occurs when $l = w$. Thus, t_f is defined by

$$w = \int_0^{t_f} \dot{l}\, dt. \tag{6.10.5}$$

The dynamic failure model given above is summarized as follows: The coalescence of tension cracks in a dynamically growing wing-crack array under high strain-rate compressive loads is modeled by (6.10.5) with $\dot{l}$ given by (6.10.4), and the corresponding static stress intensity factors given by (6.10.2b,c). To solve the problem for a given stress pulse of known rate, calculate the current value of $\dot{l}$ from (6.10.4) and then obtain t_f incrementally from (6.10.5). The failure stress, say, σ_f, is then defined by the value of the applied axial compressive stress at crack coalescence. When the stress pulse involves loading and unloading, the cracks may continue to grow and coalescence during unloading, in which case the failure stress at crack coalescence would be smaller than the failure stress when the load continues to increase at positive rates.

6.10.2. Illustrative Examples of Dynamic Brittle Failure in Compression

Using the model discussed above, Nemat-Nasser and Deng (1994) have made a series of parametric studies to illustrate various aspects of dynamic compression failure of brittle solids. For the material, these authors consider alumina, and set: Young's modulus $E = 372$GPa; Poisson's ratio $\nu = 0.22$; density $\rho = 3.9$g/cc; and the fracture toughness $K_{Ic} = 4\text{MPa}\,\text{m}^{1/2}$. For the uniaxial strain loading regimes in plane problems, $\sigma_{22}/\sigma_{11} \approx \nu/(1 - \nu) \approx 0.3$ is assumed. For $\theta = \gamma$ in Figure 6.9.11 (see page 187) and frictionless sliding, the concentrated force F is maximum when $\gamma = 0.3\pi$, as reported by Deng and Nemat-Nasser (1992b).

For a constant loading rate followed by an equal constant unloading rate, the model shows that the rate and the extent of the growth of tension cracks depend on their size and spacing. When the flaws are initially far apart, the resulting tension cracks may cease to grow soon after the unloading begins. On the other hand, during loading, or even after unloading has begun, failure can occur by crack coalescence when the pre-existing flaws are suitably spaced.

Consider now a constant strain-rate, monotonic loading to failure (defined by crack coalescence), and examine the effects of the strain rate, microstructure (defined by the initial size and spacing of the microflaws, i.e., by c and w), and the confining pressure (defined by the ratio σ_{22}/σ_{11}). These effects are illustrated in Figures 6.10.3a-d. Figure 6.10.3a shows the effect of flaw spacing on the failure load in a uniaxial stressing case, while Figure 6.10.3b shows the effect of flaw size for a fixed flaw spacing; note that the flaw spacing, w, is measured in the unit of the flaw size, c.

The results in Figures 6.10.3a,b are interesting. Figure 6.10.3a suggests that for a fixed (average) flaw size, c, the failure stress increases with increasing (average) flaw spacing, w, as should be expected, and that higher strain rates activate more closely spaced flaws. Figure 6.10.3b shows that at higher strain rates the smaller flaws interact to produce failure which then occurs at higher stresses for finer microflaws. Note the crossover of the curves in Figure 6.10.3b, as the strain rate increases.

To clarify the influence of the flaw size on the failure stress, and its relation to the strain rate, consider the critical value of the dimensionless failure stress, $\sigma_f c^{1/2}/K_{Ic}$, as a function of the strain rate and the flaw size, as shown in Figure 6.10.3c, for a fixed flaw spacing. As is seen, the model suggests that, at higher strain rates, the interaction among smaller flaws leads to failure. This is

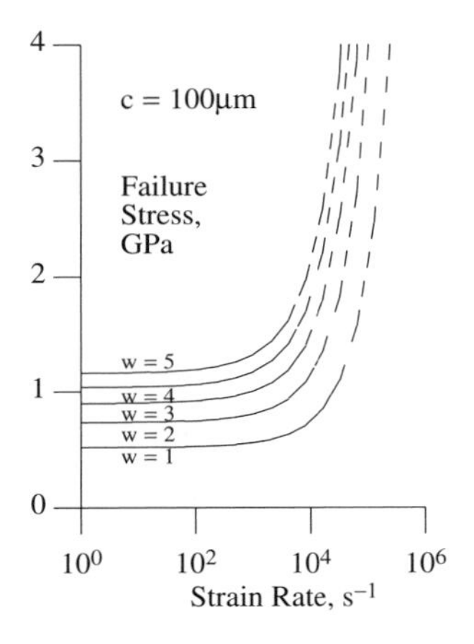

Figure 6.10.3a

A model for dynamic compression failure: effect of microflaw spacing on uniaxial failure stress (from Nemat-Nasser and Deng, 1994)

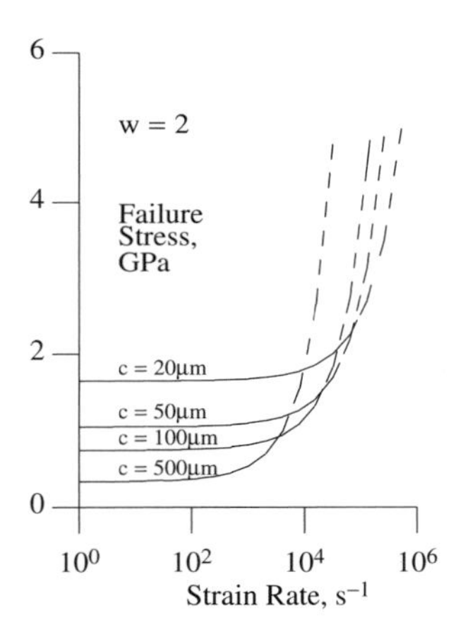

Figure 6.10.3b

A model for dynamic compression failure: effect of microflaw size on uniaxial failure stress (from Nemat-Nasser and Deng, 1994)

in accord with experimental observations.

From Figure 6.10.3c it is seen that the dimensionless normalized failure stress, $\sigma_f c^{1/2}/K_{Ic}$, for a given set of initial conditions, remains essentially independent of the strain rate until a transition strain rate, $\dot{\varepsilon}_T$, is reached, after which the failure stress increases with increasing strain rate. The interaction between the (average) flaw size, c, and the strain rate is vividly illustrated in this figure.

Consider now the effect of the confining pressure on the failure stress. This is illustrated in Figure 6.10.3d for w = c = 100μm. As is expected, the failure stress increases with increasing confining pressure. Nemat-Nasser and

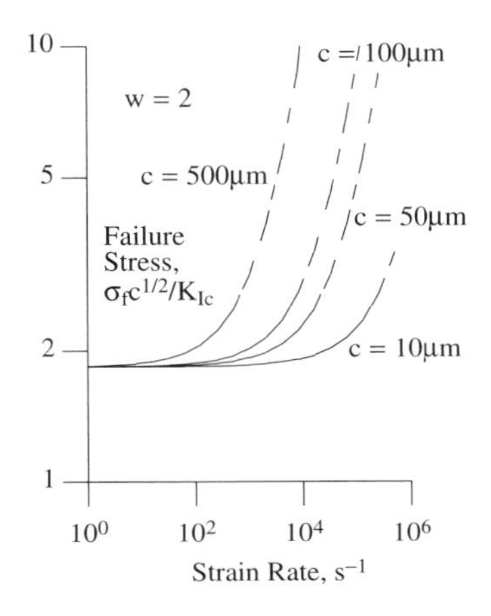

Figure 6.10.3c

A model for dynamic compression failure: effect of microflaw size on uniaxial failure stress (from Nemat-Nasser and Deng, 1994)

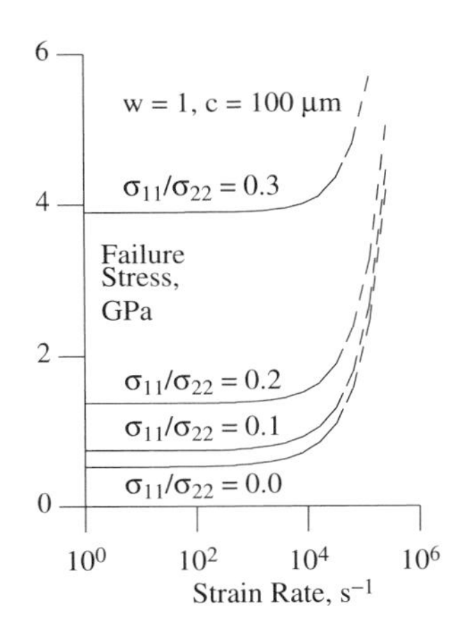

Figure 6.10.3d

A model for dynamic compression failure: effect of confining pressure on uniaxial failure stress (from Nemat-Nasser and Deng, 1994)

Deng (1994) show by additional calculations that, by increasing either the pressure or the strain rate, smaller and more closely spaced flaws interact to produce the final failure. In particular, these authors demonstrate the relation between the fragment size and the material microstructure, using the model discussed above. Under dynamic loads, the damage evolution depends on the strain rate, the microstructure, and the loading state. Hence, the fragment size at failure is also a function of the strain rate, the microstructure, and the loading condition. As the lateral confining pressure is increased, the fragment size decreases, as a result of the coalescence of tension cracks emanating from microflaws which are closer to each other; see Figure 11 of Nemat-Nasser and Deng (1994). As the strain rate increases, the fragment size decreases as a result of the coalescence of compression-induced tension cracks emanating from smaller flaws, as shown in Figure 6.10.3b.

In dynamic compression experiments, the failure may occur by microcracking or it may occur by plastic deformation of the preexisting defects; this is similar to what is considered in Subsection 6.9.6 (see Figure 6.9.14). This and other related topics are discussed by Deng and Nemat-Nasser (1994a,b). For a detailed account of other failure models see the recent book by Krajcinovic (1996).

6.11. REFERENCES

Achenbach, J.D. and Bazant, Z. P. (1975), Elastodynamic near-tip stress and displacement fields for rapidly propagating cracks in orthotropic media, *J. Appl. Mech.*, Vol. 42, 183-189.

Ashby, M. F. and Hallam, D. (1986), The failure of brittle solids containing small cracks under compressive stress states, *Acta. Metall.*, Vol. 34, 497-510.

Barsoum, M., Kangutkar, P., and Wang, A. S. D. (1992), Matrix crack initiation in ceramic matrix composites part I: experiment and test results, part II: models and simulation results, *Comp. Sci. Tech.*, Vol. 43.

Bieniawski, Z. T. (1967), Mechanics of brittle fracture of rock, *Int. Rock Mech. Min. Sci.*, Vol. 4, 395-430.

Brace, W. F. and Bombolakis, E. G. (1963). A note on brittle crack growth in compression. *J. Geophys. Res.*, Vol. 68, 3709-3713.

Bridgman, P. W. (1931), *The Physics of High Pressure*, Bell, London.

Bristow, J. R. (1960), Microcracks and the static and dynamic elastic constants of annealed and heavily cold-worked metals, *Fr. J. Appl. Phys.,* Vol. 11, 81-85.

Budiansky, B. (1965), On the elastic moduli of some heterogeneous materials, *J. Mech. Phys. Solids*, Vol. 13, 223-227.

Budiansky, B. (1986), Micromechanics II, in *Proceedings of the Tenth U.S. National Congress of Applied Mechanics,* Austin, Texas.

Budiansky, B. and O'Connell, R. J. (1976), Elastic moduli of a cracked solid, *Int. J. Solids Structures*, Vol. 12, 81-97.

Budiansky, B., Amazigo, J. C., and Evans, A. G. (1988), Small-scale crack bridging and the fracture toughness of particle reinforced ceramics, *J. Mech. Phys. Solids*, Vol. 36, 167-187.

Cherepanov, G. P. (1979), *Mechanics of brittle fracture*, translation by de Wit, R. and Cooley, W. C., McGraw-Hill, New York.

Chudnovsky, A. and Kachanov, M. (1983), Interaction of a crack with a field of microcracks, *Int. J. Eng. Sci.*, Vol. 21, 1009.

Cotterell, B. (1964), On the nature of moving cracks, *J. Appl. Mech.*, Vol. 31, 12-16.

Costin, L. S. (1985), Damage mechanics in post-failure regime, *Mech. Matr.*,

Vol. 4, 149-160.
Curtin, W. A. (1991), Theory of mechanical properties of ceramic-matrix composites, *J. Am. Ceram. Soc.*, Vol. 74, 2837-2845.
Deng, H. and Nemat-Nasser, S. (1992a), Microcrack arrays in isotropic solids, *Mech. Matr.*, Vol. 13, 15-36.
Deng, H. and Nemat-Nasser, S. (1992b), Dynamic damage evolution in brittle solids, *Mech. Mater.*, Vol. 14, 83-103.
Deng, H. and Nemat-Nasser, S. (1994a), Microcrack interaction and shear fault failure, *Int. J. Damage Mechanics*, Vol. 3, 3-37.
Deng, H. and Nemat-Nasser, S. (1994b), Dynamic damage evolution of solids in compression: Microcracking, plastic flow, and brittle-ductile transition, *J. Eng. Mat. Tech.*, Vol. 116, 286-289.
Donath, F. A., Faill, R. T., and Tobin, D. G. (1971). Deformational mode fields in experimentally deformed rock, *Bull. Geol. Soc. Am.*, Vol. 82, 1441-1461.
Dugdale, D. S. (1960), Yielding of steel sheets containing slits, *J. Mech. Phys. Solids*, Vol. 8, 100-104.
England, A. H. and Green, A. E. (1963), Some two-dimensional punch and crack problems in classical elasticity, *Proc. Camb. Phil. Soc.,* Vol. 59, 489-500.
Evans, A. G. (1990), Perspective on the development of high-toughness ceramics, *J. Am. Ceram. Soc.*, Vol. 73, 187-206.
Evans, A. G. and Fu, Y. (1985), Some effects of microcracks on the mechanical properties of brittle solids -II. microcrack toughening, *Acta Metall.* Vol. 33, 1525-1531.
Evans, A. G. and McMeeking, R. M. (1986), On the toughening of ceramics by strong reinforcements, *Acta Metall.*, Vol. 34, 2435-2441.
Fairhurst, C. and Cook, N. G. W. (1966), The phenomenon of rock splitting parallel to the direction of maximum compression in the neighbourhood of a surface, *Proc. 1st Cong. Int. Soc. Rock Mech., Lisbon*, Vol. 1, 687-692.
Freund, L.B. (1993), *Dynamic fracture mechanics,* Cambridge University Press.
Freund, L.B. and Clifton, R.J. (1974), On the uniqueness of plane elastodynamic solutions for running cracks, *Journal of Elasticity*,Vol. 4, 293-299.
Garbin, H. D. and Knopoff, L. (1973), The compressional modulus of a material permeated by a random distribution of circular cracks, *Quart. Appl. Mech.*, Vol. 30, 453-464.
Garbin, H. D. and Knopoff, L. (1975a), The shear modulus of a material permeated by a random distribution of free circular cracks, *Quart. Appl. Mech.*, Vol. 33, 296-300.
Garbin, H. D. and Knopoff, L. (1975b), Elastic moduli of a medium with liquid-filled cracks, *Quart. Appl. Mech.*, Vol. 33, 301-303.
Griggs, D. T., Jackson, D. D., Knopoff, L., and Shreve, R. L. (1975), Earthquake prediction: Modelling the anomalous v_p/v_s source region, *Science*, Vol. 187, 537-540.
Hallbauer, D. K., Wagner, H., and Cook, N. G. W. (1973), Some observations concerning the microscopic and mechanical behaviour of quartzite specimens in stiff, triaxial compression tests, *Int. J. Rock Mech. Min. Sci. Geomech. Abstr.*, Vol. 10, 713-726.
Hashin, Z. (1988), The differential scheme and its application to cracked materials, *J. Mech. Phys. Solids*, Vol. 36, 719-734.

He, M. and Hutchinson, J. W. (1989), Crack deflection at an interface between dissimilar elastic materials, *Int. J. Solids Struct.*, Vol. 25, 9.

Hill, R. (1965), A self-consistent mechanics of composite materials, *J. Mech. Phys. Solids*, Vol. 13, 213-222.

Hoek, E. and Bieniawski, Z. T. (1965), Brittle fracture propagation in rock under compression, *Int. J. Fract. Mech.*, Vol. 1, 137-155.

Hoenig, A. (1979), Elastic moduli of a non-randomly cracked body, *Int. J. Solids Struct.*, Vol. 15, 137-154.

Holzhausen, G. R. (1978), Sheet structure in rock and some related problems in rock mechanics, Ph.D. thesis, Stanford University, Stanford, California.

Holzhausen, G. R. and Johnson, A. M. (1979), Analyses of longitudinal splitting of uniaxially compressed rock cylinders, *Int. J. Rock Mech. Min. Sci. Geomech. Abstr.*, Vol. 16, 163-177.

Hori, M. and Nemat-Nasser, S. (1987), Interacting micro-cracks near the tip in the process zone of a macro-crack, *J. Mech. Phys. Solids*, Vol. 35, 601-629.

Horii, H. and Nemat-Nasser, S. (1983), Estimate of stress intensity factors for interacting cracks, in *Advances in aerospace, structures, materials and dynamics* (Yuceoglu, U., Sierakowski, R. L., and Glasgow, D. A., eds.), Vol. AD-06, ASME, New York, 111-117.

Horii, H. and Nemat-Nasser, S. (1985a), Compression-induced microcrack growth in brittle solids: Axial splitting and shear failure, *J. Geophys. Res.*, Vol. 90, B4, 3105-3125.

Horii, H. and Nemat-Nasser, S. (1985b), Elastic fields of interacting inhomogeneities, *Int. J. Solids Struct.*, Vol. 21, 731-745.

Horii, H. and Nemat-Nasser, S. (1986), Brittle failure in compression: Splitting, faulting, and brittle-ductile transition, *Phil. Trans. Roy. Soc. Lond.*, Vol. 319, 1549, 337-374.

Hutchinson, J. W. and Jensen, H. M. (1990), Models of fiber debonding and pullout in brittle composites with friction, *Mech. Matr.*, Vol. 9, 139-163.

Irwin, G. R. (1958), Fracture, *Handbuch der Physik*, Vol. VI, 551-590.

Jaeger, J. C. and Cook, N. G. W. (1963), Pinching-off and disking of rocks, *J. Geophys. Res.*, Vol. 68, 1759-1765.

Janach, W. (1977), Failure of granite under compression, *Int. J. Rock Mech. Min. Sci. Geomech. Abstr.*, Vol. 14, 209-215.

Ju, J. W. (1990), Isotropic and anisotropic damage variables in continuum damage mechanics, *J. Eng. Mech.*, Vol. 116, 2764-2770.

Ju, J.W. (1991), On two-dimensional self-consistent micromechanical damage models for brittle solids, *Int. J. Solids Structures*, Vol. 27, 227-258.

Kachanov, M. (1982), Microcrack model of rock inelasticity, Part I: frictional sliding on pre-existing microcracks, *Mech. Matr.*, Vol. 1, 3-18.

Karihaloo, B. L. and Fu, D. (1990), An anisotropic damage model for plain concrete, *Engng. Fracture Mech.*, Vol. 35, 205-209.

Karihaloo, B. L. and Huang, X. (1991), Tensile response of quasi-brittle materials, *PAGEOPH*, Vol. 137, 461-487.

Kemeny, J. and Cook, N. G. W. (1986), Effective moduli, non-linear deformation and strength of cracked elastic solids, *Int. J. Rock Mech. Mining Sci. Geomech. Abstr.*, Vol. 23, 107-118.

Kim, R. Y. and Pagano, N. J. (1991), Crack initiation in brittle matrix composites, *J. Am. Ceram. Soc.*, Vol. 74, 5.

Kirby, S. H. and Kronenberg, A. K. (1984), Deformation of clinopyroxenite: Evidence for a transition in flow mechanisms and semibrittle behaviour, *J. Geophys. Res.*, Vol. 89, 3177-3192.

Koiter, W. T. (1959), An infinite row of collinear cracks in an infinite elastic sheet, *Ingen. Arch.*, Vol. 28, 168-172.

Kotil, T., Holmes, J. W., and Comninou, M. (1990), Origin of hysteresis observed during fatigue of ceramic-matrix composites, *J. Am. Ceram. Soc.*, Vol. 73, 1879-1883.

Krajcinovic, D. (1989), Damage mechanics, *Mech. Matr.*, Vol. 8, 117-197.

Krajcinovic, D. (1996), *Damage mechanics*, Elsevier, New York.

Kranz, R. L. (1979), Crack growth and development during creep of Barre granite, *Int. J. Rock Mech. Min. Sci.*, Vol. 16, 23-36.

Kranz, R. L. (1980), The effect of confining pressure and stress difference on the static fatigue of granite, *J. Geophys. Res.*, Vol. 85, 1854-1866.

Kranz, R. L. (1983), Microcracks in rocks: a review, *Technophysics*, Vol. 100, 449-480.

Kröner, E. (1958), Berechnung der elastischen Konstanten des Vielkristalls aus den Konstanten des Einkristalls, *Z. Phys.*, Vol. 151, 504-518.

Laws, N. and Brockenbrough, J. R. (1987), The effect of micro-crack systems on the loss of stiffness of brittle solids, *Int. J. Solids Struct.*, Vol. 23, 1247-1268.

Laws, N. and Dvorak, G. J. (1987), The effect of fiber breaks and aligned penny-shaped cracks on the stiffness and energy release rates in unidirectional composites, *Int. J. Solids Struct.*, Vol. 23, 1269-1283.

Laws, N., Dvorak, G. J., and Hejazi, M. (1983), Stiffness changes in unidirectional composites caused by crack systems, *Mech. Matr.*, Vol. 2, 123-137.

Laws, N. and Lee, J. C. (1989), Microcracking in polycrystalline ceramics: elastic isotropy and thermal anisotropy, *J. Mech. Phys. Solids*, Vol. 17, 603-618.

Li, V. C., guest ed., (1992), *Micromechanical modelling of quasi-brittle materials behavior, Appl. Mech. Rev.*, Vol. 45, No. 8.

Llorca, J. and Elices, M. (1990), Fracture resistance of fiber-reinforced ceramic matrix composites, *Acta. Metall. Mater.*, Vol. 38, 2485-2492.

Mao, S. W. and Sunder, S. (1992a), Elastic anisotropy and micro-damage process in polycrystalline ice, Part I: theoretical formulation, *Int. J. Fracture*, Vol. 55, 223-243.

Mao, S. W. and Sunder, S. (1992b), Elastic anisotropy and micro-damage process in polycrystalline ice, Part II: numerical simulation, *Int. J. Fracture*, Vol. 55, 375-396.

Marshall, D. B., Cox, B. N., and Evans, A. G. (1985), The mechanics of matrix cracking in brittle matrix composites, *Acta Metall.*, Vol. 33, 2013-2021.

Marshall, D. B. and Cox, B. N. (1987), Tensile fracture of brittle matrix composites: influence of fiber strength, *Acta Metall.*, Vol. 35, 2607-2619.

McCartney, L. N. (1987), Mechanics of matrix cracking in brittle-matrix fiber-reinforced composites, *Proc. Roy. Soc. London*, Ser. A, 409.

McMeeking, R. M. and Evans, A. G. (1990), Matrix fatigue cracking in fiber composites, *Mech. Matr.*, Vol. 9, 217-227.

Mogi, K. (1966), Pressure dependence of rock strength and transition from brittle fracture to ductile flow, *Bull. Earthq. Res. Inst.*, Vol. 44, 215-232.

Myer, L. R., Kemeny, J. M., Zheng, Z., Suarez, R., Ewy, R. T., and Cook, N. G.

W. (1992), Extensile cracking in porous rock under differential compressive stress, *Appl. Mech. Rev.*, Vol. 45, 263-280.

Nemat-Nasser, S. (1987), Micromechanically based constitutive modeling of inelastic response of solids, in: *Constitutive models of deformation*, (J. Chandra, R. P. Srivastav, eds.), 120-125.

Nemat-Nasser, S. (1989), Compression-induced ductile flow of brittle materials and brittle fracturing of ductile materials, in: *Advances in fracture research*, (K. Salama, K. Ravichandran, D. M. R. Taplin, and P. Rama Rao, eds.), Vol. 1, 423-445.

Nemat-Nasser, S. and Chang, S. N. (1990), Compression-induced high strain rate void collapse, tensile cracking, and recrystallizaion in ductile single and polycrystals, *Mech. Matr.*, Vol. 10, 1-17.

Nemat-Nasser, S. and Deng, H. (1994), Strain-rate effect on brittle failure in compression, *Acta Metal. Mat.*, Vol. 42, 1013-1024.

Nemat-Nasser, S. and Hori, M. (1987), Toughening by partial or full bridging of cracks in ceramics and fiber reinforced composites, *Mech. Matr.*, Vol. 6, 245-269.

Nemat-Nasser, S. and Hori, M. (1990), Elastic solids with microdefects, in *Micromechanics and inhomogeneity - The T. Mura 65th anniversary volume*, (G. J. Weng, M. Taya, and H. Abé, eds.), Springer-Verlag, New York, 297-320.

Nemat-Nasser, S. and Horii, H. (1982), Compression-induced nonplanar crack extension with application to splitting, exfoliation, and rockburst, *J. Geophys. Res.*, Vol. 87, 6805-6821.

Nemat-Nasser, S., Isaacs, J.B. and Starrett, J.E. (1991), Hopkinson techniques for dynamic recovery experiments, *Proc. Royal Soc. London*, Vol. 435, 371-391.

Nemat-Nasser, S. and Obata, M. (1988), A microcrack model of dilatancy in brittle materials, *J. Appl. Mech.*, Vol. 110, 24-35.

Nemat-Nasser, S., Yu, N., and Hori, M. (1993), Solids with periodically distributed cracks, *Int. J. Solids Structures*, Vol. 30, No. 15, 2071-2095.

Nilsson, F. (1974), A note on the stress singularity at a nonuniformly moving crack tip, *J. Elasticity*, Vol. 4, 73-75.

O'Connell, R. J. and Budiansky, B. (1974), Seismic velocities in dry and saturated cracked solids, *J. Geophys. Res.*, Vol. 79, 5412-5426.

O'Connell, R. J. and Budiansky, B. (1977), Viscoelastic properties of fluid-saturated cracked solids, *J. Geophys. Res.*, Vol. 82, No. 36, 5719-5735.

Olsson, W. A. (1974), Microfracturing and faulting in a limestone, *Technophysics*, Vol. 24, 277-285.

Olsson, W. A. and Peng, S. S. (1976), Microcrack nucleation in marble, *Int. J. Rock Mech. Min. Sci. Geomech. Abstr.*, Vol. 13, 53-59.

Ortiz, M. (1985), A constitutive theory for the inelastic behavior of concrete, *Mech. Matr.*, Vol. 4, 67-93.

Peng, S. S. and Johnson, A. M. (1972), Crack growth and faulting in cylindrical specimens of Chelmsford granite, *Int. J. Rock Mech. Min. Sci.*, Vol. 9, 37-86.

Rice, J.R. (1968), Mathemetical analysis in the mechanics of fracture, in *Fracture*, Vol. 2, ed. H. Liebowitz, Academic Press, New York, 191-311.

Rogers, W. P. and Nemat-Nasser, S. (1990), Transformation plasticity at high strain rate in magnesia-partially-stabilized zirconia, *J. Am. Cer. Soc.*, Vol. 73, 136-139.

Rose, L. R. F. (1987), Crack reinforcement by distributed springs, *J. Mech. Phys. Solids*, Vol. 35, 383-405.

Salganik, R. L. (1973), Mechanics of bodies with many cracks, *Izv. AN SSR, Mekhanika Tverdogo Tela*, Vol. 8, 149-158.

Scholtz, C. H., Boitnott, G. and Nemat-Nasser, S. (1986), The Bridgman ring paradox revisited, *PAGEOPH*, Vol. 124, 587-599.

Sih, G. C. (1968), Some elastodynamic problems of cracks, *Int. J. Frac. Mech.*, Vol. 4, 567-574.

Sih, G. C. (1970), Dynamic aspects of crack propagation, in *Inelastic behavior of solids*, eds. Kanninen, M. F. *et al.*, McGraw-Hill series in materials science and engineering, McGraw-Hill, New York.

Sih, G. C., Paris, P. C., and Irwin, G. R. (1965), On crack in rectilinearly anisotropic bodies, *Intern. J. Fracture Mech.*, Vol. 1, 189-196.

Sneddon, I. N. and Lowengrub, M. (1969), *Crack problems in the classical theory of elasticity,* John Wiley & Sons, Inc..

Sneddon, I. N. and Srivastav, R. P. (1965), The stress in the vicinity of an infinite row of collinear cracks in an elastic body, *Proc. Roy. Soc. Edin. A*, Vol. 67, 39-49.

Sonuparlak, B. (1990), Tailoring the microstructure of ceramics and ceramic matrix composites through processing, *Composite Sci. Tech.*, Vol. 37, 299-312.

Steif, P. S. (1984), Crack extension under compressive loading, *Engng. Fract. Mech.*, Vol. 20, 463-473.

Subhash, G. (1991), Dynamic behavior of zirconia ceramics in uniaxial compression, Ph.D. Thesis, University of California, San Diego.

Subhash, G. and Nemat-Nasser, S. (1993), Dynamic stress-induced transformation and texture formation in uniaxial compression of zirconia ceramics, *J. Am. Cer. Soc.*, Vol. 76, 153-165.

Talreja, R. (1987), Continuum modeling of damage in ceramic matrix composites, *Mech. Matr.*, Vol. 12, 165-180.

Tobin, D. G. and Donath, F. A. (1971), Microscopic criteria for defining deformational modes in rock, *Bull. Geol. Soc. Am.*, Vol. 82, 1463-1476.

Tullis, J. and Yund, R. A. (1977), Experimental deformation of dry Westerly granite, *J. Geophys. Res.*, Vol. 82, 5705-5718.

Tvergaard, V. and Hutchinson, J. W. (1988), Microcracking in ceramics induced by thermal expansion or elastic anisotropy, *J. Am. Cer. Soc.*, Vol. 71, 157-166.

Walsh, J. B. (1969), New analysis of attenuation in partially melted rock, *J. Geophys. Res.*, Vol. 78, 4333-4337.

Willis, J. R. (1973), Self-similar problems in elastodynamics, *Phil. Trans. Roy. Soc. Lond.*, Vol. 274, 435-491.

Wong, T. F. (1982), Micromechanics of faulting in Westerly granite. *Int. J. Rock Mech. Min. Sci. Geomech. Abstr,*, Vol. 19, 49-64.

Wu, S. and Chudnovsky, A. (1990), The effective elastic properties of a linear elastic solid with microcracks, *J. Mater. Sci. Letters*, Vol. 4, 1457-1460.

Yoshida, H. and Horii, H. (1992), A micromechanics-based model for creep behavior of rock, *Appl. Mech. Rev.*, Vol. 45, 294-303.

Zheng, Z., Cook, N. G. W, Doyle, F. M., and Myer, L. R. (1988), Preservation of stress-induced microstructures in rock specimens, Report, Lawrence Berkeley Laboratory, University of California, Berkeley, 1-14.

CHAPTER III

ELASTIC SOLIDS WITH MICRO-INCLUSIONS

In this chapter, for an RVE with a linearly elastic and homogeneous matrix containing linearly elastic inclusions, the overall moduli are estimated in terms of the moduli of its constituents and their distribution. This RVE is used to introduce the concepts of eigenstrain and eigenstress which play key roles in estimating and bounding the overall moduli of heterogeneous solids (elastic or inelastic). The presentation is, however, general, applicable to bounded or unbounded heterogeneous solids, with inclusions having arbitrary geometries. For the special case of an infinitely extended homogeneous linearly elastic solid containing an ellipsoidal inclusion, the important results obtained by Eshelby (1957) are presented. The results of this chapter have direct application to elastic composites such as ceramics, cermets, cementitious materials, and other related heterogeneous solids. The stress and strain fields due to phase transformation or other physical processes are also discussed, e.g., a stress field induced in ceramic composites which contain partially stabilized zirconia which undergoes phase transformation under the applied stresses.

First, in Section 7, for micro-inclusions of any geometry and elasticity, exact general expressions for the overall elastic modulus and compliance tensors are established. The concepts of eigenstrain and eigenstress required to homogenize the heterogeneous RVE are introduced, Eshelby's tensor and its dual are presented, their properties are discussed, and their relations to the **H**- *and* **J**-*tensors of Section 4 are established. Homogenization on the basis of the dilute-distribution and the self-consistent models is discussed in Subsections 7.4 and 7.5, respectively, and in terms of the overall elastic energy of the RVE, in Subsection 7.6.*

Section 8 contains specific illustrative examples for elastic solids with micro-inclusions. A number of problems are worked out in detail, and numerical illustrations are presented.

Upper and lower bounds for the overall elastic moduli are presented in Section 9, based on the Hashin-Shtrikman variational principle (Subsection 9.1), leading to two functionals: one, when the overall uniform boundary tractions are prescribed, and the other, when the overall linear displacement boundary data are assigned.

The upper and lower bounds for the energy functionals are presented in Subsection 9.2, and their generalization is given in Subsection 9.3. Direct estimates of the overall moduli are presented in Subsections 9.4 and 9.5, where the Hashin-Shtrikman variational principle is generalized for boundary data other than uniform tractions and linear displacements, and the corresponding generalized bounds are obtained. With the aid of the universal theorems of Subsection 2.5.6, these bounds are then related to the bounds for the uniform traction and linear displacement boundary data. It is proved that two out of four possible approximate expressions that result are indeed rigorous bounds. Explicit, computable, exact upper and lower bounds for the overall moduli are then given, when the composite is statistically homogeneous and isotropic. Finally, it is shown in Subsection 9.6 that there are two universal bounds on two overall moduli of multi-phase composites, valid for any shape or distribution of phases. Furthermore, it is established that these universal bounds also apply to any finite elastic solid of ellipsoidal shape, consisting of any distribution of inhomogeneities of any shape and elasticity. An alternative formulation of exact computable bounds using the universal theorems of Subsection 2.5.6, is presented in Subsections 9.5.6 and 9.7.2, for the mechanical and nonmechanical properties, respectively, and in Subsection 9.8 for the coupled mechanical and nonmechanical overall parameters.

A number of averaging models are studied systematically in Section 10. This includes the dilute-distribution (Subsection 10.1) and the self-consistent methods (Subsection 10.2), as well as the differential scheme (Subsection 10.3), and the two- and three-phase models. The double-inclusion method is discussed in Subsection 10.4, together with the Mori-Tanaka result, and using the universal theorems of Section 2.5.6, it is shown that the double-inclusion model does in fact provide exact bounds. The double-inclusion model is then generalized to multi-inclusion models, where, again, all the average field quantities are estimated analytically, for a set of nested ellipsoidal regions of arbitrary aspect ratios and relative locations. The relations among various homogenization techniques are studied in Subsections 10.5 and 10.6.

The development of Eshelby's tensor in terms of the infinite-space Green function, is contained in Section 11, where the properties of this tensor and its dual are studied (Subsections 11.1 to 11.3). Then, relations among various average quantities are examined (Subsection 11.4), and the energy associated with heterogeneity and, hence, with the homogenizing eigenstrains or eigenstresses, is quantified explicitly.

SECTION 7 OVERALL ELASTIC MODULUS AND COMPLIANCE TENSORS

In this section, an RVE of volume V bounded by ∂V is considered, which consists of a uniform elastic matrix with elasticity and compliance tensors $\mathbf{C}$ and $\mathbf{D}$, containing n elastic micro-inclusions Ω_α, with elasticity and compliance tensors $\mathbf{C}^\alpha$ and $\mathbf{D}^\alpha$ ($\alpha = 1, 2, ..., n$). The micro-inclusions are perfectly bonded to the matrix. All constituents of the RVE are assumed to be linearly elastic. Hence, the overall response of the RVE is linearly elastic. The matrix and each inclusion are assumed to be uniform, but neither the matrix nor the inclusions need be isotropic. In general, the overall response of the RVE may be anisotropic, even if its constituents are isotropic. This depends on the geometry and arrangement of the micro-inclusions.

The overall elasticity and compliance tensors of the RVE are denoted by $\bar{\mathbf{C}}$ and $\bar{\mathbf{D}}$, and it is sought to estimate them in terms of the RVE's microstructural properties and geometry. As in Section 4, the cases of a prescribed macrostress and a prescribed macrostrain are considered separately. The concepts of *eigenstrain* and *eigenstress* are introduced to *homogenize* the RVE, and the corresponding *consistency conditions* are developed. Then for an ellipsoidal inhomogeneity, the Eshelby tensor and its conjugate are introduced, and they are related to the **H**- and **J**-tensors, which have been used in Sections 4, 5, and 8. These results are then employed and explicit expressions for the overall moduli are obtained by the dilute and the self-consistent methods.[1]

7.1. MACROSTRESS PRESCRIBED

For the constant macrostress $\boldsymbol{\sigma} = \boldsymbol{\sigma}^o$, the boundary tractions are

$$\mathbf{t}^o = \boldsymbol{\nu} \cdot \boldsymbol{\sigma}^o \qquad \text{on } \partial V. \tag{7.1.1}$$

Because of heterogeneity, neither the resulting stress nor the resulting strain fields in the RVE are uniform. *Define* the constant strain field $\boldsymbol{\varepsilon}^o$ by

$$\boldsymbol{\varepsilon}^o \equiv \mathbf{D} : \boldsymbol{\sigma}^o, \tag{7.1.2a}$$

and observe that the actual stress field, denoted by $\boldsymbol{\sigma}$, and strain field, denoted by $\boldsymbol{\varepsilon}$, can be expressed as

[1] Bounds on moduli are presented in Section 9, and other averaging methods are discussed and compared in Section 10.

$$\boldsymbol{\sigma} = \boldsymbol{\sigma}^{\mathrm{o}} + \boldsymbol{\sigma}^{\mathrm{d}}(\mathbf{x}), \qquad \boldsymbol{\varepsilon} = \boldsymbol{\varepsilon}^{\mathrm{o}} + \boldsymbol{\varepsilon}^{\mathrm{d}}(\mathbf{x}), \tag{7.1.2b,c}$$

where the variable stress and strain fields, $\boldsymbol{\sigma}^{\mathrm{d}}(\mathbf{x})$ and $\boldsymbol{\varepsilon}^{\mathrm{d}}(\mathbf{x})$, are the disturbances or perturbations in the prescribed uniform stress field $\boldsymbol{\sigma}^{\mathrm{o}}$ and the associated constant strain field $\boldsymbol{\varepsilon}^{\mathrm{o}}$, due to the presence of the inclusions. The total stress and strain tensors, $\boldsymbol{\sigma}$ and $\boldsymbol{\varepsilon}$, are related by Hooke's law, as follows:

$$\boldsymbol{\sigma}(\mathbf{x}) = \boldsymbol{\sigma}^{\mathrm{o}} + \boldsymbol{\sigma}^{\mathrm{d}}(\mathbf{x}) = \begin{cases} \mathbf{C} : \boldsymbol{\varepsilon}(\mathbf{x}) = \mathbf{C} : \{\boldsymbol{\varepsilon}^{\mathrm{o}} + \boldsymbol{\varepsilon}^{\mathrm{d}}(\mathbf{x})\} & \text{in } \mathrm{M} = \mathrm{V} - \Omega \\ \mathbf{C}^{\alpha} : \boldsymbol{\varepsilon}(\mathbf{x}) = \mathbf{C}^{\alpha} : \{\boldsymbol{\varepsilon}^{\mathrm{o}} + \boldsymbol{\varepsilon}^{\mathrm{d}}(\mathbf{x})\} & \text{in } \Omega_{\alpha}, \end{cases}$$

$$\boldsymbol{\varepsilon}(\mathbf{x}) = \boldsymbol{\varepsilon}^{\mathrm{o}} + \boldsymbol{\varepsilon}^{\mathrm{d}}(\mathbf{x}) = \begin{cases} \mathbf{D} : \boldsymbol{\sigma}(\mathbf{x}) = \mathbf{D} : \{\boldsymbol{\sigma}^{\mathrm{o}} + \boldsymbol{\sigma}^{\mathrm{d}}(\mathbf{x})\} & \text{in } \mathrm{M} = \mathrm{V} - \Omega \\ \mathbf{D}^{\alpha} : \boldsymbol{\sigma}(\mathbf{x}) = \mathbf{D}^{\alpha} : \{\boldsymbol{\sigma}^{\mathrm{o}} + \boldsymbol{\sigma}^{\mathrm{d}}(\mathbf{x})\} & \text{in } \Omega_{\alpha}, \end{cases} \tag{7.1.3a,b}$$

where Ω is the union of all micro-inclusions, $\Omega \equiv \bigcup_{\alpha=1}^{n} \Omega_{\alpha}$.

From the averaging theorems discussed in Section 2, and in view of (7.1.1), it follows that

$$\bar{\boldsymbol{\sigma}} \equiv <\boldsymbol{\sigma}> = \boldsymbol{\sigma}^{\mathrm{o}}. \tag{7.1.4a}$$

On the other hand, the overall average strain is given by

$$\bar{\boldsymbol{\varepsilon}} \equiv <\boldsymbol{\varepsilon}> = <\boldsymbol{\varepsilon}^{\mathrm{o}} + \boldsymbol{\varepsilon}^{\mathrm{d}}>, \tag{7.1.4b}$$

i.e., in general, $<\boldsymbol{\varepsilon}^{\mathrm{d}}> \neq \mathbf{0}$. The aim is to calculate the overall compliance $\bar{\mathbf{D}}$, such that

$$\bar{\boldsymbol{\varepsilon}} = \bar{\mathbf{D}} : \bar{\boldsymbol{\sigma}} = \bar{\mathbf{D}} : \boldsymbol{\sigma}^{\mathrm{o}}. \tag{7.1.5}$$

To this end, consider the notation

$$\bar{\boldsymbol{\varepsilon}}^{\alpha} \equiv <\boldsymbol{\varepsilon}>_{\alpha} \equiv \frac{1}{\Omega_{\alpha}} \int_{\Omega_{\alpha}} \boldsymbol{\varepsilon}(\mathbf{x})\, \mathrm{dV}, \tag{7.1.6a}$$

or, in general, for any field variable $\mathbf{T}(\mathbf{x})$, set

$$\bar{\mathbf{T}}^{\alpha} \equiv <\mathbf{T}>_{\alpha} \equiv \frac{1}{\Omega_{\alpha}} \int_{\Omega_{\alpha}} \mathbf{T}(\mathbf{x})\, \mathrm{dV}. \tag{7.1.6b}$$

Similarly, when the strain field is averaged over the matrix material of the RVE, it is convenient to write

$$\bar{\boldsymbol{\varepsilon}}^{\mathrm{M}} \equiv <\boldsymbol{\varepsilon}>_{\mathrm{M}} \equiv \frac{1}{\mathrm{M}} \int_{\mathrm{M}} \boldsymbol{\varepsilon}(\mathbf{x})\, \mathrm{dV}, \tag{7.1.6c}$$

and, for the general field variable $\mathbf{T}(\mathbf{x})$, set

$$\bar{\mathbf{T}}^{\mathrm{M}} \equiv <\mathbf{T}>_{\mathrm{M}} \equiv \frac{1}{\mathrm{M}} \int_{\mathrm{M}} \mathbf{T}(\mathbf{x})\, \mathrm{dV}. \tag{7.1.6d}$$

Thus, the volume average of (7.1.3b) over the matrix and inclusions produces

$$\bar{\boldsymbol{\varepsilon}}^{\mathrm{M}} = \mathbf{D} : \bar{\boldsymbol{\sigma}}^{\mathrm{M}}, \qquad \bar{\boldsymbol{\varepsilon}}^{\alpha} = \mathbf{D}^{\alpha} : \bar{\boldsymbol{\sigma}}^{\alpha} \qquad (\alpha \text{ not summed}). \tag{7.1.7a,b}$$

Since

$$\frac{M}{V}\,\bar{\boldsymbol{\varepsilon}}^M = \bar{\boldsymbol{\varepsilon}} - \sum_{\alpha=1}^{n} f_\alpha\,\bar{\boldsymbol{\varepsilon}}^\alpha = \bar{\mathbf{D}} : \boldsymbol{\sigma}^o - \sum_{\alpha=1}^{n} f_\alpha\,\mathbf{D}^\alpha : \bar{\boldsymbol{\sigma}}^\alpha, \tag{7.1.8a}$$

and

$$\frac{M}{V}\,\bar{\boldsymbol{\varepsilon}}^M = \frac{M}{V}\,\mathbf{D} : \bar{\boldsymbol{\sigma}}^M = \mathbf{D} : \{\boldsymbol{\sigma}^o - \sum_{\alpha=1}^{n} f_\alpha\,\bar{\boldsymbol{\sigma}}^\alpha\}, \tag{7.1.8b}$$

then

$$(\mathbf{D} - \bar{\mathbf{D}}) : \boldsymbol{\sigma}^o = \sum_{\alpha=1}^{n} f_\alpha\,(\mathbf{D} - \mathbf{D}^\alpha) : \bar{\boldsymbol{\sigma}}^\alpha = \sum_{\alpha=1}^{n} f_\alpha\,(\mathbf{D} - \mathbf{D}^\alpha) : < \boldsymbol{\sigma}^o + \boldsymbol{\sigma}^d >_\alpha, \tag{7.1.9}$$

where $f_\alpha = \Omega_\alpha/V$ is the volume fraction of the αth inclusion. This is an *exact* result. It defines the overall compliance tensor $\bar{\mathbf{D}}$ in terms of the average stresses in the inclusions. It is important to note that this result does not require knowledge of the entire field within each inclusion. Only the estimate of the *average* value of the stress in each inclusion is needed.

Since the response is linearly elastic, the disturbances or perturbations in the stress and strain fields due to the presence of inclusions, $\boldsymbol{\sigma}^d(\mathbf{x})$ and $\boldsymbol{\varepsilon}^d(\mathbf{x})$, are linear and homogeneous functions of the prescribed constant macrostress $\boldsymbol{\Sigma} = \boldsymbol{\sigma}^o$. Hence, in general,

$$(\mathbf{D}^\alpha - \mathbf{D}) : < \boldsymbol{\sigma}^o + \boldsymbol{\sigma}^d >_\alpha \equiv (\mathbf{D}^\alpha - \mathbf{D}) : \bar{\boldsymbol{\sigma}}^\alpha = \mathbf{H}^\alpha : \boldsymbol{\sigma}^o \qquad (\alpha \text{ not summed}), \tag{7.1.10}$$

where the constant fourth-order $\mathbf{H}^\alpha$-tensor is defined by

$$\bar{\boldsymbol{\varepsilon}}^\alpha - \mathbf{D} : \bar{\boldsymbol{\sigma}}^\alpha \equiv < \boldsymbol{\varepsilon}^o + \boldsymbol{\varepsilon}^d >_\alpha - \mathbf{D} : < \boldsymbol{\sigma}^o + \boldsymbol{\sigma}^d >_\alpha \equiv \mathbf{H}^\alpha : \boldsymbol{\sigma}^o. \tag{7.1.11}$$

This is the change in the average strain of Ω_α, if $\mathbf{D}^\alpha$ is replaced by $\mathbf{D}$. Note that for traction-free cavities or cracks, $\bar{\boldsymbol{\sigma}}^\alpha = \mathbf{0}$, and definition (7.1.11) is consistent with (4.3.3). Since $\boldsymbol{\sigma}^o$ is arbitrary, substitution of (7.1.10) into (7.1.9) produces

$$\bar{\mathbf{D}} = \mathbf{D} + \sum_{\alpha=1}^{n} f_\alpha\,\mathbf{H}^\alpha. \tag{7.1.12}$$

The result (7.1.12) is exact. It applies to a finite, as well as an infinitely extended RVE. There is no restriction on the geometry (i.e., shapes) or distribution of the inclusions. The only requirements are: (1) the matrix is linearly elastic and homogeneous; (2) each inclusion is linearly elastic and homogeneous; and (3) the inclusions are perfectly bonded to the matrix. Approximations and specializations are generally introduced when it is sought to estimate the constant tensors $\mathbf{H}^\alpha$ ($\alpha = 1, 2, ..., n$). To this end, the inclusions are often assumed to be ellipsoidal and other assumptions are made in order to estimate the $\mathbf{H}^\alpha$'s. This and related issues are examined later on in this chapter. Observe that, since

$$\bar{\boldsymbol{\varepsilon}}^\alpha = \frac{1}{\Omega_\alpha} \int_{\Omega_\alpha} \boldsymbol{\varepsilon}(\mathbf{x})\, dV = \frac{1}{\Omega_\alpha} \int_{\partial\Omega_\alpha} \frac{1}{2}(\mathbf{n} \otimes \mathbf{u} + \mathbf{u} \otimes \mathbf{n})\, dS, \tag{7.1.13}$$

the $\mathbf{H}^\alpha$-tensor here has the same significance as that introduced for cavities in Sections 4 and 5. The $\mathbf{u} = \mathbf{u}(\mathbf{x})$ in (7.1.13) is the displacement of the boundary of the αth inclusion. Unlike the case of an isolated cavity, the calculation of this boundary displacement field is somewhat complicated.

It is noted that (7.1.12) can be specialized to yield the Reuss (1929) estimate. Reuss assumed that the average stress of each inclusion (and hence the matrix) is equal to the applied stress $\boldsymbol{\sigma}^o$. Then, the average strain in Ω_α is given by

$$\bar{\boldsymbol{\varepsilon}}^\alpha = \mathbf{D}^\alpha : \bar{\boldsymbol{\sigma}}^\alpha = \mathbf{D}^\alpha : \boldsymbol{\sigma}^o. \tag{7.1.14a}$$

Hence, $\mathbf{H}^\alpha$ reduces to $\mathbf{D}^\alpha - \mathbf{D}$. Then, the overall compliance $\bar{\mathbf{D}}$ is estimated to be the volume average of the compliance tensor of the matrix and the inclusions, i.e.,

$$\bar{\mathbf{D}} = (1-\mathrm{f})\,\mathbf{D} + \sum_{\alpha=1}^{n} \mathrm{f}_\alpha\,\mathbf{D}^\alpha, \tag{7.1.14b}$$

where $\mathrm{f} = \sum_{\alpha=1}^{n} \mathrm{f}_\alpha$ is the total volume fraction of all inclusions.

7.2. MACROSTRAIN PRESCRIBED

When, instead of the macrostresses, the uniform macrostrains, $\mathbf{E} = \boldsymbol{\varepsilon}^o$, are prescribed, the boundary conditions for the RVE become

$$\mathbf{u}^o = \mathbf{x} \cdot \boldsymbol{\varepsilon}^o \qquad \text{on } \partial V, \tag{7.2.1}$$

and *defining*

$$\boldsymbol{\sigma}^o \equiv \mathbf{C} : \boldsymbol{\varepsilon}^o, \tag{7.2.2}$$

again observe that the presence of inclusions with different elasticity tensors introduces disturbances or changes in the uniform strain and stress fields, $\boldsymbol{\varepsilon}^o$ and $\boldsymbol{\sigma}^o$. Denoting the strain and stress disturbances by $\boldsymbol{\varepsilon}^d$ and $\boldsymbol{\sigma}^d$, respectively, express the resulting variable strain, $\boldsymbol{\varepsilon} = \boldsymbol{\varepsilon}(\mathbf{x})$, and stress, $\boldsymbol{\sigma} = \boldsymbol{\sigma}(\mathbf{x})$, as in (7.1.2c,b). The stress-strain relations are given by (7.1.3).

From the boundary conditions (7.2.1), it now follows that

$$\bar{\boldsymbol{\varepsilon}} \equiv \langle \boldsymbol{\varepsilon} \rangle = \langle \boldsymbol{\varepsilon}^o + \boldsymbol{\varepsilon}^d \rangle = \boldsymbol{\varepsilon}^o, \tag{7.2.3a}$$

but

$$\bar{\boldsymbol{\sigma}} \equiv \langle \boldsymbol{\sigma} \rangle = \langle \boldsymbol{\sigma}^o + \boldsymbol{\sigma}^d \rangle, \tag{7.2.3b}$$

in general, is not equal to $\boldsymbol{\sigma}^o$, i.e., in general, $\langle \boldsymbol{\sigma}^d \rangle \neq \mathbf{0}$. The overall elasticity tensor $\bar{\mathbf{C}}$ is *defined* through

$$\bar{\boldsymbol{\sigma}} = \bar{\mathbf{C}} : \bar{\boldsymbol{\varepsilon}} = \bar{\mathbf{C}} : \boldsymbol{\varepsilon}^o, \tag{7.2.4}$$

to be estimated in terms of the microstructure and properties of the RVE.

Following the procedure outlined in Subsection 7.1, observe that

$$\frac{M}{V}\,\bar{\boldsymbol{\sigma}}^M = \bar{\boldsymbol{\sigma}} - \sum_{\alpha=1}^{n} \mathrm{f}_\alpha\,\bar{\boldsymbol{\sigma}}^\alpha = \bar{\mathbf{C}} : \boldsymbol{\varepsilon}^o - \sum_{\alpha=1}^{n} \mathrm{f}_\alpha\,\mathbf{C}^\alpha : \bar{\boldsymbol{\varepsilon}}^\alpha, \tag{7.2.5a}$$

and that

$$\frac{M}{V}\,\bar{\boldsymbol{\sigma}}^M = \frac{M}{V}\,\mathbf{C}:\bar{\boldsymbol{\varepsilon}}^M = \mathbf{C}:(\boldsymbol{\varepsilon}^o - \sum_{\alpha=1}^{n} f_\alpha\,\bar{\boldsymbol{\varepsilon}}^\alpha). \tag{7.2.5b}$$

Hence, it follows from (7.2.4), (7.2.5a), and (7.2.5b) that

$$(\mathbf{C}-\bar{\mathbf{C}}):\boldsymbol{\varepsilon}^o = \sum_{\alpha=1}^{n} f_\alpha\,(\mathbf{C}-\mathbf{C}^\alpha):\bar{\boldsymbol{\varepsilon}}^\alpha = \sum_{\alpha=1}^{n} f_\alpha\,(\mathbf{C}-\mathbf{C}^\alpha):<\boldsymbol{\varepsilon}^o+\boldsymbol{\varepsilon}^d>_\alpha. \tag{7.2.6}$$

Again, because of linearity, the change of the average strain of Ω_α due to the homogenization associated with replacing $\mathbf{C}^\alpha$ by $\mathbf{C}$, is expressed as

$$\bar{\boldsymbol{\varepsilon}}^\alpha - \mathbf{D}:\bar{\boldsymbol{\sigma}}^\alpha \equiv \mathbf{J}^\alpha:\boldsymbol{\varepsilon}^o, \tag{7.2.7}$$

and from (7.2.6) it is deduced that (since $\boldsymbol{\varepsilon}^o$ is arbitrary)

$$\bar{\mathbf{C}} = \mathbf{C} - \sum_{\alpha=1}^{n} f_\alpha\,\mathbf{C}:\mathbf{J}^\alpha = \mathbf{C}:(\mathbf{1}^{(4s)} - \sum_{\alpha=1}^{n} f_\alpha\,\mathbf{J}^\alpha). \tag{7.2.8}$$

Definition (7.2.7) for $\mathbf{J}^\alpha$ is consistent with the corresponding definition for the case of cavities where $\bar{\boldsymbol{\sigma}}^\alpha = \mathbf{0}$; see (4.5.3). Comments which follow (7.1.12) also apply here. The constant tensors $\mathbf{H}^\alpha$ and $\mathbf{J}^\alpha$, for each inclusion, must now be estimated.

It is noted that (7.2.8) can be simplified to yield the Voigt (1889) estimate. Voigt assumed that the average strain of each inclusion (and hence the matrix) is equal to the applied strain $\boldsymbol{\varepsilon}^o$. Then, the overall elasticity tensor $\bar{\mathbf{C}}$ is given by the volume average of the elasticity tensors of the matrix and the inclusions, i.e.,

$$\bar{\mathbf{C}} = (1-f)\,\mathbf{C} + \sum_{\alpha=1}^{n} f_\alpha\,\mathbf{C}^\alpha. \tag{7.2.9}$$

7.3. EIGENSTRAIN AND EIGENSTRESS TENSORS

For clarity in presentation, a specific elasticity problem is considered and is used to introduce the concept of *eigenstrain*. Consider a *finite homogeneous linearly* elastic (not necessarily isotropic) solid with elasticity tensor $\mathbf{C}$ and compliance tensor $\mathbf{D}$, containing within it a (only one) *linearly elastic and homogeneous* (but not necessarily isotropic) inclusion Ω, of *arbitrary geometry*, with elasticity and compliance tensors $\mathbf{C}^\Omega$ and $\mathbf{D}^\Omega$. The total volume is V, bounded by ∂V, and the matrix is $M = V - \Omega$, bounded by $\partial V + \partial\Omega^M = \partial V - \partial\Omega$; see Subsection 4.1, as well as Figure 4.1.1, for a discussion of the notation. Let the solid be subjected on ∂V to either the self-equilibrating surface tractions corresponding to the uniform stress $\boldsymbol{\sigma}^o = (\boldsymbol{\sigma}^o)^T$ = constant,

$$\mathbf{t}^o = \boldsymbol{\nu}\cdot\boldsymbol{\sigma}^o \qquad \text{on } \partial V, \tag{7.3.1a}$$

or the self-compatible linear surface displacements corresponding to the uniform strain $\boldsymbol{\varepsilon}^o = (\boldsymbol{\varepsilon}^o)^T$ = constant,

$$\mathbf{u}^o = \mathbf{x} \cdot \boldsymbol{\varepsilon}^o \qquad \text{on } \partial V; \tag{7.3.1b}$$

see Figure 7.3.1a for the case when the surface tractions are prescribed. Note that (7.3.1a) and (7.3.1b) define two separate problems which are being examined simultaneously. These boundary conditions are in general *mutually exclusive* for a heterogeneous elastic solid.

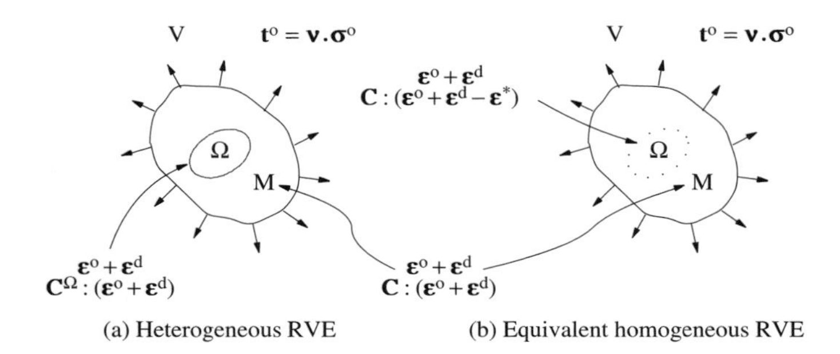

Figure 7.3.1

Equivalent homogeneous RVE and eigenstrain

If the RVE were uniform throughout its entire volume, then the stress field and hence the corresponding strain field would be uniform when tractions are prescribed on ∂V by (7.3.1a); these fields would be given by $\boldsymbol{\sigma}^o$ and $\boldsymbol{\varepsilon}^o \equiv \mathbf{D} : \boldsymbol{\sigma}^o$, respectively. Similarly, the strain field and the corresponding stress field would be uniform when displacements are prescribed on ∂V by (7.3.1b); these fields would be given by $\boldsymbol{\varepsilon}^o$ and $\boldsymbol{\sigma}^o \equiv \mathbf{C} : \boldsymbol{\varepsilon}^o$, respectively. The presence of region Ω with a different elasticity, i.e., the existence of a material mismatch, disturbs the uniform stress and strain fields in both cases. Denote the resulting variable strain and stress fields, respectively by $\boldsymbol{\varepsilon} = \boldsymbol{\varepsilon}(\mathbf{x})$ and $\boldsymbol{\sigma} = \boldsymbol{\sigma}(\mathbf{x})$, and set

$$\boldsymbol{\varepsilon} = \boldsymbol{\varepsilon}^o + \boldsymbol{\varepsilon}^d(\mathbf{x}), \qquad \boldsymbol{\sigma} = \boldsymbol{\sigma}^o + \boldsymbol{\sigma}^d(\mathbf{x}). \tag{7.3.2a,b}$$

Here $\boldsymbol{\varepsilon}^d(\mathbf{x})$ and $\boldsymbol{\sigma}^d(\mathbf{x})$ are the disturbance strain and stress fields caused by the presence of the inclusion Ω, with mismatched elasticity. From Hooke's law it follows that

$$\boldsymbol{\sigma} = \begin{cases} \mathbf{C} : (\boldsymbol{\varepsilon}^o + \boldsymbol{\varepsilon}^d(\mathbf{x})) & \text{in } M = V - \Omega \\ \mathbf{C}^\Omega : (\boldsymbol{\varepsilon}^o + \boldsymbol{\varepsilon}^d(\mathbf{x})) & \text{in } \Omega, \end{cases}$$

$$\boldsymbol{\varepsilon} = \begin{cases} \mathbf{D} : (\boldsymbol{\sigma}^o + \boldsymbol{\sigma}^d(\mathbf{x})) & \text{in } M = V - \Omega \\ \mathbf{D}^\Omega : (\boldsymbol{\sigma}^o + \boldsymbol{\sigma}^d(\mathbf{x})) & \text{in } \Omega. \end{cases} \tag{7.3.3a,b}$$

7.3.1. Eigenstrain

Instead of dealing with the above-mentioned heterogeneous solid, it is convenient and effective to consider an *equivalent homogeneous* solid which has the *uniform* elasticity tensor $\mathbf{C}$ of the matrix material *everywhere, including in* Ω. Then, in order to account for the mismatch of the material properties of the inclusion and the matrix, a suitable strain field $\boldsymbol{\varepsilon}^*(\mathbf{x})$ is introduced in Ω, such that the *equivalent homogeneous* solid has the same strain and stress fields as the *actual heterogeneous* solid under the applied tractions or displacements, whichever may be the case. The strain field $\boldsymbol{\varepsilon}^*$ necessary for this *homogenization* is called the *eigenstrain*.

Figure 7.3.1b illustrates this procedure for the case when boundary tractions corresponding to $\boldsymbol{\sigma}^{\mathrm{o}}$ are prescribed on $\partial\mathrm{V}$. In this figure the eigenstrain field is given by

$$\boldsymbol{\varepsilon}^*(\mathbf{x}) = \begin{cases} \mathbf{0} & \text{in M} \\ \boldsymbol{\varepsilon}^* & \text{in } \Omega. \end{cases} \tag{7.3.4a}$$

For this *equivalent* problem the elasticity tensor is *uniform everywhere, including in* Ω. It is given by $\mathbf{C}$. Therefore, the corresponding strain and stress fields are

$$\boldsymbol{\varepsilon}(\mathbf{x}) = \boldsymbol{\varepsilon}^{\mathrm{o}} + \boldsymbol{\varepsilon}^{\mathrm{d}}(\mathbf{x}),$$

$$\boldsymbol{\sigma}(\mathbf{x}) = \mathbf{C} : (\boldsymbol{\varepsilon}(\mathbf{x}) - \boldsymbol{\varepsilon}^*(\mathbf{x})) = \begin{cases} \mathbf{C} : (\boldsymbol{\varepsilon}^{\mathrm{o}} + \boldsymbol{\varepsilon}^{\mathrm{d}}(\mathbf{x})) & \text{in M} \\ \mathbf{C} : (\boldsymbol{\varepsilon}^{\mathrm{o}} + \boldsymbol{\varepsilon}^{\mathrm{d}}(\mathbf{x}) - \boldsymbol{\varepsilon}^*(\mathbf{x})) & \text{in } \Omega. \end{cases} \tag{7.3.4b,c}$$

As is seen, the eigenstrain field disturbs the relation between the strain and the stress. Indeed, they are no longer related through uniform elasticity $\mathbf{C}$ in Ω.

To relate the eigenstrain $\boldsymbol{\varepsilon}^*$ to the corresponding perturbation strain $\boldsymbol{\varepsilon}^{\mathrm{d}}$, consider the equivalent uniform elastic solid of volume V and *uniform* elasticity $\mathbf{C}$, and observe that, since by definition,

$$\boldsymbol{\sigma}^{\mathrm{o}} = \mathbf{C} : \boldsymbol{\varepsilon}^{\mathrm{o}} \tag{7.3.5a}$$

or

$$\boldsymbol{\varepsilon}^{\mathrm{o}} = \mathbf{D} : \boldsymbol{\sigma}^{\mathrm{o}}, \tag{7.3.5b}$$

then from (7.3.2) and (7.3.4), it follows that

$$\boldsymbol{\sigma}^{\mathrm{d}}(\mathbf{x}) = \mathbf{C} : (\boldsymbol{\varepsilon}^{\mathrm{d}}(\mathbf{x}) - \boldsymbol{\varepsilon}^*(\mathbf{x})) \quad \text{in V}. \tag{7.3.6}$$

Since the resulting stress field must be in equilibrium and must produce a compatible strain field, in general, the strain field $\boldsymbol{\varepsilon}^{\mathrm{d}}(\mathbf{x})$ is obtained in terms of an integral operator acting on the corresponding eigenstrain[2] $\boldsymbol{\varepsilon}^*(\mathbf{x})$. In the present context, it is convenient to denote this integral operator by $\mathbf{S}$, and simply set

$$\boldsymbol{\varepsilon}^{\mathrm{d}}(\mathbf{x}) \equiv \mathbf{S}(\mathbf{x}; \boldsymbol{\varepsilon}^*) \tag{7.3.7a}$$

---

[2] As will be shown in Section 9 in some detail, (7.3.6) can be solved with the aid of Green's function for V.

or

$$\varepsilon_{ij}^{d}(\mathbf{x}) \equiv S_{ij}(\mathbf{x};\, \boldsymbol{\varepsilon}^*). \tag{7.3.7b}$$

7.3.2. Eigenstress

In the above treatment, the heterogeneous finite (or infinite) linearly elastic solid consisting of a uniform matrix M and a single inclusion Ω with different elasticities (7.3.3a), is homogenized by the introduction of the eigenstrain $\boldsymbol{\varepsilon}^*(\mathbf{x})$. The homogenization can be performed by the introduction of an *eigenstress* $\boldsymbol{\sigma}^*(\mathbf{x})$, instead. To this end, set

$$\boldsymbol{\sigma}^*(\mathbf{x}) = \begin{cases} \mathbf{0} & \text{in M} \\ \boldsymbol{\sigma}^* & \text{in } \Omega. \end{cases} \tag{7.3.8a}$$

For this *alternative equivalent* problem, the elasticity tensor is again *uniform everywhere, including in* Ω, as in (7.3.4). The corresponding strain and stress fields are

$$\boldsymbol{\varepsilon}(\mathbf{x}) = \boldsymbol{\varepsilon}^{o} + \boldsymbol{\varepsilon}^{d}(\mathbf{x}),$$

$$\boldsymbol{\sigma}(\mathbf{x}) = \mathbf{C} : \boldsymbol{\varepsilon}(\mathbf{x}) + \boldsymbol{\sigma}^*(\mathbf{x}) = \begin{cases} \mathbf{C} : (\boldsymbol{\varepsilon}^{o} + \boldsymbol{\varepsilon}^{d}(\mathbf{x})) & \text{in M} \\ \mathbf{C} : (\boldsymbol{\varepsilon}^{o} + \boldsymbol{\varepsilon}^{d}(\mathbf{x})) + \boldsymbol{\sigma}^*(\mathbf{x}) & \text{in } \Omega. \end{cases} \tag{7.3.8b,c}$$

From (7.3.5), the disturbance strain and stress must satisfy

$$\boldsymbol{\sigma}^{d}(\mathbf{x}) = \mathbf{C} : \boldsymbol{\varepsilon}^{d}(\mathbf{x}) + \boldsymbol{\sigma}^*(\mathbf{x}) \qquad \text{in V}, \tag{7.3.9}$$

for the required eigenstress. As discussed in Subsection 7.3.1, in general, the stress field $\boldsymbol{\sigma}^{d}(\mathbf{x})$ is expressed in terms of an integral operator acting on the corresponding eigenstress $\boldsymbol{\sigma}^*(\mathbf{x})$. Formally, this is written as

$$\boldsymbol{\sigma}^{d}(\mathbf{x}) = \mathbf{T}(\mathbf{x};\, \boldsymbol{\sigma}^*) \tag{7.3.10a}$$

or

$$\sigma_{ij}^{d}(\mathbf{x}) = T_{ij}(\mathbf{x};\, \boldsymbol{\sigma}^*). \tag{7.3.10b}$$

7.3.3. Uniform Eigenstrain and Eigenstress

An important result[3] due to Eshelby (1957), which has played a key role in the micromechanical modeling of elastic and inelastic heterogeneous solids, as well as of nonlinear creeping fluids, is that if:

1) $V - \Omega$ is homogeneous, linearly elastic, and infinitely extended; and

[3] See an earlier similar observation in two dimensions by Hardiman (1954).

2) Ω is an ellipsoid,

then:

1) the eigenstrain $\boldsymbol{\varepsilon}^*$ necessary for homogenization is *uniform* in Ω; and

2) the resulting strain $\boldsymbol{\varepsilon}^d$ and hence, stress $\boldsymbol{\sigma}^d$, are also uniform in Ω, the former being given by

$$\boldsymbol{\varepsilon}^d = \mathbf{S}^\Omega : \boldsymbol{\varepsilon}^*, \tag{7.3.11a}$$

where the fourth-order tensor $\mathbf{S}^\Omega$ is called Eshelby's tensor, with the following properties:

a) it is symmetric with respect to the first two indices and the second two indices,

$$S^\Omega_{ijkl} = S^\Omega_{jikl} = S^\Omega_{ijlk}; \tag{7.3.11b}$$

however, it is not, in general, symmetric with respect to the exchange of ij and kl, i.e, in general, $S^\Omega_{ijkl} \neq S^\Omega_{klij}$;

b) it is independent of the material properties of the inclusion Ω;

c) it is completely defined in terms of the aspect ratios of the ellipsoidal inclusion Ω, and the elastic parameters of the surrounding matrix M; and

d) when the surrounding matrix is isotropic, then $\mathbf{S}^\Omega$ depends only on the Poisson ratio of the matrix and the aspect ratios of Ω.

In Subsection 7.3.6, the components of Eshelby's tensor are listed for several special cases. In Section 11 a detailed calculation of Eshelby's tensor is given.

When the eigenstrain $\boldsymbol{\varepsilon}^*$ and the resulting strain disturbance $\boldsymbol{\varepsilon}^d$ are uniform in Ω, then the corresponding eigenstress $\boldsymbol{\sigma}^*$ and the associated stress disturbance $\boldsymbol{\sigma}^d$ are also uniform in Ω. Therefore, a fourth-order tensor $\mathbf{T}^\Omega$, may be introduced such that

$$\boldsymbol{\sigma}^d = \mathbf{T}^\Omega : \boldsymbol{\sigma}^* \quad \text{in } \Omega. \tag{7.3.12a}$$

The tensor $\mathbf{T}^\Omega$ has symmetries similar to Eshelby's tensor $\mathbf{S}^\Omega$, i.e.,

$$T^\Omega_{ijkl} = T^\Omega_{jikl} = T^\Omega_{ijlk}, \tag{7.3.12b}$$

but in general, $T^\Omega_{ijkl} \neq T^\Omega_{klij}$.

To relate the tensor $\mathbf{T}^\Omega$ to Eshelby's tensor $\mathbf{S}^\Omega$, it is first noted from (7.3.6) and (7.3.9) that the eigenstrain and eigenstress are related by

$$\boldsymbol{\sigma}^* + \mathbf{C} : \boldsymbol{\varepsilon}^* = \mathbf{0}, \qquad \boldsymbol{\varepsilon}^* + \mathbf{D} : \boldsymbol{\sigma}^* = \mathbf{0}. \tag{7.3.13a,b}$$

From (7.3.9), (7.3.11a), and (7.3.12a), it follows that

$$\mathbf{S}^\Omega : \boldsymbol{\varepsilon}^* = \mathbf{D} : (\mathbf{T}^\Omega - \mathbf{1}^{(4s)}) : (-\mathbf{C} : \boldsymbol{\varepsilon}^*),$$

$$\mathbf{T}^\Omega : \boldsymbol{\sigma}^* = \mathbf{C} : (\mathbf{S}^\Omega - \mathbf{1}^{(4s)}) : (-\mathbf{D} : \boldsymbol{\sigma}^*). \tag{7.3.14a,b}$$

Therefore, the tensors $\mathbf{S}^\Omega$ and $\mathbf{T}^\Omega$ must satisfy

$$\mathbf{S}^{\Omega} + \mathbf{D} : \mathbf{T}^{\Omega} : \mathbf{C} = \mathbf{1}^{(4s)}, \qquad \mathbf{T}^{\Omega} + \mathbf{C} : \mathbf{S}^{\Omega} : \mathbf{D} = \mathbf{1}^{(4s)}. \tag{7.3.14c,d}$$

In component form, these are

$$S^{\Omega}_{ijkl} + D_{ijpq}\, T^{\Omega}_{pqrs}\, C_{rskl} = \frac{1}{2}(\delta_{ik}\,\delta_{jl} + \delta_{il}\,\delta_{jk}),$$

$$T^{\Omega}_{ijkl} + C_{ijpq}\, S^{\Omega}_{pqrs}\, D_{rskl} = \frac{1}{2}(\delta_{ik}\,\delta_{jl} + \delta_{il}\,\delta_{jk}). \tag{7.3.14e,f}$$

7.3.4. Consistency Conditions

For finite V, the eigenstrains or eigenstresses necessary for homogenization are, in general, nonuniform in Ω, even if Ω is ellipsoidal. Also, for a nonellipsoidal Ω, the required eigenstrains or eigenstresses are in general, variable in Ω (they are zero outside of Ω), even if V is unbounded. For the general case, the eigenstrain, $\boldsymbol{\varepsilon}^*(\mathbf{x})$, or the eigenstress, $\boldsymbol{\sigma}^*(\mathbf{x})$, is defined by the so-called *consistency conditions* which require the resulting stress field $\boldsymbol{\sigma}(\mathbf{x})$, or the strain field $\boldsymbol{\varepsilon}(\mathbf{x})$, to be the same under the applied overall loads, whether it is calculated through homogenization or directly from (7.3.3a) or from (7.3.6). Hence, the resulting stress field in Ω becomes,

$$\boldsymbol{\sigma}(\mathbf{x}) = \mathbf{C}^{\Omega} : \{\boldsymbol{\varepsilon}^{o} + \boldsymbol{\varepsilon}^{d}(\mathbf{x})\} = \mathbf{C} : \{\boldsymbol{\varepsilon}^{o} + \boldsymbol{\varepsilon}^{d}(\mathbf{x}) - \boldsymbol{\varepsilon}^{*}(\mathbf{x})\} \quad \text{in } \Omega, \tag{7.3.15a}$$

and the resulting strain field in Ω satisfies,

$$\boldsymbol{\varepsilon}(\mathbf{x}) = \mathbf{D}^{\Omega} : \{\boldsymbol{\sigma}^{o} + \boldsymbol{\sigma}^{d}(\mathbf{x})\} = \mathbf{D} : \{\boldsymbol{\sigma}^{o} + \boldsymbol{\sigma}^{d}(\mathbf{x}) - \boldsymbol{\sigma}^{*}(\mathbf{x})\} \quad \text{in } \Omega. \tag{7.3.15b}$$

Substitution into (7.3.15a) for $\boldsymbol{\varepsilon}^{d}(\mathbf{x})$ from (7.3.7) now yields an integral equation for $\boldsymbol{\varepsilon}^*(\mathbf{x})$. Similarly, substitution into (7.3.15b) from (7.3.10) yields an integral equation for $\boldsymbol{\sigma}^*(\mathbf{x})$.

It is noted that both (7.3.15a) and (7.3.15b) are valid whether uniform tractions produced by $\boldsymbol{\sigma}^{o}$ or linear displacements produced by $\boldsymbol{\varepsilon}^{o}$ are prescribed on ∂V. If the overall stress $\boldsymbol{\sigma}^{o}$ is *given*, $\boldsymbol{\varepsilon}^{o}$ is *defined* by $\mathbf{D} : \boldsymbol{\sigma}^{o}$, whereas if the overall strain $\boldsymbol{\varepsilon}^{o}$ is *given*, $\boldsymbol{\sigma}^{o}$ is *defined* by $\mathbf{C} : \boldsymbol{\varepsilon}^{o}$. In Chapter IV this procedure is detailed when V is a cuboid, representing a unit cell of an RVE with periodic microstructure. In this chapter, on the other hand, attention is confined to the case when V is unbounded and Ω is ellipsoidal, so that the homogenization eigenstrain and eigenstress are both uniform in Ω.

Whether V is bounded or not, and for any homogeneous linearly elastic inclusion Ω in a homogeneous linearly elastic matrix M, consistency conditions (7.3.15a,b) yield

$$\boldsymbol{\varepsilon}^{o} + \boldsymbol{\varepsilon}^{d}(\mathbf{x}) = \mathbf{A}^{\Omega} : \boldsymbol{\varepsilon}^{*}(\mathbf{x}), \qquad \boldsymbol{\sigma}^{o} + \boldsymbol{\sigma}^{d}(\mathbf{x}) = \mathbf{B}^{\Omega} : \boldsymbol{\sigma}^{*}(\mathbf{x}) \quad \text{in } \Omega, \tag{7.3.16a,b}$$

where

$$\mathbf{A}^{\Omega} \equiv (\mathbf{C} - \mathbf{C}^{\Omega})^{-1} : \mathbf{C}, \qquad \mathbf{B}^{\Omega} \equiv (\mathbf{D} - \mathbf{D}^{\Omega})^{-1} : \mathbf{D}. \tag{7.3.17a,b}$$

By definition, constant tensors $\mathbf{A}^{\Omega}$ and $\mathbf{B}^{\Omega}$ satisfy

$$\mathbf{D} : \mathbf{C}^{\Omega} = \mathbf{1}^{(4s)} - (\mathbf{A}^{\Omega})^{-1} = (\mathbf{1}^{(4s)} - (\mathbf{B}^{\Omega})^{-1})^{-T} \tag{7.3.17c}$$

or

$$\mathbf{C} : \mathbf{D}^{\Omega} = \mathbf{1}^{(4s)} - (\mathbf{B}^{\Omega})^{-1} = (\mathbf{1}^{(4s)} - (\mathbf{A}^{\Omega})^{-1})^{-T}, \tag{7.3.17d}$$

where the superscript -T stands for the inverse of the transpose or the transpose of the inverse.

When V is *unbounded* there is no distinction between the cases when the strain $\boldsymbol{\varepsilon}^{o}$ or the stress $\boldsymbol{\sigma}^{o}$ is prescribed.[4] Thus $\boldsymbol{\varepsilon}^{o} = \mathbf{D} : \boldsymbol{\sigma}^{o}$ or $\boldsymbol{\sigma}^{o} = \mathbf{C} : \boldsymbol{\varepsilon}^{o}$. Also, when, in addition, Ω is *ellipsoidal*, then $\boldsymbol{\varepsilon}^{d}$, $\boldsymbol{\varepsilon}^{*}$, $\boldsymbol{\sigma}^{d}$, and $\boldsymbol{\sigma}^{*}$ are all *constant* tensors in Ω. Hence, for unbounded V and ellipsoidal Ω, substitution for $\boldsymbol{\varepsilon}^{d}$ in (7.3.16a) or for $\boldsymbol{\sigma}^{d}$ in (7.3.16b) provides explicit expressions for the eigenstrain $\boldsymbol{\varepsilon}^{*}$ and eigenstress $\boldsymbol{\sigma}^{*}$ which are necessary for homogenization,

$$\boldsymbol{\varepsilon}^{*} = (\mathbf{A}^{\Omega} - \mathbf{S}^{\Omega})^{-1} : \boldsymbol{\varepsilon}^{o}, \qquad \boldsymbol{\sigma}^{*} = (\mathbf{B}^{\Omega} - \mathbf{T}^{\Omega})^{-1} : \boldsymbol{\sigma}^{o} \qquad \text{in } \Omega. \tag{7.3.18a,b}$$

These and (7.3.16a,b) now lead to

$$\boldsymbol{\varepsilon} = \boldsymbol{\varepsilon}^{o} + \boldsymbol{\varepsilon}^{d} = \mathbf{A}^{\Omega} : (\mathbf{A}^{\Omega} - \mathbf{S}^{\Omega})^{-1} : \boldsymbol{\varepsilon}^{o},$$

$$\boldsymbol{\sigma} = \boldsymbol{\sigma}^{o} + \boldsymbol{\sigma}^{d} = \mathbf{B}^{\Omega} : (\mathbf{B}^{\Omega} - \mathbf{T}^{\Omega})^{-1} : \boldsymbol{\sigma}^{o} \qquad \text{in } \Omega. \tag{7.3.19a,b}$$

Note that the strain $\boldsymbol{\varepsilon}$ and stress $\boldsymbol{\sigma}$ in Ω given by (7.3.19a) and (7.3.19b) are equivalent. From constitutive relations (7.3.3a,b), substitution of (7.3.17a,b) into (7.3.19a,b) yields

$$\boldsymbol{\sigma} = \mathbf{C}^{\Omega} : \mathbf{A}^{\Omega} : (\mathbf{A}^{\Omega} - \mathbf{S}^{\Omega})^{-1} : \boldsymbol{\varepsilon}^{o}$$

$$= \{\mathbf{C}^{\Omega} : \{\mathbf{1}^{(4s)} - \mathbf{S}^{\Omega} : (\mathbf{1}^{(4s)} - \mathbf{D} : \mathbf{C}^{\Omega})\}^{-1} : \mathbf{D}\} : \boldsymbol{\sigma}^{o},$$

$$\boldsymbol{\varepsilon} = \mathbf{D}^{\Omega} : \mathbf{B}^{\Omega} : (\mathbf{B}^{\Omega} - \mathbf{T}^{\Omega})^{-1} : \boldsymbol{\sigma}^{o}$$

$$= \{\mathbf{D}^{\Omega} : \{\mathbf{1}^{(4s)} - \mathbf{T}^{\Omega} : (\mathbf{1}^{(4s)} - \mathbf{C} : \mathbf{D}^{\Omega})\}^{-1} : \mathbf{C}\} : \boldsymbol{\varepsilon}^{o} \qquad \text{in } \Omega. \tag{7.3.20a,b}$$

Taking advantage of identities (7.3.14c,d), observe that the fourth-order tensors in the right-hand sides of (7.3.20a,b) become

$$\mathbf{C}^{\Omega} : \{\mathbf{1}^{(4s)} - \mathbf{S}^{\Omega} : (\mathbf{1}^{(4s)} - \mathbf{D} : \mathbf{C}^{\Omega})\}^{-1} : \mathbf{D} = \{\mathbf{1}^{(4s)} - \mathbf{T}^{\Omega} : (\mathbf{1}^{(4s)} - \mathbf{C} : \mathbf{D}^{\Omega})\}^{-1},$$

$$\mathbf{D}^{\Omega} : \{\mathbf{1}^{(4s)} - \mathbf{T}^{\Omega} : (\mathbf{1}^{(4s)} - \mathbf{C} : \mathbf{D}^{\Omega})\}^{-1} : \mathbf{C} = \{\mathbf{1}^{(4s)} - \mathbf{S}^{\Omega} : (\mathbf{1}^{(4s)} - \mathbf{D} : \mathbf{C}^{\Omega})\}^{-1}. \tag{7.3.20c,d}$$

Therefore, (7.3.20c,d) compared with (7.3.19a,b) yield the equivalence relations between ($\mathbf{A}^{\Omega}$, $\mathbf{S}^{\Omega}$) and ($\mathbf{B}^{\Omega}$, $\mathbf{T}^{\Omega}$), as follows:

$$\mathbf{C}^{\Omega} : \mathbf{A}^{\Omega} : (\mathbf{A}^{\Omega} - \mathbf{S}^{\Omega})^{-1} : \mathbf{D} = \mathbf{B}^{\Omega} : (\mathbf{B}^{\Omega} - \mathbf{T}^{\Omega})^{-1},$$

$$\mathbf{D}^{\Omega} : \mathbf{B}^{\Omega} : (\mathbf{B}^{\Omega} - \mathbf{T}^{\Omega})^{-1} : \mathbf{C} = \mathbf{A}^{\Omega} : (\mathbf{A}^{\Omega} - \mathbf{S}^{\Omega})^{-1}. \tag{7.3.20e,f}$$

---

[4] This is, however, not true when V is *bounded*.

7.3.5. H- and J-Tensors

Since the total strain in an ellipsoidal Ω is uniform for the unbounded V considered in Subsection 7.3.4, the corresponding $\mathbf{H}^\Omega$- and $\mathbf{J}^\Omega$-tensors defined in (7.1.11) and (7.2.7), respectively, become

$$< \boldsymbol{\varepsilon} >_\Omega - \mathbf{D} : < \boldsymbol{\sigma} >_\Omega = \bar{\boldsymbol{\varepsilon}}^\Omega - \mathbf{D} : \bar{\boldsymbol{\sigma}}^\Omega = \mathbf{H}^\Omega : \boldsymbol{\sigma}^o \quad \text{in } \Omega, \tag{7.3.21a}$$

when the overall stress $\boldsymbol{\sigma}^o$ is prescribed; and

$$< \boldsymbol{\varepsilon} >_\Omega - \mathbf{D} : < \boldsymbol{\sigma} >_\Omega = \bar{\boldsymbol{\varepsilon}}^\Omega - \mathbf{D} : \bar{\boldsymbol{\sigma}}^\Omega = \mathbf{J}^\Omega : \boldsymbol{\varepsilon}^o \quad \text{in } \Omega, \tag{7.3.21b}$$

when the overall strain $\boldsymbol{\varepsilon}^o$ is prescribed. Since the region V is unbounded, $\mathbf{H}^\Omega$ and $\mathbf{J}^\Omega$ satisfy

$$\mathbf{J}^\Omega = \mathbf{H}^\Omega : \mathbf{C}, \qquad \mathbf{H}^\Omega = \mathbf{J}^\Omega : \mathbf{D}. \tag{7.3.21c,d}$$

Comparing (7.3.20a,b) and (7.3.21a,b), note that $\mathbf{H}^\Omega$ and $\mathbf{J}^\Omega$ may be expressed in terms of Eshelby's tensor $\mathbf{S}^\Omega$ and its conjugate $\mathbf{T}^\Omega$, as

$$\mathbf{H}^\Omega = (\mathbf{D}^\Omega - \mathbf{D}) : \mathbf{B}^\Omega : (\mathbf{B}^\Omega - \mathbf{T}^\Omega)^{-1},$$

$$\mathbf{J}^\Omega = (\mathbf{D}^\Omega - \mathbf{D}) : \mathbf{C}^\Omega : \mathbf{A}^\Omega : (\mathbf{A}^\Omega - \mathbf{S}^\Omega)^{-1}, \tag{7.3.22a,b}$$

or

$$\mathbf{H}^\Omega = (\mathbf{D}^\Omega - \mathbf{D}) : \mathbf{C}^\Omega : \mathbf{A}^\Omega : (\mathbf{A}^\Omega - \mathbf{S}^\Omega)^{-1} : \mathbf{D},$$

$$\mathbf{J}^\Omega = (\mathbf{D}^\Omega - \mathbf{D}) : \mathbf{B}^\Omega : (\mathbf{B}^\Omega - \mathbf{T}^\Omega)^{-1} : \mathbf{C}. \tag{7.3.22c,d}$$

As pointed out before, the Eshelby tensor $\mathbf{S}^\Omega$ and its conjugate $\mathbf{T}^\Omega$ for a uniform ellipsoidal inclusion Ω in an unbounded uniform matrix, depend on the aspect ratios of Ω and the elastic parameters of the matrix material, but they are independent of the material properties of Ω. On the other hand, $\mathbf{H}^\Omega$ and $\mathbf{J}^\Omega$ depend on the geometry of Ω, as well as on the elasticity of both Ω and the matrix material. For cavities, on the other hand, (7.3.22c,d) reduce to

$$\mathbf{H}^\Omega = (\mathbf{1}^{(4s)} - \mathbf{S}^\Omega)^{-1} : \mathbf{D}, \qquad \mathbf{J}^\Omega = (\mathbf{1}^{(4s)} - \mathbf{S}^\Omega)^{-1}, \tag{7.3.22e,f}$$

which show that the **H**- and **J**-tensors are effective tools for homogenization of solids with cavities and cracks.

From the above equations, it is seen that the equivalence of ($\mathbf{A}^\Omega$, $\mathbf{S}^\Omega$) and ($\mathbf{B}^\Omega$, $\mathbf{T}^\Omega$), given by (7.3.20e,f), corresponds to the equivalence of $\mathbf{J}^\Omega$ and $\mathbf{H}^\Omega$, given by (7.3.21c,d). It should be kept in mind that:

1) if the solid containing an inclusion is *unbounded*, these equivalent relations always hold, since the farfield stress $\boldsymbol{\sigma}^\infty = \boldsymbol{\sigma}^o$ and strain $\boldsymbol{\varepsilon}^\infty = \boldsymbol{\varepsilon}^o$ are related by $\boldsymbol{\sigma}^o = \mathbf{C} : \boldsymbol{\varepsilon}^o$ or $\boldsymbol{\varepsilon}^o = \mathbf{D} : \boldsymbol{\sigma}^o$, and hence the response of the solid is the same whether $\boldsymbol{\sigma}^o$ or $\boldsymbol{\varepsilon}^o$ is prescribed; but

2) if the solid is *bounded*, these equivalent relations do not, in general, hold, since the response of the solid when uniform boundary tractions are prescribed is, in general, different from that when linear boundary displacements are prescribed.

As is seen from (7.3.20), the formulation in terms of (**C**, $\mathbf{A}^{\Omega}$, $\mathbf{S}^{\Omega}$) corresponds exactly to that in terms of (**D**, $\mathbf{B}^{\Omega}$, $\mathbf{T}^{\Omega}$). Furthermore, from (7.2.24), the $\mathbf{H}^{\Omega}$- and $\mathbf{J}^{\Omega}$-tensors are expressed in terms of either (**C**, $\mathbf{A}^{\Omega}$, $\mathbf{S}^{\Omega}$) or (**D**, $\mathbf{B}^{\Omega}$, $\mathbf{T}^{\Omega}$). Hence, from now on in this chapter, mainly $\mathbf{A}^{\Omega}$, $\mathbf{B}^{\Omega}$, $\mathbf{S}^{\Omega}$, and $\mathbf{T}^{\Omega}$ are used instead of $\mathbf{H}^{\Omega}$ and $\mathbf{J}^{\Omega}$.

7.3.6. Eshelby's Tensor for Special Cases

The components of Eshelby's tensor **S**, with respect to a rectangular Cartesian coordinate system are listed below when the matrix M is unbounded and isotropically elastic, and the inclusion Ω is ellipsoidal with semiprincipal axes, a_i, which coincide with the coordinate axes, x_i ($i = 1, 2, 3$); see Figure 7.3.2.

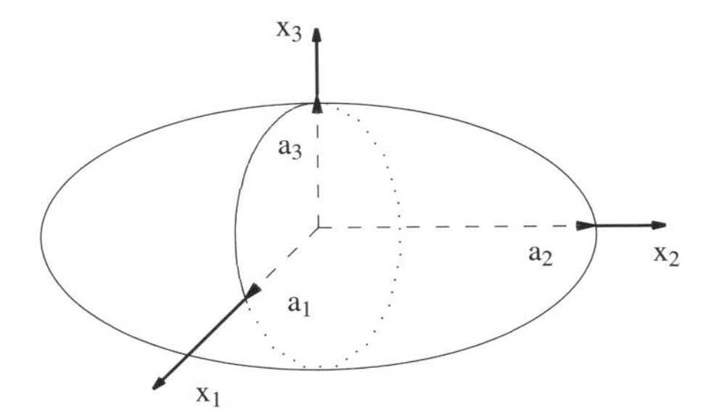

Figure 7.3.2

An ellipsoid coaxial with the Cartesian coordinates

(1) General form ($a_1 > a_2 > a_3$):

$$S_{1111} = \frac{3}{8\pi(1-\nu)}\, a_1^2\, I_{11} + \frac{1-2\nu}{8\pi(1-\nu)}\, I_1,$$

$$S_{1122} = \frac{1}{8\pi(1-\nu)}\, a_2^2\, I_{12} - \frac{1-2\nu}{8\pi(1-\nu)}\, I_1,$$

$$S_{1212} = \frac{1}{16\pi(1-\nu)}\, (a_1^2 + a_2^2)\, I_{12} + \frac{1-2\nu}{16\pi(1-\nu)}\, (I_1 + I_2). \qquad (7.3.23a\text{~}c)$$

The I_i- and I_{ij}-integrals are given by

$$I_1 = \frac{4\pi a_1 a_2 a_3}{(a_1^2 - a_2^2)(a_1^2 - a_3^2)^{1/2}}\, \{F(\theta, k) - E(\theta, k)\},$$

$$I_3 = \frac{4\pi a_1 a_2 a_3}{(a_2^2 - a_3^2)(a_1^2 - a_3^2)^{1/2}}\, \{\frac{a_2(a_1^2 - a_3^2)^{1/2}}{a_1 a_3} - E(\theta, k)\},$$

$$I_1 + I_2 + I_3 = 4\pi, \tag{7.3.23d~f}$$

and

$$3I_{11} + I_{12} + I_{13} = \frac{4\pi}{a_1^2}, \qquad 3a_1^2\, I_{11} + a_2^2\, I_{12} + a_3^2\, I_{13} = 3I_1,$$

$$I_{12} = \frac{I_2 - I_1}{a_1^2 - a_2^2}, \tag{7.3.23g~i}$$

where F and E are the elliptic integrals of the first and second kind, and

$$\theta = \arcsin\{\frac{a_1^2 - a_3^2}{a_1^2}\}^{1/2}, \qquad k = \{\frac{a_1^2 - a_2^2}{a_1^2 - a_3^2}\}^{1/2}. \tag{7.3.23j,k}$$

(2) Sphere ($a_1 = a_2 = a_3 = a$):

$$S_{ijkl} = \frac{5\nu - 1}{15(1-\nu)}\, \delta_{ij}\, \delta_{kl} + \frac{4 - 5\nu}{15(1-\nu)}\, (\delta_{ik}\, \delta_{jl} + \delta_{il}\, \delta_{jk}). \tag{7.3.24}$$

(3) Elliptic cylinder ($a_3 \to \infty$)

$$S_{1111} = \frac{1}{2(1-\nu)}\, \{\frac{a_2^2 + 2a_1a_2}{(a_1+a_2)^2} + (1-2\nu)\, \frac{a_2}{a_1+a_2}\},$$

$$S_{2222} = \frac{1}{2(1-\nu)}\, \{\frac{a_1^2 + 2a_1a_2}{(a_1+a_2)^2} + (1-2\nu)\, \frac{a_1}{a_1+a_2}\}, \qquad S_{3333} = 0,$$

$$S_{1122} = \frac{1}{2(1-\nu)}\, \{\frac{a_2^2}{(a_1+a_2)^2} - (1-2\nu)\, \frac{a_2}{a_1+a_2}\},$$

$$S_{2233} = \frac{1}{2(1-\nu)}\, \frac{2\nu a_1}{a_1+a_2},$$

$$S_{2211} = \frac{1}{2(1-\nu)}\, \{\frac{a_1^2}{(a_1+a_2)^2} - (1-2\nu)\, \frac{a_1}{a_1+a_2}\},$$

$$S_{1133} = \frac{1}{2(1-\nu)}\, \frac{2\nu a_2}{a_1+a_2}, \qquad S_{3311} = S_{3322} = 0,$$

$$S_{1212} = \frac{1}{2(1-\nu)}\, \{\frac{a_1^2 + a_2^2}{2(a_1+a_2)^2} + \frac{1-2\nu}{2}\},$$

$$S_{2323} = \frac{a_1}{2(a_1+a_2)}, \qquad S_{3131} = \frac{a_2}{2(a_1+a_2)}. \tag{7.3.25a~k}$$

(4) Penny shape ($a_1 = a_2 >> a_3$)

$$S_{1111} = S_{2222} = \frac{\pi(13-8\nu)}{32(1-\nu)}\, \frac{a_3}{a_1}, \qquad S_{3333} = 1 - \frac{\pi(1-2\nu)}{4(1-\nu)}\, \frac{a_3}{a_1},$$

$$S_{1122} = S_{2211} = \frac{\pi(8\nu - 1)}{32(1-\nu)} \frac{a_3}{a_1}, \qquad S_{1133} = S_{2233} = \frac{\pi(2\nu - 1)}{8(1-\nu)} \frac{a_3}{a_1},$$

$$S_{3311} = S_{3322} = \frac{\nu}{1-\nu} \{1 - \frac{\pi(4\nu + 1)}{8\nu} \frac{a_3}{a_1}\},$$

$$S_{1212} = \frac{\pi(7 - 8\nu)}{32(1-\nu)} \frac{a_3}{a_1}, \qquad S_{3131} = S_{2323} = \frac{1}{2}\{1 + \frac{\pi(\nu - 2)}{4(1-\nu)} \frac{a_3}{a_1}\}.$$

(7.3.26a~g)

7.3.7. Transformation Strain

As pointed out in Subsection 7.3.3, an unbounded uniform elastic solid V containing a uniform elastic ellipsoidal inhomogeneity Ω, can be homogenized by the introduction of uniform eigenstrains $\boldsymbol{\varepsilon}^*$ (or eigenstresses $\boldsymbol{\sigma}^*$) in Ω. Upon homogenization, and in view of (7.3.5), the disturbance stress and strain fields, $\boldsymbol{\sigma}^d$ and $\boldsymbol{\varepsilon}^d$, may be viewed as the stress and strain fields produced in the unbounded *homogeneous* solid (no inhomogeneity) when the region Ω undergoes a transformation which introduces in Ω *inelastic* strains[5] $\boldsymbol{\varepsilon}^*$. Figure 7.3.3 shows a portion of an unbounded uniform solid of elasticity **C**, with transformation strains $\boldsymbol{\varepsilon}^*$ defined in Ω in the absence of any applied loads. In this case, the stress field $\boldsymbol{\sigma}^d$ is self-equilibrating, with a vanishing average, i.e., $< \boldsymbol{\sigma}^d > \; = \mathbf{0}$. In general, the fields $\boldsymbol{\sigma}^d$ and $\boldsymbol{\varepsilon}^d$ are nonzero when they correspond to an actual

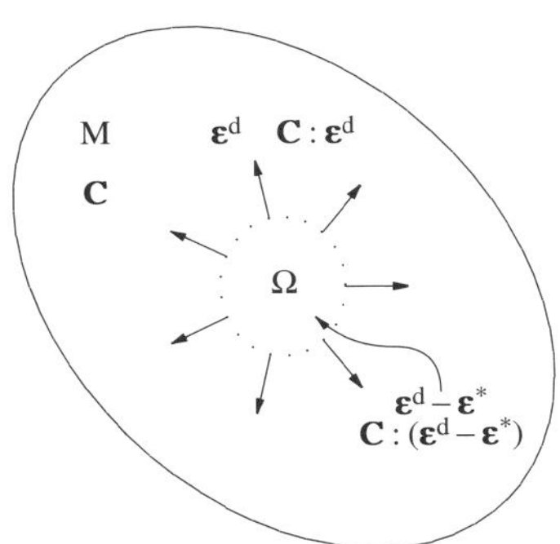

Figure 7.3.3

A portion of an unbounded uniform elastic solid of uniform elasticity **C**, in which region Ω has undergone transformation with inelastic strain $\boldsymbol{\varepsilon}^*$

[5] That is, if Ω is cut off such that the constraint imposed by its surrounding matrix is relaxed, then its strain would be $\boldsymbol{\varepsilon}^*$ which is also called the unconstrained transformation strain.

inelastic transformation strain $\boldsymbol{\varepsilon}^*$ in Ω. In the case of the inhomogeneity, on the other hand, the disturbance fields, $\boldsymbol{\sigma}^{\mathrm{d}}$ and $\boldsymbol{\varepsilon}^{\mathrm{d}}$, vanish when the applied loads are zero. In both cases, however, the strain $\boldsymbol{\varepsilon}^{\mathrm{d}}$ and the stress $\boldsymbol{\sigma}^{\mathrm{d}}$ in Ω are given by (7.3.11a) and (7.3.6), respectively. The difference between homogenization of a heterogeneous elastic solid, illustrated in Figure 7.3.1, and the strain and stress fields produced by inelastic strains $\boldsymbol{\varepsilon}^*$ in Ω, Figure 7.3.3, should nevertheless be carefully noted. In Figure 7.3.1, the eigenstrains are introduced as a *tool* to simplify the solution of the problem, whereas in Figure 7.3.3, these are the *actual* inelastic strains which may stem from shrinkage, thermal expansion, phase-transformation, or plastic deformation by slip or twinning. The basic equations are, however, quite similar, and provide a powerful technique to deal with a broad class of problems involving defects in homogeneous or inhomogeneous elastic solids.

As an illustration, assume that Ω of elasticity $\mathbf{C}^{\Omega}$ in Figure 7.3.1, actually undergoes a uniform inelastic deformation which corresponds to inelastic strains $\boldsymbol{\varepsilon}^{\mathrm{in}}$, if Ω were not constrained by its surrounding elastic matrix. In the presence of the surrounding elastic matrix M of elasticity $\mathbf{C}$, and when the farfield stresses and strains are $\boldsymbol{\sigma}^{\mathrm{o}}$ and $\boldsymbol{\varepsilon}^{\mathrm{o}}$, $\boldsymbol{\sigma}^{\mathrm{o}} = \mathbf{C} : \boldsymbol{\varepsilon}^{\mathrm{o}}$, it then follows that the actual stress field in Ω is given by

$$\boldsymbol{\sigma} = \boldsymbol{\sigma}^{\mathrm{o}} + \boldsymbol{\sigma}^{\mathrm{d}} = \mathbf{C}^{\Omega} : (\boldsymbol{\varepsilon}^{\mathrm{o}} + \boldsymbol{\varepsilon}^{\mathrm{d}} - \boldsymbol{\varepsilon}^{\mathrm{in}})$$

$$= \mathbf{C} : (\boldsymbol{\varepsilon}^{\mathrm{o}} + \boldsymbol{\varepsilon}^{\mathrm{d}} - \boldsymbol{\varepsilon}^{\mathrm{in}} - \boldsymbol{\varepsilon}^*) \quad \text{in } \Omega, \tag{7.3.27a,b}$$

where

$$\boldsymbol{\varepsilon}^{\mathrm{d}} = \mathbf{S}^{\Omega} : (\boldsymbol{\varepsilon}^{\mathrm{in}} + \boldsymbol{\varepsilon}^*) \quad \text{in } \Omega. \tag{7.3.27c}$$

In (7.3.27a,b), a part of $\boldsymbol{\varepsilon}^{\mathrm{d}}$ in Ω is due to the actual inelastic deformation of Ω, i.e., due to $\boldsymbol{\varepsilon}^{\mathrm{in}}$. This part must be defined through the relevant physical laws which govern the corresponding inelastic deformation. The remaining part of $\boldsymbol{\varepsilon}^{\mathrm{d}}$ in Ω, on the other hand, is due to eigenstrains $\boldsymbol{\varepsilon}^*$ which are introduced in order to homogenize the solid. Upon this homogenization, the uniform elasticity tensor $\mathbf{C}$ is used for the entire solid, i.e., for both the elastic matrix and, say, the elastic-plastic inclusion Ω. From (7.3.27) and definition (7.3.17), it follows that

$$\boldsymbol{\varepsilon}^* = (\mathbf{A}^{\Omega} - \mathbf{S}^{\Omega})^{-1} : \{\boldsymbol{\varepsilon}^{\mathrm{o}} + (\mathbf{S}^{\Omega} - \mathbf{1}^{(4s)}) : \boldsymbol{\varepsilon}^{\mathrm{in}}\}. \tag{7.3.27d}$$

Since in general, the relation between $\boldsymbol{\varepsilon}^{\mathrm{in}}$ and the farfield uniform strain $\boldsymbol{\varepsilon}^{\mathrm{o}}$ (or stress $\boldsymbol{\sigma}^{\mathrm{o}}$) is nonlinear and possibly history-dependent, an incremental formulation is often necessary. Because of the assumed small deformations, the corresponding rate equations in the present case[6] follow directly from (7.3.27); replace $\boldsymbol{\varepsilon}^{\mathrm{o}}$, $\boldsymbol{\varepsilon}^*$, and $\boldsymbol{\varepsilon}^{\mathrm{in}}$ by their rates.

The formulation may also be done in terms of the eigenstress. Then it is necessary to define the *decrement*, $\boldsymbol{\sigma}^{\mathrm{in}}$, in the stress due to the inelastic strain $\boldsymbol{\varepsilon}^{\mathrm{in}}$, by

---

[6] The assumption that the matrix is linearly elastic, considerably simplifies the solution of this problem. For the case when the matrix also admits inelastic deformations, the problem becomes considerably more complex; see Appendix A of Part 1.

$$\boldsymbol{\sigma}^{\mathrm{in}} = -\mathbf{C}^{\Omega} : \boldsymbol{\varepsilon}^{\mathrm{in}}, \tag{7.3.28a}$$

and obtain the consistency conditions,

$$\begin{aligned}\boldsymbol{\varepsilon} &= \boldsymbol{\varepsilon}^{\mathrm{o}} + \boldsymbol{\varepsilon}^{\mathrm{d}} \\ &= \mathbf{D}^{\Omega} : (\boldsymbol{\sigma}^{\mathrm{o}} + \boldsymbol{\sigma}^{\mathrm{d}} - \boldsymbol{\sigma}^{\mathrm{in}}) \\ &= \mathbf{D} : (\boldsymbol{\sigma}^{\mathrm{o}} + \boldsymbol{\sigma}^{\mathrm{d}} - \boldsymbol{\sigma}^{\mathrm{in}} - \boldsymbol{\sigma}^{*}) \qquad \text{in } \Omega,\end{aligned} \tag{7.3.28b}$$

where

$$\boldsymbol{\sigma}^{\mathrm{d}} = \mathbf{T}^{\Omega} : (\boldsymbol{\sigma}^{\mathrm{in}} + \boldsymbol{\sigma}^{*}). \tag{7.3.28c}$$

Hence, instead of (7.3.27d), the following equivalent relation is obtained:

$$\boldsymbol{\sigma}^{*} = (\mathbf{B}^{\Omega} - \mathbf{T}^{\Omega})^{-1} : \{\boldsymbol{\sigma}^{\mathrm{o}} + (\mathbf{T}^{\Omega} - \mathbf{1}^{(4s)}) : \boldsymbol{\sigma}^{\mathrm{in}}\}. \tag{7.3.28d}$$

7.4. ESTIMATES OF OVERALL MODULUS AND COMPLIANCE TENSORS: DILUTE DISTRIBUTION

In this subsection, the following two *exact* relations are applied to an RVE consisting of an elastic matrix and elastic micro-inclusions: from (7.2.6) when the overall strain $\boldsymbol{\varepsilon}^{\mathrm{o}}$ is prescribed,

$$(\mathbf{C} - \overline{\mathbf{C}}) : \boldsymbol{\varepsilon}^{\mathrm{o}} = \sum_{\alpha=1}^{n} f_{\alpha} (\mathbf{C} - \mathbf{C}^{\alpha}) : \bar{\boldsymbol{\varepsilon}}^{\alpha}; \tag{7.4.1a}$$

and from (7.1.9) when the overall stress $\boldsymbol{\sigma}^{\mathrm{o}}$ is prescribed,

$$(\mathbf{D} - \overline{\mathbf{D}}) : \boldsymbol{\sigma}^{\mathrm{o}} = \sum_{\alpha=1}^{n} f_{\alpha} (\mathbf{D} - \mathbf{D}^{\alpha}) : \bar{\boldsymbol{\sigma}}^{\alpha}. \tag{7.4.1b}$$

Then, the average stress $\bar{\boldsymbol{\sigma}}^{\alpha}$ or the average strain $\bar{\boldsymbol{\varepsilon}}^{\alpha}$ of each micro-inclusion Ω_{α}, is estimated in order to obtain the overall elastic parameters of the RVE. For simplicity, consider the case when all micro-inclusions are either ellipsoidal or can be approximated by ellipsoids, and are perfectly bonded to the matrix. As in the cases of microcavities and microcracks, for a dilute distribution of inclusions, interaction effects may be neglected. In the next subsection, the self-consistent method is used, which includes the interaction effects in a certain manner. Under the assumption of a dilute distribution of inhomogeneities, the overall elasticity $\overline{\mathbf{C}}$ and compliance $\overline{\mathbf{D}}$, are each other's inverse only to the first order in the volume fraction of inhomogeneities. The general case is considered first, and then results for special geometries of inclusions and their distribution are given.

As in the case of microcavities and microcracks, the case of a prescribed macrostress $\boldsymbol{\Sigma} = \boldsymbol{\sigma}^{\mathrm{o}}$, and the case of a prescribed macrostrain $\mathbf{E} = \boldsymbol{\varepsilon}^{\mathrm{o}}$, are treated separately. In either case, an infinitely extended solid under a prescribed overall stress $\boldsymbol{\sigma}^{\mathrm{o}}$ or strain $\boldsymbol{\varepsilon}^{\mathrm{o}}$, is considered with an embedded isolated inclusion Ω_{α}, in order to estimate the average strain and the average stress of a typical inclusion Ω_α. Since for these estimates the solid is assumed to be unbounded, the presence of a single inclusion does not affect the overall relations $\boldsymbol{\sigma}^o = \mathbf{C} : \boldsymbol{\varepsilon}^o$ and $\boldsymbol{\varepsilon}^o = \mathbf{D} : \boldsymbol{\sigma}^o$. Therefore, for calculating $\bar{\boldsymbol{\sigma}}^\alpha$ and $\bar{\boldsymbol{\varepsilon}}^\alpha$, on the basis of a single inclusion in an unbounded matrix, it does not make any difference whether the stress or the strain is regarded prescribed at infinity. Note, however, that the final overall elastic parameters in the "dilute estimates" (but not in the self-consistent case) depend on whether (7.4.1a) or (7.4.1b) is used.

Denote the elasticity and compliance tensors of Ω_α by $\mathbf{C}^\alpha$ and $\mathbf{D}^\alpha$, respectively, and set

$$\mathbf{A}^\alpha \equiv (\mathbf{C} - \mathbf{C}^\alpha)^{-1} : \mathbf{C}, \qquad \mathbf{B}^\alpha \equiv (\mathbf{D} - \mathbf{D}^\alpha)^{-1} : \mathbf{D}. \tag{7.4.2a,b}$$

The Eshelby tensor $\mathbf{S}$ and its conjugate $\mathbf{T}$, in general, depend on the aspect ratios of Ω_α. Hence, denote them by $\mathbf{S}^\alpha$ and $\mathbf{T}^\alpha$, respectively. From (7.3.14c,d), $\mathbf{S}^\alpha$ and $\mathbf{T}^\alpha$ satisfy

$$\mathbf{S}^\alpha + \mathbf{D} : \mathbf{T}^\alpha : \mathbf{C} = \mathbf{1}^{(4s)}, \qquad \mathbf{T}^\alpha + \mathbf{C} : \mathbf{S}^\alpha : \mathbf{D} = \mathbf{1}^{(4s)}. \tag{7.4.3a,b}$$

Then, for a single Ω_α embedded in an infinitely extended solid, under uniform stress or strain at infinity, the resulting strain and stress in Ω_α are uniform and given, in view of (7.3.20e,f), by

$$\bar{\boldsymbol{\varepsilon}}^\alpha \equiv \langle \boldsymbol{\varepsilon}^o + \boldsymbol{\varepsilon}^d \rangle_\alpha = \mathbf{A}^\alpha : (\mathbf{A}^\alpha - \mathbf{S}^\alpha)^{-1} : \boldsymbol{\varepsilon}^o,$$

$$\bar{\boldsymbol{\sigma}}^\alpha \equiv \langle \boldsymbol{\sigma}^o + \boldsymbol{\sigma}^d \rangle_\alpha = \mathbf{B}^\alpha : (\mathbf{B}^\alpha - \mathbf{T}^\alpha)^{-1} : \boldsymbol{\sigma}^o \qquad (\alpha \text{ not summed}). \tag{7.4.4a,b}$$

From (7.3.19c,d), the corresponding equivalence relations are,

$$\mathbf{C}^\alpha : \mathbf{A}^\alpha : (\mathbf{A}^\alpha - \mathbf{S}^\alpha)^{-1} : \mathbf{D} = \mathbf{B}^\alpha : (\mathbf{B}^\alpha - \mathbf{T}^\alpha)^{-1},$$

$$\mathbf{D}^\alpha : \mathbf{B}^\alpha : (\mathbf{B}^\alpha - \mathbf{T}^\alpha)^{-1} : \mathbf{C} = \mathbf{A}^\alpha : (\mathbf{A}^\alpha - \mathbf{S}^\alpha)^{-1} \qquad (\alpha \text{ not summed}). \tag{7.4.4c,d}$$

7.4.1. Macrostress Prescribed

Consider first the case when the surface tractions are regarded prescribed through (7.1.1) by the macrostress $\boldsymbol{\Sigma} = \boldsymbol{\sigma}^o$. For a dilute distribution of inclusions, the average stress of each inclusion Ω_α is approximated by the uniform stress of a single inclusion embedded in an unbounded solid with the elasticity of the matrix material, and subjected to the farfield stress $\boldsymbol{\sigma}^o$ (or strain $\boldsymbol{\varepsilon}^o \equiv \mathbf{D} : \boldsymbol{\sigma}^o$). From (7.4.4a), it then follows that

$$\bar{\boldsymbol{\sigma}}^\alpha = \mathbf{C}^\alpha : \mathbf{A}^\alpha : (\mathbf{A}^\alpha - \mathbf{S}^\alpha)^{-1} : \mathbf{D} : \boldsymbol{\sigma}^o, \tag{7.4.5a}$$

and from (7.4.4b),

$$\bar{\boldsymbol{\sigma}}^\alpha = \mathbf{B}^\alpha : (\mathbf{B}^\alpha - \mathbf{T}^\alpha)^{-1} : \boldsymbol{\sigma}^o. \tag{7.4.5b}$$

From (7.4.4c,d), it is seen that these two equations are equivalent. Now, substitution of (7.4.5a,b) into (7.1.9) yields

$$(\bar{\mathbf{D}} - \mathbf{D}) : \boldsymbol{\sigma}^o = \Big\{ \sum_{\alpha=1}^{n} f_\alpha (\mathbf{D}^\alpha - \mathbf{D}) : \mathbf{C}^\alpha : \mathbf{A}^\alpha : (\mathbf{A}^\alpha - \mathbf{S}^\alpha)^{-1} : \mathbf{D} \Big\} : \boldsymbol{\sigma}^o,$$

$$(\bar{\mathbf{D}} - \mathbf{D}) : \boldsymbol{\sigma}^o = \Big\{ \sum_{\alpha=1}^{n} f_\alpha (\mathbf{D}^\alpha - \mathbf{D}) : \mathbf{B}^\alpha : (\mathbf{B}^\alpha - \mathbf{T}^\alpha)^{-1} \Big\} : \boldsymbol{\sigma}^o. \qquad (7.4.5\text{c,d})$$

Since the resulting equations must be independent of the prescribed $\boldsymbol{\sigma}^o$, it follows that

$$\bar{\mathbf{D}} = \Big\{ \mathbf{1}^{(4s)} + \sum_{\alpha=1}^{n} f_\alpha (\mathbf{A}^\alpha - \mathbf{S}^\alpha)^{-1} \Big\} : \mathbf{D},$$

$$\bar{\mathbf{D}} = \mathbf{D} : \Big\{ \mathbf{1}^{(4s)} - \sum_{\alpha=1}^{n} f_\alpha (\mathbf{B}^\alpha - \mathbf{T}^\alpha)^{-1} \Big\}. \qquad (7.4.6\text{a,b})$$

Recall that since an unbounded region is considered, the average stress $\bar{\boldsymbol{\sigma}}^\alpha$ given by (7.4.5a) is equivalent to that given by (7.4.5b), and hence the overall compliance $\bar{\mathbf{D}}$ given by (7.4.6a) is identical to that given by (7.4.6b).

When all the inclusions are similar,[7] are all similarly aligned with respect to the fixed coordinate axes, and have the same elasticity and compliance tensors denoted, respectively, by $\mathbf{C}^I$ and $\mathbf{D}^I$, then (7.4.6a,b) reduce to

$$\bar{\mathbf{D}} = \{\mathbf{1}^{(4s)} + f\,(\mathbf{A}^I - \mathbf{S}^I)^{-1}\} : \mathbf{D}, \qquad \bar{\mathbf{D}} = \mathbf{D} : \{\mathbf{1}^{(4s)} - f\,(\mathbf{B}^I - \mathbf{T}^I)^{-1}\}, \quad (7.4.7\text{a,b})$$

where f is the volume fraction of inclusions,

$$\mathbf{A}^I \equiv (\mathbf{C} - \mathbf{C}^I)^{-1} : \mathbf{C}, \qquad \mathbf{B}^I \equiv (\mathbf{D} - \mathbf{D}^I)^{-1} : \mathbf{D}, \qquad (7.4.7\text{c,d})$$

and $\mathbf{S}^I$ and $\mathbf{T}^I$ are Eshelby's tensor and its conjugate, common to all inclusions.

7.4.2. Macrostrain Prescribed

Next, consider the case when the surface displacements are regarded prescribed through (7.2.1) by the macrostrain $\mathbf{E} = \boldsymbol{\varepsilon}^o$. The average strain in each inclusion is now approximated by the uniform strain of a single inclusion embedded in an unbounded solid with the elasticity of the matrix material, and subjected to the farfield strain $\boldsymbol{\varepsilon}^o$ (or the farfield stress $\boldsymbol{\sigma}^o \equiv \mathbf{C} : \boldsymbol{\varepsilon}^o$). Thus, from (7.4.4a,b), it follows that

$$\boldsymbol{\varepsilon}^\alpha = \mathbf{A}^\alpha : (\mathbf{A}^\alpha - \mathbf{S}^\alpha)^{-1} : \boldsymbol{\varepsilon}^o, \qquad \boldsymbol{\varepsilon}^\alpha = \mathbf{D}^\alpha : \mathbf{B}^\alpha : (\mathbf{B}^\alpha - \mathbf{T}^\alpha)^{-1} : \mathbf{C} : \boldsymbol{\varepsilon}^o. \quad (7.4.8\text{a,b})$$

Substitution of (7.4.8a,b) into the exact relation (7.4.1b) then yields

$$(\bar{\mathbf{C}} - \mathbf{C}) : \boldsymbol{\varepsilon}^o = \Big\{ \sum_{\alpha=1}^{n} f_\alpha (\mathbf{C}^\alpha - \mathbf{C}) : \mathbf{A}^\alpha : (\mathbf{A}^\alpha - \mathbf{S}^\alpha)^{-1} \Big\} : \boldsymbol{\varepsilon}^o,$$

---

[7] I.e., the inclusions have the same aspect ratios, but may not be equal in size.

$$(\bar{\mathbf{C}}-\mathbf{C}):\boldsymbol{\varepsilon}^{\mathrm{o}}=\left\{\sum_{\alpha=1}^{\mathrm{n}}\mathrm{f}_{\alpha}\,(\mathbf{C}^{\alpha}-\mathbf{C}):\mathbf{D}^{\alpha}:\mathbf{B}^{\alpha}:(\mathbf{B}^{\alpha}-\mathbf{T}^{\alpha})^{-1}:\mathbf{C}\right\}:\boldsymbol{\varepsilon}^{\mathrm{o}},\qquad(7.4.8\mathrm{c,d})$$

and since the prescribed overall strain, $\boldsymbol{\varepsilon}^{\mathrm{o}}$, is arbitrary,

$$\bar{\mathbf{C}}=\mathbf{C}:\left\{\mathbf{1}^{(4s)}-\sum_{\alpha=1}^{\mathrm{n}}\mathrm{f}_{\alpha}\,(\mathbf{A}^{\alpha}-\mathbf{S}^{\alpha})^{-1}\right\},$$

$$\bar{\mathbf{C}}=\left\{\mathbf{1}^{(4s)}+\sum_{\alpha=1}^{\mathrm{n}}\mathrm{f}_{\alpha}\,(\mathbf{B}^{\alpha}-\mathbf{T}^{\alpha})^{-1}\right\}:\mathbf{C}.\qquad(7.4.9\mathrm{a,b})$$

Again, since an unbounded region is considered, the average strain $\bar{\boldsymbol{\varepsilon}}^{\alpha}$ given by (7.4.8a) is the same as that given by (7.4.8b). Hence, the overall elasticity tensor $\bar{\mathbf{C}}$ given by (7.4.9a) is identical with that given by (7.4.9b).

If all inclusions are similar, with identical elasticity and alignment, as mentioned in Subsection 7.4.1, then $\mathbf{C}^{\alpha}\equiv\mathbf{C}^{\mathrm{I}}$, $\mathbf{D}^{\alpha}\equiv\mathbf{D}^{\mathrm{I}}$, $\mathbf{S}^{\alpha}\equiv\mathbf{S}^{\mathrm{I}}$, and $\mathbf{T}^{\alpha}\equiv\mathbf{T}^{\mathrm{I}}$ for all Ω_{α}. The overall elasticity tensor $\bar{\mathbf{C}}$ becomes

$$\bar{\mathbf{C}}=\mathbf{C}:\{\mathbf{1}^{(4s)}-\mathrm{f}\,(\mathbf{A}^{\mathrm{I}}-\mathbf{S}^{\mathrm{I}})^{-1}\},$$

$$\bar{\mathbf{C}}=\{\mathbf{1}^{(4s)}+\mathrm{f}\,(\mathbf{B}^{\mathrm{I}}-\mathbf{T}^{\mathrm{I}})^{-1}\}:\mathbf{C}.\qquad(7.4.10\mathrm{a,b})$$

7.4.3. Equivalence between Overall Compliance and Elasticity Tensors

In Subsection 7.4.1, two identical expressions are obtained for the overall compliance tensor, (7.4.6a,b). Similarly, in Subsection 7.4.2, two expressions are obtained for the overall elasticity tensor, (7.4.9a,b). A dilute distribution of micro-inclusions is assumed, in arriving at these results. As discussed in Sections 5 and 6, the overall elasticity and compliance tensors obtained by this assumption agree only up to the first order of the volume fraction of micro-inclusions. Indeed, from (7.4.6a) and (7.4.9a),

$$\bar{\mathbf{D}}:\bar{\mathbf{C}}=\mathbf{1}^{(4s)}-\sum_{\alpha=1}^{\mathrm{n}}\sum_{\beta=1}^{\mathrm{n}}\mathrm{f}_{\alpha}\,\mathrm{f}_{\beta}\,(\mathbf{A}^{\alpha}-\mathbf{S}^{\alpha})^{-1}:(\mathbf{A}^{\beta}-\mathbf{S}^{\beta})^{-1},\qquad(7.4.11\mathrm{a})$$

and from (7.4.6b) and (7.4.9b),

$$\bar{\mathbf{C}}:\bar{\mathbf{D}}=\mathbf{1}^{(4s)}-\sum_{\alpha=1}^{\mathrm{n}}\sum_{\beta=1}^{\mathrm{n}}\mathrm{f}_{\alpha}\,\mathrm{f}_{\beta}\,(\mathbf{B}^{\alpha}-\mathbf{T}^{\alpha})^{-1}:(\mathbf{B}^{\beta}-\mathbf{T}^{\beta})^{-1}.\qquad(7.4.11\mathrm{b})$$

Therefore, $\bar{\mathbf{C}}$ and $\bar{\mathbf{D}}$ are each other's inverse only to the first order in the volume fraction of micro-inclusions.

In particular, when all micro-inclusions are similar, from (7.4.7a,b) and (7.4.10a,b), it follows that

$$\bar{\mathbf{D}}:\bar{\mathbf{C}}=\mathbf{1}^{(4s)}-\mathrm{f}^{2}\,(\mathbf{A}^{\mathrm{I}}-\mathbf{S}^{\mathrm{I}})^{-1}:(\mathbf{A}^{\mathrm{I}}-\mathbf{S}^{\mathrm{I}})^{-1}=\mathbf{1}^{(4s)}+\mathrm{O}(\mathrm{f}^{2}),$$

$$\bar{\mathbf{C}}:\bar{\mathbf{D}}=\mathbf{1}^{(4s)}-\mathrm{f}^{2}\,(\mathbf{B}^{\mathrm{I}}-\mathbf{T}^{\mathrm{I}})^{-1}:(\mathbf{B}^{\mathrm{I}}-\mathbf{T}^{\mathrm{I}})^{-1}=\mathbf{1}^{(4s)}+\mathrm{O}(\mathrm{f}^{2}).\qquad(7.4.12\mathrm{a,b})$$

In terms of $\mathbf{S}^\alpha$ and $\mathbf{T}^\alpha$, the tensors $\mathbf{H}^\alpha$ and $\mathbf{H}$ can be defined from (7.4.6a,b), by

$$\mathbf{H}^\alpha \equiv (\mathbf{A}^\alpha - \mathbf{S}^\alpha)^{-1} : \mathbf{D} \equiv -\mathbf{D} : (\mathbf{B}^\alpha - \mathbf{T}^\alpha)^{-1}, \qquad \mathbf{H} \equiv \sum_{\alpha=1}^{n} f_\alpha \, \mathbf{H}^\alpha, \tag{7.4.13a,b}$$

and the tensors $\mathbf{J}^\alpha$ and $\mathbf{J}$ from (7.4.8a,b), by

$$\mathbf{J}^\alpha \equiv (\mathbf{A}^\alpha - \mathbf{S}^\alpha)^{-1} \equiv -\mathbf{D} : (\mathbf{B}^\alpha - \mathbf{T}^\alpha)^{-1} : \mathbf{C}, \qquad \mathbf{J} \equiv \sum_{\alpha=1}^{n} f_\alpha \, \mathbf{J}^\alpha. \tag{7.4.13c,d}$$

From the equivalence relations (7.4.4c,d), it follows that

$$\mathbf{J} = \mathbf{H} : \mathbf{C}, \qquad \mathbf{H} = \mathbf{J} : \mathbf{D}. \tag{7.4.13e,f}$$

Hence, (7.4.11a,b) may be rewritten in terms of the $\mathbf{H}$-tensor, as

$$\bar{\mathbf{C}} : \bar{\mathbf{D}} = \mathbf{1}^{(4s)} - \mathbf{C} : \mathbf{H} : \mathbf{C} : \mathbf{H}, \qquad \bar{\mathbf{D}} : \bar{\mathbf{C}} = \mathbf{1}^{(4s)} - \mathbf{H} : \mathbf{C} : \mathbf{H} : \mathbf{C}. \tag{7.4.14a,b}$$

These relations are exactly the same as relations (5.1.19a,b) obtained for an elastic RVE with microcavities.

7.5. ESTIMATES OF OVERALL MODULUS AND COMPLIANCE TENSORS: SELF-CONSISTENT METHOD

Consider the self-consistent method for estimating the average stress $\bar{\boldsymbol{\sigma}}^\alpha$ or the average strain $\bar{\boldsymbol{\varepsilon}}^\alpha$ for a typical micro-inclusion Ω_α. As explained in Sections 5 and 6, in the self-consistent method, one considers a typical micro-inclusion embedded in an unbounded homogeneous elastic solid which has the *yet-unknown overall moduli* of the RVE, and then calculates the average stress or strain in the embedded inclusion. Since an unbounded solid is used, computation of the average strain and stress does not depend on whether the overall strain $\boldsymbol{\varepsilon}^o$ or the overall stress $\boldsymbol{\sigma}^o$ is regarded prescribed. Moreover, these overall strains and stresses are related by the *overall elastic parameters*, i.e., $\boldsymbol{\sigma}^o = \bar{\mathbf{C}} : \boldsymbol{\varepsilon}^o$ or $\boldsymbol{\varepsilon}^o = \bar{\mathbf{D}} : \boldsymbol{\sigma}^o$, where the unknown overall elasticity and compliance tensors are denoted by $\bar{\mathbf{C}}$ and $\bar{\mathbf{D}}$, respectively. The results obtained in Subsection 7.3 now give the average strain and stress in Ω_α, to be

$$\bar{\boldsymbol{\varepsilon}}^\alpha \equiv < \boldsymbol{\varepsilon}^o + \boldsymbol{\varepsilon}^d >_\alpha = \bar{\mathbf{A}}^\alpha : (\bar{\mathbf{A}}^\alpha - \bar{\mathbf{S}}^\alpha)^{-1} : \boldsymbol{\varepsilon}^o,$$

$$\bar{\boldsymbol{\sigma}}^\alpha \equiv < \boldsymbol{\sigma}^o + \boldsymbol{\sigma}^d >_\alpha = \bar{\mathbf{B}}^\alpha : (\bar{\mathbf{B}}^\alpha - \bar{\mathbf{T}}^\alpha)^{-1} : \boldsymbol{\sigma}^o, \tag{7.5.1a,b}$$

where

$$\bar{\mathbf{A}}^\alpha \equiv (\bar{\mathbf{C}} - \mathbf{C}^\alpha)^{-1} : \bar{\mathbf{C}}, \qquad \bar{\mathbf{B}}^\alpha \equiv (\bar{\mathbf{D}} - \mathbf{D}^\alpha)^{-1} : \bar{\mathbf{D}}. \tag{7.5.1c,d}$$

In (7.5.1a,b), $\bar{\mathbf{S}}^\alpha$ and $\bar{\mathbf{T}}^\alpha$ are Eshelby's tensor and its conjugate, for the geometry of Ω_α and the *overall material properties* defined by $\bar{\mathbf{C}}$ and $\bar{\mathbf{D}}$. From (7.3.13c,d) and (7.3.19c,d), $\bar{\mathbf{S}}^\alpha$ and $\bar{\mathbf{T}}^\alpha$ satisfy

$$\bar{\mathbf{S}}^\alpha + \bar{\mathbf{D}} : \bar{\mathbf{T}}^\alpha : \bar{\mathbf{C}} = \mathbf{1}^{(4s)}, \qquad \bar{\mathbf{T}}^\alpha + \bar{\mathbf{C}} : \bar{\mathbf{S}}^\alpha : \bar{\mathbf{D}} = \mathbf{1}^{(4s)}, \tag{7.5.2a,b}$$

and tensors $\bar{\mathbf{A}}^\alpha$ and $\bar{\mathbf{B}}^\alpha$ satisfy

$$\mathbf{C}^\alpha : \bar{\mathbf{A}}^\alpha : (\bar{\mathbf{A}}^\alpha - \bar{\mathbf{S}}^\alpha)^{-1} : \bar{\mathbf{D}} = \bar{\mathbf{B}}^\alpha : (\bar{\mathbf{B}}^\alpha - \bar{\mathbf{T}}^\alpha)^{-1},$$

$$\mathbf{D}^\alpha : \bar{\mathbf{B}}^\alpha : (\bar{\mathbf{B}}^\alpha - \bar{\mathbf{T}}^\alpha)^{-1} : \bar{\mathbf{C}} = \bar{\mathbf{A}}^\alpha : (\bar{\mathbf{A}}^\alpha - \bar{\mathbf{S}}^\alpha)^{-1} \quad (\alpha \text{ not summed}). \tag{7.5.2c,d}$$

7.5.1. Macrostress Prescribed

Following the procedure in Subsection 7.4.1, consider first the case when the macrostress $\boldsymbol{\Sigma} = \boldsymbol{\sigma}^o$ is regarded prescribed. *The average stress $\bar{\boldsymbol{\sigma}}^\alpha$ of the micro-inclusion Ω_α is approximated by the uniform stress of a single inclusion embedded in an unbounded solid with the yet-unknown elasticity and compliance tensors, $\bar{\mathbf{C}}$ and $\bar{\mathbf{D}}$.* From (7.5.1a,b), the average stress of Ω_α is expressed in terms of the overall stress, $\boldsymbol{\sigma}^o$, as

$$\bar{\boldsymbol{\sigma}}^\alpha = \mathbf{C}^\alpha : \bar{\mathbf{A}}^\alpha : (\bar{\mathbf{A}}^\alpha - \bar{\mathbf{S}}^\alpha)^{-1} : \bar{\mathbf{D}} : \boldsymbol{\sigma}^o,$$

$$\bar{\boldsymbol{\sigma}}^\alpha = \bar{\mathbf{B}}^\alpha : (\bar{\mathbf{B}}^\alpha - \bar{\mathbf{T}}^\alpha)^{-1} : \boldsymbol{\sigma}^o, \tag{7.5.3a,b}$$

and substitution of (7.5.3a,b) into (7.4.1b) yields

$$(\bar{\mathbf{D}} - \mathbf{D}) : \boldsymbol{\sigma}^o = \Big\{ \sum_{\alpha=1}^{n} f_\alpha (\mathbf{D}^\alpha - \mathbf{D}) : \mathbf{C}^\alpha : \bar{\mathbf{A}}^\alpha : (\bar{\mathbf{A}}^\alpha - \bar{\mathbf{S}}^\alpha)^{-1} : \bar{\mathbf{D}} \Big\} : \boldsymbol{\sigma}^o,$$

$$(\bar{\mathbf{D}} - \mathbf{D}) : \boldsymbol{\sigma}^o = \Big\{ \sum_{\alpha=1}^{n} f_\alpha (\mathbf{D}^\alpha - \mathbf{D}) : \bar{\mathbf{B}}^\alpha : (\bar{\mathbf{B}}^\alpha - \bar{\mathbf{T}}^\alpha)^{-1} \Big\} : \boldsymbol{\sigma}^o. \tag{7.5.3c,d}$$

Since $\boldsymbol{\sigma}^o$ is arbitrary, the overall compliance $\bar{\mathbf{D}}$ becomes

$$\bar{\mathbf{D}} = \mathbf{D} + \sum_{\alpha=1}^{n} f_\alpha (\mathbf{D}^\alpha - \mathbf{D}) : \mathbf{C}^\alpha : \bar{\mathbf{A}}^\alpha : (\bar{\mathbf{A}}^\alpha - \bar{\mathbf{S}}^\alpha)^{-1} : \bar{\mathbf{D}},$$

$$\bar{\mathbf{D}} = \mathbf{D} + \sum_{\alpha=1}^{n} f_\alpha (\mathbf{D}^\alpha - \mathbf{D}) : \bar{\mathbf{B}}^\alpha : (\bar{\mathbf{B}}^\alpha - \bar{\mathbf{T}}^\alpha)^{-1}. \tag{7.5.4a,b}$$

Since the average stress $\bar{\boldsymbol{\sigma}}^\alpha$ given by (7.5.3a), is the same as that given by (7.5.3b), the overall compliance tensor $\bar{\mathbf{D}}$ given by (7.5.4a), equals that given by (7.5.4b).

When all the inclusions are similar, from (7.5.4a,b) the overall compliance $\bar{\mathbf{D}}$ becomes

$$\bar{\mathbf{D}} = \mathbf{D} + f\,(\mathbf{D}^I - \mathbf{D}) : \mathbf{C}^I : \bar{\mathbf{A}}^I : (\bar{\mathbf{A}}^I - \bar{\mathbf{S}}^I)^{-1} : \bar{\mathbf{D}},$$

$$\bar{\mathbf{D}} = \mathbf{D} + f\,(\mathbf{D}^I - \mathbf{D}) : \bar{\mathbf{B}}^I : (\bar{\mathbf{B}}^I - \bar{\mathbf{T}}^I)^{-1}, \tag{7.5.5a,b}$$

where

$$\bar{\mathbf{A}}^I = (\bar{\mathbf{C}} - \mathbf{C}^I)^{-1} : \bar{\mathbf{C}}, \qquad \bar{\mathbf{B}}^I = (\bar{\mathbf{D}} - \mathbf{D}^I)^{-1} : \bar{\mathbf{D}}, \tag{7.5.5c,d}$$

and $\bar{\mathbf{S}}^I$ and $\bar{\mathbf{T}}^I$ are Eshelby's tensor and its conjugate, common to all inclusions.

7.5.2. Macrostrain Prescribed

Next, consider the case when the macrostrain $\mathbf{E} = \boldsymbol{\varepsilon}^o$ is prescribed. In a manner similar to the preceding subsection, from (7.5.1a,b) express the average strain of the micro-inclusion Ω_α in terms of the overall strain $\boldsymbol{\varepsilon}^o$ as

$$\bar{\boldsymbol{\varepsilon}}^\alpha = \bar{\mathbf{A}}^\alpha : (\bar{\mathbf{A}}^\alpha - \bar{\mathbf{S}}^\alpha)^{-1} : \boldsymbol{\varepsilon}^o,$$

$$\bar{\boldsymbol{\varepsilon}}^\alpha = \mathbf{D}^\alpha : \bar{\mathbf{B}}^\alpha : (\bar{\mathbf{B}}^\alpha - \bar{\mathbf{T}}^\alpha)^{-1} : \bar{\mathbf{C}} : \boldsymbol{\varepsilon}^o. \qquad (7.5.6\text{a,b})$$

Substitution into (7.4.1a) now yields

$$(\bar{\mathbf{C}} - \mathbf{C}) : \boldsymbol{\varepsilon}^o = \left\{ \sum_{\alpha=1}^{n} f_\alpha (\mathbf{C}^\alpha - \mathbf{C}) : \bar{\mathbf{A}}^\alpha : (\bar{\mathbf{A}}^\alpha - \bar{\mathbf{S}}^\alpha)^{-1} \right\} : \boldsymbol{\varepsilon}^o,$$

$$(\bar{\mathbf{C}} - \mathbf{C}) : \boldsymbol{\varepsilon}^o = \left\{ \sum_{\alpha=1}^{n} f_\alpha (\mathbf{C}^\alpha - \mathbf{C}) : \mathbf{D}^\alpha : \bar{\mathbf{B}}^\alpha : (\bar{\mathbf{B}}^\alpha - \bar{\mathbf{T}}^\alpha)^{-1} : \bar{\mathbf{C}} \right\} : \boldsymbol{\varepsilon}^o. \qquad (7.5.6\text{c,d})$$

Since $\boldsymbol{\varepsilon}^o$ is arbitrary, the overall elasticity tensor $\bar{\mathbf{C}}$ is given by

$$\bar{\mathbf{C}} = \mathbf{C} + \sum_{\alpha=1}^{n} f_\alpha (\mathbf{C}^\alpha - \mathbf{C}) : \bar{\mathbf{A}}^\alpha : (\bar{\mathbf{A}}^\alpha - \bar{\mathbf{S}}^\alpha)^{-1},$$

$$\bar{\mathbf{C}} = \mathbf{C} + \sum_{\alpha=1}^{n} f_\alpha (\mathbf{C}^\alpha - \mathbf{C}) : \mathbf{D}^\alpha : \bar{\mathbf{B}}^\alpha : (\bar{\mathbf{B}}^\alpha - \bar{\mathbf{T}}^\alpha)^{-1} : \bar{\mathbf{C}}. \qquad (7.5.7\text{a,b})$$

Again, since the average strain $\bar{\boldsymbol{\varepsilon}}^\alpha$ given by (7.5.6a), is the same as that given by (7.5.6b), the overall elasticity tensor $\bar{\mathbf{C}}$ given by (7.5.7a), equals that given by (7.5.7b).

When all inclusions are similar, (7.5.7a,b) reduce to

$$\bar{\mathbf{C}} = \mathbf{C} + f\,(\mathbf{C}^I - \mathbf{C}) : \bar{\mathbf{A}}^I : (\bar{\mathbf{A}}^I - \bar{\mathbf{S}}^I)^{-1},$$

$$\bar{\mathbf{C}} = \mathbf{C} + f\,(\mathbf{C}^I - \mathbf{C}) : \mathbf{D}^I : \bar{\mathbf{B}}^I : (\bar{\mathbf{B}}^I - \bar{\mathbf{T}}^I)^{-1} : \bar{\mathbf{C}}. \qquad (7.5.8\text{a,b})$$

7.5.3. Equivalence of Overall Compliance and Elasticity Tensors Obtained by Self-Consistent Method

As discussed in Sections 5 and 6, the overall compliance and elasticity tensors obtained by the self-consistent method are each other's inverse, hence the name self-consistent. To see this, consider (7.5.7a) and multiply from the right by $\bar{\mathbf{C}}^{-1}$, and from the left by $\mathbf{C}^{-1} = \mathbf{D}$, to obtain

$$\mathbf{D} = \bar{\mathbf{C}}^{-1} + \sum_{\alpha=1}^{n} f_\alpha\, \mathbf{D} : (\mathbf{C}^\alpha - \mathbf{C}) : \bar{\mathbf{A}}^\alpha : (\bar{\mathbf{A}}^\alpha - \bar{\mathbf{S}}^\alpha)^{-1} : \bar{\mathbf{C}}^{-1}$$

$$= \bar{\mathbf{C}}^{-1} - \sum_{\alpha=1}^{n} f_\alpha (\mathbf{D}^\alpha - \mathbf{D}) : \mathbf{C}^\alpha : \bar{\mathbf{A}}^\alpha : (\bar{\mathbf{A}}^\alpha - \bar{\mathbf{S}}^\alpha)^{-1} : \bar{\mathbf{C}}^{-1}. \qquad (7.5.9\text{a})$$

Now, multiply (7.5.4b) from the right by $\bar{\mathbf{D}}^{-1}$, and from the left by $\mathbf{D}^{-1} = \mathbf{C}$, to arrive at

$$\mathbf{C} = \overline{\mathbf{D}}^{-1} + \sum_{\alpha=1}^{n} f_\alpha\, \mathbf{C} : (\mathbf{D}^\alpha - \mathbf{D}) : \overline{\mathbf{B}}^\alpha : (\overline{\mathbf{B}}^\alpha - \overline{\mathbf{T}}^\alpha)^{-1} : \overline{\mathbf{D}}^{-1}$$

$$= \overline{\mathbf{D}}^{-1} - \sum_{\alpha=1}^{n} f_\alpha\, (\mathbf{C}^\alpha - \mathbf{C}) : \mathbf{D}^\alpha : \overline{\mathbf{B}}^\alpha : (\overline{\mathbf{B}}^\alpha - \overline{\mathbf{T}}^\alpha)^{-1} : \overline{\mathbf{D}}^{-1}. \tag{7.5.9b}$$

Comparing (7.5.9a) with (7.5.4a), or (7.5.9b) with (7.5.7b), observe that

$$\overline{\mathbf{C}}^{-1} = \overline{\mathbf{D}}, \qquad \overline{\mathbf{D}}^{-1} = \overline{\mathbf{C}}. \tag{7.5.9c,d}$$

Therefore, the overall compliance tensor $\overline{\mathbf{D}}$ given by (7.5.4a,b), and the overall elasticity tensor $\overline{\mathbf{C}}$ given by (7.5.7a,b), are each other's inverse, exactly.

When all micro-inclusions are similar, a similar procedure yields, from (7.5.8a),

$$\mathbf{D} = \overline{\mathbf{C}}^{-1} + f\,(\mathbf{D} - \mathbf{D}^I) : \mathbf{C}^I : \overline{\mathbf{A}}^I : (\overline{\mathbf{A}}^I - \overline{\mathbf{S}}^I)^{-1} : \overline{\mathbf{C}}^{-1}, \tag{7.5.10a}$$

and from (7.5.5b),

$$\mathbf{C} = \overline{\mathbf{D}}^{-1} + f\,(\mathbf{C} - \mathbf{C}^I) : \mathbf{D}^I : \overline{\mathbf{B}}^I : (\overline{\mathbf{B}}^I - \overline{\mathbf{T}}^I)^{-1} : \overline{\mathbf{D}}^{-1}. \tag{7.5.10b}$$

The equivalence of (7.5.10a) and (7.5.5a), and (7.5.10b) and (7.5.8b), ensure the equivalence of $\overline{\mathbf{D}}$ and $\overline{\mathbf{C}}$.

In terms of $\overline{\mathbf{S}}^\alpha$ and $\overline{\mathbf{T}}^\alpha$, the tensor $\overline{\mathbf{H}}$ can be defined from (7.5.3c,d), by

$$\overline{\mathbf{H}} \equiv \sum_{\alpha=1}^{n} f_\alpha\, (\mathbf{D}^\alpha - \mathbf{D}) : \mathbf{C}^\alpha : \overline{\mathbf{A}}^\alpha : (\overline{\mathbf{A}}^\alpha - \overline{\mathbf{S}}^\alpha)^{-1} : \overline{\mathbf{D}}$$

$$\equiv \sum_{\alpha=1}^{n} f_\alpha\, (\mathbf{D}^\alpha - \mathbf{D}) : \overline{\mathbf{B}}^\alpha : (\overline{\mathbf{B}}^\alpha - \overline{\mathbf{T}}^\alpha)^{-1}, \tag{7.5.11a}$$

and the tensor $\overline{\mathbf{J}}$ from (7.5.6c,d), by

$$\overline{\mathbf{J}} \equiv -\sum_{\alpha=1}^{n} f_\alpha\, \mathbf{D} : (\mathbf{C}^\alpha - \mathbf{C}) : \overline{\mathbf{A}}^\alpha : (\overline{\mathbf{A}}^\alpha - \overline{\mathbf{S}}^\alpha)^{-1}$$

$$\equiv -\sum_{\alpha=1}^{n} f_\alpha\, \mathbf{D} : (\mathbf{C}^\alpha - \mathbf{C}) : \mathbf{D}^\alpha : \overline{\mathbf{B}}^\alpha : (\overline{\mathbf{B}}^\alpha - \overline{\mathbf{T}}^\alpha)^{-1} : \overline{\mathbf{C}}. \tag{7.5.11b}$$

From (7.5.11a,b) (or from the equivalence relations (7.5.2c,d)), $\overline{\mathbf{H}}$ and $\overline{\mathbf{J}}$ satisfy

$$\overline{\mathbf{J}} = \overline{\mathbf{H}} : \overline{\mathbf{C}}, \qquad \overline{\mathbf{H}} = \overline{\mathbf{J}} : \overline{\mathbf{D}}. \tag{7.5.12a,b}$$

These relations correspond to (5.1.22b,c) exactly, and hence,

$$\overline{\mathbf{D}} = \mathbf{D} + \overline{\mathbf{H}} \Longleftrightarrow \overline{\mathbf{C}} = \mathbf{C} - \mathbf{C} : \overline{\mathbf{J}}. \tag{7.5.13}$$

Formulation of effective moduli of elastic composites has been considered by a number of investigators. Related to the formulation presented in this subsection are contributions by Kerner, (1956), Hill (1963, 1965a,b), Willis (1964, 1977, 1980, 1981, 1982), Walpole (1966a,b, 1967, 1970, 1981), and Wu (1966).

7.5.4. Overall Elasticity and Compliance Tensors for Polycrystals

In certain problems it may not be feasible to distinguish between the matrix and the inclusions in an RVE. For example, if an RVE is a polycrystal, each crystal may be treated as an inclusion embedded in the remaining crystals and hence, all crystals have the same significance. In this case there is complete symmetry in treating each crystal as an inclusion. Here the concept of a matrix with embedded inclusions is no longer relevant. Nevertheless, it is possible to apply the exact relations (7.4.1a,b), in order to estimate the corresponding overall elasticity and compliance tensors. However, on physical grounds, the assumption of a dilute distribution of inclusions no longer applies, whereas the self-consistent scheme may be used. Indeed, the self-consistent method was originally proposed by Hershey (1954), Kröner (1958, 1967) and Kneer (1965), to estimate the overall moduli of polycrystals; earlier work on the elastic properties of polycrystals is due to Voigt (1910) and Reuss (1929); the self-consistent method was later applied to composites by Budiansky (1965) and Hill (1965a,b). Other related work in this area is by Hill (1952), Hashin and Shtrikman (1962), Peselnick and Meister (1965), Morris (1970, 1971), Zeller and Dederichs (1973), Korringa (1973), and Gubernatis and Krumhansl (1975); see also reviews by Hashin (1964, 1983), Watt *el al.* (1976), Kröner (1980), and Mura (1987).

For the self-consistent estimate, a single crystal is embedded in an unbounded uniform matrix which has the effective overall parameters of the polycrystal. The local average stresses and strains in the embedded crystal are then calculated and used to obtain the overall moduli. To apply (7.4.1a,b), however, the significance given to the matrix (whose elasticity and compliance tensors have been denoted by **C** and **D**, respectively) must be removed.

To this end, regard the matrix in (7.4.1a,b) as the *zeroth* inclusion, Ω_0, with the corresponding elasticity and compliance tensors $\mathbf{C}^0$ and $\mathbf{D}^0$, respectively. The volume fraction of Ω_0 is $f_0 = 1 - f$. For simplicity assume that all micro-inclusions are similar, and denote the common Eshelby tensor and its conjugate by $\bar{\mathbf{S}}^I$ and $\bar{\mathbf{T}}^I$, *where the superbar signifies that these tensors correspond to the yet-unknown overall elasticity or compliance tensors* $\bar{\mathbf{C}}$ *and* $\bar{\mathbf{D}}$. Then, rewrite (7.5.4a) and (7.5.7b), as

$$\bar{\mathbf{D}} - \mathbf{D}^0 = \sum_{\alpha=1}^{n} f_\alpha\,(\mathbf{1}^{(4s)} - \mathbf{D}^0 : \mathbf{C}^\alpha) : \{\mathbf{1}^{(4s)} + \bar{\mathbf{S}}^I : (\bar{\mathbf{D}} : \mathbf{C}^\alpha - \mathbf{1}^{(4s)})\}^{-1} : \bar{\mathbf{D}},$$

$$\bar{\mathbf{C}} - \mathbf{C}^0 = \sum_{\alpha=1}^{n} f_\alpha\,(\mathbf{1}^{(4s)} - \mathbf{C}^0 : \mathbf{D}^\alpha) : \{\mathbf{1}^{(4s)} + \bar{\mathbf{T}}^I : (\bar{\mathbf{C}} : \mathbf{D}^\alpha - \mathbf{1}^{(4s)})\}^{-1} : \bar{\mathbf{C}},$$

(7.5.14a,b)

where $\bar{\mathbf{A}}^\alpha$ and $\bar{\mathbf{B}}^\alpha$ are eliminated using (7.5.1c,d). Since

$$\sum_{\alpha=1}^{n} f_\alpha = 1, \tag{7.5.15}$$

multiplying (7.5.14a) from the right by $\bar{\mathbf{C}}$, observe that

$$\mathbf{0} = \mathbf{1}^{(4s)} - \mathbf{D}^0 : \bar{\mathbf{C}} - \sum_{\alpha=0}^{n} f_\alpha (\mathbf{1}^{(4s)} - \mathbf{D}^0 : \mathbf{C}^\alpha) : \{\mathbf{1}^{(4s)} + \mathbf{S}^I : (\bar{\mathbf{D}} : \mathbf{C}^\alpha - \mathbf{1}^{(4s)})\}^{-1}$$

$$= \mathbf{1}^{(4s)} - \sum_{\alpha=0}^{n} f_\alpha \{\mathbf{1}^{(4s)} + \mathbf{S}^I : (\bar{\mathbf{D}} : \mathbf{C}^\alpha - \mathbf{1}^{(4s)})\}^{-1}$$

$$- \mathbf{D}^0 : \bar{\mathbf{C}} : \left\{ \mathbf{1}^{(4s)} - \sum_{\alpha=0}^{n} f_\alpha \bar{\mathbf{D}} : \mathbf{C}^\alpha : \{\mathbf{1}^{(4s)} + \mathbf{S}^I : (\bar{\mathbf{D}} : \mathbf{C}^\alpha - \mathbf{1}^{(4s)})\}^{-1} \right\}$$

$$= \{\mathbf{1}^{(4s)} - \mathbf{D}^0 : \bar{\mathbf{C}} : (\mathbf{1}^{(4s)} - \bar{\mathbf{S}}^{I^{-1}})\}$$

$$: \left\{ \mathbf{1}^{(4s)} - \sum_{\alpha=0}^{n} f_\alpha \{\mathbf{1}^{(4s)} + \mathbf{S}^I : (\bar{\mathbf{D}} : \mathbf{C}^\alpha - \mathbf{1}^{(4s)})\}^{-1} \right\}. \qquad (7.5.14c)$$

In a similar manner, (7.5.14b) yields

$$\mathbf{0} = \{\mathbf{1}^{(4s)} - \mathbf{C}^0 : \bar{\mathbf{D}} : (\mathbf{1}^{(4s)} - \bar{\mathbf{T}}^I)\}$$

$$: \left\{ \mathbf{1}^{(4s)} - \sum_{\alpha=0}^{n} f_\alpha \{\mathbf{1}^{(4s)} + \mathbf{T}^I : (\bar{\mathbf{C}} : \mathbf{D}^\alpha - \mathbf{1}^{(4s)})\}^{-1} \right\}. \qquad (7.5.14d)$$

Since the tensor $(\mathbf{1}^{(4s)} - \mathbf{C}^0 : \bar{\mathbf{D}} : (\mathbf{1}^{(4s)} - \mathbf{T}^{I^{-1}}))$ is invertible,[8] the following expressions are obtained for the overall elasticity and compliance tensors $\bar{\mathbf{C}}$ and $\bar{\mathbf{D}}$, which do not distinguish between the matrix and the inclusions:

$$\sum_{\alpha=0}^{n} f_\alpha \{\mathbf{1}^{(4s)} + \bar{\mathbf{S}}^I : (\bar{\mathbf{D}} : \mathbf{C}^\alpha - \mathbf{1}^{(4s)})\}^{-1} = \mathbf{1}^{(4s)},$$

$$\sum_{\alpha=0}^{n} f_\alpha \{\mathbf{1}^{(4s)} + \bar{\mathbf{T}}^I : (\bar{\mathbf{C}} : \mathbf{D}^\alpha - \mathbf{1}^{(4s)})\}^{-1} = \mathbf{1}^{(4s)}. \qquad (7.5.16a,b)$$

It is noted that (7.5.16a) and (7.5.16b) are identical. The proof is straightforward. The equivalence relations between $\bar{\mathbf{S}}^I$ and $\bar{\mathbf{T}}^I$ are

$$\bar{\mathbf{S}}^I + \bar{\mathbf{D}} : \bar{\mathbf{T}}^I : \bar{\mathbf{C}} = \mathbf{1}^{(4s)}, \qquad \bar{\mathbf{T}}^I + \bar{\mathbf{C}} : \bar{\mathbf{S}}^I : \bar{\mathbf{D}} = \mathbf{1}^{(4s)}. \qquad (7.5.17a,b)$$

Using (7.5.15), rewrite (7.5.16a) as

$$\mathbf{0} = \sum_{\alpha=0}^{n} f_\alpha \bar{\mathbf{S}}^I : (\bar{\mathbf{D}} : \mathbf{C}^\alpha - \mathbf{1}^{(4s)}) : \{\mathbf{1}^{(4s)} + \bar{\mathbf{S}}^I : (\bar{\mathbf{D}} : \mathbf{C}^\alpha - \mathbf{1}^{(4s)})\}^{-1}. \qquad (7.5.18a)$$

With the aid of (7.5.17a), multiplication of (7.5.18a) from the right by $\bar{\mathbf{D}}$ and from the left by $\bar{\mathbf{C}}$ yields,

$$\mathbf{0} = \sum_{\alpha=0}^{n} f_\alpha (\bar{\mathbf{T}}^I - \mathbf{1}^{(4s)}) : (\bar{\mathbf{C}} : \mathbf{D}^\alpha - \mathbf{1}^{(4s)}) : \{\mathbf{1}^{(4s)} + \bar{\mathbf{T}}^I : (\bar{\mathbf{C}} : \mathbf{D}^\alpha - \mathbf{1}^{(4s)})\}^{-1}. \qquad (7.5.18b)$$

Now multiply (7.5.18b) from the left by $\bar{\mathbf{T}}^I : (\bar{\mathbf{T}}^I - \mathbf{1}^{(4s)})^{-1}$, to obtain (7.5.16b).

---

[8] Similar comments apply when $\mathbf{D}^0$ or $\mathbf{C}^0$ is replaced by $\mathbf{D}^\alpha$ or $\mathbf{C}^\alpha$.

7.6. ENERGY CONSIDERATION AND SYMMETRY OF OVERALL ELASTICITY AND COMPLIANCE TENSORS

In this section, the elastic strain energy stored in an elastic RVE is examined. The RVE contains a linearly elastic and homogeneous matrix and a set of linearly elastic and homogeneous inclusions with their possibly different elasticity tensors. From constitutive relations (7.3.3a,b), the complementary strain energy density function, $w^c(\boldsymbol{\sigma})$, and the elastic strain energy density function, $w(\boldsymbol{\varepsilon})$, are defined by

$$w^c(\boldsymbol{\sigma}) = \begin{cases} \frac{1}{2}\boldsymbol{\sigma} : \mathbf{D} : \boldsymbol{\sigma} & \text{in M} \\ \frac{1}{2}\boldsymbol{\sigma} : \mathbf{D}^\alpha : \boldsymbol{\sigma} & \text{in } \Omega_\alpha, \end{cases}$$

$$w(\boldsymbol{\varepsilon}) = \begin{cases} \frac{1}{2}\boldsymbol{\varepsilon} : \mathbf{C} : \boldsymbol{\varepsilon} & \text{in M} \\ \frac{1}{2}\boldsymbol{\varepsilon} : \mathbf{C}^\alpha : \boldsymbol{\varepsilon} & \text{in } \Omega_\alpha. \end{cases} \tag{7.6.1a,b}$$

The components of the elasticity and compliance tensors satisfy the following symmetry conditions:

$$D_{ijkl} = D_{jikl} = D_{ijlk} = D_{klij}, \qquad C_{ijkl} = C_{jikl} = C_{ijlk} = C_{klij},$$

$$D^\alpha_{ijkl} = D^\alpha_{jikl} = D^\alpha_{ijlk} = D^\alpha_{klij}, \qquad C^\alpha_{ijkl} = C^\alpha_{jikl} = C^\alpha_{ijlk} = C^\alpha_{klij}. \tag{7.6.1c~f}$$

When the macrostress $\boldsymbol{\Sigma} = \boldsymbol{\sigma}^o$ is regarded prescribed, the overall complementary strain energy function $W^c = W^c(\boldsymbol{\sigma}^o)$ is defined by

$$W^c(\boldsymbol{\sigma}^o) \equiv < w^c > = \frac{1}{2} < \boldsymbol{\sigma} : \boldsymbol{\varepsilon} > = \frac{1}{2}\boldsymbol{\sigma}^o : \bar{\boldsymbol{\varepsilon}}. \tag{7.6.2a}$$

When the macrostrain $\mathbf{E} = \boldsymbol{\varepsilon}^o$ is regarded prescribed, the overall strain energy function $W = W(\boldsymbol{\varepsilon}^o)$ becomes

$$W(\boldsymbol{\varepsilon}^o) \equiv < w > = \frac{1}{2} < \boldsymbol{\sigma} : \boldsymbol{\varepsilon} > = \frac{1}{2}\bar{\boldsymbol{\sigma}} : \boldsymbol{\varepsilon}^o. \tag{7.6.2b}$$

It is recalled that $\boldsymbol{\sigma}$ in (7.6.2a) is the stress field produced by the applied overall stress $\boldsymbol{\sigma}^o$, while $\boldsymbol{\varepsilon}$ in (7.6.2b) is the strain field produced by the overall strain $\boldsymbol{\varepsilon}^o$. In general, these resulting stress and strain fields are unrelated. Also, in general, the average strain $\bar{\boldsymbol{\varepsilon}}$ in (7.6.2a) is unrelated to the average stress $\bar{\boldsymbol{\sigma}}$ in (7.6.2b); see Subsections 2.5 and 2.6 for related comments and explanation.

The overall compliance tensor $\bar{\mathbf{D}}$ and elasticity tensor $\bar{\mathbf{C}}$ may be defined by the overall complementary and strain energy functions, W^c and W, through

$$W^c(\boldsymbol{\sigma}^o) \equiv \frac{1}{2}\boldsymbol{\sigma}^o : \bar{\mathbf{D}} : \boldsymbol{\sigma}^o, \qquad W(\boldsymbol{\varepsilon}^o) \equiv \frac{1}{2}\boldsymbol{\varepsilon}^o : \bar{\mathbf{C}} : \boldsymbol{\varepsilon}^o. \tag{7.6.3a,b}$$

The tensors $\bar{\mathbf{D}}$ and $\bar{\mathbf{C}}$ then are the compliance and elasticity tensors of a homogenized RVE which contains the same amounts of elastic energy as the heterogeneous solid, under a uniform stress $\boldsymbol{\sigma}^o$ and a uniform strain $\boldsymbol{\varepsilon}^o$, respectively. If the estimates of $\bar{\mathbf{D}}$ and $\bar{\mathbf{C}}$ are made in a consistent manner, then $\bar{\mathbf{C}}$ and $\bar{\mathbf{D}}$ must be each other's inverse. *Moreover, like* $\mathbf{D}$ *and* $\mathbf{C}$, $\bar{\mathbf{D}}$ *and* $\bar{\mathbf{C}}$ *defined by* (7.6.3a,b) *have the following symmetry properties*:

$$\bar{D}_{ijkl} = \bar{D}_{jikl} = \bar{D}_{ijlk} = \bar{D}_{klij}, \qquad \bar{C}_{ijkl} = \bar{C}_{jikl} = \bar{C}_{ijlk} = \bar{C}_{klij}. \tag{7.6.3c,d}$$

On the other hand, the $\bar{\mathbf{D}}$*- and* $\bar{\mathbf{C}}$*-tensors defined in Subsection* 7.4 *by* (7.4.6a,b) *and* (7.4.9a,b), *and in Subsection* 7.5 *by* (7.5.4a,b) *and* (7.5.8a,b), *need not and may not necessarily satisfy the symmetry conditions* $\bar{D}_{ijkl} = \bar{D}_{klij}$ *and* $\bar{C}_{ijkl} = \bar{C}_{klij}$. The aim now is to establish the corresponding energy relations, considering the prescribed macrostress and prescribed macrostrain cases separately.

7.6.1. Macrostrain Prescribed

When the prescribed macrostrain is $\mathbf{E} = \boldsymbol{\varepsilon}^o$, from the results obtained in Subsection 7.2, the average stress is *exactly* given by

$$\bar{\boldsymbol{\sigma}} = <\boldsymbol{\sigma}> = \mathbf{C} : \boldsymbol{\varepsilon}^o + \sum_{\alpha=1}^{n} f_\alpha\, (\mathbf{C}^\alpha - \mathbf{C}) : \bar{\boldsymbol{\varepsilon}}^\alpha, \tag{7.6.4}$$

where $\bar{\boldsymbol{\varepsilon}}^\alpha \equiv <\boldsymbol{\varepsilon}>_\alpha$ is the average strain of the αth micro-inclusion Ω_α. Therefore, the right-hand side of (7.6.2b) is calculated as follows:

$$\bar{\boldsymbol{\sigma}} : \boldsymbol{\varepsilon}^o = \{\mathbf{C} : \boldsymbol{\varepsilon}^o + \sum_{\alpha=1}^{n} f_\alpha\, (\mathbf{C}^\alpha - \mathbf{C}) : \bar{\boldsymbol{\varepsilon}}^\alpha\} : \boldsymbol{\varepsilon}^o. \tag{7.6.5a}$$

From (7.6.3b) and (7.6.5a) it now follows that

$$\boldsymbol{\varepsilon}^o : \bar{\mathbf{C}} : \boldsymbol{\varepsilon}^o = \{\mathbf{C} : \boldsymbol{\varepsilon}^o + \sum_{\alpha=1}^{n} f_\alpha\, (\mathbf{C}^\alpha - \mathbf{C}) : \bar{\boldsymbol{\varepsilon}}^\alpha\} : \boldsymbol{\varepsilon}^o. \tag{7.6.5b}$$

Since the response of the RVE is linearly elastic, from (7.2.7) it follows that

$$(\mathbf{C}^\alpha - \mathbf{C}) : \bar{\boldsymbol{\varepsilon}}^\alpha = -\mathbf{C} : \mathbf{J}^\alpha : \boldsymbol{\varepsilon}^o. \tag{7.6.5c}$$

Hence,

$$\boldsymbol{\varepsilon}^o : \bar{\mathbf{C}} : \boldsymbol{\varepsilon}^o = \boldsymbol{\varepsilon}^o : \{\mathbf{C} - \sum_{\alpha=1}^{n} f_\alpha\, \mathbf{C} : \mathbf{J}^\alpha\} : \boldsymbol{\varepsilon}^o. \tag{7.6.6a}$$

This is an exact result which must hold for any prescribed constant symmetric $\boldsymbol{\varepsilon}^o$. It thus follows that

$$\bar{\mathbf{C}} = sym\, \{\mathbf{C} - \sum_{\alpha=1}^{n} f_\alpha\, \mathbf{C} : \mathbf{J}^\alpha\}, \tag{7.6.6b}$$

where *sym* stands for the "symmetric part of", i.e., for a fourth-order tensor $\mathbf{T}$,

$$sym\,(T_{ijkl}) \equiv \frac{1}{2}(T_{ijkl} + T_{klij}). \tag{7.6.6c}$$

Expression (7.6.6b) *is an exact result which must hold for any linearly elastic* RVE. Approximations become necessary in order to estimate the $\mathbf{J}^\alpha$-tensor. The homogenization method discussed in Subsections 7.4 and 7.5 may be used to estimate $\mathbf{J}^\alpha$. For example, when the distribution of inclusions is dilute, the $\mathbf{J}^\alpha$-tensor is estimated by embedding Ω_α in an unbounded matrix of elasticity $\mathbf{C}$, with the overall strain $\boldsymbol{\varepsilon}^o$ or the stress $\boldsymbol{\sigma}^o = \mathbf{C} : \boldsymbol{\varepsilon}^o$ prescribed at infinity. This then yields

$$\mathbf{J}^\alpha = (\mathbf{D}^\alpha - \mathbf{D}) : \mathbf{C}^\alpha : \mathbf{A}^\alpha : (\mathbf{A}^\alpha - \mathbf{S}^\alpha)^{-1}, \tag{7.6.7a}$$

where $\mathbf{A}^\alpha$ is given by (7.4.2a). Hence,

$$\bar{\mathbf{C}} = sym\left\{ \mathbf{C} : \{\mathbf{1}^{(4s)} - \sum_{\alpha=1}^{n} f_\alpha (\mathbf{A}^\alpha - \mathbf{S}^\alpha)^{-1}\} \right\}. \tag{7.6.7b}$$

For the self-consistent estimate, on the other hand, the $\mathbf{J}^\alpha$-tensor becomes

$$\mathbf{J}^\alpha \equiv \bar{\mathbf{J}}^\alpha = (\mathbf{D}^\alpha - \mathbf{D}) : \mathbf{C}^\alpha : \bar{\mathbf{A}}^\alpha : (\bar{\mathbf{A}}^\alpha - \bar{\mathbf{S}}^\alpha)^{-1}, \tag{7.6.8a}$$

where the notation follows that in Subsection 7.5, e.g., (7.5.1c). Then, from (7.6.6b),

$$\bar{\mathbf{C}} = sym\left\{ \mathbf{C} + \sum_{\alpha=1}^{n} f_\alpha (\mathbf{C}^\alpha - \mathbf{C}) : \bar{\mathbf{A}}^\alpha : (\bar{\mathbf{A}}^\alpha - \bar{\mathbf{S}}^\alpha)^{-1} \right\}. \tag{7.6.8b}$$

Expressions (7.6.7b) and (7.6.8b) respectively are the symmetric parts of expressions (7.4.9a) and (7.5.7a). In a similar manner, one can show that the energy-based definition of the overall elasticity tensor $\bar{\mathbf{C}}$ is given in terms of $\mathbf{B}^\alpha$ and $\mathbf{T}^\alpha$ by the symmetric part of (7.4.9b) for the dilute-distribution case, and by the symmetric part of (7.5.7b) for the self-consistent case, i.e., by

$$\bar{\mathbf{C}} = sym\left\{ \{\mathbf{1}^{(4s)} + \sum_{\alpha=1}^{n} f_\alpha (\mathbf{B}^\alpha - \mathbf{T}^\alpha)^{-1}\} : \mathbf{C} \right\}, \tag{7.6.7c}$$

for the dilute case, and by

$$\bar{\mathbf{C}} = sym\left\{ \mathbf{C} - \sum_{\alpha=1}^{n} f_\alpha (\mathbf{C}^\alpha - \mathbf{C}) : \mathbf{D}^\alpha : \bar{\mathbf{B}}^\alpha : (\bar{\mathbf{B}}^\alpha - \bar{\mathbf{T}}^\alpha)^{-1} : \bar{\mathbf{C}} \right\}, \tag{7.6.8c}$$

for the self-consistent case.

7.6.2. Macrostress Prescribed

When the prescribed macrostress is $\boldsymbol{\Sigma} = \boldsymbol{\sigma}^o$, from the results obtained in Subsection 7.1, the average strain is *exactly* given by

$$\bar{\boldsymbol{\varepsilon}} = \langle \boldsymbol{\varepsilon} \rangle = \mathbf{D} : \boldsymbol{\sigma}^o + \sum_{\alpha=1}^{n} f_\alpha (\mathbf{D}^\alpha - \mathbf{D}) : \bar{\boldsymbol{\sigma}}^\alpha, \tag{7.6.9}$$

where $\bar{\boldsymbol{\sigma}}^\alpha \equiv \langle \boldsymbol{\sigma} \rangle_\alpha$ is the average stress of the αth micro-inclusion Ω_α. Then, the right-hand side of (7.6.2a) is computed as

$$\boldsymbol{\sigma}^o : \bar{\boldsymbol{\varepsilon}} = \boldsymbol{\sigma}^o : \{\mathbf{D} : \boldsymbol{\sigma}^o + \sum_{\alpha=1}^{n} f_\alpha (\mathbf{D}^\alpha - \mathbf{D}) : \bar{\boldsymbol{\sigma}}^\alpha\}. \tag{7.6.10a}$$

Hence, from (7.6.3a) and (7.6.10a), it follows that

$$\boldsymbol{\sigma}^o : \bar{\mathbf{D}} : \boldsymbol{\sigma}^o = \boldsymbol{\sigma}^o : \{\mathbf{D} : \boldsymbol{\sigma}^o + \sum_{\alpha=1}^{n} f_\alpha (\mathbf{D}^\alpha - \mathbf{D}) : \bar{\boldsymbol{\sigma}}^\alpha\}. \tag{7.6.10b}$$

Since the response of the RVE is linearly elastic, from (7.1.11) it is noted that

$$(\mathbf{D}^\alpha - \mathbf{D}) : \bar{\boldsymbol{\sigma}}^\alpha = \mathbf{H}^\alpha : \boldsymbol{\sigma}^o, \tag{7.6.10c}$$

and hence,

$$\boldsymbol{\sigma}^o : \bar{\mathbf{D}} : \boldsymbol{\sigma}^o = \boldsymbol{\sigma}^o : \{\mathbf{D} + \sum_{\alpha=1}^{n} f_\alpha \, \mathbf{H}^\alpha\} : \boldsymbol{\sigma}^o. \tag{7.6.11a}$$

This expression is an exact result which must hold for any constant and symmetric $\boldsymbol{\sigma}^o$, leading to

$$\bar{\mathbf{D}} = sym\,\{\mathbf{D} + \sum_{\alpha=1}^{n} f_\alpha \, \mathbf{H}^\alpha\}. \tag{7.6.11b}$$

Again, (7.6.11b) is an exact result. To estimate $\mathbf{H}^\alpha$, however, one is often forced to introduce simplifying approximations; see Subsections 7.4 and 7.5. For example, when the distribution of micro-inclusions is dilute,

$$(\mathbf{D}^\alpha - \mathbf{D})^{-1} : \mathbf{H}^\alpha = \mathbf{B}^\alpha : (\mathbf{B}^\alpha - \mathbf{T}^\alpha)^{-1}$$

$$= \mathbf{C}^\alpha : \mathbf{A}^\alpha : (\mathbf{A}^\alpha - \mathbf{S}^\alpha)^{-1} : \mathbf{D} \qquad (\alpha \text{ not summed}), \tag{7.6.12a}$$

where the notation follows that in Subsection 7.4. Hence, two equivalent expressions for $\bar{\mathbf{D}}$ are obtained,

$$\bar{\mathbf{D}} = sym\left\{ \mathbf{D} : \{\mathbf{1}^{(4s)} - \sum_{\alpha=1}^{n} f_\alpha \, (\mathbf{B}^\alpha - \mathbf{T}^\alpha)^{-1}\} \right\},$$

$$\bar{\mathbf{D}} = sym\left\{ \{\mathbf{1}^{(4s)} + \sum_{\alpha=1}^{n} f_\alpha \, (\mathbf{A}^\alpha - \mathbf{S}^\alpha)^{-1}\} : \mathbf{D} \right\}. \tag{7.6.12b,c}$$

For the self-consistent method,

$$(\mathbf{D}^\alpha - \mathbf{D})^{-1} : \bar{\mathbf{H}}^\alpha = \bar{\mathbf{B}}^\alpha : (\bar{\mathbf{B}}^\alpha - \bar{\mathbf{T}}^\alpha)^{-1}$$

$$= \mathbf{C}^\alpha : \bar{\mathbf{A}}^\alpha : (\bar{\mathbf{A}}^\alpha - \bar{\mathbf{S}}^\alpha)^{-1} : \bar{\mathbf{D}} \qquad (\alpha \text{ not summed}), \tag{7.6.13a}$$

where the notation follows that in Subsection 7.5. Again, two equivalent expressions are obtained,

$$\bar{\mathbf{D}} = sym\left\{ \mathbf{D} + \sum_{\alpha=1}^{n} f_\alpha \, (\mathbf{D}^\alpha - \mathbf{D}) : \bar{\mathbf{B}}^\alpha : (\bar{\mathbf{B}}^\alpha - \bar{\mathbf{T}}^\alpha)^{-1} \right\},$$

$$\bar{\mathbf{D}} = sym\left\{ \mathbf{D} + \sum_{\alpha=1}^{n} f_\alpha \, (\mathbf{D}^\alpha - \mathbf{D}) : \mathbf{C}^\alpha : \bar{\mathbf{A}}^\alpha : (\bar{\mathbf{A}}^\alpha - \bar{\mathbf{S}}^\alpha)^{-1} : \bar{\mathbf{D}} \right\}. \tag{7.6.13b,c}$$

7.6.3. Equivalence of Overall Compliance and Elasticity Tensors Obtained on the Basis of Elastic Energy

In Subsection 7.6.1, it is shown that the overall elasticity tensor defined in terms of the overall strain energy, $W = \boldsymbol{\varepsilon}^o : \bar{\mathbf{C}} : \boldsymbol{\varepsilon}^o/2$, is the symmetric part of the overall elasticity tensor defined directly in terms of the average stress and strain, $\bar{\boldsymbol{\sigma}} = \bar{\mathbf{C}} : \boldsymbol{\varepsilon}^o$; see (7.2.8) and (7.6.6b). Similarly, in Subsection 7.6.2, the overall compliance tensor defined in terms of the overall complementary strain energy, $W^c = \boldsymbol{\sigma}^o : \bar{\mathbf{D}} : \boldsymbol{\sigma}^o/2$, is the symmetric part of the overall compliance tensor defined in terms of the average strain and stress, $\bar{\boldsymbol{\varepsilon}} = \bar{\mathbf{D}} : \boldsymbol{\sigma}^o$; see (7.1.12) and (7.6.11b). *For comparison purposes, let* $\bar{\mathbf{C}}$ *and* $\bar{\mathbf{D}}$ *be the overall elasticity and compliance tensors defined for the average strain and stress, i.e., those given by* (7.2.8) *and* (7.1.12), *respectively*. As noted, these tensors may not be symmetric. Let $\bar{\mathbf{C}}^{(s)}$ and $\bar{\mathbf{D}}^{(s)}$ be defined by

$$\bar{\mathbf{C}}^{(s)} \equiv sym\,(\bar{\mathbf{C}}), \qquad \bar{\mathbf{D}}^{(s)} \equiv sym\,(\bar{\mathbf{D}}). \tag{7.6.14a,b}$$

Then, the overall elasticity and compliance tensors based on the overall elastic and complementary energies, i.e., those given by (7.6.6b) and (7.6.11b), are equal to $\bar{\mathbf{C}}^{(s)}$ and $\bar{\mathbf{D}}^{(s)}$, respectively.

As shown in Subsections 7.4.3 and 7.5.3, the $\bar{\mathbf{C}}$- and $\bar{\mathbf{D}}$-tensors satisfy certain equivalence relations: for the dilute distribution,

$$\bar{\mathbf{C}} : \bar{\mathbf{D}} = \mathbf{1}^{(4s)} + O(f^2), \qquad \bar{\mathbf{D}} : \bar{\mathbf{C}} = \mathbf{1}^{(4s)} + O(f^2); \tag{7.6.15a,b}$$

and for the self-consistent method,

$$\bar{\mathbf{C}} : \bar{\mathbf{D}} = \mathbf{1}^{(4s)}, \qquad \bar{\mathbf{D}} : \bar{\mathbf{C}} = \mathbf{1}^{(4s)}. \tag{7.6.16a,b}$$

From (7.6.14a,b), $\bar{\mathbf{C}}^{(s)} : \bar{\mathbf{D}}^{(s)}$ and $\bar{\mathbf{D}}^{(s)} : \bar{\mathbf{C}}^{(s)}$ can be calculated directly. Let $\bar{\mathbf{C}}^{(a)}$ and $\bar{\mathbf{D}}^{(a)}$ be the antisymmetric parts of $\bar{\mathbf{C}}$ and $\bar{\mathbf{D}}$, i.e., $\bar{\mathbf{C}}^{(a)} \equiv \bar{\mathbf{C}} - \bar{\mathbf{C}}^{(s)}$ and $\bar{\mathbf{D}}^{(a)} \equiv \bar{\mathbf{D}} - \bar{\mathbf{D}}^{(s)}$. Then,

$$\bar{\mathbf{C}}^{(s)} : \bar{\mathbf{D}}^{(s)} = \bar{\mathbf{C}} : \bar{\mathbf{D}} - \{\bar{\mathbf{C}}^{(a)} : \bar{\mathbf{D}}^{(s)} + \bar{\mathbf{C}}^{(s)} : \bar{\mathbf{D}}^{(a)} + \bar{\mathbf{C}}^{(a)} : \bar{\mathbf{D}}^{(a)}\},$$

$$\bar{\mathbf{D}}^{(s)} : \bar{\mathbf{C}}^{(s)} = \bar{\mathbf{D}} : \bar{\mathbf{C}} - \{\bar{\mathbf{D}}^{(a)} : \bar{\mathbf{C}}^{(s)} + \bar{\mathbf{D}}^{(s)} : \bar{\mathbf{C}}^{(a)} + \bar{\mathbf{D}}^{(a)} : \bar{\mathbf{C}}^{(a)}\}. \tag{7.6.17a,b}$$

As is seen, tensors $\bar{\mathbf{C}}^{(s)}$ and $\bar{\mathbf{D}}^{(s)}$ may not satisfy the same equivalence relations as $\bar{\mathbf{C}}$ and $\bar{\mathbf{D}}$, if $\bar{\mathbf{C}}^{(a)}$ and $\bar{\mathbf{D}}^{(a)}$ are non-zero. Therefore, in general, the overall elasticity and compliance tensors estimated from the energy-based definitions may not be each other's inverse. The inconsistency here is a direct result of the approximation involved in *estimating* the average strain or stress in the inclusions.

For the dilute distribution, however, $\bar{\mathbf{C}}^{(a)}$ and $\bar{\mathbf{D}}^{(a)}$ are zero. This is because the Eshelby tensor $\mathbf{S}^\alpha$ satisfies[9]

$$(\mathbf{C} : \mathbf{S}^\alpha)^T = \mathbf{C} : \mathbf{S}^\alpha, \qquad (\mathbf{S}^\alpha : \mathbf{D})^T = \mathbf{S}^\alpha : \mathbf{D}. \tag{7.6.18a,b}$$

Since $\mathbf{C}$ and $\mathbf{D}$ are symmetric, the tensors inside the parentheses of (7.6.7b) and (7.6.12c) are symmetric. Hence, $\bar{\mathbf{C}}^{(s)}$ and $\bar{\mathbf{D}}^{(s)}$ are equal to $\bar{\mathbf{C}}$ and $\bar{\mathbf{D}}$, and are each other's inverse up to the first order in the volume fraction of micro-inclusions.

However, the symmetric parts of the overall tensors estimated by the self-consistent method may not be each other's inverse. Even though the Eshelby tensor $\bar{\mathbf{S}}^\alpha$ is such that $\bar{\mathbf{C}} : \bar{\mathbf{S}}^\alpha$ and $\bar{\mathbf{S}}^\alpha : \bar{\mathbf{D}}$ are symmetric, as shown in (7.6.18a,b), the tensors inside the parentheses of (7.6.8b) and (7.6.13c) are not

---

[9] The proof is given in Section 11.

necessarily symmetric. Hence, from (7.6.8b) it follows that

$$\mathbf{D} - \overline{\mathbf{C}}^{(s)-1} = \mathbf{D} : sym\left\{ \sum_{\alpha=1}^{n} f_\alpha (\mathbf{C}^\alpha - \mathbf{C}) : \overline{\mathbf{A}}^\alpha : (\overline{\mathbf{A}}^\alpha - \overline{\mathbf{S}}^\alpha)^{-1} \right\} : \overline{\mathbf{C}}^{(s)-1}$$

$$\neq sym\left\{ \mathbf{D} : \sum_{\alpha=1}^{n} f_\alpha (\mathbf{C}^\alpha - \mathbf{C}) : \overline{\mathbf{A}}^\alpha : (\overline{\mathbf{A}}^\alpha - \overline{\mathbf{S}}^\alpha)^{-1} : \overline{\mathbf{C}}^{(s)-1} \right\}, \quad (7.6.19a)$$

and from (7.6.13b),

$$\mathbf{C} - \overline{\mathbf{D}}^{(s)-1} = \mathbf{C} : sym\left\{ \sum_{\alpha=1}^{n} f_\alpha (\mathbf{D}^\alpha - \mathbf{D}) : \overline{\mathbf{B}}^\alpha : (\overline{\mathbf{B}}^\alpha - \overline{\mathbf{T}}^\alpha)^{-1} \right\} : \overline{\mathbf{D}}^{(s)-1}$$

$$\neq sym\left\{ \mathbf{C} : \sum_{\alpha=1}^{n} f_\alpha (\mathbf{D}^\alpha - \mathbf{D}) : \overline{\mathbf{B}}^\alpha : (\overline{\mathbf{B}}^\alpha - \overline{\mathbf{T}}^\alpha)^{-1} : \overline{\mathbf{D}}^{(s)-1} \right\}. \quad (7.6.19b)$$

Comparison of (7.6.19a) with (7.6.13c) and (7.6.19b) with (7.6.8c), shows that $\overline{\mathbf{C}}^{(s)}$ and $\overline{\mathbf{D}}^{(s)}$ are not, in general, each other's inverse, even though $\overline{\mathbf{C}}$ and $\overline{\mathbf{D}}$ are; see Benveniste *et al.* (1991).

7.6.4. Certain Exact Identities Involving Overall Elastic Energy

In Subsections 7.6.1 and 7.6.2, the symmetric overall elasticity and compliance tensors are obtained, using the *exact* results of Subsections 7.1 and 7.2. These exact equations are derived directly from the average strain and stress in the RVE. On the other hand, from the evaluation of the average strain and complementary strain energies of the RVE, alternative *exact* equations for the overall elasticity and compliance tensors are derived, which are necessarily symmetric tensors. In this subsection, two exact identities are presented for the elastic energy of the RVE.

As a starting point, compute

$$< \boldsymbol{\sigma} : \boldsymbol{\varepsilon} >_M = < \boldsymbol{\varepsilon} : \mathbf{C} : \boldsymbol{\varepsilon} >_M = < \boldsymbol{\sigma} : \mathbf{D} : \boldsymbol{\sigma} >_M$$

$$= < \boldsymbol{\sigma} : \boldsymbol{\varepsilon} > - \sum_{\alpha=1}^{n} f_\alpha < \boldsymbol{\sigma} : \boldsymbol{\varepsilon} >_\alpha. \quad (7.6.20)$$

Then, consider the case when the macrostrain $\mathbf{E} = \boldsymbol{\varepsilon}^o$ is regarded prescribed. From (7.6.2b) and (7.6.3b), the right-hand side of (7.6.20) becomes

$$< \boldsymbol{\sigma} : \boldsymbol{\varepsilon} > - \sum_{\alpha=1}^{n} f_\alpha < \boldsymbol{\sigma} : \boldsymbol{\varepsilon} >_\alpha = \boldsymbol{\varepsilon}^o : \overline{\mathbf{C}} : \boldsymbol{\varepsilon}^o - \sum_{\alpha=1}^{n} f_\alpha < \boldsymbol{\varepsilon} : \mathbf{C}^\alpha : \boldsymbol{\varepsilon} >_\alpha, \quad (7.6.21)$$

and hence

$$\boldsymbol{\varepsilon}^o : \overline{\mathbf{C}} : \boldsymbol{\varepsilon}^o - < \boldsymbol{\varepsilon} : \mathbf{C} : \boldsymbol{\varepsilon} >_M = \sum_{\alpha=1}^{n} f_\alpha < \boldsymbol{\varepsilon} : \mathbf{C}^\alpha : \boldsymbol{\varepsilon} >_\alpha, \quad (7.6.22a)$$

where

$$< \boldsymbol{\varepsilon} : \mathbf{C} : \boldsymbol{\varepsilon} >_M = C_{ijkl} < \varepsilon_{ij}\varepsilon_{kl} >_M,$$

$$< \boldsymbol{\varepsilon} : \mathbf{C}^\alpha : \boldsymbol{\varepsilon} >_\alpha = C^\alpha_{ijkl} < \varepsilon_{ij}\varepsilon_{kl} >_\alpha. \tag{7.6.22b,c}$$

Next, consider the case when the macrostress $\boldsymbol{\Sigma} = \boldsymbol{\sigma}^o$ is prescribed. From (7.6.2a) and (7.6.3a), the right-hand side of (7.6.20) becomes

$$< \boldsymbol{\sigma} : \boldsymbol{\varepsilon} > - \sum_{\alpha=1}^{n} f_\alpha < \boldsymbol{\sigma} : \boldsymbol{\varepsilon} >_\alpha = \boldsymbol{\sigma}^o : \bar{\mathbf{D}} : \boldsymbol{\sigma}^o - \sum_{\alpha=1}^{n} f_\alpha < \boldsymbol{\sigma} : \mathbf{D}^\alpha : \boldsymbol{\sigma} >_\alpha, \tag{7.6.23}$$

and hence,

$$\boldsymbol{\sigma}^o : \bar{\mathbf{D}} : \boldsymbol{\sigma}^o - < \boldsymbol{\sigma} : \mathbf{D} : \boldsymbol{\sigma} >_M = \sum_{\alpha=1}^{n} f_\alpha < \boldsymbol{\sigma} : \mathbf{D}^\alpha : \boldsymbol{\sigma} >_\alpha, \tag{7.6.24a}$$

where

$$< \boldsymbol{\sigma} : \mathbf{D} : \boldsymbol{\sigma} >_M = D_{ijkl} < \sigma_{ij}\sigma_{kl} >_M, \qquad < \boldsymbol{\varepsilon} : \mathbf{D}^\alpha : \boldsymbol{\sigma} >_\alpha = D^\alpha_{ijkl} < \sigma_{ij}\sigma_{kl} >_\alpha. \tag{7.6.24b,c}$$

These two expressions, (7.6.21) and (7.6.23), are *exact*. As is seen from (7.6.22b,c) and (7.6.24b,c), the average quantities,[10] $< \boldsymbol{\varepsilon} \otimes \boldsymbol{\varepsilon} >_M$, $< \boldsymbol{\varepsilon} \otimes \boldsymbol{\varepsilon} >_\alpha$, $< \boldsymbol{\sigma} \otimes \boldsymbol{\sigma} >_M$, and $< \boldsymbol{\sigma} \otimes \boldsymbol{\sigma} >_\alpha$ must be estimated. For example, one may approximate as follows:

$$< \boldsymbol{\varepsilon} \otimes \boldsymbol{\varepsilon} >_\alpha = \bar{\boldsymbol{\varepsilon}}^\alpha \otimes \bar{\boldsymbol{\varepsilon}}^\alpha - < (\boldsymbol{\varepsilon} - \bar{\boldsymbol{\varepsilon}}^\alpha) \otimes (\boldsymbol{\varepsilon} - \bar{\boldsymbol{\varepsilon}}^\alpha) >_\alpha \approx \bar{\boldsymbol{\varepsilon}}^\alpha \otimes \bar{\boldsymbol{\varepsilon}}^\alpha. \tag{7.6.25a}$$

Then, for the prescribed macrostrain $\boldsymbol{\varepsilon}^o$, $< \boldsymbol{\varepsilon} \otimes \boldsymbol{\varepsilon} >_\alpha$ is estimated as

$$< \boldsymbol{\varepsilon} \otimes \boldsymbol{\varepsilon} >_\alpha \approx \{\mathbf{A}^\alpha : (\mathbf{A}^\alpha - \mathbf{S}^\alpha)^{-1} : \boldsymbol{\varepsilon}^o\} \otimes \{\mathbf{A}^\alpha : (\mathbf{A}^\alpha - \mathbf{S}^\alpha)^{-1} : \boldsymbol{\varepsilon}^o\}, \tag{7.6.25b}$$

by the dilute distribution model, or as

$$< \boldsymbol{\varepsilon} \otimes \boldsymbol{\varepsilon} >_\alpha \approx \{\bar{\mathbf{A}}^\alpha : (\bar{\mathbf{A}}^\alpha - \bar{\mathbf{S}}^\alpha)^{-1} : \boldsymbol{\varepsilon}^o\} \otimes \{\bar{\mathbf{A}}^\alpha : (\bar{\mathbf{A}}^\alpha - \bar{\mathbf{S}}^\alpha)^{-1} : \boldsymbol{\varepsilon}^o\}, \tag{7.6.25c}$$

by the self-consistent model. For the other average quantities, $< \boldsymbol{\varepsilon} \otimes \boldsymbol{\varepsilon} >_M$, $< \boldsymbol{\sigma} \otimes \boldsymbol{\sigma} >_M$, and $< \boldsymbol{\sigma} \otimes \boldsymbol{\sigma} >_\alpha$, similar approximations may be admissible.

By definition, the $\bar{\mathbf{C}}$- and $\bar{\mathbf{D}}$-tensors in the exact expressions (7.6.22a) and (7.6.24a) are symmetric, i.e., $\bar{C}_{ijkl} = \bar{C}_{klij}$, and $\bar{D}_{ijkl} = \bar{D}_{klij}$. However, these tensors may be unrelated to the symmetric overall tensors, $\bar{\mathbf{C}}^{(s)}$ and $\bar{\mathbf{D}}^{(s)}$, obtained in Subsections 7.6.1 and 7.6.2. The $\bar{\mathbf{C}}^{(s)}$- and $\bar{\mathbf{D}}^{(s)}$-tensors satisfy

$$\boldsymbol{\varepsilon}^o : (\bar{\mathbf{C}}^{(s)} : \boldsymbol{\varepsilon}^o) - \boldsymbol{\varepsilon}^o : < \mathbf{C} : \boldsymbol{\varepsilon} >_M = \sum_{\alpha=1}^{n} f_\alpha\, \boldsymbol{\varepsilon}^o : < \mathbf{C}^\alpha : \boldsymbol{\varepsilon} >_\alpha,$$

$$\boldsymbol{\sigma}^o : (\bar{\mathbf{D}}^{(s)} : \boldsymbol{\sigma}^o) - \boldsymbol{\sigma}^o : < \mathbf{D} : \boldsymbol{\sigma} >_M = \sum_{\alpha=1}^{n} f_\alpha\, \boldsymbol{\sigma}^o : < \mathbf{D}^\alpha : \boldsymbol{\sigma} >_\alpha. \tag{7.6.26a,b}$$

From a comparison of (7.6.26a) with (7.6.22a) and (7.6.26b) with (7.6.24a), it is seen that $\bar{\mathbf{C}}^{(s)}$ and $\bar{\mathbf{D}}^{(s)}$ (which are derived from the average strain and stress) are,

---

[10] In view of the linearity of the RVE, the average quantity $< \boldsymbol{\varepsilon} \otimes \boldsymbol{\varepsilon} >_M$ relates to $\boldsymbol{\varepsilon}^o \otimes \boldsymbol{\varepsilon}^o$, through a constant eighth-order tensor. Similar comments apply to $< \boldsymbol{\varepsilon} \otimes \boldsymbol{\varepsilon} >_\alpha$. Note that $< \boldsymbol{\sigma} \otimes \boldsymbol{\sigma} >_M$ and $< \boldsymbol{\sigma} \otimes \boldsymbol{\sigma} >_\alpha$ have similar relations to $\boldsymbol{\sigma}^o \otimes \boldsymbol{\sigma}^o$.

in general, different from $\bar{\mathbf{C}}$ and $\bar{\mathbf{D}}$ (which are derived from the average elastic and complementary energies). These differences are the direct consequence of the modeling approximations.

7.7. REFERENCES

Budiansky, B. (1965), On the elastic moduli of some heterogeneous materials, *J. Mech. Phys. Solids*, Vol. 13, 223-227.

Benveniste, Y., Dvorak, G. J., and Chen, T. (1991), On diagonal and elastic symmetry of the approximate effective stiffness tensor of heterogeneous media, *J. Mech. Phys. Solids*, Vol. 39, 927-946.

Eshelby, J. D. (1957), The determination of the elastic field of an ellipsoidal inclusion, and related problems, *Proc. R. Soc. London, Ser. A*, Vol. 241, 376-396.

Gubernatis, J. E. and Krumhansl, J. A. (1975), Macroscopic engineering properties of polycrystalline materials: Elastic properties, *J. Appl. Phys.*, Vol. 46, No. 5, 1875-1883.

Hardiman, N. J. (1954), Elliptic elastic inclusion in an infinite elastic plate, *Q. J. Mech. Appl. Math.*, Vol. 52, 226-230.

Hashin, Z. (1964), Theory of mechanical behaviour of heterogeneous media, *Appl. Mech. Rev.*, Vol. 17, 1-9.

Hashin, Z. (1983), Analysis of composite materials - A survey, *J. Appl. Mech.*, Vol. 50, 481-505.

Hashin, Z. and Shtrikman, S. (1962), A variational approach to the theory of the elastic behaviour of polycrystals, *J. Mech. Phys. Solids*, Vol. 10, 343-352.

Hershey, A. V. (1954), The elasticity of an isotropic aggregate of anisotropic cubic crystals, *J. Appl. Mech.*, Vol. 21, 236-241.

Hill, R. (1952), The elastic behaviour of a crystalline aggregate, *Proc. Phys. Soc., London, Sect. A*, Vol. 65, 349-354.

Hill, R. (1963), Elastic properties of reinforced solids: Some theoretical principles, *J. Mech. Phys. Solids*, Vol 11, 357-372.

Hill, R. (1965a), Theory of mechanical properties of fibre-strengthened materials - III: Self-consistent model, *J. Mech. Phys. Solids*, Vol. 13, 189-198.

Hill, R. (1965b), A self-consistent mechanics of composite materials, *J. Mech. Phys. Solids*, Vol. 13, 213-222.

Kerner, E. H. (1956), The elastic and thermo-elastic properties of composite media, *Proc. Phys. Soc.*, Vol. 69, No. 8B, 808-813.

Kneer, G. (1965), Über die Berechnung der Elastizitätsmoduln vielkristalliner Aggregate mit Textur, *Phys. Stat. Sol.*, Vol. 9, 825-838.

Korringa, J., (1973), Theory of elastic constants of heterogeneous media, *J. Math. Phys.,* Vol. 14, 509-513.

Kröner, E. (1958), Berechnung der elastischen Konstanten des Vielkristalls aus den Konstanten des Einkristalls, *Z. Phys.*, Vol. 151, 504-518.

Kröner, E. (1967), Elastic moduli of perfectly disordered composite materials, *J. Mech. Phys. Solids*, Vol. 15, No. 319.

Kröner, E. (1980), Graded and perfect disorder in random media elasticity, *J. Eng. Mech. Division*, Vol. 106, No. EM5, 889-914.

Morris, P. R. (1970), Elastic constants of polycrystals, *Int. J. Eng. Sci.*, Vol. 8, 49-61.

Morris, P. R. (1971), Iterative scheme for calculating polycrystal elastic constants, *Int. J. Engineering Science*, Vol. 9, 917-920.

Mura, T. (1987), *Micromechanics of defects in solids (2nd edition)*, Martinus Nijhoff Publishers, Dordrecht.

Peselnick, L. and Meister, R. (1965), Variational method for determining effective moduli of polycrystals: (A) hexagonal symmetry, (B) trigonal symmetry, *J. Appl. Phys.*, Vol. 36, 2879-2884.

Reuss, A. (1929), Berechnung der Fliessgrenze von Mischkristallen auf Grund der Plastizitätsbedingung für Einkristalle, *Z. Angew. Math. Mech.*, Vol. 9, 49-58.

Voigt, W. (1889), Über die Beziehung zwischen den beiden Elastizitätskonstanten isotroper Körper, *Wied. Ann. Physik*, Vol. 38, 573-587.

Voigt, W. (1910), *Lehrbuch der Kristallphysik*, Teubner, Leipzig.

Walpole, L. J. (1966a), On bounds for the overall elastic moduli of inhomogeneous systems - I, *J. Mech. Phys. Solids*, Vol. 14, 151-162.

Walpole, L. J. (1966b), On bounds for the overall elastic moduli of inhomogeneous systems - II, *J. Mech. Phys. Solids*, Vol. 14, 289-301.

Walpole, L. J. (1967), The elastic field of an inclusion in an anisotropic medium, *Proc. Roy. Soc. A*, Vol. 300, 270-289.

Walpole, L. J. (1970), Strengthening effects in elastic solids, *J. Mech. Phys. Solids*, Vol. 18, 343-358.

Walpole, L. J. (1981), Elastic behavior of composite materials: Theoretical foundations, *Advances in Applied Mechanics*, Vol. 21, 169-242.

Watt, J. P., Davies, G. F., and O'Connell, R. J. (1976), The elastic properties of composite materials, *Rev. Geophys. Space Phys.*, Vol. 14, 541-563.

Willis, J. R. (1964), Anisotropic elastic inclusion problems, *Q. J. Mech. Appl. Math.*, Vol. 17, 157-174.

Willis, J. R. (1977), Bounds and self-consistent estimates for the overall properties of anisotropic composites, *J. Mech. Phys. Solids*, Vol. 25, 185-202.

Willis, J. R. (1980), Relationships between derivations of the overall properties of composites by perturbation expansions and variational principles, in *Variational Methods in the Mechanics of Solids* (Nemat-Nasser, S., ed.), Pergamon Press, 59-66.

Willis, J. R. (1981), Variational and related methods for the overall properties of composites, *Advances in Applied Mechanics*, Vol. 21, 1-78.

Willis, J. R. (1982), Elasticity theory of composites, in *Mechanics of Solids*, The Rodney Hill 60th Anniversary Volume (Hopkins, H. G. and Sewell, M. J., eds.), Pergamon Press, Oxford, 653-686.

Wu, T. T. (1966), The effect of inclusion shape on the elastic moduli of a two-phase material, *Int. J. Solids Struc.*, Vol. 2, 1-8.

Zeller, R. and Dederichs, P. H. (1973), Elastic constant of polycrystals, *Phys. Status Solidi B*, Vol. 55, 831-842.

SECTION 8 EXAMPLES OF ELASTIC SOLIDS WITH ELASTIC MICRO-INCLUSIONS

In this section, several specific examples are worked out in some detail in order to illustrate the results of the preceding section.

8.1. RANDOM DISTRIBUTION OF SPHERICAL MICRO-INCLUSIONS

Suppose all micro-inclusions in an RVE are spherical (Figure 8.1.1), or they can be approximated as spheres. Assume that the matrix and the inclusions are both linearly elastic and isotropic, but do not have the same elastic parameters. If the distribution of the micro-inclusions is random (whether it is dilute or not), the overall response of the RVE is isotropic. Hence, in order to express the overall elasticity and compliance tensors, it suffices to obtain two independent overall elastic moduli as functions of the volume fraction of the inclusions and the elastic moduli of the matrix and the inclusions. As in Subsections 5.1, 6.4, and 6.6, a dilute distribution of inclusions is considered first, and the corresponding overall elastic moduli of the RVE are estimated. Then, the self-consistent method is used in order to take into account (in a certain approximate sense) the interaction effects, which are completely neglected by the assumption of a dilute distribution.

To this end, first observe that an isotropic fourth-order tensor can be expressed in terms of two *basic* isotropic fourth-order tensors (Hill, 1965a,b),

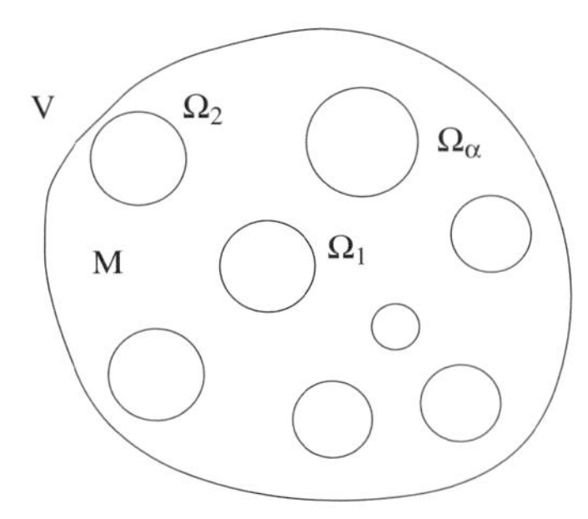

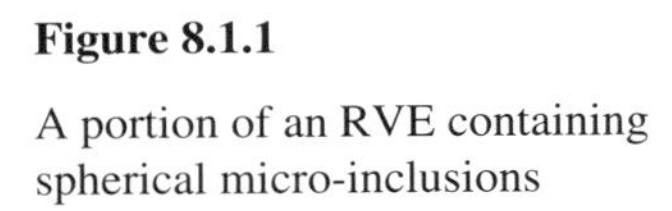
Figure 8.1.1

A portion of an RVE containing spherical micro-inclusions

$$\mathbf{E}^1 \equiv \frac{1}{3}\,\mathbf{1}^{(2)}\otimes\mathbf{1}^{(2)}, \qquad \mathbf{E}^2 \equiv -\frac{1}{3}\mathbf{1}^{(2)}\otimes\mathbf{1}^{(2)} + \mathbf{1}^{(4s)}. \tag{8.1.1a,b}$$

The tensors $\mathbf{E}^1$ and $\mathbf{E}^2$ satisfy

$$\mathbf{E}^1 : \mathbf{E}^1 = \mathbf{E}^1, \qquad \mathbf{E}^2 : \mathbf{E}^2 = \mathbf{E}^2, \qquad \mathbf{E}^1 : \mathbf{E}^2 = \mathbf{E}^2 : \mathbf{E}^1 = \mathbf{0}. \tag{8.1.1c~e}$$

Moreover, for arbitrary real numbers (a_1, a_2), (b_1, b_2), and (p, q), it follows that

$$p\,(a_1\,\mathbf{E}^1 + a_2\,\mathbf{E}^2) + q\,(b_1\,\mathbf{E}^1 + b_2\,\mathbf{E}^2) = (pa_1 + qb_1)\,\mathbf{E}^1 + (pa_2 + qb_2)\,\mathbf{E}^2,$$

$$(a_1\,\mathbf{E}^1 + a_2\,\mathbf{E}^2):(b_1\,\mathbf{E}^1 + b_2\,\mathbf{E}^2) = (a_1b_1)\,\mathbf{E}^1 + (a_2b_2)\,\mathbf{E}^2,$$

$$(a_1\,\mathbf{E}^1 + a_2\,\mathbf{E}^2)^{-1} = a_1^{-1}\,\mathbf{E}^1 + a_2^{-1}\,\mathbf{E}^2. \tag{8.1.1f~h}$$

These properties can be used to reduce certain tensorial operations on isotropic tensors to *scalar* operations on the corresponding coefficients of $\mathbf{E}^1$ and $\mathbf{E}^2$.

In terms of $\mathbf{E}^1$ and $\mathbf{E}^2$, the elasticity and compliance tensors of the matrix M and a typical micro-inclusion Ω_α, are expressed as

$$\mathbf{C} = 3K\,\mathbf{E}^1 + 2\mu\,\mathbf{E}^2, \qquad \mathbf{D} = \frac{1}{3K}\,\mathbf{E}^1 + \frac{1}{2\mu}\,\mathbf{E}^2, \tag{8.1.2a,b}$$

and

$$\mathbf{C}^\alpha = 3K^\alpha\,\mathbf{E}^1 + 2\mu^\alpha\,\mathbf{E}^2, \qquad \mathbf{D}^\alpha = \frac{1}{3K^\alpha}\,\mathbf{E}^1 + \frac{1}{2\mu^\alpha}\,\mathbf{E}^2, \tag{8.1.3a,b}$$

where K and μ are the bulk and shear moduli of M, and K^α and μ^α are the bulk and shear moduli of Ω_α. In a similar manner, the overall isotropic (for random distribution of inclusions) elasticity and compliance tensors, denoted by $\bar{\mathbf{C}}$ and $\bar{\mathbf{D}}$, are expressed as

$$\bar{\mathbf{C}} = 3\bar{K}\mathbf{E}^1 + 2\bar{\mu}\,\mathbf{E}^2, \qquad \bar{\mathbf{D}} = \frac{1}{3\bar{K}}\,\mathbf{E}^1 + \frac{1}{2\bar{\mu}}\,\mathbf{E}^2, \tag{8.1.4a,b}$$

where $\bar{K}$ and $\bar{\mu}$ are the overall bulk and shear moduli. Therefore, the tensorial equations involving $(\mathbf{C}, \mathbf{C}^\alpha, \bar{\mathbf{C}})$ or $(\mathbf{D}, \mathbf{D}^\alpha, \bar{\mathbf{D}})$ are reduced to scalar equations for $(K, K^\alpha, \bar{K})$ and $(\mu, \mu^\alpha, \bar{\mu})$.

8.1.1. Effective Moduli: Dilute Distribution of Spherical Inclusions

For a dilute distribution of micro-inclusions, a typical micro-inclusion Ω_α is embedded in an *unbounded* homogeneous solid with the elasticity and compliance tensors $\mathbf{C}$ and $\mathbf{D}$. When all micro-inclusions are spherical, the common Eshelby tensor and its conjugate are given by

$$\mathbf{S}^\alpha \equiv \mathbf{S}^I = s_1\,\mathbf{E}^1 + s_2\,\mathbf{E}^2, \qquad \mathbf{T}^\alpha \equiv \mathbf{T}^I = (1 - s_1)\,\mathbf{E}^1 + (1 - s_2)\,\mathbf{E}^2, \tag{8.1.5a,b}$$

where the coefficients s_1 and s_2 are functions of the Poisson ratio, $\nu \equiv (3K - 2\mu)/2(3K + \mu)$, of the matrix M,

$$s_1 = \frac{1 + \nu}{3(1 - \nu)}, \qquad s_2 = \frac{2(4 - 5\nu)}{15(1 - \nu)}. \tag{8.1.5c,d}$$

Since in the present case the moduli expressed in terms of $\mathbf{T}^I$ are identical with those expressed in terms of $\mathbf{S}^I$, hereinafter only the expressions in terms of $\mathbf{S}^I$ are employed.

First consider the case when the macrostress $\boldsymbol{\Sigma} = \boldsymbol{\sigma}^o$ is prescribed. From (7.4.6a) the overall compliance tensor $\overline{\mathbf{D}}$ is given by

$$\overline{\mathbf{D}} = \{\mathbf{1}^{(4s)} + \sum_{\alpha=1}^{n} f_\alpha\,(\mathbf{A}^\alpha - \mathbf{S}^I)^{-1}\} : \mathbf{D}, \tag{8.1.6a}$$

where

$$\mathbf{A}^\alpha \equiv (\mathbf{C} - \mathbf{C}^\alpha)^{-1} : \mathbf{C} = \frac{K}{K - K^\alpha}\,\mathbf{E}^1 + \frac{\mu}{\mu - \mu^\alpha}\,\mathbf{E}^2. \tag{8.1.6b}$$

Then, in terms of $\mathbf{E}^1$ and $\mathbf{E}^2$, $\overline{\mathbf{D}}$ is rewritten as

$$\overline{\mathbf{D}} = \frac{1}{3K}\,\{1 + \sum_{\alpha=1}^{n} f_\alpha\,(\frac{K}{K - K^\alpha} - s_1)^{-1}\}\,\mathbf{E}^1$$

$$+ \frac{1}{2\mu}\,\{1 + \sum_{\alpha=1}^{n} f_\alpha\,(\frac{\mu}{\mu - \mu^\alpha} - s_2)^{-1}\}\,\mathbf{E}^2. \tag{8.1.6c}$$

Hence, from (8.1.4b) and (8.1.6c), the overall bulk modulus, $\overline{K}$, and shear modulus, $\overline{\mu}$, become

$$\frac{\overline{K}}{K} = \{1 + \sum_{\alpha=1}^{n} f_\alpha\,(\frac{K}{K - K^\alpha} - s_1)^{-1}\}^{-1} = 1 - \sum_{\alpha=1}^{n} f_\alpha\,(\frac{K}{K - K^\alpha} - s_1)^{-1} + \ldots,$$

$$\frac{\overline{\mu}}{\mu} = \{1 + \sum_{\alpha=1}^{n} f_\alpha\,(\frac{\mu}{\mu - \mu^\alpha} - s_2)^{-1}\}^{-1} = 1 - \sum_{\alpha=1}^{n} f_\alpha\,(\frac{\mu}{\mu - \mu^\alpha} - s_2)^{-1} + \ldots\,. \tag{8.1.7a,b}$$

Next consider the case when the macrostrain $\mathbf{E} = \boldsymbol{\varepsilon}^o$ is prescribed. From (7.4.9a), the overall elasticity tensor $\overline{\mathbf{C}}$ is expressed in terms of $\mathbf{E}^1$ and $\mathbf{E}^2$, as

$$\overline{\mathbf{C}} = \mathbf{C} : \{\mathbf{1}^{(4s)} - \sum_{\alpha=1}^{n} f_\alpha\,(\mathbf{A}^\alpha - \mathbf{S}^I)^{-1}\}$$

$$= 3K\,\{1 - \sum_{\alpha=1}^{n} f_\alpha\,(\frac{K}{K - K^\alpha} - s_1)^{-1}\}\,\mathbf{E}^1$$

$$+ 2\mu\,\{1 - \sum_{\alpha=1}^{n} f_\alpha\,(\frac{\mu}{\mu - \mu^\alpha} - s_2)^{-1}\}\,\mathbf{E}^2. \tag{8.1.8}$$

Hence, from (8.1.4a) and (8.1.8), the overall bulk, $\overline{K}$, and shear, $\overline{\mu}$, moduli become

$$\frac{\overline{K}}{K} = 1 - \sum_{\alpha=1}^{n} f_\alpha\,(\frac{K}{K - K^\alpha} - s_1)^{-1},$$

$$\frac{\overline{\mu}}{\mu} = 1 - \sum_{\alpha=1}^{n} f_\alpha\,(\frac{\mu}{\mu - \mu^\alpha} - s_2)^{-1}. \tag{8.1.9a,b}$$

As is seen from (8.1.7a,b) and (8.1.9a,b), the overall bulk modulus given by (8.1.7a) and (8.1.9a) and shear modulus given by (8.1.7b) and (8.1.9b), agree to the first order in the volume fraction of micro-inclusions.

As a special case, let all micro-inclusions have the same elasticity, with the common bulk and shear moduli, K^I and μ^I. If the macrostress is regarded prescribed, (8.1.7a,b) yield

$$\frac{\overline{K}}{K} = \{1+f(\frac{K}{K-K^I} - s_1)^{-1}\}^{-1}, \qquad \frac{\overline{\mu}}{\mu} = \{1+f(\frac{\mu}{\mu-\mu^I} - s_2)^{-1}\}^{-1}, \tag{8.1.10a,b}$$

and if the macrostrain is regarded prescribed, (8.1.9a,b) yield

$$\frac{\overline{K}}{K} = 1-f(\frac{K}{K-K^I} - s_1)^{-1}, \qquad \frac{\overline{\mu}}{\mu} = 1-f(\frac{\mu}{\mu-\mu^I} - s_2)^{-1}, \tag{8.1.11a,b}$$

where f is the volume fraction of micro-inclusions.

8.1.2. Effective Moduli: Self-Consistent Estimates

If the distribution of micro-inclusions is random and the interaction effects are to be included to a certain extent, then the self-consistent method may be used to estimate the overall response of the RVE. Since each micro-inclusion is assumed to be embedded in an *unbounded* solid which has the unknown overall elasticity and compliance tensors, $\overline{\mathbf{C}}$ and $\overline{\mathbf{D}}$, the Eshelby tensor and its conjugate become

$$\overline{\mathbf{S}}^\alpha \equiv \overline{\mathbf{S}}^I = \overline{s}_1\,\mathbf{E}^1 + \overline{s}_2\,\mathbf{E}^2, \qquad \overline{\mathbf{T}}^\alpha \equiv \overline{\mathbf{T}}^I = (1-\overline{s}_1)\,\mathbf{E}^1 + (1-\overline{s}_2)\,\mathbf{E}^2, \tag{8.1.12a,b}$$

where $\overline{s}_1$ and $\overline{s}_2$ are defined in terms of the overall Poisson ratio, $\overline{\nu} \equiv (3\overline{K}-2\overline{\mu})/2(3\overline{K}+\overline{\mu})$, by

$$\overline{s}_1 = \frac{1+\overline{\nu}}{3(1-\overline{\nu})}, \qquad \overline{s}_2 = \frac{2(4-5\overline{\nu})}{15(1-\overline{\nu})}. \tag{8.1.12c,d}$$

Again, since the formulation in terms of $\overline{\mathbf{T}}^I$ is equivalent to that in terms of $\overline{\mathbf{S}}^I$, only the expression involving $\overline{\mathbf{S}}^I$ is employed.

Now, using the results obtained in Subsections 7.5.1 and 7.5.2, the overall elasticity and compliance tensors are obtained, which are each other's inverse. From (7.5.4a), the overall compliance tensor $\overline{\mathbf{D}}$ is given by

$$\overline{\mathbf{D}} = \mathbf{D} + \sum_{\alpha=1}^{n} f_\alpha\,(\mathbf{D}^\alpha - \mathbf{D}) : \mathbf{C}^\alpha : \overline{\mathbf{A}}^\alpha : (\overline{\mathbf{A}}^\alpha - \overline{\mathbf{S}}^I)^{-1} : \overline{\mathbf{D}}, \tag{8.1.13a}$$

where

$$\overline{\mathbf{A}}^\alpha \equiv (\overline{\mathbf{C}} - \mathbf{C}^\alpha)^{-1} : \overline{\mathbf{C}} = \frac{\overline{K}}{\overline{K}-K^\alpha}\,\mathbf{E}^1 + \frac{\overline{\mu}}{\overline{\mu}-\mu^\alpha}\,\mathbf{E}^2. \tag{8.1.13b}$$

Hence, $\overline{\mathbf{D}}$ in terms of $\mathbf{E}^1$ and $\mathbf{E}^2$ becomes

$$\overline{\mathbf{D}} = \frac{1}{3}\{\frac{1}{K} + \sum_{\alpha=1}^{n} f_\alpha\,\frac{1}{K}\,\frac{K-K^\alpha}{\overline{K}-K^\alpha}\,(\frac{\overline{K}}{\overline{K}-K^\alpha} - \overline{s}_1)^{-1}\}\,\mathbf{E}^1$$

$$+\frac{1}{2}\{\frac{1}{\mu} + \sum_{\alpha=1}^{n} f_\alpha\,\frac{1}{\mu}\,\frac{\mu-\mu^\alpha}{\overline{\mu}-\mu^\alpha}\,(\frac{\overline{\mu}}{\overline{\mu}-\mu^\alpha} - \overline{s}_2)^{-1}\}\,\mathbf{E}^2. \tag{8.1.13c}$$

Or, from (7.5.7a), the overall elasticity tensor $\overline{\mathbf{C}}$ is expressed as

$$\overline{\mathbf{C}} = \mathbf{C} + \sum_{\alpha=1}^{n} f_\alpha\,(\mathbf{C}^\alpha - \mathbf{C}) : \overline{\mathbf{A}}^\alpha : (\overline{\mathbf{A}}^\alpha - \overline{\mathbf{S}}^I)^{-1}$$

$$= 3\{K + \sum_{\alpha=1}^{n} f_\alpha\,\overline{K}\,\frac{K^\alpha - K}{\overline{K} - K^\alpha}\,(\frac{\overline{K}}{\overline{K} - K^\alpha} - \overline{s}_1)^{-1}\}\,\mathbf{E}^1$$

$$+ 2\{\mu + \sum_{\alpha=1}^{n} f_\alpha\,\overline{\mu}\,\frac{\mu^\alpha - \mu}{\overline{\mu} - \mu^\alpha}\,(\frac{\overline{\mu}}{\overline{\mu} - \mu^\alpha} - \overline{s}_2)^{-1}\}\,\mathbf{E}^2. \tag{8.1.13d}$$

Either from (8.1.13c) or (8.1.13d), the overall bulk modulus, $\overline{K}$, and shear modulus, $\overline{\mu}$, are given by

$$\frac{\overline{K}}{K} = 1 + \sum_{\alpha=1}^{n} f_\alpha\,(\frac{K^\alpha}{K} - 1)\,\{1 + (\frac{K^\alpha}{\overline{K}} - 1)\,\overline{s}_1\}^{-1},$$

$$\frac{\overline{\mu}}{\mu} = 1 + \sum_{\alpha=1}^{n} f_\alpha\,(\frac{\mu^\alpha}{\mu} - 1)\,\{1 + (\frac{\mu^\alpha}{\overline{\mu}} - 1)\,\overline{s}_2\}^{-1}. \tag{8.1.14a,b}$$

As discussed in Subsection 7.5.4, the matrix of an RVE with n spherical micro-inclusions may be regarded as an "inclusion" in the self-consistent method. Since all micro-inclusions have the common Eshelby tensor, $\mathbf{S}^I$, from (7.5.16a) it follows that

$$\sum_{\alpha=1}^{n} f_\alpha\,\{1 + \overline{s}_1\,(\frac{K^\alpha}{\overline{K}} - 1)\}^{-1}\,\mathbf{E}^1 + \sum_{\alpha=1}^{n} f_\alpha\,\{1 + \overline{s}_2\,(\frac{\mu^\alpha}{\overline{\mu}} - 1)\}^{-1}\,\mathbf{E}^2 = \mathbf{E}^1 + \mathbf{E}^2. \tag{8.1.15}$$

Therefore, the overall bulk modulus, $\overline{K}$, and shear modulus, $\overline{\mu}$, satisfy the following equations:

$$\sum_{\alpha=1}^{n} f_\alpha\,\{1 + \overline{s}_1\,(\frac{K^\alpha}{\overline{K}} - 1)\}^{-1} = 1,$$

$$\sum_{\alpha=1}^{n} f_\alpha\,\{1 + \overline{s}_2\,(\frac{\mu^\alpha}{\overline{\mu}} - 1)\}^{-1} = 1. \tag{8.1.16a,b}$$

These coincide with Budiansky's results (1965).

When all micro-inclusions consist of the same material, denote their common bulk and shear moduli by K^I and μ^I, and from (8.1.14a,b) obtain

$$\frac{\overline{K}}{K} = 1 - f\,\frac{\overline{K}(K - K^I)}{K(\overline{K} - K^I)}\,(\frac{\overline{K}}{\overline{K} - K^I} - \overline{s}_1)^{-1},$$

$$\frac{\overline{\mu}}{\mu} = 1 - f\,\frac{\overline{\mu}(\mu - \mu^I)}{\mu(\overline{\mu} - \mu^I)}\,(\frac{\overline{\mu}}{\overline{\mu} - \mu^I} - \overline{s}_2)^{-1}. \tag{8.1.17a,b}$$

It is noted that although $\overline{K}$ and $\overline{\mu}$ given by (8.1.10a,b) or (8.1.11a,b) are decoupled, $\overline{K}$ and $\overline{\mu}$ given by (8.1.17a,b) are coupled, since $\overline{s}_1$ and $\overline{s}_2$, the coefficients of Eshelby's tensor $\overline{\mathbf{S}}^I$, are determined by the unknown overall Poisson ratio, $\overline{\nu} \equiv (3\overline{K} - 2\overline{\mu})/2(3\overline{K} + \overline{\mu})$.

Figure 8.1.2 shows the graph of the overall bulk and shear moduli of an RVE containing spherical micro-inclusions of a common elasticity, where $\mu^I/\mu = 50$, and $\nu^I = \nu = 0.3$. Solid curves indicate the self-consistent estimate, given by (8.1.17a,b), and dotted curves are the results for a dilute distribution, given by (8.1.10a,b) and (8.1.11a,b). For small f, these curves coincide; (8.1.11a) and (8.1.11b), respectively, are asymptotic expressions for $\bar{K}/K$ and $\bar{\mu}/\mu$, as f goes to zero.

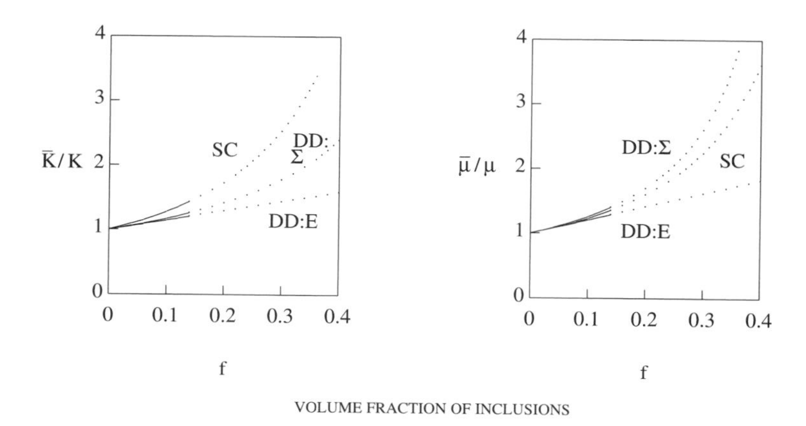

Figure 8.1.2

Overall bulk and shear moduli of an RVE with randomly distributed spherical inclusions; $K^I/K = \mu^I/\mu = 50$ and $\nu^I = \nu = 1/3$.
SC ≡ self-consistent
DD:Σ ≡ dilute distribution with macrostress prescribed
DD:E ≡ dilute distribution with macrostrain prescribed

8.2. EFFECTIVE MODULI OF AN ELASTIC PLATE CONTAINING ALIGNED REINFORCING-FIBERS

In this subsection, composite materials are considered which consist of a linearly elastic matrix reinforced by linearly elastic long and stiff microfibers. The microfibers are aligned in, say, the x_3-direction, having a random distribution in the x_1,x_2-plane. Hence, the composite is, for example, stiffer in the x_3-direction than in the x_1- and x_2-directions, when the fibers are stiffer than the matrix. The microfibers may be approximated as long circular cylinders, and, in this manner this type of composite material is modeled by an elastic RVE which contains a random distribution of infinitely long aligned microfibers. The aim then is to estimate the overall elastic parameters of this RVE in plane strain and antiplane shear. Because of the random distribution of the fibers parallel to the x_3-direction, the composite is transversely isotropic.

For simplicity, assume that both the matrix and the reinforcing microfibers are homogeneous, linearly elastic, and isotropic. Then, the analysis of the two-dimensional inplane and antiplane deformations of the composite can be based on the Eshelby tensor **S** for an isolated infinitely long circular cylindrical inclusion embedded in an unbounded homogeneous solid. In the self-consistent model, the estimate of the average stress and strain in a typical inclusion is affected by the anisotropy which results from randomly distributed but aligned fibers. Therefore, the inplane overall elastic moduli obtained by the self-consistent model are coupled with the overall moduli associated with the x_3-direction; see Subsection 5.1. As pointed out before, even for the self-consistent method, the validity of the results is limited to rather small values of the volume fraction of inhomogeneities. Therefore, in the case of the aligned but randomly distributed fibers, the induced anisotropy, because of the aligned fibers, should be rather small, unless there is considerable mismatch between the fiber and matrix stiffness. In the case of a small anisotropy, instead of a transversely isotropic overall response, one may ignore the anisotropy, and hence the coupling, treating the problem as a pseudo-isotropic two-dimensional one.

On the other hand, as in Section 5, by the introduction of the parameter κ, the three-dimensional formulation can be reduced to the corresponding two-dimensional counterpart exactly, and then the two-dimensional results may be examined appropriately by a proper interpretation of the overall effective $\bar{\kappa}$.

To this end, the necessary two-dimensional Eshelby tensor is obtained from that associated with a circular cylindrical inclusion, with its generator parallel to the x_3-axis, embedded in an infinite, *transversely isotropic,* linearly elastic material. Regarding a circular cylindrical inclusion as a limit of an ellipsoid with $a_1 = a_2$ and $a_1/a_3 \equiv a_2/a_3 \to 0$, the nonzero components of Eshelby's tensor, $\bar{\mathbf{S}}$, for this *homogenized* infinite material are as follows:

$$\bar{S}_{1111} = \bar{S}_{2222} \to \frac{2+\bar{\kappa}}{2(1+\bar{\kappa})}, \quad \bar{S}_{1122} = \bar{S}_{2211} \to \frac{2-\bar{\kappa}}{2(1+\bar{\kappa})}, \quad \bar{S}_{1212} \to \frac{\bar{\kappa}}{2(1+\bar{\kappa})},$$

$$\bar{S}_{3311} = \bar{S}_{3322} = \bar{S}_{333} \to 0, \quad \bar{S}_{1313} = \bar{S}_{2323} \to \frac{1}{4}, \qquad (8.2.1a\sim e)$$

where $\bar{\kappa} = 3 - 8\bar{\mu}(\bar{\nu}/\bar{E} + \bar{\nu}_3^2/\bar{E}_3)$ with $\bar{E}$, $\bar{\nu}$, and $\bar{\mu}$ being the effective inplane Young modulus, Poisson ratio, and shear modulus, and $\bar{E}_3$ and $\bar{\nu}_3$ being the effective Young modulus in the x_3-direction and Poisson ratio in the x_1,x_3- and x_2,x_3-directions; and $\bar{S}_{ijkl} = \bar{S}_{jikl} = \bar{S}_{ijlk}$. If the infinite body is isotropic,[1] κ is replaced by $3-4\nu$, with ν being the Poisson ratio, and these coefficients become

$$S_{1111} = S_{2222} \to \frac{5-4\nu}{8(1-\nu)}, \quad S_{1122} = S_{2211} \to \frac{1-4\nu}{8(1-\nu)}, \quad S_{1212} \to \frac{3-4\nu}{8(1-\nu)},$$

$$S_{1133} = S_{2233} = S_{3333} \to 0, \quad S_{1313} = S_{2323} \to \frac{1}{4}, \qquad (8.2.1f\sim j)$$

where $S_{ijkl} = S_{jikl} = S_{ijlk}$.

---

[1] (8.2.1f~j) can be obtained directly from the components of the Eshelby tensor for an isotropic material. In particular, $S_{1133} = S_{2233} \to 0$.

In this subsection, a plane strain state is assumed.[2] Hence, in view of (8.2.1a~d), $\bar{\mathbf{S}}^{\mathrm{I}}$ is defined by

$$\bar{S}^{\mathrm{I}}_{ijkl} = \frac{2-\bar{\kappa}}{2(1+\bar{\kappa})}\,\delta_{ij}\,\delta_{kl} + \frac{\bar{\kappa}}{1+\bar{\kappa}}\,\frac{1}{2}(\delta_{ik}\,\delta_{jl} + \delta_{il}\,\delta_{jk}) \qquad (i, j, k, l = 1, 2), \tag{8.2.2a}$$

and for antiplane shear,

$$\bar{S}^{\mathrm{I}}_{1313} = \bar{S}^{\mathrm{I}}_{3113} = \bar{S}^{\mathrm{I}}_{1331} = \bar{S}^{\mathrm{I}}_{3131} = \bar{S}^{\mathrm{I}}_{2323} = \bar{S}^{\mathrm{I}}_{3223} = \bar{S}^{\mathrm{I}}_{2332} = \bar{S}^{\mathrm{I}}_{3232} = \frac{1}{4},$$

$$\bar{S}^{\mathrm{I}}_{1323} = \bar{S}^{\mathrm{I}}_{3123} = \bar{S}^{\mathrm{I}}_{1332} = \bar{S}^{\mathrm{I}}_{3132} = \bar{S}^{\mathrm{I}}_{2313} = \bar{S}^{\mathrm{I}}_{3213} = \bar{S}^{\mathrm{I}}_{2331} = \bar{S}^{\mathrm{I}}_{3231} = 0. \tag{8.2.3a,b}$$

For an isotropic case, in view of (8.2.1f~j), define

$$S^{\mathrm{I}}_{ijkl} = \frac{2-\kappa}{2(1+\kappa)}\,\delta_{ij}\,\delta_{kl} + \frac{\kappa}{1+\kappa}\,\frac{1}{2}(\delta_{ik}\,\delta_{jl} + \delta_{il}\,\delta_{jk}) \qquad (i, j, k, l = 1, 2), \tag{8.2.2b}$$

where $\kappa = 3 - 4\nu$, and for antiplane shear, obtain

$$S^{\mathrm{I}}_{1313} = S^{\mathrm{I}}_{3113} = S^{\mathrm{I}}_{1331} = S^{\mathrm{I}}_{3131} = S^{\mathrm{I}}_{2323} = S^{\mathrm{I}}_{3223} = S^{\mathrm{I}}_{2332} = S^{\mathrm{I}}_{3232} = \frac{1}{4},$$

$$S^{\mathrm{I}}_{1323} = S^{\mathrm{I}}_{3123} = S^{\mathrm{I}}_{1332} = S^{\mathrm{I}}_{3132} = S^{\mathrm{I}}_{2313} = S^{\mathrm{I}}_{3213} = S^{\mathrm{I}}_{2331} = S^{\mathrm{I}}_{3231} = 0. \tag{8.2.3c,d}$$

In this subsection the two-dimensional overall properties of the RVE are examined, i.e., the effects of $\bar{S}^{\mathrm{I}}_{1133}$ and $\bar{S}^{\mathrm{I}}_{2233}$ or S^{I}_{1133} and S^{I}_{2233} are excluded.

Since the reinforcing microfibers are randomly distributed in the x_1,x_2-plane, the overall response of the RVE is isotropic in this plane. Hence, all the tensors (i.e., $\mathbf{C}$, $\mathbf{C}^{\mathrm{I}}$, $\bar{\mathbf{C}}$, $\mathbf{S}^{\mathrm{I}}$, etc.) involved in this particular setting are two-dimensionally isotropic. In Subsection 8.1, two base tensors, $\mathbf{E}^1$ and $\mathbf{E}^2$, are introduced in terms of $\mathbf{1}^{(2)} \otimes \mathbf{1}^{(2)}$ and $\mathbf{1}^{(4s)}$ in (8.1.1a,b), in order to express *three-dimensional isotropic tensors* and to reduce tensorial manipulations of these isotropic tensors to simple scalar operations on the coefficients of $\mathbf{E}^1$ and $\mathbf{E}^2$. Since a *two-dimensional isotropic tensor* $\mathbf{T}$ is expressed as

$$T_{ijkl} = T_1\,\delta_{ij}\,\delta_{kl} + T_2\,\frac{1}{2}(\delta_{ik}\,\delta_{jl} + \delta_{il}\,\delta_{jk}) \qquad (i, j, k, l = 1, 2), \tag{8.2.4}$$

define $\mathbf{E}^1$ and $\mathbf{E}^2$ in terms of $\delta_{ij}\,\delta_{kl}$ and $(\delta_{ik}\,\delta_{jl} + \delta_{il}\,\delta_{jk})/2$, by

$$E^1_{ijkl} \equiv \frac{1}{2}\delta_{ij}\,\delta_{kl}, \quad E^2_{ijkl} \equiv -\frac{1}{2}\delta_{ij}\,\delta_{kl} + \frac{1}{2}(\delta_{ik}\,\delta_{jl} + \delta_{il}\,\delta_{jk}). \tag{8.2.5a,b}$$

Note that the coefficients of $\delta_{ij}\,\delta_{kl}$ in $\mathbf{E}^1$ and $\mathbf{E}^2$ are $\pm 1/2$ for two-dimensional isotropic tensors, while they are $\pm 1/3$ for three-dimensional isotropic tensors. The tensors $\mathbf{E}^1$ and $\mathbf{E}^2$ satisfy

$$\mathbf{E}^1 : \mathbf{E}^1 = \mathbf{E}^1, \qquad \mathbf{E}^2 : \mathbf{E}^2 = \mathbf{E}^2, \qquad \mathbf{E}^1 : \mathbf{E}^2 = \mathbf{E}^2 : \mathbf{E}^1 = \mathbf{0}. \tag{8.2.5c~e}$$

---

[2] While a plane strain state is assumed for infinitely long fibers, the corresponding plane stress solution can be obtained by simple modification of the results; see Subsection 8.3.

In terms of $\mathbf{E}^1$ and $\mathbf{E}^2$, the two-dimensional isotropic elasticity and compliance tensors of the matrix material, denoted by $\mathbf{C}$ and $\mathbf{D}$, are expressed as

$$\mathbf{C} = 3K'\,\mathbf{E}^1 + 2\mu\mathbf{E}^2, \qquad \mathbf{D} = \frac{1}{3K'}\,\mathbf{E}^1 + \frac{1}{2\mu}\,\mathbf{E}^2, \tag{8.2.6a,b}$$

and those of the reinforcing microfibers, denoted by $\mathbf{C}^\mathrm{I}$ and $\mathbf{D}^\mathrm{I}$, as

$$\mathbf{C}^\mathrm{I} = 3K'^\mathrm{I}\,\mathbf{E}^1 + 2\mu^\mathrm{I}\,\mathbf{E}^2, \qquad \mathbf{D}^\mathrm{I} = \frac{1}{3K'^\mathrm{I}}\,\mathbf{E}^1 + \frac{1}{2\mu^\mathrm{I}}\,\mathbf{E}^2. \tag{8.2.7a,b}$$

Then $3K'/2$ is interpreted as the two-dimensional bulk modulus; the factor 3 in (8.2.6a,b) is introduced to make these equations the same as (8.1.2a,b). Here K' and K'^I are defined by

$$\frac{1}{3K'} \equiv \begin{cases} (1-\nu-2\nu^2)/E & \text{for plane strain} \\ (1-\nu)/E & \text{for plane stress,} \end{cases} \tag{8.2.6c}$$

and

$$\frac{1}{3K'^\mathrm{I}} \equiv \begin{cases} (1-\nu^\mathrm{I}-2(\nu^\mathrm{I})^2)/E^\mathrm{I} & \text{for plane strain} \\ (1-\nu^\mathrm{I})/E^\mathrm{I} & \text{for plane stress.} \end{cases} \tag{8.2.7c}$$

Note that the corresponding shear modulus does not change under plane strain or plane stress, i.e., $\mu = E/2(1+\nu)$ and $\mu^\mathrm{I} = E^\mathrm{I}/2(1+\nu^\mathrm{I})$.

In a similar manner, the two-dimensional overall elasticity and compliance tensors, denoted by $\bar{\mathbf{C}}$ and $\bar{\mathbf{D}}$, can be expressed in terms of $\mathbf{E}^1$ and $\mathbf{E}^2$, as

$$\bar{\mathbf{C}} = 3\bar{K}'\,\mathbf{E}^1 + 2\bar{\mu}\,\mathbf{E}^2, \qquad \bar{\mathbf{D}} = \frac{1}{3\bar{K}'}\,\mathbf{E}^1 + \frac{1}{2\bar{\mu}}\,\mathbf{E}^2, \tag{8.2.8a,b}$$

where $\bar{K}'$ is defined by

$$\frac{1}{3\bar{K}'} \equiv \begin{cases} (1-\bar{\nu})/\bar{E} - 2\bar{\nu}_3^2/\bar{E}_3 & \text{for plane strain} \\ (1-\bar{\nu})/\bar{E} & \text{for plane stress.} \end{cases} \tag{8.2.8c}$$

Note that $\bar{\mu}$ remains the same for plane strain or plane stress.

Since the Eshelby tensor $\bar{\mathbf{S}}^\mathrm{I}$ or $\mathbf{S}^\mathrm{I}$, given by (8.2.3), is two-dimensionally isotropic, its conjugate $\bar{\mathbf{T}}^\mathrm{I}$ or $\mathbf{T}^\mathrm{I}$ is also two-dimensionally isotropic. In terms of $\mathbf{E}^1$ and $\mathbf{E}^2$, $\bar{\mathbf{S}}^\mathrm{I}$ and $\bar{\mathbf{T}}^\mathrm{I}$ are expressed as

$$\bar{\mathbf{S}}^\mathrm{I} = \bar{s}_1\,\mathbf{E}^1 + \bar{s}_2\,\mathbf{E}^2, \qquad \bar{\mathbf{T}}^\mathrm{I} = (1-\bar{s}_1)\,\mathbf{E}^1 + (1-\bar{s}_2)\,\mathbf{E}^2, \tag{8.2.9a,b}$$

and $\mathbf{S}^\mathrm{I}$ and $\mathbf{T}^\mathrm{I}$ as

$$\mathbf{S}^\mathrm{I} = s_1\,\mathbf{E}^1 + s_2\,\mathbf{E}^2, \qquad \mathbf{T}^\mathrm{I} = (1-s_1)\,\mathbf{E}^1 + (1-s_2)\,\mathbf{E}^2, \tag{8.2.9c,d}$$

where

$$\bar{s}_1 = \frac{2}{1+\bar{\kappa}}, \qquad \bar{s}_2 = \frac{\bar{\kappa}}{1+\bar{\kappa}}, \qquad s_1 = \frac{2}{1+\kappa}, \qquad s_2 = \frac{\kappa}{1+\kappa}, \tag{8.2.9e~h}$$

where $\bar{\kappa} = 3 - 8\bar{\mu}(\bar{\nu}/\bar{E} + \bar{\nu}_3^2/\bar{E}_3)$ and $\kappa = 3 - 4\nu$ for plane strain.

8.2.1. Effective Moduli: Dilute Distribution of Fibers

As in Subsection 8.1, first consider an RVE which contains a random and dilute distribution of aligned reinforcing microfibers, and assume that the interaction between neighboring fibers is negligible. Hence, estimate the average strain and stress of each fiber by the uniform strain and stress in an isolated microfiber embedded in an unbounded homogeneous solid which has the matrix elasticity and compliance tensors $\mathbf{C}$ and $\mathbf{D}$. When the farfield stress and strain $\boldsymbol{\sigma}^o$ and $\boldsymbol{\varepsilon}^o$ (which satisfy $\boldsymbol{\sigma}^o = \mathbf{C} : \boldsymbol{\varepsilon}^o$ or $\boldsymbol{\varepsilon}^o = \mathbf{D} : \boldsymbol{\sigma}^o$) are prescribed for this unbounded solid, the exact uniform strain and stress in the isolated microfiber can be calculated in terms of the corresponding Eshelby tensor $\mathbf{S}^I$. Hence, the average stress and strain of a typical microfiber in the composite, denoted by $\bar{\boldsymbol{\varepsilon}}^I$ and $\bar{\boldsymbol{\sigma}}^I$, are approximated by

$$\bar{\boldsymbol{\varepsilon}}^I = \mathbf{A}^I : (\mathbf{A}^I - \mathbf{S}^I)^{-1} : \boldsymbol{\varepsilon}^o, \qquad \bar{\boldsymbol{\sigma}}^I = \mathbf{C}^I : \mathbf{A}^I : (\mathbf{A}^I - \mathbf{S}^I)^{-1} : \mathbf{D} : \boldsymbol{\sigma}^o, \qquad (8.2.10a,b)$$

where $\mathbf{A}^I$ is the two-dimensional fourth-order tensor

$$\mathbf{A}^I = (\mathbf{C} - \mathbf{C}^I)^{-1} : \mathbf{C}. \qquad (8.2.10c)$$

The overall compliance and elasticity tensors, $\bar{\mathbf{D}}$ and $\bar{\mathbf{C}}$, are determined from (8.2.10): when the macrostress $\boldsymbol{\sigma} = \boldsymbol{\sigma}^o$ is prescribed, from (7.4.7a),

$$\bar{\mathbf{D}} = \{\mathbf{1}^{(4s)} + f\,(\mathbf{A}^I - \mathbf{S}^I)^{-1}\} : \mathbf{D}; \qquad (8.2.11a)$$

and when the macrostrain $\mathbf{E} = \boldsymbol{\varepsilon}^o$ is prescribed, from (7.4.10a),

$$\bar{\mathbf{C}} = \mathbf{C} : \{\mathbf{1}^{(4s)} - f\,(\mathbf{A}^I - \mathbf{S}^I)^{-1}\}, \qquad (8.2.12a)$$

where f is the volume fraction of the microfibers, and the two-dimensional fourth-order identity tensor, $(\delta_{ik}\,\delta_{jl} + \delta_{il}\,\delta_{jk})/2$, is denoted by $\mathbf{1}^{(4s)}$.

In terms of $\mathbf{E}^1$ and $\mathbf{E}^2$, the tensorial equation (8.2.11a) reduces to

$$\frac{1}{3\bar{K}'}\,\mathbf{E}^1 + \frac{1}{2\bar{\mu}}\,\mathbf{E}^2 = \frac{1}{3K'}\,\{1 + f\,(\frac{K'}{K' - K'^I} - s_1)^{-1}\}\,\mathbf{E}^1$$

$$+ \frac{1}{2\mu}\,\{1 + f\,(\frac{\mu}{\mu - \mu^I} - s_2)^{-1}\}\,\mathbf{E}^2, \qquad (8.2.11b)$$

and (8.2.12a), becomes

$$3\bar{K}'\,\mathbf{E}^1 + 2\bar{\mu}\,\mathbf{E}^2 = 3K'\,\{1 - f\,(\frac{K'}{K' - K'^I} - s_1)^{-1}\}\,\mathbf{E}^1$$

$$+ 2\mu\,\{1 - f\,(\frac{\mu}{\mu - \mu^I} - s_2)^{-1}\}\,\mathbf{E}^2. \qquad (8.2.12b)$$

Therefore, the overall elastic moduli $\bar{K}'$ and $\bar{\mu}$ are obtained: when $\boldsymbol{\Sigma} = \boldsymbol{\sigma}^o$ is prescribed,

$$\frac{\bar{K}'}{K'} = \{1 + f\,(\frac{K'}{K' - K'^I} - s_1)^{-1}\}^{-1} = 1 - f\,(\frac{K'}{K' - K'^I} - s_1)^{-1} + O(f^2),$$

$$\frac{\bar{\mu}}{\mu} = \{1 + f\,(\frac{\mu}{\mu - \mu^I} - s_2)^{-1}\}^{-1} = 1 - f\,(\frac{\mu}{\mu - \mu^I} - s_2)^{-1} + O(f^2); \qquad (8.2.13a,b)$$

and when $\mathbf{E} = \boldsymbol{\varepsilon}^o$ is prescribed,

$$\frac{\bar{K}'}{K'} = 1 - f\,(\frac{K'}{K' - K'^I} - s_1)^{-1},$$

$$\frac{\bar{\mu}}{\mu} = 1 - f\,(\frac{\mu}{\mu - \mu^I} - s_2)^{-1}. \tag{8.2.14a,b}$$

8.2.2. Effective Moduli: Self-Consistent Estimates

Next consider the case when the distribution of aligned reinforcing microfibers in the x_1,x_2-plane is random, and the interaction between the microfibers is to be included using the self-consistent method. From (8.2.9) and the overall Poisson ratio $\bar{\nu}$, the Eshelby tensor $\bar{\mathbf{S}}^I$ is defined for an infinitely long circular cylinder embedded in an unbounded solid, by

$$\bar{\mathbf{S}}^I = \bar{s}_1\,\mathbf{E}^1 + \bar{s}_2\,\mathbf{E}^2, \tag{8.2.15a}$$

where

$$\bar{s}_1 = \frac{2}{1+\bar{\kappa}}, \qquad \bar{s}_2 = \frac{\bar{\kappa}}{1+\bar{\kappa}}. \tag{8.2.15b,c}$$

The overall compliance and elasticity tensors $\bar{\mathbf{C}}$ and $\bar{\mathbf{D}}$ are determined from (8.2.16): when the macrostress $\mathbf{\Sigma} = \boldsymbol{\sigma}^o$ is prescribed, from (7.5.5a),

$$\bar{\mathbf{D}} = \mathbf{D} + f\,(\mathbf{D}^I - \mathbf{D}) : \mathbf{C}^I : \bar{\mathbf{A}}^I : (\bar{\mathbf{A}}^I - \bar{\mathbf{S}}^I)^{-1} : \bar{\mathbf{D}}$$

$$= \frac{1}{3}\,\{\frac{1}{K'} + f\,\frac{1}{\bar{K}'}\,\frac{K' - K'^I}{\bar{K}' - K'^I}\,(\frac{\bar{K}'}{\bar{K}' - K'^I} - \bar{s}_1)^{-1}\}\,\mathbf{E}^1$$

$$+ \frac{1}{2}\{\frac{1}{\mu} + f\,\frac{1}{\mu}\,\frac{\mu - \mu^I}{\bar{\mu} - \mu^I}\,(\frac{\bar{\mu}}{\bar{\mu} - \mu^I} - \bar{s}_2)^{-1}\}\,\mathbf{E}^2; \tag{8.2.16}$$

and when the macrostrain $\mathbf{E} = \boldsymbol{\varepsilon}^o$ is prescribed, from (7.5.8a),

$$\bar{\mathbf{C}} = \mathbf{C} + f\,(\mathbf{C}^I - \mathbf{C}) : \bar{\mathbf{A}}^I : (\bar{\mathbf{A}}^I - \bar{\mathbf{S}}^I)^{-1}$$

$$= 3\,\{K' + f\,\bar{K}'\,\frac{K'^I - K'}{\bar{K}' - K'^I}\,(\frac{\bar{K}'}{\bar{K}' - K'^I} - \bar{s}_1)^{-1}\}\,\mathbf{E}^1$$

$$+ 2\,\{\mu + f\,\bar{\mu}\,\frac{\mu^I - \mu}{\bar{\mu} - \mu^I}\,(\frac{\bar{\mu}}{\bar{\mu} - \mu^I} - \bar{s}_2)^{-1}\}\,\mathbf{E}^2. \tag{8.2.17}$$

As is shown in Subsection 7.5, the above two equations, (8.2.16) and (8.2.17), are identical, giving the same overall moduli $\bar{K}'$ and $\bar{\mu}$. Hence,

$$\frac{\bar{K}'}{K'} = 1 - f\,\frac{\bar{K}'(K' - K'^I)}{K'(\bar{K}' - K'^I)}\,(\frac{\bar{K}'}{\bar{K}' - K'^I} - \bar{s}_1)^{-1} = 1 - f\,(\frac{K'}{K' - K'^I} - s_1)^{-1} + O(f^2),$$

$$\frac{\bar{\mu}}{\mu} = 1 - f\,\frac{\bar{\mu}(\mu - \mu^I)}{\mu(\bar{\mu} - \mu^I)}\,(\frac{\bar{\mu}}{\bar{\mu} - \mu^I} - \bar{s}_2)^{-1} = 1 - f\,(\frac{\mu}{\mu - \mu^I} - s_2)^{-1} + O(f^2). \tag{8.2.18a,b}$$

Figure 8.2.1 shows the graphs of the overall moduli $\bar{K}'$ and $\bar{\mu}$ as functions of the volume fraction of the microfibers f, obtained by the assumption of a dilute distribution, (8.2.13a,b) and (8.2.14a,b), and by the self-consistent method, (8.2.18a,b). As is seen, for small f, the three estimates agree.

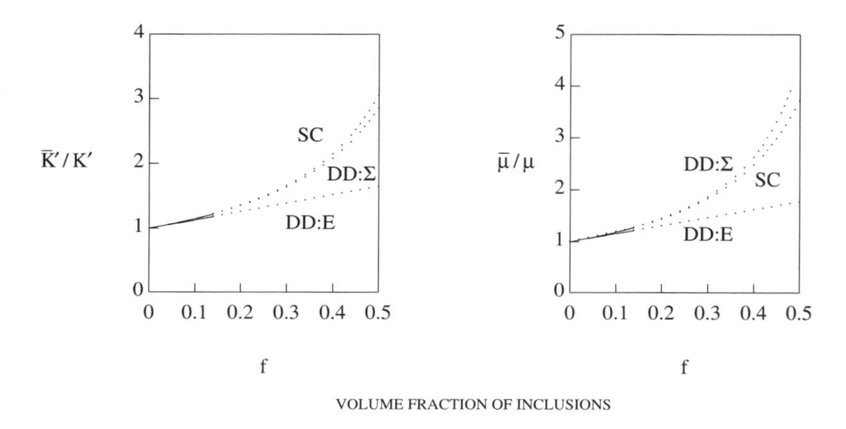

Figure 8.2.1

Two-dimensional overall bulk and shear moduli of an RVE with aligned circular cylindrical inclusions; $K'^{I}/K' = \mu^{I}/\mu = 50$ and $\nu^{I} = \nu = 1/3$
SC ≡ self-consistent
DD:Σ ≡ dilute distribution with macrostress prescribed
DD:E ≡ dilute distribution with macrostrain prescribed

8.2.3. Effective Moduli in Antiplane Shear: Dilute-Distribution and Self-Consistent Estimates

In this subsection, the overall shear moduli of the RVE in the x_1,x_3- and x_2,x_3-planes are estimated for antiplane shearing. Due to the isotropy of the matrix and the reinforcing microfibers, and the random distribution of the microfibers in the x_1,x_2-plane, the overall shear moduli in the x_1,x_3- and x_2,x_3-planes are the same, i.e.,

$$\bar{\mu}_{13} = \bar{\mu}_{23} \equiv \bar{\mu}_3. \tag{8.2.19}$$

Furthermore, the fourth-order tensors involved in this antiplane shear problem (i.e., **C**, $\bar{\mathbf{C}}$, **S**, etc.) are expressed by means of the two by two identity matrix

$$[I^{(4s)}_{ab}] = \begin{bmatrix} 1 & 0 \\ 0 & 1 \end{bmatrix}, \tag{8.2.20}$$

so that all tensorial equations are reduced to the corresponding scalar equations for the coefficients of this matrix; see Subsection 6.4.3.

From this observation, the four tensorial equations, (7.4.7a) and (7.5.10a) for the dilute distribution, and (7.5.5a) and (7.5.10b) for the self-consistent method, are reduced to four scalar equations, similar to (8.2.11) and (8.2.12) obtained in Subsection 8.2.1, and (8.2.16) and (8.2.17) obtained in Subsection 8.2.2, respectively. Hence, the overall shear modulus $\bar{\mu}_3$ for the antiplane shear is given by the same equation that gives the overall shear modulus $\bar{\mu}$ for the plane strain case: from the assumption of a dilute distribution, when σ^o_{13} or σ^o_{23} is prescribed,

$$\frac{\bar{\mu}_3}{\mu} = \{1 + f\,(\frac{\mu}{\mu - \mu^I} - s_3)^{-1}\}^{-1} = 1 - f\,(\frac{\mu}{\mu - \mu^I} - s_3)^{-1} + O(f^2); \tag{8.2.21}$$

and when ε^o_{13} or ε^o_{23} is prescribed,

$$\frac{\bar{\mu}_3}{\mu} = 1 - f\,(\frac{\mu}{\mu - \mu^I} - s_3)^{-1}; \tag{8.2.22}$$

and by the self-consistent estimate, when either, say, σ^o_{13} or, say, ε^o_{13} is prescribed,

$$\frac{\bar{\mu}_3}{\mu} = 1 - f\,\frac{\bar{\mu}_3(\mu - \mu^I)}{\mu(\bar{\mu}_3 - \mu^I)}\,(\frac{\bar{\mu}_3}{\bar{\mu}_3 - \mu^I} - \bar{s}_3)^{-1} = 1 - f\,(\frac{\mu}{\mu - \mu^I} - s_3)^{-1} + O(f^2), \tag{8.2.23}$$

where, from (8.2.3),

$$S^I_{1313} = ... = S^I_{2323} = ... \equiv s_3 = \frac{1}{4},$$

$$\bar{S}^I_{1313} = ... = \bar{S}^I_{2323} = ... \equiv \bar{s}_3 = \frac{1}{4}. \tag{8.2.24}$$

It is instructive to compare the overall shear modulus $\bar{\mu}_3$, for the antiplane shear, given by (8.2.21) and (8.2.22) for the dilute distribution, and by (8.2.23) for the self-consistent method, with the overall shear modulus $\bar{\mu}$ for the plane problem, given by (8.2.13b) and (8.2.14b) for the dilute distribution, and by (8.2.18b) for the self-consistent method. When the Poisson ratio of the matrix and that of the composite, ν and $\bar{\nu}$, is 1/2, the $\mathbf{E}^2$-coefficients of the Eshelby tensors $\mathbf{S}^I$ and $\bar{\mathbf{S}}^I$, i.e., s_2 and $\bar{s}_2$, reduce to 1/2. In this case, $\bar{\mu}_3$ coincides with $\bar{\mu}$. In general, $0 \leq \nu < 1/2$ and $0 \leq \bar{\nu} \leq 1/2$, and hence

$$s_2 > s_3, \qquad \bar{s}_2 > \bar{s}_3. \tag{8.2.25a,b}$$

If the reinforcing microfibers are stiffer than the matrix, i.e., $\mu^I > \mu$, it follows that $\bar{\mu}_3 > \bar{\mu}$, according to the assumption of a dilute distribution; compare (8.2.21) with (8.2.13b), and (8.2.22) with (8.2.14b). Furthermore, expecting $\bar{\mu} > \mu$ in this case, note that $\bar{\mu}_3 > \bar{\mu}$ according to the self-consistent estimate; compare (8.2.23) with (8.2.17b). This stiffer response of the RVE in the x_3-direction is reasonable on physical grounds.

Figure 8.2.2 shows the graph of $\bar{\mu}_3$ with respect to the volume fraction of microfibers f. As is seen, for small f, the three results, (8.2.21), (8.2.22), and (8.2.23) agree with each other. Note that these estimates are valid only for small values of f. In this connection, recall that the results of the dilute-distribution model may actually violate the exact theorems of Subsection 2.6, further revealing the limitation of these estimates.

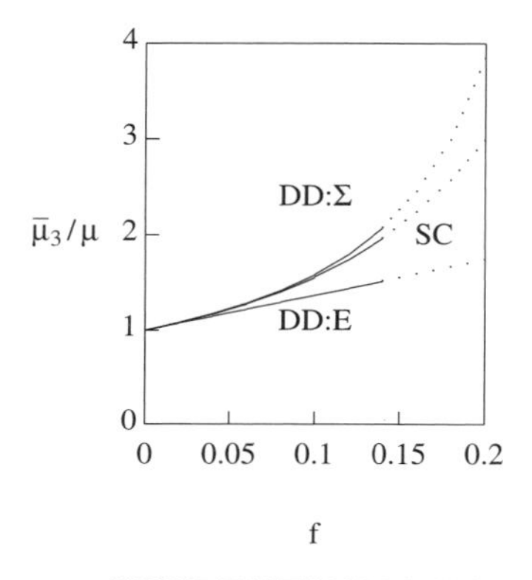

Figure 8.2.2

Two-dimensional overall anti-shear moduli of an RVE with aligned circular cylindrical inclusions; $K'^{I}/K' = \mu^{I}/\mu = 50$ and $\nu^{I} = \nu = 1/3$.
SC ≡ self-consistent
DD:Σ ≡ dilute distribution with macrostress prescribed
DD:E ≡ dilute distribution with macrostrain prescribed

Early work on elastic properties of reinforced elastic materials is by Dewey (1947), Hashin (1959, 1962, 1965, 1970, 1979), Paul (1960), Hill (1963, 1964, 1965a,b), Hashin and Rosen (1964), Budiansky (1965), Adams *et al.* (1967), Walpole (1970, 1981), Christensen and Waals (1972), Korringa (1973), Chu *et al.* (1980), and Willis (1983). An interesting observation by Hill (1964) for any two-phase composite with aligned cylindrical fibers, is that not all the corresponding effective moduli are independent. For isotropic constituents, Hill provides two relations which connect the overall parameters, leading to only three independent overall moduli for this kind of transversely isotropic composite. The procedure has other potential applications, some of which have been pursued by Dvorak (1990) and Dvorak and Chen (1989).

8.3. THREE-DIMENSIONAL ANALYSIS OF PLANE STRAIN AND PLANE STRESS STATES

In the preceding subsection, a *plane strain* state is assumed, and an infinitely long fiber is used as a model for analyzing the two-dimensional inplane problem. The results, however, can also be used for analyzing an inplane problem in *plane stress*. This and related issues are examined in this subsection.

8.3.1. Reduction of Three-Dimensional Moduli to Two-Dimensional Moduli

For simplicity consider a case when the macrostress, $\mathbf{\Sigma} = \boldsymbol{\sigma}^{o}$, is prescribed. As shown in (7.4.7a), the overall compliance tensor, $\bar{\mathbf{D}}$, is given by

$$\bar{\mathbf{D}} = (\mathbf{1}^{(4s)} + f\,\hat{\mathbf{A}}) : \mathbf{D}, \tag{8.3.1a}$$

where

$$\hat{\mathbf{A}} = \begin{cases} \{(\mathbf{C} - \mathbf{C}^{\mathrm{I}})^{-1} : \mathbf{C} - \mathbf{S}^{\mathrm{I}}\}^{-1} & \text{dilute distribution} \\ \{(\bar{\mathbf{C}} - \mathbf{C}^{\mathrm{I}})^{-1} : \bar{\mathbf{C}} - \bar{\mathbf{S}}^{\mathrm{I}}\}^{-1} & \text{self–consistent.} \end{cases} \tag{8.3.1b}$$

The matrix form of these tensors is

$$[(...)_{ab}] = \begin{bmatrix} [(...)^{(1)}_{ab}] & [0_{ab}] \\ [0_{ab}] & [(...)^{(2)}_{ab}] \end{bmatrix}, \tag{8.3.2}$$

where $[(...)^{(1)}_{ab}]$ and $[(...)^{(2)}_{ab}]$ are three by three matrices. Hence, normal stresses (normal strains) and shear stresses (shear strains) are decoupled, and it suffices to examine only $[(...)^{(1)}_{ab}]$ for plane strain or plane stress.

In terms of $[(...)^{(1)}_{ab}]$, the average strain, $\bar{\boldsymbol{\varepsilon}} = \bar{\mathbf{D}} : \boldsymbol{\sigma}^{o}$, is given by

$$[\bar{\gamma}^{(1)}_{a}] = [\bar{D}^{(1)}_{ab}]\,[\tau^{o(1)}_{b}] = ([1^{(1)}_{ac}] + f\,[\hat{A}^{(1)}_{ac}])\,[D^{(1)}_{cb}]\,[\tau^{o(1)}_{b}], \tag{8.3.3}$$

where $[\bar{\gamma}^{(1)}_{a}] = [\bar{\varepsilon}_{11}, \bar{\varepsilon}_{22}, \bar{\varepsilon}_{33}]^{T}$, $[\tau^{o(1)}_{a}] = [\sigma^{o}_{11}, \sigma^{o}_{22}, \sigma^{o}_{33}]^{T}$, and $[1^{(1)}_{ab}]$ is a three by three identity matrix.

As mentioned in Subsection 5.1, the nominal two-dimensional matrix which relates the normal inplane strains, $[\bar{\gamma}_1, \bar{\gamma}_2]^{T}$, to the normal inplane stresses, $[\tau^{o}_1, \tau^{o}_2]^{T}$, is given by

$$[P_{ap}]^{T}\,[\bar{D}^{(1)}_{pq}]\,[P'_{qb}] = [P_{ap}]^{T}\,([1^{(1)}_{pq}] + f\,[\hat{A}^{(1)}_{pq}])\,[D^{(1)}_{qr}]\,[P'_{rb}], \tag{8.3.4a}$$

for plane strain, and by

$$[P_{ap}]^{T}\,[\bar{D}^{(1)}_{pq}]\,[P_{qb}] = [P_{ap}]^{T}\,([1^{(1)}_{pq}] + f\,[\hat{A}^{(1)}_{pq}])\,[D^{(1)}_{qr}]\,[P_{rb}], \tag{8.3.4b}$$

for plane stress; see (5.1.5a,b) and (5.1.6a,b). Note that $[P'_{ab}]$ is now defined by

$$[P'_{ab}] = [P'_{ab}](\bar{D}_{cd}) \equiv \begin{bmatrix} 1 & 0 \\ 0 & 1 \\ -\bar{D}_{31}/\bar{D}_{33} & -\bar{D}_{32}/\bar{D}_{33} \end{bmatrix}, \tag{8.3.5}$$

where the quantity in parentheses denotes the argument of $[P'_{ab}]$. Therefore, if the three-dimensional $\hat{\mathbf{A}}$ is obtained from a three-dimensional analysis of an RVE, then, the nominal two-dimensional compliance tensor can be obtained using (8.3.4a) for plane strain and (8.3.4b) for plane stress. However, in a two-dimensional analysis, only the nominal part of $\hat{\mathbf{A}}$, i.e., $[P_{ap}]^T ([1^{(1)}_{pq}] + f\,[\hat{A}^{(1)}_{pq}])\,[D^{(1)}_{qr}]\,[P'_{rb}]$ for plane strain and $[P_{ap}]^T ([1^{(1)}_{pq}] + f\,[\hat{A}^{(1)}_{pq}])\,[D^{(1)}_{qr}]\,[P_{rb}]$ for plane stress, is estimated. Moreover, this nominal part is often given in terms of the nominal compliance tensor, that is,

$$\begin{Bmatrix} [P_{ap}]^T ([1^{(1)}_{pq}] + f\,[\hat{A}^{(1)}_{pq}])\,[D^{(1)}_{qr}]\,[P'_{rb}] \\ [P_{ap}]^T ([1^{(1)}_{pq}] + f\,[\hat{A}^{(1)}_{pq}])\,[D^{(1)}_{qr}]\,[P_{rb}] \end{Bmatrix}$$

$$= \begin{Bmatrix} \text{function of } [P_{ap}][D_{pq}][P'_{qb}] \text{ or } [P_{ap}][\bar{D}_{pq}][P'_{qb}] & \text{plane strain} \\ \text{function of } [P_{ap}][D_{pq}][P_{qb}] \text{ or } [P_{ap}][\bar{D}_{pq}][P_{qb}] & \text{plane stress.} \end{Bmatrix} \tag{8.3.6}$$

Therefore, the estimate of the overall compliance tensor changes, depending on whether plane strain or plane stress conditions are assumed.[3]

8.3.2. Two-Dimensional Nominal Eshelby Tensor

Consider now a two-dimensional nominal Eshelby tensor, denoted by $\mathbf{S}'$, which is obtained from the three-dimensional Eshelby tensor. Note that, in matrix form similar to (8.3.3), the strain field produced by an eigenstrain tensor, $\boldsymbol{\varepsilon} = \mathbf{S} : \boldsymbol{\varepsilon}^*$, is expressed as

$$[\gamma^{(1)}_a] = [S^{(1)}_{ab}]\,[\gamma^{*(1)}_b], \tag{8.3.7}$$

where $[\gamma^{(1)}_a] = [\varepsilon_{11}, \varepsilon_{22}, \varepsilon_{33}]^T$ and $[\gamma^{*(1)}_a] = [\varepsilon^*_{11}, \varepsilon^*_{22}, \varepsilon^*_{33}]^T$. It is *assumed* that ε_{33} vanishes in plane strain. Hence, with the aid of $[P_{ab}]$ and $[P'_{ab}](S_{cd})$, a two by two matrix for the two-dimensional nominal Eshelby tensor is given by[4]

$$[S'_{ab}] = [P_{ap}]^T\,[S^{(1)}_{pq}]\,[P'_{qb}](S_{cd}). \tag{8.3.8a}$$

In particular, if components $S_{i3}(= S_{ii33})$ (i not summed) of Eshelby's tensor vanish as in (8.2.1), then (8.3.8a) becomes

$$[S'_{ab}] = [P_{ap}]^T\,[S^{(1)}_{pq}]\,[P_{qb}]. \tag{8.3.8b}$$

The nominal Eshelby tensor used in Subsection 8.2 is obtained in this manner.

Now, consider the stress produced by an eigenstrain tensor, $\boldsymbol{\sigma} = \mathbf{C} : (\mathbf{S} - \mathbf{1}^{(4s)}) : \boldsymbol{\varepsilon}^*$. In matrix form, $\boldsymbol{\sigma}$ is expressed as

---

[3] Similar comments apply when the overall strains are prescribed, $\mathbf{E} = \boldsymbol{\varepsilon}^o$.

[4] Note that the argument of $[P'_{ab}]$ now is the relevant component of $[S_{ab}]$.

$$[\tau_a^{(1)}] = [C_{ab}^{(1)}]\,([S_{bc}^{(1)}] - [1_{bc}^{(1)}])\,[\gamma_c^{*(1)}], \tag{8.3.9}$$

where $[\tau_a^{(1)}] = [\sigma_{11}, \sigma_{22}, \sigma_{33}]^T$. If it is *assumed* that σ_{33} vanishes in plane stress, then, with the aid of $[P_{ab}]$ and $[P'_{ab}](S_{cd})$, a two by two matrix for the two-dimensional nominal Eshelby tensor, $\mathbf{S}'$, is obtained to be,

$$[S'_{ab}] = [P_{ap}]^T\,[S_{pq}^{(1)}]\,[P'_{qb}]((\mathbf{C} : (\mathbf{S} - \mathbf{1}^{(4s)}))_{cd}). \tag{8.3.10}$$

Some attention must be paid to the argument of the matrix $[P'_{ab}]$. Since $\tau_3^{(1)}$ in (8.3.9) vanishes, the argument must therefore be the relevant components of $\mathbf{C} : (\mathbf{S} - \mathbf{1}^{(4s)})$, as indicated.

It is seen that the two-dimensional nominal Eshelby tensor for plane strain (or plane stress) is obtained by *assuming* that ε_{33} (or σ_{33}) produced by $\boldsymbol{\varepsilon}^*$ vanishes. However, the resulting two-dimensional Eshelby tensor is not the same as that obtained by reducing $(\mathbf{1}^{(4s)} + f\,\hat{\mathbf{A}})$ with $\hat{\mathbf{A}} = \{(\mathbf{C} - \mathbf{C}^I)^{-1} : \mathbf{C} - \mathbf{S}^I\}^{-1}$ or $\hat{\mathbf{A}} = \{(\bar{\mathbf{C}} - \mathbf{C}^I)^{-1} : \bar{\mathbf{C}} - \bar{\mathbf{S}}^I\}^{-1}$, to the corresponding two-dimensional tensor; see Subsection 8.3.1. Indeed, the overall moduli obtained with the aid of the nominal Eshelby tensor (8.3.8) for plane strain, differ from those obtained through the nominal Eshelby tensor (8.3.10) for plane stress. This inconsistency does not occur if the overall moduli are obtained directly by using the nominal part of $(\mathbf{1}^{(4s)} + f\,\hat{\mathbf{A}})$. However, especially for an anisotropic case, the computation of $(\mathbf{1}^{(4s)} + f\,\hat{\mathbf{A}})$ (or the Eshelby tensor itself) is more complex. In this case, plane approximations may be used , as discussed above.[5]

8.3.3. Computation of Nominal Eshelby Tensor for Plane Stress

As shown in the previous subsection, the two-dimensional Eshelby tensor used in Subsection 8.2 is the nominal Eshelby tensor for plane strain, obtained from the Eshelby tensor for a cylindrical fiber in a transversely isotropic material. In this subsection, the two-dimensional nominal Eshelby tensor for plane stress is computed.

The matrix form of $\bar{\mathbf{C}}$ for a transversely isotropic material is

$$[\bar{C}_{ab}^{(1)}] = \begin{bmatrix} \bar{C}_{11} & \bar{C}_{12} & \bar{C}_{13} \\ \bar{C}_{12} & \bar{C}_{11} & \bar{C}_{13} \\ \bar{C}_{13} & \bar{C}_{13} & \bar{C}_{33} \end{bmatrix}, \tag{8.3.11}$$

and $\bar{C}_{ab}$'s are given by (3.1.13h~l), with E, E_3, ν, and ν_3 replaced by $\bar{E}$, $\bar{E}_3$, $\bar{\nu}$, and $\bar{\nu}_3$, respectively. Hence,

---

[5] The inconsistency is due to the plane approximation, and is derived from the *noncommutativity of certain matrix operations*; for example, $[P_{ap}]^T[\hat{A}_{pq}^{(1)}][P_{qb}] = [P_{ap}]^T\{[(C - C^I)_{ap}^{(1)}]^{-1}[C_{qr}^{(1)}] - [S_{pr}^{(1)}]\}^{-1}[P_{rb}] \neq \{([P_{ap}]^T([C_{pq}^{(1)}] - [C_{pq}^{I(1)}])^{-1}[P_{qr}])([P_{rs}]^T[C_{st}^{(1)}][P_{tb}]) - ([P_{pq}]^T[S_{qr}^{(1)}][P_{rb}])\}^{-1}$.

$$[\bar{C}^{(1)}_{ab}]\,([\bar{S}^{(1)}_{bc}] - [1^{(1)}_{bc}]) = \frac{1}{2(1+\bar{\kappa})}$$

$$\times \begin{bmatrix} 2\bar{C}_{12} - \bar{\kappa}(\bar{C}_{11}+\bar{C}_{12}) & 2\bar{C}_{11} - \bar{\kappa}(\bar{C}_{11}+\bar{C}_{12}) & \bar{S}_{13}(\bar{C}_{11}+\bar{C}_{12}) - 2(1+\bar{\kappa})\bar{C}_{13} \\ 2\bar{C}_{11} - \bar{\kappa}(\bar{C}_{11}+\bar{C}_{12}) & 2\bar{C}_{12} - \bar{\kappa}(\bar{C}_{11}+\bar{C}_{12}) & \bar{S}_{13}(\bar{C}_{11}+\bar{C}_{12}) - 2(1+\bar{\kappa})\bar{C}_{13} \\ 2(1-\bar{\kappa})\bar{C}_{13} & 2(1-\bar{\kappa})\bar{C}_{13} & 2\bar{S}_{13}\bar{C}_{13} - 2(1+\bar{\kappa})\bar{C}_{33} \end{bmatrix}. \tag{8.3.12}$$

Although $\bar{S}_{1133} = \bar{S}_{2233}$ is not given in (8.2.1), due to the symmetry of $\bar{\mathbf{C}} : \bar{\mathbf{S}}$, it is obtained to be

$$\bar{S}_{1133} = \frac{4\bar{C}_{1133}}{C_{1111}+C_{1122}}, \qquad \bar{S}_{13} = \frac{4\bar{C}_{13}}{C_{11}+C_{12}}. \tag{8.3.13a,b}$$

Hence, the two by three matrix $[P'_{ab}]$ for plane stress is

$$[P'_{ab}]((\bar{\mathbf{C}} : (\bar{\mathbf{S}} - \mathbf{1}^{(4s)}))^{(1)}_{ab}) = \begin{bmatrix} 1 & 0 \\ 0 & 1 \\ -\dfrac{(1-\kappa)\bar{C}_{13}}{\bar{S}_{13}\bar{C}_{13} - (1+\kappa)\bar{C}_{33}} & -\dfrac{(1-\kappa)\bar{C}_{13}}{\bar{S}_{13}\bar{C}_{13} - (1+\kappa)\bar{C}_{33}} \end{bmatrix}. \tag{8.3.14}$$

Since $\kappa = 3 - 8\bar{\mu}(\bar{\nu}/\bar{E} + \bar{\nu}_3^2/\bar{E}_3)$, the nominal Eshelby tensor is obtained from (8.3.10), to be

$$[\bar{S}'_{ab}] = \begin{bmatrix} \dfrac{3+\bar{\nu}}{4} - \dfrac{1+\bar{\nu}}{8}(1-\bar{\nu}_3^2\dfrac{\bar{E}}{\bar{E}_3})^{-1} & -\dfrac{1-2\bar{\nu}}{4} + \dfrac{1+\bar{\nu}}{8}(1-\bar{\nu}_3^2\dfrac{\bar{E}}{\bar{E}_3})^{-1} \\ -\dfrac{1-2\bar{\nu}}{4} + \dfrac{1+\bar{\nu}}{8}(1-\bar{\nu}_3^2\dfrac{\bar{E}}{\bar{E}_3})^{-1} & \dfrac{3+\bar{\nu}}{4} - \dfrac{1+\bar{\nu}}{8}(1-\bar{\nu}_3^2\dfrac{\bar{E}}{\bar{E}_3})^{-1} \end{bmatrix}. \tag{8.3.15a}$$

In particular, if the material is isotropic, (8.3.15a) becomes

$$[S'_{ab}] = \frac{1}{8(1-\nu)} \begin{bmatrix} 5-4\nu-2\nu^2 & -1+2\nu-2\nu^2 \\ -1+2\nu-2\nu^2 & 5-4\nu-2\nu^2 \end{bmatrix}. \tag{8.3.15b}$$

8.4. REFERENCES

Adams, D. F., Dooner, D. R., and Thomas, R. L. (1967), Mechanical behavior of fiber-reinforced composite materials, *J. Compos. Mat.*, Vol.1, 4-17, 152-165.

Budiansky, B. (1965), On the elastic moduli of some heterogeneous materials, *J. Mech. Phys. Solids*, Vol. 13, 223-227.

Christensen, R. M. and Waals, F. M. (1972), Effective stiffness of randomly oriented fiber-reinforced composites, *J. Compos. Mat.*, Vol. 6, 518-532.

Chu, T. W. W., Nomura, S., and Taya, M. (1980), Self-consistent approach to elastic stiffness of short-fiber composites, *J. Compos. Mat.*, Vol. 14, 178-188.

Dewey, J. M. (1947), The elastic constants of materials loaded with non-rigid fillers, *J. Appl. Phys.*, Vol. 18, 578-581.

Dvorak, G. J. (1990), On uniform fields in heterogeneous media, *Proc. R. Soc. Lond.* A, Vol. 431, 89-110.

Dvorak, G. J. and Chen, T. (1989), Thermal expansion of three-phase composite materials, *J. Appl. Mech.*, Vol. 56, 418-422.

Hashin, Z. (1959), The moduli of an elastic solid containing spherical particles of another elastic material, in *Nonhomogeneity in Elasticity and Plasticity* (Olszak, W., ed.), Pergamon Press, 463-478.

Hashin, Z. (1962), The elastic moduli of heterogeneous materials, *J. Appl. Mech.*, Vol. 29, 143-150.

Hashin, Z. (1965), On elastic behaviour of fiber reinforced materials of arbitrary transverse phase geometry, *J. Mech. Phys. Solids*, Vol. 13, 119-134.

Hashin, Z. (1970), *Theory of fiber reinforced materials*, NASA CR-1974.

Hashin, Z. (1979), Analysis of properties of fiber composites with anisotropic constituents,

Hashin, Z. and Rosen, B. W. (1964), The elastic moduli of fiber reinforced materials, *J. Appl. Mech.*, Vol. 31, 223-232.

Hill, R. (1963), Elastic properties of reinforced solids: Some theoretical principles, *J. Mech. Phys. Solids*, Vol 11, 357-372.

Hill, R. (1964), Theory of mechanical properties of fibre-strengthened materials - I: Elastic behaviour, *J. Mech. Phys. Solids*, Vol. 12, 199-212.

Hill, R. (1965a), Theory of mechanical properties of fibre-strengthened materials - III: Self-consistent model, *J. Mech. Phys. Solids*, Vol. 13, 189-198.

Hill, R. (1965b), A self-consistent mechanics of composite materials, *J. Mech. Phys. Solids*, Vol. 13, 213-222.

Korringa, J. (1973), Theory of elastic constants of heterogeneous media *J. Math. Phys.*, Vol. 14, 509-513. *J. Appl. Mech.*, Vol. 46, 543-550.

Paul, B. (1960), Prediction of the elastic constants of multiphase materials, *Trans. Am . Inst. Min. Metall. Pet. Eng.*, Vol. 218, 36-41.

Walpole, L. J. (1970), Strengthening effects in elastic solids, *J. Mech. Phys. Solids*, Vol. 18, 343-358.

Walpole, L. J. (1981), Elastic behavior of composite materials: Theoretical foundations, *Advances in Applied Mechanics*, Vol. 21, 169-242.

Willis, J. R. (1983), The overall elastic response of composite materials, *J. Appl. Mech.*, Vol. 50, 1202-1209.

SECTION 9 UPPER AND LOWER BOUNDS FOR OVERALL ELASTIC MODULI

In this section the focus is again on an RVE which consists of a linearly elastic matrix and linearly elastic micro-inclusions. In Section 7, eigenstrains or eigenstresses are introduced to define an equivalent homogeneous solid for a heterogeneous RVE. The stress and strain fields in the equivalent homogeneous solid depend on the distribution of the corresponding eigenstrains (or eigenstresses). For the exact eigenstrain (or eigenstress) field which satisfies the consistency conditions, the resulting stress and strain fields in the equivalent homogeneous solid coincide with the actual stress and strain fields of the original heterogeneous RVE. It is easier to seek to solve this equivalent homogeneous solid problem than the original heterogeneous one. Furthermore, the strain and complementary strain energy functionals of the equivalent solid, when regarded as functionals of the eigenstrain (or eigenstress), are stationary for the exact eigenstrain (or eigenstress). Depending on the heterogeneity of the original RVE, these functionals provide global maximum or minimum values for the actual total strain and complementary energy functionals. This remarkable result relating to the elastic energy of the equivalent solid was obtained by Hashin and Shtrikman (1962a,b), and is called the Hashin-Shtrikman variational principle. Its formulation in terms of both eigenstress (also called polarization stress) and eigenstrain (also called polarization strain) is given by Hill (1963) who develops these principles from the classical variational theorems of elasticity; see Section 19, Part 2, and also Subsection 13.5 of Chapter IV.

It is still difficult to obtain the exact eigenstrain and eigenstress fields which produce in the equivalent homogeneous solid the actual stress and strain fields of the original heterogeneous RVE. However, with the aid of the variational principle, approximate eigenstrain (or eigenstress) fields which yield strict upper and lower bounds for the overall parameters of the RVE can be constructed. Willis (1977) has generalized the Hashin-Shtrikman variational principle using the Green function of an unbounded equivalent homogeneous solid. He has sought to construct upper and lower bounds for the overall parameters of the heterogeneous solid. Willis (1981, p.18) comments that the approximations involved in using the Green function of the unbounded domain may render the results, more as plausible estimates than rigorous bounds. It turns out that the general theorems of Subsection 2.5.6, i.e., Theorem I and II, can be used to show rigorously that two out of four possible bounding expressions that result from the generalized Hashin-Shtrikman bounds, as obtained with the aid of the approximate Green function, remain rigorous bounds. These new results were reported by the authors in the first edition of the present book, and are discussed in Subsection 9.5; see also Nemat-Nasser and Hori (1995). Other related contributions are by Hashin (1965, 1967), Walpole (1966a,b, 1969, 1981), Korringa (1973), Willis and Acton (1976), Kröner (1977), and Wu and McCullough

(1977).[1]

First, the Hashin-Shtrikman variational principle, as generalized by Willis, is presented in this section. Two integral operators, $\mathbf{\Lambda}$ and $\mathbf{\Gamma}$, are defined which determine the stress and strain fields in the equivalent homogeneous solid produced by a prescribed eigenstrain or eigenstress, respectively associated with the uniform traction and linear displacement boundary data. In terms of $\mathbf{\Lambda}$ and $\mathbf{\Gamma}$, the strain and complementary strain energy of the equivalent solid are defined as functionals of the eigenstrain and eigenstress, respectively. Then, Willis' formulation is followed, the Green function of the equivalent solid is introduced, and the exact expressions for the stress and strain fields produced by the eigenstrains or eigenstresses are formulated. The upper and lower bounds are obtained in terms of integral operators $\mathbf{\Lambda}$ and $\mathbf{\Gamma}$. Then, approximations of these integral operators are introduced, and a solution method is outlined. In Subsection 9.5, the Hashin-Shtrikman variational principle is generalized for boundary data other than uniform tractions and linear displacements, and the corresponding generalized bounds are obtained. With the aid of Theorems I and II of Subsection 2.5.6, these bounds are then related to the bounds for the uniform traction and linear displacement boundary data. It is proved that two out of four possible approximate expressions that result are indeed rigorous bounds. Explicit, computable, exact upper and lower bounds for the overall moduli are then given when the composite is statistically homogeneous and isotropic. Finally, it is shown in Subsection 9.6 that these new observations (originally reported by the authors in the first edition of the present book) lead to *universal* bounds on two overall moduli of multi-phase composites, valid for any shape or distribution of phases. Furthermore, it is established that the bounds are valid for any *finite* elastic solid of ellipsoidal shape, consisting of any distribution of inhomogeneities of any shape and elasticity. In Section 13, it is proved that the same bounds emerge for multi-phase composites with periodic, but otherwise completely arbitrary, microstructure (see Subsection 13.5).

For historical reasons, the bounds on the overall properties in this section are based on the Hashin-Shtrikman variational principle. An alternative formulation of *exact computable bounds* is to use the universal Theorems I and II of Subsection 2.5.6, together with proper choices of the reference elasticity or compliance tensors; this is presented in Subsection 9.5.6. These bounds are valid for any *finite* elastic solid of ellipsoidal shape, consisting of any distribution of inhomogeneities of any shape and elasticity. Theorems I and II of Subsection 2.7 are used in Subsection 9.7.2 to formulate bounds on parameters which define nonmechanical properties (e.g., conductivity and resistivity tensors) of composites, and in Subsection 9.8 to obtain exact upper and lower bounds for the coupled mechanical and nonmechanical moduli, using piezoelectricity as a specific case.

---

[1] Accorsi and Nemat-Nasser (1986) have used the Hashin-Shtrikman variational principle to obtain bounds on the overall elasticity and instantaneous elastic-plastic moduli of composites with periodic microstructures; see Sections 12, 13, and 14 for detailed accounts of some basic results for heterogeneous solids with periodic microstructure.

9.1. HASHIN-SHTRIKMAN VARIATIONAL PRINCIPLE

The Hashin-Shtrikman variational principle is formulated for a general linearly elastic heterogeneous RVE. The elasticity and compliance tensors are denoted by $\mathbf{C}'$ and $\mathbf{D}'$, respectively, where $\mathbf{C}' = (\mathbf{D}')^{-1}$ or $\mathbf{D}' = (\mathbf{C}')^{-1}$. They are, in general, functions of the position vector $\mathbf{x}$, i.e.,

$$\mathbf{C}' = \mathbf{C}'(\mathbf{x}), \qquad \mathbf{D}' = \mathbf{D}'(\mathbf{x}). \tag{9.1.1a,b}$$

These tensors satisfy the following symmetry conditions:

$$C'_{ijkl} = C'_{jikl} = C'_{ijlk} = C'_{klij}, \qquad D'_{ijkl} = D'_{jikl} = D'_{ijlk} = D'_{klij}. \tag{9.1.1c,d}$$

In particular, when the RVE consists of a linearly elastic homogeneous matrix and n distinct linearly elastic homogeneous micro-inclusions, $\mathbf{C}'$ and $\mathbf{D}'$ become

$$\mathbf{C}' = \begin{cases} \mathbf{C} & \text{in M} \\ \mathbf{C}^\alpha & \text{in } \Omega_\alpha, \end{cases} \qquad \mathbf{D}' = \begin{cases} \mathbf{D} & \text{in M} \\ \mathbf{D}^\alpha & \text{in } \Omega_\alpha, \end{cases} \tag{9.1.2a,b}$$

where $\mathbf{C}$ and $\mathbf{D}$, and $\mathbf{C}^\alpha$ and $\mathbf{D}^\alpha$ are the elasticity and compliance tensors of the matrix M and the micro-inclusions Ω_α ($\alpha = 1, 2, ..., n$), respectively.

9.1.1. Macrostress Prescribed

First consider the case when the RVE is subjected to uniform tractions produced by the constant macrostress $\boldsymbol{\Sigma} = \boldsymbol{\sigma}^o$; see Figure 9.1.1a. From the averaging theorem,

$$\bar{\boldsymbol{\sigma}} \equiv < \boldsymbol{\sigma} > = \boldsymbol{\sigma}^o, \tag{9.1.3a}$$

and the overall compliance tensor of the RVE, denoted by $\bar{\mathbf{D}}$, is *defined* by

$$\bar{\boldsymbol{\varepsilon}} \equiv < \boldsymbol{\varepsilon} > \equiv \bar{\mathbf{D}} : \boldsymbol{\sigma}^o. \tag{9.1.3b}$$

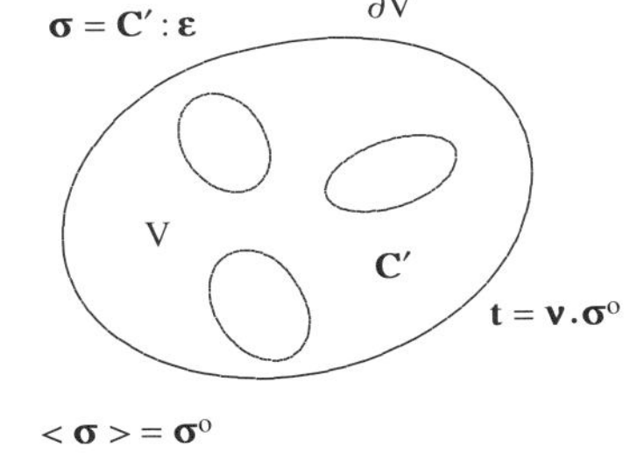

Figure 9.1.1a

Heterogeneous RVE of volume V, bounded by ∂V, subjected to uniform tractions $\mathbf{t} = \boldsymbol{\nu}.\boldsymbol{\sigma}^o$

Note that $\mathbf{C}'$ and $\mathbf{D}'$ are variable, and there may exist material discontinuity surfaces in V. However, the tractions and displacements must remain continuous across these surfaces. Hence, their presence does not affect the averaging theorems which are derived with the aid of the Gauss theorem; see Subsection 2.4. In view of this observation, the existence of material discontinuity surfaces does not require any special treatment in the present context.

Now consider an *equivalent homogeneous* solid with an overall geometry identical to that of the RVE, and introduce the eigenstrain field necessary for this homogenization. Let the homogenized solid consist of a *comparison material,* with constant elasticity and compliance tensors **C** and **D**. Note that in (9.1.2a,b), **C** and **D** are the elasticity and compliance tensors of the homogeneous matrix material. Here **C** and **D** are associated with an *arbitrary elastic comparison material* which is used for the homogenization of the original heterogeneous RVE. Introduce the eigenstrain field $\boldsymbol{\varepsilon}^* = \boldsymbol{\varepsilon}^*(\mathbf{x})$ such that, in the equivalent homogeneous solid with elasticity and compliance tensors **C** and **D**, the same stress and strain fields as exist in the original heterogeneous RVE are produced. Hence, there follow the consistency conditions,

$$\boldsymbol{\sigma}(\mathbf{x}) = \mathbf{C}'(\mathbf{x}) : \boldsymbol{\varepsilon}(\mathbf{x}) = \mathbf{C} : \{\boldsymbol{\varepsilon}(\mathbf{x}) - \boldsymbol{\varepsilon}^*(\mathbf{x})\},$$

$$\boldsymbol{\varepsilon}(\mathbf{x}) = \mathbf{D}'(\mathbf{x}) : \boldsymbol{\sigma}(\mathbf{x}) = \mathbf{D} : \boldsymbol{\sigma}(\mathbf{x}) + \boldsymbol{\varepsilon}^*(\mathbf{x}). \tag{9.1.4a,b}$$

Then, in terms of the eigenstrain $\boldsymbol{\varepsilon}^*(\mathbf{x})$, the stress field in the equivalent homogeneous solid, $\boldsymbol{\sigma}(\mathbf{x})$, which coincides with that in the original heterogeneous RVE, is given by

$$\boldsymbol{\sigma}(\mathbf{x}) = \{\mathbf{D}'(\mathbf{x}) - \mathbf{D}\}^{-1} : \boldsymbol{\varepsilon}^*(\mathbf{x}). \tag{9.1.4c}$$

Now consider the fact that the boundary tractions are uniform. If the perturbation stress $\boldsymbol{\sigma}^{\rm d}(\mathbf{x})$ is defined in the equivalent homogeneous solid through

$$\boldsymbol{\sigma}^{\rm d}(\mathbf{x}) \equiv \boldsymbol{\sigma}(\mathbf{x}) - \boldsymbol{\sigma}^{\rm o}, \tag{9.1.5a}$$

then,

$$\boldsymbol{\nu} \cdot \boldsymbol{\sigma}^{\rm d}(\mathbf{x}) = \mathbf{0} \quad \text{on } \partial{\rm V}; \tag{9.1.5b}$$

see Figure 9.1.1b. Because of linearity, the perturbation stress, $\boldsymbol{\sigma}^{\rm d}(\mathbf{x})$, in

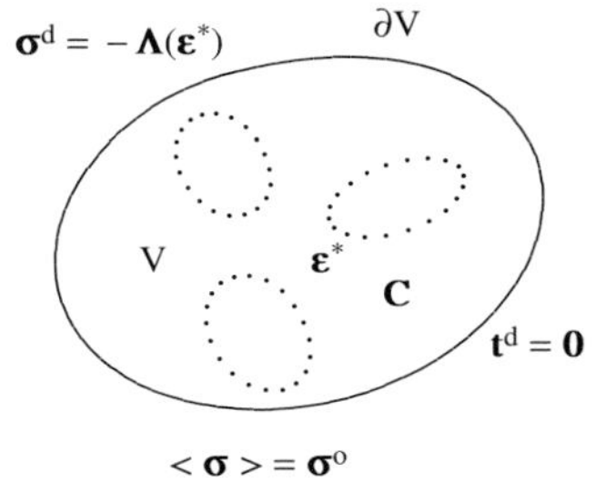

Figure 9.1.1b

Homogenizing eigenstrain field $\boldsymbol{\varepsilon}^*(\mathbf{x})$, distributed in homogeneous V of elasticity **C**, produces disturbance fields which leave boundary ∂V of V traction-free, $\mathbf{t}^{\rm d} = \boldsymbol{\nu} \cdot \boldsymbol{\sigma}^{\rm d} = \mathbf{0}$

general, is a linear functional of the distributed eigenstrains. Then, *formally* define an integral operator $\mathbf{\Lambda}(\mathbf{x}; ...)$ which determines $\boldsymbol{\sigma}^d(\mathbf{x})$ in terms of $\boldsymbol{\varepsilon}^*(\mathbf{x})$, through

$$\boldsymbol{\sigma}^d(\mathbf{x}) \equiv -\mathbf{\Lambda}(\mathbf{x}; \boldsymbol{\varepsilon}^*). \tag{9.1.6a}$$

The operator $\mathbf{\Lambda}(\mathbf{x}; ...)$ (denoted by $\mathbf{\Lambda}$ for short) has the following properties:

1) $\mathbf{\Lambda}$ depends on the geometry and material properties of the equivalent solid, but not on the prescribed (or required) eigenstrains;
2) the stress and strain fields determined by $\mathbf{\Lambda}$ are statically and kinematically admissible;
3) the stress field produced by $\mathbf{\Lambda}$ satisfies the zero-traction boundary conditions (9.1.5b).

In terms of Green's function $\mathbf{G}(\mathbf{x}, \mathbf{y})$ *which satisfies traction-free boundary conditions on* ∂V, integral operator $\mathbf{\Lambda}$ is expressed as

$$\mathbf{\Lambda}(\mathbf{x}; \boldsymbol{\varepsilon}^*) = -\mathbf{C} : \Big\{ \int_V \nabla_x \otimes \mathbf{G}(\mathbf{x}, \mathbf{y}) \cdot \{\nabla_y \cdot (\mathbf{C} : \boldsymbol{\varepsilon}^*(\mathbf{y}))\} \, dV_y - \boldsymbol{\varepsilon}^*(\mathbf{x}) \Big\}, \tag{9.1.6b}$$

or in component form,

$$\Lambda_{ij}(\mathbf{x}; \boldsymbol{\varepsilon}^*) = -C_{ijpq} \Big\{ \int_V \frac{\partial}{\partial x_p} G_{qs}(\mathbf{x}, \mathbf{y}) \, \{\frac{\partial}{\partial y_r}(C_{rskl}\, \varepsilon^*_{kl})\} \, dV_y - \varepsilon^*_{pq}(\mathbf{x}) \Big\}. \tag{9.1.6c}$$

The integral operator $\mathbf{\Lambda}$ is related to the integral operator $\mathbf{S}$ which has been introduced by (7.3.7) in Subsection 7.3. While $\mathbf{S}$ yields the perturbation strain field $\boldsymbol{\varepsilon}^d$ associated with a given eigenstrain field $\boldsymbol{\varepsilon}^*$, operator $-\mathbf{\Lambda}$ yields the corresponding stress field $\boldsymbol{\sigma}^d$. Hence,

$$\mathbf{\Lambda}(\mathbf{x}; \boldsymbol{\varepsilon}^*) = -\mathbf{C} : \{\mathbf{S}(\mathbf{x}; \boldsymbol{\varepsilon}^*) - \boldsymbol{\varepsilon}^*(\mathbf{x})\},$$

$$\mathbf{S}(\mathbf{x}; \boldsymbol{\varepsilon}^*) = -\mathbf{D} : \mathbf{\Lambda}(\mathbf{x}; \boldsymbol{\varepsilon}^*) + \boldsymbol{\varepsilon}^*(\mathbf{x}). \tag{9.1.6d,e}$$

In general, with respect to the averaging operator $< \;\; >$, the integral operator $\mathbf{\Lambda}$ is self-adjoint, i.e., for arbitrary eigenstrain fields $\mathbf{e}^{*(1)} = \mathbf{e}^{*(1)}(\mathbf{x})$ and $\mathbf{e}^{*(2)} = \mathbf{e}^{*(2)}(\mathbf{x})$ in the equivalent homogeneous solid,

$$< \mathbf{e}^{*(1)}(\mathbf{x}) : \mathbf{\Lambda}(\mathbf{x}; \mathbf{e}^{*(2)}) > = < \mathbf{e}^{*(2)}(\mathbf{x}) : \mathbf{\Lambda}(\mathbf{x}; \mathbf{e}^{*(1)}) >. \tag{9.1.7a}$$

The proof is straightforward. Let $\boldsymbol{\sigma}^{d(\alpha)} = \boldsymbol{\sigma}^{d(\alpha)}(\mathbf{x})$ and $\boldsymbol{\varepsilon}^{d(\alpha)} = \boldsymbol{\varepsilon}^{d(\alpha)}(\mathbf{x})$ be the stress and strain fields produced by the eigenstrain $\mathbf{e}^{*(\alpha)}$. Then from (9.1.4b) and (9.1.6a),

$$\boldsymbol{\varepsilon}^{d(\alpha)}(\mathbf{x}) \equiv \mathbf{D} : \boldsymbol{\sigma}^{d(\alpha)}(\mathbf{x}) + \mathbf{e}^{*(\alpha)}(\mathbf{x}),$$

$$\boldsymbol{\sigma}^{d(\alpha)}(\mathbf{x}) \equiv -\mathbf{\Lambda}(\mathbf{x}; \mathbf{e}^{*(\alpha)}) \qquad (\alpha = 1, 2). \tag{9.1.7b,c}$$

The eigenstrain $\mathbf{e}^{*(\alpha)}$ produces self-equilibrating body forces through $\nabla \cdot (\mathbf{C} : \mathbf{e}^{*(\alpha)})$. Hence, $\boldsymbol{\sigma}^{d(\alpha)} = -\mathbf{\Lambda}(\mathbf{x}; \mathbf{e}^{*(\alpha)})$ is statically admissible. Since $\boldsymbol{\sigma}^{d(\alpha)}$ and $\boldsymbol{\varepsilon}^{d(\alpha)}$ are statically and kinematically admissible, by the averaging theorem,

$$< \boldsymbol{\sigma}^{d(\alpha)} : \boldsymbol{\varepsilon}^{d(\beta)} > = \frac{1}{V} \int_{\partial V} \mathbf{t}^{d(\alpha)} \boldsymbol{\cdot} \mathbf{u}^{d(\beta)} \, dS, \tag{9.1.7d}$$

where $\mathbf{t}^{d(\alpha)}$ and $\mathbf{u}^{d(\beta)}$ are the tractions and displacements on ∂V relating to $\boldsymbol{\sigma}^{d(\alpha)}$ and $\boldsymbol{\varepsilon}^{d(\beta)}$, respectively. From the boundary conditions associated with $\boldsymbol{\Lambda}$, (9.1.5b), the surface integral in the right-hand side of (9.1.7d) vanishes. Hence, from (9.1.7b~d),

$$\begin{aligned} < \mathbf{e}^{*(\alpha)} : \boldsymbol{\Lambda}(\mathbf{e}^{*(\beta)}) > &= < (\boldsymbol{\varepsilon}^{d(\alpha)} - \mathbf{D} : \boldsymbol{\sigma}^{d(\alpha)}) : (-\, \boldsymbol{\sigma}^{d(\beta)}) > \\ &= < \boldsymbol{\sigma}^{d(\alpha)} : \mathbf{D} : \boldsymbol{\sigma}^{d(\beta)} > \\ &= < (\boldsymbol{\Lambda} : \mathbf{e}^{*(\alpha)}) : \mathbf{D} : (\boldsymbol{\Lambda} : \mathbf{e}^{*(\beta)}) >. \end{aligned} \tag{9.1.7e}$$

The symmetry of $\mathbf{D}$ with respect to the first and second pairs of its indices now implies (9.1.7a). The symmetry embedded in (9.1.7a,e) is displayed by the notation

$$\boldsymbol{\Lambda}(\mathbf{x}; \boldsymbol{\varepsilon}^*) \equiv (\boldsymbol{\Lambda} : \boldsymbol{\varepsilon}^*)(\mathbf{x}), \tag{9.1.6f}$$

and (9.1.7a) is rewritten as follows:

$$< \mathbf{e}^{*(\alpha)} : (\boldsymbol{\Lambda} : \mathbf{e}^{*(\beta)}) > \equiv < \mathbf{e}^{*(\alpha)} : \boldsymbol{\Lambda} : \mathbf{e}^{*(\beta)} >. \tag{9.1.7f}$$

In this notation, the operation $\boldsymbol{\Lambda} : \mathbf{e}^{*(\beta)}$ is to be understood in the sense of (9.1.6b,c,f), i.e., $\boldsymbol{\Lambda} : \mathbf{e}^{*(\beta)} \equiv (\boldsymbol{\Lambda} : \mathbf{e}^{*(\beta)})(\mathbf{x}) \equiv \boldsymbol{\Lambda}(\mathbf{x}; \mathbf{e}^{*(\beta)})$. Notation (9.1.7f) is used in the sequel. Note from (9.1.7e) that

$$< \; : \boldsymbol{\Lambda} : \; > \equiv < \; : (\boldsymbol{\Lambda} : \mathbf{D} : \boldsymbol{\Lambda}) : \; >. \tag{9.1.7g}$$

From (9.1.4c) and (9.1.6a), the consistency condition (9.1.4a) is replaced by the following consistency condition written for the eigenstrain field:

$$(\mathbf{D}'(\mathbf{x}) - \mathbf{D})^{-1} : \boldsymbol{\varepsilon}^*(\mathbf{x}) + (\boldsymbol{\Lambda} : \boldsymbol{\varepsilon}^*)(\mathbf{x}) - \boldsymbol{\sigma}^o = \mathbf{0}. \tag{9.1.8}$$

This consistency condition is a linear integral equation which defines the eigenstrain field necessary for homogenization. The solution to this integral equation is the *exact* eigenstrain field which gives the actual stress and strain fields of the original RVE.

Now a functional is constructed for the eigenstrain field in the equivalent homogeneous solid such that the exact eigenstrain that satisfies the consistency condition (9.1.8) renders this functional stationary. Using the symmetry of $\boldsymbol{\Lambda}$, define a functional, I, for an arbitrary eigenstrain, $\mathbf{e}^*$, by

$$I(\mathbf{e}^*; \boldsymbol{\sigma}^o) \equiv \frac{1}{2} < \mathbf{e}^* : \{(\mathbf{D}' - \mathbf{D})^{-1} + \boldsymbol{\Lambda}\} : \mathbf{e}^* > - < \boldsymbol{\sigma}^o : \mathbf{e}^* >, \tag{9.1.9a}$$

where $< \mathbf{e}^* : \boldsymbol{\Lambda} : \mathbf{e}^* > \equiv < \mathbf{e}^* : (\boldsymbol{\Lambda} : \mathbf{e}^*) > = < (\boldsymbol{\Lambda} : \mathbf{e}^*) : \mathbf{e}^* >$. The overall stress $\boldsymbol{\sigma}^o$ is regarded as fixed. The boundary conditions imposed on $\boldsymbol{\Lambda}$ then lead to $< (\boldsymbol{\Lambda} : \mathbf{e}^*) > = \mathbf{0}$. The first variation of functional I is given by

$$\delta I(\mathbf{e}^*; \boldsymbol{\sigma}^o) = < \delta\mathbf{e}^* : \left\{ \{(\mathbf{D}' - \mathbf{D})^{-1} + \boldsymbol{\Lambda}\} : \mathbf{e}^* - \boldsymbol{\sigma}^o \right\} >, \tag{9.1.9b}$$

where $\delta\mathbf{e}^*$ is an arbitrary variation in the eigenstrain field. As is seen, for the *exact* eigenstrain $\boldsymbol{\varepsilon}^*$ which satisfies (9.1.8), i.e., the eigenstrain field which produces in the equivalent homogeneous solid the same stress and strain fields as in the original heterogeneous RVE, $\delta \mathrm{I}(\boldsymbol{\varepsilon}^*; \boldsymbol{\sigma}^{\mathrm{o}}) = 0$. Hence $\mathrm{I}(\boldsymbol{\varepsilon}^*; \boldsymbol{\sigma}^{\mathrm{o}})$ is the stationary value of (9.1.9a). Moreover, the vanishing of the first variation of $\mathrm{I}(\boldsymbol{\varepsilon}^*; \boldsymbol{\sigma}^{\mathrm{o}})$, for arbitrary variation of the eigenstrain, yields the consistency condition (9.1.8).

Using the exact eigenstrain $\boldsymbol{\varepsilon}^*$, rewrite the functional I, (9.1.9a), as

$$\mathrm{I}(\mathbf{e}^*; \boldsymbol{\sigma}^{\mathrm{o}}) = \frac{1}{2} < (\mathbf{e}^* - \boldsymbol{\varepsilon}^*) : \{(\mathbf{D}' - \mathbf{D})^{-1} + \boldsymbol{\Lambda}\} : (\mathbf{e}^* - \boldsymbol{\varepsilon}^*) > + \mathrm{I}(\boldsymbol{\varepsilon}^*; \boldsymbol{\sigma}^{\mathrm{o}}) \tag{9.1.10a}$$

which attains its stationary value for $\mathbf{e}^* = \boldsymbol{\varepsilon}^*$. Note that $\mathrm{I}(\boldsymbol{\varepsilon}^*; \boldsymbol{\sigma}^{\mathrm{o}})$ is the change of the complementary strain energy associated with the difference between the reference and the overall compliance tensors, $\mathbf{D} - \bar{\mathbf{D}}$. Indeed, direct substitution of $\boldsymbol{\varepsilon}^*$ into (9.1.9a) yields

$$\mathrm{I}(\boldsymbol{\varepsilon}^*; \boldsymbol{\sigma}^{\mathrm{o}}) = -\frac{1}{2} < \boldsymbol{\varepsilon}^* : \boldsymbol{\sigma}^{\mathrm{o}} > = \frac{1}{2} \boldsymbol{\sigma}^{\mathrm{o}} : (\mathbf{D} - \bar{\mathbf{D}}) : \boldsymbol{\sigma}^{\mathrm{o}}, \tag{9.1.10b}$$

where $\bar{\mathbf{D}}$ is the overall compliance tensor defined by (9.1.3b). In Subsection 9.2, representations (9.1.10a) and (9.1.10b) are used to establish extremum principles for calculating the eigenstrain field $\boldsymbol{\varepsilon}^*$ and for obtaining bounds on the overall moduli $\bar{\mathbf{D}}$.

For uniform traction boundary data, (9.1.9) and (9.1.10) define the Hashin-Shtrikman variational principle, in the sense that the Euler equation associated with functional $\mathrm{I}(\mathbf{e}^*; \boldsymbol{\sigma}^{\mathrm{o}})$ is the corresponding consistency condition, (9.1.8).

9.1.2. Macrostrain Prescribed

Next, consider the case when the RVE is subjected to linear displacements produced by the constant macrostrain $\mathbf{E} = \boldsymbol{\varepsilon}^{\mathrm{o}}$; see Figure 9.1.2a. From the averaging theorem,

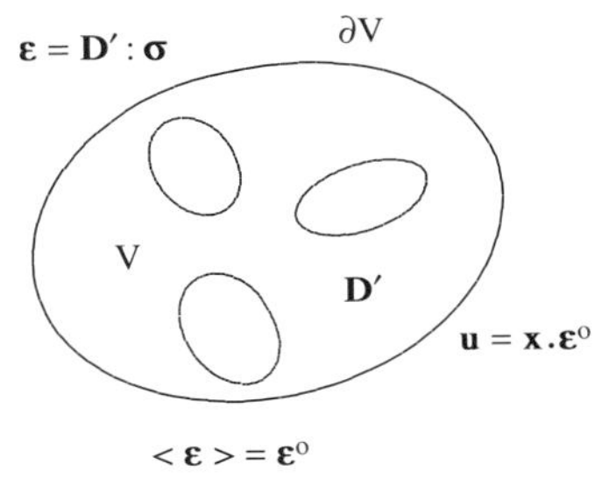

Figure 9.1.2a

Heterogeneous RVE of volume V, bounded by ∂V, subjected to linear displacements $\mathbf{u} = \mathbf{x} \cdot \boldsymbol{\varepsilon}^{\mathrm{o}}$

$$\bar{\boldsymbol{\varepsilon}} \equiv \langle \boldsymbol{\varepsilon} \rangle = \boldsymbol{\varepsilon}^{o}, \tag{9.1.11a}$$

and the overall elasticity tensor of the RVE, denoted by $\bar{\mathbf{C}}$, is *defined* by

$$\bar{\boldsymbol{\sigma}} \equiv \langle \boldsymbol{\sigma} \rangle \equiv \bar{\mathbf{C}} : \boldsymbol{\varepsilon}^{o}. \tag{9.1.11b}$$

Now, instead of the eigenstrain $\boldsymbol{\varepsilon}^*$, introduce an *eigenstress field,* $\boldsymbol{\sigma}^*$, in the equivalent homogeneous solid, such that the final stress and strain fields coincide with those of the original heterogeneous RVE. The consistency condition for the eigenstress $\boldsymbol{\sigma}^*$ is given by

$$\boldsymbol{\varepsilon}(\mathbf{x}) = \mathbf{D}'(\mathbf{x}) : \boldsymbol{\sigma}(\mathbf{x}) = \mathbf{D} : \{\boldsymbol{\sigma}(\mathbf{x}) - \boldsymbol{\sigma}^*(\mathbf{x})\},$$

$$\boldsymbol{\sigma}(\mathbf{x}) = \mathbf{C}'(\mathbf{x}) : \boldsymbol{\varepsilon}(\mathbf{x}) = \mathbf{C} : \boldsymbol{\varepsilon}(\mathbf{x}) + \boldsymbol{\sigma}^*(\mathbf{x}), \tag{9.1.12a,b}$$

where $\mathbf{D}$ and $\mathbf{C}$ are the compliance and elasticity tensors of the equivalent homogeneous solid. Then, in terms of the eigenstress $\boldsymbol{\sigma}^*$, the strain field in the homogeneous solid is expressed as

$$\boldsymbol{\varepsilon}(\mathbf{x}) = \{\mathbf{C}'(\mathbf{x}) - \mathbf{C}\}^{-1} : \boldsymbol{\sigma}^*(\mathbf{x}), \tag{9.1.12c}$$

with the corresponding stress field defined by (9.1.12b). These stress and strain fields in the equivalent homogeneous solid are identical with the actual stress and strain fields in the original heterogeneous RVE.

Define the perturbation strain $\boldsymbol{\varepsilon}^{d}(\mathbf{x})$ in the equivalent homogeneous solid by

$$\boldsymbol{\varepsilon}^{d}(\mathbf{x}) \equiv \boldsymbol{\varepsilon}(\mathbf{x}) - \boldsymbol{\varepsilon}^{o}. \tag{9.1.13a}$$

Then, the displacement field associated with $\boldsymbol{\varepsilon}^{d}(\mathbf{x})$ satisfies

$$\mathbf{u}^{d} = \mathbf{0} \qquad \text{on } \partial V; \tag{9.1.13b}$$

see Figure 9.1.2b. The perturbation strain field $\boldsymbol{\varepsilon}^{d}(\mathbf{x})$, produced by the eigenstress $\boldsymbol{\sigma}^*(\mathbf{x})$, can be expressed in terms of an integral operator $\boldsymbol{\Gamma}(\mathbf{x}; ...)$ (denoted by $\boldsymbol{\Gamma}$ for short), in a manner similar to that which led to the introduction of the operator $\boldsymbol{\Lambda}$. Hence, write

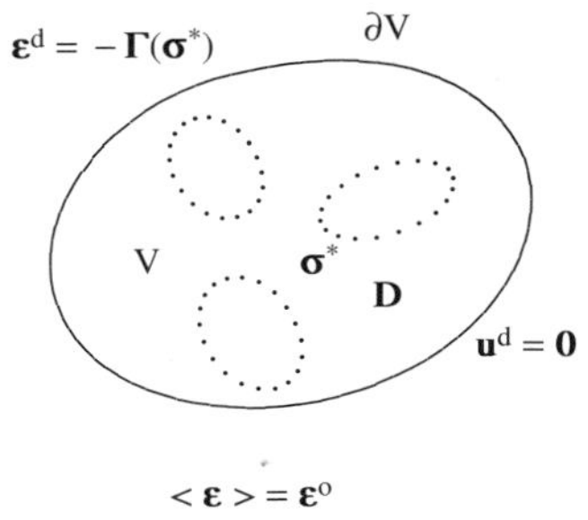

Figure 9.1.2b

Homogenizing eigenstress field $\boldsymbol{\sigma}^*(\mathbf{x})$, distributed in homogeneous V of compliance $\mathbf{D}$, produces disturbance fields, such that displacements on boundary ∂V of V vanish; $\mathbf{u}^{d} = \mathbf{0}$

$$\boldsymbol{\varepsilon}^{\rm d}(\mathbf{x}) \equiv -\boldsymbol{\Gamma}(\mathbf{x};\, \boldsymbol{\sigma}^*), \tag{9.1.14a}$$

or, following the notation introduced for $\boldsymbol{\Lambda}$ in (9.1.6f), set

$$\boldsymbol{\varepsilon}^{\rm d}(\mathbf{x}) = -(\boldsymbol{\Gamma} : \boldsymbol{\sigma}^*)(\mathbf{x}). \tag{9.1.14b}$$

In terms of Green's function $\mathbf{G}(\mathbf{x}, \mathbf{y})$ which *satisfies zero surface displacement boundary conditions on* ∂V, integral operator $\boldsymbol{\Gamma}$ is expressed as

$$\boldsymbol{\Gamma}(\mathbf{x};\, \boldsymbol{\sigma}^*) = -\int_{\rm V} \frac{1}{2}\Big\{ \{\nabla_{\mathbf{x}} \otimes \mathbf{G}(\mathbf{x}, \mathbf{y}) \cdot (\nabla_{\mathbf{y}} \cdot \boldsymbol{\sigma}^*(\mathbf{y}))\} + \{\nabla_{\mathbf{x}} \otimes \mathbf{G}(\mathbf{x}, \mathbf{y}) \cdot (\nabla_{\mathbf{y}} \cdot \boldsymbol{\sigma}^*(\mathbf{y}))\}^{\rm T} \Big\}\, {\rm dV_y}, \tag{9.1.14c}$$

or in component form,

$$\Gamma_{ij}(\mathbf{x};\, \boldsymbol{\sigma}^*) = -\int_{\rm V} \frac{1}{2}\{ \frac{\partial}{\partial x_i} G_{jl}(\mathbf{x}, \mathbf{y}) \frac{\partial}{\partial y_k} \sigma^*_{kl}(\mathbf{y}) + \frac{\partial}{\partial x_j} G_{il}(\mathbf{x}, \mathbf{y}) \frac{\partial}{\partial y_k} \sigma^*_{kl}(\mathbf{y}) \}\, {\rm dV_y}. \tag{9.1.14d}$$

The integral operator $\boldsymbol{\Gamma}$ is given by the integral operator $\mathbf{T}(\mathbf{x};\, ...)$ introduced by (7.3.10) in Subsection 7.3. Indeed, $\boldsymbol{\Gamma}$ and $\mathbf{T}$ respectively give the strain and stress fields due to a given eigenstress field,

$$\boldsymbol{\Gamma}(\mathbf{x};\, \boldsymbol{\sigma}^*) = -\mathbf{D} : \{\mathbf{T}(\mathbf{x};\, \boldsymbol{\sigma}^*) - \boldsymbol{\sigma}^*(\mathbf{x})\},$$

$$\mathbf{T}(\mathbf{x};\, \boldsymbol{\sigma}^*) = -\mathbf{C} : \boldsymbol{\Gamma}(\mathbf{x};\, \boldsymbol{\sigma}) + \boldsymbol{\sigma}^*(\mathbf{x}). \tag{9.1.14e,f}$$

Table 9.1.1 summarizes the relations among integral operators $\mathbf{S}$, $\mathbf{T}$, $\boldsymbol{\Lambda}$, and $\boldsymbol{\Gamma}$.

The integral operator $\boldsymbol{\Gamma}$ satisfies properties 1) and 2) which are stated for $\boldsymbol{\Lambda}$, after (9.1.6a). However, while $\boldsymbol{\Lambda}$ produces zero tractions on ∂V, $\boldsymbol{\Gamma}$ produces zero displacements on ∂V. Hence, instead of property 3) of $\boldsymbol{\Lambda}$, the operator $\boldsymbol{\Gamma}$ satisfies the following condition:

3′) the displacement field produced by $\boldsymbol{\Gamma}$ vanishes on the boundary ∂V, i.e., (9.1.13b) is satisfied identically.

In general, similarly to $\boldsymbol{\Lambda}$, the integral operator $\boldsymbol{\Gamma}$ is self-adjoint, i.e., for arbitrary eigenstress fields $\mathbf{s}^{*(1)}$ and $\mathbf{s}^{*(2)}$ in the equivalent homogeneous solid,

$$< \mathbf{s}^{*(1)} : (\boldsymbol{\Gamma} : \mathbf{s}^{*(2)}) > = < \mathbf{s}^{*(2)} : (\boldsymbol{\Gamma} : \mathbf{s}^{*(1)}) > = < \mathbf{s}^{*(1)} : \boldsymbol{\Gamma} : \mathbf{s}^{*(2)} >; \tag{9.1.15a}$$

here, again, the same notation as in (9.1.7f) is used. The proof is similar to that for $\boldsymbol{\Lambda}$. Taking advantage of the boundary conditions (9.1.13b), observe that

$$< \mathbf{s}^{*(\alpha)} : (\boldsymbol{\Gamma} : \mathbf{s}^{*(\beta)}) > = < (\boldsymbol{\Gamma} : \mathbf{s}^{*(\alpha)}) : \mathbf{C} : (\boldsymbol{\Gamma} : \mathbf{s}^{*(\beta)}) >, \tag{9.1.15b}$$

where $-(\boldsymbol{\Gamma} : \mathbf{s}^{*(\alpha)})(\mathbf{x})$ is the strain field associated with $\mathbf{s}^{*(\alpha)}$. From the symmetry of $\mathbf{C}$ with respect to the first and second pairs of its indices, (9.1.15a) now follows. Note from (9.1.15b), that

Table 9.1.1

Relation among integral operators $\mathbf{S}$, $\mathbf{T}$, $\mathbf{\Lambda}$, and $\mathbf{\Gamma}$

uniform traction boundary conditions: $\mathbf{S}$ and $\mathbf{\Lambda}$ for $\boldsymbol{\varepsilon}^*$		
disturbance strain	$\mathbf{S}(\mathbf{x};\, \boldsymbol{\varepsilon}^*)$	$-\mathbf{D} : \mathbf{\Lambda}(\mathbf{x};\, \boldsymbol{\varepsilon}^*) + \boldsymbol{\varepsilon}^*$
disturbance stress	$\mathbf{C} : (\mathbf{S}(\mathbf{x};\, \boldsymbol{\varepsilon}^*) - \boldsymbol{\varepsilon}^*)$	$-\mathbf{\Lambda}(\mathbf{x};\, \boldsymbol{\varepsilon}^*)$
linear displacement boundary conditions: $\mathbf{T}$ and $\mathbf{\Gamma}$ for $\boldsymbol{\sigma}^*$		
disturbance strain	$\mathbf{D} : (\mathbf{T}(\mathbf{x};\, \boldsymbol{\sigma}^*) - \boldsymbol{\sigma}^*)$	$-\mathbf{\Gamma}(\mathbf{x};\, \boldsymbol{\sigma}^*)$
disturbance stress	$\mathbf{T}(\mathbf{x};\, \boldsymbol{\sigma}^*)$	$-\mathbf{C} : \mathbf{\Gamma}(\mathbf{x};\, \boldsymbol{\sigma}^*) + \boldsymbol{\sigma}^*$

$$< :\mathbf{\Gamma}: > \equiv < :(\mathbf{\Gamma} : \mathbf{C} : \mathbf{\Gamma}) : >. \tag{9.1.15c}$$

From (9.1.12b) and (9.1.14a), the following consistency condition for the eigenstress $\boldsymbol{\sigma}^*$ is obtained,

$$(\mathbf{C}'(\mathbf{x}) - \mathbf{C})^{-1} : \boldsymbol{\sigma}^*(\mathbf{x}) + (\mathbf{\Gamma} : \boldsymbol{\sigma}^*)(\mathbf{x}) - \boldsymbol{\varepsilon}^{\mathrm{o}} = \mathbf{0}. \tag{9.1.16}$$

For a prescribed $\boldsymbol{\varepsilon}^{\mathrm{o}}$, this is a linear integral equation which defines $\boldsymbol{\sigma}^*$ in the equivalent homogeneous solid, such that the corresponding final stress and strain fields are identical with those of the original heterogeneous RVE.

Consider now the following functional, J, similar to the functional I, which is stationary for the eigenstress field which satisfies the consistency condition (9.1.16):

$$\mathrm{J}(\mathbf{s}^*;\, \boldsymbol{\varepsilon}^{\mathrm{o}}) \equiv \frac{1}{2} < \mathbf{s}^* : \{(\mathbf{C}' - \mathbf{C})^{-1} + \mathbf{\Gamma}\} : \mathbf{s}^* > - < \boldsymbol{\varepsilon}^{\mathrm{o}} : \mathbf{s}^* >, \tag{9.1.17a}$$

where the fact that $\mathbf{\Gamma}$ is self-adjoint is also used. In (9.1.17a), the overall strain $\boldsymbol{\varepsilon}^{\mathrm{o}}$ is fixed. The boundary conditions imposed on $\mathbf{\Gamma}$ lead to $< (\mathbf{\Gamma} : \mathbf{s}^*) > = \mathbf{0}$. Thus, the first variation of J, for an arbitrary variation $\delta\mathbf{s}^*$ of the eigenstress field, yields

$$\delta\mathrm{J}(\mathbf{s}^*;\, \boldsymbol{\varepsilon}^{\mathrm{o}}) = < \delta\mathbf{s}^* : \left\{ (\mathbf{C}' - \mathbf{C})^{-1} + \mathbf{\Gamma}\} : \mathbf{s}^* - \boldsymbol{\varepsilon}^{\mathrm{o}} \right\} >. \tag{9.1.17b}$$

Therefore, for the *exact* eigenstress $\boldsymbol{\sigma}^*$ which satisfies (9.1.16), $\delta\mathrm{J}(\boldsymbol{\sigma}^*;\, \boldsymbol{\varepsilon}^{\mathrm{o}}) = 0$. Hence, for the eigenstress field which produces in the equivalent homogeneous solid the exact stress and strain fields of the original heterogeneous RVE, the functional $\mathrm{J}(\boldsymbol{\sigma}^*;\, \boldsymbol{\varepsilon}^{\mathrm{o}})$ is stationary. Furthermore, the vanishing of the first variation of $\mathrm{J}(\boldsymbol{\sigma}^*;\, \boldsymbol{\varepsilon}^{\mathrm{o}})$, for arbitrary variation of the eigenstress, leads to the consistency condition (9.1.16).

In a manner similar to (9.1.10a), the functional J can be rewritten as

$$\mathrm{J}(\mathbf{s}^*;\, \boldsymbol{\varepsilon}^o) = \frac{1}{2} < (\mathbf{s}^* - \boldsymbol{\sigma}^*) : \{(\mathbf{C}' - \mathbf{C})^{-1} + \boldsymbol{\Gamma}\} : (\mathbf{s}^* - \boldsymbol{\sigma}^*) > + \,\mathrm{J}(\boldsymbol{\sigma}^*;\, \boldsymbol{\varepsilon}^o) \tag{9.1.18a}$$

which, for $\mathbf{s}^* = \boldsymbol{\sigma}^*$, has the stationary value $\mathrm{J}(\boldsymbol{\sigma}^*;\, \boldsymbol{\varepsilon}^o)$, and can be used to establish bounds for the overall moduli. Direct substitution of $\boldsymbol{\sigma}^*$ into (9.1.17a) yields

$$\mathrm{J}(\boldsymbol{\sigma}^*;\, \boldsymbol{\varepsilon}^o) = -\frac{1}{2} < \boldsymbol{\sigma}^* : \boldsymbol{\varepsilon}^o > = \frac{1}{2} \boldsymbol{\varepsilon}^o : (\mathbf{C} - \bar{\mathbf{C}}) : \boldsymbol{\varepsilon}^o, \tag{9.1.18b}$$

where $\bar{\mathbf{C}}$ is the overall elasticity tensor defined by (9.1.11b). Therefore, similarly to $\mathrm{I}(\boldsymbol{\varepsilon}^*;\, \boldsymbol{\sigma}^o)$ which gives the change in the complementary strain energy due to the inhomogeneity of the RVE, $\mathrm{J}(\boldsymbol{\sigma}^*;\, \boldsymbol{\varepsilon}^o)$ gives the change in the corresponding strain energy. It should be noted that the functional I is defined with the overall stress $\boldsymbol{\sigma}^o$ prescribed, while the functional J is defined with the overall strain $\boldsymbol{\varepsilon}^o$ fixed. Hence, for *an arbitrary heterogeneous elastic solid,* $\mathrm{I}(\boldsymbol{\varepsilon}^*;\, \boldsymbol{\sigma}^o)$ and $\mathrm{J}(\boldsymbol{\sigma}^*;\, \boldsymbol{\varepsilon}^o)$ may not be related, even if $\boldsymbol{\sigma}^o = \mathbf{C} : \boldsymbol{\varepsilon}^o$ or $\boldsymbol{\varepsilon}^o = \mathbf{D} : \boldsymbol{\sigma}^o$, with $\mathbf{C} = \mathbf{D}^{-1}$ and $\mathbf{D} = \mathbf{C}^{-1}$.

9.2. UPPER AND LOWER BOUNDS FOR ENERGY FUNCTIONALS

In Subsection 9.1, functionals I and J are introduced for prescribed eigenstrain and eigenstress fields in the equivalent homogeneous solid. It is shown that the eigenstrain and eigenstress fields, $\boldsymbol{\varepsilon}^*$ and $\boldsymbol{\sigma}^*$, which satisfy their corresponding consistency conditions, respectively render I and J stationary. Under certain conditions, these stationary values become the extremum values of these functionals.

To establish this, first note that a fourth-order symmetric tensor $\mathbf{A}$ is positive-definite (negative-definite) when, for every symmetric second-order tensor $\mathbf{t}$ of nonzero magnitude, $\mathbf{t} : \mathbf{t} \neq 0$, the following inequality holds:

$$\mathbf{t} : \mathbf{A} : \mathbf{t} > (<)\, 0. \tag{9.2.1a}$$

$\mathbf{A}$ is called positive-semi-definite (negative-semi-definite), when under the same conditions,

$$\mathbf{t} : \mathbf{A} : \mathbf{t} \geq (\leq)\, 0. \tag{9.2.1b}$$

Hence, the elasticity and compliance tensors, $\mathbf{C}'$ and $\mathbf{D}'$, are positive-definite, since the energy required for any elastic deformation is always positive.

The *integral operators*, $\boldsymbol{\Lambda}$ and $\boldsymbol{\Gamma}$, associated with the averaging operator $< \;>$, namely the operators $< : \{(\mathbf{D}' - \mathbf{D})^{-1} + \boldsymbol{\Lambda}\} : >$ and $< : \{(\mathbf{C}' - \mathbf{C})^{-1} + \boldsymbol{\Gamma}\} : >$, are called positive-definite (negative-definite) or positive-semi-definite (negative-semi-definite) if, for every eigenstrain field $\mathbf{e}^*$ and eigenstress field $\mathbf{s}^*$ with nonzero norms, $< \mathbf{e}^* : \mathbf{e}^* > \neq 0$ and $< \mathbf{s}^* : \mathbf{s}^* > \neq 0$, it follows that

$$< \mathbf{e}^* : \{(\mathbf{D}' - \mathbf{D})^{-1} + \mathbf{\Lambda}\} : \mathbf{e}^* > \quad > (<)\, 0 \quad \text{or} \quad \geq (\leq)\, 0,$$

$$< \mathbf{s}^* : \{(\mathbf{C}' - \mathbf{C})^{-1} + \mathbf{\Gamma}\} : \mathbf{s}^* > \quad > (<)\, 0, \quad \text{or} \quad \geq (\leq)\, 0. \tag{9.2.2a,b}$$

From (9.1.10a) it is seen that if the integral operator $< \, : \{(\mathbf{D}' - \mathbf{D})^{-1} + \mathbf{\Lambda}\} : \, >$ is positive-definite (negative-definite), then the local stationary value of the functional I, i.e., $\mathrm{I}(\boldsymbol{\varepsilon}^*; \boldsymbol{\sigma}^o)$, is the corresponding global minimum (maximum), i.e., for any eigenstrain field $\mathbf{e}^*$,

$$\mathrm{I}(\boldsymbol{\varepsilon}^*; \boldsymbol{\sigma}^o) \leq (\geq)\, \mathrm{I}(\mathbf{e}^*; \boldsymbol{\sigma}^o), \tag{9.2.2c}$$

where $\boldsymbol{\varepsilon}^*$ is the eigenstrain field which satisfies the consistency condition (9.1.8). Similarly, from (9.1.18a), if $< \, : \{(\mathbf{C}' - \mathbf{C})^{-1} + \mathbf{\Gamma}\} : \, >$ is positive-definite (negative-definite), then for any eigenstress field $\mathbf{s}^*$,

$$\mathrm{J}(\boldsymbol{\sigma}^*; \boldsymbol{\varepsilon}^o) \leq (\geq)\, \mathrm{J}(\mathbf{s}^*; \boldsymbol{\varepsilon}^o), \tag{9.2.2d}$$

where $\boldsymbol{\sigma}^*$ is the eigenstress field which satisfies the consistency condition (9.1.16).

In this subsection it is shown that the positive-definiteness (negative-definiteness) of, e.g., the tensor $\mathbf{C}' - \mathbf{C}$ is equivalent to the positive-definiteness (negative-definiteness) of the integral operator $< \, : \{(\mathbf{C}' - \mathbf{C})^{-1} + \mathbf{\Gamma}\} : \, >$ *and* the negative-definiteness (positive-definiteness) of the integral operator $< \, : \{(\mathbf{D}' - \mathbf{D})^{-1} + \mathbf{\Lambda}\} : \, >$. In view of (9.2.2c), minimization (maximization) of $\mathrm{I}(\mathbf{e}^*; \boldsymbol{\sigma}^o)$ with respect to $\mathbf{e}^*$, over a suitable class of approximating functions, results in an optimal estimate of $\boldsymbol{\varepsilon}^*$ within the considered function-space, when $\mathbf{C}' - \mathbf{C}$ is positive-definite (negative-definite). Similar comments apply to $\mathrm{J}(\mathbf{s}^*; \boldsymbol{\varepsilon}^o)$, viewed as a functional of the eigenstress field $\mathbf{s}^*(\mathbf{x})$.

From the identities

$$\mathbf{C}' - \mathbf{C} = (\mathbf{C}' - \mathbf{C}) : (\mathbf{C}' - \mathbf{C})^{-1} : (\mathbf{C}' - \mathbf{C}),$$

$$\mathbf{D}' - \mathbf{D} = (\mathbf{D}' - \mathbf{D}) : (\mathbf{D}' - \mathbf{D})^{-1} : (\mathbf{D}' - \mathbf{D}), \tag{9.2.3a,b}$$

it is seen that: (1) if $\mathbf{C}' - \mathbf{C}$ is positive-definite (negative-definite), then $(\mathbf{C}' - \mathbf{C})^{-1}$ is positive-definite (negative-definite); and (2) if $\mathbf{D}' - \mathbf{D}$ is positive-definite (negative-definite), then $(\mathbf{D}' - \mathbf{D})^{-1}$ is positive-definite (negative-definite), i.e.,

$$\mathbf{C}' - \mathbf{C} \text{ is p.d. (n.d.)} \Longleftrightarrow (\mathbf{C}' - \mathbf{C})^{-1} \text{ is p.d. (n.d.)},$$

$$\mathbf{D}' - \mathbf{D} \text{ is p.d. (n.d.)} \Longleftrightarrow (\mathbf{D}' - \mathbf{D})^{-1} \text{ is p.d. (n.d.)}, \tag{9.2.3c,d}$$

where p.d. and n.d. stand for positive-definite and negative-definite, respectively.

9.2.1. Stiff Micro-Inclusions

First, assume $\mathbf{C}' - \mathbf{C}$ is positive-definite, i.e., choose the comparison elasticity tensor $\mathbf{C}$ such that $\mathbf{C}' - \mathbf{C}$ is positive-definite[2]. For example, when an RVE contains micro-inclusions which are stiffer than the surrounding uniform matrix and there are suitable symmetries, identify $\mathbf{C}$ with the elastic tensor of the matrix material, in which case, the class of eigenstresses (eigenstrains) is restricted to vanish within matrix M. From (9.1.15b), for any eigenstress field $\mathbf{s}^*$, $\mathbf{\Gamma}$ satisfies

$$< \mathbf{s}^* : \mathbf{\Gamma} : \mathbf{s}^* > = < (\mathbf{\Gamma} : \mathbf{s}^*) : \mathbf{C} : (\mathbf{\Gamma} : \mathbf{s}^*) >. \tag{9.2.4}$$

Since $\mathbf{C}$ is a positive-definite tensor, $< : \mathbf{\Gamma} : >$ is a positive-definite operator. Hence, the positive-definiteness of the tensor $\mathbf{C}' - \mathbf{C}$ implies the positive-definiteness of the operator $< : \{(\mathbf{C}' - \mathbf{C})^{-1} + \mathbf{\Gamma}\} : >$.

Furthermore, the positive-definiteness of the tensor $\mathbf{C}' - \mathbf{C}$ implies the negative-definiteness of the operator $< : \{(\mathbf{D}' - \mathbf{D})^{-1} + \mathbf{\Lambda}\} : >$. To see this, let $\boldsymbol{\sigma}^{\mathrm{d}}$ and $\boldsymbol{\varepsilon}^{\mathrm{d}}$ be the strain and stress fields produced by an arbitrary eigenstrain field $\mathbf{e}^*$, i.e.,

$$\boldsymbol{\varepsilon}^{\mathrm{d}} \equiv \mathbf{D} : \boldsymbol{\sigma}^{\mathrm{d}} + \mathbf{e}^*, \qquad \boldsymbol{\sigma}^{\mathrm{d}} \equiv -(\mathbf{\Lambda} : \mathbf{e}^*). \tag{9.2.5a,b}$$

Then, $< \mathbf{e}^* : \mathbf{\Lambda} : \mathbf{e}^* >$ becomes

$$\begin{aligned} < \mathbf{e}^* : \mathbf{\Lambda} : \mathbf{e}^* > &= < \mathbf{e}^* : \mathbf{C} : \mathbf{e}^* > - < \mathbf{e}^* : \mathbf{C} : \boldsymbol{\varepsilon}^{\mathrm{d}} > \\ &= < \mathbf{e}^* : \mathbf{C} : \mathbf{e}^* > - < \boldsymbol{\varepsilon}^{\mathrm{d}} : \mathbf{C} : \boldsymbol{\varepsilon}^{\mathrm{d}} >. \end{aligned} \tag{9.2.5c}$$

From identity

$$(\mathbf{D}' - \mathbf{D})^{-1} = -\mathbf{C} : (\mathbf{C}' - \mathbf{C})^{-1} : \mathbf{C} - \mathbf{C}, \tag{9.2.5d}$$

it follows that

$$\begin{aligned} < \mathbf{e}^* : (\mathbf{D}' - \mathbf{D})^{-1} : \mathbf{e}^* > = &- < (\mathbf{C} : \mathbf{e}^*) : (\mathbf{C}' - \mathbf{C})^{-1} : (\mathbf{C} : \mathbf{e}^*) > \\ &- < \mathbf{e}^* : \mathbf{C} : \mathbf{e}^* >. \end{aligned} \tag{9.2.5e}$$

Hence, adding (9.2.5c) and (9.2.5e), obtain

$$\begin{aligned} < \mathbf{e}^* : \{(\mathbf{D}' - \mathbf{D})^{-1} + \mathbf{\Lambda}\} : \mathbf{e}^* > = &- < (\mathbf{C} : \mathbf{e}^*) : (\mathbf{C}' - \mathbf{C})^{-1} : (\mathbf{C} : \mathbf{e}^*) > \\ &- < \boldsymbol{\varepsilon}^{\mathrm{d}} : \mathbf{C} : \boldsymbol{\varepsilon}^{\mathrm{d}} >. \end{aligned} \tag{9.2.5f}$$

Therefore, if $\mathbf{C}' - \mathbf{C}$ is positive-definite, then the tensors $(\mathbf{D}' - \mathbf{D})^{-1}$ and $\mathbf{D}' - \mathbf{D}$, and the integral operator $< : \{(\mathbf{D}' - \mathbf{D})^{-1} + \mathbf{\Lambda}\} : >$ are all negative-definite.

---

[2] The same comments apply even when $\mathbf{C}' - \mathbf{C}$ is positive-semi-definite. However, the positive-semi-definiteness of $\mathbf{C}' - \mathbf{C}$ implies that some eigenvalues of $\mathbf{C}$ coincide with the corresponding eigenvalues of $\mathbf{C}'$ at some points within V, where, then, $(\mathbf{C}' - \mathbf{C})^{-1}$ is not defined. In such a case, restrict the appropriate components of the eigenstrain field to vanish where the corresponding eigenvalues of $\mathbf{C}' - \mathbf{C}$ vanish. This restriction is mandatory, otherwise functional J will not be well defined. Thus, the eigenstrains must always vanish wherever the reference elasticity tensor equals the actual material elasticity tensor, as discussed in Section 7.

9.2.2. Compliant Micro-Inclusions

Next, assume $\mathbf{D}' - \mathbf{D}$ is positive-definite, i.e., choose a stiff comparison material of uniform elasticity $\mathbf{C} = \mathbf{D}^{-1}$. This may correspond to the case when an RVE contains micro-inclusions which have suitable symmetry and are more compliant than the surrounding uniform matrix with elasticity $\mathbf{C} = \mathbf{D}^{-1}$. Then the integral operators $< : \{(\mathbf{D}' - \mathbf{D})^{-1} + \mathbf{\Lambda}\} : >$ and $< : \{(\mathbf{C}' - \mathbf{C})^{-1} + \mathbf{\Gamma}\} : >$ are positive-definite and negative-definite, respectively. From (9.1.7e), for any eigenstrain $\mathbf{e}^*$,

$$< \mathbf{e}^* : \mathbf{\Lambda} : \mathbf{e}^* > = < (\mathbf{\Lambda} : \mathbf{e}^*) : \mathbf{D} : (\mathbf{\Lambda} : \mathbf{e}^*) >. \tag{9.2.6}$$

Hence, similarly to operator $< : \mathbf{\Gamma} : >$, the integral operator $< : \mathbf{\Lambda} : >$ is positive-definite. Thus, when $\mathbf{D}' - \mathbf{D}$ is positive-definite, then the integral operator $< : \{(\mathbf{D}' - \mathbf{D})^{-1} + \mathbf{\Lambda}\} : >$ is positive-definite.

As in (9.2.5), the negative-definiteness of the integral operator $< : \{(\mathbf{C}' - \mathbf{C})^{-1} + \mathbf{\Gamma}\} : >$ follows from the positive-definiteness of the tensor $\mathbf{D}' - \mathbf{D}$. Using (9.1.15b), observe that

$$< \mathbf{s}^* : \mathbf{\Gamma} : \mathbf{s}^* > = < \mathbf{s}^* : \mathbf{D} : \mathbf{s}^* > - < \boldsymbol{\sigma}^{\mathrm{d}} : \mathbf{D} : \boldsymbol{\sigma}^{\mathrm{d}} >, \tag{9.2.7a}$$

where $\boldsymbol{\sigma}^{\mathrm{d}}$ is the stress field produced by the eigenstress $\mathbf{s}^*$ through

$$\boldsymbol{\sigma}^{\mathrm{d}} = -\mathbf{C} : (\mathbf{\Gamma} : \mathbf{s}^*) + \mathbf{s}^*. \tag{9.2.7b}$$

Using identities similar to those in (9.2.5d), obtain

$$< \mathbf{s}^* : (\mathbf{C}' - \mathbf{C})^{-1} : \mathbf{s}^* > = -< (\mathbf{D} : \mathbf{s}^*) : (\mathbf{D}' - \mathbf{D})^{-1} : (\mathbf{D} : \mathbf{s}^*) > \\ - < \mathbf{s}^* : \mathbf{D} : \mathbf{s}^* >. \tag{9.2.7c}$$

The addition of (9.2.7a) and (9.2.7c) yields

$$< \mathbf{s}^* : \{(\mathbf{C}' - \mathbf{C})^{-1} + \mathbf{\Gamma}\} : \mathbf{s}^* > = -< (\mathbf{D} : \mathbf{s}^*) : (\mathbf{D}' - \mathbf{D})^{-1} : (\mathbf{D} : \mathbf{s}^*) > \\ - < \boldsymbol{\sigma}^{\mathrm{d}} : \mathbf{D} : \boldsymbol{\sigma}^{\mathrm{d}} >. \tag{9.2.7d}$$

If $\mathbf{D}' - \mathbf{D}$ is positive-definite, then the tensors $(\mathbf{C}' - \mathbf{C})^{-1}$ and $\mathbf{C}' - \mathbf{C}$, and the integral operator $< : \{(\mathbf{C}' - \mathbf{C})^{-1} + \mathbf{\Gamma}\} : >$ are all negative-definite.

9.2.3. Bounds for Elastic Strain and Complementary Elastic Energies

Using the overall compliance and elasticity tensors $\overline{\mathbf{D}}$ and $\overline{\mathbf{C}}$, given by (9.1.3b) and (9.1.11b), define the average complementary strain energy and the average strain energy, respectively, as follows: the average complementary strain energy is

$$\mathrm{W}^{\mathrm{c}}(\boldsymbol{\sigma}^{\mathrm{o}}) \equiv \frac{1}{2} \boldsymbol{\sigma}^{\mathrm{o}} : \overline{\mathbf{D}} : \boldsymbol{\sigma}^{\mathrm{o}}, \tag{9.2.8a}$$

when $\boldsymbol{\sigma}^{\mathrm{o}}$ is prescribed; and the average strain energy is

$$\mathrm{W}(\boldsymbol{\varepsilon}^{\mathrm{o}}) \equiv \frac{1}{2}\boldsymbol{\varepsilon}^{\mathrm{o}} : \bar{\mathbf{C}} : \boldsymbol{\varepsilon}^{\mathrm{o}}, \tag{9.2.8b}$$

when $\boldsymbol{\varepsilon}^{\mathrm{o}}$ is prescribed.[3] Let W^{co} and W^{o}, respectively, be the comparison average complementary strain energy and strain energy, i.e.,

$$\mathrm{W}^{\mathrm{co}}(\boldsymbol{\sigma}^{\mathrm{o}}) \equiv \frac{1}{2}\boldsymbol{\sigma}^{\mathrm{o}} : \mathbf{D} : \boldsymbol{\sigma}^{\mathrm{o}}, \qquad \mathrm{W}^{\mathrm{o}}(\boldsymbol{\varepsilon}^{\mathrm{o}}) \equiv \frac{1}{2}\boldsymbol{\varepsilon}^{\mathrm{o}} : \mathbf{C} : \boldsymbol{\varepsilon}^{\mathrm{o}}. \tag{9.2.8c,d}$$

From (9.1.10b) and (9.1.18b), the values of the functionals $\mathrm{I}(\boldsymbol{\varepsilon}^*; \boldsymbol{\sigma}^{\mathrm{o}})$ and $\mathrm{J}(\boldsymbol{\sigma}^*; \boldsymbol{\varepsilon}^{\mathrm{o}})$ are

$$\mathrm{I}(\boldsymbol{\varepsilon}^*; \boldsymbol{\sigma}^{\mathrm{o}}) = \mathrm{W}^{\mathrm{co}}(\boldsymbol{\sigma}^{\mathrm{o}}) - \mathrm{W}^{\mathrm{c}}(\boldsymbol{\sigma}^{\mathrm{o}}), \qquad \mathrm{J}(\boldsymbol{\sigma}^*; \boldsymbol{\varepsilon}^{\mathrm{o}}) = \mathrm{W}^{\mathrm{o}}(\boldsymbol{\varepsilon}^{\mathrm{o}}) - \mathrm{W}(\boldsymbol{\varepsilon}^{\mathrm{o}}). \tag{9.2.8e,f}$$

Then, from the results obtained in the preceding subsections, the positive-definiteness or negative-definiteness of the integral operators $< : \{(\mathbf{D}' - \mathbf{D})^{-1} + \boldsymbol{\Lambda}\} : >$ and $< : \{(\mathbf{C}' - \mathbf{C})^{-1} + \boldsymbol{\Gamma}\} : >$ depends on whether $\mathbf{D}' - \mathbf{D}$ or $\mathbf{C}' - \mathbf{C}$ is positive-definite, i.e.,

$$\{\mathbf{C}' - \mathbf{C} \text{ is p.d.} \Longleftrightarrow \mathbf{D}' - \mathbf{D} \text{ is n.d.}\} \Rightarrow \left\{ \begin{array}{l} < : \{(\mathbf{D}' - \mathbf{D})^{-1} + \boldsymbol{\Lambda}\} : > \text{is n.d.} \\ < : \{(\mathbf{C}' - \mathbf{C})^{-1} + \boldsymbol{\Gamma}\} : > \text{is p.d.} \end{array} \right.$$

$$\{\mathbf{D}' - \mathbf{D} \text{ is p.d.} \Longleftrightarrow \mathbf{C}' - \mathbf{C} \text{ is n.d.}\} \Rightarrow \left\{ \begin{array}{l} < : \{(\mathbf{D}' - \mathbf{D})^{-1} + \boldsymbol{\Lambda}\} : > \text{is p.d.} \\ < : \{(\mathbf{C}' - \mathbf{C})^{-1} + \boldsymbol{\Gamma}\} : > \text{is n.d.} \end{array} \right. \tag{9.2.9a,b}$$

Therefore, if $\mathbf{C}' - \mathbf{C}$ is positive-definite (if $\mathbf{D}' - \mathbf{D}$ is positive-definite), then, for arbitrary eigenstrain and eigenstress fields $\mathbf{e}^*$ and $\mathbf{s}^*$, $\mathrm{W}^{\mathrm{c}}(\boldsymbol{\sigma}^{\mathrm{o}})$ and $\mathrm{W}(\boldsymbol{\varepsilon}^{\mathrm{o}})$ satisfy

$$\mathrm{W}^{\mathrm{co}}(\boldsymbol{\sigma}^{\mathrm{o}}) - \mathrm{W}^{\mathrm{c}}(\boldsymbol{\sigma}^{\mathrm{o}}) \geq (\leq)\ \frac{1}{2} < \mathbf{e}^* : \{(\mathbf{D}' - \mathbf{D})^{-1} + \boldsymbol{\Lambda}\} : \mathbf{e}^* > - \boldsymbol{\sigma}^{\mathrm{o}} : < \mathbf{e}^* >,$$

$$\mathrm{W}^{\mathrm{o}}(\boldsymbol{\varepsilon}^{\mathrm{o}}) - \mathrm{W}(\boldsymbol{\varepsilon}^{\mathrm{o}}) \leq (\geq)\ \frac{1}{2} < \mathbf{s}^* : \{(\mathbf{C}' - \mathbf{C})^{-1} + \boldsymbol{\Gamma}\} : \mathbf{s}^* > - < \mathbf{s}^* > : \boldsymbol{\varepsilon}^{\mathrm{o}}, \tag{9.2.10a,b}$$

where $\boldsymbol{\sigma}^{\mathrm{o}}$ is the prescribed overall stress for W^{co} and W^{c}, and $\boldsymbol{\varepsilon}^{\mathrm{o}}$ is the prescribed overall strain for W^{o} and W. The equality in each case holds only when $\mathbf{e}^* = \boldsymbol{\varepsilon}^*$ and $\mathbf{s}^* = \boldsymbol{\sigma}^*$, respectively. It should be recalled that for a bounded V, the stress and strain fields associated with a prescribed $\boldsymbol{\sigma}^{\mathrm{o}}$, in general, are different from those corresponding to a given $\boldsymbol{\varepsilon}^{\mathrm{o}}$. Hence $\bar{\mathbf{C}}$ and $\bar{\mathbf{D}}$, in general, need not be related. However, as pointed out before, if the RVE is indeed statistically representative, and if a consistent averaging scheme is used, then one expects that $\bar{\mathbf{C}}$ and $\bar{\mathbf{D}}$ be each other's inverse.

Consider now the particular case when the RVE consists of a homogeneous matrix and n distinct but homogeneous micro-inclusions. The volume average over V in the functionals I and J is decomposed into the volume average over the matrix M and the sum of those over each micro-inclusion Ω_α. Choose the matrix with elasticity and compliance tensors $\mathbf{C}$ and $\mathbf{D}$, as the comparison material. Since $(\mathbf{C}' - \mathbf{C})^{-1}$ and $(\mathbf{D}' - \mathbf{D})^{-1}$ are not defined in M, restrict eigenstrain and eigenstress fields to vanish in M. The functionals I and J then are

[3] These are two separate problems.

well defined, with finite values. From (9.2.10a,b) it follows that if $\mathbf{C}^\alpha - \mathbf{C}$ is positive-definite (if $\mathbf{D}^\alpha - \mathbf{D}$ is positive-definite) for all $\alpha = 1, 2, ..., n$, then, for any arbitrary eigenstress and eigenstrain fields, $\mathbf{s}^*$ and $\mathbf{e}^*$, vanishing in M,

$$W^{co}(\boldsymbol{\sigma}^o) - W^c(\boldsymbol{\sigma}^o)$$

$$\geq (\leq) \sum_{\alpha=1}^{n} f_\alpha \frac{1}{2} < \mathbf{e}^* : \{(\mathbf{D}^\alpha - \mathbf{D})^{-1} + \boldsymbol{\Lambda}\} : \mathbf{e}^* >_\alpha - \boldsymbol{\sigma}^o : < \mathbf{e}^* >,$$

$$W^o(\boldsymbol{\varepsilon}^o) - W(\boldsymbol{\varepsilon}^o)$$

$$\leq (\geq) \sum_{\alpha=1}^{n} f_\alpha \frac{1}{2} < \mathbf{s}^* : \{(\mathbf{C}^\alpha - \mathbf{C})^{-1} + \boldsymbol{\Gamma}\} : \mathbf{s}^* >_\alpha - < \mathbf{s}^* > : \boldsymbol{\varepsilon}^o, \qquad (9.2.11a,b)$$

where

$$< \mathbf{e}^* > = \sum_{\alpha=1}^{n} f_\alpha < \mathbf{e}^* >_\alpha, \qquad < \mathbf{s}^* > = \sum_{\alpha=1}^{n} f_\alpha < \mathbf{s}^* >_\alpha, \qquad (9.2.11c,d)$$

and, as before, f_α is the volume average of Ω_α. Again, the equality in each case holds only when $\mathbf{e}^* = \boldsymbol{\varepsilon}^*$ and $\mathbf{s}^* = \boldsymbol{\sigma}^*$. It should be noted that if $\mathbf{D}' - \mathbf{D} = \mathbf{0}$ and $\mathbf{C}' - \mathbf{C} = \mathbf{0}$ in M, then the exact eigenstress and eigenstrain fields in the equivalent homogeneous solid that produce the same stress and strain fields as in the original heterogeneous RVE, vanish identically over the matrix M. Hence, they belong to the group of restricted trial eigenstrain and eigenstress fields which vanish over M, rendering I and J finite.

9.3. GENERALIZED BOUNDS ON OVERALL ENERGIES

Consider an RVE of volume V which contains linearly elastic and homogeneous micro-inclusions of possibly different elasticity tensors, embedded in a linearly elastic and homogeneous matrix. From (9.1.2a,b), the elasticity and compliance tensors $\mathbf{C}'(\mathbf{x})$ and $\mathbf{D}'(\mathbf{x})$ of the RVE are given by

$$\mathbf{C}'(\mathbf{x}) = H_M(\mathbf{x})\, \mathbf{C} + \sum_{\alpha=1}^{n} H_\alpha(\mathbf{x})\, \mathbf{C}^\alpha,$$

$$\mathbf{D}'(\mathbf{x}) = H_M(\mathbf{x})\, \mathbf{D} + \sum_{\alpha=1}^{n} H_\alpha(\mathbf{x})\, \mathbf{D}^\alpha, \qquad (9.3.1a,b)$$

where $H_\alpha(\mathbf{x}) = H(\mathbf{x}; \Omega_\alpha)$ (or $H_M(\mathbf{x}) = H(\mathbf{x}; M)$) is the Heaviside step function which takes on the value 1 if $\mathbf{x}$ is in Ω_α (or M) and is 0 otherwise.

In Subsection 9.2.3, particular eigenstrain and eigenstress fields are chosen in the equivalent homogeneous solid, which vanish in the matrix, since the reference elasticity and compliance tensors are set to coincide with those of the matrix. Hereinafter, in order to consider a more general case, do not impose any such restriction on the reference elasticity and compliance tensors; they may, for example, coincide with those of the αth inclusion[4]. For simplicity, treat the matrix phase as the 0th inclusion phase, Ω_0, and denote its elasticity and compliance by $\mathbf{C}^0$ and $\mathbf{D}^0$.

Now, in order to consider Hashin-Shtrikman bounds on the overall elastic moduli, choose particular eigenstrain and eigenstress fields which take on *distinct constant* values in each micro-inclusion, i.e.,

$$\mathbf{e}^*(\mathbf{x}) = \sum_{\alpha=0}^{n} H_\alpha(\mathbf{x})\, \mathbf{e}^{*\alpha}, \qquad \mathbf{s}^*(\mathbf{x}) = \sum_{\alpha=0}^{n} H_\alpha(\mathbf{x})\, \mathbf{s}^{*\alpha}. \tag{9.3.2a,b}$$

Here $\mathbf{e}^{*\alpha}$ and $\mathbf{s}^{*\alpha}$ ($\alpha = 0, 1, 2, ..., n$) are *constant tensors.* From these fields, bounds on the overall elasticity and compliance tensors are computed by optimizing I and J, in accordance with (9.2.2c,d). The technique is expected to yield good bounds when the volume fraction of each inclusion is suitably small. On the other hand, when an inclusion is large so that the assumption of a uniform eigenstrain or eigenstress in it appears inappropriate, one may subdivide this inclusion into several subregions and use the uniform eigenstrain and eigenstress in each *subregion.* Such an approach is mandatory for estimating the instantaneous (or incrementally linear) effective moduli of an RVE with nonlinear constituents, e.g., an RVE with elastic-plastic materials; see Accorsi and Nemat-Nasser (1986), and Nemat-Nasser *et al.* (1986) for illustrations.

9.3.1. Correlation Tensors

With piecewise constant eigenstrain and eigenstress fields, $\mathbf{e}^*(\mathbf{x})$ and $\mathbf{s}^*(\mathbf{x})$, (9.3.2a,b), the integral operators $\mathbf{\Lambda}$ and $\mathbf{\Gamma}$ defined by (9.1.6) and (9.1.14) reduce to tensor operators, $\mathbf{\Lambda}^{\alpha\beta}$ and $\mathbf{\Gamma}^{\alpha\beta}$, acting on the constant eigenstrain and eigenstress tensors, $\mathbf{e}^{*\beta}$ and $\mathbf{s}^{*\beta}$. These tensor operators determine, for example, the average stress, $<\boldsymbol{\sigma}>_\alpha$, and strain, $<\boldsymbol{\varepsilon}>_\alpha$, in micro-inclusion Ω_α, for the constant eigenstrain and eigenstress, $\mathbf{e}^{*\beta}$ and $\mathbf{s}^{*\beta}$, prescribed on Ω_β, ($\alpha, \beta = 0, 1, 2, ..., n$). The tensors $\mathbf{\Lambda}^{\alpha\beta}$ and $\mathbf{\Gamma}^{\alpha\beta}$ are called the *correlation tensors.* In this subsection, these correlation tensors are defined explicitly in terms of integral operators $\mathbf{\Lambda}$ and $\mathbf{\Gamma}$, for piecewise constant trial eigenstrain and eigenstress fields.

First, consider the integral operator $\mathbf{\Lambda}$, introduced in (9.1.6a,b), and note again that

$$-(\mathbf{\Lambda} : \boldsymbol{\varepsilon}^*)(\mathbf{x}) \equiv \mathbf{C} : (\mathbf{S}(\mathbf{x};\, \boldsymbol{\varepsilon}^*) - \boldsymbol{\varepsilon}^*(\mathbf{x})), \tag{9.3.3a}$$

where the integral operator $\mathbf{S}(\mathbf{x};\, \boldsymbol{\varepsilon}^*)$ is defined by (7.3.7) in terms of the Green function of *the homogeneous linearly elastic solid* V *bounded by* ∂V, *with zero tractions prescribed on* ∂V. As pointed out in Subsection 9.1, the actual calculation of the Green function for a bounded region of arbitrary shape is, in general, not feasible. Nevertheless, the properties of such a Green function can be used effectively to establish general expressions which, in many important applications, lend themselves to accurate estimates. The Hashin-Shtrikman bounds, as generalized by Willis, provide an illustration of this procedure.

---

[4] In this case, the eigenstrains and eigenstresses must vanish there.

The integral operator $\mathbf{\Gamma}$ relates to the integral operator $\mathbf{T}(\mathbf{x};\, \boldsymbol{\sigma}^*)$, as follows:

$$-(\mathbf{\Gamma} : \boldsymbol{\sigma}^*)(\mathbf{x}) \equiv \mathbf{D} : (\mathbf{T}(\mathbf{x};\, \boldsymbol{\sigma}^*) - \boldsymbol{\sigma}^*(\mathbf{x})), \tag{9.3.3b}$$

where $\mathbf{T}(\mathbf{x};\, \boldsymbol{\sigma}^*)$ is defined by (7.3.10) in terms of the Green function of the bounded V, which satisfies *zero displacement boundary conditions on* ∂V.

Using the piecewise constant eigenstrain field (9.3.2a), first observe that

$$< \mathbf{e}^*(\mathbf{x}) : \mathbf{\Lambda}(\mathbf{x};\, \mathbf{e}^*) >_\alpha = \mathbf{e}^{*\alpha} : < \mathbf{\Lambda}(\mathbf{x};\, \mathbf{e}^*) >_\alpha \qquad (\alpha \text{ not summed}). \tag{9.3.4a}$$

Similarly, using the piecewise constant eigenstress field (9.3.2b), obtain

$$< \mathbf{s}^*(\mathbf{x}) : \mathbf{\Gamma}(\mathbf{x};\, \mathbf{s}^*) >_\alpha = \mathbf{s}^{*\alpha} : < \mathbf{\Gamma}(\mathbf{x};\, \mathbf{s}^*) >_\alpha \qquad (\alpha \text{ not summed}). \tag{9.3.4b}$$

For the volume average of the stress $< \mathbf{\Lambda} : \mathbf{e}^* >_\alpha$ and strain $< \mathbf{\Gamma} : \mathbf{s}^* >_\alpha$ in (9.3.4a) and (9.3.4b), introduce the fourth-order constant correlation tensors, $\mathbf{\Lambda}^{\alpha\beta}$ and $\mathbf{\Gamma}^{\alpha\beta}$ (α, β = 0, 1, 2, ..., n), such that for arbitrary $\mathbf{e}^{*\beta}$ and $\mathbf{s}^{*\beta}$,

$$f_\beta\, \mathbf{\Lambda}^{\alpha\beta} : \mathbf{e}^{*\beta} \equiv < \mathbf{\Lambda}(\mathbf{x};\, H_\beta\, \mathbf{e}^{*\beta}) >_\alpha \qquad \text{or} \qquad f_\beta\, \Lambda^{\alpha\beta}_{ijkl}\, e^{*\beta}_{kl} \equiv < \Lambda_{ij}(\mathbf{x};\, H_\beta\, \mathbf{e}^{*\beta}) >_\alpha$$

$$(\beta \text{ not summed}), \tag{9.3.5a,b}$$

and

$$f_\beta\, \mathbf{\Gamma}^{\alpha\beta} : \mathbf{s}^{*\beta} \equiv < \mathbf{\Gamma}(\mathbf{x};\, H_\beta\, \mathbf{s}^{*\beta}) >_\alpha \qquad \text{or} \qquad f_\beta\, \Gamma^{\alpha\beta}_{ijkl}\, s^{*\beta}_{kl} \equiv < \Gamma_{ij}(\mathbf{x};\, H_\beta\, \mathbf{s}^{*\beta}) >_\alpha$$

$$(\beta \text{ not summed}), \tag{9.3.5c,d}$$

where H_β is the Heaviside step function which equals 1 in Ω_β and 0 elsewhere. The fourth-order correlation tensor $\mathbf{\Lambda}^{\alpha\beta}$ (tensor $\mathbf{\Gamma}^{\alpha\beta}$) represents the influence of the eigenstrain $\mathbf{e}^{*\beta}$ (eigenstress $\mathbf{s}^{*\beta}$) in the βth micro-inclusion Ω_β on the αth micro-inclusion Ω_α, for the class of piecewise constant eigenstrain (eigenstress) fields. *These tensors depend only on the geometries of the inclusions and the* RVE, *as well as the elasticity tensor* $\mathbf{C} = \mathbf{D}^{-1}$ *of the reference material, but not on the eigenstrains and eigenstresses, nor on the elasticity of the corresponding inclusions.*

From the self-adjointness of the integral operator $\mathbf{\Lambda}$ or $\mathbf{\Gamma}$, i.e., from $< H_\alpha\, \mathbf{e}^{*\alpha} : \mathbf{\Lambda} : H_\beta\, \mathbf{e}^{*\beta} > \; = \; < H_\beta\, \mathbf{e}^{*\beta} : \mathbf{\Lambda} : H_\alpha\, \mathbf{e}^{*\alpha} >$ for any $\mathbf{e}^{*\alpha}$ and $\mathbf{e}^{*\beta}$, or from $< H_\alpha\, \mathbf{s}^{*\alpha} : \mathbf{\Gamma} : H_\beta\, \mathbf{s}^{*\beta} > \; = \; < H_\beta\, \mathbf{s}^{*\beta} : \mathbf{\Gamma} : H_\alpha\, \mathbf{s}^{*\alpha} >$ for any $\mathbf{s}^{*\alpha}$ and $\mathbf{s}^{*\beta}$, the correlation tensors satisfy

$$\mathbf{\Lambda}^{\alpha\beta} = (\mathbf{\Lambda}^{\beta\alpha})^T \qquad \text{or} \qquad \Lambda^{\alpha\beta}_{ijkl} = \Lambda^{\beta\alpha}_{klij} \tag{9.3.5e,f}$$

and

$$\mathbf{\Gamma}^{\alpha\beta} = (\mathbf{\Gamma}^{\beta\alpha})^T \qquad \text{or} \qquad \Gamma^{\alpha\beta}_{ijkl} = \Gamma^{\beta\alpha}_{klij}. \tag{9.3.5g,h}$$

In general, however, these correlation tensors are not symmetric with respect to their superscripts α and β.[5]

---

[5] As shown in (9.1.6) and (9.1.14), integral operators $\mathbf{\Lambda}$ and $\mathbf{\Gamma}$ can be expressed in terms of Green function $\mathbf{G}$. While (9.3.5e~h) are directly derived from (3.2.7), i.e., $G_{ij}(\mathbf{x},\, \mathbf{y}) = G_{ji}(\mathbf{y},\, \mathbf{x})$, the correlation tensors are symmetric with respect to their superscripts, if Green's function is of the form

Using the correlation tensors $\mathbf{\Lambda}^{\alpha\beta}$ and $\mathbf{\Gamma}^{\alpha\beta}$, rewrite (9.3.4a) and (9.3.4b), as

$$\mathbf{e}^{*\alpha} : < \mathbf{\Lambda}(\mathbf{x};\, \mathbf{e}^*) >_\alpha = \sum_{\beta=0}^{n} f_\beta\, \mathbf{e}^{*\alpha} : \mathbf{\Lambda}^{\alpha\beta} : \mathbf{e}^{*\beta},$$

$$\mathbf{s}^{*\alpha} : < \mathbf{\Gamma}(\mathbf{x};\, \mathbf{s}^*) >_\alpha = \sum_{\beta=0}^{n} f_\beta\, \mathbf{s}^{*\alpha} : \mathbf{\Gamma}^{\alpha\beta} : \mathbf{s}^{*\beta} \qquad (\alpha \text{ not summed}), \qquad (9.3.6a,b)$$

and hence obtain

$$\sum_{\alpha=0}^{n} f_\alpha\, \mathbf{e}^{*\alpha} : < \mathbf{\Lambda}(\mathbf{x};\, \mathbf{e}^*) >_\alpha = \sum_{\alpha=0}^{n} \sum_{\beta=0}^{n} f_\alpha f_\beta\, \mathbf{e}^{*\alpha} : \mathbf{\Lambda}^{\alpha\beta} : \mathbf{e}^{*\beta},$$

$$\sum_{\alpha=0}^{n} f_\alpha\, \mathbf{s}^{*\alpha} : < \mathbf{\Gamma}(\mathbf{x};\, \mathbf{s}^*) >_\alpha = \sum_{\alpha=0}^{n} \sum_{\beta=0}^{n} f_\alpha f_\beta\, \mathbf{s}^{*\alpha} : \mathbf{\Gamma}^{\alpha\beta} : \mathbf{s}^{*\beta}. \qquad (9.3.6c,d)$$

9.3.2. Upper and Lower Bounds on Overall Energies

With the aid of the correlation tensors, $\mathbf{\Lambda}^{\alpha\beta}$ and $\mathbf{\Gamma}^{\alpha\beta}$, for the piecewise constant trial eigenstrain and eigenstress fields, $\mathbf{e}^*(\mathbf{x})$ and $\mathbf{s}^*(\mathbf{x})$, functionals $I(\mathbf{e}^*;\, \boldsymbol{\sigma}^o)$ and $J(\mathbf{s}^*;\, \boldsymbol{\varepsilon}^o)$ defined respectively by (9.1.9a) and (9.1.17a), reduce to quadratic forms (with linear terms) in the constant eigenstrains $\mathbf{e}^{*\alpha}$ and eigenstresses $\mathbf{s}^{*\alpha}$, ($\alpha = 0, 1, 2, ..., n$), respectively. These quadratic forms then provide upper and lower bounds on the overall complementary elastic and elastic energies, $W^c(\boldsymbol{\sigma}^o)$ and $W(\boldsymbol{\varepsilon}^o)$. The aim now is to formally calculate these quadratic expressions, and obtain their optimal values which determine bounds on the overall energies.

For prescribed overall constant stress and strain, $\boldsymbol{\sigma}^o$ and $\boldsymbol{\varepsilon}^o$, the upper and lower bounds on the overall complementary elastic energy $W^c(\boldsymbol{\sigma}^o)$ and the overall elastic energy $W(\boldsymbol{\varepsilon}^o)$ are given by (9.2.11a) and (9.2.11b), respectively. Substituting (9.3.5) and (9.3.6) into these bounds, obtain the following bounds on $W^c(\boldsymbol{\sigma}^o)$ and $W(\boldsymbol{\varepsilon}^o)$, when $\mathbf{C}^\alpha - \mathbf{C}$ is positive-definite (when $\mathbf{D}^\alpha - \mathbf{D}$ is positive-definite) for all α's:

$$W^{co}(\boldsymbol{\sigma}^o) - W^c(\boldsymbol{\sigma}^o) \geq (\leq)\; I(\mathbf{e}^*;\, \boldsymbol{\sigma}^o),$$

$$W^o(\boldsymbol{\varepsilon}^o) - W(\boldsymbol{\varepsilon}^o) \leq (\geq)\; J(\mathbf{s}^*;\, \boldsymbol{\varepsilon}^o). \qquad (9.3.7a,b)$$

In general, the exact eigenstrain and eigenstress fields which produce the same stress and strain fields in the original heterogeneous RVE, are not piecewise constant but vary within each micro-inclusion. Hence, in most cases, the inequalities $<$ and $>$ instead of $\leq$ and $\geq$ apply in (9.3.7a,b). From *functionals* $I(\mathbf{e}^*;\, \boldsymbol{\sigma}^o)$ and $J(\mathbf{s}^*;\, \boldsymbol{\varepsilon}^o)$, define *functions* I' and J', using the piecewise constant eigenstrain and eigenstress fields. Let $\{\mathbf{e}^{*\alpha}\}$ and $\{\mathbf{s}^{*\alpha}\}$ ($\alpha = 0, 1, 2, ..., n$) stand

---

$\mathbf{G}(\mathbf{x}, \mathbf{y}) = \mathbf{G}(\mathbf{x} - \mathbf{y})$, as shown in Subsection 9.4. Note that the transpose in (9.3.5e~h) is with respect to the first and second pairs of the *subscripts*.

for the set of constant eigenstrains and eigenstresses, and set

$$\mathrm{I}(\mathbf{e}^*;\, \boldsymbol{\sigma}^o) \equiv \mathrm{I}'(\{\mathbf{e}^{*\alpha}\};\, \boldsymbol{\sigma}^o), \qquad \mathrm{J}(\mathbf{s}^*;\, \boldsymbol{\varepsilon}^o) \equiv \mathrm{J}'(\{\mathbf{s}^{*\alpha}\};\, \boldsymbol{\varepsilon}^o), \tag{9.3.8a,b}$$

where

$$\mathrm{I}'(\{\mathbf{e}^{*\alpha}\};\, \boldsymbol{\sigma}^o) \equiv \sum_{\alpha=0}^{n} \sum_{\beta=0}^{n} \frac{1}{2}\mathbf{e}^{*\alpha} : \mathbf{I}^{\alpha\beta} : \mathbf{e}^{*\beta} - \boldsymbol{\sigma}^o : \bar{\mathbf{e}}^*,$$

$$\mathrm{J}'(\{\mathbf{s}^{*\alpha}\};\, \boldsymbol{\varepsilon}^o) \equiv \sum_{\alpha=0}^{n} \sum_{\beta=0}^{n} \frac{1}{2}\mathbf{s}^{*\alpha} : \mathbf{J}^{\alpha\beta} : \mathbf{s}^{*\beta} - \bar{\mathbf{s}}^* : \boldsymbol{\varepsilon}^o, \tag{9.3.8c,d}$$

with the fourth-order tensors $\mathbf{I}^{\alpha\beta}$ and $\mathbf{J}^{\alpha\beta}$ being defined by

$$\mathbf{I}^{\alpha\beta} \equiv \mathrm{f}_\alpha\, \delta_{\alpha\beta}\, (\mathbf{D}^\beta - \mathbf{D})^{-1} + \mathrm{f}_\alpha\, \mathrm{f}_\beta\, \boldsymbol{\Lambda}^{\alpha\beta},$$

$$\mathbf{J}^{\alpha\beta} \equiv \mathrm{f}_\alpha\, \delta_{\alpha\beta}\, (\mathbf{C}^\beta - \mathbf{C})^{-1} + \mathrm{f}_\alpha\, \mathrm{f}_\beta\, \boldsymbol{\Gamma}^{\alpha\beta} \qquad (\alpha,\ \beta \text{ not summed}). \tag{9.3.8e,f}$$

Note that $\mathbf{I}^{\alpha\beta}$ and $\mathbf{J}^{\alpha\beta}$ are symmetric with respect to the first and second pairs of their subscript indices, i.e., $\mathrm{I}_{ijkl}^{\alpha\beta} = \mathrm{I}_{klij}^{\alpha\beta}$ and $\mathrm{J}_{ijkl}^{\alpha\beta} = \mathrm{J}_{klij}^{\alpha\beta}$.

The optimal (or stationary) values of the quadratic expressions I′ and J′ are computed by setting equal to zero the corresponding derivative with respect to $\mathbf{e}^{*\beta}$ and $\mathbf{s}^{*\beta}$, respectively, i.e., from

$$\frac{\partial \mathrm{I}'}{\partial \mathbf{e}^{*\beta}}(\{\mathbf{e}^{*\alpha}\};\, \boldsymbol{\sigma}^o) = \mathbf{0}, \qquad \frac{\partial \mathrm{J}'}{\partial \mathbf{s}^{*\beta}}(\{\mathbf{s}^{*\alpha}\};\, \boldsymbol{\varepsilon}^o) = \mathbf{0}. \tag{9.3.9a,b}$$

These are systems of linear equations for the unknowns $\{\mathbf{e}^{*\beta}\}$ and $\{\mathbf{s}^{*\beta}\}$, respectively. Let $\{\boldsymbol{\varepsilon}^{*\alpha}\}$ and $\{\boldsymbol{\sigma}^{*\alpha}\}$ be the corresponding solutions, i.e., the solutions to the following sets of n linear tensorial equations:

$$\sum_{\beta=0}^{n} \mathbf{I}^{\alpha\beta} : \boldsymbol{\varepsilon}^{*\beta} - \mathrm{f}_\alpha\, \boldsymbol{\sigma}^o = \mathbf{0},$$

$$\sum_{\beta=0}^{n} \mathbf{J}^{\alpha\beta} : \boldsymbol{\sigma}^{*\beta} - \mathrm{f}_\alpha\, \boldsymbol{\varepsilon}^o = \mathbf{0} \qquad (\alpha = 0,\, 1,\, 2,\, \ldots,\, n). \tag{9.3.10a,b}$$

In general, when the basic problem is well-posed, both (9.3.10a) and (9.3.10b) have a unique solution. This is assumed to be the case in the following.

If $\mathbf{C}' - \mathbf{C}$ is positive-definite (if $\mathbf{D}' - \mathbf{D}$ is positive-definite) everywhere in V, the functionals I and J are negative-definite and positive-definite (positive-definite and negative-definite), and hence they have the global maximum and minimum (minimum and maximum), respectively. Since the functions I′ and J′ are defined by substituting piecewise constant eigenstrains and eigenstresses into I and J, I′ and J′ must have the global maximum or minimum, when I and J have the global maximum or minimum, respectively. On the other hand, when the set of linear tensorial equations (9.3.10a,b) has a unique solution, both I′ and J′ have one and only one stationary value. Therefore, if $\mathbf{C}' - \mathbf{C}$ is positive-definite (if $\mathbf{D}' - \mathbf{D}$ is positive-definite), and if the solution of (9.3.10a,b) is unique, then the unique stationary values of I′ and J′ that are given by the solution of (9.3.10a,b), are the corresponding global maximum or minimum. Actually, the uniqueness of the solution of (9.3.10a,b) is guaranteed, when $\mathbf{C}' - \mathbf{C}$ is positive-definite (when $\mathbf{D}' - \mathbf{D}$ is positive-definite);[6] for a general proof, see Willis (1989). The simple proof is as follows: the integral operators in the functionals I and J, $< : \{(\mathbf{D}' - \mathbf{D})^{-1} + \boldsymbol{\Lambda}\} : >$ and $< : \{(\mathbf{C}' - \mathbf{C})^{-1} + \boldsymbol{\Gamma}\} : >$, are definite when either $\mathbf{C}' - \mathbf{C}$ or $\mathbf{D}' - \mathbf{D}$ is definite, which means that only $\mathbf{e}^* = \mathbf{0}$ and $\mathbf{s}^* = \mathbf{0}$ lead to

$$< \mathbf{e}^* : \{(\mathbf{D}' - \mathbf{D})^{-1} + \boldsymbol{\Lambda}\} : \mathbf{e}^* > = \sum_{\alpha=0}^{n} \sum_{\beta=0}^{n} \mathbf{e}^{*\alpha} : \mathbf{I}^{\alpha\beta} : \mathbf{e}^{*\beta} = 0,$$

$$< \mathbf{s}^* : \{(\mathbf{C}' - \mathbf{C})^{-1} + \boldsymbol{\Gamma}\} : \mathbf{s}^* > = \sum_{\alpha=0}^{n} \sum_{\beta=0}^{n} \mathbf{s}^{*\alpha} : \mathbf{J}^{\alpha\beta} : \mathbf{s}^{*\beta} = 0, \qquad (9.3.11a,b)$$

and the left-hand side of (9.3.11a,b) cannot be zero for nonzero piecewise constant eigenstrains or eigenstresses. Hence, if $\{\boldsymbol{\varepsilon}^{*\alpha(K)}\}$ and $\{\boldsymbol{\sigma}^{*\alpha(K)}\}$ (K = 1, 2) are two solutions of (9.3.10a,b), then, $\{\boldsymbol{\varepsilon}^{*\alpha(1)} - \boldsymbol{\varepsilon}^{*\alpha(2)}\}$ and $\{\boldsymbol{\sigma}^{*\alpha(1)} - \boldsymbol{\sigma}^{*\alpha(2)}\}$ satisfy

$$\sum_{\alpha=0}^{n} \sum_{\beta=0}^{n} (\boldsymbol{\varepsilon}^{*\alpha(1)} - \boldsymbol{\varepsilon}^{*\alpha(2)}) : \mathbf{I}^{\alpha\beta} : (\boldsymbol{\varepsilon}^{*\beta(1)} - \boldsymbol{\varepsilon}^{*\beta(2)}) = 0,$$

$$\sum_{\alpha=0}^{n} \sum_{\beta=0}^{n} (\boldsymbol{\sigma}^{*\alpha(1)} - \boldsymbol{\sigma}^{*\alpha(2)}) : \mathbf{J}^{\alpha\beta} : (\boldsymbol{\sigma}^{*\beta(1)} - \boldsymbol{\sigma}^{*\beta(2)}) = 0. \qquad (9.3.11c,d)$$

From (9.3.11a,b), it follows that $\boldsymbol{\varepsilon}^{*\alpha(1)} = \boldsymbol{\varepsilon}^{*\alpha(2)}$ and $\boldsymbol{\sigma}^{*\alpha(1)} = \boldsymbol{\sigma}^{*\alpha(2)}$, for $\alpha = 0, 1, 2, ..., n$. Therefore, the solution of (9.3.10a,b) is unique, if $\mathbf{C}' - \mathbf{C}$ is positive-definite (if $\mathbf{D}' - \mathbf{D}$ is positive-definite).[7] In this case, the stationary values of I′ and J′ are the corresponding global maximum and minimum. In summary, whenever operator $< : \{(\mathbf{D}' - \mathbf{D})^{-1} + \boldsymbol{\Lambda}\} : >$ or $< : \{(\mathbf{C}' - \mathbf{C})^{-1} + \boldsymbol{\Gamma}\} : >$ is positive-definite (negative-definite), the corresponding matrix, $\{\mathbf{I}^{\alpha\beta}\}$ or $\{\mathbf{J}^{\alpha\beta}\}$, is definite and therefore invertible.

In terms of the optimal constant eigenstrains and eigenstresses, $\{\boldsymbol{\varepsilon}^{*\alpha}\}$ and $\{\boldsymbol{\sigma}^{*\alpha}\}$, which are the solutions of (9.3.10a,b), the functions I′ and J′ are written as

$$I'(\{\mathbf{e}^{*\alpha}\}; \boldsymbol{\sigma}^o) = \sum_{\alpha=0}^{n} \sum_{\beta=0}^{n} \frac{1}{2} (\mathbf{e}^{*\alpha} - \boldsymbol{\varepsilon}^{*\alpha}) : \mathbf{I}^{\alpha\beta} : (\mathbf{e}^{*\beta} - \boldsymbol{\varepsilon}^{*\beta}) + I'(\{\boldsymbol{\varepsilon}^{*\alpha}\}; \boldsymbol{\sigma}^o),$$

$$J'(\{\mathbf{s}^{*\alpha}\}; \boldsymbol{\varepsilon}^o) = \sum_{\alpha=0}^{n} \sum_{\beta=0}^{n} \frac{1}{2} (\mathbf{s}^{*\alpha} - \boldsymbol{\sigma}^{*\alpha}) : \mathbf{J}^{\alpha\beta} : (\mathbf{s}^{*\beta} - \boldsymbol{\sigma}^{*\beta}) + J'(\{\boldsymbol{\sigma}^{*\alpha}\}; \boldsymbol{\varepsilon}^o). \qquad (9.3.12a,b)$$

Therefore, for any constant eigenstrains and eigenstresses, $\{\mathbf{e}^{*\alpha}\}$ and $\{\mathbf{s}^{*\alpha}\}$,

$$I'(\{\boldsymbol{\varepsilon}^{*\alpha}\}; \boldsymbol{\sigma}^o) \geq (\leq)\, I'(\{\mathbf{e}^{*\alpha}\}; \boldsymbol{\sigma}^o),$$

---

[6] The following proof is essentially the same as that for the uniqueness of the solution of linearly elastic problems; see Part 2, Subsection 18.3.

[7] The above proof also implies that if the operator $< : \{(\mathbf{D}' - \mathbf{D})^{-1} + \boldsymbol{\Lambda}\} : >$ or $< : \{(\mathbf{C}' - \mathbf{C})^{-1} + \boldsymbol{\Gamma}\} : >$ is positive-definite or negative-definite, then the corresponding tensors $\mathbf{I}^{\alpha\beta}$ or $\mathbf{J}^{\alpha\beta}$ are positive-definite or negative-definite, respectively.

$$J'(\{\boldsymbol{\sigma}^{*\alpha}\}; \boldsymbol{\varepsilon}^o) \leq (\geq) \; J'(\{\mathbf{s}^{*\alpha}\}; \boldsymbol{\varepsilon}^o), \tag{9.3.12c,d}$$

if $\mathbf{C}^\alpha - \mathbf{C}$ is positive-definite, (if $\mathbf{D}^\alpha - \mathbf{D}$ is positive-definite), for $\alpha = 0, 1, 2, ..., n$.

From the governing sets of $n+1$ linear equations (9.3.10a,b), compute the values of $I'(\{\boldsymbol{\varepsilon}^{*\alpha}\}; \boldsymbol{\sigma}^o)$ and $J'(\{\boldsymbol{\sigma}^{*\alpha}\}; \boldsymbol{\varepsilon}^o)$, as

$$I'(\{\boldsymbol{\varepsilon}^{*\alpha}\}; \boldsymbol{\sigma}^o) = -\frac{1}{2}\boldsymbol{\sigma}^o : \bar{\boldsymbol{\varepsilon}}^*,$$

$$J'(\{\boldsymbol{\sigma}^{*\alpha}\}; \boldsymbol{\varepsilon}^o) = -\frac{1}{2}\bar{\boldsymbol{\sigma}}^* : \boldsymbol{\varepsilon}^o, \tag{9.3.13a,b}$$

where

$$\bar{\boldsymbol{\varepsilon}}^* \equiv \sum_{\alpha=0}^{n} f_\alpha \, \boldsymbol{\varepsilon}^{*\alpha}, \qquad \bar{\boldsymbol{\sigma}}^* \equiv \sum_{\alpha=0}^{n} f_\alpha \, \boldsymbol{\sigma}^{*\alpha}. \tag{9.3.13c,d}$$

Therefore, from (9.3.7), (9.3.8), (9.3.12), and (9.3.13), the upper and lower bounds on the overall elastic energies are given by

$$W^{co}(\boldsymbol{\sigma}^o) - W^c(\boldsymbol{\sigma}^o) \geq (\leq) \; -\frac{1}{2}\boldsymbol{\sigma}^o : \bar{\boldsymbol{\varepsilon}}^*,$$

$$W^o(\boldsymbol{\varepsilon}^o) - W(\boldsymbol{\varepsilon}^o) \leq (\geq) \; -\frac{1}{2}\bar{\boldsymbol{\sigma}}^* : \boldsymbol{\varepsilon}^o, \tag{9.3.14a,b}$$

when $\mathbf{C}^\alpha - \mathbf{C}$ is positive-definite (when $\mathbf{D}^\alpha - \mathbf{D}$ is positive-definite).

9.3.3. Subregion Approximation Method

As mentioned above, the correlation tensors for the trial piecewise constant eigenstrain and eigenstress fields depend only on the geometries of the equivalent homogeneous solid, the reference elasticity, and the geometries of the inclusions and the RVE. Hence, in principle, the trial correlation tensors can be defined for any pair of micro-inclusions, Ω_α and Ω_β. Since an inclusion can be divided into a number of subregions, with each subregion being viewed as an inclusion, more and more accurate estimates of the correlation tensors may be expected by further subdivision.

To improve the approximation method, divide each micro-inclusion Ω_α into a set of several subregions. Ordering these subregions from 0 to N (> n), redefine Ω_α to be the αth subregion, with $\alpha = 0, 1, 2, ..., N$. Several subregions now have common uniform elasticity and compliance tensors. However, each such subregion may have its own uniform eigenstrain and eigenstress, unequal to those of other subregions, even though all these subregions may belong to the same original inclusion which has been subdivided for the purpose of calculation. The procedure of Subsection 9.3.2 may now be followed to calculate the trial correlation tensors $\boldsymbol{\Lambda}^{\alpha\beta}$ and $\boldsymbol{\Gamma}^{\alpha\beta}$, the quadratic expressions I' and J', and their optimal values. Then, for a properly posed problem, it is expected that the exact eigenstrain and eigenstress fields, $\boldsymbol{\varepsilon}^*(\mathbf{x})$ and $\boldsymbol{\sigma}^*(\mathbf{x})$, are given by the limit of the optimal piecewise constant eigenstrains and eigenstresses, $\{\boldsymbol{\varepsilon}^{*\alpha}\}$ and

$\{\boldsymbol{\sigma}^{*\alpha}\}$, i.e.,

$$\lim_{\Omega_\alpha \to \mathbf{x}} \boldsymbol{\varepsilon}^{*\alpha} = \boldsymbol{\varepsilon}^*(\mathbf{x}), \qquad \lim_{\Omega_\alpha \to \mathbf{x}} \boldsymbol{\sigma}^{*\alpha} = \boldsymbol{\sigma}^*(\mathbf{x}). \tag{9.3.15a,b}$$

Therefore, as N increases, it is expected that more and more accurate solutions to the consistency conditions result.

Using this approximating method with a piecewise constant trial eigenstrain field, Nemat-Nasser and Taya (1981, 1985) evaluated the overall elastic energies of elastic solids with periodic structures; see also Nemat-Nasser *et al.* (1982) and Iwakuma and Nemat-Nasser (1983). In this case, the trial correlation tensors, $\boldsymbol{\Lambda}^{\alpha\beta}$ and $\boldsymbol{\Gamma}^{\alpha\beta}$, can be expressed in Fourier series, and the optimal values of the quadratic expressions, I' and J', can be computed analytically. These authors show that, when the volume fraction of micro-inclusions is small, even a small number of subregions yields good estimates of the exact average elastic energies. These and related topics are discussed in Sections 12 and 13.

9.4. DIRECT ESTIMATES OF OVERALL MODULI

Expressions for the exact upper and lower bounds on the average complementary and elastic energies are obtained in the preceding subsections. However, it is rather difficult to compute these bounds, since: (1) for a bounded RVE, the exact integral operators $\boldsymbol{\Lambda}$ and $\boldsymbol{\Gamma}$ are very complicated; and (2) the exact correlation tensors $\boldsymbol{\Lambda}^{\alpha\beta}$ and $\boldsymbol{\Gamma}^{\alpha\beta}$ cannot, in general, be calculated explicitly. In order to obtain explicit expressions for the bounds, it is necessary to estimate the integral operators and the corresponding correlation tensors. Willis (1977) proposed an asymptotic method to explicitly determine estimates of the correlation tensors through simple integral operators which are defined in terms of the Green function of an infinite homogeneous elastic solid. For a heterogeneous finite elastic solid, the integral operators and the corresponding correlation tensors derived in this approximate manner may not (since errors are introduced by the approximation) produce bounds on the average elastic energies, even if the Hashin-Shtrikman variational principle is applied. In simple cases, such as when the microgeometry is *statistically isotropic*, the results in such an approximation depend only on the volume fraction of micro-inclusions and are independent of other geometrical properties of the RVE.[8] Since the volume fraction of micro-inclusions is a geometrical quantity which can be measured easily, results of this kind are generally regarded as useful. Numerical computations seem to support this. Furthermore, the approach has been justified by considering the RVE as part of a very large heterogeneous body with mean stress and strain in common with the RVE. The resulting tractions and displacements on boundary ∂V of the RVE then fluctuate about the corresponding average stress

[8] This should be contrasted with the results presented in Section 13 for periodic microstructures where the Green function and the bounds can be calculated to any desired degree of accuracy.

and strain tensors. It is suggested (Willis, 1977) that the boundary contributions to the overall average elastic strain and complementary elastic strain energies, by these fluctuations, may be neglected. This then allows use of the simple translationally invariant Green function of an unbounded domain to estimate the operators $\mathbf{\Lambda}$ and $\mathbf{\Gamma}$. It is shown in Subsection 9.5 that the error in such an approximation may actually affect the corresponding bounds. However, based on the universal Theorems I and II of Subsection 2.5.6, the effects of this approximation are established in Subsection 9.5.3, and *computable rigorous* bounds are obtained in Subsection 9.5.4.

In this subsection, expressions which approximate the exact integral operators, $\mathbf{\Lambda}$ and $\mathbf{\Gamma}$, are obtained. First, in terms of the Green function $\mathbf{G}^{\infty}$ for an infinitely extended homogeneous domain, exact expressions for $\mathbf{\Lambda}$ and $\mathbf{\Gamma}$ are developed, and then these are approximated with the aid of several assumptions. For piecewise constant trial eigenstress and eigenstrain fields, the correlation tensors are then computed, and explicit bounds on the overall elastic energies are estimated according to the Hashin-Shtrikman variational principle. Finally, from these energy bounds, the bounds on the overall elasticity and compliance tensors of the RVE are obtained.

9.4.1. Boundary-Value Problems for Equivalent Homogeneous Solid

Consider the boundary-value problem for the displacement field $\mathbf{u}(\mathbf{x})$ of the equivalent homogeneous bounded solid V with uniform elasticity tensor $\mathbf{C}$ ($= \mathbf{D}^{-1}$). As shown in Section 11, for tractions $\mathbf{t}$ prescribed on ∂V, and body forces $\boldsymbol{\nabla}\cdot\mathbf{T}$ distributed in V, the displacement field $\mathbf{u}$ is given by the solution of the following boundary-value problem:

$$\boldsymbol{\nabla}\cdot(\mathbf{C} : \boldsymbol{\nabla}\otimes\mathbf{u}(\mathbf{x})) + \boldsymbol{\nabla}\cdot\mathbf{T}(\mathbf{x}) = \mathbf{0} \qquad \mathbf{x} \text{ in V},$$

$$\boldsymbol{\nu}(\mathbf{x})\cdot(\mathbf{C} : \boldsymbol{\nabla}\otimes\mathbf{u}(\mathbf{x})) = \mathbf{t}(\mathbf{x}) \qquad \mathbf{x} \text{ on } \partial\text{V}, \tag{9.4.1a,b}$$

where $\mathbf{T}(\mathbf{x})$ is some second-order tensor field, and $\boldsymbol{\nu}(\mathbf{x})$ is the outward unit normal at $\mathbf{x}$ on ∂V. In (9.4.1), the tractions $\mathbf{t}$ are regarded as arbitrary, except that $\boldsymbol{\nabla}\cdot\mathbf{T}$ and $\mathbf{t}$ must be in equilibrium.

To formulate boundary-value problem (9.4.1a,b) *in terms of the infinite-body Green function, regard the finite homogeneous domain* V *as part of an infinitely extended homogeneous solid.* Then, for arbitrary $\mathbf{T}(\mathbf{x})$ defined in V, tractions acting on the boundary ∂V are such that the continuity of displacements and tractions is satisfied. The required solution can be expressed in terms of the Green function $\mathbf{G}^{\infty}(\mathbf{z})$ of the unbounded homogeneous solid, where, instead of $\mathbf{t}$, some suitable tractions $\hat{\mathbf{t}}$ are distributed on ∂V such that the resulting tractions due to both $\mathbf{T}$ and $\hat{\mathbf{t}}$ equal $\mathbf{t}$. Then, the solution $\mathbf{u}$ is exactly given by[9]

---

[9] Note that the solutions corresponding to the two tensor fields, $\mathbf{T}$ in V and $\hat{\mathbf{t}}$ on ∂V, are independent of each other. Hence, the displacement field produced by both fields is given by superposition of the one produced by $\mathbf{T}$ and that produced by $\hat{\mathbf{t}}$.

$$\mathbf{u}(\mathbf{x}) = \mathbf{U}(\mathbf{x};\, \mathbf{T};\, \hat{\mathbf{t}};\, \hat{\mathbf{u}})$$

$$\equiv -\int_V \mathbf{T}(\mathbf{y}) : (\nabla_y \otimes \mathbf{G}^{\infty T}(\mathbf{y}-\mathbf{x}))\, dV_y$$

$$+\int_{\partial V} \{\boldsymbol{\nu}(\mathbf{y}) \cdot \mathbf{T}(\mathbf{y}) + \hat{\mathbf{t}}(\mathbf{y})\} \cdot \mathbf{G}^{\infty T}(\mathbf{y}-\mathbf{x})\, dS_y$$

$$-\int_{\partial V} \hat{\mathbf{u}}(\mathbf{y}) \cdot (\boldsymbol{\nu}(\mathbf{y}) \cdot \mathbf{C} : (\nabla_y \otimes \mathbf{G}^{\infty T}(\mathbf{y}-\mathbf{x})))\, dS_y, \qquad (9.4.2a)$$

or in component form,

$$U_i(\mathbf{x};\, \mathbf{T};\, \hat{\mathbf{t}};\, \hat{\mathbf{u}}) \equiv -\int_V G^{\infty}_{ji,k}(\mathbf{y}-\mathbf{x})\, T_{kj}(\mathbf{y})\, dV_y$$

$$+\int_{\partial V} \{\nu_k(\mathbf{y})\, T_{kj}(\mathbf{y}) + t_j(\mathbf{y})\}\, G^{\infty}_{ji}(\mathbf{y}-\mathbf{x})\, dS_y$$

$$-\int_{\partial V} \nu_m(\mathbf{y})\, C_{mnkl}\, G^{\infty}_{ki,l}(\mathbf{y}-\mathbf{x})\, \hat{u}_n(\mathbf{y})\, dS_y. \qquad (9.4.2b)$$

In (9.4.2a), $\hat{\mathbf{u}}$ is the resulting boundary displacements which is determined by setting[10] $\frac{1}{2}\mathbf{U}(\mathbf{x};\, \mathbf{T};\, \hat{\mathbf{t}};\, \hat{\mathbf{u}}) = \hat{\mathbf{u}}$, for $\mathbf{x}$ on ∂V. Subscript $\mathbf{y}$ indicates that the corresponding operation is with respect to the variable $\mathbf{y}$; see Subsection 11.1. The quantity $\mathbf{U}(\mathbf{x};\, \mathbf{T};\, \hat{\mathbf{t}};\, \hat{\mathbf{u}})$ is an integral operator which acts on the first- and second-order tensor fields, $\hat{\mathbf{t}}$ and $\mathbf{T}$, once the boundary displacements $\hat{\mathbf{u}}$ are determined; quantities with a caret, ^, are boundary data.

In particular, let the displacement fields of the equivalent homogeneous solid, respectively produced by an eigenstrain field $\mathbf{e}^*$ and an eigenstress field $\mathbf{s}^*$, be denoted by

$$\mathbf{u}^{\Sigma}(\mathbf{x}) = \mathbf{U}(\mathbf{x};\; -\mathbf{C} : \mathbf{e}^*;\, \mathbf{0};\, \hat{\mathbf{u}}^{\Sigma}), \qquad (9.4.3a)$$

and

$$\mathbf{u}^{E}(\mathbf{x}) = \mathbf{U}(\mathbf{x};\, \mathbf{s}^*;\, \hat{\mathbf{t}}^{E};\, \mathbf{0}) \qquad \mathbf{x} \text{ in } V, \qquad (9.4.3b)$$

where $\hat{\mathbf{u}}^{\Sigma}$ or $\hat{\mathbf{t}}^{E}$ is introduced on ∂V to satisfy zero-traction or zero-displacement boundary conditions[11] on ∂V. The unknown displacements or tractions are determined from the following integral equations:

$$\boldsymbol{\nu}(\mathbf{x}) \cdot \mathbf{C} : \{\nabla \otimes \mathbf{U}(\mathbf{x};\; -\mathbf{C} : \mathbf{e}^*;\, \mathbf{0};\, \hat{\mathbf{u}}^{\Sigma}) - \mathbf{e}^*(\mathbf{x})\} = \mathbf{0} \qquad \mathbf{x} \text{ on } \partial V, \qquad (9.4.4a)$$

and

$$\mathbf{U}(\mathbf{x};\, \mathbf{s}^*;\, \hat{\mathbf{t}}^{E};\, \mathbf{0}) = \mathbf{0} \qquad \mathbf{x} \text{ on } \partial V; \qquad (9.4.4b)$$

see (9.1.5b) and (9.1.13b). For the bounded equivalent homogeneous solid, the boundary-value problems associated with the eigenstrain $\mathbf{e}^*$ and eigenstress $\mathbf{s}^*$

---

[10] A coefficient 1/2 appears as $\mathbf{x}$ is on the boundary, assumed to be smooth.

[11] Note that two separate boundary-value problems are considered simultaneously for the same RVE, one corresponding to zero displacements (denoted by superscript E) and the other corresponding to zero tractions (denoted by superscript Σ) on ∂V. These are two separate problems, and the two boundary data are, in general, independent.

are solved exactly by (9.4.3), if the unknown boundary data, $\hat{\mathbf{u}}^{\Sigma}$ and $\hat{\mathbf{t}}^{\mathrm{E}}$, can be determined. But, in general, this is a very difficult problem, even if the Green function $\mathbf{G}^{\infty}$ is known explicitly.

9.4.2. Simplified Integral Operators

Using (9.4.3a,b), express the integral operators $\boldsymbol{\Lambda}$ and $\boldsymbol{\Gamma}$ in terms of the integral operator $\mathbf{U}$, as

$$-\boldsymbol{\Lambda}(\mathbf{x};\,\mathbf{e}^*) = \mathbf{C} : (\boldsymbol{\nabla} \otimes \mathbf{U}(\mathbf{x};\, -\mathbf{C} : \mathbf{e}^*;\, \mathbf{0};\, \hat{\mathbf{u}}^{\Sigma})) - \mathbf{C} : \mathbf{e}^*(\mathbf{x}),$$

$$-\boldsymbol{\Gamma}(\mathbf{x};\,\mathbf{s}^*) = \frac{1}{2}\{(\boldsymbol{\nabla} \otimes \mathbf{U}(\mathbf{x};\,\mathbf{s}^*;\,\hat{\mathbf{t}}^{\mathrm{E}};\,\mathbf{0})) + (\boldsymbol{\nabla} \otimes \mathbf{U}(\mathbf{x};\,\mathbf{s}^*;\,\hat{\mathbf{t}}^{\mathrm{E}};\,\mathbf{0}))^{\mathrm{T}}\}. \qquad (9.4.5a,b)$$

These formal expressions are *exact*. From (9.4.5a,b) now seek to obtain simple integral operators which approximate the stress and strain fields produced by the corresponding eigenstrain and eigenstress fields.

To this end, first rearrange the integral operator $\mathbf{U}$. From the Gauss theorem, rewrite (9.4.2b) as

$$\begin{aligned}\mathbf{U}(\mathbf{x};\,\mathbf{T};\,\hat{\mathbf{t}};\,\hat{\mathbf{u}}) = &-\int_{\mathrm{V}} (\mathbf{T}(\mathbf{y}) - \overline{\mathbf{T}}) : (\boldsymbol{\nabla}_{\mathbf{y}} \otimes \mathbf{G}^{\infty\mathrm{T}}(\mathbf{y}-\mathbf{x}))\, d\mathrm{V}_{\mathbf{y}} \\ &+\int_{\partial\mathrm{V}} \{\boldsymbol{\nu}(\mathbf{y}) \cdot (\mathbf{T}(\mathbf{y}) - \overline{\mathbf{T}}) + \hat{\mathbf{t}}(\mathbf{y})\} \cdot \mathbf{G}^{\infty\mathrm{T}}(\mathbf{y}-\mathbf{x})\, d\mathrm{S}_{\mathbf{y}} \\ &-\int_{\partial\mathrm{V}} \hat{\mathbf{u}}(\mathbf{y}) \cdot (\boldsymbol{\nu}(\mathbf{y}) \cdot \mathbf{C} : (\boldsymbol{\nabla}_{\mathbf{y}} \otimes \mathbf{G}^{\infty\mathrm{T}}(\mathbf{y}-\mathbf{x})))\, d\mathrm{S}_{\mathbf{y}}, \\ = &\ \mathbf{U}(\mathbf{x};\,\mathbf{T} - \overline{\mathbf{T}};\,\hat{\mathbf{t}};\,\hat{\mathbf{u}}),\end{aligned} \qquad (9.4.6)$$

where $\overline{\mathbf{T}}$ is the (constant) average of $\mathbf{T}$ over V, i.e., $\overline{\mathbf{T}} \equiv \langle \mathrm{T} \rangle$. It has been pointed out in the literature (Willis, 1977, 1981) that if the required boundary conditions are either zero tractions or zero displacements, it is then expected that the integrand of the surface integrals should have the following two properties: (1) $\mathbf{T} - \overline{\mathbf{T}}$ has an oscillatory spatial variation about zero on ∂V; and (2) the unknown traction $\hat{\mathbf{t}}$ also has an oscillatory spatial variation about zero on ∂V, provided that the effect of $\hat{\mathbf{u}}$ is negligibly small. The approximation is then based on the expectation that the contribution to the displacements at the interior points of a very large (relative to the size of the inhomogeneities) V by the surface integrals in (9.4.6) may be neglected, except when the displacement of points within a thin "boundary layer" close to ∂V is of concern. Hence, for interior points $\mathbf{x}$ in V, the displacement $\mathbf{u}$ is approximated by[12]

$$\mathbf{u}(\mathbf{x}) \approx \mathbf{U}^{\mathrm{A}}(\mathbf{x};\,\mathbf{T} - \overline{\mathbf{T}}), \qquad (9.4.7a)$$

where

---

[12] The consequences of this assumption are rigorously established in Subsection 9.5.

$$\mathbf{U}^{\mathrm{A}}(\mathbf{x};\, \mathbf{T}) \equiv -\int_{\mathrm{V}} \mathbf{T}(\mathbf{y}) : (\boldsymbol{\nabla}_{\mathbf{y}} \otimes \mathbf{G}^{\infty\mathrm{T}}(\mathbf{y}-\mathbf{x}))\, \mathrm{dV}_{\mathbf{y}}. \tag{9.4.7b}$$

If the exact integral operator $\mathbf{U}$ is replaced by the integral operator $\mathbf{U}^{\mathrm{A}}$ in (9.4.3a,b), the displacement fields, $\mathbf{u}^{\Sigma}$ and $\mathbf{u}^{\mathrm{E}}$, produced by the eigenstrain and eigenstress fields, $\mathbf{e}^*$ and $\mathbf{s}^*$, are respectively approximated by (two separate problems)

$$\mathbf{u}^{\Sigma}(\mathbf{x}) \approx \mathbf{U}^{\mathrm{A}}(\mathbf{x};\, -\mathbf{C} : (\mathbf{e}^* - \overline{\mathbf{e}}^*)), \qquad \mathbf{u}^{\mathrm{E}}(\mathbf{x}) \approx \mathbf{U}^{\mathrm{A}}(\mathbf{x};\, \mathbf{s}^* - \overline{\mathbf{s}}^*), \tag{9.4.8a,b}$$

where $\overline{\mathbf{e}}^* \equiv \langle \mathbf{e}^* \rangle$ and $\overline{\mathbf{s}}^* \equiv \langle \mathbf{s}^* \rangle$. Hence, substitution of (9.4.8a,b) into (9.4.5a,b) yields the following approximation of the integral operators $\boldsymbol{\Lambda}$ and $\boldsymbol{\Gamma}$:

$$\boldsymbol{\Lambda}(\mathbf{x};\, \mathbf{e}^*) \equiv (\boldsymbol{\Lambda} : \mathbf{e}^*)(\mathbf{x}) \approx (\boldsymbol{\Lambda}^{\mathrm{A}} : \mathbf{e}^*)(\mathbf{x}),$$

$$\boldsymbol{\Gamma}(\mathbf{x};\, \mathbf{s}^*) \equiv (\boldsymbol{\Gamma} : \mathbf{s}^*)(\mathbf{x}) \approx (\boldsymbol{\Gamma}^{\mathrm{A}} : \mathbf{s}^*)(\mathbf{x}), \tag{9.4.9a,b}$$

where

$$-(\boldsymbol{\Lambda}^{\mathrm{A}} : \mathbf{e}^*)(\mathbf{x}) \equiv \mathbf{C} : \{\boldsymbol{\nabla} \otimes \mathbf{U}^{\mathrm{A}}(\mathbf{x};\, -\mathbf{C} : (\mathbf{e}^* - \overline{\mathbf{e}}^*))\} - \mathbf{C} : (\mathbf{e}^*(\mathbf{x}) - \overline{\mathbf{e}}^*),$$

$$-(\boldsymbol{\Gamma}^{\mathrm{A}} : \mathbf{s}^*)(\mathbf{x}) \equiv \frac{1}{2}\{(\boldsymbol{\nabla} \otimes \mathbf{U}^{\mathrm{A}}(\mathbf{x};\, \mathbf{s}^* - \overline{\mathbf{s}}^*)) + (\boldsymbol{\nabla} \otimes \mathbf{U}^{\mathrm{A}}(\mathbf{x};\, \mathbf{s}^* - \overline{\mathbf{s}}^*))^{\mathrm{T}}\}, \tag{9.4.9c,d}$$

are the approximate forms of the corresponding original operators. By their construction, *the volume averages of* $(\boldsymbol{\Lambda}^{\mathrm{A}} : \mathbf{e}^*)(\mathbf{x})$ *and* $(\boldsymbol{\Gamma}^{\mathrm{A}} : \mathbf{s}^*)(\mathbf{x})$ *over* V *are identically zero,*

$$\langle \boldsymbol{\Lambda}^{\mathrm{A}} : \mathbf{e}^* \rangle = \mathbf{0}, \qquad \langle \boldsymbol{\Gamma}^{\mathrm{A}} : \mathbf{s}^* \rangle = \mathbf{0}, \tag{9.4.9e,f}$$

when V *is an ellipsoid*; see Subsection 11.3.3 and Equation (11.3.18b). This observation is used in Subsections 9.5 and 9.6, to obtain exact computable bounds for the overall moduli.

9.4.3. Approximate Correlation Tensors

Using the (approximate) integral operators $\boldsymbol{\Lambda}^{\mathrm{A}}$ and $\boldsymbol{\Gamma}^{\mathrm{A}}$ given by (9.4.9a,b), now approximate the correlation tensors, $\boldsymbol{\Lambda}^{\alpha\beta}$ and $\boldsymbol{\Gamma}^{\alpha\beta}$, for piecewise constant trial eigenstrain and eigenstress fields. Assuming a suitable reference elasticity tensor, $\mathbf{C}$, and following definitions (9.3.5a,b) for $\boldsymbol{\Lambda}^{\alpha\beta}$ and $\boldsymbol{\Gamma}^{\alpha\beta}$, consider

$$\mathrm{f}_\beta\, \boldsymbol{\Lambda}^{\alpha\beta} : \mathbf{e}^{*\beta} \approx \langle \boldsymbol{\Lambda}^{\mathrm{A}} : \mathrm{H}_\beta \mathbf{e}^{*\beta} \rangle_\alpha$$

$$= \left\{ -\frac{1}{\Omega_\alpha} \int_{\Omega_\alpha} \{ \int_{\Omega_\beta} \mathbf{C} : \boldsymbol{\Gamma}^{\infty}(\mathbf{y}-\mathbf{x}) : \mathbf{C}\, \mathrm{dV}_{\mathbf{y}} \}\, \mathrm{dV}_{\mathbf{x}} + \mathbf{C} \right\} : (\delta_{\alpha\beta} - \mathrm{f}_\beta)\, \mathbf{e}^{*\beta} \tag{9.4.10a}$$

and

$$\mathrm{f}_\beta\, \boldsymbol{\Gamma}^{\alpha\beta} : \mathbf{s}^{*\beta} \approx \langle \boldsymbol{\Gamma}^{\mathrm{A}} : \mathrm{H}_\beta \mathbf{s}^{*\beta} \rangle_\alpha$$

$$= \left\{ \frac{1}{\Omega_\alpha} \int_{\Omega_\alpha} \{ \int_{\Omega_\beta} \mathbf{\Gamma}^\infty(\mathbf{y}-\mathbf{x})\, dV_\mathbf{y} \}\, dV_\mathbf{x} \right\} : (\delta_{\alpha\beta} - f_\beta)\, \mathbf{s}^{*\beta}$$

$$(\alpha,\ \beta \text{ not summed}), \qquad (9.4.10b)$$

where $H_\beta = H(\mathbf{x};\ \Omega_\beta)$, and $\mathbf{e}^{*\beta}$ and $\mathbf{s}^{*\beta}$ are the arbitrary constant eigenstrain and eigenstress; $\mathbf{x}$ and $\mathbf{y}$ are in the αth and βth micro-inclusions, Ω_α and Ω_β, respectively; and the fourth-order tensor field $\mathbf{\Gamma}^\infty(\mathbf{z})$ is defined by

$$\Gamma^\infty_{ijkl} = -\frac{1}{4} \{ G^\infty_{ik,jl} + G^\infty_{jk,il} + G^\infty_{il,jk} + G^\infty_{jl,ik} \}. \qquad (9.4.11)$$

It is still difficult to compute the above double integral analytically, even though the Green function $\mathbf{G}^\infty$ can be obtained explicitly, at least when an isotropic reference elasticity $\mathbf{C}$ is used. From the properties of $\mathbf{G}^\infty$, however, it is possible to estimate these integrals and obtain an explicit approximation for the *expected values* of the correlation tensors $\mathbf{\Lambda}^{\alpha\beta}$ and $\mathbf{\Gamma}^{\alpha\beta}$, based on *statistical homogeneity and isotropy of the distribution of the micro-inclusions.* As shown by Willis (1977) for a medium with statistically homogeneous and isotropic microgeometry, the expected, or the average, value of the double integral of $\mathbf{\Gamma}^\infty$ for all possible arrangements of Ω_α and Ω_β can be expressed as

$$\left[\frac{1}{\Omega_\alpha} \int_{\Omega_\alpha} \int_{\Omega_\beta} \mathbf{\Gamma}^\infty(\mathbf{y}-\mathbf{x})\, dV_\mathbf{y}\, dV_\mathbf{x} \right]_{\text{expected}} \approx \mathbf{\Gamma}^{A\alpha\beta}, \qquad (9.4.12a)$$

where the expected value of the correlation tensor $\mathbf{\Gamma}^{A\alpha\beta}$ is defined by

$$\mathbf{\Gamma}^{A\alpha\beta} \equiv (\delta_{\alpha\beta} \frac{1}{f_\beta} - 1)\, \mathbf{P} \qquad (\beta \text{ not summed}), \qquad (9.4.12b)$$

with the fourth-order tensor $\mathbf{P}$ given by

$$\mathbf{P} \equiv \int_{|\mathbf{z}|<a} \mathbf{\Gamma}^\infty(\mathbf{z})\, dV_\mathbf{z} \qquad (a > 0). \qquad (9.4.13a)$$

As will be shown in Section 11, tensor $\mathbf{P}$ satisfies the symmetry properties:

$$P_{ijkl} = P_{klij}\ (= P_{jikl} = P_{ijlk}). \qquad (9.4.13b)$$

Since the contribution of $H_\beta(\mathbf{x})\, \mathbf{e}^{*\beta}$ to the average eigenstrain $\bar{\mathbf{e}}^*$ is $f_\beta\, \mathbf{e}^{*\beta}$, subtract it from $H_\beta(\mathbf{x})\, \mathbf{e}^{*\beta}$ (β not summed), in order to compute the approximate correlation tensor $\mathbf{\Lambda}^{A\alpha\beta}$; see (9.4.9c). Then, direct substitution of (9.4.12) into (9.4.10a) yields $\mathbf{\Lambda}^{A\alpha\beta}$, as

$$\mathbf{\Lambda}^{A\alpha\beta} \equiv -(\delta_{\alpha\beta} \frac{1}{f_\beta} - 1)\, \mathbf{C} : \mathbf{P} : \mathbf{C} + \delta_{\alpha\beta} \frac{1}{f_\beta}\, \mathbf{C} - \mathbf{C} \qquad (\beta \text{ not summed}). \qquad (9.4.14)$$

It is seen from the symmetry of $\mathbf{P}$, (9.4.13b), that tensors $\mathbf{\Lambda}^{A\alpha\beta}$ and $\mathbf{\Gamma}^{A\alpha\beta}$ are symmetric with respect to the first and last pairs of their subscript indices, i.e.,

$$\Lambda^{A\alpha\beta}_{ijkl} = \Lambda^{A\alpha\beta}_{klij}\ (= \Lambda^{A\alpha\beta}_{jikl} = \Lambda^{A\alpha\beta}_{ijlk}),$$

$$\Gamma^{A\alpha\beta}_{ijkl} = \Gamma^{A\alpha\beta}_{klij} \; (= \Gamma^{A\alpha\beta}_{jikl} = \Gamma^{A\alpha\beta}_{ijlk}), \tag{9.4.15a,b}$$

for $\alpha, \beta = 0, 1, 2, ..., n$.

A brief derivation of (9.4.12) and (9.4.13) is as follows; see Willis (1977) for details. Since $\mathbf{\Gamma}^\infty(\mathbf{s}^*)$ is defined by

$$\mathbf{\Gamma}^\infty(\mathbf{x}; \mathbf{s}^*) \equiv \int \mathbf{\Gamma}^\infty(\mathbf{x}-\mathbf{y}) : (\mathbf{s}^* - \overline{\mathbf{s}}^*) \, dV_y, \tag{9.4.16a}$$

for a piecewise constant distribution of eigenstresses, $< \mathbf{s}^* : \mathbf{\Gamma}^\infty(\mathbf{s}^*) >$ becomes

$$< \mathbf{s}^* : \mathbf{\Gamma}^\infty(\mathbf{s}^*) > = \sum_{\alpha=0}^{n} \sum_{\beta=0}^{n} \frac{1}{V} \iint (H_\alpha(\mathbf{x})\, \mathbf{s}^{*\alpha})$$

$$: \mathbf{\Gamma}^\infty(\mathbf{x}-\mathbf{y}) : (H_\beta(\mathbf{y})\, \mathbf{s}^{*\beta} - \overline{\mathbf{s}}^*) dV_x \, dV_y. \tag{9.4.16b}$$

Based on the assumption of statistical homogeneity, interpret the above volume average as the expected value or the ensemble average taken at a fixed point. Then, noting that the expected value of finding phase α at $\mathbf{x}$ is given by f_α, the right side of (9.4.16b) becomes

$$\sum_{\alpha=0}^{n} \sum_{\beta=0}^{n} \mathbf{s}^{*\alpha} : \left\{ \int (\phi_{\alpha\beta}(|\mathbf{x}-\mathbf{y}|) - f_\alpha f_\beta)\, \mathbf{\Gamma}^\infty(\mathbf{x}-\mathbf{y}) \, dV_y \right\} : \mathbf{s}^{*\beta},$$

where $\phi_{\alpha\beta}$ is the two-point correlation function that represents the probability of finding phases α and β at $\mathbf{x}$ and $\mathbf{y}$, respectively. Note that the two-point correlation function must depend only on $|\mathbf{x}-\mathbf{y}|$, in view of statistical homogeneity (translation invariance) and isotropy (rotation invariance).

In view of the assumed disorder, $\phi_{\alpha\beta}$ tends to $f_\alpha f_\beta$ when $|\mathbf{x}-\mathbf{y}|$ becomes large. Furthermore, $\phi_{\alpha\beta}(0) = f_\alpha$ when $\alpha = \beta$ and $\phi_{\alpha\beta}(0) = 0$ when $\alpha \neq \beta$; that is, the probability of finding the αth phase at any point is given by f_α, and the probability of finding different αth and βth phases at one point is zero. Hence, the *contribution* to the volume average, $< \mathbf{s}^* : \mathbf{\Gamma}^\infty(\mathbf{s}^*) >$, by the αth and βth phases may be estimated, as

$$\int (\phi_{\alpha\beta}(|\mathbf{x}-\mathbf{y}|) - f_\alpha f_\beta)\, \mathbf{\Gamma}^\infty(\mathbf{x}-\mathbf{y}) \, dV_y = (\phi_{\alpha\beta}(0) - f_\alpha f_\beta)\, \mathbf{P}, \tag{9.4.16c}$$

where identity

$$\int_{a<|z|<b} \mathbf{\Gamma}^\infty(\mathbf{z}) \, dV_z = \mathbf{0} \qquad (a > b > 0) \tag{9.4.16d}$$

is also used. Hence, (9.4.12) and (9.4.13) are obtained.

Since $\mathbf{\Gamma}^\infty$ is given by the gradient of the Green function $\mathbf{G}^\infty$, it depends on the elastic properties of the comparison material, i.e., on $\mathbf{C}$ (or $\mathbf{D}$). For the case of an isotropic elastic material, Eshelby (1957) obtained an explicit form for $\mathbf{P}$,

$$\mathbf{P} = \frac{1}{2\mu(1-\nu)} \left\{ -\frac{1}{15} \mathbf{1}^{(2)} \otimes \mathbf{1}^{(2)} + \frac{2(4-5\nu)}{15} \mathbf{1}^{(4s)} \right\}, \tag{9.4.17}$$

where μ and ν are the shear modulus and Poisson ratio of the isotropic material. For the case of an anisotropic elastic material, $\mathbf{P}$ has been obtained for several cases; for example, Kneer (1965), Willis (1970), and Kinoshita and Mura (1971).

9.4.4. Optimal Eigenstrains and Eigenstresses

Using the simplified integral operators which are derived from the Green function for an infinite domain, approximate the functionals I and J, by

$$\mathrm{I}(\mathbf{e}^*;\,\boldsymbol{\sigma}^o) \approx \mathrm{I}^A(\mathbf{e}^*;\,\boldsymbol{\sigma}^o), \qquad \mathrm{J}(\mathbf{s}^*;\,\boldsymbol{\varepsilon}^o) \approx \mathrm{J}^A(\mathbf{s}^*;\,\boldsymbol{\varepsilon}^o), \tag{9.4.18a,b}$$

where I^A and J^A are obtained by replacing the integral operators $\boldsymbol{\Lambda}$ and $\boldsymbol{\Gamma}$ in I and J, with the integral operators $\boldsymbol{\Lambda}^A$ and $\boldsymbol{\Gamma}^A$ which are defined by (9.4.9c,d).

For piecewise constant trial eigenstrain and eigenstress fields, $\mathbf{e}^* = \sum_{\alpha=0}^{n} \mathrm{H}_\alpha\, \mathbf{e}^{*\alpha}$ and $\mathbf{s}^* = \sum_{\alpha=0}^{n} \mathrm{H}_\alpha\, \mathbf{s}^{*\alpha}$, construct the quadratic expressions I′ and J′ in (9.3.8a,b) from I and J. In a similar manner, for such piecewise constant fields, define quadratic expressions $\mathrm{I}^{A\prime}$ and $\mathrm{J}^{A\prime}$ from I^A and J^A, as

$$\mathrm{I}^A(\mathbf{e}^*;\,\boldsymbol{\sigma}^o) \equiv \mathrm{I}^{A\prime}(\{\mathbf{e}^{*\alpha}\};\,\boldsymbol{\sigma}^o), \qquad \mathrm{J}^A(\mathbf{s}^*;\,\boldsymbol{\varepsilon}^o) \equiv \mathrm{J}^{A\prime}(\{\mathbf{s}^{*\alpha}\};\,\boldsymbol{\varepsilon}^o), \tag{9.4.19a,b}$$

where

$$\mathrm{I}^{A\prime}(\{\mathbf{e}^{*\alpha}\};\,\boldsymbol{\sigma}^o) \equiv \sum_{\alpha=0}^{n}\sum_{\beta=0}^{n} \frac{1}{2}\mathbf{e}^{*\alpha} : \mathbf{I}^{A\alpha\beta} : \mathbf{e}^{*\beta} - \boldsymbol{\sigma}^o : \bar{\mathbf{e}}^*,$$

$$\mathrm{J}^{A\prime}(\{\mathbf{s}^{*\alpha}\};\,\boldsymbol{\varepsilon}^o) \equiv \sum_{\alpha=0}^{n}\sum_{\beta=0}^{n} \frac{1}{2}\mathbf{s}^{*\alpha} : \mathbf{J}^{A\alpha\beta} : \mathbf{s}^{*\beta} - \bar{\mathbf{s}}^* : \boldsymbol{\varepsilon}^o, \tag{9.4.19c,d}$$

with the fourth-order tensors $\mathbf{I}^{A\alpha\beta}$ and $\mathbf{J}^{A\alpha\beta}$ being obtained by replacing the trial correlation tensors $\boldsymbol{\Lambda}^{\alpha\beta}$ and $\boldsymbol{\Gamma}^{\alpha\beta}$ in $\mathbf{I}^{\alpha\beta}$ and $\mathbf{J}^{\alpha\beta}$ with the approximate correlation tensors $\boldsymbol{\Lambda}^{A\alpha\beta}$ and $\boldsymbol{\Gamma}^{A\alpha\beta}$ which are defined by (9.4.12) and (9.4.14), for $\alpha, \beta = 0, 1, 2, ..., n$. Hence, from (9.4.18a,b), estimate I′ and J′ by

$$\mathrm{I}'(\{\mathbf{e}^{*\alpha}\};\,\boldsymbol{\sigma}^o) \approx \mathrm{I}^{A\prime}(\{\mathbf{e}^{*\alpha}\};\,\boldsymbol{\sigma}^o), \qquad \mathrm{J}'(\{\mathbf{s}^{*\alpha}\};\,\boldsymbol{\varepsilon}^o) \approx \mathrm{J}^{A\prime}(\{\mathbf{s}^{*\alpha}\};\,\boldsymbol{\varepsilon}^o). \tag{9.4.18c,d}$$

Substituting (9.4.14) into (9.4.19a), write the quadratic expression $\mathrm{I}^{A\prime}$, as

$$\begin{aligned}
&\mathrm{I}^{A\prime}(\{\mathbf{e}^{*\alpha}\};\,\boldsymbol{\sigma}^o) \\
&= \sum_{\alpha=0}^{n} \mathrm{f}_\alpha \frac{1}{2}\mathbf{e}^{*\alpha} : \{(\mathbf{D}^\alpha - \mathbf{D})^{-1} - \mathbf{C} : \mathbf{P} : \mathbf{C} + \mathbf{C}\} : \mathbf{e}^{*\alpha} \\
&\qquad + \frac{1}{2}\bar{\mathbf{e}}^* : \mathbf{C} : (\mathbf{P} - \mathbf{D}) : \mathbf{C} : \bar{\mathbf{e}}^* - \boldsymbol{\sigma}^o : \bar{\mathbf{e}}^* \\
&= - \sum_{\alpha=0}^{n} \mathrm{f}_\alpha \frac{1}{2}(-\,\mathbf{C} : \mathbf{e}^{*\alpha}) : \{(\mathbf{C}^\alpha - \mathbf{C})^{-1} + \mathbf{P}\} : (-\,\mathbf{C} : \mathbf{e}^{*\alpha}) \\
&\qquad + \frac{1}{2}(-\,\mathbf{C} : \bar{\mathbf{e}}^*) : (\mathbf{P} - \mathbf{D}) : (-\,\mathbf{C} : \bar{\mathbf{e}}^*) + (\mathbf{D} : \boldsymbol{\sigma}^o) : (-\,\mathbf{C} : \bar{\mathbf{e}}^*),
\end{aligned} \tag{9.4.20a}$$

where

$$(\mathbf{D}^\alpha - \mathbf{D})^{-1} = -\,\mathbf{C} : (\mathbf{C}^\alpha - \mathbf{C})^{-1} : \mathbf{C}^\alpha = -\,\mathbf{C} - \mathbf{C} : (\mathbf{C}^\alpha - \mathbf{C})^{-1} : \mathbf{C} \tag{9.4.21a}$$

and

$$(\mathbf{D}^\alpha - \mathbf{D})^{-1} - \mathbf{C} : \mathbf{P} : \mathbf{C} + \mathbf{C} = -\,\mathbf{C} : \{(\mathbf{C}^\alpha - \mathbf{C})^{-1} + \mathbf{P}\} : \mathbf{C} \tag{9.4.21b}$$

are used. In a similar manner, substituting (9.4.12b) into (9.4.19b), write the quadratic form $J^{A\prime}$, as

$$J^{A\prime}(\{\mathbf{s}^{*\alpha}\};\boldsymbol{\varepsilon}^{o}) = \sum_{\alpha=0}^{n} f_\alpha \frac{1}{2}\mathbf{s}^{*\alpha}:\{(\mathbf{C}^{\alpha}-\mathbf{C})^{-1}+\mathbf{P}\}:\mathbf{s}^{*\alpha} - \frac{1}{2}\bar{\mathbf{s}}^{*}:\mathbf{P}:\bar{\mathbf{s}}^{*} - \bar{\mathbf{s}}^{*}:\boldsymbol{\varepsilon}^{o}. \tag{9.4.20b}$$

Note that the symmetry properties, $P_{ijkl} = P_{klij}$ and $(\mathbf{C}:\mathbf{P}:\mathbf{C})_{ijkl} = (\mathbf{C}:\mathbf{P}:\mathbf{C})_{klij}$, are used to derive (9.4.20a,b).

The optimal eigenstrains and eigenstresses, $\{\boldsymbol{\varepsilon}^{*\alpha}\}$ and $\{\boldsymbol{\sigma}^{*\alpha}\}$, are determined such that the derivatives of $I^{A\prime}(\{\mathbf{e}^{*\alpha}\};\boldsymbol{\sigma}^{o})$ and $J^{A\prime}(\{\mathbf{s}^{*\alpha}\};\boldsymbol{\varepsilon}^{o})$ with respect to $\mathbf{e}^{*\alpha}$ and $\mathbf{s}^{*\alpha}$ vanish for all α. Since $\mathbf{P}$ is a constant tensor, the governing sets of $n+1$ linear tensorial equations are given by

$$\{(\mathbf{C}^{\alpha}-\mathbf{C})^{-1}+\mathbf{P}\}:(-\,\mathbf{C}:\boldsymbol{\varepsilon}^{*\alpha}) - \mathbf{P}:(-\,\mathbf{C}:\bar{\boldsymbol{\varepsilon}}^{*}) - (\mathbf{D}:\boldsymbol{\sigma}^{o}+\bar{\boldsymbol{\varepsilon}}^{*}) = \mathbf{0},$$

$$\{(\mathbf{C}^{\alpha}-\mathbf{C})^{-1}+\mathbf{P}\}:\boldsymbol{\sigma}^{*\alpha} - \mathbf{P}:\bar{\boldsymbol{\sigma}}^{*} - \boldsymbol{\varepsilon}^{o} = \mathbf{0},$$

$$(\alpha = 0, 1, 2, ..., n). \tag{9.4.22a,b}$$

The above two sets of tensorial equations are essentially the same, and if $-\mathbf{C}:\boldsymbol{\varepsilon}^{*\alpha}$ and $\mathbf{D}:\boldsymbol{\sigma}^{o}+\bar{\boldsymbol{\varepsilon}}^{*}$ in (9.4.22a) are replaced by $\boldsymbol{\sigma}^{*\alpha}$ and $\boldsymbol{\varepsilon}^{o}$, respectively, one obtains (9.4.22b). Furthermore, solving each set of equations explicitly, obtain the solution of, say, (9.4.22b), to be $\boldsymbol{\sigma}^{*\alpha} = \{(\mathbf{C}^{\alpha}-\mathbf{C})^{-1}+\mathbf{P}\}^{-1}:\{\boldsymbol{\varepsilon}^{o} + \mathbf{P}:\bar{\boldsymbol{\sigma}}^{*}\}$. Then the average eigenstress, $\bar{\boldsymbol{\sigma}}^{*}$, is given by

$$\bar{\boldsymbol{\sigma}}^{*} = (\mathbf{1}^{(4s)} - \bar{\boldsymbol{\Sigma}}:\mathbf{P})^{-1}:\bar{\boldsymbol{\Sigma}}:\boldsymbol{\varepsilon}^{o}, \tag{9.4.23a}$$

where

$$\bar{\boldsymbol{\Sigma}} = \sum_{\alpha=0}^{n} f_\alpha \{(\mathbf{C}^{\alpha}-\mathbf{C})^{-1}+\mathbf{P}\}^{-1}$$

$$= \sum_{\alpha=0}^{n} f_\alpha (\mathbf{C}^{\alpha}-\mathbf{C}):\{\mathbf{1}^{(4s)}+\mathbf{P}:(\mathbf{C}^{\alpha}-\mathbf{C})\}^{-1}. \tag{9.4.23b}$$

In a similar manner, from (9.4.22a) obtain

$$\bar{\boldsymbol{\varepsilon}}^{*} = -\mathbf{D}:\{\mathbf{1}^{(4s)} - \bar{\boldsymbol{\Sigma}}:(\mathbf{P}-\mathbf{D})\}^{-1}:\bar{\boldsymbol{\Sigma}}:(\mathbf{D}:\boldsymbol{\sigma}^{o}). \tag{9.4.23c}$$

As shown in (9.3.13a,b), the optimal values of $I^{A\prime}(\{\boldsymbol{\varepsilon}^{*\alpha}\};\boldsymbol{\sigma}^{o})$ and $J^{A\prime}(\{\boldsymbol{\sigma}^{*\alpha}\};\boldsymbol{\varepsilon}^{o})$ are given by half of the inner product of $\boldsymbol{\sigma}^{o}:\bar{\boldsymbol{\varepsilon}}^{*}$ and $\bar{\boldsymbol{\sigma}}^{*}:\boldsymbol{\varepsilon}^{o}$, respectively. Indeed, using (9.4.20~23), obtain

$$I^{A\prime}(\{\boldsymbol{\varepsilon}^{*\alpha}\};\boldsymbol{\sigma}^{o}) = -\,\frac{1}{2}\boldsymbol{\sigma}^{o}:\bar{\boldsymbol{\varepsilon}}^{*A},$$

$$J^{A\prime}(\{\boldsymbol{\sigma}^{*\alpha}\};\boldsymbol{\varepsilon}^{o}) = -\,\frac{1}{2}\bar{\boldsymbol{\sigma}}^{*A}:\boldsymbol{\varepsilon}^{o}, \tag{9.4.24a,b}$$

where, to distinguish the optimal average eigenstrain and eigenstress of the exact functions I' and J' from those of the approximated functions $I^{A\prime}$ and $J^{A\prime}$, the superscript A is used for the latter, i.e.,

$$\bar{\boldsymbol{\varepsilon}}^{*\mathrm{A}} \equiv -\mathbf{D} : \{\, \mathbf{1}^{(4s)} - \bar{\boldsymbol{\Sigma}} : (\mathbf{P} - \mathbf{D})\,\}^{-1} : \bar{\boldsymbol{\Sigma}} : \mathbf{D} : \boldsymbol{\sigma}^{\mathrm{o}},$$

$$\bar{\boldsymbol{\sigma}}^{*\mathrm{A}} \equiv (\mathbf{1}^{(4s)} - \bar{\boldsymbol{\Sigma}} : \mathbf{P})^{-1} : \bar{\boldsymbol{\Sigma}} : \boldsymbol{\varepsilon}^{\mathrm{o}}. \qquad (9.4.24\mathrm{c,d})$$

9.5. GENERALIZED VARIATIONAL PRINCIPLES; EXACT BOUNDS

It is established in this subsection that the approximate expressions $\boldsymbol{\Lambda}^{\mathrm{A}}$ and $\boldsymbol{\Gamma}^{\mathrm{A}}$ given by (9.4.9c,d), can actually be used to obtain *rigorous* bounds for the overall moduli of the heterogeneous *finite* solid. This is done with the aid of the universal Theorems I and II of Subsection 2.5.6. The proof given in the present subsection does not depend on the statistical nature of the RVE, although this is an important feature which has been considered, in order to validate the approximations used to estimate these bounds; Willis (1977, 1981).

The results obtained here apply to any heterogeneous linearly elastic solid of overall ellipsoidal shape, independently of its size, and the distribution and elasticity tensors of its heterogeneities. The elastic inclusions may be distributed in the considered finite *ellipsoidal* RVE in an arbitrary manner, and their elasticity may deviate from the average properties of the composite by orders of magnitude, without affecting the proof. Since the initial RVE can be regarded as a suitably large part of an infinitely extended heterogeneous solid with the same overall properties, it is always possible to choose an RVE in an ellipsoidal shape. Indeed, from the discussion of Section 2, it is clear that the overall properties must not depend on the shape and the size of the RVE, as long as the RVE is suitably large. The choice of an ellipsoidal RVE, therefore, is for convenience of analysis only. While the ellipsoidal shape of the RVE is essential for establishing the final proof (Subsection 9.5.4), *general results developed in Subsections* 9.5.1, 9.5.2, *and* 9.5.3 *hold for* RVE's *of any shape or size, containing any number of phases with arbitrary elastic moduli.*

9.5.1. Generalization of Energy Functionals and Bounds

Integral operator $\boldsymbol{\Lambda} = \boldsymbol{\Lambda}(\mathbf{x}; \mathbf{e}^*)$ defined by (9.1.6), yields the disturbance stress field, $-\boldsymbol{\sigma}^{\mathrm{d}}$, due to eigenstrain field $\mathbf{e}^*$ in the homogeneous solid[13] V of uniform elasticity $\mathbf{C} = \mathbf{D}^{-1}$, such that the boundary ∂V is left traction-free; see Figure 9.5.1a. If the corresponding disturbance strain field is denoted by $\boldsymbol{\varepsilon}^{\mathrm{d}} \equiv \mathbf{D} : \boldsymbol{\sigma}^{\mathrm{d}} + \mathbf{e}^{*\mathrm{d}}$ with[14] $\mathbf{e}^{*\mathrm{d}} \equiv \mathbf{e}^* - < \mathbf{e}^* >$, then it follows that

---

[13] V is not necessarily an ellipsoid.

[14] The disturbance strain associated with the disturbance stress is defined by $\boldsymbol{\varepsilon}^{\mathrm{d}} \equiv \mathbf{D} : \boldsymbol{\sigma}^{\mathrm{d}} + \mathbf{e}^*$ in Subsection 9.1. The volume average of $\boldsymbol{\varepsilon}^{\mathrm{d}}$ is nonzero (given by $< \mathbf{e}^* >$) in this definition. In the present subsection, the disturbance strain is defined such that its volume average vanishes.

$$< \boldsymbol{\sigma}^{\mathrm{d}} : \boldsymbol{\varepsilon}^{\mathrm{d}} > = \frac{1}{\mathrm{V}} \int_{\partial \mathrm{V}} \mathbf{t}^{\mathrm{d}} \cdot \mathbf{u}^{\mathrm{d}} \, \mathrm{dS} = 0, \tag{9.5.1a}$$

where $\mathbf{t}^{\mathrm{d}}$ ($\equiv \mathbf{0}$) and $\mathbf{u}^{\mathrm{d}}$ are the associated boundary tractions and displacements, respectively. As is shown in Subsections 9.1 and 9.2, the integral operator $< : \{(\mathbf{D}' - \mathbf{D})^{-1} + \boldsymbol{\Lambda}\} : >$ is self-adjoint, and it is positive-definite (negative-definite) when $\mathbf{D}' - \mathbf{D}$ is positive-definite (negative-definite).

Similarly, integral operator $\boldsymbol{\Gamma} = \boldsymbol{\Gamma}(\mathbf{x}; \mathbf{s}^*)$ defined by (9.1.14), yields the disturbance strain field, $-\boldsymbol{\varepsilon}^{\mathrm{d}}$, due to the eigenstress field $\mathbf{s}^*$, in the homogeneous solid V of uniform elasticity $\mathbf{C}$, such that the corresponding boundary displacements on ∂V vanish identically (a different problem); see Figure 9.5.1b. Hence, if the corresponding disturbance stress field is denoted by $\boldsymbol{\sigma}^{\mathrm{d}} \equiv \mathbf{C} : \boldsymbol{\varepsilon}^{\mathrm{d}} + \mathbf{s}^{*\mathrm{d}}$ with $\mathbf{s}^{*\mathrm{d}} \equiv \mathbf{s}^* - < \mathbf{s}^* >$, then (9.5.1a) also holds for this problem, and the corresponding integral operator $< : \{(\mathbf{C}' - \mathbf{C})^{-1} + \boldsymbol{\Gamma}\} : >$ is self-adjoint, as well as positive-definite (negative-definite), when $\mathbf{C}' - \mathbf{C}$ is positive-definite (negative-definite).

For both of the above stated problems, the Hashin-Shtrikman variational principle holds, and (9.2.9) and (9.2.10), therefore, follow directly from the properties of operators $\boldsymbol{\Lambda}$ and $\boldsymbol{\Gamma}$. Note that (9.2.10a) corresponds to the uniform traction boundary data with the average stress $\boldsymbol{\sigma}^{\mathrm{o}}$ prescribed, whereas (9.2.10b) is for the linear displacement boundary data with the average strain $\boldsymbol{\varepsilon}^{\mathrm{o}}$ prescribed. The key to the establishment of the Hashin-Shtrikman variational

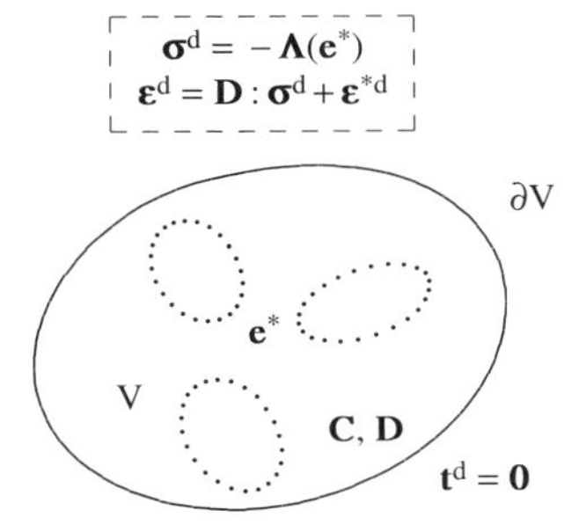

Figure 9.5.1a

Eigenstrain field $\mathbf{e}^*(\mathbf{x})$ in homogeneous V produces disturbance fields which leave ∂V traction-free, $\mathbf{t}^{\mathrm{d}} = \boldsymbol{\nu} \cdot \boldsymbol{\sigma}^{\mathrm{d}} = \mathbf{0}$ on ∂V

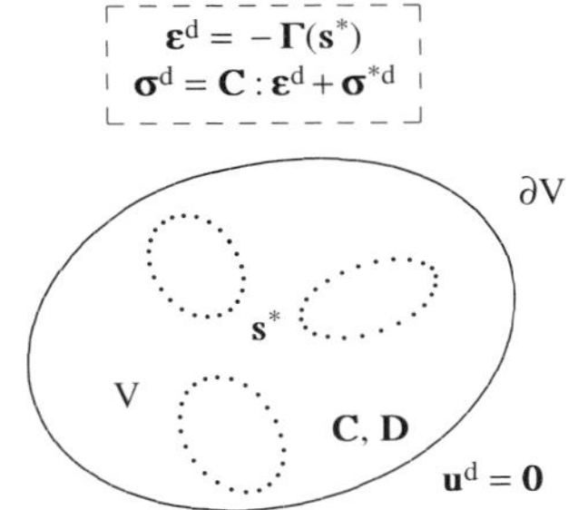

Figure 9.5.1b

Eigenstress field $\mathbf{s}^*(\mathbf{x})$ in homogeneous V produces disturbance fields, such that displacements on ∂V vanish; $\mathbf{u}^{\mathrm{d}} = \mathbf{0}$ on ∂V

principle, and hence (9.2.10), the bounds for these cases, is the fact that for any eigenstrain or eigenstress field, the disturbance stress or strain field satisfies (9.5.1a) identically.

For the Hashin-Shtrikman variational principle to apply, it is not necessary to restrict operators $\mathbf{\Lambda}$ and $\mathbf{\Gamma}$ to correspond, respectively, to traction-free and zero-displacement boundary conditions on boundary ∂V of homogeneous V. The only requirement is that (9.5.1a) holds for the associated disturbance stress and strain fields. Then, the corresponding (total) stress and strain fields, $\boldsymbol{\sigma}$ and $\boldsymbol{\varepsilon}$, satisfy

$$< \boldsymbol{\sigma} : \boldsymbol{\varepsilon} > = < \boldsymbol{\sigma} > : < \boldsymbol{\varepsilon} >. \tag{9.5.2}$$

This ensures that the overall moduli defined in terms of the elastic or complementary elastic energies coincide with those defined in terms of the average stress-strain relations; see Subsection 2.3. As mentioned, the volume average of the disturbance stresses and the disturbance strains is zero, i.e.,

$$< \boldsymbol{\sigma}^{\rm d} > = \frac{1}{V} \int_{\partial V} \mathbf{x} \otimes \mathbf{t}^{\rm d} \, dS = \mathbf{0},$$

$$< \boldsymbol{\varepsilon}^{\rm d} > = \frac{1}{V} \int_{\partial V} sym \, \{\mathbf{v} \otimes \mathbf{u}^{\rm d}\} \, dS = \mathbf{0}; \tag{9.5.1b,c}$$

see Subsections 2.1 and 2.2. Conditions (9.5.1b) and (9.5.1c) are equivalent, when the disturbance stress and strain fields are related through $\boldsymbol{\varepsilon}^{\rm d} - \mathbf{D} : \boldsymbol{\sigma}^{\rm d} = \mathbf{e}^* - < \mathbf{e}^* >$, if $\mathbf{e}^*$ is prescribed, or through $\boldsymbol{\sigma}^{\rm d} - \mathbf{C} : \boldsymbol{\varepsilon}^{\rm d} = \mathbf{s}^* - < \mathbf{s}^* >$, if $\mathbf{s}^*$ is prescribed.

Consider now a general class of boundary data for homogeneous solid V of uniform elasticity $\mathbf{C} = \mathbf{D}^{-1}$, such that neither the disturbance traction nor the disturbance displacement field, on boundary ∂V, is identically zero, but, instead, at every point on ∂V, these disturbance fields satisfy

$$\mathbf{t}^{\rm d} \boldsymbol{\cdot} \mathbf{u}^{\rm d} = 0 \qquad \text{on } \partial V, \tag{9.5.3}$$

where $\mathbf{t}^{\rm d} = \mathbf{v} \boldsymbol{\cdot} (\boldsymbol{\sigma} - < \boldsymbol{\sigma} >)$ and $\mathbf{u}^{\rm d} = \mathbf{u} - \mathbf{x} \boldsymbol{\cdot} < \boldsymbol{\varepsilon} >$ on ∂V. The disturbance traction field and disturbance displacement field are associated with the disturbance stress field and the disturbance strain field, respectively.

For this class of boundary data, consider first the case when the overall stress is prescribed to be $< \boldsymbol{\sigma} > = \boldsymbol{\sigma}^{\rm o}$. Denote by $\mathbf{\Lambda}^{\rm G}$ the corresponding integral operator which, for any eigenstrain field $\mathbf{e}^*(\mathbf{x})$, yields the disturbance stress field, $\boldsymbol{\sigma}^{\rm d}(\mathbf{x}) = - \mathbf{\Lambda}^{\rm G}(\mathbf{x}; \mathbf{e}^*)$, and the corresponding disturbance strain and hence displacement fields, $\boldsymbol{\varepsilon}^{\rm d}$ and $\mathbf{u}^{\rm d}$, such that (9.5.3) is satisfied identically;[15] see Figure 9.5.1c. Define for $\mathbf{\Lambda}^{\rm G}$ the following functional:

$$\hat{I}(\mathbf{e}^*; \mathbf{\Lambda}^{\rm G}; \boldsymbol{\sigma}^{\rm o}) \equiv \frac{1}{2} < \mathbf{e}^* : \{(\mathbf{D}' - \mathbf{D})^{-1} + \mathbf{\Lambda}^{\rm G}\} : \mathbf{e}^* > - < \boldsymbol{\sigma}^{\rm o} : \mathbf{e}^* >. \tag{9.5.4a}$$

The Hashin-Shtrikman variational principle is applicable to this class of functionals, since, by definition, $\mathbf{\Lambda}^{\rm G}$ is self-adjoint and satisfies

---

[15] The class of operators $\mathbf{\Lambda}^{\rm G}$ includes operator $\mathbf{\Lambda}$ associated with the traction-free boundary conditions, i.e., (9.1.5b).

$$< \mathbf{e}^* : \mathbf{\Lambda}^{\mathrm{G}} : \mathbf{e}^* > = < (\mathbf{\Lambda}^{\mathrm{G}} : \mathbf{e}^*) : \mathbf{D} : (\mathbf{\Lambda}^{\mathrm{G}} : \mathbf{e}^*) >$$

$$= < \mathbf{e}^* : \mathbf{C} : \mathbf{e}^* > - < \boldsymbol{\varepsilon}^{\mathrm{d}} : \mathbf{C} : \boldsymbol{\varepsilon}^{\mathrm{d}} >. \tag{9.5.5a}$$

These properties of $\mathbf{\Lambda}^{\mathrm{G}}$ ensure the positive-definiteness (negative-definiteness) of operator $< : \{(\mathbf{D}' - \mathbf{D})^{-1} + \mathbf{\Lambda}^{\mathrm{G}}\} : >$, when $\mathbf{D}' - \mathbf{D}$ is positive-definite (negative-definite); see Subsection 9.2.[16]

Let $\boldsymbol{\varepsilon}^{*\mathrm{G}}$ be the eigenstrain field that renders $\hat{\mathrm{I}}(\mathbf{e}^*; \mathbf{\Lambda}^{\mathrm{G}}; \boldsymbol{\sigma}^{\mathrm{o}})$ stationary. Due to the self-adjointness of $\mathbf{\Lambda}^{\mathrm{G}}$, this eigenstrain, $\boldsymbol{\varepsilon}^{*\mathrm{G}}$, satisfies

$$\delta\hat{\mathrm{I}}(\boldsymbol{\varepsilon}^{*\mathrm{G}}; \mathbf{\Lambda}^{\mathrm{G}}; \boldsymbol{\sigma}^{\mathrm{o}}) = < \delta\mathbf{e}^* : \left\{ \{(\mathbf{D}' - \mathbf{D})^{-1} + \mathbf{\Lambda}^{\mathrm{G}}\} : \boldsymbol{\varepsilon}^{*\mathrm{G}} - \boldsymbol{\sigma}^{\mathrm{o}} \right\} > = 0. \tag{9.5.6a}$$

The Euler equation of $\hat{\mathrm{I}}$ coincides with the consistency condition under the considered general boundary conditions, and hence the stationary value of $\hat{\mathrm{I}}$ is given by

$$\hat{\mathrm{I}}(\boldsymbol{\varepsilon}^{*\mathrm{G}}; \mathbf{\Lambda}^{\mathrm{G}}; \boldsymbol{\sigma}^{\mathrm{o}}) = -\frac{1}{2}\,\bar{\boldsymbol{\varepsilon}}^{*\mathrm{G}} : \boldsymbol{\sigma}^{\mathrm{o}} = \frac{1}{2}\,\boldsymbol{\sigma}^{\mathrm{o}} : (\mathbf{D} - \bar{\mathbf{D}}^{\mathrm{G}}) : \boldsymbol{\sigma}^{\mathrm{o}}, \tag{9.5.7a}$$

where $\bar{\mathbf{D}}^{\mathrm{G}}$ is the overall compliance tensor under the prescribed, possibly mixed, boundary conditions. This is the *generalized Hashin-Shtrikman variational principle*. Furthermore, it follows from (9.5.5a) that for any eigenstrain field $\mathbf{e}^*$,

$$\hat{\mathrm{I}}(\boldsymbol{\varepsilon}^{*\mathrm{G}}; \mathbf{\Lambda}^{\mathrm{G}}; \boldsymbol{\sigma}^{\mathrm{o}}) \leq (\geq)\; \hat{\mathrm{I}}(\mathbf{e}^*; \mathbf{\Lambda}^{\mathrm{G}}; \boldsymbol{\sigma}^{\mathrm{o}}), \tag{9.5.8a}$$

if $\mathbf{D}' - \mathbf{D}$ is positive-definite (negative-definite). These are the *generalized Hashin-Shtrikman bounds.* Note that, *since condition* (9.5.3) *is satisfied for the general boundary data considered above, the energy definition of* $\bar{\mathbf{D}}^{\mathrm{G}}$ *and its definition based on the average stress-strain relations are identical.*

The case when the overall strain is prescribed to be $< \boldsymbol{\varepsilon} > = \boldsymbol{\varepsilon}^{\mathrm{o}}$, is treated in a similar manner. Denote by $\mathbf{\Gamma}^{\mathrm{G}}$ the integral operator which, for any eigenstresses $\mathbf{s}^*(\mathbf{x})$, yields the disturbance strain field, $\boldsymbol{\varepsilon}^{\mathrm{d}}(\mathbf{x}) = -\mathbf{\Gamma}^{\mathrm{G}}(\mathbf{x}; \mathbf{s}^*)$, and the

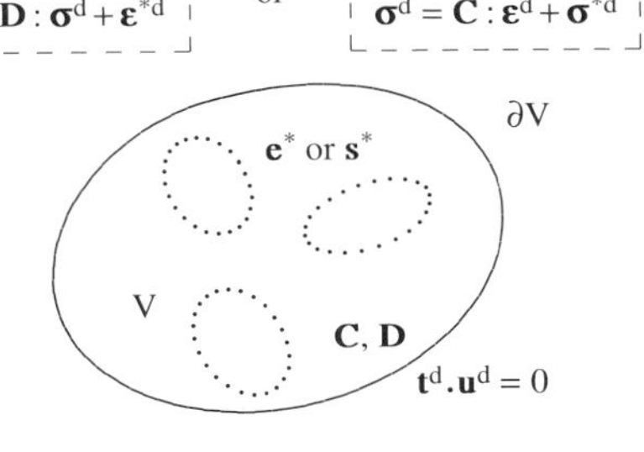

Figure 9.5.1c

Eigenstrain field $\mathbf{e}^*(\mathbf{x})$ or eigenstress field $\mathbf{s}^*(\mathbf{x})$ in homogeneous V produces disturbance fields, such that although neither tractions nor displacements are identically zero on ∂V, their inner product vanishes there

---

[16] Equation (9.5.5a) holds whether the disturbance strain field is defined by $\boldsymbol{\varepsilon}^{\mathrm{d}} = \mathbf{D} : \boldsymbol{\sigma}^{\mathrm{d}} + \mathbf{e}^*$ or by $\boldsymbol{\varepsilon}^{\mathrm{d}} = \mathbf{D} : \boldsymbol{\sigma}^{\mathrm{d}} + (\mathbf{e}^* - < \mathbf{e}^* >)$.

corresponding disturbance stress and displacement fields, such that (9.5.3) is satisfied identically.[17] The Hashin-Shtrikman variational principle is then applicable. Define the following functional for $\mathbf{\Gamma}^G$:

$$\hat{J}(\mathbf{s}^*;\, \mathbf{\Gamma}^G;\, \boldsymbol{\varepsilon}^o) \equiv \frac{1}{2} < \mathbf{s}^* : \{(\mathbf{C}' - \mathbf{C})^{-1} + \mathbf{\Gamma}^G\} : \mathbf{s}^* > - < \boldsymbol{\varepsilon}^o : \mathbf{s}^* >, \tag{9.5.4b}$$

where $< \,:\mathbf{\Gamma}^G\,: >$ satisfies[18]

$$< \mathbf{s}^* : \mathbf{\Gamma}^G : \mathbf{s}^* > = < (\mathbf{\Gamma}^G : \mathbf{s}^*) : \mathbf{C} : (\mathbf{\Gamma}^G : \mathbf{s}^*) >$$

$$= < \mathbf{s}^* : \mathbf{D} : \mathbf{s}^* > - < \boldsymbol{\sigma}^d : \mathbf{D} : \boldsymbol{\sigma}^d >. \tag{9.5.5b}$$

These properties of $\mathbf{\Gamma}^G$ ensure the positive-definiteness (negative-definiteness) of operator $< \,: \{(\mathbf{C}' - \mathbf{C})^{-1} + \mathbf{\Gamma}^G\} :\, >$, when $\mathbf{C}' - \mathbf{C}$ is positive-definite (negative-definite); see Subsection 9.2. Then, observe that $\boldsymbol{\sigma}^{*G}$ which solves

$$\delta\hat{J}(\boldsymbol{\sigma}^{*G};\, \mathbf{\Gamma}^G;\, \boldsymbol{\varepsilon}^o) = < \delta\mathbf{s}^* : \left\{ \{(\mathbf{C}' - \mathbf{C})^{-1} + \mathbf{\Gamma}^G\} : \boldsymbol{\sigma}^{*G} - \boldsymbol{\varepsilon}^o \right\} > = 0, \tag{9.5.6b}$$

renders $\hat{J}$ stationary, and satisfies the consistency condition under the prescribed boundary data. Then, the stationary value of $\hat{J}$ is given by

$$\hat{J}(\boldsymbol{\sigma}^{*G};\, \mathbf{\Gamma}^G;\, \boldsymbol{\varepsilon}^o) = -\frac{1}{2}\, \bar{\boldsymbol{\sigma}}^{*G} : \boldsymbol{\varepsilon}^o = \frac{1}{2}\, \boldsymbol{\varepsilon}^o : (\mathbf{C} - \bar{\mathbf{C}}^G) : \boldsymbol{\varepsilon}^o, \tag{9.5.7b}$$

where $\bar{\mathbf{C}}^G$ is the overall elasticity tensor under the prescribed, possibly mixed, boundary conditions. From (9.5.5b) it follows that, for any eigenstress field $\mathbf{s}^*$,

$$\hat{J}(\boldsymbol{\sigma}^{*G};\, \mathbf{\Gamma}^G;\, \boldsymbol{\varepsilon}^o) \leq (\geq)\, \hat{J}(\mathbf{s}^*;\, \mathbf{\Gamma}^G;\, \boldsymbol{\varepsilon}^o), \tag{9.5.8b}$$

if $\mathbf{C}' - \mathbf{C}$ is positive-definite (negative-definite). Equations (9.5.4b) to (9.5.7b) define the *second generalized Hashin-Shtrikman variational principle*, and (9.5.8b) gives the *corresponding generalized bounds*. It should be noted that the results obtained in this subsection, namely (9.5.7a,b) and (9.5.8a,b), are valid for any *finite* isolated V of any shape or size, i.e., V is *not* necessarily an ellipsoid. Note also that *since condition* (9.5.3) *is satisfied, the energy definition of* $\bar{\mathbf{C}}^G$ *and its definition through the average stress-strain relations are identical.*

The class of functionals $\hat{I}(\mathbf{e}^*;\, \mathbf{\Lambda}^G;\, \boldsymbol{\sigma}^o)$ includes the functional $I(\mathbf{e}^*;\, \boldsymbol{\sigma}^o)$ which corresponds to uniform boundary tractions, and which has been considered in the preceding subsections. Similarly, the class of functionals $\hat{J}(\mathbf{e}^*;\, \mathbf{\Lambda}^G;\, \boldsymbol{\varepsilon}^o)$ includes the functional $J(\mathbf{s}^*;\, \boldsymbol{\varepsilon}^o)$ for linear displacement boundary data. To emphasize the corresponding boundary data, the operator $\mathbf{\Lambda}$ associated with the *uniform traction boundary data* will be denoted by $\mathbf{\Lambda}^\Sigma$, and the operator $\mathbf{\Gamma}$ associated with the *linear displacement boundary data* will be denoted by $\mathbf{\Gamma}^E$. The corresponding functionals will be displayed as

---

[17] The class of operators $\mathbf{\Gamma}^G$ includes operator $\mathbf{\Gamma}$ associated with the zero-displacement boundary conditions, i.e., (9.1.13b).

[18] Similarly to (9.5.5a), (9.5.5b) holds whether $\boldsymbol{\sigma}^d$ is defined by $\boldsymbol{\sigma}^d = \mathbf{C} : \mathbf{\Gamma}^G + \mathbf{s}^*$ or by $\boldsymbol{\sigma}^d = \mathbf{C} : \mathbf{\Gamma}^G + (\mathbf{s}^* - < \mathbf{s}^* >)$.

$$\mathrm{I}(\mathbf{e}^*;\, \boldsymbol{\sigma}^o) \equiv \hat{\mathrm{I}}(\mathbf{e}^*;\, \boldsymbol{\Lambda}^{\Sigma};\, \boldsymbol{\sigma}^o), \qquad \mathrm{J}(\mathbf{s}^*;\, \boldsymbol{\varepsilon}^o) \equiv \hat{\mathrm{J}}(\mathbf{s}^*;\, \boldsymbol{\Gamma}^{E};\, \boldsymbol{\varepsilon}^o). \tag{9.5.9a,b}$$

Table 9.5.1 summarizes the functionals and the corresponding bounds for various boundary data.

It may be instructive to note that when the boundary data of the original heterogeneous RVE consist of either uniform tractions or linear displacements, the classical principles of the minimum potential and complementary potential

Table 9.5.1

Finite-space operators and their properties

boundary data	trial field	Hashin-Shtrikman functional
uniform tractions	$\mathbf{e}^* = \mathbf{e}^*(\mathbf{x})$	$\mathrm{I}(\mathbf{e}^*;\, \boldsymbol{\sigma}^o) \equiv \hat{\mathrm{I}}(\mathbf{e}^*;\, \boldsymbol{\Lambda}^{\Sigma};\, \boldsymbol{\sigma}^o)$
linear displacement	$\mathbf{s}^* = \mathbf{s}^*(\mathbf{x})$	$\mathrm{J}(\mathbf{s}^*;\, \boldsymbol{\varepsilon}^o) \equiv \hat{\mathrm{J}}(\mathbf{s}^*;\, \boldsymbol{\Gamma}^{E};\, \boldsymbol{\varepsilon}^o)$
general: $<\boldsymbol{\sigma}> = \boldsymbol{\sigma}^o$	$\mathbf{e}^* = \mathbf{e}^*(\mathbf{x})$	$\hat{\mathrm{I}}(\mathbf{e}^*;\, \boldsymbol{\Lambda}^{G};\, \boldsymbol{\sigma}^o)$
general: $<\boldsymbol{\varepsilon}> = \boldsymbol{\varepsilon}^o$	$\mathbf{s}^* = \mathbf{s}^*(\mathbf{x})$	$\hat{\mathrm{J}}(\mathbf{s}^*;\, \boldsymbol{\Gamma}^{G};\, \boldsymbol{\varepsilon}^o)$
boundary data	**exact field**	**optimal value**
uniform tractions	$\boldsymbol{\varepsilon}^* = \boldsymbol{\varepsilon}^*(\mathbf{x})$	$\mathrm{I}(\boldsymbol{\varepsilon}^*;\, \boldsymbol{\sigma}^o) = \boldsymbol{\sigma}^o : (\mathbf{D} - \overline{\mathbf{D}}^{\Sigma}) : \boldsymbol{\sigma}^o / 2$
linear displacement	$\boldsymbol{\sigma}^* = \boldsymbol{\sigma}^*(\mathbf{x})$	$\mathrm{J}(\boldsymbol{\sigma}^*;\, \boldsymbol{\varepsilon}^o) = \boldsymbol{\varepsilon}^o : (\mathbf{C} - \overline{\mathbf{C}}^{E}) : \boldsymbol{\varepsilon}^o / 2$
general: $<\boldsymbol{\sigma}> = \boldsymbol{\sigma}^o$	$\boldsymbol{\varepsilon}^{*G} = \boldsymbol{\varepsilon}^{*G}(\mathbf{x})$	$\hat{\mathrm{I}}(\boldsymbol{\varepsilon}^{*G};\, \boldsymbol{\Lambda}^{G};\, \boldsymbol{\sigma}^o) = \boldsymbol{\sigma}^o : (\mathbf{D} - \overline{\mathbf{D}}^{G}) : \boldsymbol{\sigma}^o / 2$
general: $<\boldsymbol{\varepsilon}> = \boldsymbol{\varepsilon}^o$	$\boldsymbol{\sigma}^{*G} = \boldsymbol{\sigma}^{*G}(\mathbf{x})$	$\hat{\mathrm{J}}(\boldsymbol{\sigma}^{*G};\, \boldsymbol{\Gamma}^{G};\, \boldsymbol{\varepsilon}^o) = \boldsymbol{\varepsilon}^o : (\mathbf{C} - \overline{\mathbf{C}}^{G}) : \boldsymbol{\varepsilon}^o / 2$
restrictions		**bounds**
$\mathbf{D}' - \mathbf{D}$ is p.d.		$\mathrm{I}(\boldsymbol{\varepsilon}^*;\, \boldsymbol{\sigma}^o) \le \mathrm{I}(\mathbf{e}^*;\, \boldsymbol{\sigma}^o)$ $\hat{\mathrm{I}}(\boldsymbol{\varepsilon}^{*G};\, \boldsymbol{\Lambda}^{G};\, \boldsymbol{\sigma}^o) \le \hat{\mathrm{I}}(\mathbf{e}^*;\, \boldsymbol{\Lambda}^{G};\, \boldsymbol{\sigma}^o)$
$\mathbf{D}' - \mathbf{D}$ is n.d.		$\mathrm{I}(\boldsymbol{\varepsilon}^*;\, \boldsymbol{\sigma}^o) \ge \mathrm{I}(\mathbf{e}^*;\, \boldsymbol{\sigma}^o)$ $\hat{\mathrm{I}}(\boldsymbol{\varepsilon}^{*G};\, \boldsymbol{\Lambda}^{G};\, \boldsymbol{\sigma}^o) \ge \hat{\mathrm{I}}(\mathbf{e}^*;\, \boldsymbol{\Lambda}^{G};\, \boldsymbol{\sigma}^o)$
$\mathbf{C}' - \mathbf{C}$ is p.d.		$\mathrm{J}(\boldsymbol{\sigma}^*;\, \boldsymbol{\varepsilon}^o) \le \mathrm{J}(\mathbf{s}^*;\, \boldsymbol{\varepsilon}^o)$ $\hat{\mathrm{J}}(\boldsymbol{\sigma}^{*G};\, \boldsymbol{\Lambda}^{G};\, \boldsymbol{\varepsilon}^o) \le \hat{\mathrm{J}}(\mathbf{s}^*;\, \boldsymbol{\Lambda}^{G};\, \boldsymbol{\varepsilon}^o)$
$\mathbf{C}' - \mathbf{C}$ is n.d.		$\mathrm{J}(\boldsymbol{\sigma}^*;\, \boldsymbol{\varepsilon}^o) \ge \mathrm{J}(\mathbf{s}^*;\, \boldsymbol{\varepsilon}^o)$ $\hat{\mathrm{J}}(\boldsymbol{\sigma}^{*G};\, \boldsymbol{\Lambda}^{G};\, \boldsymbol{\varepsilon}^o) \ge \hat{\mathrm{J}}(\mathbf{s}^*;\, \boldsymbol{\Lambda}^{G};\, \boldsymbol{\varepsilon}^o)$

energies can be used to establish the Hashin-Shtrikman variational principle, as is shown by Hill (1963). For the general boundary data considered in this subsection, on the other hand, the universal Theorems I and II of Subsection 2.5.6 provide the necessary inequalities which compare the energies associated with different boundary data which produce either the same overall stresses or the same overall strains.

9.5.2. Inequalities among Generalized Energy Functionals

Now, consider Theorems I and II of Subsection 2.5.6, in order to obtain inequalities that hold among the members of each of the classes of functionals $\hat{\mathrm{I}}$ and $\hat{\mathrm{J}}$. As is shown by (9.5.7a,b), the stationary value of $\hat{\mathrm{I}}$ (of $\hat{\mathrm{J}}$) gives the overall compliance (elasticity) tensor of the original heterogeneous RVE for the corresponding boundary conditions. For the same overall strains, or the same overall stresses, the elastic and complementary energies associated with these or any other general boundary data can be compared with the aid of universal Theorems I and II of Subsection 2.5.6.

According to Theorem I, among all boundary conditions[19] that yield the same average strains, $\bar{\boldsymbol{\varepsilon}}$, the uniform traction boundary conditions render the strain energy an absolute minimum. This yields

$$\frac{1}{2}\,\bar{\boldsymbol{\varepsilon}} : (\bar{\mathbf{C}}^{\mathrm{G}}_{E} - \bar{\mathbf{C}}^{\Sigma}) : \bar{\boldsymbol{\varepsilon}} \geq 0, \tag{9.5.10a}$$

where $\bar{\mathbf{C}}^{\Sigma}$ is the overall elasticity tensor of the RVE when subjected to the uniform traction boundary conditions (signified by the superscript Σ), and $\bar{\mathbf{C}}^{\mathrm{G}}_{E}$ is the overall elasticity tensor defined through the average strain energy[20] of the same RVE under some other general boundary conditions which produce the same overall average strain, $\bar{\boldsymbol{\varepsilon}}$. It thus follows that $\bar{\mathbf{C}}^{\mathrm{G}}_{E} - \bar{\mathbf{C}}^{\Sigma}$ is positive-semi-definite. Then, in view of identity (9.2.5d), $(\bar{\mathbf{C}}^{\mathrm{G}}_{E})^{-1}$ and $\bar{\mathbf{D}}^{\Sigma} = (\bar{\mathbf{C}}^{\Sigma})^{-1}$ satisfy

$$\frac{1}{2}\,\boldsymbol{\sigma}^{\mathrm{o}} : ((\bar{\mathbf{C}}^{\mathrm{G}}_{E})^{-1} - \bar{\mathbf{D}}^{\Sigma}) : \boldsymbol{\sigma}^{\mathrm{o}} \leq 0, \tag{9.5.11a}$$

for any constant $\boldsymbol{\sigma}^{\mathrm{o}}$. Hence, $(\bar{\mathbf{C}}^{\mathrm{G}}_{E})^{-1} - \bar{\mathbf{D}}^{\Sigma}$ is negative-semi-definite.

Similarly, according to Theorem II of Subsection 2.5.6, among all boundary conditions that yield the same average stresses, $\bar{\boldsymbol{\sigma}}$, the linear displacement boundary conditions render the complementary elastic strain energy an absolute minimum, i.e.,

---

[19] These may or may not satisfy (9.5.3).

[20] Since the considered general boundary data need *not* satisfy (9.5.3), and hence (9.5.2) may not hold, the energy definition of the overall elasticity tensor, $\bar{\mathbf{C}}^{\mathrm{G}}_{E}$, and its definition by the average stress-strain relations may *not*, in general, be the same. Subscript E emphasizes that this quantity is defined through the *average energy* of the RVE. Note that $\bar{\mathbf{C}}^{\Sigma}$ is the same for both definitions, since it corresponds to uniform traction boundary data.

$$\frac{1}{2}\,\bar{\boldsymbol{\sigma}} : (\bar{\mathbf{D}}^{\mathrm{G}}_{E} - \bar{\mathbf{D}}^{\mathrm{E}}) : \bar{\boldsymbol{\sigma}} \geq 0, \tag{9.5.10b}$$

where $\bar{\mathbf{D}}^{\mathrm{E}}$ is the overall compliance tensor of the RVE when subjected to the linear displacement boundary conditions (signified by the superscript E), and $\bar{\mathbf{D}}^{\mathrm{G}}_{E}$ is the overall compliance tensor defined through the average complementary strain energy[21] of the same RVE under some other general boundary conditions, both producing the same overall average stress, $\bar{\boldsymbol{\sigma}}$. Hence, $\bar{\mathbf{C}}^{\mathrm{E}} = (\bar{\mathbf{D}}^{\mathrm{E}})^{-1}$ satisfies

$$\frac{1}{2}\,\boldsymbol{\varepsilon}^{\mathrm{o}} : ((\bar{\mathbf{D}}^{\mathrm{G}}_{E})^{-1} - \bar{\mathbf{C}}^{\mathrm{E}}) : \boldsymbol{\varepsilon}^{\mathrm{o}} \leq 0, \tag{9.5.11b}$$

for any $\boldsymbol{\varepsilon}^{\mathrm{o}}$. Note that even under the same boundary conditions, $\bar{\mathbf{C}}^{\mathrm{G}}_{E}$ and $\bar{\mathbf{D}}^{\mathrm{G}}_{E}$ are not, in general, each other's inverse, unless (9.5.2) is satisfied.

Now, if the general boundary data considered above are restricted such that they satisfy (9.5.3), then the energy definition of the overall moduli and their definition based on the average stress-strain relations coincide. Then $(\bar{\mathbf{C}}^{\mathrm{G}}_{E})^{-1}$ in (9.5.11a) is expressed as $\bar{\mathbf{D}}^{\mathrm{G}}_{E} = \bar{\mathbf{D}}^{\mathrm{G}}$, and with the aid of (9.5.7a), the following inequality for the stationary value of functional $\hat{\mathrm{I}}$ is obtained:

$$\hat{\mathrm{I}}(\boldsymbol{\varepsilon}^{*\Sigma};\, \boldsymbol{\Lambda}^{\Sigma};\, \boldsymbol{\sigma}^{\mathrm{o}}) \leq \hat{\mathrm{I}}(\boldsymbol{\varepsilon}^{*\mathrm{G}};\, \boldsymbol{\Lambda}^{\mathrm{G}};\, \boldsymbol{\sigma}^{\mathrm{o}}), \tag{9.5.12a}$$

where $\boldsymbol{\varepsilon}^{*\Sigma}$ is the eigenstrain field that satisfies the same consistency condition (9.5.6a), with $\boldsymbol{\Lambda}^{\mathrm{G}}$ replaced by $\boldsymbol{\Lambda}^{\Sigma}$ corresponding to uniform traction data. In a similar manner, with the aid of (9.5.7b) and (9.5.11b), the following inequality for the stationary value of functional $\hat{\mathrm{J}}$ is obtained:

$$\hat{\mathrm{J}}(\boldsymbol{\sigma}^{*\mathrm{E}};\, \boldsymbol{\Gamma}^{\mathrm{E}};\, \boldsymbol{\varepsilon}^{\mathrm{o}}) \leq \hat{\mathrm{J}}(\boldsymbol{\sigma}^{*\mathrm{G}};\, \boldsymbol{\Gamma}^{\mathrm{G}};\, \boldsymbol{\varepsilon}^{\mathrm{o}}), \tag{9.5.12b}$$

where $\boldsymbol{\sigma}^{*\mathrm{E}}$ is the eigenstress field that satisfies (9.5.6b), with $\boldsymbol{\Gamma}^{\mathrm{G}}$ replaced by $\boldsymbol{\Gamma}^{\mathrm{E}}$. Note that for these cases, the general boundary conditions are restricted to satisfy (9.5.3), so that (9.5.2) holds.

Observe that the left side of (9.5.12a) is given by $\mathrm{I}(\boldsymbol{\varepsilon}^{*};\, \boldsymbol{\sigma}^{\mathrm{o}})$ with $\boldsymbol{\varepsilon}^{*} \equiv \boldsymbol{\varepsilon}^{*\Sigma}$, since functional I belongs to the class of $\hat{\mathrm{I}}$'s; see (9.5.9a). Similarly, the left side of (9.5.12b) is given by $\mathrm{J}(\boldsymbol{\sigma}^{*};\, \boldsymbol{\varepsilon}^{\mathrm{o}})$, with $\boldsymbol{\sigma}^{*} \equiv \boldsymbol{\sigma}^{*\mathrm{E}}$; see (9.5.9b).

Figure 9.5.2 summarizes the results of this subsection in the form of a flow chart. Note that the results above the two middle dashed boxes correspond to any general boundary data (denoted by superscript G), whereas those below these dashed boxes are for boundary data which satisfy (9.5.3).

9.5.3. Functionals with Simplified Integral Operators

Disturbance stress and strain fields produced by integral operators $\boldsymbol{\Lambda}^{\mathrm{G}}$ or $\boldsymbol{\Gamma}^{\mathrm{G}}$, introduced in Subsection 9.5.1, satisfy particular prescribed boundary conditions on ∂V, such that (9.5.3) is satisfied. In Subsection 9.4.2, simplified integral operators $\boldsymbol{\Lambda}^{\mathrm{A}}$ and $\boldsymbol{\Gamma}^{\mathrm{A}}$ are introduced to approximate $\boldsymbol{\Lambda}$ and $\boldsymbol{\Gamma}$. This approximation is equivalent to embedding the uniform isolated V of elasticity $\mathbf{C} = \mathbf{D}^{-1}$ in

---

[21] Comments similar to the preceding footnote about $\bar{\mathbf{C}}^{\mathrm{G}}_{E}$ also apply to $\bar{\mathbf{D}}^{\mathrm{G}}_{E}$.

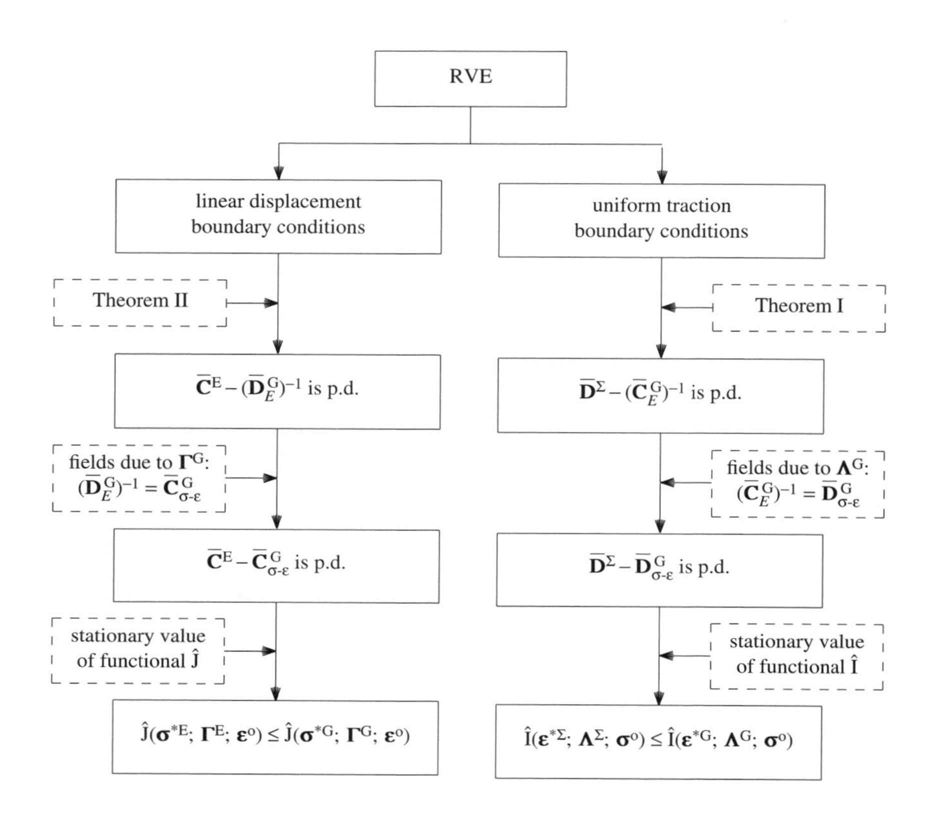

Figure 9.5.2

Flow chart of exact inequalities for different boundary data with the same average strain $<\boldsymbol{\varepsilon}> = \boldsymbol{\varepsilon}^{\mathrm{o}}$ (left half) and the same average stress $<\boldsymbol{\sigma}> = \boldsymbol{\sigma}^{\mathrm{o}}$ (right half); results above the middle dashed boxes are for boundary data which need not satisfy (9.5.3), whereas those below these boxes satisfy (9.5.3)

an unbounded homogeneous domain of the same uniform elasticity, and considering its response under overall stresses or strains, when eigenstrains or eigenstresses are distributed within V. In this manner, V is regarded as a portion, V′, of a uniform infinite region of uniform elasticity $\mathbf{C} = \mathbf{D}^{-1}$, within which suitable eigenstrains or eigenstresses are distributed.

Note that *constant* eigenstrains, $\mathbf{e}^{*\mathrm{o}}$, or eigenstresses, $\mathbf{s}^{*\mathrm{o}}$, uniformly distributed throughout the infinite domain, produce no disturbance fields in any subdomain, since the divergences of $\mathbf{C} : \mathbf{e}^{*\mathrm{o}}$ and $\mathbf{s}^{*\mathrm{o}}$ are identically zero everywhere. Hence, for any arbitrary eigenstrain field, $\mathbf{e}^{*}$, defined within V′, it is only necessary to consider the strain and stress fields produced in V′ by the *disturbance eigenstrain field*, $\mathbf{e}^{*\mathrm{d}} \equiv \mathbf{e}^{*} - <\mathbf{e}^{*}>$, and then add the average strains and stresses corresponding to the uniform eigenstrains, $< \mathbf{e}^* >$. Similar comments apply to any eigenstress field, $\mathbf{s}^*$, prescribed within V′, so that only the fields produced by the disturbance eigenstresses, $\mathbf{s}^{*d} \equiv \mathbf{s}^* - < \mathbf{s}^* >$, need to be examined and superimposed on the average fields produced in V′ by the uniform eigenstresses, $< \mathbf{s}^* >$, applied over the infinite region; see Figure 9.5.3. Therefore, only the effects of the disturbance eigenstrains and eigenstresses, $\mathbf{e}^{*d}$ and $\mathbf{s}^{*d}$, are considered in the following.

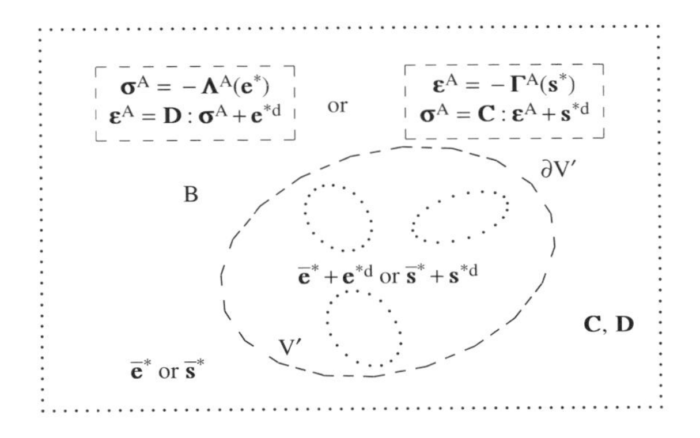

Figure 9.5.3

Eigenstrain field $\mathbf{e}^*$ or eigenstress field $\mathbf{s}^*$ in infinite homogeneous region B; $\mathbf{e}^{*d}$ or $\mathbf{s}^{*d}$ is distributed in an arbitrary subregion V′, and $\bar{\mathbf{e}}^*$ or $\bar{\mathbf{s}}^*$ is distributed throughout B; for ellipsoidal V′, $< \boldsymbol{\sigma}^A > = \mathbf{0}$, $< \boldsymbol{\varepsilon}^A > = \mathbf{0}$, and $< \boldsymbol{\sigma}^A : \boldsymbol{\varepsilon}^A > \leq 0$

The approximate operators, $\boldsymbol{\Lambda}^A$ and $\boldsymbol{\Gamma}^A$, are not associated with any particular prescribed boundary conditions on $\partial V \equiv \partial V'$, since the disturbance stress and strain fields that they yield change the corresponding boundary data on $\partial V'$, as the eigenstrains or eigenstresses are changed. Thus, (9.5.3) may not necessarily be satisfied, and the Hashin-Shtrikman variational principle may not be applicable to the functionals defined for these operators, namely to functionals $I^A(\mathbf{e}^*; \boldsymbol{\Lambda}^A; \boldsymbol{\sigma}^o)$ and $J^A(\mathbf{s}^*; \boldsymbol{\Gamma}^A; \boldsymbol{\varepsilon}^o)$, which are given by

$$I^A(\mathbf{e}^*; \boldsymbol{\Lambda}^A; \boldsymbol{\sigma}^o) \equiv \frac{1}{2} < \mathbf{e}^* : \{(\mathbf{D}' - \mathbf{D})^{-1} + \boldsymbol{\Lambda}^A\} : \mathbf{e}^* > - < \boldsymbol{\sigma}^o : \mathbf{e}^* >,$$

$$J^A(\mathbf{s}^*; \boldsymbol{\Gamma}^A; \boldsymbol{\sigma}^o) \equiv \frac{1}{2} < \mathbf{s}^* : \{(\mathbf{C}' - \mathbf{C})^{-1} + \boldsymbol{\Gamma}^A\} : \mathbf{s}^* > - < \boldsymbol{\varepsilon}^o : \mathbf{s}^* >; \quad (9.5.13\text{a,b})$$

see Subsection 9.4.4.[22]

Although the fields due to simplified operators $\boldsymbol{\Lambda}^A$ and $\boldsymbol{\Gamma}^A$ may not satisfy (9.5.3) and hence, (9.5.2) may not survive the involved approximation, the

[22] In Subsection 9.4.4, these functionals are denoted by $I^A(\mathbf{e}^*; \boldsymbol{\sigma}^o)$ and $J^A(\mathbf{s}^*; \boldsymbol{\varepsilon}^o)$. In this and the following subsections, the dependence on integral operators $\boldsymbol{\Lambda}^A$ and $\boldsymbol{\Gamma}^A$ is displayed in the corresponding arguments.

following result is always true for any eigenstrain and eigenstress field, $\mathbf{e}^*$ and $\mathbf{s}^*$, defined in any finite subdomain V′ (of any shape) of a uniform infinite domain of constant elasticity $\mathbf{C} = \mathbf{D}^{-1}$:

1) operators $< \, : \boldsymbol{\Lambda}^{A} : \, >$ and $< \, : \boldsymbol{\Gamma}^{A} : \, >$ are self-adjoint and positive-definite.

This property is proved at the end of this subsection.

From the self-adjointness and positive-definiteness of operators $\boldsymbol{\Lambda}^A$ and $\boldsymbol{\Gamma}^A$, it follows that the stationary value of I^A (of J^A) is its absolute minimum, if $\mathbf{D}' - \mathbf{D}$ (if $\mathbf{C}' - \mathbf{C}$) is positive-definite. That is

$$I^A(\boldsymbol{\varepsilon}^{*A}; \boldsymbol{\Lambda}^A; \boldsymbol{\sigma}^o) \leq I^A(\mathbf{e}^*; \boldsymbol{\Lambda}^A; \boldsymbol{\sigma}^o) \qquad \text{for any } \mathbf{e}^* \text{ when } \mathbf{D}' - \mathbf{D} \text{ is p.d.,}$$

$$J^A(\boldsymbol{\sigma}^{*A}; \boldsymbol{\Gamma}^A; \boldsymbol{\varepsilon}^o) \leq J^A(\mathbf{s}^*; \boldsymbol{\Gamma}^A; \boldsymbol{\varepsilon}^o) \qquad \text{for any } \mathbf{s}^* \text{ when } \mathbf{C}' - \mathbf{C} \text{ is p.d.,}$$

(9.5.14a,b)

where $\boldsymbol{\varepsilon}^{*A}$ and $\boldsymbol{\sigma}^{*A}$ are the solutions of

$$\delta I^A(\boldsymbol{\varepsilon}^{*A}; \boldsymbol{\Lambda}^A; \boldsymbol{\sigma}^o) = < \delta\mathbf{e}^* : \left\{ \{(\mathbf{D}' - \mathbf{D})^{-1} + \boldsymbol{\Lambda}^A\} : \boldsymbol{\varepsilon}^{*A} - \boldsymbol{\sigma}^o \right\} > = 0,$$

$$\delta J^A(\boldsymbol{\sigma}^{*A}; \boldsymbol{\Gamma}^A; \boldsymbol{\varepsilon}^o) = < \delta\mathbf{s}^* : \left\{ \{(\mathbf{C}' - \mathbf{C})^{-1} + \boldsymbol{\Gamma}^A\} : \boldsymbol{\sigma}^{*A} - \boldsymbol{\varepsilon}^o \right\} > = 0.$$

(9.5.15a,b)

The Euler equations in (9.5.15a) and (9.5.15b) are the same as those in (9.5.6a) and (9.5.6b), respectively, except for the corresponding integral operators and hence for the boundary data on ∂V′.

In addition to the self-adjointness and the positive-definiteness, operators $\boldsymbol{\Lambda}^A$ and $\boldsymbol{\Gamma}^A$ have the following property, independently of the shape of V′:

2) operators $< \, : (\boldsymbol{\Lambda}^A : \mathbf{D} : \boldsymbol{\Lambda}^A - \boldsymbol{\Lambda}^A) : \, >$ and $< \, : (\boldsymbol{\Gamma}^A : \mathbf{C} : \boldsymbol{\Gamma}^A - \boldsymbol{\Gamma}^A) : \, >$ are negative-definite.

Since $< \boldsymbol{\Lambda}^A >$ and $< \boldsymbol{\Gamma}^A >$ vanish[23] for any eigenstrains and eigenstresses, if V′ is an *ellipsoid*, property 2) can be used to obtain inequalities which relate the elastic and complementary elastic strain energies in the ellipsoidal V′ to its overall elasticity and compliance tensors which are defined by means of the average stress-strain relations. Property 2) is also proved at the end of this subsection.

Now, consider an arbitrary eigenstrain field, $\mathbf{e}^*$, in *ellipsoidal* V′. Denote the strain and stress fields produced by the disturbance eigenstrains through $\boldsymbol{\Lambda}^A$, by[24] $\boldsymbol{\sigma}^A(\mathbf{x}) = -\boldsymbol{\Lambda}^A(\mathbf{x}; \mathbf{e}^*)$ and $\boldsymbol{\varepsilon}^A(\mathbf{x}) = \mathbf{D} : \boldsymbol{\sigma}^A(\mathbf{x}) + \mathbf{e}^{*d}(\mathbf{x})$. Since V′ is ellipsoidal, and since operator $< \, : (\boldsymbol{\Lambda}^A : \mathbf{D} : \boldsymbol{\Lambda}^A - \boldsymbol{\Lambda}^A) : \, >$ is negative-definite, these strain and stress fields satisfy

[23] This is proved in Subsection 11.3.3.

[24] As mentioned, constant eigenstrains do not produce stresses, if they are distributed throughout the infinite domain. Hence, $\boldsymbol{\sigma}^A = -\boldsymbol{\Lambda}^A(\mathbf{e}^{*d})$ is written as $-\boldsymbol{\Lambda}^A(\mathbf{e}^*)$.

$$<\boldsymbol{\sigma}^{\mathrm{A}}> = < -\boldsymbol{\Lambda}^{\mathrm{A}}(\mathbf{e}^{*\mathrm{d}})> = \mathbf{0},$$

$$<\boldsymbol{\varepsilon}^{\mathrm{A}}> = < -\mathbf{D}:\boldsymbol{\Lambda}^{\mathrm{A}}(\mathbf{e}^{*\mathrm{d}}) + \mathbf{e}^{*\mathrm{d}}> = \mathbf{0}, \qquad \text{(9.5.16a~c)}$$

$$<\boldsymbol{\sigma}^{\mathrm{A}}:\boldsymbol{\varepsilon}^{\mathrm{A}}> = <\mathbf{e}^{*\mathrm{d}}:(\boldsymbol{\Lambda}^{\mathrm{A}}:\mathbf{D}:\boldsymbol{\Lambda}^{\mathrm{A}} - \boldsymbol{\Lambda}^{\mathrm{A}}):\mathbf{e}^{*\mathrm{d}}> \leq 0.$$

Under uniform farfield stresses $\boldsymbol{\sigma}^{\mathrm{o}}$, the resulting (total) stress and strain fields in the infinite uniform domain which contains V′, are given by

$$\boldsymbol{\sigma}(\mathbf{x}) = \boldsymbol{\sigma}^{\mathrm{o}} + \boldsymbol{\sigma}^{\mathrm{A}}(\mathbf{x}), \qquad \boldsymbol{\varepsilon}(\mathbf{x}) = \mathbf{D}:\boldsymbol{\sigma}^{\mathrm{o}} + <\mathbf{e}^{*}> + \boldsymbol{\varepsilon}^{\mathrm{A}}(\mathbf{x}), \qquad \text{(9.5.17a,b)}$$

and these satisfy

$$<\boldsymbol{\sigma}:\boldsymbol{\varepsilon}> - <\boldsymbol{\sigma}>:<\boldsymbol{\varepsilon}> = <\boldsymbol{\sigma}^{\mathrm{A}}:\boldsymbol{\varepsilon}^{\mathrm{A}}> \leq 0, \qquad \text{(9.5.18a)}$$

where $<\boldsymbol{\sigma}> = \boldsymbol{\sigma}^{\mathrm{o}}$ and $<\boldsymbol{\varepsilon}> = \bar{\boldsymbol{\varepsilon}} = \mathbf{D}:\boldsymbol{\sigma}^{\mathrm{o}} + <\mathbf{e}^{*}>$. From (9.5.18a), the following inequality is obtained:

$$\frac{1}{2}\,\bar{\boldsymbol{\varepsilon}}:(\bar{\mathbf{C}}_{E} - \bar{\mathbf{C}}_{\sigma\text{-}\varepsilon}):\bar{\boldsymbol{\varepsilon}} \leq 0, \qquad \text{(9.5.19a)}$$

where $\bar{\mathbf{C}}_{E}$ is the overall elasticity tensor of V′ defined through the *average elastic strain energy*, $<\boldsymbol{\sigma}:\boldsymbol{\varepsilon}>/2 \equiv \bar{\boldsymbol{\varepsilon}}:\bar{\mathbf{C}}_{E}:\bar{\boldsymbol{\varepsilon}}/2$; and $\bar{\mathbf{C}}_{\sigma\text{-}\varepsilon}$ is the overall elasticity tensor of V′ defined through the *average stress-strain relations*, $<\boldsymbol{\sigma}> \equiv \bar{\mathbf{C}}_{\sigma\text{-}\varepsilon}:\bar{\boldsymbol{\varepsilon}}$. Thus, $(\bar{\mathbf{C}}_{E})^{-1}$ and $\bar{\mathbf{D}}_{\sigma\text{-}\varepsilon} \equiv (\bar{\mathbf{C}}_{\sigma\text{-}\varepsilon})^{-1}$ satisfy

$$\frac{1}{2}\,\boldsymbol{\sigma}^{\mathrm{o}}:(\bar{\mathbf{D}}_{\sigma\text{-}\varepsilon} - (\bar{\mathbf{C}}_{E})^{-1}):\boldsymbol{\sigma}^{\mathrm{o}} \leq 0, \qquad \text{(9.5.20a)}$$

for any $\boldsymbol{\sigma}^{\mathrm{o}}$.

In a similar manner, consider an arbitrary eigenstress field $\mathbf{s}^{*}$, defined in *ellipsoidal region* V′, and let $\boldsymbol{\varepsilon}^{\mathrm{A}}(\mathbf{x}) = -\boldsymbol{\Gamma}^{\mathrm{A}}(\mathbf{x};\,\mathbf{s}^{*})$ and $\boldsymbol{\sigma}^{\mathrm{A}}(\mathbf{x}) = \mathbf{C}:\boldsymbol{\varepsilon}^{\mathrm{A}}(\mathbf{x}) + \mathbf{s}^{*\mathrm{d}}(\mathbf{x})$ be the corresponding strain and stress fields[25] produced through $\boldsymbol{\Gamma}^{\mathrm{A}}$. Since V′ is ellipsoidal, and since operator $<\ :(\boldsymbol{\Gamma}^{\mathrm{A}}:\mathbf{C}:\boldsymbol{\Gamma}^{\mathrm{A}} - \boldsymbol{\Gamma}^{\mathrm{A}}):\ >$ is negative-definite, $\boldsymbol{\varepsilon}^{\mathrm{A}}$ and $\boldsymbol{\sigma}^{\mathrm{A}}$ satisfy

$$<\boldsymbol{\varepsilon}^{\mathrm{A}}> = < -\boldsymbol{\Gamma}^{\mathrm{A}}(\mathbf{s}^{*\mathrm{d}})> = \mathbf{0},$$

$$<\boldsymbol{\sigma}^{\mathrm{A}}> = < -\mathbf{C}:\boldsymbol{\Gamma}^{\mathrm{A}}(\mathbf{s}^{*\mathrm{d}}) + \mathbf{s}^{*\mathrm{d}}> = \mathbf{0}, \qquad \text{(9.5.16d~f)}$$

$$<\boldsymbol{\sigma}^{\mathrm{A}}:\boldsymbol{\varepsilon}^{\mathrm{A}}> = <\mathbf{s}^{*\mathrm{d}}:(\boldsymbol{\Gamma}^{\mathrm{A}}:\mathbf{C}:\boldsymbol{\Gamma}^{\mathrm{A}} - \boldsymbol{\Gamma}^{\mathrm{A}}):\mathbf{s}^{*\mathrm{d}}> \leq 0.$$

Under uniform farfield strains $\boldsymbol{\varepsilon}^{\mathrm{o}}$, the resulting (total) stress and strain fields which are given by

$$\boldsymbol{\varepsilon}(\mathbf{x}) = \boldsymbol{\varepsilon}^{\mathrm{o}} + \boldsymbol{\varepsilon}^{\mathrm{A}}(\mathbf{x}), \qquad \boldsymbol{\sigma}(\mathbf{x}) = \mathbf{C}:\boldsymbol{\varepsilon}^{\mathrm{o}} + <\mathbf{s}^{*}> + \boldsymbol{\sigma}^{\mathrm{A}}(\mathbf{x}), \qquad \text{(9.5.17c,d)}$$

satisfy

$$<\boldsymbol{\sigma}:\boldsymbol{\varepsilon}> - <\boldsymbol{\sigma}>:<\boldsymbol{\varepsilon}> = <\boldsymbol{\sigma}^{\mathrm{A}}:\boldsymbol{\varepsilon}^{\mathrm{A}}> \leq 0, \qquad \text{(9.5.18b)}$$

and hence the following inequality is obtained:

---

[25] Comments similar to the preceding footnote about $\boldsymbol{\Lambda}^{\mathrm{A}}$ also apply to $\boldsymbol{\Gamma}^{\mathrm{A}}$.

$$\frac{1}{2}\,\bar{\boldsymbol{\sigma}} : (\bar{\mathbf{D}}_E - \bar{\mathbf{D}}_{\sigma\text{-}\varepsilon}) : \bar{\boldsymbol{\sigma}} \leq 0, \tag{9.5.19b}$$

where $\bar{\boldsymbol{\sigma}} = <\boldsymbol{\sigma}> = \mathbf{C} : \boldsymbol{\varepsilon}^o + <\mathrm{s}^*>$, and $\bar{\mathbf{D}}_E$ and $\bar{\mathbf{D}}_{\sigma\text{-}\varepsilon}$ are the overall compliance tensors defined for V′ through the *average elastic strain energy*, and the *average stress-strain relations*, respectively. Thus, $(\bar{\mathbf{D}}_E)^{-1}$ and $\bar{\mathbf{C}}_{\sigma\text{-}\varepsilon} \equiv (\bar{\mathbf{D}}_{\sigma\text{-}\varepsilon})^{-1}$ satisfy

$$\frac{1}{2}\,\boldsymbol{\varepsilon}^o : (\bar{\mathbf{C}}_{\sigma\text{-}\varepsilon} - (\bar{\mathbf{D}}_E)^{-1}) : \boldsymbol{\varepsilon}^o \leq 0, \tag{9.5.20b}$$

for any $\boldsymbol{\varepsilon}^o$.

Table 9.5.2 summarizes the results of this subsection. The superscript A denotes the results obtained through operators $\boldsymbol{\Lambda}^A$ or $\boldsymbol{\Gamma}^A$ associated with the infinite-space Green function.

It will now be proved that operators $< :\boldsymbol{\Lambda}^A: >$ and $< :\boldsymbol{\Gamma}^A: >$ are self-adjoint and positive-definite for any eigenstrain and eigenstress field, respectively, and that $< :(\boldsymbol{\Lambda}^A : \mathbf{D} : \boldsymbol{\Lambda}^A - \boldsymbol{\Lambda}^A): >$ and $< :(\boldsymbol{\Gamma}^A : \mathbf{C} : \boldsymbol{\Lambda}^A - \boldsymbol{\Gamma}^A): >$ are negative-definite. For illustration, the proof is given only for operator $\boldsymbol{\Gamma}^A$ here. Consider an infinite domain B of uniform elasticity $\mathbf{C}$. By definition, integral operator $\boldsymbol{\Gamma}^A$ determines disturbance strains in B for eigenstresses distributed in B. Let V′ be a region in B, within which eigenstresses $\mathbf{s}^*(\mathbf{x})$ are distributed. As is discussed in Subsection 12.4, the Fourier transform representation of $\boldsymbol{\Gamma}^A(\mathbf{s}^*)$ is given by

$$\boldsymbol{\Gamma}^A(\mathbf{x};\,\mathbf{s}^*) = \int_{-\infty}^{+\infty} F\boldsymbol{\Gamma}^A(\boldsymbol{\xi}) : F\mathbf{s}^*(\boldsymbol{\xi}) \exp(\iota\boldsymbol{\xi}\cdot\mathbf{x})\, dV_\xi, \tag{9.5.21a}$$

where

$$F\mathbf{s}^*(\boldsymbol{\xi}) = \frac{1}{(2\pi)^3}\int_{V'} \mathbf{s}^*(\mathbf{x}) \exp(-\iota\boldsymbol{\xi}\cdot\mathbf{x})\, dV_{\mathbf{x}},$$

$$F\boldsymbol{\Gamma}^A(\boldsymbol{\xi}) = \begin{cases} sym\,(\boldsymbol{\xi}\otimes(\boldsymbol{\xi}\cdot\mathbf{C}\cdot\boldsymbol{\xi})^{-1}\otimes\boldsymbol{\xi}) & \boldsymbol{\xi} \neq \mathbf{0} \\ \mathbf{0} & \boldsymbol{\xi} = \mathbf{0}, \end{cases} \tag{9.5.21b,c}$$

with *sym* standing for the symmetric part of the fourth-order tensor.

Since eigenstresses uniformly distributed in B do not produce disturbance strains, it suffices to consider an eigenstress field which vanishes outside of V′ and has zero volume average over V′. Then, $<\mathrm{s}^* : \boldsymbol{\Gamma}^A : \mathbf{s}^*>$ is given by

$$\begin{aligned} <\mathrm{s}^* : \boldsymbol{\Gamma}^A : \mathbf{s}^*> &= \frac{1}{V'}\int_{V'} \mathbf{s}^*(\mathbf{x}) : \boldsymbol{\Gamma}^A(\mathbf{x};\,\mathbf{s}^*)\, dV_{\mathbf{x}} \\ &= \frac{1}{V'}\int_{B} \mathbf{s}^*(\mathbf{x}) : \boldsymbol{\Gamma}^A(\mathbf{x};\,\mathbf{s}^*)\, dV_{\mathbf{x}} \\ &= \frac{1}{V'}\int_{-\infty}^{+\infty} F\mathbf{s}^*(-\boldsymbol{\xi}) : F\boldsymbol{\Gamma}^A(\boldsymbol{\xi}) : F\mathbf{s}^*(\boldsymbol{\xi})\, dV_\xi. \end{aligned} \tag{9.5.22}$$

Since tensor $F\boldsymbol{\Gamma}^A(\boldsymbol{\xi})$ is symmetric with respect to the first and last pairs of its indices for any $\boldsymbol{\xi}$, operator $< :\boldsymbol{\Gamma}^A: >$ is self-adjoint.

Table 9.5.2

Infinite-space operators and their properties

properties of operators $\boldsymbol{\Lambda}^{\mathrm{A}}(\mathbf{x};\, \mathbf{e}^*)$ and $\boldsymbol{\Gamma}^{\mathrm{A}}(\mathbf{x};\, \mathbf{s}^*)$ associated with $<\,>$ (eigenstrains or eigenstresses vanish outside of V′)		
self-adjointness	$<\mathbf{e}^{*1} : \boldsymbol{\Lambda}^{\mathrm{A}} : \mathbf{e}^{*2}> \; = \; <\mathbf{e}^{*2} : \boldsymbol{\Lambda}^{\mathrm{A}} : \mathbf{e}^{*1}>$	$<\mathbf{s}^{*1} : \boldsymbol{\Gamma}^{\mathrm{A}} : \mathbf{s}^{*2}> \; = \; <\mathbf{s}^{*2} : \boldsymbol{\Gamma}^{\mathrm{A}} : \mathbf{s}^{*1}>$
positive-definiteness	$<\mathbf{e}^* : \boldsymbol{\Lambda}^{\mathrm{A}} : \mathbf{e}^*> \, \geq 0$	$<\mathbf{s}^* : \boldsymbol{\Gamma}^{\mathrm{A}} : \mathbf{s}^*> \, \geq 0$
negative-definiteness	$<\mathbf{e}^* : (\boldsymbol{\Lambda}^{\mathrm{A}} : \mathbf{D} : \boldsymbol{\Lambda}^{\mathrm{A}} - \boldsymbol{\Lambda}^{\mathrm{A}}) : \mathbf{e}^*> \, \leq 0$	$<\mathbf{s}^* : (\boldsymbol{\Gamma}^{\mathrm{A}} : \mathbf{C} : \boldsymbol{\Gamma}^{\mathrm{A}} - \boldsymbol{\Gamma}^{\mathrm{A}}) : \mathbf{s}^*> \, \leq 0$
fields produced by operators for ellipsoidal V′		
volume average	$<\boldsymbol{\Lambda}^{\mathrm{A}}(\mathbf{e}^*)> \, = \mathbf{0}$	$<\boldsymbol{\Gamma}^{\mathrm{A}}(\mathbf{s}^*)> \, = \mathbf{0}$
disturbance fields	$\boldsymbol{\sigma}^{\mathrm{A}}(\mathbf{x}) = -\boldsymbol{\Lambda}^{\mathrm{A}}(\mathbf{x};\, \mathbf{e}^*)$ $\boldsymbol{\varepsilon}^{\mathrm{A}}(\mathbf{x}) = \mathbf{D} : \boldsymbol{\sigma}^{\mathrm{A}}(\mathbf{x}) + \mathbf{e}^{*\mathrm{d}}(\mathbf{x})$	$\boldsymbol{\varepsilon}^{\mathrm{A}}(\mathbf{x}) = -\boldsymbol{\Gamma}^{\mathrm{A}}(\mathbf{x};\, \mathbf{s}^*)$ $\boldsymbol{\sigma}^{\mathrm{A}}(\mathbf{x}) = \mathbf{C} : \boldsymbol{\varepsilon}^{\mathrm{A}}(\mathbf{x}) + \mathbf{s}^{*\mathrm{d}}(\mathbf{x})$
total fields	$\boldsymbol{\sigma}(\mathbf{x}) = \boldsymbol{\sigma}^{\mathrm{o}} + \boldsymbol{\sigma}^{\mathrm{A}}(\mathbf{x})$ $\boldsymbol{\varepsilon}(\mathbf{x}) = \mathbf{D} : \boldsymbol{\sigma}^{\mathrm{o}} + <\mathbf{e}^*> + \boldsymbol{\varepsilon}^{\mathrm{A}}(\mathbf{x})$	$\boldsymbol{\varepsilon}(\mathbf{x}) = \boldsymbol{\varepsilon}^{\mathrm{o}} + \boldsymbol{\varepsilon}^{\mathrm{A}}(\mathbf{x})$ $\boldsymbol{\varepsilon}(\mathbf{x}) = \mathbf{C} : \boldsymbol{\varepsilon}^{\mathrm{o}} + <\mathbf{s}^*> + \boldsymbol{\sigma}^{\mathrm{A}}(\mathbf{x})$
relations among fields produced by operators for ellipsoidal V′		
$<\boldsymbol{\sigma} : \boldsymbol{\varepsilon}>$	$\bar{\boldsymbol{\varepsilon}} : \overline{\mathbf{C}}{}_{E}^{\Sigma'} : \bar{\boldsymbol{\varepsilon}}$	$\bar{\boldsymbol{\sigma}} : \mathbf{D}_{E}^{\mathrm{E}'} : \bar{\boldsymbol{\sigma}}$
$<\boldsymbol{\sigma}>, <\boldsymbol{\varepsilon}>$ -relations	$\boldsymbol{\sigma}^{\mathrm{o}} = \overline{\mathbf{C}}{}_{\sigma\text{-}\varepsilon}^{\Sigma'} : \bar{\boldsymbol{\varepsilon}}$	$\boldsymbol{\varepsilon}^{\mathrm{o}} = \mathbf{D}_{\sigma\text{-}\varepsilon}^{\mathrm{E}'} : \bar{\boldsymbol{\sigma}}$
$<\boldsymbol{\sigma} : \boldsymbol{\varepsilon}> - <\boldsymbol{\sigma}> : <\boldsymbol{\varepsilon}>$	$<\boldsymbol{\sigma}^{\mathrm{A}} : \boldsymbol{\varepsilon}^{\mathrm{A}}> \, \leq 0$	$<\boldsymbol{\sigma}^{\mathrm{A}} : \boldsymbol{\varepsilon}^{\mathrm{A}}> \, \leq 0$
overall moduli	$\bar{\boldsymbol{\varepsilon}} : (\overline{\mathbf{C}}{}_{E}^{\Sigma'} - \overline{\mathbf{C}}{}_{\sigma\text{-}\varepsilon}^{\Sigma'}) : \bar{\boldsymbol{\varepsilon}} \leq 0$ $\boldsymbol{\sigma}^{\mathrm{o}} : (\mathbf{D}_{\sigma\text{-}\varepsilon}^{\Sigma'} - (\overline{\mathbf{C}}{}_{E}^{\Sigma'})^{-1}) : \boldsymbol{\sigma}^{\mathrm{o}} \leq 0$	$\bar{\boldsymbol{\sigma}} : (\mathbf{D}_{E}^{\mathrm{E}'} - \mathbf{D}_{\sigma\text{-}\varepsilon}^{\mathrm{E}'}) : \bar{\boldsymbol{\sigma}} \leq 0$ $\boldsymbol{\varepsilon}^{\mathrm{o}} : (\overline{\mathbf{C}}{}_{\sigma\text{-}\varepsilon}^{\mathrm{E}'} - (\mathbf{D}_{E}^{\mathrm{E}'})^{-1}) : \boldsymbol{\varepsilon}^{\mathrm{o}} \leq 0$

Since $\mathbf{s}^*$ is real-valued, $F\mathbf{s}^*(-\boldsymbol{\xi})$ is the complex conjugate of $F\mathbf{s}^*(\boldsymbol{\xi})$; see (9.5.21b). From definition (9.5.21c) of $F\boldsymbol{\Gamma}^{\mathrm{A}}$, and the symmetry of $F\mathbf{s}^*$, $F\mathbf{s}^*(-\boldsymbol{\xi}) : F\boldsymbol{\Gamma}^{\mathrm{A}}(\boldsymbol{\xi}) : F\mathbf{s}^*(\boldsymbol{\xi})$ becomes

$$F\mathbf{s}^*(-\boldsymbol{\xi}) : F\boldsymbol{\Gamma}^{\mathrm{A}}(\boldsymbol{\xi}) : F\mathbf{s}^*(\boldsymbol{\xi}) = \{\boldsymbol{\xi} \cdot F\mathbf{s}^*(-\boldsymbol{\xi})\} \cdot (\boldsymbol{\xi} \cdot \mathbf{C} \cdot \boldsymbol{\xi})^{-1} \cdot \{\boldsymbol{\xi} \cdot F\mathbf{s}^*(\boldsymbol{\xi})\}$$

$$= \mathbf{Z} \cdot (\boldsymbol{\xi} \cdot \mathbf{C} \cdot \boldsymbol{\xi}) \cdot \mathbf{Z}, \tag{9.5.23a}$$

where $\mathbf{Z} = (\boldsymbol{\xi}.\mathbf{C}.\boldsymbol{\xi})^{-1}.\{\boldsymbol{\xi}.F\mathbf{s}^*(\boldsymbol{\xi})\}$. Because of the symmetry and positive-definiteness of **C**, the following relations always holds:

$$\mathbf{Z}.(\boldsymbol{\xi}.\mathbf{C}.\boldsymbol{\xi}).\mathbf{Z} = sym\,\{\mathbf{Z}\otimes\boldsymbol{\xi}\} : \mathbf{C} : sym\,\{\mathbf{Z}\otimes\boldsymbol{\xi}\}$$

$$= sym\,\{(Re\mathbf{Z})\otimes\boldsymbol{\xi}\} : \mathbf{C} : sym\,\{(Re\mathbf{Z})\otimes\boldsymbol{\xi}\}$$

$$+ sym\,\{(Im\mathbf{Z})\otimes\boldsymbol{\xi}\} : \mathbf{C} : sym\,\{(Im\mathbf{Z})\otimes\boldsymbol{\xi}\} \geq 0. \tag{9.5.23b}$$

It follows from (9.5.23a) and (9.5.23b) that $< : \boldsymbol{\Gamma}^A : >$ is positive-definite. The self-adjointness and positive-definiteness of $< : \boldsymbol{\Lambda}^A : >$ can be proved in essentially the same manner. Note that the above results hold independently of the shape and size of V′.

With the aid of the Fourier transformation, it can be proved that $< : (\boldsymbol{\Gamma}^A : \mathbf{C} : \boldsymbol{\Gamma}^A - \boldsymbol{\Gamma}^A) : > \leq 0$. Since **C** is positive-definite and $\mathbf{s}^*$ vanishes outside of V′, the following inequality holds:

$$< \mathbf{s}^* : (\boldsymbol{\Gamma}^A : \mathbf{C} : \boldsymbol{\Gamma}^A - \boldsymbol{\Gamma}^A) : \mathbf{s}^* >$$

$$= \frac{1}{V'} \int_{V'} \boldsymbol{\Gamma}^A(\mathbf{x};\, \mathbf{s}^*) : \{\mathbf{C} : \boldsymbol{\Gamma}^A(\mathbf{x};\, \mathbf{s}^*) - \mathbf{s}^*(\mathbf{x})\}\, dV_\mathbf{x}$$

$$\leq \frac{1}{V'} \int_{B} \boldsymbol{\Gamma}^A(\mathbf{x};\, \mathbf{s}^*) : \{\mathbf{C} : \boldsymbol{\Gamma}^A(\mathbf{x};\, \mathbf{s}^*) - \mathbf{s}^*(\mathbf{x})\}\, dV_\mathbf{x}. \tag{9.5.24}$$

In terms of $F\mathbf{s}^*$ and $F\boldsymbol{\Gamma}^A$, the right side of (9.5.24) is given by

$$\frac{1}{V'} \int_{-\infty}^{+\infty} F\mathbf{s}^*(-\boldsymbol{\xi}) : \{F\boldsymbol{\Gamma}^A(-\boldsymbol{\xi}) : \mathbf{C} : F\boldsymbol{\Gamma}^A(\boldsymbol{\xi}) - F\boldsymbol{\Gamma}^A(-\boldsymbol{\xi})\} : F\mathbf{s}^*(\boldsymbol{\xi})\, dV_\xi.$$

From definition (9.5.21c) of $F\boldsymbol{\Gamma}^A$,

$$F\boldsymbol{\Gamma}^A(-\boldsymbol{\xi}) : \mathbf{C} : F\boldsymbol{\Gamma}^A(\boldsymbol{\xi}) = F\boldsymbol{\Gamma}^A(\boldsymbol{\xi}) : \mathbf{C} : F\boldsymbol{\Gamma}^A(\boldsymbol{\xi}) = F\boldsymbol{\Gamma}^A(\boldsymbol{\xi}), \tag{9.5.25}$$

for $\boldsymbol{\xi} \neq \mathbf{0}$, and the term in the curly brackets vanishes. Therefore, it is proved that for any eigenstresses which vanish outside of V′, $< : (\boldsymbol{\Gamma}^A : \mathbf{C} : \boldsymbol{\Gamma}^A - \boldsymbol{\Gamma}^A) : > \leq 0$. In essentially the same manner, it is proved that $< : (\boldsymbol{\Lambda}^A : \mathbf{D} : \boldsymbol{\Gamma}^A - \boldsymbol{\Lambda}^A) : > \leq 0$, for any eigenstrains which vanish outside of V′. Note again that V′ *does not need to be an ellipsoid.* Hence, if eigenstrains $\mathbf{e}^*$ or eigenstresses $\mathbf{s}^*$ satisfy $< \boldsymbol{\Lambda}^A(\mathbf{e}^*) > = \mathbf{0}$ or $< \boldsymbol{\Gamma}^A(\mathbf{s}^*) > = \mathbf{0}$, then, inequalities (9.5.18) to (9.5.20) hold. These latter conditions are necessarily satisfied when V′ is ellipsoidal.

9.5.4. Exact Bounds Based on Simplified Functionals

In Subsection 9.5.2, two exact inequalities, (9.5.11a) and (9.5.11b), are obtained for the overall compliance and elasticity tensors of the *original* heterogeneous RVE (denoted by V with boundary ∂V) subjected to any arbitrary boundary data, using Theorems I and II of Subsection 2.5.6; see Figure 9.5.2. In Subsection 9.5.3, two other exact inequalities, (9.5.14a) and (9.5.14b), are obtained for functionals I^A and J^A, for any eigenstrains and eigenstresses prescribed in an arbitrary uniform region, V′, of an infinite uniform domain, B.

Furthermore, for an *ellipsoidal* RVE, two additional exact inequalities, (9.5.20a) and (9.5.20b), relate the overall moduli defined through the *average energy* and the *average stress-strain relations*, when the disturbance fields are given by approximate operators $\mathbf{\Lambda}^{\mathrm{A}}$ and $\mathbf{\Gamma}^{\mathrm{A}}$; see Table 9.5.2. Two sets of exact inequalities, namely, {(9.5.11a), (9.5.14a), (9.5.20a)} and {(9.5.11b), (9.5.14b), (9.5.20b)}, are employed in this subsection, and exact *computable* bounds for the overall moduli of the original heterogeneous RVE are obtained. The case of traction boundary data is examined first.

To this end, choose V′ and V to be identical *ellipsoids*. The first is part of the uniform infinite B, and the second represents the original RVE. For any eigenstrain field $\mathbf{e}^*$ defined in V′, $< \mathbf{\Lambda}^{\mathrm{A}}(\mathbf{e}^*) > \equiv \mathbf{0}$, where the average is over V′; see (9.4.9e). It then follows that the left side of (9.5.14a) can be expressed as

$$\mathrm{I}^{\mathrm{A}}(\boldsymbol{\varepsilon}^{*\mathrm{A}};\ \mathbf{\Lambda}^{\mathrm{A}};\ \boldsymbol{\sigma}^{\mathrm{o}}) = -\frac{1}{2}\,\bar{\boldsymbol{\varepsilon}}^{*\mathrm{A}} : \boldsymbol{\sigma}^{\mathrm{o}} = \frac{1}{2}\,\boldsymbol{\sigma}^{\mathrm{o}} : (\mathbf{D} - \bar{\mathbf{D}}^{\Sigma'}_{\sigma\text{-}\varepsilon}) : \boldsymbol{\sigma}^{\mathrm{o}}, \tag{9.5.26a}$$

where $\bar{\mathbf{D}}^{\Sigma'}_{\sigma\text{-}\varepsilon}$ is the overall compliance tensor of the original RVE which is homogenized by the eigenstrain field $\boldsymbol{\varepsilon}^{*\mathrm{A}}$, and which is subjected to the following boundary conditions:

$$\begin{aligned} \mathbf{u}(\mathbf{x}) &= \mathbf{x}\boldsymbol{\cdot}(\mathbf{D} : \boldsymbol{\sigma}^{\mathrm{o}}) + \mathbf{U}(\mathbf{x};\ -\mathbf{C} : (\boldsymbol{\varepsilon}^{*\mathrm{A}} - \bar{\boldsymbol{\varepsilon}}^{*\mathrm{A}})), \\ \mathbf{t}(\mathbf{x}) &= \boldsymbol{\nu}(\mathbf{x})\boldsymbol{\cdot}(\boldsymbol{\sigma}^{\mathrm{o}} - \mathbf{\Lambda}^{\mathrm{A}}(\mathbf{x};\ \boldsymbol{\varepsilon}^{*\mathrm{A}})), \end{aligned} \tag{9.5.27a}$$

where $\mathbf{U}$ is defined by (9.4.7). The eigenstrain field $\boldsymbol{\varepsilon}^{*\mathrm{A}}$ solves the consistency conditions for boundary data (9.5.27a) *exactly*, and hence homogenizes the original RVE for these boundary data *exactly*.

The definition of the overall compliance, $\bar{\mathbf{D}}^{\Sigma'}_{\sigma\text{-}\varepsilon}$ given by (9.5.26a), is based on the *average stress-strain relations* associated with the special boundary data (9.5.27a).[26] From (9.5.20a), the following inequality now follows:

$$\frac{1}{2}\,\boldsymbol{\sigma}^{\mathrm{o}} : (\bar{\mathbf{D}}^{\Sigma'}_{\sigma\text{-}\varepsilon} - (\bar{\mathbf{C}}^{\Sigma'}_{E})^{-1}) : \boldsymbol{\sigma}^{\mathrm{o}} \le 0, \tag{9.5.28a}$$

where $\bar{\mathbf{C}}^{\Sigma'}_{E}$ is the the overall elasticity tensor which is defined by the *average strain energy* of the RVE for boundary data (9.5.27a). Inequality (9.5.11a) therefore applies to this $\bar{\mathbf{C}}^{\Sigma'}_{E}$, i.e.,

$$\frac{1}{2}\,\boldsymbol{\sigma}^{\mathrm{o}} : ((\bar{\mathbf{C}}^{\Sigma'}_{E})^{-1} - \bar{\mathbf{D}}^{\Sigma}) : \boldsymbol{\sigma}^{\mathrm{o}} \le 0. \tag{9.5.29a}$$

Finally, from (9.5.28a) and (9.5.29a) it follows that

$$\frac{1}{2}\,\boldsymbol{\sigma}^{\mathrm{o}} : (\bar{\mathbf{D}}^{\Sigma'}_{\sigma\text{-}\varepsilon} - \bar{\mathbf{D}}^{\Sigma}) : \boldsymbol{\sigma}^{\mathrm{o}} \le 0, \tag{9.5.30a}$$

for any constant $\boldsymbol{\sigma}^{\mathrm{o}}$. Hence, $\bar{\mathbf{D}}^{\Sigma'}_{\sigma\text{-}\varepsilon} - \bar{\mathbf{D}}^{\Sigma}$ is negative-semi-definite.

It is still difficult to compute $\bar{\mathbf{D}}^{\Sigma'}_{\sigma\text{-}\varepsilon}$ exactly, since it requires an exact value of $\bar{\boldsymbol{\varepsilon}}^{*\mathrm{A}}$. Inequality (9.5.14a) may now be invoked and, instead of the exact

---

[26] Subscript σ-ε emphasizes that the moduli are obtained through the average stress-strain relations. Superscript Σ denotes that the uniform traction data are prescribed, and superscript Σ' signifies that $\mathrm{V} \equiv \mathrm{V}'$ is embedded in an infinite solid. Note that the boundary conditions (9.5.27a) or (9.5.27b) cannot be prescribed *a priori*; all primed quantities correspond to these boundary data.

eigenstrain field, $\boldsymbol{\varepsilon}^{*A}$, a piecewise constant eigenstrain field may be used, together with the approximate operator $\boldsymbol{\Lambda}^A$. As in Subsection 9.4, denote the value of the functional I^A for a piecewise constant distribution of eigenstrains, $\{\mathbf{e}^{*\alpha}\}$, by $I^{A\prime}(\{\mathbf{e}^{*\alpha}\}; \boldsymbol{\sigma}^o)$, where $\mathbf{e}^{*\alpha}$ is the constant eigenstrain field defined on Ω_α ($\alpha = 1, 2, ..., n$). Let the stationary value of $I^{A\prime}$ be attained by $\{\boldsymbol{\varepsilon}^{*\alpha}\}$ which is the solution of $\partial I^{A\prime}/\partial \mathbf{e}^{*\alpha} = \mathbf{0}$ ($\alpha = 1, 2, ..., n$). For the class of piecewise constant eigenstrains, the optimal value $I^{A\prime}(\{\boldsymbol{\varepsilon}^{*\alpha}\}; \boldsymbol{\sigma}^o)$ obtained in this manner is now expressed as

$$I^{A\prime}(\{\boldsymbol{\varepsilon}^{*\alpha}\}; \boldsymbol{\sigma}^o) \equiv \frac{1}{2}\, \boldsymbol{\sigma}^o : (\mathbf{D} - \overline{\mathbf{D}}^{\Sigma A}) : \boldsymbol{\sigma}^o, \tag{9.5.31a}$$

where $\overline{\mathbf{D}}^{\Sigma A}$, defined from $I^{A\prime}(\{\boldsymbol{\varepsilon}^{*\alpha}\}; \boldsymbol{\sigma}^o)$, is the final estimate of the overall compliance tensor. From inequality (9.5.14a) it now follows that

$$\frac{1}{2}\, \boldsymbol{\sigma}^o : (\overline{\mathbf{D}}^{\Sigma A} - \overline{\mathbf{D}}^{\Sigma'}_{\sigma\text{-}\varepsilon}) : \boldsymbol{\sigma}^o \leq 0, \tag{9.5.32a}$$

if $\mathbf{D}' - \mathbf{D}$ is positive-definite. Together with (9.5.30a), a computable bound on the overall compliance tensor $\overline{\mathbf{D}}^{\Sigma}$ is obtained as follows:

$$\boldsymbol{\sigma}^o : \overline{\mathbf{D}}^{\Sigma} : \boldsymbol{\sigma}^o \geq \boldsymbol{\sigma}^o : \overline{\mathbf{D}}^{\Sigma A} : \boldsymbol{\sigma}^o, \tag{9.5.33a}$$

for any $\boldsymbol{\sigma}^o$, if $\mathbf{D}' - \mathbf{D}$ is positive-definite. In the right half of Figure 9.5.4, the above results are summarized in the form of a flow chart.

The case corresponding to the class of functionals $\hat{J}(\mathbf{s}^*; \boldsymbol{\Gamma}^G; \boldsymbol{\varepsilon}^o)$ with approximation $J^A(\mathbf{s}^*; \boldsymbol{\Gamma}^A; \boldsymbol{\varepsilon}^o)$, is treated in the same manner. First the overall elasticity for the functional $J^A(\boldsymbol{\sigma}^{*A}; \boldsymbol{\Gamma}^A; \boldsymbol{\varepsilon}^o)$ is defined by

$$J^A(\boldsymbol{\sigma}^{*A}; \boldsymbol{\Gamma}^A; \boldsymbol{\varepsilon}^o) = -\frac{1}{2}\, \bar{\boldsymbol{\sigma}}^{*A} : \boldsymbol{\varepsilon}^o = \frac{1}{2}\, \boldsymbol{\varepsilon}^o : (\mathbf{C} - \overline{\mathbf{C}}^{E'}_{\sigma\text{-}\varepsilon}) : \boldsymbol{\varepsilon}^o, \tag{9.5.26b}$$

where $\boldsymbol{\sigma}^{*A}$ is the eigenstress field which *exactly* homogenizes the original RVE under the following boundary data:

$$\begin{aligned} \mathbf{u}(\mathbf{x}) &= \mathbf{x} \cdot \boldsymbol{\varepsilon}^o + \mathbf{U}(\mathbf{x}; (\boldsymbol{\sigma}^{*A} - \bar{\boldsymbol{\sigma}}^{*A})), \\ \mathbf{t}(\mathbf{x}) &= \boldsymbol{\nu}(\mathbf{x}) \cdot (\mathbf{C} : \boldsymbol{\varepsilon}^o - \boldsymbol{\Gamma}^A(\mathbf{x}; \boldsymbol{\sigma}^{*A})). \end{aligned} \tag{9.5.27b}$$

Note that $\overline{\mathbf{C}}^{E'}_{\sigma\text{-}\varepsilon}$ is the overall elasticity tensor defined through the *average stress-strain relations*. Hence, inequality (9.5.20b) applies, i.e.,

$$\frac{1}{2}\, \boldsymbol{\varepsilon}^o : (\overline{\mathbf{C}}^{E'}_{\sigma\text{-}\varepsilon} - (\overline{\mathbf{D}}^{E'}_{E})^{-1}) : \boldsymbol{\varepsilon}^o \leq 0, \tag{9.5.28b}$$

where $\overline{\mathbf{D}}^{E'}_{E}$ is the overall compliance tensor which is defined by the *average complementary elastic strain energy* of the RVE under the special boundary data (9.5.27b). Since inequality (9.5.11b) applies to this $\overline{\mathbf{C}}^{E'}_{E}$,

$$\frac{1}{2}\, \boldsymbol{\varepsilon}^o : ((\overline{\mathbf{D}}^{E'}_{E})^{-1} - \overline{\mathbf{C}}^{E}) : \boldsymbol{\varepsilon}^o \leq 0, \tag{9.5.29b}$$

and in view of (9.5.28b), the following inequality holds:

$$\frac{1}{2}\, \boldsymbol{\varepsilon}^o : (\overline{\mathbf{C}}^{E'}_{\sigma\text{-}\varepsilon} - \overline{\mathbf{C}}^{E}) : \boldsymbol{\varepsilon}^o \leq 0, \tag{9.5.30b}$$

for any constant $\boldsymbol{\varepsilon}^o$. Hence, $\overline{\mathbf{C}}^{E'}_{\sigma\text{-}\varepsilon} - \overline{\mathbf{C}}^{E}$ is negative-semi-definite.

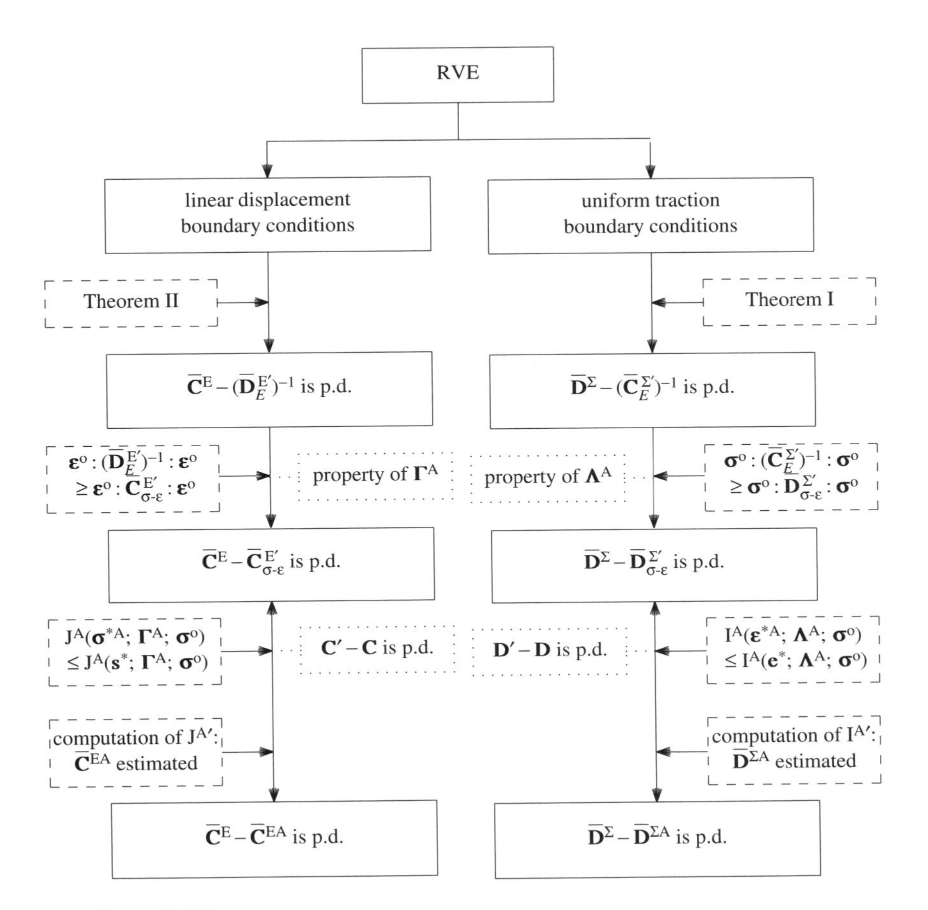

Figure 9.5.4

Flow chart of exact inequalities when homogenized V is regarded as part of an infinite domain with approximate operators $\boldsymbol{\Gamma}^{A}$ and $\boldsymbol{\Lambda}^{A}$; $\bar{\mathbf{D}}_{E}^{E'}$ is the overall compliance tensor when V is subjected to boundary conditions produced by $\boldsymbol{\Gamma}^{A}$ (symbolized by E′), and $\bar{\mathbf{C}}_{E}^{\Sigma'}$ is the overall elasticity tensor when V is subjected to boundary conditions produced by $\boldsymbol{\Lambda}^{A}$ (symbolized by Σ′)

Again, based on inequality (9.5.14b), $J^{A}(\mathbf{s}^{*}; \boldsymbol{\Gamma}^{A}; \boldsymbol{\varepsilon}^{o})$ is replaced by the computable function, $J^{A'}(\{\mathbf{s}^{*\alpha}\}; \boldsymbol{\varepsilon}^{o})$, for piecewise constant eigenstresses, $\{\mathbf{s}^{*\alpha}\}$. The optimal value of this function is $J^{A'}(\{\boldsymbol{\sigma}^{*\alpha}\}; \boldsymbol{\varepsilon}^{o})$, where $\{\boldsymbol{\sigma}^{*\alpha}\}$ is the solution of $\partial J^{A'}/\partial \mathbf{s}^{*\alpha} = \mathbf{0}$ ($\alpha = 1, 2, ..., n$). The optimal value $J^{A'}(\{\boldsymbol{\sigma}^{*\alpha}\}; \boldsymbol{\varepsilon}^{o})$ now becomes

$$J^{A'}(\{\boldsymbol{\sigma}^{*\alpha}\}; \boldsymbol{\varepsilon}^{o}) \equiv \frac{1}{2}\, \boldsymbol{\varepsilon}^{o} : (\mathbf{C} - \bar{\mathbf{C}}^{EA}) : \boldsymbol{\varepsilon}^{o}, \tag{9.5.31b}$$

where $\bar{\mathbf{C}}^{EA}$, defined from $J^{A'}(\{\boldsymbol{\sigma}^{*\alpha}\}; \boldsymbol{\varepsilon}^{o})$, is the final estimate of the overall elasticity tensor. Inequality (9.5.14b) now yields

$$\frac{1}{2}\,\boldsymbol{\varepsilon}^{\mathrm{o}} : (\overline{\mathbf{C}}^{\mathrm{EA}} - \overline{\mathbf{C}}^{\mathrm{E}'}_{\sigma\text{-}\varepsilon}) : \boldsymbol{\varepsilon}^{\mathrm{o}} \leq 0, \tag{9.5.32b}$$

if $\mathbf{C}' - \mathbf{C}$ is positive-definite. In view of (9.5.30b), a computable bound on the overall elasticity tensor $\overline{\mathbf{C}}^{\mathrm{E}}$ is obtained as follows:

$$\boldsymbol{\varepsilon}^{\mathrm{o}} : \overline{\mathbf{C}}^{\mathrm{E}} : \boldsymbol{\varepsilon}^{\mathrm{o}} \geq \boldsymbol{\varepsilon}^{\mathrm{o}} : \overline{\mathbf{C}}^{\mathrm{EA}} : \boldsymbol{\varepsilon}^{\mathrm{o}}, \tag{9.5.33b}$$

for any $\boldsymbol{\varepsilon}^{\mathrm{o}}$, if $\mathbf{C}' - \mathbf{C}$ is positive-definite. In the left half of Figure 9.5.4, these results are summarized in the form of a flow chart.

From (9.5.33a,b) the following exact relations between the energy functionals and their approximations are obtained:

$$\hat{\mathrm{I}}(\boldsymbol{\varepsilon}^*; \boldsymbol{\Lambda}^{\Sigma}; \boldsymbol{\sigma}^{\mathrm{o}}) \leq \mathrm{I}^{\mathrm{A}'}(\{\boldsymbol{\varepsilon}^{*\alpha}\}; \boldsymbol{\sigma}^{\mathrm{o}}), \tag{9.5.34a}$$

when $\mathbf{D}' - \mathbf{D}$ is positive-definite; and

$$\hat{\mathrm{J}}(\boldsymbol{\sigma}^*; \boldsymbol{\Gamma}^{\mathrm{E}}; \boldsymbol{\varepsilon}^{\mathrm{o}}) \leq \mathrm{J}^{\mathrm{A}'}(\{\boldsymbol{\varepsilon}^{*\alpha}\}; \boldsymbol{\varepsilon}^{\mathrm{o}}), \tag{9.5.34b}$$

when $\mathbf{C}' - \mathbf{C}$ is positive-definite.

From (9.5.33a,b), it is therefore possible to state the following rigorously established, *computable* bounds for the overall moduli of an arbitrary heterogeneous, linearly elastic solid:

$$\begin{aligned} &\overline{\mathbf{D}}^{\Sigma} - \overline{\mathbf{D}}^{\Sigma\mathrm{A}} \quad \text{is p.d.} \qquad \text{when } \mathbf{D}' - \mathbf{D} \text{ is p.d.,} \\ &\overline{\mathbf{C}}^{\mathrm{E}} - \overline{\mathbf{C}}^{\mathrm{EA}} \quad \text{is p.d.} \qquad \text{when } \mathbf{C}' - \mathbf{C} \text{ is p.d..} \end{aligned} \tag{9.5.35a,b}$$

These results are exact and provide lower and upper bounds for the overall moduli. Note that the considered boundary data for (9.5.35a) are *uniform tractions*, whereas the boundary data which lead to (9.5.35b) are *linear displacements*; these are signified by superscripts Σ and E, respectively. The approximation of replacing $\boldsymbol{\Lambda}$ by $\boldsymbol{\Lambda}^{\mathrm{A}}$ precludes the possibility of obtaining bounds under uniform tractions when $\mathbf{D}' - \mathbf{D}$ is negative-definite. Similarly, when $\boldsymbol{\Gamma}$ is replaced by $\boldsymbol{\Gamma}^{\mathrm{A}}$, then the possibility of obtaining bounds from linear displacement boundary data when $\mathbf{C}' - \mathbf{C}$ is negative-definite is precluded. In view of these observations, bounds which are obtained under these conditions should be regarded as *estimates* of the overall moduli rather than rigorous bounds. Only inequalities (9.5.35a,b) survive with certainty the errors introduced by approximating the Green function $\mathbf{G}(\mathbf{x}, \mathbf{y})$ by that of the infinite domain, in the manner discussed in Subsection 9.4.

Figure 9.5.4 summarizes the results of this subsection in the form of a flow chart.

9.5.5. Calculation of Bounds

As shown in Subsection 9.5.4, exact inequalities (9.5.35a,b) provide computable bounds for the overall tensors $\overline{\mathbf{D}}^{\Sigma}$ and $\overline{\mathbf{C}}^{\mathrm{E}}$, in terms of $\overline{\mathbf{D}}^{\Sigma\mathrm{A}}$ and $\overline{\mathbf{C}}^{\mathrm{EA}}$, respectively. Indeed, for statistically homogeneous and isotropic RVE's, substitution of (9.4.24c) into (9.4.24a) and (9.4.24d) into (9.4.24b), yields $\overline{\mathbf{D}}^{\Sigma\mathrm{A}}$ and $\overline{\mathbf{C}}^{\mathrm{EA}}$, as follows:

$$\overline{\mathbf{D}}^{\Sigma A} \equiv \mathbf{D} - \mathbf{D} : \{\mathbf{1}^{(4s)} - \overline{\mathbf{\Sigma}} : (\mathbf{P} - \mathbf{D})\}^{-1} : \overline{\mathbf{\Sigma}} : \mathbf{D}$$

$$\overline{\mathbf{C}}^{EA} = \mathbf{C} + \{\mathbf{1}^{(4s)} - \overline{\mathbf{\Sigma}} : \mathbf{P}\}^{-1} : \overline{\mathbf{\Sigma}}, \tag{9.5.36a,b}$$

where $\overline{\mathbf{\Sigma}}$ is given by (9.4.23b). After some tensor manipulation, $\overline{\mathbf{D}}^{\Sigma A}$ and $\overline{\mathbf{C}}^{EA}$ are expressed in terms of $\mathbf{P}$ which can be computed explicitly. In this manner, $\overline{\mathbf{D}}^{\Sigma A}$ becomes

$$\overline{\mathbf{D}}^{\Sigma A} = \mathbf{D} : \left\{ \mathbf{C} - \{\mathbf{1}^{(4s)} - \overline{\mathbf{\Sigma}} : (\mathbf{P} - \mathbf{D})\}^{-1} : \overline{\mathbf{\Sigma}} \right\} : \mathbf{D}$$

$$= \mathbf{D} : \left\{ \mathbf{C} - \overline{\mathbf{\Sigma}} : \{\mathbf{1}^{(4s)} - (\mathbf{P} - \mathbf{D}) : \overline{\mathbf{\Sigma}}\}^{-1} \right\} : \mathbf{D}$$

$$= \mathbf{D} : \left\{ \mathbf{C} : \{\mathbf{1}^{(4s)} - (\mathbf{P} - \mathbf{D}) : \overline{\mathbf{\Sigma}}\} - \overline{\mathbf{\Sigma}} \right\} : \{\mathbf{1}^{(4s)} - (\mathbf{P} - \mathbf{D}) : \overline{\mathbf{\Sigma}}\}^{-1} : \mathbf{D}$$

$$= \{\mathbf{1}^{(4s)} - \mathbf{P} : \overline{\mathbf{\Sigma}}\} : \{\mathbf{1}^{(4s)} - (\mathbf{P} - \mathbf{D}) : \overline{\mathbf{\Sigma}}\}^{-1} : \mathbf{D}, \tag{9.5.37a}$$

and $\overline{\mathbf{C}}^{EA}$ takes on the form

$$\overline{\mathbf{C}}^{EA} = \mathbf{C} + \overline{\mathbf{\Sigma}} : \{\mathbf{1}^{(4s)} - \mathbf{P} : \overline{\mathbf{\Sigma}}\}^{-1}$$

$$= \left\{ \mathbf{C} : \{\mathbf{1}^{(4s)} - \mathbf{P} : \overline{\mathbf{\Sigma}}\} + \overline{\mathbf{\Sigma}} \right\} : \{\mathbf{1}^{(4s)} - \mathbf{P} : \overline{\mathbf{\Sigma}}\}^{-1}$$

$$= \mathbf{C} : \{\mathbf{1}^{(4s)} - (\mathbf{P} - \mathbf{D}) : \overline{\mathbf{\Sigma}}\} : \{\mathbf{1}^{(4s)} - \mathbf{P} : \overline{\mathbf{\Sigma}}\}^{-1}. \tag{9.5.37b}$$

Since substitution of (9.4.23b) into $\overline{\mathbf{\Sigma}}$ in $\{\mathbf{1}^{(4s)} - (\mathbf{P} - \mathbf{D}) : \overline{\mathbf{\Sigma}}\}$ and $\{\mathbf{1}^{(4s)} - \mathbf{P} : \overline{\mathbf{\Sigma}}\}$ yields

$$\mathbf{1}^{(4s)} - (\mathbf{P} - \mathbf{D}) : \overline{\mathbf{\Sigma}} = \mathbf{D} : \left\{ \sum_{\alpha=0}^{n} f_\alpha \mathbf{C}^\alpha : \{\mathbf{1}^{(4s)} + \mathbf{P} : (\mathbf{C}^\alpha - \mathbf{C})\}^{-1} \right\},$$

$$\mathbf{1}^{(4s)} - \mathbf{P} : \overline{\mathbf{\Sigma}} = \sum_{\alpha=0}^{n} f_\alpha \{\mathbf{1}^{(4s)} + \mathbf{P} : (\mathbf{C}^\alpha - \mathbf{C})\}^{-1}, \tag{9.5.38a,b}$$

$\overline{\mathbf{D}}^{\Sigma A}$ and $\overline{\mathbf{C}}^{EA}$ are finally expressed in terms of $\mathbf{P}$, as

$$\overline{\mathbf{D}}^{\Sigma A} = \left\{ \sum_{\alpha=0}^{n} f_\alpha \{\mathbf{1}^{(4s)} + \mathbf{P} : (\mathbf{C}^\alpha - \mathbf{C})\}^{-1} \right\}$$

$$: \left\{ \sum_{\beta=0}^{n} f_\beta \mathbf{C}^\beta : \{\mathbf{1}^{(4s)} + \mathbf{P} : (\mathbf{C}^\beta - \mathbf{C})\}^{-1} \right\}^{-1},$$

$$\overline{\mathbf{C}}^{EA} = \left\{ \sum_{\alpha=0}^{n} f_\alpha \mathbf{C}^\alpha : \{\mathbf{1}^{(4s)} + \mathbf{P} : (\mathbf{C}^\alpha - \mathbf{C})\}^{-1} \right\}$$

$$: \left\{ \sum_{\beta=0}^{n} f_\beta \, \{ \mathbf{1}^{(4s)} + \mathbf{P} : (\mathbf{C}^\beta - \mathbf{C}) \}^{-1} \right\}^{-1}. \tag{9.5.39a,b}$$

Therefore, most remarkably, but as should be expected from their derivation, these bounding tensors are each other's inverse,

$$\overline{\mathbf{C}}^{EA} = (\overline{\mathbf{D}}^{\Sigma A})^{-1} \quad \text{or} \quad \overline{\mathbf{D}}^{\Sigma A} = (\overline{\mathbf{C}}^{EA})^{-1}. \tag{9.5.39c,d}$$

Inequalities (9.5.35a) and (9.5.35b) provide the upper bound on the overall compliance and elasticity tensors, $\overline{\mathbf{D}}^{\Sigma}$ and $\overline{\mathbf{C}}^{E}$, in such a manner that the tensors $\overline{\mathbf{D}}^{\Sigma} - \overline{\mathbf{D}}^{\Sigma A}$ and $\overline{\mathbf{C}}^{E} - \overline{\mathbf{C}}^{EA}$ are positive-definite, when $\mathbf{D}^\alpha - \mathbf{D}$ and $\mathbf{C}^\alpha - \mathbf{C}$ are positive-definite.[27]

The tensor $\mathbf{P}$ can be computed explicitly. Hence, explicit bounds on the overall elastic energies and tensors are obtained. Hashin and Shtrikman (1963) apply the Fourier transform to the stress and strain fields and estimate the values of the correlation tensors for an isotropic case. Walpole (1966b) generalizes the bounds for an anisotropic case. Willis (1977) and Kröner (1977) examine the general properties of the Hashin-Shtrikman bounds; for additional comments and references, see Walpole (1981), Willis (1981), and Mura (1987). Recently, Milton and Kohn (1988) have addressed certain mathematical aspects of these bounds; see also Kantor and Bergman (1984), Francfort and Murat (1986), and Torquato and Lado (1992). The generalized variational principles and the corresponding generalized bounds, as well as the accompanying inequalities which lead to the computable exact bounds (9.5.35a,b) are new observations. These observations are used in Subsection 9.6 to obtain universal bounds on two overall moduli, which apply to *any* heterogeneous elastic solid.

9.5.6. Alternative Formulation of Exact Inequalities: Direct Evaluation of Exact Bounds

A number of exact inequalities can be derived based on Theorems I and II of Subsection 2.5.6 and a proper choice of the reference elasticity and compliance tensors. To this end, the following results are considered: 1) generalization of Theorems I and II of Subsection 2.5.6; 2) consequences of negative-definiteness of $\mathbf{D}' - \mathbf{D}$ or $\mathbf{C}' - \mathbf{C}$; and 3) consequences of the vanishing of $< \mathbf{\Lambda}^A >$ or $< \mathbf{\Gamma}^A >$ when V is ellipsoidal. In this manner, exact computable upper and lower bounds for the overall moduli and compliances are established directly without invoking the variational principle;[28] i.e., *bounds on the overall elasticity and compliance tensors of a general heterogeneous elastic solid consisting of any number of phases of any shape, size, and distribution are obtained directly by proper choices of the reference elasticity and compliance tensors and by the use of Theorems* I *and* II. These results are then related to the corresponding bounds obtained in the preceding subsections.

---

[27] These are two mutually exclusive cases.

[28] For historical reasons, bounds developed up to this point are based on the Hashin-Shtrikman variational principle.

First, observe that the proof of Theorem II of Subsection 2.5.6 can be directly applied to establish the following result: for *any* disturbance stress field $\boldsymbol{\sigma}^{\mathrm{d}}$ which satisfies the equations of equilibrium and has zero volume average,

$$\frac{1}{2} < \boldsymbol{\sigma}^{\mathrm{E}} : \mathbf{D}' : \boldsymbol{\sigma}^{\mathrm{E}} > \leq \frac{1}{2} < (\boldsymbol{\sigma}^{\mathrm{o}} + \boldsymbol{\sigma}^{\mathrm{d}}) : \mathbf{D}' : (\boldsymbol{\sigma}^{\mathrm{o}} + \boldsymbol{\sigma}^{\mathrm{d}}) >, \tag{9.5.40a}$$

where $\boldsymbol{\sigma}^{\mathrm{E}}$ is the stress field of the linear displacement boundary data, with $< \boldsymbol{\sigma}^{\mathrm{E}} > = \boldsymbol{\sigma}^{\mathrm{o}}$. Similarly, the proof of Theorem I of Subsection 2.5.6 can be directly applied to show that: for *any* disturbance strain field $\boldsymbol{\varepsilon}^{\mathrm{d}}$ which is compatible with a disturbance displacement field (i.e., is the symmetric part of the gradient of a displacement field), and has zero volume average,

$$\frac{1}{2} < \boldsymbol{\varepsilon}^{\Sigma} : \mathbf{C}' : \boldsymbol{\varepsilon}^{\Sigma} > \leq \frac{1}{2} < (\boldsymbol{\varepsilon}^{\mathrm{o}} + \boldsymbol{\varepsilon}^{\mathrm{d}}) : \mathbf{C}' : (\boldsymbol{\varepsilon}^{\mathrm{o}} + \boldsymbol{\varepsilon}^{\mathrm{d}}) >, \tag{9.5.40b}$$

where $\boldsymbol{\varepsilon}^{\Sigma}$ is the strain field of the uniform traction boundary data, with $< \boldsymbol{\varepsilon}^{\Sigma} > = \boldsymbol{\varepsilon}^{\mathrm{o}}$.

Note that $\boldsymbol{\sigma}^{\mathrm{d}}$ in (9.5.40a) is not necessarily associated with a compatible strain field, and that $\boldsymbol{\varepsilon}^{\mathrm{d}}$ in (9.5.40b) is not necessarily associated with an equilibrating stress field. In Subsection 2.5.6, Theorems I and II are stated for cases when the considered fields are associated with compatible strains and equilibrating stresses, such that they can be actually produced in the RVE when subjected to suitable boundary conditions. Here, on the other hand, these theorems are generalized by (9.5.40a,b) to include cases when the strains are compatible but the stresses need not be in equilibrium, or cases when the stresses are in equilibrium but the strains need not be compatible; Willis (1992).

Next, *let* $\mathbf{D}' - \mathbf{D}$ be *negative-semi-definite*. Then, for *any arbitrary* pair of fields $\boldsymbol{\sigma}$ and $\mathbf{s}$, the following inequality holds:

$$\begin{aligned} 0 &\geq (\boldsymbol{\sigma} - \mathbf{s}) : (\mathbf{D}' - \mathbf{D}) : (\boldsymbol{\sigma} - \mathbf{s}) \\ &= \boldsymbol{\sigma} : (\mathbf{D}' - \mathbf{D}) : \boldsymbol{\sigma} - 2\,\boldsymbol{\sigma} : (\mathbf{D}' - \mathbf{D}) : \mathbf{s} + \mathbf{s} : (\mathbf{D}' - \mathbf{D}) : \mathbf{s} \\ &= \boldsymbol{\sigma} : \mathbf{D}' : \boldsymbol{\sigma} - \boldsymbol{\sigma} : \mathbf{D} : \boldsymbol{\sigma} - 2\,\mathbf{e}^* : \boldsymbol{\sigma} + \mathbf{e}^* : (\mathbf{D}' - \mathbf{D})^{-1} : \mathbf{e}^*, \end{aligned} \tag{9.5.41a}$$

where[29] $\mathbf{e}^* \equiv (\mathbf{D}' - \mathbf{D}) : \mathbf{s}$. Thus, taking the volume average over V, and dividing both sides by 2, obtain

$$\frac{1}{2} < \boldsymbol{\sigma} : \mathbf{D}' : \boldsymbol{\sigma} > \leq \frac{1}{2} < \boldsymbol{\sigma} : \mathbf{D} : \boldsymbol{\sigma} + 2\,\boldsymbol{\sigma} : \mathbf{e}^* - \mathbf{e}^* : (\mathbf{D}' - \mathbf{D})^{-1} : \mathbf{e}^* >. \tag{9.5.42a}$$

This inequality holds for *any pair* of $\boldsymbol{\sigma}$ and $\mathbf{e}^*$.

Similarly, *let* $\mathbf{C}' - \mathbf{C}$ *be negative-semi-definite*. Then, for *any arbitrary* pair of fields $\boldsymbol{\varepsilon}$ and $\mathbf{e}$,

$$0 \geq (\boldsymbol{\varepsilon} - \mathbf{e}) : (\mathbf{C}' - \mathbf{C}) : (\boldsymbol{\varepsilon} - \mathbf{e})$$

---

[29] Note that $\mathbf{s} = (\mathbf{D}' - \mathbf{D})^{-1} : \mathbf{e}^*$ may be viewed as the stress field associated with the homogenizing eigenstrain field $\mathbf{e}^*$; compare with (9.1.4c).

$$= \boldsymbol{\varepsilon} : \mathbf{C}' : \boldsymbol{\varepsilon} - \boldsymbol{\varepsilon} : \mathbf{C} : \boldsymbol{\varepsilon} - 2\,\mathbf{s}^* : \boldsymbol{\varepsilon} + \mathbf{s}^* : (\mathbf{C}' - \mathbf{C})^{-1} : \mathbf{s}^*, \tag{9.5.41b}$$

where[30] $\mathbf{s}^* \equiv (\mathbf{C}' - \mathbf{C}) : \mathbf{e}$. It now follows that

$$\frac{1}{2} < \boldsymbol{\varepsilon} : \mathbf{C}' : \boldsymbol{\varepsilon} > \leq \frac{1}{2} < \boldsymbol{\varepsilon} : \mathbf{C} : \boldsymbol{\varepsilon} + 2\,\mathbf{s}^* : \boldsymbol{\varepsilon} - \mathbf{s}^* : (\mathbf{C}' - \mathbf{C})^{-1} : \mathbf{s}^* >. \tag{9.5.42b}$$

This inequality holds for *any pair* of $\boldsymbol{\varepsilon}$ and $\mathbf{s}^*$.

Finally, consider a stress field which is produced by an arbitrary eigenstrain field, say, $\mathbf{e}^*$, i.e., $\boldsymbol{\sigma}^{\mathrm{A}} = -\boldsymbol{\Lambda}^{\mathrm{A}}(\mathbf{e}^*)$, and apply (9.5.40a) and (9.5.42a). By definition, $\boldsymbol{\sigma}^{\mathrm{A}}$ is in equilibrium, and has zero volume average when V is an *ellipsoid*. Thus, consider an ellipsoidal V, replace $\boldsymbol{\sigma}^{\mathrm{d}}$ by $\boldsymbol{\sigma}^{\mathrm{A}}$ in (9.5.40a), $\boldsymbol{\sigma}$ by $\boldsymbol{\sigma}^{\mathrm{o}} + \boldsymbol{\sigma}^{\mathrm{A}}$ in (9.5.42a), and obtain

$$\frac{1}{2} < \boldsymbol{\sigma}^{\mathrm{E}} : \mathbf{D}' : \boldsymbol{\sigma}^{\mathrm{E}} > \leq \frac{1}{2} < (\boldsymbol{\sigma}^{\mathrm{o}} + \boldsymbol{\sigma}^{\mathrm{A}}) : \mathbf{D}' : (\boldsymbol{\sigma}^{\mathrm{o}} + \boldsymbol{\sigma}^{\mathrm{A}}) >$$

$$\leq \frac{1}{2} < \{\, (\boldsymbol{\sigma}^{\mathrm{o}} + \boldsymbol{\sigma}^{\mathrm{A}}) : \mathbf{D} : (\boldsymbol{\sigma}^{\mathrm{o}} + \boldsymbol{\sigma}^{\mathrm{A}}) + 2\,(\boldsymbol{\sigma}^{\mathrm{o}} + \boldsymbol{\sigma}^{\mathrm{A}}) : \mathbf{e}^*$$

$$- \mathbf{e}^* : (\mathbf{D}' - \mathbf{D})^{-1} : \mathbf{e}^* \,\} >. \tag{9.5.43a}$$

Inequality (9.5.43a) holds for any $\mathbf{e}^*$, as long as $\mathbf{D}' - \mathbf{D}$ is negative-semi-definite and V is ellipsoidal. From $< \boldsymbol{\sigma}^{\mathrm{A}} > = \mathbf{0}$, the first two terms in the right side of (9.5.43a) become

$$< (\boldsymbol{\sigma}^{\mathrm{o}} + \boldsymbol{\sigma}^{\mathrm{A}}) : \mathbf{D} : (\boldsymbol{\sigma}^{\mathrm{o}} + \boldsymbol{\sigma}^{\mathrm{A}}) + 2\,(\boldsymbol{\sigma}^{\mathrm{o}} + \boldsymbol{\sigma}^{\mathrm{A}}) : \mathbf{e}^* >$$

$$= \boldsymbol{\sigma}^{\mathrm{o}} : \mathbf{D} : \boldsymbol{\sigma}^{\mathrm{o}} + < \boldsymbol{\sigma}^{\mathrm{A}} : \mathbf{D} : \boldsymbol{\sigma}^{\mathrm{A}} + \boldsymbol{\sigma}^{\mathrm{A}} : \mathbf{e}^* + (\boldsymbol{\sigma}^{\mathrm{A}} + 2\,\boldsymbol{\sigma}^{\mathrm{o}}) : \mathbf{e}^* >$$

$$= \boldsymbol{\sigma}^{\mathrm{o}} : \mathbf{D} : \boldsymbol{\sigma}^{\mathrm{o}} + < \mathrm{e}^* : (\boldsymbol{\Lambda}^{\mathrm{A}} : \mathbf{D} : \boldsymbol{\Lambda}^{\mathrm{A}} - \boldsymbol{\Lambda}^{\mathrm{A}}) : \mathbf{e}^* - \mathbf{e}^* : (\boldsymbol{\Lambda}^{\mathrm{A}}(\mathbf{e}^*) - 2\,\boldsymbol{\sigma}^{\mathrm{o}}) >. \tag{9.5.44a}$$

In this manner, inequality (9.5.43a) is rewritten as

$$\frac{1}{2} < \boldsymbol{\sigma}^{\mathrm{E}} : \mathbf{D}' : \boldsymbol{\sigma}^{\mathrm{E}} > \leq \frac{1}{2}\, \boldsymbol{\sigma}^{\mathrm{o}} : \mathbf{D} : \boldsymbol{\sigma}^{\mathrm{o}} - \left\{ \frac{1}{2} < \mathrm{e}^* : \{(\mathbf{D}' - \mathbf{D})^{-1} + \boldsymbol{\Lambda}^{\mathrm{A}}\} : \mathbf{e}^* > \right.$$

$$\left. - < \mathrm{e}^* > : \boldsymbol{\sigma}^{\mathrm{o}} \right\} + \frac{1}{2} < \mathrm{e}^* : (\boldsymbol{\Lambda}^{\mathrm{A}} : \mathbf{D} : \boldsymbol{\Lambda}^{\mathrm{A}} - \boldsymbol{\Lambda}^{\mathrm{A}}) : \mathbf{e}^* >. \tag{9.5.45a}$$

By definition, the left side of (9.5.45a) is given by $\boldsymbol{\sigma}^{\mathrm{o}} : \overline{\mathbf{D}}^{\mathrm{E}} : \boldsymbol{\sigma}^{\mathrm{o}} / 2$, and the terms in the curly brackets of the right side of (9.5.45a) equal functional $\mathrm{I}^{\mathrm{A}}(\mathbf{e}^*;\, \boldsymbol{\Lambda}^{\mathrm{A}};\, \boldsymbol{\sigma}^{\mathrm{o}})$. Since the last expression of (9.5.45a) is negative due to the property of $\boldsymbol{\Lambda}^{\mathrm{A}}$ when V is ellipsoidal, the following inequality is obtained:

$$\frac{1}{2}\, \boldsymbol{\sigma}^{\mathrm{o}} : (\overline{\mathbf{D}}^{\mathrm{E}} - \mathbf{D}) : \boldsymbol{\sigma}^{\mathrm{o}} \leq -\mathrm{I}^{\mathrm{A}}(\mathbf{e}^*;\, \boldsymbol{\Lambda}^{\mathrm{A}};\, \boldsymbol{\sigma}^{\mathrm{o}}), \tag{9.5.46a}$$

when $\mathbf{D}' - \mathbf{D}$ is negative-semi-definite. This inequality is *exact*. It is a direct

---

[30] Note that $\mathbf{e} = (\mathbf{C}' - \mathbf{C})^{-1} : \mathbf{s}^*$ may be viewed as the strain field associated with the homogenizing eigenstress field $\mathbf{s}^*$; compare with (9.1.12c).

result of Theorem II of Subsection 2.5.6, and the choice of the reference compliance tensor, **D**.

A similar exact inequality is obtained for J^A directly from (9.5.40b) and (9.5.42b). Since $\boldsymbol{\varepsilon}^A = -\boldsymbol{\Gamma}^A(\mathbf{s}^*)$ is compatible and has zero volume average when V is ellipsoidal, replace $\boldsymbol{\varepsilon}^d$ by $\boldsymbol{\varepsilon}^A$ in (9.5.40b), $\boldsymbol{\varepsilon}$ by $\boldsymbol{\varepsilon}^o + \boldsymbol{\varepsilon}^A$ in (9.5.42b), and obtain

$$\frac{1}{2} < \boldsymbol{\varepsilon}^\Sigma : \mathbf{C}' : \boldsymbol{\varepsilon}^\Sigma > \leq \frac{1}{2} < (\boldsymbol{\varepsilon}^o + \boldsymbol{\varepsilon}^A) : \mathbf{C}' : (\boldsymbol{\varepsilon}^o + \boldsymbol{\varepsilon}^A) >$$

$$\leq \frac{1}{2} < \{ (\boldsymbol{\varepsilon}^o + \boldsymbol{\varepsilon}^A) : \mathbf{C} : (\boldsymbol{\varepsilon}^o + \boldsymbol{\varepsilon}^A) + 2 (\boldsymbol{\varepsilon}^o + \boldsymbol{\varepsilon}^A) : \mathbf{s}^*$$

$$- \mathbf{s}^* : (\mathbf{C}' - \mathbf{C})^{-1} : \mathbf{s}^* \} >. \tag{9.5.43b}$$

Inequality (9.5.43b) holds for any $\mathbf{s}^*$, as long as $\mathbf{C}' - \mathbf{C}$ is negative-semi-definite and V is ellipsoidal. From $< \boldsymbol{\sigma}^A > = \mathbf{0}$, the first two terms in the right side of (9.5.43b) become

$$< (\boldsymbol{\varepsilon}^o + \boldsymbol{\varepsilon}^A) : \mathbf{C} : (\boldsymbol{\varepsilon}^o + \boldsymbol{\varepsilon}^A) + 2 (\boldsymbol{\varepsilon}^o + \boldsymbol{\varepsilon}^A) : \mathbf{s}^* >$$

$$= \boldsymbol{\varepsilon}^o : \mathbf{C} : \boldsymbol{\varepsilon}^o + < \mathbf{s}^* : (\boldsymbol{\Gamma}^A : \mathbf{C} : \boldsymbol{\Gamma}^A - \boldsymbol{\Gamma}^A) : \mathbf{s}^* - \mathbf{s}^* : (\boldsymbol{\Gamma}^A(\mathbf{s}^*) - 2\, \boldsymbol{\varepsilon}^o) >, \tag{9.5.44b}$$

and inequality (9.5.43b) is rewritten as

$$\frac{1}{2} < \boldsymbol{\varepsilon}^\Sigma : \bar{\mathbf{C}}' : \boldsymbol{\varepsilon}^\Sigma > \leq \frac{1}{2}\, \boldsymbol{\varepsilon}^o : \mathbf{C} : \boldsymbol{\varepsilon}^o - \left\{ \frac{1}{2} < \mathbf{s}^* : \{(\mathbf{C}' - \mathbf{C})^{-1} + \boldsymbol{\Gamma}^A\} : \mathbf{s}^* \right.$$

$$\left. - < \mathbf{s}^* > : \boldsymbol{\varepsilon}^o \right\} + \frac{1}{2} < \mathbf{s}^* : (\boldsymbol{\Gamma}^A : \mathbf{C} : \boldsymbol{\Gamma}^A - \boldsymbol{\Gamma}^A) : \mathbf{s}^* >. \tag{9.5.45b}$$

The left side of (9.5.45b) is equal to $\boldsymbol{\varepsilon}^o : \bar{\mathbf{C}}^\Sigma : \boldsymbol{\varepsilon}^o / 2$ by definition, and the terms in the curly brackets in the right side of (9.5.45b) equal $J^A(\mathbf{s}^*;\, \boldsymbol{\Gamma}^A : \boldsymbol{\varepsilon}^o)$. The last expression in (9.5.45b) is negative due to the properties of $\boldsymbol{\Gamma}^A$ when V is ellipsoidal. Hence, the following inequality is obtained:

$$\frac{1}{2}\, \boldsymbol{\varepsilon}^o : (\bar{\mathbf{C}}^\Sigma - \mathbf{C}) : \boldsymbol{\varepsilon}^o \leq - J^A(\mathbf{s}^*;\, \boldsymbol{\Gamma}^A;\, \boldsymbol{\varepsilon}^o), \tag{9.5.46b}$$

when $\mathbf{C}' - \mathbf{C}$ is negative-semi-definite.[31] Again, this inequality is *exact*, and follows directly from Theorem I of Subsection 2.5.6 and the choice of the reference elasticity tensor, **C**.

Functionals $I^A(\mathbf{e}^*;\, \boldsymbol{\Lambda}^A;\, \boldsymbol{\sigma}^o)$ and $J^A(\mathbf{s}^*;\, \boldsymbol{\Gamma}^A;\, \boldsymbol{\varepsilon}^o)$ in the right side of (9.5.46a,b) can be computed for piecewise constant eigenstrains and eigenstresses, respectively. As shown in Subsection 9.5.4, the values of the functionals I^A and J^A for such piecewise constant eigenstrains and eigenstresses are given by functions $I^{A\prime}$ and $J^{A\prime}$, and their optimal values are $I^{A\prime}(\{\boldsymbol{\varepsilon}^{*\alpha}\};\, \boldsymbol{\sigma}^o) = \boldsymbol{\sigma}^o : (\mathbf{D} - \bar{\mathbf{D}}^{\Sigma A}) : \boldsymbol{\sigma}^o / 2$ and $J^{A\prime}(\{\boldsymbol{\sigma}^{*\alpha}\};\, \boldsymbol{\varepsilon}^o) = \boldsymbol{\varepsilon}^o : (\mathbf{C} - \bar{\mathbf{C}}^{EA}) : \boldsymbol{\varepsilon}^o / 2$, where $\bar{\mathbf{D}}^{\Sigma A}$ and

[31] In terms of functionals defined in Subsection 9.5.2, inequalities (9.5.46a) and (9.5.46b) are written as $\hat{I}(\boldsymbol{\varepsilon}^{*E};\, \boldsymbol{\Lambda}^E;\, \boldsymbol{\sigma}^o) \geq I^A(\mathbf{e}^*;\, \boldsymbol{\Lambda}^A;\, \boldsymbol{\sigma}^o)$ and $\hat{J}(\boldsymbol{\sigma}^{*\Sigma};\, \boldsymbol{\Gamma}^\Sigma;\, \boldsymbol{\varepsilon}^o) \geq J^A(\mathbf{s}^*;\, \boldsymbol{\Gamma}^A;\, \boldsymbol{\varepsilon}^o)$.

$\bar{\mathbf{C}}^{EA}$ are estimated by (9.5.39a,b) of Subsection 9.5.5. Thus, (9.5.46a) and (9.5.46b) yield

$$\boldsymbol{\sigma}^o : \bar{\mathbf{D}}^E : \boldsymbol{\sigma}^o \leq \boldsymbol{\sigma}^o : \bar{\mathbf{D}}^{\Sigma A} : \boldsymbol{\sigma}^o,$$

$$\boldsymbol{\varepsilon}^o : \bar{\mathbf{C}}^{\Sigma} : \boldsymbol{\varepsilon}^o \leq \boldsymbol{\varepsilon}^o : \bar{\mathbf{C}}^{EA} : \boldsymbol{\varepsilon}^o, \tag{9.5.47a,b}$$

if $\mathbf{D}' - \mathbf{D}$ and $\mathbf{C}' - \mathbf{C}$ are negative-semi-definite,[32] respectively. Again, inequalities (9.5.47a,b) are *exact*.

Inequalities (9.5.47a) and (9.5.47b) lead to the following conclusions: for the overall moduli of an arbitrary heterogeneous, linearly elastic solid, containing any number of different phases of arbitrary distribution and shape,

$$\bar{\mathbf{D}}^E - \bar{\mathbf{D}}^{\Sigma A} \quad \text{is n.d.} \qquad \text{when } \mathbf{D}' - \mathbf{D} \text{ is n.d.,}$$

$$\bar{\mathbf{C}}^{\Sigma} - \bar{\mathbf{C}}^{EA} \quad \text{is n.d.} \qquad \text{when } \mathbf{C}' - \mathbf{C} \text{ is n.d.,} \tag{9.5.48a,b}$$

where $\bar{\mathbf{D}}^{\Sigma A}$ and $\bar{\mathbf{C}}^{EA}$ are *computable*. The negative-definiteness of $\mathbf{D}' - \mathbf{D}$ is equivalent to the positive-definiteness of $\mathbf{C}' - \mathbf{C}$. Hence, (9.5.48a,b) can be rewritten as

$$\bar{\mathbf{D}}^{\Sigma} - \bar{\mathbf{D}}^{EA} \quad \text{is p.d.} \qquad \text{when } \mathbf{D}' - \mathbf{D} \text{ is p.d.,}$$

$$\bar{\mathbf{C}}^{E} - \bar{\mathbf{C}}^{\Sigma A} \quad \text{is p.d.} \qquad \text{when } \mathbf{C}' - \mathbf{C} \text{ is p.d..} \tag{9.5.49a,b}$$

Since $\bar{\mathbf{D}}^{EA} = \bar{\mathbf{D}}^{\Sigma A}$ and $\bar{\mathbf{C}}^{\Sigma A} = \bar{\mathbf{C}}^{EA}$, (9.5.49a) and (9.5.49b) are equivalent[33] to (9.5.35a) and (9.5.35b).

9.6. UNIVERSAL BOUNDS FOR OVERALL MODULI

In the preceding subsections, exact inequalities are established between the average strain energy of an ellipsoidal RVE and approximate functionals I^A and J^A, by combining the Hashin-Shtrikman variational principle[34] with Theorems I and II of Subsection 2.5.6. An estimate of the exact bounds for the overall moduli of the RVE is obtained based on the assumption of statistical homogeneity and isotropy. However, *two exact bounds for certain combinations of the components of the overall elasticity tensor can be deduced rigorously, without any additional assumptions other than that the* RVE *be ellipsoidal.* These bounds apply to a general heterogeneous elastic solid consisting

---

[32] These are two mutually exclusive cases which are being examined simultaneously, where each case can be realized by a proper choice of the corresponding reference tensor.

[33] As shown in Subsection 9.5.5, $\bar{\mathbf{D}}^{\Sigma A}$ and $\bar{\mathbf{C}}^{EA}$ are computed from functionals I^A and J^A with the assumption of statistical homogeneity and isotropy, and are each other's inverse. If these functionals are computed in a different manner, $\bar{\mathbf{D}}^{\Sigma A}$ and $\bar{\mathbf{C}}^{EA}$ may not be each other's inverse, and bounds (9.5.35a,b) and (9.5.49a,b) then need not be identical.

[34] Or directly, by a proper choice of the reference elasticity tensor.

of any number of phases with any arbitrary shape and elasticity tensor.

In this subsection, the equivalence between I^A and J^A is first established. Then, using J^A only, exact inequalities which relate the average strain energy in the RVE to J^A are summarized. From the values of J^A for piecewise constant eigenstresses, two universal bounds are deduced, one for $\overline{C}_{iijj}$ and the other for $\overline{C}_{ijij}$. Then, from the equivalence relation, other bounds for $\overline{\mathbf{D}}$ are obtained. Finally, it is shown that when all phases of the RVE are isotropic, these universal bounds coincide with the conventional bounds which are obtained by assuming the statistical homogeneity and isotropy of the microstructure.

9.6.1. Equivalence of Two Approximate Functionals

In an infinite homogeneous domain B, fields produced by eigenstrains $\mathbf{e}^*$ through operator $\mathbf{\Lambda}^A$ coincide with those produced by eigenstresses $\mathbf{s}^*$ through operator $\mathbf{\Gamma}^A$, if $\mathbf{s}^* = -\mathbf{C} : \mathbf{e}^*$ or $\mathbf{e}^* = -\mathbf{D} : \mathbf{s}^*$. This is because both $\mathbf{\Lambda}^A$ and $\mathbf{\Gamma}^A$ are obtained from the Green function of the infinite B. Indeed, the following relations hold between $\mathbf{\Lambda}^A$ and $\mathbf{\Gamma}^A$, for any arbitrary $\mathbf{e}^*$ and $\mathbf{s}^*$:

$$\mathbf{\Lambda}^A(\mathbf{x};\, \mathbf{e}^{*d}) = \mathbf{C} : \{\mathbf{\Gamma}^A(\mathbf{x};\, -\mathbf{C} : \mathbf{e}^{*d}) + \mathbf{e}^{*d}(\mathbf{x})\},$$

$$\mathbf{\Gamma}^A(\mathbf{x};\, \mathbf{s}^{*d}) = \mathbf{D} : \{\mathbf{\Lambda}^A(\mathbf{x};\, -\mathbf{D} : \mathbf{s}^{*d}) + \mathbf{s}^{*d}(\mathbf{x})\}, \tag{9.6.1a,b}$$

where $\mathbf{e}^{*d} \equiv \mathbf{e}^* - < \mathbf{e}^* >$ and $\mathrm{s}^{*d} \equiv \mathbf{s}^* - < \mathrm{s}^* >$ are the disturbance eigenstrains and eigenstresses.

As shown in Subsection 9.3, the following identity holds:

$$(\mathbf{C}' - \mathbf{C})^{-1} = -\mathbf{D} - \mathbf{D} : (\mathbf{D}' - \mathbf{D})^{-1} : \mathbf{D}. \tag{9.6.2}$$

Hence, if functional J^A is written as

$$J^A(\mathbf{s}^*;\, \boldsymbol{\varepsilon}^o) = -\frac{1}{2} < (-\mathbf{D} : \mathbf{s}^*) : \{(\mathbf{D}' - \mathbf{D})^{-1} + \mathbf{\Lambda}^A\} : (-\mathbf{D} : \mathbf{s}^*) >$$

$$-\frac{1}{2} < (-\mathbf{D} : \mathbf{s}^*) : \{- < \mathrm{s}^* > - 2\,(\mathbf{C} : \boldsymbol{\varepsilon}^o)\} >, \tag{9.6.3}$$

then it follows that I^A and J^A are related through

$$I^A(-\mathbf{D} : \mathbf{s}^*;\, \mathbf{C} : \boldsymbol{\varepsilon}^o + < \mathrm{s}^* >) = -J^A(\mathbf{s}^*;\, \boldsymbol{\varepsilon}^o) + \frac{1}{2} < \mathrm{s}^* > : \mathbf{D} : < \mathrm{s}^* >, \tag{9.6.4a}$$

or

$$I^A(-\mathbf{D} : \mathbf{s}^*;\, \mathbf{C} : \boldsymbol{\varepsilon}^o) = -J^A(\mathbf{s}^*;\, \boldsymbol{\varepsilon}^o) - \frac{1}{2} < \mathrm{s}^* > : \mathbf{D} : < \mathrm{s}^* >. \tag{9.6.4b}$$

This is the equivalence relation between functionals I^A and J^A.

From equivalence relation (9.6.4a), it is seen that the eigenstress field which renders $J^A(\mathbf{s}^*;\, \boldsymbol{\varepsilon}^o)$ stationary, gives an eigenstress field which renders $I^A(\mathbf{e}^*;\, \boldsymbol{\sigma}^o)$ stationary, when the average eigenstress is fixed and $\mathbf{C} : \boldsymbol{\varepsilon}^o + < \mathrm{s}^* >$ is denoted[35] by $\boldsymbol{\sigma}^{o}$. Furthermore, if $\bar{\mathbf{C}}^{A}$ is determined from the stationary value of J^{A}, say, $J^{A}(\boldsymbol{\sigma}^{*}; \boldsymbol{\varepsilon}^{o})$, as

$$J^{A}(\boldsymbol{\sigma}^{*}; \boldsymbol{\varepsilon}^{o}) \equiv \frac{1}{2}\, \boldsymbol{\varepsilon}^{o} : (\mathbf{C} - \bar{\mathbf{C}}^{A}) : \boldsymbol{\varepsilon}^{o}, \tag{9.6.5a}$$

then the stationary value of the corresponding I^{A} is

$$I^{A}(\boldsymbol{\varepsilon}^{*}; \boldsymbol{\sigma}^{o}) = \frac{1}{2}\, \boldsymbol{\sigma}^{o} : (\mathbf{D} - (\bar{\mathbf{C}}^{A})^{-1}) : \boldsymbol{\sigma}^{o}, \tag{9.6.5b}$$

where $\boldsymbol{\varepsilon}^{*} = -\mathbf{D} : \boldsymbol{\sigma}^{*}$ and $\boldsymbol{\sigma}^{o} = \mathbf{C} : \boldsymbol{\varepsilon}^{o} + < \boldsymbol{\sigma}^{*} >$. The proof is straightforward: since $\boldsymbol{\sigma}^{*}$ renders J^{A} stationary, $J^{A}(\boldsymbol{\sigma}^{*}; \boldsymbol{\varepsilon}^{o})$ equals $-< \boldsymbol{\sigma}^{*} > : \boldsymbol{\varepsilon}^{o}/2$. Hence, $< \boldsymbol{\sigma}^{*} >$ is expressed in terms of $\bar{\mathbf{C}}^{A}$, as

$$< \boldsymbol{\sigma}^{*} > = (\bar{\mathbf{C}}^{A} - \mathbf{C}) : \boldsymbol{\varepsilon}^{o}. \tag{9.6.5c}$$

Substitution of (9.6.5a) and (9.6.5c) into (9.6.4a) yields

$$\begin{aligned} I^{A}(\boldsymbol{\varepsilon}^{*}; \boldsymbol{\sigma}^{o}) &= -J^{A}(\boldsymbol{\sigma}^{*}; \boldsymbol{\varepsilon}^{o}) + \frac{1}{2} < \boldsymbol{\sigma}^{*} > : \mathbf{D} : < \boldsymbol{\sigma}^{*} > \\ &= \frac{1}{2}\, \boldsymbol{\varepsilon}^{o} : \{\bar{\mathbf{C}}^{A} - \mathbf{C} + (\bar{\mathbf{C}}^{A} - \mathbf{C}) : \mathbf{D} : (\bar{\mathbf{C}}^{A} - \mathbf{C})\} : \boldsymbol{\varepsilon}^{o} \\ &= \frac{1}{2}\, (\bar{\mathbf{C}}^{A} : \boldsymbol{\varepsilon}^{o}) : (\mathbf{D} - (\bar{\mathbf{C}}^{A})^{-1}) : (\bar{\mathbf{C}}^{A} : \boldsymbol{\varepsilon}^{o}). \end{aligned} \tag{9.6.5d}$$

Since $\boldsymbol{\sigma}^{o} = \mathbf{C} : \boldsymbol{\varepsilon}^{o} + < \boldsymbol{\sigma}^{*} >$ equals $\bar{\mathbf{C}}^{A} : \boldsymbol{\varepsilon}^{o}$, (9.6.5d) yields (9.6.5b).

Therefore, functional $I^{A}(\mathbf{e}^{*}; \boldsymbol{\sigma}^{o})$ or its stationary value can be computed from functional $J^{A}(\mathbf{s}^{*}; \boldsymbol{\varepsilon}^{o})$, by replacing $\mathbf{e}^{*}$ with $-\mathbf{D} : \mathbf{s}^{*}$.[36] Hence, it suffices to consider either I^{A} or J^{A}. In this subsection, only J^{A} is studied, since operator $\boldsymbol{\Gamma}^{A}$ in J^{A} is of simpler form than operator $\boldsymbol{\Lambda}^{A}$ in I^{A}.

9.6.2. Summary of Exact Inequalities

The results obtained in Subsections 9.5.1 to 9.5.4 can be summarized as follows:

1) According to Theorem II of Subsection 2.5.6, for any arbitrary RVE, $\bar{\mathbf{C}}^{E} - (\bar{\mathbf{D}}^{G}_{E})^{-1}$ is positive-semi-definite, where $\bar{\mathbf{C}}^{E}$ and $\bar{\mathbf{D}}^{G}_{E}$ are the overall elasticity and compliance tensors, defined through the strain energy of the RVE, respectively for the linear displacement boundary data and for other general boundary data, both yielding the same average stresses.

---

[35] For notational convenience, only in this subsection, $\mathbf{C} : \boldsymbol{\varepsilon}^{o} + < \mathbf{s}^{*} >$ is designated by $\boldsymbol{\sigma}^{o}$. In all other subsections of Section 9, $\boldsymbol{\sigma}^{o}$ equals $\mathbf{C} : \boldsymbol{\varepsilon}^{o}$.

[36] For approximation, a certain class of eigenstress fields is usually used in computing functional J^{A}. If a stationary value of J^{A} is given by (9.6.5a) on such eigenstress fields, then a stationary value of I^{A} on the corresponding class of eigenstrain fields is *always* given by (9.6.5b). Indeed, for a piecewise constant distribution of eigenstresses and eigenstrains, the bounds produced by J^{A} and I^{A} (or $J^{A\prime}$ and $I^{A\prime}$), $\bar{\mathbf{C}}^{EA}$ and $\bar{\mathbf{D}}^{\Sigma A}$, are each other's inverse; see Subsection 9.5.5.

2) In an ellipsoidal RVE, the integral operator $\mathbf{\Gamma}^A$ satisfies $< \mathbf{\Gamma}^A > = \mathbf{0}$ for any eigenstress field. Moreover, operator $< : (\mathbf{\Gamma}^A : \mathbf{C} : \mathbf{\Gamma}^A - \mathbf{\Gamma}^A) : >$ is negative-semi-definite. It follows that $(\overline{\mathbf{D}}^G_E)^{-1} - \overline{\mathbf{C}}^G_{\sigma\text{-}\varepsilon}$ is positive-semi-definite, where $\overline{\mathbf{C}}^G_{\sigma\text{-}\varepsilon}$ is the overall elasticity tensor defined through the average stress-strain relations.

3) Since $\mathbf{\Gamma}^A$ is positive-definite, when $\mathbf{C}' - \mathbf{C}$ is positive-semi-definite, then, it follows that $J^A(\boldsymbol{\sigma}^*; \boldsymbol{\varepsilon}^o) = \frac{1}{2}\, \boldsymbol{\varepsilon}^o : (\mathbf{C} - \overline{\mathbf{C}}^G_{\sigma\text{-}\varepsilon}) : \boldsymbol{\varepsilon}^o$ is the minimum of $J^A(\mathbf{s}^*; \boldsymbol{\varepsilon}^o)$, where $\boldsymbol{\sigma}^*$ is the eigenstress field that minimizes J^A and hence homogenizes the RVE when it is subjected to *some* boundary conditions which cannot be prescribed *a priori*.

From 1) to 3), the following inequality holds for any $\mathbf{s}^*$ and any $\boldsymbol{\varepsilon}^o$:

$$\frac{1}{2}\, \boldsymbol{\varepsilon}^o : (\mathbf{C} - \overline{\mathbf{C}}^E) : \boldsymbol{\varepsilon}^o \leq J^A(\mathbf{s}^*; \boldsymbol{\varepsilon}^o) \qquad \text{if } \mathbf{C}' - \mathbf{C} \text{ is p.d..} \tag{9.6.6}$$

Besides the positive-definiteness of $\mathbf{C}' - \mathbf{C}$, a sufficient condition for the validity of (9.6.6) is that the RVE be ellipsoidal.

In the following subsection, functional $J^A(\mathbf{s}^*; \boldsymbol{\varepsilon}^o)$ is computed without any special assumptions other than the overall shape of the RVE, which must be ellipsoidal. The key to the computation is the special properties of operator $\mathbf{\Gamma}^A$. The results do not depend on the aspect ratios or the orientation of the ellipsoidal RVE, i.e., the same universal expressions for the bounds are obtained for any ellipsoidal RVE, or for that matter, for any heterogeneous elastic solid (with any heterogeneities) of overall ellipsoidal shape.

9.6.3. Universal Bounds for Overall Moduli of Ellipsoidal RVE (1)

Now, consider an ellipsoidal RVE consisting of $n+1$ distinct subregions, Ω_α, with elasticity $\mathbf{C}^\alpha$ and volume fraction f_α ($\alpha = 0, ..., n$), where the subregion corresponding to the matrix is denoted by Ω_0, with elasticity $\mathbf{C}^0$. To homogenize this RVE, consider a subregion, V′, in an infinite homogeneous solid, B, with the same geometry as that of the RVE. Let B be isotropic, with elasticity tensor $\mathbf{C} = 2\mu\nu/(1-2\nu)\mathbf{1}^{(2)} \otimes \mathbf{1}^{(2)} + 2\mu\mathbf{1}^{(4s)}$, such that $\mathbf{C}^\alpha - \mathbf{C}$ is positive-definite for $\alpha = 0, ..., n$. First, distribute piecewise constant eigenstresses, $\mathbf{s}^{*\alpha}$, in each Ω_α within V′, and then distribute the average eigenstress, $\overline{\mathbf{s}}^* \equiv \sum_{\alpha=0}^{n} f_\alpha \mathbf{s}^{*\alpha}$, uniformly in B − V′, such that the disturbance strain and stress fields in V′ are due only to the disturbance eigenstress field defined by

$$\mathbf{s}^{*d}(\mathbf{x}) = \sum_{\alpha=0}^{n} H(\mathbf{x}; \Omega_\alpha)\,(\mathbf{s}^{*\alpha} - \overline{\mathbf{s}}^*), \tag{9.6.7}$$

where $H(\mathbf{x}; \Omega_\alpha) = 1$ for $\mathbf{x}$ in Ω_α, and 0 otherwise. Finally, consider uniform strains $\boldsymbol{\varepsilon}^o$ prescribed in B.

Since V′ is ellipsoidal and $\overline{\mathbf{s}}^*$ is distributed throughout B, $< \mathbf{s}^* : \mathbf{\Gamma}^A : \mathbf{s}^* >$ equals $< \mathbf{s}^{*d} : \mathbf{\Gamma}^A : \mathbf{s}^{*d} >$, and functional J^A is given by

$$J^A(\mathbf{s}^*; \boldsymbol{\varepsilon}^o) = \frac{1}{2} < \mathbf{s}^* : (\mathbf{C}' - \mathbf{C})^{-1} : \mathbf{s}^* > - < \boldsymbol{\varepsilon}^o : \mathbf{s}^* >$$

$$+ \frac{1}{2} < \mathbf{s}^{*d} : \boldsymbol{\Gamma}^A : \mathbf{s}^{*d} >. \tag{9.6.8a}$$

For piecewise constant eigenstresses prescribed in V′, functional J^A is given by function $J^{A\prime}$ which, in view of (9.6.8a), becomes

$$J^{A\prime}(\{\mathbf{s}^{*\alpha}\}; \boldsymbol{\varepsilon}^o) = \frac{1}{2} \sum_{\alpha=0}^{n} f_\alpha \, \mathbf{s}^{*\alpha} : \{(\mathbf{C}^\alpha - \mathbf{C})^{-1} : \mathbf{s}^{*\alpha} - 2\,\boldsymbol{\varepsilon}^o\}$$

$$+ \frac{1}{2} \sum_{\alpha=0}^{n} \sum_{\beta=0}^{n} < \{H(\Omega_\alpha)\,(\mathbf{s}^{*\alpha} - \bar{\mathbf{s}}^*)\} : \boldsymbol{\Gamma}^A : \{H(\Omega_\beta)\,(\mathbf{s}^{*\beta} - \bar{\mathbf{s}}^*)\} >. \tag{9.6.8b}$$

To compute $J^{A\prime}$, integral $< \{H(\Omega_\alpha)\,(\mathbf{s}^{*\alpha} - \bar{\mathbf{s}}^*)\} : \boldsymbol{\Gamma}^A : \{H(\Omega_\beta)\,(\mathbf{s}^{*\beta} - \bar{\mathbf{s}}^*)\} >$ needs to be evaluated.

To this end, consider the Fourier transform of operator $\boldsymbol{\Gamma}^A$, given by (9.5.21). Since $\mathbf{C}$ is isotropic, $F\boldsymbol{\Gamma}^A$ defined by (9.5.21c) becomes

$$F\boldsymbol{\Gamma}^A(\boldsymbol{\xi}) = \frac{1}{\mu} \{-\frac{1}{2(1-\nu)} \bar{\boldsymbol{\xi}} \otimes \bar{\boldsymbol{\xi}} \otimes \bar{\boldsymbol{\xi}} \otimes \bar{\boldsymbol{\xi}} + sym\,(\bar{\boldsymbol{\xi}} \otimes \mathbf{1}^{(2)} \otimes \bar{\boldsymbol{\xi}})\}, \tag{9.6.9a}$$

where $\bar{\boldsymbol{\xi}} \equiv \boldsymbol{\xi} / |\boldsymbol{\xi}|$. From $\bar{\xi}_i \bar{\xi}_i = 1$, the components of $F\Gamma^A_{ijkl}$ satisfy

$$F\Gamma^A_{iijj} = \frac{1}{\mu} \frac{1-2\nu}{2(1-\nu)},$$

$$F\Gamma^A_{ijij} = \frac{1}{\mu} \frac{3-4\nu}{2(1-\nu)}, \tag{9.6.9b,c}$$

and hence the inverse Fourier transforms of $F\Gamma^A_{iijj}$ and $F\Gamma^A_{ijij}$ satisfy

$$\frac{1}{(2\pi)^3} \int_{-\infty}^{\infty} F\Gamma^A_{iijj}(\boldsymbol{\xi}) \exp(\iota \boldsymbol{\xi} \cdot \mathbf{x})\, dV_\xi = \frac{1}{\mu} \frac{1-2\nu}{2(1-\nu)} \delta(\mathbf{x}),$$

$$\frac{1}{(2\pi)^3} \int_{-\infty}^{\infty} F\Gamma^A_{ijij}(\boldsymbol{\xi}) \exp(\iota \boldsymbol{\xi} \cdot \mathbf{x})\, dV_\xi = \frac{1}{\mu} \frac{3-4\nu}{2(1-\nu)} \delta(\mathbf{x}), \tag{9.6.10a,b}$$

where $\delta(\mathbf{x})$ is the delta function at the origin.

From (9.6.10a,b), the following *exact* relations are obtained for $< H(\Omega_\alpha)\, \Gamma^A_{ijkl}\, H(\Omega_\beta) >$:

$$< H(\Omega_\alpha)\, \Gamma^A_{iijj}\, H(\Omega_\beta) > = \begin{cases} f_\alpha \dfrac{1}{\mu} \dfrac{1-2\nu}{2(1-\nu)} & \text{if } \alpha = \beta \\ 0 & \text{otherwise,} \end{cases} \tag{9.6.11a}$$

and

$$< H(\Omega_\alpha)\, \Gamma^A_{ijij}\, H(\Omega_\beta) > = \begin{cases} f_\alpha \dfrac{1}{\mu} \dfrac{3-4\nu}{2(1-\nu)} & \text{if } \alpha = \beta \\ 0 & \text{otherwise.} \end{cases} \tag{9.6.11b}$$

Since the isotropic tensor **P** defined by (9.4.13) satisfies $P_{iijj} = (1-2\nu)/2\mu(1-\nu)$ and $P_{ijij} = (3-4\nu)/2\mu(1-\nu)$, (9.6.11a,b) are rewritten as

$$< H(\Omega_\alpha)\, \Gamma^A_{iijj}\, H(\Omega_\beta) > = \begin{cases} f_\alpha\, P_{iijj} & \text{if } \alpha = \beta \\ 0 & \text{otherwise,} \end{cases}$$

$$< H(\Omega_\alpha)\, \Gamma^A_{ijij}\, H(\Omega_\beta) > = \begin{cases} f_\alpha\, P_{ijij} & \text{if } \alpha = \beta \\ 0 & \text{otherwise.} \end{cases} \tag{9.6.11c,d}$$

Note that *exact* expressions (9.6.11a,b) and (9.6.11c,d) *hold for any* Ω_α *and* Ω_β *of arbitrary shape, orientation, and relative location.*

Taking advantage of (9.6.11), function $J^{A\prime}$ given by (9.6.8) can be computed exactly by choosing special forms for $\boldsymbol{\varepsilon}^o$ and $\mathbf{s}^{*\alpha}$'s. First, suppose that the uniform overall strain is dilatational, $\boldsymbol{\varepsilon}^o = \varepsilon^o\, \mathbf{1}^{(2)}$. Then, setting $\mathbf{s}^{*\alpha} = s^{*\alpha}\, \mathbf{1}^{(2)}$, and using (9.6.11c), compute $J^{A\prime}$ as

$$J^{A\prime}(\{s^{*\alpha}\, \mathbf{1}^{(2)}\};\, \varepsilon^o\, \mathbf{1}^{(2)})$$

$$= \frac{1}{2} \sum_{\alpha=0}^{n} f_\alpha\, s^{*\alpha}\, (\mathbf{C}^\alpha - \mathbf{C})^{-1}_{iijj}\, s^{*\alpha} - 3\,\bar{s}^*\, \varepsilon^o + \frac{1}{2} \sum_{\alpha=0}^{n} f_\alpha\, (s^{*\alpha} - \bar{s}^*)\, P_{iijj}\, (s^{*\alpha} - \bar{s}^*)$$

$$= \frac{1}{2} \sum_{\alpha=0}^{n} f_\alpha\, s^{*\alpha}\, \{(\mathbf{C}^\alpha - \mathbf{C})^{-1}_{iijj} + P_{iijj}\}\, s^{*\alpha} - \frac{1}{2}\, \bar{s}^*\, P_{iijj}\, \bar{s}^* - 3\,\bar{s}^*\, \varepsilon^o, \tag{9.6.12a}$$

where $\bar{s}^* \equiv \sum_{\alpha=0}^{n} f_\alpha\, s^{*\alpha}$. This computation of $J^{A\prime}$ does not involve any assumptions except that V' is ellipsoidal. It applies to any RVE, consisting of any arbitrary multi-phase microstructure.

Next, consider three cases of overall biaxial shearing with nonzero strain components, $\varepsilon^o(\mathbf{e}_1 \otimes \mathbf{e}_1 - \mathbf{e}_2 \otimes \mathbf{e}_2)$, $\varepsilon^o(\mathbf{e}_2 \otimes \mathbf{e}_2 - \mathbf{e}_3 \otimes \mathbf{e}_3)$, and $\varepsilon^o(\mathbf{e}_3 \otimes \mathbf{e}_3 - \mathbf{e}_1 \otimes \mathbf{e}_1)$, and three cases of overall pure shearing with nonzero strain components, $\varepsilon^o(\mathbf{e}_2 \otimes \mathbf{e}_3 + \mathbf{e}_3 \otimes \mathbf{e}_2)$, $\varepsilon^o(\mathbf{e}_3 \otimes \mathbf{e}_1 + \mathbf{e}_1 \otimes \mathbf{e}_3)$, and $\varepsilon^o(\mathbf{e}_1 \otimes \mathbf{e}_2 + \mathbf{e}_2 \otimes \mathbf{e}_1)$. For simplicity, denote the tensor products of the unit base vectors associated with these six cases by $\mathbf{s}_i$ ($i = 1, 2, ..., 6$), i.e., $\mathbf{s}_1 = \mathbf{e}_1 \otimes \mathbf{e}_1 - \mathbf{e}_2 \otimes \mathbf{e}_2$, ..., $\mathbf{s}_4 = \mathbf{e}_2 \otimes \mathbf{e}_3 + \mathbf{e}_3 \otimes \mathbf{e}_2$, ... For each $\boldsymbol{\varepsilon}^o = \varepsilon^o \mathbf{s}_i$, set $\mathbf{s}^{*\alpha} = s^{*\alpha} \mathbf{s}_i$, and write the corresponding energy function $J^{A\prime}$ as $J^{A\prime}(\{s^{*\alpha}\mathbf{s}_i\};\, \varepsilon^o \mathbf{s}_i)$. Using (9.6.11c) and (9.6.11d), compute the following function:

$$\frac{1}{3}\left[\, J^{A\prime}(\{s^{*\alpha}\mathbf{s}_1\};\, \varepsilon^o\mathbf{s}_1) + J^{A\prime}(\{s^{*\alpha}\mathbf{s}_2\};\, \varepsilon^o\mathbf{s}_2) + J^{A\prime}(\{s^{*\alpha}\mathbf{s}_3\};\, \varepsilon^o\mathbf{s}_3)\, \right]$$

$$+ \frac{1}{2}\left[\, J^{A\prime}(\{s^{*\alpha}\mathbf{s}_4\};\, \varepsilon^o\mathbf{s}_4) + J^{A\prime}(\{s^{*\alpha}\mathbf{s}_5\};\, \varepsilon^o\mathbf{s}_5) + J^{A\prime}(\{s^{*\alpha}\mathbf{s}_6\};\, \varepsilon^o\mathbf{s}_6)\, \right]$$

$$= \frac{1}{2} \sum_{\alpha=0}^{n} f_\alpha \, s^{*\alpha} \{(\mathbf{C}^\alpha - \mathbf{C})^{-1}_{ijij} - \frac{1}{3}(\mathbf{C}^\alpha - \mathbf{C})^{-1}_{iijj}\} \, s^{*\alpha} - 5\,\bar{s}^*\,\varepsilon^o$$

$$+ \frac{1}{2} \sum_{\alpha=0}^{n} f_\alpha \, (s^{*\alpha} - \bar{s}^*) \{P_{ijij} - \frac{1}{3}P_{iijj}\} (s^{*\alpha} - \bar{s}^*)$$

$$= \frac{1}{2} \sum_{\alpha=0}^{n} f_\alpha \, s^{*\alpha} \{(\mathbf{C}^\alpha - \mathbf{C})^{-1}_{ijij} - \frac{1}{3}(\mathbf{C}^\alpha - \mathbf{C})^{-1}_{iijj} + P_{ijij} - \frac{1}{3}P_{iijj}\} \, s^{*\alpha}$$

$$- \frac{1}{2} \bar{s}^* \{P_{ijij} - \frac{1}{3}P_{iijj}\} \bar{s}^* - 5\,\bar{s}^*\,\varepsilon^o. \tag{9.6.12b}$$

Again, this computation of $J^{A\prime}$ does not involve any assumptions except that V' is ellipsoidal. The result applies to any RVE consisting of any arbitrary microstructure.

The right sides of (9.6.12a) and (9.6.12b) can be rewritten in a unified manner, as follows:

$$N_I \left\{ \frac{1}{2} \sum_{\alpha=0}^{n} f_\alpha \, s^{*\alpha} \{(\mathbf{C}^\alpha - \mathbf{C})^{-1}_I + P_I\} \, s^{*\alpha} - \frac{1}{2} \, \bar{s}^* P_I \bar{s}^* - \bar{s}^* \, \varepsilon^o \right\} \qquad (I = 1, 2),$$

where $N_1 = 3$ and $N_2 = 5$; and where $P_1 = P_{iijj}/N_1$ and $P_2 = (P_{ijij} - P_{iijj}/3)/N_2$, with similar expressions for $(\mathbf{C}^\alpha - \mathbf{C})^{-1}_I$ or any other fourth-order tensor. The above quadratic form (with linear terms) can be optimized with respect to $\{s^{*\alpha}\}$. Let $\{\sigma^{*\alpha}\}$ be the eigenstresses that render the corresponding functions stationary. Then, the corresponding optimal values of $J^{A\prime}$ are given by

$$J^{A\prime}(\{\sigma^{*\alpha}\,\mathbf{1}^{(2)}\};\,\varepsilon^o\,\mathbf{1}^{(2)}) = \frac{1}{2}\,N_1\,\varepsilon^o\,(C_1 - \bar{C}_1)\,\varepsilon^o,$$

$$\frac{1}{3}\left[J^{A\prime}(\{\sigma^{*\alpha}\,\mathbf{s}_1\};\,\varepsilon^o\,\mathbf{s}_1) + \dots \right] + \frac{1}{2}\left[J^{A\prime}(\{\sigma^{*\alpha}\,\mathbf{s}_4\};\,\varepsilon^o\,\mathbf{s}_4) + \dots \right]$$

$$= \frac{1}{2}\,N_2\,\varepsilon^o\,(C_2 - \bar{C}_2)\,\varepsilon^o, \tag{9.6.13a,b}$$

where

$$\bar{C}_I \equiv \sum_{\alpha=0}^{n} f_\alpha \left[C_I + \frac{1}{(\mathbf{C}^\alpha - \mathbf{C})^{-1}_I} \right] \left[1 + \frac{P_I}{(\mathbf{C}^\alpha - \mathbf{C})^{-1}_I} \right]^{-1}$$

$$\times \left\{ \sum_{\beta=0}^{n} f_\beta \left[1 + \frac{P_I}{(\mathbf{C}^\beta - \mathbf{C})^{-1}_I} \right]^{-1} \right\}^{-1}, \tag{9.6.13c}$$

for $I = 1, 2$. In view of (9.6.6), the following bounds for $\bar{\mathbf{C}}^E$ are obtained:

$$\bar{C}^E_I \geq \bar{C}_I \qquad (I = 1, 2) \qquad \text{if } \mathbf{C}' - \mathbf{C} \text{ is p.d.}. \tag{9.6.14a,b}$$

While the results obtained in Subsection 9.5 are only valid for the case when statistical homogeneity and isotropy exist, inequalities (9.6.14a) and (9.6.14b), corresponding to $I = 1$ and $I = 2$, respectively, hold for any microstructure of the RVE. Hence, these are indeed *universal bounds*. In fact, these bounds are valid for any finite elastic composite with overall ellipsoidal shape, independently of the shape, distribution, and elasticity of its multi-phase inclusions.

9.6.4. Universal Bounds for Overall Moduli of Ellipsoidal RVE (2)

Exact inequalities (9.6.14a,b), obtained in Subsection 9.6.3, are for the overall elasticity tensor, $\overline{\mathbf{C}}^{\mathrm{E}}$, of an ellipsoidal RVE under linear displacement boundary conditions. In essentially the same manner, similar exact inequalities are obtained for the overall compliance tensor, $\overline{\mathbf{D}}^{\Sigma}$, of the same ellipsoidal RVE under uniform traction boundary conditions. Indeed, from the exact inequality

$$\frac{1}{2}\,\boldsymbol{\sigma}^{\mathrm{o}} : (\mathbf{D} - \overline{\mathbf{D}}^{\Sigma}) : \boldsymbol{\sigma}^{\mathrm{o}} \leq \mathrm{I}^{\mathrm{A}}(\mathbf{e}^{*};\, \boldsymbol{\sigma}^{\mathrm{o}}) \qquad \text{if } \mathbf{D}' - \mathbf{D} \text{ is p.d.,} \tag{9.6.15}$$

inequalities similar to (9.6.14a,b) are deduced by computing the right side of (9.6.15) for suitable piecewise constant eigenstrains. In this computation, the properties of operator $\boldsymbol{\Lambda}^{\mathrm{A}}$, similar to (9.6.11) of operator $\boldsymbol{\Gamma}^{\mathrm{A}}$, are used.

Without considering the properties of $\boldsymbol{\Lambda}^{\mathrm{A}}$, the right side of (9.6.15) can be directly computed from the results obtained in Subsection 9.6.3, if the equivalence relation established in Subsection 9.6.1 is applied. Indeed, equivalence relations similar to (9.6.4a) hold between functions $\mathrm{J}^{\mathrm{A}\prime}$ and $\mathrm{I}^{\mathrm{A}\prime}$, and it can be easily shown that if $\{\boldsymbol{\sigma}^{*\alpha}\}$ renders $\mathrm{I}^{\mathrm{A}\prime}(\{\mathrm{s}^{*\alpha}\};\, \boldsymbol{\varepsilon}^{\mathrm{o}})$ stationary, then, $\{\boldsymbol{\varepsilon}^{*\alpha}\} = \{-\mathbf{D} : \boldsymbol{\sigma}^{*\alpha}\}$ renders $\mathrm{J}^{\mathrm{A}\prime}(\{\mathbf{e}^{*\alpha}\};\, \boldsymbol{\sigma}^{\mathrm{o}})$ stationary, when $\boldsymbol{\sigma}^{\mathrm{o}}$ is set to be $\mathbf{C} : \boldsymbol{\varepsilon}^{\mathrm{o}} + \overline{\boldsymbol{\sigma}}^{*}$. Furthermore, if the stationary value $\mathrm{J}^{\mathrm{A}\prime}(\{\boldsymbol{\sigma}^{*\alpha}\};\, \boldsymbol{\varepsilon}^{\mathrm{o}})$ is expressed in terms of $\overline{\mathbf{C}}^{\mathrm{A}}$, as[37]

$$\mathrm{J}^{\mathrm{A}\prime}(\{\boldsymbol{\sigma}^{*\alpha}\};\, \boldsymbol{\varepsilon}^{\mathrm{o}}) = \frac{1}{2}\,\boldsymbol{\varepsilon}^{\mathrm{o}} : (\mathbf{C} - \overline{\mathbf{C}}^{\mathrm{A}}) : \boldsymbol{\varepsilon}^{\mathrm{o}}, \tag{9.6.16a}$$

then, the stationary value $\mathrm{I}^{\mathrm{A}\prime}(\{\boldsymbol{\varepsilon}^{*\alpha}\};\, \boldsymbol{\sigma}^{\mathrm{o}})$ is given by

$$\mathrm{I}^{\mathrm{A}\prime}(\{\boldsymbol{\varepsilon}^{*\alpha}\};\, \boldsymbol{\sigma}^{\mathrm{o}}) = \frac{1}{2}\,\boldsymbol{\sigma}^{\mathrm{o}} : (\mathbf{D} - (\overline{\mathbf{C}}^{\mathrm{A}})^{-1}) : \boldsymbol{\sigma}^{\mathrm{o}}. \tag{9.6.16b}$$

Taking advantage of (9.6.16), consider the case when the applied uniform strain is dilatational, and assume that the homogenizing eigenstress is also dilatational, i.e., $\boldsymbol{\varepsilon}^{\mathrm{o}} = \varepsilon^{\mathrm{o}}\,\mathbf{1}^{(2)}$ and $\mathbf{s}^{*\alpha} = \mathrm{s}^{*\alpha}\,\mathbf{1}^{(2)}$. The stationary value of $\mathrm{J}^{\mathrm{A}\prime}$ for these uniform strain and eigenstress fields is given by (9.6.13a). It is rewritten as

$$\mathrm{J}^{\mathrm{A}\prime}(\{\sigma^{*\alpha}\,\mathbf{1}^{(2)}\};\, \varepsilon^{\mathrm{o}}\,\mathbf{1}^{(2)}) = \frac{1}{2}\,(\varepsilon^{\mathrm{o}}\,\mathbf{1}^{(2)}) : (\mathbf{C} - \overline{C}_1\,\mathbf{E}^1) : (\varepsilon^{\mathrm{o}}\,\mathbf{1}^{(2)}), \tag{9.6.17a}$$

where $\mathbf{E}^1 \equiv \mathbf{1}^{(2)} \otimes \mathbf{1}^{(2)}/3$; see (8.1.1a). Hence, it follows from (9.6.16) that the stationary value of $\mathrm{I}^{\mathrm{A}\prime}(\{\mathrm{e}^{*\alpha}\,\mathbf{1}\};\, \sigma^{\mathrm{o}}\,\mathbf{1})$ is

$$\mathrm{I}^{\mathrm{A}\prime}(\{\varepsilon^{*\alpha}\,\mathbf{1}\};\, \sigma^{\mathrm{o}}\,\mathbf{1}) = \frac{1}{2}\,(\sigma^{\mathrm{o}}\,\mathbf{1}) : (\mathbf{D} - \frac{1}{\overline{C}_1}\,\mathbf{E}^1) : (\sigma^{\mathrm{o}}\,\mathbf{1}). \tag{9.6.17b}$$

In view of (9.6.15), the following bound for $\overline{\mathbf{D}}^{\Sigma}$ is obtained:

$$\overline{\mathrm{D}}_1^{\Sigma} \geq \frac{1}{\overline{C}_1} \qquad \text{if } \mathbf{D}' - \mathbf{D} \text{ is p.d.,} \tag{9.6.18a}$$

where $\overline{\mathrm{D}}_1^{\Sigma} = \overline{\mathrm{D}}_{iijj}^{\Sigma}/3$. This bound is valid for any finite elastic composite

---

[37] As commented on in Subsection 9.6.1, if a stationary value of $\mathrm{J}^{\mathrm{A}\prime}$ is given by (9.6.16a) on a class of eigenstress fields, the stationary value of $\mathrm{I}^{\mathrm{A}\prime}$ on the corresponding class of eigenstrain fields is *always* given by (9.6.16b).

consisting of arbitrary microstructure, as long as its overall shape is ellipsoidal.

Manipulations similar to the above, yield a bound for $\bar{D}_2^\Sigma = \bar{D}_{ijij}^\Sigma - \bar{D}_{iijj}^\Sigma/3$ which can be expressed in terms of $\bar{C}_2$. Indeed, overall pure shear strains and the homogenizing shear eigenstresses, $\boldsymbol{\varepsilon}^o = \varepsilon^o\,\mathbf{s}_i$ and $\mathbf{s}^{*\alpha} = s^{*\alpha}\,\mathbf{s}_i$, correspond to the overall pure shear stresses and the homogenizing shear eigenstrains, $\boldsymbol{\sigma}^o = \varepsilon^o\,\mathbf{s}_i$ and $\mathbf{e}^{*\alpha} = e^{*\alpha}\,\mathbf{s}_i$, for i = 1, 2, ..., 6. Thus, since the stationary value of the sum of $J^{A\prime}$'s given by (9.6.13b), is

$$\frac{1}{3}\Big[\,J^{A\prime}(\{\sigma^{*\alpha}\,\mathbf{s}_1\};\,\varepsilon^o\,\mathbf{s}_1) + \ldots\Big] + \frac{1}{2}\Big[\,J^{A\prime}(\{\sigma^{*\alpha}\,\mathbf{s}_4\};\,\varepsilon^o\,\mathbf{s}_4) + \ldots\Big]$$

$$= \frac{1}{3}\Big[\,(\varepsilon^o\,\mathbf{s}_1) : (\mathbf{C} - \bar{C}_2\,\mathbf{E}^2) : (\varepsilon^o\,\mathbf{s}_1) + \ldots\Big]$$

$$+\frac{1}{2}\Big[\,(\varepsilon^o\,\mathbf{s}_4) : (\mathbf{C} - \bar{C}_2\,\mathbf{E}^2) : (\varepsilon^o\,\mathbf{s}_4) + \ldots\Big], \tag{9.6.19a}$$

where $\mathbf{E}^2 \equiv \mathbf{1}^{(4s)} - \mathbf{E}^2$, then the stationary value of the corresponding sum of $I^{A\prime}$'s is given by

$$\frac{1}{3}\Big[\,I^{A\prime}(\{\varepsilon^{*\alpha}\,\mathbf{s}_1\};\,\sigma^o\,\mathbf{s}_1) + \ldots\Big] + \frac{1}{2}\Big[\,I^{A\prime}(\{\varepsilon^{*\alpha}\,\mathbf{s}_4\};\,\sigma^o\,\mathbf{s}_4) + \ldots\Big]$$

$$= \frac{1}{3}\Big[\,(\sigma^o\,\mathbf{s}_1) : (\mathbf{D} - \frac{1}{\bar{C}_2}\,\mathbf{E}^2) : (\sigma^o\,\mathbf{s}_1) + \ldots\Big]$$

$$+\frac{1}{2}\Big[\,(\sigma^o\,\mathbf{s}_4) : (\mathbf{D} - \frac{1}{\bar{C}_2}\,\mathbf{E}^2) : (\sigma^o\,\mathbf{s}_4) + \ldots\Big]. \tag{9.6.19b}$$

In view of (9.6.15), the following bound for $\bar{\mathbf{D}}^\Sigma$ is obtained:

$$\bar{D}_2^\Sigma \geq \frac{1}{\bar{C}_2} \qquad \text{if } \mathbf{D}' - \mathbf{D} \text{ is p.d..} \tag{9.6.18b}$$

Again, this bound is valid for any finite elastic composite consisting of arbitrary microstructure.

9.6.5. Relation between Universal Bounds and Estimated Bounds

Now, consider relations between the universal bounds, $\bar{C}_1$ and $\bar{C}_2$, and the estimate of the overall elasticity tensor, $\bar{\mathbf{C}}^{EA}$, based on the assumption of statistical homogeneity and isotropy. For simplicity, let all Ω_α's be isotropic, having a distribution such that $\bar{\mathbf{C}}^{EA}$ is isotropic. In this case, $\bar{\mathbf{C}}^{EA}$ can be expressed in terms of the basic isotropic tensors, $\mathbf{E}^1$ and $\mathbf{E}^2$, as

$$\bar{\mathbf{C}}^{EA} = \bar{C}_1^{EA}\,\mathbf{E}^1 + \bar{C}_2^{EA}\,\mathbf{E}^2; \tag{9.6.20a}$$

see Subsection 8.1. In this manner, definition (9.5.39b) of $\bar{\mathbf{C}}^{EA}$ is reduced to two algebraic relations involving the coefficients of these basic tensors. Hence, coefficient $\bar{C}_1^{EA}$ is given by

$$\bar{C}_I^{EA} = \left\{ \sum_{\alpha=0}^{n} f_\alpha C_I^\alpha \{1_I^{(4s)} + P_I(C_I^\alpha - C_I)\}^{-1} \right\}$$

$$\times \left\{ \sum_{\beta=0}^{n} f_\beta \{1_I^{(4s)} + P_I(C_I^\beta - C_I)\}^{-1} \right\}^{-1}, \qquad (9.6.20b)$$

for I = 1, 2. Note that C_I, C_I^α, P_I, and $1_I^{(4s)}$ are the corresponding coefficients of the isotropic tensors, $\mathbf{C}$, $\mathbf{C}^\alpha$, $\mathbf{P}$, and $\mathbf{1}^{(4s)}$, respectively.

Coefficient $\bar{C}_I^{EA}$ given by (9.6.20b) is related to the universal bound $\bar{C}_1$. To prove this, first observe that the coefficient for $\mathbf{E}^1$ is determined by the fourth-order contraction with $\mathbf{E}^1$, i.e.,

$$C_1 = \frac{1}{N_1} C_{iijj}, \qquad C_1^\alpha = \frac{1}{N_1} C_{iijj}^\alpha, \qquad P_1 = \frac{1}{N_1} P_{iijj},$$

$$1_1^{(4s)} = \{\frac{1}{2}(\delta_{ik}\delta_{jl} + \delta_{il}\delta_{jk})\}\{\frac{1}{3}\delta_{ij}\delta_{kl}\} = 1. \qquad (9.6.21a\text{~}d)$$

Next, since $\mathbf{C}^\alpha$ and $\mathbf{C}$ are isotropic, $(\mathbf{C}^\alpha - \mathbf{C})^{-1}$ is given by

$$(\mathbf{C}^\alpha - \mathbf{C})^{-1} = \frac{1}{C_1^\alpha - C_1}\mathbf{E}^1 + \frac{1}{C_2^\alpha - C_2}\mathbf{E}^2. \qquad (9.6.22)$$

Finally, taking advantage of (9.6.21) and (9.6.22), rewrite $\bar{C}_1$, defined by (9.6.14c), as

$$\bar{C}_1 = \left\{ \sum_{\alpha=0}^{n} f_\alpha C_1^\alpha \{1 + P_1(C_1^\alpha - C_1)\}^{-1} \right\}$$

$$\times \left\{ \sum_{\beta=0}^{n} f_\beta \{1 + P_1(C_1^\beta - C_1))\}^{-1} \right\}^{-1}. \qquad (9.6.23)$$

The right side of (9.6.23) is given by (9.6.16b) for I = 1. Hence,

$$\bar{C}_1 = \bar{C}_1^{EA} = \frac{1}{N_1}\bar{C}_{iijj}^{EA}. \qquad (9.6.24a)$$

This is the relation between $\bar{C}_1$ and $\bar{\mathbf{C}}^{EA}$ for the isotropic case.

In a similar manner, it is shown that universal bound (9.6.18a) given by $1/\bar{C}_1$ is related to $\bar{\mathbf{D}}^{\Sigma A}$. Since $\bar{\mathbf{D}}^{\Sigma A}$ is the inverse of $\bar{\mathbf{C}}^{EA}$, it is given by

$$\bar{\mathbf{D}}^{\Sigma A} = \bar{D}_1^{\Sigma A}\mathbf{E}^1 + \bar{D}_2^{\Sigma A}\mathbf{E}^2 = \frac{1}{\bar{C}_1^{EA}}\mathbf{E}^1 + \frac{1}{\bar{C}_2^{EA}}\mathbf{E}^2. \qquad (9.6.25)$$

Hence, it follows from (9.6.23) that

$$\frac{1}{\bar{C}_1} = \frac{1}{\bar{C}_1^{EA}} = \bar{D}_1^{\Sigma A} = \frac{1}{N_1}\bar{D}_{iijj}^{\Sigma A}. \qquad (9.6.26a)$$

Therefore, it is concluded from (9.6.24a) and (9.6.26a) that the upper bounds for $\bar{C}_{iijj}^{E}$ and $\bar{D}_{iijj}^{\Sigma}$, determined by $\bar{C}_1$, remain unchanged whether or not Ω_α's and the overall RVE are isotropic.

In essentially the same manner, $\bar{C}_2^{EA}$ and $\bar{D}_2^{\Sigma A}$ can be related to $\bar{C}_2$, as follows:

$$\bar{C}_2 = \bar{C}_2^{EA} = \frac{1}{N_2} (\bar{C}_{ijij}^{EA} - \frac{1}{3} \bar{C}_{iijj}^{EA}) \tag{9.6.24b}$$

and

$$\frac{1}{\bar{C}_2} = \frac{1}{\bar{C}_2^{EA}} = \bar{D}_2^{\Sigma A} = \frac{1}{N_2} (\bar{D}_{ijij}^{\Sigma A} - \frac{1}{3} \bar{D}_{iijj}^{\Sigma A}). \tag{9.6.26b}$$

9.7. BOUNDS FOR OVERALL NONMECHANICAL MODULI

In Subsection 2.7, certain nonmechanical properties, such as electrostatic, magnetostatic, thermal, and diffusional properties, are briefly reviewed, and it is shown that universal Theorems I and II also apply to this class of problems. To treat this class of nonmechanical properties in a unified manner, a generalized force field $\mathbf{p} = \mathbf{p}(\mathbf{x})$, with its conjugate flux field $\mathbf{q} = \mathbf{q}(\mathbf{x})$ is introduced. The force field is the gradient of a scalar potential field $u = u(\mathbf{x})$, i.e., $\mathbf{p} = -\nabla u$. In the absence of sources, the flux field is divergence-free, i.e., $\nabla \cdot \mathbf{q} = 0$. Moreover, these conjugate fields are related through constitutive relations, $\mathbf{q} = \mathbf{K}' \cdot \mathbf{p}$ and $\mathbf{p} = \mathbf{R}' \cdot \mathbf{q}$, where, for ease in referencing, $\mathbf{K}'$ and $\mathbf{R}'$ are referred to as *conductivity* and *resistivity* tensors, respectively. These are second-order, positive-definite, symmetric tensors, which may stand for other properties; e.g., $\mathbf{K}'$ is the dielectric tensor in electrostatics. For heterogeneous materials, $\mathbf{K}' = \mathbf{K}'(\mathbf{x})$ and $\mathbf{R}' = \mathbf{R}'(\mathbf{x})$.

In this subsection, it is shown that the application of the generalized Hashin-Shtrikman variational principles and the universal Theorems I and II of Subsection 2.7 leads to exact bounds for the overall conductivity or resistivity tensor, $\bar{\mathbf{K}}$ or $\bar{\mathbf{R}}$. In particular, it is demonstrated that the resulting bounds on $\bar{K}_{ii}$ and $\bar{R}_{ii}$ are *universal* in the sense that they are independent of the shape and distribution of the phases within the composite, i.e., they apply to any composites with any number and distribution of phases. Indeed, the same bounds are obtained for the unit cell of composites with periodic microstructure; see Section 13.

The presentation in this subsection is structured such that the results of Subsections 9.5 and 9.6 directly apply to the considered cases of nonmechanical material parameters. First, the corresponding generalized Hashin-Shtrikman variational principles are developed. Then, these principles are combined with universal Theorems I and II of Subsection 2.7, and computable bounds for the overall conductivity tensor $\bar{\mathbf{K}}$ or resistivity tensor $\bar{\mathbf{R}}$ are obtained in terms of the Green function of the infinite homogeneous body. Finally, the universal bounds valid for any heterogeneous medium of any number and distribution of phases are obtained.

In view of the mathematical similarity in the mechanical and nonmechanical field equations, the procedure[38] for obtaining bounds on the mechanical properties can easily be followed to produce bounds on the nonmechanical parameters; see, e.g., Hashin and Shtrikman (1962c) for bounds for magnetic permeability, Schapery (1968) and Willis (1977) for bounds on thermal properties, and Milton (1990) who reviews the mathematical fundamentals, and treats the overall mechanical and nonmechanical properties of composites.[39]

9.7.1. Generalized Hashin-Shtrikman Variational Principle

Consider an RVE whose force and flux fields, $\mathbf{p}(\mathbf{x})$ and $\mathbf{q}(\mathbf{x})$, are related through the following linear constitutive relation:

$$\mathbf{q}(\mathbf{x}) = \mathbf{K}'(\mathbf{x}) \cdot \mathbf{p}(\mathbf{x}), \tag{9.7.1a}$$

where the conductivity tensor $\mathbf{K}'(\mathbf{x})$ stands for any other nonmechanical (or mechanical) material tensor discussed in Subsection 2.7. Note that the force field is given by the gradient of a potential field, u, and the flux field is divergence-free,

$$\mathbf{p}(\mathbf{x}) = -\boldsymbol{\nabla}\, \mathrm{u}(\mathbf{x}), \qquad \boldsymbol{\nabla} \cdot \mathbf{q}(\mathbf{x}) = \mathbf{0} \qquad \text{in V}. \tag{9.7.1b,c}$$

Moreover, on the boundary ∂V, either the potential or the flux is prescribed, such that the disturbance potential field, $\mathrm{u}^{\mathrm{d}} \equiv \mathbf{u} - \mathbf{x} \cdot < \mathbf{p} >$, and the disturbance flux field, $\mathbf{q}^{\mathrm{d}} = \mathbf{q} - < \mathbf{q} >$, satisfy

$$(\mathbf{v}(\mathbf{x}) \cdot \mathbf{q}^{\mathrm{d}}(\mathbf{x}))\, \mathrm{u}^{\mathrm{d}}(\mathbf{x}) = 0 \qquad \text{on } \partial\mathrm{V}. \tag{9.7.2a}$$

Note that u^{d} is associated with the disturbance force field, $\mathbf{p}^{\mathrm{d}} = \mathbf{p} - < \mathbf{p} >$. When the boundary conditions that satisfy (9.7.2a) are prescribed, the average force and flux satisfy

$$< \mathbf{q} \cdot \mathbf{p} > = < \mathbf{q} > \cdot < \mathbf{p} >; \tag{9.7.2b}$$

see Subsection 2.7.

To obtain a general variational principle similar to (9.5.7a), introduce an equivalent homogeneous RVE with *constant* reference conductivity tensor, $\mathbf{K}$, and prescribe a suitable homogenizing *eigenflux field*,[40] $\mathbf{q}^* = \mathbf{q}^*(\mathbf{x})$. Flux and force fields in the equivalent homogeneous RVE are related through

$$\mathbf{q}(\mathbf{x}) = \mathbf{K} \cdot \mathbf{p}(\mathbf{x}) + \mathbf{q}^*(\mathbf{x}). \tag{9.7.3}$$

Since these fields in the equivalent homogeneous RVE coincide with those in

---

[38] In addition to a variational method similar to that presented in this section, Milton (1990) summarizes another technique to obtain bounds on the overall properties, and discusses the relation between the two methods; see also Milton and Kohn (1988).

[39] While the bounds presented in this section involve the volume fractions of inhomogeneities and the corresponding two-point correlation tensors as the only required geometric information about the microstructure, improved bounds are obtained when higher-order correlation tensors are available; see, e.g., Torquato (1991) for discussions and references.

[40] An eigenflux field here plays a role similar to the eigenstress field in preceding subsections; see, for example, Subsection 7.3.

the original heterogeneous RVE, they must satisfy (9.7.1a), and hence

$$\mathbf{K}'(\mathbf{x})\cdot\mathbf{p}(\mathbf{x}) = \mathbf{K}\cdot\mathbf{p}(\mathbf{x}) + \mathbf{q}^*(\mathbf{x}) \tag{9.7.4a}$$

or

$$(\mathbf{K}'(\mathbf{x}) - \mathbf{K})^{-1}\cdot\mathbf{q}^*(\mathbf{x}) - \mathbf{p}(\mathbf{x}) = \mathbf{0}. \tag{9.7.4b}$$

This is the *consistency condition* for the eigenflux field.

The governing field equation for a potential field u in the equivalent homogeneous RVE is deduced from (9.7.1a,b) and (9.7.3),

$$\boldsymbol{\nabla}\cdot\{\mathbf{K}\cdot(-\boldsymbol{\nabla}\mathrm{u}(\mathbf{x})) + \mathbf{q}^*(\mathbf{x})\} = 0, \tag{9.7.5}$$

subjected to the general boundary conditions satisfying (9.7.2a). This boundary-value problem for u can be solved for given $\mathbf{q}^*$, and the resulting force field is expressed in terms of $\mathbf{q}^*$, as

$$\mathbf{p}(\mathbf{x}) = \mathbf{p}^{\mathrm{o}} - \boldsymbol{\Gamma}^{\mathrm{G}}(\mathbf{x};\, \mathbf{q}^*), \tag{9.7.6a}$$

where $\boldsymbol{\Gamma}^{\mathrm{G}}$ is an integral operator which, for a given $\mathbf{q}^*$, produces the disturbance force field, $\mathbf{p}^{\mathrm{d}} \equiv \mathbf{p} - <\mathbf{p}>$. *Superscript* G *here stands for general boundary conditions that satisfy* (9.7.2a). On account of (9.7.2a), operator $\boldsymbol{\Gamma}^{\mathrm{G}}$ is self-adjoint, and operator $<\cdot\{(\mathbf{K}' - \mathbf{K})^{-1} + \boldsymbol{\Gamma}^{\mathrm{G}}\}\cdot>$ is positive-semi-definite (negative-semi-definite) when $\mathbf{K}' - \mathbf{K}$ is positive-semi-definite (negative-semi-definite).[41]

When $<\mathbf{p}>$ is prescribed to be $\mathbf{p}^{\mathrm{o}}$, consistency condition (9.7.4b) is rewritten in terms of the integral operator $\boldsymbol{\Gamma}^{\mathrm{G}}$, as follows:

$$(\mathbf{K}'(\mathbf{x}) - \mathbf{K})^{-1}\cdot\mathbf{q}^*(\mathbf{x}) + \boldsymbol{\Gamma}^{\mathrm{G}}(\mathbf{x};\, \mathbf{q}^*) - \mathbf{p}^{\mathrm{o}} = \mathbf{0}. \tag{9.7.7}$$

In view of (9.7.7), for an arbitrary eigenflux field, $\boldsymbol{\psi}^* = \boldsymbol{\psi}^*(\mathbf{x})$, introduce a functional defined by

$$\hat{\mathrm{J}}(\boldsymbol{\psi}^*;\, \boldsymbol{\Gamma}^{\mathrm{G}};\, \mathbf{p}^{\mathrm{o}}) \equiv \frac{1}{2} <\boldsymbol{\psi}^*\cdot\{(\mathbf{K}' - \mathbf{K})^{-1} + \boldsymbol{\Gamma}^{\mathrm{G}}\}\cdot\boldsymbol{\psi}^*> - <\boldsymbol{\psi}^*>\cdot\mathbf{p}^{\mathrm{o}}. \tag{9.7.8a}$$

Since the first variation of $\hat{\mathrm{J}}$ is

$$\delta\hat{\mathrm{J}}(\boldsymbol{\psi}^*;\, \boldsymbol{\Gamma}^{\mathrm{G}};\, \mathbf{p}^{\mathrm{o}}) = <\delta\boldsymbol{\psi}^*\cdot\Big\{\, \{(\mathbf{K}' - \mathbf{K})^{-1} + \boldsymbol{\Gamma}^{\mathrm{G}}\}\cdot\boldsymbol{\psi}^* - \mathbf{p}^{\mathrm{o}} \Big\}>, \tag{9.7.8b}$$

the exact eigenflux field, $\mathbf{q}^*$, which satisfies (9.7.7) renders $\hat{\mathrm{J}}$ stationary. The stationary value of $\hat{\mathrm{J}}$ is given by

$$\hat{\mathrm{J}}(\mathbf{q}^*;\, \boldsymbol{\Gamma}^{\mathrm{G}};\, \mathbf{p}^{\mathrm{o}}) = -\frac{1}{2}\,\mathbf{q}^*\cdot\mathbf{p}^{\mathrm{o}} = \frac{1}{2}\,\mathbf{p}^{\mathrm{o}}\cdot(\mathbf{K} - \overline{\mathbf{K}}^{\mathrm{G}})\cdot\mathbf{p}^{\mathrm{o}}, \tag{9.7.8c}$$

where $\overline{\mathbf{K}}^{\mathrm{G}}$ is the overall conductivity tensor for the prescribed *general boundary data*.[42] This is one of the *generalized Hashin-Shtrikman variational principles for nonmechanical properties* of any heterogeneous RVE under general

---

[41] The proof of these properties of the integral operator is essentially the same as that given in Subsections 9.1 and 9.2.

[42] From (9.7.2b), it follows that $<\mathbf{q}> = \overline{\mathbf{K}}^{\mathrm{G}}.\mathbf{p}^{\mathrm{o}}$ and $<\mathbf{q}.\mathbf{p}> = \mathbf{p}^{\mathrm{o}}.\overline{\mathbf{K}}^{\mathrm{G}}.\mathbf{p}^{\mathrm{o}}$.

boundary data which satisfy (9.7.2a). Moreover, from the positive-semi-definiteness (negative-semi-definiteness) of the operator $<\bullet\{(\mathbf{K}' - \mathbf{K})^{-1} + \mathbf{\Gamma}^{\mathrm{G}}\}\bullet>$, it immediately follows that, for any eigenflux field, $\boldsymbol{\psi}^*$,

$$\frac{1}{2}\mathbf{p}^{\mathrm{o}}\bullet(\mathbf{K} - \overline{\mathbf{K}}^{\mathrm{G}})\bullet\mathbf{p}^{\mathrm{o}} = \hat{\mathrm{J}}(\mathbf{p}^*;\ \mathbf{\Gamma}^{\mathrm{G}} : \mathbf{p}^{\mathrm{o}}) \leq (\geq)\ \hat{\mathrm{J}}(\boldsymbol{\psi}^*;\ \mathbf{\Gamma}^{\mathrm{G}} : \mathbf{p}^{\mathrm{o}}), \tag{9.7.9a}$$

when $\mathbf{K}' - \mathbf{K}$ is positive-semi-definite (negative-semi-definite). These are the *generalized Hashin-Shtrikman bounds* corresponding to general boundary data (9.7.2a).

As in the case of mechanical fields, the formulation here may also be implemented in terms of the *eigenforce* field, say, $\boldsymbol{\phi}^* = \boldsymbol{\phi}^*(\mathbf{x})$, conjugate to the eigenflux field $\boldsymbol{\psi}^* = \boldsymbol{\psi}^*(\mathbf{x})$. If the reference resistivity is $\mathbf{R} = \mathbf{K}^{-1}$, then the eigenfields are related by $\mathbf{R}\bullet\boldsymbol{\psi}^* + \boldsymbol{\phi}^* = \mathbf{0}$ or $\mathbf{K}\bullet\boldsymbol{\phi}^* + \boldsymbol{\psi}^* = \mathbf{0}$; compare with (7.3.13a,b). Then, in terms of $\mathbf{\Gamma}^{\mathrm{G}}$, an integral operator for $\boldsymbol{\phi}^*$ is defined by

$$\mathbf{\Lambda}^{\mathrm{G}}(\mathbf{x};\ \boldsymbol{\phi}^*) \equiv \mathbf{K}\bullet\{\mathbf{\Gamma}^{\mathrm{G}}(\mathbf{x};\ -\mathbf{K}\bullet\boldsymbol{\phi}^*) + (\boldsymbol{\phi}^* - <\boldsymbol{\phi}^*>)\}, \tag{9.7.6b}$$

which produces a disturbance flux field, $\mathbf{q}^{\mathrm{d}} \equiv \mathbf{q} - <\mathbf{q}>$. In terms of integral operator $\mathbf{\Lambda}^{\mathrm{G}}$, functional $\hat{\mathrm{I}}$ conjugate to the functional $\hat{\mathrm{J}}$, is defined as

$$\hat{\mathrm{I}}(\boldsymbol{\phi}^*;\ \mathbf{\Lambda}^{\mathrm{G}};\ \mathbf{q}^{\mathrm{o}}) \equiv \frac{1}{2} <\boldsymbol{\phi}^*\bullet\{(\mathbf{R}' - \mathbf{R})^{-1} + \mathbf{\Lambda}^{\mathrm{G}}\}\bullet\boldsymbol{\phi}^*> - <\boldsymbol{\phi}^*>\bullet\mathbf{q}^{\mathrm{o}}, \tag{9.7.8d}$$

for a given $\mathbf{q}^{\mathrm{o}}$. Since the first variation of $\hat{\mathrm{I}}$ is

$$\delta\hat{\mathrm{I}}(\boldsymbol{\phi}^*;\ \mathbf{\Lambda}^{\mathrm{G}};\ \mathbf{q}^{\mathrm{o}}) \equiv <\delta\boldsymbol{\phi}^*\bullet\Big\{\{(\mathbf{R}' - \mathbf{R})^{-1} + \mathbf{\Lambda}^{\mathrm{G}}\}\bullet\boldsymbol{\phi}^* - \mathbf{q}^{\mathrm{o}}\Big\}>, \tag{9.7.8e}$$

the exact homogenizing eigenforce field, $\mathbf{p}^* \equiv -\mathbf{R}\bullet\mathbf{q}^*$, renders $\hat{\mathrm{I}}$ stationary. The stationary value is given by

$$\hat{\mathrm{I}}(\mathbf{p}^*;\ \mathbf{\Lambda}^{\mathrm{G}};\ \mathbf{q}^{\mathrm{o}}) = -\frac{1}{2}\ \mathbf{p}^*\bullet\mathbf{q}^{\mathrm{o}} = \frac{1}{2}\ \mathbf{q}^{\mathrm{o}}\bullet(\mathbf{R} - \overline{\mathbf{R}}^{\mathrm{G}})\bullet\mathbf{q}^{\mathrm{o}}, \tag{9.7.8f}$$

where $\overline{\mathbf{R}}^{\mathrm{G}}$ is the overall resistivity tensor for the prescribed *general boundary data*[43] that satisfy (9.7.2a). Then, the second *generalized Hashin-Shtrikman variational principle applied to nonmechanical fields*, takes on the following form:

$$\frac{1}{2}\mathbf{p}^{\mathrm{o}}\bullet(\mathbf{R} - \overline{\mathbf{R}}^{\mathrm{G}})\bullet\mathbf{q}^{\mathrm{o}} = \hat{\mathrm{I}}(\mathbf{p}^*;\ \mathbf{\Lambda}^{\mathrm{G}};\ \mathbf{q}^{\mathrm{o}}) \leq (\geq)\ \hat{\mathrm{I}}(\boldsymbol{\phi}^*;\ \mathbf{\Lambda}^{\mathrm{G}} : \mathbf{q}^{\mathrm{o}}), \tag{9.7.9b}$$

when $\mathbf{R}' - \mathbf{R}$ is positive-semi-definite (negative-semi-definite).

9.7.2. Consequence of Universal Theorems

Although (9.7.9a,b) is exact, it is not easy to determine the integral operator $\mathbf{\Gamma}^{\mathrm{G}}$ which produces disturbance fields satisfying the prescribed boundary conditions. In order to use an integral operator which can be explicitly determined, combine the Hashin-Shtrikman variational principles with Theorems I

[43] Again, from (9.7.2b), it follows that $<\mathbf{p}> = \overline{\mathbf{R}}^{\mathrm{G}}.\mathbf{q}^{\mathrm{o}}$ and $<\mathbf{p}.\mathbf{q}> = \mathbf{q}^{\mathrm{o}}.\overline{\mathbf{R}}^{\mathrm{G}}.\mathbf{q}^{\mathrm{o}}$.

and II of Subsection 2.7. First, observe the following inequality which is obtained directly from the proof of Theorem II:

$$\frac{1}{2} < \mathbf{p}^{\mathrm{P}} \cdot \mathbf{K}' \cdot \mathbf{p}^{\mathrm{P}} > \leq \frac{1}{2} < (\mathbf{p}^{\mathrm{o}} + \mathbf{p}^{\mathrm{d}}) \cdot \mathbf{K}' \cdot (\mathbf{p}^{\mathrm{o}} + \mathbf{p}^{\mathrm{d}}) >, \tag{9.7.10a}$$

where $\mathbf{p}^{\mathrm{P}} = \mathbf{p}^{\mathrm{P}}(\mathbf{x})$ is the force field when the RVE is subjected to the linear potential boundary conditions which yield the average force $\mathbf{p}^{\mathrm{o}}$; and $\mathbf{p}^{\mathrm{d}} = \mathbf{p}^{\mathrm{d}}(\mathbf{x})$ is an arbitrary disturbance force field which is *compatible*,[44] i.e., is given by the gradient of a certain disturbance potential u^{d}, $\mathbf{p}^{\mathrm{d}} \equiv -\nabla \mathrm{u}^{\mathrm{d}}$, and has zero volume average, $< \mathbf{p}^{\mathrm{d}} > = \mathbf{0}$.

Let $\mathbf{K}$ be a constant conductivity tensor such that $\mathbf{K}' - \mathbf{K}$ is negative-semi-definite, i.e., for any arbitrary pair of fields $\mathbf{p}$ and $\mathbf{p}'$,

$$0 \geq (\mathbf{p} - \mathbf{p}') \cdot (\mathbf{K}' - \mathbf{K}) \cdot (\mathbf{p} - \mathbf{p}'). \tag{9.7.11}$$

Replacing $\mathbf{p}'$ by $\mathbf{p}' \equiv (\mathbf{K}' - \mathbf{K})^{-1} \cdot \boldsymbol{\psi}^*$, and taking the volume average of the resulting inequality, obtain

$$0 \geq \frac{1}{2} < \mathbf{p} \cdot (\mathbf{K}' - \mathbf{K}) \cdot \mathbf{p} - 2\,\boldsymbol{\psi}^* \cdot \mathbf{p} + \boldsymbol{\psi}^* \cdot (\mathbf{K}' - \mathbf{K})^{-1} \cdot \boldsymbol{\psi}^* >. \tag{9.7.12}$$

Then, taking advantage of inequality (9.7.12), evaluate the right side of (9.7.10a) and obtain

$$\frac{1}{2} < \mathbf{p}^{\mathrm{P}} \cdot \mathbf{K}' \cdot \mathbf{p}^{\mathrm{P}} > \leq \frac{1}{2} \mathbf{p}^{\mathrm{o}} \cdot \mathbf{K} \cdot \mathbf{p}^{\mathrm{o}}$$

$$- \frac{1}{2} < \boldsymbol{\psi}^* \cdot (\mathbf{K}' - \mathbf{K})^{-1} \cdot \boldsymbol{\psi}^* - \boldsymbol{\psi}^* \cdot (\mathbf{p}^{\mathrm{d}} + 2\mathbf{p}^{\mathrm{o}}) > + \frac{1}{2} < \mathbf{p}^{\mathrm{d}} \cdot \mathbf{q}^{\mathrm{d}} >, \tag{9.7.13}$$

where $\mathbf{q}^{\mathrm{d}} = \mathbf{K} \cdot \mathbf{p}^{\mathrm{d}} + \boldsymbol{\psi}^{*\mathrm{d}}$. Note that inequality (9.7.13) holds for any pair of $\boldsymbol{\psi}^*$ and $\mathbf{p}^{\mathrm{d}}$, as long as $\mathbf{K}' - \mathbf{K}$ is negative-semi-definite.

Now, relate $\mathbf{p}^{\mathrm{d}}$ and $\boldsymbol{\psi}^{*\mathrm{d}}$ through an integral operator which can be obtained from the Green function of an infinite homogeneous solid B with uniform conductivity $\mathbf{K}$. As in the case of mechanical fields, this integral operator is denoted by $\boldsymbol{\Gamma}^{\mathrm{A}}$, i.e., $\mathbf{p}^{\mathrm{d}}(\mathbf{x}) = -\boldsymbol{\Gamma}^{\mathrm{A}}(\mathbf{x};\, \boldsymbol{\psi}^{*\mathrm{d}})$. The explicit expressions of $\boldsymbol{\Gamma}^{\mathrm{A}}$ and the Green function are given at the end of this subsection.[45] When the subregion V′ is *ellipsoidal*,[46] the following three identities hold:

$$< \mathbf{p}^{\mathrm{A}} > = < -\boldsymbol{\Gamma}^{\mathrm{A}}(\boldsymbol{\psi}^*) > = \mathbf{0},$$

$$< \mathbf{q}^{\mathrm{A}} > = < -\mathbf{K} \cdot \boldsymbol{\Gamma}^{\mathrm{A}}(\boldsymbol{\psi}^*) + \boldsymbol{\psi}^{*\mathrm{d}} > = \mathbf{0}, \tag{9.7.14a~c}$$

$$< \mathbf{q}^{\mathrm{A}} \cdot \mathbf{p}^{\mathrm{A}} > = < \boldsymbol{\psi}^{*\mathrm{d}} \cdot \{\boldsymbol{\Gamma}^{\mathrm{A}} \cdot \mathbf{K} \cdot \boldsymbol{\Gamma}^{\mathrm{A}} - \boldsymbol{\Gamma}^{\mathrm{A}}\} \cdot \boldsymbol{\psi}^* > \leq 0,$$

where $\mathbf{q}^{\mathrm{A}} = \mathbf{K} \cdot \mathbf{p}^{\mathrm{A}} + \boldsymbol{\psi}^{*\mathrm{d}}$ is the corresponding flux field; the proof of (9.7.14a~c)

---

[44] The disturbance force field $\mathbf{p}^{\mathrm{d}}$ is not necessarily associated with an "equilibrating" flux field, i.e., $\mathbf{K}' \cdot \mathbf{p}^{\mathrm{d}}$ need not necessarily be divergence-free.

[45] See (9.7.24) and (9.7.25a~c).

is given at the end of this subsection.

In view of (9.7.14a), replace $\mathbf{p}^d$ in (9.7.13) by $\mathbf{p}^A$, and from (9.7.14c), note that the last term in the right side of (9.7.13) is negative. Hence, inequality (9.7.13) is rewritten as

$$\frac{1}{2} < \mathbf{p}^P \cdot \mathbf{K}' \cdot \mathbf{p}^P > \leq \frac{1}{2} \mathbf{p}^o \cdot \mathbf{K} \cdot \mathbf{p}^o - \hat{J}(\boldsymbol{\psi}^*; \boldsymbol{\Gamma}^A; \mathbf{p}^o). \tag{9.7.15}$$

Furthermore, if $\overline{\mathbf{K}}^P$ is defined as the overall conductivity of the RVE when subjected to the linear potential boundary conditions, then $< \mathbf{p}^P \cdot \mathbf{K}' \mathbf{p}^P >$ is given by $\mathbf{p}^o \cdot \overline{\mathbf{K}}^P \cdot \mathbf{p}^o$, and (9.7.15) yields

$$\frac{1}{2} \mathbf{p}^o \cdot (\overline{\mathbf{K}}^P - \mathbf{K}) \cdot \mathbf{p}^o \leq -\hat{J}(\boldsymbol{\psi}^*; \boldsymbol{\Gamma}^A; \mathbf{p}^o), \tag{9.7.16a}$$

if $\mathbf{K}' - \mathbf{K}$ is negative-definite. This is an *exact* bound for $\overline{\mathbf{K}}^P$, directly expressed in terms of the simple and known integral operator $\boldsymbol{\Gamma}^A$ of the infinite uniform body B. Note that this formulation is implemented directly in terms of the universal Theorem II and the negative-definiteness of $\mathbf{K}' - \mathbf{K}$. The only assumption is that (9.7.14a,b) hold, which is necessarily ensured when V′ is ellipsoidal.

In essentially the same manner, an *exact* bound for the overall resistivity is obtained. First, Theorem I of Subsection 2.7 is generalized as

$$\frac{1}{2} < \mathbf{q}^Q \cdot \mathbf{R}' \cdot \mathbf{q}^Q > \leq \frac{1}{2} < (\mathbf{q}^o + \mathbf{q}^d) \cdot \mathbf{R}' \cdot (\mathbf{q}^o + \mathbf{q}^d) >, \tag{9.7.10b}$$

for any equilibrating $\mathbf{q}^d$ with zero average, where $\mathbf{q}^Q$ is the divergence-free flux field when the RVE is subjected to the uniform flux boundary conditions which yield $< \mathbf{q}^Q > = \mathbf{q}^o$. Then, $\mathbf{q}^d(\mathbf{x})$ in (9.7.10b) is set equal to $-\boldsymbol{\Lambda}^A(\mathbf{x}; \boldsymbol{\phi}^*)$, where integral operator $\boldsymbol{\Lambda}^A$ is defined by $\boldsymbol{\Lambda}^A(\mathbf{x}; \boldsymbol{\phi}^*) \equiv -\mathbf{K} \cdot \boldsymbol{\Gamma}^A(\mathbf{x}; \boldsymbol{\psi}^*) - \boldsymbol{\psi}^{*d}(\mathbf{x})$ with $\boldsymbol{\phi}^* \equiv -\mathbf{R} \cdot \boldsymbol{\psi}^*$. In this manner, the following *exact* inequality is obtained:

$$\frac{1}{2} \mathbf{q}^o \cdot (\overline{\mathbf{R}}^Q - \mathbf{R}) \cdot \mathbf{q}^o \leq -\hat{I}(\boldsymbol{\phi}^*; \boldsymbol{\Lambda}^A; \mathbf{q}^o), \tag{9.7.16b}$$

when $\mathbf{R}' - \mathbf{R}$ is negative-definite and the RVE is ellipsoidal.[47] Note that $\overline{\mathbf{R}}^Q$ is the overall resistivity of the RVE when it is subjected to the uniform flux boundary conditions.

9.7.3. Universal Bounds for Overall Conductivity

Since $\boldsymbol{\Gamma}^A$ is explicitly given, functional $\hat{J}(\boldsymbol{\psi}^*; \boldsymbol{\Gamma}^A; \mathbf{p}^o)$ can be computed exactly for certain classes of suitable eigenflux fields. Suppose that an ellipsoidal RVE contains n distinct microconstituents embedded in a matrix. Let the matrix and microconstituents be denoted by Ω_0 and $\Omega_1, \Omega_2, ..., \Omega_n$. The conductivity and volume fraction of Ω_α are $\mathbf{K}^\alpha$ and f_α ($\alpha = 0, 1, ..., n$). To homogenize this RVE, consider an ellipsoidal subregion V′ of an infinite homogeneous body B. The geometry of V′ is the same as that of the RVE. Let the body B be

---

[46] When (9.7.14a,b) hold, then (9.7.14c) follows whether or not V′ is ellipsoidal.

[47] The properties of $\boldsymbol{\Lambda}^A$ which are the same as those of $\boldsymbol{\Gamma}^A$, (9.7.14a~c), must be used to derive

isotropic, with conductivity tensor $\mathbf{K} = K\,\mathbf{1}^{(2)}$, such that $\mathbf{K}^{\alpha} - \mathbf{K}$ is negative-definite for all α. Assume that B is subjected to farfield uniform forces $\mathbf{p}^{o}$. Then, distribute piecewise constant eigenfluxes in B in the following manner: 1) $\boldsymbol{\psi}^{*\alpha}$ in each Ω_{α} within V′; and 2) $\bar{\boldsymbol{\psi}}^{*} = \sum_{\alpha=0}^{n} f_{\alpha}\,\boldsymbol{\psi}^{*\alpha}$ uniformly in B − V′. Thus, in V′, the disturbance eigenflux field is

$$\boldsymbol{\psi}^{*d}(\mathbf{x}) = \sum_{\alpha=0}^{n} H(\mathbf{x};\,\Omega_{\alpha})\,(\boldsymbol{\psi}^{*\alpha} - \bar{\boldsymbol{\psi}}^{*}), \tag{9.7.17}$$

where $H(\mathbf{x};\,\Omega_{\alpha}) = 1$ for $\mathbf{x}$ in Ω_{α}, and 0 otherwise.

The average eigenflux $\bar{\boldsymbol{\psi}}^{*}$ does not produce disturbance fields, since it is uniformly distributed throughout B. Moreover, $< \boldsymbol{\Gamma}^{A}(\boldsymbol{\psi}^{*d}) >$ vanishes, since V′ is ellipsoidal. Hence, in this setting, functional $\hat{J}$ reduces to the function $J^{A\prime}$ of $\boldsymbol{\psi}^{*\alpha}$'s, defined by

$$J^{A\prime}(\{\boldsymbol{\psi}^{*\alpha}\};\,\mathbf{p}^{o}) = \frac{1}{2}\sum_{\alpha=0}^{n} f_{\alpha}\,\boldsymbol{\psi}^{*\alpha}\cdot\{(\mathbf{K}^{\alpha} - \mathbf{K})^{-1}\cdot\boldsymbol{\psi}^{*\alpha} - 2\mathbf{p}^{o}\}$$

$$+\frac{1}{2}\sum_{\alpha=0}^{n}\sum_{\beta=0}^{n} < \{H(\Omega_{\alpha})(\boldsymbol{\psi}^{*\alpha} - \bar{\boldsymbol{\psi}}^{*})\}\cdot\boldsymbol{\Gamma}^{A}\cdot\{H(\Omega_{\beta})(\boldsymbol{\psi}^{*\beta} - \bar{\boldsymbol{\psi}}^{*})\} >. \tag{9.7.18}$$

To compute $J^{A\prime}$, $< \{H(\Omega_{\alpha})(\boldsymbol{\psi}^{*\alpha} - \bar{\boldsymbol{\psi}}^{*})\}\cdot\boldsymbol{\Gamma}^{A}\cdot\{H(\Omega_{\beta})(\boldsymbol{\psi}^{*\beta} - \bar{\boldsymbol{\psi}}^{*})\} >$ needs to be evaluated.

Since $\mathbf{K}$ is isotropic, the integral operator $\boldsymbol{\Gamma}^{A}$ has the following property:

$$< H(\Omega_{\alpha})\,\Gamma^{A}_{ii}\,H(\Omega_{\beta}) > = \begin{cases} f_{\alpha}\,\dfrac{1}{K} & \text{if } \alpha = \beta \\ 0 & \text{otherwise.} \end{cases} \tag{9.7.19a}$$

The proof of this property is given at the end of this subsection. Introduce an isotropic second-order tensor $\mathbf{P} \equiv (1/3K)\,\mathbf{1}^{(2)}$, and rewrite (9.7.19a) as

$$< H(\Omega_{\alpha})\,\Gamma^{A}_{ii}\,H(\Omega_{\beta}) > = \begin{cases} f_{\alpha}\,P_{ii} & \text{if } \alpha = \beta \\ 0 & \text{otherwise.} \end{cases} \tag{9.7.19b}$$

Note that the left side of (9.7.19) can be explicitly evaluated in terms of $\mathbf{P}$, if statistical homogeneity and isotropy are assumed; see Subsection 9.4 for the case of mechanical fields.

Taking advantage of exact relation (9.7.19), proceed as follows: let only one component of the uniform force $\mathbf{p}^{o}$, say, p^{o}_{1}, be nonzero, and let the corresponding nonzero component of $\boldsymbol{\psi}^{*\alpha}$ be $\psi^{*\alpha}_{1}$ for each Ω_{α}. The function $J^{A\prime}$ for this case is denoted by $J^{A\prime}(\{\psi^{*\alpha}_{1}\};\,p^{o}_{1})$. For three possible cases of this kind, identify the nonzero components of $\mathbf{p}^{o}$ and $\boldsymbol{\psi}^{*\alpha}$ by p^{o} and $\psi^{*\alpha}$. Then, consider the following sum of the resulting three cases:

(9.7.16b) but no other assumptions are required. Hence, (9.7.16b) holds for any ellipsoidal RVE.

$$J^{A\prime}(\{\psi_1^{*\alpha}\};\ p_1^o) + J^{A\prime}(\{\psi_2^{*\alpha}\};\ p_2^o) + J^{A\prime}(\{\psi_3^{*\alpha}\};\ p_3^o)$$

$$= \frac{1}{2}\sum_{\alpha=0}^{n} f_\alpha\, \psi^{*\alpha}\, \{(\mathbf{K}^\alpha - \mathbf{K})_{ii}^{-1} + P_{ii}\}\, \psi^{*\alpha} - \frac{1}{2}\overline{\psi}^*\, P_{ii}\, \overline{\psi}^* - 3\overline{\psi}^*\, p^o. \tag{9.7.20}$$

Since the right side of (9.7.20) is a quadratic form (with linear terms) for $\{\psi^{*\alpha}\}$, it can be optimized. Let $\{q^{*\alpha}\}$ be the eigenflux that renders the right side of (9.7.20) stationary, and compute

$$J^{A\prime}(\{q_1^{*\alpha}\};\ p_1^o) + J^{A\prime}(\{q_2^{*\alpha}\};\ p_2^o) + J^{A\prime}(\{q_3^{*\alpha}\};\ p_3^o) = \frac{1}{2}\, p^o\, (K_{ii} - \overline{K})\, p^o, \tag{9.7.21a}$$

where

$$\overline{K} \equiv \left\{ \sum_{\alpha=0}^{n} f_\alpha \left[K_{ii} + \frac{9}{(\mathbf{K}^\alpha - \mathbf{K})_{ii}^{-1}} \right] \left[1 + \frac{P_{ii}}{(\mathbf{K}^\alpha - \mathbf{K})_{ii}^{-1}} \right]^{-1} \right\}$$

$$\times \left\{ \sum_{\beta=0}^{n} f_\beta \left[1 + \frac{P_{ii}}{(\mathbf{K}^\beta - \mathbf{K})_{ii}^{-1}} \right]^{-1} \right\}^{-1}. \tag{9.7.21b}$$

In view of (9.7.16a), the following bound for $\overline{\mathbf{K}}^P$ is obtained:

$$\overline{K}_{ii}^P \geq \overline{K} \qquad \text{if } \mathbf{K}' - \mathbf{K} \text{ is negative–definite.} \tag{9.7.22a}$$

By definition of the integral operator $\mathbf{\Gamma}^A$, the following equivalence relation holds between $\hat{I}(\boldsymbol{\phi}^*;\ \boldsymbol{\Lambda}^A;\ \mathbf{q}^o)$ and $\hat{J}(\boldsymbol{\psi}^*;\ \mathbf{\Gamma}^A;\ \mathbf{p}^o)$:

$$I^A(\boldsymbol{\phi}^*;\ \boldsymbol{\Lambda}^A;\ \mathbf{q}^o) + J^A(\boldsymbol{\psi}^*;\ \mathbf{\Gamma}^A;\ \mathbf{p}^o) = \frac{1}{2} < \boldsymbol{\psi}^* > : \mathbf{R} : < \boldsymbol{\psi}^* >, \tag{9.7.23}$$

where $\boldsymbol{\phi}^* = -\mathbf{R} \cdot \boldsymbol{\psi}^*$ and $\mathbf{q}^o = \mathbf{K} \cdot \mathbf{p}^o + < \boldsymbol{\psi}^* >$. From (9.7.23), I^A which corresponds to J^A for the above special eigenflux field, is computed exactly. Hence, in view of inequality (9.7.16b), the following bound for $\overline{\mathbf{R}}^Q$ is obtained:

$$\overline{R}_{ii}^Q \geq \frac{9}{\overline{K}} \qquad \text{if } \mathbf{R}' - \mathbf{R} \text{ is negative–definite,} \tag{9.7.22b}$$

where $\overline{K}$ is given by (9.7.21b).

Now, the explicit form of the Green function of the infinite homogeneous B and the associated integral operator $\mathbf{\Gamma}^A$ is presented, and the properties of $\mathbf{\Gamma}^A$ are proved. Since, in the presence of a source b, the governing equation for a potential u in B is $\boldsymbol{\nabla} \cdot \{-\mathbf{K} \cdot (\boldsymbol{\nabla} u)\} + b = 0$, the Green function of B is

$$g^\infty(\mathbf{x}) = \frac{-1}{(2\pi)^3} \int_{-\infty}^{+\infty} (\boldsymbol{\xi} \cdot \mathbf{K} \cdot \boldsymbol{\xi})^{-1} \exp(\iota \boldsymbol{\xi} \cdot \mathbf{x})\, dV_\xi. \tag{9.7.24}$$

Hence, when an eigenflux field $\boldsymbol{\psi}^*$ is prescribed in B, the resulting force field is given by

$$-\mathbf{p}^A(\mathbf{x}) = \mathbf{\Gamma}^A(\mathbf{x};\ \boldsymbol{\psi}^*)$$

$$= \int_{-\infty}^{+\infty} F\mathbf{\Gamma}^{A}(\boldsymbol{\xi}) \cdot F\boldsymbol{\psi}^{*}(\boldsymbol{\xi}) \exp(\iota\boldsymbol{\xi}\cdot\mathbf{x})\, dV_{\xi}, \tag{9.7.25a}$$

where

$$F\boldsymbol{\psi}^{*}(\boldsymbol{\xi}) = \frac{1}{(2\pi)^3} \int_{V'} \boldsymbol{\psi}^{*}(\mathbf{x}) \exp(-\iota\boldsymbol{\xi}\cdot\mathbf{x})\, dV_{\mathbf{x}},$$

$$F\mathbf{\Gamma}^{A}(\boldsymbol{\xi}) = \begin{cases} \bar{\boldsymbol{\xi}}\otimes\bar{\boldsymbol{\xi}} / \bar{\boldsymbol{\xi}}\cdot\mathbf{K}\cdot\bar{\boldsymbol{\xi}} & \boldsymbol{\xi} \neq \mathbf{0} \\ 0 & \boldsymbol{\xi} = \mathbf{0}, \end{cases} \tag{9.7.25b,c}$$

and where $\bar{\boldsymbol{\xi}} \equiv \boldsymbol{\xi} / |\boldsymbol{\xi}|$.

The proof of (9.7.14a~c) and (9.7.19a) is as follows: 1) For any arbitrary eigenflux distributed in an ellipsoidal region, the volume average of the resulting forces is given by the product of a certain second-order tensor and the average of the eigenflux. Hence, (9.7.14a) and (9.7.14b) hold as long as V′ is ellipsoidal and $< \boldsymbol{\psi}^{*d} > = \mathbf{0}$. 2) Due to the positive-definiteness of $\mathbf{K}$, the following inequality holds:

$$< \mathbf{q}^{d}\cdot\mathbf{p}^{d} > = < \boldsymbol{\psi}^{*d}\cdot(\mathbf{\Gamma}^{A}\cdot\mathbf{K}^{A}\cdot\mathbf{\Gamma} - \mathbf{\Gamma}^{A})\cdot\boldsymbol{\psi}^{*d} >$$

$$\leq \frac{1}{V'} \int_{B} \mathbf{\Gamma}^{A}(\mathbf{x};\, \boldsymbol{\psi}^{*d})\cdot(\mathbf{K}\cdot\mathbf{\Gamma}^{A}(\mathbf{x};\, \boldsymbol{\psi}^{*d}) - \boldsymbol{\psi}^{*d})\, dV_{\mathbf{x}}. \tag{9.7.26a}$$

The right side of (9.7.26a) is expressed in terms of $F\boldsymbol{\psi}^{*d}$ and $F\mathbf{\Gamma}$, as

$$\frac{1}{V'} \int_{-\infty}^{+\infty} F\boldsymbol{\psi}^{*d}(-\boldsymbol{\xi})\cdot\{F\mathbf{\Gamma}^{A}(-\boldsymbol{\xi})\cdot\mathbf{K}\cdot F\mathbf{\Gamma}^{A}(\boldsymbol{\xi}) - F\mathbf{\Gamma}^{A}(-\boldsymbol{\xi})\}\cdot F\boldsymbol{\psi}^{*d}(\boldsymbol{\xi})\, dV_{\xi}.$$

From (9.7.16c), $F\mathbf{\Gamma}^{A}$ satisfies

$$F\mathbf{\Gamma}^{A}(-\boldsymbol{\xi})\cdot\mathbf{K}\cdot F\mathbf{\Gamma}^{A}(\boldsymbol{\xi}) = F\mathbf{\Gamma}^{A}(\boldsymbol{\xi})\cdot\mathbf{K}\cdot F\mathbf{\Gamma}^{A}(\boldsymbol{\xi}) = F\mathbf{\Gamma}^{A}(\boldsymbol{\xi}), \tag{9.7.26b}$$

for $\boldsymbol{\xi} \neq \mathbf{0}$, and the term in the curly brackets vanishes. Hence, (9.7.14c) holds. 3) For isotropic $\mathbf{K}$, the Fourier transform of $\mathbf{\Gamma}^{A}$ is

$$F\mathbf{\Gamma}^{A}(\boldsymbol{\xi}) = \frac{1}{K}\, \bar{\boldsymbol{\xi}}\otimes\bar{\boldsymbol{\xi}}, \tag{9.7.27a}$$

where $\bar{\boldsymbol{\xi}} \equiv \boldsymbol{\xi} / |\boldsymbol{\xi}|$; see definition (9.7.25a,b) of $\mathbf{\Gamma}^{A}$. From $\bar{\xi}_i \bar{\xi}_i = 1$, the components of $F\mathbf{\Gamma}^{A}$ satisfy

$$F\Gamma^{A}_{ii} = \frac{1}{K}, \tag{9.7.27b}$$

from which the inverse Fourier transform of $F\Gamma^{A}_{ii}$ becomes

$$\frac{1}{(2\pi)^3} \int_{-\infty}^{+\infty} F\Gamma^{A}_{ii}(\boldsymbol{\xi}) \exp(\iota\boldsymbol{\xi}\cdot\mathbf{x})\, dV_{\xi} = \frac{1}{K}\, \delta(\mathbf{x}), \tag{9.7.27c}$$

where $\delta(\mathbf{x})$ is the delta function at the origin. Hence, (9.7.19a) is obtained.

9.8. BOUNDS FOR OVERALL MODULI OF PIEZOELECTRIC RVE'S

Consider a linear RVE of heterogeneous piezoelectric materials whose properties are defined by spatially variable moduli $(\mathbf{C}', \mathbf{h}', \mathbf{R}') = (\mathbf{C}'(\mathbf{x}), \mathbf{h}'(\mathbf{x}), \mathbf{R}'(\mathbf{x}))$. To simplify the expressions, denote the inverse relations corresponding to these moduli by $(\mathbf{D}', \mathbf{i}', \mathbf{K}') = (\mathbf{D}'(\mathbf{x}), \mathbf{i}'(\mathbf{x}), \mathbf{K}'(\mathbf{x}))$, instead of $((\mathbf{D}^{\mathrm{P}})', (\mathbf{i}^{\mathrm{P}})', (\mathbf{K}^{\mathrm{P}})')$ which are defined by (2.8.26a~c), keeping in mind, however, that $(\mathbf{D}', \mathbf{i}', \mathbf{K}')$ now stand for $((\mathbf{D}^{\mathrm{P}})', (\mathbf{i}^{\mathrm{P}})', (\mathbf{K}^{\mathrm{P}})')$ *in all equations that follow*.

In this subsection, first the Hashin-Shtrikman variational principles are generalized and applied to the coupled piezoelectricity problems. Then these variational principles are used to obtain exact bounds on the corresponding moduli. Except for the notation, the presentation here follows the recent work by Hori and Nemat-Nasser (1998).

9.8.1. Generalized Hashin-Shtrikman Variational Principles

First, apply the equivalent inclusion method to coupled problems of piezoelectricity. The constitutive relations of the heterogeneous RVE are

$$\boldsymbol{\sigma}(\mathbf{x}) = \mathbf{C}'(\mathbf{x}) : \boldsymbol{\varepsilon}(\mathbf{x}) + \mathbf{h}'(\mathbf{x}) \cdot \mathbf{q}(\mathbf{x}),$$

$$\mathbf{p}(\mathbf{x}) = (\mathbf{h}')^{\mathrm{T}}(\mathbf{x}) : \boldsymbol{\varepsilon}(\mathbf{x}) + \mathbf{R}'(\mathbf{x}) \cdot \mathbf{q}(\mathbf{x}), \tag{9.8.1a,b}$$

or, in index notation,

$$\sigma_{ij}(\mathbf{x}) = C_{ijkl}{}'(\mathbf{x})\, \varepsilon_{kl}(\mathbf{x}) + h_{ijk}{}'(\mathbf{x})\, q_k(\mathbf{x}),$$

$$p_i(\mathbf{x}) = h_{kli}{}'(\mathbf{x})\, \varepsilon_{kl}(\mathbf{x}) + R_{ik}{}'(\mathbf{x})\, q_k(\mathbf{x}). \tag{9.8.1c,d}$$

This RVE is subjected to suitable boundary conditions which satisfy

$$\mathbf{t}^{\mathrm{d}}(\mathbf{x}) \cdot \mathbf{u}^{\mathrm{d}}(\mathbf{x}) + q^{\mathrm{d}}(\mathbf{x})\, u^{\mathrm{d}}(\mathbf{x}) = 0 \qquad \text{on } \partial V, \tag{9.8.2a}$$

where

$$\mathbf{t}^{\mathrm{d}} = \boldsymbol{\nu} \cdot (\boldsymbol{\sigma} - <\boldsymbol{\sigma}>), \qquad \mathbf{u}^{\mathrm{d}} = \mathbf{u} - \mathbf{x} \cdot <\boldsymbol{\varepsilon}>,$$

$$q^{\mathrm{d}} = \boldsymbol{\nu} \cdot (\mathbf{q} - <\mathbf{q}>), \qquad u^{\mathrm{d}} = u - \mathbf{x} \cdot <\mathbf{p}>.$$

Condition (9.8.2a) leads to

$$<\boldsymbol{\sigma} : \boldsymbol{\varepsilon}> + <\mathbf{q} \cdot \mathbf{p}> = <\boldsymbol{\sigma}> : <\boldsymbol{\varepsilon}> + <\mathbf{q}> \cdot <\mathbf{p}>. \tag{9.8.2b}$$

Hence, the boundary data are consistent, ensuring that the energy definition of the corresponding moduli coincides with that obtained through the averaging of the field equations.

Introduce a homogeneous RVE with constant reference material properties $(\mathbf{C}, \mathbf{h}, \mathbf{R})$, and prescribe within it suitable eigenstress and eigen-electric fields, $\boldsymbol{\sigma}^* = \boldsymbol{\sigma}^*(\mathbf{x})$ and $\mathbf{p}^* = \mathbf{p}^*(\mathbf{x})$, such that the resulting coupled fields are the same as those of the original heterogeneous RVE. The constitutive relations of this homogeneous RVE are

$$\boldsymbol{\sigma}(\mathbf{x}) = \mathbf{C} : \boldsymbol{\varepsilon}(\mathbf{x}) + \mathbf{h}\cdot\mathbf{q}(\mathbf{x}) + \boldsymbol{\sigma}^*(\mathbf{x}),$$

$$\mathbf{p}(\mathbf{x}) = \mathbf{h}^{\mathrm{T}} : \boldsymbol{\varepsilon}(\mathbf{x}) + \mathbf{R}\cdot\mathbf{q}(\mathbf{x}) + \mathbf{p}^*(\mathbf{x}), \tag{9.8.3a,b}$$

or, in index notation,

$$\sigma_{ij}(\mathbf{x}) = C_{ijkl}\,\varepsilon_{kl}(\mathbf{x}) + h_{ijk}\,q_k(\mathbf{x}) + \sigma^*_{ij}(\mathbf{x}),$$

$$p_i(\mathbf{x}) = h_{kli}\,\varepsilon_{kl}(\mathbf{x}) + R_{ik}\,q_k(\mathbf{x}) + p^*_i(\mathbf{x}). \tag{9.8.3c,d}$$

Since these must be the same as those of the original problem, the consistency conditions for $\boldsymbol{\sigma}^*$ and $\mathbf{p}^*$ are obtained as follows:

$$\mathbf{C}'(\mathbf{x}) : \boldsymbol{\varepsilon}(\mathbf{x}) + \mathbf{h}'(\mathbf{x})\cdot\mathbf{q}(\mathbf{x}) = \mathbf{C} : \boldsymbol{\varepsilon}(\mathbf{x}) + \mathbf{h}\cdot\mathbf{q}(\mathbf{x}) + \boldsymbol{\sigma}^*(\mathbf{x}),$$

$$(\mathbf{h}')^{\mathrm{T}}(\mathbf{x}) : \boldsymbol{\varepsilon}(\mathbf{x}) + \mathbf{R}'(\mathbf{x})\cdot\mathbf{q}(\mathbf{x}) = \mathbf{h}^{\mathrm{T}} : \boldsymbol{\varepsilon}(\mathbf{x}) + \mathbf{R}\cdot\mathbf{q}(\mathbf{x}) + \mathbf{p}^*(\mathbf{x}), \tag{9.8.4a,b}$$

or

$$(\mathbf{C}'(\mathbf{x}) - \mathbf{C}) : \boldsymbol{\varepsilon}(\mathbf{x}) + (\mathbf{h}'(\mathbf{x}) - \mathbf{h})\cdot\mathbf{q}(\mathbf{x}) = \boldsymbol{\sigma}^*(\mathbf{x}),$$

$$(\mathbf{h}'(\mathbf{x}) - \mathbf{h})^{\mathrm{T}} : \boldsymbol{\varepsilon}(\mathbf{x}) + (\mathbf{R}'(\mathbf{x}) - \mathbf{R})\cdot\mathbf{q}(\mathbf{x}) = \mathbf{p}^*(\mathbf{x}). \tag{9.8.4c,d}$$

In view of (2.8.6a,b), the governing equations for the mechanical displacement and the electric potential fields in the homogeneous RVE become

$$\boldsymbol{\nabla}\cdot(\mathbf{C}^{\mathrm{P}} : (\boldsymbol{\nabla}\otimes\mathbf{u}(\mathbf{x}))) - \boldsymbol{\nabla}\cdot(\mathbf{i}\cdot(\boldsymbol{\nabla}\mathrm{u}(\mathbf{x}))) + \boldsymbol{\nabla}\cdot\boldsymbol{\sigma}^*(\mathbf{x}) - \boldsymbol{\nabla}\cdot(\mathbf{i}\cdot\mathbf{p}^*(\mathbf{x})) = \mathbf{0},$$

$$-\,\boldsymbol{\nabla}\cdot(\mathbf{i}^{\mathrm{T}} : (\boldsymbol{\nabla}\otimes\mathbf{u}(\mathbf{x}))) - \boldsymbol{\nabla}\cdot(\mathbf{K}\cdot(\boldsymbol{\nabla}\mathrm{u}(\mathbf{x}))) - \boldsymbol{\nabla}\cdot(\mathbf{K}\cdot\mathbf{p}^*(\mathbf{x})) = 0, \tag{9.8.5a,b}$$

or, in index form,

$$C^{\mathrm{P}}_{ijkl}\,u_{k,li}(\mathbf{x}) - i_{ijk}\,\mathrm{u}_{,ki}(\mathbf{x}) + \sigma^*_{ij,i}(\mathbf{x}) - i_{ijk}\,p^*_{k,i}(\mathbf{x}) = 0,$$

$$-\,i_{kli}\,u_{k,li}(\mathbf{x}) - K_{ik}\,\mathrm{u}_{,ik}(\mathbf{x}) - K_{ik}\,p^*_{k,i}(\mathbf{x}) = 0, \tag{9.8.5c,d}$$

where $\mathbf{K} = \mathbf{R}^{-1}$ and $\mathbf{C}^{\mathrm{P}}$ and $\mathbf{i}$ are given by (2.8.5a,b). Solve this boundary-value problem for $\mathbf{u}$ and u with the boundary conditions which satisfy (9.8.2), when $\boldsymbol{\sigma}^*$ and $\mathbf{p}^*$ are given and the average strain and electric displacement are prescribed, i.e., $<\boldsymbol{\varepsilon}> = \boldsymbol{\varepsilon}^{\mathrm{o}}$ and $<\mathrm{q}> = \mathbf{q}^{\mathrm{o}}$. Hence, formally express the mechanical strain and the electric displacement in terms of suitable integral operators which are defined by the Green function of the boundary-value problem, as follows:

$$\boldsymbol{\varepsilon}(\mathbf{x}) = \boldsymbol{\varepsilon}^{\mathrm{o}} - \boldsymbol{\Gamma}^{\mathrm{G}}(\mathbf{x};\, \boldsymbol{\sigma}^*,\, \mathbf{p}^*), \qquad \mathbf{q}(\mathbf{x}) = \mathbf{q}^{\mathrm{o}} - \boldsymbol{\lambda}^{\mathrm{G}}(\mathbf{x};\, \boldsymbol{\sigma}^*,\, \mathbf{p}^*). \tag{9.8.6a,b}$$

Here, $\boldsymbol{\Gamma}^{\mathrm{G}}$ and $\boldsymbol{\lambda}^{\mathrm{G}}$ are integral operators defined for any given geometry and zero boundary data, in a manner similar to the corresponding integral operators introduced in Subsections 9.1 and 9.7.

Making use of (9.8.6a,b), integral equations for $\boldsymbol{\sigma}^*$ and $\mathbf{p}^*$ are now obtained from the consistency conditions (9.8.4c,d). Indeed, (9.8.4c,d) can be rewritten such that $\boldsymbol{\varepsilon}$ and $\mathbf{q}$ are explicitly expressed as[48]

$$\boldsymbol{\varepsilon}(\mathbf{x}) = \Delta\mathbf{C}^{-1}(\mathbf{x}) : \boldsymbol{\sigma}^*(\mathbf{x}) + \Delta\mathbf{h}^{-1}(\mathbf{x})\cdot\mathbf{p}^*(\mathbf{x}),$$

$$\mathbf{q}(\mathbf{x}) = \Delta\mathbf{h}^{-\mathrm{T}}(\mathbf{x}) : \boldsymbol{\sigma}^*(\mathbf{x}) + \Delta\mathbf{R}^{-1}(\mathbf{x}) \cdot \mathbf{p}^*(\mathbf{x}), \tag{9.8.4e,f}$$

where

$$\Delta\mathbf{C}(\mathbf{x}) = (\mathbf{C}'(\mathbf{x}) - \mathbf{C}) - (\mathbf{h}'(\mathbf{x}) - \mathbf{h}) \cdot (\mathbf{R}'(\mathbf{x}) - \mathbf{R})^{-1} \cdot (\mathbf{h}'(\mathbf{x}) - \mathbf{h}),$$

$$\Delta\mathbf{h}^{-1}(\mathbf{x}) = -\Delta\mathbf{C}^{-1}(\mathbf{x}) : (\mathbf{h}'(\mathbf{x}) - \mathbf{h}) \cdot (\mathbf{R}'(\mathbf{x}) - \mathbf{R})^{-1},$$

$$\Delta\mathbf{R}(\mathbf{x}) = (\mathbf{R}'(\mathbf{x}) - \mathbf{R}) - (\mathbf{h}'(\mathbf{x}) - \mathbf{h})^{\mathrm{T}} : (\mathbf{C}'(\mathbf{x}) - \mathbf{C})^{-1} : (\mathbf{h}'(\mathbf{x}) - \mathbf{h}). \tag{9.8.4g~i}$$

Hence, the integral equations for $\boldsymbol{\sigma}^*$ and $\mathbf{p}^*$ are

$$\Delta\mathbf{C}^{-1}(\mathbf{x}) : \boldsymbol{\sigma}^*(\mathbf{x}) + \Delta\mathbf{h}^{-1}(\mathbf{x}) \cdot \mathbf{p}^*(\mathbf{x}) + \boldsymbol{\Gamma}^{\mathrm{G}}(\mathbf{x};\, \boldsymbol{\sigma}^*,\, \mathbf{p}^*) - \boldsymbol{\varepsilon}^{\mathrm{o}} = \mathbf{0},$$

$$\Delta\mathbf{h}^{-\mathrm{T}}(\mathbf{x}) : \boldsymbol{\sigma}^*(\mathbf{x}) + \Delta\mathbf{R}^{-1}(\mathbf{x}) \cdot \mathbf{p}^*(\mathbf{x}) + \boldsymbol{\lambda}^{\mathrm{G}}(\mathbf{x};\, \boldsymbol{\sigma}^*,\, \mathbf{p}^*) - \mathbf{q}^{\mathrm{o}} = \mathbf{0}. \tag{9.8.7a,b}$$

Now define a functional for arbitrary $(\mathbf{s}^*, \boldsymbol{\phi}^*)$, whose Euler equations coincide with (9.8.7a,b). In view of (9.8.2b), this functional is given by

$$\mathrm{J}(\mathbf{s}^*, \boldsymbol{\phi}^*; \boldsymbol{\Gamma}^{\mathrm{G}}, \boldsymbol{\lambda}^{\mathrm{G}}; \boldsymbol{\varepsilon}^{\mathrm{o}}, \mathbf{q}^{\mathrm{o}}) = \frac{1}{2} < \mathbf{s}^* : (\Delta\mathbf{C}^{-1} : \mathbf{s}^* + \Delta\mathbf{h}^{-1} \cdot \boldsymbol{\phi}^* + \boldsymbol{\Gamma}^{\mathrm{G}}(\mathbf{s}^*, \boldsymbol{\phi}^*))$$

$$+ \boldsymbol{\phi}^* \cdot (\Delta\mathbf{h}^{-\mathrm{T}} : \mathbf{s}^* + \Delta\mathbf{R}^{-1} \cdot \boldsymbol{\phi}^* + \boldsymbol{\lambda}^{\mathrm{G}}(\mathbf{s}^*, \boldsymbol{\phi}^*)) > - (< \mathbf{s}^* > : \boldsymbol{\varepsilon}^{\mathrm{o}} + < \boldsymbol{\phi}^* > \cdot \mathbf{q}^{\mathrm{o}}). \tag{9.8.8a}$$

The first variation of J is

$$\delta\mathrm{J}(\mathbf{s}^*, \boldsymbol{\phi}^*; \boldsymbol{\Gamma}^{\mathrm{G}}, \boldsymbol{\lambda}^{\mathrm{G}}; \boldsymbol{\varepsilon}^{\mathrm{o}}, \mathbf{q}^{\mathrm{o}})$$

$$= < \delta\mathbf{s}^* : (\Delta\mathbf{C}^{-1} : \mathbf{s}^* + \Delta\mathbf{h}^{-1} \cdot \boldsymbol{\phi}^* + \boldsymbol{\Gamma}^{\mathrm{G}}(\mathbf{s}^*, \boldsymbol{\phi}^*) - \boldsymbol{\varepsilon}^{\mathrm{o}})$$

$$+ \delta\boldsymbol{\phi}^* \cdot (\Delta\mathbf{h}^{-\mathrm{T}} : \mathbf{s}^* + \Delta\mathbf{R}^{-1} \cdot \boldsymbol{\phi}^* + \boldsymbol{\lambda}^{\mathrm{G}}(\mathbf{s}^*, \boldsymbol{\phi}^*) - \mathbf{q}^{\mathrm{o}}) >. \tag{9.8.8b}$$

As is seen, the Euler equations coincide with the consistency conditions (9.8.4e,f). The exact eigenstress and eigen-electric fields, $\boldsymbol{\sigma}^*$ and $\mathbf{p}^*$, which satisfy (9.8.7a,b) render J stationary. The stationary value is

$$\mathrm{J}(\boldsymbol{\sigma}^*, \mathbf{p}^*; \boldsymbol{\Gamma}^{\mathrm{G}}, \boldsymbol{\lambda}^{\mathrm{G}}; \boldsymbol{\varepsilon}^{\mathrm{o}}, \mathbf{q}^{\mathrm{o}})$$

$$= -\frac{1}{2} (\boldsymbol{\sigma}^* : \boldsymbol{\varepsilon}^{\mathrm{o}} + \mathbf{p}^* \cdot \mathbf{q}^{\mathrm{o}})$$

$$= \frac{1}{2} (\boldsymbol{\varepsilon}^{\mathrm{o}} : (\mathbf{C} - \overline{\mathbf{C}}^{\mathrm{G}}) : \boldsymbol{\varepsilon}^{\mathrm{o}} + 2\boldsymbol{\varepsilon}^{\mathrm{o}} : (\mathbf{h} - \overline{\mathbf{h}}^{\mathrm{G}}) \cdot \mathbf{q}^{\mathrm{o}} + \mathbf{q}^{\mathrm{o}} \cdot (\mathbf{R} - \overline{\mathbf{R}}^{\mathrm{G}}) \cdot \mathbf{q}^{\mathrm{o}}), \tag{9.8.8c}$$

where $(\overline{\mathbf{C}}^{\mathrm{G}}, \overline{\mathbf{h}}^{\mathrm{G}}, \overline{\mathbf{R}}^{\mathrm{G}})$ are the overall material properties for the prescribed boundary data which satisfy (9.8.2a).

If the reference material properties are chosen such that $(\Delta\mathbf{C}, \Delta\mathbf{h}, \Delta\mathbf{R})$ are positive-semi-definite (negative-semi-definite)[49], then the integral operators

[48] In index notation, these equations are given as $\varepsilon_{ij}(\mathbf{x}) = \Delta C^{-1}_{ijkl}(\mathbf{x})\, \sigma^*_{kl}(\mathbf{x}) + \Delta h^{-1}_{ijk}(\mathbf{x})\, p^*_k(\mathbf{x})$ and $q_i(\mathbf{x}) = \Delta h^{-1}_{ikl}(\mathbf{x})\, \sigma^*_{kl}(\mathbf{x}) + \Delta R^{-1}_{ik}(\mathbf{x})\, p^*_k(\mathbf{x})$, respectively.

$(\boldsymbol{\Gamma}^G, \boldsymbol{\lambda}^G)$ associated with the averaging operator <...> for any pair of suitably smooth eigenstress and eigen-electric fields, $\mathbf{s}^*$ and $\boldsymbol{\phi}^*$, produce the following positive-semi-definite (negative-semi-definite) scalar quantity:

$$L(\mathbf{s}^*, \boldsymbol{\phi}^*) = < \mathbf{s}^* : (\Delta\mathbf{C}^{-1} : \mathbf{s}^* + \Delta\mathbf{h}^{-1} \cdot \boldsymbol{\phi}^* + \boldsymbol{\Gamma}^G(\mathbf{s}^*, \boldsymbol{\phi}^*))$$

$$+ \boldsymbol{\phi}^* \cdot (\Delta\mathbf{h}^{-T} : \mathbf{s}^* + \Delta\mathbf{R}^{-1} \cdot \boldsymbol{\phi}^* + \boldsymbol{\lambda}^G(\mathbf{s}^*, \boldsymbol{\phi}^*)) >.$$

In such a case, the operator $L(\mathbf{s}^*, \boldsymbol{\phi}^*)$ is called positive-semi-definite (negative-semi-definite). Hence, it follows that

$$\frac{1}{2}\,(\boldsymbol{\varepsilon}^o : (\mathbf{C} - \overline{\mathbf{C}}^G) : \boldsymbol{\varepsilon}^o + 2\boldsymbol{\varepsilon}^o : (\mathbf{h} - \overline{\mathbf{h}}^G) \cdot \mathbf{q}^o + \mathbf{q}^o \cdot (\mathbf{R} - \overline{\mathbf{R}}^G) \cdot \mathbf{q}^o)$$

$$= \mathrm{J}(\boldsymbol{\sigma}^*, \mathbf{p}^*; \boldsymbol{\Gamma}^G, \boldsymbol{\lambda}^G; \boldsymbol{\varepsilon}^o, \mathbf{q}^o)$$

$$\leq (\geq)\ \mathrm{J}(\mathbf{s}^*, \boldsymbol{\phi}^*; \boldsymbol{\Gamma}^G, \boldsymbol{\lambda}^G; \boldsymbol{\varepsilon}^o, \mathbf{q}^o), \tag{9.8.9a}$$

when $(\Delta\mathbf{C}, \Delta\mathbf{h}, \Delta\mathbf{R})$ are positive-semi-definite (negative-semi-definite).

An alternative functional and the related inequalities can be obtained using the dual formulation discussed in the preceding subsections. Instead of $\mathbf{s}^*$ and $\boldsymbol{\phi}^*$, prescribe $\mathbf{e}^*$ and $\boldsymbol{\psi}^*$ in the homogeneous RVE. Setting $\mathbf{s}^* = -\mathbf{C} : \mathbf{e}^*$ and $\boldsymbol{\phi}^* = -\mathbf{R} \cdot \boldsymbol{\psi}^*$, define the integral operators $\boldsymbol{\Lambda}^G$ and $\boldsymbol{\gamma}^G$ as follows:

$$\boldsymbol{\Lambda}^G(\mathbf{x}; \mathbf{e}^*, \boldsymbol{\psi}^*) = \mathbf{C} : (\boldsymbol{\Gamma}^G(\mathbf{x}; -\mathbf{C} : \mathbf{e}^*, -\mathbf{R} \cdot \boldsymbol{\psi}^*) - (\mathbf{e}^* - < \mathbf{e}^* >)$$

$$+ \mathbf{h} \cdot \boldsymbol{\lambda}^G(\mathbf{x}; -\mathbf{C} : \mathbf{e}^*, -\mathbf{R} \cdot \boldsymbol{\psi}^*), \tag{9.8.6c}$$

and

$$\boldsymbol{\gamma}^G(\mathbf{x}; \mathbf{e}^*, \boldsymbol{\psi}^*) = \mathbf{h}^T : \boldsymbol{\Gamma}^G(\mathbf{x}; -\mathbf{C} : \mathbf{e}^*, -\mathbf{R} \cdot \boldsymbol{\psi}^*)$$

$$+ \mathbf{R} \cdot (\boldsymbol{\lambda}^G(\mathbf{x}; -\mathbf{C} : \mathbf{e}^*, -\mathbf{R} \cdot \boldsymbol{\psi}^*) - (\boldsymbol{\psi}^* - < \boldsymbol{\psi}^* >)). \tag{9.8.6d}$$

These are the expressions for the resulting disturbance stress and electric field, respectively, i.e., $\boldsymbol{\sigma}^d = \boldsymbol{\sigma} - < \boldsymbol{\sigma} > = -\boldsymbol{\Lambda}^G$ and $\mathbf{p}^d = \mathbf{p} - < \mathbf{p} > = -\boldsymbol{\gamma}^G$. The consistency conditions now are[50]

$$\boldsymbol{\varepsilon}(\mathbf{x}) = \Delta\mathbf{D}^{-1}(\mathbf{x}) : \boldsymbol{\varepsilon}^*(\mathbf{x}) + \Delta\mathbf{i}^{-1}(\mathbf{x}) \cdot \mathbf{q}^*(\mathbf{x}),$$

$$\mathbf{p}(\mathbf{x}) = \Delta\mathbf{i}^{-T}(\mathbf{x}) : \boldsymbol{\varepsilon}^*(\mathbf{x}) + \Delta\mathbf{K}^{-1}(\mathbf{x}) \cdot \mathbf{q}^*(\mathbf{x}), \tag{9.8.4j,k}$$

where

$$\Delta\mathbf{D}(\mathbf{x}) = (\mathbf{D}'(\mathbf{x}) - \mathbf{D}) - (\mathbf{i}'(\mathbf{x}) - \mathbf{i}) \cdot (\mathbf{K}'(\mathbf{x}) - \mathbf{K})^{-1} \cdot (\mathbf{i}'(\mathbf{x}) - \mathbf{i}),$$

$$\Delta\mathbf{i}^{-1}(\mathbf{x}) = -\Delta\mathbf{D}^{-1}(\mathbf{x}) : (\mathbf{i}'(\mathbf{x}) - \mathbf{i}) \cdot (\mathbf{K}'(\mathbf{x}) - \mathbf{K})^{-1},$$

---

[49] When, for any pair of nonzero strain tensor, $\boldsymbol{\varepsilon}$, and electric displacement vector, $\mathbf{q}$, the scalar quantity, $\boldsymbol{\varepsilon} : \mathbf{C} : \boldsymbol{\varepsilon} + 2\boldsymbol{\varepsilon} : \mathbf{h}.\mathbf{q} + \mathbf{q}.\mathrm{R}.\mathbf{q} \geq (\leq)\ 0$, is non-negative (non-positive), then $(\mathbf{C}, \mathbf{h}, \mathbf{R})$ are called positive-semi-definite (negative-semi-definite).

$$\Delta\mathbf{K}(\mathbf{x}) = (\mathbf{K}'(\mathbf{x}) - \mathbf{K}) - (\mathbf{i}'(\mathbf{x}) - \mathbf{i})^{\mathrm{T}} : (\mathbf{D}'(\mathbf{x}) - \mathbf{D})^{-1} : (\mathbf{i}'(\mathbf{x}) - \mathbf{i}). \qquad (9.8.4l\sim n)$$

A functional for $(\mathbf{e}^*, \boldsymbol{\psi}^*)$ is then defined as

$$\mathrm{I}(\mathbf{e}^*, \boldsymbol{\psi}^*; \boldsymbol{\Lambda}^{\mathrm{G}}, \boldsymbol{\gamma}^{\mathrm{G}}; \boldsymbol{\sigma}^{\mathrm{o}}, \mathbf{p}^{\mathrm{o}}) = \frac{1}{2} < \mathbf{e}^* : (\Delta\mathbf{D}^{-1} : \mathbf{e}^* + \Delta\mathbf{i}^{-1} \cdot \boldsymbol{\psi}^* + \boldsymbol{\Lambda}^{\mathrm{G}}(\mathbf{e}^*, \boldsymbol{\psi}^*))$$

$$+ \boldsymbol{\psi}^* \cdot (\Delta\mathbf{i}^{-\mathrm{T}} : \mathbf{e}^* + \Delta\mathbf{K}^{-1} \cdot \boldsymbol{\psi}^* + \boldsymbol{\lambda}^{\mathrm{G}}(\mathbf{e}^*, \boldsymbol{\psi}^*)) > - (< \mathbf{e}^* > : \boldsymbol{\sigma}^{\mathrm{o}} + < \boldsymbol{\psi}^* > \cdot \mathbf{p}^{\mathrm{o}}),$$

$$(9.8.8\mathrm{d})$$

whose first variation is

$$\delta\mathrm{I}(\mathbf{e}^*, \boldsymbol{\psi}^*; \boldsymbol{\Lambda}^{\mathrm{G}}, \boldsymbol{\gamma}^{\mathrm{G}}; \boldsymbol{\sigma}^{\mathrm{o}}, \mathbf{p}^{\mathrm{o}})$$

$$= < \delta\mathbf{e}^* : (\Delta\mathbf{D}^{-1} : \mathbf{e}^* + \Delta\mathbf{i}^{-1} \cdot \boldsymbol{\psi}^* + \boldsymbol{\Lambda}^{\mathrm{G}}(\mathbf{e}^*, \boldsymbol{\psi}^*) - \boldsymbol{\sigma}^{\mathrm{o}})$$

$$+ \delta\boldsymbol{\psi}^* \cdot (\Delta\mathbf{i}^{-\mathrm{T}} : \mathbf{e}^* + \Delta\mathbf{K}^{-1} \cdot \boldsymbol{\psi}^* + \boldsymbol{\gamma}^{\mathrm{G}}(\mathbf{e}^*, \boldsymbol{\psi}^*) - \mathbf{p}^{\mathrm{o}}) >. \qquad (9.8.8\mathrm{e})$$

The exact eigenstress tensor and eigen-electric-displacement vector, $\boldsymbol{\varepsilon}^*$ and $\mathbf{q}^*$, which satisfy consistency conditions (9.8.4j,k), render I stationary. The stationary value is

$$\mathrm{I}(\boldsymbol{\varepsilon}^*, \mathbf{q}^*; \boldsymbol{\Lambda}^{\mathrm{G}}, \boldsymbol{\gamma}^{\mathrm{G}}; \boldsymbol{\sigma}^{\mathrm{o}}, \mathbf{p}^{\mathrm{o}})$$

$$= -\frac{1}{2} (\boldsymbol{\varepsilon}^* : \boldsymbol{\sigma}^{\mathrm{o}} + \mathbf{q}^* \cdot \mathbf{p}^{\mathrm{o}})$$

$$= \frac{1}{2} (\boldsymbol{\sigma}^{\mathrm{o}} : (\mathbf{D} - \overline{\mathbf{D}}^{\mathrm{G}}) : \boldsymbol{\sigma}^{\mathrm{o}} + 2\boldsymbol{\sigma}^{\mathrm{o}} : (\mathbf{i} - \overline{\mathbf{i}}^{\mathrm{G}}) \cdot \mathbf{p}^{\mathrm{o}} + \mathbf{p}^{\mathrm{o}} \cdot (\mathbf{K} - \overline{\mathbf{K}}^{\mathrm{G}}) \cdot \mathbf{p}^{\mathrm{o}}). \quad (9.8.8\mathrm{f})$$

Finally, when $(\mathbf{D}' - \mathbf{D}, \mathbf{i}' - \mathbf{i}, \mathbf{K}' - \mathbf{K})$ are positive-semi-definite (negative-semi-definite), the following inequality holds:

$$\frac{1}{2} (\boldsymbol{\sigma}^{\mathrm{o}} : (\mathbf{D} - \overline{\mathbf{D}}^{\mathrm{G}}) : \boldsymbol{\sigma}^{\mathrm{o}} + 2\boldsymbol{\sigma}^{\mathrm{o}} : (\mathbf{i} - \overline{\mathbf{i}}^{\mathrm{G}}) \cdot \mathbf{p}^{\mathrm{o}} + \mathbf{p}^{\mathrm{o}} \cdot (\mathbf{K} - \overline{\mathbf{K}}^{\mathrm{G}}) \cdot \mathbf{p}^{\mathrm{o}})$$

$$= \mathrm{I}(\boldsymbol{\varepsilon}^*, \mathbf{q}^*; \boldsymbol{\Lambda}^{\mathrm{G}}, \boldsymbol{\gamma}^{\mathrm{G}}; \boldsymbol{\sigma}^{\mathrm{o}}, \mathbf{p}^{\mathrm{o}})$$

$$\leq (\geq)\ \mathrm{I}(\mathbf{e}^*, \boldsymbol{\psi}^*; \boldsymbol{\Lambda}^{\mathrm{G}}, \boldsymbol{\gamma}^{\mathrm{G}}; \boldsymbol{\sigma}^{\mathrm{o}}, \mathbf{p}^{\mathrm{o}}). \qquad (9.8.9\mathrm{b})$$

9.8.2. Consequence of Universal Theorems

As in the uncoupled mechanical and non-mechanical cases, it is not easy to explicitly determine integral operators $(\boldsymbol{\Gamma}^{\mathrm{G}}, \boldsymbol{\lambda}^{\mathrm{G}})$ or $(\boldsymbol{\Lambda}^{\mathrm{G}}, \boldsymbol{\gamma}^{\mathrm{G}})$. However, as in Subsection 9.5, Theorems I and II can be used to obtain computable bounds based on the integral operators $(\boldsymbol{\Gamma}^{\mathrm{A}}, \boldsymbol{\lambda}^{\mathrm{A}})$ or $(\boldsymbol{\Lambda}^{\mathrm{A}}, \boldsymbol{\gamma}^{\mathrm{A}})$ which are defined for an

---

[50] Note that $\mathbf{D}$ here and in the sequel stands for $\mathbf{D}^{\mathrm{P}} = (\mathbf{C} - \mathbf{h}.\mathbf{R}^{-1}.\mathbf{h}^{\mathrm{T}})^{-1}$ and $\mathbf{K}$ stands for $\mathbf{K}^{\mathrm{P}} = (\mathbf{R} - \mathbf{h}^{\mathrm{T}} : \mathbf{C}^{-1} : \mathbf{h})^{-1}$. Similar comments apply to $\mathbf{D}'$ and $\mathbf{K}'$; see definitions (2.8.26).

infinitely extended homogeneous domain. *As before, this is done in two steps: first consider uniform boundary data and use Theorems I and II, and then, relate the corresponding results to those obtained using the integral operators of the infinite homogeneous domain.*

Hence, taking advantage of Theorem II, and using uniform boundary conditions, obtain

$$\frac{1}{2} < \boldsymbol{\varepsilon}^{\mathrm{EQ}} : \mathbf{C}' : \boldsymbol{\varepsilon}^{\mathrm{EQ}} + 2\mathbf{q}^{\mathrm{EQ}} : \mathbf{h}' : \boldsymbol{\varepsilon}^{\mathrm{EQ}} + \mathbf{q}^{\mathrm{EQ}} : \mathbf{R}' : \mathbf{q}^{\mathrm{EQ}} >$$

$$\leq \frac{1}{2} < (\boldsymbol{\varepsilon}^{\mathrm{o}} + \boldsymbol{\varepsilon}^{\mathrm{d}}) : \mathbf{C}' : (\boldsymbol{\varepsilon}^{\mathrm{o}} + \boldsymbol{\varepsilon}^{\mathrm{d}}) + 2(\mathbf{q}^{\mathrm{o}} + \mathbf{q}^{\mathrm{d}}) : \mathbf{h}' : (\boldsymbol{\varepsilon}^{\mathrm{o}} + \boldsymbol{\varepsilon}^{\mathrm{d}})$$

$$+ (\mathbf{q}^{\mathrm{o}} + \mathbf{q}^{\mathrm{d}}) : \mathbf{R}' : (\mathbf{q}^{\mathrm{o}} + \mathbf{q}^{\mathrm{d}}) >, \tag{9.8.10a}$$

where $\boldsymbol{\varepsilon}^{\mathrm{EQ}} = \boldsymbol{\varepsilon}^{\mathrm{EQ}}(\mathbf{x})$ and $\mathbf{q}^{\mathrm{EQ}} = \mathbf{q}^{\mathrm{EQ}}(\mathbf{x})$ are the strain tensor and electric displacement field vector which result when the RVE is subjected to suitable uniform boundary conditions corresponding to the constant strain and the constant electric displacement, $(\boldsymbol{\varepsilon}^{\mathrm{o}}, \mathbf{q}^{\mathrm{o}})$, i.e., when $(\mathbf{u}, \mathrm{q}) = (\mathbf{x} \cdot \boldsymbol{\varepsilon}^{\mathrm{o}}, \boldsymbol{\nu} \cdot \mathbf{q}^{\mathrm{o}})$ on $\partial \mathrm{V}$, such that the overall average strain and electric displacement are $\boldsymbol{\varepsilon}^{\mathrm{o}}$ and $\mathbf{q}^{\mathrm{o}}$, respectively. A computable bound is obtained by combining this inequality with (9.8.9a).

An alternative formulation is possible, as in the case of the uncoupled mechanical and non-mechanical problems. Let $(\mathbf{C}, \mathbf{h}, \mathbf{R})$ be constant material properties such that $(\Delta\mathbf{C}, \Delta\mathbf{h}, \Delta\mathbf{R})$ are negative-semi-definite, i.e., for any pair of $(\boldsymbol{\varepsilon}, \mathbf{q})$ and $(\boldsymbol{\varepsilon}', \mathbf{q}')$,

$$0 \geq (\boldsymbol{\varepsilon} - \boldsymbol{\varepsilon}') : (\mathbf{C}' - \mathbf{C}) : (\boldsymbol{\varepsilon} - \boldsymbol{\varepsilon}') + 2(\mathbf{q} - \mathbf{q}') \cdot (\mathbf{h}' - \mathbf{h}) : (\boldsymbol{\varepsilon} - \boldsymbol{\varepsilon}')$$

$$+ (\mathbf{q} - \mathbf{q}') \cdot (\mathbf{R}' - \mathbf{R}) \cdot (\mathbf{q} - \mathbf{q}'). \tag{9.8.11}$$

Replacing $\boldsymbol{\varepsilon}'$ and $\mathbf{q}'$ by $\boldsymbol{\varepsilon}' = \Delta\mathbf{C}^{-1} : \mathbf{s}^* + \Delta\mathbf{h}^{-1} \cdot \boldsymbol{\phi}^*$ and $\mathbf{q}' = \Delta\mathbf{h}^{-\mathrm{T}} : \mathbf{s}^* + \Delta\mathbf{R}^{-1} \cdot \boldsymbol{\phi}^*$, and taking average of the resulting inequality, obtain

$$0 \geq < \boldsymbol{\varepsilon} : (\mathbf{C}' - \mathbf{C}) : \boldsymbol{\varepsilon} + 2\boldsymbol{\varepsilon} : (\mathbf{h}' - \mathbf{h}) \cdot \mathbf{q} + \mathbf{q} \cdot (\mathbf{R}' - \mathbf{R}) \cdot \mathbf{q}$$

$$- 2(\mathbf{s}^* : \boldsymbol{\varepsilon} + \boldsymbol{\phi}^* \cdot \mathbf{q}) + \mathbf{s}^* : \Delta\mathbf{C}^{-1} : \mathbf{s}^* + 2\mathbf{s}^* : \Delta\mathbf{h}^{-1} \cdot \mathbf{q}^* + \mathbf{q}^* \cdot \Delta\mathbf{R}^{-1} \cdot \mathbf{q}^* >. \tag{9.8.12}$$

Now compare the right-hand side of (9.8.10a) with (9.8.12), and obtain

$$\frac{1}{2} < \boldsymbol{\varepsilon}^{\mathrm{EQ}} : \mathbf{C}' : \boldsymbol{\varepsilon}^{\mathrm{EQ}} + 2\mathbf{q}^{\mathrm{EQ}} : \mathbf{h}' : \boldsymbol{\varepsilon}^{\mathrm{EQ}} + \mathbf{q}^{\mathrm{EQ}} : \mathbf{R}' : \mathbf{q}^{\mathrm{EQ}} >$$

$$\leq \frac{1}{2} (\boldsymbol{\varepsilon}^{\mathrm{o}} : \mathbf{C} : \boldsymbol{\varepsilon}^{\mathrm{o}} + 2\boldsymbol{\varepsilon}^{\mathrm{o}} : \mathbf{h} \cdot \mathbf{q}^{\mathrm{o}} + \mathbf{q}^{\mathrm{o}} \cdot \mathbf{R} \cdot \mathbf{q}^{\mathrm{o}})$$

$$- \frac{1}{2} < \Big\{ \mathbf{s}^* : (\Delta\mathbf{C}^{-1} : \mathbf{s}^* + \Delta\mathbf{h}^{-1} \cdot \boldsymbol{\phi}^*) - \mathbf{s}^* : (\boldsymbol{\varepsilon}^{\mathrm{d}} + 2\boldsymbol{\varepsilon}^{\mathrm{o}})$$

$$+ \boldsymbol{\phi}^* \cdot (\Delta\mathbf{h}^{-T} : \mathbf{s}^* + \Delta\mathbf{R}^{-1} \cdot \boldsymbol{\phi}^*) - \boldsymbol{\phi}^* \cdot (\mathbf{q}^d + 2\mathbf{q}^o) \Big\} >$$

$$+ \frac{1}{2} < \boldsymbol{\sigma}^d : \boldsymbol{\varepsilon}^d + \mathbf{q}^d \cdot \mathbf{p}^d >. \tag{9.8.13}$$

As in the case of the mechanical and non-mechanical problems, it can be proved that the following relations[51] hold for any arbitrary ellipsoidal RVE:

$$< \boldsymbol{\varepsilon}^A > = < -\boldsymbol{\Gamma}^A(\mathbf{s}^*, \boldsymbol{\psi}^*) > = \mathbf{0},$$

$$< \mathbf{q}^A > = < -\boldsymbol{\lambda}^A(\mathbf{s}^*, \boldsymbol{\psi}^*) > = \mathbf{0},$$

$$< \boldsymbol{\sigma}^A > = < -\mathbf{C} : \boldsymbol{\Gamma}^A(\mathbf{s}^*, \boldsymbol{\psi}^*) - \mathbf{h} : \boldsymbol{\lambda}^A(\mathbf{s}^*, \boldsymbol{\psi}^*) + \mathbf{s}^{*d} > = \mathbf{0},$$

$$< \mathbf{p}^A > = < -\mathbf{h}^T : \boldsymbol{\Gamma}^A(\mathbf{s}^*, \boldsymbol{\psi}^*) - \mathbf{R} : \boldsymbol{\lambda}^A(\mathbf{s}^*, \boldsymbol{\psi}^*) + \boldsymbol{\psi}^{*d} > = \mathbf{0},$$

$$< \boldsymbol{\sigma}^A : \boldsymbol{\varepsilon}^A + \mathbf{q}^A \cdot \mathbf{p}^A > = < (\boldsymbol{\Gamma}^A : (\mathbf{C} : \boldsymbol{\Gamma}^A + \mathbf{h} \cdot \boldsymbol{\lambda}^A) + \boldsymbol{\lambda}^A \cdot (\mathbf{h}^T : \boldsymbol{\Gamma}^A + \mathbf{R} \cdot \boldsymbol{\lambda}^A))$$

$$- (\mathbf{s}^* : (\mathbf{C} : \boldsymbol{\Gamma}^A + \mathbf{h} \cdot \boldsymbol{\lambda}^A) + \boldsymbol{\phi}^* \cdot (\mathbf{h}^T : \boldsymbol{\Gamma}^A + \mathbf{R} \cdot \boldsymbol{\lambda}^A)) > \leq 0.$$

(9.8.14a~e)

Here, *superscript* A *stands for the quantities which are computed using the integral operators that are determined from Green's function of an infinite homogeneous body.*

In view of (9.8.14a~e), replacing $(\boldsymbol{\varepsilon}^d, \mathbf{q}^d)$ by $(\boldsymbol{\varepsilon}^A, \mathbf{q}^A)$, obtain the following inequality from (9.8.13):

$$\frac{1}{2} < \boldsymbol{\varepsilon}^{EQ} : \mathbf{C}' : \boldsymbol{\varepsilon}^{EQ} + 2\mathbf{q}^{EQ} : \mathbf{h}' : \boldsymbol{\varepsilon}^{EQ} + \mathbf{q}^{EQ} : \mathbf{R}' : \mathbf{q}^{EQ} >$$

$$\leq \frac{1}{2} (\boldsymbol{\varepsilon}^o : \mathbf{C} : \boldsymbol{\varepsilon}^o + 2\boldsymbol{\varepsilon}^o : \mathbf{h} \cdot \mathbf{q}^o + \mathbf{q}^o \cdot \mathbf{R} \cdot \mathbf{q}^o) - J(\mathbf{s}^*, \boldsymbol{\phi}^*; \boldsymbol{\Gamma}^A, \boldsymbol{\lambda}^A; \boldsymbol{\varepsilon}^o, \mathbf{q}^o). \tag{9.8.15}$$

Furthermore, if $(\overline{\mathbf{C}}^{EQ}, \overline{\mathbf{h}}^{EQ}, \overline{\mathbf{R}}^{EQ})$ are defined as the overall material properties of the RVE subjected to the uniform boundary conditions of $(\boldsymbol{\varepsilon}^o, \mathbf{q}^o)$, (9.8.15) yields

$$\frac{1}{2} (\boldsymbol{\varepsilon}^o : (\overline{\mathbf{C}}^{EQ} - \mathbf{C}) : \boldsymbol{\varepsilon}^o + 2\boldsymbol{\varepsilon}^o : (\overline{\mathbf{h}}^{EQ} - \mathbf{h}) \cdot \mathbf{q}^o + \mathbf{q}^o \cdot (\overline{\mathbf{R}}^{EQ} - \mathbf{R}) \cdot \mathbf{q}^o)$$

$$\leq -J(\mathbf{s}^*, \boldsymbol{\phi}^*; \boldsymbol{\Gamma}^A, \boldsymbol{\lambda}^A; \boldsymbol{\varepsilon}^o, \mathbf{q}^o). \tag{9.8.16a}$$

---

[51] The proof is essentially the same as in the case of the uncoupled mechanical and non-mechanical problems. Furthermore, Eshelby's results can be extended to the coupled problem, i.e., it can be shown that constant eigenstress and eigen-electric fields or constant eigenstrain and eigen-electric-displacement fields, distributed in any arbitrary ellipsoidal region, produce constant fields within the region; this has been shown by Deeg (1980). However, the computation of the corresponding Eshelby's tensors in the coupled mechanical and non-mechanical problem, requires tedious manipulations.

In this expression, the functional J is computable in terms of the Green functions of the homogeneous infinite domain. Therefore, this quantity can be computed in terms of the corresponding Eshelby tensors, as outlined in Subsection 9.8.3.

In essentially the same manner, obtain the following inequality from Theorem I:

$$\frac{1}{2} < \boldsymbol{\sigma}^{\Sigma P} : \mathbf{D}' : \boldsymbol{\sigma}^{\Sigma P} + 2\mathbf{p}^{\Sigma P} : \mathbf{i}' : \boldsymbol{\sigma}^{\Sigma P} + \mathbf{p}^{\Sigma P} : \mathbf{K}' : \mathbf{p}^{\Sigma P} >$$

$$\leq \frac{1}{2} < (\boldsymbol{\sigma}^{o} + \boldsymbol{\sigma}^{d}) : \mathbf{D}' : (\boldsymbol{\sigma}^{o} + \boldsymbol{\sigma}^{d}) + 2(\mathbf{p}^{o} + \mathbf{p}^{d}) : \mathbf{i}' : (\boldsymbol{\sigma}^{o} + \boldsymbol{\sigma}^{d})$$

$$+ (\mathbf{p}^{o} + \mathbf{p}^{d}) : \mathbf{K}' : (\mathbf{p}^{o} + \mathbf{p}^{d}) >. \tag{9.8.10b}$$

When $(\mathbf{D}, \mathbf{i}, \mathbf{K})$ are chosen such that $(\mathbf{D}' - \mathbf{D}, \mathbf{i}' - \mathbf{i}, \mathbf{K}' - \mathbf{K})$ are negative-semi-definite, the right side of (9.8.10b) can be evaluated, leading to

$$\frac{1}{2} (\boldsymbol{\sigma}^{o} : (\overline{\mathbf{D}}^{\Sigma P} - \mathbf{D}) : \boldsymbol{\sigma}^{o} + 2\boldsymbol{\sigma}^{o} : (\overline{\mathbf{i}}^{\Sigma P} - \mathbf{i}) \boldsymbol{\cdot} \mathbf{p}^{o} + \mathbf{p}^{o} \boldsymbol{\cdot} (\overline{\mathbf{K}}^{\Sigma P} - \mathbf{K}) \boldsymbol{\cdot} \mathbf{p}^{o})$$

$$\leq -\mathrm{I}(\mathbf{e}^{*}, \boldsymbol{\psi}^{*}; \boldsymbol{\Lambda}^{A}, \boldsymbol{\gamma}^{A}; \boldsymbol{\sigma}^{o}, \mathbf{p}^{o}). \tag{9.8.16b}$$

As in the case of (9.8.16a), $(\overline{\mathbf{D}}^{\Sigma P}, \overline{\mathbf{i}}^{\Sigma P}, \overline{\mathbf{K}}^{\Sigma P})$ are the overall material properties of the RVE subjected to the uniform boundary conditions of $(\boldsymbol{\sigma}^{o}, \mathbf{p}^{o})$, i.e., $(\mathbf{t}, \mathrm{u}) = (\boldsymbol{\nu} \boldsymbol{\cdot} \boldsymbol{\sigma}^{o}, \mathbf{x} \boldsymbol{\cdot} \mathbf{p}^{o})$.

9.8.3. Comments on Computing Bounds for Overall Moduli

Now, seek to compute the values of the bounds for the piezoelectric overall moduli which are formally given by (9.8.16a,b). Note that the existence of Eshelby-type (constant) tensors can easily be established for the coupled piezoelectricity problems, and, indeed, general formal expressions can be obtained for these tensors. Once the existence of Eshelby's tensor for piezoelectricity is established, explicit Fourier transform expressions of these tensors are given in this subsection. These expressions can be numerically computed in a straightforward manner.

To simplify the expressions, use the following governing equations for the mechanical displacement and electric potential, $\mathbf{u}$ and u, instead of (9.8.5a,b):

$$\boldsymbol{\nabla} \boldsymbol{\cdot} (\mathbf{C}^{P} : (\boldsymbol{\nabla} \otimes \mathbf{u}(\mathbf{x}))) - \boldsymbol{\nabla} \boldsymbol{\cdot} (\mathbf{i} : (\boldsymbol{\nabla} \mathrm{u}(\mathbf{x}))) + \boldsymbol{\nabla} \boldsymbol{\cdot} \boldsymbol{\sigma}^{*}(\mathbf{x}) = \mathbf{0},$$

$$-\boldsymbol{\nabla} \boldsymbol{\cdot} (\mathbf{K} \boldsymbol{\cdot} (\boldsymbol{\nabla} \mathrm{u}(\mathbf{x}))) - \boldsymbol{\nabla} \boldsymbol{\cdot} (\mathbf{i}^{T} : (\boldsymbol{\nabla} \otimes \mathbf{u}(\mathbf{x}))) + \boldsymbol{\nabla} \boldsymbol{\cdot} \mathbf{q}^{*}(\mathbf{x}) = 0, \tag{9.8.17a,b}$$

where $\boldsymbol{\sigma}^{*}$ and $\mathbf{q}^{*}$ are eigenstress and eigen-electric-displacement fields.

Consider an unbounded uniform body, B, and assume[52] that $\boldsymbol{\sigma}^{*}$ and $\mathbf{q}^{*}$ are uniformly distributed within an ellipsoidal domain, Ω, contained in B.

---

[52] Rewrite the right-hand sides of equations (9.8.17a,b) as $\boldsymbol{\nabla} \boldsymbol{\cdot} (\mathbf{C}^{P} : (\boldsymbol{\nabla} \otimes \mathbf{u})) + \mathbf{f}$ and $\boldsymbol{\nabla} \boldsymbol{\cdot} (\mathbf{K} \boldsymbol{\cdot} (\boldsymbol{\nabla} \mathrm{u})) + \mathrm{U}$, where $\mathbf{f} = \boldsymbol{\nabla} \boldsymbol{\cdot} \boldsymbol{\sigma}^{*} - \boldsymbol{\nabla} \boldsymbol{\cdot} (\mathbf{i} \boldsymbol{\cdot} (\boldsymbol{\nabla} \mathrm{u}))$ and $\mathrm{U} = \boldsymbol{\nabla} \boldsymbol{\cdot} \mathbf{q}^{*} - \boldsymbol{\nabla} \boldsymbol{\cdot} (\mathbf{i}^{T} : (\boldsymbol{\nabla} \otimes \mathbf{u}))$ are regarded as body forces and electric charges. By definition, $\boldsymbol{\nabla} \boldsymbol{\cdot} \boldsymbol{\sigma}^{*}$ and $\boldsymbol{\nabla} \boldsymbol{\cdot} \mathbf{q}^{*}$ produce body forces and electric charges

The existence of Eshelby's tensors for the coupled piezoelectricity problem[53] can be established, without explicitly obtaining these tensors. For simplicity, assume transverse isotropy, with the axis of symmetry being the x_3-axis. The constitutive relations are

$$\begin{bmatrix} [\sigma] \\ [q] \end{bmatrix} = \begin{bmatrix} [C^P] & [i] \\ -[i]^T & [1^{(3)}] \end{bmatrix} \begin{bmatrix} [W] & [0] \\ [0]^T & [1^3] \end{bmatrix} \begin{bmatrix} [\varepsilon] \\ [p] \end{bmatrix}, \tag{9.8.18a}$$

where $[\sigma] = [\sigma_{11}, \sigma_{22}, \sigma_{33}, \sigma_{23}, \sigma_{31}, \sigma_{12}]^T$ and $[q] = [q_1, q_2, q_2]^T$, $[\varepsilon]$ and $[p]$ are defined in the same manner as $[\sigma]$ and $[q]$, respectively, $[W]$ is a diagonal matrix consisting of (1, 1, 1, 2, 2, 2) as defined by (3.1.6c), $[1^3]$ is a three by three unit matrix, and $([C^P], [i], [K])$ are given by

$$[C^P] = \begin{bmatrix} C^P_{1111} & C^P_{1111} - 2C^P_{1212} & C^P_{1133} & 0 & 0 & 0 \\ C^P_{1111} - 2C^P_{1212} & C^P_{1111} & C^P_{1133} & 0 & 0 & 0 \\ C^P_{1133} & C^P_{1133} & C^P_{3333} & 0 & 0 & 0 \\ 0 & 0 & 0 & C^P_{1313} & 0 & 0 \\ 0 & 0 & 0 & 0 & C^P_{1313} & 0 \\ 0 & 0 & 0 & 0 & 0 & C^P_{1212} \end{bmatrix},$$

$$[i]^T = \begin{bmatrix} 0 & 0 & 0 & 0 & i^P_{131} & 0 \\ 0 & 0 & 0 & i^P_{131} & 0 & 0 \\ i^P_{113} & i^P_{113} & i^P_{333} & 0 & 0 & 0 \end{bmatrix}, \qquad [K] = \begin{bmatrix} K_{11} & 0 & 0 \\ 0 & K_{11} & 0 \\ 0 & 0 & K_{33} \end{bmatrix}. \tag{9.8.18b~d}$$

Dunn and Wienecke (1996) have obtained Green's function,[54] $\mathbf{G}^P$, for the unbounded transversely isotropic body, in terms of the weighted distance parameters defined by

$$R_\alpha^2 = x_1^2 + x_2^2 + (\nu_\alpha x_3)^2, \tag{9.8.19}$$

for $\alpha = 0, 1, 2, 3$, where ν_0 is determined from the above moduli and $\nu_{1,2,3}$ are the roots of a third-order polynomial equation whose coefficients are given by the above moduli. In this derivation, it is shown that if the second-order derivative of Green's function, $G^P_{ik,jl}(\mathbf{x}-\mathbf{y})$, is integrated with respect to $\mathbf{y}$ in the ellipsoidal Ω, then, the integral can be expressed in terms of polyharmonic functions of R_α's. Then, according to Walpole (1967), the integral is constant when $\mathbf{x}$ lies within Ω. The resulting constant forth-order tensor yields Eshelby's tensor for the piezoelectricity problem.

---

which behave like delta functions across the boundary $\partial\Omega$ of Ω. The remaining body forces and electric charges, $-\boldsymbol{\nabla}.(\mathbf{i}.(\boldsymbol{\nabla}u))$ and $\boldsymbol{\nabla}.(\mathbf{i}^T:(\boldsymbol{\nabla}\otimes\mathbf{u}))$, vanish in Ω, but smoothly decay outside of Ω, even though they behave like delta functions on $\partial\Omega$. Therefore, Eshelby's tensors for the coupled piezoelectricity problem are different from those for the mechanical and non-mechanical problems, even though the same ellipsoidal domain with the same (uncoupled) moduli are considered.

[53] The fact that Eshelby's tensor for the piezoelectricity problem is constant in an ellipsoidal region, has been discussed by Deeg (1980), Benveniste (1992), Dunn and Taya (1993), and Li and Dunn (1998) who consider coupled piezoelectromagnetic media.

[54] In index notation, $\mathbf{G}^P$ is expressed as G^P_{ik}, where i = 4 corresponds to an electric potential and k = 4 corresponds to an electric charge, while i = 1, 2,3 and k = 1, 2, 3 correspond to displace-

Once the existence of (constant) Eshelby's tensor for the piezoelectricity problem is established, the tensor can be computed numerically using the Fourier transform. The required computation is straightforward.

The governing equations (9.8.17a,b) can be solved by taking the Fourier transform of **u** and u. The associated strain and electric field are expressed as

$$\boldsymbol{\varepsilon}(\mathbf{x}) = \frac{1}{(2\pi)^3}\int dV_\xi \int_\Omega dV_y \exp(\iota\boldsymbol{\xi}\cdot(\mathbf{x}-\mathbf{y}))$$

$$\left\{ \mathrm{sym}^4\left\{ \boldsymbol{\xi}\otimes\mathbf{A}^2(\boldsymbol{\xi})\otimes\boldsymbol{\xi} \right\} : \boldsymbol{\sigma}^* - \mathrm{sym}^3\left\{ \boldsymbol{\xi}\otimes\mathbf{A}^1(\boldsymbol{\xi})\otimes\boldsymbol{\xi} \right\}\cdot\mathbf{q}^*, \right\}$$

$$\mathbf{p}(\mathbf{x}) = \frac{1}{(2\pi)^3}\int dV_\xi \int_\Omega dV_y \exp(\iota\boldsymbol{\xi}\cdot(\mathbf{x}-\mathbf{y}))$$

$$\left\{ \mathrm{sym}^3\left\{ \boldsymbol{\xi}\otimes\mathbf{A}^1(\boldsymbol{\xi})\otimes\boldsymbol{\xi} \right\}^{\mathrm{T}} : \boldsymbol{\sigma}^* + \left\{ \mathrm{A}^0(\boldsymbol{\xi})\,\boldsymbol{\xi}\otimes\boldsymbol{\xi} \right\}\cdot\mathbf{q}^* \right\}, \tag{9.8.19a,b}$$

where $\mathrm{sym}^4(\,)_{ijkl} = ((\,)_{ijkl} + (\,)_{jikl} + (\,)_{ijlk} + (\,)_{jilk})/4$, $\mathrm{sym}^3(\,)_{ijk} = ((\,)_{ijk} + (\,)_{jik})/2$, and $(\mathbf{A}^2, \mathbf{A}^1, \mathrm{A}^0)$ are

$$\mathrm{A}^0 = \frac{1}{\tilde{\mathrm{K}} + \tilde{\mathbf{i}}^{\mathrm{T}}\cdot(\tilde{\mathbf{C}}^{\mathrm{P}})^{-1}\cdot\tilde{\mathbf{i}}}, \qquad \mathrm{A}^1 = \mathrm{A}^0\,(\boldsymbol{\xi}\cdot(\tilde{\mathbf{C}}^{\mathrm{P}})^{-1}), \qquad \mathbf{A}^2 = (\tilde{\mathbf{C}}^{\mathrm{P}} + \frac{1}{\tilde{\mathrm{K}}}\boldsymbol{\xi}\otimes\boldsymbol{\xi})^{-1}, \tag{9.8.19c~e}$$

with $(\tilde{\mathbf{C}}^{\mathrm{P}}, \tilde{\mathbf{i}}, \tilde{\mathrm{K}})$ being defined as

$$\tilde{\mathbf{C}}^{\mathrm{P}} = \boldsymbol{\xi}\cdot\mathbf{C}^{\mathrm{P}}\cdot\boldsymbol{\xi}, \qquad \tilde{\mathbf{i}} = \boldsymbol{\xi}\cdot\mathbf{i}\cdot\boldsymbol{\xi}, \qquad \tilde{\mathrm{K}} = \boldsymbol{\xi}\cdot\mathbf{K}\cdot\boldsymbol{\xi}. \tag{9.8.19f~h}$$

$\boldsymbol{\varepsilon}$ and **p** do not depend on **x** when **x** is within Ω, i.e.,

$$\boldsymbol{\varepsilon}(\mathbf{x}) = \boldsymbol{\Gamma}^4 : \boldsymbol{\sigma}^* - \boldsymbol{\Gamma}^3\cdot\mathbf{q}^*, \qquad \mathbf{p}(\mathbf{x}) = (\boldsymbol{\Gamma}^3)^{\mathrm{T}} : \boldsymbol{\sigma}^* + \boldsymbol{\Gamma}^2\cdot\mathbf{q}^*, \tag{9.8.20a,b}$$

for any **x** in Ω. Here, $\boldsymbol{\Gamma}^{4,\,3,\,2}$ are the fourth-, third-, and second-order tensors, which correspond to Eshelby's tensor for the piezoelectricity problem. These tensors can be computed easily, by taking the volume average of (9.8.19a,b) over Ω, as follows:

$$\begin{bmatrix} \boldsymbol{\Gamma}^4 \\ \boldsymbol{\Gamma}^3 \\ \boldsymbol{\Gamma}^2 \end{bmatrix} = \frac{\Omega}{(2\pi)^3}\int dV_\xi\, g(\boldsymbol{\xi})\, g(-\boldsymbol{\xi}) \begin{bmatrix} \mathrm{sym}^4\left\{ \boldsymbol{\xi}\otimes\mathbf{A}^2(\boldsymbol{\xi})\otimes\boldsymbol{\xi} \right\} \\ \mathrm{sym}^3\left\{ \boldsymbol{\xi}\otimes\mathbf{A}^1(\boldsymbol{\xi})\otimes\boldsymbol{\xi} \right\} \\ \mathrm{A}^0\,\boldsymbol{\xi}\otimes\boldsymbol{\xi} \end{bmatrix}, \tag{9.8.21}$$

where

$$g(\boldsymbol{\xi}) = \frac{1}{\Omega}\int_\Omega \exp(\iota\boldsymbol{\xi}\cdot\mathbf{x})\, dV. \tag{9.8.22}$$

As in Section 12, this is called the g-integral[55] and can be computed explicitly

---

ment and force in the x_i- and x_k-direction, respectively.

for an ellipsoidal Ω.

For the transversely isotropic case given by (9.8.18), $(\tilde{\mathbf{C}}^{\mathrm{P}}, \tilde{\mathbf{i}}, \tilde{\mathrm{K}})$ appearing in (9.8.21) are expressed in matrix form, as follows:

$$[\tilde{\mathrm{C}}^{\mathrm{P}}] = \begin{bmatrix} \mathrm{C}^{\mathrm{P}}_{1111}\xi_1^2 + \mathrm{C}^{\mathrm{P}}_{1212}\xi_2^2 + \mathrm{C}^{\mathrm{P}}_{1313}\xi_3^2 & (\mathrm{C}^{\mathrm{P}}_{1122} - \mathrm{C}^{\mathrm{P}}_{1212})\xi_1\xi_2 & (\mathrm{C}^{\mathrm{P}}_{1133} + \mathrm{C}^{\mathrm{P}}_{1313})\xi_1\xi_3 \\ (\mathrm{C}^{\mathrm{P}}_{1122} - \mathrm{C}^{\mathrm{P}}_{1212})\xi_1\xi_2 & \mathrm{C}^{\mathrm{P}}_{1212}\xi_1^2 + \mathrm{C}^{\mathrm{P}}_{1111}\xi_2^2 + \mathrm{C}^{\mathrm{P}}_{1313}\xi_3^2 & (\mathrm{C}^{\mathrm{P}}_{1133} + \mathrm{C}^{\mathrm{P}}_{1313})\xi_2\xi_3 \\ (\mathrm{C}^{\mathrm{P}}_{1133} + \mathrm{C}^{\mathrm{P}}_{1313})\xi_1\xi_3 & (\mathrm{C}^{\mathrm{P}}_{1133} + \mathrm{C}^{\mathrm{P}}_{1313})\xi_1\xi_3 & \mathrm{C}^{\mathrm{P}}_{1313}(\xi_1^2 + \xi_2^2) + \mathrm{C}^{\mathrm{P}}_{3333}\xi_3^2 \end{bmatrix}, \quad [\tilde{\xi}] = \begin{bmatrix} (\mathrm{i}_{113} + \mathrm{i}_{131})\xi_1\xi_3 \\ (\mathrm{i}_{113} + \mathrm{i}_{131})\xi_2\xi_3 \\ \mathrm{i}_{131}(\xi_1^2 + \xi_2^2) + \mathrm{i}_{333}\xi_3^2 \end{bmatrix},$$

$$\tilde{\mathrm{K}} = \mathrm{K}_{11}(\xi_1^2 + \xi_2^2) + \mathrm{K}_{33}\xi_3^2. \qquad \text{(9.8.23a~c)}$$

Hence, Eshelby's tensors for the piezoelectricity problem, namely, $(\mathbf{\Gamma}^4, \mathbf{\Gamma}^3, \mathbf{\Gamma}^2)$ can be determined by using $(\mathbf{A}^2, \mathbf{A}^1, \mathrm{A}^0)$ which are computed from (9.8.19c~e) with the above $(\tilde{\mathbf{C}}^{\mathrm{P}}, \tilde{\mathbf{i}}, \tilde{\mathrm{K}})$.

9.9. REFERENCES

Accorsi, M. L. and Nemat-Nasser, S. (1986), Bounds on the overall elastic and instantaneous elastoplastic moduli of periodic composites, *Mech. Matr.*, Vol. 5, No. 3, 209-220.

Benveniste, Y. (1992), The determination of the elastic and electric fields in a piezoelectric inhomogeneity, *J. Appl. Phys.*, Vol. 72, No.1, 1086-1095.

Deeg, W. F. (1980), The analysis of dislocation, crack, and inclusion problems in piezoelectric solids, PhD dissertation, Stanford University.

Dunn, M. L. and Taya, M. (1993), An analysis of piezoelectric composite materials containing ellipsoidal inhomogeneities, *Proc. R. Soc. London, A*, Vol. 443, 265-287.

Dunn, M. L. and Wienecke, H. A. (1996), Green's functions for transversely isotropic piezoelectric solids, *Int. J. Solids Struct.*, Vol. 33, No.30, 4571-4581.

Eshelby, J. D. (1957), The determination of the elastic field of an ellipsoidal inclusion, and related problems, *Proc. R. Soc. London, A*, Vol. 241, 376-396.

Francfort, G. A. and Murat, F. (1986), Homogenization and optimal bounds in linear elasticity, *Archive Rat. Mech. and Analysis*, Vol. 94, 307-334.

Hashin, Z. (1965), On elastic behaviour of fiber reinforced materials of arbitrary transverse phase geometry, *J. Mech. Phys. Solids*, Vol. 13, 119-134.

Hashin, Z. (1967), Variational principles of elasticity in terms of polarization tensors, *J. Engng. Sci.*, Vol. 5, 213-223.

---

[55] This g is essentially the same as the Fourier transform of the characteristic function for Ω.

Hashin, Z. and Shtrikman, S. (1962a), On some variational principles in anisotropic and nonhomogeneous elasticity, *J. Mech. Phys. Solids*, Vol. 10, 335-342.

Hashin, Z. and Shtrikman, S. (1962b), A variational approach to the theory of the elastic behaviour of polycrystals, *J. Mech. Phys. Solids*, Vol. 10, 343-352.

Hashin, Z. and Shtrikman, S. (1962c), A variational approach to the theory of the magnetic permeability of multiphase materials, *J. Appl. Phys.*, Vol. 33, 3125-3131.

Hashin, Z. and Shtrikman, S. (1963), A variational approach to the theory of the elastic behaviour of multiphase materials, *J. Mech. Phys. Solids*, Vol. 11, 127-140.

Hill, R. (1963), Elastic properties of reinforced solids: Some theoretical principles, *J. Mech. Phys. Solids*, Vol. 11, 357-372.

Hori, M. and Nemat-Nasser, S. (1998), Universal bounds for effective piezoelectric moduli, *Mech. Mat.*, Vol. 30, No. 1, 1-19.

Iwakuma, T. and Nemat-Nasser, S. (1983), Composites with periodic microstructure, *Advances and Trends in Structural and Solid Mechanics*, Pergamon Press, 13-19 or *Computers and Structures*, Vol. 16, Nos. 1-4, 13-19.

Kantor, Y. and Bergman, D. J. (1984), Improved rigorous bounds on the effective elastic moduli of a composite material, *J. Mech. Phys. Solids*, Vol. 32, 41-61.

Kinoshita, N. and Mura, T. (1971), Elastic fields of inclusion in anisotropic media, *Phys. Stat. Sol. A*, Vol. 5, 759-768.

Kneer, G. (1965), Über die Berechnung der Elastizitätsmoduln Vielkristalliner Aggregate mit Textur, *Phys. Stat. Sol.*, Vol. 9, 825-838.

Korringa, J. (1973), Theory of elastic constants of heterogeneous media, *J. Math. Phys.,* Vol. 14, 509-513.

Kröner, E. (1977), Bounds for effective elastic moduli of disordered materials, *J. Mech. Phys. Solids*, Vol. 25, 137-155.

Li, J. Y. and Dunn, M. L. (1998), Anisotropic couple-field inclusion and inhomogeneity problems, *Phil. Mag.*, Vol. 77, No. 5, 1341-1350.

Milton, G. W. (1990), On characterizing the set of possible effective tensors of composites: The variational method and the translation method, *Communications on Pure and Applied Mathematics*, Vol. 43, 63-125.

Milton, G. W. and Kohn, R. (1988), Variational bounds on the effective moduli of anisotropic composites, *J. Mech. Phys. Solids*, Vol. 36, 597-629.

Mura, T. (1987), *Micromechanics of defects in solids (2nd Edition)*, Martinus Nijhoff Publishers, Dordrecht.

Nemat-Nasser, S. and Hori, M. (1995), Universal bounds for overall properties of linear and nonlinear heterogeneous solids, *J. Engineering Mat. Tech.*, Vol. 117, 412-432.

Nemat-Nasser, S., Iwakuma, T., and Accorsi, M. (1986), Cavity growth and grain boundary sliding in polycrystalline solids, *Mech. Matr.*, Vol. 5, No. 4, 317-329.

Nemat-Nasser, S., Iwakuma, T., and Hejazi, M. (1982), On composites with periodic structure, *Mech. Matr.*, Vol. 1, No. 2, 239-267.

Nemat-Nasser, S. and Taya, M. (1981), On effective moduli of an elastic body containing periodically distributed voids, *Quarterly of Applied Mathematics*,

Vol. 71, 335-362.

Nemat-Nasser, S. and Taya, M. (1985), On effective moduli of an elastic body containing periodically distributed voids: Comments and corrections, *Quarterly of Applied Mathematics,* Vol. 43, No. 2, 187-188.

Schapery, R. A. (1968), Thermal expansion coefficients of composite materials based on energy principles, *J. Comp. Mat.*, Vol. 2, No. 3, 380-404.

Torquato, S. (1991), Random heterogeneous media: microstructure and improved bounds on effective properties, *Appl. Mech. Rev.*, Vol. 42, No. 2, 37-76.

Torquato, S. and Lado, F. (1992), Improved bounds on the effective elastic moduli of random arrays of cylinders, *J. Appl. Mech.*, Vol. 59, 1-6.

Walpole, L. J. (1966a), On bounds for the overall elastic moduli of inhomogeneous systems - I, *J. Mech. Phys. Solids*, Vol. 14, 151-162.

Walpole, L. J. (1966b), On bounds for the overall elastic moduli of inhomogeneous systems - II, *J. Mech. Phys. Solids*, Vol. 14, 289-301.

Walpole, L. J. (1967), The elastic field of an inclusion in an anisotropic medium, *Proc. Roy. Soc. A*, Vol. 300, 270-289.

Walpole, L. J. (1969), On the overall elastic moduli of composite materials, *J. Mech. Phys. Solids*, Vol. 17, 235-251.

Walpole, L. J. (1981), Elastic behavior of composite materials: Theoretical foundations, *Advances in Applied Mechanics*, Vol. 21, 169-242.

Willis, J. R. (1970), *Asymmetric problems of elasticity*, Adams Prize Essay, Cambridge University.

Willis, J. R. (1977), Bounds and self-consistent estimates for the overall properties of anisotropic composites, *J. Mech. Phys. Solids*, Vol. 25, 185-202.

Willis, J. R. (1981), Variational and related methods for the overall properties of composites, *Advances in Applied Mechanics*, Vol. 21, 1-78.

Willis, J. R. (1989), *Private communication.*

Willis, J. R. (1992), *Private communication.*

Willis, J. R. and Acton, J. R. (1976), The overall elastic moduli of a dilute suspension of spheres, *Q.J. Mech. Appl. Math.*, Vol. 29, 163-177.

Wu, C. T. D. and McCullough, R. L. (1977), Constitutive relationships for heterogeneous materials, in *Developments in Composite Materials* (Holister, C. D., ed.), 119-187.

SECTION 10 SELF-CONSISTENT, DIFFERENTIAL, AND RELATED AVERAGING METHODS

In the preceding sections, the overall moduli of a linearly elastic RVE are evaluated by: (1) the assumption of a dilute distribution of inhomogeneities, and (2) the self-consistent method. In the first case, the interaction effects are ignored, and, in the second case, this interaction is included in a certain sense. Both methods, however, are valid only when the volume fraction of inhomogeneities is rather small, although the self-consistent estimate may apply over a wider range. There is an alternative method, called the *differential scheme*, which applies over a much wider range of the volume fraction of inhomogeneities. In this scheme, differential equations for the overall moduli are derived by evaluating the change in the moduli of a homogenized RVE due to the introduction of an infinitesimally small amount of microconstituents. The overall moduli are then determined as suitable solutions of these differential equations. In this section the self-consistent and differential schemes are related to the results obtained by the dilute-distribution assumption, and their relations are discussed in some detail. Other related averaging schemes are also considered, including the two-phase, double-inclusion (or the three-phase), and the multi-inclusion methods. In addition, it is shown that the average strain within each annulus in a nested set of ellipsoidal regions of arbitrary aspect ratios and relative orientations and positions, embedded in an infinite uniform elastic solid, can be computed *exactly*, when each annulus undergoes arbitrary transformation with uniform but distinct (i.e., different from annulus to annulus) eigenstrains; the eigenstrain of the innermost ellipsoid need not be uniform. Explicit results are presented for this problem, and used to obtain estimates of the overall moduli of composites with several layers of coatings of different elasticities.

10.1. SUMMARY OF EXACT RELATIONS BETWEEN AVERAGE QUANTITIES

For a linearly elastic RVE, with matrix M and micro-elements Ω_α, the following two *exact* equations define the overall moduli: when the constant macrostrain $\mathbf{E} = \boldsymbol{\varepsilon}^o$ is prescribed,

$$(\bar{\mathbf{C}} - \mathbf{C}) : \boldsymbol{\varepsilon}^o = \sum_{\alpha=1}^{n} f_\alpha \, (\mathbf{C}^\alpha - \mathbf{C}) : \bar{\boldsymbol{\varepsilon}}^\alpha; \qquad (10.1.1a)$$

and when the constant macrostress $\boldsymbol{\Sigma} = \boldsymbol{\sigma}^o$ is prescribed,

$$(\overline{\mathbf{D}} - \mathbf{D}) : \boldsymbol{\sigma}^o = \sum_{\alpha=1}^{n} f_\alpha \, (\mathbf{D}^\alpha - \mathbf{D}) : \bar{\boldsymbol{\sigma}}^\alpha. \qquad (10.1.1b)$$

In general, $\overline{\mathbf{C}}$ given by (10.1.1a) and $\overline{\mathbf{D}}$ given by (10.1.1b), may not be each other's inverse. Since the RVE is linearly elastic, the average strain and stress in each inclusion relate linearly to the uniform boundary data: when the macrostrain $\mathbf{E} = \boldsymbol{\varepsilon}^o$ is prescribed,

$$\bar{\boldsymbol{\varepsilon}}^\alpha = \bar{\boldsymbol{\varepsilon}}^\alpha(\boldsymbol{\varepsilon}^o) \equiv \mathbf{E}^\alpha : \boldsymbol{\varepsilon}^o, \qquad (10.1.2a)$$

and when the macrostress $\boldsymbol{\Sigma} = \boldsymbol{\sigma}^o$ is prescribed,

$$\bar{\boldsymbol{\sigma}}^\alpha = \bar{\boldsymbol{\sigma}}^\alpha(\boldsymbol{\sigma}^o) \equiv \mathbf{F}^\alpha : \boldsymbol{\sigma}^o, \qquad (10.1.2b)$$

where the fourth-order tensors $\mathbf{E}^\alpha$ and $\mathbf{F}^\alpha$ depend on the material properties and geometry of all constituents:

$$\mathbf{E}^\alpha = \mathbf{E}^\alpha(\mathbf{C}^\alpha, \Omega_\alpha; \mathbf{C}, \mathrm{M}; \mathbf{C}^1, \Omega_1; ...; \mathbf{C}^\beta, \Omega_\beta; ...; \mathbf{C}^n, \Omega_n),$$

$$\mathbf{F}^\alpha = \mathbf{F}^\alpha(\mathbf{D}^\alpha, \Omega_\alpha; \mathbf{D}, \mathrm{M}; \mathbf{D}^1, \Omega_1; ...; \mathbf{D}^\beta, \Omega_\beta; ...; \mathbf{D}^n, \Omega_n), \qquad (\beta \neq \alpha), \qquad (10.1.2c,d)$$

where M and Ω_β stand for the geometry (shapes, locations, etc.) of the matrix and the βth micro-element. Note that $\mathbf{E}^\alpha$ and $\mathbf{F}^\alpha$ relate to tensors $\mathbf{h}^\alpha$ and $\mathbf{j}^\alpha$ of Sections 4 and 7, by

$$(\mathbf{C}^\alpha - \mathbf{C}) : \mathbf{E}^\alpha = \mathbf{j}^\alpha, \qquad (\mathbf{D}^\alpha - \mathbf{D}) : \mathbf{F}^\alpha = \mathbf{h}^\alpha \qquad (\alpha \text{ not summed}). \qquad (10.1.3a,b)$$

Here, it is more convenient to use $\mathbf{E}^\alpha$ and $\mathbf{F}^\alpha$, as defined by (10.1.2a,b).

From (10.1.1a,b) and (10.1.2a,b),

$$\overline{\mathbf{C}} = \mathbf{C} + \sum_{\alpha=1}^{n} f_\alpha \, (\mathbf{C}^\alpha - \mathbf{C}) : \mathbf{E}^\alpha \qquad (\boldsymbol{\varepsilon}^o \text{ prescribed}),$$

$$\overline{\mathbf{D}} = \mathbf{D} + \sum_{\alpha=1}^{n} f_\alpha \, (\mathbf{D}^\alpha - \mathbf{D}) : \mathbf{F}^\alpha \qquad (\boldsymbol{\sigma}^o \text{ prescribed}). \qquad (10.1.4a,b)$$

In general, it is difficult to obtain exact expressions for $\mathbf{E}^\alpha$ and $\mathbf{F}^\alpha$. With proper estimates of $\mathbf{E}^\alpha$ and $\mathbf{F}^\alpha$, however, reasonable estimates of $\overline{\mathbf{C}}$ and $\overline{\mathbf{D}}$ can be obtained. Note that, in general, $\mathbf{E}^\alpha$ and $\mathbf{F}^\alpha$ are not related to each other, since they represent the response of the αth micro-element, Ω_α, in a finite RVE of volume V, under different boundary conditions.

10.1.1. Assumptions in Dilute-Distribution Model

Since it is difficult to obtain the tensors $\mathbf{E}^\alpha$ and $\mathbf{F}^\alpha$ exactly, approximate estimates are often sought, based on simple assumptions. The simplest estimates are obtained from the Reuss and Voigt approximations which produce bounds; see Subsections 2.5 and 7.1, and Equations (7.1.14) and (7.2.9), where $\mathbf{E}^\alpha$ and $\mathbf{F}^\alpha$ are taken to be $\mathbf{1}^{(4s)}$, the identity tensor.

The next simplest assumption is that of a dilute distribution of inhomogeneities, where interaction among the inhomogeneities is neglected. This assumption yields reasonable estimates of $\mathbf{E}^\alpha$ and $\mathbf{F}^\alpha$, when the volume fraction of micro-elements is relatively small and the micro-elements are far apart. With the assumption of a dilute distribution of inhomogeneities, consider a fictitious unbounded homogeneous solid, denoted by B, whose moduli are those of the matrix material and in which an isolated micro-element, Ω_α, is embedded. Denote the corresponding average strain and stress in Ω_α due to farfield uniform strain $\boldsymbol{\varepsilon}^\infty = \boldsymbol{\varepsilon}^o$ and stress $\boldsymbol{\sigma}^\infty = \boldsymbol{\sigma}^o$, respectively, by

$$\bar{\boldsymbol{\varepsilon}}^\alpha(\boldsymbol{\varepsilon}^o) \equiv \mathbf{E}^\infty : \boldsymbol{\varepsilon}^o, \qquad \bar{\boldsymbol{\sigma}}^\alpha(\boldsymbol{\sigma}^o) \equiv \mathbf{F}^\infty : \boldsymbol{\sigma}^o, \tag{10.1.5a,b}$$

where

$$\mathbf{E}^\infty = \mathbf{E}^\infty(\mathbf{C}^\alpha, \Omega_\alpha; \mathbf{C}), \qquad \mathbf{F}^\infty = \mathbf{F}^\infty(\mathbf{D}^\alpha, \Omega_\alpha; \mathbf{D}). \tag{10.1.5c,d}$$

Here, Ω_α stands for the shape and orientation of the αth micro-element. Unlike for the bounded V, in the present case, for either $\boldsymbol{\sigma}^o = \mathbf{C} : \boldsymbol{\varepsilon}^o$ or $\boldsymbol{\varepsilon}^o = \mathbf{D} : \boldsymbol{\sigma}^o$,

$$\mathbf{F}^\infty = \mathbf{C}^\alpha : \mathbf{E}^\infty : \mathbf{D}, \qquad \mathbf{E}^\infty = \mathbf{D}^\alpha : \mathbf{F}^\infty : \mathbf{C}. \tag{10.1.6a,b}$$

This is because the farfield strain and stress are related through $\boldsymbol{\sigma}^o = \mathbf{C} : \boldsymbol{\varepsilon}^o$ or $\boldsymbol{\varepsilon}^o = \mathbf{D} : \boldsymbol{\sigma}^o$, and hence produce identical fields in B. Depending on $\mathbf{C}$, $\mathbf{C}^\alpha$ (or $\mathbf{D}$, $\mathbf{D}^\alpha$), and Ω_α, it may be possible to calculate $\mathbf{E}^\infty$ and $\mathbf{F}^\infty$ directly. In the sequel, denote the estimate of $\mathbf{E}^\alpha$ and $\mathbf{F}^\alpha$ obtained by means of the dilute-distribution assumption (i.e., by embedding an isolated Ω_α in an unbounded matrix with elasticity $\mathbf{C}$) by superposed *DD* (for dilute distribution), i.e., set

$$\mathbf{E}^\alpha \approx \mathbf{E}^{DD\alpha} \equiv \mathbf{E}^\infty(\mathbf{C}^\alpha, \Omega_\alpha; \mathbf{C}), \qquad \mathbf{F}^\alpha \approx \mathbf{F}^{DD\alpha} \equiv \mathbf{F}^\infty(\mathbf{D}^\alpha, \Omega_\alpha; \mathbf{D}). \tag{10.1.7a,b}$$

Denoting by $\bar{\mathbf{C}}^{DD}$ the corresponding estimated overall elasticity tensor when the RVE is subjected to $\boldsymbol{\varepsilon}^o$, and by $\bar{\mathbf{D}}^{DD}$ the corresponding estimated overall compliance tensor when the RVE is subjected to $\boldsymbol{\sigma}^o$, obtain

$$\bar{\mathbf{C}}^{DD} \equiv \mathbf{C} + \sum_{\alpha=1}^{n} f_\alpha\, (\mathbf{C}^\alpha - \mathbf{C}) : \mathbf{E}^{DD\alpha}$$

$$\bar{\mathbf{D}}^{DD} \equiv \mathbf{D} + \sum_{\alpha=1}^{n} f_\alpha\, (\mathbf{D}^\alpha - \mathbf{D}) : \mathbf{F}^{DD\alpha}. \tag{10.1.8a,b}$$

From the equivalence relations (10.1.6a,b) for $\mathbf{E}^\infty$ and $\mathbf{F}^\infty$,

$$(\mathbf{C}^\alpha - \mathbf{C}) : \mathbf{E}^{DD\alpha} = -\mathbf{C} : (\mathbf{D}^\alpha - \mathbf{D}) : \mathbf{F}^{DD\alpha} : \mathbf{C},$$

$$(\mathbf{D}^\alpha - \mathbf{D}) : \mathbf{F}^{DD\alpha} = -\mathbf{D} : (\mathbf{C}^\alpha - \mathbf{C}) : \mathbf{E}^{DD\alpha} : \mathbf{D} \quad (\alpha \text{ not summed}), \tag{10.1.9a,b}$$

and the estimated overall tensors $\bar{\mathbf{C}}^{DD}$ and $\bar{\mathbf{D}}^{DD}$ satisfy

$$\frac{1}{2}\,(\mathbf{D} : \bar{\mathbf{C}}^{DD} + \bar{\mathbf{D}}^{DD} : \mathbf{C}) = \mathbf{1}^{(4s)},$$

$$\frac{1}{2}\,(\mathbf{C} : \bar{\mathbf{D}}^{DD} + \bar{\mathbf{C}}^{DD} : \mathbf{D}) = \mathbf{1}^{(4s)}. \tag{10.1.9c,d}$$

Hence, from $(\mathbf{D} - \bar{\mathbf{D}}^{DD}) : (\mathbf{C} - \bar{\mathbf{C}}^{DD})$ or $(\mathbf{C} - \bar{\mathbf{C}}^{DD}) : (\mathbf{D} - \bar{\mathbf{D}}^{DD})$, it is seen that the deviation of $\bar{\mathbf{D}}^{DD} : \bar{\mathbf{C}}^{DD}$ and $\bar{\mathbf{C}}^{DD} : \bar{\mathbf{D}}^{DD}$ from $\mathbf{1}^{(4s)}$ is second-order in the volume fraction of inhomogeneities,

$$\bar{\mathbf{D}}^{DD} : \bar{\mathbf{C}}^{DD} - \mathbf{1}^{(4s)} = \sum_{\alpha=1}^{n} \sum_{\beta=1}^{n} f_\alpha f_\beta \{(\mathbf{C}^\alpha - \mathbf{C}) : \mathbf{E}^{DD\alpha}\} : \{(\mathbf{D}^\beta - \mathbf{D}) : \mathbf{F}^{DD\beta}\},$$

$$\bar{\mathbf{C}}^{DD} : \bar{\mathbf{D}}^{DD} - \mathbf{1}^{(4s)} = \sum_{\alpha=1}^{n} \sum_{\beta=1}^{n} f_\alpha f_\beta \{(\mathbf{D}^\alpha - \mathbf{D}) : \mathbf{F}^{DD\alpha}\} : \{(\mathbf{C}^\beta - \mathbf{C}) : \mathbf{E}^{DD\beta}\}. \tag{10.1.10a,b}$$

Although $\bar{\mathbf{C}}^{DD}$ and $\bar{\mathbf{D}}^{DD}$ are the overall tensors for different boundary conditions, they are each other's inverse to the first order in the volume fraction of micro-elements.

10.1.2. Dilute Distribution: Modeling Approximation

As commented on in Subsections 5.1.1, the effective overall moduli estimated on the basis of the dilute-distribution assumption contradict the exact inequalities (2.5.44) of Subsection 2.5, which are based on fundamental energy theorems in linear elasticity. Here, this issue is examined in some detail and it is shown that the contradiction stems from the approximation used in the dilute-distribution modeling to estimate the concentration tensors, $\mathbf{E}$ and $\mathbf{F}$.

For simplicity, consider a two-phase composite, and let the corresponding RVE consist of the matrix phase with elasticity $\mathbf{C} = \mathbf{D}^{-1}$ and the inclusion phase with elasticity $\mathbf{C}^I = (\mathbf{D}^I)^{-1}$. The overall elasticity when the RVE is subjected to the *linear displacement boundary conditions*, and the overall compliance when the RVE is subjected to the *uniform traction boundary conditions* are exactly given by

$$\bar{\mathbf{C}}^E = \mathbf{C} + f\,(\mathbf{C}^I - \mathbf{C}) : \mathbf{E}^E, \qquad \bar{\mathbf{D}}^\Sigma = \mathbf{D} + f\,(\mathbf{D}^I - \mathbf{D}) : \mathbf{F}^\Sigma, \tag{10.1.11a,b}$$

where superscript E or Σ emphasizes that the corresponding quantity is obtained for prescribed macrostrain, $\mathbf{E} = \boldsymbol{\varepsilon}^o$, or prescribed macrostress, $\boldsymbol{\Sigma} = \boldsymbol{\sigma}^o$, respectively; hence, $\mathbf{E}^E$ is the concentration tensor for *the average strain* in the inclusions under the overall strain $\mathbf{E}$, while $\mathbf{F}^\Sigma$ is the concentration tensor for *the average stress* in the inclusion under the overall stress $\boldsymbol{\Sigma}$.

Taking advantage of identities

$$\bar{\mathbf{D}}^\Sigma - \mathbf{D} = -\mathbf{D} : (\bar{\mathbf{C}}^\Sigma - \mathbf{C}) : \bar{\mathbf{D}}^\Sigma,$$

$$\mathbf{D}^I - \mathbf{D} = -\mathbf{D} : (\mathbf{C}^I - \mathbf{C}) : \mathbf{D}^I, \tag{10.1.12a,b}$$

with $\bar{\mathbf{C}}^\Sigma = (\bar{\mathbf{D}}^\Sigma)^{-1}$, rewrite (10.1.11b) as

$$\bar{\mathbf{C}}^\Sigma = \mathbf{C} + f\,(\mathbf{C}^I - \mathbf{C}) : (\mathbf{D}^I : \mathbf{F}^\Sigma : \bar{\mathbf{C}}^\Sigma). \tag{10.1.11c}$$

Subtracting (10.1.11c) from (10.1.11a), compute the difference $\bar{\mathbf{C}}^E - \bar{\mathbf{C}}^\Sigma$, as

$$\bar{\mathbf{C}}^E - \bar{\mathbf{C}}^\Sigma = f\,(\mathbf{C}^I - \mathbf{C}) : \{\mathbf{E}^E - \mathbf{D}^I : \mathbf{F}^\Sigma : \bar{\mathbf{C}}^\Sigma\}. \tag{10.1.13}$$

Since the overall elasticity tensors $\bar{\mathbf{C}}^E$ and $\bar{\mathbf{C}}^\Sigma$ are defined in terms of two different boundary-value problems, it is clear that they need not coincide. Indeed, the following inequality holds between $\bar{\mathbf{C}}^E$ and $\bar{\mathbf{C}}^\Sigma$:

$$\boldsymbol{\varepsilon}^o : (\bar{\mathbf{C}}^E - \bar{\mathbf{C}}^\Sigma) : \boldsymbol{\varepsilon}^o \geq 0, \qquad \text{for any } \boldsymbol{\varepsilon}^o; \tag{10.1.14a}$$

see Subsection[1] 2.5. For illustration, assume that $\mathbf{C}$, $\mathbf{C}^I$, and $\bar{\mathbf{C}}$, are all isotropic. Then, (10.1.14a) reduces to the following scalar equation:

$$(\bar{C}^E)^\alpha - (\bar{C}^\Sigma)^\alpha \geq 0, \qquad \text{for } \alpha = 1, 2, \tag{10.1.14b}$$

where $(\;)^\alpha$'s are the coefficients of the unit isotropic tensors, $\mathbf{E}^1$ and $\mathbf{E}^2$, defined in Section 8; see (8.1.1).

In the dilute-distribution model, concentration tensors $\mathbf{E}^E$ and $\mathbf{F}^\Sigma$ are *approximated* by $\mathbf{E}^\infty = \mathbf{E}^\infty(\Omega, \mathbf{C}^I; \mathbf{C})$ and $\mathbf{F}^\infty = \mathbf{E}^\infty(\Omega, \mathbf{D}^I; \mathbf{D})$, which, respectively, give the average strain and the average stress in an inclusion of elasticity $\mathbf{C}^I = (\mathbf{D}^I)^{-1}$, embedded in an infinite homogeneous solid of elasticity $\mathbf{C} = \mathbf{D}^{-1}$, when farfield strains and stresses satisfying $\boldsymbol{\varepsilon}^o = \mathbf{D} : \boldsymbol{\sigma}^o$ or $\boldsymbol{\sigma}^o = \mathbf{C} : \boldsymbol{\varepsilon}^o$ are prescribed. Therefore, these concentration tensors are related through $\mathbf{F}^\infty = \mathbf{C}^I : \mathbf{E}^\infty : \mathbf{D}$; see (10.1.6). The scalar equation corresponding to (10.1.13) then becomes

$$(\bar{C}^{EDD})^\alpha - (\bar{C}^{\Sigma DD})^\alpha = -f\,\{(C^I)^\alpha - (C)^\alpha\}\,(E^\infty)^\alpha\,(D)^\alpha\,\{(\bar{C}^{\Sigma DD})^\alpha - (C)^\alpha\},$$
$$\text{for } \alpha = 1, 2 \qquad (\alpha \text{ not summed}). \tag{10.1.15a}$$

If the inclusions are stiffer than the matrix, $(C^I)^\alpha$ and $(\bar{C}^{\Sigma DD})^\alpha$ will be greater than $(C)^\alpha$. On the other hand, if the inclusions are more compliant than the matrix, $(C^I)^\alpha$ and $(\bar{C}^{\Sigma DD})^\alpha$ will be smaller than $(C)^\alpha$. Since $(E^\infty)^\alpha$ is positive, the right-hand side of (10.1.15a) is always negative. This implies that

$$(\bar{C}^{EDD})^\alpha - (\bar{C}^{\Sigma DD})^\alpha < 0, \qquad \text{for } \alpha = 1, 2. \tag{10.1.15b}$$

This inequality is a direct result of the *approximation* used in the dilute-distribution model to calculate the average strains and stresses in the inhomogeneities. Inequality (10.1.15b), therefore, should *not* be viewed as *contradicting* the general result embedded in inequality (10.1.14a) and displayed by (10.1.14b) for an isotropic inclusion and matrix. If the concentration tensors, $\mathbf{E}^E$ and $\mathbf{F}^\Sigma$, are calculated *exactly*, then (10.1.14b) is obtained instead of (10.1.15b). Indeed, this dichotomy should be viewed as a measure of the limitation of the dilute-distribution *modeling approximation*.

10.2. SELF-CONSISTENT METHOD

As already explained in Sections 5 and 7, to estimate the average stress or strain in a typical inhomogeneity, the self-consistent method embeds this inhomogeneity in a fictitious unbounded homogeneous solid B which has the yet-unknown overall properties of the RVE, instead of those of the matrix material

[1] It should be kept in mind that inequality (10.1.14a) is derived on the basis of the average strain energy, and that for the linear displacement and uniform traction boundary conditions, the overall moduli obtained from the average stress-strain relation are identical with the overall moduli obtained from the corresponding average strain energy.

(used in the dilute-distribution approach). The resulting elasticity and compliance tensors obtained for prescribed overall strains and stresses, respectively, are then each other's inverse. To distinguish these unbounded homogeneous solids, use the notation,

$$B = B(\hat{\mathbf{C}}) = B(\hat{\mathbf{D}}), \tag{10.2.1}$$

when the elasticity of the material of B is $\hat{\mathbf{C}} = \hat{\mathbf{D}}^{-1}$. With this notation, the unbounded homogeneous solid used in the dilute-distribution assumption is denoted by $B(\mathbf{C})$ or $B(\mathbf{D})$, while that used in the self-consistent method is denoted by $B(\bar{\mathbf{C}})$ or $B(\bar{\mathbf{D}})$.

Consider a successive approximation to estimate the overall tensors, and regard the self-consistent method as its limiting case. Let $\bar{\mathbf{C}}^{(N)}$ be the overall elasticity tensor in the Nth approximation, with $\bar{\mathbf{C}}^{(1)} = \mathbf{C}$, the matrix elasticity tensor. Denote by $\bar{\mathbf{D}}^{(N)}$ the overall compliance tensor (not necessarily the inverse of $\bar{\mathbf{C}}^{(N)}$) for the Nth approximation. Set

$$\bar{\mathbf{C}}^{(N+1)} = \mathbf{C} + \sum_{\alpha=1}^{n} f_\alpha \, (\mathbf{C}^\alpha - \mathbf{C}) : \mathbf{E}^{\alpha(N)},$$

$$\bar{\mathbf{D}}^{(N+1)} = \mathbf{D} + \sum_{\alpha=1}^{n} f_\alpha \, (\mathbf{D}^\alpha - \mathbf{D}) : \mathbf{F}^{\alpha(N)}, \tag{10.2.2a,b}$$

where

$$\mathbf{E}^{\alpha(N)} = \mathbf{E}^\infty(\mathbf{C}^\alpha, \Omega_\alpha; \bar{\mathbf{C}}^{(N)}), \qquad \mathbf{F}^{\alpha(N)} = \mathbf{F}^\infty(\mathbf{D}^\alpha, \Omega_\alpha; \bar{\mathbf{D}}^{(N)}). \tag{10.2.2c,d}$$

Thus, $\mathbf{E}^{\alpha(N)}$ and $\mathbf{F}^{\alpha(N)}$ are the estimates of the concentration tensors $\mathbf{E}^\alpha$ and $\mathbf{F}^\alpha$, on the basis of embedding Ω_α in infinitely extended homogeneous linearly elastic solids with elasticity and compliance tensors $\bar{\mathbf{C}}^{(N)}$ and $\bar{\mathbf{D}}^{(N)}$, respectively.

With N = 2, the results of the dilute-distribution assumption are obtained,

$$\bar{\mathbf{C}}^{(2)} = \mathbf{C} + \sum_{\alpha=1}^{n} f_\alpha \, (\mathbf{C}^\alpha - \mathbf{C}) : \mathbf{E}^{\alpha(1)} = \mathbf{C} + \sum_{\alpha=1}^{n} f_\alpha \, (\mathbf{C}^\alpha - \mathbf{C}) : \mathbf{E}^{DD\alpha},$$

$$\bar{\mathbf{D}}^{(2)} = \mathbf{D} + \sum_{\alpha=1}^{n} f_\alpha \, (\mathbf{D}^\alpha - \mathbf{D}) : \mathbf{F}^{\alpha(1)} = \mathbf{D} + \sum_{\alpha=1}^{n} f_\alpha \, (\mathbf{D}^\alpha - \mathbf{D}) : \mathbf{F}^{DD\alpha}, \tag{10.2.3a,b}$$

and hence $\bar{\mathbf{C}}^{(2)} = \bar{\mathbf{C}}^{DD}$ and $\bar{\mathbf{D}}^{(2)} = \bar{\mathbf{D}}^{DD}$.

Note that to obtain $\bar{\mathbf{C}}^{(N+1)}$, the fictitious unbounded homogeneous solid $B(\bar{\mathbf{C}}^{(N)})$ is used, whereas to obtain $\bar{\mathbf{D}}^{(N+1)}$, $B(\bar{\mathbf{D}}^{(N)})$ is used; see (10.2.2c,d). Hence, $\bar{\mathbf{C}}^{(N+1)}$ and $\bar{\mathbf{D}}^{(N+1)}$ are not necessarily each other's inverse. Since $B(\bar{\mathbf{C}}^{(N)})$ and $B(\bar{\mathbf{D}}^{(N)})$ are influenced by all microconstituents in the RVE, $\bar{\mathbf{C}}^{(N+1)}$ and $\bar{\mathbf{D}}^{(N+1)}$ take into account their interaction in a certain sense. It is plausible that as N increases, $\bar{\mathbf{C}}^{(N)}$ and $\bar{\mathbf{D}}^{(N)}$ become better estimates of the overall elasticity and compliance tensors. The limits $\bar{\mathbf{C}}^{(\infty)}$ and $\bar{\mathbf{D}}^{(\infty)}$ are defined by

$$\bar{\mathbf{C}}^{(\infty)} \equiv \lim_{N\to\infty} \bar{\mathbf{C}}^{(N)}, \qquad \bar{\mathbf{D}}^{(\infty)} \equiv \lim_{N\to\infty} \bar{\mathbf{D}}^{(N)}. \tag{10.2.4a,b}$$

If these limits exist, then they may be considered the best estimates of the overall elasticity and compliance tensors, $\bar{\mathbf{C}}$ and $\bar{\mathbf{D}}$, *for this class of approximate solutions.* From the recurrence formulae (10.2.2a,b),

$$\bar{\mathbf{C}}^{(\infty)} = \mathbf{C} + \sum_{\alpha=1}^{n} f_\alpha\, (\mathbf{C}^\alpha - \mathbf{C}) : \mathbf{E}^{\alpha(\infty)},$$

$$\bar{\mathbf{D}}^{(\infty)} = \mathbf{D} + \sum_{\alpha=1}^{n} f_\alpha\, (\mathbf{D}^\alpha - \mathbf{D}) : \mathbf{F}^{\alpha(\infty)}, \tag{10.2.5a,b}$$

where

$$\mathbf{E}^{\alpha(\infty)} = \mathbf{E}^\infty(\mathbf{C}^\alpha,\, \Omega_\alpha;\, \bar{\mathbf{C}}^{(\infty)}), \qquad \mathbf{F}^{\alpha(\infty)} = \mathbf{F}^\infty(\mathbf{D}^\alpha,\, \Omega_\alpha;\, \bar{\mathbf{D}}^{(\infty)}). \tag{10.2.5c,d}$$

Hence, in this method, the original exact tensors $\mathbf{E}^\alpha$ and $\mathbf{F}^\alpha$ are approximated by

$$\mathbf{E}^\alpha \approx \mathbf{E}^{\alpha(\infty)}, \qquad \mathbf{F}^\alpha \approx \mathbf{F}^{\alpha(\infty)}, \tag{10.2.6a,b}$$

which are the *self-consistent estimates.* The corresponding overall elasticity and compliance tensors, $\bar{\mathbf{C}}^{SC}$ and $\bar{\mathbf{D}}^{SC}$, are derived from the following equations:

$$\bar{\mathbf{C}}^{SC} = \mathbf{C} + \sum_{\alpha=1}^{n} f_\alpha\, (\mathbf{C}^\alpha - \mathbf{C}) : \mathbf{E}^{SC\alpha},$$

$$\bar{\mathbf{D}}^{SC} = \mathbf{D} + \sum_{\alpha=1}^{n} f_\alpha\, (\mathbf{D}^\alpha - \mathbf{D}) : \mathbf{F}^{SC\alpha}, \tag{10.2.7a,b}$$

where

$$\mathbf{E}^{SC\alpha} = \mathbf{E}^\infty(\mathbf{C}^\alpha,\, \Omega_\alpha;\, \bar{\mathbf{C}}^{SC}), \qquad \mathbf{F}^{SC\alpha} = \mathbf{F}^\infty(\mathbf{D}^\alpha,\, \Omega_\alpha;\, \bar{\mathbf{D}}^{SC}). \tag{10.2.7c,d}$$

Therefore, from comparison of (10.2.5a~d) with (10.2.7a~d), $\bar{\mathbf{C}}^{SC}$ and $\bar{\mathbf{D}}^{SC}$ agree with the limits $\bar{\mathbf{C}}^{(\infty)}$ and $\bar{\mathbf{D}}^{(\infty)}$, i.e.,

$$\bar{\mathbf{C}}^{SC} \equiv \bar{\mathbf{C}}^{(\infty)}, \qquad \bar{\mathbf{D}}^{SC} \equiv \bar{\mathbf{D}}^{(\infty)}. \tag{10.2.8a,b}$$

Since $\bar{\mathbf{C}}^{(N)}$ and $\bar{\mathbf{D}}^{(N)}$ correspond to the RVE subjected to different boundary conditions, they may not necessarily be each other's inverse. Indeed, from the recurrence formulae (10.2.2a,b), it is seen that $\bar{\mathbf{C}}^{(N+1)}$ and $\bar{\mathbf{D}}^{(N+1)}$ are each other's inverse only up to the first order in the volume fraction of micro-elements. The proof is straightforward. If $\mathbf{C} - \bar{\mathbf{C}}^{(N)}$ and $\mathbf{D} - \bar{\mathbf{D}}^{(N)}$ are O(f), with $f = \sum_{\alpha=1}^{n} f_\alpha$, then $\bar{\mathbf{C}}^{(N+1)}$ and $\bar{\mathbf{D}}^{(N+1)}$ become

$$\bar{\mathbf{C}}^{(N+1)} = \mathbf{C} + O(f), \qquad \bar{\mathbf{D}}^{(N+1)} = \mathbf{D} + O(f), \tag{10.2.9a,b}$$

and hence,

$$\bar{\mathbf{C}}^{(N+1)} : \bar{\mathbf{D}}^{(N+1)} = \mathbf{1}^{(4s)} + O(f^2), \qquad \bar{\mathbf{D}}^{(N+1)} : \bar{\mathbf{C}}^{(N+1)} = \mathbf{1}^{(4s)} + O(f^2). \tag{10.2.9c,d}$$

Now examine the equivalence of $\bar{\mathbf{C}}^{(N)}$ and $\bar{\mathbf{D}}^{(N)}$ in greater detail. Since $B(\bar{\mathbf{C}}^{(N)}) = B((\bar{\mathbf{C}}^{(N)})^{-1})$ and $B(\bar{\mathbf{D}}^{(N)}) = B((\bar{\mathbf{D}}^{(N)})^{-1})$, from (10.1.6a,b), it follows that

$$\mathbf{E}^\infty(\mathbf{C}^\alpha,\, \Omega_\alpha;\, \bar{\mathbf{C}}^{(N)}) = \mathbf{D}^\alpha : \mathbf{F}^\infty(\mathbf{D}^\alpha,\, \Omega_\alpha;\, (\bar{\mathbf{C}}^{(N)})^{-1}) : \bar{\mathbf{C}}^{(N)},$$

$$\mathbf{F}^\infty(\mathbf{D}^\alpha,\, \Omega_\alpha;\, \bar{\mathbf{D}}^{(N)}) = \mathbf{C}^\alpha : \mathbf{E}^\infty(\mathbf{C}^\alpha,\, \Omega_\alpha;\, (\bar{\mathbf{D}}^{(N)})^{-1}) : \bar{\mathbf{D}}^{(N)} \quad (\alpha \text{ not summed}). \tag{10.2.10a,b}$$

Using (10.2.10a,b), and multiplying (10.2.2a) by $\mathbf{D}$ from the left and by $(\bar{\mathbf{C}}^{(N)})^{-1}$

from the right, obtain

$$(\bar{\mathbf{C}}^{(N)})^{-1} = \mathbf{D} : \{\bar{\mathbf{C}}^{(N+1)} : (\bar{\mathbf{C}}^{(N)})^{-1}\} + \sum_{\alpha=1}^{n} f_\alpha\, (\mathbf{D}^\alpha - \mathbf{D}) : \mathbf{F}^\infty(\mathbf{D}^\alpha, \Omega_\alpha; (\bar{\mathbf{C}}^{(N)})^{-1}). \tag{10.2.11a}$$

Similarly, multiplying (10.2.2b) by $\mathbf{C}$ from the left and by $\mathbf{D}^{(N)-1}$ from the right, obtain

$$(\bar{\mathbf{D}}^{(N)})^{-1} = \mathbf{C} : \{\bar{\mathbf{D}}^{(N+1)} : (\bar{\mathbf{D}}^{(N)})^{-1}\} + \sum_{\alpha=1}^{n} f_\alpha\, (\mathbf{C}^\alpha - \mathbf{C}) : \mathbf{E}^\infty(\mathbf{C}^\alpha, \Omega_\alpha; (\bar{\mathbf{D}}^{(N)})^{-1}). \tag{10.2.11b}$$

Comparing (10.2.11a) with (10.2.2b), and (10.2.11b) with (10.2.2a), note that

$$\bar{\mathbf{D}}^{(N+1)} - (\bar{\mathbf{C}}^{(N)})^{-1} = \mathbf{D} : \{\mathbf{1}^{(4s)} - \bar{\mathbf{C}}^{(N+1)} : (\bar{\mathbf{C}}^{(N)})^{-1}\}$$

$$+ \sum_{\alpha=1}^{n} f_\alpha\, (\mathbf{D}^\alpha - \mathbf{D}) : \{\mathbf{F}^\infty(\mathbf{D}^\alpha, \Omega_\alpha; \bar{\mathbf{D}}^{(N)}) - \mathbf{F}^\infty(\mathbf{D}^\alpha, \Omega_\alpha; (\bar{\mathbf{C}}^{(N)})^{-1})\},$$

$$\bar{\mathbf{C}}^{(N+1)} - (\bar{\mathbf{D}}^{(N)})^{-1} = \mathbf{C} : \{\mathbf{1}^{(4s)} - \bar{\mathbf{D}}^{(N+1)} : (\bar{\mathbf{D}}^{(N)})^{-1}\}$$

$$+ \sum_{\alpha=1}^{n} f_\alpha\, (\mathbf{C}^\alpha - \mathbf{C}) : \{\mathbf{E}^\infty(\mathbf{C}^\alpha, \Omega_\alpha; \bar{\mathbf{C}}^{(N)}) - \mathbf{E}^\infty(\mathbf{C}^\alpha, \Omega_\alpha; (\bar{\mathbf{D}}^{(N)})^{-1})\}. \tag{10.2.12a,b}$$

As N goes to infinity, the left-hand sides of (10.2.12a,b) approach

$$\lim_{N\to\infty} \{\bar{\mathbf{D}}^{(N+1)} - (\bar{\mathbf{C}}^{(N)})^{-1}\} = \bar{\mathbf{D}}^{(\infty)} - (\bar{\mathbf{C}}^{(\infty)})^{-1},$$

$$\lim_{N\to\infty} \{\bar{\mathbf{C}}^{(N+1)} - (\bar{\mathbf{D}}^{(N)})^{-1}\} = \bar{\mathbf{C}}^{(\infty)} - (\bar{\mathbf{D}}^{(\infty)})^{-1}, \tag{10.2.12c,d}$$

and the terms in the summation in the right-hand sides of (10.2.12a,b) approach

$$\lim_{N\to\infty} \{\mathbf{F}^\infty(\mathbf{D}^\alpha, \Omega_\alpha; \bar{\mathbf{D}}^{(N)}) - \mathbf{F}^\infty(\mathbf{D}^\alpha, \Omega_\alpha; (\bar{\mathbf{C}}^{(N)})^{-1})\}$$

$$= \mathbf{F}^\infty(\mathbf{D}^\alpha, \Omega_\alpha; \bar{\mathbf{D}}^{(\infty)}) - \mathbf{F}^\infty(\mathbf{D}^\alpha, \Omega_\alpha; (\bar{\mathbf{C}}^{(\infty)})^{-1}),$$

$$\lim_{N\to\infty} \{\mathbf{E}^\infty(\mathbf{C}^\alpha, \Omega_\alpha; \bar{\mathbf{C}}^{(N)}) - \mathbf{E}^\infty(\mathbf{C}^\alpha, \Omega_\alpha; (\bar{\mathbf{D}}^{(N)})^{-1})\}$$

$$= \mathbf{E}^\infty(\mathbf{C}^\alpha, \Omega_\alpha; \bar{\mathbf{C}}^{(\infty)}) - \mathbf{E}^\infty(\mathbf{C}^\alpha, \Omega_\alpha; (\bar{\mathbf{D}}^{(\infty)})^{-1}). \tag{10.2.12e,f}$$

Since

$$\lim_{N\to\infty} \bar{\mathbf{C}}^{(N+1)} : (\bar{\mathbf{C}}^{(N)})^{-1} = \mathbf{1}^{(4s)}, \qquad \lim_{N\to\infty} \bar{\mathbf{D}}^{(N+1)} : (\bar{\mathbf{D}}^{(N)})^{-1} = \mathbf{1}^{(4s)}, \tag{10.2.12g,h}$$

$\bar{\mathbf{C}}^{(\infty)}$ and $\bar{\mathbf{D}}^{(\infty)}$, and consequently $\bar{\mathbf{C}}^{SC}$ and $\bar{\mathbf{D}}^{SC}$, are each other's exact inverse,

$$\bar{\mathbf{D}}^{SC} = (\bar{\mathbf{C}}^{SC})^{-1}, \qquad \bar{\mathbf{C}}^{SC} = (\bar{\mathbf{D}}^{SC})^{-1}. \tag{10.2.13a,b}$$

As pointed out in Sections 6, 7, and 8, the overall elasticity or compliance tensor, $\bar{\mathbf{C}}^{SC}$ or $\bar{\mathbf{D}}^{SC}$, estimated by the self-consistent method, is given by the solution of a (nonlinear) tensorial equation. However, this tensorial equation cannot, in general, be solved directly, especially when the geometries and material properties of the microconstituents are complicated. In this case, direct computation of the limit of the sequence $\{\bar{\mathbf{C}}^{(N)}\}$ or $\{\bar{\mathbf{D}}^{(N)}\}$ is an effective alternative algorithm for obtaining $\bar{\mathbf{C}}^{SC}$ or $\bar{\mathbf{D}}^{SC}$. In three examples of Section 8, namely, Figures 8.1.2, 8.2.1, and 8.2.2, $\bar{\mathbf{C}}^{SC}$ is obtained by directly computing $\{\bar{\mathbf{C}}^{(N)}\}$ such that $\bar{\mathbf{C}}^{(N)} - \bar{\mathbf{C}}^{(N-1)}$ is essentially zero. Numerically, such direct evaluation of the limit of this sequence appears stable and efficient. For the problems of Section 8, it produces physically meaningful solutions of the nonlinear equations.

Since V is bounded, in principle $\bar{\mathbf{C}}$ and $\bar{\mathbf{D}}$ may not be each other's inverse. However, an RVE represents one macroscopic material point at which the microstructure must be characterized in a statistical sense. It therefore suffices to obtain a statistical estimate of the overall response of this RVE under various macroscopic conditions. This may be more significant than the exact response of the RVE under particular microscopic boundary conditions. In this sense, the overall tensors estimated by the self-consistent method, $\bar{\mathbf{C}}^{SC}$ and $\bar{\mathbf{D}}^{SC}$, which are each other's inverse, may be a more suitable representation than the exact overall elasticity and compliance tensors calculated for the boundary conditions of the linear displacements and uniform tractions. It is in this sense that the results of various models discussed in this book should be examined.

10.3. DIFFERENTIAL SCHEME

As pointed out before, estimates of the overall moduli of heterogeneous linearly elastic solids by the assumption of a dilute distribution of microelements, as well as by the self-consistent method, have a limited range of applicability. While it is plausible that the self-consistent method may yield reasonable estimates for greater values of the volume fraction of inhomogeneities than the dilute-distribution assumption, its range is still quite limited. Indeed, for porous elastic solids and for elastic solids with microcracks, the self-consistent method gives zero stiffness (zero values for the overall shear and bulk moduli) at unreasonably small values of the void volume fraction or the crack density parameter; see Sections 5 and 6, and Figures 5.1.5, 6.4.2, 6.4.3, and 6.6.2. In this book several alternatives which apply over a broad range of densities of inhomogeneities are considered. They are: (1) estimates obtained by the dilute-distribution assumption; (2) estimates obtained by the self-consistent method; (3) estimates obtained by the so-called *differential scheme;* (4) estimates obtained by assuming periodically distributed inhomogeneities; and (5) estimates obtained by other averaging schemes. The fourth alternative is discussed and illustrated in Chapter IV. In this subsection the differential scheme is examined. Subsection 10.4 deals with other averaging schemes, including the two-phase model (Benveniste, 1987), and the double- and multi-inclusion methods (Hori and Nemat-Nasser, 1993).

As an introduction, consider an RVE which contains a uniform elastic matrix of elasticity $\mathbf{C} = \mathbf{D}^{-1}$, and micro-inclusions with common elasticity $\mathbf{C}^1 =$

$(\mathbf{D}^1)^{-1}$, i.e., a two-phase RVE. To estimate the overall elasticity $\overline{\mathbf{C}}$, when the volume fraction of the inclusion is f, the differential scheme begins with a uniform RVE of elasticity $\mathbf{C}$, containing an infinitesimally small volume fraction of inhomogeneities, δf. The overall moduli are estimated by the dilute-distribution assumption which, since $\delta f \ll 1$, yields accurate results. Then a new homogeneous solid whose elasticity tensor is the uniform elasticity which has just been calculated, i.e., the overall elasticity obtained in the previous step, is considered, and an infinitesimally small increment, δf, of the inhomogeneities is added. Again, the dilute-distribution assumption is used to obtain the new overall elasticity tensor. This process is continued until the final volume fraction of inhomogeneities is obtained. The mathematical formulation of this procedure leads to an ordinary differential equation for the overall elasticity tensor as a function of the volume fraction of inhomogeneities, f. This differential equation is integrated and the overall moduli for any value of f are obtained. The method may be formulated in terms of the elasticity tensor or the compliance tensor which will be each other's inverse.

In the sequel first the differential scheme for a two-phase RVE is formulated, and then the results are extended to cases involving inhomogeneities of several phases. Note at the outset that, inasmuch as only the volume fractions of inhomogeneities are prescribed, the final solution is not unique, depending on the chosen integration paths. However, one may fix the integration path on physical grounds and obtain reasonable results.

The differential scheme for estimating the overall properties of heterogeneous media was used by Roscoe (1952, 1973) who examined the viscosity of suspensions of rigid spheres and properties of composites with elastic and viscoelastic constituents. The application of the concept to composites and solids with microcracks is discussed by Boucher (1974), McLaughlin (1977), Cleary *et al.* (1980), Norris (1985), Hashin (1988), and Nemat-Nasser and Hori (1990). Other related contributions are Sen *et al.* (1981) and Sheng and Callegari (1984) who consider geophysical applications, and Henyey and Pomphrey (1982) who use an iterative scheme. Milton (1984) has used the concept of embedding in defined proportions, dilute concentrations of phases of a heterogeneous body within a sequentially homogenized medium, and has established the corresponding relation to other averaging methods, especially to realization of bounds through sequential packing; see also Milton (1990) who presents a broad framework to obtain possible effective tensors for composites, and Torquato (1991) who reviews the literature in the general area of random heterogeneous media with improved bounds for effective parameters, and provides useful comments and references.

10.3.1. Two-Phase RVE

Here the uniform matrix of elasticity $\mathbf{C}$ contains only one kind of inclusion of common elasticity $\mathbf{C}^1$; the corresponding compliance tensors are $\mathbf{D}$ and $\mathbf{D}^1$, respectively. For conceptual simplicity, let there be only one inclusion Ω_1 with volume fraction f_1 in the RVE. Regard the overall elasticity and compliance tensors of the RVE as functions of f_1, i.e.,

$$\bar{\mathbf{C}}\ (= \bar{\mathbf{C}}(\mathbf{C}^1,\ \Omega_1;\ \mathbf{C},\ \mathrm{M})) \equiv \bar{\mathbf{C}}(f_1),$$

$$\bar{\mathbf{D}}\ (= \bar{\mathbf{D}}(\mathbf{D}^1,\ \Omega_1;\ \mathbf{D},\ \mathrm{M})) \equiv \bar{\mathbf{D}}(f_1). \tag{10.3.1a,b}$$

Seek to obtain an expression which governs the variations of $\bar{\mathbf{C}}$ and $\bar{\mathbf{D}}$ in terms of f_1, *with the shape of the inclusion, Ω_1, kept fixed.*

From the *exact* equation (10.1.4a), $\bar{\mathbf{C}}(f_1)$ satisfies

$$\bar{\mathbf{C}}(f_1) - \mathbf{C} = f_1(\mathbf{C}^1 - \mathbf{C}) : \mathbf{E}^1(\mathbf{C}^1,\ \Omega_1;\ \mathbf{C},\ \mathrm{M}), \tag{10.3.1c}$$

when the macrostrain $\mathbf{E} = \boldsymbol{\varepsilon}^{\mathrm{o}}$ is regarded prescribed, and in view of the *exact* equation (10.1.4b), $\bar{\mathbf{D}}(f_1)$ satisfies

$$\bar{\mathbf{D}}(f_1) - \mathbf{D} = f_1(\mathbf{D}^1 - \mathbf{D}) : \mathbf{F}^1(\mathbf{D}^1,\ \Omega_1;\ \mathbf{D},\ \mathrm{M}), \tag{10.3.1d}$$

when the macrostress $\boldsymbol{\Sigma} = \boldsymbol{\sigma}^{\mathrm{o}}$ is regarded prescribed. The tensors $\mathbf{E}^1$ and $\mathbf{F}^1$ determine the average strain and stress in Ω_1, i.e., $\bar{\boldsymbol{\varepsilon}}^1(\boldsymbol{\varepsilon}^{\mathrm{o}})$ and $\bar{\boldsymbol{\sigma}}^1(\boldsymbol{\sigma}^{\mathrm{o}})$, respectively. Consider the change in $\mathbf{E}^1$ and $\mathbf{F}^1$, and hence the change in $\bar{\mathbf{C}}$ and $\bar{\mathbf{D}}$, caused by an infinitesimally small change in f_1.

First, to examine the change in $\bar{\mathbf{C}}$, construct the following procedure:

1) Consider an RVE of volume V which consists of a uniform matrix M of volume v and an inclusion Ω_1 of volume v_1. Let the overall elasticity tensor be $\bar{\mathbf{C}}(f_1)$, where $f_1 = v_1/(v + v_1)$.
2) Replace this RVE by a uniform solid, M′, of volume $v + v_1$, and elasticity tensor $\bar{\mathbf{C}}(f_1)$.
3) Add an infinitesimally small inclusion $\delta\Omega_1$ of volume δv_1 and elasticity tensor $\mathbf{C}^1$, and calculate the resulting new overall elasticity tensor by the dilute-distribution assumption.

The geometry of $\delta\Omega_1$ is assumed to be similar to that of Ω_1. The volume fractions f_1 and $f_1 + \delta f_1$ are given by

$$f_1 = \frac{v_1}{v + v_1}, \qquad f_1 + \delta f_1 = \frac{v_1 + \delta v_1}{v + v_1 + \delta v_1}, \tag{10.3.2a,b}$$

where $\delta f_1 << 1$. Hence, the volume fraction δf_1 of the additional $\delta\Omega_1$ in the composite of total volume $v + v_1 + \delta v_1$ is given by

$$\frac{\delta v_1}{v + v_1 + \delta v_1} = \frac{\delta f_1}{1 - f_1}; \tag{10.3.2c}$$

subtract (10.3.2a) from (10.3.2b). Let the overall elasticity tensor of this imaginary composite of volume $v + v_1 + \delta v_1$ be denoted by $\bar{\mathbf{C}}(f_1 + \delta f_1)$. Since the composite consists of a uniform matrix M′ of volume $v + v_1$ and elasticity tensor $\bar{\mathbf{C}}(f_1)$, and an inclusion $\delta\Omega_1$ of volume δv_1 and elasticity tensor $\mathbf{C}^1$, the exact expression for $\bar{\mathbf{C}}(f_1 + \delta f_1)$ is, from (10.3.1c),

$$\bar{\mathbf{C}}(f_1 + \delta f_1) - \bar{\mathbf{C}}(f_1) = \frac{\delta f_1}{1 - f_1}\ \{\mathbf{C}^1 - \bar{\mathbf{C}}(f_1)\} : \mathbf{E}^1(\mathbf{C}^1,\ \delta\Omega_1;\ \bar{\mathbf{C}}(f_1),\ \mathrm{M}'). \tag{10.3.3}$$

Now take the limit as the volume fraction of the micro-element $\delta\Omega_1$ in the composite $M' + \delta\Omega_1$ approaches zero. As $\delta\Omega_1$ approaches zero, the response of $\delta\Omega_1$ embedded in the finite M' approaches that of an isolated micro-element embedded in $B(\bar{\mathbf{C}})$ which is an unbounded homogeneous solid of elasticity tensor $\bar{\mathbf{C}}(f_1)$. This is the *limiting case* of the assumption of a dilute distribution of inhomogeneities. Hence,

$$\lim_{\delta\Omega_1 \to 0} \mathbf{E}^1(\mathbf{C}^1, \delta\Omega_1; \bar{\mathbf{C}}(f_1), M') = \mathbf{E}^\infty(\mathbf{C}^1, \Omega_1; \bar{\mathbf{C}}(f_1)). \tag{10.3.4}$$

Here $\mathbf{E}^\infty$ is the tensor that *exactly* determines the average strain of an isolated micro-element in an unbounded homogeneous solid of uniform elasticity $\bar{\mathbf{C}}$; see Subsection 10.1. Therefore, taking the limit $\delta f_1 \to 0$ in (10.3.3), obtain a differential equation for $\bar{\mathbf{C}}$ with respect to f_1,

$$\frac{d}{df_1} \bar{\mathbf{C}}(f_1) = \frac{1}{1-f_1} \{\mathbf{C}^1 - \bar{\mathbf{C}}(f_1)\} : \mathbf{E}^{DS}, \tag{10.3.5a}$$

where

$$\mathbf{E}^{DS} = \mathbf{E}^{DS}(f_1) = \mathbf{E}^\infty(\mathbf{C}^1, \Omega_1; \bar{\mathbf{C}}(f_1)). \tag{10.3.5b}$$

Since $\bar{\mathbf{C}}$ is equal to $\mathbf{C}$ at $f_1 = 0$, the initial condition is[2]

$$\bar{\mathbf{C}}(0) = \mathbf{C}. \tag{10.3.5c}$$

Hence, one has an initial-value problem for $\bar{\mathbf{C}}$.

Next examine the effect of δf_1 on the overall compliance tensor $\bar{\mathbf{D}}(f_1)$. Following the same procedure as above, arrive at

$$\frac{d}{df_1} \bar{\mathbf{D}}(f_1) = \frac{1}{1-f_1} \{\mathbf{D}^1 - \bar{\mathbf{D}}(f_1)\} : \mathbf{F}^{DS}, \tag{10.3.6a}$$

where

$$\mathbf{F}^{DS} = \mathbf{F}^{DS}(f_1) = \mathbf{F}^\infty(\mathbf{D}^1, \Omega_1; \bar{\mathbf{D}}(f_1)), \tag{10.3.6b}$$

with the initial condition[3]

$$\bar{\mathbf{D}}(0) = \mathbf{D}. \tag{10.3.6c}$$

10.3.2. Multi-Phase RVE

Using the technique of Subsection 10.3.1, examine a multi-phase RVE, containing n distinct linearly elastic and homogeneous micro-elements embedded in its linearly elastic and homogeneous matrix. Let the volume fraction of each micro-element Ω_α be f_α, and regard the overall elasticity and compliance tensors as functions of all f_α, i.e.,

---

[2] It should be noted that in the differential equation (10.3.5a), $\bar{\mathbf{C}}$ approaches $\mathbf{C}^1$ as f_1 approaches 1, such that $\mathbf{C}^1 - \bar{\mathbf{C}}$ vanishes at $f_1 = 1$, and the right-hand side of (10.3.5a) remains bounded. Hence, $\bar{\mathbf{C}}(1) = \mathbf{C}^1$ need not be imposed as an additional condition.

[3] As in (10.3.5), in the differential equation (10.3.6a), $\bar{\mathbf{D}}$ approaches $\mathbf{D}^1$ as f_1 goes to 1. Hence, $\bar{\mathbf{D}}(1) = \mathbf{D}^1$ need not be imposed.

$$\bar{\mathbf{C}}\ (= \bar{\mathbf{C}}(\mathbf{C}^1, \Omega_1; ...; \mathbf{C}^n, \Omega_n; \mathbf{C}, \mathrm{M})) \equiv \bar{\mathbf{C}}(f_1, ..., f_n),$$

$$\bar{\mathbf{D}}\ (= \bar{\mathbf{D}}(\mathbf{D}^1, \Omega_1; ...; \mathbf{D}^n, \Omega_n; \mathbf{D}, \mathrm{M})) \equiv \bar{\mathbf{D}}(f_1, ..., f_n). \tag{10.3.7a,b}$$

All micro-inclusions need not have the same shape, but each micro-element Ω_α is assumed to have a distinct *similar* shape for any value of f_α, and hence only the effects of the volume fractions f_α on $\bar{\mathbf{C}}$ and $\bar{\mathbf{D}}$ are considered.

Consider an infinitesimally small change in the volume fraction of one typical micro-element, say, Ω_α, and examine its effects on the overall moduli using the procedure outlined in the preceding subsection. In this manner, arrive at

$$\bar{\mathbf{C}}(..., f_\alpha + \delta f_\alpha, ...) - \bar{\mathbf{C}}(..., f_\alpha, ...)$$

$$= \frac{\delta f_\alpha}{1 - f_\alpha} \{\mathbf{C}^\alpha - \bar{\mathbf{C}}(..., f_\alpha, ...)\} : \mathbf{E}^\alpha(\mathbf{C}^\alpha, \delta\Omega_\alpha; \bar{\mathbf{C}}(..., f_\alpha, ...), \mathrm{M}')$$
$$(\alpha \text{ not summed; } \alpha = 1, ..., n), \tag{10.3.8a}$$

when f_α is changed to $f_\alpha + \delta f_\alpha$ for a fixed α. Similarly, obtain for the overall compliance,

$$\bar{\mathbf{D}}(..., f_\alpha + \delta f_\alpha, ...) - \bar{\mathbf{D}}(..., f_\alpha, ...)$$

$$= \frac{\delta f_\alpha}{1 - f_\alpha} \{\mathbf{D}^\alpha - \bar{\mathbf{D}}(..., f_\alpha, ...)\} : \mathbf{F}^\alpha(\mathbf{D}^\alpha, \delta\Omega_\alpha; \bar{\mathbf{D}}(..., f_\alpha, ...), \mathrm{M}')$$
$$(\alpha \text{ not summed; } \alpha = 1, ..., n). \tag{10.3.8b}$$

Note that in (10.3.8a,b) the homogenized matrix is M′ which consists of the original matrix, M, and all inhomogeneities, Ω_β ($\beta = 1, ..., n$), and that the infinitesimally small inhomogeneity $\delta\Omega_\alpha$ has a shape similar to that of Ω_α.

Taking the limit $\delta f_\alpha \to 0$, obtain partial differential equations from (10.3.8a,b) for the overall elasticity and compliance tensors estimated by the *differential scheme*. Denote these by $\bar{\mathbf{C}}^{DS} = \bar{\mathbf{C}}^{DS}(f_1, ..., f_n)$ and $\bar{\mathbf{D}}^{DS} = \bar{\mathbf{D}}^{DS}(f_1, ..., f_n)$, respectively, and observe that

$$\frac{\partial}{\partial f_\alpha} \bar{\mathbf{C}}^{DS}(f_1, ..., f_n) = \frac{1}{1 - f_\alpha} (\mathbf{C}^\alpha - \bar{\mathbf{C}}^{DS}) : \mathbf{E}^{DS\alpha},$$

$$\frac{\partial}{\partial f_\alpha} \bar{\mathbf{D}}^{DS}(f_1, ..., f_n) = \frac{1}{1 - f_\alpha} (\mathbf{D}^\alpha - \bar{\mathbf{D}}^{DS}) : \mathbf{F}^{DS\alpha}$$
$$(\alpha \text{ not summed; } \alpha = 1, ..., n), \tag{10.3.9a,b}$$

where

$$\mathbf{E}^{DS\alpha} = \mathbf{E}^{DS\alpha}(f_1, ..., f_n) = \mathbf{E}^\infty(\mathbf{C}^\alpha, \Omega_\alpha; \bar{\mathbf{C}}^{DS}(f_1, ..., f_n)),$$

$$\mathbf{F}^{DS\alpha} = \mathbf{F}^{DS\alpha}(f_1, ..., f_n) = \mathbf{F}^\infty(\mathbf{D}^\alpha, \Omega_\alpha; \bar{\mathbf{D}}^{DS}(f_1, ..., f_n)). \tag{10.3.9c,d}$$

The initial conditions of the sets of partial differential equations (10.3.9a,b) are

$$\bar{\mathbf{C}}^{DS}(0, ..., 0) = \mathbf{C}, \qquad \bar{\mathbf{D}}^{DS}(0, ..., 0) = \mathbf{D}. \tag{10.3.9e,f}$$

It is clear that the solution to the system of partial differential equations (10.3.9a) or (10.3.9b), with the corresponding initial conditions (10.3.9e) or (10.3.9f), is not unique, because these partial differential equations give the gradient of $\bar{\mathbf{C}}^{DS}$ or $\bar{\mathbf{D}}^{DS}$ with respect to only the volume fraction of each inhomogeneity. From a physical point of view, this nonuniqueness is not acceptable: the overall response of an elastic composite should not depend on the manner by which the solution is constructed. That is, even if there are infinitely many ways that the construction of a finite composite can be theoretically interpreted (i.e., by different sequential additions of small amounts of inhomogeneities), the response of the final composite to given boundary data should be unique. Therefore, the possible solutions must be restricted by introducing reasonable simplifying assumptions.

One technique of constructing the final *solution* incrementally is to start with the initial step of a uniform matrix, M, containing infinitesimally small inhomogeneities, $\delta\Omega_\alpha$ ($\alpha = 1, ..., n$), each centered at the centroid of the corresponding Ω_α, and each having a shape *similar* to the corresponding Ω_α. Furthermore, fix the relative fraction of inhomogeneities such that $\delta\Omega_\alpha/\delta\Omega_\beta = \Omega_\alpha/\Omega_\beta$ for all α and β. Thus, set

$$f_\alpha = \rho_\alpha f, \qquad \delta f_\alpha = \rho_\alpha \delta f, \qquad \sum_{\alpha=1}^{n} \rho_\alpha = 1, \tag{10.3.10a~c}$$

where ρ_α, $\alpha = 1, 2, ..., n$, are fixed numbers giving the relative volume fractions of inhomogeneities, and note that

$$\bar{\mathbf{C}}^{DS}(\rho_1 f, \rho_2 f, ..., \rho_n f) = \bar{\mathbf{C}}^{DS}(f),$$

$$\bar{\mathbf{D}}^{DS}(\rho_1 f, \rho_2 f, ..., \rho_n f) = \bar{\mathbf{D}}^{DS}(f). \tag{10.3.10d,e}$$

The parameter f then varies from zero to a final value, as the sizes of the inhomogeneities are increased while their respective shapes remain self-similar. The overall moduli $\bar{\mathbf{C}}^{DS}$ and $\bar{\mathbf{D}}^{DS}$ are now regarded as functions of only the total volume fraction f of the entire set of micro-elements, Ω, in the RVE. In this manner, the partial differential equations (10.3.9a,b) are reduced to the following two ordinary differential equations:

$$\frac{1}{\rho_\alpha}\frac{d}{df}\bar{\mathbf{C}}^{DS}(f) = \frac{1}{1-\rho_\alpha f}(\mathbf{C}^\alpha - \bar{\mathbf{C}}^{DS}) : \mathbf{E}^{DS\alpha},$$

$$\frac{1}{\rho_\alpha}\frac{d}{df}\bar{\mathbf{D}}^{DS}(f) = \frac{1}{1-\rho_\alpha f}(\mathbf{D}^\alpha - \bar{\mathbf{D}}^{DS}) : \mathbf{F}^{DS\alpha} \quad (\alpha \text{ not summed}), \tag{10.3.11a,b}$$

where $\mathbf{E}^{DS\alpha}$ and $\mathbf{F}^{DS\alpha}$ are now regarded as functions of f.

Multiplying by ρ_α^2 and then summing (10.3.11a,b) over all α's, obtain the following ordinary differential equations for $\bar{\mathbf{C}}^{DS}$ and $\bar{\mathbf{D}}^{DS}$:

$$\frac{d}{df}\bar{\mathbf{C}}^{DS}(f) = \sum_{\alpha=1}^{n}\frac{\rho_\alpha^2}{1-\rho_\alpha f}(\mathbf{C}^\alpha - \bar{\mathbf{C}}^{DS}) : \mathbf{E}^{DS\alpha}, \qquad \bar{\mathbf{C}}^{DS}(0) = \mathbf{C}, \tag{10.3.12a,c}$$

and

$$\frac{d}{df}\,\bar{\mathbf{D}}^{DS}(f) = \sum_{\alpha=1}^{n} \frac{\rho_\alpha^2}{1-\rho_\alpha f}\,(\mathbf{D}^\alpha - \bar{\mathbf{D}}^{DS}) : \mathbf{F}^{DS\alpha}, \qquad \bar{\mathbf{D}}^{DS}(0) = \mathbf{D}. \quad (10.3.12b,d)$$

10.3.3. Equivalence between Overall Elasticity and Compliance Tensors

The initial-value problem (10.3.9a,e) gives the overall elasticity tensor $\bar{\mathbf{C}}^{DS}$ of the RVE subjected to a constant macrostrain, while the initial-value problem (10.3.9b,f) gives the overall compliance tensor $\bar{\mathbf{D}}^{DS}$ of the RVE subjected to a constant macrostress. In principle, $\bar{\mathbf{C}}^{DS}$ and $\bar{\mathbf{D}}^{DS}$ need not be each other's inverse. However, in order to obtain physically admissible solutions from the partial differential equations (10.3.9a,b), one may assign the equivalence of $\bar{\mathbf{C}}^{DS}$ and $\bar{\mathbf{D}}^{DS}$ as a necessary condition. As discussed before, such an equivalence is more desirable for estimating the *expected* response of an RVE, than the exact solution of the corresponding initial-value problem.

To this end, observe that: 1) by definition, $\mathbf{F}^\infty$ and $\mathbf{E}^\infty$ are

$$\mathbf{F}^\infty(\mathbf{D}^\alpha, \Omega_\alpha; (\bar{\mathbf{C}}^{DS})^{-1}) = \mathbf{C}^\alpha : \mathbf{E}^\infty(\mathbf{C}^\alpha, \Omega_\alpha; \bar{\mathbf{C}}^{DS}) : (\bar{\mathbf{C}}^{DS})^{-1},$$

$$\mathbf{E}^\infty(\mathbf{C}^\alpha, \Omega_\alpha; (\bar{\mathbf{D}}^{DS})^{-1}) = \mathbf{D}^\alpha : \mathbf{F}^\infty(\mathbf{D}^\alpha, \Omega_\alpha; \bar{\mathbf{D}}^{DS}) : (\bar{\mathbf{D}}^{DS})^{-1}$$
$$(\alpha \text{ not summed}); \qquad (10.3.13a,b)$$

and 2) differentiation of $\bar{\mathbf{C}}^{DS} : (\bar{\mathbf{C}}^{DS})^{-1} = \mathbf{1}^{(4s)}$ and $\bar{\mathbf{D}}^{DS} : (\bar{\mathbf{D}}^{DS})^{-1} = \mathbf{1}^{(4s)}$ with respect to f_α yields

$$\{\frac{\partial}{\partial f_\alpha}\,\bar{\mathbf{C}}^{DS}\} : (\bar{\mathbf{C}}^{DS})^{-1} + \bar{\mathbf{C}}^{DS} : \{\frac{\partial}{\partial f_\alpha}\,(\bar{\mathbf{C}}^{DS})^{-1}\} = \mathbf{0},$$

$$\{\frac{\partial}{\partial f_\alpha}\,\bar{\mathbf{D}}^{DS}\} : (\bar{\mathbf{D}}^{DS})^{-1} + \bar{\mathbf{D}}^{DS} : \{\frac{\partial}{\partial f_\alpha}\,(\bar{\mathbf{D}}^{DS})^{-1}\} = \mathbf{0}, \qquad (10.3.14a,b)$$

and hence,

$$\frac{\partial}{\partial f_\alpha}\,(\bar{\mathbf{C}}^{DS})^{-1} = -\,(\bar{\mathbf{C}}^{DS})^{-1} : \{\frac{\partial}{\partial f_\alpha}\,\bar{\mathbf{C}}^{DS}\} : (\bar{\mathbf{C}}^{DS})^{-1},$$

$$\frac{\partial}{\partial f_\alpha}\,(\bar{\mathbf{D}}^{DS})^{-1} = -\,(\bar{\mathbf{D}}^{DS})^{-1} : \{\frac{\partial}{\partial f_\alpha}\,\bar{\mathbf{D}}^{DS}\} : (\bar{\mathbf{D}}^{DS})^{-1}. \qquad (10.3.14c,d)$$

Then, from (10.3.13a,b) and (10.3.14c,d), the partial differential equations (10.3.9a,b) for $\bar{\mathbf{C}}^{DS}$ and $\bar{\mathbf{D}}^{DS}$ can be transformed into those for $(\bar{\mathbf{C}}^{DS})^{-1}$ and $(\bar{\mathbf{D}}^{DS})^{-1}$, respectively, as follows:

$$\frac{\partial}{\partial f_\alpha}\,(\bar{\mathbf{C}}^{DS})^{-1} = \frac{1}{1-f_\alpha}\,\{\mathbf{D}^\alpha - (\bar{\mathbf{C}}^{DS})^{-1}\} : \mathbf{F}^\infty(\mathbf{D}^\alpha, \Omega_\alpha; (\bar{\mathbf{C}}^{DS})^{-1}),$$

$$\frac{\partial}{\partial f_\alpha}\,(\bar{\mathbf{C}}^{DS})^{-1} = \frac{1}{1-f_\alpha}\,\{\mathbf{C}^\alpha - (\bar{\mathbf{D}}^{DS})^{-1}\} : \mathbf{E}^\infty(\mathbf{C}^\alpha, \Omega_\alpha; (\bar{\mathbf{D}}^{DS})^{-1})$$
$$(\alpha \text{ not summed}), \qquad (10.3.15a,b)$$

for $\alpha = 1, ..., n$, and the initial conditions (10.3.9c,d) are transformed into

$$(\overline{\mathbf{C}}^{DS})^{-1}(0, ..., 0) = \mathbf{D}, \qquad (\overline{\mathbf{D}}^{DS})^{-1}(0, ..., 0) = \mathbf{C}. \tag{10.3.15c,d}$$

Here, in deriving (10.3.15a,b), the following identities are used:

$$(\mathbf{C}^{\alpha} - \overline{\mathbf{C}}^{DS}) : \mathbf{D}^{\alpha} = -\overline{\mathbf{C}}^{DS} : \{\mathbf{D}^{\alpha} - (\overline{\mathbf{C}}^{DS})^{-1}\},$$

$$(\mathbf{D}^{\alpha} - \overline{\mathbf{D}}^{DS}) : \mathbf{C}^{\alpha} = -\overline{\mathbf{D}}^{DS} : \{\mathbf{C}^{\alpha} - (\overline{\mathbf{D}}^{DS})^{-1}\}. \tag{10.3.16a,b}$$

From comparison of (10.3.15a) with (10.3.9b) and (10.3.15b) with (10.3.9a), respectively, it is seen that $(\overline{\mathbf{C}}^{DS})^{-1}$ satisfies the same initial-value problem as $\overline{\mathbf{D}}^{DS}$. And from comparison of (10.3.15c) with (10.3.9f) and (10.3.15d) with (10.3.9e), it is seen that $(\overline{\mathbf{D}}^{DS})^{-1}$ satisfies the same initial value problem as $\overline{\mathbf{C}}^{DS}$. Therefore, if the solutions of the differential scheme are uniquely chosen,[4] then it follows that

$$(\overline{\mathbf{C}}^{DS})^{-1} = \overline{\mathbf{D}}^{DS}, \qquad (\overline{\mathbf{D}}^{DS})^{-1} = \overline{\mathbf{C}}^{DS} \tag{10.3.17a,b}$$

or

$$\overline{\mathbf{C}}^{DS} : \overline{\mathbf{D}}^{DS} = \mathbf{1}^{(4s)}, \qquad \overline{\mathbf{D}}^{DS} : \overline{\mathbf{C}}^{DS} = \mathbf{1}^{(4s)}. \tag{10.3.17c,d}$$

Thus, the two partial differential equations, (10.3.12a) and (10.3.12b), are *equivalent*.

10.4. TWO-PHASE MODEL AND DOUBLE-INCLUSION METHOD

A major shortcoming of the self-consistent method is that the interaction between the inclusions and their immediate surrounding matrix material of different elasticity is not directly included in the model. When the elasticity of the matrix deviates considerably from that of the inclusions, one may use the *double-inclusion method* which does take this fact into account directly. In this method, the average strain or stress in a typical inclusion is estimated by embedding a typical inclusion Ω in a finite ellipsoidal region V *of matrix elasticity*, and then this double-inclusion is embedded in an infinite uniform solid. When the elasticity of the infinite solid is that of the matrix, one has the two-phase model (Benveniste, 1987). On the other hand, when this elasticity is the yet-unknown overall elasticity of the composite, one has the three-phase or double-inclusion model (Hori and Nemat-Nasser, 1993); see Subsection 10.6. For prescribed overall strains or stresses, the average strains and stresses in the ellipsoidal region V are then obtained, and used to estimate its overall moduli, using the exact expressions (10.1.1a,b). This procedure for the double-inclusion model is significantly simplified by the use of the observation by Tanaka and Mori (1972), based on the properties of the Eshelby tensor; this and its generalizations are discussed in Subsection 11.3.3. The result exploits Eshelby's solution of the

[4] The partial differential equations (10.3.15) can be reduced to ordinary differential equations, in the same manner as (10.3.9) is reduced to (10.3.11) or (10.3.12). In this case, the solution is unique.

single-inclusion problem considered in some detail in Section 11.

The Tanaka-Mori result addresses the following problem: Suppose that uniform eigenstrains are distributed in an ellipsoidal domain Ω embedded in an infinite homogeneous domain, and let V be an arbitrary ellipsoidal domain which includes Ω. Then, the volume average of the strains and stresses produced by the eigenstrains in Ω, when taken over $V - \Omega$, is identically zero, if V and Ω are *coaxial and similar*[5].

From this exact result, the two-phase averaging scheme is derived and then generalized to the double-inclusion case which may involve noncoaxial and dissimilar ellipsoids of different elasticities, embedded in an infinite solid of yet different elasticity. Note the difference between this kind of averaging scheme and the three averaging methods discussed in this section. These latter methods consider a single inclusion embedded in an unbounded homogeneous elastic solid, and use this model to estimate the average strain or stress in a typical inclusion of a finite RVE consisting of many inclusions within a matrix material. On the other hand, in the double-inclusion method, a finite subregion of an infinite domain, which contains an inclusion is considered as an RVE, and the exact average strain and stress in this subregion are computed for uniform farfield stresses or strains. This method and its extensions to three-phase and multi-inclusion models are examined in this subsection.

10.4.1. Basic Formulation: Two-Phase Model

Consider an infinite homogeneous domain, B, of matrix elasticity $\mathbf{C}$ and compliance $\mathbf{D}$, in which an isolated inclusion Ω of elasticity $\mathbf{C}^{\Omega}$ and compliance $\mathbf{D}^{\Omega}$ is embedded. Let V contain Ω, and be a subregion of the infinite domain. Denote by M the part of V which is outside of Ω, $M = V - \Omega$. The elasticity and compliance fields of this infinite composite, $\mathbf{C}'$ and $\mathbf{D}'$, are

$$\mathbf{C}' = \mathbf{C}'(\mathbf{x}) = \begin{cases} \mathbf{C}^{\Omega} & \mathbf{x} \text{ in } \Omega \\ \mathbf{C} & \mathbf{x} \text{ in M} \\ \mathbf{C} & \mathbf{x} \text{ outside of V} \end{cases} \tag{10.4.1a}$$

and

$$\mathbf{D}' = \mathbf{D}'(\mathbf{x}) = \begin{cases} \mathbf{D}^{\Omega} & \mathbf{x} \text{ in } \Omega \\ \mathbf{D} & \mathbf{x} \text{ in M} \\ \mathbf{D} & \mathbf{x} \text{ outside of V;} \end{cases} \tag{10.4.1b}$$

see Figure 10.4.1.

The following exact average relations hold among the strains and stresses in V, Ω, and M:

[5] It is shown in Subsection 11.3.3 that the average strain in the ellipsoidal domain V is exactly defined by the Eshelby tensor for V and the average eigenstrain in Ω, even when Ω is not ellipsoidal and the eigenstrain distribution in Ω is not uniform; see (11.3.18b).

Figure 10.4.1

A double inclusion $V = M + \Omega$ is embedded in a uniform infinite domain B; elasticity of Ω is $\mathbf{C}^\Omega$, that of $M = V - \Omega$, and the infinite domain is $\mathbf{C}$; this is a two-phase model

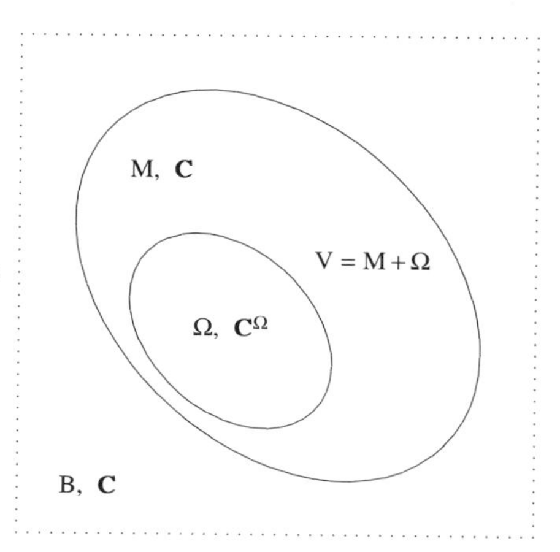

$$< \boldsymbol{\varepsilon} >_V = f < \boldsymbol{\varepsilon} >_\Omega + (1 - f) < \boldsymbol{\varepsilon} >_M,$$

$$< \boldsymbol{\sigma} >_V = f < \boldsymbol{\sigma} >_\Omega + (1 - f) < \boldsymbol{\sigma} >_M, \qquad (10.4.2a,b)$$

where $f = \Omega/V$. Let the farfield strains and stresses be denoted by $\boldsymbol{\varepsilon}^\infty$ and $\boldsymbol{\sigma}^\infty$, and note that $\boldsymbol{\sigma}^\infty = \mathbf{C} : \boldsymbol{\varepsilon}^\infty$ or $\boldsymbol{\varepsilon}^\infty = \mathbf{D} : \boldsymbol{\sigma}^\infty$. Let the homogenizing eigenstrain be $\boldsymbol{\varepsilon}^*$. Then, if Ω is ellipsoidal, its strain and stress fields are *exactly* given by

$$\boldsymbol{\varepsilon}(\mathbf{x}) = \boldsymbol{\varepsilon}^\infty + \mathbf{S}^\Omega : \boldsymbol{\varepsilon}^*,$$

$$\boldsymbol{\sigma}(\mathbf{x}) = \mathbf{C} : \{\boldsymbol{\varepsilon}^\infty + (\mathbf{S}^\Omega - \mathbf{1}^{(4s)}) : \boldsymbol{\varepsilon}^*\} \qquad \mathbf{x} \text{ in } \Omega. \qquad (10.4.3a,b)$$

Here the eigenstrain $\boldsymbol{\varepsilon}^*$ is determined from consistency condition (7.3.15a), to be

$$\boldsymbol{\varepsilon}^* = (\mathbf{A}^\Omega - \mathbf{S}^\Omega)^{-1} : \boldsymbol{\varepsilon}^\infty, \qquad (10.4.3c)$$

where $\mathbf{A}^\Omega = (\mathbf{C} - \mathbf{C}^\Omega)^{-1} : \mathbf{C}$, and $\mathbf{S}^\Omega$ is Eshelby's tensor for Ω; see Subsection 7.3. Since the strain and stress fields are uniform in Ω, their volume averages over Ω are given by (10.4.3), i.e.,

$$< \boldsymbol{\varepsilon} >_\Omega = \boldsymbol{\varepsilon}^\infty + \mathbf{S}^\Omega : \boldsymbol{\varepsilon}^*, \qquad < \boldsymbol{\sigma} >_\Omega = \mathbf{C} : \{\boldsymbol{\varepsilon}^\infty + (\mathbf{S}^\Omega - \mathbf{1}^{(4s)}) : \boldsymbol{\varepsilon}^*\}. \quad (10.4.4a,b)$$

The eigenstrain defined by (10.4.3c) disturbs the strain and stress fields outside of Ω. Hence, these fields are not uniform. Let V be an ellipsoid coaxial with and similar to Ω. Then, according to the Tanaka-Mori result, the volume average of these disturbances, taken over $M = V - \Omega$, vanishes. Hence,

$$< \boldsymbol{\varepsilon} >_M = \boldsymbol{\varepsilon}^\infty, \qquad < \boldsymbol{\sigma} >_M = \mathbf{C} : \boldsymbol{\varepsilon}^\infty. \qquad (10.4.5a,b)$$

Substituting (10.4.4a,b) and (10.4.5a,b) into (10.4.2a,b), obtain the *exact* average strain and stress over V, as follows:

$$< \boldsymbol{\varepsilon} >_V = \boldsymbol{\varepsilon}^\infty + f\, \mathbf{S}^\Omega : \boldsymbol{\varepsilon}^*,$$

$$< \boldsymbol{\sigma} >_V = \mathbf{C} : \{\boldsymbol{\varepsilon}^\infty + f\, (\mathbf{S}^\Omega - \mathbf{1}^{(4s)}) : \boldsymbol{\varepsilon}^*\}. \qquad (10.4.6a,b)$$

Therefore, the overall elasticity, denoted by $\bar{\mathbf{C}}^{TP}$ for this two-phase model, is defined through $< \boldsymbol{\sigma} >_V = \bar{\mathbf{C}}^{TP} : < \boldsymbol{\varepsilon} >_V$, and is given by

$$\bar{\mathbf{C}}^{TP} = \mathbf{C} : \{\, \mathbf{1}^{(4s)} + f\, (\mathbf{S}^\Omega - \mathbf{1}^{(4s)}) : (\mathbf{A}^\Omega - \mathbf{S}^\Omega)^{-1} \,\}$$

$$: \{\, \mathbf{1}^{(4s)} + f\, \mathbf{S}^\Omega : (\mathbf{A}^\Omega - \mathbf{S}^\Omega)^{-1} \,\}^{-1}. \qquad (10.4.7)$$

In deriving (10.4.7), one assumption, that the overall moduli of the two-phase material are given by the average response of the above specified V, is made. The mathematical derivation, however, is *exact*. Furthermore, the overall elasticity and compliance defined on the basis of (10.4.6a,b):

1) are each other's inverse;

2) do not depend on the surface data on ∂V; and

3) do not depend on the location of Ω relative to V.

Except for special cases,[6] the surface data on ∂V are unknown. Because of this, it cannot be proven that the same overall elasticity is obtained by considering the average strain energy of the composite. However, from (10.4.3a,b),

$$(\boldsymbol{\sigma} : \boldsymbol{\varepsilon})(\mathbf{x}) = \{\boldsymbol{\varepsilon}^\infty + (\mathbf{S}^\Omega - \mathbf{1}^{(4s)}) : \boldsymbol{\varepsilon}^*\} : \mathbf{C} : \{\boldsymbol{\varepsilon}^\infty + \mathbf{S}^\Omega : \boldsymbol{\varepsilon}^*\} \qquad \mathbf{x} \text{ in } \Omega, \qquad (10.4.3d)$$

and its volume average over Ω is given by

$$< \boldsymbol{\sigma} : \boldsymbol{\varepsilon} >_\Omega = \boldsymbol{\varepsilon}^\infty : \{\mathbf{1}^{(4s)} + (\mathbf{S}^\Omega - \mathbf{1}^{(4s)}) : (\mathbf{A}^\Omega - \mathbf{S}^\Omega)^{-1}\}^T : \mathbf{C}$$

$$: \{\mathbf{1}^{(4s)} + \mathbf{S}^\Omega : (\mathbf{A}^\Omega - \mathbf{S}^\Omega)^{-1}\} : \boldsymbol{\varepsilon}^\infty. \qquad (10.4.4c)$$

Hence, assuming that the average strain energy over M, $< \boldsymbol{\sigma} : \boldsymbol{\varepsilon} >_M$, is approximated by $\boldsymbol{\varepsilon}^\infty : \mathbf{C} : \boldsymbol{\varepsilon}^\infty$, and using the exact average relation

$$< \boldsymbol{\sigma} : \boldsymbol{\varepsilon} >_V = f < \boldsymbol{\sigma} : \boldsymbol{\varepsilon} >_\Omega + (1 - f) < \boldsymbol{\sigma} : \boldsymbol{\varepsilon} >_M, \qquad (10.4.2c)$$

estimate the average strain energy over V, as

$$< \boldsymbol{\sigma} : \boldsymbol{\varepsilon} >_V = \boldsymbol{\varepsilon}^\infty : \{\mathbf{1}^{(4s)} + f(\mathbf{S}^\Omega - \mathbf{1}^{(4s)}) : (\mathbf{A}^\Omega - \mathbf{S}^\Omega)^{-1}\}^T : \mathbf{C}$$

$$: \{\mathbf{1}^{(4s)} + f\mathbf{S}^\Omega : (\mathbf{A}^\Omega - \mathbf{S}^\Omega)^{-1}\} : \boldsymbol{\varepsilon}^\infty$$

$$+ f(1 - f)\, \boldsymbol{\varepsilon}^\infty : \{(\mathbf{S}^\Omega - \mathbf{1}^{(4s)}) : (\mathbf{A}^\Omega - \mathbf{S}^\Omega)^{-1}\}^T : \mathbf{C} : \{\mathbf{S}^\Omega : (\mathbf{A}^\Omega - \mathbf{S}^\Omega)^{-1}\} : \boldsymbol{\varepsilon}^\infty$$

$$= < \boldsymbol{\varepsilon} >_V : \bar{\mathbf{C}}^{TP} : < \boldsymbol{\varepsilon} >_V$$

---

[6] For example, if 1) V and Ω are cocentered spheres, 2) the farfield stresses or strains are isotropic, i.e., of the form $\sigma^\infty \mathbf{1}^{(2)}$ or $\varepsilon^\infty \mathbf{1}^{(2)}$, and 3) $\mathbf{C}$ and $\mathbf{C}^\Omega$ are isotropic, then the field variables are spherically symmetric, and the surface tractions and displacements become uniform and linear on ∂V.

$$+ f(1-f)\, \boldsymbol{\varepsilon}^\infty : \{(\mathbf{S}^\Omega - \mathbf{1}^{(4s)}) : (\mathbf{A}^\Omega - \mathbf{S}^\Omega)^{-1}\}^T : \mathbf{C} : \{\mathbf{S}^\Omega : (\mathbf{A}^\Omega - \mathbf{S}^\Omega)^{-1}\} : \boldsymbol{\varepsilon}^\infty. \tag{10.4.6c}$$

If f is close to 0 or 1, the second term in the right-hand side of (10.4.6c) may be neglected. Therefore, for a relatively small or large volume fraction, the symmetric part of the overall elasticity defined by (10.4.7) may be used to obtain a good estimate of the average strain energy[7] in V.

The above procedure can be directly applied to a multi-phase RVE. Suppose that a finite ellipsoidal subregion V of the infinite domain contains several micro-inclusions Ω_α, ($\alpha = 1, 2, ..., n$). Instead of (10.4.2a,b), then

$$< \boldsymbol{\varepsilon} >_V = \sum_{\alpha=1}^{n} f_\alpha < \boldsymbol{\varepsilon} >_\alpha + (1-f) < \boldsymbol{\varepsilon} >_M,$$

$$< \boldsymbol{\sigma} >_V = \sum_{\alpha=1}^{n} f_\alpha < \boldsymbol{\sigma} >_\alpha + (1-f) < \boldsymbol{\sigma} >_M, \tag{10.4.8a,b}$$

where f_α is the volume fraction of Ω_α, and $f = \sum_{\alpha=1}^{n} f_\alpha$. For the same method to work, each Ω_α of distinct elasticity $\mathbf{C}^\alpha$, must be an ellipsoid coaxial and similar to V. Neglecting the interaction among the micro-inclusions, set

$$< \boldsymbol{\varepsilon} >_\alpha = \boldsymbol{\varepsilon}^\infty + \mathbf{S}^\Omega : \boldsymbol{\varepsilon}^{*\alpha},$$

$$< \boldsymbol{\sigma} >_\alpha = \mathbf{C} : \{\boldsymbol{\varepsilon}^\infty + (\mathbf{S}^\Omega - \mathbf{1}^{(4s)}) : \boldsymbol{\varepsilon}^{*\alpha}\}, \tag{10.4.9a,b}$$

where

$$\boldsymbol{\varepsilon}^{*\alpha} = (\mathbf{A}^\alpha - \mathbf{S}^\Omega)^{-1} : \boldsymbol{\varepsilon}^\infty, \tag{10.4.9c}$$

with $\mathbf{A}^\alpha = (\mathbf{C} - \mathbf{C}^\alpha)^{-1} : \mathbf{C}$, $\mathbf{S}^\Omega$ being the common Eshelby tensor for all Ω_α's. The average strain and stress over M are still given by (10.4.5a,b). Hence, substitution of (10.4.9a~c) and (10.4.5a,b) into (10.4.8a,b) yields

$$< \boldsymbol{\varepsilon} >_V = \boldsymbol{\varepsilon}^\infty + \sum_{\alpha=1}^{n} f_\alpha\, \mathbf{S}^\Omega : (\mathbf{A}^\alpha - \mathbf{S}^\Omega)^{-1} : \boldsymbol{\varepsilon}^\infty,$$

$$< \boldsymbol{\sigma} >_V = \mathbf{C} : \{\boldsymbol{\varepsilon}^\infty + \sum_{\alpha=1}^{n} f_\alpha\, (\mathbf{S}^\Omega - \mathbf{1}^{(4s)}) : (\mathbf{A}^\alpha - \mathbf{S}^\Omega)^{-1} : \boldsymbol{\varepsilon}^\infty\}. \tag{10.4.10a,b}$$

Therefore, the overall elasticity $\bar{\mathbf{C}}^{MP}$ of this multi-phase composite becomes

$$\bar{\mathbf{C}}^{MP} = \mathbf{C} : \{\mathbf{1}^{(4s)} + \sum_{\alpha=1}^{n} f_\alpha\, (\mathbf{S}^\Omega - \mathbf{1}^{(4s)}) : (\mathbf{A}^\alpha - \mathbf{S}^\Omega)^{-1}\}$$

$$: \{\mathbf{1}^{(4s)} + \sum_{\alpha=1}^{n} f_\alpha\, \mathbf{S}^\Omega : (\mathbf{A}^\alpha - \mathbf{S}^\Omega)^{-1}\}^{-1}. \tag{10.4.11}$$

The overall compliance $\bar{\mathbf{D}}^{MP}$ is given by the inverse of $\bar{\mathbf{C}}^{MP}$.

---

[7] Note that $< \boldsymbol{\sigma} : \boldsymbol{\varepsilon} >_M = \boldsymbol{\varepsilon}^\infty : \mathbf{C} : \boldsymbol{\varepsilon}^\infty$ is assumed in deriving (10.4.6c). This implies small volume fractions of the inclusion phase.

10.4.2. Comments on Two-Phase Model

The averaging scheme presented in Subsection 10.4.1 is proposed by Benveniste (1987) on the basis of the following model,[8] due to Mori and Tanaka (1973):

Suppose a number of micro-inclusions are randomly distributed in an RVE *subjected to linear displacement boundary conditions. If an additional inclusion* Ω *is embedded in this* RVE*, then the average strain over* Ω *becomes*

$$< \boldsymbol{\varepsilon} >_{\Omega} = \Delta\boldsymbol{\varepsilon} + < \boldsymbol{\varepsilon} >_{\mathrm{M}}, \tag{10.4.12a}$$

where M *denotes the matrix phase surrounding* Ω*, and* $\Delta\boldsymbol{\varepsilon}$ *is given by the strain which is produced in an isolated inclusion when it is embedded in an infinite uniform domain.*[9] *Since the number of inclusions is large, the average strain over the newly embedded inclusion* Ω *must be the same as that of the other pre-existing ones. Hence,* (10.4.12a) *gives a relation between the average strains in the inclusion and the matrix phases.*

To obtain the relation between $\Delta\boldsymbol{\varepsilon}$ and $< \boldsymbol{\varepsilon} >_{\mathrm{M}}$, regard $< \boldsymbol{\varepsilon} >_{\mathrm{M}}$ as the farfield strain $\boldsymbol{\varepsilon}^{\infty}$, introduced in Subsection 10.4.1, and compute the average strain in Ω in terms of $< \boldsymbol{\varepsilon} >_{\mathrm{M}}$. From (10.4.3a) and (10.4.4a),

$$\Delta\boldsymbol{\varepsilon} = \mathbf{S}^{\Omega} : (\mathbf{A}^{\Omega} - \mathbf{S}^{\Omega})^{-1} : < \boldsymbol{\varepsilon} >_{\mathrm{M}}. \tag{10.4.12b}$$

The average strain in the inclusion phase is obtained by substituting (10.4.12b) into (10.4.12a). The average strain and hence the corresponding average stress in the inclusion and matrix phases are now expressed in terms of $< \boldsymbol{\varepsilon} >_{\mathrm{M}}$. Then, with the aid of (10.4.2a,b), the average strain and stress in the RVE are also expressed in terms of $< \boldsymbol{\varepsilon} >_{\mathrm{M}}$, and the overall moduli of the RVE are computed. In this manner, Benveniste (1987) derives the overall compliance under uniform tractions, and the overall elasticity under linear displacements. The final result is the same as (10.4.7).[10]

The Mori-Tanaka model and the differential scheme consider the change in the overall moduli when a small increment of the inclusion phase is introduced in a homogeneous composite. The corresponding change in the field variables is neglected by the former, whereas in the latter scheme the change in the field variables is related to the change of the volume fraction of the inclusion. It is shown in Subsection 10.5 that, as the volume fraction goes to zero, the two methods agree asymptotically, although their specific predictions may be different, depending on the problem.

---

[8] This is called the Mori-Tanaka model. A similar result is obtained when the RVE is subjected to uniform traction boundary conditions.

[9] Mori and Tanaka (1973) also address the local field variables in the embedded inclusions, not only their volume averages; see also Tanaka and Mori (1972).

[10] Mori and Wakashima (1990) have obtained (10.4.7), using (10.4.12) and successive iterations. The derivation in Subsection 10.4.1 is terse and transparent, requiring *no* iteration nor any restriction on the boundary data on ∂V.

10.4.3. Relation with Hashin-Shtrikman Bounds

When V and Ω are spherical, the overall elasticity given by the two-phase model coincides with either the upper or the lower bound obtained from the Hashin-Shtrikman variational principle. Assume that the matrix is isotropic and all inclusions are spherical. If the reference elasticity is set to be the matrix elasticity, a bound on the overall elasticity, $\bar{\mathbf{C}}^{HS}$, is given by (9.5.39b), i.e.,

$$\bar{\mathbf{C}}^{HS} = \Big\{(1-f)\,\mathbf{C} + \sum_{\alpha=1}^{n} f_\alpha\, \mathbf{C}^\alpha : \{\mathbf{1}^{(4s)} + \mathbf{S}^S : \mathbf{D} : (\mathbf{C}^\alpha - \mathbf{C})\}^{-1}\Big\}$$

$$:\Big\{(1-f)\,\mathbf{1}^{(4s)} + \sum_{\alpha=1}^{n} f_\alpha\, \{\mathbf{1}^{(4s)} + \mathbf{S}^S : \mathbf{D} : (\mathbf{C}^\alpha - \mathbf{C})\}^{-1}\Big\}^{-1}, \qquad (10.4.13)$$

where $\mathbf{P}$ in (9.5.28b) is replaced by $\mathbf{S}^S : \mathbf{D}$, with $\mathbf{S}^S$ being Eshelby's tensor for a sphere in an isotropic solid[11]; see Subsection 9.5. On the other hand, from

$$(\mathbf{A}^\alpha - \mathbf{S}^S)^{-1} = -\mathbf{D} : (\mathbf{C}^\alpha - \mathbf{C}) : \{\mathbf{1}^{(4s)} + \mathbf{S}^S : \mathbf{D} : (\mathbf{C}^\alpha - \mathbf{C})\}^{-1} \qquad (10.4.14a)$$

and

$$\mathbf{1}^{(4s)} = (1-f)\mathbf{1}^{(4s)} + \sum_{\alpha=1}^{n} f_\alpha\, \mathbf{1}^{(4s)}, \qquad (10.4.14b)$$

rewrite (10.4.11) as

$$\bar{\mathbf{C}}^{MP} = \mathbf{C} : \Big\{(1-f)\,\mathbf{1}^{(4s)} + \sum_{\alpha=1}^{n} f_\alpha\, \mathbf{D} : \mathbf{C}^\alpha : \{\mathbf{1}^{(4s)} + \mathbf{S}^S : \mathbf{D} : (\mathbf{C}^\alpha - \mathbf{C})\}^{-1}\Big\}$$

$$:\Big\{(1-f)\,\mathbf{1}^{(4s)} + \sum_{\alpha=1}^{n} f_\alpha\, \{\mathbf{1}^{(4s)} + \mathbf{S}^S : \mathbf{D} : (\mathbf{C}^\alpha - \mathbf{C})\}^{-1}\Big\}^{-1}. \qquad (10.4.14c)$$

As is seen, the right-hand sides of (10.4.13) and (10.4.14c) are the same. Hence, $\bar{\mathbf{C}}^{MP}$ coincides with $\bar{\mathbf{C}}^{HS}$, if the inclusions are all spherical and the matrix is isotropic. Weng (1990) shows the relation between the above two estimates of the overall moduli, $\bar{\mathbf{C}}^{HS}$ and $\bar{\mathbf{C}}^{MP}$.

The agreement of the overall moduli obtained by the multi-phase model and the corresponding Hashin-Shtrikman bounds is revealing. The multi-phase model includes the interaction between the matrix and an embedded inclusion exactly, but it does not take into account interaction among inclusions (which may not be negligible for large volume fractions of inclusions). Since the Hashin-Shtrikman variational principle considers the average strain energy stored in an RVE, it is seen that if the lower (upper) bound of the Hashin-Shtrikman variational principle coincides with the multi-phase result, then the interaction among inclusions *always* makes a positive (negative) contribution to the average strain energy, in such a manner that the corresponding bound remains valid. Indeed, if $\mathbf{C}' - \mathbf{C}$ is positive-definite (if $\mathbf{D}' - \mathbf{D}$ is positive-definite), then

[11] The matrix phase may be regarded as the 0th inclusion phase, i.e., $M = \Omega_0$ and $\mathbf{C} = \mathbf{C}^0$. In this case, $f_0\, \mathbf{C}^0 : \{\mathbf{1}^{(4s)} + \mathbf{S}^S : \mathbf{D} : (\mathbf{C}^0 - \mathbf{C})\}^{-1}$ and $f_0\, \{\mathbf{1}^{(4s)} + \mathbf{S}^S : \mathbf{D} : (\mathbf{C}^0 - \mathbf{C})\}^{-1}$ reduce to $(1-f)\,\mathbf{C}$ and $(1-f)\,\mathbf{1}^{(4s)}$, respectively.

$$\frac{1}{2}\boldsymbol{\varepsilon}^{\mathrm{o}} : \overline{\mathbf{C}} : \boldsymbol{\varepsilon}^{\mathrm{o}} \geq (\leq) \; \frac{1}{2}\boldsymbol{\varepsilon}^{\mathrm{o}} : \overline{\mathbf{C}}^{HS} : \boldsymbol{\varepsilon}^{\mathrm{o}} \equiv \frac{1}{2}\boldsymbol{\varepsilon}^{\mathrm{o}} : \overline{\mathbf{C}}^{MP} : \boldsymbol{\varepsilon}^{\mathrm{o}}, \tag{10.4.15}$$

where $\overline{\mathbf{C}}$ is the *exact* overall elasticity of the composite that gives the exact overall strain energy. Hence, if the inclusions are more stiff (compliant) than the matrix, their interaction *always* results in an increase (decrease) in the overall strain energy.

10.4.4. Generalization of Eshelby's Results

The two-phase model discussed in the preceding subsections is based on Eshelby's results. The double-inclusion method discussed in this and the following subsections is based on the following generalization of Eshebly's results. Consider average strains and stresses in an ellipsoidal region V of an unbounded uniform elastic solid B with elasticity $\mathbf{C}$, when V undergoes phase transformations in the following manner: 1) an ellipsoidal subregion Ω of V undergoes transformation corresponding to eigenstrains $\boldsymbol{\varepsilon}^{*1}(\mathbf{x})$ (not necessarily constant); and 2) the remaining part of V, $\Gamma \equiv \mathrm{V} - \Omega$, undergoes transformations with uniform eigenstrains $\boldsymbol{\varepsilon}^{*2}$; see Figure 10.4.2. Ellipsoids Ω and V need not be either coaxial or similar. Although the resulting strain and stress fields, $\boldsymbol{\varepsilon}^{\mathrm{d}}$ and $\boldsymbol{\sigma}^{\mathrm{d}}$, may not be constant in V and Ω, the average strain and stress in Ω are *exactly* given by

$$< \boldsymbol{\varepsilon}^{\mathrm{d}} >_{\Omega} = \mathbf{S}^{\Omega} : < \boldsymbol{\varepsilon}^{*1} >_{\Omega} + (\mathbf{S}^{\mathrm{V}} - \mathbf{S}^{\Omega}) : \boldsymbol{\varepsilon}^{*2},$$

$$< \boldsymbol{\sigma}^{\mathrm{d}} >_{\Omega} = \mathbf{C} : (\mathbf{S}^{\Omega} - \mathbf{1}^{(4s)}) : < \boldsymbol{\varepsilon}^{*1} >_{\Omega} + \mathbf{C} : (\mathbf{S}^{\mathrm{V}} - \mathbf{S}^{\Omega}) : \boldsymbol{\varepsilon}^{*2}, \tag{10.4.16a,b}$$

where $\mathbf{S}^{\mathrm{V}}$ and $\mathbf{S}^{\Omega}$ are Eshelby's tensors for ellipsoids V and Ω, respectively.

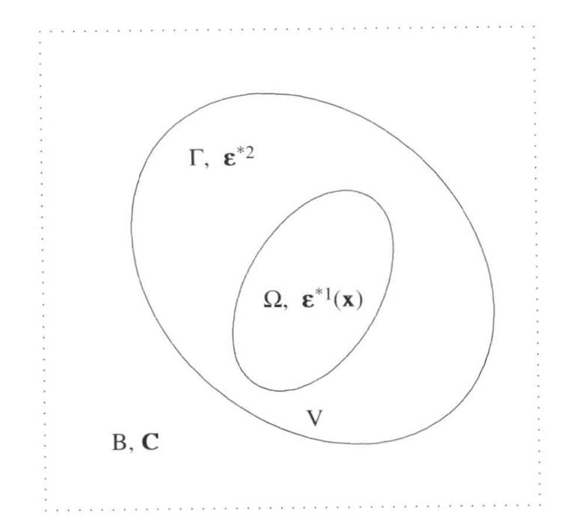

Figure 10.4.2

An unbounded uniform solid B contains two ellipsoidal regions, V and Ω ($\Omega \subset \mathrm{V}$), with eigenstrains $\boldsymbol{\varepsilon}^{*1}(\mathbf{x})$ in Ω and uniform eigenstrains $\boldsymbol{\varepsilon}^{*2}$ in $\Gamma = \mathrm{V} - \Omega$

Similarly, the average strain and stress over V are *exactly* given by

$$< \boldsymbol{\varepsilon}^d >_V = \mathbf{S}^V : \{f < \boldsymbol{\varepsilon}^{*1} >_\Omega + (1-f)\,\boldsymbol{\varepsilon}^{*2}\},$$

$$< \boldsymbol{\sigma}^d >_V = \mathbf{C} : (\mathbf{S}^V - \mathbf{1}^{(4s)}) : \{f < \boldsymbol{\varepsilon}^{*1} >_\Omega + (1-f)\,\boldsymbol{\varepsilon}^{*2}\}, \qquad (10.4.16c,d)$$

where f is the volume fraction of Ω in V, $f = \Omega / V$. Finally, in view of (10.4.16a~d), the average strain and stress over Γ are given by

$$< \boldsymbol{\varepsilon}^d >_\Gamma = \mathbf{S}^V : \boldsymbol{\varepsilon}^{*2} + \frac{f}{1-f}\,(\mathbf{S}^V - \mathbf{S}^\Omega) : (< \boldsymbol{\varepsilon}^{*1} >_\Omega - \boldsymbol{\varepsilon}^{*2}),$$

$$< \boldsymbol{\sigma}^d >_\Gamma = \mathbf{C} : (\mathbf{S}^V - \mathbf{1}^{(4s)}) : \boldsymbol{\varepsilon}^{*2} + \frac{f}{1-f}\,\mathbf{C} : (\mathbf{S}^V - \mathbf{S}^\Omega) : (< \boldsymbol{\varepsilon}^{*1} >_\Omega - \boldsymbol{\varepsilon}^{*2}). \qquad (10.4.16e,f)$$

Again, these equations are *exact*.

The proof of (10.4.16a~f) directly follows from the Tanaka-Mori result presented in Subsection 11.3. Indeed, the volume average of the strain produced by $\boldsymbol{\varepsilon}^*(\mathbf{x})$ in Ω is[12]

$$< (\text{strain due to } \boldsymbol{\varepsilon}^{*1}) >_D = \begin{cases} \mathbf{S}^\Omega : < \boldsymbol{\varepsilon}^{*1} >_\Omega & D = \Omega \\ f\,\mathbf{S}^V : < \boldsymbol{\varepsilon}^{*1} >_\Omega & D = V \end{cases} \qquad (10.4.17a)$$

and the volume average of the corresponding stress field is

$$< (\text{stress due to } \boldsymbol{\varepsilon}^{*1}) >_D = \begin{cases} \mathbf{C} : (\mathbf{S}^\Omega - \mathbf{1}^{(4s)}) : < \boldsymbol{\varepsilon}^{*1} >_\Omega & D = \Omega \\ f\,\mathbf{C} : (\mathbf{S}^V - \mathbf{1}^{(4s)}) : < \boldsymbol{\varepsilon}^{*1} >_\Omega & D = V. \end{cases} \qquad (10.4.17b)$$

Fields due to constant $\boldsymbol{\varepsilon}^{*2}$ in Γ can be obtained by superposing the fields due to $-\boldsymbol{\varepsilon}^{*2}$ distributed over the entire Ω and the fields due to $\boldsymbol{\varepsilon}^{*2}$ distributed over the entire V. Hence, the volume average of the strain and stress fields produced by $\boldsymbol{\varepsilon}^{*2}$ in Γ are computed by applying the Tanaka-Mori result separately to the fields due to $-\boldsymbol{\varepsilon}^{*2}$ in Ω and $\boldsymbol{\varepsilon}^{*2}$ in V, i.e.,

$$< (\text{strain due to } \boldsymbol{\varepsilon}^{*2}) >_D = \begin{cases} (\mathbf{S}^V - \mathbf{S}^\Omega) : \boldsymbol{\varepsilon}^{*2} & D = \Omega \\ (1-f)\,\mathbf{S}^V : \boldsymbol{\varepsilon}^{*2} & D = V, \end{cases}$$

$$< (\text{stress due to } \boldsymbol{\varepsilon}^{*2}) >_D = \begin{cases} \mathbf{C} : (\mathbf{S}^V - \mathbf{S}^\Omega) : \boldsymbol{\varepsilon}^{*2} & D = \Omega \\ (1-f)\,\mathbf{C} : (\mathbf{S}^V - \mathbf{1}^{(4s)}) : \boldsymbol{\varepsilon}^{*2} & D = V. \end{cases} \qquad (10.4.17c,d)$$

The volume averages of $\boldsymbol{\varepsilon}^d$ and $\boldsymbol{\sigma}^d$ taken over Ω and V, (10.4.16a~d), are obtained directly from (10.4.17a~d).

The above results can be generalized to the case when V consists of a series of annular subregions, each of which has distinct constant eigenstrains. To be specific, consider a nested series of ellipsoidal regions, Ω_α ($\alpha = 1, 2, ..., m$), with $\Omega_m \equiv V$, which satisfy $\Omega_1 \subset \Omega_2 \subset ... \subset \Omega_m$, and denote the annulus

---

[12] While Eshelby's technique determines local field quantities inside (and outside) an ellipsoidal region of constant eigenstrain, the Tanaka-Mori result gives only the average field quantities over the ellipsoidal regions.

between Ω_α and $\Omega_{\alpha-1}$ by $\Gamma_\alpha \equiv \Omega_\alpha - \Omega_{\alpha-1}$ ($\alpha = 2, 3, ..., m$); see Figure 10.4.3. Then, consider the following distribution of eigenstrains:

$$\boldsymbol{\varepsilon}^*(\mathbf{x}) = \mathrm{H}(\mathbf{x};\, \Omega_1)\, \boldsymbol{\varepsilon}^{*1}(\mathbf{x}) + \sum_{\alpha=2}^{m} \mathrm{H}(\mathbf{x};\, \Gamma_\alpha)\, \boldsymbol{\varepsilon}^{*\alpha}, \tag{10.4.18}$$

where each $\boldsymbol{\varepsilon}^{*\alpha}$ ($\alpha = 2, 3, ..., m$) is constant, and $\mathrm{H}(\mathbf{x};\, \mathrm{D})$ is the Heaviside step function, taking on the value 1 when $\mathbf{x}$ is in D, and 0 otherwise; note that $\boldsymbol{\varepsilon}^{*1}(\mathbf{x})$ need not be constant, and that each annulus has different but constant eigenstrains. The resulting strain and stress fields are denoted by $\boldsymbol{\varepsilon}^{\mathrm{d}}$ and $\boldsymbol{\sigma}^{\mathrm{d}}$.

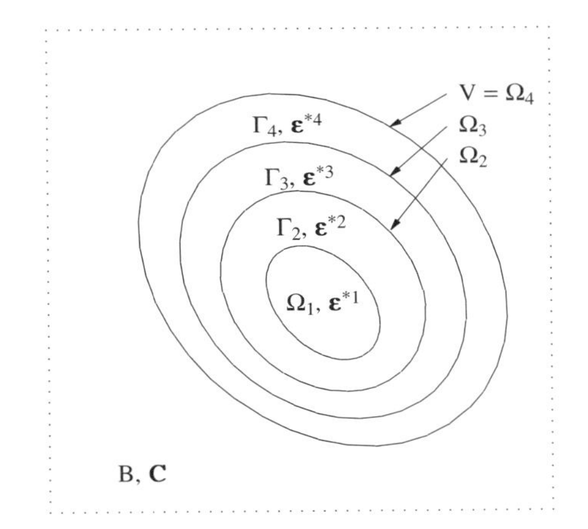

Figure 10.4.3

A nested sequence of 4 inclusions embedded in infinite domain B; Ω_α with eigenstrain $\boldsymbol{\varepsilon}^{*\alpha}$ ($\alpha = 1, 2, 3, 4$); B with elasticity $\mathbf{C}$

Apply the Tanaka-Mori result to the fields produced by $\boldsymbol{\varepsilon}^*(\mathbf{x})$ in Ω_1. The volume average of the resulting strain field over any $\Omega_\alpha \supseteq \Omega_1$ then is

$$< \text{strain due to } \boldsymbol{\varepsilon}^{*1} >_\alpha = \frac{\Omega_1}{\Omega_\alpha}\, \mathbf{S}^\alpha : < \boldsymbol{\varepsilon}^{*1} >_1 \qquad (\alpha \text{ not summed}), \tag{10.4.19a}$$

where $\mathbf{S}^\alpha$ is Eshelby's tensor for Ω_α, and subscript α or 1 on $< >$ emphasizes that the volume average is taken over Ω_α or Ω_1. The strain field due to $\boldsymbol{\varepsilon}^{*\beta}$ in Γ_β ($\beta = 2, 3, ..., m$) is obtained by superposing the strain field due to $-\boldsymbol{\varepsilon}^{*\beta}$ distributed in $\Omega_{\beta-1}$, and the strain field due to $\boldsymbol{\varepsilon}^{*\beta}$ distributed in Ω_β. Hence, the average strain over Ω_α due to $\boldsymbol{\varepsilon}^{*\beta}$ is

$$< \text{strain due to } \boldsymbol{\varepsilon}^{*\beta} >_\alpha = \begin{cases} (\mathbf{S}^\beta - \mathbf{S}^{\beta-1}) : \boldsymbol{\varepsilon}^{*\beta} & \Omega_\alpha \subset \Omega_\beta \\ \{(\Omega_\beta - \Omega_{\beta-1})/\Omega_\alpha\}\, \mathbf{S}^\alpha : \boldsymbol{\varepsilon}^{*\beta} & \Omega_\alpha \supseteq \Omega_\beta \end{cases} \qquad (\beta \text{ not summed}). \tag{10.4.19b}$$

Since all subregions have the same elasticity, the resulting average stresses are obtained directly from the corresponding average elastic strains.

The volume average of the strain field $\boldsymbol{\varepsilon}^{\mathrm{d}}$, over Ω_α is obtained by superposition of (10.4.19a) and (10.4.19b) for $\beta = 2, 3, ..., m$, and then the volume average of $\boldsymbol{\varepsilon}^{\mathrm{d}}$ over annular region Γ_α is computed. This leads to

$$<\boldsymbol{\varepsilon}^{d}>_{1} = \mathbf{S}^{1} : <\boldsymbol{\varepsilon}^{*1}>_{1} + \sum_{\beta=2}^{m} (\mathbf{S}^{\beta} - \mathbf{S}^{\beta-1}) : \boldsymbol{\varepsilon}^{*\beta} \tag{10.4.20a}$$

and

$$<\boldsymbol{\varepsilon}^{d}>_{\alpha}' = \frac{1}{F_{\alpha} - F_{\alpha-1}} (\mathbf{S}^{\alpha} - \mathbf{S}^{\alpha-1}) : \left\{ F_{1} <\boldsymbol{\varepsilon}^{*1}>_{1} + \sum_{\beta=2}^{\alpha-1} (F_{\beta} - F_{\beta-1})\, \boldsymbol{\varepsilon}^{*\beta} \right\}$$

$$+ \frac{1}{F_{\alpha} - F_{\alpha-1}} \{ (F_{\alpha} - 2F_{\alpha-1})\, \mathbf{S}^{\alpha} + F_{\alpha-1}\, \mathbf{S}^{\alpha-1} \} : \boldsymbol{\varepsilon}^{*\alpha}$$

$$+ \sum_{\beta=\alpha+1}^{m} (\mathbf{S}^{\beta} - \mathbf{S}^{\beta-1}) : \boldsymbol{\varepsilon}^{*\beta}, \tag{10.4.20b}$$

for [13] $\alpha = 2, 3, ..., m$, where $< >_{\alpha}'$ denotes the volume average taken over the annular region Γ_{α}, and F_{α} is the volume fraction of Ω_{α} relative to $\Omega_{m} \equiv V$, i.e.,

$$< >_{\alpha}' \equiv \frac{1}{F_{\alpha} - F_{\alpha-1}} (F_{\alpha} < >_{\alpha} - F_{\alpha-1} < >_{\alpha-1}), \qquad F_{\alpha} \equiv \frac{\Omega_{\alpha}}{V}. \tag{10.4.21a,b}$$

The corresponding average stresses are computed in the same manner, and are expressed in terms of $<\boldsymbol{\varepsilon}^{d}>$, as

$$<\boldsymbol{\sigma}^{d}>_{1} = \mathbf{C} : (<\boldsymbol{\varepsilon}^{d}>_{1} - <\boldsymbol{\varepsilon}^{*1}>_{1}),$$

$$<\boldsymbol{\sigma}^{d}>_{\alpha}' = \mathbf{C} : (<\boldsymbol{\varepsilon}^{d}>_{\alpha} - \boldsymbol{\varepsilon}^{*\alpha}), \tag{10.4.20c,d}$$

for $\alpha = 2, 3, ..., m$. In particular, if all Ω_{α}'s are similar and coaxial, then, (10.4.20a) and (10.4.20b) become

$$<\boldsymbol{\varepsilon}^{d}>_{1} = \mathbf{S} : <\boldsymbol{\varepsilon}^{*1}>_{1}, \qquad <\boldsymbol{\varepsilon}^{d}>_{\alpha}' = \mathbf{S} : \boldsymbol{\varepsilon}^{*\alpha}, \tag{10.4.22a,b}$$

for $\alpha = 2, 3, ..., m$, where $\mathbf{S}$ is the Eshelby tensor common to all Ω_{α}'s.

10.4.5. Double-Inclusion Method

The exact results presented in the preceding subsection can serve as a basis for a new averaging scheme, called the *double-inclusion* method. To illustrate this method, consider an ellipsoidal inclusion V which includes another ellipsoidal inclusion Ω in it and is embedded in an unbounded region of elasticity $\mathbf{C}$. The elasticity of Ω and the remaining part of V, $M \equiv V - \Omega$, is $\mathbf{C}^{\Omega}$ and $\mathbf{C}^{M}$, respectively. The elasticity tensor field then is

$$\mathbf{C}' = \mathbf{C}'(\mathbf{x}) = \begin{cases} \mathbf{C}^{\Omega} & \text{if } \mathbf{x} \text{ in } \Omega \\ \mathbf{C}^{M} & \text{if } \mathbf{x} \text{ in M} \\ \mathbf{C} & \text{otherwise.} \end{cases} \tag{10.4.23}$$

Hence, M may be viewed as the matrix phase which contains an inclusion Ω, and which is surrounded by a homogenized infinite solid of uniform elasticity $\mathbf{C}$;

[13] Note that in (10.4.20b), the first summation is omitted for $\alpha = 2$, and the second summation is omitted for $\alpha = m$.

see Figure 10.4.4.

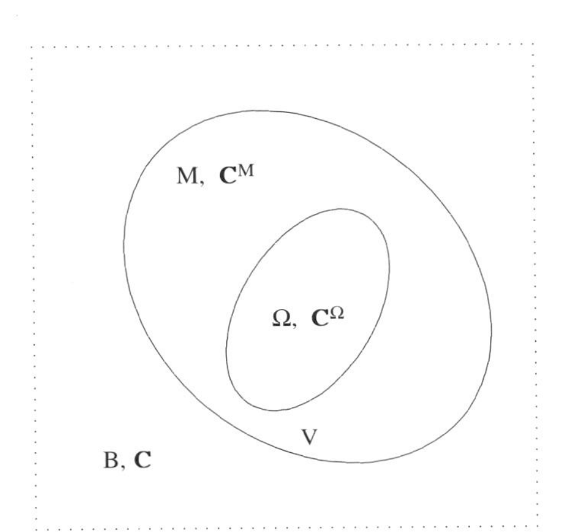

Figure 10.4.4

An unbounded uniform solid B of elasticity $\mathbf{C}$, contains two ellipsoidal regions, V and Ω ($\Omega \subset \mathrm{V}$), of elasticity tensors $\mathbf{C}^{\Omega}$ in Ω and $\mathbf{C}^{\mathrm{M}}$ in $\Gamma = \mathrm{V} - \Omega$

The fields produced in the double-inclusion V when the infinite domain is subjected to farfield strains $\boldsymbol{\varepsilon}^{\infty}$, are not uniform, due to the existence of the inclusion Ω within V. Hence, the homogenizing eigenstrain field in V is no longer constant (unlike the single-inclusion case), and may, indeed, suffer a jump across $\partial\Omega$. Though it is difficult to compute the exact homogenizing eigenstrain field, the volume average of the eigenstrains taken over Ω and M can be estimated by applying the generalized Eshelby results of the preceding subsection.

Since the average strains and stresses, taken over the inclusion phase Ω and the matrix phase M, are given by (10.4.16a,b) and (10.4.16e,f), the following average consistency conditions are obtained (Hori and Nemat-Nasser, 1993):

$$\mathbf{C}^{\Omega} : \{\boldsymbol{\varepsilon}^{\infty} + \mathbf{S}^{\Omega} : \bar{\boldsymbol{\varepsilon}}^{*\Omega} + (\mathbf{S}^{\mathrm{V}} - \mathbf{S}^{\Omega}) : \bar{\boldsymbol{\varepsilon}}^{*\mathrm{M}}\}$$

$$= \mathbf{C} : \{\boldsymbol{\varepsilon}^{\infty} + (\mathbf{S}^{\Omega} - \mathbf{1}^{(4s)}) : \bar{\boldsymbol{\varepsilon}}^{*\Omega} + (\mathbf{S}^{\mathrm{V}} - \mathbf{S}^{\Omega}) : \bar{\boldsymbol{\varepsilon}}^{*\mathrm{M}}\} \tag{10.4.24a}$$

and

$$\mathbf{C}^{\mathrm{M}} : \{\boldsymbol{\varepsilon}^{\infty} + \mathbf{S}^{\mathrm{V}} : \bar{\boldsymbol{\varepsilon}}^{*\mathrm{M}} + \frac{\mathrm{f}}{1-\mathrm{f}}(\mathbf{S}^{\mathrm{V}} - \mathbf{S}^{\Omega}) : (\bar{\boldsymbol{\varepsilon}}^{*\Omega} - \bar{\boldsymbol{\varepsilon}}^{*\mathrm{M}})\}$$

$$= \mathbf{C} : \{\boldsymbol{\varepsilon}^{\infty} + (\mathbf{S}^{\mathrm{V}} - \mathbf{1}^{(4s)}) : \bar{\boldsymbol{\varepsilon}}^{*\mathrm{M}} + \frac{\mathrm{f}}{1-\mathrm{f}}(\mathbf{S}^{\mathrm{V}} - \mathbf{S}^{\Omega}) : (\bar{\boldsymbol{\varepsilon}}^{*\Omega} - \bar{\boldsymbol{\varepsilon}}^{*\mathrm{M}})\}, \tag{10.4.24b}$$

where $\bar{\boldsymbol{\varepsilon}}^{*\Omega}$ and $\bar{\boldsymbol{\varepsilon}}^{*\mathrm{M}}$ are estimates of the average eigenstrains[14] over Ω and M. Solving the set of tensorial equations, (10.4.24a,b), for $\bar{\boldsymbol{\varepsilon}}^{*\Omega}$ and $\bar{\boldsymbol{\varepsilon}}^{*\mathrm{M}}$, and substituting the results into (10.4.16c,d), compute the average strain and stress over V.

---

[14] Although $\bar{\boldsymbol{\varepsilon}}^{*\Omega}$ is the volume average of the exact eigenstrains over Ω, $\bar{\boldsymbol{\varepsilon}}^{*\mathrm{M}}$ is not the corresponding exact quantity over M. The error due to this approximation is examined by Hori and Nemat-Nasser (1993).

Then, the overall elasticity of the double-inclusion is given by

$$\overline{\mathbf{C}} = \mathbf{C} : \{\mathbf{1}^{(4s)} + (\mathbf{S}^{V} - \mathbf{1}^{(4s)}) : \boldsymbol{A}\} : \{\mathbf{1}^{(4s)} + \mathbf{S}^{V} : \boldsymbol{A}\}^{-1}, \tag{10.4.25a}$$

where $\boldsymbol{A}$ is defined by

$$\mathrm{f}\,\overline{\boldsymbol{\varepsilon}}^{*\Omega} + (1-\mathrm{f})\,\overline{\boldsymbol{\varepsilon}}^{*M} \equiv \boldsymbol{A} : \boldsymbol{\varepsilon}^{\infty}. \tag{10.4.25b}$$

Note that the left side of (10.4.25b) is the volume average of the eigenstrains over V.

As an example, consider the case when V and Ω are similar and coaxial ellipsoids. In this case, $\overline{\boldsymbol{\varepsilon}}^{*\Omega}$ and $\overline{\boldsymbol{\varepsilon}}^{*M}$ are given by

$$\overline{\boldsymbol{\varepsilon}}^{*\Omega} = (\mathbf{A}^{\Omega} - \mathbf{S})^{-1} : \boldsymbol{\varepsilon}^{\infty}, \qquad \overline{\boldsymbol{\varepsilon}}^{*M} = (\mathbf{A}^{M} - \mathbf{S})^{-1} : \boldsymbol{\varepsilon}^{\infty}, \tag{10.4.26a,b}$$

where $\mathbf{S}$ is Eshelby's tensor common to Ω and M, $\mathbf{A}^{\Omega} \equiv (\mathbf{C} - \mathbf{C}^{\Omega})^{-1} : \mathbf{C}$, and $\mathbf{A}^{M} \equiv (\mathbf{C} - \mathbf{C}^{M})^{-1} : \mathbf{C}$. Hence, the overall elasticity $\overline{\mathbf{C}}$ is obtained by substituting the following $\boldsymbol{A}$ into (10.4.25a):

$$\boldsymbol{A} = \mathrm{f}\,(\mathbf{A}^{\Omega} - \mathbf{S})^{-1} + (1-\mathrm{f})\,(\mathbf{A}^{M} - \mathbf{S})^{-1}. \tag{10.4.26c}$$

If $\mathbf{C}$ is set equal to $\mathbf{C}^{M}$, then $(\mathbf{A}^{M} - \mathbf{S})^{-1}$ vanishes, and the resulting overall elasticity tensor coincides with that obtained by the two-phase model; see (10.4.7). On the other hand, $\mathbf{C}$ of the infinite body may be set equal to the yet-unknown overall elasticity tensor of the double-inclusion, $\overline{\mathbf{C}}$. From (10.4.25a), $\boldsymbol{A}$ given by (10.4.26c) then vanishes, and hence

$$\mathrm{f}\,(\overline{\mathbf{A}}^{\Omega} - \overline{\mathbf{S}})^{-1} + (1-\mathrm{f})\,(\overline{\mathbf{A}}^{M} - \overline{\mathbf{S}})^{-1} = \mathbf{0}, \tag{10.4.27}$$

where $\overline{\mathbf{S}}$ is Eshelby's tensor for the elasticity $\overline{\mathbf{C}}$, $\overline{\mathbf{A}}^{M} \equiv (\overline{\mathbf{C}} - \mathbf{C}^{M})^{-1} : \overline{\mathbf{C}}$, and $\overline{\mathbf{A}}^{\Omega} \equiv (\overline{\mathbf{C}} - \mathbf{C}^{\Omega})^{-1} : \overline{\mathbf{C}}$. *The overall moduli* $\overline{\mathbf{C}}$ *satisfying* (10.4.27), *remarkably, coincide with the estimate by the self-consistent method.* The proof is straightforward, although the required algebraic manipulation is rather tedious. First, multiplying (10.4.27) by $\overline{\mathbf{A}}^{\Omega} - \overline{\mathbf{S}}$ from the left and by $\overline{\mathbf{A}}^{M} - \overline{\mathbf{S}}$ from the right, obtain

$$\{\mathrm{f}\,(\overline{\mathbf{C}} - \mathbf{C}^{M})^{-1} + (1-\mathrm{f})\,(\overline{\mathbf{C}} - \mathbf{C}^{\Omega})^{-1}\} : \overline{\mathbf{C}} - \overline{\mathbf{S}} = \mathbf{0}. \tag{10.4.28a}$$

Next, multiplying (10.4.28a) by $\overline{\mathbf{C}} - \mathbf{C}^{M}$ from the left, and taking advantage of

$$(\overline{\mathbf{C}} - \mathbf{C}^{M}) : (\overline{\mathbf{C}} - \mathbf{C}^{\Omega})^{-1} = (\mathbf{C}^{\Omega} - \mathbf{C}^{M}) : (\overline{\mathbf{C}} - \mathbf{C}^{\Omega})^{-1} + \mathbf{1}^{(4s)}, \tag{10.4.28b}$$

arrive at

$$(\overline{\mathbf{C}} - \mathbf{C}^{M}) : \{(\overline{\mathbf{C}} - \mathbf{C}^{\Omega})^{-1} : \overline{\mathbf{C}} - \overline{\mathbf{S}}\} = \mathrm{f}\,(\mathbf{C}^{\Omega} - \mathbf{C}^{M}) : (\overline{\mathbf{C}} - \mathbf{C}^{\Omega})^{-1} : \overline{\mathbf{C}}. \tag{10.4.28c}$$

Finally, using

$$(\overline{\mathbf{C}} - \mathbf{C}^{\Omega})^{-1} : \overline{\mathbf{C}} : \{(\overline{\mathbf{C}} - \mathbf{C}^{\Omega})^{-1} : \overline{\mathbf{C}} - \overline{\mathbf{S}}\}^{-1}$$

$$= \mathbf{1}^{(4s)} + \overline{\mathbf{S}} : \{(\overline{\mathbf{C}} - \mathbf{C}^{\Omega})^{-1} : \overline{\mathbf{C}} - \overline{\mathbf{S}}\}^{-1}, \tag{10.4.28d}$$

and multiplying (10.4.28c) by $\{(\overline{\mathbf{C}} - \mathbf{C}^{\Omega})^{-1} : \overline{\mathbf{C}} - \overline{\mathbf{S}}\}^{-1}$ from the left, obtain

$$\overline{\mathbf{C}} - \mathbf{C}^{M} = \mathrm{f}\,(\mathbf{C}^{\Omega} - \mathbf{C}^{M}) : \{\mathbf{1}^{(4s)} + \overline{\mathbf{S}} : \{(\overline{\mathbf{C}} - \mathbf{C}^{\Omega})^{-1} : \overline{\mathbf{C}} - \overline{\mathbf{S}}\}^{-1}\}. \tag{10.4.28e}$$

Since $\mathbf{1}^{(4s)} + \overline{\mathbf{S}} : \{(\overline{\mathbf{C}} - \mathbf{C}^{\Omega})^{-1} : \overline{\mathbf{C}} - \overline{\mathbf{S}}\}^{-1}$ is $\mathbf{E}^{\infty}(\mathbf{C}^{\Omega}, \Omega; \overline{\mathbf{C}})$, the overall moduli satisfying (10.4.27) coincide with the overall moduli estimated by the self-consistent method; see Subsection 10.2.

If it is assumed that V (= Ω_2) is not similar to and coaxial with Ω (= Ω_1), then a different estimate for the overall elasticity tensor results; see Hori and Nemat-Nasser (1993). In this sense, this three-phase model is more general than the self-consistent model. It should be noted that, as in the double-inclusion model, no approximation is made in this case, except for the assumption of a piecewise constant eigenstrain field. The location of Ω relative to V is arbitrary.

10.4.6. Multi-Inclusion Method

The double-inclusion method considered in the preceding subsection can be generalized to a multi-inclusion method. The multi-inclusion is an ellipsoid, V, which contains a nested series of ellipsoids, Ω_α ($\alpha = 1, 2, ..., m$), such that $\Omega_1 \subset \Omega_2 \subset ... \subset \Omega_m \equiv V$; see Figure 10.4.5. The smallest ellipsoid Ω_1 and each annular region $\Gamma_\alpha = \Omega_\alpha - \Omega_{\alpha-1}$ have uniform elasticities $\mathbf{C}^1$ and $\mathbf{C}^\alpha$ ($\alpha = 2, 3, ..., m$), and the multi-inclusion V is embedded in an infinite domain of uniform elasticity $\mathbf{C}$. Hence, the elasticity tensor field is

$$\mathbf{C}' = \mathbf{C}'(\mathbf{x}) = \begin{cases} \mathbf{C}^1 & \text{if } \mathbf{x} \text{ in } \Omega_1 \\ \mathbf{C}^\alpha & \text{if } \mathbf{x} \text{ in } \Gamma_\alpha \quad (\alpha = 2, 3,...,m) \\ \mathbf{C} & \text{otherwise.} \end{cases} \tag{10.4.29}$$

To compute the average field quantities over the multi-inclusion when the farfield strains $\boldsymbol{\varepsilon}^\infty$ are prescribed, use the results obtained in Subsection 10.4.4. Then, estimate the volume average of the eigenstrain field through the following average consistency conditions:

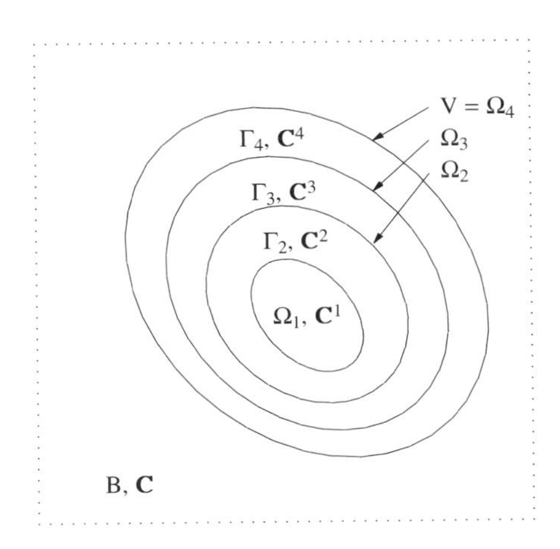

Figure 10.4.5

A five-phase inclusion embedded in infinite domain B; Ω_1 and Γ_α with elasticities $\mathbf{C}^1$ and $\mathbf{C}^\alpha$ ($\alpha = 2, 3, 4$); and B with elasticity $\mathbf{C}$

$$\mathbf{C}^1 : (\boldsymbol{\varepsilon}^\infty + <\boldsymbol{\varepsilon}^d>_1) = \mathbf{C} : (\boldsymbol{\varepsilon}^\infty + <\boldsymbol{\varepsilon}^d>_1 - \bar{\boldsymbol{\varepsilon}}^{*1}) \tag{10.4.30a}$$

and

$$\mathbf{C}^\alpha : (\boldsymbol{\varepsilon}^\infty + <\boldsymbol{\varepsilon}^d>'_\alpha) = \mathbf{C} : (\boldsymbol{\varepsilon}^\infty + <\boldsymbol{\varepsilon}^d>'_\alpha - \bar{\boldsymbol{\varepsilon}}^{*\alpha}) \quad (\alpha \text{ not summed}), \tag{10.4.30b}$$

for $\alpha = 2, 3, ..., m$, where $<\boldsymbol{\varepsilon}^d>_1$ and $<\boldsymbol{\varepsilon}^d>'_\alpha$ are the averages of the strain field over Ω_1 and Γ_α, respectively, computed from (10.4.19a) and (10.4.19b), by substituting the average eigenstrains over Ω_1 and Γ_α, $\bar{\boldsymbol{\varepsilon}}^{*1}$ and $\bar{\boldsymbol{\varepsilon}}^{*\alpha}$, into $<\boldsymbol{\varepsilon}^*>_1$ and $\boldsymbol{\varepsilon}^{*\alpha}$ $(\alpha = 2, 3, ..., m)$. Now solve the set of m tensorial equations, (10.4.30), for $\bar{\boldsymbol{\varepsilon}}^{*1}$ and $\bar{\boldsymbol{\varepsilon}}^{*\alpha}$, and obtain the overall elasticity as

$$\bar{\mathbf{C}} = \mathbf{C} : \{\mathbf{1}^{(4s)} + (\mathbf{S}^V - \mathbf{1}^{(4s)}) : \boldsymbol{A}\} : \{\mathbf{1}^{(4s)} + \mathbf{S}^V : \boldsymbol{A}\}^{-1}, \tag{10.4.31a}$$

where $\mathbf{S}^V$ is the Eshelby tensor for $V \equiv \Omega_m$, and $\boldsymbol{A}$ is defined by

$$\sum_{\alpha=1}^{m} f_\alpha \bar{\boldsymbol{\varepsilon}}^{*\alpha} \equiv \boldsymbol{A} : \boldsymbol{\varepsilon}^\infty. \tag{10.4.31b}$$

Again, the left side of (10.4.31b) is the volume average of the eigenstrains taken over V.

If Ω_α's are similar and coaxial, the average consistency conditions are reduced to

$$\mathbf{C}^\alpha : (\boldsymbol{\varepsilon}^\infty + \mathbf{S} : \bar{\boldsymbol{\varepsilon}}^{*\alpha}) = \mathbf{C} : \{\boldsymbol{\varepsilon}^\infty + (\mathbf{S} - \mathbf{1}^{(4s)}) : \bar{\boldsymbol{\varepsilon}}^{*\alpha}\} \quad (\alpha \text{ not summed}), \tag{10.4.32}$$

where $\mathbf{S}$ is the Eshelby tensor common to all Ω_α's. The solution of (10.4.32) then is

$$\bar{\boldsymbol{\varepsilon}}^{*\alpha} = (\mathbf{A}^\alpha - \mathbf{S})^{-1} : \boldsymbol{\varepsilon}^\infty, \tag{10.4.33a}$$

where $\mathbf{A}^\alpha \equiv (\mathbf{C} - \mathbf{C}^\alpha)^{-1} : \mathbf{C}$, and $\boldsymbol{A}$ in (10.4.31b) is given by

$$\boldsymbol{A} = \sum_{\alpha=1}^{m} f_\alpha (\mathbf{A}^\alpha - \mathbf{S})^{-1}, \tag{10.4.33b}$$

where $f_1 \equiv \Omega_1 / V$ and $f_\alpha \equiv \Gamma_\alpha / V$ $(\alpha = 1, 2, ..., m)$. Hence, the overall elasticity tensor is given by

$$\bar{\mathbf{C}} = \mathbf{C} : \{\mathbf{1}^{(4s)} + \sum_{\alpha=1}^{m} f_\alpha (\mathbf{S} - \mathbf{1}^{(4s)}) : (\mathbf{A}^\alpha - \mathbf{S})^{-1}\}$$

$$: \{\mathbf{1}^{(4s)} + \sum_{\alpha=1}^{m} f_\alpha \mathbf{S} : (\mathbf{A}^\alpha - \mathbf{S})^{-1}\}^{-1}. \tag{10.4.33c}$$

This overall elasticity tensor is of a form similar to that estimated by the multi-phase inclusion; see (10.4.11).

10.4.7. Multi-Phase Composite Model

With the aid of the generalized Eshelby results obtained in Subsection 10.4.4, the interaction among adjacent inclusions can be estimated through the interaction between V and the surrounding homogenized infinite domain. Hence, in a composite consisting of inclusions of different materials, the interaction among various constituents may be accounted for, in an approximate manner, by choosing a suitable arrangement of different inclusions in a multi-phase composite model. This type of multi-phase composite model can also be used to analyze heterogeneous materials which do not have a matrix phase, such as polycrystals.

To illustrate this multi-phase composite model in a specific manner, suppose that the ellipsoid V contains $n-1$ ellipsoidal heterogeneities Ω_α ($\alpha = 1, 2, ..., n-1$). The remaining part of V is denoted by Γ; see Figure 10.4.6a. The elasticities of Ω_α and Γ are $\mathbf{C}^\alpha$ and $\mathbf{C}^n$, respectively, and the volume fractions of Ω_α and Γ are defined by $f_\alpha = \Omega_\alpha / V$ ($\alpha = 1, 2, ..., n-1$) and $f_n = \Gamma / V$. When this multi-phase composite is embedded in an infinite domain with uniform elasticity $\mathbf{C}$, the elasticity tensor field is

$$\mathbf{C}(\mathbf{x}) = \begin{cases} \mathbf{C}^\alpha & \text{if } \mathbf{x} \text{ in } \Omega_\alpha \quad (\alpha = 1, 2, ..., n-1) \\ \mathbf{C}^n & \text{if } \mathbf{x} \text{ in } \Gamma \\ \mathbf{C} & \text{otherwise.} \end{cases} \tag{10.4.34}$$

Apply the generalized Eshelby results to this multi-phase composite V subjected to farfield strains $\boldsymbol{\varepsilon}^\infty$. In view of (10.4.34), the average consistency conditions are

$$\mathbf{C}^\alpha : (\boldsymbol{\varepsilon}^\infty + <\boldsymbol{\varepsilon}^d>_\alpha) = \mathbf{C} : (\boldsymbol{\varepsilon}^\infty + <\boldsymbol{\varepsilon}^d>_\alpha - \bar{\boldsymbol{\varepsilon}}^{*\alpha}) \qquad (\alpha \text{ not summed}), \tag{10.4.35}$$

for $\alpha = 1, 2, ..., n$, where $\bar{\boldsymbol{\varepsilon}}^{*\alpha}$ and $\boldsymbol{\varepsilon}^{*n}$ are the average eigenstrains, and $< >_\alpha$ and

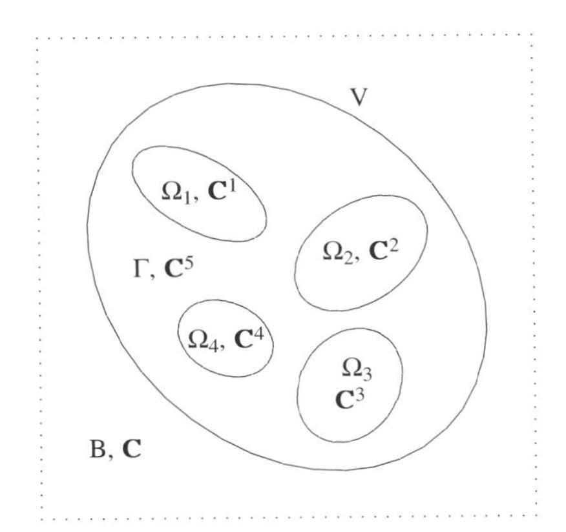

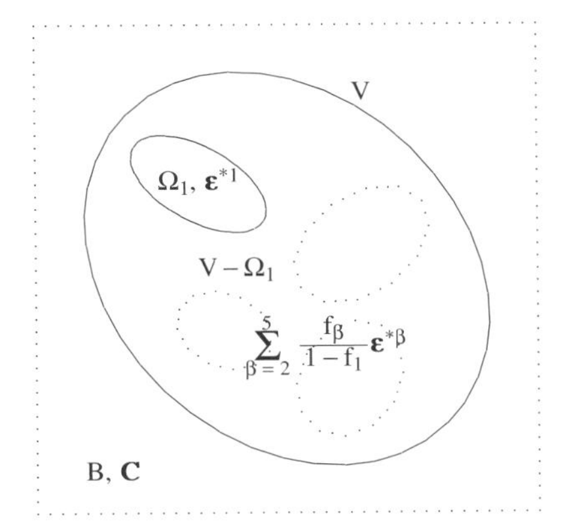

Figure 10.4.6

(a) A 5-phase composite is embedded in infinite domain B; Ω_α with elasticity $\mathbf{C}^\alpha$ ($\alpha = 1, ..., 4$) ; Γ with elasticity $\mathbf{C}^5$; and B with elasticity $\mathbf{C}$

(b) To estimate $<\boldsymbol{\varepsilon}^d>_1$, eigenstrains $\boldsymbol{\varepsilon}^{*1}$ are distributed in Ω_1, and eigenstrains $\sum_{\beta=2}^{5} \frac{f_\beta}{1-f_1} \boldsymbol{\varepsilon}^{*\beta}$ are *uniformly* distributed in $V - \Omega_1$

$< >_n$ are the volume averages taken over Ω_α ($\alpha = 1, 2, ..., n-1$) and Γ.

The volume averages of the strains produced by the eigenstrain field are obtained in terms of the average eigenstrains. From

$$< (.) >_{V-\Omega_\alpha} = \sum_{\beta \neq \alpha}^{n} \frac{f_\beta}{1-f_\alpha} < (.) >_\beta, \tag{10.4.36}$$

for each Ω_α, *assume that* $V - \Omega_\alpha$ *is a region where the average eigenstrains,* $\sum_{\beta \neq \alpha}^{n} \{f_\beta/(1-f_\alpha)\}\bar{\boldsymbol{\varepsilon}}^{*\beta}$, *are uniformly distributed*; see Figure 10.4.6b. Then, the average strain over Ω_α is estimated by

$$< \boldsymbol{\varepsilon}^d >_\alpha = \mathbf{S}^\alpha : \bar{\boldsymbol{\varepsilon}}^{*\alpha} + (\mathbf{S}^V - \mathbf{S}^\alpha) : \left\{ \sum_{\beta \neq \alpha}^{n} \frac{f_\beta}{1-f_\alpha} \bar{\boldsymbol{\varepsilon}}^{*\beta} \right\}, \tag{10.4.37a}$$

for $\alpha = 1, 2, ..., n-1$, where $\mathbf{S}^\alpha$ and $\mathbf{S}^V$ are Eshelby's tensors for Ω_α and V, respectively. For Γ, from $< \boldsymbol{\varepsilon}^d >_V = \mathbf{S}^V : < \boldsymbol{\varepsilon}^* >_V$ and (10.4.37a), it follows that

$$< \boldsymbol{\varepsilon}^d >_n = \mathbf{S}^V : \bar{\boldsymbol{\varepsilon}}^{*n} + \sum_{\alpha=1}^{n-1} \frac{f_\alpha}{f_n(1-f_\alpha)} (\mathbf{S}^V - \mathbf{S}^\alpha) : \left\{ \bar{\boldsymbol{\varepsilon}}^{*\alpha} - \sum_{\beta=1}^{n} f_\beta \bar{\boldsymbol{\varepsilon}}^{*\beta} \right\}. \tag{10.4.37b}$$

Substitution of (10.4.37a,b) into (10.4.36) yields a set of n tensorial equations for n average eigenstrains, from which $\bar{\boldsymbol{\varepsilon}}^{*\alpha}$'s are computed, and the average response of each inclusion is evaluated. Finally, the overall elasticity tensor $\bar{\mathbf{C}}$ is obtained, which is identical with (10.4.31). Note that (10.4.37a) still holds, even when the matrix phase is absent, in which case f_n is zero.

Furthermore, if a multi-phase composite consisting of $n-1$ different kinds of inclusions with similar shapes but distinct elasticities, is embedded in a matrix M, then the consistency conditions are (10.4.35), and the average eigenstrains are given by (10.4.36). As shown in the preceding subsection, it is easy to prove that the estimates of the overall moduli by the self-consistent and the Mori-Tanaka models are obtained from this multi-phase composite model, by setting $\mathbf{C} = \bar{\mathbf{C}}$ and $\mathbf{C} = \mathbf{C}^M$, respectively.

10.4.8. Bounds on Overall Moduli by Double-Inclusion Method

The overall moduli obtained by means of the double-inclusion method, depend on the location and orientation of the inner inclusion, relative to the outer one. Bounds for these moduli can be computed by applying the universal inequalities; see Hori and Nemat-Nasser (1994) and Nemat-Nasser and Hori (1995). The statistical homogeneity, used with the Hashin-Shtrikman variational principle, is *not* required in the double-inclusion model.

Referring to Figure 10.4.4, observe that, if the elasticity $\mathbf{C}$ of B is such that $\mathbf{C}'(\mathbf{x}) - \mathbf{C}$ is negative-semi-definite, then

$$0 \geq \frac{1}{2} \boldsymbol{\varepsilon}^o : (\bar{\mathbf{C}} - \mathbf{C}) : \boldsymbol{\varepsilon}^o - J(\boldsymbol{\varepsilon}^*; \boldsymbol{\varepsilon}^o), \tag{10.4.38a}$$

where J is a functional of the eigenstrain $\boldsymbol{\varepsilon}^*$, defined by

$$\mathrm{J}(\boldsymbol{\varepsilon}^*;\,\boldsymbol{\varepsilon}^o) \equiv \frac{1}{2} < (-\mathbf{C}:\boldsymbol{\varepsilon}^*) : \{(\mathbf{C}'-\mathbf{C})^{-1} : (-\mathbf{C}:\boldsymbol{\varepsilon}^*)$$

$$-(\boldsymbol{\varepsilon}^d - < \boldsymbol{\varepsilon}^d >_V) - 2\boldsymbol{\varepsilon}^o \} >_V. \qquad (10.4.38b)$$

Inequality (10.4.38a) holds for any arbitrary $\boldsymbol{\varepsilon}^*(\mathbf{x})$.

Since $< \boldsymbol{\varepsilon}^* : \mathbf{C} : \boldsymbol{\varepsilon}^* >_V$ is self-adjoint for $\boldsymbol{\varepsilon}^*$ which vanishes outside of V (see Figure 10.4.4), the Euler equation of J coincides with the consistency condition, if $< \boldsymbol{\varepsilon}^d >_V + \boldsymbol{\varepsilon}^o$ is regarded as $\boldsymbol{\varepsilon}^\infty$. Furthermore, J can be computed exactly for a piecewise constant eigenstrain field, i.e., $\boldsymbol{\varepsilon}^*(\mathbf{x}) = \boldsymbol{\varepsilon}^{*1}$ in Ω and $\boldsymbol{\varepsilon}^*(\mathbf{x}) = \boldsymbol{\varepsilon}^{*2}$ in $\mathrm{M} = \mathrm{V} - \Omega$. Then, J is reduced to a quadratic form (with linear terms) for $\boldsymbol{\varepsilon}^{*1}$ and $\boldsymbol{\varepsilon}^{*2}$. The condition that this quadratic form be stationary, yields the average consistency condition. The stationary value of this quadratic form is then given by $\boldsymbol{\varepsilon}^o : (\mathbf{C} - \overline{\mathbf{C}}^{DI}) : \boldsymbol{\varepsilon}^o/2$, where $\overline{\mathbf{C}}^{DI}$ is given by (10.4.25a), i.e.,

$$\overline{\mathbf{C}}^{DI} = \mathbf{C} : \{\mathbf{1}^{(4s)} + (\mathbf{S}^V - \mathbf{1}^{(4s)}) : \boldsymbol{A}\} : \{\mathbf{1}^{(4s)} + \mathbf{S}^V : \boldsymbol{A}\}^{-1},$$

where $\boldsymbol{A}$ is defined by (10.4.25b). This $\overline{\mathbf{C}}^{DI}$ is an upper bound for the exact overall modulus tensor, $\overline{\mathbf{C}}$, which corresponds to the exact overall elastic strain energy, i.e.,

$$0 \geq \frac{1}{2}\,\boldsymbol{\varepsilon}^o : (\overline{\mathbf{C}} - \overline{\mathbf{C}}^{DI}) : \boldsymbol{\varepsilon}^o. \qquad (10.4.39a)$$

The double-inclusion model can be formulated in terms of the corresponding compliance tensors with the farfield stress, $\boldsymbol{\sigma}^\infty$, being prescribed. Then, it can easily be deduced that, when $\mathbf{D}' - \mathbf{D}$ is negative-semi-definite, an upper bound for the exact overall compliance tensor, $\overline{\mathbf{D}}$, results; here, $\mathbf{D}' = (\mathbf{C}')^{-1}$, and $\mathbf{D}$ is the compliance of B in Figure 10.4.4. This upper bound is given by

$$0 \geq \frac{1}{2}\,\boldsymbol{\sigma}^o : (\overline{\mathbf{D}} - (\overline{\mathbf{C}}^{DI})^{-1}) : \boldsymbol{\sigma}^o. \qquad (10.4.39b)$$

Note that $\overline{\mathbf{C}}$ in (10.4.39a) may not, in general, be the inverse of $\overline{\mathbf{D}}$ in (10.4.39b), because, $\overline{\mathbf{D}}$ and $\overline{\mathbf{C}}$ are the overall elasticity and compliance tensors when the finite solid V is subjected to uniform traction and linear displacement boundary conditions, respectively. Also, observe that, to compute $\overline{\mathbf{C}}^{DI}$ in (10.4.39a), the elasticity tensor $\mathbf{C}$ is chosen such that $\mathbf{C}' - \mathbf{C}$ is rendered negative-semi-definite, while to obtain $\overline{\mathbf{C}}^{DI}$ in (10.4.39b), the elasticity tensor $\mathbf{C} = \mathbf{D}^{-1}$ must be chosen such that it renders $\mathbf{C}' - \mathbf{C}$ positive-semi-definite and, hence, makes $\mathbf{D}' - \mathbf{D}$ negative-semi-definite.

These bounds for the overall modulus and compliance tensors apply to double inclusions of any arbitrary configuration, i.e., the elasticity of each phase, the relative location and orientation of Ω in V, and the aspect ratios of ellipsoids V and Ω, are arbitrary. In particular, when V and Ω are spheres and $\mathbf{C}$ is either $\mathbf{C}^M$ or $\mathbf{C}^\Omega$, then the resulting bounds coincide with the conventional Hashin-Shtrikman bounds for the two-phase composite materials. For $\mathrm{V} = \Omega$ and $\mathbf{C} = \mathbf{C}^M$, the results of the Mori-Tanaka method are obtained. Hence, the overall moduli estimated by the Mori-Tanaka method are always bounds for this case, as shown by Weng (1990).

The universal inequalities can be used to obtain conditions under which the results, (10.4.33c), of the multi-inclusion model, are bounds for the overall moduli. Indeed, inequality (10.4.38a) applies to a nested set of ellipsoidal inclusions of any arbitrary orientations and aspect ratios, provided that the elasticity tensor $\mathbf{C}$ of the infinite solid B (see Figure 10.4.5) is chosen such that $\mathbf{C}'(\mathbf{x}) - \mathbf{C}$ is negative-semi-definite. It should be noted that the only difference between the multi-inclusion (nested composite) in Figure 10.4.5 and the double-inclusion in Figure 10.4.4, is the elasticity tensor field in V.

When a piecewise constant eigenstrain field is used in functional J of inequality (10.4.38a), then J is reduced to a quadratic form (with linear terms) for these constant eigenstrains. As in the case of the double-inclusion, this reduction is exact, and the stationary value of this quadratic form is given by $\boldsymbol{\varepsilon}^{\mathrm{o}} : (\mathbf{C} - \bar{\mathbf{C}}^{MI}) : \boldsymbol{\varepsilon}^{\mathrm{o}}/2$, where $\bar{\mathbf{C}}^{MI}$ is given by the right-hand side of (10.4.33c), i.e.,

$$\bar{\mathbf{C}}^{MI} = \mathbf{C} : \{\mathbf{1}^{(4s)} + \sum_{\alpha=1}^{m} f_{\alpha}\, (\mathbf{S} - \mathbf{1}^{(4s)}) : (\mathbf{A}^{\alpha} - \mathbf{S})^{-1}\}$$

$$: \{\mathbf{1}^{(4s)} + \sum_{\alpha=1}^{n} f_{\alpha}\, \mathbf{S} : (\mathbf{A}^{\alpha} - \mathbf{S})^{-1}\}^{-1}.$$

Therefore, this $\bar{\mathbf{C}}^{MI}$ provides an upper bound for the exact overall elasticity tensor when $\mathbf{C}$ is chosen such that $\mathbf{C}^{\alpha} - \mathbf{C}$ is negative-semi-definite, for all α's. Similarly, an upper bound for the exact $\bar{\mathbf{D}}$ can be constructed, using uniform boundary tractions on V and choosing $\mathbf{C}$ such that $\mathbf{C}' - \mathbf{C}$ is positive-semi-definite. The result is similar to (10.4.39b).

10.5. EQUIVALENCE AMONG ESTIMATES BY DILUTE DISTRIBUTION, SELF-CONSISTENT, DIFFERENTIAL, AND DOUBLE-INCLUSION METHODS

In Subsections 10.1, 10.2, 10.3, and 10.4, the overall elasticity and compliance tensors, $\bar{\mathbf{C}}$ and $\bar{\mathbf{D}}$, are estimated by the dilute-distribution assumption (*DD*), the self-consistent method (*SC*), the differential scheme (*DS*), and the double-inclusion method (*DI*). In these approximations, the exact $\mathbf{E}^{\alpha}$- and $\mathbf{F}^{\alpha}$-tensors which determine the average strain and stress of each micro-element for the prescribed macrostrain and macrostress, are replaced by the approximate $\mathbf{E}^{\infty}$- and $\mathbf{F}^{\infty}$-tensors which determine the average strain and stress of one isolated micro-element embedded in an unbounded homogeneous solid of suitable elasticity, and subjected to a constant farfield strain and stress, respectively. The overall moduli estimated by these approximations are: by the dilute-distribution assumption,

$$\bar{\mathbf{C}}^{DD} = \mathbf{C} + \sum_{\alpha=1}^{n} f_{\alpha}\, (\mathbf{C}^{\alpha} - \mathbf{C}) : \mathbf{E}^{\infty}(\mathbf{C}^{\alpha}, \Omega_{\alpha};\ \mathbf{C}),$$

$$\bar{\mathbf{D}}^{DD} = \mathbf{D} + \sum_{\alpha=1}^{n} f_\alpha\,(\mathbf{D}^\alpha - \mathbf{D}) : \mathbf{F}^\infty(\mathbf{D}^\alpha,\ \Omega_\alpha;\ \mathbf{D}); \tag{10.5.1a,b}$$

by the self-consistent method,

$$\bar{\mathbf{C}}^{SC} = \mathbf{C} + \sum_{\alpha=1}^{n} f_\alpha\,(\mathbf{C}^\alpha - \mathbf{C}) : \mathbf{E}^\infty(\mathbf{C}^\alpha,\ \Omega_\alpha;\ \bar{\mathbf{C}}^{SC}); \tag{10.5.2}$$

by the differential scheme,

$$\frac{\partial}{\partial f_\alpha}\,\bar{\mathbf{C}}^{DS} = \frac{1}{1-f_\alpha}\,(\mathbf{C}^\alpha - \bar{\mathbf{C}}^{DS}) : \mathbf{E}^\infty(\mathbf{C}^\alpha,\ \Omega_\alpha;\ \bar{\mathbf{C}}^{DS}),$$

$$\bar{\mathbf{C}}^{DS}(0) = \mathbf{C}; \tag{10.5.3a,b}$$

and by the double-inclusion method,

$$\bar{\mathbf{C}}^{DI} = \mathbf{C} : \{\mathbf{1}^{(4s)} + \sum_{\alpha=1}^{n} f_\alpha\,(\mathbf{S}^\Omega - \mathbf{1}^{(4s)}) : (\mathbf{A}^\alpha - \mathbf{S}^\Omega)^{-1}\}$$

$$: \{\mathbf{1}^{(4s)} + \sum_{\alpha=1}^{n} f_\alpha\,\mathbf{S}^\Omega : (\mathbf{A}^\alpha - \mathbf{S}^\Omega)\}^{-1}. \tag{10.5.4}$$

Now consider the asymptotic behavior of these tensors, $\bar{\mathbf{C}}^{DD}$, $\bar{\mathbf{D}}^{DD}$, $\bar{\mathbf{C}}^{SC}$, $\bar{\mathbf{C}}^{DS}$, and $\bar{\mathbf{C}}^{MS}$, as f_α approaches zero, when $f_\beta = 0$ for $\beta \neq \alpha$. The zeroth-order asymptotic term of all four estimates is the matrix elasticity, $\mathbf{C}$. From (10.5.1a), $\bar{\mathbf{C}}^{DD}$ behaves as

$$\bar{\mathbf{C}}^{DD} \approx \mathbf{C} + f_\alpha\,(\mathbf{C}^\alpha - \mathbf{C}) : \mathbf{E}^\infty(\mathbf{C}^\alpha,\ \Omega_\alpha;\ \mathbf{C}) \qquad (\alpha \text{ not summed}). \tag{10.5.5a}$$

As shown in Subsection 10.1, from the equivalence of $\mathbf{E}^\infty$ and $\mathbf{F}^\infty$, $\bar{\mathbf{D}}^{DD}$ behaves as

$$(\bar{\mathbf{D}}^{DD})^{-1} \approx \{\mathbf{D} : \{\mathbf{1}^{(4s)} - f_\alpha\,(\mathbf{C}^\alpha - \mathbf{C}) : \mathbf{E}^\infty(\mathbf{C}^\alpha,\ \Omega_\alpha;\ \mathbf{C}) : \mathbf{D}\}\}^{-1}$$

$$\approx \mathbf{C} + f_\alpha\,(\mathbf{C}^\alpha - \mathbf{C}) : \mathbf{E}^\infty(\mathbf{C}^\alpha,\ \Omega_\alpha;\ \mathbf{C}) \qquad (\alpha \text{ not summed}). \tag{10.5.5b}$$

For the self-consistent method or the differential scheme, with initial conditions $\bar{\mathbf{C}}^{SC}(0) = \bar{\mathbf{C}}^{DS}(0) = \mathbf{C}$, the behavior as f_α approaches zero is given by

$$\bar{\mathbf{C}}^{SC} \approx \mathbf{C} + f_\alpha\,(\mathbf{C}^\alpha - \mathbf{C}) : \mathbf{E}^\infty(\bar{\mathbf{C}}^{SC}(0), \mathbf{C}^\alpha;\ \Omega_\alpha)$$

$$= \mathbf{C} + f_\alpha\,(\mathbf{C}^\alpha - \mathbf{C}) : \mathbf{E}^\infty(\mathbf{C}^\alpha,\ \Omega_\alpha;\ \mathbf{C}),$$

$$\bar{\mathbf{C}}^{DS} \approx \bar{\mathbf{C}}^{DS}(0) + f_\alpha\,(\mathbf{C}^\alpha - \bar{\mathbf{C}}^{DS}(0)) : \mathbf{E}^\infty(\bar{\mathbf{C}}^{DS}(0), \mathbf{C}^\alpha;\ \Omega_\alpha)$$

$$= \mathbf{C} + f_\alpha\,(\mathbf{C}^\alpha - \mathbf{C}) : \mathbf{E}^\infty(\mathbf{C}^\alpha,\ \Omega_\alpha;\ \mathbf{C}) \qquad (\alpha \text{ not summed}), \tag{10.5.5c,d}$$

where the third argument of $\mathbf{E}^\infty$ or $\mathbf{F}^\infty$ is replaced by $\mathbf{C}$. For the double-inclusion method, from

$$\mathbf{S}^\alpha : (\mathbf{A}^\alpha - \mathbf{S}^\alpha)^{-1} = \mathbf{E}^\infty(\mathbf{C}^\alpha,\ \Omega_\alpha;\ \mathbf{C}) - \mathbf{1}^{(4s)},$$

$$\mathbf{C} : (\mathbf{S}^{\alpha} - \mathbf{1}^{(4s)}) : (\mathbf{A}^{\alpha} - \mathbf{S}^{\alpha})^{-1} = \mathbf{C}^{\alpha} : \mathbf{E}^{\infty}(\mathbf{C}^{\alpha}, \Omega_{\alpha};\ \mathbf{C}) - \mathbf{C}$$

$$(\alpha \text{ not summed}), \qquad (10.5.6a,b)$$

it follows that

$$\bar{\mathbf{C}}^{DI} \approx \{\mathbf{C} + \mathrm{f}_{\alpha}\,(\mathbf{C}^{\alpha} : \mathbf{E}^{\infty}(\mathbf{C}^{\alpha}, \Omega_{\alpha};\ \mathbf{C}) - \mathbf{C})\} : \{\mathbf{1}^{(4s)} + \mathrm{f}_{\alpha}\,(\mathbf{E}^{\infty}(\mathbf{C}^{\alpha}, \Omega_{\alpha};\ \mathbf{C}) - \mathbf{1}^{(4s)})\}^{-1}$$

$$= \mathbf{C} + \mathrm{f}_{\alpha}\,(\mathbf{C}^{\alpha} - \mathbf{C}) : \mathbf{E}^{\infty}(\mathbf{C}^{\alpha}, \Omega_{\alpha};\ \mathbf{C}). \qquad (10.5.5e)$$

Hence, as is seen from (10.5.5a~e), $\bar{\mathbf{C}}^{DD}$, $\bar{\mathbf{D}}^{DD}$, $\bar{\mathbf{C}}^{SC}$, $\bar{\mathbf{C}}^{DS}$, and $\bar{\mathbf{C}}^{DI}$, give the same estimate of the overall response of the RVE for small volume fractions of micro-elements. Asymptotically, therefore, these methods are all consistent at small volume fractions of inhomogeneities.

10.6. OTHER AVERAGING SCHEMES

The three averaging schemes, *DD*, *SC*, and *DS*, studied in this section share a common model for computing the average strain (or stress) in a typical inhomogeneity. *The common model may be called the two-phase model, since it is based on an isolated inclusion embedded in an infinite homogeneous domain.* The model yields explicit results with the aid of the Eshelby tensor. There are other models which can be used for certain heterogeneous materials, and require relatively simple analytic estimates. The double- and multi-inclusion models are two examples. Two other such models are: 1) the composite-spheres model, considered by Hashin (1962), and 2) a three-phase model. Neither is directly related to Eshelby's solution. The three-phase model was initially proposed by Fröhlich and Sack (1946) who modeled the overall properties of viscous fluids containing elastic or rigid spheres, by solving the problem of a concentric elastic (or rigid) sphere embedded in a fluid shell which in turn is embedded in a homogenized medium of the effective overall properties. For the rigid spherical suspension problem, Fröhlich and Sack obtain the effective viscosity using their three-phase model. For the elastic spheres dispersed in a viscous fluid, an ordinary linear differential equation is obtained for the shearing. Mackenzie (1950) follows Fröhlich and Sack, and considers a similar three-phase model consisting of a spherical cavity embedded in a shell of the matrix material which in turn is embedded in an infinite domain of the homogenized material, in order to estimate the overall properties of an elastic solid containing spherical cavities. His final results turn out to match the self-consistent estimates, as is shown in Subsection (10.4.4); see (10.4.25b). Smith (1974) works out the three-phase model in some detail and obtains a quadratic equation for the effective shear modulus. The same result is reported by Christensen and Lo (1979) who have reexamined the three-phase model and provide comments and discussion, showing that a term neglected by Smith is in fact negligible. In addition, these authors examine the plane problem of solids with cylindrical inclusions, using the same three-phase model. The three-phase model is briefly reviewed in Subsection 10.6.2.

10.6.1. Composite-Spheres Model

The composite-spheres model assumes that an RVE may be represented by a finite body V consisting of a matrix M and a single inclusion Ω in it, where these are cocentered spheres of radii b and a. Both M and Ω have distinct elasticities and distinct compliances, $\mathbf{C}$ and $\mathbf{C}^\Omega$, and $\mathbf{D}$ and $\mathbf{D}^\Omega$. Hence,

$$\mathbf{C}' = \mathbf{C}'(\mathbf{x}) = \begin{cases} \mathbf{C}^\Omega & \Omega\ (0 < r < a) \\ \mathbf{C} & M\ (a < r < b), \end{cases}$$

$$\mathbf{D}' = \mathbf{D}'(\mathbf{x}) = \begin{cases} \mathbf{D}^\Omega & \Omega\ (0 < r < a) \\ \mathbf{D} & M\ (a < r < b), \end{cases} \tag{10.6.1a,b}$$

where $r = |\mathbf{x}|$, with the origin at the common center of the spheres. The volume fraction of Ω is $f = (a/b)^3$; see Figure 10.6.1. Both M and Ω are isotropic.

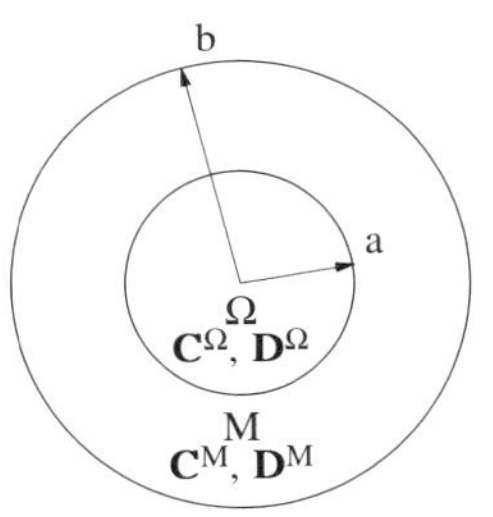

Figure 10.6.1

Composite-spheres model

Under spherically symmetric loading, the deformation of V is spherically symmetric. For example, in hydrostatic loading ($\mathbf{t} = \boldsymbol{\nu}\cdot(\sigma^o \mathbf{1}^{(2)})$ with $\boldsymbol{\nu}$ being the outer unit normal on ∂V), the displacement field becomes $\mathbf{u}(\mathbf{x}) = u_r(r)\, \mathbf{x}/|\mathbf{x}|$. The radial displacement u_r then is

$$u_r(r) = \begin{cases} A^\Omega r & 0 < r < a \\ A r + B r^{-2} & a < r < b, \end{cases} \tag{10.6.2a}$$

where

$$A = L\,\frac{\sigma^o}{4\mu}, \qquad B = b^3\,(L\,\frac{3K}{4\mu} - 1)\,\frac{\sigma^o}{4\mu}, \qquad A^\Omega = A + \frac{B}{a^3}, \tag{10.6.2b~d}$$

with

$$L = \frac{4}{3}\,\frac{3K^\Omega + 4\mu}{4(K^\Omega - K)\,f + 3(K\,K^\Omega/\mu) + 4K}. \tag{10.6.2e}$$

As usual, K and μ are the bulk and shear moduli for the matrix, while K^Ω and

μ^Ω are those for the inclusion. From the boundary condition, the average stress over V is $\bar{\boldsymbol{\sigma}} = \sigma^o \, \mathbf{1}^{(2)}$, and from (10.6.2), the average strain over V becomes

$$\bar{\boldsymbol{\varepsilon}} = \frac{1}{V} \int_{\partial V} u_r(b) \, \mathbf{v} \otimes \mathbf{v} \, dS = \sigma^o \left\{ K + f \frac{(K^\Omega - K)(3K + 4\mu)}{3K + 4\mu + 3(1-f)(K^\Omega - K)} \right\}^{-1} \mathbf{1}^{(2)}, \tag{10.6.3}$$

since $\mathbf{v} = \mathbf{x} / |\mathbf{x}|$ on the boundary. Therefore, the overall bulk modulus $\bar{K}$ is given by

$$\frac{\bar{K}}{K} = 1 + f \frac{(K^\Omega - K)(3K + 4\mu)}{K\{3K + 4\mu + 3(1-f)(K^\Omega - K)\}}. \tag{10.6.4}$$

This estimate of the overall bulk modulus also gives the average strain energy due to hydrostatic loading σ^o. The resulting radial displacement field is linear in $\mathbf{x}$. Thus, the same $\bar{K}$ is obtained for a linear displacement boundary condition. The overall bulk modulus is therefore *uniquely* given by (10.6.4), since in this case the upper and lower bounds coincide.[15] The stress and strain fields within these composite spheres remain unchanged when the composite is embedded in an infinite homogeneous elastic solid of bulk modulus $\bar{K}$, (10.6.4), under farfield hydrostatic stress σ^o, or when the infinite solid is filled with composite spheres of suitable dimensions but a common volume fraction, leading to an assemblage of spherical composites; Hashin (1962). Similar comments apply to an assemblage of aligned composite cylinders of circular cross sections, as discussed by Hashin and Rosen (1964).

In order to estimate the overall shear modulus $\bar{\mu}$, other loading conditions must be considered. The solution then is more complicated than (10.6.2). For large volume fractions of inclusions, Hashin (1962) obtains the following estimate of $\bar{\mu}$:

$$\frac{\bar{\mu}}{\mu^\Omega} = 1 - (1-f) \frac{(\mu^\Omega - \mu)\{(7 - 5\nu)\,\mu + 2(4 - 5\nu)\,\mu^\Omega\}}{15(1-\nu)\,\mu\,\mu^\Omega}, \tag{10.6.5}$$

where ν is the Poisson ratio of the matrix.

10.6.2. Three-Phase Model

A three-phase model is shown in Figure 10.6.2. An isolated double-inclusion V which consists of a matrix M and a single inclusion Ω within it, is embedded in an infinite homogeneous domain subjected to farfield uniform stresses or strains, $\boldsymbol{\sigma}^\infty$ or $\boldsymbol{\varepsilon}^\infty$. The elasticity and compliance of M and Ω are $\mathbf{C}$ and $\mathbf{C}^\Omega$, and $\mathbf{D}$ and $\mathbf{D}^\Omega$, respectively, whereas those of the surrounding infinite domain are the yet-unknown overall elasticity and compliance of the composite $V = M + \Omega$, i.e., $\bar{\mathbf{C}}$ and $\bar{\mathbf{D}}$.

---

[15] As shown in Section 2, the overall moduli of a finite body depend on the prescribed boundary conditions, and the uniform traction and linear displacement boundary conditions provide bounds for the moduli corresponding to any general boundary data; see Subsection 2.5.6.

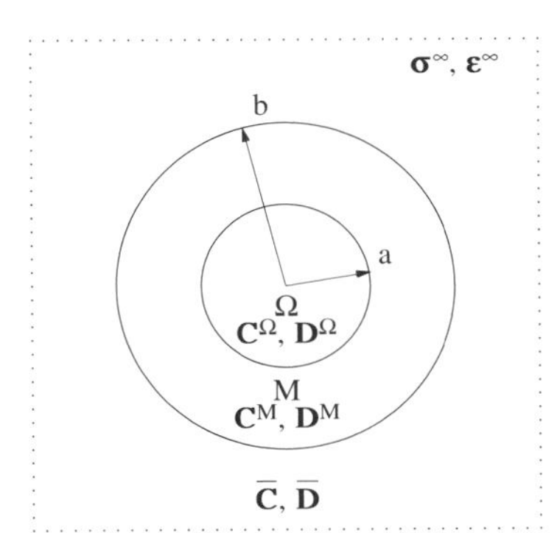

Figure 10.6.2

Three-phase model

Since three distinct materials, M, Ω, and the surrounding infinite body, are considered, this model may be called the three-phase model. It may also be called a generalized self-consistent method, since, in addition to including the interaction of an inclusion with its immediate surrounding matrix, it considers the overall effects of the composite on each double-inclusion. While the self-consistent method seeks to predict the interaction of an inclusion and its neighboring microstructure (the combined effect of the matrix and other inclusions), this model includes (in a certain approximate sense) the interaction between the inclusion and the surrounding matrix, as well as the neighboring microstructure.

Unlike the single-inclusion problem solved by Eshelby, there is no simple analytic solution for the general double-inclusion case. Hence, the shape and arrangement of the double inclusion must be specified such that simple analytic results are obtained. Usually, V and Ω are taken to be cocentered spheres or cylinders. As the simplest example, a solution of cocentered spheres is presented. Similarly to the composite-spheres model of Subsection 10.6.1, it is assumed that: 1) the radii of V and Ω are b and a, respectively, i.e., the elasticity and compliance tensors are

$$\mathbf{C}' = \mathbf{C}'(\mathbf{x}) = \begin{cases} \mathbf{C}^{\Omega} & 0 < \mathrm{r} < \mathrm{a} \\ \mathbf{C} & \mathrm{a} < \mathrm{r} < \mathrm{b} \\ \overline{\mathbf{C}} & \mathrm{b} < \mathrm{r}, \end{cases}$$

$$\mathbf{D}' = \mathbf{D}'(\mathbf{x}) = \begin{cases} \mathbf{D}^{\Omega} & 0 < \mathrm{r} < \mathrm{a} \\ \mathbf{D} & \mathrm{a} < \mathrm{r} < \mathrm{b} \\ \overline{\mathbf{D}} & \mathrm{b} < \mathrm{r}; \end{cases} \tag{10.6.6a,b}$$

and 2) $\mathbf{C}$ and $\mathbf{C}^{\Omega}$ are isotropic, and hence $\overline{\mathbf{C}}$ becomes isotropic.

When the farfield stress is hydrostatic, $\boldsymbol{\sigma}^{\infty} = \sigma^{\mathrm{o}}\mathbf{1}^{(2)}$, or the farfield strain is dilatational, $\boldsymbol{\varepsilon}^{\infty} = \sigma^{\mathrm{o}}\mathbf{1}^{(2)}$, the resulting stress and strain fields become spherically symmetric, and only radial displacement $\mathrm{u_r = u_r(r)}$ is produced,

$$u_r(r) = \begin{cases} A^{\Omega} r & 0 < r < a \\ A\, r + B\, r^{-2} & a < r < b \\ \bar{A}\, r + \bar{B}\, r^{-2} & b < r, \end{cases} \tag{10.6.7a}$$

where A, B, and A^{Ω} are given by (10.6.2b~d), and

$$\bar{A} = \frac{\sigma^o}{3K}, \qquad \frac{\bar{B}}{b^2} = (A - \bar{A})\, b + \frac{B}{b^2}. \tag{10.6.7b,c}$$

The overall bulk modulus $\bar{K}$ of this three-phase model coincides with that of the composite-spheres model. Note that the resulting tractions and displacements on the boundary of V are uniform and linear, respectively. Hence, $\bar{K}$ given by (10.6.4) is also the *unique* overall bulk modulus of this model.

If farfield shearing is prescribed, the overall shear modulus of the composite, $\bar{\mu}$, can be estimated. However, unlike the case of the spherically symmetric loading, surface tractions and displacements on ∂V now fluctuate. Hence, it is seen that: 1) the overall shear modulus may not be the same when it is defined through the relation between the average strain and stress over V, compared with that defined through the average strain energy; and 2) the predicted overall shear modulus is not unique, unless its upper and lower bounds coincide. Following Smith (1974), $\bar{\mu}$ is computed from the average elastic energy, leading to

$$A_2 \{\frac{\bar{\mu}}{\mu}\}^2 + 2A_1 \{\frac{\bar{\mu}}{\mu}\} + A_0 = 0, \tag{10.6.8a}$$

where

$$A_2 = 8f^{10/3}d(4-5\nu)\eta_1 - 2f^{7/3}\{63d\eta_2 + 2\eta_1\eta_3\} + 252f^{5/3}d\eta_2$$
$$- 25fd(7 - 12\nu + 8\nu^2)\eta_2 + 4(7-10\nu)\eta_2\eta_3,$$

$$A_1 = -2f^{10/3}d(1-5\nu)\eta_1 + 2f^{7/3}\{63d\eta_2 + 2\eta_1\eta_3\} - 252f^{5/3}d\eta_2$$
$$+ 75fd(3-\nu)\nu\eta_2 + \frac{3}{2}(15\nu - 7)\eta_2\eta_3,$$

$$A_0 = 4f^{10/3}d(5\nu - 7)\eta_1 - 2f^{7/3}\{63d\eta_2 + 2\eta_1\eta_3\}$$
$$- 252f^{5/3}d\eta_2 + 25fd(\nu^2 - 1)\eta_2 - 4(7+5\nu)\eta_2\eta_3, \tag{10.6.8b~d}$$

with $d = \mu^{\Omega}/\mu - 1$, and

$$\eta_1 = (49 - 50\nu\nu^{\Omega})d + 35(d+1)(\nu^{\Omega} - 2\nu) + 35(2\nu^{\Omega} - \nu),$$

$$\eta_2 = 5\nu^{\Omega}(d-3) + 7(d+5), \qquad \eta_3 = (d+1)(8-10\nu) + (7-5\nu). \tag{10.6.8e~g}$$

Christensen (1990) has examined the results of several models for extreme conditions of composites containing rather high concentrations of rigid inclusions, in an effort to provide insight into the validity and limitations of these models. Referring to the above three-phase model as *the generalized self-consistent method* (GSCM), Christensen compares the overall shear modulus obtained by this model with those given by the differential scheme (DS) and the

Mori-Tanaka method (MT) as proposed by Benveniste (1987). In addition, experimental data collected by Thomas (1965)[16] on viscosity of suspensions of particles, e.g., glass spheres in liquids, are used to check the effectiveness of various model estimates. Christensen finds that the GSCM results correlate better with the experimental data; these and additonal results are summarized in Figure 10.6.3, and are discussed below.

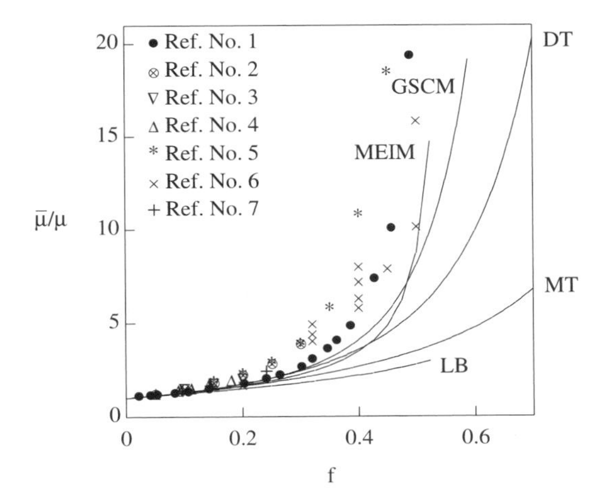

Figure 10.6.3

Comparison of estimates of overall shear modulus; LB: lower bound; GSCM: generalized self-consistent method; DT: differential scheme; MT: Mori-Tanaka method; MEIM: modified equivalent inclusion method; Data taken from following references: No. 1: Ting and Luebbers (1957), No. 2: Ward and Whitmore (1950), No. 3: Broughton and Windebank (1938), No. 4: Manley and Mason (1954), No. 5: Vand (1948), No. 6: Sweeny and Geckler (1954), No. 7: Saunders (1961)

Yu (1992) has examined the sources of data used by Thomas (1965), and has observed that the reported experimental data depend on particle size distribution, shearing rate, and even on the type of the viscometer used for measurements, as has been pointed out by Jeffrey and Acrivos (1976). The overall response at high concentrations of suspensions is generally non-Newtonian, especially at high shearing rates. Thus, while such comparisons provide better understanding of the results of various models, caution must be exercised in drawing definite conclusions on the effectiveness of various models. In a recent work, Nemat-Nasser *et al.* (1993) have re-examined the results presented by Christensen (1990), and have provided an alternative method based on the

[16] An earlier effort to collect and correlate similar data is by Rutgers (1962a,b). These and related issues are reviewed by Jeffrey and Acrivos (1976).

periodic microstructure, which seems to also produce results in good accord with experimental data used by Christensen. These authors also have re-examined the sources of data used by Thomas (1965), and provide useful critique and comments. In addition to Christensen's results, Figure 10.6.3 of Nemat-Nasser *et al.* (1993), provides new results based on a periodic structure which Nemat-Nasser *et al.* call the *modified equivalent inclusion method* (MEIM). This method uses a periodic microstructure in which the unit cell is homogenized using the yet-unknown overall moduli for the reference elasticity tensor; see Section 12.

10.7. REFERENCES

Benveniste, Y. (1987), A new approach to the application of Mori-Tanaka's theory in composite materials, *Mech. Matr.*, Vol. 6, No. 2, 147-157.

Boucher, S. (1974), On the effective moduli of isotropic two-phase elastic composites, *J. Composite Materials*, Vol. 8, 82-89.

Broughton, G. and Windebank, C. S. (1938), Agglomeration and viscosity in dilute suspensions, *Industrial and Engineering Chem.*, Vol. 4, 407-409.

Christensen, R. M. (1990), A critical evaluation for a class of micromechanics models, *J. Mech. Phys. Solids*, Vol. 38, No. 3, 379-404.

Christensen, R. M. and Lo, K. H. (1979), Solutions for effective shear properties in three phase sphere and cylinder models, *J. Mech. Phys. Solids,* Vol. 27, 315-330.

Cleary, M. P., Chen, I.-W., and Lee, S.-M. (1980), Self-consistent techniques of heterogeneous media, *J. Engrn. Mech. Div. ASCE*, Vol. 106, 861.

Fröhlich, H. and Sack, R. (1946), Theory of the rheological properties of dispersions, *Proc. Roy. Soc. Lond.,* Ser. A, Vol. 185, 415-430.

Hashin, Z. (1962), The elastic moduli of heterogeneous materials, *ASME J. Appl. Mech.*, Vol. 29, 143-150.

Hashin, Z. (1988), The differential scheme and its application to cracked materials, *J. Mech. Phys. Solids*, Vol. 21, 236-241.

Hashin, Z. and Rosen, B. W. (1964), The elastic moduli of fiber reinforced materials, *J. Appl. Mech.*, Vol. 31, 223-232.

Henyey, F. S., and Pomphrey, N. (1982), Self-consistent elastic moduli of a cracked solid, *Geopys. Res. Lett.*, Vol. 9, 903-906.

Hori, M. and Nemat-Nasser, S. (1993), Double-inclusion model and overall moduli of materials with microstructure, *Mech. Matr.*, Vol. 14, No. 3, 189-206.

Hori, M. and Nemat-Nasser, S. (1994), Double-inclusion model and overall moduli of multi-phase composites, *J. Engineering Mat. Tech.*, Vol. 116, 305-309.

Jeffrey, D. J. and A. Acrivos (1976), The rheological properties of suspensions of rigid particles, *Am. Inst. Chem. Engineers*, Vol. 22, 417-432.

Mackenzie, J. K. (1950), The elastic constants of a solid containing spherical holes, *Proc. Phys. Soc.,* Vol. B.63, 2-11.

Manley, R. St. J. and Mason, S. G. (1954), The viscosity of suspensions of spheres: A note on the particle interaction coefficient, *Canadian J. Chem.*, Vol. 32, 763-767.

McLaughlin, R. (1977), A study of the differential scheme for composite materials, *Int. J. Engng. Sci.*, Vol. 15, 237-244.

Milton, G. W. (1984), Microgeometries corresponding exactly with effective medium theories, in: *Physics and Chemistry of Porous Media*, AIP Conference Proceedings, American Institute of Physics, New York, Vol. 107.

Milton, G. W. (1990), On characterizing the set of possible effective tensors of composites: The variational method and the traslation method, *Communications on Pure and Applied Mathematics*, Vol. 43, 63-125.

Mori, T. and Tanaka, K. (1973), Average stress in matrix and average elastic energy of materials with misfitting inclusions, *Acta Met.*, Vol. 21, 571-574.

Mori, T. and Wakashima, K. (1990), Successive iteration method in the evaluation of average fields in elastically inhomogeneous materials, in *Micromechanics and inhomogeneity - The T. Mura 65th anniversary volume,* Springer-Verlag, New York, 269-282.

Nemat-Nasser, S. and Hori, M. (1990), Elastic solids with microdefects, in *Micromechanics and inhomogeneity - The T. Mura 65th anniversary volume,* Springer-Verlag, New York, 297-320.

Nemat-Nasser, S., Yu, N., and Hori, M. (1993), Bounds and estimates of overall moduli of composites with periodic microstructure, *Mech. Matr.*, Vol. 15, No. 3, 163-181.

Nemat-Nasser, S. and Hori, M. (1995), Universal bounds for overall properties of linear and nonlinear heterogeneous solids, *J. Engineering Mat. Tech.*, Vol. 117, 412-432.

Norris, A. N. (1985), A differential scheme for the effective moduli of composites, *Mech. Matr.*, Vol. 4, No. 1, 1-16.

Roscoe, R. (1952) The viscosity of suspensions of rigid spheres,*Brit. J. Appl. Phys.,*, Vol. 3, 267-269.

Roscoe, R. (1973), Isotropic composites with elastic or viscoelastic phases: General bounds for the moduli and solutions for special geometries, *Rheol. Acta*, Vol. 12, 404-411.

Rutgers, R. (1962a), Relative viscosity of suspensions of rigid spheres in Newtonian liquids, *Rheol. Acta*, Vol. 2, 202-210.

Rutgers, R. (1962b), Relative viscosity and concentration, *Reol. Acta*, Vol. 2, 305-348 .

Saunders, F. L. (1961), Rheological properties of monodisperse latex systems I. Concentration dependence of relative viscosity, *J. Colloid Science*, Vol. 16, 13-22.

Sen, P. N., Scala, C., and Cohen, M. H. (1981), A self similar model for sedimentary rocks with application to the dielectric constant of fused glass beads, *Geophysics*, Vol. 46, 781-795.

Sheng, P. and Callegari, A. J. (1984), Differential effective medium theory for sedimentary rocks, in: *Physics and Chemistry of Porous Media*, AIP Conference Proceedings, American Institute of Physics, New York, Vol. 107.

Smith, J. C. (1974), Correction and extension of van der Poel's method for calculating the shear modulus of a particulate composite, *J. Res. Natl. Bur. Stand. Sect. A,* Vol. 78, 355-361.

Sweeny, K. H. and Geckler, R. D. (1954), The rheology of suspensions, *J. Appl. Phys.*, Vol. 25, 1135-1144.

Tanaka, K. and Mori, T. (1972), Note on volume integrals of the elastic field around an ellipsoidal inclusion, *J. Elasticity*, Vol. 2, 199-200.

Thomas, D. G. (1965), Transport characteristics of suspension VIII. A note of the viscosity of Newtonian suspensions of uniform spherical particles, *J. Colloid Sci*, Vol. 20, 267-277.

Ting, A. P. and Luebbers, R. H. (1957), Viscosity of suspensions of spherical and other isodimensional particles in liquids, *Am. Inst. Chem. Engineers*, Vol. 1, 111-116.

Torquato, S. (1991), Random heterogeneous media: microstructure and improved bounds on effective properties, *Appl. Mech. Rev.*, Vol. 42, No. 2, 37-76.

Vand, V. (1948), Viscosity of solutions and suspensions. II Experimental determination of the vicosity-concentration function of spherical suspensions, *J. Phys. Colloid Chem.*, Vol. 52, 277-299.

Ward, S. G. and Whitmore, R. L. (1950), Studies of the viscosity and sedimentation of suspensions Part 1.-The viscosity of suspension of spherical particles, *British J. Appl. Phys.*, Vol. 1, 286-290.

Weng, G. J. (1990), The theoretical connection between Mori-Tanaka's theory and the Hashin-Shtrikman-Walpole variational bounds, *Int. J. Engng. Sci.*, Vol. 28, No. 11, 1111-1120.

Yu, Niann-i. (1992), *Overall properties of composite materials: a general model based on periodic microstructure*, Ph.D. Thesis, University of California, San Diego.

SECTION 11 ESHELBY'S TENSOR AND RELATED TOPICS

In Section 7, Eshelby's tensor $\mathbf{S}^{\Omega}$ and its conjugate $\mathbf{T}^{\Omega}$ are introduced for an infinitely extended homogeneous linearly elastic solid and are used to determine the constant strain and stress fields in an ellipsoidal subdomain Ω, in which a uniform eigenstrain or eigenstress is distributed. As shown in Sections 8 and 10, Eshelby's tensor can be used to estimate the average strains and stresses in the micro-inclusions embedded in an elastic matrix. These tensors can be computed analytically and therefore, they provide an effective means for estimating the overall moduli of a heterogeneous RVE.

In this section an infinitely extended homogeneous linearly elastic solid is considered within a portion of which either eigenstrains or eigenstresses are distributed (not necessarily uniformly). The Green function for the unbounded solid is then used to formulate the resulting displacement field in terms of two integral operators, $\mathbf{S}^{\infty}(\mathbf{x}; \boldsymbol{\varepsilon}^*)$ and $\mathbf{T}^{\infty}(\mathbf{x}; \boldsymbol{\sigma}^*)$, where $\boldsymbol{\varepsilon}^* = \boldsymbol{\varepsilon}^*(\mathbf{x})$ and $\boldsymbol{\sigma}^* = \boldsymbol{\sigma}^*(\mathbf{x})$ are the corresponding prescribed eigenstrain and eigenstress.[1] The integral operators reduce to tensor operators, when the distribution of eigenstrains or eigenstresses is uniform and the domain Ω in which they are distributed is an isolated ellipsoid in an unbounded uniform medium, resulting in the Eshelby tensor, $\mathbf{S}^{\Omega}$, and its conjugate, $\mathbf{T}^{\Omega}$. Explicit expressions are derived for these tensors in the case of an isotropic matrix, and the anisotropic case is briefly discussed. Then a number of interesting properties of the Eshelby tensor are examined, including: (1) its symmetry properties; (2) the Tanaka-Mori result which is then generalized and used to estimate the interaction among inclusions and to solve the double-inclusion problem; and (3) the disturbances in the average field values produced by prescribed eigenstrains or eigenstresses, where exact expressions are obtained for the average strain, stress, and strain energy and complementary strain energy densities.

11.1. EIGENSTRAIN AND EIGENSTRESS PROBLEMS

Consider an infinitely extended domain, denoted by V^{∞}, consisting of a homogeneous linearly elastic solid of elasticity $\mathbf{C}$ (not necessarily isotropic). The displacement field produced by distributed body forces, $\mathbf{f}$, is formulated in

---

[1] Note that only the eigenstrain or the eigenstress but not both can be prescribed arbitrarily in a given region. Hence, when both eigenstrain and eigenstress are mentioned, either two separate problems are considered, or the two fields are mutually dependent. The intended alternative should be clear from the context.

terms of the corresponding Green function. Then, eigenstrains or eigenstresses distributed within a certain finite domain in V^∞, are expressed as equivalent body forces, and the corresponding displacement, strain, and stress fields are obtained in terms of the integral operator $\mathbf{S}^\infty$ and $\mathbf{T}^\infty$.

11.1.1. Green's Function for Infinite Domain

First consider the Green function for an infinitely extended homogeneous linearly elastic solid, V^∞, with elasticity and compliance tensors $\mathbf{C}$ and $\mathbf{D}$. As pointed out before, the Green function $G^\infty_{ij}(\mathbf{x}, \mathbf{y})$ gives the displacement in the x_i-direction at point $\mathbf{x}$, produced by a unit point force applied in the x_j-direction at point $\mathbf{y}$. More precisely, the Green function is the vector-valued *fundamental*[2] *solution of the operator* $\mathbf{L} \equiv \boldsymbol{\nabla} \boldsymbol{.} \{\mathbf{C} : (\boldsymbol{\nabla} \otimes)\}$,

$$\boldsymbol{\nabla} \boldsymbol{.} \{\mathbf{C} : (\boldsymbol{\nabla} \otimes \mathbf{G}^\infty(\mathbf{x}, \mathbf{y}))\} + \delta(\mathbf{x} - \mathbf{y})\, \mathbf{1}^{(2)} = \mathbf{0} \qquad \mathbf{x} \text{ in } V^\infty, \tag{11.1.1a}$$

where $\delta(\mathbf{x} - \mathbf{y})$ is the delta function, and

$$\mathbf{G}^\infty(\mathbf{x}, \mathbf{y}) \rightarrow \mathbf{0} \qquad \text{as } |\mathbf{x}| \rightarrow \infty. \tag{11.1.1b}$$

In component form, (11.1.1a) is

$$C_{ijkl}\, G^\infty_{lm,ik}(\mathbf{x}, \mathbf{y}) + \delta(\mathbf{x} - \mathbf{y})\, \delta_{jm} = 0 \qquad \mathbf{x} \text{ in } V^\infty. \tag{11.1.1c}$$

Since the domain V^∞ is homogeneous and unbounded, only the difference between the $\mathbf{x}$-point (where the displacement is measured) and the $\mathbf{y}$-point (where the unit force is applied) determines the Green function,

$$\mathbf{G}^\infty(\mathbf{x}, \mathbf{y}) = \mathbf{G}^\infty(\mathbf{x} - \mathbf{y}). \tag{11.1.2}$$

The farfield condition (11.1.1b) now becomes

$$\mathbf{G}^\infty(\mathbf{z}) \rightarrow 0 \qquad \text{as } |\mathbf{z}| \rightarrow \infty. \tag{11.1.3a}$$

The gradient of $\mathbf{G}^\infty$ also vanishes for large values of $\mathbf{z}$, and hence the farfield strains and stresses are zero. For an arbitrary finite domain W within V^∞ to be in equilibrium, the resultant tractions on the surface ∂W of W must satisfy

$$\int_{\partial W} \mathbf{n}(\mathbf{x}) \boldsymbol{.} \{\mathbf{C} : (\boldsymbol{\nabla}_\mathbf{x} \otimes \mathbf{G}^\infty(\mathbf{x} - \mathbf{y})\}\, dS_\mathbf{x} = \begin{cases} -\mathbf{1}^{(2)} & \text{if } \mathbf{y} \text{ in W} \\ \mathbf{0} & \text{otherwise,} \end{cases} \tag{11.1.3b}$$

where $\mathbf{n}$ is the outer unit normal of ∂W, and subscript $\mathbf{x}$ stands for differentiation (integration) with respect to $\mathbf{x}$. Indeed, with the aid of the Gauss theorem, (11.1.3b) is obtained by integrating the governing equation (11.1.1a) over W.

The Green function has an important symmetry property, derived from the reciprocal theorem. Let $\mathbf{u}^{(\alpha)}(\mathbf{x})$ be the displacement field produced by a point force $\mathbf{f}^{(\alpha)}$ applied at $\mathbf{x}^{(\alpha)}$, for $\alpha = 1, 2$; note that these are two different displacement fields, each associated with its own point loading. Using the reciprocal theorem, and noting that $\mathbf{f}^{(\alpha)}$ is a point force, obtain

---

[2] See, e.g., Morse and Feshbach (1953), Stakgold (1967), and Roach (1982).

$$\mathbf{f}^{(1)} \cdot \mathbf{u}^{(2)}(\mathbf{x}^{(1)}) = \mathbf{f}^{(2)} \cdot \mathbf{u}^{(1)}(\mathbf{x}^{(2)}). \tag{11.1.4a}$$

Since $\mathbf{u}^{(\alpha)}(\mathbf{x})$ is given by $\mathbf{G}^{\infty}(\mathbf{x} - \mathbf{x}^{(\alpha)}) \cdot \mathbf{f}^{(\alpha)}$, this leads to

$$\mathbf{f}^{(1)} \cdot \mathbf{G}^{\infty}(\mathbf{x}^{(1)} - \mathbf{x}^{(2)}) \cdot \mathbf{f}^{(2)} = \mathbf{f}^{(2)} \cdot \mathbf{G}^{\infty}(\mathbf{x}^{(2)} - \mathbf{x}^{(1)}) \cdot \mathbf{f}^{(1)}. \tag{11.1.4b}$$

The above equation must hold for any $\mathbf{f}^{(\alpha)}$ applied at any point $\mathbf{x}^{(\alpha)}$. Therefore, the Green function $\mathbf{G}^{\infty}$ has the following symmetry property:

$$\mathbf{G}^{\infty}(\mathbf{z}) = \mathbf{G}^{\infty T}(-\mathbf{z}) \qquad \text{or} \qquad G^{\infty}_{ij}(\mathbf{z}) = G^{\infty}_{ji}(-\mathbf{z}). \tag{11.1.5a,b}$$

It is apparent from the governing equations (11.1.1), that $\mathbf{G}$ is an even function of $\mathbf{z}$. Thus, in view of (11.1.5), it follows that

$$\mathbf{G}^{\infty}(\mathbf{z}) = \mathbf{G}^{\infty T}(\mathbf{z}) \qquad \text{or} \qquad G^{\infty}_{ij}(\mathbf{z}) = G^{\infty}_{ji}(\mathbf{z}). \tag{11.1.6a,b}$$

11.1.2. The Body-Force Problem

Using the Green function $\mathbf{G}^{\infty}$, consider the displacement field $\mathbf{u}$ produced by distributed body forces $\mathbf{f}$. The boundary-value problem for $\mathbf{u}$ is,

$$\nabla \cdot (\mathbf{C} : \nabla \otimes \mathbf{u}(\mathbf{x})) + \mathbf{f}(\mathbf{x}) = \mathbf{0} \qquad \mathbf{x} \text{ in } V^{\infty}, \tag{11.1.7a}$$

with

$$\mathbf{u}(\mathbf{x}) \to \mathbf{0} \qquad \text{as } |\mathbf{x}| \to \infty. \tag{11.1.7b}$$

From linearity, the solution is given by integrating $\mathbf{G}^{\infty}$ for the prescribed body forces $\mathbf{f}$. Assuming that $\mathbf{f}$ vanishes sufficiently quickly toward infinity, obtain

$$\mathbf{u}(\mathbf{x}) = \int_{V^{\infty}} \mathbf{G}^{\infty}(\mathbf{x} - \mathbf{y}) \cdot \mathbf{f}(\mathbf{y})\, dV_{\mathbf{y}}. \tag{11.1.8}$$

This is the *unique* solution of the boundary-value problem (11.1.7), valid for any $\mathbf{f}$ (discontinuous or not) which renders the integral finite.

Now suppose the body forces, $\mathbf{f}$, are given by the divergence of a tensor field $\mathbf{T}(\mathbf{x})$ which is sufficiently smooth in W but suffers a jump to $\mathbf{0}$ across ∂W, i.e.,

$$\mathbf{T}(\mathbf{x}) = \begin{cases} \mathbf{T}(\mathbf{x}) \neq \mathbf{0} & \mathbf{x} \text{ in W} \\ \mathbf{0} & \text{otherwise,} \end{cases} \tag{11.1.9a}$$

and

$$\mathbf{f}(\mathbf{x}) = \nabla \cdot \mathbf{T}(\mathbf{x}) \qquad \mathbf{x} \text{ in W}. \tag{11.1.9b}$$

Then, while $\mathbf{f}(\mathbf{x})$ is finite within W, it behaves as a delta function across ∂W. This behavior is represented by concentrated forces distributed within a thin layer about ∂W, representing the jump in $\mathbf{T}(\mathbf{x})$ across this boundary; the overall effect, therefore, is represented by additional tractions acting on W over its boundary ∂W. Denote these tractions by $[\mathbf{t}](\mathbf{x})$ for $\mathbf{x}$ on ∂W, and obtain

$$[\mathbf{t}](\mathbf{x}) \equiv \mathbf{n}(\mathbf{x}) \cdot [\mathbf{T}](\mathbf{x}) = -\mathbf{n}(\mathbf{x}) \cdot \{\lim_{\mathbf{x}^+ \to \mathbf{x}} \mathbf{T}(\mathbf{x}^+)\} \qquad \mathbf{x} \text{ on } \partial\text{W}, \tag{11.1.9c}$$

where the minus sign is due to the fact that the unit outer normal $\mathbf{n}$ points from the inside toward the outside of W, and $\mathbf{x}^+$ is a point inside W. The resulting displacement field produced by body forces $\mathbf{f}(\mathbf{x})$ distributed within W, and tractions $[\mathbf{t}](\mathbf{x})$ acting on ∂W, i.e., the displacement field corresponding to (11.1.9b) and (11.1.9c), is given by

$$\mathbf{u}(\mathbf{x}) = \int_W \mathbf{G}^{\infty T}(\mathbf{y}-\mathbf{x}) \cdot \mathbf{f}(\mathbf{y})\, dV_y + \int_{\partial W} \mathbf{G}^{\infty T}(\mathbf{y}-\mathbf{x}) \cdot [\mathbf{t}](\mathbf{y})\, dS_y, \tag{11.1.10}$$

where the symmetry of $\mathbf{G}^\infty$, (11.1.5), is used. In view of (11.1.9b,c), use the Gauss theorem to rewrite (11.1.10) as

$$\mathbf{u}(\mathbf{x}) = -\int_W \mathbf{T}(\mathbf{y}) : (\nabla_y \otimes \mathbf{G}^{\infty T}(\mathbf{y}-\mathbf{x}))\, dV_y \tag{11.1.11a}$$

or, in component form,

$$u_i(\mathbf{x}) = -\int_W G^\infty_{ki,j}(\mathbf{y}-\mathbf{x})\, T_{jk}(\mathbf{y})\, dV_y, \tag{11.1.11b}$$

where subscript j following a comma denotes derivative with respect to y_j. Note, since (11.1.10) is valid for any $\mathbf{x}$ in V^∞, the finite domain problem can also be solved in a similar manner, if $\mathbf{x}$ is restricted to remain within W. In this case, $[\mathbf{t}]$ is replaced by suitable surface tractions on ∂W, in order to satisfy the prescribed boundary conditions; see Subsection 9.4.

11.1.3. The Eigenstrain- or Eigenstress-Problem

Using the general results obtained in Subsection 11.1.2, consider the problem of an eigenstrain field $\boldsymbol{\varepsilon}^*$ (or eigenstress field $\boldsymbol{\sigma}^*$) prescribed in W and vanishing identically outside of W,

$$\boldsymbol{\varepsilon}^*(\mathbf{x}) = \begin{cases} \boldsymbol{\varepsilon}^*(\mathbf{x}) & \mathbf{x} \text{ in W} \\ \mathbf{0} & \text{otherwise} \end{cases} \tag{11.1.12a}$$

or

$$\boldsymbol{\sigma}^*(\mathbf{x}) = \begin{cases} \boldsymbol{\sigma}^*(\mathbf{x}) & \mathbf{x} \text{ in W} \\ \mathbf{0} & \text{otherwise.} \end{cases} \tag{11.1.12b}$$

Let $\mathbf{u}^e$ (let $\mathbf{u}^s$) be the displacement field produced by $\boldsymbol{\varepsilon}^*$ (by $\boldsymbol{\sigma}^*$) and denote by $\boldsymbol{\varepsilon}^e$ and $\boldsymbol{\sigma}^e$ (by $\boldsymbol{\varepsilon}^s$ and $\boldsymbol{\sigma}^s$) the corresponding strain and stress fields. These strain and stress fields are expressed as

$$\boldsymbol{\varepsilon}^e = \frac{1}{2}\{(\nabla \otimes \mathbf{u}^e) + (\nabla \otimes \mathbf{u}^e)^T\}, \qquad \boldsymbol{\sigma}^e = \mathbf{C} : \boldsymbol{\varepsilon}^e - \mathbf{C} : \boldsymbol{\varepsilon}^*, \tag{11.1.13a,b}$$

and

$$\boldsymbol{\varepsilon}^s = \frac{1}{2}\{(\nabla \otimes \mathbf{u}^s) + (\nabla \otimes \mathbf{u}^s)^T\}, \qquad \boldsymbol{\sigma}^s = \mathbf{C} : \boldsymbol{\varepsilon}^s + \boldsymbol{\sigma}^*; \tag{11.1.13c,d}$$

see Section 7.

As explained in Subsection 9.3, the divergence of the eigenstrain $\boldsymbol{\varepsilon}^*$ and eigenstress $\boldsymbol{\sigma}^*$ can be regarded as *equivalent body forces*. Indeed, direct substitution of the constitutive relations (11.1.13b) and (11.1.13d) into the equations of equilibrium yields

$$\boldsymbol{\nabla}\boldsymbol{\cdot}\boldsymbol{\sigma}^e = \boldsymbol{\nabla}\boldsymbol{\cdot}\{\mathbf{C} : (\boldsymbol{\nabla} \otimes \mathbf{u}^e)\} + \boldsymbol{\nabla}\boldsymbol{\cdot}(-\,\mathbf{C} : \boldsymbol{\varepsilon}^*) = \mathbf{0} \tag{11.1.13e,f}$$

and

$$\boldsymbol{\nabla}\boldsymbol{\cdot}\boldsymbol{\sigma}^s = \boldsymbol{\nabla}\boldsymbol{\cdot}\{\mathbf{C} : (\boldsymbol{\nabla} \otimes \mathbf{u}^s)\} + \boldsymbol{\nabla}\boldsymbol{\cdot}\boldsymbol{\sigma}^* = \mathbf{0}. \tag{11.1.13g,h}$$

The divergence of $-\,\mathbf{C} : \boldsymbol{\varepsilon}^*$ (of $\boldsymbol{\sigma}^*$) appears like a distribution of body forces in the governing equations for the displacement field $\mathbf{u}^e$ (field $\mathbf{u}^s$); see (11.1.7a). Hence, the displacement field $\mathbf{u}^e$ (field $\mathbf{u}^s$) may be expressed in terms of the Green function $\mathbf{G}^\infty$, as

$$\mathbf{u}^e(\mathbf{x}) = \int_W \{\mathbf{C} : \boldsymbol{\varepsilon}^*(\mathbf{y})\} : \{\boldsymbol{\nabla}_\mathbf{y} \otimes \mathbf{G}^{\infty T}(\mathbf{y}-\mathbf{x})\}\, dV_\mathbf{y} \tag{11.1.14a}$$

and

$$\mathbf{u}^s(\mathbf{x}) = -\int_W \boldsymbol{\sigma}^*(\mathbf{y}) : \{\boldsymbol{\nabla}_\mathbf{y} \otimes \mathbf{G}^{\infty T}(\mathbf{y}-\mathbf{x})\}\, dV_\mathbf{y} \tag{11.1.14b}$$

or, in component form,

$$u_i^e(\mathbf{x}) = \int_W G^\infty_{ki,l}(\mathbf{y}-\mathbf{x})\, C_{klmn}\, \varepsilon^*_{mn}(\mathbf{y})\, dV_\mathbf{y} \tag{11.1.14c}$$

and

$$u_i^s(\mathbf{x}) = -\int_W G^\infty_{ki,l}(\mathbf{y}-\mathbf{x})\, \sigma^*_{kl}(\mathbf{y})\, dV_\mathbf{y}. \tag{11.1.14d}$$

The displacement field $\mathbf{u}^e$ (field $\mathbf{u}^s$) obtained above, satisfies the condition, $\mathbf{u}^e(\mathbf{z}) \rightarrow \mathbf{0}$, ($\mathbf{u}^s(\mathbf{z}) \rightarrow \mathbf{0}$), as $|\,\mathbf{z}\,| \rightarrow \infty$, and is continuous across ∂W.

In general, the strain and stress fields $\boldsymbol{\varepsilon}^e$ and $\boldsymbol{\sigma}^e$ (field $\boldsymbol{\varepsilon}^s$ and $\boldsymbol{\sigma}^s$) are discontinuous across surfaces where the eigenstrain (eigenstress) admits finite discontinuities; the displacement field $\mathbf{u}^e$ (field $\mathbf{u}^s$) is, of course, continuous, as are the tractions $\boldsymbol{\nu}\boldsymbol{\cdot}\boldsymbol{\sigma}^e$ (traction $\boldsymbol{\nu}\boldsymbol{\cdot}\boldsymbol{\sigma}^s$).

Using the strain-displacement and the constitutive relations given by (11.1.13), define the integral operator $\mathbf{S}^\infty$ (operator $\mathbf{T}^\infty$) which determines $\boldsymbol{\varepsilon}^e$ (determines $\boldsymbol{\sigma}^s$), in terms of $\boldsymbol{\varepsilon}^*$ (of $\boldsymbol{\sigma}^*$), as follows:

$$\mathbf{S}^\infty(\mathbf{x};\, \boldsymbol{\varepsilon}^*) \equiv \int_W \boldsymbol{\Gamma}^\infty(\mathbf{y}-\mathbf{x}) : \mathbf{C} : \boldsymbol{\varepsilon}^*(\mathbf{y})\, dV_\mathbf{y} \tag{11.1.15a}$$

and

$$\mathbf{T}^\infty(\mathbf{x};\, \boldsymbol{\sigma}^*) \equiv -\,\mathbf{C} : \{\int_W \boldsymbol{\Gamma}^\infty(\mathbf{y}-\mathbf{x}) : \boldsymbol{\sigma}^*(\mathbf{y})\, dV_\mathbf{y}\} + \boldsymbol{\sigma}^*(\mathbf{x}), \tag{11.1.15b}$$

or, in component form,

$$S^\infty_{ij}(\mathbf{x};\, \varepsilon^*_{kl}) \equiv \int_W \Gamma^\infty_{ijmn}(\mathbf{y}-\mathbf{x})\, C_{mnkl}\, \varepsilon^*_{kl}(\mathbf{y})\, dV_\mathbf{y} \tag{11.1.15c}$$

and

$$T^\infty_{ij}(\mathbf{x};\, \sigma^*_{kl}) \equiv -\,C_{ijmn}\, \{\int_W \Gamma^\infty_{mnkl}(\mathbf{y}-\mathbf{x})\, \sigma^*_{kl}(\mathbf{y})\, dV_\mathbf{y}\} + \sigma^*_{ij}(\mathbf{x}), \tag{11.1.15d}$$

where, as shown in (9.4.10c), the tensor field $\boldsymbol{\Gamma}^\infty(\mathbf{z})$ is defined by

$$\Gamma^{\infty}_{ijkl} = -\frac{1}{4}\,\{G^{\infty}_{ik,jl} + G^{\infty}_{jk,il} + G^{\infty}_{il,jk} + G^{\infty}_{jl,ik}\}. \tag{11.1.15e}$$

If the prescribed eigenstrains and eigenstresses *correspond to each other* in the sense that

$$\boldsymbol{\sigma}^*(\mathbf{x}) = -\mathbf{C} : \boldsymbol{\varepsilon}^*(\mathbf{x}) \qquad \text{or} \qquad \boldsymbol{\varepsilon}^*(\mathbf{x}) = -\mathbf{D} : \boldsymbol{\sigma}^*(\mathbf{x}) \qquad \mathbf{x} \text{ in W}, \tag{11.1.16a,b}$$

then the resulting displacement fields $\mathbf{u}^e$ and $\mathbf{u}^s$ are the same, as are the resulting strain and stress fields. Then (11.1.13a) and (11.1.13c) agree, and

$$\boldsymbol{\sigma}^e = \mathbf{C} : (\boldsymbol{\varepsilon}^e - \boldsymbol{\varepsilon}^*) = \mathbf{C} : \boldsymbol{\varepsilon}^s + \boldsymbol{\sigma}^* = \boldsymbol{\sigma}^s. \tag{11.1.17}$$

Therefore, the integral operators $\mathbf{S}^{\infty}$ and $\mathbf{T}^{\infty}$ satisfy

$$\mathbf{T}^{\infty}(\mathbf{x};\, -\mathbf{C} : \boldsymbol{\varepsilon}^*) = \mathbf{C} : \{\mathbf{S}^{\infty}(\mathbf{x};\, \boldsymbol{\varepsilon}^*) - \boldsymbol{\varepsilon}^*(\mathbf{x})\} \tag{11.1.18a}$$

and

$$\mathbf{S}^{\infty}(\mathbf{x};\, -\mathbf{D} : \boldsymbol{\sigma}^*) = \mathbf{D} : \{\mathbf{T}^{\infty}(\mathbf{x};\, \boldsymbol{\sigma}^*) - \boldsymbol{\sigma}^*(\mathbf{x})\}, \tag{11.1.18b}$$

provided that $\boldsymbol{\sigma}^*$ and $\boldsymbol{\varepsilon}^*$ are related by (11.1.16a,b). Equations (11.1.18a,b) are the *equivalence relations* between the integral operators $\mathbf{S}^{\infty}$ and $\mathbf{T}^{\infty}$.

11.2. ESHELBY'S TENSOR

Since the integral operators $\mathbf{S}^{\infty}$ and $\mathbf{T}^{\infty}$ are equivalent, only $\mathbf{S}^{\infty}$ is computed. Here this is done when: (1) the eigenstrains are *uniformly* distributed in an *ellipsoidal* domain, and (2) the unbounded homogeneous solid V^{∞} is *isotropic.* The resulting tensor which relates the strain field in the ellipsoid to the prescribed uniform eigenstrain is derived. This is *Eshelby's tensor.* The results for an anisotropic V^{∞} are also briefly examined. Hereinafter in this subsection, the components of tensors are used, mainly to avoid any possible confusion.

11.2.1. Uniform Eigenstrains in an Ellipsoidal Domain

First consider the case when a constant eigenstrain ε^{*o}_{ij} is uniformly distributed in a certain finite domain W (not necessarily ellipsoidal) within V^{∞}. In terms of the Heaviside step function, the eigenstrain field $\varepsilon^*_{ij}(\mathbf{x})$ is expressed as

$$\varepsilon^*_{ij}(\mathbf{x}) = H(\mathbf{x};\, W)\, \varepsilon^{*o}_{ij} \qquad \mathbf{x} \text{ in } V^{\infty}, \tag{11.2.1}$$

where $H(\mathbf{x};\, W)$ is 1 for $\mathbf{x}$ in W, and 0 otherwise. Note that ε^{*o}_{ij} is the magnitude of the uniform eigenstrain field $\varepsilon^*_{ij}(\mathbf{x})$. From (11.1.14a), the resulting displacement field $u_i(\mathbf{x})$ is given by

$$u_i(\mathbf{x}) = \left\{\int_W G^{\infty}_{ki,l}(\mathbf{y} - \mathbf{x})\, dV_{\mathbf{y}}\right\} C_{klmn}\, \varepsilon^{*o}_{mn}, \tag{11.2.2}$$

and from (11.1.15a), the strain field $\varepsilon_{ij}(\mathbf{x})$ becomes

$$\varepsilon_{ij}(\mathbf{x}) = S^{\infty}_{ijkl}(\mathbf{x};\, W)\, \varepsilon^{*o}_{kl}, \tag{11.2.3a}$$

where the tensor field $S^{\infty}_{ijkl}(\mathbf{x};\, W)$ is

$$S^{\infty}_{ijkl}(\mathbf{x};\, W) \equiv \{\int_W \Gamma^{\infty}_{ijmn}(\mathbf{x}-\mathbf{y})\, dV_y\}\, C_{mnkl}. \tag{11.2.3b}$$

Since the partial derivatives of the uniform eigenstrain field $\varepsilon^*_{ij}(\mathbf{x}) \equiv \varepsilon^{*o}_{ij}$ are zero in W, the equivalent body forces vanish in W. However, across the boundary ∂W, the divergence of $\varepsilon^*_{ij}(\mathbf{x})$ varies as a delta function. This boundary then is a discontinuity surface for the strain and stress fields. To examine this discontinuity, denote the inside and outside of the domain W by superscripts + and $-$, respectively, and let n_i be the unit normal vector pointing from the inside to the outside of W (from W^+ to W^-). Although the displacements $u_i(\mathbf{x})$ and tractions $n_i(\mathbf{x})\,\sigma_{ij}(\mathbf{x})$ are continuous across ∂W, the quantity (called pseudo-tractions)

$$[t_j](\mathbf{x}) \equiv n_i(\mathbf{x})\{\lim_{\mathbf{x}^+\to\mathbf{x}} C_{ijkl}\,\varepsilon_{kl}(\mathbf{x}^+) - \lim_{\mathbf{x}^-\to\mathbf{x}} C_{ijkl}\,\varepsilon_{kl}(\mathbf{x}^-)\}, \tag{11.2.4a}$$

is discontinuous for $\mathbf{x}$ on ∂W, and from the continuity of tractions,

$$[t_j](\mathbf{x}) = [C_{ijkl}\,\varepsilon_{kl}\, n_i](\mathbf{x}) = n_i(\mathbf{x}) \lim_{\mathbf{x}^+\to\mathbf{x}} C_{ijkl}\,\varepsilon^*_{kl}(\mathbf{x}^+) \qquad \mathbf{x} \text{ on } \partial W \tag{11.2.4b}$$

or

$$[t_j](\mathbf{x}) = n_i(\mathbf{x})\,(C_{ijkl}\,\varepsilon^{*o}_{kl}) \qquad \mathbf{x} \text{ on } \partial W. \tag{11.2.5a}$$

The strain field in W, given by (11.2.3a),

$$\varepsilon_{ij}(\mathbf{x}) = S^{\infty}_{ijkl}(\mathbf{x};\, W)\, \varepsilon^{*o}_{kl}, \tag{11.2.5b}$$

and the corresponding stress field,

$$\sigma_{ij}(\mathbf{x}) = C_{ijkl}\, \{S^{\infty}_{klmn}(\mathbf{x};\, W) - 1^{(4s)}_{klmn}\}\, \varepsilon^{*o}_{mn} \qquad \mathbf{x} \text{ in } W, \tag{11.2.5c}$$

are, in general, nonuniform within W, unless the integral in the right side of (11.2.3b) turns out to be independent of $\mathbf{x}$. When W is ellipsoidal in shape, Eshelby (1957) has shown that this integral is constant, and hence, a uniform eigenstrain distributed in W produces uniform strain and stress fields in W, with continuous displacements and tractions across ∂W; see also Hardiman[3] (1954), Eshelby (1961), Hill (1965), and Bacon *et al.* (1979).

11.2.2. Eshelby's Tensor for an Isotropic Solid

Eshelby (1957) shows that if the domain where the *uniform* eigenstrain ε^*_{ij} is distributed is an *ellipsoid,* say, Ω, then the resulting strain and stress fields in Ω are uniform. For an isotropic V^{∞}, Eshelby gave explicit expressions for the tensor which relates the resulting uniform strain in the ellipsoid to the prescribed

[3] It is pointed out that Hardiman in a Ph.D. thesis (1951) has shown that stress and strain fields in an elliptic inclusion in an infinite plate, are uniform when the farfield strains or stresses are uniform.

eigenstrain ε^*_{ij}. From the results of Subsection 11.2.1, it is seen that the integral (11.2.3b) is independent of **x** in W, only if W has a suitable shape.

To compute the tensor field $S^{\infty}_{ijkl}(\mathbf{x};\,\Omega)$ *explicitly*, it is assumed that:

1) Ω is an ellipsoid, with the semi-axes a_i parallel to the x_i-direction (i = 1, 2, 3); and

2) the linearly elastic homogeneous solid V^{∞} is isotropic,

$$C_{ijkl} = 2\mu\{\frac{\nu}{1-2\nu}\delta_{ij}\,\delta_{kl} + \frac{1}{2}(\delta_{ik}\,\delta_{jl} + \delta_{il}\,\delta_{jk})\}, \tag{11.2.6}$$

where μ and ν are the shear modulus and Poisson ratio.

The Green function $G^{\infty}_{ij}(\mathbf{z})$ for an infinitely extended body with isotropic elasticity is given by (see Part 2)

$$G^{\infty}_{ij}(\mathbf{z}) = \frac{1}{16\pi\mu(1-\nu)\,|\,\mathbf{z}\,|}\,\{(3-4\nu)\,\delta_{ij} + \frac{z_i z_j}{|\,\mathbf{z}\,|^2}\}, \tag{11.2.7}$$

where $|\,\mathbf{z}\,| = \sqrt{z_i z_i}$. Substitute (11.2.7) into (11.1.14), to obtain

$$u_i = -\frac{\varepsilon^*_{kl}}{8\pi(1-\nu)}\,\{\psi_{,ikl} - 2\nu\,\delta_{kl}\,\phi_{,i} - 2(1-\nu)(\delta_{ik}\,\phi_{,l} + \delta_{il}\,\phi_{,k})\}, \tag{11.2.8a}$$

where a subscript following a comma stands for differentiation with respect to the corresponding argument, and

$$\psi \equiv \psi(\mathbf{x}) \equiv \int_{\Omega} |\,\mathbf{x}-\mathbf{y}\,|\,dV_{\mathbf{y}}, \qquad \phi \equiv \phi(\mathbf{x}) \equiv \int_{\Omega} \frac{1}{|\,\mathbf{x}-\mathbf{y}\,|}\,dV_{\mathbf{y}}. \tag{11.2.8b,c}$$

The corresponding strain field now becomes

$$\varepsilon_{ij} = -\frac{1}{8\pi(1-\nu)}\,\{\psi_{,ijkl} - 2\nu\,\phi_{,ij}\delta_{kl} - (1-\nu)\,(\delta_{ik}\phi_{,jl} + \delta_{jk}\phi_{,il} + \delta_{il}\phi_{,jk} + \delta_{jl}\phi_{,ik})\}\,\varepsilon^*_{kl}. \tag{11.2.8d}$$

From (11.2.8d), define the following fourth-order tensor:

$$S^{\infty}_{ijkl}(\mathbf{x};\,\Omega) = -\frac{1}{8\pi(1-\nu)}\,\{\psi_{,ijkl} - 2\nu\,\phi_{,ij}\delta_{kl} - (1-\nu)\,(\delta_{ik}\phi_{,jl} + \delta_{jk}\phi_{,il} + \delta_{il}\phi_{,jk} + \delta_{jl}\phi_{,ik})\} \tag{11.2.9}$$

which determines the strains at any point **x**, produced by constant unit eigenstrains distributed in Ω. To obtain tensor field S^{∞}_{ijkl}, it is necessary to evaluate the second derivatives of $\phi(\mathbf{x})$ and the fourth derivatives of $\psi(\mathbf{x})$, with respect to the coordinate variables, x_1, x_2, and x_3. These derivatives may be conveniently expressed in terms of a set of integrals over Ω, often referred to as the I-integrals, as follows:

$$\phi_{,ij} = -\,\delta_{ij}\,I_i - x_i\,I_{ij}, \qquad \psi_{,ijkl} = \delta_{ij}\,(x_k\,J_{ik})_{,l} + (x_i\,x_j\,J_{ij})_{,kl}, \tag{11.2.10a,b}$$

where i, j, k, and l are *not* summed, and I_i and I_{ij} are

$$I_i = I_i(\lambda) \equiv 2\pi\,a_1\,a_2\,a_3 \int_{\lambda}^{\infty} (A_i + s)^{-1}\,\frac{ds}{\Delta(s)},$$

$$I_{ij} = I_{ij}(\lambda) \equiv 2\pi\, a_1\, a_2\, a_3 \int_{\lambda}^{\infty} (A_i + s)^{-1} (A_j + s)^{-1} \frac{ds}{\Delta(s)}, \tag{11.2.10c,d}$$

with $A_i = a_i^2$ and $\Delta(s) = \sqrt{(A_1 + s)(A_2 + s)(A_3 + s)}$; and J_{ij} is

$$J_{ij} = J_{ij}(\lambda) \equiv A_i\, I_{ij}(\lambda) - I_j(\lambda) \qquad \text{(j not summed)}. \tag{11.2.10e}$$

The argument λ in the above expressions is equal to zero when $\mathbf{x}$ is in Ω, and for $\mathbf{x}$ outside of Ω, it takes on the largest positive root of the following equation:

$$x_i\, (A_i + \lambda)^{-1} = \sum_{i=1}^{3} x_i^2\, (a_i^2 + \lambda)^{-1} = 1. \tag{11.2.10f}$$

Substitution of (11.2.10a~f) into (11.2.9) yields

$$\begin{aligned} 8\pi(1-\nu)\, S_{ijkl}^{\infty} &= 8\pi(1-\nu)\, S_{ijkl}^{\Omega} + 2\nu\, \delta_{kl}\, x_i\, I_{i,j} \\ &+ (1-\nu)\, \{\delta_{il}\, x_k\, I_{k,j} + \delta_{jl}\, x_k\, I_{k,i} + \delta_{ik}\, x_l\, I_{l,j} + \delta_{jk}\, x_l\, I_{l,i}\} \\ &+ \delta_{ij}\, x_k\, J_{ik,l} + (\delta_{ik}\, x_j + \delta_{jk}\, x_i)\, J_{ij,l} + (\delta_{il}\, x_j + \delta_{jl}\, x_i)\, J_{ij,k} + x_i\, x_j\, J_{ij,kl}, \end{aligned} \tag{11.2.11a}$$

where S_{ijkl}^{Ω} is

$$8\pi(1-\nu) S_{ijkl}^{\Omega} \equiv \delta_{ij}\, \delta_{kl}\, (2\nu I_i + J_{ik}) + (\delta_{ik}\, \delta_{jl} + \delta_{jk}\, \delta_{il})\, \{(1-\nu)(I_k + I_l) + J_{ij}\}, \tag{11.2.11b}$$

and repeated indices are *not* summed. This tensor is constant, and is completely defined in terms of the I-integrals.

As an illustration, assume Ω is a sphere of radius a. From $A_i = a^2$ (i = 1, 2, 3), the I-integrals, (11.2.10c,d), become

$$I_i(\lambda) = \frac{4\pi}{3}\, a^3\, (a^2 + \lambda)^{-3/2}, \qquad I_{ij}(\lambda) = \frac{4\pi}{5}\, a^3\, (a^2 + \lambda)^{-5/2}, \tag{11.2.12a,b}$$

and J_i reduces to

$$J_i(\lambda) = -\frac{4\pi}{15}\, a^3\, (a^2 + \lambda)^{-5/2}\, (2a^2 - 5\lambda), \tag{11.2.12c}$$

for i, j = 1, 2, 3. Note that $\lambda = 0$ for $|\mathbf{x}| \le a$, and $\lambda = |\mathbf{x}|^2 - a^2$ for $|\mathbf{x}| > a$. Hence, the derivatives of the I-integrals with respect to x_j are given by

$$I_{i,j}(\lambda) = \begin{cases} -4\pi\, a^3\, (a^2 + \lambda)^{-5/2}\, x_j, & \text{for } |\mathbf{x}| \le a \\ 0 & \text{for } |\mathbf{x}| > a \end{cases} \tag{11.2.12d}$$

and

$$I_{ij,k}(\lambda) = \begin{cases} -4\pi\, a^3\, (a^2 + \lambda)^{-7/2}\, x_k, & \text{for } |\mathbf{x}| \le a \\ 0 & \text{for } |\mathbf{x}| > a, \end{cases} \tag{11.2.12e}$$

for i, j, k = 1, 2, 3. Substitute (11.2.12a~e) into (11.2.11a), to compute S_{ijkl}^{∞} for this case explicitly. In particular, for S_{ijkl}^{Ω} given by (11.2.11b), obtain

$$S_{ijkl}^{\Omega} \equiv \delta_{ij}\, \delta_{kl}\, \frac{5\nu - 1}{15(1-\nu)} + (\delta_{ik}\, \delta_{jl} + \delta_{jk}\, \delta_{il})\, \frac{2(4 - 5\nu)}{15(1-\nu)}. \tag{11.2.12f}$$

Suppose $\mathbf{x}$ is inside Ω. The parameter λ is then zero[4] and all derivatives of the I-integrals vanish identically. The tensor field $S^{\infty}_{ijkl}(\mathbf{x};\,\Omega)$ equals $S^{\Omega}_{ijkl}(\mathbf{x})$ at $\lambda = 0$, and is uniform inside Ω. This is *Eshelby's tensor*. The uniform strain and stress fields in Ω then are

$$\varepsilon_{ij} = S^{\Omega}_{ijkl}\,\varepsilon^*_{kl}, \qquad \sigma_{ij} = C_{ijkl}\,(S^{\Omega}_{klmn} - 1^{(4s)}_{klmn})\,\varepsilon^*_{mn}. \tag{11.2.13a,b}$$

Therefore, the uniform eigenstrain ε^*_{ij} prescribed in an ellipsoidal domain Ω, produces uniform tractions on $\partial\Omega$,

$$t_j^+ = n_i\,C_{ijkl}\{\,S^{\Omega}_{klmn}(\Omega) - 1^{(4s)}_{klmn}\}\,\varepsilon^*_{mn}. \tag{11.2.13c}$$

11.2.3. Eshelby's Tensor for Anisotropic Media

When the unbounded linearly elastic homogeneous solid V^{∞} is anisotropic, a procedure similar to that for the isotropic case shows that if the prescribed eigenstrain is uniform in the ellipsoidal domain, then the strain and stress fields inside this ellipsoidal domain are also uniform. However, the derivation of the Eshelby tensor is more tedious compared with the isotropic case.

Excluding the mathematical details, consider the reason why the tensor field $S^{\infty}_{ijkl}(\mathbf{x};\,\Omega)$ is constant in Ω. In general, the Green function for the unbounded domain is expressed in terms of *polyharmonic potentials,* Φ^s, which are defined by

$$\Phi^s(\mathbf{x};\,\Omega) = -\,\frac{1}{4\pi(2s-2)!}\int_{\Omega} |\,\mathbf{x}-\mathbf{y}\,|^{2s-3}\,dV_y. \tag{11.2.14a}$$

Indeed, for the isotropic case, the functions ϕ and ψ are given by Φ^2 and Φ^1, respectively,

$$\psi = -\,8\pi\,\Phi^2, \qquad \phi = -\,4\pi\,\Phi^1. \tag{11.2.14b,c}$$

The two potentials Φ^1 and Φ^2 are called harmonic and biharmonic potentials.

Let ∇^2 be the Laplacian with respect to $\mathbf{x}$, i.e., $\nabla^2 \equiv \partial_i\partial_i$. Since $\partial_i\,|\,\mathbf{x}-\mathbf{y}\,| = (x_i - y_i)/\,|\,\mathbf{x}-\mathbf{y}\,|$, then

$$\nabla^2\,|\,\mathbf{x}-\mathbf{y}\,|^{2s-3} = (2s-2)(2s-3)\,|\,\mathbf{x}-\mathbf{y}\,|^{2s-5}. \tag{11.2.15a}$$

Therefore, the polyharmonic potentials satisfy $\nabla^2\Phi^s = \Phi^{s-1}$ for $s > 1$, and

$$\nabla^{2s}\Phi^s(\mathbf{x};\,\Omega) = \begin{cases} 1 & \text{for } \mathbf{x} \text{ in } \Omega \\ 0 & \text{otherwise.} \end{cases} \tag{11.2.15b}$$

This is because the harmonic potential $\Phi^1(\mathbf{x})$ is such that $\nabla^2\Phi^1(\mathbf{x})$ is 1 for $\mathbf{x}$ inside, and 0 for $\mathbf{x}$ outside Ω.

In general, the tensor field $S^{\infty}_{ijkl}(\mathbf{x};\,\Omega)$ is given by $\nabla^{2s}\Phi^s$'s, whether the unbounded solid is isotropic or not. Then, from (11.2.15b), $S^{\infty}_{ijkl}(\mathbf{x};\,\Omega)$ is constant for $\mathbf{x}$ in Ω. Walpole (1967) has given detailed results for the polyharmonic

[4] The values of the I-integrals at $\lambda = 0$ are explicitly given by the complete elliptic integrals.

potentials relating to the eigenstrain problem in the anisotropic case; he solves the corresponding integral equations for the eigenstrain field by successive approximations, using polyharmonic potentials. Other related references are:[5] Bhargava and Radhakrishna (1963, 1964) who present results for orthotropic media; Willis (1964) who considers cubic symmetry; Chen (1967) who examines a solid with one plane of symmetry; and Yang and Chen (1976) who discuss general anisotropic plane problems and give results for orthorhombic anisotropy. These and related results are summarized by Mura (1987) who also provides additional references.

11.3. SOME BASIC PROPERTIES OF ESHELBY'S TENSOR

11.3.1. Symmetry of the Eshelby Tensor

In the preceding subsections, the integral operator $S_{ij}^{\infty}(\mathbf{x};\, \varepsilon_{kl}^{*})$ and the tensor field $S_{ijkl}^{\infty}(\mathbf{x};\, W)$ are explicitly expressed in terms of the tensor field $\Gamma_{ijkl}^{\infty}(\mathbf{z})$ which is given by the gradient of the Green function $G_{ij}^{\infty}(\mathbf{z})$ for the unbounded domain V^{∞}; see (11.1.15a) and (11.2.3b). The integration of $\Gamma_{ijkl}^{\infty}(\mathbf{z})$ essentially determines these operators, and hence yields the Eshelby tensor S_{ijkl}^{Ω} when the domain W is an ellipsoid. Define the tensor field

$$P_{ijkl}(\mathbf{x};\, W) \equiv \int_W \Gamma_{ijkl}^{\infty}(\mathbf{y}-\mathbf{x})\, dV_y, \tag{11.3.1}$$

where W is an arbitrary finite domain (not necessarily ellipsoidal). *If* W *is ellipsoidal, it is denoted by* Ω, *and the corresponding Eshelby tensor, by* S_{ijkl}^{Ω}. Then, in terms of $P_{ijkl}(\mathbf{x};\, \Omega)$, this tensor becomes,

$$S_{ijkl}^{\Omega} = P_{ijmn}^{\Omega}\, C_{mnkl}, \tag{11.3.2a}$$

where

$$P_{ijkl}^{\Omega}(\mathbf{x};\, \Omega) \equiv P_{ijkl}^{\Omega} = \text{constant} \qquad \text{for } \mathbf{x} \text{ in } \Omega. \tag{11.3.2b}$$

The properties of the Green function, $G_{ij}^{\infty}(\mathbf{z})$, will now be used to study the tensor field $P_{ijkl}(\mathbf{x};\, W)$, and to obtain the symmetry properties of the Eshelby tensor S_{ijkl}^{Ω}. It should be noted that in this section, the infinite homogeneous domain V^{∞} is not assumed to be isotropic nor has it any special symmetries. *The following results are valid for a general anisotropic linearly elastic* V^{∞}.

From the symmetry properties (11.1.5) and (11.1.6) of the Green function, $G_{ij}^{\infty}(\mathbf{z}) = G_{ji}^{\infty}(-\mathbf{z})$ and $G_{ij}^{\infty}(\mathbf{z}) = G_{ji}^{\infty}(\mathbf{z})$, it follows that

[5] See also Stroh (1958, 1962), Barnett (1972), Barnett and Lothe (1973), Ting and Yan (1991), and Ting (1992). Lubarda and Markenscoff (1998) have shown that the constancy of Eshlbey's tensor, $S_{ijkl}^{\infty}(\mathbf{x};\, \Omega)$, is essentially limited to ellipsoidal Ω.

$$G^{\infty}_{ij,k}(\mathbf{z}) = -G^{\infty}_{ji,k}(-\mathbf{z}), \qquad G^{\infty}_{ij,kl}(\mathbf{z}) = G^{\infty}_{ji,kl}(-\mathbf{z}),$$
$$G^{\infty}_{ij,k}(\mathbf{z}) = G^{\infty}_{ji,k}(\mathbf{z}), \qquad G^{\infty}_{ij,kl}(\mathbf{z}) = G^{\infty}_{ji,kl}(\mathbf{z}). \tag{11.3.3a~d}$$

Therefore, in view of (11.1.15e), $\Gamma^{\infty}_{ijkl}(\mathbf{z})$ is symmetric with respect to the first and last pairs of its indices, i.e.,

$$\Gamma^{\infty}_{ijkl}(\mathbf{z}) = -\frac{1}{4}\{G^{\infty}_{ki,jl}(-\mathbf{z}) + G^{\infty}_{kj,il}(-\mathbf{z}) + G^{\infty}_{li,jk}(-\mathbf{z}) + G^{\infty}_{lj,ik}(-\mathbf{z})\}$$
$$= \Gamma^{\infty}_{klij}(-\mathbf{z}) = \Gamma^{\infty}_{klij}(\mathbf{z}). \tag{11.3.4}$$

Based on (11.3.4), it is now proved that $P_{ijkl}(\mathbf{x};\, W)$ is also symmetric with respect to the first and second pairs of its indices. From (11.3.1),

$$P_{ijkl}(\mathbf{x};\, W) = \int_{W-\mathbf{x}} \Gamma^{\infty}_{ijkl}(\mathbf{z})\, dV_z, \tag{11.3.5a}$$

where $\mathbf{z} = \mathbf{y} - \mathbf{x}$, and $W - \mathbf{x}$ denotes the rigid-body translation of region W by $\mathbf{x}$. Then, from (11.3.4), $P^{T}_{ijkl}(\mathbf{x};\, W) = P_{klij}(\mathbf{x};\, W)$ becomes

$$P_{klij}(\mathbf{x};\, W) = \int_{W-\mathbf{x}} \Gamma^{\infty}_{klij}(\mathbf{z})\, dV_z$$
$$= \int_{W-\mathbf{x}} \Gamma^{\infty}_{ijkl}(\mathbf{z})\, dV_z = P_{ijkl}(\mathbf{x};\, W). \tag{11.3.5b}$$

As is seen, $P_{ijkl}(\mathbf{x};\, W)$ is symmetric with respect to the first and second pairs of its indices, for any $\mathbf{x}$ in V^{∞}. Note that (11.3.5c) holds whether or not W is ellipsoidal.

For an ellipsoidal Ω, $P_{ijkl}(\mathbf{x};\, \Omega)$ is constant, denoted by $P_{ijkl}(\Omega)$, for $\mathbf{x}$ in Ω, (11.3.2b). From (11.3.5c),

$$P^{\Omega}_{ijkl} = P^{\Omega}_{klij}. \tag{11.3.6}$$

Therefore, the Eshelby tensor S^{Ω}_{ijkl} satisfies the following two symmetry properties:

$$S^{\Omega}_{ijmn}\, D_{mnkl} = S^{\Omega}_{klmn}\, D_{mnij} \;(= P_{ijkl}), \tag{11.3.7}$$

with $D_{ijkl} = C^{-1}_{ijkl}$ being the compliance tensor, and

$$C_{ijmn}\, S^{\Omega}_{mnkl} = C_{klmn}\, S^{\Omega}_{mnij} \;(= C_{ijmn}\, P^{\Omega}_{mnpq}\, C_{pqkl}). \tag{11.3.8}$$

From these properties of the Eshelby tensor S^{Ω}_{ijkl}, and its conjugate $T^{\Omega}_{ijkl} = 1^{(4s)}_{ijkl} - D_{ijmn}S^{\Omega}_{mnpq}C_{pqkl}$, it follows that the overall elasticity and compliance tensors, $\bar{C}_{ijkl}$ and $\bar{D}_{ijkl}$, estimated by the dilute assumption, are always symmetric; see Subsection 7.6.

11.3.2. Conjugate Eshelby Tensor

Using the equivalence relations (11.1.18a,b), define the tensor field $T^{\infty}_{ijkl}(\mathbf{x};\, W)$ for a finite domain W, as follows:

$$T^{\infty}_{ijkl}(\mathbf{x};\, W) \equiv -C_{ijmn}\int_W \Gamma^{\infty}_{mnkl}(\mathbf{y}-\mathbf{x})\, dV_\mathbf{y} + H(\mathbf{x};\, W)\, 1^{(4s)}_{ijkl} \tag{11.3.9a}$$

which, in terms of $P_{ijkl}(\mathbf{x};\, W)$, becomes

$$T^{\infty}_{ijkl}(\mathbf{x};\, W) = -C_{ijmn}\, P_{mnkl}(\mathbf{x};\, W) + H(\mathbf{x};\, W)\, 1^{(4s)}_{ijkl}. \tag{11.3.9b}$$

When W is an ellipsoid, Ω, $T^{\infty}_{ijkl}(\mathbf{x};\, \Omega)$ is constant, denoted by T^{Ω}_{ijkl}, for $\mathbf{x}$ in Ω. Indeed, from (11.3.2b),

$$T^{\Omega}_{ijkl}(\Omega) = -C_{ijmn}\, P^{\Omega}_{mnkl}(\Omega) + 1^{(4s)}_{ijkl}. \tag{11.3.10}$$

The tensor T^{Ω}_{ijkl} is *conjugate* to the Eshelby tensor S^{Ω}_{ijkl}. The equivalence relations between S^{Ω}_{ijkl} and T^{Ω}_{ijkl} are

$$S^{\Omega}_{ijkl} + D_{ijkl}\, T^{\Omega}_{klmn}\, C_{mnkl} = 1^{(4s)}_{ijkl},$$

$$T^{\Omega}_{ijkl} + C_{ijkl}\, S^{\Omega}_{klmn}\, D_{mnkl} = 1^{(4s)}_{ijkl}. \tag{11.3.11a,b}$$

From the symmetry of P^{Ω}_{ijkl}, the symmetry of T^{Ω}_{ijkl} follows,

$$D_{ijmn}\, T^{\Omega}_{mnkl}(\Omega) = D_{klmn}\, T^{\Omega}_{mnij}(\Omega),$$

$$T^{\Omega}_{ijmn}(\Omega)\, C_{mnkl} = T^{\Omega}_{klmn}(\Omega)\, C_{mnij}. \tag{11.3.12a,b}$$

11.3.3. Evaluation of Average Quantities

An interesting application of Eshelby's fundamental results is the *Tanaka-Mori observation* (Tanaka and Mori, 1972):

> *Suppose that an eigenstrain is uniformly distributed in a certain finite domain* W *of arbitrary shape. Let* Ω_1 *and* Ω_2 *be two arbitrary ellipsoidal domains such that* W *is totally contained within* Ω_1, *and* Ω_1 *is totally contained within* Ω_2; see Figure 11.3.1. *Then, the average strain in subregion* $\Omega_2 - \Omega_1$ *is completely determined by Eshelby's tensors corresponding to* Ω_1 *and* Ω_2.

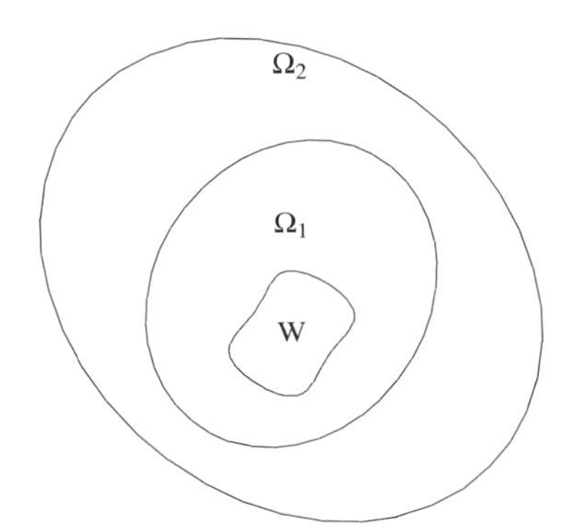

Figure 11.3.1

Tanaka-Mori result: ellipsoidal Ω_1 and Ω_2 and finite domain W

Consider the proof of the Tanaka-Mori result, using the tensor field $P_{ijkl}(\mathbf{x}; W)$.

For a uniform eigenstrain ε^*_{ij} in the finite domain W within an unbounded uniform region V^∞, the strain field is

$$\varepsilon_{ij}(\mathbf{x}) = P_{ijkl}(\mathbf{x}; W)\, C_{klmn}\, \varepsilon^*_{mn} \qquad \mathbf{x} \text{ in } V^\infty. \tag{11.3.13a}$$

Therefore, the average strain in the region between two ellipsoidal domains, Ω_1 and Ω_2, becomes

$$\bar{\varepsilon}^{\Omega_2-\Omega_1}_{ij} \equiv \frac{1}{\Omega_2-\Omega_1}\, \{\int_{\Omega_2-\Omega_1} P_{ijkl}(\mathbf{x}; W)\, dV_{\mathbf{x}}\}\, C_{klmn}\, \varepsilon^*_{mn}$$

$$= \frac{1}{\Omega_2-\Omega_1}\, \{\int_{\Omega_2-\Omega_1} \{\int_W \Gamma^\infty_{ijkl}(\mathbf{y}-\mathbf{x})\, dV_{\mathbf{y}}\}\, dV_{\mathbf{x}}\}\, C_{klmn}\, \varepsilon^*_{mn}, \tag{11.3.13b}$$

where $\mathbf{x}$ and $\mathbf{y}$ are in $\Omega_2-\Omega_1$ and W, respectively.

Since the domains W and $\Omega_2-\Omega_1$ do not intersect, the integrand $\Gamma^\infty_{ijkl}(\mathbf{x}-\mathbf{y})$ in (11.3.13b) is not singular. Hence, the order of integration can be changed,

$$\int_{\Omega_2-\Omega_1} \{\int_W \Gamma^\infty_{ijkl}(\mathbf{y}-\mathbf{x})\, dV_{\mathbf{y}}\}\, dV_{\mathbf{x}} = \int_W \{\int_{\Omega_2-\Omega_1} \Gamma^\infty_{ijkl}(\mathbf{y}-\mathbf{x})\, dV_{\mathbf{x}}\}\, dV_{\mathbf{y}}. \tag{11.3.14a}$$

As shown in (11.3.6), the first integral in the right-hand side is given by

$$\int_{\Omega_2-\Omega_1} \Gamma^\infty_{ijkl}(\mathbf{y}-\mathbf{x})\, dV_{\mathbf{x}} = P^\Omega_{ijkl}(\Omega_2) - P^\Omega_{ijkl}(\Omega_1) \tag{11.3.14b}$$

which does not depend on $\mathbf{y}$, if $\mathbf{y}$ is in $W \subset \Omega_1 \subset \Omega_2$.

Therefore, from (11.3.2), (11.3.13), and (11.3.14), the average strain $\bar{\varepsilon}^{\Omega_2-\Omega_1}_{ij}$ is given in terms of the Eshelby tensor by

$$\bar{\varepsilon}^{\Omega_2-\Omega_1}_{ij} = \frac{W}{\Omega_2-\Omega_1}\, \{S^\Omega_{ijkl}(\Omega_2) - S^\Omega_{ijkl}(\Omega_1)\}\, \varepsilon^*_{kl}, \tag{11.3.15}$$

where $S^\Omega_{ijkl}(\Omega_1)$ and $S^\Omega_{ijkl}(\Omega_2)$ are the Eshelby tensors corresponding to the ellipsoidal domains Ω_1 and Ω_2, respectively. Equation (11.3.15) expresses the Tanaka-Mori result. The average stress in $\Omega_2-\Omega_1$ is given by $C_{ijkl}\, \bar{\varepsilon}^{\Omega_2-\Omega_1}_{ij}$, since $\sigma_{ij} = C_{ijkl}\, \varepsilon_{kl}$ outside W. Next, examine two consequences of the Tanaka-Mori result.

As is shown in Subsection 11.2.2 for the isotropic case, the Eshelby tensor depends on both the shape and the orientation of the ellipsoid, but only the ratios of the semi-axes enter the components of this tensor in a coordinate system coincident with the directions of the principal semi-axes of the ellipsoid. Therefore, if the two ellipsoidal regions Ω_1 and Ω_2 have the same shape and orientation, i.e., if the corresponding semi-axes have common ratios and directions, then

$$S^\Omega_{ijkl}(\Omega_2) = S^\Omega_{ijkl}(\Omega_1), \tag{11.3.16a}$$

and the average strain (and, hence, the average stress) in $\Omega_2-\Omega_1$ vanishes,

$$\bar{\varepsilon}_{ij}^{\Omega_2-\Omega_1} = 0. \tag{11.3.16b}$$

This result holds for any W with any arbitrary shape.

When the eigenstrain ε^*_{ij} is distributed in an ellipsoidal domain, say, Ω, let Ω_1 coincide with Ω; see Figure 11.3.2. Then, from the Tanaka-Mori result, the average strain in any ellipsoidal region Ω' which includes Ω $(\Omega \subset \Omega')$ is given by

$$\bar{\varepsilon}_{ij}^{\Omega'} = \frac{\Omega}{\Omega'}\, S_{ijkl}^{\Omega}(\Omega')\, \varepsilon^*_{kl}. \tag{11.3.17a}$$

As is seen, (11.3.17a) does not depend on the shape or the orientation of Ω. Indeed, Ω need not be ellipsoidal. The average stress is determined by

$$\bar{\sigma}_{ij}^{\Omega'} = \frac{\Omega}{\Omega'}\, C_{ijmn}\, \{S_{mnkl}^{\Omega}(\Omega') - 1_{mnkl}^{(4s)}\}\, \varepsilon^*_{kl}, \tag{11.3.17b}$$

since $\sigma_{ij} = C_{ijkl}\,(\varepsilon_{kl} - \varepsilon^*_{kl})$ in Ω.

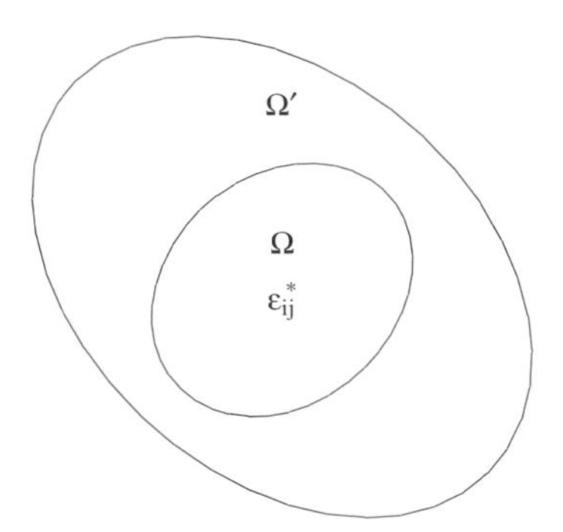

Figure 11.3.2

Ellipsoidal domains Ω and Ω', where eigenstrains ε^*_{ij} are distributed in Ω

The Tanaka-Mori result can be further generalized to include *nonuniform distributions of eigenstrains in the region,* W, *of any arbitrary shape.* Suppose $\boldsymbol{\varepsilon}^* = \boldsymbol{\varepsilon}^*(\mathbf{x})$ is integrable on W, having the average $\bar{\boldsymbol{\varepsilon}}^* = <\boldsymbol{\varepsilon}^*>_W$. Then (11.3.13b) becomes

$$\bar{\varepsilon}_{ij}^{\Omega_2-\Omega_1} \equiv \frac{1}{\Omega_2-\Omega_1}\, \{\int_{\Omega_2-\Omega_1} \{\int_W \Gamma_{ijkl}(\mathbf{y}-\mathbf{x})\, C_{klmn}\, \varepsilon^*_{mn}(\mathbf{y})\}\, dV_{\mathbf{y}}\}\, dV_{\mathbf{x}}$$

$$= \frac{1}{\Omega_2-\Omega_1} \int_W \{\int_{\Omega_2-\Omega_1} \Gamma^{\infty}_{ijkl}(\mathbf{y}-\mathbf{x})\, dV_{\mathbf{x}}\}\, C_{klmn}\, \varepsilon^*_{mn}(\mathbf{y})\, dV_{\mathbf{y}}$$

$$= \frac{W}{\Omega_2-\Omega_1}\, \{S_{ijkl}^{\Omega}(\Omega_2) - S_{ijkl}^{\Omega}(\Omega_1)\} < \varepsilon^*_{kl} >_W. \tag{11.3.18a}$$

Thus the average strain over $\Omega_2 - \Omega_1$ vanishes: (1) when Ω_1 and Ω_2 are similar and coaxial ellipsoids, (in this case, the distribution of $\boldsymbol{\varepsilon}^*(\mathbf{x})$ in W may be arbitrary); and (2) when $\boldsymbol{\varepsilon}^*(\mathbf{x})$ has a zero average over W (in this case, Ω_1 and Ω_2

need not be similar or coaxial, but both must be ellipsoids). Expression (11.3.18a) remains valid, as Ω_1 coincides with W. *Then the average strain in any ellipsoidal region Ω' which includes an arbitrary domain,* W, *in which variable eigenstrains $\boldsymbol{\varepsilon}^*(\mathbf{x})$ are distributed, is given by*

$$\bar{\varepsilon}_{ij}^{\Omega'} = \frac{W}{\Omega'} < \varepsilon_{kl}^{*} >_W. \tag{11.3.18b}$$

Then the corresponding average stress is

$$\bar{\sigma}_{ij}^{\Omega'} = \frac{W}{\Omega'} D_{ijmn} \{S_{mnke}^{\Omega}(\Omega') - 1_{mnkl}^{(4s)}\} < \varepsilon_{kl}^{*} >_W. \tag{11.3.18c}$$

Thus the average strain and stress induced in an ellipsoidal region Ω' by arbitrary eigenstrains $\boldsymbol{\varepsilon}^*(\mathbf{x})$ distributed in a region $W \subset \Omega'$ of arbitrary shape, are always zero if the eigenstrains have zero average over W. All these and all other results presented in this section are *exact*.

11.4. RELATIONS AMONG AVERAGE QUANTITIES

As shown in Sections 7 and 9, a heterogeneous solid can be homogenized by distributing suitable eigenstrains or eigenstresses within its inhomogeneities. The stress and strain fields produced in the homogenized solid by external loads in the presence of such eigenstrains or eigenstresses then coincide with the corresponding actual fields. Moreover, the estimate of the overall moduli of a heterogeneous RVE depends only on the average field quantities within its inhomogeneities. In this subsection, these average field quantities are expressed in terms of the corresponding eigenstrains or eigenstresses of the homogenized solid.

11.4.1. General Relations

Choose a certain finite region V in an unbounded homogeneous domain V^∞, and consider relations among various average field quantities in V. It is assumed that the eigenstrain $\boldsymbol{\varepsilon}^*$ or eigenstress $\boldsymbol{\sigma}^*$ may be distributed in V, in order to produce arbitrary displacements or tractions on ∂V. Since a jump in the eigenstrains (or eigenstresses) across ∂V is equivalent to a layer of body forces applied there, this body force layer (or the corresponding surface displacements) on ∂V may be chosen such that the strain and stress fields due to the eigenstrains and eigenstresses coincide with those of a heterogeneous bounded RVE with prescribed boundary conditions.

The stress and strain fields produced by the eigenstrain $\boldsymbol{\varepsilon}^*$ and eigenstress $\boldsymbol{\sigma}^*$ are now examined at the same time, but as two separate problems for the same region V. Following the notation of Section 7.3, let $\boldsymbol{\varepsilon}^d$ and $\boldsymbol{\sigma}^d$ be the corresponding (disturbance) strain and stress fields. These fields satisfy the strain-displacement relations and the equations of equilibrium,

$$\boldsymbol{\varepsilon}^{\mathrm{d}} = \frac{1}{2}\{(\boldsymbol{\nabla}\otimes\mathbf{u}^{\mathrm{d}}) + (\boldsymbol{\nabla}\otimes\mathbf{u}^{\mathrm{d}})^{\mathrm{T}}\}, \qquad \boldsymbol{\nabla}\boldsymbol{\cdot}\boldsymbol{\sigma}^{\mathrm{d}} = \mathbf{0}, \tag{11.4.1a,b}$$

where $\mathbf{u}^{\mathrm{d}}$ is the displacement field. Hence, $\boldsymbol{\varepsilon}^{\mathrm{d}}$ and $\boldsymbol{\sigma}^{\mathrm{d}}$ are kinematically and statically admissible fields, related by

$$\boldsymbol{\sigma}^{\mathrm{d}} = \begin{cases} \mathbf{C} : (\boldsymbol{\varepsilon}^{\mathrm{d}} - \boldsymbol{\varepsilon}^{*}) \\ \mathbf{C} : \boldsymbol{\varepsilon}^{\mathrm{d}} + \boldsymbol{\sigma}^{*}, \end{cases} \qquad \boldsymbol{\varepsilon}^{\mathrm{d}} = \begin{cases} \mathbf{D} : \boldsymbol{\sigma}^{\mathrm{d}} + \boldsymbol{\varepsilon}^{*} \\ \mathbf{D} : (\boldsymbol{\sigma}^{\mathrm{d}} - \boldsymbol{\sigma}^{*}) \end{cases} \tag{11.4.2a,b}$$

for a prescribed eigenstrain $\boldsymbol{\varepsilon}^{*}$, or a prescribed eigenstress $\boldsymbol{\sigma}^{*}$, respectively. Note that either (but not both) $\boldsymbol{\varepsilon}^{*}$ or $\boldsymbol{\sigma}^{*}$ is prescribed. These are two separate problems which are being examined simultaneously. If $\boldsymbol{\sigma}^{*} = -\mathbf{C} : \boldsymbol{\varepsilon}^{*}$, then the two problems are identical, otherwise, they are two separate problems.

From the kinematical admissibility of the strain field and the statical admissibility of the stress field, the average strain and stress are given by the surface data on $\partial\mathrm{V}$,

$$< \boldsymbol{\varepsilon}^{\mathrm{d}} > = \frac{1}{\mathrm{V}} \int_{\partial \mathrm{V}} \frac{1}{2}(\mathbf{v}\otimes\mathbf{u}^{\mathrm{d}} + \mathbf{u}^{\mathrm{d}}\otimes\mathbf{v})\, \mathrm{dS},$$

$$< \boldsymbol{\sigma}^{\mathrm{d}} > = \frac{1}{\mathrm{V}} \int_{\partial \mathrm{V}} (\mathbf{v}\boldsymbol{\cdot}\boldsymbol{\sigma}^{\mathrm{d}})\otimes\mathbf{x}\, \mathrm{dS}, \tag{11.4.3a,b}$$

where the displacements and tractions are assumed to be continuous everywhere in V. The continuity of displacements and tractions does not imply the continuity of the corresponding strain and stress fields. Indeed, if the eigenstrains and eigenstresses are discontinuous across some surfaces, the strain or stress fields are also discontinuous there. Taking the volume average of (11.4.2a,b), note the following relations for the above defined average quantities:

$$< \boldsymbol{\sigma}^{\mathrm{d}} > = \begin{cases} \mathbf{C} : (< \boldsymbol{\varepsilon}^{\mathrm{d}} > - < \boldsymbol{\varepsilon}^{*} >) \\ \mathbf{C} : < \boldsymbol{\varepsilon}^{\mathrm{d}} > + < \boldsymbol{\sigma}^{*} >, \end{cases}$$

$$< \boldsymbol{\varepsilon}^{\mathrm{d}} > = \begin{cases} \mathbf{D} : < \boldsymbol{\sigma}^{\mathrm{d}} > + < \boldsymbol{\varepsilon}^{*} > \\ \mathbf{D} : (< \boldsymbol{\sigma}^{\mathrm{d}} > - < \boldsymbol{\sigma}^{*} >). \end{cases} \tag{11.4.4a,b}$$

These relations are *exact.* They are valid for a finite V with arbitrary shape, and for any variable (admissible) eigenstrain or eigenstress field prescribed on any region within V. These relations always hold whether or not the displacements or tractions are discontinuous. When the boundary $\partial\mathrm{V}$ of V is traction-free, it follows from (11.4.3b) and (11.4.4a) that $< \boldsymbol{\varepsilon}^{\mathrm{d}} > = < \boldsymbol{\varepsilon}^{*} >$, and when the boundary displacements on $\partial\mathrm{V}$ vanish, (11.4.3a) and (11.4.4b) show that $< \boldsymbol{\sigma}^{\mathrm{d}} > = < \boldsymbol{\sigma}^{*} >$. Again, these results are not restricted to any specific geometry for V or any specific distribution of eigenstrains and eigenstresses, nor are they restricted to any specific shape of the regions within V, where these eigenstrains and eigenstresses are distributed.

Next consider the volume average of the strain energy density produced by prescribed eigenstrains or eigenstresses. *Define the stress-work* W^{*} by

$$\mathrm{W}^{*} \equiv \frac{1}{2} < \boldsymbol{\sigma}^{\mathrm{d}} : \boldsymbol{\varepsilon}^{\mathrm{d}} >. \tag{11.4.5a}$$

Again, from kinematical and statical admissibility of the strain and stress fields, it follows that

$$W^* = \frac{1}{2} < \boldsymbol{\sigma}^d > : < \boldsymbol{\varepsilon}^d > + \text{(B.C.)}, \tag{11.4.5b}$$

where (B.C.) stands for the "boundary contributions", given by

$$\text{(B.C.)} \equiv \frac{1}{2V} \int_{\partial V} \{\mathbf{v} \cdot (\boldsymbol{\sigma}^d - < \boldsymbol{\sigma}^d >)\} \cdot \{\mathbf{u}^d - \mathbf{x} \cdot < \boldsymbol{\varepsilon}^d >\} \, dS. \tag{11.4.5c}$$

As in (11.4.3a,b), the continuity of displacements and tractions is also assumed.

From the constitutive relations (11.4.2a,b), define the *elastic strain,*

$$\boldsymbol{\varepsilon}^{de} \equiv \mathbf{D} : \boldsymbol{\sigma}^d \equiv \begin{cases} \boldsymbol{\varepsilon}^d - \boldsymbol{\varepsilon}^*, \\ \boldsymbol{\varepsilon}^d + \mathbf{D} : \boldsymbol{\sigma}^*, \end{cases} \tag{11.4.6}$$

such that $\mathbf{C} : \boldsymbol{\varepsilon}^{de}$ is the resulting stress $\boldsymbol{\sigma}^d$. Then, the average *elastic strain energy,* E^*, produced by the eigenstrains or eigenstresses is

$$E^* \equiv \frac{1}{2} < \boldsymbol{\sigma}^d : \boldsymbol{\varepsilon}^{de} >. \tag{11.4.7a}$$

The total strain $\boldsymbol{\varepsilon}^d$ is the sum of the elastic strain $\boldsymbol{\varepsilon}^{de}$ and the inelastic strain $(\boldsymbol{\varepsilon}^d - \boldsymbol{\varepsilon}^{de})$. Using (11.4.2a,b) and (11.4.5a), express the average elastic strain energy E^*, as

$$2E^* = \begin{cases} 2W^* - < \boldsymbol{\sigma}^d : \boldsymbol{\varepsilon}^* > \\ 2W^* + < \boldsymbol{\sigma}^d : \mathbf{D} : \boldsymbol{\sigma}^* >. \end{cases} \tag{11.4.7b}$$

11.4.2. Superposition of Uniform Strain and Stress Fields

Consider a uniform V^∞, within which eigenstrain $\boldsymbol{\varepsilon}^*$ (eigenstress $\boldsymbol{\sigma}^*$) is prescribed in a subregion V. Let the farfield strain and stress be $\boldsymbol{\varepsilon}^\infty$ and $\boldsymbol{\sigma}^\infty$. Then, it follows that

$$\boldsymbol{\varepsilon}^\infty \equiv \lim_{|\mathbf{x}| \to \infty} \boldsymbol{\varepsilon}(\mathbf{x}), \qquad \boldsymbol{\sigma}^\infty \equiv \lim_{|\mathbf{x}| \to \infty} \boldsymbol{\sigma}(\mathbf{x}). \tag{11.4.8a,b}$$

The limiting procedure does not depend on the direction of $\mathbf{x}$. The farfield strain and stress are related by

$$\boldsymbol{\sigma}^\infty = \mathbf{C} : \boldsymbol{\varepsilon}^\infty, \qquad \boldsymbol{\varepsilon}^\infty = \mathbf{D} : \boldsymbol{\sigma}^\infty, \tag{11.4.8c,d}$$

since $\boldsymbol{\varepsilon}^*$ and $\boldsymbol{\sigma}^*$ are prescribed on a finite region V within V^∞.

If the farfield strain or stress is prescribed to be $\boldsymbol{\varepsilon}^\infty = \boldsymbol{\varepsilon}^o$ or $\boldsymbol{\sigma}^\infty = \boldsymbol{\sigma}^o$ $(\boldsymbol{\sigma}^o = \mathbf{C} : \boldsymbol{\varepsilon}^o$ or $\boldsymbol{\varepsilon}^o = \mathbf{D} : \boldsymbol{\sigma}^o)$, in the absence of eigenstrains or eigenstresses, the resulting strain and stress fields are constant and given by $\boldsymbol{\sigma}^o$ and $\boldsymbol{\varepsilon}^o$ everywhere in a homogeneous V^∞. Superimpose these uniform fields and the disturbance fields, $\boldsymbol{\sigma}^d$ and $\boldsymbol{\varepsilon}^d$, which are produced by the eigenstrains or eigenstresses. Since $\boldsymbol{\sigma}^d$ and $\boldsymbol{\varepsilon}^d$ vanish at infinity, the prescribed farfield boundary conditions are unchanged by this superposition.

Denoting the superposed strain and stress fields by $\boldsymbol{\sigma}$ and $\boldsymbol{\varepsilon}$ (as in Section 7), define these fields in terms of the uniform and disturbance fields, as follows:

$$\boldsymbol{\varepsilon} \equiv \boldsymbol{\varepsilon}^o + \boldsymbol{\varepsilon}^d, \qquad \boldsymbol{\sigma} \equiv \boldsymbol{\sigma}^o + \boldsymbol{\sigma}^d. \tag{11.4.9a,b}$$

Then, the exact relations (11.4.4a,b) become

$$< \boldsymbol{\sigma} > = \boldsymbol{\sigma}^o + < \boldsymbol{\sigma}^d > = \begin{cases} \mathbf{C} : (\boldsymbol{\varepsilon}^o + < \boldsymbol{\varepsilon}^d > - < \boldsymbol{\varepsilon}^* >) \\ \mathbf{C} : (\boldsymbol{\varepsilon}^o + < \boldsymbol{\varepsilon}^d >) + < \boldsymbol{\sigma}^* >, \end{cases}$$

$$< \boldsymbol{\varepsilon} > = \boldsymbol{\varepsilon}^o + < \boldsymbol{\varepsilon}^d > = \begin{cases} \mathbf{D} : (\boldsymbol{\sigma}^o + < \boldsymbol{\sigma}^d >) - < \boldsymbol{\varepsilon}^* > \\ \mathbf{D} : (\boldsymbol{\sigma}^o + < \boldsymbol{\sigma}^d > - < \boldsymbol{\sigma}^* >). \end{cases} \tag{11.4.10a,b}$$

Now compute the change in the stress-work due to the existence of eigenstrains or eigenstresses. Since the elastic strain energy in the absence of any eigenstrain or eigenstress is given by $\boldsymbol{\sigma}^o : \boldsymbol{\varepsilon}^o/2$, define this change in the stress work by

$$\Delta W^* \equiv \frac{1}{2} < \boldsymbol{\sigma} : \boldsymbol{\varepsilon} > - \frac{1}{2} \boldsymbol{\sigma}^o : \boldsymbol{\varepsilon}^o, \tag{11.4.11a}$$

and from (11.4.9a,b), rewrite (11.4.11a) as

$$\Delta W^* = \frac{1}{2} (\boldsymbol{\sigma}^o : < \boldsymbol{\varepsilon}^d > + \boldsymbol{\varepsilon}^o : < \boldsymbol{\sigma}^d >) + W^*, \tag{11.4.11b}$$

where W^* is given by (11.4.5a). In a similar manner, for the average elastic strain energy, consider the change due to the existence of eigenstrains or eigenstresses, defined by

$$\Delta E^* \equiv \frac{1}{2} < \boldsymbol{\sigma} : \boldsymbol{\varepsilon}^e > - \frac{1}{2} \boldsymbol{\sigma}^o : \boldsymbol{\varepsilon}^o, \tag{11.4.12}$$

where $\boldsymbol{\varepsilon}^e$ is the elastic part of the total strain, $\boldsymbol{\varepsilon}^e = \boldsymbol{\varepsilon}^o + \boldsymbol{\varepsilon}^{de}$. From (11.4.8a,b), (11.4.12) becomes

$$2\Delta E^* = \begin{cases} 2\Delta W^* - \boldsymbol{\sigma}^o : < \boldsymbol{\varepsilon}^* > - < \boldsymbol{\sigma}^d : \boldsymbol{\varepsilon}^* > \\ 2\Delta W^* + < \boldsymbol{\sigma}^* > : \boldsymbol{\varepsilon}^o + < \boldsymbol{\sigma}^d : \mathbf{D} : \boldsymbol{\sigma}^* > \end{cases} \tag{11.4.13a}$$

or in terms of W^* and E^*, from (11.4.7b),

$$2\Delta E^* = \begin{cases} 2\Delta W^* - \boldsymbol{\sigma}^o : < \boldsymbol{\varepsilon}^* > + 2E^* - 2W^* \\ 2\Delta W^* + < \boldsymbol{\sigma}^* > : \boldsymbol{\varepsilon}^o + 2E^* - 2W^*. \end{cases} \tag{11.4.13b}$$

11.4.3. Prescribed Boundary Conditions

Using the general results obtained in Subsections 11.4.1 and 11.4.2, consider relations for the average quantities associated with particular boundary conditions prescribed on ∂V, i.e., uniform surface tractions or linear surface displacements. In either case, the tractions or displacements on ∂V are determined by certain prescribed constant symmetric tensors, $\boldsymbol{\sigma}^o$ or $\boldsymbol{\varepsilon}^o$. The uniform tractions are prescribed by

$$\boldsymbol{\nu} \cdot \boldsymbol{\sigma} = \boldsymbol{\nu} \cdot \boldsymbol{\sigma}^o \quad \text{on } \partial V, \tag{11.4.14a}$$

and the linear displacements are prescribed by

$$\mathbf{u} = \mathbf{x} \cdot \boldsymbol{\varepsilon}^{\circ} \qquad \text{on } \partial V. \tag{11.4.15a}$$

Note that, in general, $\boldsymbol{\sigma}^{\circ}$ and $\boldsymbol{\varepsilon}^{\circ}$ are not related; the boundary conditions (11.4.14a) and (11.4.15a) are mutually exclusive. However, one can always define $\boldsymbol{\varepsilon}^{\circ} \equiv \mathbf{D} : \boldsymbol{\sigma}^{\circ}$ in the first case, and $\boldsymbol{\sigma}^{\circ} \equiv \mathbf{C} : \boldsymbol{\varepsilon}^{\circ}$ in the second-case, that is, $\boldsymbol{\varepsilon}^{\circ}$ can be *defined* to stand for $\mathbf{D} : \boldsymbol{\sigma}^{\circ}$ if $\boldsymbol{\sigma}^{\circ}$ is *prescribed*, and $\boldsymbol{\sigma}^{\circ}$ can be *defined* to represent $\mathbf{C} : \boldsymbol{\varepsilon}^{\circ}$ if $\boldsymbol{\varepsilon}^{\circ}$ is *prescribed*.

Since the uniform strain and stress fields satisfy the above boundary conditions automatically, the disturbance strain and stress fields, $\boldsymbol{\sigma}^{d}$ and $\boldsymbol{\varepsilon}^{d}$, produced by the eigenstrains and eigenstresses, satisfy either zero tractions or zero displacements, depending on whether (11.4.14a) or (11.4.15a) is prescribed. Under (11.4.14a),

$$\mathbf{v} \cdot \boldsymbol{\sigma}^{d} = \mathbf{0} \qquad \text{on } \partial V \qquad (\mathbf{u}^{d} \neq \mathbf{0} \text{ on } \partial V), \tag{11.4.14b}$$

and under (11.4.15a),

$$\mathbf{u}^{d} = \mathbf{0} \qquad \text{on } \partial V \qquad (\mathbf{v} \cdot \boldsymbol{\sigma}^{d} \neq \mathbf{0} \text{ on } \partial V). \tag{11.4.15b}$$

The (average) constitutive relations (11.4.3a,b), together with (11.4.14b), now yield

$$< \boldsymbol{\sigma}^{d} > = \mathbf{0}, \qquad < \boldsymbol{\varepsilon}^{d} > = \begin{cases} < \boldsymbol{\varepsilon}^{*} > \\ -\mathbf{D} : < \boldsymbol{\sigma}^{*} >, \end{cases} \tag{11.4.16a,b}$$

whereas (11.4.3a,b) and (11.4.15b) give

$$< \boldsymbol{\varepsilon}^{d} > = \mathbf{0}, \qquad < \boldsymbol{\sigma}^{d} > = \begin{cases} -\mathbf{C} : < \boldsymbol{\varepsilon}^{*} > \\ < \boldsymbol{\sigma}^{*} >. \end{cases} \tag{11.4.17a,b}$$

In either case, the stress-work W^{*} is zero. Hence, from (11.4.7b), the elastic strain energy E^{*} is given by

$$2E^{*} = \begin{cases} -< \boldsymbol{\sigma}^{d} : \boldsymbol{\varepsilon}^{*} > \\ < \boldsymbol{\sigma}^{d} : \mathbf{D} : \boldsymbol{\sigma}^{*} >. \end{cases} \tag{11.4.18}$$

From (11.4.16), (11.4.17), and (11.4.18), the average total strain and stress are: when uniform boundary tractions are prescribed by (11.4.14a),

$$< \boldsymbol{\sigma} > = \boldsymbol{\sigma}^{\circ}, \qquad < \boldsymbol{\varepsilon} > = \begin{cases} \boldsymbol{\varepsilon}^{\circ} + < \boldsymbol{\varepsilon}^{*} > \\ \boldsymbol{\varepsilon}^{\circ} - \mathbf{D} : < \boldsymbol{\sigma}^{*} >; \end{cases} \tag{11.4.19a,b}$$

and when linear boundary displacements (11.4.15a) are given,

$$< \boldsymbol{\varepsilon} > = \boldsymbol{\varepsilon}^{\circ}, \qquad < \boldsymbol{\sigma} > = \begin{cases} \boldsymbol{\sigma}^{\circ} + < \boldsymbol{\sigma}^{*} > \\ \boldsymbol{\sigma}^{\circ} - \mathbf{C} : < \boldsymbol{\varepsilon}^{*} >. \end{cases} \tag{11.4.20a,b}$$

Under either boundary conditions, from (11.4.11b) and (11.4.13b), the change in the stress-work, ΔW^{*}, and that of the elastic strain energy, ΔE^{*}, are: when uniform boundary tractions are prescribed,

$$2\Delta W^{*} = \begin{cases} \boldsymbol{\sigma}^{\circ} : < \boldsymbol{\varepsilon}^{*} > \\ -< \boldsymbol{\sigma}^{*} > : \boldsymbol{\varepsilon}^{\circ}, \end{cases} \qquad 2\Delta E^{*} = 2E^{*} = \begin{cases} -< \boldsymbol{\sigma}^{d} : \boldsymbol{\varepsilon}^{*} > \\ < \boldsymbol{\sigma}^{d} : \mathbf{D} : \boldsymbol{\sigma}^{*} >; \end{cases} \tag{11.4.21a,b}$$

and when linear boundary displacements are prescribed,

$$2\Delta W^* = \begin{cases} -\boldsymbol{\sigma}^o : <\boldsymbol{\varepsilon}^*> \\ <\boldsymbol{\sigma}^*> : \boldsymbol{\varepsilon}^o, \end{cases} \qquad 2\Delta E^* = 2E^* = \begin{cases} -<\boldsymbol{\sigma}^d : \boldsymbol{\varepsilon}^*> \\ <\boldsymbol{\sigma}^d : \mathbf{D} : \boldsymbol{\sigma}^*>. \end{cases} \tag{11.4.22a,b}$$

When (11.4.21) and (11.4.22) describe the same problem, so that $\boldsymbol{\sigma}^o$ and $\boldsymbol{\varepsilon}^o$ are related through $\boldsymbol{\sigma}^o = \mathbf{C} : \boldsymbol{\varepsilon}^o$ or $\boldsymbol{\varepsilon}^o = \mathbf{D} : \boldsymbol{\sigma}^o$, it is seen that in (11.4.21a) and (11.4.22a), the absolute value of ΔW^* is the same but its sign changes, while ΔE^* in (11.4.21b) and (11.4.22b) coincides and is given by E^*.

Finally, suppose an RVE of volume V and arbitrary elasticity $\mathbf{C}' = \mathbf{C}'(\mathbf{x})$, is subjected on its boundary ∂V to *self-equilibrating tractions of arbitrary variation*, and denote the corresponding stress and strain fields by $\boldsymbol{\sigma}(\mathbf{x})$ and $\boldsymbol{\varepsilon}(\mathbf{x})$, respectively. Consider, in addition, a *homogeneous* solid of volume V′ with the *same geometry* as that of V, but of *uniform elasticity* **C**, subjected on its boundary $\partial V'$ to the *same tractions* as those acting on ∂V, and denote the corresponding stress and strain fields by $\boldsymbol{\sigma}^H(\mathbf{x})$ and $\boldsymbol{\varepsilon}^H(\mathbf{x})$, respectively. Let $\boldsymbol{\varepsilon}^*(\mathbf{x})$ be the eigenstrain field which homogenizes V of variable elasticity $\mathbf{C}'(\mathbf{x})$ to V′ of constant elasticity **C**. Define the (average) *energy of inhomogeneity* by

$$\Delta W^* \equiv \frac{1}{2} <\boldsymbol{\sigma} : \boldsymbol{\varepsilon}> - \frac{1}{2} <\boldsymbol{\sigma}^H : \boldsymbol{\varepsilon}^H>. \tag{11.4.23a}$$

It then follows that

$$2\Delta W^* = \begin{cases} <\boldsymbol{\sigma}^H : \boldsymbol{\varepsilon}^*> \\ -<\boldsymbol{\sigma}^* : \boldsymbol{\varepsilon}^H>, \end{cases} \tag{11.4.23b}$$

as may be verified by direct calculation, where $\boldsymbol{\sigma}^* = -\mathbf{C} : \boldsymbol{\varepsilon}^*$. Similarly, when arbitrary self-compatible displacements are prescribed on ∂V, the average energy of the inhomogeneity becomes

$$2\Delta W^* = \begin{cases} -<\boldsymbol{\sigma}^H : \boldsymbol{\varepsilon}^*> \\ <\boldsymbol{\sigma}^* : \boldsymbol{\varepsilon}^H>. \end{cases} \tag{11.4.24}$$

11.5. REFERENCES

Bacon, D. J., Barnett, D. M., and Scattergood, R. O. (1979), Anisotropic continuum theory of lattice defects, *Prog. in Mat. Sci.*, Vol. 23, 51-262.

Barnett, D. M. (1972), The precise evaluation of derivatives of the anisotropic elastic Green's function, *Phys. stat. sol. (b)*, Vol. 49, 741-748.

Barnett, D. M. and Lothe, J. (1973), Synthesis of the sextic and the integral formalism for dislocations, Green's functions, and surface waves in anisotropic elastic solids, *Physica Norveg.*, Vol. 7, 13-19.

Bhargava, R. D. and Radhakrishna, H. C. (1963), Elliptic inclusions in a stressed matrix, *Proc. Camb. Phil. Soc.*, Vol. 59, 821-832.

Bhargava, R. D. and Radhakrishna, H. C. (1964), Elliptic inclusion in orthotropic medium, *Phys. Soc. Japan*, No. 19, 396-405.

Chen, W. T. (1967), On an elliptic elastic inclusion in an anisotropic medium, *Q. J. Mech. Appl. Math.*, No. 20, 307-313.

Eshelby, J. D. (1957), The determination of the elastic field of an ellipsoidal

inclusion, and related problems, *Proc. Roy. Soc.*, Vol. A241, 376-396.
Eshelby, J. D. (1961), Elastic inclusions and inhomogeneities, in *Prog. in Solid Mech.*, Snedden, I. N. and Hill, R. eds., Vol. II, North-Holland, 89-140.
Hardiman, N. J. (1954), Elliptic elastic inclusion in an infinite elastic plate, *Q. J. Mech. Appl. Math.*, Vol. 52, 226-230.
Hill, R. (1965), A self-consistent mechanics of composite materials, *J. Mech. Phys. Solids*, Vol. 13, 213-222.
Lubarda, V. A. and Markenscoff, X. (1998), On the absence of Eshelby property for non-ellipsoidal inclusions, *Int. J. Solids Structure*, Vol. 35, No. 25, 3405-3411.
Morse, P. M. and Feshbach, H. (1953), *Methods of theoretical physics*, McGraw-Hill, New York.
Mura, T.(1987), *Micromechanics of Defects in Solids*, Martinus Nijhoff Publishers, Dordrecht.
Roach, G. F. (1982), *Green's functions* (second edition), Cambridge University Press.
Stakgold, I. (1967), *Boundary value problems of mathematical physics*, Vol. I, Macmillan, New York.
Stroh, A. N. (1958), Dislocations and cracks in anisotropic elasticity, *Phil. Mag.*, Vol. 3, 625-646.
Stroh, A. N. (1962), Steady state problems in anisotropic elasticity, *J. Mech. Phys. Solids*, Vol. 41, 77-103.
Tanaka, K. and Mori, T. (1972), Note on volume integrals of the elastic field around an ellipsoidal inclusion, *J. Elasticity*, Vol. 2, 199-200.
Ting, T. C. T. and Yan, G. (1991), The anisotropic elastic solids with an elliptic hole or rigid inclusion, *Int. J. Solids Structure*, Vol. 27, 1879-1894.
Ting, T. C. T. (1992), Image singularities of Green's functions for anisotropic elastic half-spaces and bimaterials, *Q. J. Mech. Appl. Math.*, Vol. 45, 119-139.
Walpole, L. J. (1967), The elastic field of an inclusion in an anisotropic medium, *Proc. Roy. Soc. London, Ser. A*, Vol. 300, 270-289.
Willis, J. R. (1964), Anisotropic elastic inclusion problems, *Q. J. Mech. Appl. Math.*, Vol. 17, 157-174.
Yang, H. C. and Chen, Y. T. (1976), Generalized plane problems of elastic inclusions in anisotropic solids, *J. Appl. Mech.*, Vol. 98, 424-430.

CHAPTER IV

SOLIDS WITH PERIODIC MICROSTRUCTURE

In the preceding sections attention is focused on solids with irregular microheterogeneities. The random distribution of inhomogeneities is the corresponding limiting case. At the other extremum is the limiting case of perfect regularity. This may be modeled by an infinitely extended solid with periodic structure. In this chapter the general properties of solids with periodic microstructure are examined, i.e., solids with periodically distributed cavities, cracks, or inclusions. Many advanced composites can be modeled in this manner. Furthermore, the results provide useful limiting values for the overall properties of solids with microheterogeneities.

For a periodic structure, the concept of a unit cell is introduced in Subsections 12.1 and 12.2, and, using Fourier series, the general solution is obtained in Subsection 12.3. Homogenization of a unit cell with the aid of variable eigenstrains or eigenstresses, is introduced in Subsection 12.4, and the corresponding consistency conditions are obtained. Then, specific classes of problems are solved for illustration, providing detailed results for two-phase microstructure (Subsection 12.5), inclusions and cavities (Subsection 12.6), and microcracks (Subsection 12.7). Finally, Section 12 is completed by considering a unit cell with rate-dependent or rate-independent nonlinear constituents (Subsection 12.8).

The focus of Section 13 is on the overall response of linearly elastic solids containing periodically distributed uniform micro-inclusions. For this, an equivalent homogeneous solid with suitable periodic eigenstrain or eigenstress fields, is defined (Subsection 13.1), the Hashin-Shtrikman variational principles are developed, exact computable bounds on the overall moduli are obtained (Subsections 13.2 and 13.3), and the results are exemplified (Subsections 13.4 to 13.6). In Subsection 13.5, exact bounds for two overall elastic parameters are obtained, which are valid for inclusions of any shape or distribution.

The concept of mirror images of points and vectors is introduced in Section 14, and then used to decompose tensor-valued functions defined on the unit cell, to their symmetric and antisymmetric parts. The decomposition is applied to Fourier series representations of tensor-valued field quantities such as strain, stress, and elastic moduli, resulting in considerable economy in numerical computation and clarity in restrictions which must be imposed on the boundary data.

SECTION 12 GENERAL PROPERTIES AND FIELD EQUATIONS

This section addresses a number of general physical and mathematical properties of solids with periodic microstructure. First, attention is focused on the essential differences between an RVE with a random distribution of microheterogeneities and an RVE of a solid with periodic microstructure. Then the Fourier series expansion technique is used to solve the problem of a linearly elastic solid with periodically distributed cavities or linearly elastic inclusions. To this end, a unit cell is specified which encompasses the structure of the solid and which is used to reduce the solution of an infinitely extended periodic structure to that of a finite unit cell. The boundary conditions for the unit cell are examined with some care and the uniqueness of the periodic field variables is discussed. A series of illustrative examples are worked out, and application of the results to periodically distributed inelastic inhomogeneities in an inelastic matrix is also briefly examined.

12.1. PERIODIC MICROSTRUCTURE AND RVE

As has been discussed before, the distribution of microdefects may, in many cases, be regarded to be random and therefore, be described statistically. The concept of an RVE with random distribution of defects has been introduced to deal with such situations. In many problems the material may possess a regular microstructure which lends itself to a completely different type of modeling, namely, a solid with periodically distributed defects or inclusions. Such a perfectly regular distribution, of course, does not exist in actual cases. However, the model can be quite useful, since it provides limiting values for various overall material properties.

Whereas the boundary data for an RVE with irregular heterogeneities are defined by the local values of the macrostress or macrostrain in the corresponding continuum at the corresponding material element, the same interpretation cannot be applied directly to the model of a solid with a periodic structure. In this latter model the periodicity extends to infinity in all directions and therefore, the boundary conditions need to be prescribed differently in a precise and useful manner. In particular, these boundary conditions must be such that they lead to a periodic distribution of the field quantities, namely, displacement, strain, and stress. Hence, a representative unit cell is considered, which encompasses the periodic geometry and material properties, as well as the periodicity of the field variables, and the overall properties of the solid with periodic microstructure are studied in terms of the overall properties of this representative cell. This representative cell is called the *unit cell.* By definition the solid with a periodic structure must consist of an infinite monolithic collection of such unit cells, satisfying the continuity of displacements and tractions across all cell boundaries. For the unit cell, the average strains and stresses can be calculated in terms of the geometry and properties of its constituents, and the overall average elasticity and compliance tensors can be defined. In this manner, instead of an RVE, a unit cell is employed, in order to estimate the constitutive properties of the continuum material. Note that, in many advanced composites, the microstructure can indeed be accurately modeled as a periodic one and therefore, the present results directly apply to such structures. However, an estimate of the overall properties of solids on the basis of an *assumed* periodic microstructure has broader application and, indeed, provides limiting values for cases where actual periodicity may not exist. Note that when the shapes and the arrangement of the inclusions within the unit cell are regular, following a certain pattern, then the overall response of the solid will entail some corresponding symmetries. On the other hand, when the arrangement within the unit cell is irregular and more or less random, the overall response may become isotropic even for the periodically arranged unit cells. The general results presented in this section and the following two sections apply to both cases.

12.2. PERIODICITY AND UNIT CELL

A model of an infinitely extended linearly elastic solid with periodically varying inhomogeneities is now examined. The elastic and compliance tensors are periodic functions of the position. For an arbitrary loading, the displacement, strain, and stress fields may not, in general, be periodic. However, special idealized prescribed deformations or stresses may be considered which produce periodically varying field quantities and also lead to practically useful results.

Assume that the infinitely extended periodic structure is obtained from a unit cell which is repeated indefinitely in all directions. The unit cell does not need to be in the shape of a parallelepiped. But, for a broad class of periodic structures, a unit cell of that kind can always be chosen. The parallelepiped unit cell lends itself to simpler analysis and, therefore, is considered in this chapter.

Denote by $2a_i$ ($i = 1, 2, 3$), the dimensions of the parallelepiped unit cell in the x_i-direction. Then the elasticity tensor $\mathbf{C}' = \mathbf{C}'(\mathbf{x})$ satisfies the following periodic condition:

$$\mathbf{C}'(\mathbf{x} + \mathbf{d}) = \mathbf{C}'(\mathbf{x}), \tag{12.2.1}$$

where

$$\mathbf{d} = \sum_{i=1}^{3} 2m_i\, a_i\, \mathbf{e}_i, \tag{12.2.2a}$$

with m_i ($i = 1, 2, 3$) arbitrary integers. It is convenient to introduce

$$\mathbf{a}^1 = 2a_1\,\mathbf{e}_1, \qquad \mathbf{a}^2 = 2a_2\,\mathbf{e}_2, \qquad \mathbf{a}^3 = 2a_3\,\mathbf{e}_3, \tag{12.2.2b~d}$$

to denote the edges of the unit cell. The vectors $\mathbf{a}^\alpha$ ($\alpha = 1, 2, 3$), now define the regular structure of the periodic solid. Note that from one set of regularity vectors $\mathbf{a}^\alpha$ ($\alpha = 1, 2, 3$), other sets can be constructed. For example, $\mathbf{a}^1 + \mathbf{a}^2$ and $\mathbf{a}^2 - \mathbf{a}^1$ can replace $\mathbf{a}^1$ and $\mathbf{a}^2$. As shown in Figure 12.2.1, several parallelepiped unit cells can be constructed for one periodic structure. Therefore, depending on the problem, different suitable regularity vectors may be chosen. Note also that the regularity vectors need not be mutually orthogonal. Here, however, orthogonality is assumed for the sake of simplicity in presentation.

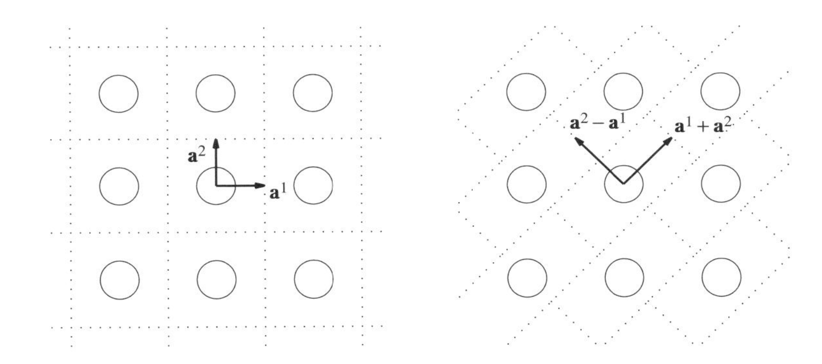

Figure 12.2.1

Two unit cells for one periodic structure

Let $\mathbf{u} = \mathbf{u}(\mathbf{x})$, $\boldsymbol{\varepsilon} = \boldsymbol{\varepsilon}(\mathbf{x})$, and $\boldsymbol{\sigma} = \boldsymbol{\sigma}(\mathbf{x})$ be the displacement, strain, and stress fields. They must satisfy the three governing field equations,

$$\boldsymbol{\varepsilon}(\mathbf{x}) = \frac{1}{2}\{\boldsymbol{\nabla}\otimes\mathbf{u}(\mathbf{x}) + (\boldsymbol{\nabla}\otimes\mathbf{u}(\mathbf{x}))^T\},$$

$$\boldsymbol{\nabla}\boldsymbol{\cdot}\boldsymbol{\sigma}(\mathbf{x}) = \mathbf{0}, \qquad \boldsymbol{\sigma}(\mathbf{x}) = \mathbf{C}'(\mathbf{x}) : \boldsymbol{\varepsilon}(\mathbf{x}), \tag{12.2.3a~c}$$

for any $\mathbf{x}$. Consider periodic solutions,

$$\boldsymbol{\sigma}(\mathbf{x}+\mathbf{d}) = \boldsymbol{\sigma}(\mathbf{x}), \qquad \boldsymbol{\varepsilon}(\mathbf{x}+\mathbf{d}) = \boldsymbol{\varepsilon}(\mathbf{x}). \tag{12.2.4a,b}$$

From these and (12.2.1), it follows that

$$\boldsymbol{\sigma}(\mathbf{x}+\mathbf{d}) = \mathbf{C}'(\mathbf{x}) : \boldsymbol{\varepsilon}(\mathbf{x}). \tag{12.2.4c}$$

If the displacement field $\mathbf{u}$ satisfies the periodicity, its gradient and hence the strain field become periodic. The converse, however, may not be the case. That is, even if the strain field satisfies periodicity, the corresponding displacement field may not be periodic. This is because a uniform strain field, which satisfies periodicity,

$$\boldsymbol{\varepsilon}(\mathbf{x}+\mathbf{d}) = \boldsymbol{\varepsilon}(\mathbf{x}) = \boldsymbol{\varepsilon}^{\mathrm{o}}, \tag{12.2.5a}$$

produces a linear displacement field,

$$\mathbf{u}(\mathbf{x}) = \mathbf{x} \cdot \boldsymbol{\varepsilon}^{\mathrm{o}}, \tag{12.2.5b}$$

which is not periodic; this notwithstanding, a computational scheme with piecewise linear (hence discontinuous) periodic displacement field, is proposed by Aboudi (1991). In order for the corresponding displacement field to be periodic, subtract from the total strain field the corresponding uniform part (if it exists). In this manner, it is required that the average strain for a periodic strain field vanish. The uniform part (if it exists) will then be dealt with separately. In Subsection 12.3, the periodicity of the displacement field is examined in some detail.

In general, besides periodicity properties, the field variables possess certain *symmetry and antisymmetry properties.* From these properties, zero traction or zero displacement conditions can be assigned on certain planes in a periodic structure; see Section 14. These conditions may be used to simplify and considerably reduce the actual required computation.

12.3. FOURIER SERIES

Since the field variables in a solid with a periodic structure and suitable boundary data satisfy periodicity, it is useful to consider the Fourier series representation of these variables. Recall

$$\exp(\iota x) = \cos x + \iota \sin x \qquad (\iota = \sqrt{-1}). \tag{12.3.1}$$

Define the domain of a typical unit cell, U, by

$$\mathrm{U} = \{\mathbf{x};\, -a_i < x_i < a_i \ (i = 1, 2, 3)\}, \tag{12.3.2}$$

and introduce new variables ξ_i, by

$$\xi_i \equiv \xi_i(n_i) \equiv \frac{n_i \pi}{a_i} \qquad (n_i = 0, \pm 1, ...;\ i \text{ not summed; } i = 1, 2, 3). \tag{12.3.3a}$$

For simplicity, *treat* ξ_i as components of a vector, say, $\boldsymbol{\xi}$, and write

$$\boldsymbol{\xi} = \xi_i \mathbf{e}_i, \qquad x_i \xi_i = \mathbf{x} \cdot \boldsymbol{\xi}. \tag{12.3.3b,c}$$

Note that the set of complex functions $\exp(\iota \mathbf{x} \cdot \boldsymbol{\xi})$ is periodic,

$$\exp\{\iota(\mathbf{x}+\mathbf{d}) \cdot \boldsymbol{\xi}\} = \exp(\iota \mathbf{x} \cdot \boldsymbol{\xi}), \tag{12.3.4a}$$

with $\mathbf{d}$ defined by (12.2.2a), and orthonormal,

$$\frac{1}{U}\int_U \exp(\iota\mathbf{x}\cdot\boldsymbol{\xi})\exp(-\iota\mathbf{x}\cdot\boldsymbol{\zeta})\,dV_x = \begin{cases} 1 & \text{if } \boldsymbol{\xi} = \boldsymbol{\zeta} \\ 0 & \text{otherwise,} \end{cases} \tag{12.3.4b}$$

where $\zeta_i = m_i\pi/a_i$ ($m_i = 0, \pm 1, ..., $; i not summed; $i = 1, 2, 3$), and

$$U = 8\,a_1\,a_2\,a_3. \tag{12.3.4c}$$

Note also that

$$\frac{\partial}{\partial x_i}\exp(\iota\mathbf{x}\cdot\boldsymbol{\xi}) = \iota\,\xi_i\exp(\iota\mathbf{x}\cdot\boldsymbol{\xi}) \qquad (i = 1, 2, 3) \tag{12.3.4d}$$

or

$$\boldsymbol{\nabla}\{\exp(\iota\mathbf{x}\cdot\boldsymbol{\xi})\} = \iota\,\boldsymbol{\xi}\exp(\iota\mathbf{x}\cdot\boldsymbol{\xi}). \tag{12.3.4e}$$

The set of functions, $\exp(\iota\mathbf{x}\cdot\boldsymbol{\xi})$, $\xi_i = n_i\pi/a_i$, ($n_i = 0, \pm 1, ...$), defines an orthogonal basis such that a suitably smooth arbitrary periodic function $f(\mathbf{x})$, with periodicity $f(\mathbf{x}+\mathbf{d}) = f(\mathbf{x})$, can be expressed as

$$f(\mathbf{x}) = \sum_{\xi} \mathcal{F}f(\boldsymbol{\xi})\exp(\iota\mathbf{x}\cdot\boldsymbol{\xi}), \tag{12.3.5a}$$

where the summation is taken for all integers n_i ($i = 1, 2, 3$). The coefficients $\mathcal{F}f(\boldsymbol{\xi})$ are defined by

$$\mathcal{F}f(\boldsymbol{\xi}) \equiv \frac{1}{U}\int_U f(\mathbf{x})\exp(-\iota\mathbf{x}\cdot\boldsymbol{\xi})\,dV_x. \tag{12.3.5b}$$

These coefficients are unique, and the infinite sum converges at every point in the unit cell if $f(\mathbf{x})$ is sufficiently smooth; see Sneddon (1951).

12.3.1. Displacement and Strain Fields

Now, seek to express the periodic field variables in the unit cell in terms of the Fourier series. First consider the displacement field $\mathbf{u}(\mathbf{x})$. From (12.3.5),

$$\mathbf{u}(\mathbf{x}) = \sum_{\xi}{}' \mathcal{F}\mathbf{u}(\boldsymbol{\xi})\exp(\iota\mathbf{x}\cdot\boldsymbol{\xi}), \tag{12.3.6a}$$

where

$$\mathcal{F}\mathbf{u}(\boldsymbol{\xi}) = \frac{1}{U}\int_U \mathbf{u}(\mathbf{x})\exp(-\iota\mathbf{x}\cdot\boldsymbol{\xi})\,dV_x \tag{12.3.6b}$$

or, in component form,

$$\mathcal{F}u_i(\boldsymbol{\xi}) = \frac{1}{U}\int_U u_i(\mathbf{x})\exp(-\iota\mathbf{x}\cdot\boldsymbol{\xi})\,dV_x. \tag{12.3.6c}$$

In (12.3.6a) and in the sequel, a prime on Σ indicates that the summation is taken over all integers n_i in (12.3.3a), such that $\boldsymbol{\xi} \neq \mathbf{0}$. The term corresponding to $\boldsymbol{\xi} = \mathbf{0}$, i.e., $\mathcal{F}\mathbf{u}(\mathbf{0})$, should be omitted, since it corresponds to a rigid-body translation.

Using (12.3.6b), express the coefficients of the Fourier expansion of $\boldsymbol{\nabla}\otimes\mathbf{u}(\mathbf{x})$, as

$$F(\nabla\otimes\mathbf{u})(\boldsymbol{\xi}) = \frac{1}{U}\int_U \nabla\otimes\mathbf{u}(\mathbf{x})\exp(-\iota\mathbf{x}\cdot\boldsymbol{\xi})\,dV_{\mathbf{x}}. \tag{12.3.7a}$$

From the Gauss theorem, (12.3.7a) is rewritten as

$$F(\nabla\otimes\mathbf{u})(\boldsymbol{\xi}) = \frac{1}{U}\int_{\partial U} \boldsymbol{\nu}(\mathbf{x})\otimes\mathbf{u}(\mathbf{x})\exp(-\iota\mathbf{x}\cdot\boldsymbol{\xi})\,dS_{\mathbf{x}}$$

$$-\frac{1}{U}\int_U \nabla\{\exp(-\iota\mathbf{x}\cdot\boldsymbol{\xi})\}\otimes\mathbf{u}(\mathbf{x})\,dV_{\mathbf{x}}. \tag{12.3.7b}$$

Because of periodicity,

$$\mathbf{u}(\mathbf{x}^+) = \mathbf{u}(\mathbf{x}^-), \qquad \exp(\iota\mathbf{x}^+\cdot\boldsymbol{\xi}) = \exp(\iota\mathbf{x}^-\cdot\boldsymbol{\xi}) \qquad \text{on } \partial U, \tag{12.3.8a,b}$$

where $\mathbf{x}^+$ is on $x_i = +a_i$ and $\mathbf{x}^-$ is on $x_i = -a_i$, for i = 1, 2, 3. Since the unit external normal vector on ∂U, $\boldsymbol{\nu}(\mathbf{x})$, satisfies

$$\boldsymbol{\nu}(\mathbf{x}^+) = -\boldsymbol{\nu}(\mathbf{x}^-), \tag{12.3.8c}$$

the surface integral in the right-hand side of (12.3.7b) vanishes,

$$\int_{\partial U} \boldsymbol{\nu}(\mathbf{x})\otimes\mathbf{u}(\mathbf{x})\exp(-\iota\mathbf{x}\cdot\boldsymbol{\xi})\,dS_{\mathbf{x}} = \mathbf{0}. \tag{12.3.8d}$$

The displacement gradient then becomes

$$\nabla\otimes\mathbf{u}(\mathbf{x}) = \sum_{\boldsymbol{\xi}}{}' F(\nabla\otimes\mathbf{u})(\boldsymbol{\xi})\exp(\iota\mathbf{x}\cdot\boldsymbol{\xi}), \tag{12.3.9a}$$

where

$$F(\nabla\otimes\mathbf{u})(\boldsymbol{\xi}) = \iota\boldsymbol{\xi}\otimes F\mathbf{u}(\boldsymbol{\xi}). \tag{12.3.9b}$$

Hence, the strain field $\boldsymbol{\varepsilon}$ has the following Fourier series expansion:

$$\boldsymbol{\varepsilon}(\mathbf{x}) = \sum_{\boldsymbol{\xi}}{}' F\boldsymbol{\varepsilon}(\boldsymbol{\xi})\exp(\iota\mathbf{x}\cdot\boldsymbol{\xi}), \tag{12.3.10a}$$

where

$$F\boldsymbol{\varepsilon}(\boldsymbol{\xi}) = \frac{\iota}{2}\,\{\boldsymbol{\xi}\otimes F\mathbf{u}(\boldsymbol{\xi}) + F\mathbf{u}(\boldsymbol{\xi})\otimes\boldsymbol{\xi}\}. \tag{12.3.10b}$$

Assuming that the order of differentiation and summation can be exchanged in the Fourier series expansions, obtain (12.3.9a) directly from the gradient of the right-hand side of (12.3.6a). This formal exchange of the operators is possible only if $\mathbf{u}(\mathbf{x})$ is periodic, suitably smooth, and the surface integral corresponding to (12.3.7a) vanishes for all $\boldsymbol{\xi}$, i.e., (12.3.8d) holds.

From (12.3.8d), it can be proved that the average strain over the unit cell must vanish identically if the displacement field is periodic. For $\xi_i = 0$ (or $n_i = 0$), (12.3.8d) becomes

$$\frac{1}{U}\int_{\partial U} \boldsymbol{\nu}(\mathbf{x})\otimes\mathbf{u}(\mathbf{x})\,dS_{\mathbf{x}} = <\nabla\otimes\mathbf{u}> = \mathbf{0}, \tag{12.3.11}$$

where $U = 8a_1a_2a_3$, and < > stands for the volume average over U. *Therefore, for a periodic displacement field, the average strain must vanish.*

12.3.2. Stress Field

The periodic elasticity tensor $\mathbf{C}'(\mathbf{x})$ has the Fourier series expansion,

$$\mathbf{C}'(\mathbf{x}) = \sum_{\boldsymbol{\xi}} F\mathbf{C}' \exp(\iota\mathbf{x}\cdot\boldsymbol{\xi}), \tag{12.3.12a}$$

where

$$F\mathbf{C}'(\boldsymbol{\xi}) = \frac{1}{U}\int_U \mathbf{C}'(\mathbf{x}) \exp(-\iota\mathbf{x}\cdot\boldsymbol{\xi})\, dV_x, \tag{12.3.12b}$$

with $FC'_{ijkl} = FC'_{jikl} = FC'_{ijlk} = FC'_{klij}$. Then,[1]

$$\boldsymbol{\sigma}(\mathbf{x}) = \mathbf{C}'(\mathbf{x}) : \boldsymbol{\varepsilon}(\mathbf{x})$$

$$= \{\sum_{\boldsymbol{\xi}} F\mathbf{C}'(\boldsymbol{\xi}) \exp(\iota\mathbf{x}\cdot\boldsymbol{\xi})\} : \{\sum_{\boldsymbol{\zeta}}{}' F\boldsymbol{\varepsilon}(\boldsymbol{\zeta}) \exp(\iota\mathbf{x}\cdot\boldsymbol{\zeta})\}. \tag{12.3.13a}$$

If the order of the two infinite summations can be rearranged, then

$$\boldsymbol{\sigma}(\mathbf{x}) = \sum_{\boldsymbol{\xi}} \{\sum_{\boldsymbol{\zeta}}{}' F\mathbf{C}'(\boldsymbol{\xi}-\boldsymbol{\zeta}) : F\boldsymbol{\varepsilon}(\boldsymbol{\zeta})\} \exp(\iota\mathbf{x}\cdot\boldsymbol{\xi}). \tag{12.3.13b}$$

Note from (12.3.13b), that

$$F\boldsymbol{\sigma}(\boldsymbol{\xi}) = \frac{1}{U}\int_U \boldsymbol{\sigma}(\mathbf{x}) \exp(-\iota\mathbf{x}\cdot\boldsymbol{\xi})\, dV_x$$

$$= \iota \sum_{\boldsymbol{\zeta}}{}' F\mathbf{C}'(\boldsymbol{\xi}-\boldsymbol{\zeta}) : \{(\boldsymbol{\zeta})\otimes F\mathbf{u}(\boldsymbol{\zeta})\}. \tag{12.3.14a}$$

Note also that, if $\mathbf{C}'$ is a constant, say, $\mathbf{C}$, then

$$F\boldsymbol{\sigma}(\boldsymbol{\xi}) = \iota\,\mathbf{C} : \{\boldsymbol{\xi}\otimes F\mathbf{u}(\boldsymbol{\xi})\} \tag{12.3.14b}$$

which is linear in $\boldsymbol{\xi}$ and $F\mathbf{u}(\boldsymbol{\xi})$.

Whatever may be the relation between $F\mathbf{u}(\boldsymbol{\xi})$ and $F\boldsymbol{\sigma}(\boldsymbol{\xi})$, the divergence of the stress field satisfies

$$F(\boldsymbol{\nabla}\cdot\boldsymbol{\sigma})(\boldsymbol{\xi}) = \frac{1}{U}\int_{\partial U} \boldsymbol{\nu}(\mathbf{x})\cdot\boldsymbol{\sigma}(\mathbf{x}) \exp(-\iota\mathbf{x}\cdot\boldsymbol{\xi})\, dS_x$$

$$+ \frac{1}{U}\int_U \iota\boldsymbol{\xi}\cdot\boldsymbol{\sigma}(\mathbf{x}) \exp(-\iota\mathbf{x}\cdot\boldsymbol{\xi})\, dV_x$$

$$= \iota\boldsymbol{\xi}\cdot F\boldsymbol{\sigma}(\boldsymbol{\xi}). \tag{12.3.15}$$

As pointed out before, this is because the stress field $\boldsymbol{\sigma}(\mathbf{x})$ is periodic and the surface integral on ∂U vanishes for all $\boldsymbol{\xi}$, i.e.,

$$\int_{\partial U} \boldsymbol{\nu}(\mathbf{x})\cdot\boldsymbol{\sigma}(\mathbf{x}) \exp(-\iota\mathbf{x}\cdot\boldsymbol{\xi})\, dS_x = \mathbf{0}. \tag{12.3.16a}$$

In particular, taking $\boldsymbol{\xi} = \mathbf{0}$, observe that

---

[1] In (12.3.13a), the uniform part of the stress field is included.

$$\int_{\partial U} \boldsymbol{\nu}(\mathbf{x}) \cdot \boldsymbol{\sigma}(\mathbf{x}) \, dS_{\mathbf{x}} = \mathbf{0} \tag{12.3.16b}$$

which stems from the equilibrium of the unit cell.

12.4. HOMOGENIZATION

An elastic solid with periodically distributed inhomogeneities can be homogenized by the introduction of suitable periodically distributed eigenstrains or eigenstresses. In this approach, the periodic elasticity tensor $\mathbf{C}'$ is replaced by a reference constant elasticity tensor $\mathbf{C}$ and a suitable periodic eigenstrain (or eigenstress) field. The uniform solid with constant elasticity tensor $\mathbf{C}$, is referred to as the *equivalent homogeneous solid.* In this section, this general problem is formulated by the Fourier series approach.[2]

12.4.1. Periodic Eigenstrain and Eigenstress Fields

In the equivalent homogeneous solid of reference constant elasticity $\mathbf{C}$, the required eigenstrain (or eigenstress) field is periodic, i.e.,

$$\boldsymbol{\varepsilon}^*(\mathbf{x}+\mathbf{d}) = \boldsymbol{\varepsilon}^*(\mathbf{x}), \qquad \boldsymbol{\sigma}^*(\mathbf{x}+\mathbf{d}) = \boldsymbol{\sigma}^*(\mathbf{x}), \tag{12.4.1a,b}$$

where $\mathbf{d}$ is given by (12.2.2a). The Fourier series representations of the eigenstrain and eigenstress fields are

$$\boldsymbol{\varepsilon}^*(\mathbf{x}) = \sum_{\boldsymbol{\xi}}{}' \, F\boldsymbol{\varepsilon}^*(\boldsymbol{\xi}) \exp(\iota \mathbf{x} \cdot \boldsymbol{\xi}),$$

$$\boldsymbol{\sigma}^*(\mathbf{x}) = \sum_{\boldsymbol{\xi}}{}' \, F\boldsymbol{\sigma}^*(\boldsymbol{\xi}) \exp(\iota \mathbf{x} \cdot \boldsymbol{\xi}), \tag{12.4.2a,3a}$$

where

$$F\boldsymbol{\varepsilon}^*(\boldsymbol{\xi}) = \frac{1}{U} \int_U \boldsymbol{\varepsilon}^*(\mathbf{x}) \exp(-\iota \mathbf{x} \cdot \boldsymbol{\xi}) \, dV_{\mathbf{x}},$$

$$F\boldsymbol{\sigma}^*(\boldsymbol{\xi}) = \frac{1}{U} \int_U \boldsymbol{\sigma}^*(\mathbf{x}) \exp(-\iota \mathbf{x} \cdot \boldsymbol{\xi}) \, dV_{\mathbf{x}}, \tag{12.4.2b,3b}$$

with $(F\boldsymbol{\varepsilon}^*)^T = F\boldsymbol{\varepsilon}^*$ and $(F\boldsymbol{\sigma}^*)^T = F\boldsymbol{\sigma}^*$. The terms $F\boldsymbol{\varepsilon}^*(\mathbf{0})$ and $F\boldsymbol{\sigma}^*(\mathbf{0})$ need not be considered in this representation,[3] since they correspond to uniform eigenstrains

---

[2] See also Bakhvalov and Panasenko (1984) who apply an asymptotic expansion as well as the Fourier series expansion to fields in a solid with periodic microstructure.

[3] Unlike periodic strain or stress fields, periodic eigenstress and eigenstrain fields are allowed to have nonzero volume averages, $< \boldsymbol{\varepsilon}^* >$ and $< \boldsymbol{\sigma}^* >$, i.e., homogeneous parts. Although these homogeneous parts produce no *periodic* strain or stress fields, they are required for a complete representation of the final solution; note that these constant eigenstrains and eigenstresses do affect the *homogeneous* strains and stresses, through $\boldsymbol{\sigma}^o = \mathbf{C} : (\boldsymbol{\varepsilon}^o - < \boldsymbol{\varepsilon}^* >)$ or $\boldsymbol{\varepsilon}^o = \mathbf{D} : (\boldsymbol{\sigma}^o - < \boldsymbol{\sigma}^* >)$.

and eigenstresses, and hence do not produce (equivalent) distributed body forces, i.e., $\nabla \cdot \boldsymbol{\varepsilon}^* = \mathbf{0}$ and $\nabla \cdot \boldsymbol{\sigma}^* = \mathbf{0}$ for these terms. As pointed out before, uniform strain and stress fields are dealt with separately; see Subsection 12.4.5.

To relate the field variables in the equivalent homogenized solid to those in the original heterogeneous solid with periodic structure, use the following *consistency conditions* for the eigenstrain (or eigenstress) field:

$$\mathbf{C}'(\mathbf{x}) : \boldsymbol{\varepsilon}(\mathbf{x}) = \mathbf{C} : \{\boldsymbol{\varepsilon}(\mathbf{x}) - \boldsymbol{\varepsilon}^*(\mathbf{x})\},$$

$$\mathbf{D}'(\mathbf{x}) : \boldsymbol{\sigma}(\mathbf{x}) = \mathbf{D} : \{\boldsymbol{\sigma}(\mathbf{x}) - \boldsymbol{\sigma}^*(\mathbf{x})\} \qquad \text{for } \mathbf{x} \in \mathrm{U}, \tag{12.4.4a,b}$$

where $\mathbf{D}'$ and $\mathbf{D}$ are the actual and the reference compliance tensors, respectively,[4] i.e., $\mathbf{D}'(\mathbf{x}) = \mathbf{C}'^{-1}(\mathbf{x})$ and $\mathbf{D} = \mathbf{C}^{-1}$; see Subsection 12.4.5. It is important to note that the homogeneous (uniform) parts of the field variables $\boldsymbol{\varepsilon}(\mathbf{x})$, $\boldsymbol{\sigma}(\mathbf{x})$, $\boldsymbol{\varepsilon}^*(\mathbf{x})$, and $\boldsymbol{\sigma}^*(\mathbf{x})$ are *included* in (12.4.4). When the average stress and strain tensors, $< \boldsymbol{\sigma}(\mathbf{x}) > = \boldsymbol{\sigma}^o$ and $< \boldsymbol{\varepsilon}(\mathbf{x}) > = \boldsymbol{\varepsilon}^o$, are nonzero and are included in the consistency conditions, the corresponding average eigenfields $< \boldsymbol{\sigma}^*(\mathbf{x}) >$ and $< \boldsymbol{\varepsilon}^*(\mathbf{x}) >$ must also be included in $\boldsymbol{\sigma}^*(\mathbf{x})$ and $\boldsymbol{\varepsilon}^*(\mathbf{x})$; see (12.4.22a,b). Hence, consistency conditions (12.4.4a,b) may be interpreted to be identical with (12.4.22a,b), given later on, if quantities $\boldsymbol{\varepsilon}(\mathbf{x})$, $\boldsymbol{\sigma}(\mathbf{x})$, $\boldsymbol{\varepsilon}^*(\mathbf{x})$, and $\boldsymbol{\sigma}^*(\mathbf{x})$ include their corresponding averages.

12.4.2. Governing Equations

First consider the periodic field variables produced by a prescribed periodic eigenstress field, $\boldsymbol{\sigma}^*$. The stress field $\boldsymbol{\sigma}$ is given by[5] $\mathbf{C} : \boldsymbol{\varepsilon} + \boldsymbol{\sigma}^*$. Equilibrium then requires that

$$\nabla \cdot \{\mathbf{C} : (\nabla \otimes \mathbf{u}(\mathbf{x}))\} + \nabla \cdot \boldsymbol{\sigma}^*(\mathbf{x}) = \mathbf{0}, \qquad \text{for } \mathbf{x} \in \mathrm{U}. \tag{12.4.5a}$$

Hence, $\nabla \cdot \boldsymbol{\sigma}^*$ can be regarded as the distributed body forces. Noting that both $\nabla \otimes \mathbf{u}$ and $\boldsymbol{\sigma}^*$ are periodic, use a Fourier series expansion and arrive at

$$-\boldsymbol{\xi} \cdot \mathbf{C} : \{\boldsymbol{\xi} \otimes \mathcal{F}\mathbf{u}(\boldsymbol{\xi})\} + \iota \boldsymbol{\xi} \cdot \mathcal{F}\boldsymbol{\sigma}^*(\boldsymbol{\xi}) = \mathbf{0}, \qquad \text{for } \boldsymbol{\xi} \neq \mathbf{0}. \tag{12.4.5b}$$

From (12.4.5b), the coefficients $\mathcal{F}\mathbf{u}(\boldsymbol{\xi})$ are obtained *uniquely* in terms of those of the eigenstress field. If $(\boldsymbol{\xi} \cdot \mathbf{C} \cdot \boldsymbol{\xi})$ has an inverse (choose $\mathbf{C}$ with a unique inverse), then

$$\mathcal{F}\mathbf{u}(\boldsymbol{\xi}) = \iota\, (\boldsymbol{\xi} \cdot \mathbf{C} \cdot \boldsymbol{\xi})^{-1} \cdot \{\boldsymbol{\xi} \cdot \mathcal{F}\boldsymbol{\sigma}^*(\boldsymbol{\xi})\} \tag{12.4.6a}$$

or in component form,

$$\mathcal{F}u_i(\boldsymbol{\xi}) = \iota\, (\boldsymbol{\xi} \cdot \mathbf{C} \cdot \boldsymbol{\xi})_{il}^{-1} \{\xi_k\, \mathcal{F}\sigma_{kl}^*(\boldsymbol{\xi})\}. \tag{12.4.6b}$$

---

[4] As mentioned in the previous footnote, the homogeneous part of $\boldsymbol{\varepsilon}^*$ or $\boldsymbol{\sigma}^*$ must be included in consistency condition (12.4.4a,b).

[5] Superscript p is used later on for periodic strain and stress fields to emphasize their periodicity. Here, however, the superscript p is omitted.

Note that (12.4.6a) necessarily requires that $\boldsymbol{\xi} \neq \mathbf{0}$. From (12.3.10b) and (12.3.14b), it then follows that

$$F\boldsymbol{\varepsilon}(\boldsymbol{\xi}) = -\,sym\,\{\boldsymbol{\xi}\otimes(\boldsymbol{\xi}.\mathbf{C}.\boldsymbol{\xi})^{-1}\otimes\boldsymbol{\xi}\} : F\boldsymbol{\sigma}^*(\boldsymbol{\xi}),$$

$$F\boldsymbol{\sigma}(\boldsymbol{\xi}) = -\,\mathbf{C} : sym\,\{\boldsymbol{\xi}\otimes(\boldsymbol{\xi}.\mathbf{C}.\boldsymbol{\xi})^{-1}\otimes\boldsymbol{\xi}\} : F\boldsymbol{\sigma}^*(\boldsymbol{\xi}) + F\boldsymbol{\sigma}^*(\boldsymbol{\xi}), \qquad (12.4.7a,b)$$

where *sym* stands for the symmetric part of the corresponding fourth-order tensor, i.e., $sym\,T_{ijkl} = (T_{ijkl} + T_{jikl} + T_{ijlk} + T_{jilk})/4$. In component form, these become

$$F\varepsilon_{ij}(\boldsymbol{\xi}) = -\,sym\,\{\xi_i(\boldsymbol{\xi}.\mathbf{C}.\boldsymbol{\xi})^{-1}_{jk}\,\xi_l\}\,F\sigma^*_{kl}(\boldsymbol{\xi}),$$

$$F\sigma_{ij}(\boldsymbol{\xi}) = -\,C_{ijmn}\,sym\,\{\xi_m(\boldsymbol{\xi}.\mathbf{C}.\boldsymbol{\xi})^{-1}_{nk}\xi_l\}\,F\sigma^*_{kl}(\boldsymbol{\xi}) + F\sigma^*_{ij}(\boldsymbol{\xi}). \qquad (12.4.7c,d)$$

In a similar manner, the field variables associated with a prescribed periodic eigenstrain field $\boldsymbol{\varepsilon}^*$ are obtained. This leads to

$$F\mathbf{u}(\boldsymbol{\xi}) = -\,\iota\,(\boldsymbol{\xi}.\mathbf{C}.\boldsymbol{\xi})^{-1}.\{\boldsymbol{\xi}.(\mathbf{C} : F\boldsymbol{\varepsilon}^*(\boldsymbol{\xi}))\} \qquad (12.4.8a)$$

or in component form,

$$Fu_i(\boldsymbol{\xi}) = -\,\iota\,(\boldsymbol{\xi}.\mathbf{C}.\boldsymbol{\xi})^{-1}_{in}\,\{\xi_m\,C_{mnkl}\,F\varepsilon^*_{kl}(\boldsymbol{\xi})\}. \qquad (12.4.8b)$$

The strain and stress fields are then defined by

$$F\boldsymbol{\varepsilon}(\boldsymbol{\xi}) = sym\,\{\boldsymbol{\xi}\otimes(\boldsymbol{\xi}.\mathbf{C}.\boldsymbol{\xi})^{-1}\otimes\boldsymbol{\xi}\} : \mathbf{C} : F\boldsymbol{\varepsilon}^*(\boldsymbol{\xi}),$$

$$F\boldsymbol{\sigma}(\boldsymbol{\xi}) = \mathbf{C} : \left[\,sym\,\{\boldsymbol{\xi}\otimes(\boldsymbol{\xi}.\mathbf{C}.\boldsymbol{\xi})^{-1}\otimes\boldsymbol{\xi}\} : \mathbf{C} - \mathbf{1}^{(4s)}\right] : F\boldsymbol{\varepsilon}^*(\boldsymbol{\xi}) \qquad (12.4.9a,b)$$

or in component form,

$$F\varepsilon_{ij}(\boldsymbol{\xi}) = sym\,\{\xi_i\,(\boldsymbol{\xi}.\mathbf{C}.\boldsymbol{\xi})^{-1}_{jm}\,\xi_n\}\,C_{mnkl}\,F\varepsilon^*_{kl}(\boldsymbol{\xi}),$$

$$F\sigma_{ij}(\boldsymbol{\xi}) = C_{ijpq}\left[\,sym\,\{\xi_p\,(\boldsymbol{\xi}.\mathbf{C}.\boldsymbol{\xi})^{-1}_{qm}\,\xi_n\}\,C_{mnkl} - 1^{(4s)}_{pqkl}\right] F\varepsilon^*_{kl}(\boldsymbol{\xi}). \qquad (12.4.9c,d)$$

12.4.3. Periodic Integral Operators

Let $\mathbf{S}^P(\mathbf{x};\,\boldsymbol{\varepsilon}^*)$ $(= \boldsymbol{\varepsilon}(\mathbf{x}))$ and $\mathbf{T}^P(\mathbf{x};\,\boldsymbol{\sigma}^*)$ $(= \boldsymbol{\sigma}(\mathbf{x}))$ be the integral operators for $\boldsymbol{\varepsilon}^*$ and $\boldsymbol{\sigma}^*$, defined in the manner[6] of Section 7, i.e., $\mathbf{S}^P$ gives a periodic strain field produced by $\boldsymbol{\varepsilon}^*$, and $\mathbf{T}^P$ gives a periodic stress field produced by $\boldsymbol{\sigma}^*$. Then, these integral operators reduce to fourth-order tensor operators, $F\mathbf{S}^P = F\mathbf{S}^P(\boldsymbol{\xi})$ and $F\mathbf{T}^P = F\mathbf{T}^P(\boldsymbol{\xi})$, for the coefficients of the Fourier series expansions of the strain and stress fields. These now determine the coefficients of Fourier series representations of the strain and stress fields in terms of those of the eigenstrain

---

[6] In Section 7, integral operators **S** and **T** are introduced to obtain the disturbance stress and strain fields produced by prescribed eigenstrain and eigenstress fields, respectively. While **S** and **T** give the fields which do not violate the prescribed boundary conditions, $\mathbf{S}^P$ and $\mathbf{T}^P$ give periodic fields whose volume averages vanish.

and eigenstress fields. From (12.4.9a) and (12.4.7b),

$$F\boldsymbol{\varepsilon}(\boldsymbol{\xi}) = F\mathbf{S}^{P}(\boldsymbol{\xi}) : F\boldsymbol{\varepsilon}^{*}(\boldsymbol{\xi}), \qquad F\boldsymbol{\sigma}(\boldsymbol{\xi}) = F\mathbf{T}^{P}(\boldsymbol{\xi}) : F\boldsymbol{\sigma}^{*}(\boldsymbol{\xi}), \tag{12.4.10a,b}$$

where

$$F\mathbf{S}^{P}(\boldsymbol{\xi}) = sym\,\{\boldsymbol{\xi}\otimes(\boldsymbol{\xi}\cdot\mathbf{C}\cdot\boldsymbol{\xi})^{-1}\otimes\boldsymbol{\xi}\} : \mathbf{C},$$

$$F\mathbf{T}^{P}(\boldsymbol{\xi}) = -\mathbf{C} : sym\,\{\boldsymbol{\xi}\otimes(\boldsymbol{\xi}\cdot\mathbf{C}\cdot\boldsymbol{\xi})^{-1}\otimes\boldsymbol{\xi}\} + \mathbf{1}^{(4s)} \tag{12.4.10c,d}$$

or in component form,

$$FS^{P}_{ijkl}(\boldsymbol{\xi}) = sym\,\{\xi_i\,(\boldsymbol{\xi}\cdot\mathbf{C}\cdot\boldsymbol{\xi})^{-1}_{jm}\,\xi_n\}\,C_{mnkl},$$

$$FT^{P}_{ijkl}(\boldsymbol{\xi}) = -C_{ijmn}\,sym\,\{\xi_m\,(\boldsymbol{\xi}\cdot\mathbf{C}\cdot\boldsymbol{\xi})^{-1}_{nk}\,\xi_l\} + 1^{(4s)}_{ijkl}. \tag{12.4.10e,f}$$

As is seen, $F\mathbf{S}^{P}$ and $F\mathbf{T}^{P}$ satisfy the following equivalence relations:

$$F\mathbf{S}^{P} + \mathbf{D} : F\mathbf{T}^{P} : \mathbf{C} = \mathbf{1}^{(4s)}, \qquad F\mathbf{T}^{P} + \mathbf{C} : F\mathbf{S}^{P} : \mathbf{D} = \mathbf{1}^{(4s)}. \tag{12.4.11a,b}$$

In a similar manner, define integral operators $\boldsymbol{\Lambda}^{P}(\mathbf{x};\,\boldsymbol{\varepsilon}^{*})$ $(= -\boldsymbol{\sigma})$ and $\boldsymbol{\Gamma}^{P}(\mathbf{x};\,\boldsymbol{\sigma}^{*})$ $(= -\boldsymbol{\varepsilon})$, where $\boldsymbol{\sigma}$ and $\boldsymbol{\varepsilon}$ are periodic stress and strain fields[7] due to $\boldsymbol{\varepsilon}^{*}$ and $\boldsymbol{\sigma}^{*}$, respectively. These integral operators reduce to the fourth-order tensors $F\boldsymbol{\Lambda}^{P} = F\boldsymbol{\Lambda}^{P}(\boldsymbol{\xi})$ and $F\boldsymbol{\Gamma}^{P} = F\boldsymbol{\Gamma}^{P}(\boldsymbol{\xi})$ which correspond to the tensor fields $\boldsymbol{\Lambda}^{A}(\mathbf{x})$ and $\boldsymbol{\Gamma}^{A}(\mathbf{x})$ introduced in Subsection 9.4; see Subsection 9.4 or 11.1. From (12.4.7a) and (12.4.9b),

$$F\boldsymbol{\sigma}(\boldsymbol{\xi}) = -F\boldsymbol{\Lambda}^{P}(\boldsymbol{\xi}) : F\boldsymbol{\varepsilon}^{*}(\boldsymbol{\xi}), \qquad F\boldsymbol{\varepsilon}(\boldsymbol{\xi}) = -F\boldsymbol{\Gamma}^{P}(\boldsymbol{\xi}) : F\boldsymbol{\sigma}^{*}(\boldsymbol{\xi}), \tag{12.4.12a,b}$$

where

$$F\boldsymbol{\Lambda}^{P}(\boldsymbol{\xi}) = -\mathbf{C} : sym\,\{\boldsymbol{\xi}\otimes(\boldsymbol{\xi}\cdot\mathbf{C}\cdot\boldsymbol{\xi})^{-1}\otimes\boldsymbol{\xi}\} : \mathbf{C} + \mathbf{C},$$

$$F\boldsymbol{\Gamma}^{P}(\boldsymbol{\xi}) = sym\,\{\boldsymbol{\xi}\otimes(\boldsymbol{\xi}\cdot\mathbf{C}\cdot\boldsymbol{\xi})^{-1}\otimes\boldsymbol{\xi}\} \tag{12.4.12c,d}$$

or in component form,

$$F\Lambda^{P}_{ijkl}(\boldsymbol{\xi}) = -C_{ijpq}\,sym\,\{\xi_p\,(\boldsymbol{\xi}\cdot\mathbf{C}\cdot\boldsymbol{\xi})^{-1}_{qr}\,\xi_s\}\,C_{rskl} + C_{ijkl},$$

$$F\Gamma^{P}_{ijkl}(\boldsymbol{\xi}) = sym\,\{\xi_i\,(\boldsymbol{\xi}\cdot\mathbf{C}\cdot\boldsymbol{\xi})^{-1}_{jk}\,\xi_l\}. \tag{12.4.12e,f}$$

From the symmetry of $\mathbf{C}$, $C_{ijkl} = C_{jikl} = C_{ijlk} = C_{klij}$, the second-order tensor $(\boldsymbol{\xi}\cdot\mathbf{C}\cdot\boldsymbol{\xi})$ and its inverse are symmetric, i.e., $(\boldsymbol{\xi}\cdot\mathbf{C}\cdot\boldsymbol{\xi})^{-1}_{ij} = (\boldsymbol{\xi}\cdot\mathbf{C}\cdot\boldsymbol{\xi})^{-1}_{ji}$. Hence, $F\boldsymbol{\Lambda}^{P}$ and $F\boldsymbol{\Gamma}^{P}$ have the following symmetry properties:

$$F\Lambda^{P}_{ijkl} = F\Lambda^{P}_{jikl} = F\Lambda^{P}_{ijlk} = F\Lambda^{P}_{klij},$$

$$F\Gamma^{P}_{ijkl} = F\Gamma^{P}_{jikl} = F\Gamma^{P}_{ijlk} = F\Gamma^{P}_{klij}; \tag{12.4.12g,h}$$

compare (12.4.12d) with (11.3.4).

---

[7] Similarly to $\mathbf{S}$ and $\mathbf{S}^{P}$, or $\mathbf{T}$ and $\mathbf{T}^{P}$, the integral operator $\boldsymbol{\Lambda}$ or $\boldsymbol{\Gamma}$ introduced in Section 9 gives a strain field which does not violate prescribed boundary conditions, while $\boldsymbol{\Lambda}^{P}$ or $\boldsymbol{\Gamma}^{P}$ gives a periodic stress or strain field whose volume average vanishes.

The fourth-order tensors ${}^F\mathbf{S}^P$ and ${}^F\mathbf{T}^P$ are given in terms of ${}^F\boldsymbol{\Gamma}^P$, by

$$ {}^F\mathbf{S}^P(\boldsymbol{\xi}) = -{}^F\boldsymbol{\Gamma}^P(\boldsymbol{\xi}) : \mathbf{C}, \qquad {}^F\mathbf{T}^P(\boldsymbol{\xi}) = -\mathbf{C} : {}^F\boldsymbol{\Gamma}^P(\boldsymbol{\xi}) + \mathbf{1}^{(4s)}. \tag{12.4.13a,b} $$

Therefore, using ${}^F\boldsymbol{\Gamma}^P$, express the integral operators $\mathbf{S}^P$ and $\mathbf{T}^P$ for the eigenstrain, $\boldsymbol{\varepsilon}^*$, and the eigenstress, $\boldsymbol{\sigma}^*$, respectively, as

$$ \mathbf{S}^P(\mathbf{x};\, \boldsymbol{\varepsilon}^*) = \sum_{\xi}{}' \, {}^F\boldsymbol{\Gamma}^P(\boldsymbol{\xi}) : \mathbf{C} : \{\frac{1}{U}\int_U \boldsymbol{\varepsilon}^*(\mathbf{y}) \exp(\iota\boldsymbol{\xi}\boldsymbol{\cdot}(\mathbf{x}-\mathbf{y}))\, dV_y\}, $$

$$ \mathbf{T}^P(\mathbf{x};\, \boldsymbol{\sigma}^*) = -\sum_{\xi}{}' \, \mathbf{C} : {}^F\boldsymbol{\Gamma}^P(\boldsymbol{\xi}) : \{\frac{1}{U}\int_U \boldsymbol{\sigma}^*(\mathbf{y}) \exp(\iota\boldsymbol{\xi}\boldsymbol{\cdot}(\mathbf{x}-\mathbf{y}))\, dV_y\} + \boldsymbol{\sigma}^*(\mathbf{x}) \tag{12.4.14a,b} $$

or, in component form,

$$ S^P_{ij}(\mathbf{x};\, \boldsymbol{\varepsilon}^*) = \sum_{\xi}{}' \, {}^F\Gamma^P_{ijmn}(\boldsymbol{\xi})\, C_{mnkl}\, \{\frac{1}{U}\int_U \varepsilon^*_{kl}(\mathbf{y}) \exp(\iota\boldsymbol{\xi}\boldsymbol{\cdot}(\mathbf{x}-\mathbf{y}))\, dV_y\}, $$

$$ T^P_{ij}(\mathbf{x};\, \boldsymbol{\sigma}^*) = -\sum_{\xi}{}' \, C_{ijmn}\, {}^F\Gamma^P_{mnkl}\, \{\frac{1}{U}\int_U \sigma^*_{kl}(\mathbf{y}) \exp(\iota\boldsymbol{\xi}\boldsymbol{\cdot}(\mathbf{x}-\mathbf{y}))\, dV_y\} + \sigma^*_{ij}(\mathbf{x}). \tag{12.4.14c,d} $$

Note that (12.4.14a~d) exclude terms associated with $\boldsymbol{\xi} = \mathbf{0}$; this is signified by a prime on the summation sign.

12.4.4. Isotropic Matrix

The periodic microstructure may be homogenized by an isotropic reference constant elasticity tensor, i.e., an isotropic uniform equivalent solid. This may be an effective approach when the matrix material is isotropic and the volume fraction of inhomogeneities is not very large. In principle, one may always choose an isotropic constant reference elasticity tensor for the equivalent homogenized solid. However, the nature of the solution to the final set of equations depends on how closely the elasticity of the homogenized solid represents the final overall elasticity tensor.

Let the constant *reference* elasticity tensor $\mathbf{C}$ be isotropic,

$$ \mathbf{C} = \lambda\, \mathbf{1}^{(2)} \otimes \mathbf{1}^{(2)} + 2\mu\, \mathbf{1}^{(4s)}. \tag{12.4.15} $$

Direct substitution into $\boldsymbol{\xi}\boldsymbol{\cdot}\mathbf{C}\boldsymbol{\cdot}\boldsymbol{\xi}$ yields

$$ \boldsymbol{\xi}\boldsymbol{\cdot}\mathbf{C}\boldsymbol{\cdot}\boldsymbol{\xi} = (\lambda+\mu)\, \boldsymbol{\xi}\otimes\boldsymbol{\xi} + \mu\, \xi^2 \mathbf{1}^{(2)}, \tag{12.4.16a} $$

where

$$ \xi = |\boldsymbol{\xi}| = \sqrt{\xi_i\, \xi_i}. \tag{12.4.16b} $$

Set $(\boldsymbol{\xi}\boldsymbol{\cdot}\mathbf{C}\boldsymbol{\cdot}\boldsymbol{\xi})^{-1} = A\, \boldsymbol{\xi}\otimes\boldsymbol{\xi} + B\, \mathbf{1}^{(2)}$. Then, $(\boldsymbol{\xi}\boldsymbol{\cdot}\mathbf{C}\boldsymbol{\cdot}\boldsymbol{\xi})\boldsymbol{\cdot}(\boldsymbol{\xi}\boldsymbol{\cdot}\mathbf{C}\boldsymbol{\cdot}\boldsymbol{\xi})^{-1} = \mathbf{1}^{(2)}$ becomes

$$ \{A(\lambda+2\mu)\xi^2 + B(\lambda+\mu)\}\, \boldsymbol{\xi}\otimes\boldsymbol{\xi} + B\mu\xi^2\, \mathbf{1}^{(2)} = \mathbf{1}^{(2)}, \tag{12.4.17a} $$

from which A and B are obtained,

$$A = -\frac{\lambda+\mu}{\mu(\lambda+2\mu)}\xi^{-4}, \qquad B = \frac{1}{\mu}\xi^{-2}. \tag{12.4.17b,c}$$

Therefore, the inverse of $(\boldsymbol{\xi}\cdot\mathbf{C}\cdot\boldsymbol{\xi})$ is

$$(\boldsymbol{\xi}\cdot\mathbf{C}\cdot\boldsymbol{\xi})^{-1} = \frac{1}{\mu}\,\xi^{-2}\,\{-\frac{\lambda+\mu}{\lambda+2\mu}\xi^{-2}\boldsymbol{\xi}\otimes\boldsymbol{\xi} + \mathbf{1}^{(2)}\} \qquad (\boldsymbol{\xi} \neq \mathbf{0}). \tag{12.4.17d}$$

To simplify notation, define $\bar{\boldsymbol{\xi}} = \boldsymbol{\xi}/\xi$ with $\xi = |\boldsymbol{\xi}|$. Then, (12.4.17d) becomes

$$(\boldsymbol{\xi}\cdot\mathbf{C}\cdot\boldsymbol{\xi})^{-1} = \frac{1}{\mu}\,\xi^{-2}\{-\frac{\lambda+\mu}{\lambda+2\mu}\,\bar{\boldsymbol{\xi}}\otimes\bar{\boldsymbol{\xi}} + \mathbf{1}^{(2)}\} \tag{12.4.18a}$$

or in component form,

$$(\boldsymbol{\xi}\cdot\mathbf{C}\cdot\boldsymbol{\xi})^{-1}_{ij} = \frac{1}{\mu}\,\xi^{-2}\,\{-\frac{\lambda+\mu}{\lambda+2\mu}\bar{\xi}_i\,\bar{\xi}_j + \delta_{ij}\}. \tag{12.4.18b}$$

Using (12.4.18), compute the fourth-order tensor $_F\mathbf{S}^P$ when the reference equivalent homogeneous solid is isotropic,

$$\begin{aligned} {}_F\mathbf{S}^P(\boldsymbol{\xi}) &= 2sym\,(\bar{\boldsymbol{\xi}}\otimes\mathbf{1}^{(2)}\otimes\bar{\boldsymbol{\xi}}) - \frac{2(\lambda+\mu)}{\lambda+2\mu}\,\bar{\boldsymbol{\xi}}\otimes\bar{\boldsymbol{\xi}}\otimes\bar{\boldsymbol{\xi}}\otimes\bar{\boldsymbol{\xi}} + \frac{\lambda}{\lambda+2\mu}\,\bar{\boldsymbol{\xi}}\otimes\bar{\boldsymbol{\xi}}\otimes\mathbf{1}^{(2)} \\ &= 2sym\,(\bar{\boldsymbol{\xi}}\otimes\mathbf{1}^{(2)}\otimes\bar{\boldsymbol{\xi}}) - \frac{1}{1-\nu}\,\bar{\boldsymbol{\xi}}\otimes\bar{\boldsymbol{\xi}}\otimes\bar{\boldsymbol{\xi}}\otimes\bar{\boldsymbol{\xi}} + \frac{\nu}{1-\nu}\,\bar{\boldsymbol{\xi}}\otimes\bar{\boldsymbol{\xi}}\otimes\mathbf{1}^{(2)}, \end{aligned} \tag{12.4.19}$$

where ν is the Poisson ratio, $\nu = \lambda/2(\lambda+\mu)$. Similarly, $_F\mathbf{T}^P$ becomes

$$\begin{aligned} {}_F\mathbf{T}^P(\boldsymbol{\xi}) &= \mathbf{1}^{(4s)} - 2sym\,(\bar{\boldsymbol{\xi}}\otimes\mathbf{1}^{(2)}\otimes\bar{\boldsymbol{\xi}}) - \frac{\lambda}{\lambda+2\mu}\,\mathbf{1}^{(2)}\otimes\bar{\boldsymbol{\xi}}\otimes\bar{\boldsymbol{\xi}} \\ &\quad + \frac{2(\lambda+\mu)}{\lambda+2\mu}\,\bar{\boldsymbol{\xi}}\otimes\bar{\boldsymbol{\xi}}\otimes\bar{\boldsymbol{\xi}}\otimes\bar{\boldsymbol{\xi}} \\ &= \mathbf{1}^{(4s)} - 2sym\,(\bar{\boldsymbol{\xi}}\otimes\mathbf{1}^{(2)}\otimes\bar{\boldsymbol{\xi}}) - \frac{\nu}{1-\nu}\,\mathbf{1}^{(2)}\otimes\bar{\boldsymbol{\xi}}\otimes\bar{\boldsymbol{\xi}} + \frac{1}{1-\nu}\,\bar{\boldsymbol{\xi}}\otimes\bar{\boldsymbol{\xi}}\otimes\bar{\boldsymbol{\xi}}\otimes\bar{\boldsymbol{\xi}}, \end{aligned} \tag{12.4.20}$$

and $_F\boldsymbol{\Gamma}^P$ becomes

$$\begin{aligned} {}_F\boldsymbol{\Gamma}^P(\boldsymbol{\xi}) &= \frac{1}{\mu}\{-\frac{\lambda+\mu}{\lambda+2\mu}\bar{\boldsymbol{\xi}}\otimes\bar{\boldsymbol{\xi}}\otimes\bar{\boldsymbol{\xi}}\otimes\bar{\boldsymbol{\xi}} + sym\,(\bar{\boldsymbol{\xi}}\otimes\mathbf{1}^{(2)}\otimes\bar{\boldsymbol{\xi}})\} \\ &= \frac{1}{\mu}\{-\frac{1}{2(1-\nu)}\,\bar{\boldsymbol{\xi}}\otimes\bar{\boldsymbol{\xi}}\otimes\bar{\boldsymbol{\xi}}\otimes\bar{\boldsymbol{\xi}} + sym\,(\bar{\boldsymbol{\xi}}\otimes\mathbf{1}^{(2)}\otimes\bar{\boldsymbol{\xi}})\}. \end{aligned} \tag{12.4.21}$$

As is seen, $_F\Gamma^P_{ijkl} = {}_F\Gamma^P_{klij}$, whereas this symmetry does not hold for $_F\mathbf{S}^P$ and $_F\mathbf{T}^P$.

12.4.5. Consistency Conditions

In Subsection 12.4.3, the periodic strain and stress fields produced by a prescribed periodic eigenstrain (or eigenstress) field are obtained . As mentioned in Subsection 12.3, a periodic displacement field cannot include a linear part which corresponds to a uniform strain field.

Uniform stress and strain fields, however, are periodic, and satisfy all the governing field equations. Therefore, the consistency conditions can be rewritten for the eigenstrain and eigenstress, as

$$\mathbf{C}'(\mathbf{x}) : (\boldsymbol{\varepsilon}^{\mathrm{o}} + \boldsymbol{\varepsilon}^{\mathrm{p}}(\mathbf{x})) = \mathbf{C} : (\boldsymbol{\varepsilon}^{\mathrm{o}} + \boldsymbol{\varepsilon}^{\mathrm{p}}(\mathbf{x}) - \boldsymbol{\varepsilon}^{*}(\mathbf{x})),$$

$$\mathbf{D}'(\mathbf{x}) : (\boldsymbol{\sigma}^{\mathrm{o}} + \boldsymbol{\sigma}^{\mathrm{p}}(\mathbf{x})) = \mathbf{D} : (\boldsymbol{\sigma}^{\mathrm{o}} + \boldsymbol{\sigma}^{\mathrm{p}}(\mathbf{x}) - \boldsymbol{\sigma}^{*}(\mathbf{x})), \qquad (12.4.22\mathrm{a,b})$$

where $\boldsymbol{\varepsilon}^{\mathrm{o}}$ and $\boldsymbol{\sigma}^{\mathrm{o}}$ are constant symmetric tensors, and $\boldsymbol{\varepsilon}^{\mathrm{p}}$ and $\boldsymbol{\sigma}^{\mathrm{p}}$ are the periodic strain and stress fields produced by $\boldsymbol{\varepsilon}^{*}$ and $\boldsymbol{\sigma}^{*}$, respectively.[8] From (12.4.10a,b), $\boldsymbol{\varepsilon}^{\mathrm{p}}$ and $\boldsymbol{\sigma}^{\mathrm{p}}$ are given by

$$\boldsymbol{\varepsilon}^{\mathrm{p}}(\mathbf{x}) = \sum_{\boldsymbol{\xi}}{}' F\mathbf{S}^{\mathrm{P}}(\boldsymbol{\xi}) : \{\frac{1}{\mathrm{U}}\int_{\mathrm{U}} \boldsymbol{\varepsilon}^{*}(\mathbf{y}) \exp(\iota\boldsymbol{\xi}\cdot(\mathbf{x}-\mathbf{y}))\, \mathrm{dV_y}\},$$

$$\boldsymbol{\sigma}^{\mathrm{p}}(\mathbf{x}) = \sum_{\boldsymbol{\xi}}{}' F\mathbf{T}^{\mathrm{P}}(\boldsymbol{\xi}) : \{\frac{1}{\mathrm{U}}\int_{\mathrm{U}} \boldsymbol{\sigma}^{*}(\mathbf{y}) \exp(\iota\boldsymbol{\xi}\cdot(\mathbf{x}-\mathbf{y}))\, \mathrm{dV_y}\}. \qquad (12.4.22\mathrm{c,d})$$

With the aid of (12.4.22c), consistency condition (12.4.22a) yields the following integral equation for the eigenstrain field:

$$(\mathbf{C}'(\mathbf{x}) - \mathbf{C}) : \left[\, \boldsymbol{\varepsilon}^{\mathrm{o}} + \sum_{\boldsymbol{\xi}}{}' F\mathbf{S}^{\mathrm{P}}(\boldsymbol{\xi}) : \{\frac{1}{\mathrm{U}}\int_{\mathrm{U}} \boldsymbol{\varepsilon}^{*}(\mathbf{y}) \exp(\iota\boldsymbol{\xi}\cdot(\mathbf{x}-\mathbf{y}))\, \mathrm{dV_y}\} \right]$$

$$+ \mathbf{C} : \boldsymbol{\varepsilon}^{*}(\mathbf{x}) = \mathbf{0}. \qquad (12.4.22\mathrm{e})$$

Similarly, from (12.4.22d) and the consistency condition (12.4.22b), the following integral equation for the eigenstress field is obtained:

$$(\mathbf{D}'(\mathbf{x}) - \mathbf{D}) : \left[\, \boldsymbol{\sigma}^{\mathrm{o}} - \sum_{\boldsymbol{\xi}}{}' F\mathbf{T}^{\mathrm{P}}(\boldsymbol{\xi}) : \{\frac{1}{\mathrm{U}}\int_{\mathrm{U}} \boldsymbol{\sigma}^{*}(\mathbf{y}) \exp(\iota\boldsymbol{\xi}\cdot(\mathbf{x}-\mathbf{y}))\, \mathrm{dV_y}\} \right]$$

$$+ \mathbf{D} : \boldsymbol{\sigma}^{*}(\mathbf{x}) = \mathbf{0}. \qquad (12.4.22\mathrm{f})$$

Integral equations (12.4.22e,f) can be solved with the aid of Fourier series. Using (12.3.13), obtain, from (12.4.22e),

$$F\mathbf{C}'(\boldsymbol{\xi}) : \boldsymbol{\varepsilon}^{\mathrm{o}} + \sum_{\boldsymbol{\zeta}}{}' F\mathbf{C}'(\boldsymbol{\xi} - \boldsymbol{\zeta}) : F\mathbf{S}^{\mathrm{P}}(\boldsymbol{\zeta}) : F\boldsymbol{\varepsilon}^{*}(\boldsymbol{\zeta})$$

$$= \mathbf{C} : \{\boldsymbol{\varepsilon}^{\mathrm{o}} + F\mathbf{S}^{\mathrm{P}}(\boldsymbol{\xi}) : F\boldsymbol{\varepsilon}^{*}(\boldsymbol{\xi}) - F\boldsymbol{\varepsilon}^{*}(\boldsymbol{\xi})\}, \qquad (12.4.23\mathrm{a})$$

and from (12.4.22f),

$$F\mathbf{D}'(\boldsymbol{\xi}) : \boldsymbol{\sigma}^{\mathrm{o}} + \sum_{\boldsymbol{\zeta}}{}' F\mathbf{D}'(\boldsymbol{\xi} - \boldsymbol{\zeta}) : F\mathbf{T}^{\mathrm{P}}(\boldsymbol{\zeta}) : F\boldsymbol{\sigma}^{*}(\boldsymbol{\zeta})$$

$$= \mathbf{D} : \{\boldsymbol{\sigma}^{\mathrm{o}} + F\mathbf{T}^{\mathrm{P}}(\boldsymbol{\xi}) : F\boldsymbol{\sigma}^{*}(\boldsymbol{\xi}) - F\boldsymbol{\sigma}^{*}(\boldsymbol{\xi})\}. \qquad (12.4.23\mathrm{b})$$

Here, *the uniform eigenstrain and eigenstress fields associated with the Fourier*

---

[8] Superscript p emphasizes that these fields satisfy periodicity, and hence have zero volume averages.

coefficients $F\boldsymbol{\varepsilon}^*(\mathbf{0})$ *and* $F\boldsymbol{\sigma}^*(\mathbf{0})$ *must be included in* (12.4.23a,b), *although these terms are not included in* (12.4.22c,d). This is because the uniform fields $\boldsymbol{\varepsilon}^o$ and $\boldsymbol{\sigma}^o$ are related through

$$\boldsymbol{\sigma}^o = \mathbf{C} : (\boldsymbol{\varepsilon}^o - F\boldsymbol{\varepsilon}^*(\mathbf{0})) \qquad \text{or} \qquad \boldsymbol{\varepsilon}^o = \mathbf{D} : (\boldsymbol{\sigma}^o - F\boldsymbol{\sigma}^*(\mathbf{0})), \tag{12.4.24a,b}$$

as well as through the overall relations $\boldsymbol{\sigma}^o = \bar{\mathbf{C}} : \boldsymbol{\varepsilon}^o$ or $\boldsymbol{\varepsilon}^o = \bar{\mathbf{D}} : \boldsymbol{\sigma}^o$. If the infinite sums in (12.4.23a,b) are truncated, i.e., if n_i is restricted to, $-N < n_i < N$ ($i = 1, 2, 3$), then these equations are reduced to two sets of $(2N + 1)^3$ linear equations for the complex-valued tensors $F\boldsymbol{\varepsilon}^*(\boldsymbol{\xi})$ and $F\boldsymbol{\sigma}^*(\boldsymbol{\xi})$. In this manner, approximate solutions for the eigenstrains and eigenstresses are obtained. When the elasticity tensor $\mathbf{C}'(\mathbf{x})$ is suitably smooth, e.g., it is piecewise constant, then it can be shown that the infinite series in (12.4.23a,b) are convergent, and the exact solutions of (12.4.23a,b) can be estimated to any desired degree of accuracy, by choosing N suitably large.

It is emphasized that substitution for $\boldsymbol{\varepsilon}^p(\mathbf{x})$ from (12.4.22c) into (12.4.22a), yields the integral equation (12.4.22e) for the eigenstrain field $\boldsymbol{\varepsilon}^*(\mathbf{x})$. Similarly, from (12.4.22d) and (12.4.22b), the integral equation (12.4.22f) is obtained for the eigenstress field $\boldsymbol{\sigma}^*(\mathbf{x})$. The solution of these integral equations then provides the required exact eigenstrain (eigenstress) field which produces, for a chosen reference uniform elasticity tensor $\mathbf{C}$, a homogenized solid equivalent to the original periodically heterogeneous solid. In the latter part of this section, this technique is illustrated by means of several examples.

12.4.6. Alternative Formulation

The formulation of a heterogeneous elastic solid with periodic microstructure in terms of a representative unit cell, using Fourier series expansions of the field quantities, is effective, straightforward, and allows for an evaluation of the overall properties of the heterogeneous solid in a systematic manner. In addition, as pointed out at the beginning of Section 13, and elsewhere in Chapter IV, the results also directly apply to a finite heterogeneous elastic solid, in the shape of the considered unit cell subjected to suitable overall boundary data.

There is an alternative but essentially equivalent method of formulating the problem of a heterogeneous solid with periodic microstructure, which directly deals with the infinitely extended solid consisting of identical unit cells repeated in all three directions to infinity, and employs Green's function for an infinite *homogeneous* elastic solid having the reference uniform elasticity tensor, $\mathbf{C}$. Furthermore, this Green function takes on a particularly simple form when the reference elasticity tensor, $\mathbf{C}$, is isotropic; see Subsection 11.1. This technique has been discussed and illustrated by Walker *et al.* (1990, 1991). In this subsection the equivalence of the two alternative formulations is established.

As is pointed out in Subsection 12.4.3, there are four equivalent integral operators which can be used to establish the overall response and properties of a homogenized unit cell of periodic structure. These are: $-\boldsymbol{\Gamma}^P(\mathbf{x}; \boldsymbol{\sigma}^*)$ and $\mathbf{T}^P(\mathbf{x}; \boldsymbol{\sigma}^*)$, which, respectively, give the periodic strain and stress fields in the homogenized unit cell, due to the prescribed eigenstresses $\boldsymbol{\sigma}^*$; and $-\boldsymbol{\Lambda}^P(\mathbf{x}; \boldsymbol{\varepsilon}^*)$

and $\mathbf{S}^{\mathrm{P}}(\mathbf{x}; \boldsymbol{\varepsilon}^*)$, which, respectively, yield the periodic stress and strain fields in the homogenized unit cell, produced by the prescribed eigenstrains $\boldsymbol{\varepsilon}^*$. Because of the equivalence of these operators, in the sequel, only $\boldsymbol{\varepsilon}^{\mathrm{p}} = -\boldsymbol{\Gamma}^{\mathrm{P}}(\mathbf{x}; \boldsymbol{\sigma}^*)$ is considered, where homogenization of the unit cell is accomplished by the introduction of the appropriate eigenstress field $\boldsymbol{\sigma}^*(\mathbf{x})$.

As detailed in the previous subsections, the integral operator $\boldsymbol{\Gamma}^{\mathrm{P}}(\mathbf{x}; \boldsymbol{\sigma}^*)$ has the following Fourier series representation:

$$\boldsymbol{\Gamma}^{\mathrm{P}}(\mathbf{x}; \boldsymbol{\sigma}^*) = \sum_{\boldsymbol{\xi}}{}' \, F\boldsymbol{\Gamma}^{\mathrm{P}}(\boldsymbol{\xi}) : < \boldsymbol{\sigma}^*(\mathbf{y}) \exp(\iota\boldsymbol{\xi}\cdot(\mathbf{x}-\mathbf{y})) >_{\mathrm{U}} \qquad \text{for } \mathbf{x} \text{ in U}, \tag{12.4.25}$$

where the Fourier coefficients, $F\boldsymbol{\Gamma}^{\mathrm{P}}(\boldsymbol{\xi})$, are defined by (12.4.12d). In (12.4.25), the discrete variable, $\boldsymbol{\xi}$, has the components, $\xi_i = \pi n_i/a_i$ (i not summed), and the triple summation is with respect to $n_i = 0, \pm 1, ..., \pm\infty$, with $\boldsymbol{\xi} \neq \mathbf{0}$. Since the transformation from Fourier series to Fourier transform representation requires transformation from discrete variables in the triple infinite summation, to continuous variables in the corresponding triple infinite integration, it proves convenient to introduce a three by three *diagonal* matrix, $\mathbf{A}$, with the following nonzero components:

$$A_{11} = 2a_1, \qquad A_{22} = 2a_2, \qquad A_{33} = 2a_3, \tag{12.4.26a~c}$$

and express $\boldsymbol{\xi}$ in terms of $\mathbf{n}$, with components n_i (i = 1, 2, 3), as

$$\boldsymbol{\xi} = 2\pi\, \mathbf{A}^{-1}\cdot\mathbf{n}. \tag{12.4.26d}$$

Consider now the infinitely extended homogenized linearly elastic solid of reference elasticity $\mathbf{C}$, and let $\boldsymbol{\sigma}^*(\mathbf{x})$ be an eigenstress field prescribed in a finite region B, such that it admits a Fourier transform and an inverse Fourier transform, as follows:

$$F\boldsymbol{\sigma}^*(\boldsymbol{\zeta}) = \frac{1}{(2\pi)^3} \int_{-\infty}^{+\infty} \boldsymbol{\sigma}^*(\mathbf{x}) \exp(-\iota\boldsymbol{\zeta}\cdot\mathbf{x})\, dV_{\mathbf{x}},$$

$$\boldsymbol{\sigma}^*(\mathbf{x}) = \int_{-\infty}^{+\infty} F\boldsymbol{\sigma}^*(\boldsymbol{\zeta}) \exp(\iota\boldsymbol{\zeta}\cdot\mathbf{x})\, dV_{\zeta}. \tag{12.4.27a,b}$$

This prescribed eigenstress field produces the strain field, $\boldsymbol{\varepsilon}(\mathbf{x})$, which may be expressed as

$$\boldsymbol{\varepsilon}(\mathbf{x}) = -\boldsymbol{\Gamma}(\mathbf{x}; \boldsymbol{\sigma}^*), \tag{12.4.28a}$$

where

$$\boldsymbol{\Gamma}(\mathbf{x}; \boldsymbol{\sigma}^*) = \frac{1}{(2\pi)^3} \int_{-\infty}^{+\infty} \{\int_{\mathrm{B}} F\boldsymbol{\Gamma}(\boldsymbol{\zeta}) : \boldsymbol{\sigma}^*(\mathbf{y}) \exp(\iota\boldsymbol{\zeta}\cdot(\mathbf{x}-\mathbf{y}))\, dV_{\mathbf{y}}\}\, dV_{\zeta}. \tag{12.4.28b}$$

In (12.4.28b) the integration with respect to $\mathbf{y}$ is taken over the region where $\boldsymbol{\sigma}^*$ is prescribed. This eigenstress field must be appropriately restricted so that the corresponding infinite integrals are meaningful and finite.

Under suitable restrictions, it may be expected that the Green function for the homogeneous infinite domain should also apply to the equivalent homogenized periodic structure, even if the field variables are not assumed to be periodic.[9] The form of the Fourier transform and the form of the Fourier expansion are essentially the same, except for the following two points:

1) the variables in the Fourier transform are *continuous* and real, and range from minus infinity to plus infinity, while those in the Fourier series are *discrete* integers which also range from minus infinity to plus infinity; and

2) the Fourier transform uses $\int_{-\infty}^{+\infty} \exp(\iota\boldsymbol{\xi}.\mathbf{x})\, dV_{\mathbf{x}} = (2\pi)^3$ for $\boldsymbol{\xi} = \mathbf{0}$ and is zero otherwise, while the Fourier series uses $\int_U \exp(\iota\boldsymbol{\xi}.\mathbf{x})\, dV_{\mathbf{x}} = U$ for $\boldsymbol{\xi} = \mathbf{0}$ and is zero otherwise, where $\boldsymbol{\xi} = 2\pi\mathbf{A}^{-1}.\mathbf{n}$.

The periodic structure consists of an infinite number of unit cells, each of which contains the same eigenstress. Hence, the field variables in the periodic structure are periodic. It therefore may be expected that *if the integral operator* $\boldsymbol{\Gamma}$ *is applied to the eigenstresses in the infinite set of unit cells, it should produce a strain field identical with the one defined by* $\boldsymbol{\Gamma}^P$ *over a typical unit cell.* That is, if all unit cells are identified by $U(\mathbf{m})$, with $\mathbf{m}$ being a set of three integers, $\mathbf{m} \equiv (m_1, m_2, m_3)$, then the following indentity is expected to hold:

$$\boldsymbol{\Gamma}^P(\mathbf{x}; \boldsymbol{\sigma}^*) = \sum_{\mathbf{m}} \boldsymbol{\Gamma}(\mathbf{x}; (\boldsymbol{\sigma}^* \text{ in } U(\mathbf{m}))\,) \qquad \text{for } \mathbf{x} \text{ in } U(\mathbf{0}), \tag{12.4.29}$$

where $\boldsymbol{\sigma}^*$ in $\boldsymbol{\Gamma}^P$ is the eigenstress field in the original unit cell, $U = U(\mathbf{0})$, and the same eigenstress field is distributed in each cell. Although (12.4.29) is intuitively acceptable, it is not certain whether the infinite sum of $\boldsymbol{\Gamma}(\mathbf{x}; (\boldsymbol{\sigma}^* \text{ in } U(\mathbf{m}))\,)$ in the right-hand side is actually convergent. On the other hand, the left-hand side of (12.4.29) is a well-defined quantity for a rather general class of eigenfunctions $\boldsymbol{\sigma}^*$.

With the aid of *the Poisson sum formula*, (12.4.29) can be established formally.[10] To simplify notation, define the term which appears in $\boldsymbol{\Gamma}^P$ by

$$\mathbf{h}(\mathbf{x}; \mathbf{n}) \equiv F\boldsymbol{\Gamma}^P(2\pi\, \mathbf{A}^{-1}.\mathbf{n}) < \exp(\iota(2\pi\, \mathbf{A}^{-1}.\mathbf{n}).(\mathbf{x}-\mathbf{y})) >, \tag{12.4.30a}$$

where the volume average is taken with respect to $\mathbf{y}$. Then, the Poisson sum formula transforms the infinite summation of $\mathbf{h}(\mathbf{x}; \mathbf{n})$ with respect to $\mathbf{n}$, into the following alternative form:

$$\sum_{\mathbf{n}}{}' \mathbf{h}(\mathbf{x}; \mathbf{n}) = \sum_{\mathbf{m}} \frac{1}{(2\pi)^3}\, U \int_{-\infty}^{+\infty} \mathbf{h}(\mathbf{x}; \boldsymbol{\zeta}') \exp(\iota\boldsymbol{\zeta}.\mathbf{A}.\mathbf{m})\, dV_{\zeta}, \tag{12.4.31}$$

where $\boldsymbol{\zeta}' = \mathbf{A}.\boldsymbol{\zeta}/2\pi$ and $U = 8a_1a_2a_3$. Hence, from

$$\mathbf{h}(\mathbf{x}; \boldsymbol{\zeta}') = F\boldsymbol{\Gamma}^P(2\pi\, \mathbf{A}^{-1}.\boldsymbol{\zeta}') < \exp(\iota(2\pi\, \mathbf{A}^{-1}.\boldsymbol{\zeta}').(\mathbf{x}-\mathbf{y})) >_U$$

---

[9] Note that the Fourier transform cannot be applied directly to a periodic field, since the infinite integral of the periodic function, which does not vanish at infinity, may not converge.

[10] The convergence of the infinite summation is discussed by several authors; see Furuhashi *et al.* (1981) and Mura (1987) for periodically distributed eigenstrains, and Horii and Sahasakmontri (1990) for a problem of periodically distributed cracks; see also Jeffrey (1973) and McCoy and Beran (1976) for nonmechanical problems.

$$= F\mathbf{\Gamma}^{\mathrm{P}}(\boldsymbol{\zeta})\, \frac{1}{\mathrm{U}} \int_{\mathrm{U}} \exp(\iota \boldsymbol{\zeta} \cdot (\mathbf{x} - \mathbf{y}))\, \mathrm{dV}_{\mathbf{y}}, \tag{12.4.30b}$$

the following expression for $\mathbf{\Gamma}^{\mathrm{P}}(\mathbf{x};\, \boldsymbol{\sigma}^*)$ is obtained:

$$\mathbf{\Gamma}^{\mathrm{P}}(\mathbf{x};\, \boldsymbol{\sigma}^*) = \sum_{\mathbf{m}} \frac{1}{(2\pi)^3} \int_{-\infty}^{+\infty} F\mathbf{\Gamma}^{\mathrm{P}}(\boldsymbol{\zeta}) \exp(\iota \boldsymbol{\zeta} \cdot \mathbf{x})$$

$$\times \{ \int_{\mathrm{U}} \boldsymbol{\sigma}^*(\mathbf{y}) \exp(-\iota \boldsymbol{\zeta} \cdot (\mathbf{y} - \mathbf{A} \cdot \mathbf{m}))\, \mathrm{dV}_{\mathbf{y}} \}\, \mathrm{dV}_{\zeta}. \tag{12.4.32a}$$

The eigenstress field in the periodic structure satisfies the periodicity, $\boldsymbol{\sigma}^*(\mathbf{y} + \mathbf{A} \cdot \mathbf{m}) = \boldsymbol{\sigma}^*(\mathbf{y})$, for any $\mathbf{m}$. Hence, since $\mathrm{U}(\mathbf{m})$ may be obtained through a rigid-body translation of $\mathrm{U}(\mathbf{0})$ by $\mathbf{A} \cdot \mathbf{m}$, the sum of the integral of $\boldsymbol{\sigma}^*(\mathbf{y}) \exp(-\iota \boldsymbol{\zeta} \cdot (\mathbf{y} - \mathbf{A} \cdot \mathbf{m}))$ for different $\mathbf{m}$'s taken over $\mathrm{U} = \mathrm{U}(\mathbf{0})$, can be replaced by the sum of the integral of $\boldsymbol{\sigma}^*(\mathbf{y}) \exp(-\iota \boldsymbol{\zeta} \cdot \mathbf{y})$ taken over $\mathrm{U}(\mathbf{m})$'s, as follows:

$$\sum_{\mathbf{m}} \int_{\mathrm{U}} \boldsymbol{\sigma}^*(\mathbf{y}) \exp(-\iota \boldsymbol{\zeta} \cdot (\mathbf{y} - \mathbf{A} \cdot \mathbf{m}))\, \mathrm{dV}_{\mathbf{y}} = \sum_{\mathbf{m}} \int_{\mathrm{U}(\mathbf{m})} \boldsymbol{\sigma}^*(\mathbf{y}) \exp(-\iota \boldsymbol{\zeta} \cdot \mathbf{y})\, \mathrm{dV}_{\mathbf{y}}. \tag{12.4.33}$$

Hence, (12.4.32a) becomes

$$\mathbf{\Gamma}^{\mathrm{P}}(\mathbf{x};\, \boldsymbol{\sigma}^*) = \frac{1}{(2\pi)^3} \int_{-\infty}^{+\infty} F\mathbf{\Gamma}^{\mathrm{P}}(\boldsymbol{\zeta}) \exp(\iota \boldsymbol{\zeta} \cdot \mathbf{x})$$

$$\times \{ \sum_{\mathbf{m}} \int_{\mathrm{U}(\mathbf{m})} \boldsymbol{\sigma}^*(\mathbf{y}) \exp(-\iota \boldsymbol{\zeta} \cdot \mathbf{y})\, \mathrm{dV}_{\mathbf{y}} \}\, \mathrm{dV}_{\zeta}$$

$$= \sum_{\mathbf{m}} \frac{1}{(2\pi)^3} \int_{-\infty}^{+\infty} \{ \int_{\mathrm{U}(\mathbf{m})} F\mathbf{\Gamma}(\boldsymbol{\zeta}) : \boldsymbol{\sigma}^*(\mathbf{y}) \exp(\iota \boldsymbol{\zeta} \cdot (\mathbf{x} - \mathbf{y}))\, \mathrm{dV}_{\mathbf{y}} \}\, \mathrm{dV}_{\zeta}. \tag{12.4.32b}$$

Since the second integral in the right-hand side of (12.4.32b) is $\mathbf{\Gamma}(\mathbf{x};\, \boldsymbol{\sigma}^*)$ for $\mathbf{x}$ in $\mathrm{U}(\mathbf{0})$ and $\boldsymbol{\sigma}^*$ in $\mathrm{U}(\mathbf{m})$, identity (12.4.29) is proved.

In the above discussion, the infinite summation in (12.4.29) is the limit of a corresponding finite sum as the number of terms goes to infinity. Hence, (12.4.29) should read

$$\mathbf{\Gamma}^{\mathrm{P}}(\mathbf{x};\, \boldsymbol{\sigma}^*) = \lim_{\mathbf{N} \to \infty} \sum_{\mathbf{n}}^{\mathbf{N}}{}' \, \mathbf{h}(\mathbf{x};\, \mathbf{n})$$

$$= \lim_{\mathbf{M} \to \infty} \sum_{\mathbf{m}}^{\mathbf{M}} \mathbf{\Gamma}(\mathbf{x};\, (\boldsymbol{\sigma}^* \text{ in } \mathrm{U}(\mathbf{m}))\,). \tag{12.4.34}$$

For certain problems, the convergence of the sum of $\mathbf{h}(\mathbf{x};\, \mathbf{n})$ up to $\pm\mathbf{N}$, as $\mathbf{N}$ goes to infinity, may be slower than the convergence of the sum of $\mathbf{\Gamma}(\mathbf{x};\, (\boldsymbol{\sigma}^* \text{ in } \mathrm{U}(\mathbf{m}))\,)$ up to $\pm\mathbf{M}$, as $\mathbf{M}$ goes to infinity. In cases of this kind, it may be better to use a Green function formulation similar to (12.4.28), instead of the Fourier series formulation (12.4.27), to compute the field variables in the periodic structure model.[11]

12.5. TWO-PHASE PERIODIC MICROSTRUCTURE

Several illustrative examples of estimating the overall moduli of solids with periodic microstructures are presented in this subsection. For simplicity, only two-phase periodic microstructures are considered, where one phase is the matrix, M, with elasticity $\mathbf{C}$, and the other phase is an inclusion, Ω, with elasticity $\mathbf{C}^{\Omega}$, i.e.,

$$\mathbf{C}'(\mathbf{x}) = \mathrm{H}(\mathbf{x};\, \mathrm{M})\, \mathbf{C} + \mathrm{H}(\mathbf{x};\, \Omega)\, \mathbf{C}^{\Omega}, \tag{12.5.1a}$$

where $\mathrm{H}(\mathbf{x};\, \mathrm{M})$ and $\mathrm{H}(\mathbf{x};\, \Omega)$ are the Heaviside step functions associated with points in M and Ω, respectively. For the equivalent homogeneous solid, the matrix elasticity $\mathbf{C}$ is used. Then the eigenstrains are nonzero only in Ω, i.e.,

$$\boldsymbol{\varepsilon}^*(\mathbf{x}) = \mathrm{H}(\mathbf{x};\, \Omega)\, \boldsymbol{\varepsilon}^*(\mathbf{x}). \tag{12.5.1b}$$

12.5.1. Average Eigenstrain Formulation

From (12.4.22a), the consistency condition is

$$\{\mathrm{H}(\mathbf{x};\, \mathrm{M})\, \mathbf{C} + \mathrm{H}(\mathbf{x};\, \Omega)\, \mathbf{C}^{\Omega}\} : (\boldsymbol{\varepsilon}^{\mathrm{o}} + \boldsymbol{\varepsilon}^{\mathrm{p}}(\mathbf{x})) = \mathbf{C} : \{\boldsymbol{\varepsilon}^{\mathrm{o}} + \boldsymbol{\varepsilon}^{\mathrm{p}}(\mathbf{x}) - \mathrm{H}(\mathbf{x};\, \Omega)\, \boldsymbol{\varepsilon}^*(\mathbf{x})\},$$

$$\mathbf{x} \text{ in } \mathrm{U} = \mathrm{M} + \Omega, \tag{12.5.2a}$$

where the disturbance (periodic) strain field $\boldsymbol{\varepsilon}^{\mathrm{p}}(\mathbf{x})$ is determined by the eigenstrain field $\mathrm{H}(\mathbf{x};\, \Omega)\, \boldsymbol{\varepsilon}^*(\mathbf{x})$. Equivalently, (12.5.2a) can be rewritten as

$$\mathbf{C}^{\Omega} : (\boldsymbol{\varepsilon}^{\mathrm{o}} + \boldsymbol{\varepsilon}^{\mathrm{p}}(\mathbf{x})) = \mathbf{C} : (\boldsymbol{\varepsilon}^{\mathrm{o}} + \boldsymbol{\varepsilon}^{\mathrm{p}}(\mathbf{x}) - \boldsymbol{\varepsilon}^*(\mathbf{x})), \qquad \mathbf{x} \text{ in } \Omega, \tag{12.5.2b}$$

since the eigenstrains vanish in M because of the choice of the reference elasticity tensor.

Assume that the matrix (and hence the reference equivalent homogeneous solid) is isotropic. As shown in Subsection 12.4, the disturbance strain $\boldsymbol{\varepsilon}^{\mathrm{p}}$ is determined from

$$\boldsymbol{\varepsilon}^{\mathrm{p}}(\mathbf{x}) = \sum_{\boldsymbol{\xi}}{}' \, {}^F\mathbf{S}^{\mathrm{P}}(\boldsymbol{\xi}) : \{\frac{1}{\mathrm{U}} \int_{\mathrm{U}} \mathrm{H}(\mathbf{y};\, \Omega)\, \boldsymbol{\varepsilon}^*(\mathbf{y}) \exp(\iota\boldsymbol{\xi} \boldsymbol{\cdot} (\mathbf{x} - \mathbf{y}))\, \mathrm{dV}_{\mathbf{y}}\}$$

$$= \sum_{\boldsymbol{\xi}}{}' \, {}^F\mathbf{S}^{\mathrm{P}}(\boldsymbol{\xi}) : \{\frac{1}{\mathrm{U}} \int_{\Omega} \boldsymbol{\varepsilon}^*(\mathbf{y}) \exp(\iota\boldsymbol{\xi} \boldsymbol{\cdot} (\mathbf{x} - \mathbf{y}))\, \mathrm{dV}_{\mathbf{y}}\}, \tag{12.5.3}$$

where the fourth-order tensor ${}^F\mathbf{S}^{\mathrm{P}}$ is given by (12.4.13a). When $\mathbf{C}$ is isotropic,

[11] Expressions (12.4.29) and (12.4.34) are formally introduced; see Mura (1987) for discussions on the convergence of these infinite sums.

from (12.4.19) $F\mathbf{S}^{\mathrm{P}}$ can be rewritten as (Nemat-Nasser *et al.*, 1982; and Iwakuma and Nemat-Nasser, 1983)

$$F\mathbf{S}^{\mathrm{P}}(\boldsymbol{\xi}) = F\mathbf{S}^{1}(\boldsymbol{\xi}) - \frac{1}{1-\nu}\, F\mathbf{S}^{2}(\boldsymbol{\xi}) + \frac{\nu}{1-\nu}\, F\mathbf{S}^{3}(\boldsymbol{\xi}), \tag{12.5.4a}$$

where

$$F\mathbf{S}^{1}(\boldsymbol{\xi}) = 2sym\,(\bar{\boldsymbol{\xi}} \otimes \mathbf{1}^{(2)} \otimes \bar{\boldsymbol{\xi}}), \qquad F\mathbf{S}^{2}(\boldsymbol{\xi}) = \bar{\boldsymbol{\xi}} \otimes \bar{\boldsymbol{\xi}} \otimes \bar{\boldsymbol{\xi}} \otimes \bar{\boldsymbol{\xi}},$$

$$F\mathbf{S}^{3}(\boldsymbol{\xi}) = \bar{\boldsymbol{\xi}} \otimes \bar{\boldsymbol{\xi}} \otimes \mathbf{1}^{(2)}, \tag{12.5.4b~d}$$

with $\bar{\boldsymbol{\xi}} = \boldsymbol{\xi}/\xi$. As shown in earlier sections, when an infinitely extended isotropic elastic solid contains an isolated ellipsoidal inclusion, then the tensor $\mathbf{S}^{\mathrm{P}}$ reduces to Eshelby's constant tensor $\mathbf{S}$ which only depends on the Poisson ratio ν of the infinite solid. Tensors $F\mathbf{S}^{1}$, $F\mathbf{S}^{2}$, and $F\mathbf{S}^{3}$ are also independent of the material properties. Equations (12.5.4a~d), therefore, provide a powerful tool to study the overall response of solids with a periodic distribution of *arbitrary* inclusions and defects in an *isotropic* elastic matrix.

The exact eigenstrain field $\boldsymbol{\varepsilon}^{*}$ is defined by the solution of the set of linear equations (12.4.23a) for the Fourier coefficients, $F\boldsymbol{\varepsilon}^{*}$. In general, this is a difficult task; see Nemat-Nasser and Taya (1981, 1985) and Nunan and Keller (1984) for discussions of alternative methods of solution. A good approximation may be obtained if the eigenstrain $\mathrm{H}(\mathbf{x};\,\Omega)\,\boldsymbol{\varepsilon}^{*}(\mathbf{x})$ is replaced by its average value, $\mathrm{H}(\mathbf{x};\,\Omega)\,\bar{\boldsymbol{\varepsilon}}^{*}$, in (12.5.3), and the result is entered into the consistency condition (12.5.2b).[12] In this manner, the average eigenstrain $\bar{\boldsymbol{\varepsilon}}^{*}$ is estimated from

$$\boldsymbol{\varepsilon}^{\mathrm{p}}(\mathbf{x}) = \{\sum_{\boldsymbol{\xi}}{}'\, \mathrm{f}\,\mathrm{g}(-\,\boldsymbol{\xi})\, F\mathbf{S}^{\mathrm{P}}(\boldsymbol{\xi})\, \exp(\iota\boldsymbol{\xi}\cdot\mathbf{x})\} : \bar{\boldsymbol{\varepsilon}}^{*}, \tag{12.5.5a}$$

where $\mathrm{f} = \Omega/\mathrm{U}$, and

$$\mathrm{g}(\boldsymbol{\xi}) \equiv \frac{1}{\Omega} \int_{\Omega} \exp(\iota\boldsymbol{\xi}\cdot\mathbf{x})\, \mathrm{dV}_{\mathbf{x}}. \tag{12.5.5b}$$

Now the average value of the periodic strain in Ω becomes

$$\frac{1}{\Omega} \int_{\Omega} \boldsymbol{\varepsilon}^{\mathrm{p}}(\mathbf{x})\, \mathrm{dV}_{\mathrm{x}} = \mathbf{S}^{\mathrm{P}} : \bar{\boldsymbol{\varepsilon}}^{*}, \tag{12.5.6a}$$

where

$$\mathbf{S}^{\mathrm{P}} = \sum_{\boldsymbol{\xi}}{}'\, \mathrm{f}\,\mathrm{g}(-\boldsymbol{\xi})\, \mathrm{g}(\boldsymbol{\xi})\, F\mathbf{S}^{\mathrm{P}}(\boldsymbol{\xi}). \tag{12.5.6b}$$

It should be noted that since $\mathrm{g}(\boldsymbol{\xi})$ and $\mathrm{g}(-\boldsymbol{\xi})$ are complex conjugates, their product $\mathrm{g}(-\boldsymbol{\xi})\,\mathrm{g}(\boldsymbol{\xi})$ is real. Therefore, $\mathbf{S}^{\mathrm{P}}$ is a real-valued tensor, since $F\mathbf{S}^{\mathrm{P}}(\boldsymbol{\xi})$ is also real-valued.

Using (12.5.2b) and (12.5.6b), obtain an *average consistency condition* for the *average eigenstrain* $\bar{\boldsymbol{\varepsilon}}^{*}$, as follows:

---

[12] The reason for the effectiveness of this approximation is established in Section 13, with the aid of the Hashin-Shtrikman variational principle; see also Nunan and Keller (1984) where this is verified computationally.

$$\mathbf{C}^{\Omega} : (\boldsymbol{\varepsilon}^{\mathrm{o}} + \mathbf{S}^{\mathrm{P}} : \bar{\boldsymbol{\varepsilon}}^{*}) = \mathbf{C} : \{\boldsymbol{\varepsilon}^{\mathrm{o}} + (\mathbf{S}^{\mathrm{P}} - \mathbf{1}^{(4s)}) : \bar{\boldsymbol{\varepsilon}}^{*}\}. \tag{12.5.7a}$$

Note that the effect of the geometry of the inclusion Ω is represented in $\mathbf{S}^{\mathrm{P}}$ through the g-integral. For a given geometry of Ω, the g-integral, the three infinite series in (12.5.4a), and hence $\mathbf{S}^{\mathrm{P}}$ can be computed once and for all, and the results can then be used to solve the inclusion problem for various different material properties determined by $\mathbf{C}^{\Omega}$. Since the average eigenstrain $\bar{\boldsymbol{\varepsilon}}^{*}$ is obtained through (12.5.7a), the overall elasticity $\bar{\mathbf{C}}$ is computed from

$$\bar{\mathbf{C}} : \boldsymbol{\varepsilon}^{\mathrm{o}} = \mathbf{C} : (\boldsymbol{\varepsilon}^{\mathrm{o}} - \mathrm{f}\, \bar{\boldsymbol{\varepsilon}}^{*}) \tag{12.5.8a}$$

which equals the uniform overall stress $\boldsymbol{\sigma}^{\mathrm{o}}$.

Substituting from (12.5.7a) for $\bar{\boldsymbol{\varepsilon}}^{*}$ into (12.5.8a), and in view of the fact that $\boldsymbol{\varepsilon}^{\mathrm{o}}$ is arbitrary, obtain an explicit expression for the overall elasticity tensor,

$$\bar{\mathbf{C}} = \mathbf{C} - \mathrm{f}\, \{(\mathbf{C} - \mathbf{C}^{\Omega})^{-1} - \mathbf{S}^{\mathrm{P}} : \mathbf{C}^{-1}\}^{-1}. \tag{12.5.9a}$$

The form of (12.5.9a) appears slightly different from that of the equations derived for an RVE model; see, for example, Section 10. However, taking advantage of the fact that $\{(\mathbf{C} - \mathbf{C}^{\Omega})^{-1} - \mathbf{S}^{\mathrm{P}} : \mathbf{C}^{-1}\}^{-1}$ can be rewritten as $\mathbf{C} : \{(\mathbf{C} - \mathbf{C}^{\Omega})^{-1} : \mathbf{C} - \mathbf{S}^{\mathrm{P}}\}^{-1}$, rewrite (12.5.9a) as

$$\bar{\mathbf{C}} = \mathbf{C} : \left\{ \mathbf{1}^{(4s)} - \mathrm{f}\, \{(\mathbf{C} - \mathbf{C}^{\Omega})^{-1} : \mathbf{C} - \mathbf{S}^{\mathrm{P}}\}^{-1} \right\} \tag{12.5.9b}$$

which is of the same form as that obtained for a typical RVE,[13] except for the difference between $\mathbf{S}$, the Eshelby tensor for the infinite domain, and $\mathbf{S}^{\mathrm{P}}$, the tensor for the periodic structure defined by (12.5.4b). Note that the overall compliance tensor in this case becomes

$$\bar{\mathbf{D}} = \mathbf{D} : \left\{ \mathbf{1}^{(4s)} - \mathrm{f}\, \{(\mathbf{D} - \mathbf{D}^{\Omega})^{-1} : \mathbf{D} - \mathbf{T}^{\mathrm{P}}\}^{-1} \right\}, \tag{12.5.9c}$$

where $\mathbf{T}^{\mathrm{P}}$ is defined by $\mathbf{C} : (\mathbf{S}^{\mathrm{P}} - \mathbf{1}^{(4s)})$.

As the spacing of the inclusions increases, or as the volume fraction of Ω approaches zero, it is expected that the tensor $\mathbf{S}^{\mathrm{P}}$ defined by (12.5.6b) should satisfy, for an ellipsoidal Ω,

$$\lim_{\mathrm{f}\to 0} \mathbf{S}^{\mathrm{P}} = \lim_{\mathrm{f}\to 0} \{\sum_{\boldsymbol{\xi}}{}' \, \mathrm{f}\, g(-\boldsymbol{\xi})\, g(\boldsymbol{\xi})\, F\mathbf{S}^{\mathrm{P}}(\boldsymbol{\xi})\} = \mathbf{S}, \tag{12.5.10}$$

where $\mathbf{S}$ is the Eshelby tensor corresponding to Ω. Indeed, the discrete Fourier series can be replaced by a continuous Fourier integral, as the dimensions of the unit cell become very large. For an ellipsoidal Ω, one can then retrieve Eshelby's tensor. It should be kept in mind that only for an ellipsoidal Ω, does the distribution of the exact eigenstrain become uniform in Ω as f goes to zero, and (12.5.10) reduces to Eshelby's tensor. For other shapes of Ω, the distribution of the exact eigenstrain may not be uniform as f goes to zero, or as the dimensions of the unit cell become very large.

---

[13] See, e.g., (7.4.10a) with $\mathbf{A}^{\mathrm{I}} = (\mathbf{C} - \mathbf{C}^{\mathrm{I}})^{-1}$, of Subsection 7.4.

12.5.2. Modification for Multi-Phase Periodic Microstructure

When each unit cell contains several inclusions with different elastic moduli and geometries, (12.5.5b), (12.5.6b), and (12.5.7a) of the preceding subsection require minor modifications. Denote the elasticity and the average eigenstrain in the αth inclusion, Ω_α, by $\mathbf{C}^\alpha$ and $\bar{\boldsymbol{\varepsilon}}^{*\alpha}$ ($\alpha = 1, 2, ..., n$), respectively. Consistency condition (12.5.2b) for a two-phase periodic microstructure is replaced by

$$\mathbf{C}^\alpha : (\boldsymbol{\varepsilon}^o + \boldsymbol{\varepsilon}^p(\mathbf{x})) = \mathbf{C} : (\boldsymbol{\varepsilon}^o + \boldsymbol{\varepsilon}^p(\mathbf{x}) - \boldsymbol{\varepsilon}^*(\mathbf{x})), \qquad \mathbf{x} \text{ in } \Omega_\alpha, \tag{12.5.2c}$$

and the g-integral and $\mathbf{S}^P$, defined by (12.5.5b) and (12.5.6b), are replaced by the g_α-integral[14] and $\mathbf{S}^P(\Omega_\alpha, \Omega_\beta)$, which are defined by

$$g_\alpha(\boldsymbol{\xi}) \equiv \frac{1}{\Omega_\alpha} \int_{\Omega_\alpha} \exp(\iota \boldsymbol{\xi} \cdot \mathbf{x}) \, dV_\mathbf{x} \tag{12.5.5c}$$

and

$$\mathbf{S}^P(\Omega_\alpha, \Omega_\beta) \equiv \sum_{\xi}{}' f_\beta \, g_\alpha(\boldsymbol{\xi}) \, g_\beta(-\boldsymbol{\xi}) \, \mathcal{F}\mathbf{S}^P(\boldsymbol{\xi}). \tag{12.5.6c}$$

Hence, average consistency condition (12.5.7a) is replaced by the following average consistency condition over Ω_α:

$$\mathbf{C}^\alpha : \{\boldsymbol{\varepsilon}^o + \sum_{\beta=1}^{n} \mathbf{S}^P(\Omega_\alpha, \Omega_\beta) : \bar{\boldsymbol{\varepsilon}}^{*\beta}\}$$

$$= \mathbf{C} : \{\boldsymbol{\varepsilon}^o + \sum_{\beta=1}^{n} (\mathbf{S}^P(\Omega_\alpha, \Omega_\beta) - \delta_{\alpha\beta} \mathbf{1}^{(4s)}) : \bar{\boldsymbol{\varepsilon}}^{*\beta}\}, \tag{12.5.7b}$$

for $\alpha = 1, 2, ..., n$, where[15] $f_\alpha \equiv \Omega_\alpha / U$. Then, the overall uniform stress is given by

$$\bar{\mathbf{C}} : \boldsymbol{\varepsilon}^o = \mathbf{C} : \{\boldsymbol{\varepsilon}^o - \sum_{\alpha=1}^{n} f_\alpha \bar{\boldsymbol{\varepsilon}}^{*\alpha}\}, \tag{12.5.8b}$$

the overall elasticity $\bar{\mathbf{C}}$ is computed from the solution of the set of linear tensorial equations (12.5.7b), for $\{\bar{\boldsymbol{\varepsilon}}^{*\alpha}\}$, and hence $\bar{\boldsymbol{\varepsilon}}^* = \sum_{\alpha=1}^{n} f_\alpha \bar{\boldsymbol{\varepsilon}}^{*\alpha}$.

12.5.3. Properties of the g-Integral

The geometry of Ω is represented by $g(\boldsymbol{\xi})$, the volume integral of $\exp(\iota \boldsymbol{\xi} \cdot \mathbf{x})$ over Ω. The tensor $\mathbf{S}^P$ therefore applies to any anisotropic inclusion with any arbitrary shape. The g-integral can be related to the Fourier series coefficients of the Heaviside step function, i.e.,

---

[14] Note that, like $\mathbf{S}^P$, the correlation tensor $\mathbf{S}^P(\Omega_\alpha, \Omega_\beta)$ does not change if both Ω_α and Ω_β are moved by the same rigid-body translation; see Subsection 12.5.3 for a more detailed explanation.

[15] Since $\mathcal{F}\mathbf{S}^P(-\boldsymbol{\xi}) = \mathcal{F}\mathbf{S}^P(\boldsymbol{\xi})$ and $g_\alpha(-\boldsymbol{\xi})$ is the complex conjugate of $g_\alpha(\boldsymbol{\xi})$, $\mathbf{S}^P(\Omega_\alpha, \Omega_\beta)$ defined by (12.5.6c), is real-valued.

$$H(\mathbf{x};\Omega)=\sum_{\xi}\frac{1}{U}\int_{\Omega}\exp(\iota\boldsymbol{\xi}\cdot(\mathbf{x}-\mathbf{y}))\,dV_y$$

$$=\sum_{\xi} f\,g(-\boldsymbol{\xi})\exp(\iota\boldsymbol{\xi}\cdot\mathbf{x}). \tag{12.5.11a}$$

Hence, the following identity holds (multiply both sides of (12.5.11a) by $\exp(\iota\boldsymbol{\zeta}\cdot\mathbf{x})$, and average over Ω):

$$g(\boldsymbol{\zeta})=\sum_{\xi} f\,g(-\boldsymbol{\xi})\,g(\boldsymbol{\zeta}+\boldsymbol{\xi}). \tag{12.5.11b}$$

In particular, for $\boldsymbol{\zeta}=\mathbf{0}$,

$$g(\mathbf{0})=1=\sum_{\xi} f\,g(-\boldsymbol{\xi})\,g(\boldsymbol{\xi}). \tag{12.5.11c}$$

For illustration, consider an ellipsoidal Ω with the axis b_i parallel to the x_i-direction, i = 1, 2, 3. Then, for $\boldsymbol{\xi}\neq\mathbf{0}$, $g(\boldsymbol{\xi})$ is given by[16]

$$g(\boldsymbol{\xi})=\frac{3}{\eta^3}(\sin\eta-\eta\cos\eta)\qquad(\eta\neq 0), \tag{12.5.12a}$$

where

$$\eta=\pi\left[\{n_1\frac{b_1}{a_1}\}^2+\{n_2\frac{b_2}{a_2}\}^2+\{n_3\frac{b_3}{a_3}\}^2\right]^{1/2}. \tag{12.5.12b}$$

The g-integral is real-valued in (12.5.12a), satisfying $g(-\boldsymbol{\xi})=g(\boldsymbol{\xi})$.

For the numerical computation, follow Nemat-Nasser *et al.* (1982) and Iwakuma and Nemat-Nasser (1983), and decompose $\mathbf{S}^P$ into several parts. To this end, first define

$$h_1(\boldsymbol{\xi})=(\bar{\xi}_1)^2,\qquad h_2(\boldsymbol{\xi})=(\bar{\xi}_2)^2,\qquad h_3(\boldsymbol{\xi})=(\bar{\xi}_3)^2,$$

$$h_4(\boldsymbol{\xi})=\bar{\xi}_2\,\bar{\xi}_3,\qquad h_5(\boldsymbol{\xi})=\bar{\xi}_3\,\bar{\xi}_1,\qquad h_6(\boldsymbol{\xi})=\bar{\xi}_1\,\bar{\xi}_2, \tag{12.5.13a~f}$$

and

$$h_{IJ}(\boldsymbol{\xi})=h_I(\boldsymbol{\xi})\,h_J(\boldsymbol{\xi}),\qquad \text{for } I, J=1, 2, ..., 6. \tag{12.5.13g}$$

Then, introduce infinite sums S_I and S_{IJ}, as follows:

$$S_I\equiv\sum_{\xi}{}' f\,g(-\boldsymbol{\xi})\,g(\boldsymbol{\xi})\,h_I(\boldsymbol{\xi}),\qquad S_{IJ}\equiv\sum_{\xi}{}' f\,g(-\boldsymbol{\xi})\,g(\boldsymbol{\xi})\,h_{IJ}(\boldsymbol{\xi}). \tag{12.5.14a,b}$$

Hence, $\mathbf{S}^P$ for an isotropic reference elasticity tensor, becomes

$$\mathbf{S}^P_{ijkl}=\frac{1}{2}\left[\delta_{il}\,S_{I(j,\,k)}+\delta_{ik}\,S_{I(j,\,l)}+\delta_{jl}\,S_{I(i,\,k)}+\delta_{jk}\,S_{I(i,\,l)}\right]$$

---

[16] Computation of (12.5.12) is straightforward, if spherical coordinates, (r,ϕ,θ), are used instead of Cartesian coordinates, (x_1, x_2, x_3), where $r^2=\sum_{i=1}^{3}(x_i/b_i)^2$, ϕ is the angle between the $\mathbf{x}$- and $\boldsymbol{\xi}$-directions, and θ is the angle of the projection of the $\mathbf{x}$-direction onto the plane perpendicular to the $\boldsymbol{\xi}$-direction, measured from a certain line on the projection plane which passes through the origin.

$$- \frac{1}{1-\nu} S_{I(i,j)J(k,l)} + \frac{\nu}{1-\nu} \delta_{kl} S_{I(i,j)}, \tag{12.5.15}$$

where $I(i, j) = I(j, i) = 1, 2, 3, 4, 5$, or 6 for $(i, j) = (1, 1), (2, 2), (3, 3), (2, 3), (3, 1)$, or $(1, 2)$, respectively.

Note that the product $g(-\boldsymbol{\xi})\, g(\boldsymbol{\xi})$ is expressed by the double volume integral,

$$g(-\boldsymbol{\xi})\, g(\boldsymbol{\xi}) = \left\{ \frac{1}{\Omega} \int_\Omega \exp(-\iota\boldsymbol{\xi}\cdot\mathbf{y})\, dV_y \right\} \left\{ \frac{1}{\Omega} \int_\Omega \exp(\iota\boldsymbol{\xi}\cdot\mathbf{x})\, dV_x \right\}$$

$$= \frac{1}{\Omega^2} \int_\Omega \int_\Omega \exp(\iota\boldsymbol{\xi}\cdot(\mathbf{x}-\mathbf{y}))\, dV_x\, dV_y \tag{12.5.16}$$

which is unchanged by an arbitrary rigid-body translation[17] of the region Ω. On the other hand, a rigid-body rotation changes the value of $g(-\boldsymbol{\xi})\, g(\boldsymbol{\xi})$. Hence, $g(-\boldsymbol{\xi})\, g(\boldsymbol{\xi})$ is independent of the location of Ω but not of its orientation; this property is also shared by the $\mathbf{S}^P$-tensor. Self-similar variations of Ω change $g(\boldsymbol{\xi})$, and hence, $\mathbf{S}^P$. Therefore, like the Eshelby tensor $\mathbf{S}$, tensor $\mathbf{S}^P$ depends on the aspect ratios and the orientation of the inclusion when this inclusion is ellipsoidal; however, unlike the Eshelby tensor, $\mathbf{S}^P$ *does depend on the size of* Ω *relative to the unit cell.*

12.6. ELASTIC INCLUSIONS AND CAVITIES

As shown in Subsection 12.5, the geometrical effects of the periodically distributed inhomogeneities can be separated from the effects of their material properties, when an isotropic reference elasticity tensor is used. With the aid of the g-integral, the $\mathbf{S}^P$-tensor is defined, which plays a role in the periodic case similar to that of the Eshelby tensor in the homogeneous unbounded solid containing an ellipsoidal inclusion. For the periodic case, however, it is rather difficult to compute the displacements and tractions on the boundary of the inhomogeneity. Hence, the $\mathbf{S}^P$-tensor is used to obtain the overall parameters of solids with periodic microstructures. In this section, periodic distributions of elastic inclusions are considered, including cavities which can be regarded as elastic inclusions with zero stiffness, and cracks which can be regarded as suitable limiting cases of cavities.

---

[17] This is quite reasonable, since $\mathbf{S}^P$ expresses interaction effects among all Ω's in the periodic structure. Such interactions are determined by the shape, size, and orientation of Ω, and should not depend on the relative position of Ω within the unit cell which can be defined somewhat arbitrarily within the (equivalent) homogeneous periodic structure. If Ω is moved by a rigid-body translation, say, $\mathbf{d}$, then $g(\boldsymbol{\xi})$ changes to $\exp(\iota\boldsymbol{\xi}.\mathbf{d})\, g(\boldsymbol{\xi})$. However, the product $g(-\boldsymbol{\xi})\, g(\boldsymbol{\xi})$ remains the same, and the sum in the right-hand side of (12.5.6b) does not change for such rigid-body translation of Ω.

12.6.1. Elastic Spherical Inclusions

First, consider the case of equally spaced spherical inclusions in an isotropic homogeneous elastic matrix. The unit cell U is then a cube of dimension a, containing a spherical inclusion Ω of radius b at its center; see Figure 12.6.1. For simplicity, it is assumed that both the matrix and the inclusion are isotropic, having the shear moduli μ and μ^{Ω}, respectively, and a common Poisson ratio ν. Then,

$$(\mathbf{C}-\mathbf{C}^{\Omega})^{-1} : \mathbf{C} = \frac{\mu}{\mu-\mu^{\Omega}}\,\mathbf{1}^{(4s)}. \tag{12.6.1}$$

The overall elasticity $\bar{\mathbf{C}}$ has cubic symmetry. Its components can be expressed in terms of three moduli, $\bar{\lambda}$, $\bar{\mu}$, and $\bar{\mu}'$, as follows:

$$\bar{C}_{1111} = \bar{C}_{2222} = \bar{C}_{3333} = \bar{\lambda} + 2\bar{\mu} + \bar{\mu}',$$

$$\bar{C}_{1122} = \bar{C}_{1133} = \bar{C}_{2211} = \cdots = \bar{\lambda},$$

$$\bar{C}_{2323} = \bar{C}_{3131} = \bar{C}_{1212} = \cdots = \bar{\mu}, \tag{12.6.2a~c}$$

with all other $\bar{C}_{ijkl}$'s vanishing. The overall bulk modulus $\bar{K}$ is given by

$$\bar{K} = \frac{1}{9}\bar{C}_{iijj} = \frac{1}{3}(3\,\bar{\lambda} + 2\,\bar{\mu} + \bar{\mu}'). \tag{12.6.3a}$$

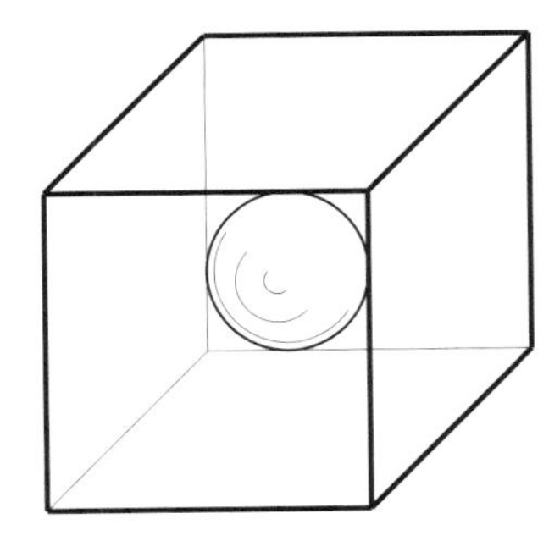

Figure 12.6.1

Cubic unit cell with spherical inclusion at center

The parameter $\bar{\mu}'$ measures the degree of cubic anisotropy; it is zero for isotropic materials (see Section 3). From (12.6.2), it follows that

$$\bar{\mu}' = \bar{C}_{1111} - 2\bar{C}_{2323} - \bar{C}_{1122}. \tag{12.6.3b}$$

For a cubic unit cell, $a_1 = a_2 = a_3 = a$, and for a spherical inclusion, $b_1 = b_2 = b_3 = b$. The g-integral, (12.5.12), then becomes

$$g(\boldsymbol{\xi}) = \frac{3}{(b\xi)^3}\{\sin(b\xi) - (b\xi)\cos(b\xi)\}, \tag{12.6.4}$$

where $\xi = |\boldsymbol{\xi}|$. In this special case, the infinite sums $\bar{S}_I$ and $\bar{S}_{IJ}$ reduce to

$$\bar{S}_1 = \bar{S}_2 = \bar{S}_3 = \sum_{\xi}{}' f\, g(-\xi)\, g(\xi)\, (\bar{\xi}_1)^2,$$

$$\bar{S}_{11} = \bar{S}_{22} = \bar{S}_{33} = \sum_{\xi}{}' f\, g(-\xi)\, g(\xi)\, (\bar{\xi}_1)^4,$$

$$\bar{S}_{44} = \bar{S}_{55} = \bar{S}_{66} = \bar{S}_{12} = \bar{S}_{23} = \bar{S}_{31} = \sum_{\xi}{}' f\, g(-\xi)\, g(\xi)\, (\bar{\xi}_2\, \bar{\xi}_3)^2, \qquad (12.6.5a\sim c)$$

with all other terms vanishing. For a given b/a, or a given f, these infinite sums can be computed once and for all, and their values then used for various computations.

Consider now two problems, one, a composite containing elastic inclusions with $\mu^\Omega/\mu = 3$, and the other, a solid with voids for which $\mu^\Omega/\mu = 0$; the Poisson ratio is 0.3. Figure 12.6.2 shows the overall moduli $\bar{K}$, $\bar{\lambda}$, $\bar{\mu}$, and $\bar{C}_{1111} = \bar{\lambda} + 2\bar{\mu} + \bar{\mu}'$ for various f, when Ω is spherical. Results for indicated ellipsoidal voids and inclusions are also plotted in Figure 12.6.2.

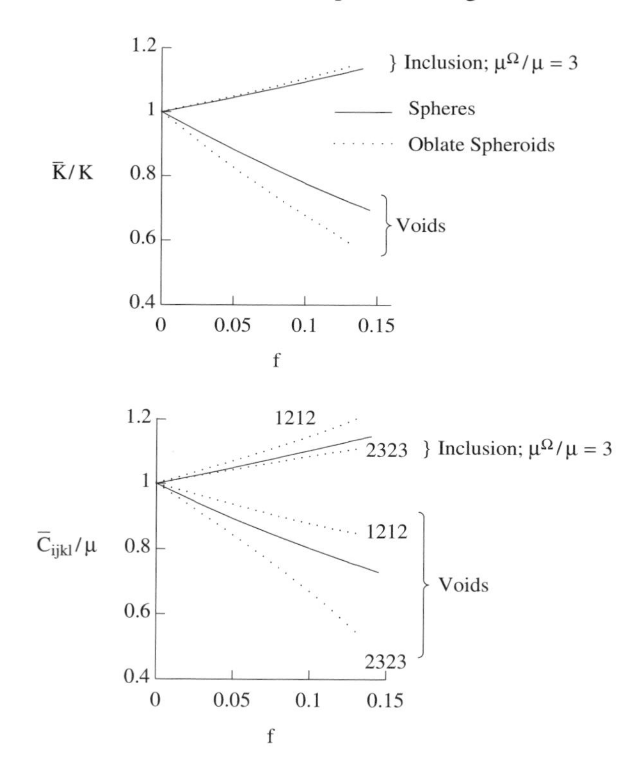

Figure 12.6.2

Estimate of overall elastic moduli of a body with spherical or ellipsoidal (aspect ratios $b_2/b_1 = 1$, $b_3/b_1 = 0.25$) voids or inclusions; $\nu = 0.3$

For spherical cavities, $\mathbf{C}^\Omega = \mathbf{0}$, and the overall bulk modulus, $\bar{K}$, defined by (12.6.3a), becomes

$$\bar{K} = \frac{1}{9}\{\mathbf{C} : (\mathbf{1}^{(4s)} - f\,(\mathbf{1}^{(4s)} - \mathbf{S}^P)^{-1})\}_{iijj}. \tag{12.6.6a}$$

Since $\mathbf{S}^P$ has cubic symmetry, (12.6.6a) yields

$$\frac{\bar{K}}{K} = 1 - f\,\frac{1}{3}\,(\mathbf{1}^{(4s)} - \mathbf{S}^P)^{-1}_{iijj} = 1 - f\,\frac{1}{1 - \frac{1}{3}S^P_{iijj}}, \tag{12.6.6b}$$

and in view of $\bar{\xi}_i\,\xi_i = 1$, it follows that

$$S^P_{iijj} = \frac{1+\nu}{1-\nu}\sum_{\xi}{}'\, f\,g(-\xi)\,g(\xi) = \frac{1+\nu}{1-\nu}\,\{\sum_{\xi} f\,g(-\xi)\,g(\xi) - f\}. \tag{12.6.7a}$$

From identity (12.5.11c),

$$S^P_{iijj} = \frac{1+\nu}{1-\nu}(1-f). \tag{12.6.7b}$$

Hence, the overall bulk modulus $\bar{K}$ is

$$\frac{\bar{K}}{K} = 1 - \frac{3(1-\nu)\,f}{2(1-2\nu) + (1+\nu)\,f}. \tag{12.6.6c}$$

Note that this coincides with the Hashin-Shtrikman upper bound discussed in Section 9. Indeed, as is shown in Section 13, in certain cases, estimates based on piecewise constant eigenstress (or eigenstrain) fields in a periodic solid and the Hashin-Shtrikman variational principle, provide the same set of equations for the unknown values of the eigenstresses, when a suitable reference elasticity (or compliance) tensor is used. Moreover, bounds for $\bar{C}_{iijj}$ are universal, and hence valid for any number and distribution of phases; see Subsections 9.6, 9.7, and 13.5. In the present illustration, this universal upper bound for voids is given by (12.6.6c), and *is actually attained by spherical voids*, as can be seen from the results in Figure 12.6.2 for the effective bulk modulus. While these are also upper bounds for oblate spheroidal voids (as seen from the results in Figure 12.6.2), sharper upper bounds are obtained when the effect of geometry is included. Similar comments apply to the case of inclusions; see Subsection 12.6.2, below. In this case, the results for the bulk modulus are the lower bounds, since the elasticity of the matrix is used as the reference one, and the inclusions are stiffer in this example; see Subsection 13.5 for more discussion.

12.6.2. Elastic Ellipsoidal Inclusions

Next, consider general ellipsoidal inhomogeneities, periodically embedded in an elastic homogeneous isotropic matrix. The unit cell U is a cube of dimension a, containing at its center an ellipsoidal inclusion Ω, with its major axis b_i parallel to the coordinate direction x_i ($i = 1, 2, 3$). Only the infinite sums, S_I, S_{II} (I not summed), and S_{IJ} ($I \neq J$) are nonzero, with all other infinite sums vanishing. The value of these sums depends on the volume fraction and the aspect ratios of the inclusion, i.e., on f, b_2/b_1, and b_3/b_1. Different geometries have been considered by Iwakuma and Nemat-Nasser (1983), where the required sums are tabulated, and several illustrative examples are given.

As an illustration, consider a composite whose matrix is isotropically elastic and contains periodically distributed isotropic elastic inclusions. Let the common Poisson ratio be ν. When $b_2/b_1 = 1$, it follows that

$$\bar{C}_{1111} = \bar{C}_{2222}, \qquad \bar{C}_{3311} = \bar{C}_{3322}, \qquad \bar{C}_{3131} = \bar{C}_{2323}, \qquad \text{etc.} \tag{12.6.8}$$

The relevant moduli for this case with $b_3/b_1 = 0.25$, are plotted in Figure 12.6.2, for voids, $\mu^{\Omega}/\mu = 0$, and inclusions with $\mu^{\Omega}/\mu = 3$. For voids, the results are the upper bounds, and for inclusions, they are the lower bounds.

12.6.3. Cylindrical Voids

Consider now an isotropic elastic body which contains periodically distributed circular-cylindrical voids of common radius b and length l, with their common generator in the x_1-direction. The unit cell is a cube of dimension a; see Figure 12.6.3. From the definition of $g(\boldsymbol{\xi})$, (12.5.5b), obtain

$$g(\boldsymbol{\xi}) = \frac{2}{y} J_1(y) \frac{\sin x}{x}, \tag{12.6.9a}$$

where J_1 is the order 1 Bessel function of the first kind, and

$$x = \pi n_1 \frac{l}{a}, \qquad y = 2\pi (n_2^2 + n_3^2)^{1/2} \frac{b}{a}, \tag{12.6.9b,c}$$

and the volume fraction f is given by[18] $\pi b^2 l/a^3$.

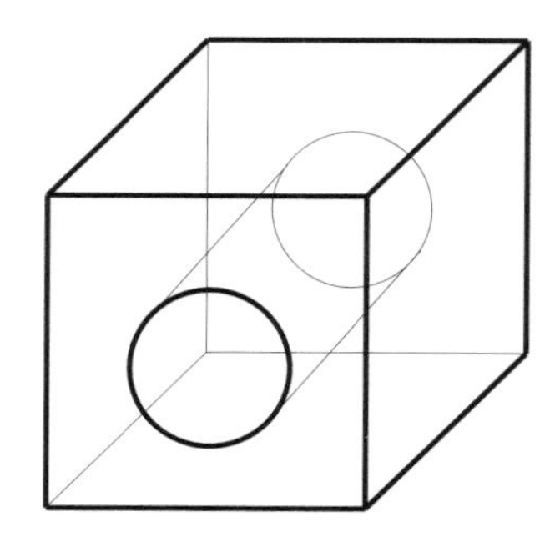

Figure 12.6.3

Cubic unit cell with circular-cylindrical void at center

Using (12.6.9), compute nonzero infinite sums, S_1, S_2, S_{11}, S_{22}, S_{12} and S_{23}, and note that

$$S_3 = S_2, \qquad S_{33} = S_{22}, \qquad S_{13} = S_{55} = S_{66} = S_{12}, \qquad S_{44} = S_{23}. \tag{12.6.10}$$

---

[18] Similarly to (12.5.12), computation of (12.6.9) is straightforward, if cylindrical coordinates, (r, θ, x_3), are used, where $r^2 = \sum_{i=1}^{2} (x_i/a)^2$, and $\sum_{i=1}^{2} \xi_i x_i = r\xi\cos\theta$, with $\xi^2 = \sum_{i=1}^{2} (a\xi_i)^2$.

The results for short cylindrical voids, $l/\mathrm{a} = \mathrm{b}/l = 0.5$, are plotted in Figure 12.6.4.

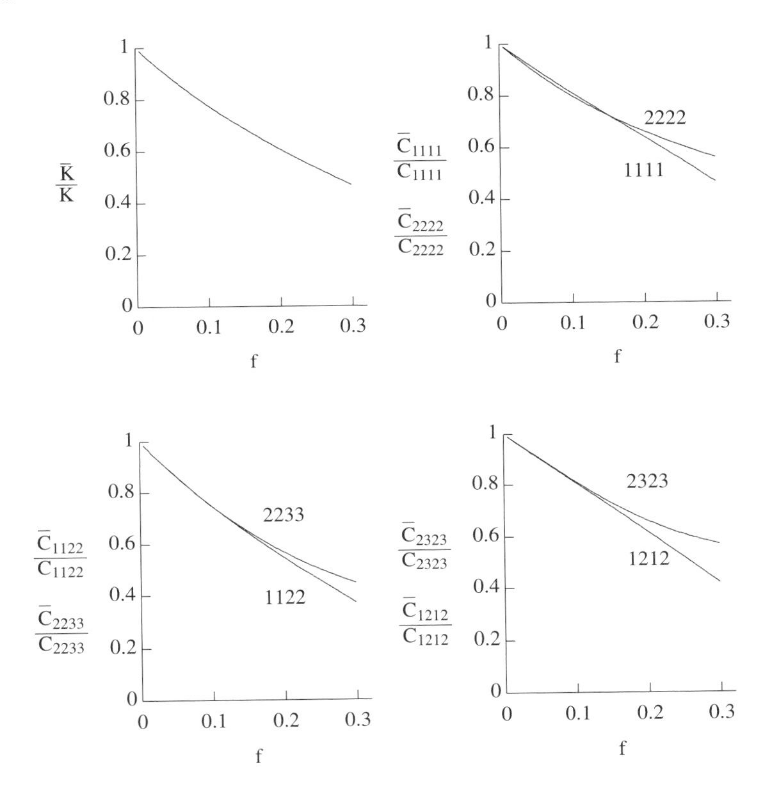

Figure 12.6.4

Estimate of effective moduli for periodically distributed short circular-cylindrical voids; $\nu = 0.3$, $l/\mathrm{a} = \mathrm{b}/l = 0.5$

In the case when circular-cylindrical voids are infinitely long in the x_1-direction, i.e.,

$$\frac{l}{\mathrm{a}_1} = 1, \quad \frac{\mathrm{b}}{l} = 0 \quad (\mathrm{a}_1 >> \mathrm{a}_2 = \mathrm{a}_3 > \mathrm{b}), \tag{12.6.11a,b}$$

the problem is reduced to the two-dimensional plane strain problem with circular holes in the x_2,x_3-plane. Then, it is necessary to evaluate only the quantities S_2 (= S_3), S_{22} (= S_{33}), and S_{23}, which are related to the inplane quantities, and S_{12} (= S_{13}) which is related to the antiplane quantities. For simplicity, consider the inplane quantities. Figure 12.6.5 shows the corresponding overall elastic moduli. For comparison, results for randomly distributed voids are also plotted. Figure 12.6.5 also gives the graph of the quantity

$$\rho = 2\overline{C}_{2323}\,(\overline{C}_{2222} - \overline{C}_{2233})^{-1} \tag{12.6.12}$$

which is the ratio of the shear moduli for simple shearing of the x_2,x_3-directions, and for simple shearing at 45° with these directions. The parameter ρ represents the degree of cubic anisotropy. For the isotropic case, $\rho = 1$.

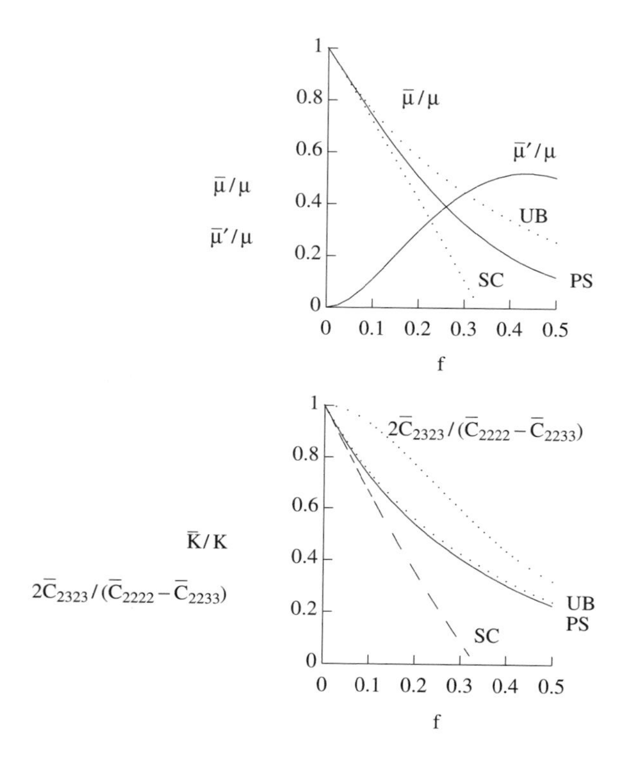

Figure 12.6.5

Estimate of effective moduli of a body with circular-cylindrical holes; plane-strain problem; PS: periodic structure, UB: upper bound, and SC: self-consistent method

12.7. PERIODICALLY DISTRIBUTED MICROCRACKS

In the preceding subsection, solids containing periodically distributed cavities are considered. A crack with traction-free surfaces can be treated as a flat cavity whose thickness becomes infinitesimally small. As discussed in Subsections 12.5, the geometrical effects of cavities or inclusions in a periodic structure are represented only by the g-integral, the volume average of $\exp(\iota\boldsymbol{\xi}\cdot\mathbf{x})$ over the cavity or inclusion. In this subsection, the limiting value of the g-integral is examined, as a flat cavity tends to become a microcrack.

12.7.1. Limit of Eshelby's Solution

Before considering a periodic structure, examine an infinite elastic body containing an ellipsoidal cavity, and take the limit as the cavity approaches a penny-shaped crack; Willis (1968). The limiting procedure must result in a measure for the crack density which then replaces the void volume fraction; see Section 6. This technique is illustrated in terms of the limiting values of the Eshelby tensor and the corresponding homogenizing eigenstrain field. Note that an infinite body problem may be viewed as a limiting case of a periodic structure, as the size of the inclusions relative to the size of the unit cell becomes very small. Therefore, the limiting procedure mentioned for the infinite body problem is applicable directly to the periodic problem.

For simplicity, consider an infinitely extended homogeneous solid with isotropic elasticity $\mathbf{C}$, containing an oblate cavity Ω', whose major axis a_i is parallel to the x_i-direction. Assume $a_1 = a_2 >> a_3$. Set

$$h \equiv \frac{a_3}{a_1}, \tag{12.7.1}$$

and note that as h goes to zero, Ω' approaches a penny-shaped crack of radius a_1, which is denoted by Ω. The solid is subjected to the farfield uniform stress $\boldsymbol{\sigma}^o$. Solve this problem by introducing a suitable homogenizing uniform eigenstrain $\boldsymbol{\varepsilon}^*$ in Ω'. Since Ω' has zero stiffness, the consistency condition becomes

$$\mathbf{0} = \boldsymbol{\sigma}^o + \mathbf{C} : \{\mathbf{S} : \boldsymbol{\varepsilon}^* - \boldsymbol{\varepsilon}^*\} \tag{12.7.2a}$$

or

$$(\mathbf{1}^{(4s)} - \mathbf{S}) : \boldsymbol{\varepsilon}^* = \mathbf{D} : \boldsymbol{\sigma}^o, \tag{12.7.2b}$$

where $\mathbf{S}$ is the Eshelby tensor for Ω'. As has been explained in Section 11, $\mathbf{C} : \mathbf{S}$ is symmetric with respect to the first and last pairs of its indices, although $\mathbf{S}$ is not symmetric. For simplicity, rewrite (12.7.2b) as

$$(\mathbf{C} - \mathbf{C} : \mathbf{S}) : \boldsymbol{\varepsilon}^* = \boldsymbol{\sigma}^o. \tag{12.7.2c}$$

Except for the symmetry of $\mathbf{C} : \mathbf{S}$, (12.7.2c) is a better form for the crack problem than (12.7.2b); see Subsections 12.7.2 and 12.7.3.

The product of elasticity, $\mathbf{C}$, and Eshelby tensor, $\mathbf{S}$, for the oblate spheroidal Ω' is expressed as

$$\mathbf{C} : \mathbf{S} = \mathbf{C} : \mathbf{S}^{(0)} + h\mathbf{C} : \mathbf{S}^{(1)} + O(h^2). \tag{12.7.3}$$

Here, in matrix form,[19] $\mathbf{C} : \mathbf{S}^{(0)}$ and $\mathbf{C} : \mathbf{S}^{(1)}$ are expressed as

---

[19] The matrix form of the product of $\mathbf{C}$ with $\mathbf{S}$ or similar tensors is the same as that of $\mathbf{C}$, i.e., the components of the matrix coincide with the corresponding components of the tensor; see Sections 3 and 15.

$$[(\mathbf{C}:\mathbf{S}^{(0)})_{ab}] = \begin{bmatrix} [(\mathbf{C}:\mathbf{S}^{(0)})_{ab}^{(1)}] & [0_{ab}] \\ [0_{ab}] & [(\mathbf{C}:\mathbf{S}^{(0)})_{ab}^{(2)}] \end{bmatrix}, \tag{12.7.4a}$$

and

$$[(\mathbf{C}:\mathbf{S}^{(1)})_{ab}] = \begin{bmatrix} [(\mathbf{C}:\mathbf{S}^{(1)})_{ab}^{(1)}] & [0_{ab}] \\ [0_{ab}] & [(\mathbf{C}:\mathbf{S}^{(1)})_{ab}^{(2)}] \end{bmatrix}, \tag{12.7.5a}$$

where three by three matrices $[(\mathbf{C}:\mathbf{S}^{(0)})_{ab}^{(1)}]$ and $[(\mathbf{C}:\mathbf{S}^{(0)})_{ab}^{(2)}]$ are

$$[(\mathbf{C}:\mathbf{S}^{(0)})_{ab}^{(1)}] = \frac{E}{(1-\nu^2)(1-2\nu)} \begin{bmatrix} \nu^2 & \nu^2 & \nu(1-\nu) \\ \nu^2 & \nu^2 & \nu(1-\nu) \\ \nu(1-\nu) & \nu(1-\nu) & (1-\nu)^2 \end{bmatrix},$$

$$[(\mathbf{C}:\mathbf{S}^{(0)})_{ab}^{(2)}] = \frac{E}{2(1+\nu)} \begin{bmatrix} 1 & 0 & 0 \\ 0 & 1 & 0 \\ 0 & 0 & 0 \end{bmatrix}, \tag{12.7.4b,c}$$

and $[(\mathbf{C}:\mathbf{S}^{(1)})_{ab}^{(1)}]$ and $[(\mathbf{C}:\mathbf{S}^{(1)})_{ab}^{(2)}]$ are

$$[(\mathbf{C}:\mathbf{S}^{(1)})_{ab}^{(1)}] = \frac{\pi E}{32(1-\nu^2)} \begin{bmatrix} 13 & -(1-16\nu) & -4(1+2\nu) \\ -(1-16\nu) & 13 & -4(1+2\nu) \\ -4(1+2\nu) & -4(1+2\nu) & -8 \end{bmatrix},$$

$$[(\mathbf{C}:\mathbf{S}^{(1)})_{ab}^{(2)}] = \frac{\pi E}{32(1-\nu^2)} \begin{bmatrix} -4(2-\nu) & 0 & 0 \\ 0 & -4(2-\nu) & 0 \\ 0 & 0 & 7-8\nu \end{bmatrix}. \tag{12.7.5b,c}$$

Substituting (12.7.3), (12.7.4), and (12.7.5) into (12.7.2c), obtain

$$\boldsymbol{\varepsilon}^* = (\mathbf{C}-\mathbf{C}:\mathbf{S})^{-1}:\boldsymbol{\sigma}^o = \frac{1}{h}\,\mathbf{U}^{(-1)}:\boldsymbol{\sigma}^o + O(h^0), \tag{12.7.6}$$

where $\mathbf{U}^{(-1)}$ has the following three nonzero components:

$$U_{3333}^{(-1)} = \frac{4(1-\nu^2)}{\pi E}, \qquad U_{2323}^{(-1)} = U_{3131}^{(-1)} = \frac{2(1-\nu^2)}{(2-\nu)\pi E}, \tag{12.7.7a,b}$$

with $U_{ijkl}^{(-1)} = U_{jikl}^{(-1)} = U_{ijlk}^{(-1)} = U_{klij}^{(-1)}$.

If the limit of (12.7.6) as h goes to zero is taken, it is seen that the eigenstrain $\boldsymbol{\varepsilon}^*$ for the penny-shaped crack Ω becomes unbounded, with the leading term given by

$$\lim_{h\to 0} \boldsymbol{\varepsilon}^* = \lim_{h\to 0} \frac{1}{h}\,\mathbf{U}^{(-1)}:\boldsymbol{\sigma}^o. \tag{12.7.8a}$$

The volume integral (not the volume average) of $\boldsymbol{\varepsilon}^*$ over Ω, however, gives

$$\lim_{h\to 0} \Omega'\,\boldsymbol{\varepsilon}^* = \overline{\Omega}\,\mathbf{U}^{(-1)}:\boldsymbol{\sigma}^o, \tag{12.7.8b}$$

where $\overline{\Omega} = 4\pi a_1^3/3$. From these results, it is seen that, as h goes to zero:

1) although the limit of Eshelby's tensor, $\mathbf{S}$, exists, the inverse of $\mathbf{C}-\mathbf{C}:\mathbf{S}$ becomes unbounded; and

2) although the uniform homogenizing eigenstrain $\boldsymbol{\varepsilon}^*$ becomes unbounded in Ω, its volume integral over Ω remains finite.

Applying (12.7.8b) to analyze the RVE, write the volume average of the eigenstrain over V, as

$$\frac{1}{V}\int_{\Omega} \boldsymbol{\varepsilon}^* \, dV = \frac{\bar{\Omega}}{V} \, \mathbf{U}^{(-1)} : \boldsymbol{\sigma}^{o} \tag{12.7.9}$$

which is consistent with the results discussed in Section 6; in particular, use $a_1^3/V = 3\bar{\Omega}/4\pi V$ as the measure of the cavity density; see Section 6.

12.7.2. The g-Integral for a Crack

Following the above limiting procedure, consider the limit of the g-integral, as an ellipsoidal cavity reduces to a crack. As a simple example, consider an oblate spheroid Ω' at the center of a unit cell, with its major axis b_i being parallel to the x_i-direction, and assume $b_1 = b_2 = b$ and $b_3 \equiv hb << b_1$.

From definition (12.5.5b), the g-integral for Ω', denoted by g', becomes

$$g'(\boldsymbol{\xi}) \equiv \frac{1}{\Omega'}\int_{\Omega} \exp(\iota(\xi_1 x_1 + \xi_2 x_2)) \, \{\int_{-h'}^{h'} \exp(\iota \xi_3 x_3) \, dx_3\} \, dx_1 dx_2$$

$$= \frac{1}{\bar{\Omega}}\int_{\Omega} \exp(\iota(\xi_1 x_1 + \xi_2 x_2)) \, \{\frac{1}{h} \, \frac{2\sin(\xi_3 h')}{\xi_3}\} \, dx_1 dx_2, \tag{12.7.10a}$$

where Ω is a circle of radius b on the plane $x_3 = 0$ (a penny-shaped crack), $\bar{\Omega} \equiv 4\pi b^3/3 = \Omega'/h$, and

$$h' = h'(x_1, x_2) = hb \left[1 - \{\frac{x_1}{b}\}^2 - \{\frac{x_2}{b}\}^2\right]^{1/2}. \tag{12.7.10b}$$

Since

$$\lim_{h \to 0} \frac{\sin(\xi_3 h')}{\xi_3 h'} = 1, \tag{12.7.11}$$

for any x_1 and x_2, the limit of the g'-integral becomes

$$g(\boldsymbol{\xi}) \equiv \lim_{h \to 0} g'(\boldsymbol{\xi})$$

$$= \frac{2b}{\bar{\Omega}}\int_{\Omega} \left[1 - \{\frac{x_1}{b}\}^2 - \{\frac{x_2}{b}\}^2\right]^{1/2} \exp(\iota(\xi_1 x_1 + \xi_2 x_2)) \, dx_1 dx_2 \tag{12.7.12a}$$

which is bounded. Indeed, with the aid of (12.5.12b), explicitly compute g defined by (12.7.12a), as follows: defining $\eta' \equiv \pi(n_1^2 + n_2^2)^{1/2} \, b/a$, note that

$$g(\boldsymbol{\xi}) = \lim_{h \to 0} \frac{3}{\eta^3}(\sin\eta - \eta\cos\eta)$$

$$= \begin{cases} 3\eta'^{-3} \, (\sin\eta' - \eta'\cos\eta'), & \text{for } \eta' > 0 \\ 1 & \text{for } \eta' = 0, \end{cases} \tag{12.7.12b}$$

where η is given by (12.5.12b), i.e., $\eta \equiv \pi(n_1^2 + n_2^2 + h^2 n_3^2)^{1/2} \, b/a$.

It should be noted that, in the above limiting case, a penny-shaped crack Ω is regarded as an oblate spheroid of an infinitely small thickness. If the limits of other flat cavities are considered, the corresponding g-integral will be different. For example, for a circular-cylinder of radius b and height hb (<< b) and generator along the x_3-axis, the limit of the corresponding g-integral, denoted by g'', becomes

$$\lim_{h\to 0} g''(\boldsymbol{\xi}) = \frac{1}{\pi b^2}\int_{\Omega} \exp(\iota(\xi_1 x_1 + \xi_2 x_2))\, dx_1 dx_2. \tag{12.7.13a}$$

This, however, may not correspond to a penny-shaped crack. Indeed, with the aid of (12.6.9), explicitly compute g'' as

$$\lim_{h\to 0} g''(\boldsymbol{\xi}) = \lim_{h\to 0} \frac{2}{y} J_1(y) \frac{\sin x}{x} = \frac{2}{y} J_1(y), \tag{12.7.13b}$$

where $x = \pi n_3 hb/a$ and $y = \pi(n_1^2 + n_2^2)^{1/2} b/a$; see (12.6.9b,c). As is seen, (12.7.13b) is different from (12.7.12b).

12.7.3. Piecewise Constant Distribution of Eigenstrain

Now, consider an isotropic homogeneous cubic unit cell, containing a penny-shaped crack Ω of radius b, at its center, on the plane $x_3 = 0$; see Figure 12.7.1. As in Subsection 12.7.2, consider the limit of an oblate spheroidal void Ω' of radius b and thickness hb, and obtain the penny-shaped crack as h goes to zero.

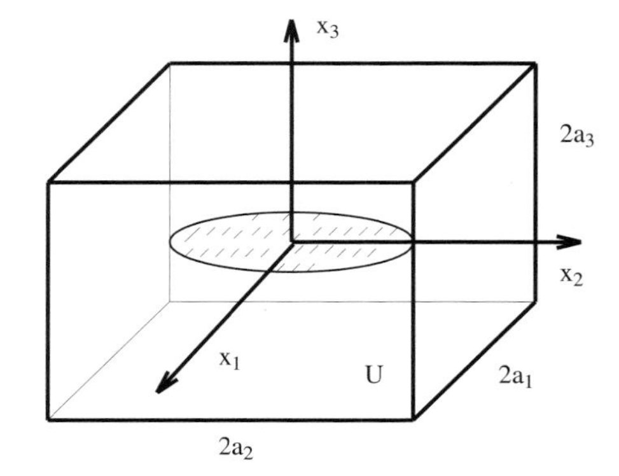

Figure 12.7.1

Cubic unit cell with penny-shaped crack at center

For illustration, choose a *constant* distribution of eigenstrains, $\bar{\boldsymbol{\varepsilon}}^*$ in Ω', i.e., $H(\mathbf{x}; \Omega')\, \bar{\boldsymbol{\varepsilon}}^*$. According to definition (12.5.6), define $\mathbf{S}^{P\prime}$ for this Ω' by

$$\mathbf{S}^{P\prime} \equiv \sum_{\xi}{}' f' g'(-\boldsymbol{\xi})\, g'(\boldsymbol{\xi})\, \mathcal{F}\mathbf{S}^P(\boldsymbol{\xi}), \tag{12.7.14a}$$

where $f' = \Omega'/U$. The volume average of the periodic strains taken over Ω' is given by $\mathbf{S}^{P\prime} : \bar{\boldsymbol{\varepsilon}}^*$. Although $\mathcal{F}\mathbf{S}^P$ is not symmetric with respect to the first and second pairs of its indices, the following tensor is symmetric:

$$\mathbf{C} : \mathbf{S}^{\mathrm{P}\prime} = \sum_{\boldsymbol{\xi}}{}' \,(\mathrm{hf})\, \mathrm{g}'(-\boldsymbol{\xi})\, \mathrm{g}'(\boldsymbol{\xi})\, \mathbf{C} : {}_F\mathbf{S}^{\mathrm{P}}(\boldsymbol{\xi}). \tag{12.7.14b}$$

Since the periodic part of the eigenstrain field $\mathrm{H}(\Omega')\bar{\boldsymbol{\varepsilon}}^*$ is $(\mathrm{H}(\Omega') - \mathrm{f}')\,\bar{\boldsymbol{\varepsilon}}^*$, the volume average of the periodic stresses taken over Ω', is given by $\mathbf{C} : \mathbf{S}^{\mathrm{P}\prime} : \bar{\boldsymbol{\varepsilon}}^* - (1-\mathrm{f}')\,\mathbf{C} : \bar{\boldsymbol{\varepsilon}}^*$.

Suppose that a homogeneous stress $\boldsymbol{\sigma}^{\mathrm{o}}$ is prescribed. Then, the average consistency condition for $\bar{\boldsymbol{\varepsilon}}^*$ is

$$\boldsymbol{\sigma}^{\mathrm{o}} + \mathbf{C} : \mathbf{S}^{\mathrm{P}\prime} : \bar{\boldsymbol{\varepsilon}}^* - (1-\mathrm{f}')\,\mathbf{C} : \bar{\boldsymbol{\varepsilon}}^* = \mathbf{0} \tag{12.7.15a}$$

or

$$\{(1-\mathrm{f}')\,\mathbf{C} - \mathbf{C} : \mathbf{S}^{\mathrm{P}\prime}\} : \bar{\boldsymbol{\varepsilon}}^* = \boldsymbol{\sigma}^{\mathrm{o}}. \tag{12.7.15b}$$

As h goes to zero, the inverse of $(1-\mathrm{f}')\,\mathbf{C} - \mathbf{C} : \mathbf{S}^{\mathrm{P}\prime}$ diverges, though $\mathbf{S}^{\mathrm{P}\prime}$ remains finite, as in the case of $\mathbf{S}$ (Eshelby tensor for the oblate spheroid); see Subsection 12.7.1. The proof is straightforward. First, taking advantage of $|\bar{\boldsymbol{\xi}}| = 1$, decompose $\mathbf{C} : {}_F\mathbf{S}^{\mathrm{P}\prime}$, as

$$\mathbf{C} : {}_F\mathbf{S}^{\mathrm{P}\prime} = \mathbf{C} : {}_F\mathbf{S}^{\mathrm{P}(0)\prime} + \mathbf{C} : {}_F\mathbf{S}^{\mathrm{P}(1)\prime}. \tag{12.7.16}$$

Here, in a manner similar to (12.7.4,5), $\mathbf{C} : {}_F\mathbf{S}^{\mathrm{P}(0)\prime}$ and $\mathbf{C} : {}_F\mathbf{S}^{\mathrm{P}(1)\prime}$ are expressed in matrix form as

$$[(\mathbf{C} : {}_F\mathbf{S}^{\mathrm{P}(0)\prime})_{\mathrm{ab}}] = \begin{bmatrix} [(\mathbf{C} : {}_F\mathbf{S}^{\mathrm{P}(0)\prime})^{(1)}_{\mathrm{ab}}] & [0_{\mathrm{ab}}] \\ [0_{\mathrm{ab}}] & [(\mathbf{C} : {}_F\mathbf{S}^{\mathrm{P}(0)\prime})^{(2)}_{\mathrm{ab}}] \end{bmatrix} \tag{12.7.17a}$$

and

$$[(\mathbf{C} : {}_F\mathbf{S}^{\mathrm{P}(1)\prime})_{\mathrm{ab}}] = \begin{bmatrix} [(\mathbf{C} : {}_F\mathbf{S}^{\mathrm{P}(1)\prime})^{(1)}_{\mathrm{ab}}] & [(\mathbf{C} : {}_F\mathbf{S}^{\mathrm{P}(1)\prime})^{(3)}_{\mathrm{ab}}] \\ [(\mathbf{C} : {}_F\mathbf{S}^{\mathrm{P}(1)\prime})^{(3)}_{\mathrm{ab}}]^{\mathrm{T}} & [(\mathbf{C} : {}_F\mathbf{S}^{\mathrm{P}(1)\prime})^{(2)}_{\mathrm{ab}}] \end{bmatrix}, \tag{12.7.18a}$$

where the three by three matrices $[(\mathbf{C} : {}_F\mathbf{S}^{\mathrm{P}(0)\prime})^{(1)}_{\mathrm{ab}}]$ and $[(\mathbf{C} : {}_F\mathbf{S}^{\mathrm{P}(0)\prime})^{(2)}_{\mathrm{ab}}]$ are

$$[(\mathbf{C} : {}_F\mathbf{S}^{\mathrm{P}(0)\prime})^{(1)}_{\mathrm{ab}}] = \frac{\mathrm{E}}{(1-\nu^2)(1-2\nu)} \begin{bmatrix} \nu^2 & \nu^2 & \nu(1-\nu) \\ \nu^2 & \nu^2 & \nu(1-\nu) \\ \nu(1-\nu) & \nu(1-\nu) & (1-\nu)^2 \end{bmatrix},$$

$$[(\mathbf{C} : {}_F\mathbf{S}^{\mathrm{P}(0)\prime})^{(2)}_{\mathrm{ab}}] = \frac{\mathrm{E}}{2(1+\nu)} \begin{bmatrix} 1 & 0 & 0 \\ 0 & 1 & 0 \\ 0 & 0 & 0 \end{bmatrix}, \tag{12.7.17b,c}$$

and $[(\mathbf{C} : {}_F\mathbf{S}^{\mathrm{P}(1)\prime})^{(1)}_{\mathrm{ab}}]$, $[(\mathbf{C} : {}_F\mathbf{S}^{\mathrm{P}(1)\prime})^{(2)}_{\mathrm{ab}}]$, and $[(\mathbf{C} : {}_F\mathbf{S}^{\mathrm{P}(1)\prime})^{(3)}_{\mathrm{ab}}]$ are

$$[(\mathbf{C} : {}_F\mathbf{S}^{\mathrm{P}(1)\prime})^{(1)}_{\mathrm{ab}}] = \frac{\mathrm{E}}{(1-\nu^2)}$$

$$\times \begin{bmatrix} 2\hat{\xi}_1 - \hat{\xi}_1^2 & \nu(\hat{\xi}_1 + \hat{\xi}_2) - \hat{\xi}_1\hat{\xi}_2 & -\nu\hat{\xi}_2 - \hat{\xi}_1\hat{\xi}_3 \\ \nu(\hat{\xi}_1 + \hat{\xi}_2) - \hat{\xi}_1\hat{\xi}_2 & 2\hat{\xi}_2 - \hat{\xi}_2^2 & -\nu\hat{\xi}_1 - \hat{\xi}_2\hat{\xi}_3 \\ -\nu\hat{\xi}_2 - \hat{\xi}_1\hat{\xi}_3 & -\nu\hat{\xi}_1 - \hat{\xi}_2\hat{\xi}_3 & -(\hat{\xi}_1 + \hat{\xi}_2)^2 \end{bmatrix},$$

$$[(\mathbf{C} : {}_F\mathbf{S}^{\mathrm{P}(1)\prime})^{(2)}_{\mathrm{ab}}] = \frac{\mathrm{E}}{(1-\nu^2)}$$

$$\times \begin{bmatrix} -(1-\nu)\hat{\xi}_1/2-\hat{\xi}_4^2 & (1-\nu)\hat{\xi}_6/2-\hat{\xi}_4\hat{\xi}_5 & (1-\nu)\hat{\xi}_5/2-\hat{\xi}_4\hat{\xi}_6 \\ (1-\nu)\hat{\xi}_6/2-\hat{\xi}_4\hat{\xi}_5 & -(1-\nu)\hat{\xi}_2/2-\hat{\xi}_5^2 & (1-\nu)\hat{\xi}_4/2-\hat{\xi}_5\hat{\xi}_6 \\ (1-\nu)\hat{\xi}_5/2-\hat{\xi}_4\hat{\xi}_6 & (1-\nu)\hat{\xi}_4/2-\hat{\xi}_5\hat{\xi}_6 & (1-\nu)(\hat{\xi}_1+\hat{\xi}_2)/2-\hat{\xi}_6^2 \end{bmatrix},$$

$$[(\mathbf{C} : {}_F\mathbf{S}^{P(1)\prime})^{(3)}_{ab}] = \frac{E}{(1-\nu^2)}$$

$$\times \begin{bmatrix} \nu\hat{\xi}_4-\hat{\xi}_5\hat{\xi}_6 & \hat{\xi}_5-\hat{\xi}_1\hat{\xi}_5 & \hat{\xi}_6-\hat{\xi}_1\hat{\xi}_6 \\ \hat{\xi}_4-\hat{\xi}_2\hat{\xi}_4 & \nu\hat{\xi}_5-\hat{\xi}_4\hat{\xi}_6 & \hat{\xi}_6-\hat{\xi}_2\hat{\xi}_6 \\ \hat{\xi}_4-\hat{\xi}_3\hat{\xi}_4 & \hat{\xi}_5-\hat{\xi}_3\hat{\xi}_5 & \nu\hat{\xi}_6-\hat{\xi}_4\hat{\xi}_5 \end{bmatrix}. \qquad (12.7.18b\sim d)$$

Here, $\hat{\xi}_1 \equiv \bar{\xi}_1^2$, $\hat{\xi}_2 \equiv \bar{\xi}_2^2$, $\hat{\xi}_3 \equiv \bar{\xi}_3^2$, $\hat{\xi}_4 \equiv \bar{\xi}_2\bar{\xi}_3$, $\hat{\xi}_5 \equiv \bar{\xi}_3\bar{\xi}_1$, and $\hat{\xi}_6 \equiv \bar{\xi}_1\bar{\xi}_2$, with $\bar{\boldsymbol{\xi}}$ being given by[20] $\boldsymbol{\xi}/|\boldsymbol{\xi}|$.

From $\sum_{\xi}' f'g'(-\boldsymbol{\xi})g(\boldsymbol{\xi}) = 1 - f'$ and from ${}_F\mathbf{S}^{P(0)\prime}(\boldsymbol{\xi}) = \mathbf{S}^{(0)}$ for $\boldsymbol{\xi} \neq \mathbf{0}$, it follows that

$$\mathbf{C} : \mathbf{S}^{P\prime} = (1-f')\,\mathbf{C} : \mathbf{S}^{(0)} + f' \sum_{\xi}{}' g'(-\boldsymbol{\xi})\, g'(\boldsymbol{\xi})\, \mathbf{C} : {}_F\mathbf{S}^{P(1)\prime}(\boldsymbol{\xi}). \qquad (12.7.19)$$

In the limit of $h \rightarrow 0$, g' becomes independent of ξ_3. Hence, the triple infinite summation for $\mathbf{C} : \mathbf{S}^{P(1)}$ with respect to ξ_1, ξ_2, and ξ_3, reduces to the double infinite summation with respect to ξ_1 and ξ_2, based on the following identities:

$$\sum_{m=-\infty}^{+\infty} \frac{1}{a^2+m^2} = \frac{\pi}{a} \coth(\pi a),$$

$$\sum_{m=-\infty}^{+\infty} \frac{1}{(a^2+m^2)^2} = \frac{1}{2a^4} \{\pi a \coth(\pi a) + (\pi a)^2 \mathrm{cosech}^2(\pi a)\}; \qquad (12.7.20a,b)$$

for example,

$$\lim_{h\rightarrow 0} \sum_{\xi} g'(-\boldsymbol{\xi})\, g'(\boldsymbol{\xi})\, \bar{\xi}_1^2 = \sum_{\xi_1,\,\xi_2}{}' g(-\boldsymbol{\xi})\, g(\boldsymbol{\xi})\, \xi_1^2 \left\{ \sum_{\xi_3} \frac{1}{(\xi_1^2+\xi_2^2)+\xi_3^2} \right\}, \qquad (12.7.20c)$$

since g is independent of ξ_3. Substituting (12.7.19) into (12.7.15b), and using (12.7.18a~d), obtain

$$\bar{\boldsymbol{\varepsilon}}^* = \{(1-f')\,\mathbf{C} - \mathbf{C} : \mathbf{S}^{P\prime}\}^{-1} : \boldsymbol{\sigma}^o$$

$$= \frac{1}{h}\, \mathbf{U}^{P(-1)} : \boldsymbol{\sigma}^o + O(h^0), \qquad (12.7.21)$$

where $\mathbf{U}^{P(-1)}$ has the following three nonzero components:

$$U^{P(-1)}_{3333} = \frac{1}{f}\, \frac{1-\nu^2}{E}\, \frac{1}{s_2},$$

[20] In (12.7.17) and (12.7.18), terms which involve odd powers of $\bar{\xi}_i$ for i = 1, 2, or 3, may be omitted on account of the existing symmetry. This is discussed in Section 14.

$$U^{P(-1)}_{2323} = U^{P(-1)}_{3131} = \frac{1}{f}\,\frac{1-\nu^2}{2E}\,\frac{1}{(3-\nu)\,s_1 - s_2}, \tag{12.7.22a,b}$$

with $U^{P(-1)}_{ijkl} = U^{P(-1)}_{jikl} = U^{P(-1)}_{ijlk} = U^{P(-1)}_{klij}$, and $f = \pi(b/a)^3/6$. From (12.7.20), s_1 and s_2 are given by

$$s_1 = \sum_{\xi_1,\,\xi_2}{}' \frac{1}{2}\, g(-\boldsymbol{\xi})\, g(\boldsymbol{\xi})\, (\pi/\xi')\coth(\pi\xi'),$$

$$s_2 = \sum_{\xi_1,\,\xi_2}{}' \frac{1}{2}\, g(-\boldsymbol{\xi})\, g(\boldsymbol{\xi})\, \{(\pi/\xi')\coth(\pi\xi') + (\pi/\xi')^2 \mathrm{cosech}^2(\pi\xi')\}, \tag{12.7.22c,d}$$

where $\boldsymbol{\xi}'$ is given by $\xi'_1 = \xi_1$, $\xi'_2 = \xi_2$, and $\xi'_3 = 0$, with $\xi' = |\boldsymbol{\xi}'|$ $(= (\xi_1^2 + \xi_2^2)^{1/2})$, and $\bar{\boldsymbol{\xi}}' = \boldsymbol{\xi}'/|\boldsymbol{\xi}'|$. Note that the following relations are used:

$$\sum_{\xi} g'(-\boldsymbol{\xi})\, g'(\boldsymbol{\xi})\, \bar{\xi}_1^2 = \sum_{\xi} g'(-\boldsymbol{\xi})\, g'(\boldsymbol{\xi})\, \bar{\xi}_2^2,$$

$$\sum_{\xi} g'(-\boldsymbol{\xi})\, g'(\boldsymbol{\xi})\, \bar{\xi}_1^4 = \sum_{\xi} g'(-\boldsymbol{\xi})\, g'(\boldsymbol{\xi})\, \bar{\xi}_2^4; \tag{12.7.23a,b}$$

see (12.7.12b).

As in Subsection 12.7.1, observe that

$$\lim_{h\to 0} \bar{\boldsymbol{\varepsilon}}^* = \lim_{h\to 0} \frac{1}{h}\, \mathbf{U}^{P(-1)} : \boldsymbol{\sigma}^o, \qquad \lim_{h\to 0} f'\,\bar{\boldsymbol{\varepsilon}}^* = f\,\mathbf{U}^{P(-1)} : \boldsymbol{\sigma}^o. \tag{12.7.24a,b}$$

Hence, at the limit:

1) although $\mathbf{C} : \mathbf{S}^{P'}$ is bounded, the inverse of $(1-f')\,\mathbf{C} - \mathbf{C} : \mathbf{S}^{P'}$ is unbounded; and
2) although $\bar{\boldsymbol{\varepsilon}}^*$ is unbounded, its volume integral $f'\,\bar{\boldsymbol{\varepsilon}}^*$ remains finite;

see Subsection 12.7.1.

As the oblate cavity Ω' approaches the penny-shaped crack Ω, the average strain over the unit cell becomes

$$\mathbf{D} : \boldsymbol{\sigma}^o + \lim_{h\to 0} f'\,\bar{\boldsymbol{\varepsilon}}^* = \mathbf{D} : \boldsymbol{\sigma}^o + f\,\{\lim_{h\to 0} h\,\bar{\boldsymbol{\varepsilon}}^*\}. \tag{12.7.25a}$$

Hence, using (12.7.24b), estimate the overall compliance $\bar{\mathbf{D}}$, as

$$\bar{\mathbf{D}} = \mathbf{D} + f\,\mathbf{U}^{P(-1)}. \tag{12.7.25b}$$

The limit in (12.7.25a) is finite.

12.7.4. Stress Intensity Factor of Periodic Cracks

It is interesting to estimate the stress intensity factor of the periodically distributed cracks. Here, the stress intensity factor is a measure of the singularity of the stress field near the crack tip; if deformation is small and quasi-static and the material is linearly elastic, the stress field near the crack tip diverges proportionally to the square root of the distance from the crack tip, and the coefficient of that proportionality determines the stress intensity factor; see Section 21.

As an example, consider the isolated penny-shaped crack with radius a_1 in an unbounded solid, studied in Subsection 12.7.1. When subjected to a farfield uniform tension σ^o_{33}, σ_{33} near the crack tip becomes

$$\sigma_{33}(r) = K_I r^{-1/2}, \tag{12.7.26a}$$

where K_I defined by

$$K_I \equiv 2\left[\frac{a_1}{\pi}\right]^{1/2} \sigma^o_{33}, \tag{12.7.26b}$$

is called the *Mode I stress intensity factor,* and r measures the distance from the crack tip to points on the plane containing the crack. The crack-opening displacement, denoted by $[u_3]$, vanishes near the crack tip, since

$$[u_3](R) = K_I R^{1/2}, \tag{12.7.27}$$

where R is the distance to the crack tip from points on the crack. Due to the linearity of the problem, the coefficient for $[u_3]$ is proportional to the stress intensity factor, K_I.

From the uniform but diverging eigenstrains given by (12.7.8a), (12.7.27) is obtained. For the oblate spheroidal cavity Ω', uniform eigenstrains produce uniform strains in Ω', and hence the displacement field in Ω becomes

$$\mathbf{u}(\mathbf{x}) = \mathbf{x} \cdot (\boldsymbol{\varepsilon}^o + \mathbf{S} : \boldsymbol{\varepsilon}^*), \tag{12.7.28a}$$

where $\boldsymbol{\varepsilon}^o = \mathbf{D} : \boldsymbol{\sigma}^o$. As Ω' approaches the penny-shaped crack, the difference of the displacements of the upper and lower surfaces of the cavity gives the crack-opening displacement, $[\mathbf{u}]$. Taking advantage of $x_3 = \pm bh\,\{1 - (x_1/b)^2 - (x_2/b)^2\}^{1/2}$ on the cavity surfaces, obtain

$$\begin{aligned}
[u_3] &= \lim_{h \to 0} [x_3]\,(\varepsilon^o_{33} + (\mathbf{S} : \boldsymbol{\varepsilon}^*)_{33}) \\
&= 2b\sqrt{1 - (x_1/b)^2 - (x_2/b)^2}\,(\mathbf{S}^{(0)} : \mathbf{U}^{(-1)} : \boldsymbol{\sigma}^o)_{33} \\
&= 2b\sqrt{1 - (x_1/b)^2 - (x_2/b)^2}\, S^{(0)}_{3333}\, U^{(-1)}_{3333}\, \sigma^o_{33}.
\end{aligned} \tag{12.7.28b}$$

Hence, (12.7.28b) coincides with (12.7.27).

The above technique may be applied to the periodically distributed penny-shaped cracks, studied in Subsection 12.7.3. For diverging average eigenstrains given by (12.7.24a), the crack-opening displacement[21] under uniform tension σ^o_{33} is estimated by

$$[u_3] = 2b\sqrt{1 - (x_1/b)^2 - (x_2/b)^2}\, S^{P(0)}_{3333}\, U^{P(-1)}_{3333}\, \sigma^o_{33}. \tag{12.7.29}$$

Taking advantage of linearity, now estimate the stress intensity factor for the periodic cracks. Since $\bar{\boldsymbol{\varepsilon}}^*$ obtained in Subsection 12.7.3 is an average value

[21] It should be noted that, unlike an isolated crack, a uniform distribution of eigenstrains does not produce the exact solution for periodically distributed cracks. Hence, the exact crack-opening displacement (and hence the singularity of the stress field) varies along the crack edge.

of the eigenstrains, denote by $\bar{K}_I^P$ the *average of the stress intensity factor taken along the crack edge*, and obtain

$$\frac{\bar{K}_I^P}{K_I} = \frac{U_{3333}^{P(-1)}}{U_{3333}^{(-1)}}, \tag{12.7.30}$$

where K_I is given by (12.7.26b).

12.7.5. Illustrative Examples

The method of this subsection has been employed by Nemat-Nasser *et al.* (1992) to study stiffness degradation of elastic solids containing periodically distributed cracks. For illustration, some results obtained by these authors are briefly reviewed here.

As the first example, consider a unit cell which contains a penny-shaped crack normal to the x_3-axis, at its center; see Figure 12.7.1. The unit cell is a general parallelepiped with dimensions $2a_i$ ($i = 1, 2, 3$), and the radius of the penny-shaped crack is b.

The overall compliance tensor $\bar{\mathbf{D}}$ of the unit cell is given by (12.7.25b). The presence of the crack increases $\bar{D}_{3333}$ (the inverse of the Young modulus in the x_3–direction, $\bar{E}_3$), and $\bar{D}_{2323} = \bar{D}_{3131}$ (the inverse of the shear modulus for the x_2,x_3- or x_3,x_1-plane, $\bar{\mu}_3$), since the crack is normal to the x_3-axis. As an illustration, Figure 12.7.2 shows the graph of $\bar{\mu}_3$ in terms of the crack density parameter $f = 4\pi b^3/3U$. Estimates based on the differential scheme and the self-consistent method are also shown for comparison.

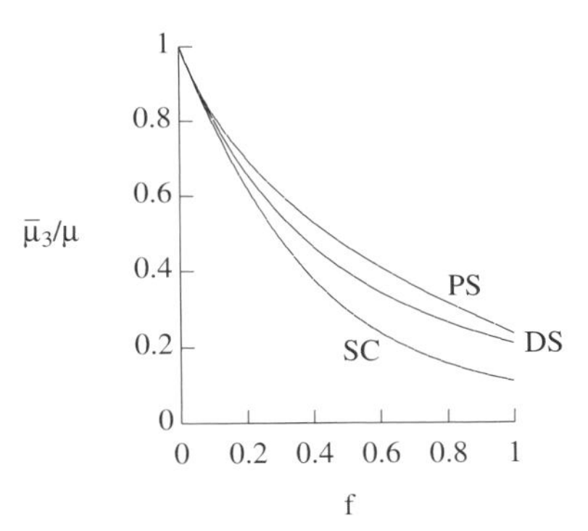

Figure 12.7.2

Estimate of shear modulus for solids with penny-shaped cracks:
SC: self-consistent
DS: differential scheme
PS: periodic structure

For the periodic calculation, a constant eigenstrain is used within the entire crack. The result, therefore, is an upper bound, as discussed in Section 13. It turns out, however, that remarkably accurate estimates are obtained by such a crude approximation; this is similar to the case of periodic voids (see Nunan and Keller, 1984).

As the second example, consider a unit cell which contains a flat slit crack normal to the x_3-axis with crack tips along the x_1-axis; see Figure 12.7.3. This infinitely extended slit crack can be treated as a two-dimensional line crack under plane strain conditions.

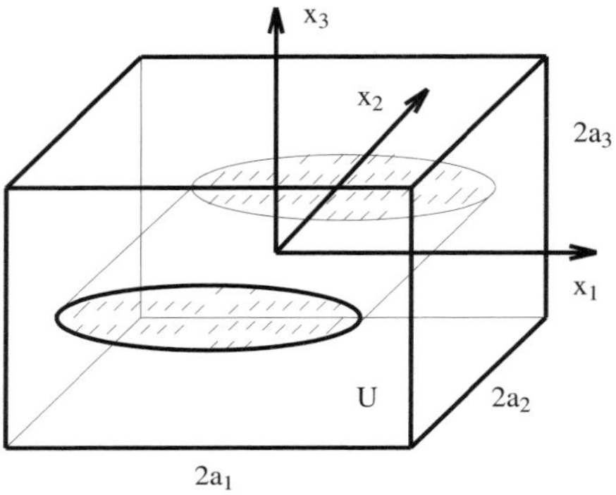

Figure 12.7.3

Cubic unit cell with infinitely long slit crack normal to the x_3-axis

The corresponding g-integral is given by the limit of an elliptical-cylindrical cavity as its thickness vanishes. i.e.,

$$g(\boldsymbol{\xi}) = \lim_{h \to 0} \frac{2}{y'} J_1(y') \frac{\sin x}{x}$$

$$= \frac{2}{y} J_1(y) \frac{\sin x}{x}, \tag{12.7.31}$$

where $x = a_2\xi_2$, $y' = b(\xi_1^2 + h^2\xi_3^2)^{1/2}$, and $y = b\xi_1$, with hb being the thickness of the cavity.[22] In essentially the same manner as shown in Subsection 12.7.4, $\mathbf{U}^{P(-1)}$, the limit of $h\,\{(1-f')\,\mathbf{C} - \mathbf{C} : \mathbf{S}^{P'}\}^{-1}$ as h goes to zero, is computed to be

$$U^{P(-1)}_{3333} = \frac{1}{f}\frac{1-\nu^2}{E}\frac{1}{\hat{s}_2}, \qquad U^{P(-1)}_{2323} = \frac{1}{f}\frac{1+\nu}{2E}\frac{1}{\hat{s}_1},$$

$$U^{P(-1)}_{3131} = \frac{1}{f}\frac{1-\nu^2}{4E}\frac{1}{\hat{s}_1 - \hat{s}_2}, \tag{9.4.32a~c}$$

where $\hat{s}_1$ and $\hat{s}_2$ are

$$\hat{s}_1 = \sum_{\xi_1}{}' \, g(-\boldsymbol{\xi})\, g(\boldsymbol{\xi})\, (a_1\xi_1) \coth(a_1\xi_1),$$

[22] Note that under plane strain conditions, the Fourier series expansion in the x_2-direction is not performed, and hence $\sin x/x = 1$ for $\xi_2 = 0$.

$$\hat{s}_2 = \sum_{\xi_1}' \frac{g(-\xi)\,g(\xi)}{2} \{(a_1\xi_1)\coth(a_1\xi_1) + (a_1\xi_1)^2 \mathrm{cosech}^2(a_1\xi_1)\}. \qquad (12.7.32d,e)$$

Note that $\hat{s}_1$ and $\hat{s}_2$ are given by the infinite summation with respect to ξ_1 only. As an illustration, Figure 12.7.4 shows the graph of $\bar{\mu}_3$ in terms of the crack density parameter $f = (b/a)^2$. The results are again for a constant eigenstrain field within the entire crack. To check the accuracy of this upper-bound estimate, Nemat-Nasser *et al.* (1992) subdivide the slit crack into 200 elements, and using a piecewise constant eigenstrain field, obtain improved bounds which are shown by heavy dots in Figure 12.7.4. As is seen, the crudest approximation with a single uniform eigenstrain field provides rather accurate upper bounds.

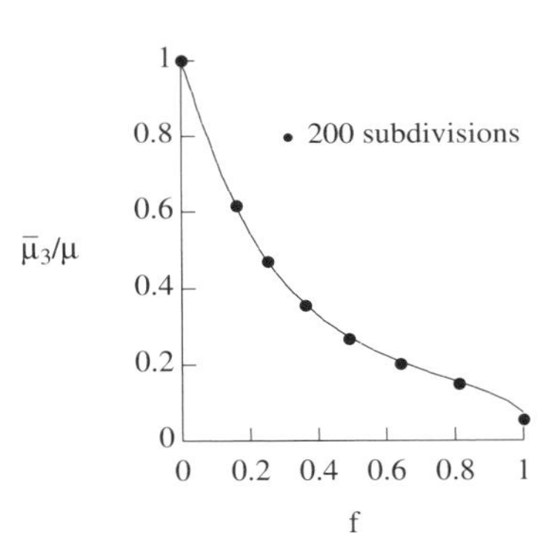

Figure 12.7.4

Estimate of shear modulus for a solid with periodically distributed slit cracks

12.8. APPLICATION TO NONLINEAR COMPOSITES

Suppose neither the matrix nor the inhomogeneities in a composite with periodic microstructure are linearly elastic. For illustration, consider only small strains and rotations. In the following, *both* rate-dependent and rate-independent cases are examined simultaneously. *For the rate-dependent materials, let* $\boldsymbol{\varepsilon}$ *stand for the strain rate, whereas for the rate-independent cases,* $\boldsymbol{\varepsilon}$ *continues to denote the strain.* Consider an incremental formulation, and denote the increment of the strain (strain rate) and stress, respectively, by $d\boldsymbol{\varepsilon}$ and $d\boldsymbol{\sigma}$. Consider the constitutive relations,

$$d\boldsymbol{\varepsilon} = \mathbf{D}' : d\boldsymbol{\sigma}, \qquad d\boldsymbol{\sigma} = \mathbf{C}' : d\boldsymbol{\varepsilon}, \qquad (12.8.1a,b)$$

where $\mathbf{D}' = \mathbf{D}'(\mathbf{x};\, \boldsymbol{\sigma})$ is the instantaneous compliance tensor field, which, in general, depends on the location within the unit cell, as well as on the stress and possibly on other material parameters; see Appendix A. For example, for the J_2-plasticity model, $\mathbf{D}'$ is given by (A.2.9a), and its inverse, $\mathbf{C}'$, is given by (A.2.9b). For the rate-dependent case, various models may be considered.

As a simple illustration, consider the nonlinear viscoplastic model of Subsection A.3.3, where the *strain-rate increment*[23] relates to the stress increment through the instantaneous compliance given by (A.3.5b). Now, the entire formulation of this section applies to this rate-independent (rate-dependent) elastoplastic (viscoplastic) unit cell. For example, in the case of metal-matrix composites, say, aluminum matrix-alumina inclusions, the matrix may be modeled by rate-independent J_2-plasticity, so that

$$\boldsymbol{D}^{\mathrm{M}} = \mathbf{D}^{\mathrm{M}} + \frac{\boldsymbol{\sigma}' \otimes \boldsymbol{\sigma}'}{4\,\tau^2\,\mathrm{H}^{\mathrm{M}}},$$

$$\boldsymbol{C}^{\mathrm{M}} = \mathbf{C}^{\mathrm{M}} - \frac{(\mathbf{C}^{\mathrm{M}} : \boldsymbol{\sigma}') \otimes (\mathbf{C}^{\mathrm{M}} : \boldsymbol{\sigma}')}{4\,\tau^2\,\mathrm{H}^{\mathrm{M}} + \boldsymbol{\sigma}' : \mathbf{C}^{\mathrm{M}} : \boldsymbol{\sigma}'}, \tag{12.8.2a,b}$$

where $\boldsymbol{\sigma}' = \boldsymbol{\sigma} - \sigma_{ii}/3\,\mathbf{1}^{(2)}$ is the deviatoric stress; $\tau - (\boldsymbol{\sigma}' : \boldsymbol{\sigma}'/2)^{1/2}$ is the work-hardening parameter; and H^{M} may be constant for linear hardening (Accorsi and Nemat-Nasser, 1986), or it may be fixed by considering a power-law fit for the uniaxial stress-strain relation of the material, i.e.,

$$\mathrm{H}^{\mathrm{M}} = \sigma_{\mathrm{Y}}^{\mathrm{o}}\,\mathrm{N}\left(\frac{\gamma}{\gamma_{\mathrm{o}}}\right)^{\mathrm{N}-1}, \tag{12.8.2c}$$

where $\sigma_{\mathrm{Y}}^{\mathrm{o}}$ is the initial yield stress, N is a material parameter, and γ_{o} is the initial yield strain. When N = 1, the uniaxial stress-strain relation is bilinear. In general, N is small, less than 0.1 for certain metals (about 0.08 for 4340 steels).

As a further simplification, let the matrix elasticity tensor be isotropic. Then, (12.8.2b), for example, reduces to

$$\boldsymbol{C}^{\mathrm{M}} = \lambda^{\mathrm{M}}\,\mathbf{1}^{(2)} \otimes \mathbf{1}^{(2)} + 2\mu^{\mathrm{M}}\,\mathbf{1}^{(4s)} - \frac{(\mu^{\mathrm{M}})^2}{(\mu^{\mathrm{M}} + \mathrm{H}^{\mathrm{M}})\,\tau^2}\,\boldsymbol{\sigma}' \otimes \boldsymbol{\sigma}'$$

$$= 2\mu^{\mathrm{M}}\left\{\frac{\nu^{\mathrm{M}}}{1 - 2\nu^{\mathrm{M}}}\,\mathbf{1}^{(2)} \otimes \mathbf{1}^{(2)} + \mathbf{1}^{(4s)} - \frac{\mu^{\mathrm{M}}}{(1 + \beta^{\mathrm{M}})\,\tau^2}\,\boldsymbol{\sigma}' \otimes \boldsymbol{\sigma}'\right\}, \tag{12.8.2d}$$

where $\beta^{\mathrm{M}} = \mathrm{H}^{\mathrm{M}}/\mu^{\mathrm{M}}$. Further, let the inclusions be isotropic and linearly elastic,

$$\mathbf{C}^{\mathrm{I}} = 2\mu^{\mathrm{I}}\left\{\frac{\nu^{\mathrm{I}}}{1 - 2\nu^{\mathrm{I}}}\,\mathbf{1}^{(2)} \otimes \mathbf{1}^{(2)} + \mathbf{1}^{(4s)}\right\}. \tag{12.8.3}$$

Since the instantaneous moduli of the matrix material, (12.8.2d), depend on the state of stress at each material point, and since this stress field is not, in general, uniform throughout the matrix, for accurate calculation it may be necessary to subdivide the matrix into finite elements and use the average stress within each element to determine the effective instantaneous modulus tensor of the element. Then the homogenizing eigenstrain (or eigenstress) field within the matrix may be assumed to be uniform, and the average stress taken over each element is computed in the same manner as for the linearly elastic case. It is also possible to subdivide the inclusion into suitably small regions and use the constant eigenstrain in each subregion.

---

[23] The strain-rate increment is denoted in Appendix A by $d\dot{\boldsymbol{\varepsilon}}$, but here it is denoted by $d\boldsymbol{\varepsilon}$.

In general, at each stage of incremental loading, the instantaneous modulus tensor field, $\mathbf{C}'$, of the unit cell is given by

$$\mathbf{C}'(\mathbf{x}) = \begin{cases} \mathcal{C}^{\mathrm{M}}(\boldsymbol{\sigma}') & \text{for } \mathbf{x} \text{ in M} \\ \mathbf{C}^{\mathrm{I}} & \text{for } \mathbf{x} \text{ in I.} \end{cases} \tag{12.8.4}$$

In this manner, each subdivision is assigned its instantaneous modulus tensor, depending on its location within the matrix or inclusion, where in the latter case the modulus tensor remains constant throughout the loading history. Denote by $\mathbf{C}^{\alpha}$ ($\alpha = 1, 2, ..., n$) the current values of the modulus tensor for the αth element. Then, with $\mathbf{C}$ as the reference modulus tensor, the results of Subsection 12.5.2 are applied to obtain the overall instantaneous elastoplastic modulus tensor of the composite. From the system of linear equations (12.5.7b), the average eigenstrain in each element, $\bar{\boldsymbol{\varepsilon}}^{*\alpha}$, is obtained. Then, (12.5.8b) yields the corresponding overall modulus tensor $\bar{\mathbf{C}}$.

As discussed in Section 13, this method provides upper and lower bounds on the moduli, provided that the reference elasticity tensor, $\mathbf{C}$, is properly chosen. Since for each element, constant moduli are assumed at each stage of the incremental loading, approximations which may violate the bounds are necessarily introduced. Yet, bounds can be established if the instantaneous moduli of the elastoplastically deforming elements are suitably chosen.

For the rate-dependent viscoplastic case, a similar analysis applies; see Nemat-Nasser *et al.* (1986). In this case, the *strain-rate* increment in each element is related to the stress increment by (12.8.1a,b). For example, for the power-law model, $\mathbf{D}'$ is given in terms of the deviatoric stress tensor, $\boldsymbol{\sigma}'$, by (A.3.5.b), i.e.,

$$\mathbf{D}' = \eta\, \tau^{n} \{\mathbf{1}^{(4s)} + \frac{n}{2\tau^2} \boldsymbol{\sigma}' \otimes \boldsymbol{\sigma}'\}, \tag{12.8.5}$$

where η and n are material parameters. Hence, the same procedure can be used to obtain the instantaneous overall modulus tensor which relates the overall (e.g., prescribed) *strain-rate* increment $\mathrm{d}\boldsymbol{\varepsilon}^{\mathrm{o}}$ to the corresponding overall stress increment $< \mathrm{d}\boldsymbol{\sigma} >$; see Nemat-Nasser *et al.* (1986) for illustrative examples.

Note that it is possible to treat the entire unit cell as an inhomogeneous cell, and directly apply the Fourier analysis to solve field equations, without the use of homogenization approaches, i.e., without the use of eigenstrains or eigenstresses. In such an approach, a large number of linear equations must be solved at each incremental loading. Moreover, even for the linearly elastic case, the benefit of automatically bounding the overall parameters is generally lost in such a direct approach. With the use of homogenization, on the other hand, accurate bounds are obtained with rather crude approximations. Indeed, it is possible to treat the entire matrix or the entire inclusion as a single element with uniform eigenstrains, and yet obtain reasonable bounds; see Accorsi and Nemat-Nasser (1986). This procedure provides an effective tool for design, where crude approximations are made at early stages of the material development, say, using a single-element homogenization, and then, once a final design is obtained, a rather refined homogenization is employed to calculate the corresponding response. Indeed, this procedure can be used in large-scale finite-element codes, when micromechanics with crude approximations is employed to obtain constitutive relations of the composite in each finite element, and then, when necessary, e.g., for damage evaluation, refined calculations are considered, in order to assess local material responses.

Even for highly nonlinear elastoplastic metal-matrix-ceramic composites, the use of a single region with an average homogenization eigenstrain field leads to a closed-form result for the overall properties of the composite. In this case, it follows from (12.5.9b) that

$$\bar{\mathbf{C}} = \mathbf{C}^{\mathrm{I}} : \left\{ \mathbf{1}^{(4s)} - f\,\{(\mathbf{C}^{\mathrm{I}} - \boldsymbol{C}^{\mathrm{M}})^{-1} : \mathbf{C}^{\mathrm{I}} - \mathbf{S}^{\mathrm{P}}\}^{-1} \right\}, \tag{12.8.6}$$

when the matrix is homogenized and the inclusion elasticity tensor is used for the reference modulus tensor. The result is an upper bound. Note that $\mathbf{S}^{\mathrm{P}}$ in this case is given by (12.5.6b) and (12.5.4a~d), where the g-integral over the matrix is simply given by

$$g_{\mathrm{M}}(\boldsymbol{\xi}) \equiv < \exp(\iota\boldsymbol{\xi}\boldsymbol{.}\mathbf{x})>_{\mathrm{M}} = (1-f)\, g_{\mathrm{U}}(\boldsymbol{\xi}) - \frac{f}{1-f}\, g_{\mathrm{I}}(\boldsymbol{\xi}), \tag{12.8.7}$$

with $g_{\mathrm{U}}(\boldsymbol{\xi}) = 1$ for $\boldsymbol{\xi} = \mathbf{0}$, and $= 0$ otherwise. In this manner, $\mathbf{S}^{\mathrm{P}}$ and $\mathbf{C}^{\mathrm{I}}$ remain unchanged in (12.8.6), and only $\boldsymbol{C}^{\mathrm{M}}$ is changed with continued loading. The results are explicit. Note that $\mathbf{C}^{\mathrm{I}}$ need not be used for the reference elasticity, and one may use $\mathbf{C}^{\mathrm{M}}$, instead. Indeed, it is also possible to use the yet-unknown overall modulus tensor $\bar{\mathbf{C}}$ for the reference one, and obtain estimates which are expected to be closer to the exact solution than the corresponding upper bounds; see Subsection 13.5 for further comments.

12.9. REFERENCES

Aboudi, J. (1991), *Mechanics of composite materials - A unified micromechanical approach*, Elsevier, Amsterdam.

Accorsi, M. L. and Nemat-Nasser, S. (1986), Bounds on the overall elastic and instantaneous elastoplastic moduli of periodic composites, *Mech. Matr.*, Vol. 5, 209-220.

Bakhvalov, N. S. and Panasenko, G. (1984), *Homogenization: averaging processes in periodic media*, Kluwer Academic Publ., Dordrecht.

Furuhashi, R., Kinoshita, N., and Mura, T. (1981), Periodic distribution of inclusions, *Int. J. Eng. Sci.*, Vol. 19, 231-236.

Horii, H. and Sahasakmontri, K. (1990), Mechanical properties of cracked solids: Validity of the self-consistent method, in: *Micromechanics and inhomogeneity - The T. Mura 65th anniversary volume*, (G. J. Weng, M. Taya, and H. Abé, eds.), Springer-Verlag, New York, 137-159.

Iwakuma, T. and Nemat-Nasser, S. (1983), Composites with periodic microstructure, *Advances and Trends in Structural and Solid Mechanics*, Pergamon Press, 13-19 or *Computers and Structures*, Vol. 16, Nos. 1-4, 13 -19.

Jeffrey, D. J. (1973), Conduction through a random suspension of spheres, *Proc. Roy. Soc. Lond. Ser. A.*, Vol. 335, 355-367.

McCoy, J. J. and Beran, M. J. (1976), On the effective thermal conductivity of a random suspension of spheres, *Int. J. Eng. Sci.*, Vol. 14, 7-18.

Mura, T. (1987), *Micromechanics of defects in solids (2nd Edition)*, Martinus Nijhoff Publishers, Dordrecht.

Nemat-Nasser, S., Iwakuma, T., and Hejazi, M. (1982), On composites with periodic structure, *Mech. Matr.*, Vol. 1, 239-267.

Nemat-Nasser, S., Iwakuma, T., and Accorsi, M. (1986), Cavity growth and grain boundary sliding in polycrystalline solids, *Mech. Matr.*, Vol. 5, 317-329.

Nemat-Nasser, S. and Taya, M. (1981), On effective moduli of an elastic body containing periodically distributed voids, *Quarterly of Applied Mathematics*, Vol. 39, 43-59.

Nemat-Nasser, S. and Taya, M. (1985), On effective moduli of an elastic body containing periodically distributed voids: Comments and corrections, *Quarterly of Applied Mathematics,* Vol. 43, 187-188.

Nemat-Nasser, S., Yu, N., and Hori, M. (1993), Solids with periodically distributed cracks, *Int. J. Solids Struct.*, Vol. 30, No. 15, 2071-2095.

Nunan, K. C. and Keller, J. B. (1984), Effective elasticity tensor of a periodic composite, *J. Mech. Phys. Solids*, Vol. 32, 259-280.

Sneddon, I. N. (1951), *Progress in solid mechanics* (I. N. Sneddon and R. Hill, eds.), Vol. 2, North Holland, Amsterdam.

Walker, K. P., Jordan, E. H., and Freed, A. D. (1990), Equivalence of Green's function and the Fourier series representation of composites with periodic structure, in *Micromechanics and inhomogeneity - The T. Mura 65th anniversary volume* (G. J. Weng, M. Taya, and H. Abé, eds.), Springer-Verlag, New York, 535-558.

Walker, K. P., Freed, A. D., and Jordan, E. H. (1991), Microstress analysis of periodic composites, *Composites Engineering*, Vol. 1, 29-40.

Willis, J. R. (1968), The stress field around an elliptical crack in an anisotropic medium, *Int. J. Eng. Sci.*, Vol. 6, 253-263.

SECTION 13 OVERALL PROPERTIES OF SOLIDS WITH PERIODIC MICROSTRUCTURE

In Section 12, an elastic solid with periodic microstructure is considered, and the representation of the corresponding field variables in terms of Fourier series is examined. The governing field (partial differential) equations are reduced to sets of linear algebraic equations for the Fourier coefficients. These can be solved exactly. Hence, the overall elasticity and compliance tensors of an elastic solid with periodic microstructure can be estimated to any desired degree of accuracy, and, at least in principle, the problem admits an exact (unique) solution. The actual evaluation of the moduli, however, in general entails considerable numerical effort, even though this can be reduced through the use of various symmetry and antisymmetry properties of the periodic structure, as discussed in Section 14. Furthermore, often, instead of an exact distribution of the field quantities, the overall constitutive response associated with the average field variables may be of greater interest. The overall response of a periodically heterogeneous solid of this kind is defined in terms of the volume averages of the stress and strain taken over a typical unit cell.

In this section, estimates of the overall response of solids with periodic microstructure, consisting of simple arrangements of linearly elastic uniform micro-inclusions (or defects) embedded in a linearly elastic uniform matrix are sought. To this end, first, relations between the average field quantities in the equivalent homogeneous solid are established by the introduction of suitable periodic eigenstrain or eigenstress fields. Then, the Hashin-Shtrikman variational principle,[1] discussed for an RVE in Section 9, is applied to solids with periodic microstructure, and bounds on the overall moduli are obtained by defining energy functionals for the eigenstrain or eigenstress fields in the equivalent homogeneous solid. Finally, this application is considered in some detail, bounds for the overall elastic energy of the unit cell, and, hence, bounds for the overall elasticity parameters of elastic solids with periodic microstructure are obtained. These bounds are *exact* and can be computed to any desired degree of accuracy. In Subsection 13.5 it is shown that there are always two overall elastic parameters (e.g., $\bar{K}$ and $\bar{\mu}$ for the isotropic case) whose bounds obtained on the basis of the periodic and random (RVE) microstructure, are identical, and hence exact. Moreover, these bounds are valid for any number of inclusions of any shape, distribution, and elasticity.

In this section, bounds on the overall moduli and compliances are developed with the aid of the Hashin-Shtrikman variational principles. As is pointed out in Subsections 9.5.6 and 9.7.2, these bounds for an RVE can be

[1] See Hashin and Shtrikman (1962), and Willis (1977).

obtained *directly* with the aid of Theorems I and II of Subsection 2.5.6, and by a suitable choice of the *reference elasticity or compliance tensor.* A similar approach can be used for the periodic case. In this approach, a reference, say, elasticity tensor $\mathbf{C}$ is chosen such that the difference $\mathbf{C}'(\mathbf{x}) - \mathbf{C}$ is positive-semi-definite (negative-semi-definite), where $\mathbf{C}'(\mathbf{x})$ now is the elasticity tensor of the unit cell, U. Then, with the aid of the theorem of the minimum potential energy,[2] computable bounds result from the properties of $\mathbf{C}' - \mathbf{C}$ and the properties of the integral operators for the periodic structure. Similar comments apply when the formulation is cast in terms of the compliance tensor. This formulation is presented in Subsection 13.2.4.

13.1. GENERAL EQUIVALENT HOMOGENEOUS SOLID

Consider an elastic solid consisting of a periodically arranged set of identical unit cells. Each cell contains n distinct linearly elastic micro-inclusions, Ω_α, embedded in a linearly elastic matrix, M. Consider a typical unit cell, and as before, denote the elasticity and compliance tensors of the matrix and micro-inclusions by $\mathbf{C}^M$ and $\mathbf{D}^M$, and $\mathbf{C}^\alpha$ and $\mathbf{D}^\alpha$, respectively. Note that $\mathbf{C}^M = (\mathbf{D}^M)^{-1}$ and $\mathbf{C}^\alpha = (\mathbf{D}^\alpha)^{-1}$, for[2] $\alpha = 1, 2, ..., n$.

13.1.1. Notation and Introductory Comments

The notation used throughout this section is essentially the same as that used in Section 12. It is summarized as follows: the domain of a unit cell is denoted by

$$U = M + \Omega = M + \sum_{\alpha=1}^{n} \Omega_\alpha; \tag{13.1.1}$$

the elasticity and compliance tensor fields, $\mathbf{C}'$ and $\mathbf{D}'$, in U are

$$\mathbf{C}' = \mathbf{C}'(\mathbf{x}) = H(\mathbf{x};\, M)\,\mathbf{C}^M + \sum_{\alpha=1}^{n} H(\mathbf{x};\, \Omega_\alpha)\,\mathbf{C}^\alpha,$$

$$\mathbf{D}' = \mathbf{D}'(\mathbf{x}) = H(\mathbf{x};\, M)\,\mathbf{D}^M + \sum_{\alpha=1}^{n} H(\mathbf{x};\, \Omega_\alpha)\,\mathbf{D}^\alpha, \tag{13.1.2a,b}$$

where $H(\mathbf{x};\, M)$ and $H(\mathbf{x};\, \Omega_\alpha)$ are the Heaviside step functions. The volume fraction[3] of the αth micro-inclusion Ω_α and that of all inclusions Ω are defined,

[2] The superscript M for $\mathbf{C}^M$ and $\mathbf{D}^M$ emphasizes that the elasticity and compliance tensors are for the matrix, which may *not* necessarily be the same as those for the *reference* homogeneous solid, which will be denoted by $\mathbf{C}$ and $\mathbf{D}$.

[3] The volume fraction for the unit cell expresses the corresponding volume fraction for the unbounded periodic solid, as well as for a *finite* solid represented by the corresponding unit cell.

respectively, by

$$f_\alpha \equiv \frac{\Omega_\alpha}{U}, \qquad f \equiv \frac{\Omega}{U} = 1 - \frac{M}{U} = \sum_{\alpha=1}^{n} f_\alpha. \tag{13.1.3a,b}$$

As before, the volume average of a periodic tensor field **T** taken over the unit cell U of a typical micro-inclusion Ω_α is given by

$$< \mathbf{T} > \equiv \frac{1}{U}\int_U \mathbf{T}(\mathbf{x})\, dV, \qquad < \mathbf{T} >_\alpha \equiv \frac{1}{\Omega_\alpha}\int_{\Omega_\alpha} \mathbf{T}(\mathbf{x})\, dV. \tag{13.1.4a,b}$$

These are the corresponding averages of **T** in the unbounded periodic solid; $< \mathbf{T} >$ represents the volume average in the entire solid, and $< \mathbf{T} >_\alpha$ is that over the Ω_α-inclusion. Hence, these average quantities can be viewed either as corresponding to an unbounded periodic solid or to a *finite* body represented by the bounded unit cell.

The results obtained in Section 12 may be applied to a periodic solid with a general distribution of micro-inclusions of arbitrary shapes. In this section, however, rather simple distributions of micro-inclusions with simple shapes are considered, where the application of the basic results for the general case is illustrated. In particular, in all illustrations, it is assumed that:

(*) the unit cell remains unchanged upon reflection with respect to the plane $x_i = 0$, for i = 1, 2, and 3, i.e., the unit cell is elastically and geometrically completely symmetric with respect to all three coordinate planes.

In other words, the elasticity and compliance tensor fields, $\mathbf{C}'$ and $\mathbf{D}'$, satisfy three mirror-image reflections with respect to the coordinate planes,[4] i.e.,

$$\mathbf{C}'(\mathbf{x}^1) = \mathbf{C}'(\mathbf{x}^2) = \mathbf{C}'(\mathbf{x}^3) = \mathbf{C}'(\mathbf{x}),$$

$$\mathbf{D}'(\mathbf{x}^1) = \mathbf{D}'(\mathbf{x}^2) = \mathbf{D}'(\mathbf{x}^3) = \mathbf{D}'(\mathbf{x}), \tag{13.1.5a,b}$$

where $\mathbf{x}^1$, $\mathbf{x}^2$, and $\mathbf{x}^3$ are the mirror images of point $\mathbf{x}$ with respect to the planes $x_1 = 0$, $x_2 = 0$, and $x_3 = 0$, respectively. A unit cell with an isotropic matrix and an isotropic spherical micro-inclusion at its center is an example which satisfies (*) and (13.1.5a,b); see Figure 13.1.1.

13.1.2. Macrofield Variables and Homogeneous Solutions

As explained in Subsection 2.5, the macrofield quantities are given by the volume averages of the corresponding microfield variables. For an RVE, uniform boundary conditions completely determine the volume averages of the microfield variables: for example, uniform boundary tractions, $\mathbf{t} = \mathbf{v} \cdot \boldsymbol{\sigma}^o$, determine the average stress, $\boldsymbol{\sigma}^o$, and linear boundary displacements, $\mathbf{u} = \mathbf{x} \cdot \boldsymbol{\varepsilon}^o$, determine the average strain, $\boldsymbol{\varepsilon}^o$. The macrostress $\boldsymbol{\Sigma}$ and/or macrostrain $\mathbf{E}$, which are the averages of the microstress and microstrain fields, completely fix these

[4] This condition is referred to as the fourth MI sym/ant condition in Section 14; see Subsection 14.1 for explanation and details.

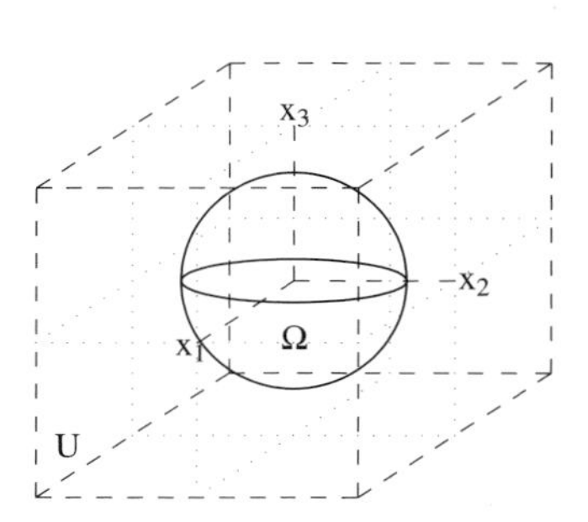

Figure 13.1.1

Example of fully symmetric periodic structure: an isotropic matrix and an isotropic spherical inclusion at its center

simple boundary conditions for the corresponding RVE. Furthermore, with either boundary data, $<\boldsymbol{\sigma}:\boldsymbol{\varepsilon}> = <\boldsymbol{\sigma}>:<\boldsymbol{\varepsilon}>$ which, in general, may not hold for an RVE subjected to general fluctuating boundary data; see comments in Subsection 2.6, as well as discussions in Subsection 9.5.1.

Similar notions *cannot* be applied directly to unbounded periodic solids, since the boundary data for a typical unit cell must be such that the periodicity of the field variables is ensured. The average microfield variables in a periodic solid cannot be defined through simple boundary conditions. On the other hand, the periodic field variables are unique for a prescribed homogeneous strain or stress; see Subsection 14.4. Each of these homogeneous field quantities is the volume average of the corresponding field variable in the unit cell; note that the volume average of the strain and stress fields associated with a *continuous* periodic displacement field is zero. Hence, the macrofield quantities for periodic solids are the associated homogeneous strain and stress fields: *one to be prescribed, and the other to be determined.*

For example, suppose that the homogeneous strain field $\boldsymbol{\varepsilon}^{\rm o}$ is prescribed. The associated average stress field then is given by

$$\bar{\boldsymbol{\sigma}} = <\boldsymbol{\sigma}> = <\mathbf{C}':(\boldsymbol{\varepsilon}^{\rm o}+\boldsymbol{\varepsilon}^{\rm p})> \equiv \bar{\mathbf{C}}:\boldsymbol{\varepsilon}^{\rm o}, \tag{13.1.6a}$$

where $\boldsymbol{\varepsilon}^{\rm p}$ is the periodic strain field (with zero average), and $\bar{\mathbf{C}}$ is the overall elasticity tensor. Similarly, when the homogeneous stress field $\boldsymbol{\sigma}^{\rm o}$ is prescribed, the associated average strain field becomes

$$\bar{\boldsymbol{\varepsilon}} = <\boldsymbol{\varepsilon}> = <\mathbf{D}':(\boldsymbol{\sigma}^{\rm o}+\boldsymbol{\sigma}^{\rm p})> \equiv \bar{\mathbf{D}}:\boldsymbol{\sigma}^{\rm o}, \tag{13.1.6b}$$

where $\boldsymbol{\sigma}^{\rm p}$ is the periodic stress field (with zero average), and $\bar{\mathbf{D}}$ is the overall compliance tensor.

From the uniqueness of the periodic strain $\boldsymbol{\varepsilon}^{\rm p}$ and stress $\boldsymbol{\sigma}^{\rm p}$, both with zero average corresponding to prescribed $\boldsymbol{\varepsilon}^{\rm o}$ or $\boldsymbol{\sigma}^{\rm o}$, it follows that the overall elasticity and compliance tensors are each other's inverse, i.e.,

$$\bar{\mathbf{C}}:\bar{\mathbf{D}} = \bar{\mathbf{D}}:\bar{\mathbf{C}} = \mathbf{1}^{(4s)}. \tag{13.1.7}$$

The macrostress and macrostrain tensors are then related through

$$\boldsymbol{\Sigma} = \overline{\mathbf{C}} : \mathbf{E}, \qquad \mathbf{E} = \overline{\mathbf{D}} : \boldsymbol{\Sigma}. \tag{13.1.8a,b}$$

It is emphasized here that for an elastic solid with periodic microstructure, the microscopic and hence macroscopic responses are the same whether $\boldsymbol{\Sigma} = \boldsymbol{\sigma}^o$ or $\mathbf{E} = \boldsymbol{\varepsilon}^o$ is regarded prescribed, whereas, for an arbitrary finite RVE, the *microscopic* response is different depending on whether $\boldsymbol{\Sigma} = \boldsymbol{\sigma}^o$, $\mathbf{E} = \boldsymbol{\varepsilon}^o$, or general fluctuating boundary data are regarded prescribed; see Subsections 2.5, 2.6, and 9.5.

13.1.3. Periodic Microstructure versus RVE

It may be helpful to pause for a moment and review some important differences between the model of a heterogeneous solid with periodic microstructure, and a heterogeneous solid with more or less randomly distributed microheterogeneities, modeled by an RVE.

The most obvious difference is a geometric one. The microstructure in the periodic case can be described exactly, through a representative unit cell, whereas for an RVE, in general, only a statistical description is possible. As pointed out before, a totally structured microheterogeneity and a totally random one can be viewed as extreme cases which provide limits and, therefore, useful information in actual applications. Moreover, while the unit cell in the periodic case is repeated in a regular manner, the structure and arrangement within the unit cell can be irregular, leading to an isotropic[5] overall response.

The second important difference between the concept of an RVE and the periodic model is that, through the application of a Fourier series representation, the periodic (elasticity) model can be solved essentially exactly in many important cases, whereas in the case of an RVE, only estimates based on specialized models (e.g., the dilute distribution, the self-consistent, and the differential models) are possible.

The third important difference relates to the boundary data and the uniqueness of the solution. For the periodic case, the unit cell is homogenized by the introduction of a unique eigenstress (or eigenstrain) field, resulting in: 1) a unique overall elasticity tensor for a prescribed overall uniform strain field; and 2) a unique overall compliance tensor for a prescribed overall uniform stress field. These overall elasticity and compliance tensors have an exact inverse relation and do not depend on whether the overall stress or strain field is regarded prescribed. On the other hand, the overall elasticity and compliance tensors for an RVE, in general, depend on the assumed boundary data and on the approximating model that is used to estimate them. In fact, as discussed in Subsection 2.5, the uniform boundary tractions for a prescribed overall strain and the linear boundary displacements for a prescribed overall stress, yield the smallest values for the total macro-stress and -strain potentials, respectively, for

[5] Complete randomness of the microstructure is a sufficient condition for the overall modulus tensor to be isotropic. Necessary and sufficient conditions for the isotropy of this tensor are discussed in Section 15.

a fixed finite RVE. Indeed, for an RVE, the energy-based definition of the overall moduli, in general, does not lead to the same quantities as those obtained through the average stress-strain relations, unless special boundary data are considered, whereas for the periodic case, the two definitions produce identical results. Furthermore, not all approximating techniques yield elasticity (for prescribed strain) and compliance (for prescribed stress) tensors which are each other's inverse[6].

Finally, there is a difference between the periodic case and the RVE problem when bounds on the overall moduli are sought, using the Hashin-Shtrikman variational principle. For the periodic case, as is shown, one can use a unit cell with a well-defined structure and obtain exact bounds. For the RVE model, on the other hand, certain approximations are often required to obtain computable bounds, and, as shown in Subsections 9.5, 9.6, and 9.7, it is still necessary to establish whether or not the involved approximations invalidate the bounds; as is shown in those subsections, with the aid of Theorems I and II of Subsection 2.5.6, exact computable bounds are, nevertheless, obtained in the case of an ellipsoidal RVE.

13.1.4. Unit Cell as a Bounded Body

Before considering the Hashin-Shtrikman variational principle for a solid with periodic microstructure, the field variables in an RVE used in the Hashin-Shtrikman variational principle are summarized, as follows:

- $\boldsymbol{\varepsilon}^{\mathrm{o}}$: a given uniform strain field associated with prescribed linear displacement boundary data
- $\mathbf{e}^*$, $\mathbf{s}^*$: homogenization eigenstrain, eigenstress fields for which the functionals I, J are respectively defined
- $\boldsymbol{\varepsilon}^{\mathrm{d}}$, $\boldsymbol{\sigma}^{\mathrm{d}}$: strain, stress fields produced by $\mathbf{e}^*$ or $\mathbf{s}^*$ in the homogenized solid.

The homogenized unit cell in the periodic problem may be regarded as a homogenized bounded equivalent solid, with the corresponding eigenstrain, eigenstress fields as the homogenizing field. In this case, the basic fields then are:

- $\boldsymbol{\varepsilon}^{\mathrm{o}}$: a homogeneous strain representing the overall uniform strain
- $\mathbf{e}^*$, $\mathbf{s}^*$: homogenization eigenstrain, eigenstress fields with homogeneous parts $<\mathbf{e}^*>$, $<\mathbf{s}^*>$ and periodic parts $\mathbf{e}^* - <\mathbf{e}^*>$, $\mathbf{s}^* - <\mathbf{s}^*>$

---

[6] Since the global response of an RVE depends on the considered surface data, one may expect that the overall moduli are similarly dependent on the assumed surface data. However, in view of the limited information which can be had on the geometry, microstructure, and boundary data of an RVE, it is assumed that the results from two limiting sets of boundary data, i.e., uniform tractions and linear displacements (see Subsection 2.5), provide adequate information about the overall response of the RVE, especially since these, in general, provide bounds on the moduli associated with any other general boundary data; see Theorems I and II of Subsection 2.5.6.

$\boldsymbol{\varepsilon}^p$, $\boldsymbol{\sigma}^p$: periodic strain, stress fields produced in the homogenized solid by the periodic parts, $\mathbf{e}^* - < \mathbf{e}^* >$, $\mathbf{s}^* - < \mathbf{s}^* >$, of $\mathbf{e}^*$, $\mathbf{s}^*$.

For the bounded solid, $\boldsymbol{\varepsilon}^o$ prescribes linear displacement boundary conditions, and hence the solution is unique; for the periodic problem, $\boldsymbol{\varepsilon}^o$ is accompanied by a unique set of periodic field variables, and hence the solution is unique. The uniqueness of the solution guarantees that, for a given homogeneous strain $\boldsymbol{\varepsilon}^o$, there is only one exact eigenstress (or eigenstrain) field that gives the same field variables in the equivalent homogeneous solid as those in the original heterogeneous solid. Similar comments apply when the macrostress, $\boldsymbol{\sigma}^o$, is regarded prescribed. Therefore, the unit cell of a periodic microstructure may be regarded as a bounded solid which is homogenized through the introduction of the corresponding eigenstress (or eigenstrain) field, even though the boundary conditions of the unit cell as an element of an infinite periodic solid, may not be explicitly defined *a priori*; see Subsection 14.4.

13.1.5. Equivalent Homogeneous Solid for Periodic Microstructure

Now consider a *homogeneous* solid which represents the original solid of periodic microstructure through a *periodic distribution of eigenstrains or eigenstresses*. The elasticity and compliance of the solid are denoted by $\mathbf{C}$ and $\mathbf{D}$, satisfying $\mathbf{C} : \mathbf{D} = \mathbf{1}^{(4s)}$ or $\mathbf{D} : \mathbf{C} = \mathbf{1}^{(4s)}$. They can be the elasticity and compliance of the matrix phase, $\mathbf{C}^M$ and $\mathbf{D}^M$, or those of the αth inclusion phase, $\mathbf{C}^\alpha$ and $\mathbf{D}^\alpha$, or any other positive-definite reference elasticity and compliance tensors. In order to obtain the most general results, leave $\mathbf{C}$ and $\mathbf{D}$ unspecified,[7] to be defined in each case for the convenience of the analysis.

As in Section 9, an eigenstrain or eigenstress field is denoted in the equivalent homogeneous solid by $\mathbf{e}^* = \mathbf{e}^*(\mathbf{x})$ or $\mathbf{s}^* = \mathbf{s}^*(\mathbf{x})$, respectively. The homogeneous and periodic parts of $\mathbf{e}^*$ or $\mathbf{s}^*$ are denoted by $< \mathbf{e}^* >$ and $\mathbf{e}^* - < \mathbf{e}^* >$, or $< \mathbf{s}^* >$ and $\mathbf{s}^* - < \mathbf{s}^* >$. Let $\boldsymbol{\varepsilon}^p$ and $\boldsymbol{\sigma}^p$ be the periodic strain and stress fields due to $\mathbf{s}^*$. Then,

$$\boldsymbol{\varepsilon}^p = \mathbf{D} : \{\boldsymbol{\sigma}^p - (\mathbf{s}^* - < \mathbf{s}^* >)\} \tag{13.1.9a}$$

or

$$\boldsymbol{\sigma}^p = \mathbf{C} : \boldsymbol{\varepsilon}^p + (\mathbf{s}^* - < \mathbf{s}^* >). \tag{13.1.9b}$$

As shown in Subsection 12.4, the integral operator $\boldsymbol{\Gamma}^P$ determines $\boldsymbol{\varepsilon}^p$ for given $\mathbf{s}^*$, as follows:

$$-\boldsymbol{\varepsilon}^p(\mathbf{x}) = \boldsymbol{\Gamma}^P(\mathbf{x}; \mathbf{s}^*)$$

$$= \sum_{\boldsymbol{\xi}}{}' \, F\boldsymbol{\Gamma}^P(\boldsymbol{\xi}) : \{\frac{1}{U}\int_U \mathbf{s}^*(\mathbf{y}) \exp(\iota\boldsymbol{\xi}\cdot(\mathbf{x}-\mathbf{y}))\, dV_y\}, \tag{13.1.10a}$$

where the fourth-order tensor $F\boldsymbol{\Gamma}^P = F\boldsymbol{\Gamma}^P(\boldsymbol{\xi})$ is given by (12.4.12d). Then, the

---

[7] One requirement for $\mathbf{C}$ or $\mathbf{D}$ is that any arbitrary distribution of eigenstrains or eigenstresses should result in *unique* field variables in the equivalent homogenized solid.

associated periodic stress field is,

$$\boldsymbol{\sigma}^{p}(\mathbf{x}) = -\,\mathbf{C} : \boldsymbol{\Gamma}^{P}(\mathbf{x};\, \mathbf{s}^{*}) + (\mathbf{s}^{*}(\mathbf{x}) - <\mathbf{s}^{*}>). \tag{13.1.10b}$$

In a similar manner, define the conjugate integral operator $\boldsymbol{\Lambda}^{P}$ for an eigenstrain field, $\mathbf{e}^{*}$. Again, denoting the periodic stress and strain fields due to $\mathbf{e}^{*}$ by $\boldsymbol{\sigma}^{p}$ and $\boldsymbol{\varepsilon}^{p}$, obtain

$$\boldsymbol{\sigma}^{p} = \mathbf{C} : \{\boldsymbol{\varepsilon}^{p} - (\mathbf{e}^{*} - <\mathbf{e}^{*}>)\} \tag{13.1.11a}$$

or

$$\boldsymbol{\varepsilon}^{p} = \mathbf{D} : \boldsymbol{\sigma}^{p} + (\mathbf{e}^{*} - <\mathbf{e}^{*}>). \tag{13.1.11b}$$

Define the integral operator which determines $\boldsymbol{\sigma}^{p}$ due to $\mathbf{e}^{*}$, as

$$-\,\boldsymbol{\sigma}^{p}(\mathbf{x}) = \boldsymbol{\Lambda}^{P}(\mathbf{x};\, \mathbf{e}^{*})$$

$$= \sum_{\boldsymbol{\xi}}{}' \, F\boldsymbol{\Lambda}^{P}(\boldsymbol{\xi}) : \{\frac{1}{U}\int_{U} \mathbf{e}^{*}(\mathbf{y}) \exp(\iota\boldsymbol{\xi}\boldsymbol{\cdot}(\mathbf{x}-\mathbf{y}))\, dV_{y}\}. \tag{13.1.12a}$$

Then, the associated periodic strain field becomes[8]

$$\boldsymbol{\varepsilon}^{p}(\mathbf{x}) = -\,\mathbf{D} : \boldsymbol{\Lambda}^{P}(\mathbf{x};\, \mathbf{e}^{*}) + (\mathbf{e}^{*}(\mathbf{x}) - <\mathbf{e}^{*}>). \tag{13.1.12b}$$

Since the periodic fields due to $\mathbf{e}^{*}$ are the same as those due to $\mathbf{s}^{*} = -\,\mathbf{C} : \mathbf{e}^{*}$, the fourth-order tensor $F\boldsymbol{\Lambda}^{P} = F\boldsymbol{\Lambda}^{P}(\boldsymbol{\xi})$ in (13.1.12a) can be expressed in terms of $F\boldsymbol{\Gamma}^{P}$, as

$$F\boldsymbol{\Lambda}^{P}(\boldsymbol{\xi}) = -\,\mathbf{C} : F\boldsymbol{\Gamma}^{P}(\boldsymbol{\xi}) : \mathbf{C} + \mathbf{C} \qquad \text{for } \boldsymbol{\xi} \neq \mathbf{0}; \tag{13.1.13}$$

see Subsection 12.4. It should be noted that, similarly to the corresponding integral operators, $\boldsymbol{\Lambda}^{A}$ and $\boldsymbol{\Gamma}^{A}$, for an infinite homogeneous solid, the two integral operators, $\boldsymbol{\Lambda}^{P}$ and $\boldsymbol{\Gamma}^{P}$, for the unit cell are related to each other; see Table 13.1.1.

Although the homogeneous parts of the eigenstress or eigenstrain field, $<\mathbf{s}^{*}>$ or $<\mathbf{e}^{*}>$, do not produce periodic strain and stress fields, they are related to the (prescribed) homogeneous strain and stress,[9] $\boldsymbol{\varepsilon}^{o}$ and $\boldsymbol{\sigma}^{o}$, in the homogeneous solid. Indeed, corresponding to (13.1.9a,b),

$$\boldsymbol{\varepsilon}^{o} = \mathbf{D} : (\boldsymbol{\sigma}^{o} - <\mathbf{s}^{*}>) \qquad \text{or} \qquad \boldsymbol{\sigma}^{o} = \mathbf{C} : \boldsymbol{\varepsilon}^{o} + <\mathbf{s}^{*}>, \tag{13.1.9c,d}$$

and corresponding to (13.1.11a,b),

---

[8] If the periodic parts of the eigenstress and eigenstrain are denoted by $\mathbf{s}^{*p}$ and $\mathbf{e}^{*p}$, then, (13.1.10a,b) are rewritten as $-\,\boldsymbol{\varepsilon}^{p}(\mathbf{x}) = \boldsymbol{\Gamma}^{P}(\mathbf{x};\, \mathbf{s}^{*p})$ and $\boldsymbol{\sigma}^{p}(\mathbf{x}) = -\,\mathbf{C} : \boldsymbol{\Gamma}^{P}(\mathbf{x};\, \mathbf{s}^{*p}) + \mathbf{s}^{*p}(\mathbf{x})$, and (13.1.12a,b), as $-\,\boldsymbol{\sigma}^{p}(\mathbf{x}) = \boldsymbol{\Lambda}^{P}(\mathbf{x};\, \mathbf{e}^{*p})$ and $\boldsymbol{\varepsilon}^{p}(\mathbf{x}) = -\,\mathbf{D} : \boldsymbol{\Lambda}^{P}(\mathbf{x};\, \mathbf{e}^{*p}) + \mathbf{e}^{*p}(\mathbf{x})$. These are similar to the presentations used in Subsections 9.5 and 9.6.

[9] Special attention must be paid to the difference between the homogeneous fields in the equivalent homogeneous periodic structure and those in the original heterogeneous periodic structure; in the former, the homogeneous fields are related through (13.1.9c,d) when $\mathbf{s}^{*}$ is prescribed or (13.1.11c,d) when $\mathbf{e}^{*}$ is prescribed, while, in the latter, they are related through the exact overall elasticity or compliance tensor. i.e., $\boldsymbol{\sigma}^{o} = \overline{\mathbf{C}} : \boldsymbol{\varepsilon}^{o}$ or $\boldsymbol{\varepsilon}^{o} = \overline{\mathbf{D}} : \boldsymbol{\sigma}^{o}$.

Table 13.1.1

Relation between integral operators $\mathbf{\Lambda}^{\mathrm{P}}$ and $\mathbf{\Gamma}^{\mathrm{P}}$ and between $\mathbf{\Lambda}^{\mathrm{A}}$ and $\mathbf{\Gamma}^{\mathrm{A}}$

periodic fields	$\mathbf{\Lambda}^{\mathrm{P}}$	$\mathbf{\Gamma}^{\mathrm{P}}$
$\boldsymbol{\varepsilon}^{\mathrm{p}}$	$-\mathbf{D}:\mathbf{\Lambda}^{\mathrm{P}}(\mathbf{e}^*)+\mathbf{e}^{*\mathrm{p}}$	$-\mathbf{\Gamma}^{\mathrm{P}}(\mathbf{s}^*)$
$\boldsymbol{\sigma}^{\mathrm{p}}$	$-\mathbf{\Lambda}^{\mathrm{P}}(\mathbf{e}^*)$	$-\mathbf{C}:\mathbf{\Gamma}^{\mathrm{P}}(\mathbf{s}^*)+\mathbf{s}^{*\mathrm{p}}$
disturbance fields	$\mathbf{\Lambda}^{\mathrm{A}}$	$\mathbf{\Gamma}^{\mathrm{A}}$
$\boldsymbol{\varepsilon}^{\mathrm{A}}$	$-\mathbf{D}:\mathbf{\Lambda}^{\mathrm{A}}(\mathbf{e}^*)+\mathbf{e}^{*\mathrm{d}}$	$-\mathbf{\Gamma}^{\mathrm{A}}(\mathbf{s}^*)$
$\boldsymbol{\sigma}^{\mathrm{A}}$	$-\mathbf{\Lambda}^{\mathrm{A}}(\mathbf{e}^*)$	$-\mathbf{C}:\mathbf{\Gamma}^{\mathrm{A}}(\mathbf{s}^*)+\mathbf{s}^{*\mathrm{d}}$

$$\boldsymbol{\sigma}^{\mathrm{o}}=\mathbf{C}:(\boldsymbol{\varepsilon}^{\mathrm{o}}-<\mathbf{e}^*>) \qquad \text{or} \qquad \boldsymbol{\varepsilon}^{\mathrm{o}}=\mathbf{D}:\boldsymbol{\sigma}^{\mathrm{o}}+<\mathbf{e}^*>. \tag{13.1.11c,d}$$

As is shown in Section 12, the exact eigenstrain or eigenstress fields, $\boldsymbol{\varepsilon}^*=\boldsymbol{\varepsilon}^*(\mathbf{x})$ and $\boldsymbol{\sigma}^*=\boldsymbol{\sigma}^*(\mathbf{x})$, produce the same field variables in the equivalent homogeneous solid as those in the original heterogeneous periodic one. From (13.1.9a~d) and (13.1.11a~d), it can be concluded that 1) the periodic parts of $\boldsymbol{\varepsilon}^*$ and $\boldsymbol{\sigma}^*$ produce the periodic fields, and 2) their homogeneous parts correspond to the homogeneous fields. In the Hashin-Shtrikman variational principle, it is convenient to use an eigenstrain field if a homogeneous stress is prescribed, and an eigenstress field if a homogeneous strain is prescribed. Therefore, throughout this section, use the following consistency conditions:[10]

$$(\mathbf{D}'-\mathbf{D})^{-1}:\boldsymbol{\varepsilon}^*+\mathbf{\Lambda}^{\mathrm{P}}(\boldsymbol{\varepsilon}^*)-\boldsymbol{\sigma}^{\mathrm{o}}=\mathbf{0},$$

$$(\mathbf{C}'-\mathbf{C})^{-1}:\boldsymbol{\sigma}^*+\mathbf{\Gamma}^{\mathrm{P}}(\boldsymbol{\sigma}^*)-\boldsymbol{\varepsilon}^{\mathrm{o}}=\mathbf{0}, \tag{13.1.14a,b}$$

where, to simplify notation, the argument $\mathbf{x}$ is not displayed explicitly. The exact eigenstrain and eigenstress fields depend on the elasticity or compliance of the equivalent homogeneous solid, $\mathbf{C}$ or $\mathbf{D}$, which *is* the *reference* elasticity or compliance tensor. It is emphasized that, unlike for an RVE, for a prescribed microstructure, (13.1.14a) and (13.1.14b) are exactly equivalent, and hence only one of them needs to be considered. In general, however, both alternative representations are examined simultaneously, in order to display the elegant duality of this formulation.

---

[10] Consistency conditions (13.1.14a,b) are essentially the same as (12.4.22a,b). In this section, only (13.1.14a,b) are used, since they are the Euler equations of the energy functionals in the Hashin-Shtrikman variational principle.

Note that when $\mathbf{s}^*$ and $\mathbf{e}^*$ are the exact eigenstress and eigenstrain fields, $\boldsymbol{\sigma}^*$ and $\boldsymbol{\varepsilon}^*$, then the homogeneous stress and strain fields prescribed in the equivalent homogeneous periodic structure are related by the exact overall elasticity and compliance tensors, $\bar{\mathbf{C}}$ and $\bar{\mathbf{D}}$, of the original heterogeneous periodic structure. For arbitrary eigenstress and eigenstrain fields, $\mathbf{s}^*$ and $\mathbf{e}^*$, the overall elasticity and compliance tensors given through (13.1.9c,d) and (13.1.11c,d) do not coincide with $\bar{\mathbf{C}}$ and $\bar{\mathbf{D}}$. However, these overall tensors are each other's inverse when $\mathbf{s}^* = -\mathbf{C} : \mathbf{e}^*$ or $\mathbf{e}^* = -\mathbf{D} : \mathbf{s}^*$.

13.2. HASHIN-SHTRIKMAN VARIATIONAL PRINCIPLE APPLIED TO PERIODIC STRUCTURES

In this subsection, it is proved that the Hashin-Shtrikman variational principle can be applied to a periodic distribution of eigenstrains or eigenstresses, prescribed in the equivalent homogeneous solid.[11] Using the integral operators introduced in Subsection 13.1, two functionals, I and J, are defined for arbitrary eigenstrain and eigenstress fields *in the unit cell*, respectively as

$$\mathrm{I}(\mathbf{e}^*; \boldsymbol{\sigma}^o) \equiv \frac{1}{2} < \mathbf{e}^* : \{(\mathbf{D}' - \mathbf{D})^{-1} : \mathbf{e}^* + \boldsymbol{\Lambda}^{\mathrm{P}}(\mathbf{e}^*) - 2\,\boldsymbol{\sigma}^o\} >,$$

$$\mathrm{J}(\mathbf{s}^*; \boldsymbol{\varepsilon}^o) \equiv \frac{1}{2} < \mathbf{s}^* : \{(\mathbf{C}' - \mathbf{C})^{-1} : \mathbf{s}^* + \boldsymbol{\Gamma}^{\mathrm{P}}(\mathbf{s}^*) - 2\,\boldsymbol{\varepsilon}^o\} >, \qquad (13.2.1\mathrm{a,b})$$

where $\boldsymbol{\sigma}^o$ and $\boldsymbol{\varepsilon}^o$ are the prescribed homogeneous stress and strain.

13.2.1. Self-Adjointness

Since $\mathbf{D}'$ and $\mathbf{D}$ or $\mathbf{C}'$ and $\mathbf{C}$ are symmetric with respect to the first and last pairs of their indices, $(\mathbf{D}' - \mathbf{D})^{-1}$ or $(\mathbf{C}' - \mathbf{C})^{-1}$ has a similar symmetry. Hence, if $\boldsymbol{\Lambda}^{\mathrm{P}}(\mathbf{x};\, \mathbf{e}^*)$ and $\boldsymbol{\Gamma}^{\mathrm{P}}(\mathbf{x};\, \mathbf{s}^*)$, associated with the averaging operator $< >$, are self-adjoint, the integral operators in the functionals I and J become self-adjoint.

To prove the self-adjointness of $< : \boldsymbol{\Lambda}^{\mathrm{P}}(\ ) >$ or $< : \boldsymbol{\Gamma}^{\mathrm{P}}(\ ) >$, consider the volume average of the inner product of a periodic stress field and a periodic strain field. Let $\boldsymbol{\sigma}^{\mathrm{p}} = \boldsymbol{\sigma}^{\mathrm{p}}(\mathbf{x})$ and $\boldsymbol{\varepsilon}^{\mathrm{p}} = \boldsymbol{\varepsilon}^{\mathrm{p}}(\mathbf{x})$ be arbitrary periodic strain and stress fields, *which may be unrelated, but respectively are statically and kinematically admissible*. According to the averaging theorem (Section 2),

$$< \boldsymbol{\sigma}^{\mathrm{p}} : \boldsymbol{\varepsilon}^{\mathrm{p}} > = \frac{1}{\mathrm{U}} \int_{\partial \mathrm{U}} \mathbf{t}^{\mathrm{p}} \cdot \mathbf{u}^{\mathrm{p}}\, d\mathrm{S}, \qquad (13.2.2\mathrm{a})$$

where the tractions $\mathbf{t}^{\mathrm{p}}$ are given by $\boldsymbol{\nu} \cdot \boldsymbol{\sigma}^{\mathrm{p}}$, and $\mathbf{u}^{\mathrm{p}}$ is the displacement field

[11] As pointed out before and discussed in Subsection 13.2.4, the final bounds can be obtained directly by a proper choice of the reference elasticity or compliance tensor.

associated with $\boldsymbol{\varepsilon}^p$. Since $\boldsymbol{\sigma}^p$ and $\mathbf{u}^p$ are periodic, the surface integral of $\mathbf{t}^p \cdot \mathbf{u}^p$ over ∂U must vanish, i.e.,[12]

$$\int_{\partial U} \mathbf{t}^p \cdot \mathbf{u}^p \, dS = 0. \tag{13.2.2b}$$

The proof is straightforward. As shown in Subsection 12.3, the periodicity of $\mathbf{u}^p$ and $\boldsymbol{\sigma}^p$ implies

$$\mathbf{u}^p(\mathbf{x}^+) = \mathbf{u}^p(\mathbf{x}^-) \tag{13.2.3a}$$

and

$$(\boldsymbol{\nu} \cdot \boldsymbol{\sigma}^p)(\mathbf{x}^+) = -(\boldsymbol{\nu} \cdot \boldsymbol{\sigma}^p)(\mathbf{x}^-), \tag{13.2.3b}$$

for $\mathbf{x}^+$ and $\mathbf{x}^-$ on $x_i = \pm a_i$ (i = 1, 2, 3), since $\boldsymbol{\sigma}(\mathbf{x}^+) = \boldsymbol{\sigma}(\mathbf{x}^-)$ and $\boldsymbol{\nu}(\mathbf{x}^+) = -\boldsymbol{\nu}(\mathbf{x}^-)$. Hence,

$$\mathbf{t}^p(\mathbf{x}^+) \cdot \mathbf{u}^p(\mathbf{x}^+) = -\mathbf{t}^p(\mathbf{x}^-) \cdot \mathbf{u}^p(\mathbf{x}^-). \tag{13.2.3c}$$

On the boundary surfaces, say, the $(x_i = +a_i)$-plane and the $(x_i = -a_i)$-plane, (13.2.3c) yields

$$\int_{x_i = +a_i} \mathbf{t}^p(\mathbf{x}) \cdot \mathbf{u}^p(\mathbf{x}) \, dS = -\int_{x_i = -a_i} \mathbf{t}^p(\mathbf{x}) \cdot \mathbf{u}^p(\mathbf{x}) \, dS. \tag{13.2.4}$$

Hence, (13.2.2b) is proved.

Use (13.2.2b) to prove the self-adjointness of the $\boldsymbol{\Lambda}^P$-integral operator associated with the averaging operator < >. Now, let $\boldsymbol{\sigma}^p$ and $\boldsymbol{\varepsilon}^p$ be the periodic stress and strain fields due to[13] $\mathbf{e}^*$, $\boldsymbol{\sigma}^p = -\boldsymbol{\Lambda}^P(\mathbf{e}^*)$ and $\boldsymbol{\varepsilon}^p = \mathbf{D} : \boldsymbol{\sigma}^p + (\mathbf{e}^* - < \mathbf{e}^* >)$. Then,

$$\begin{aligned} < \mathbf{e}^* : \boldsymbol{\Lambda}^P(\mathbf{e}^*) > &= < (\boldsymbol{\varepsilon} - \mathbf{D} : \boldsymbol{\sigma}) : (-\boldsymbol{\sigma}^p) > \\ &= < (\boldsymbol{\varepsilon}^p - \mathbf{D} : \boldsymbol{\sigma}^p) : (-\boldsymbol{\sigma}^p) > \\ &= < \boldsymbol{\sigma}^p : \mathbf{D} : \boldsymbol{\sigma}^p >, \end{aligned} \tag{13.2.5a}$$

where $\boldsymbol{\varepsilon}$ and $\boldsymbol{\sigma}$ are the total strain and stress fields, given by the sum of the homogeneous part and the periodic part, and the fact that the volume average of the periodic fields vanishes, is used. Hence, $< : \boldsymbol{\Lambda}^P(\;) >$ is self-adjoint. In a similar manner, if $\boldsymbol{\varepsilon}^p$ is the periodic strain field due to $\mathbf{s}^*$, $\boldsymbol{\varepsilon}^p = -\boldsymbol{\Gamma}^P(\mathbf{s}^*)$, with the associated periodic stress field, $\boldsymbol{\sigma}^p = \mathbf{C} : \boldsymbol{\varepsilon}^p + \mathbf{s}^* - < \mathbf{s}^* >$, then

$$< \mathbf{s}^* : \boldsymbol{\Gamma}^P(\mathbf{s}^*) > = < \boldsymbol{\varepsilon}^p : \mathbf{C} : \boldsymbol{\varepsilon}^p >. \tag{13.2.5b}$$

Hence, $< : \boldsymbol{\Gamma}^P(\;) >$ is also self-adjoint. In view of the self-adjointness, notation $< \mathbf{e}^* : \boldsymbol{\Lambda}^P(\mathbf{e}^*) > \equiv < \mathbf{e}^* : \boldsymbol{\Lambda}^P : \mathbf{e}^* >$ and $< \mathbf{s}^* : \boldsymbol{\Gamma}^P(\mathbf{s}^*) > \equiv < \mathbf{s}^* : \boldsymbol{\Gamma}^P : \mathbf{s}^* >$ is used; see (9.1.7f) and (9.1.15a) of Subsection 9.1.

---

[12] In view of this fact, the energy-based and the stress, strain-based definitions of $\overline{\mathbf{C}} = \overline{\mathbf{D}}^{-1}$ are identical in the periodic model, i.e., in this case, $< \boldsymbol{\sigma} : \boldsymbol{\varepsilon} > = \boldsymbol{\sigma}^o : \boldsymbol{\varepsilon}^o$.

[13] Here and in the sequel, the spatial variable $\mathbf{x}$ in $\boldsymbol{\Lambda}^P(\mathbf{x}; \mathbf{e}^*)$ or $\boldsymbol{\Gamma}^P(\mathbf{x}; \mathbf{s}^*)$ is not displayed (but implied), whenever this variable is implied but not displayed in $\boldsymbol{\sigma}^p(\mathbf{x})$ or $\boldsymbol{\varepsilon}^p(\mathbf{x})$.

The self-adjointness of $< \,:\boldsymbol{\Lambda}^{\mathrm{P}}:\, >$ and $< \,:\boldsymbol{\Gamma}^{\mathrm{P}}:\, >$ can also be proved using the Fourier series. From (13.1.9) and (13.1.11), for the periodic strain and stress fields, $\boldsymbol{\varepsilon}^{\mathrm{p}}$ and $\boldsymbol{\sigma}^{\mathrm{p}}$, produced by an arbitrary eigenstrain, $\mathbf{e}^*$, or an arbitrary eigenstress, $\mathbf{s}^*$,

$$< \boldsymbol{\sigma}^{\mathrm{p}} : \boldsymbol{\varepsilon}^{\mathrm{p}} > = \begin{cases} \displaystyle\sum_{\boldsymbol{\xi}}{}' \, F\mathbf{e}^*(-\boldsymbol{\xi}) : (F\boldsymbol{\Lambda}^{\mathrm{P}})^{\mathrm{T}}(-\boldsymbol{\xi}) : \{\mathbf{D} : F\boldsymbol{\Lambda}^{\mathrm{P}}(\boldsymbol{\xi}) - \mathbf{1}^{(4s)}\} : F\mathbf{e}^*(\boldsymbol{\xi}) \\ \displaystyle\sum_{\boldsymbol{\xi}}{}' \, F\mathbf{s}^*(-\boldsymbol{\xi}) : (F\boldsymbol{\Gamma}^{\mathrm{P}})^{\mathrm{T}}(-\boldsymbol{\xi}) : \{\mathbf{C} : F\boldsymbol{\Gamma}^{\mathrm{P}}(\boldsymbol{\xi}) - \mathbf{1}^{(4s)}\} : F\mathbf{s}^*(\boldsymbol{\xi}), \end{cases} \tag{13.2.6}$$

where $F\mathbf{e}^*$ and $F\mathbf{s}^*$ are Fourier series coefficients of $\mathbf{e}^*$ and $\mathbf{s}^*$; in deriving (13.2.6), the fact that $< \exp(\iota\boldsymbol{\xi}\cdot\mathbf{x}) > = 0$ for $\boldsymbol{\xi} \neq \mathbf{0}$, is used. Since $F\boldsymbol{\Gamma}^{\mathrm{P}}(\boldsymbol{\xi}) = sym\,\{\boldsymbol{\xi}\otimes(\boldsymbol{\xi}\cdot\mathbf{C}\cdot\boldsymbol{\xi})^{-1}\otimes\boldsymbol{\xi}\}$ for $\boldsymbol{\xi} \neq \mathbf{0}$ and $C_{ijkl} = C_{jikl} = C_{ijlk}$, it follows that $(F\boldsymbol{\Gamma}^{\mathrm{P}})^{\mathrm{T}}(-\boldsymbol{\xi}) = F\boldsymbol{\Gamma}^{\mathrm{P}}(\boldsymbol{\xi})$, and[14]

$$\begin{aligned} F\boldsymbol{\Gamma}^{\mathrm{P}}(\boldsymbol{\xi}) : \mathbf{C} : F\boldsymbol{\Gamma}^{\mathrm{P}}(\boldsymbol{\xi}) &= sym\,\{\boldsymbol{\xi}\otimes(\boldsymbol{\xi}\cdot\mathbf{C}\cdot\boldsymbol{\xi})^{-1}\otimes\boldsymbol{\xi}\} : \mathbf{C} : \{\boldsymbol{\xi}\otimes(\boldsymbol{\xi}\cdot\mathbf{C}\cdot\boldsymbol{\xi})^{-1}\otimes\boldsymbol{\xi}\} \\ &= sym\,\{\boldsymbol{\xi}\otimes(\boldsymbol{\xi}\cdot\mathbf{C}\cdot\boldsymbol{\xi})^{-1}\otimes\boldsymbol{\xi}\} \\ &= F\boldsymbol{\Gamma}^{\mathrm{P}}(\boldsymbol{\xi}). \end{aligned} \tag{13.2.7a}$$

In a similar manner, $(F\boldsymbol{\Lambda}^{\mathrm{P}})^{\mathrm{T}}(-\boldsymbol{\xi}) = F\boldsymbol{\Lambda}^{\mathrm{P}}(\boldsymbol{\xi})$, and

$$F\boldsymbol{\Lambda}^{\mathrm{P}}(\boldsymbol{\xi}) : \mathbf{D} : F\boldsymbol{\Lambda}^{\mathrm{P}}(\boldsymbol{\xi}) = F\boldsymbol{\Lambda}^{\mathrm{P}}(\boldsymbol{\xi}). \tag{13.2.7b}$$

Hence, each element of the infinite sums in (13.2.6) vanishes for all $\boldsymbol{\xi}$'s, and the self-adjointness of the integral operators associated with the averaging operator is proved.

13.2.2. Hashin-Shtrikman Variational Principle and Bounds on Overall Moduli

The operators $< \,:\boldsymbol{\Lambda}^{\mathrm{P}}:\, >$ and $< \,:\boldsymbol{\Gamma}^{\mathrm{P}}:\, >$ in the energy functionals I and J are self-adjoint. From this, three basic results follow:

1) The variations of I and J are given by

$$\delta \mathrm{I}(\mathbf{e}^*; \boldsymbol{\sigma}^{\mathrm{o}}) = < \delta\mathbf{e}^* : \left\{ (\mathbf{D}' - \mathbf{D})^{-1} : \mathbf{e}^* + \boldsymbol{\Lambda}^{\mathrm{P}} : \mathbf{e}^* - \boldsymbol{\sigma}^{\mathrm{o}} \right\} >,$$

$$\delta \mathrm{J}(\mathbf{s}^*; \boldsymbol{\varepsilon}^{\mathrm{o}}) = < \delta\mathbf{s}^* : \left\{ (\mathbf{C}' - \mathbf{C})^{-1} : \mathbf{s}^* + \boldsymbol{\Gamma}^{\mathrm{P}} : \mathbf{s}^* - \boldsymbol{\varepsilon}^{\mathrm{o}} \right\} >. \tag{13.2.8a,b}$$

The Euler equations of I and J are consistency conditions (13.1.14a,b) which yield the *exact* eigenstrain field $\boldsymbol{\varepsilon}^*$ and the *exact* eigenstress field $\boldsymbol{\sigma}^*$, respectively.

---

[14] Compare with (9.1.15c) and (9.1.7g).

2) The stationary values of the functionals $I(\boldsymbol{\varepsilon}^*; \boldsymbol{\sigma}^o)$ and $J(\boldsymbol{\sigma}^*; \boldsymbol{\varepsilon}^o)$ are given by

$$I(\boldsymbol{\varepsilon}^*; \boldsymbol{\sigma}^o) = -\frac{1}{2} < \boldsymbol{\varepsilon}^* > : \boldsymbol{\sigma}^o = \frac{1}{2} \boldsymbol{\sigma}^o : (\mathbf{D} - \bar{\mathbf{D}}) : \boldsymbol{\sigma}^o,$$

$$J(\boldsymbol{\sigma}^*; \boldsymbol{\varepsilon}^o) = -\frac{1}{2} < \boldsymbol{\sigma}^* > : \boldsymbol{\varepsilon}^o = \frac{1}{2} \boldsymbol{\varepsilon}^o : (\mathbf{C} - \bar{\mathbf{C}}) : \boldsymbol{\varepsilon}^o, \qquad (13.2.9a,b)$$

where $\bar{\mathbf{D}}$ and $\bar{\mathbf{C}}$ are the overall elasticity and compliance tensors of the unit cell.

3) If the tensor field $\mathbf{C}' - \mathbf{C}$ (tensor field $\mathbf{D}' - \mathbf{D}$) is positive-definite in the entire unit cell, then the functional I has the global maximum (the global minimum) and the functional J has the global minimum (the global maximum), i.e., for any arbitrary eigenstrain $\mathbf{e}^*$ and eigenstress $\mathbf{s}^*$, and when $\mathbf{C}' - \mathbf{C}$ (when $\mathbf{D}' - \mathbf{D}$) is positive-definite,

$$I(\boldsymbol{\varepsilon}^*; \boldsymbol{\sigma}^o) \geq (\leq)\, I(\mathbf{e}^*; \boldsymbol{\sigma}^o), \qquad J(\boldsymbol{\sigma}^*; \boldsymbol{\varepsilon}^o) \leq (\geq)\, J(\mathbf{s}^*; \boldsymbol{\varepsilon}^o), \qquad (13.2.10a,b)$$

where equality holds only when $\mathbf{e}^* = \boldsymbol{\varepsilon}^*$ and $\mathbf{s}^* = \boldsymbol{\sigma}^*$, respectively; see Section 9 for a detailed derivation.

Table 13.2.1 provides a comparison between the Hashin-Shtrikman variational principle as applied to an RVE and as applied to a periodic solid.

From 2) and 3) and the corresponding functionals, bounds are obtained for the overall elasticity and compliance tensors of the unit cell. Since $\bar{\mathbf{C}}$ and $\bar{\mathbf{D}}$ are each other's inverse, (13.2.10a) and (13.2.10b) give two pairs of upper and lower bounds for the overall moduli. These bounds can be computed *exactly,* since the exact integral operators $\boldsymbol{\Lambda}^P$ and $\boldsymbol{\Gamma}^P$ are given by the corresponding Fourier series. These two pairs of bounds are exactly equivalent, and hence there is only one upper and one lower bound obtained from (13.2.10a,b), as shown in the sequel.

13.2.3. Equivalence of Two Energy Functionals

The two energy functionals defined for an RVE may not be related, in the sense that they are associated with field variables which are produced by different boundary conditions.[15] For the periodic structure, however, both the local and global responses of the unit cell remain the same whether homogeneous strains or stresses are prescribed. The two energy functionals I and J are related, and from (13.2.9a,b), they yield the overall elasticity and compliance tensors which are each other's inverse. Indeed, the two functionals are equivalent if $\mathbf{e}^* = -\mathbf{D} : \mathbf{s}^*$, with $\boldsymbol{\sigma}^o = \mathbf{C} : \boldsymbol{\varepsilon}^o + < \mathbf{s}^* >$.

[15] When a homogenized RVE is embedded in an unbounded uniform solid of reference elasticity, and the infinite-body Green function is used, then, as shown in Subsection 9.5, the corresponding functionals, I^A and J^A, become equivalent if equivalent eigenstrains and eigenstresses are used. In this case, the boundary data of the embedded RVE depend on the homogenizing eigenstrains (or eigenstresses) and cannot be prescribed *a priori*. This is similar to the periodic case.

Table 13.2.1

Hashin-Shtrikman variational principle as applied to RVE and as applied to periodic structure

	RVE	periodic structure
operator	$\boldsymbol{\Lambda}$ (uniform tractions) $\boldsymbol{\Gamma}$ (linear displacements)	$\boldsymbol{\Lambda}^{\mathrm{P}}$ (periodic) $\boldsymbol{\Gamma}^{\mathrm{P}}$ (periodic)
$I(\mathbf{e}^*; \boldsymbol{\sigma}^{\mathrm{o}})$	$\frac{1}{2} < \mathbf{e}^* : \{(\mathbf{D}' - \mathbf{D})^{-1} + \boldsymbol{\Lambda}\} : \mathbf{e}^* > - < \mathbf{e}^* > : \boldsymbol{\sigma}^{\mathrm{o}}$	$\frac{1}{2} < \mathbf{e}^* : \{(\mathbf{D}' - \mathbf{D})^{-1} + \boldsymbol{\Lambda}^{\mathrm{P}}\} : \mathbf{e}^* > - < \mathbf{e}^* > : \boldsymbol{\sigma}^{\mathrm{o}}$
$J(\mathbf{s}^*; \boldsymbol{\varepsilon}^{\mathrm{o}})$	$\frac{1}{2} < \mathbf{s}^* : \{(\mathbf{C}' - \mathbf{C})^{-1} + \boldsymbol{\Gamma}\} : \mathbf{s}^* > - < \mathbf{s}^* > : \boldsymbol{\varepsilon}^{\mathrm{o}}$	$\frac{1}{2} < \mathbf{s}^* : \{(\mathbf{C}' - \mathbf{C})^{-1} + \boldsymbol{\Gamma}^{\mathrm{P}}\} : \mathbf{s}^* > - < \mathbf{s}^* > : \boldsymbol{\varepsilon}^{\mathrm{o}}$
$\delta I(\mathbf{e}^*; \boldsymbol{\sigma}^{\mathrm{o}})$	$< \delta\mathbf{e}^* : \{(\mathbf{D}' - \mathbf{D})^{-1} : \mathbf{e}^* + \boldsymbol{\Lambda} : \mathbf{e}^* - \boldsymbol{\sigma}^{\mathrm{o}}\} >$	$< \delta\mathbf{e}^* : \{(\mathbf{D}' - \mathbf{D})^{-1} : \mathbf{e}^* + \boldsymbol{\Lambda}^{\mathrm{P}} : \mathbf{e}^* - \boldsymbol{\sigma}^{\mathrm{o}}\} >$
$\delta J(\mathbf{s}^*; \boldsymbol{\varepsilon}^{\mathrm{o}})$	$< \delta\mathbf{s}^* : \{(\mathbf{C}' - \mathbf{C})^{-1} : \mathbf{s}^* + \boldsymbol{\Gamma} : \mathbf{s}^* - \boldsymbol{\varepsilon}^{\mathrm{o}}\} >$	$< \delta\mathbf{s}^* : \{(\mathbf{C}' - \mathbf{C})^{-1} : \mathbf{s}^* + \boldsymbol{\Gamma}^{\mathrm{P}} : \mathbf{s}^* - \boldsymbol{\varepsilon}^{\mathrm{o}}\} >$
$I(\boldsymbol{\varepsilon}^*; \boldsymbol{\sigma}^{\mathrm{o}})$	$-\frac{1}{2} < \boldsymbol{\varepsilon}^* > : \boldsymbol{\sigma}^{\mathrm{o}} = \frac{1}{2} \boldsymbol{\sigma}^{\mathrm{o}} : (\mathbf{D} - \bar{\mathbf{D}}) : \boldsymbol{\sigma}^{\mathrm{o}}$	
$J(\boldsymbol{\sigma}^*; \boldsymbol{\varepsilon}^{\mathrm{o}})$	$-\frac{1}{2} < \boldsymbol{\sigma}^* > : \boldsymbol{\varepsilon}^{\mathrm{o}} = \frac{1}{2} \boldsymbol{\varepsilon}^{\mathrm{o}} : (\mathbf{C} - \bar{\mathbf{C}}) : \boldsymbol{\varepsilon}^{\mathrm{o}}$	
$\mathbf{C}' - \mathbf{C}$ is p.d.	$I(\boldsymbol{\varepsilon}^*; \boldsymbol{\sigma}^{\mathrm{o}}) \geq I(\mathbf{e}^*; \boldsymbol{\sigma}^{\mathrm{o}})$ $J(\boldsymbol{\sigma}^*; \boldsymbol{\varepsilon}^{\mathrm{o}}) \leq J(\mathbf{s}^*; \boldsymbol{\varepsilon}^{\mathrm{o}})$	
$\mathbf{D}' - \mathbf{D}$ is p.d.	$I(\boldsymbol{\varepsilon}^*; \boldsymbol{\sigma}^{\mathrm{o}}) \leq I(\mathbf{e}^*; \boldsymbol{\sigma}^{\mathrm{o}})$ $J(\boldsymbol{\sigma}^*; \boldsymbol{\varepsilon}^{\mathrm{o}}) \geq J(\mathbf{s}^*; \boldsymbol{\varepsilon}^{\mathrm{o}})$	

To prove this, recall the relation between the two corresponding integral operators, $\boldsymbol{\Lambda}^{\mathrm{P}}$ and $\boldsymbol{\Gamma}^{\mathrm{P}}$; see Subsection 13.1.5. For an arbitrary eigenstress field $\mathbf{s}^*$,

$$\boldsymbol{\Gamma}^{\mathrm{P}}(\mathbf{s}^*) = \mathbf{D} : \{\boldsymbol{\Lambda}^{\mathrm{P}}(-\mathbf{D} : \mathbf{s}^*) + (\mathbf{s}^* - < \mathbf{s}^* >)\}. \tag{13.2.11a}$$

Furthermore, the following identity holds:

$$(\mathbf{C}' - \mathbf{C})^{-1} = -\mathbf{D} : (\mathbf{D}' - \mathbf{D})^{-1} : \mathbf{D}' = -\mathbf{D} - \mathbf{D} : (\mathbf{D}' - \mathbf{D})^{-1} : \mathbf{D}; \tag{13.2.11b}$$

see Subsection 9.3. From (13.2.11a,b),

$$(\mathbf{C}' - \mathbf{C})^{-1} : \mathbf{s}^* + \boldsymbol{\Gamma}^{\mathrm{P}}(\mathbf{s}^*) = \mathbf{D} : \{(\mathbf{D}' - \mathbf{D})^{-1} : (-\mathbf{D} : \mathbf{s}^*) + \boldsymbol{\Lambda}^{\mathrm{P}}(-\mathbf{D} : \mathbf{s}^*)\}$$

$$-\mathbf{D} : < \mathbf{s}^* >, \tag{13.2.11c}$$

and $\mathrm{J}(\mathbf{s}^*; \boldsymbol{\varepsilon}^o)$ is computed as

$$\mathrm{J}(\mathbf{s}^*; \boldsymbol{\varepsilon}^o) = -\frac{1}{2} < (-\mathbf{D} : \mathbf{s}^*) : \{(\mathbf{D}' - \mathbf{D})^{-1} : (-\mathbf{D} : \mathbf{s}^*) + \boldsymbol{\Lambda}^P : (-\mathbf{D} : \mathbf{s}^*)$$

$$- < \mathbf{s}^* > - 2(\mathbf{C} : \boldsymbol{\varepsilon}^o)\} >. \tag{13.2.11d}$$

Therefore, I and J are related through

$$\mathrm{I}(-\mathbf{D} : \mathbf{s}^*; \mathbf{C} : \boldsymbol{\varepsilon}^o + < \mathbf{s}^* >) = -\mathrm{J}(\mathbf{s}^*; \boldsymbol{\varepsilon}^o) + \frac{1}{2} < \mathbf{s}^* > : \mathbf{D} : < \mathbf{s}^* > \tag{13.2.12a}$$

or

$$\mathrm{I}(-\mathbf{D} : \mathbf{s}^*; \mathbf{C} : \boldsymbol{\varepsilon}^o) = -\mathrm{J}(\mathbf{s}^*; \boldsymbol{\varepsilon}^o) - \frac{1}{2} < \mathbf{s}^* > : \mathbf{D} : < \mathbf{s}^* >. \tag{13.2.12b}$$

This is the equivalence relation between the functionals I and J. Note that, in (13.2.12a), the second argument of I is given by $\mathbf{C} : \boldsymbol{\varepsilon}^o + < \mathbf{s}^* >$, so that if s^* is the exact homogenizing eigenstress, $\boldsymbol{\sigma}^*$, then, $\mathbf{C} : \boldsymbol{\varepsilon}^o + < \boldsymbol{\sigma}^* >$ is the homogeneous stress, $\boldsymbol{\sigma}^o$, given by $\boldsymbol{\sigma}^o = \bar{\mathbf{C}} : \boldsymbol{\varepsilon}^o$. This equivalence relation, (13.2.12a), is considered in greater detail in the following.[16]

It is seen that, if the average eigenstress is fixed, then I in the left side of (13.2.12a) is stationary when J in the right side of (13.2.12a) is stationary. More precisely, consider a suitable class of eigenstress fields, and let $\boldsymbol{\sigma}^*$ be the eigenstress field which renders $\mathrm{J}(\mathbf{s}^*; \boldsymbol{\varepsilon}^o)$ stationary. Then, in the corresponding class of eigenstrain fields, the eigenstrain $\boldsymbol{\varepsilon}^* = -\mathbf{D} : \boldsymbol{\sigma}^*$ renders $\mathrm{I}(\mathbf{e}^*; \boldsymbol{\sigma}^o)$ stationary, where $\boldsymbol{\sigma}^o$ is given by $\mathbf{C} : \boldsymbol{\varepsilon}^o + < \boldsymbol{\sigma}^* >$. This can be proved directly from the consistency condition. With the aid of (13.2.11a) and (13.2.11b), the left side of consistency condition (13.1.14b) is rewritten as

$$(\mathbf{D}' - \mathbf{D})^{-1} : (-\mathbf{D} : \boldsymbol{\sigma}^*) + \boldsymbol{\Lambda}^P(-\mathbf{D} : \boldsymbol{\sigma}^*) - (\mathbf{C} : \boldsymbol{\varepsilon}^o + < \boldsymbol{\sigma}^* >).$$

Hence, $\boldsymbol{\varepsilon}^* = -\mathbf{D} : \boldsymbol{\sigma}^*$ is the exact eigenstrain that satisfies (13.1.14a) when $\boldsymbol{\sigma}^o$ is given by $\mathbf{C} : \boldsymbol{\varepsilon}^o + < \boldsymbol{\sigma}^* >$.

Since the class of eigenstresses used to obtain the optimal $\boldsymbol{\sigma}^*$ may not include the exact homogenizing eigenstress, the stationary value of J yields a bound for the overall elasticity tensor, $\bar{\mathbf{C}}$. Denote this bound by $\bar{\mathbf{C}}^P$, i.e., in terms of the stationary value of J, $\bar{\mathbf{C}}^P$ is defined by

$$\mathrm{J}(\boldsymbol{\sigma}^*; \boldsymbol{\varepsilon}^o) \equiv \frac{1}{2}\, \boldsymbol{\varepsilon}^o : (\mathbf{C} - \bar{\mathbf{C}}^P) : \boldsymbol{\varepsilon}^o. \tag{13.2.13a}$$

Then, the stationary value of the corresponding I is given by

$$\mathrm{I}(\boldsymbol{\varepsilon}^*; \boldsymbol{\sigma}^o) = \frac{1}{2}\, \boldsymbol{\sigma}^o : (\mathbf{D} - (\bar{\mathbf{C}}^P)^{-1}) : \boldsymbol{\sigma}^o. \tag{13.2.13b}$$

Hence, $(\bar{\mathbf{C}}^P)^{-1}$ is a bound for the overall compliance tensor $\bar{\mathbf{D}}$. For the periodic structure, the overall compliance tensor $\bar{\mathbf{D}}$ is the inverse of the overall elasticity

[16] See Subsection 9.6.1 for the case of an RVE.

tensor $\bar{\mathbf{C}}$, and hence bounds given by $\bar{\mathbf{C}}^{\mathrm{P}}$ and $(\bar{\mathbf{C}}^{\mathrm{P}})^{-1}$ are the same. The proof is essentially the same as that presented in Subsection 9.6.1. Indeed, taking advantage of $<\boldsymbol{\sigma}^*> = (\bar{\mathbf{C}}^{\mathrm{P}} - \mathbf{C}) : \boldsymbol{\varepsilon}^{\mathrm{o}}$, compute $\mathrm{I}(\boldsymbol{\varepsilon}^*; \boldsymbol{\sigma}^{\mathrm{o}})$ as

$$\mathrm{I}(\boldsymbol{\varepsilon}^*; \boldsymbol{\sigma}^{\mathrm{o}}) = -\,\mathrm{J}(\boldsymbol{\sigma}^*; \boldsymbol{\varepsilon}^{\mathrm{o}}) + \frac{1}{2} < \boldsymbol{\sigma}^* > : \mathrm{D} : < \boldsymbol{\sigma}^* >$$

$$= \frac{1}{2}\,(\bar{\mathbf{C}}^{\mathrm{P}} : \boldsymbol{\varepsilon}^{\mathrm{o}}) : (\mathbf{D} - (\bar{\mathbf{C}}^{\mathrm{P}})^{-1}) : (\bar{\mathbf{C}}^{\mathrm{P}} : \boldsymbol{\varepsilon}^{\mathrm{o}}). \tag{13.2.14}$$

Since $\bar{\mathbf{C}}^{\mathrm{P}} : \boldsymbol{\varepsilon}^{\mathrm{o}}$ equals $\mathbf{C} : \boldsymbol{\varepsilon}^{\mathrm{o}} + < \boldsymbol{\sigma}^* > = \boldsymbol{\sigma}^{\mathrm{o}}$, (13.2.13b) is obtained.

13.2.4. Alternative Formulation of Exact Bounds

Exact computable bounds for the overall elasticity or compliance tensor of a solid with a periodic microstructure can be obtained *without invoking the variational principle*, in essentially the same manner as those obtained for an ellipsoidal RVE; see Subsection 9.5.6.[17] While the overall response, and hence the overall elasticity and compliance tensors of an RVE depend on the nature of the prescribed boundary data, the overall elasticity and compliance tensors in the periodic case are unique, since the stress and strain fields in the corresponding unit cell are uniquely determined when periodic boundary conditions are prescribed. Theorems I and II of Subsection 2.5.6 are, therefore, not needed for formulating bounds on the overall parameters in the periodic case. Instead, the theorems of the minimum potential energy and the minimum complementary potential energy are employed for this case.

Now, obtain an exact inequality which yields bounds for the overall elasticity tensor, $\bar{\mathbf{C}}$, in the periodic case. To this end, a reference elasticity tensor $\mathbf{C}$ is chosen such that $\mathbf{C}' - \mathbf{C}$ is negative-semi-definite. In addition, the properties of integral operators $\boldsymbol{\Gamma}^{\mathrm{P}}$ and the theorem of the minimum potential energy[18] are used.

First, let $\mathbf{C}' - \mathbf{C}$ be negative-semi-definite. Then, for any two strain fields, $\boldsymbol{\varepsilon}$ and $\mathbf{e}$,

$$0 \geq (\boldsymbol{\varepsilon} - \mathbf{e}) : (\mathbf{C}' - \mathbf{C}) : (\boldsymbol{\varepsilon} - \mathbf{e}). \tag{13.2.15a}$$

Replacing $\mathbf{e}$ in (13.2.15a) by $(\mathbf{C}' - \mathbf{C})^{-1} : \mathbf{s}^*$ for some eigenstress field $\mathbf{s}^*$ defined in the unit cell, and taking the volume average of the resulting inequality over the unit cell, obtain

$$0 \geq \frac{1}{2} < \boldsymbol{\varepsilon} : (\mathbf{C}' - \mathbf{C}) : \boldsymbol{\varepsilon} - 2\,\mathbf{s}^* : \boldsymbol{\varepsilon} + \mathbf{s}^* : (\mathbf{C}' - \mathbf{C})^{-1} : \mathbf{s}^* >, \tag{13.2.16a}$$

where equality holds if and only if $\boldsymbol{\varepsilon} = (\mathbf{C}' - \mathbf{C})^{-1} : \mathbf{s}^*$.

Next, let $\boldsymbol{\varepsilon}$ be given by

---

[17] See also Subsection 9.7 for exact bounds for nonmechanical overall properties.

[18] See Subsection 2.5 or Subsection 19.4 of Part 2.

$$\boldsymbol{\varepsilon} = \boldsymbol{\varepsilon}^{\mathrm{o}} + \boldsymbol{\varepsilon}^{\mathrm{p}} = \boldsymbol{\varepsilon}^{\mathrm{o}} - \boldsymbol{\Gamma}^{\mathrm{P}}(\mathbf{s}^*), \tag{13.2.17a}$$

where $\boldsymbol{\varepsilon}^{\mathrm{o}}$ and $\boldsymbol{\varepsilon}^{\mathrm{p}}$ are homogeneous and periodic strain fields. Since inequality (13.2.16a) holds for any pair of $\boldsymbol{\varepsilon}$ and $\mathbf{s}^*$, obtain

$$0 \geq \frac{1}{2}(<\boldsymbol{\varepsilon} : \mathbf{C}' : \boldsymbol{\varepsilon}> - \boldsymbol{\varepsilon}^{\mathrm{o}} : \mathbf{C} : \boldsymbol{\varepsilon}^{\mathrm{o}})$$
$$+ \frac{1}{2} < \mathbf{s}^* : (\mathbf{C}' - \mathbf{C})^{-1} : \mathbf{s}^* - \mathbf{s}^* : (\boldsymbol{\varepsilon}^{\mathrm{p}} + 2\boldsymbol{\varepsilon}^{\mathrm{o}}) > + \frac{1}{2\mathrm{U}} \int_{\partial \mathrm{U}} (\boldsymbol{\nu} \cdot \boldsymbol{\sigma}^{\mathrm{p}}) \cdot \mathbf{u}^{\mathrm{p}} \, d\mathrm{S}, \tag{13.2.18a}$$

where $\mathbf{u}^{\mathrm{p}}$ and $\boldsymbol{\sigma}^{\mathrm{p}}$ are the periodic displacement and stress fields associated with $\boldsymbol{\varepsilon}^{\mathrm{p}} = -\boldsymbol{\Gamma}^{\mathrm{P}}(\mathbf{s}^*)$, through $\boldsymbol{\varepsilon}^{\mathrm{p}} = sym\,\{\boldsymbol{\nabla} \otimes \mathbf{u}^{\mathrm{p}}\}$ and $\boldsymbol{\sigma}^{\mathrm{p}} = \mathbf{C} : \boldsymbol{\varepsilon}^{\mathrm{p}} + \mathbf{s}^* - < \mathbf{s}^* >$, respectively. Due to the periodicity of $\boldsymbol{\sigma}^{\mathrm{p}}$ and $\mathbf{u}^{\mathrm{p}}$, the surface integral in the right side of (13.2.18a) vanishes. Hence, (13.2.18a) becomes

$$0 \geq \frac{1}{2}(<\boldsymbol{\varepsilon} : \mathbf{C}' : \boldsymbol{\varepsilon}> - \boldsymbol{\varepsilon}^{\mathrm{o}} : \mathbf{C} : \boldsymbol{\varepsilon}^{\mathrm{o}}) + \mathrm{J}(\mathbf{s}^*;\, \boldsymbol{\varepsilon}^{\mathrm{o}}), \tag{13.2.19a}$$

where J is the energy functional defined by (13.2.1b).

Finally, evaluate $<\boldsymbol{\varepsilon} : \mathbf{C}' : \boldsymbol{\varepsilon}>$, using the theorem of the minimum potential energy.[19] Since $\mathbf{u}^{\mathrm{p}}$ due to $\mathbf{s}^*$ satisfies the periodic boundary conditions on the boundary of the unit cell, $\mathbf{u} = \mathbf{x} \cdot \boldsymbol{\varepsilon}^{\mathrm{o}} + \mathbf{u}^{\mathrm{p}}$ is a kinematically admissible displacement field. Hence, $<\boldsymbol{\varepsilon} : \mathbf{C}' : \boldsymbol{\varepsilon}>/2$ is the volume average of the strain energy of the unit cell evaluated for this $\mathbf{u}$. According to the theorem of the minimum potential energy, among all kinematically admissible displacement fields, $\mathbf{u}$, $<\boldsymbol{\varepsilon} : \mathbf{C}' : \boldsymbol{\varepsilon}>/2$ takes on the minimum value when $\mathbf{C}' : (\boldsymbol{\nabla} \otimes \mathbf{u})$ satisfies the equations of equilibrium. This minimum corresponds to the volume average of the exact strain energy when the unit cell is subjected to the homogeneous strain field $\boldsymbol{\varepsilon}^{\mathrm{o}}$. Therefore, for any $\boldsymbol{\varepsilon} = \boldsymbol{\varepsilon}^{\mathrm{o}} - \boldsymbol{\Gamma}^{\mathrm{P}}(\mathbf{s}^*)$,

$$0 \geq \frac{1}{2}\boldsymbol{\varepsilon}^{\mathrm{o}} : \overline{\mathbf{C}} : \boldsymbol{\varepsilon}^{\mathrm{o}} - \frac{1}{2} <\boldsymbol{\varepsilon} : \mathbf{C}' : \boldsymbol{\varepsilon}>, \tag{13.2.20a}$$

where $\overline{\mathbf{C}}$ is the overall elasticity tensor.

From (13.2.19a) and (13.2.20a), it follows that

$$0 \geq \frac{1}{2}\boldsymbol{\varepsilon}^{\mathrm{o}} : (\overline{\mathbf{C}} - \mathbf{C}) : \boldsymbol{\varepsilon}^{\mathrm{o}} + \mathrm{J}(\mathbf{s}^*;\, \boldsymbol{\varepsilon}^{\mathrm{o}}). \tag{13.2.21a}$$

This is an exact inequality which holds for any arbitrary heterogeneous elastic solid with a periodic microstructure. Note that equality in (13.2.19a) holds if $\boldsymbol{\varepsilon} = (\mathbf{C}' - \mathbf{C})^{-1} : \mathbf{s}^*$, and equality in (13.2.20a) holds if $\mathbf{C}' : \boldsymbol{\varepsilon}$ satisfies the equations of equilibrium. Hence, equality in (13.2.21a) holds for the exact homogenizing eigenstress field, $\boldsymbol{\sigma}^*$, for which $\boldsymbol{\varepsilon} = \boldsymbol{\varepsilon}^{\mathrm{o}} - \boldsymbol{\Gamma}^{\mathrm{P}}(\boldsymbol{\sigma}^*)$ satisfies both conditions.

In a similar manner, an exact inequality which yields bounds for $\overline{\mathbf{D}}$, the overall compliance tensor of a solid with periodic microstructure, is derived

---

[19] Hill (1963) uses the principle of the minimum potential energy to obtain an exact inequality similar to (13.2.20a) for an RVE, subjected to linear displacement boundary conditions. When an RVE is subjected to general boundary data, on the other hand, Theorem II of Subsection 2.5.6 is necessary, as discussed in Subsection 9.5.

from: 1) the negative-semi-definiteness of $\mathbf{D}' - \mathbf{D}$; 2) the properties of integral operators $\mathbf{\Lambda}^{\mathrm{p}}$; and 3) the theorem of the minimum complementary potential energy. Indeed, if a reference compliance tensor, $\mathbf{D}$, is chosen such that $\mathbf{D}' - \mathbf{D}$ is negative-semi-definite, then for any two stress fields, $\boldsymbol{\sigma}$ and $\mathbf{s}$,

$$0 \geq (\boldsymbol{\sigma} - \mathbf{s}) : (\mathbf{D}' - \mathbf{D}) : (\boldsymbol{\sigma} - \mathbf{s}). \tag{13.2.15b}$$

If $\mathbf{s}$ is replaced by $(\mathbf{D}' - \mathbf{D})^{-1} : \mathbf{e}^*$ for some eigenstrain field $\mathbf{e}^*$ defined in the unit cell, the volume average of (13.2.15b) becomes

$$0 \geq \frac{1}{2} < \boldsymbol{\sigma} : (\mathbf{D}' - \mathbf{D}) : \boldsymbol{\sigma} - 2\,\mathbf{s}^* : \boldsymbol{\sigma} + \mathbf{e}^* : (\mathbf{D}' - \mathbf{D})^{-1} : \mathbf{e}^* >. \tag{13.2.16b}$$

Then, introducing homogeneous and periodic stress fields, $\boldsymbol{\sigma}^{\mathrm{o}}$ and $\boldsymbol{\sigma}^{\mathrm{p}} = -\mathbf{\Lambda}^{\mathrm{p}}(\mathbf{e}^*)$, define $\boldsymbol{\sigma}$ by

$$\boldsymbol{\sigma} = \boldsymbol{\sigma}^{\mathrm{o}} + \boldsymbol{\sigma}^{\mathrm{p}} = \boldsymbol{\sigma}^{\mathrm{o}} - \mathbf{\Lambda}^{\mathrm{p}}(\mathbf{e}^*). \tag{13.2.17b}$$

Substituting (13.2.17b) into (13.2.16b), obtain

$$0 \geq \frac{1}{2} (< \boldsymbol{\sigma} : \mathbf{D}' : \boldsymbol{\sigma} > - \boldsymbol{\sigma}^{\mathrm{o}} : \mathbf{D} : \boldsymbol{\sigma}^{\mathrm{o}})$$

$$+ \frac{1}{2} < \mathbf{e}^* : (\mathbf{D}' - \mathbf{D})^{-1} : \mathbf{e}^* - \mathbf{e}^* : (\boldsymbol{\sigma}^{\mathrm{p}} + 2\boldsymbol{\sigma}^{\mathrm{o}}) > + \frac{1}{2U} \int_{\partial U} (\boldsymbol{\nu} \cdot \boldsymbol{\sigma}^{\mathrm{p}}) \cdot \mathbf{u}^{\mathrm{p}} \, dS, \tag{13.2.18b}$$

where $\mathbf{u}^{\mathrm{p}}$ is the periodic displacement field associated with $\boldsymbol{\sigma}^{\mathrm{p}}$. Since $\boldsymbol{\sigma}^{\mathrm{p}}$ and $\mathbf{u}^{\mathrm{p}}$ are periodic, the surface integral in the right side of (13.2.18b) vanishes, and (13.2.18b) becomes

$$0 \geq \frac{1}{2} (< \boldsymbol{\sigma} : \mathbf{C}' : \boldsymbol{\sigma} > - \boldsymbol{\sigma}^{\mathrm{o}} : \mathbf{D} : \boldsymbol{\sigma}^{\mathrm{o}}) + I(\mathbf{e}^*;\, \boldsymbol{\sigma}^{\mathrm{o}}), \tag{13.2.19b}$$

where I is defined by (13.2.1a). Since $\boldsymbol{\sigma}^{\mathrm{p}}$ produced by $\mathbf{e}^*$ satisfies the equation of equilibrium and the periodic boundary conditions, $\boldsymbol{\sigma} = \boldsymbol{\sigma}^{\mathrm{o}} + \boldsymbol{\sigma}^{\mathrm{p}}$ is statically admissible. According to the theorem of the minimum complementary potential energy,[20] among all statically admissible stress fields, $< \boldsymbol{\sigma} : \mathbf{D}' : \boldsymbol{\sigma} >/2$ is minimized by the stress field which renders $\mathbf{D}' : \boldsymbol{\sigma}$ compatible. The minimum corresponds to the volume average of the exact value of the complementary potential energy of the unit cell under the overall stress $\boldsymbol{\sigma}^{\mathrm{o}}$, i.e., $\boldsymbol{\sigma}^{\mathrm{o}} : \overline{\mathbf{D}} : \boldsymbol{\sigma}^{\mathrm{o}}/2$. Therefore, for any $\boldsymbol{\sigma} = \boldsymbol{\sigma}^{\mathrm{o}} - \mathbf{\Lambda}^{\mathrm{p}}(\mathbf{e}^*)$,

$$0 \geq \frac{1}{2} \boldsymbol{\sigma}^{\mathrm{o}} : \overline{\mathbf{D}} : \boldsymbol{\varepsilon}^{\mathrm{o}} - \frac{1}{2} < \boldsymbol{\sigma} : \mathbf{D}' : \boldsymbol{\sigma} >. \tag{13.2.20b}$$

From (13.2.19b) and (13.2.20b), it follows that

$$0 \geq \frac{1}{2} \boldsymbol{\sigma}^{\mathrm{o}} : (\overline{\mathbf{D}} - \mathbf{D}) : \boldsymbol{\sigma}^{\mathrm{o}} + I(\mathbf{e}^*;\, \boldsymbol{\sigma}^{\mathrm{o}}). \tag{13.2.21b}$$

This inequality is exact, and holds for any arbitrary elastic solid with a periodic

[20] The principle of the minimum complementary potential energy is used to obtain an inequality similar to (13.2.20b) for the case of an RVE subjected to uniform traction boundary data; see Hill (1963). When general boundary data are considered, Theorem I of Subsection 2.5.6 is necessary; see Subsection 9.5.

microstructure. Note that equality in (13.2.21b) holds, if the exact homogenizing eigenstrain field $\boldsymbol{\varepsilon}^*$ is used.

13.3. APPLICATION OF FOURIER SERIES EXPANSION TO ENERGY FUNCTIONALS

Since the Hashin-Shtrikman variational principle applies to solids with periodic microstructure, it can be used to obtain *exact* bounds for the overall elasticity or compliance tensor of the unit cell. Moreover, *one of the bounds always approaches the exact solution, as the estimate of the corresponding eigenstrain or eigenstress field is improved.* In this section, the Fourier series expansion is applied, and the results are illustrated and discussed. Since functionals I and J in (13.2.1a,b) are equivalent, only functional J is used. To reduce algebraic manipulation, rewrite the J-functional using (13.1.2a), as

$$J(\mathbf{s}^*; \boldsymbol{\varepsilon}) = \frac{1}{2}\Big\{ (1-f) < \mathbf{s}^* : (\mathbf{C}^M - \mathbf{C})^{-1} : \mathbf{s}^* >_M + \sum_{\alpha=1}^{n} f_\alpha < \mathbf{s}^* : (\mathbf{C}^\alpha - \mathbf{C})^{-1} : \mathbf{s}^* >_\alpha \Big\} + \frac{1}{2} < \mathbf{s}^* : \boldsymbol{\Gamma}^P : \mathbf{s}^* > - < \mathbf{s}^* : \boldsymbol{\varepsilon}^o >. \tag{13.3.1}$$

In this and the following subsections, the homogeneous parts of the eigenstress and eigenstrain fields are denoted by $< \mathbf{s}^* > = \overline{\mathbf{s}}^*$ and $< \mathbf{e}^* > = \overline{\mathbf{e}}^*$.

13.3.1. Fourier Series Representation of Eigenstress

As shown in Subsection 13.2.2, the Euler equation of the functional J obtained from (13.2.8b) is consistency condition[21] (13.1.14b). Hence, another form of the Fourier series expansion of the consistency condition can be obtained, if the functional J given by (13.3.1) is used, yielding

$$\sum_{\boldsymbol{\zeta}} \Big\{ (1-f)\, g_M(\boldsymbol{\xi}+\boldsymbol{\zeta})\, (\mathbf{C}^M - \mathbf{C})^{-1} + \sum_{\alpha=1}^{n} f_\alpha\, g_\alpha(\boldsymbol{\xi}+\boldsymbol{\zeta})\, (\mathbf{C}^\alpha - \mathbf{C})^{-1} \Big\} : F\boldsymbol{\sigma}^*(\boldsymbol{\zeta}) + F\boldsymbol{\Gamma}^P(-\boldsymbol{\xi}) : F\boldsymbol{\sigma}^*(-\boldsymbol{\xi}) - g_U(\boldsymbol{\xi})\, \boldsymbol{\varepsilon}^o = \mathbf{0}, \tag{13.3.2}$$

where

[21] In Subsection 12.4, a general Fourier series expansion of the consistency condition is given; see (12.4.23b).

$$g_U(\boldsymbol{\xi}) \equiv \, < \exp(\iota\boldsymbol{\xi}\cdot\mathbf{x}) > \, = \begin{cases} 1 & \text{if } \boldsymbol{\xi} = \mathbf{0} \\ 0 & \text{otherwise,} \end{cases} \tag{13.3.3a}$$

and

$$g_M(\boldsymbol{\xi}) \equiv \, < \exp(\iota\boldsymbol{\xi}\cdot\mathbf{x}) >_M, \qquad g_\alpha(\boldsymbol{\xi}) \equiv \, < \exp(\iota\boldsymbol{\xi}\cdot\mathbf{x}) >_\alpha; \tag{13.3.3b,c}$$

see Subsection 12.5.[22]

Now substitute the Fourier series representation of the eigenstress field into the *functional*[23] J to define the following *function* J for the Fourier series coefficients of an eigenstress field, $F\mathbf{s}^*(\boldsymbol{\xi})$'s:

$$\begin{aligned} J(\{F\mathbf{s}^*\}; \boldsymbol{\varepsilon}^o) &\equiv \mathrm{J}(\{\sum_{\boldsymbol{\xi}} F\mathbf{s}^*(\boldsymbol{\xi}) \exp(\iota\boldsymbol{\xi}\cdot\mathbf{x})\}; \boldsymbol{\varepsilon}^o) \\ &= \frac{1}{2} \sum_{\boldsymbol{\xi}} \sum_{\boldsymbol{\zeta}} F\mathbf{s}^*(\boldsymbol{\xi}) : \Big\{ (1-f)\, g_M(\boldsymbol{\xi}+\boldsymbol{\zeta})\, (\mathbf{C}^M - \mathbf{C})^{-1} \\ &\qquad + \sum_{\alpha=1}^{n} f_\alpha\, g_\alpha(\boldsymbol{\xi}+\boldsymbol{\zeta})\, (\mathbf{C}^\alpha - \mathbf{C})^{-1} \Big\} : F\mathbf{s}^*(\boldsymbol{\zeta}) \\ &\quad + \frac{1}{2} \sum_{\boldsymbol{\xi}} F\mathbf{s}^*(\boldsymbol{\xi}) : F\boldsymbol{\Gamma}^P(-\boldsymbol{\xi}) : F\mathbf{s}^*(-\boldsymbol{\xi}) - F\mathbf{s}^*(\mathbf{0}) : \boldsymbol{\varepsilon}^o, \end{aligned} \tag{13.3.4a}$$

where the homogeneous part $\bar{\mathbf{s}}^* = \, < \mathrm{s}^* >$ of the eigenstress is given by $F\mathbf{s}^*(\mathbf{0})$, and correspondingly, $F\boldsymbol{\Gamma}^P(\mathbf{0}) = \mathbf{0}$. As is seen, J is quadratic (with linear terms) in the Fourier series coefficients $F\mathbf{s}^*(\boldsymbol{\xi})$.

The derivative of J with respect to $F\mathbf{s}^*(\boldsymbol{\xi})$ is

$$\begin{aligned} \frac{\partial J}{\partial F\mathbf{s}^*(\boldsymbol{\xi})}(\{F\mathbf{s}^*\}; \boldsymbol{\varepsilon}^o) &= \sum_{\boldsymbol{\zeta}} \Big\{ (1-f)\, g_M(\boldsymbol{\xi}+\boldsymbol{\zeta})\, (\mathbf{C}^M - \mathbf{C})^{-1} \\ &\qquad + \sum_{\alpha=1}^{n} f_\alpha\, g_\alpha(\boldsymbol{\xi}+\boldsymbol{\zeta})\, (\mathbf{C}^\alpha - \mathbf{C})^{-1} \Big\} : F\mathbf{s}^*(\boldsymbol{\zeta}) \\ &\quad + F\boldsymbol{\Gamma}^P(-\boldsymbol{\xi}) : F\mathbf{s}^*(-\boldsymbol{\xi}) - g_U(\boldsymbol{\xi})\, \boldsymbol{\varepsilon}^o. \end{aligned} \tag{13.3.4b}$$

The set of linear tensorial equations, $\partial J/\partial F\mathbf{s}^* = \mathbf{0}$, corresponds to the Fourier series expansion of the Euler equations of the functional J, that is, to consistency conditions[24] (13.3.2). In terms of $\partial J/\partial \mathbf{s}^*$, express function J as

---

[22] Note that (13.3.3a) holds for all $\boldsymbol{\xi}$, including $\mathbf{0}$, if $F\boldsymbol{\Gamma}^P(\mathbf{0})$ is defined as $\mathbf{0}$.

[23] The *functional* is denoted by Latin J, and the corresponding *function* by *italic J*.

[24] Since the Euler equations of J are the consistency equations, it is expected that the Euler equations of J coincide with the corresponding Fourier series expansion. In other words, the operation of taking the variation of functional J is *commutable* with the operation of expanding its argument function $\mathbf{s}^*$ in the Fourier series.

$$J(\{F\mathbf{s}^*\};\boldsymbol{\varepsilon}^{\mathrm{o}})=\frac{1}{2}\sum_{\boldsymbol{\xi}}F\mathbf{s}^*(\boldsymbol{\xi}):\{\frac{\partial J}{\partial F\mathbf{s}^*(-\boldsymbol{\xi})}\}-\frac{1}{2}F\mathbf{s}^*(\mathbf{0}):\boldsymbol{\varepsilon}^{\mathrm{o}}. \tag{13.3.4c}$$

Hence, for the exact eigenstress field $\boldsymbol{\sigma}^*$ whose Fourier series coefficients satisfy $\partial J/\partial F\mathbf{s}^*(\{F\boldsymbol{\sigma}^*\};\boldsymbol{\varepsilon}^{\mathrm{o}})=\mathbf{0}$, the value of the J-function is given by $-\bar{\boldsymbol{\sigma}}^*:\boldsymbol{\varepsilon}^{\mathrm{o}}/2$, since $F\boldsymbol{\sigma}^*(\mathbf{0})=<\boldsymbol{\sigma}^*>=\bar{\boldsymbol{\sigma}}^*$.

13.3.2. Truncated Fourier Series of Eigenstress Field

As shown in Section 9, the Hashin-Shtrikman variational principle holds for any sufficiently smooth distribution of eigenstresses. Hence, if a Fourier series representation of the eigenstress field up to, say, N is truncated, i.e., if a trial eigenstress field of the form

$$\mathbf{s}^*(\mathbf{x})=\sum_{\boldsymbol{\xi}}^{\pm\mathrm{N}}F\mathbf{s}^*(\boldsymbol{\xi})\exp(\iota\boldsymbol{\xi}\cdot\mathbf{x}), \tag{13.3.5}$$

where $-\pi\mathrm{N}/a_i\leq\xi_i\leq\pi\mathrm{N}/a_i$ $(i=1,\ 2,\ 3)$, is used, then the most suitable corresponding Fourier coefficients are obtained by optimizing the associated quadratic form. Indeed, if $\mathbf{C}'-\mathbf{C}$ (if $\mathbf{D}'-\mathbf{D}$) is positive-definite, the sharpest upper or lower bounds on the overall moduli that can be computed for the class of eigenstress fields defined by (13.3.5), are obtained.

Now, define a new quadratic form, $\hat{J}$, for the truncated eigenstress field, by

$$\hat{J}(\{F\mathbf{s}^*\};\boldsymbol{\varepsilon}^{\mathrm{o}})\equiv\mathrm{J}(\{\sum_{\boldsymbol{\xi}}^{\pm\mathrm{N}}F\mathbf{s}^*(\boldsymbol{\xi})\exp(\iota\boldsymbol{\xi}\cdot\mathbf{x})\};\boldsymbol{\varepsilon}^{\mathrm{o}})$$

$$=\frac{1}{2}\sum_{\boldsymbol{\xi}}^{\pm\mathrm{N}}\sum_{\boldsymbol{\zeta}}^{\pm\mathrm{N}}F\mathbf{s}^*(\boldsymbol{\xi}):\Big\{(1-\mathrm{f})\,g_{\mathrm{M}}(\boldsymbol{\xi}+\boldsymbol{\zeta})\,(\mathbf{C}^{\mathrm{M}}-\mathbf{C})^{-1}$$

$$+\sum_{\alpha=1}^{n}\mathrm{f}_{\alpha}\,g_{\alpha}(\boldsymbol{\xi}+\boldsymbol{\zeta})\,(\mathbf{C}^{\alpha}-\mathbf{C})^{-1}\Big\}:F\mathbf{s}^*(\boldsymbol{\zeta})$$

$$+\frac{1}{2}\sum_{\boldsymbol{\xi}}^{\pm\mathrm{N}}F\mathbf{s}^*(\boldsymbol{\xi}):F\boldsymbol{\Gamma}^{\mathrm{P}}(-\boldsymbol{\xi}):F\mathbf{s}^*(-\boldsymbol{\xi})-F\mathbf{s}^*(\mathbf{0}):\boldsymbol{\varepsilon}^{\mathrm{o}}. \tag{13.3.6a}$$

It should be noted that if instead of (13.3.1), (13.2.1b) is used as the energy functional, then it would be necessary to expand $(\mathbf{C}'-\mathbf{C})^{-1}$ into the Fourier series. The truncation of the Fourier series expansion of $(\mathbf{C}'-\mathbf{C})^{-1}$, however, alters the elasticity of the periodic structure. In the functional (13.3.1), on the other hand, $(\mathbf{C}'-\mathbf{C})^{-1}$ is piecewise constant, and (13.3.6a) represents this exactly.

The derivative of $\hat{J}$ with respect to $F\mathbf{s}^*(\boldsymbol{\xi})$ is given by

$$\frac{\partial\hat{J}}{\partial F\mathbf{s}^*(\boldsymbol{\xi})}(\{F\mathbf{s}^*\};\boldsymbol{\varepsilon}^{\mathrm{o}})=\sum_{\boldsymbol{\zeta}}^{\pm\mathrm{N}}\Big\{(1-\mathrm{f})\,g_{\mathrm{M}}(\boldsymbol{\xi}+\boldsymbol{\zeta})\,(\mathbf{C}^{\mathrm{M}}-\mathbf{C})^{-1}$$

$$+ \sum_{\alpha=1}^{n} f_\alpha\, g_\alpha(\boldsymbol{\xi}+\boldsymbol{\zeta})\,(\mathbf{C}^\alpha-\mathbf{C})^{-1} \Big\} : F\mathbf{s}^*(\boldsymbol{\zeta})$$

$$+ F\boldsymbol{\Gamma}^{\mathrm{P}}(-\boldsymbol{\xi}) : F\mathbf{s}^*(-\boldsymbol{\xi}) - g_U(\boldsymbol{\xi})\,\boldsymbol{\varepsilon}^{\mathrm{o}}. \tag{13.3.6b}$$

And $\hat{J}$ is expressed as

$$\hat{J}(\{F\mathbf{s}^*\}; \boldsymbol{\varepsilon}^{\mathrm{o}}) = \frac{1}{2} \sum_{\boldsymbol{\xi}}^{\pm N} F\mathbf{s}^*(\boldsymbol{\xi}) : \{\frac{\partial J}{\partial F\mathbf{s}^*(-\boldsymbol{\xi})}\} - \frac{1}{2} F\mathbf{s}^*(\mathbf{0}) : \boldsymbol{\varepsilon}^{\mathrm{o}}. \tag{13.3.6c}$$

Note that the set of linear tensorial equations, $\partial\hat{J}/\partial F\mathbf{s}^*(\boldsymbol{\xi}) = \mathbf{0}$ for $(2N+1)^3$ $\boldsymbol{\xi}$'s, coincides with the Fourier series expansion of the consistency condition, truncated up to N, i.e., with

$$\sum_{\boldsymbol{\zeta}}^{\pm N} \Big\{ (1-f)\, g_M(\boldsymbol{\xi}+\boldsymbol{\zeta})\,(\mathbf{C}^M-\mathbf{C})^{-1} + \sum_{\alpha=1}^{n} f_\alpha\, g_\alpha(\boldsymbol{\xi}+\boldsymbol{\zeta})\,(\mathbf{C}^\alpha-\mathbf{C})^{-1} \Big\} : F\boldsymbol{\sigma}^*(\boldsymbol{\zeta})$$

$$+ F\boldsymbol{\Gamma}^{\mathrm{P}}(-\boldsymbol{\xi}) : F\boldsymbol{\sigma}^*(-\boldsymbol{\xi}) - g_U(\boldsymbol{\xi})\,\boldsymbol{\varepsilon}^{\mathrm{o}} = \mathbf{0}. \tag{13.3.7}$$

The solution of this set of equations produces the optimal value of $\hat{J}$. Furthermore, if $\mathbf{C}'-\mathbf{C}$ (if $\mathbf{D}'-\mathbf{D}$) is positive-definite, $\hat{J}(\{F\boldsymbol{\sigma}^*\}; \boldsymbol{\varepsilon}^{\mathrm{o}})$ gives the sharpest upper and lower bounds of the overall elasticity and compliance that can be obtained, on the class of eigenstress fields given by (13.3.5).

13.3.3. Matrix Representation of Euler Equations

The set of linear tensorial equations (13.3.7) for the Fourier coefficients of the eigenstress field optimizes the function $\hat{J}$ on the class (13.3.5). As mentioned in Subsection 13.1, the exact distribution of the periodic eigenstress field, $\boldsymbol{\sigma}^* = \boldsymbol{\sigma}^*(\mathbf{x})$ is not needed, but its average, $\bar{\boldsymbol{\sigma}}^* = <\boldsymbol{\sigma}^*>$ is, in order to estimate the overall response of the solid with periodic microstructure. Hence, (13.3.7) *need not be solved for* $F\boldsymbol{\sigma}^* = F\boldsymbol{\sigma}^*(\boldsymbol{\xi})$ *for all* $\boldsymbol{\xi}$'s, *but only for* $F\boldsymbol{\sigma}^*(\mathbf{0})$ which equals $<\boldsymbol{\sigma}^*>$. Instead of solving (13.3.7) exactly, estimate $F\boldsymbol{\sigma}^*(\mathbf{0})$ from (13.3.7).

From (13.1.3), the g-integral for M can be expressed in terms of the g-integrals for U and Ω_α's, as

$$(1-f)\, g_M(\boldsymbol{\xi}) = g_U(\boldsymbol{\xi}) - \sum_{\alpha=1}^{n} f_\alpha\, g_\alpha(\boldsymbol{\xi}). \tag{13.3.8}$$

Hence, (13.3.7) becomes

$$\sum_{\boldsymbol{\zeta}}^{\pm N} \Big\{ \sum_{\alpha=1}^{n} f_\alpha\, g_\alpha(\boldsymbol{\xi}+\boldsymbol{\zeta})\, \{(\mathbf{C}^\alpha-\mathbf{C})^{-1} - (\mathbf{C}^M-\mathbf{C})^{-1}\} \Big\} : F\boldsymbol{\sigma}^*(\boldsymbol{\zeta})$$

$$+ \{F\boldsymbol{\Gamma}^{\mathrm{P}}(-\boldsymbol{\xi}) - (\mathbf{C}^M-\mathbf{C})^{-1}\} : F\boldsymbol{\sigma}^*(-\boldsymbol{\xi}) = g_U(-\boldsymbol{\xi})\,\boldsymbol{\varepsilon}^{\mathrm{o}}. \tag{13.3.9}$$

As will be shown later, this form of the consistency conditions makes it possible to estimate the asymptotic behavior of $\boldsymbol{\sigma}^*(\mathbf{0})$, as f_α's go to zero.

For actual calculations, it is convenient to use the matrix form of second- and fourth-order symmetric tensors. Let $[F\sigma_a^*] = [F\boldsymbol{\sigma}_a^*(\boldsymbol{\xi})] = [F\boldsymbol{\sigma}_a^*](n_i)$ be a six by one column matrix corresponding to $F\boldsymbol{\sigma}^*(\boldsymbol{\xi})$, where a = 1, 2, ..., 6, and $n_i = 0, \pm 1, ..., \pm N$ for i = 1, 2, 3. Note that the index n_i relates to the Fourier variable ξ_i by $\xi_i = \pi n_i/a_i$, (i = 1, 2, 3 ; i not summed). By arranging $[F\boldsymbol{\sigma}_a^*](n_1, n_2, n_3)$ in rows in accordance with the indexes (n_1, n_2, n_3), a $6(2N+1)^3$ by one column matrix $[F\sigma^*]$ is formed,

$$[F\sigma^*] \equiv [[F\sigma_a^*]^T(0, 0, 0), [F\sigma_a^*]^T(1, 0, 0), ...]^T. \tag{13.3.10a}$$

For the right-hand side of (13.3.9), form another $6(2N+1)^3$ by one column matrix

$$[\varepsilon^o] \equiv [[\varepsilon_a^o]^T, [0]^T, ...]^T \tag{13.3.10b}$$

which is nonzero only in its first six rows. In a similar manner, form six by six matrices

$$[A_{ab}](n_1, n_2, n_3; m_1, m_2, m_3) = g_U(\boldsymbol{\xi}+\boldsymbol{\zeta})\, [F\Gamma_{ab}^P(-\boldsymbol{\xi}) + (\mathbf{C}^M - \mathbf{C})_{ab}^{-1}]$$

$$= \begin{cases} [F\Gamma_{ab}^P(-\boldsymbol{\xi}) + (\mathbf{C}^M - \mathbf{C})_{ab}^{-1}] & \text{if } \boldsymbol{\zeta} = -\boldsymbol{\xi} \\ [0_{ab}] & \text{otherwise,} \end{cases}$$

$$[B_{ab}](n_1, n_2, n_3; m_1, m_2, m_3) = \sum_{\alpha=1}^{n} f_\alpha\, g_\alpha(\boldsymbol{\xi}+\boldsymbol{\zeta})$$

$$\times\, [(\mathbf{C}^\alpha - \mathbf{C})_{ab}^{-1} - (\mathbf{C}^M - \mathbf{C})_{ab}^{-1}], \tag{13.3.10c,d}$$

from which the following $6(2N+1)^3$ by $6(2N+1)^3$ square matrices, $[A]$ and $[B]$, are constructed:

$$[A] \equiv \begin{bmatrix} [A_{ab}](0,0,0;0,0,0) & [A_{ab}](0,0,0;1,0,0) & \dots \\ [A_{ab}](1,0,0;0,0,0) & [A_{ab}](1,0,0;1,0,0) & \dots \\ \dots & \dots & \dots \end{bmatrix}$$

$$= \begin{bmatrix} [A_{ab}](0,0,0;0,0,0) & [0_{ab}] & \dots \\ [0_{ab}] & [A_{ab}](1,0,0;1,0,0) & \dots \\ \dots & \dots & \dots \end{bmatrix},$$

$$[B] \equiv \begin{bmatrix} [B_{ab}](0,0,0;0,0,0) & [B_{ab}](0,0,0;1,0,0) & \dots \\ [B_{ab}](1,0,0;0,0,0) & [B_{ab}](1,0,0;1,0,0) & \dots \\ \dots & \dots & \dots \end{bmatrix}, \tag{13.3.10e,f}$$

where the index m_i relates to the Fourier variable ζ_i by $\zeta_i = \pi m_i/a_i$ (i = 1, 2, 3 ; i not summed). In (13.3.10e,f), $[0_{ab}]$ stands for a six by six zero matrix. As is seen, matrix $[A]$ contains the effect of the material properties of the original heterogeneous solid through $\boldsymbol{\Gamma}^P$, while matrix $[B]$ brings in the material properties and the geometry of the micro-inclusions through $\mathbf{C}^\alpha$ and the integral g_Ω.

Using (13.3.10a~f), consider the following matrix form of the consistency condition:

$$([A] + [B])[F\sigma^*] = [\varepsilon^o]. \tag{13.3.11}$$

In the absence of Ω_α's, the volume fraction f_α is zero for all α's, and (13.3.11) yields

$$[F\sigma^*] = [A]^{-1}\,[\varepsilon^o], \tag{13.3.12a}$$

where, from definition (13.3.10c), the inverse of matrix $[A]$ is given by

$$[A]^{-1} = \begin{bmatrix} [A_{ab}]^{-1}(0,0,0;0,0,0) & [0_{ab}] & \cdots \\ [0_{ab}] & [A_{ab}]^{-1}(1,0,0;1,0,0) & \cdots \\ \cdots & \cdots & \cdots \end{bmatrix}. \tag{13.3.12b}$$

Here, $[A_{ab}]^{-1}$ is the inverse of $[A_{ab}]$, since the square $6(2N+1)^3$ by $6(2N+1)^3$ matrix $[A](n_i;\,m_i)$ is *diagonal* with respect to (n_i) and (m_i).

Taking advantage of (13.3.12b), compute the inverse of the matrix in the left-hand side of (13.3.11), as

$$\begin{aligned}([A] + [B])^{-1} &= \{[A]\,([1] + [A]^{-1}[B])\}^{-1} \\ &= ([1] + [A]^{-1}[B])^{-1}\,[A]^{-1},\end{aligned} \tag{13.3.13a}$$

where $[1]$ is the $6(2N+1)^3$ by $6(2N+1)^3$ unit matrix. Therefore, for $\det([A]^{-1}[B]) < 1$,

$$([1] + [A]^{-1}[B])^{-1} = \sum_{k=0}^{\infty} \{-[A]^{-1}[B]\}^k. \tag{13.3.13b}$$

Since only the first six elements of the column matrix $[\varepsilon^o]$ are nonzero, only the corresponding terms in the inverse matrix $([A] + [B])^{-1}$ contribute to the value of $[F\sigma^*]$. Hence, $[F\sigma_a^*](\mathbf{0})$ is given by

$$\begin{aligned}[F\sigma_a^*](0,0,0) &= \left\{\sum_{k=0}^{\infty} \{-[A_{ab}]^{-1}[B_{bc}]\}^k(0,0,0;0,0,0)\right\} \\ &\quad\times [A_{cd}]^{-1}(0,0,0;0,0,0)\,[\varepsilon_d^o](0,0,0).\end{aligned} \tag{13.3.14}$$

Estimate (13.3.14) holds for any arbitrary $\mathbf{C}$ for which $(\mathbf{C}^M - \mathbf{C})^{-1}$ and $(\mathbf{C}^\alpha - \mathbf{C})^{-1}$ exist.

In particular, if the summation in (13.3.14) is truncated up to $k = 1$, or if the Fourier series expansion of the eigenstress for $\boldsymbol{\xi} = \mathbf{0}$ is taken, (13.3.14) or (13.3.9) immediately yields

$$\begin{aligned}&\{[A_{ab}](0,0,0;0,0,0) + [B_{ab}](0,0,0;0,0,0)\}\,[F\sigma_b^*](0,0,0) \\ &\quad= [\varepsilon_a^o](0,0,0).\end{aligned} \tag{13.3.15a}$$

Using (13.3.8) again, observe that the tensor form of (13.3.15a) is

$$\{(1-f)\,(\mathbf{C}^M - \mathbf{C})^{-1} + \sum_{\alpha=1}^{n} f_\alpha\,(\mathbf{C}^\alpha - \mathbf{C})^{-1}\} : F\mathbf{s}^*(\mathbf{0}) = \boldsymbol{\varepsilon}^o, \tag{13.3.15b}$$

where $F\boldsymbol{\Gamma}^{\mathrm{P}}(\mathbf{0}) = \mathbf{0}$ and $g_\alpha(\mathbf{0}) = 1$ are used. This is the first-order asymptotic expansion of the exact eigenstress field, which is a good estimate if the f_α's are suitably small. Since (13.3.15b) gives

$$F\mathbf{s}^*(\mathbf{0}) = \{(1-f)\,(\mathbf{C}^{\mathrm{M}} - \mathbf{C})^{-1} + \sum_{\alpha=1}^{n} f_\alpha\,(\mathbf{C}^{\alpha} - \mathbf{C})^{-1}\}^{-1} : \boldsymbol{\varepsilon}^{\mathrm{o}}, \tag{13.3.15c}$$

the overall elasticity tensor $\bar{\mathbf{C}}$ is estimated by

$$\bar{\mathbf{C}} \approx \mathbf{C} + \{(1-f)\,(\mathbf{C}^{\mathrm{M}} - \mathbf{C})^{-1} + \sum_{\alpha=1}^{n} f_\alpha\,(\mathbf{C}^{\alpha} - \mathbf{C})^{-1}\}^{-1}. \tag{13.3.15d}$$

Note that *the one-term approximations* (13.3.15d) are valid only if f_α's are suitably small. If $\mathbf{C}' - \mathbf{C}$ (if $\mathbf{D}' - \mathbf{D}$) is positive-definite, however, the right-hand side of (13.3.15d) is the *exact* lower (upper) bound, respectively. Furthermore, as shown in Subsection 13.2.3, the inverse of the right side of (13.3.15d) is the *exact* upper (lower) bound for the overall compliance tensor, $\bar{\mathbf{D}}$, if $\mathbf{C}' - \mathbf{C}$ (if $\mathbf{D}' - \mathbf{D}$) is positive-definite. For finite values of f_α's, these bounds are not expected to be sharp. They can be improved by including additional terms in the truncated Fourier series representation (13.3.14).

13.4. EXAMPLE: ONE-DIMENSIONAL PERIODIC MICROSTRUCTURE

In order to illustrate the application of the Hashin-Shtrikman variational principle to solids with periodic microstructure, consider a simple one-dimensional two-phase example of a *string* with a periodic structure. All tensorial expressions reduce to scalar ones, while the form of the equations remains unchanged. Furthermore, the problem admits an exact solution which can be used as a reference point. Moreover, if $u = u(x)$ is identified with temperature, $\sigma = \sigma(x)$ with the heat flux, and $C' = C'(x)$ with the conductivity, the formulation then corresponds to steady heat conduction in a periodic infinite space, with conductivity varying periodically in the x-direction only. The objective then is to estimate the effective conductivity $\bar{C}$, together with the bounds.[25]

13.4.1. Exact Solution

Consider an infinitely long straight string (or an infinite space, in the heat conduction case) consisting of a periodic arrangement of a matrix M of elasticity (conductivity) C^{M} and an inclusion Ω of elasticity (conductivity) C^{Ω}; see Figure 13.4.1. The string lies along the x-coordinate line, and is viewed as an infinite set of identical unit cells, $U \equiv \{x \mid |x| \leq a\}$, with the elasticity field,

[25] Since the analogy with heat conductivity is obvious, it is not explicitly stated at all appropriate occasions.

$$C'(x) = H(x;\, M)\, C^M + H(x;\, \Omega)\, C^\Omega, \tag{13.4.1a}$$

where the Heaviside step functions $H(x;\, M)$ and $H(x;\, \Omega)$ are

$$H(x;\, M) = \begin{cases} 1 & \text{if } b < |x| < a \\ 0 & \text{otherwise,} \end{cases}$$

$$H(x;\, \Omega) = \begin{cases} 1 & \text{if } |x| < b \\ 0 & \text{otherwise.} \end{cases} \tag{13.4.1b,c}$$

Hence, the volume fraction of the inclusion is

$$f \equiv \frac{\Omega}{U} \equiv \frac{b}{a}. \tag{13.4.2}$$

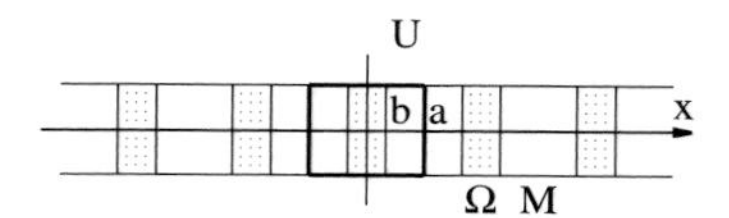

Figure 13.4.1
An infinitely long straight string consisting of periodic arrangement of a matrix and an inclusion

The governing equations for the displacement, strain, and stress fields, denoted by u, ε, and σ, are

$$\varepsilon(x) = \nabla u(x), \qquad \nabla\sigma(x) = 0, \qquad \sigma(x) = C'(x)\, \varepsilon(x), \tag{13.4.3a~c}$$

where $\nabla = d/dx$. These fields are divided into uniform and periodic parts, $\{u^o, \varepsilon^o, \sigma^o\}$ and $\{u^p, \varepsilon^p, \sigma^p\}$, respectively, where ε^o and σ^o are constant, $u^o = \varepsilon^o x$ is linear, and u^p, ε^p, and σ^p are periodic, satisfying

$$u^p(x+2a) = u^p(x), \qquad \varepsilon^p(x+2a) = \varepsilon^p(x), \qquad \sigma^p(x+2a) = \sigma^p(x). \tag{13.4.4a~c}$$

When a uniform overall strain ε^o is prescribed, the equilibrium equation is

$$\nabla\{C'(x)\,(\nabla u^p(x) + \varepsilon^o)\} = 0 \qquad \text{for x in U} \tag{13.4.5a}$$

or, equivalently,

$$\nabla\{C'(x)\,(\varepsilon^p(x) + \varepsilon^o)\} = 0 \qquad \text{for x in U.} \tag{13.4.5b}$$

The compatibility of the strain field is always satisfied in one-dimensional problems.

A general solution of (13.4.5b) is $C'(x)\{\varepsilon^p(x) + \varepsilon^o\} = \text{constant}$. The boundary conditions for $u^p(x)$ at $x = \pm a$ need not be defined explicitly. From the vanishing of the volume average of the periodic disturbance strain,

$$<\varepsilon^p> = \frac{1}{U}\int_U \varepsilon^p(x)\, dx = 0, \tag{13.4.6}$$

the *exact* solution for (13.4.5) is obtained, as

$$\varepsilon^p(x) = H(x;\, M)\, \varepsilon^{pM} + H(x;\, \Omega)\, \varepsilon^{p\Omega}, \tag{13.4.7a}$$

where

$$\varepsilon^{pM} = \{(C^M)^{-1}\,\bar{C} - 1\}\, \varepsilon^o, \qquad \varepsilon^{p\Omega} = \{(C^\Omega)^{-1}\,\bar{C} - 1\}\, \varepsilon^o, \tag{13.4.7b,c}$$

with

$$\bar{C} \equiv \{(1-f)\,(C^M)^{-1} + f\,(C^\Omega)^{-1}\}^{-1}. \tag{13.4.7d}$$

In this one-dimensional example, the uniform strain ε^o produces only a uniform stress,

$$\sigma^o = \bar{C}\, \varepsilon^o; \tag{13.4.8}$$

the periodic disturbance stress σ^p is identically zero. This is the exact solution of the problem.

The overall elasticity of this heterogeneous solid, $\bar{C}$, is the same as that obtained by the Reuss assumption; see Subsections 7.1 and 7.5. This is the consequence of the required continuity of stress, which leads to a uniform stress field. If the strain is assumed to remain continuous, then the Voigt model would result, but this does not satisfy the compatibility of deformation, and hence, is not an acceptable solution for the present problem.

13.4.2. Equivalent Homogeneous Solid with Periodic Eigenstress Field

An equivalent homogeneous one-dimensional solid of elasticity C, with a periodically distributed eigenstress σ^* is introduced. Instead of solving (13.4.5) directly, consider the following consistency condition:

$$(C'(x) - C)^{-1}\, \sigma^*(x) + \Gamma^P(x;\, \sigma^*) - \varepsilon^o = 0 \qquad x \text{ in } U \tag{13.4.9a}$$

or

$$\begin{cases} (C^M - C)^{-1}\, \sigma^*(x) + \Gamma^P(x;\, \sigma^*) - \varepsilon^o = 0 & x \text{ in } M \\ (C^\Omega - C)^{-1}\, \sigma^*(x) + \Gamma^P(x;\, \sigma^*) - \varepsilon^o = 0 & x \text{ in } \Omega, \end{cases} \tag{13.4.9b}$$

where $\Gamma^P(x;\, \sigma^*)$, operating on the eigenstress field σ^*, determines the corresponding periodic strain field,

$$-\,\varepsilon^p(x) \equiv \Gamma^P(x;\, \sigma^*)$$

$$\equiv \sum_{\xi}{}'\, {}_F\Gamma^P(\xi)\, \{\frac{1}{U}\int_U \sigma^*(y)\, \exp(\iota\xi(x-y))\, dy\}. \tag{13.4.10a}$$

From $\nabla(C\,\varepsilon^p + \sigma^*) = 0$, the coefficient ${}_F\Gamma^P$ is

$${}_F\Gamma^P(\xi) = C^{-1}, \qquad \text{for } \xi \neq 0. \tag{13.4.10b}$$

Hence, $\Gamma^P(x;\, \sigma^*)$ is explicitly given by

$$\Gamma^P(x;\, \sigma^*) = C^{-1}\{\sigma^*(x) - <\sigma^*>\}. \tag{13.4.10c}$$

The uniform part of σ^* produces no periodic strains and stresses, and is only related to the homogeneous strain and stress through $\sigma^o = C\,\varepsilon^o + <\sigma^*>$.

Substituting (13.4.10c) into (13.4.9b), obtain the following equation for the exact eigenstress field:

$$\begin{cases} \{(C^M - C)^{-1} + C^{-1}\}\,\sigma^*(x) - (\varepsilon^o + C^{-1}<\sigma^*>) = 0 & \text{x in M} \\ \{(C^\Omega - C)^{-1} + C^{-1}\}\,\sigma^*(x) - (\varepsilon^o + C^{-1}<\sigma^*>) = 0 & \text{x in } \Omega. \end{cases} \tag{13.4.11}$$

Hence, in terms of $<\sigma^*>$, the exact eigenstress field is expressed as

$$\sigma^*(x) = H(x;\,M)\,\sigma^{*M} + H(x;\,\Omega)\,\sigma^{*\Omega}, \tag{13.4.12a}$$

where

$$\sigma^{*M} = \{(C^M - C)^{-1} + C^{-1}\}^{-1}\,(\varepsilon^o + C^{-1}<\sigma^*>),$$

$$\sigma^{*\Omega} = \{(C^\Omega - C)^{-1} + C^{-1}\}^{-1}\,(\varepsilon^o + C^{-1}<\sigma^*>). \tag{13.4.12b,c}$$

Since $<\sigma^*> = (1-f)\sigma^{*M} + f\sigma^{*\Omega}$, it follows from (13.4.12b,c) that

$$<\sigma^*> = \left[\,1 - C\{(1-f)(C^M)^{-1} + f(C^\Omega)^{-1}\}\right]\,\{(1-f)(C^M)^{-1} + f(C^\Omega)^{-1}\}^{-1}\,\varepsilon^o$$

$$= (\bar{C} - C)\,\varepsilon^o. \tag{13.4.12d}$$

Substituting (13.4.12d) into (13.4.12b,c), obtain σ^*, as

$$\sigma^*(x) = H(x;\,M)\,(1 - (C^M)^{-1}\,C)\,(\bar{C}\,\varepsilon^o) + H(x;\,\Omega)\,(1 - (C^\Omega)^{-1}\,C)\,(\bar{C}\,\varepsilon^o). \tag{13.4.12e}$$

This is the solution of consistency condition (13.4.11). Note that it is easily shown that the exact eigenstress field produces the same fields as in the original heterogeneous periodic structure. Compute a periodic strain field as an example. In terms of ε^{pM} and $\varepsilon^{p\Omega}$ given by (13.4.7b,c), σ^* is expressed as

$$\sigma^*(x) = <\sigma^*> - H(x;\,M)(C\,\varepsilon^{pM}) - H(x;\,\Omega)(C\,\varepsilon^{p\Omega}). \tag{13.4.12f}$$

Hence, from $\varepsilon^p = -\Gamma^P(\sigma^*) = -(\sigma^* - <\sigma^*>)\,C^{-1}$, it is shown that the periodic strain due to σ^* coincides with that given by (13.4.7a~d).

13.4.3. Hashin-Shtrikman Variational Principle

Following the procedure discussed in the preceding subsections, apply the Hashin-Shtrikman variational principle to this one-dimensional periodic structure. The functional J is now given by

$$J(s^*;\,\varepsilon^o) = \frac{1}{2} < s^*\,\{(C' - C)^{-1}\,s^* + \Gamma^P(s^*) - 2\,\varepsilon^o\} >, \tag{13.4.13a}$$

where C is any (constant) reference elasticity and Γ^P is defined for this reference elasticity by (13.4.10c). The variation of J becomes

$$\delta J(s^*; \varepsilon^o) = < \delta s^* \{(C' - C)^{-1} s^* + \Gamma^P(s^*) - \varepsilon^o\} >, \tag{13.4.13b}$$

and the Euler equation of (13.4.13b) coincides with consistency condition (13.4.9a).

As an exercise, it is shown that the exact eigenstress field indeed renders the energy functional J stationary. To this end, rewrite (13.4.13a) using (13.4.1), as

$$J(s^*; \varepsilon^o) = \frac{1}{2} \left\{ (1-f) < s^* (C^M - C)^{-1} s^* >_M + f < s^* (C^\Omega - C)^{-1} s^* >_\Omega \right\}$$

$$+ \frac{1}{2} < s^* (\Gamma^P(s^*) - 2\,\varepsilon^o) >, \tag{13.4.14}$$

and define the g_Ω- and g_M-integrals as

$$g_\Omega(\xi) \equiv < \exp(\iota\xi x) >_\Omega, \qquad g_M(\xi) \equiv < \exp(\iota\xi x) >_M. \tag{13.4.15a,b}$$

Also, define the g_U-integral as

$$g_U(\xi) \equiv < \exp(\iota\xi x) > \equiv \begin{cases} 1 & \xi = 0 \\ 0 & \text{otherwise;} \end{cases} \tag{13.4.15c}$$

see Subsection 12.5.

First, consider the Fourier series expansion of an arbitrary eigenstress field, s^*,

$$s^*(x) = \sum_\xi \mathcal{F}s^*(\xi) \exp(\iota\xi x) \tag{13.4.16a}$$

which includes a uniform part $< s^* > = \mathcal{F}s^*(0)$. Then, apply the operator Γ^P to this eigenstress field,

$$\Gamma^P(x; s^*) = \sum_\xi \mathcal{F}\Gamma^P(\xi)\, \mathcal{F}s^*(\xi) \exp(\iota\xi x), \tag{13.4.16b}$$

where $\mathcal{F}\Gamma^P(0)$ is set equal to zero for any C, since the uniform eigenstress does not cause any deformation in the homogenized infinite solid. Substituting (13.4.15a,b) into (13.4.14), define J as a function of the Fourier coefficients of the eigenstress, $\mathcal{F}s^*$,

$$J(\{\mathcal{F}s^*\}; \varepsilon^o) \equiv J(\{\sum_\xi \mathcal{F}s^*(\xi) \exp(\iota\xi x)\}; \varepsilon^o)$$

$$= \frac{1}{2} \sum_\xi \sum_\zeta \mathcal{F}s^*(\xi) \left\{ (1-f)\, g_M(\xi+\zeta)\, (C^M - C)^{-1} \right.$$

$$\left. + f\, g_\Omega(\xi+\zeta)\, (C^\Omega - C)^{-1} \right\} \mathcal{F}s^*(\zeta)$$

$$+ \frac{1}{2} \sum_\xi \mathcal{F}s^*(\xi)\, \mathcal{F}\Gamma^P(-\xi)\, \mathcal{F}s^*(-\xi) - \mathcal{F}s^*(0)\, \varepsilon^o. \tag{13.4.17a}$$

As is seen, J is a quadratic (with linear terms) form in the Fourier series coefficients, $\mathcal{F}s^*(\xi)$.

The extremum of J is obtained by taking partial differentiation with respect to $F\mathrm{s}^*(\xi)$, i.e.,

$$\frac{\partial J}{\partial F\mathrm{s}^*(\xi)}(\{F\mathrm{s}^*\};\,\varepsilon^{\mathrm{o}}) = \sum_{\zeta}\Big\{(1-\mathrm{f})\,\mathrm{g_M}(\xi+\zeta)\,(\mathrm{C^M}-\mathrm{C})^{-1} + \mathrm{f}\,\mathrm{g}_\Omega(\xi+\zeta)\,(\mathrm{C}^\Omega-\mathrm{C})^{-1}\Big\}\,F\mathrm{s}^*(\zeta)$$

$$+\,F\Gamma^{\mathrm{P}}(-\xi)\,F\mathrm{s}^*(-\xi) - \mathrm{g_U}(\xi)\,\varepsilon^{\mathrm{o}} = 0, \qquad (13.4.17\mathrm{b})$$

for all ξ. Then, J is expressed as

$$J(\{F\mathrm{s}^*\};\,\varepsilon^{\mathrm{o}}) = \frac{1}{2}\sum_{\xi} F\mathrm{s}^*(\xi)\,\{\frac{\partial J}{\partial F\mathrm{s}^*(-\xi)}\} - \frac{1}{2}F\mathrm{s}^*(0)\,\varepsilon^{\mathrm{o}}. \qquad (13.4.17\mathrm{c})$$

As mentioned in Subsection 13.3, the Fourier series representation of the consistency condition is identical with

$$\frac{\partial J}{\partial F\mathrm{s}^*(\xi)}(\{F\mathrm{s}^*\};\,\varepsilon^{\mathrm{o}}) = 0 \qquad \text{for any } \xi, \qquad (13.4.17\mathrm{d})$$

and the solution of this set of equations, $\{F\sigma^*(\xi)\}$, is the Fourier series coefficient of the exact eigenstress field.

It is shown that the exact eigenstress field given by (13.4.12e), σ^*, indeed satisfies (13.4.17d). Since σ^* given by (13.4.12e) is rewritten as

$$\sigma^*(\mathrm{x}) = \mathrm{H}(\mathrm{x};\,\mathrm{M})\,\sigma^{*\mathrm{M}} + \mathrm{H}(\mathrm{x};\,\Omega)\,\sigma^{*\Omega}, \qquad (13.4.18\mathrm{a})$$

where $\sigma^{*\mathrm{M}} = (1-(\mathrm{C^M})^{-1}\,\mathrm{C})\,(\bar{\mathrm{C}}\,\varepsilon^{\mathrm{o}})$ and $\sigma^{*\Omega} = (1-(\mathrm{C}^\Omega)^{-1}\,\mathrm{C})\,(\bar{\mathrm{C}}\,\varepsilon^{\mathrm{o}})$, the Fourier series representation of σ^* is

$$\sigma^*(\mathrm{x}) = \sum_{\zeta}\Big\{(1-\mathrm{f})\,\mathrm{g_M}(-\zeta)\,\sigma^{*\mathrm{M}} + \mathrm{f}\,\mathrm{g}_\Omega(-\zeta)\,\sigma^{*\Omega}\Big\}\exp(\iota\zeta\mathrm{x}). \qquad (13.4.18\mathrm{b})$$

As shown in Subsection 12.5.3, the g-integrals satisfy

$$\mathrm{g}_\Omega(\xi) = \sum_{\zeta}\mathrm{f}\,\mathrm{g}_\Omega(\zeta+\xi)\,\mathrm{g}_\Omega(-\zeta), \qquad \mathrm{g_M}(\xi) = \sum_{\zeta}(1-\mathrm{f})\,\mathrm{g_M}(\zeta+\xi)\,\mathrm{g_M}(-\zeta),$$

$$0 = \sum_{\zeta}(1-\mathrm{f})\,\mathrm{g}_\Omega(\zeta+\xi)\,\mathrm{g_M}(-\zeta) = \sum_{\zeta}\mathrm{f}\,\mathrm{g_M}(\zeta+\xi)\,\mathrm{g}_\Omega(-\zeta). \qquad (13.4.19\mathrm{a}\sim\mathrm{c})$$

Hence, the value of $\partial J/\partial F\mathrm{s}^*(\xi)$ can be computed for these $F\sigma^*$'s, as follows:

$$\frac{\partial J}{\partial F\mathrm{s}^*(\xi)}(\{F\sigma^*\};\,\varepsilon^{\mathrm{o}}) = (1-\mathrm{f})(\mathrm{C^M}-\mathrm{C})^{-1}\,\mathrm{g_M}(\xi)\,\sigma^{*\mathrm{M}} + \mathrm{f}\,(\mathrm{C}^\Omega-\mathrm{C})^{-1}\,\mathrm{g}_\Omega(\xi)\,\sigma^{*\Omega}$$

$$+\,F\Gamma^{\mathrm{P}}(-\xi)\,\{(1-\mathrm{f})\,\mathrm{g_M}(\xi)\,\sigma^{*\mathrm{M}} + \mathrm{f}\,\mathrm{g}_\Omega(\xi)\sigma^{*\Omega}\} - \mathrm{g_U}(\xi)\,\varepsilon^{\mathrm{o}}. \qquad (13.4.20\mathrm{a})$$

From $(\mathrm{C^M}-\mathrm{C})^{-1}\,\sigma^{*\mathrm{M}} = (\mathrm{C^M})^{-1}\,(\bar{\mathrm{C}}\,\varepsilon^{\mathrm{o}})$ and $(\mathrm{C}^\Omega-\mathrm{C})^{-1}\,\sigma^{*\Omega} = (\mathrm{C}^\Omega)^{-1}\,(\bar{\mathrm{C}}\,\varepsilon^{\mathrm{o}})$, it follows that the right side of (13.4.20a) becomes

$$\{(1-\mathrm{f})\,(\mathrm{C^M})^{-1}\mathrm{g_M}(\xi) + \mathrm{f}\,(\mathrm{C}^\Omega)^{-1}\mathrm{g}_\Omega(\xi)\}\,(\bar{\mathrm{C}}\,\varepsilon^{\mathrm{o}})$$

$$+\,\{(1-\mathrm{f})\,(\mathrm{C}^{-1}-(\mathrm{C^M})^{-1})\mathrm{g_M}(\xi) + \mathrm{f}\,(\mathrm{C}^{-1}-(\mathrm{C}^\Omega)^{-1})\mathrm{g}_\Omega(\xi)\}\,(\bar{\mathrm{C}}\,\varepsilon^{\mathrm{o}}) + \mathrm{g_U}(\xi)\varepsilon^{\mathrm{o}}$$

$$= \{(1-f)\,g_M(\xi) + f\,g_\Omega(\xi)\}\,C^{-1}\,\bar{C}\,\varepsilon^o - g_U(\xi)\,C^{-1}\,\bar{C}\,\varepsilon^o, \tag{13.4.20b}$$

where $F\Gamma^P(\xi) = (1 - g_U(\xi))\,C^{-1}$ and $g_U(0) = g_\Omega(0) = 1$ are used. From $g_U = (1-f)\,g_M + f\,g_\Omega$, since $\bar{C}$ is given by (13.4.7d), the right side of (13.4.20b) vanishes for any ξ. Hence, it is shown that for the Fourier series coefficients of the exact eigenstress σ^*, the extremum conditions of J are satisfied identically. It is seen that for any choice of the reference elasticity, the energy functional collapses into the exact solution, for this one-dimensional case.

13.5. PIECEWISE CONSTANT APPROXIMATION AND UNIVERSAL BOUNDS

In Subsection 12.6, the overall bulk modulus for the unit cell in a periodic microstructure containing spherical cavities has been estimated, arriving at (12.6.6c). As is pointed out, *the expression is actually an exact upper bound for a solid with cavities, independently of the shape and structure of the cavities within the unit cell.* Universal upper and lower bounds of this kind can be obtained for solids containing periodically distributed inclusions of arbitrary geometry and elasticity. Moreover, a similar result can also be extracted for at least one other overall modulus, as will be discussed in the sequel. The important point to emphasize is that, for a periodic microstructure, these bounds are exact, and can be evaluated to any desired degree of accuracy. As shown in Subsections 9.5.5, 9.6.3, and 9.6.4, similar *exact computable* bounds are obtained for any linearly elastic solid of any heterogeneities, as long as the overall geometry of the RVE is ellipsoidal.

13.5.1. Piecewise Constant Approximation of Eigenstress Field

As discussed in Subsection 9.4, for a general finite RVE, approximations are required in order to calculate the correlation tensors in terms of the Green function for the unbounded homogeneous elastic solid with the chosen reference elasticity tensor. However, if an ellipsoidal RVE is embedded in the unbounded homogeneous solid, exact bounds for the overall moduli can be obtained with the aid of Theorems I and II of Subsection 2.5.6.

The remarkable result presented below is that the exact bounds for two overall elastic moduli obtained on the basis of the periodic microstructure, i.e., those obtained by considering an RVE and using Willis' approximation, and those obtained for an ellipsoidal RVE in the infinite homogeneous solid: (1) actually coincide; and (2) are valid for inclusions of any geometry and elasticity, as long as the matrix and the inclusions are both piecewise uniform.[26] Since the relation between the last two bounds is shown in Subsection 9.6, the relation

[26] This is not a major restriction, since any variable elasticity can be approximated by a piecewise constant tensor field, to any desired degree of accuracy.

between the first two bounds is shown in this subsection.[27]

To this end, and referring to Subsection 9.5 for the RVE and Subsection 13.2 for the periodic microstructure, compare the corresponding energy functional associated with a trial homogenizing eigenstress field, $\mathbf{s}^*$, i.e.,

$$J^A(\mathbf{s}^*;\, \boldsymbol{\varepsilon}^o) = \frac{1}{2} < \mathbf{s}^* : \{(\mathbf{C}' - \mathbf{C})^{-1} : \mathbf{s}^* + \boldsymbol{\Gamma}^A(\mathbf{s}^*) - 2\,\boldsymbol{\varepsilon}^o\} >_V, \qquad (13.5.1a)$$

for the RVE, and

$$J^P(\mathbf{s}^*;\, \boldsymbol{\varepsilon}^o) = \frac{1}{2} < \mathbf{s}^* : \{(\mathbf{C}' - \mathbf{C})^{-1} : \mathbf{s}^* + \boldsymbol{\Gamma}^P(\mathbf{s}^*) - 2\,\boldsymbol{\varepsilon}^o\} >_U, \qquad (13.5.2a)$$

for the periodic microstructure, where superscript A or P emphasizes whether the corresponding quantity is for the RVE (as *approximated* by using the infinite body Green function) or the periodic structure. Use a common reference elasticity $\mathbf{C}$, for both J^A and J^P. The integral operator $\boldsymbol{\Gamma}^A(\mathbf{s}^*)$ is defined in terms of the Green function[28] for an infinite homogeneous body with elasticity $\mathbf{C}$, while the integral operator $\boldsymbol{\Gamma}^P(\mathbf{s}^*)$ is defined for the equivalent homogeneous unit cell with elasticity $\mathbf{C}$, associated with the original periodic microstructure.

Now, assume that the geometry and material properties of the RVE and the periodic solid have the following in common:

1) $n + 1$ distinct inclusions, Ω_α, with elasticity $\mathbf{C}^\alpha$ and volume fraction f_α, for $\alpha = 0$ (1, ..., n);

2) $n + 1$ constant eigenstresses, $\mathbf{s}^{*\alpha}$, respectively, distributed in Ω_α, for $\alpha = 0$, 1, ..., n; and

3) the reference elasticity $\mathbf{C}$ is isotropic, $\mathbf{C} = \{2\mu\nu/(1 - 2\nu)\}\, \mathbf{1}^{(2)} \otimes \mathbf{1}^{(2)} + 2\mu\, \mathbf{1}^{(4s)}$, with the shear modulus μ and the Poisson ratio ν.

Here, for simplicity, the matrix is regarded as the 0th micro-inclusion with elasticity $\mathbf{C}^0 = \mathbf{C}^M$ and volume fraction $f_0 = M/V$; see Subsections 9.3 and 9.4. From 2), $\mathbf{C}'$ and $\mathbf{s}^*$ in J^A and J^P are given by

$$\mathbf{C}'(\mathbf{x}) = \sum_{\alpha=0}^{n} H(\mathbf{x};\, \Omega_\alpha)\, \mathbf{C}^\alpha, \qquad \mathbf{s}^*(\mathbf{x}) = \sum_{\alpha=0}^{n} H(\mathbf{x};\, \Omega_\alpha)\, \mathbf{s}^{*\alpha}, \qquad (13.5.3a,b)$$

where $H(\mathbf{x};\, \Omega)$ is the Heaviside step function with the value one when $\mathbf{x}$ is in Ω, and zero otherwise.

---

[27] As is shown, the bounds for the periodic case are essentially the same as those for the ellipsoidal RVE, except for the difference in the integral operators for the eigenfield; the former is given by the Fourier series expansion, while the latter is given by the Fourier transform.

[28] As shown in Subsection 9.4, the Green function for the infinite domain *approximates* the Green function for V, except for the region near the boundary, if V is a sufficiently large finite domain subjected to linear displacement boundary conditions.

13.5.2. Computation of Energy Functions and Universal Bounds

As discussed in Subsection 9.4, substitution of (13.5.3a,b) into (13.5.1a) yields energy function $J^{A'}$ for piecewise constant $\mathbf{s}^{*\alpha}$'s,

$$J^{A'}(\{\mathbf{s}^{*\alpha}\}; \boldsymbol{\varepsilon}^o) = \frac{1}{2} \sum_{\alpha=0}^{n} f_\alpha \, \mathbf{s}^{*\alpha} : \{(\mathbf{C}^\alpha - \mathbf{C})^{-1} + \mathbf{P}\} : \mathbf{s}^{*\alpha}$$

$$- \frac{1}{2} \bar{\mathbf{s}}^* : \mathbf{P} : \bar{\mathbf{s}}^* - \bar{\mathbf{s}}^* : \boldsymbol{\varepsilon}^o, \tag{13.5.1b}$$

where $\bar{\mathbf{s}}^* = \sum_{\alpha=0}^{n} f_\alpha \, \mathbf{s}^{*\alpha}$, and $\mathbf{P}$ is defined by

$$\mathbf{P} \equiv \frac{1}{2\mu} \frac{1}{1-\nu} \{ -\frac{1}{15} \mathbf{1}^{(2)} \otimes \mathbf{1}^{(2)} + \frac{2(4-5\nu)}{15} \mathbf{1}^{(4s)} \}; \tag{13.5.4}$$

see Subsection 9.4 for detailed derivation of $\mathbf{P}$.

Similarly, substitution of (13.5.3a,b) into (13.5.2a) yields

$$J^{P'}(\{\mathbf{s}^{*\alpha}\}; \boldsymbol{\varepsilon}^o) = \frac{1}{2} \sum_{\alpha=0}^{n} f_\alpha \, \mathbf{s}^{*\alpha} : \{(\mathbf{C}^\alpha - \mathbf{C})^{-1} : \mathbf{s}^{*\alpha} + \sum_{\beta=0}^{n} f_\beta \, \boldsymbol{\Gamma}^{P\alpha\beta} : \mathbf{s}^{*\beta} - 2\boldsymbol{\varepsilon}^o\}, \tag{13.5.2b}$$

where

$$\boldsymbol{\Gamma}^{P\alpha\beta} \equiv \sum_{\boldsymbol{\xi}}{}' \, g_\alpha(-\boldsymbol{\xi}) \, g_\beta(\boldsymbol{\xi}) \, F\boldsymbol{\Gamma}^P(\boldsymbol{\xi}),$$

$$= \sum_{\boldsymbol{\xi}}{}' \, g_\alpha(-\boldsymbol{\xi}) \, g_\beta(\boldsymbol{\xi}) \, \frac{1}{\mu} \{ -\frac{1}{2(1-\nu)} \bar{\boldsymbol{\xi}} \otimes \bar{\boldsymbol{\xi}} \otimes \bar{\boldsymbol{\xi}} \otimes \bar{\boldsymbol{\xi}} + sym\,(\bar{\boldsymbol{\xi}} \otimes \mathbf{1}^{(2)} \otimes \bar{\boldsymbol{\xi}}) \}, \tag{13.5.5}$$

with $\bar{\boldsymbol{\xi}} = \boldsymbol{\xi}/|\boldsymbol{\xi}|$; see Subsection 12.4.4 for detailed derivation of $F\boldsymbol{\Gamma}^P$.

As mentioned in Subsection 12.5.3, the g-integral, $g_\alpha(\boldsymbol{\xi})$, corresponds to the coefficients of the Fourier expansion of the Heaviside step function associated with the region Ω_α ($\alpha = 0, 1, ..., n$). Hence,

$$\sum_{\boldsymbol{\xi}}{}' \, f_\alpha \, g_\alpha(-\boldsymbol{\xi}) \, g_\beta(\boldsymbol{\xi}) = \begin{cases} 1 - f_\alpha & \text{if } \alpha = \beta \\ -f_\alpha & \text{otherwise,} \end{cases} \tag{13.5.6}$$

for any Ω_α and Ω_β. Therefore, taking advantage of identities,

$$F\Gamma^P_{iijj} = \frac{1}{\mu} \frac{1-2\nu}{2(1-\nu)}, \qquad F\Gamma^P_{ijij} = \frac{1}{\mu} \frac{3-4\nu}{2(1-\nu)}, \tag{13.5.7a,b}$$

obtain the following *exact* results:

$$f_\alpha \, \Gamma^{P\alpha\beta}_{iijj} = \begin{cases} (1-f_\alpha) \dfrac{1}{\mu} \dfrac{1-2\nu}{2(1-\nu)} & \text{if } \alpha = \beta \\ -f_\alpha \dfrac{1}{\mu} \dfrac{1-2\nu}{2(1-\nu)} & \text{otherwise} \end{cases} \tag{13.5.8a}$$

and

$$f_\alpha \Gamma^{P\alpha\beta}_{ijij} = \begin{cases} (1 - f_\alpha) \dfrac{1}{\mu} \dfrac{3 - 4\nu}{2(1 - \nu)} & \text{if } \alpha = \beta \\ - f_\alpha \dfrac{1}{\mu} \dfrac{3 - 4\nu}{2(1 - \nu)} & \text{otherwise.} \end{cases} \tag{13.5.8b}$$

Since $\mathbf{P}$ satisfies $P_{iijj} = (1 - 2\nu)/2\mu(1 - \nu)$ and $P_{ijij} = (3 - 4\nu)/2\mu(1 - \nu)$, rewrite[29] (13.5.8a,b) as

$$f_\alpha \Gamma^{P\alpha\beta}_{iijj} = \begin{cases} (1 - f_\alpha)\, P_{iijj} & \text{if } \alpha = \beta \\ - f_\alpha\, P_{iijj} & \text{otherwise} \end{cases} \tag{13.5.8c}$$

and

$$f_\alpha \Gamma^{P\alpha\beta}_{ijij} = \begin{cases} (1 - f_\alpha)\, P_{ijij} & \text{if } \alpha = \beta \\ - f_\alpha\, P_{ijij} & \text{otherwise.} \end{cases} \tag{13.5.8d}$$

Since the above properties of $\mathbf{\Gamma}^P$ are essentially the same as those of the integral operator for the infinite solid, $\mathbf{\Gamma}^A$, the procedure used in Subsection 9.6.3 can be applied to the periodic case. First, suppose that the uniform overall strain is dilatational, $\boldsymbol{\varepsilon}^o = \varepsilon^o \mathbf{1}$. Then, setting $\mathbf{s}^{*\alpha} = s^{*\alpha}\mathbf{1}$, compute $J^{A\prime}$ as

$$J^{A\prime}(\{s^{*\alpha}\mathbf{1}\}; \varepsilon^o\mathbf{1}) = \frac{1}{2} \sum_{\alpha=0}^{n} f_\alpha\, s^{*\alpha} \{(\mathbf{C}^\alpha - \mathbf{C})^{-1}_{iijj} + P_{iijj}\}\, s^{*\alpha}$$

$$- \frac{1}{2} \overline{s}^*\, P_{iijj}\, \overline{s}^* - 3\overline{s}^*\, \varepsilon^o, \tag{13.5.9a}$$

and taking advantage of (13.5.8c), compute $J^{P\prime}$ as

$$J^{P\prime}(\{s^{*\alpha}\mathbf{1}\}; \varepsilon^o\mathbf{1})$$

$$= \frac{1}{2} \sum_{\alpha=0}^{n} f_\alpha\, s^{*\alpha} \{(\mathbf{C}^\alpha - \mathbf{C})^{-1}_{iijj}\, s^{*\alpha} + \sum_{\beta=0}^{n} \Gamma^{P\alpha\beta}_{iijj}\, s^{*\beta} - 2\delta_{ij}\delta_{ij}\, \varepsilon^o\}$$

$$= \frac{1}{2} \sum_{\alpha=0}^{n} f_\alpha\, s^{*\alpha} \{(\mathbf{C}^\alpha - \mathbf{C})^{-1}_{iijj} + P_{iijj}\}\, s^{*\alpha} - \frac{1}{2} \overline{s}^*\, P_{iijj}\, \overline{s}^* - 3\overline{s}^* \varepsilon^o, \tag{13.5.10a}$$

where $\overline{s}^* = \sum_{\alpha=0}^{n} f_\alpha\, s^{*\alpha}$. As is seen here, the value of the energy function $J^{P\prime}$ coincides with that of the energy function $J^{A\prime}$. Hence, the optimal value of $J^{P\prime}$ also coincides with that of $J^{A\prime}$. The overall elasticity $\overline{\mathbf{C}}$ for prescribed $\boldsymbol{\varepsilon}^o = \varepsilon^o\,\mathbf{1}$ is $\overline{C}_{iijj}$ which is proportional to the overall bulk modulus, $\overline{K}$; see (12.6.3a). *Therefore, the bound for* $\overline{K}$ *estimated from the periodic microstructure is the same as that from an* RVE. These results are valid for inclusions Ω_α with arbitrary shape, elasticity, and relative location. For the periodic case, the bound involves no approximation.[30]

---

[29] It should be noted that (13.5.8c,d) are *exact* expressions for the correlation tensors for Ω_α and Ω_β of any *shape* and *location* within the unit cell, whereas $\mathbf{P}$ in (13.5.9a,b) only *approximately* determines the correlation tensors in an RVE, unless *statistical homogeneity and isotropy* hold; see Subsection 9.5.

[30] The fact that bounds on the bulk modulus are universal for two-phase composites with *statistically homogeneous and isotropic microstructure*, has been known; see, e.g. Milton and Kohn (1988). Here, and in Subsections 9.6 and 9.7, it is shown that this result is valid for *any microstruc-*

Next, consider three cases of overall biaxial shearing with nonzero strain components, $\varepsilon^o(\mathbf{e}_1\otimes\mathbf{e}_1 - \mathbf{e}_2\otimes\mathbf{e}_2)$, $\varepsilon^o(\mathbf{e}_2\otimes\mathbf{e}_2 - \mathbf{e}_3\otimes\mathbf{e}_3)$, and $\varepsilon^o(\mathbf{e}_3\otimes\mathbf{e}_3 - \mathbf{e}_1\otimes\mathbf{e}_1)$, and three cases of overall pure shearing with nonzero strain components, $\varepsilon^o(\mathbf{e}_2\otimes\mathbf{e}_3 + \mathbf{e}_3\otimes\mathbf{e}_2)$, $\varepsilon^o(\mathbf{e}_3\otimes\mathbf{e}_1 + \mathbf{e}_1\otimes\mathbf{e}_3)$, and $\varepsilon^o(\mathbf{e}_1\otimes\mathbf{e}_2 + \mathbf{e}_2\otimes\mathbf{e}_1)$. In the same manner as in Subsection 9.6.3, the tensor products of the unit base vectors associated with these six cases are denoted by $\mathbf{s}_1 = \mathbf{e}_1\otimes\mathbf{e}_1 - \mathbf{e}_2\otimes\mathbf{e}_2$, ..., $\mathbf{s}_4 = \mathbf{e}_2\otimes\mathbf{e}_3 + \mathbf{e}_3\otimes\mathbf{e}_2$, ... For each $\boldsymbol{\varepsilon}^o = \varepsilon^o\mathbf{s}_i$, set $\mathbf{s}^{*\alpha} = s^{*\alpha}\mathbf{s}_i$, and write the corresponding energy function $J^{A\prime}$ as $J^{A\prime}(\{s^{*\alpha}\mathbf{s}_i\}; \varepsilon^o\mathbf{s}_i)$. Then, compute the following functions:

$$
\begin{aligned}
&\frac{1}{3}\Big[\, J^{A\prime}(\{s^{*\alpha}\mathbf{s}_1\}; \varepsilon^o\mathbf{s}_1) + J^{A\prime}(\{s^{*\alpha}\mathbf{s}_2\}; \varepsilon^o\mathbf{s}_2) + J^{A\prime}(\{s^{*\alpha}\mathbf{s}_3\}; \varepsilon^o\mathbf{s}_3)\Big] \\
&\qquad + \frac{1}{2}\Big[\, J^{A\prime}(\{s^{*\alpha}\mathbf{s}_4\}; \varepsilon^o\mathbf{s}_4) + J^{A\prime}(\{s^{*\alpha}\mathbf{s}_5\}; \varepsilon^o\mathbf{s}_5) + J^{A\prime}(\{s^{*\alpha}\mathbf{s}_6\}; \varepsilon^o\mathbf{s}_6)\Big] \\
&= \frac{1}{2}\sum_{\alpha=0}^{n} f_\alpha\, s^{*\alpha}\, \{(\mathbf{C}^\alpha - \mathbf{C})^{-1}_{ijij} - \frac{1}{3}(\mathbf{C}^\alpha - \mathbf{C})^{-1}_{iijj}\}\, s^{*\alpha} - 5\,\bar{s}^*\,\varepsilon^o \\
&\qquad + \frac{1}{2}\sum_{\alpha=0}^{n} f_\alpha\, (s^{*\alpha} - \bar{s}^*)\, \{P_{ijij} - \frac{1}{3}P_{iijj}\}\, (s^{*\alpha} - \bar{s}^*) \\
&= \frac{1}{2}\sum_{\alpha=0}^{n} f_\alpha\, s^{*\alpha}\, \{(\mathbf{C}^\alpha - \mathbf{C})^{-1}_{ijij} - \frac{1}{3}(\mathbf{C}^\alpha - \mathbf{C})^{-1}_{iijj} + P_{ijij} - \frac{1}{3}P_{iijj}\}\, s^{*\alpha} \\
&\qquad - \frac{1}{2}\,\bar{s}^*\, \{P_{ijij} - \frac{1}{3}P_{iijj}\}\, \bar{s}^* - 5\,\bar{s}^*\,\varepsilon^o
\end{aligned} \tag{13.5.9b}
$$

and

$$
\begin{aligned}
&\frac{1}{3}\Big[\, J^{P\prime}(\{s^{*\alpha}\mathbf{s}_1\}; \varepsilon^o\mathbf{s}_1) + J^{P\prime}(\{s^{*\alpha}\mathbf{s}_2\}; \varepsilon^o\mathbf{s}_2) + J^{P\prime}(\{s^{*\alpha}\mathbf{s}_3\}; \varepsilon^o\mathbf{s}_3)\Big] \\
&\qquad + \frac{1}{2}\Big[\, J^{P\prime}(\{s^{*\alpha}\mathbf{s}_4\}; \varepsilon^o\mathbf{s}_4) + J^{P\prime}(\{s^{*\alpha}\mathbf{s}_5\}; \varepsilon^o\mathbf{s}_5) + J^{P\prime}(\{s^{*\alpha}\mathbf{s}_6\}; \varepsilon^o\mathbf{s}_6)\Big] \\
&= \frac{1}{2}\sum_{\alpha=0}^{n} f_\alpha\, s^{*\alpha}\, \{(\mathbf{C}^\alpha - \mathbf{C})^{-1}_{ijij} - \frac{1}{3}(\mathbf{C}^\alpha - \mathbf{C})^{-1}_{iijj}\}\, s^{*\alpha} - 5\,\bar{s}^*\,\varepsilon^o \\
&\qquad + \frac{1}{2}\sum_{\alpha=0}^{n} f_\alpha\, (s^{*\alpha} - \bar{s}^*)\, \{P_{ijij} - \frac{1}{3}P_{iijj}\}\, (s^{*\alpha} - \bar{s}^*) \\
&= \frac{1}{2}\sum_{\alpha=0}^{n} f_\alpha\, s^{*\alpha}\, \{(\mathbf{C}^\alpha - \mathbf{C})^{-1}_{ijij} - \frac{1}{3}(\mathbf{C}^\alpha - \mathbf{C})^{-1}_{iijj} + P_{ijij} - \frac{1}{3}P_{iijj}\}\, s^{*\alpha} \\
&\qquad - \frac{1}{2}\,\bar{s}^*\, \{P_{ijij} - \frac{1}{3}P_{iijj}\}\, \bar{s}^* - 5\,\bar{s}^*\,\varepsilon^o.
\end{aligned} \tag{13.5.10b}
$$

As is seen, here again, the value of the sum of the energy functions $J^{P\prime}$ coincides

---

ture with any number of phases as long as the RVE is ellipsoidal or it is a unit cell of a periodic microstructure. These observations and the result that $\bar{C}_{iijj}$ is also universal, enjoying similar attributes as $\bar{C}_{ijij} - \bar{C}_{iijj}/3$, are proved by Nemat-Nasser and Hori (1993).

with that of the sum of the energy functions $J^{A'}$. Hence, the optimal value of the sum for the periodic case coincides with that for the RVE. These sums correspond to the value of the overall elasticity parameter, $\bar{C}_{ijij} - \bar{C}_{iijj}/3$, which yields the overall shear modulus, $\bar{\mu}$, if $\bar{\mathbf{C}}$ is isotropic. *Therefore, the universal bound for* $\bar{\mu}$, *estimated from the periodic microstructure is the same as that from the* RVE *when the latter is statistically both homogeneous and isotropic.*[31] The expression for $\bar{C}_{ijij} - \bar{C}_{iijj}/3$ is valid for any Ω_α, with any shape or elasticity.

With the aid of the equivalence relation between the functionals J and I, it is shown that the universal bounds for the overall compliance tensor $\bar{\mathbf{D}}$ obtained through I^A, coincide with those obtained in Subsection 9.5 through $I^{A'}$, where I^A is a function obtained from I for piecewise constant eigenstrains; see Subsection 9.6 for detailed derivations to obtain the bounds for $\bar{\mathbf{D}}$ from the equivalence relation between the functionals.

Note that, in the periodic case, improved bounds are obtained when the optimal eigenstrain (or eigenstress) tensor field is used. These improved bounds then include the effect of the geometry and relative locations of the inhomogeneities. They are always within the universal bounds; see Subsections 12.6 and 13.6 for illustration.

13.5.3. General Piecewise Constant Approximation of Eigenstress Field

Now, consider the piecewise constant approximation of the eigenstress field, in a more general manner. As shown in Subsection 13.5.1, such an eigenstress field is given by

$$\mathbf{s}^*(\mathbf{x}) = \sum_{\alpha=0}^{n} H_\alpha(\mathbf{x})\, \mathbf{s}^{*\alpha}, \tag{13.5.11a}$$

where the $\mathbf{s}^{*\alpha}$'s are constants, and $H_\alpha(\mathbf{x})$ is the Heaviside step function for Ω_α, i.e., $H(\mathbf{x}) \equiv H(\mathbf{x}; \Omega_\alpha)$. For this eigenstress field, functional J, given by (13.3.1), becomes

$$J(\{\sum_{\alpha=0}^{n} H_\alpha\, \mathbf{s}^{*\alpha}\}; \boldsymbol{\varepsilon}^o)$$

$$= \frac{1}{2} \sum_{\alpha=0}^{n} f_\alpha\, \mathbf{s}^{*\alpha} : \left\{ (\mathbf{C}^\alpha - \mathbf{C})^{-1} : \mathbf{s}^{*\alpha} + < \boldsymbol{\Gamma}^P(\{\sum_{\beta=0}^{n} H_\beta\, \mathbf{s}^{*\beta}\}) >_\alpha - 2\boldsymbol{\varepsilon}^o \right\}, \tag{13.5.11b}$$

and its variation with respect to $\mathbf{s}^{*\alpha}$ is given by

$$\delta J(\{\sum_{\alpha=0}^{n} H_\alpha\, \mathbf{s}^{*\alpha}\}; \boldsymbol{\varepsilon}^o)$$

---

[31] As mentioned, the results obtained for the periodic microstructure and for the ellipsoidal RVE are *exact*.

$$= \sum_{\alpha=0}^{n} f_\alpha \, \delta \mathbf{s}^{*\alpha} : \left\{ (\mathbf{C}^\alpha - \mathbf{C})^{-1} : \mathbf{s}^{*\alpha} + < \boldsymbol{\Gamma}^P(\{\sum_{\beta=0}^{n} H_\beta \, \mathbf{s}^{*\beta}\}) >_\alpha - \boldsymbol{\varepsilon}^o \right\}. \tag{13.5.11c}$$

The Euler equation of (13.5.11c) coincides with the volume average of consistency condition (13.1.14b) taken over Ω_α.

In (13.5.11a~c), the volume average of the periodic strain produced by $\mathbf{s}^{*\alpha}$'s needs to be computed over Ω_β. The correlation tensor $\boldsymbol{\Gamma}^{P\alpha\beta}$ determines the average strain over Ω_α produced by the eigenstress field which takes on a constant value, $\mathbf{s}^{*\beta}$, in Ω_β. Since integral operator $\boldsymbol{\Gamma}^P$ is explicitly given by the Fourier series expansion, this correlation tensor becomes

$$< \boldsymbol{\Gamma}^P(H_\beta \, \mathbf{s}^{*\beta}) >_\alpha \equiv f_\beta \, \boldsymbol{\Gamma}^{P\alpha\beta} : \mathbf{s}^{*\beta} \qquad (\beta \text{ not summed}), \tag{13.5.12a}$$

where H_β is the Heaviside step function for Ω_β; see Subsection 9.3. In terms of $\mathcal{F}\boldsymbol{\Gamma}^P$, correlation tensor $\boldsymbol{\Gamma}^{P\alpha\beta}$ can be *exactly* expressed as

$$\boldsymbol{\Gamma}^{P\alpha\beta} = \sum_{\boldsymbol{\xi}} g_\alpha(\boldsymbol{\xi}) \, g_\beta(-\boldsymbol{\xi}) \, \mathcal{F}\boldsymbol{\Gamma}^P(\boldsymbol{\xi}). \tag{13.5.12b}$$

Note that $\boldsymbol{\Gamma}^{P\alpha\beta}$ is real-valued, since $g_\alpha(-\boldsymbol{\xi})$ is the complex conjugate of $g_\alpha(\boldsymbol{\xi})$ and $\mathcal{F}\boldsymbol{\Gamma}^P(-\boldsymbol{\xi})$ is equal to $\mathcal{F}\boldsymbol{\Gamma}^P(\boldsymbol{\xi})$[32]. Since $\mathcal{F}\boldsymbol{\Gamma}^P(\boldsymbol{\xi})$ is symmetric with respect to the first and last pairs of its indices, correlation tensor $\boldsymbol{\Gamma}^{P\alpha\beta}$ satisfies

$$\boldsymbol{\Gamma}^{P\alpha\beta} = (\boldsymbol{\Gamma}^{P\beta\alpha})^T = \boldsymbol{\Gamma}^{P\beta\alpha} \tag{13.5.12c}$$

or

$$\Gamma^{P\alpha\beta}_{ijkl} = \Gamma^{P\beta\alpha}_{klij} = \Gamma^{P\beta\alpha}_{ijkl}; \tag{13.5.12d}$$

compare these with (9.3.5).[33] It should be noted that tensor $\mathbf{S}^P(\Omega_\alpha; \Omega_\beta)$ defined by (12.5.6) is given by

$$\mathbf{S}^P(\Omega_\alpha; \Omega_\beta) = f_\beta \, \boldsymbol{\Gamma}^{P\alpha\beta} : \mathbf{C}, \tag{13.5.12e}$$

since $\mathcal{F}\mathbf{S}^P(\boldsymbol{\xi}) = \mathcal{F}\boldsymbol{\Gamma}^P(\boldsymbol{\xi}) : \mathbf{C}$; see (12.4.13a).

Taking advantage of the above-defined correlation tensors, use a piecewise constant approximation of the eigenstress field, where $\mathbf{s}^{*\alpha}$ is constant (tensor) in region Ω_α which has constant elasticity tensor $\mathbf{C}^\alpha$. This region is viewed as an inclusion, Ω_α.[34] Expressing (13.5.11b) in terms of the correlation tensors $\Gamma^{P\alpha\beta}$, define J' as a function of the constant eigenstresses, $\mathbf{s}^{*\alpha}$,

---

[32] As mentioned in Subsection 12.5, the correlation tensor $\boldsymbol{\Gamma}^{P\alpha\beta}$ does not change, if Ω_α and Ω_β are moved by the same rigid-body translation.

[33] Another correlation tensor, $\boldsymbol{\Lambda}^{P\alpha\beta}$ can be defined through the $\boldsymbol{\Lambda}^P$-integral operator. Since $\boldsymbol{\Lambda}^P$ is essentially the same as $\boldsymbol{\Gamma}^P$, tensor $\boldsymbol{\Lambda}^{P\alpha\beta}$ satisfies the same symmetry conditions as (13.5.12c,d).

[34] Note that each actual inclusion can be divided into a finite number of subregions, and then each subregion treated as an inclusion; see Iwakuma and Nemat-Nasser (1983), Accorsi and Nemat-Nasser (1986), and Nemat-Nasser *et al.* (1986). This method is practically useful when an incremental formulation for composites with nonlinear material properties is considered; see Subsection 12.8 and Appendix A of Part 1.

$$J'(\{\mathbf{s}^{*\alpha}\};\boldsymbol{\varepsilon}^{o}) \equiv \frac{1}{2}\sum_{\alpha=0}^{n}\sum_{\beta=0}^{n}\mathbf{s}^{*\alpha}:\mathbf{J}^{P\alpha\beta}:\mathbf{s}^{*\beta}-\bar{\mathbf{s}}^{*}:\boldsymbol{\varepsilon}^{o}, \tag{13.5.13a}$$

where

$$\bar{\mathbf{s}}^{*} = \sum_{\alpha=0}^{n} f_\alpha\,\mathbf{s}^{*\alpha}, \tag{13.5.14a}$$

and $\mathbf{J}^{P\alpha\beta}$ is defined by

$$\mathbf{J}^{P\alpha\beta} \equiv f_\alpha\,\delta_{\alpha\beta}\,(\mathbf{C}^{\alpha}-\mathbf{C})^{-1}+f_\alpha\,f_\beta\,\boldsymbol{\Gamma}^{P\alpha\beta} \qquad (\alpha \text{ not summed}). \tag{13.5.14b}$$

Note that $\mathbf{J}^{P\alpha\beta}$ is symmetric with respect to the superscripts and the first and second pairs of its subscripts, i.e., $\mathbf{J}^{P\alpha\beta} = \mathbf{J}^{P\beta\alpha} = (\mathbf{J}^{P\alpha\beta})^{T}$, where transpose, T, refers to the subscripts.

As in Subsections 9.3 and 9.4, J' defined by (13.5.13a) is quadratic (with linear terms) in $\{\mathbf{s}^{*\alpha}\}$. The optimal (or stationary) value of this quadratic form is computed by setting the corresponding derivative with respect to $\mathbf{s}^{*\beta}$ equal to zero, i.e.,

$$\frac{\partial J'}{\partial \mathbf{s}^{*\beta}}(\{\mathbf{s}^{*\alpha}\};\boldsymbol{\varepsilon}^{o}) = \sum_{\gamma=0}^{n}\mathbf{J}^{P\beta\gamma}:\mathbf{s}^{*\gamma}-f_\beta\,\boldsymbol{\varepsilon}^{o}$$

$$= f_\beta\left\{\sum_{\gamma=0}^{n}\{\delta_{\beta\gamma}(\mathbf{C}^{\beta}-\mathbf{C})^{-1}+f_\gamma\,\boldsymbol{\Gamma}^{P\beta\gamma}\}:\mathbf{s}^{*\gamma}-\boldsymbol{\varepsilon}^{o}\right\}. \tag{13.5.13b}$$

Hence, a set of linear equations, $\partial J'/\partial \mathbf{s}^{*\alpha} = \mathbf{0}$, is obtained, for the $\mathbf{s}^{*\alpha}$'s. The quadratic form J' is now expressed in terms of $\partial J'/\partial \mathbf{s}^{*\alpha}$, as

$$J'(\{\mathbf{s}^{*\alpha}\};\boldsymbol{\varepsilon}^{o}) = \frac{1}{2}\sum_{\beta=0}^{n}\mathbf{s}^{*\beta}:\{\frac{\partial J'}{\partial \mathbf{s}^{*\beta}}\}-\frac{1}{2}\bar{\mathbf{s}}^{*}:\boldsymbol{\varepsilon}^{o}. \tag{13.5.13c}$$

Denote the solution of the set of linear tensorial equations, $\partial J'/\partial \mathbf{s}^{*\beta}(\{\mathbf{s}^{*\alpha}\};\boldsymbol{\varepsilon}^{o}) = \mathbf{0}$, by $\{\boldsymbol{\sigma}^{*\alpha}\}$, and observe that the optimal value of J' is $-\bar{\boldsymbol{\sigma}}^{*}:\boldsymbol{\varepsilon}^{o}/2$, with $\bar{\boldsymbol{\sigma}}^{*}$ given by (13.5.14a), with $\mathbf{s}^{*\alpha} = \boldsymbol{\sigma}^{*\alpha}$.

Similarly to all the previous cases, the set of linear equations obtained from (13.5.13b), is the same as the averaged consistency conditions. Indeed, the right side of (13.5.13b) corresponds to (12.5.7b) which is derived by taking the volume average of the consistency condition over Ω_α[35]. However, (13.5.13b) is an approximation of the averaged consistency condition with piecewise constant eigenstresses. That is, the volume average of the consistency condition taken over Ω_α is

$$(\mathbf{C}^{\alpha}-\mathbf{C})^{-1}:<\boldsymbol{\sigma}^{*}>_\alpha+<\boldsymbol{\Gamma}(\boldsymbol{\sigma}^{*})>_\alpha-\boldsymbol{\varepsilon}^{o}=\mathbf{0} \qquad (\alpha \text{ not summed}), \tag{13.5.15a}$$

and the approximation of $\boldsymbol{\sigma}^{*}$ by $\sum_{\alpha=0}^{n} H_\alpha\,\boldsymbol{\sigma}^{*\alpha}$ yields

[35] Note that $\mathbf{S}^{P}(\Omega_\alpha;\,\Omega_\beta)$ in (12.5.7b) exactly corresponds to the correlation tensor $\boldsymbol{\Gamma}^{P\alpha\beta}$; see (13.5.12b).

$$(\mathbf{C}^{\alpha}-\mathbf{C})^{-1}:\boldsymbol{\sigma}^{*\alpha}+\sum_{\beta=0}^{n} f_{\beta}\,\boldsymbol{\Gamma}^{P\alpha\beta}:\boldsymbol{\sigma}^{*\beta}-\boldsymbol{\varepsilon}^{o}=\mathbf{0} \qquad (\alpha \text{ not summed}) \tag{13.5.15b}$$

which coincides with the equations obtained from (13.5.13b).

13.6. EXAMPLES

This subsection includes two examples which illustrate the application of a piecewise constant eigenstress field to the energy functional of the Hashin-Shtrikman variational principle. The first example is a one-dimensional periodic structure examined in Subsection 13.4. The exact solution is obtained by using a piecewise constant distribution of eigenstresses, as expected from the results of Subsection 13.4. The second example is a three-dimensional periodic structure with a unit cell which contains an ellipsoidal inclusion. The geometry of the inclusion varies from a sphere to an oblate or prolate ellipsoid. The upper and lower bounds of the overall moduli are computed.

13.6.1. Example (1): One-Dimensional Periodic Structure

As a simple illustrative example, consider the one-dimensional periodic structure examined in Subsection 13.4. The one-dimensional elasticity field is given by

$$C'(x)=H(x;\,M)\,C^{M}+H(x;\,\Omega)\,C^{\Omega}, \tag{13.6.1}$$

where C^M and C^Ω are the elasticity tensors of the matrix and inclusion phase; see Figure 13.4.1. The energy functional for this periodic structure is

$$J(\sigma^{*},\varepsilon^{o})=\frac{1}{2}<s^{*}\{(C'-C)^{-1}s^{*}+\Gamma^{P}(s^{*})-2\,\varepsilon^{o}\}>, \tag{13.6.2a}$$

where Γ^P is given by (13.5.10a), i.e.,

$$\begin{aligned}\Gamma^{P}(x;\,s^{*})&=\sum_{\xi}{}'\,\mathcal{F}\Gamma^{P}(\xi)\,\{\frac{1}{U}\int_{U}s^{*}(y)\exp(\iota\xi(x-y))\,dy\}\\&=C^{-1}\{s^{*}(x)-<s^{*}>\}.\end{aligned} \tag{13.6.2b}$$

As shown in Subsection 13.4.2, the exact eigenstress is piecewise constant. Hence, if a piecewise constant distribution of the eigenstresses is used for the energy functional J given by (13.5.17), then the exact solution results. Indeed, for a piecewise constant eigenstress of the following form:

$$s^{*}(x)=H(x;\,M)\,s^{*M}+H(x;\,\Omega)\,s^{*\Omega}, \tag{13.6.3}$$

with s^{*M} and $s^{*\Omega}$ constants, the energy functional becomes

$$J(s^*; \varepsilon^o) = \frac{1}{2}(1-f)\, s^{*M} \Big\{ (C^M - C)^{-1} s^{*M} + < \Gamma^P(s^*) >_M - 2\,\varepsilon^o \Big\}$$

$$+ \frac{1}{2} f\, s^{*\Omega} \Big\{ (C^\Omega - C)^{-1} s^{*\Omega} + < \Gamma^P(s^*) >_\Omega - 2\,\varepsilon^o \Big\}, \tag{13.6.4}$$

where $< \Gamma^P(s^*) >_M$ and $< \Gamma^P(s^*) >_\Omega$ are

$$< \Gamma^P(s^*) >_M = f\, C^{-1} s^{*M} - f\, C^{-1} s^{*\Omega},$$

$$< \Gamma^P(s^*) >_\Omega = -(1-f)\, C^{-1} s^{*M} + (1-f)\, C^{-1} s^{*\Omega}. \tag{13.6.5b,c}$$

The optimal conditions of this function are given by

$$\frac{\partial J}{\partial s^{*M}}(s^*; \varepsilon^o) = (1-f) \Big\{ (C^M - C)^{-1} s^{*M} + < \Gamma^P(s^*) >_M - \varepsilon^o \Big\} = 0,$$

$$\frac{\partial J}{\partial s^{*\Omega}}(s^*; \varepsilon^o) = f \Big\{ (C^\Omega - C)^{-1} s^{*\Omega} + < \Gamma^P(s^*) >_\Omega - \varepsilon^o \Big\} = 0. \tag{13.6.6a,b}$$

These two equations are the averages of the consistency condition taken over M and Ω, respectively; see (13.4.9b). Indeed, in view of (13.6.5b,c), (13.6.6a,b) yield

$$\{(C^M - C)^{-1} + C^{-1}\}\, s^{*M} - (\varepsilon^o + C^{-1} < s^* >) = 0,$$

$$\{(C^\Omega - C)^{-1} + C^{-1}\}\, s^{*\Omega} - (\varepsilon^o + C^{-1} < s^* >) = 0. \tag{13.6.7a,b}$$

These two equations are identical with (13.4.11). Hence, the solutions of (13.6.6a,b) coincide with the exact eigenstresses, σ^{*M} and $\sigma^{*\Omega}$, given by (13.4.12).

13.6.2. Example (2): Three-Dimensional Periodic Structure

Next, consider a three-dimensional two-phase composite which consists of matrix and inclusion phases, M and Ω, with elasticity tensors $\mathbf{C}^M$ and $\mathbf{C}^\Omega$. In order to approximately homogenize this composite, apply a piecewise constant eigenstress field, which equals $\mathbf{s}^{*M}$ in M and $\mathbf{s}^{*\Omega}$ in Ω. Hence, the elasticity and eigenstress fields are

$$\mathbf{C}'(\mathbf{x}) = H(\mathbf{x}; M)\, \mathbf{C}^M + H(\mathbf{x}; \Omega)\, \mathbf{C}^\Omega,$$

$$\mathbf{s}^*(\mathbf{x}) = H(\mathbf{x}; M)\, \mathbf{s}^{*M} + H(\mathbf{x}; \Omega)\, \mathbf{s}^{*\Omega}, \tag{13.6.8a,b}$$

where $H(\mathbf{x}; M)$ and $H(\mathbf{x}; \Omega)$ are the Heaviside step functions for M and Ω, respectively.

For $\mathbf{C}'$ and $\mathbf{s}^*$ given by (13.6.8a,b), the energy functional J given by (13.3.1) becomes an energy function J′,

$$J(\mathbf{s}^*; \boldsymbol{\varepsilon}^o) = J'(\mathbf{s}^{*M}, \mathbf{s}^{*\Omega}; \boldsymbol{\varepsilon}^o)$$

$$= \frac{1}{2} \{\mathbf{s}^{*M} : \mathbf{J}^{PMM} : \mathbf{s}^{*M} + 2\, \mathbf{s}^{*M} : \mathbf{J}^{PM\Omega} : \mathbf{s}^{*\Omega} + \mathbf{s}^{*\Omega} : \mathbf{J}^{P\Omega\Omega} : \mathbf{s}^{*\Omega}\}$$

$$-\bar{\mathbf{s}}^{*} : \boldsymbol{\varepsilon}^{o}, \qquad (13.6.9a)$$

and the stationary value of J′ is attained when

$$\frac{\partial J'}{\partial \mathbf{s}^{*M}} = \mathbf{J}^{PMM} : \mathbf{s}^{*M} + \mathbf{J}^{PM\Omega} : \mathbf{s}^{*\Omega} - (1-f)\, \boldsymbol{\varepsilon}^{o} = \mathbf{0},$$
$$\frac{\partial J'}{\partial \mathbf{s}^{*\Omega}} = \mathbf{J}^{P\Omega\Omega} : \mathbf{s}^{*\Omega} + \mathbf{J}^{PM\Omega} : \mathbf{s}^{*M} - f\, \boldsymbol{\varepsilon}^{o} = \mathbf{0}. \qquad (13.6.9b,c)$$

In (13.6.9a~c), $\bar{\mathbf{s}}^{*}$ and $\mathbf{J}^{P\alpha\beta}$'s are given by

$$\bar{\mathbf{s}}^{*} = (1-f)\, \mathbf{s}^{*M} + f\, \mathbf{s}^{*\Omega}, \qquad (13.6.10a)$$

and

$$\mathbf{J}^{PMM} = (1-f)\, (\mathbf{C}^{M} - \mathbf{C})^{-1} + (1-f)^2\, \boldsymbol{\Gamma}^{PMM},$$

$$\mathbf{J}^{P\Omega\Omega} = f\, (\mathbf{C}^{\Omega} - \mathbf{C})^{-1} + f^2\, \boldsymbol{\Gamma}^{P\Omega\Omega},$$

$$\mathbf{J}^{PM\Omega} = f\, (1-f)\, \boldsymbol{\Gamma}^{PM\Omega}, \qquad (13.6.10b\sim d)$$

where the correlation tensors, $\boldsymbol{\Gamma}^{P\alpha\beta}$'s, are defined by

$$\boldsymbol{\Gamma}^{P\alpha\beta} \equiv \sum_{\xi} g_{\alpha}(\boldsymbol{\xi})\, g_{\beta}(-\boldsymbol{\xi})\, F\boldsymbol{\Gamma}^{P}(\boldsymbol{\xi}), \qquad (13.6.11)$$

for α, β = M, Ω. In terms of $\partial J'/\partial \mathbf{s}^{*M}$ and $\partial J'/\partial \mathbf{s}^{*\Omega}$, the quadratic form J′ is given by

$$J'(\mathbf{s}^{*M}, \mathbf{s}^{*\Omega}; \boldsymbol{\varepsilon}^{o}) = \frac{1}{2} \left\{ \mathbf{s}^{*M} : \frac{\partial J'}{\partial \mathbf{s}^{*M}} + \mathbf{s}^{*\Omega} : \frac{\partial J'}{\partial \mathbf{s}^{*\Omega}} \right\} - \frac{1}{2} \bar{\mathbf{s}}^{*} : \boldsymbol{\varepsilon}^{o}; \qquad (13.6.9d)$$

for detailed derivations, see (13.5.12), (13.5.13), and (13.5.14) of Subsection 13.5.3. Since the correlation tensors can be computed, the set of tensorial equations (13.6.9b,c) can be solved directly.

It should be noted that for a two-phase composite, the correlation tensors given by (13.6.11) are related. Since the g-integrals, g_M and g_Ω, satisfy

$$(1-f)\, g_M(\boldsymbol{\xi}) + f\, g_{\Omega}(\boldsymbol{\xi}) = 0 \qquad \text{for } \boldsymbol{\xi} \neq \mathbf{0}, \qquad (13.6.12)$$

$\boldsymbol{\Gamma}^{PMM}$ and $\boldsymbol{\Gamma}^{PM\Omega}$ are expressed in terms of $\boldsymbol{\Gamma}^{P\Omega\Omega}$, as

$$\boldsymbol{\Gamma}^{PMM} = \left[\frac{f}{1-f} \right]^2 \boldsymbol{\Gamma}^{P\Omega\Omega}, \qquad \boldsymbol{\Gamma}^{PM\Omega} = -\frac{f}{1-f}\, \boldsymbol{\Gamma}^{P\Omega\Omega}. \qquad (13.6.13a,b)$$

If $\mathbf{C}^{M} - \mathbf{C}$ and $\mathbf{C}^{\Omega} - \mathbf{C}$ are positive-definite (negative-definite), the solution of (13.6.9b,c) gives lower bounds (upper bounds) for the overall elasticity tensor $\bar{\mathbf{C}}$. For simplicity, assume that the inclusions are stiffer than the matrix, i.e., $\mathbf{C}^{\Omega} - \mathbf{C}^{M}$ is positive-definite. Then, the reference elasticity $\mathbf{C}$ is set to be either $\mathbf{C}^{M}$ or $\mathbf{C}^{\Omega}$. Computing the stationary value of J′ for these cases, obtain the upper and lower bounds for $\bar{\mathbf{C}}$, as follows:

Case (1), $\mathbf{C} = \mathbf{C}^M$: since $\mathbf{s}^{*M}$ is zero, the set of tensorial equations (13.6.9b,c) reduces to the following tensorial equation for $\mathbf{s}^{*\Omega}$:

$$(\mathbf{C}^\Omega - \mathbf{C}^M)^{-1} : \mathbf{s}^{*\Omega} + f\,\mathbf{\Gamma}^{P\Omega\Omega}(\mathbf{C}^M) : \mathbf{s}^{*\Omega} = \boldsymbol{\varepsilon}^o, \tag{13.6.14a}$$

where $\mathbf{C}^M$ in the argument of $\mathbf{\Gamma}^{P\Omega\Omega}$ emphasizes that $\mathbf{\Gamma}^{P\Omega\Omega}$ is computed for the reference elasticity $\mathbf{C} = \mathbf{C}^M$. From (13.6.14a), $\mathbf{s}^{*\Omega}$ is obtained, as

$$\mathbf{s}^{*\Omega} = \{(\mathbf{C}^\Omega - \mathbf{C}^M)^{-1} + f\,\mathbf{\Gamma}^{P\Omega\Omega}(\mathbf{C}^M)\}^{-1} : \boldsymbol{\varepsilon}^o. \tag{13.6.14b}$$

Thus, the lower bound for the overall elasticity, denoted by $\overline{\mathbf{C}}^-$, is

$$\overline{\mathbf{C}}^- \equiv \mathbf{C}^M + f\,\{(\mathbf{C}^\Omega - \mathbf{C}^M)^{-1} + f\,\mathbf{\Gamma}^{P\Omega\Omega}(\mathbf{C}^M)\}^{-1}. \tag{13.6.14c}$$

Case (2), $\mathbf{C} = \mathbf{C}^\Omega$: since $\mathbf{s}^{*\Omega}$ is zero, (13.6.9b,c) reduces to

$$(\mathbf{C}^M - \mathbf{C}^\Omega)^{-1} : \mathbf{s}^{*M} + (1-f)\,\mathbf{\Gamma}^{PMM}(\mathbf{C}^\Omega) : \mathbf{s}^{*M} = \boldsymbol{\varepsilon}^o, \tag{13.6.15a}$$

from which $\mathbf{s}^{*M}$ is

$$\mathbf{s}^{*M} = \{(\mathbf{C}^M - \mathbf{C}^\Omega)^{-1} + (1-f)\,\mathbf{\Gamma}^{PMM}(\mathbf{C}^\Omega)\}^{-1} : \boldsymbol{\varepsilon}^o, \tag{13.6.15b}$$

and the upper bound for the overall elasticity, denoted by $\overline{\mathbf{C}}^+$, is

$$\overline{\mathbf{C}}^+ \equiv \mathbf{C}^\Omega + (1-f)\,\{(\mathbf{C}^M - \mathbf{C}^\Omega)^{-1} + (1-f)\,\mathbf{\Gamma}^{PMM}(\mathbf{C}^\Omega)\}^{-1}. \tag{13.6.15c}$$

From (13.6.14c) and (13.6.15c), the bounds of the overall elasticity are explicitly obtained, as follows:

$$\boldsymbol{\varepsilon}^o : \overline{\mathbf{C}}^- : \boldsymbol{\varepsilon}^o \le \boldsymbol{\varepsilon}^o : \overline{\mathbf{C}} : \boldsymbol{\varepsilon}^o \le \boldsymbol{\varepsilon}^o : \overline{\mathbf{C}}^+ : \boldsymbol{\varepsilon}^o. \tag{13.6.16}$$

Now, as a simple example, consider a cubic unit cell which contains a sphere, an oblate spheroid, and a prolate spheroid at its center. To obtain results for different volume fractions of the inclusion phase, the size of the inclusion is increased until it touches the edges of the cube, while the shape of the inclusion remains self-similar: for an oblate spheroid, $b_2/b_1 = 1.00$ and $b_3/b_1 = 0.25$; and for a prolate spheroid, $b_2/b_1 = 0.25$ and $b_3/b_1 = 0.25$; see Figure 13.6.1.

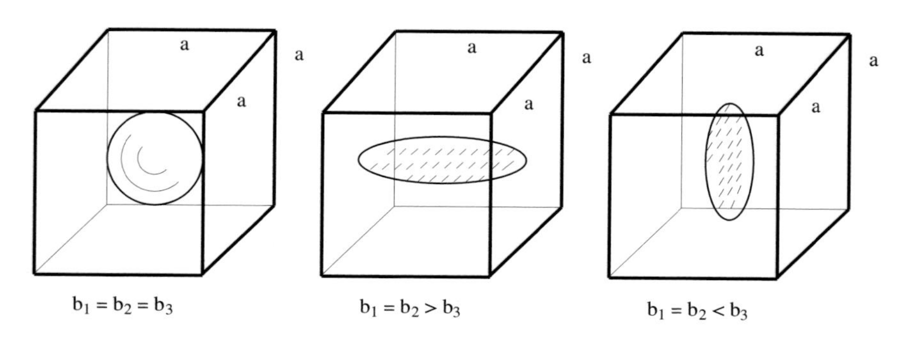

Figure 13.6.1

Cubic unit cell with sphere, oblate spheroid, and prolate spheroid

Figures 13.6.2 and 13.6.3 show the corresponding bounds for $\bar{C}_{iijj}$ and $\bar{C}_{ijij}$. The universal bounds for $\bar{C}_{iijj}$ obtained in Subsection 13.5 coincide with the bounds obtained for the sphere, since, due to the symmetry with respect to the $(x_1 = 0)$-, $(x_2 = 0)$-, and $(x_3 = 0)$-planes, the optimal eigenstress given by (13.6.14b) and (13.6.15b) satisfies $\sigma_{11}^{*\Omega} = \sigma_{22}^{*\Omega} = \sigma_{33}^{*\Omega}$ and $\sigma_{11}^{*M} = \sigma_{22}^{*M} = \sigma_{33}^{*M}$.

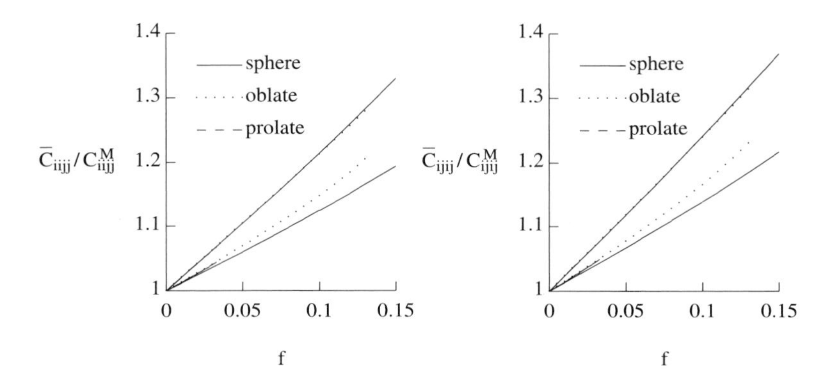

Figure 13.6.2

Estimate of upper and lower bounds for $\bar{C}_{iijj}$ and $\bar{C}_{ijij}$ of periodic structure containing sphere, oblate spheroid ($b_2/b_1 = 1$, $b_3/b_1 = 0.25$), and prolate spheroid ($b_2/b_1 = 0.25$, $b_3/b_1 = 0.25$)

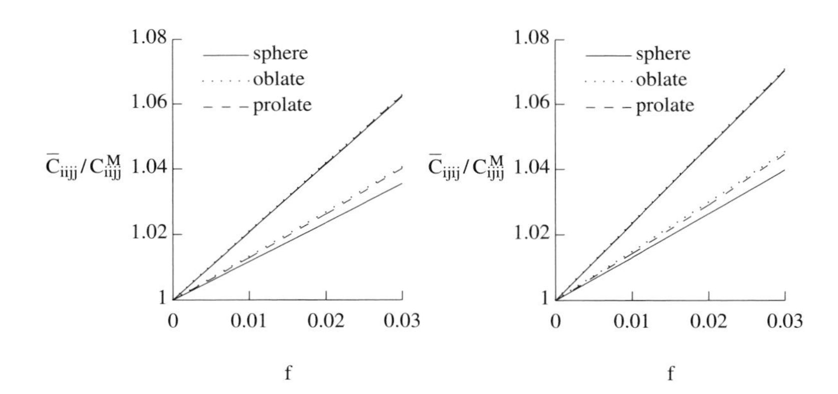

Figure 13.6.3

Estimate of upper and lower bounds for $\bar{C}_{iijj}$ and $\bar{C}_{ijij}$ of periodic structure containing sphere, oblate spheroid ($b_2/b_1 = 1$, $b_3/b_1 = 0.25$), and prolate spheroid ($b_2/b_1 = 0.25$, $b_3/b_1 = 0.25$)

The maximum value of f for the oblate spheroidal inclusion is around 0.03, which corresponds to the case when the inclusion just touches the walls of the unit cell. Figure 13.6.3 shows more details for this and other cases, when f is rather small; for additional results and illustrations, see Nemat-Nasser *et al.* (1993). Note that, as discussed in Section 10, when the volume fraction of the inclusion goes to zero, the overall moduli tend to approach the solution of the dilute distribution, which, in this case, coincides with the corresponding lower bound.

13.7. REFERENCES

Accorsi, M. L. and Nemat-Nasser, S. (1986), Bounds on the overall elastic and instantaneous elastoplastic moduli of periodic composites, *Mech. Matr.*, Vol. 5, No. 3, 209-220.

Hashin, Z. and Shtrikman, S. (1962), On some variational principles in anisotropic and nonhomogeneous elasticity, *J. Mech. Phys. Solids*, Vol. 10, 335-342.

Hill, R. (1963), Elastic properties of reinforced solids: Some theoretical principles, *J. Mech. Phys. Solids*, Vol. 11, 357-372.

Iwakuma, T. and Nemat-Nasser, S. (1983), Composites with periodic microstructure, *Advances and Trends in Structural and Solid Mechanics*, Pergamon Press, 13-19 or *Computers and Structures*, Vol. 16, Nos. 1-4, 13-19.

Milton, G. W. and Kohn, R. (1988), Variational bounds on the effective moduli of anisotropic composites, *J. Mech. Phys. Solids*, Vol. 36, 597-629.

Nemat-Nasser, S., Iwakuma, T., and Accorsi, M. (1986), Cavity growth and grain boundary sliding in polycrystalline solids, *Mech. Matr.*, Vol. 5, No. 4, 317-32.

Nemat-Nasser, S., Yu, N., and Hori, M. (1993), Bounds and estimates of overall moduli of composites with periodic microstructure, *Mech. Mat.*, Vol. 15, No. 3, 163-181.

Nemat-Nasser, S. and Hori, M. (1993), *Micromechanics: Overall properties of heterogeneous elastic solids*, Elsevier, Amsterdam.

Willis, J. R. (1977), Bounds and self-consistent estimates for the overall properties of anisotropic composites, *J. Mech. Phys. Solids*, Vol. 25, 185-202.

SECTION 14 MIRROR-IMAGE DECOMPOSITION OF PERIODIC FIELDS

As is illustrated, the general formulation of Sections 12 and 13 can be applied to solve a variety of problems based on periodic microstructure. In many of these problems, the appropriate boundary conditions which ensure periodicity and uniqueness of solution, can be established intuitively by inspection. Furthermore, in the actual calculation, the exponential representation can be used directly and, essentially, with impunity.

The fields associated with a solid of periodic microstructure, and the corresponding boundary conditions for the representative unit cell can be systematically decomposed into symmetric and antisymmetric fields, using a sequence of mirror-image reflections with respect to the rectangular Cartesian coordinate planes. This will then allow: (1) clear identification of the restrictions on the boundary data that must be prescribed for the unit cell, to ensure uniqueness of the solution; and (2) considerable reduction in the actual calculation, by exploiting various existing symmetry/antisymmetry properties of the Fourier series representation of the field quantities; see Fotiu and Nemat-Nasser (1995) for illustrations.

In this section, first the concept of mirror images of points and vectors is introduced, and then this concept is used to decompose tensor-valued functions defined on the unit cell, into their symmetric and antisymmetric parts. The decomposition is then applied to Fourier series representations of tensor-valued field quantities such as strain, stress, and elastic moduli, resulting in considerable economy in numerical computation and clarity in restrictions which must be imposed on the boundary data.

In this section the term *mirror image,* abbreviated to MI, is used to designate quantities obtained by means of reflection with respect to one, two, or three coordinate planes, in a well-defined manner. Furthermore, the term mirror image symmetry/antisymmetry is used to collectively refer to various decompositions of field quantities into parts which possess symmetry or antisymmetry properties with respect to reflections about various coordinate planes. For simplicity, the abbreviation MI sym/ant is used to designate this.

14.1. MIRROR IMAGES OF POSITION VECTORS AND VECTORS

The unit cell will continue to be defined by a parallelepiped with sides parallel to the rectangular Cartesian coordinate planes, as in (12.3.2), where the origin of the coordinate system is at the center of the parallelepiped. With respect to the coordinate planes, to any point within the unit cell, except the origin of the coordinates and points on the coordinate axes, can be associated a unique set of seven images obtained by a sequence of reflections with respect to the coordinate planes. Together with the initial point, a unique set of eight points, which are mirror images of each other, is obtained.

Superscripts ±1, ±2, ±3, and ±4 are introduced, to denote these mirror images; see Figure 14.1.1. Consider a typical point $\mathbf{x}$ and, for simplicity, assume (unless otherwise stated to the contrary) all its components are nonnegative. Identify this point by $\mathbf{x}^4$, i.e., set

$$\mathbf{x}^4 \equiv \mathbf{x}. \tag{14.1.1}$$

Then define three mirror images of $\mathbf{x}^4$, denoted by $\mathbf{x}^i$ which are obtained by the reflection of $\mathbf{x}^4$ with respect to the plane $x_i = 0$, for i = 1, 2, 3, i.e., set

$$(x_1^1, x_2^1, x_3^1) \equiv (-x_1^4, +x_2^4, +x_3^4),$$

$$(x_1^2, x_2^2, x_3^2) \equiv (+x_1^4, -x_2^4, +x_3^4),$$

$$(x_1^3, x_2^3, x_3^3) \equiv (+x_1^4, +x_2^4, -x_3^4). \tag{14.1.2~4}$$

In this notation, the superscript, i, ranging over positive integers 1, 2, 3, denotes the plane of reflection, so that all components of $\mathbf{x}^i$ are identical with the corresponding components of the original point $\mathbf{x} = \mathbf{x}^4$, except the ith component which is equal in magnitude to the ith component of $\mathbf{x}^4$, but has a reversed sign, i.e., $x_j^i = x_j^4$ for $j \neq i$, and $x_i^i = -x_i^4$ (i not summed).

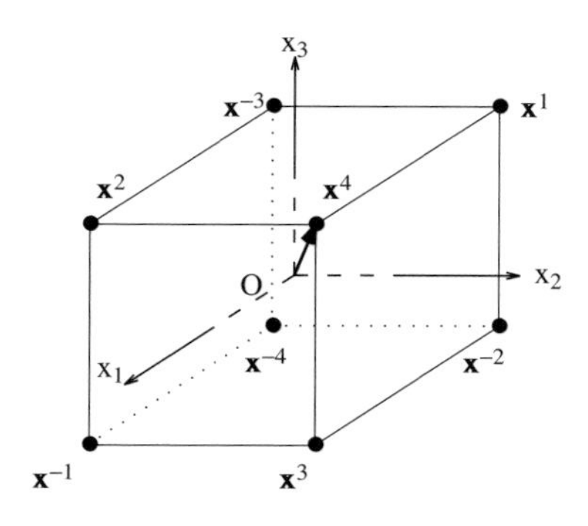

Figure 14.1.1

Eight mirror images $\mathbf{x}^i$ of position vector $\mathbf{x}$

Next consider images of the point $\mathbf{x}^4$, obtained by double reflection with respect to the coordinate planes. To this end, it is convenient to denote the point $-\mathbf{x}^m$ by $\mathbf{x}^{-m}$, and let $\mathbf{x}^{-1}$, $\mathbf{x}^{-2}$, or $\mathbf{x}^{-3}$ denote points obtained by double reflection of $\mathbf{x}^4$ with respect to the two planes $x_2 = 0$ and $x_3 = 0$, $x_3 = 0$ and $x_1 = 0$, or $x_1 = 0$ and $x_2 = 0$, respectively. In this notation, the negative superscript indicates the coordinate plane *excluded* in the double reflection; the double reflection refers to consecutive reflections with respect to two *distinct* coordinate planes. Thus, except for the ith component which is identical with the ith component of $\mathbf{x}^4$, the sign of the two other components of $\mathbf{x}^{-i}$ is reversed by the double reflection, i.e., $x_j^{-i} = -x_j^4$ for $j \neq i$, and $x_j^{-i} = x_j^4$ for $j = i$.

To complete the set, denote by $\mathbf{x}^{-4}$ the image obtained by three consecutive reflections with respect to the three coordinate planes $x_1 = 0$, $x_2 = 0$, and $x_3 = 0$, arriving at the diagonally symmetric image of $\mathbf{x}^4$. In this manner, for any given point, $\mathbf{x}$, in the unit cell, a unique set of eight points, $\mathbf{x}^i$ ($i = \pm1$, ..., ±4), is defined, which are mirror images of each other, with the original point denoted by $\mathbf{x}^4$. As shown in Figure 14.1.1, this set of eight points forms a parallelepiped, with each point at one of its corners. Points $\mathbf{x}^m$ and $\mathbf{x}^{-m}$ ($m = 1$, ..., 4) are centrally symmetric and located at the ends of the four diagonals of the parallelepiped. It should be noted that this way, a set of eight mirror-image points can also be formally defined for the origin of the coordinates or for a point on a coordinate axis, even though the eight images are not distinct.

The process of mirror imaging of points, described above, can conveniently be expressed in terms of a transformation induced by an eight by eight matrix with coefficients α^{ij} which take on values $+1$ or -1; see Table 14.1.1. For $j = 1, 2, 3$, α^{ij} is determined by the relation between the jth component of $\mathbf{x}^i$ and the jth component of $\mathbf{x}^4$. Definitions (14.1.1~4) then are collectively expressed as

$$x_j^i = \alpha^{ij}\, x_j^4 \qquad (i = \pm1, \pm2, \pm3, \pm4;\ j \text{ not summed};\ j = 1, 2, 3). \tag{14.1.5a}$$

Table 14.1.1

Mirror-image coefficients α^{ij}

	i = 4	3	2	1	−1	−2	−3	−4
j = 4	+1	+1	+1	+1	+1	+1	+1	+1
3	+1	−1	+1	+1	−1	−1	+1	−1
2	+1	+1	−1	+1	−1	+1	−1	−1
1	+1	+1	+1	−1	+1	−1	−1	−1
−1	+1	−1	−1	+1	+1	−1	−1	+1
−2	+1	−1	+1	−1	−1	+1	−1	+1
−3	+1	+1	−1	−1	−1	−1	+1	+1
−4	+1	−1	−1	−1	+1	+1	+1	−1

Hence, the components of α^{ij} are the identity transformation (i = 4), the single reflection (i = 1, 2, 3), the double reflection (i = − 1, − 2, − 3), and the triple reflection (i = − 4) for j = 1, 2, 3.

The definition of α^{ij} is now sought to be expanded to include elements for both i *and* j over the range ±1, ±2, ±3, and ±4; note that (14.1.5a) defines α^{ij} for j = 1, 2, 3, only. To this end, define α^{ij} for the remaining j = 4, − 1, − 2, − 3, and − 4, as follows:

$$\alpha^{i4} \equiv 1,$$

$$\alpha^{i-1} \equiv \alpha^{i2}\,\alpha^{i3}, \qquad \alpha^{i-2} \equiv \alpha^{i3}\,\alpha^{i1}, \qquad \alpha^{i-3} \equiv \alpha^{i1}\,\alpha^{i2},$$

$$\alpha^{i-4} \equiv \alpha^{i1}\,\alpha^{i2}\,\alpha^{i3} \qquad \text{(i not summed).} \qquad (14.1.5\text{b}{\sim}\text{f})$$

As is seen, α^{ij} defined by (14.1.5a~f) is symmetric,

$$\alpha^{ij} = \alpha^{ji} \qquad (i, j = \pm 1, ..., \pm 4). \qquad (14.1.6)$$

α^{ij} are called the *reflection coefficients.* Note again that, in (14.1.5a), j takes on only the values 1, 2, and 3. It is only in the definition of the reflection coefficients, α^{ij}, that i and j are allowed to range over ±1 to ±4, rendering the corresponding matrix symmetric.

The mirror image of a vector $\mathbf{v}$ is defined in exactly the same manner. Indeed, for a given vector $\mathbf{v}$, set

$$\mathbf{v}^4 \equiv \mathbf{v}, \qquad (14.1.7\text{a})$$

and using the reflection coefficients α^{ij}, introduce its reflections $\mathbf{v}^i$ by

$$v_j^i \equiv \alpha^{ij}\, v_j^4 \qquad \text{(j not summed),} \qquad (14.1.7\text{b})$$

for i = ±1, ±2, ±3, and ±4.

Attach the vector $\mathbf{v}^i$ to the point $\mathbf{x}^i$, and observe that it can be regarded as the mirror image of the vector $\mathbf{v}^4$ attached to the point $\mathbf{x}^4$, where *the corresponding reflections for both the vector and the point to which it is attached* (i.e., its position vector) *are made by the same set of reflection coefficients*. For example, the mirror image of $\mathbf{v}^4$ at $\mathbf{x}^4$ with respect to $x_1 = 0$ is $\mathbf{v}^1$ at $\mathbf{x}^1$; the double mirror image of $\mathbf{v}^4$ at $\mathbf{x}^4$ with respect to the planes $x_2 = 0$ and $x_3 = 0$ is $\mathbf{v}^{-1}$ at $\mathbf{x}^{-1}$; and the triple mirror image with respect to $x_1 = 0$, $x_2 = 0$, and $x_3 = 0$ is $\mathbf{v}^{-4}$ at $\mathbf{x}^{-4}$. Figure 14.1.2 shows the eight vectors $\mathbf{v}^i$ at the corresponding eight position vectors $\mathbf{x}^i$. Like $\mathbf{x}^i$, the vector $\mathbf{v}^i$ is called the ith mirror image of $\mathbf{v}$. It should be noted that the set of the above defined eight mirror images of a position vector (a vector) is *closed* in the sense that the ith mirror image of the jth mirror image is one of the eight mirror images for any i and j.

This subsection is closed by briefly commenting on mirror-image construction in two dimensions. In this case, the subscripts take on values 1, 2, and the superscripts, the values ±1, ±2, ±4. Denoting the original point $\mathbf{x}$, by $\mathbf{x}^4$, and using the same reflection coefficients defined by (14.1.5), observe that $x_j^i = \alpha^{ij}\, x_j^4$ for i = ±1, ±2, ±4 and j = 1, 2 (j not summed). Then, $\mathbf{x}^{-1}$ and $\mathbf{x}^{-2}$ coincide with

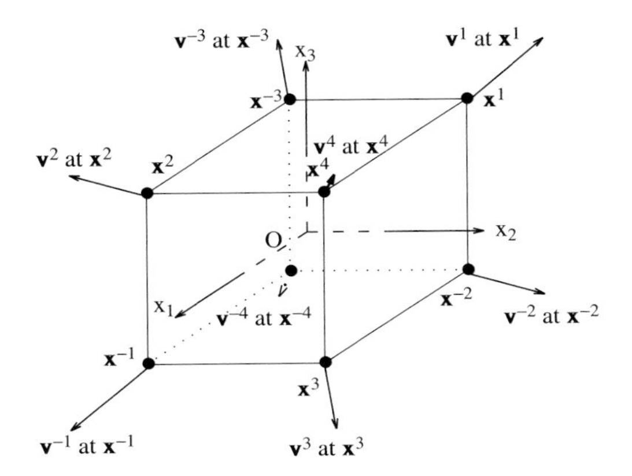

Figure 14.1.2

Eight vectors $\mathbf{v}^i$ at the corresponding eight position vectors $\mathbf{x}^i$

$\mathbf{x}^2$ and $\mathbf{x}^1$, respectively. Properties of the reflection coefficients for three dimensions, α^{ij} ($i, j = \pm 1, ..., \pm 4$), then do not change when one goes from three to two dimensions, for which $i, j = \pm 1, \pm 2$, and ± 4. Indeed, α^{ij} is defined for $j = 4, -1, -2, -4$ by $\alpha^{i4} = 1$, $\alpha^{i-1} = \alpha^{i2}$, $\alpha^{i-2} = \alpha^{i1}$, and $\alpha^{i-4} = \alpha^{i1}\,\alpha^{i2}$ (i not summed), and α^{ij} remains symmetric. Then, the identity transformation results for $i = 4$, and α^{-4i} defines the double reflection. There are four mirror images for any point (except for the origin and for points on the coordinate axes) one at each corner of a corresponding rectangular parallelogram.[1]

In a similar manner, mirror images of two-dimensional vectors attached to points in a plane are defined. Here again, set $\mathbf{v}^4 \equiv \mathbf{v}$ and $v_j^i = \alpha^{ij}\, v_j^4$ for $i = \pm 1, \pm 2$, and ± 4, $j = 1, 2$ (j not summed). Figure 14.1.3 shows the four mirror images of $\mathbf{v}^4 = \mathbf{v}$, each attached to the corresponding mirror image of $\mathbf{x}^4 = \mathbf{x}$.

[1] In general, 2^n mirror images can be obtained for an n-dimensional vector, $(v_1, v_2, ..., v_n)$, by changing the sign of each component, i.e., $(\pm v_1, \pm v_2, ..., \pm v_n)$. The mirror image can be written $(\hat{\alpha}_1 v_1, \hat{\alpha}_2 v_2, ..., \hat{\alpha}_n v_n)$, with $\hat{\alpha}_p = \pm 1$ for $p = 1, 2, ..., n$. These $\hat{\alpha}_p$'s are the reflection coefficients of an n-dimensional vector; $\hat{\alpha}^p = +1$ or $\hat{\alpha}^p = -1$ corresponds to the MI sym or MI ant with respect to $x_p = 0$ for each p. Indeed, in the three-dimensional case, α^{ij} can be rewritten as $\hat{\alpha}_1\,\hat{\alpha}_2\,\hat{\alpha}_3$. However, in the three-dimensional or two-dimensional cases considered here, the notation α^{ij} tends to reduce the number of required superscripts or subscripts for denoting the mirror images.

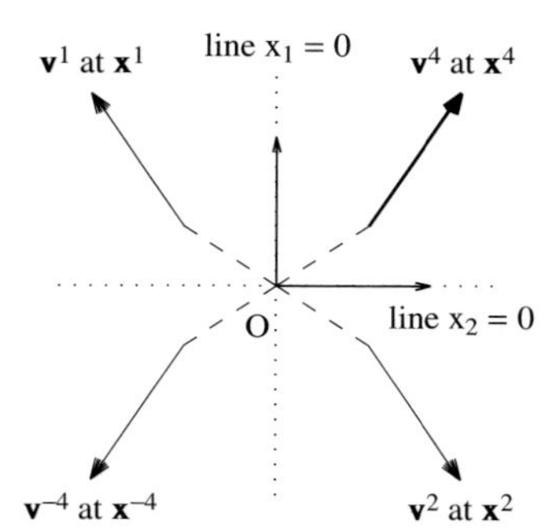

Figure 14.1.3

Mirror images of a vector $\mathbf{v} \equiv \mathbf{v}^4$ attached at the point $\mathbf{x} \equiv \mathbf{x}^4$

14.2. MIRROR-IMAGE SYMMETRY/ANTISYMMETRY OF TENSOR FIELDS

Now consider in a unit cell, a general tensor field that satisfies particular symmetry/antisymmetry conditions with respect to positions and associated vectors. The basic concept is similar to that for a scalar-valued function of a one-dimensional spatial argument. Let f(x) be such a scalar-valued function. Then, its symmetric and antisymmetric parts with respect to the origin $x = 0$ are defined by

$$f^{sym}(x) \equiv \frac{1}{2}\{f(x)+f(-x)\}, \qquad f^{ant}(x) \equiv \frac{1}{2}\{f(x)-f(-x)\}. \tag{14.2.1a,b}$$

These satisfy

$$f^{sym}(-x) = +f^{sym}(x), \qquad f^{ant}(-x) = -f^{ant}(x), \tag{14.2.1c,d}$$

and therefore,

$$f(x) = f^{sym}(x) + f^{ant}(x). \tag{14.2.1e}$$

For a tensor field in a multi-dimensional space, a similar decomposition involves both the position argument and an associated set of vector arguments, as is discussed below.

14.2.1. Mirror-Image (MI) Sym/Ant of Tensor Fields

Let $\mathbf{T} = \mathbf{T}(\mathbf{x})$ be an nth-order tensor field defined on a unit cell. Regard $\mathbf{T}$ as a linear operator which maps sets of n vectors, $\mathbf{v}$, ..., $\mathbf{w}$, into real numbers,

$$T_{q_1 \cdots q_n}(\mathbf{x})\, v_{q_1} \cdots w_{q_n} \rightarrow R, \tag{14.2.2}$$

where R is the set of real numbers. Replacing both the position and the vector arguments by their mirror images with respect to the plane $x_1 = 0$, define the following two tensor fields, $\mathbf{T}^{sym}$ and $\mathbf{T}^{ant}$:

$$T^{sym}_{q_1 \dots q_n}(\mathbf{x})\, v_{q_1} \dots w_{q_n} \equiv \frac{1}{2}\{T_{q_1 \dots q_n}(\mathbf{x}^4)\, v^4_{q_1} \dots w^4_{q_n} + T_{q_1 \dots q_n}(\mathbf{x}^1)\, v^1_{q_1} \dots w^1_{q_n}\},$$

$$T^{ant}_{q_1 \dots q_n}(\mathbf{x})\, v_{q_1} \dots w_{q_n} \equiv \frac{1}{2}\{T_{q_1 \dots q_n}(\mathbf{x}^4)\, v^4_{q_1} \dots w^4_{q_n} - T_{q_1 \dots q_n}(\mathbf{x}^1)\, v^1_{q_1} \dots w^1_{q_n}\}. \tag{14.2.3a,b}$$

The tensor fields $\mathbf{T}^{sym}$ and $\mathbf{T}^{ant}$ satisfy

$$T^{sym}_{q_1 \dots q_n}(\mathbf{x}^1)\, v^1_{q_1} \dots w^1_{q_n} = +\, T^{sym}_{q_1 \dots q_n}(\mathbf{x}^4)\, v^4_{q_1} \dots w^4_{q_n},$$

$$T^{ant}_{q_1 \dots q_n}(\mathbf{x}^1)\, v^1_{q_1} \dots w^1_{q_n} = -\, T^{ant}_{q_1 \dots q_n}(\mathbf{x}^4)\, v^4_{q_1} \dots w^4_{q_n}, \tag{14.2.3c,d}$$

for any $\mathbf{x}$ and $\mathbf{v}$, ..., $\mathbf{w}$, and therefore

$$\mathbf{T}(\mathbf{x}) = \mathbf{T}^{sym}(\mathbf{x}) + \mathbf{T}^{ant}(\mathbf{x}). \tag{14.2.3e}$$

The tensor fields $\mathbf{T}^{sym}$ and $\mathbf{T}^{ant}$ obtained in this manner are respectively, symmetric and antisymmetric with respect to the plane $x_1 = 0$.

The above example is followed and the mirror-image symmetry and antisymmetry for a tensor field $\mathbf{T}$ are defined. If $\mathbf{T}$ satisfies

$$T_{q_1 \dots q_n}(\mathbf{x}^i)\, v^i_{q_1} \dots w^i_{q_n} = \alpha\, T_{q_1 \dots q_n}(\mathbf{x}^4)\, v^4_{q_1} \dots w^4_{q_n} \quad \text{(i not summed)}, \tag{14.2.4}$$

then $\mathbf{T}$ is called *mirror-image symmetric* (*MI sym*) when $\alpha = +1$, and *mirror-image antisymmetric* (*MI ant*) when $\alpha = -1$, with respect to the plane $x_i = 0$, for $i = 1, 2, 3$. Note that, here, MI sym/ant is introduced for a tensor field and not for its components; this will be examined further below. In definition (14.2.4), the reflections of both the position and the vectors attached to that position are involved. The range of the superscript i in (14.2.4) can be easily extended to include double and triple reflections in the manner discussed in Subsection 14.1. It is this extended definition of symmetry and antisymmetry that is considered in the following.

14.2.2. MI Sym/Ant Decomposition of Tensor Fields

Using the reflection coefficients α^{ij}, introduce eight tensor fields, $\mathbf{T}^i$ ($i = \pm 1, \dots, \pm 4$), for any given tensor field $\mathbf{T}$, as follows:

$$T^i_{q_1 \dots q_n}(\mathbf{x})\, v_{q_1} \dots w_{q_n} \equiv \frac{1}{8} \sum_{j=-4}^{4} \alpha^{ij}\, T_{q_1 \dots q_n}(\mathbf{x}^j)\, v^j_{q_1} \dots w^j_{q_n}. \tag{14.2.5a}$$

Then, each $\mathbf{T}^i$ satisfies the MI sym/ant relation,

$$T^i_{q_1 \dots q_n}(\mathbf{x}^k)\, v^k_{q_1} \dots w^k_{q_n} \equiv \alpha^{ik}\, T^i_{q_1 \dots q_n}(\mathbf{x}^4)\, v^4_{q_1} \dots w^4_{q_n}, \quad \text{(i not summed)}, \tag{14.2.5b}$$

for $k = \pm 1, \pm 2, \pm 3$, and ± 4.

The proof is straightforward. As pointed out in Subsection 14.1, the jth mirror image of the kth mirror image of a position vector belongs to the set of its eight mirror images. Hence, there exists a certain j' in $\pm1, ..., \pm4$, such that

$$(\mathbf{x}^k)^j = \mathbf{x}^{j'} \qquad \text{for } k, j = \pm1, ..., \pm4. \tag{14.2.6a}$$

Then, from (14.1.5a),

$$\alpha^{ji}(\alpha^{ki}x_i^4) = \alpha^{j'i}x_i^4 \qquad \text{(i not summed)}. \tag{14.2.6b}$$

Since x_i^4 is arbitrary, $\alpha^{ji}\alpha^{ki} = \alpha^{j'i}$ ($i = \pm1, ..., \pm4$; i not summed). Therefore, from the symmetry of $\alpha^{ij} = \alpha^{ji}$, define j' in terms of j and k by

$$j' = j'(j;\, k); \;\; \alpha^{ij'} \equiv \alpha^{ij}\,\alpha^{ik} \qquad \text{(i not summed)}. \tag{14.2.6c}$$

From (14.2.6c), the reflection coefficient α^{ij} becomes

$$\alpha^{ij} = \alpha^{ij}(\alpha^{ik})^2 = \alpha^{ik}(\alpha^{ij}\alpha^{ik}) = \alpha^{ik}\alpha^{ij'}. \tag{14.2.7a}$$

Hence, in view of (14.2.5a), the left-hand side of (14.2.5b) becomes

$$\begin{aligned} T^i_{q_1 \cdots q_n}(\mathbf{x}^k)\, v^k_{q_1} \cdots w^k_{q_n} &= \frac{1}{8} \sum_{j=-4}^{4} \alpha^{ij}\, T_{q_1 \cdots q_n}((\mathbf{x}^k)^j)\, (\mathbf{v}^k)^j_{q_1} \cdots (\mathbf{w}^k)^j_{q_n} \\ &= \alpha^{ik} \left\{ \frac{1}{8} \sum_{j'=-4}^{4} \alpha^{ij'}\, T_{q_1 \cdots q_n}(\mathbf{x}^{j'})\, v^{j'}_{q_1} \cdots w^{j'}_{q_n} \right\} \\ &= \alpha^{ik}\, T^i_{q_1 \cdots q_n}(\mathbf{x}^4)\, v^4_{q_1} \cdots w^4_{q_n}. \end{aligned} \tag{14.2.7b}$$

Since $\mathbf{x}^4 = \mathbf{x}$, the MI sym/ant condition (14.2.5b) is proved.

Since MI sym/ant is defined for each plane $x_i = 0$ ($i = 1, 2, 3$), there are eight combinations of MI sym/ant. Indeed, for $\mathbf{T}^i$,

	$\mathbf{T}^4$	$\mathbf{T}^3$	$\mathbf{T}^2$	$\mathbf{T}^1$	$\mathbf{T}^{-1}$	$\mathbf{T}^{-2}$	$\mathbf{T}^{-3}$	$\mathbf{T}^{-4}$
$x_1 = 0$	sym	sym	sym	ant	sym	ant	ant	ant
$x_2 = 0$	sym	sym	ant	sym	ant	sym	ant	ant
$x_3 = 0$	sym	ant	sym	sym	ant	ant	sym	ant

These correspond to α^{ij} for $i = 1, 2, 3$, where $\alpha^{ij} = +1$ is regarded as MI sym and $\alpha^{ij} = -1$ is regarded as MI ant. Hence, the elements of the set $\{\mathbf{T}^i\}$ are linearly independent in the sense that

$$\sum_{i=-4}^{4} p_i\, \mathbf{T}^i(\mathbf{x}) = 0 \quad \Longleftrightarrow \quad p_{-4} = p_{-3} = ... = p_4 = 0. \tag{14.2.8a}$$

Therefore, the decomposition of the tensor field $\mathbf{T}$ into $\mathbf{T}^i$,

$$\mathbf{T}(\mathbf{x}) = \sum_{i=-4}^{4} \mathbf{T}^i(\mathbf{x}), \tag{14.2.8b}$$

is unique. Conversely, the set $\mathbf{T}^i$ determines a unique $\mathbf{T}$, and thus

$$\mathbf{T}(\mathbf{x}) \Longleftrightarrow \{\mathbf{T}^i(\mathbf{x})\}. \tag{14.2.8c}$$

The tensor field $\mathbf{T}^i$ is called the *ith MI sym/ant part* of $\mathbf{T}$, satisfying the *ith MI sym/ant condition* defined by the above.

14.2.3. Components of MI Sym/Ant Parts

From definition (14.2.5a), the components of the ith MI sym/ant part, $\mathbf{T}^i$, of the tensor field $\mathbf{T}$ in the x_i-coordinates are given by

$$T^i_{q_1 \dots q_n}(\mathbf{x}) \equiv \frac{1}{8} \sum_{j=-4}^{4} \alpha^{ij} T_{q_1 \dots q_n}(\mathbf{x})\, \alpha^{q_1 j} \dots \alpha^{q_n j}. \tag{14.2.9}$$

Note that these components are different from the components of the following tensor field:

$$\frac{1}{8} \sum_{j=-4}^{4} \alpha^{ij}\, \mathbf{T}(\mathbf{x}^j) \equiv \frac{1}{8} \sum_{j=-4}^{4} \alpha^{ij} T_{q_1 \dots q_n}(\mathbf{x}^j)\, \mathbf{e}_{q_1} \otimes \dots \otimes \mathbf{e}_{q_n} \tag{14.2.10}$$

which satisfies the symmetry or antisymmetry condition only for the position argument, but not for the corresponding vector arguments.

As discussed in Subsection 14.2.2, the product of α^{ij} with $\alpha^{q_1 j}$, ..., and $\alpha^{q_n j}$ corresponds to the n+1 sequence of the ith, the q_1th, ... , and the q_nth mirror image operations. Hence, i' and $\alpha^{i'j}$ are expressed as

$$i' = i'(i;\, q_1, \dots, q_n); \quad \alpha^{i'j} \equiv \alpha^{ij} \alpha^{q_1 j} \dots \alpha^{q_n j} \quad \text{(j not summed)}. \tag{14.2.11}$$

In terms of the above-defined i', (14.2.9) is rewritten as

$$T^i_{q_1 \dots q_n}(\mathbf{x}) = \frac{1}{8} \sum_{j=-4}^{4} \alpha^{i'j} T_{q_1 \dots q_n}(\mathbf{x}^j). \tag{14.2.12a}$$

Hence the components satisfy their own symmetry or antisymmetry conditions which are different from the MI sym/ant defined for $\mathbf{T}^i$. The symmetry or antisymmetry properties of the components are now given by

$$T^i_{q_1 \dots q_n}(\mathbf{x}^k) = \alpha^{i'k} T^i_{q_1 \dots q_n}(\mathbf{x}^4). \tag{14.2.12b}$$

The proof for (14.2.12b) is exactly the same as that for (14.2.5b).

From the definition of i', (14.2.11), the symmetry/antisymmetry of the components $T^i_{q_1 \dots q_n}$ depends on the superscript i as well as subscripts q_1, ..., q_n. For example, the fourth MI sym/ant parts, $\mathbf{a}^4$ and $\mathbf{A}^4$, of the first-order tensor field $\mathbf{a}$ and the second-order symmetric tensor field $\mathbf{A}$ ($A_{ij} = A_{ji}$) have the following components:

	a^4_1	a^4_2	a^4_3	A^4_{11}	A^4_{22}	A^4_{33}	A^4_{23}	A^4_{31}	A^4_{12}
$x_1 = 0$	ant	sym	sym	sym	sym	sym	sym	ant	ant
$x_2 = 0$	sym	ant	sym	sym	sym	sym	ant	sym	ant
$x_3 = 0$	sym	sym	ant	sym	sym	sym	ant	ant	sym

and the first MI sym/ant parts, $\mathbf{a}^1$ and $\mathbf{A}^1$, of $\mathbf{a}$ and $\mathbf{A}$ have the following components:

	a^1_1	a^1_2	a^1_3	A^1_{11}	A^1_{22}	A^1_{33}	A^1_{23}	A^1_{31}	A^1_{12}
$x_1 = 0$	sym	ant	ant	ant	ant	ant	ant	sym	sym
$x_2 = 0$	sym	ant	sym	sym	sym	sym	ant	sym	ant
$x_3 = 0$	sym	sym	ant	sym	sym	sym	ant	ant	sym

Note that with respect to the plane $x_1 = 0$, the respective "sym" and "ant" of a_i^1 and A_{ij}^1 are the reverse of those of a_i^4 and A_{ij}^4, while with respect to the planes $x_2 = 0$ and $x_3 = 0$, the respective "sym" and "ant" of a_i^1 and A_{ij}^1 are the same as those of a_i^4 and A_{ij}^4.

14.2.4. Operations on MI Sym/Ant Parts of Tensor Fields

From a physical point of view, the above-defined MI sym/ant parts, $\mathbf{T}^i$ ($i = \pm 1, ..., \pm 4$), of tensor field $\mathbf{T}$ have an intuitive appeal. Moreover, the gradient of $\mathbf{T}^i$ also satisfies the same MI sym/ant conditions. This allows for decomposing a physical field variable, defined in a unit cell, into eight independent parts, as shown later in Subsection[2] 14.4. Each part may be treated independently. The final results may then be combined to construct a complete solution.

From definition (14.2.9), the gradients of the components of $\mathbf{T}^i$ are computed as

$$\frac{\partial}{\partial x_k} T^i_{q_1 \ldots q_n}(\mathbf{x}) = \frac{1}{8} \sum_{j=-4}^{4} \alpha^{ij} \frac{\partial}{\partial x_k} T_{q_1 \ldots q_n}(\mathbf{x}^j)\, \alpha^{q_1 j} \ldots \alpha^{q_n j}. \tag{14.2.13a}$$

Since $\partial x_l^j / \partial x_k = \delta_{lk}\, \alpha^{kj}$ (k not summed),

$$\frac{\partial}{\partial x_k} T^i_{q_1 \ldots q_n}(\mathbf{x}) = \frac{1}{8} \sum_{j=-4}^{4} \alpha^{ij} \{\alpha^{kj} \frac{\partial}{\partial x_k} T_{q_1 \ldots q_n}(\mathbf{x}^j)\}\, \alpha^{q_1 j} \ldots \alpha^{q_n j}$$

$$= \{\frac{\partial}{\partial x_k} T_{q_1 \ldots q_n}\}^i(\mathbf{x}), \tag{14.2.13b}$$

where $\partial T_{q_1 \ldots q_n} / \partial x_k$ is regarded as an $(n+1)$th-order tensor field.

From (14.2.13b), it follows that the gradient, divergence, and curl of the ith MI sym/ant part of $\mathbf{T}(\mathbf{x})$, respectively, are the ith MI sym/ant part of the gradient, divergence, and curl of $\mathbf{T}(\mathbf{x})$. That is,

$$\boldsymbol{\nabla} \cdot (\mathbf{T}^i) = (\boldsymbol{\nabla} \cdot \mathbf{T})^i,$$

$$\boldsymbol{\nabla} \otimes (\mathbf{T}^i) = (\boldsymbol{\nabla} \otimes \mathbf{T})^i,$$

$$\boldsymbol{\nabla} \times (\mathbf{T}^i) = (\boldsymbol{\nabla} \times \mathbf{T})^i, \tag{14.2.14a~c}$$

for $i = \pm 1, \pm 2, \pm 3$, and ± 4.

For two arbitrary tensor fields, $\mathbf{S}(\mathbf{x})$ and $\mathbf{T}(\mathbf{x})$, the following identities hold with respect to the tensor product or the partial or full contraction of indices:

$$\mathbf{S}^i(\mathbf{x}^k) * \mathbf{T}^j(\mathbf{x}^k) = (\alpha^{ik}\, \alpha^{jk})\, \mathbf{S}^i(\mathbf{x}^4) * \mathbf{T}^j(\mathbf{x}^4), \tag{14.2.15a}$$

---

[2] Other definitions of symmetric or antisymmetric parts of a tensor field, say, (14.2.10), may not yield mathematical properties of this kind.

where $*$ denotes a tensorial operation. As is seen, $\mathbf{S}^i(\mathbf{x})*\mathbf{T}^j(\mathbf{x})$ satisfies the i′-th MI sym/ant condition; i′ is determined by the relation $\alpha^{ik}\alpha^{kj} = \alpha^{i'k}$. In particular, for i = 4,

$$\mathbf{S}^4(\mathbf{x}^k)*\mathbf{T}^j(\mathbf{x}^k) = \alpha^{jk}\,\mathbf{S}^4(\mathbf{x}^4)*\mathbf{T}^j(\mathbf{x}^4). \tag{14.2.15b}$$

Hence, the MI sym/ant condition of a tensor field $\mathbf{T}(\mathbf{x})$ is preserved with respect to the tensor product or contraction with the tensor, satisfying the fourth MI sym/ant condition, $\mathbf{S}^4(\mathbf{x})$, i.e.,

$$\mathbf{S}^4(\mathbf{x})*\mathbf{T}^j(\mathbf{x}) = (\mathbf{S}^4*\mathbf{T})^j(\mathbf{x}). \tag{14.2.15c}$$

A tensor field which satisfies the fourth MI sym/ant condition is fully symmetric with respect to reflection about all three coordinate planes.

14.3. MIRROR-IMAGE SYMMETRY AND ANTISYMMETRY OF FOURIER SERIES

In Subsection 14.2, the mirror-image symmetry and antisymmetry (MI sym/ant) of tensor fields have been defined, and the decomposition of a tensor field into eight linearly independent parts has been examined. Similar operations can be performed on the coefficients of Fourier series of tensor fields. This is examined in the present subsection.

14.3.1. MI Sym/Ant of Complex Kernel

The Fourier series expansion of periodic field variables has been expressed in terms of the complex kernel, $\exp(\iota\boldsymbol{\xi}\cdot\mathbf{x})$. In terms of cosine and sine functions, $\exp(\iota\boldsymbol{\xi}\cdot\mathbf{x})$ becomes

$$\exp(\iota\boldsymbol{\xi}\cdot\mathbf{x}) = c_1c_2c_3 - c_1s_2s_3 - s_1c_2s_3 - s_1s_2c_3$$
$$- \iota\{s_1s_2s_3 - s_1c_2c_3 - c_1s_2c_3 - c_1c_2s_3\}, \tag{14.3.1}$$

where c_i and s_i are

$$c_i = \cos(P_i), \qquad s_i = \sin(P_i), \qquad P_i \equiv \xi_i x_i \qquad \text{(i not summed)}. \tag{14.3.2a~c}$$

The functions c_i and s_i are respectively, symmetric and antisymmetric with respect to $P_i = 0$. Hence, (14.3.1) is the decomposition of $\exp(\iota\boldsymbol{\xi}\cdot\mathbf{x}) = \exp(\iota(P_1 + P_2 + P_3))$ into eight symmetric/antisymmetric parts with respect to $P_i = 0$, for i = 1, 2, 3.

Based on the above observation, using the reflection coefficients α^{ij}, define the MI sym/ant part of the kernel $\exp(\iota\boldsymbol{\xi}\cdot\mathbf{x})$ by

$$\exp^i(\iota\boldsymbol{\xi}\cdot\mathbf{x}) \equiv \frac{1}{8}\sum_{j=-4}^{4}\alpha^{ij}\exp(\iota\boldsymbol{\xi}\cdot\mathbf{x}^j); \tag{14.3.3a}$$

from (14.3.1), $\exp^i(\iota\boldsymbol{\xi}\cdot\mathbf{x})$'s are given by

$$\begin{aligned}
&\exp^4(\iota\boldsymbol{\xi}\cdot\mathbf{x}) = c_1c_2c_3, && \exp^{-4}(\iota\boldsymbol{\xi}\cdot\mathbf{x}) = -\iota\, s_1s_2s_3,\\
&\exp^3(\iota\boldsymbol{\xi}\cdot\mathbf{x}) = \iota\, c_1c_2s_3, && \exp^{-3}(\iota\boldsymbol{\xi}\cdot\mathbf{x}) = -\, s_1s_2c_3,\\
&\exp^2(\iota\boldsymbol{\xi}\cdot\mathbf{x}) = \iota\, c_1s_2c_3, && \exp^{-2}(\iota\boldsymbol{\xi}\cdot\mathbf{x}) = -\, s_1c_2s_3,\\
&\exp^1(\iota\boldsymbol{\xi}\cdot\mathbf{x}) = \iota\, s_1c_2c_3, && \exp^{-1}(\iota\boldsymbol{\xi}\cdot\mathbf{x}) = -\, c_1s_2s_3.
\end{aligned} \tag{14.3.3b~i}$$

The eight mirror images of the Fourier variable $\boldsymbol{\xi}$ can be defined in the ξ_1,ξ_2,ξ_3-space (or $\boldsymbol{\xi}$-space) with respect to the three planes, $\xi_1 = 0$, $\xi_2 = 0$, and $\xi_3 = 0$, in exactly the same manner as is done for the point $\mathbf{x}$ in the x_1,x_2,x_3-space (or $\mathbf{x}$-space). For simplicity, set

$$\boldsymbol{\xi}^4 = \boldsymbol{\xi}, \tag{14.3.4a}$$

and using the reflection coefficients α^{ij}, define $\boldsymbol{\xi}^i$ by

$$\xi^i_j = \alpha^{ij}\,\xi^4_j \qquad \text{(j not summed; j = 1, 2,3)}, \tag{14.3.4b}$$

for[3] $i = \pm1, \pm2, \pm3$, and ±4. Since $P_i = \xi_i x_i$ (i not summed), $\exp^i(\iota\boldsymbol{\xi}\cdot\mathbf{x})$ can be regarded as the ith MI sym/ant part of $\exp(\iota\boldsymbol{\xi}\cdot\mathbf{x})$ in the $\boldsymbol{\xi}$-space, instead of the $\mathbf{x}$-space, i.e.,

$$\exp^i(\iota\boldsymbol{\xi}\cdot\mathbf{x}) = \frac{1}{8}\sum_{j=-4}^{4} \alpha^{ij}\exp(\iota\boldsymbol{\xi}^j\cdot\mathbf{x}); \tag{14.3.3j}$$

compare with (14.3.3a). Hence, $\exp^i(\iota\boldsymbol{\xi}\cdot\mathbf{x})$ can be regarded as the MI sym/ant part of $\exp(\iota\boldsymbol{\xi}\cdot\mathbf{x})$ with respect to either $\mathbf{x}$ or $\boldsymbol{\xi}$.

Using the definition of the kth mirror image, $\mathbf{x}^k$, when the argument of $\exp^i$ is changed from $\iota\boldsymbol{\xi}\cdot\mathbf{x} = \iota\boldsymbol{\xi}^4\cdot\mathbf{x}^4$ to $\iota\boldsymbol{\xi}^k\cdot\mathbf{x}^4$, obtain

$$\begin{aligned}
\exp^i(\iota\boldsymbol{\xi}^k\cdot\mathbf{x}) &= \frac{1}{8}\sum_{j=-4}^{4} \alpha^{ij}\exp(\iota(\boldsymbol{\xi}^k)^j\cdot\mathbf{x})\\
&= \frac{1}{8}\sum_{j=-4}^{4} \alpha^{ik}\,\alpha^{ij}\exp(\iota\boldsymbol{\xi}^j\cdot\mathbf{x}).
\end{aligned} \tag{14.3.4c}$$

Equation (14.3.4c) remains valid if $\boldsymbol{\xi}$ and $\mathbf{x}$ are interchanged. Indeed, the following identity for $\exp^i(\iota\boldsymbol{\xi}\cdot\mathbf{x})$ follows from (14.3.4) (or from the general result (14.2.7b)):

$$\exp^i(\iota\boldsymbol{\xi}^k\cdot\mathbf{x}) = \exp^i(\iota\boldsymbol{\xi}\cdot\mathbf{x}^k) = \alpha^{ik}\exp^i(\iota\boldsymbol{\xi}\cdot\mathbf{x}) \qquad \text{(i not summed)}. \tag{14.3.5}$$

Identity (14.3.5) plays a key role in calculating the MI sym/ant of the Fourier coefficients in the next subsection.

14.3.2. MI Sym/Ant of Fourier Series

First, consider the Fourier series coefficients of the ith MI sym/ant part of a tensor field with respect to the kernel function $\exp(\iota\boldsymbol{\xi}\cdot\mathbf{x})$. As in the previous

[3] Unless stated otherwise, assume $\xi_i > 0$ for $\boldsymbol{\xi} = \boldsymbol{\xi}^4$ (i = 1, 2, 3).

subsection, let $\mathbf{T} = \mathbf{T}(\mathbf{x})$ be an nth-order tensor field, and let $\mathcal{F}\mathbf{T} = \mathcal{F}\mathbf{T}(\boldsymbol{\xi})$ be its Fourier series coefficients. In the same manner as $\mathcal{F}\mathbf{T}$ is defined for a given $\mathbf{T}$, the Fourier series coefficient of $\mathbf{T}^i = \mathbf{T}^i(\mathbf{x})$, denoted by $\mathcal{F}(\mathbf{T}^i) = \mathcal{F}(\mathbf{T}^i)(\boldsymbol{\xi})$, is defined by

$$\mathcal{F}(\mathbf{T}^i)(\boldsymbol{\xi}) \equiv \frac{1}{U}\int_U \mathbf{T}^i(\mathbf{x}) \exp(-\iota\boldsymbol{\xi}\cdot\mathbf{x})\, dV. \tag{14.3.6a}$$

Using the definition of $\mathbf{T}^i$, and taking advantage of $\boldsymbol{\xi}\cdot\mathbf{x} = \boldsymbol{\xi}^j\cdot\mathbf{x}^j$ (j not summed), compute the component of $\mathcal{F}(\mathbf{T}^i)$, as

$$\mathcal{F}(T^i_{q_1 \dots q_n})(\boldsymbol{\xi}) = \frac{1}{U}\int_U \{\frac{1}{8}\sum_{j=-4}^{4} \alpha^{i'j}\, T_{q_1 \dots q_n}(\mathbf{x}^j)\} \exp(-\iota\boldsymbol{\xi}\cdot\mathbf{x})\, dV_x,$$

$$= \frac{1}{8}\sum_{j=-4}^{4} \alpha^{i'j} \{\frac{1}{U}\int_U T_{q_1 \dots q_n}(\mathbf{x}^j) \exp(-\iota\boldsymbol{\xi}^j\cdot\mathbf{x}^j)\, dV_{\mathbf{x}}\}, \tag{14.3.6b}$$

where i' is defined by (14.2.11), and subscript $\mathbf{x}$ for dV emphasizes that the integral is taken with respect to $\mathbf{x}$, and not $\mathbf{x}^j$. Since the volume integral over U does not change if the integral argument $\mathbf{x}$ is replaced by $\mathbf{x}^j$, then

$$\mathcal{F}(T^i_{q_1 \dots q_n})(\boldsymbol{\xi}) = \frac{1}{8}\sum_{j=-4}^{4} \alpha^{i'j} \{\frac{1}{U}\int_U T_{q_1 \dots q_n}(\mathbf{x}) \exp(-\iota\boldsymbol{\xi}^j\cdot\mathbf{x})\, dV_{\mathbf{x}}\},$$

$$= \frac{1}{8}\sum_{j=-4}^{4} \alpha^{i'j}\, \mathcal{F}T_{q_1 \dots q_n}(\boldsymbol{\xi}^j), \tag{14.3.6c}$$

for $i = \pm1, \pm2, \pm3$, and ±4.

In view of (14.3.6c), define the ith MI sym/ant part of the Fourier series coefficient, denoted by $(\mathcal{F}\mathbf{T})^i(\boldsymbol{\xi})$, with respect to $\boldsymbol{\xi}$, as follows: regarding $\mathcal{F}\mathbf{T}(\boldsymbol{\xi})$ as an nth-order tensor field in the $\boldsymbol{\xi}$-space,[4] construct its eight MI sym/ant parts, $(\mathcal{F}\mathbf{T})^i(\boldsymbol{\xi})$, as

$$(\mathcal{F}T)^i_{q_1 \dots q_n}(\boldsymbol{\xi}) \equiv \frac{1}{8}\sum_{j=-4}^{4} \alpha^{i'j}\, \mathcal{F}T_{q_1 \dots q_n}(\boldsymbol{\xi}^j), \tag{14.3.7a}$$

for[5] $i = \pm1, \pm2, \pm3$, and ±4. Here, i' is also defined by (14.2.11), and hence, this definition is essentially the same as $\mathbf{T}^i(\mathbf{x})$ for $\mathbf{T}(\mathbf{x})$; see (14.2.12a). From definition (14.3.7a), $(\mathcal{F}\mathbf{T})^i$ satisfies

$$(\mathcal{F}\mathbf{T})^i_{q_1 \dots q_n}(\boldsymbol{\xi}^k) = \alpha^{i'k}\, (\mathcal{F}\mathbf{T})^i_{q_1 \dots q_n}(\boldsymbol{\xi}^4), \tag{14.3.7b}$$

in the same manner as $\mathbf{T}^i$ satisfies (14.2.12b).

From comparison of (14.3.6c) and (14.3.7a), it is seen that the Fourier series coefficient of the ith MI sym/ant part of the tensor field $\mathbf{T}(\mathbf{x})$ with respect to $\mathbf{x}$ coincides with the ith MI sym/ant part of the Fourier series coefficient $\mathcal{F}\mathbf{T}(\boldsymbol{\xi})$ with respect to $\boldsymbol{\xi}$. For simplicity, both $\mathcal{F}(\mathbf{T}^i)$ and $(\mathcal{F}\mathbf{T})^i$ are designated by

[4] Note that $\mathcal{F}\mathbf{T}(\boldsymbol{\xi})$ is not continuous in the $\boldsymbol{\xi}$-space, since its arguments take on discrete values, i.e., $\xi_p = n_p\pi/a_p$ (p not summed).

[5] Although $\mathbf{T}$ as a linear operator maps n given vectors to a real number, $\mathcal{F}\mathbf{T}$ as a linear operator maps n given vectors to a complex number, since the components of $\mathcal{F}\mathbf{T}$ are complex-valued.

$\mathcal{F}\mathbf{T}^i$, i.e.,

$$\mathcal{F}\mathbf{T}^i(\boldsymbol{\xi}) \equiv \mathcal{F}(\mathbf{T}^i)(\boldsymbol{\xi}) \equiv (\mathcal{F}\mathbf{T})^i(\boldsymbol{\xi}), \tag{14.3.8}$$

for i = ±1, ±2, ±3, and ±4.

Next, consider the Fourier series expansion in terms of the MI sym/ant parts of the kernel function, $\exp^j(\iota\boldsymbol{\xi}\cdot\mathbf{x})$. In essentially the same manner as the ith MI sym/ant part of the tensor field $\mathbf{T}^i$ is expanded in terms of $\exp(\iota\boldsymbol{\xi}\cdot\mathbf{x})$ in the above discussion, $\mathbf{T}^i$ can be expanded in terms of $\exp^j(\iota\boldsymbol{\xi}\cdot\mathbf{x})$. To demonstrate this, first expand the original tensor field $\mathbf{T}$ in terms of $\exp(\iota\boldsymbol{\xi}\cdot\mathbf{x})$, as

$$T_{q_1 \dots q_n}(\mathbf{x}) = \sum_{\boldsymbol{\xi}} \mathcal{F}T_{q_1 \dots q_n}(\boldsymbol{\xi}) \exp(\iota\boldsymbol{\xi}\cdot\mathbf{x}). \tag{14.3.9a}$$

Then, applying the MI sym/ant decomposition to the kernel function, $\exp(\iota\boldsymbol{\xi}\cdot\mathbf{x})$, obtain the expansion of the ith MI sym/ant part, $\mathbf{T}^i$, as

$$T^i_{q_1 \dots q_n}(\mathbf{x}) = \sum_{\boldsymbol{\xi}} \mathcal{F}T_{q_1 \dots q_n}(\boldsymbol{\xi}) \left\{\frac{1}{8} \sum_{j=-4}^{4} \alpha^{i'j} \exp(\iota\boldsymbol{\xi}\cdot\mathbf{x}^j)\right\}, \tag{14.3.9b}$$

where i' is defined by (14.2.11).

In (14.3.9b), the infinite summation with respect to $\boldsymbol{\xi}$ may be performed as follows: Select all $\boldsymbol{\xi}$'s with nonnegative components, i.e., $\xi_I \geq 0$, and order them. For simplicity, this $\boldsymbol{\xi}$ is denoted by $\boldsymbol{\xi}^4$. Although eight mirror images, $\boldsymbol{\xi}^k$ (k = ±1, ..., ±4), can be defined for $\boldsymbol{\xi}^4$, some of them may coincide with each other. The number of distinct mirror images is at most eight and at least one which corresponds to the origin, and is given by 2^K, where K is the number of nonzero components of $\boldsymbol{\xi}^4$, i.e., K = 0, 1, 2, or 3. Then, for each $\boldsymbol{\xi}^4$, sum all corresponding distinct mirror images, $\boldsymbol{\xi}^k$'s, to obtain a contribution associated with that $\boldsymbol{\xi}^4$, and sum all these terms over all $\boldsymbol{\xi}^4$'s. To represent the summation in accordance with the above procedure, write

$$\sum_{\boldsymbol{\xi}} (\dots)(\boldsymbol{\xi}) = \sum_{\boldsymbol{\xi}^4}{}^{+} \sum_{k=-4}^{4}{}^{*} (\dots)(\boldsymbol{\xi}^k), \tag{14.3.10a}$$

where superscripts + and * on the summation Σ denote that the sum is taken only for $\boldsymbol{\xi}^4$'s with nonnegative components, and that the sum is taken only for distinct $\boldsymbol{\xi}^k$'s. Furthermore, the sum for distinct $\boldsymbol{\xi}^k$'s is given by dividing the sum for all eight mirror images by $2^{3-K} = 8/2^K$, i.e.,

$$\sum_{k=-4}^{4}{}^{*} (\dots)(\boldsymbol{\xi}^k) = 2^K \left\{\frac{1}{8} \sum_{k=-4}^{4} (\dots)(\boldsymbol{\xi}^k)\right\}. \tag{14.3.10b}$$

Using (14.3.10a,b), and taking advantage of $\alpha^{ij} = \pm 1$ for any i and j and $\boldsymbol{\xi}^k\cdot\mathbf{x}^j = \boldsymbol{\xi}\cdot(\mathbf{x}^j)^k$, rewrite the Fourier series expansion (14.3.9b) as

$$T^i_{q_1 \dots q_n}(\mathbf{x}) = \sum_{\boldsymbol{\xi}^4}{}^{+} 2^K \left\{ \frac{1}{8} \sum_{k=-4}^{4} \mathcal{F}T_{q_1 \dots q_n}(\boldsymbol{\xi}^k) \left\{\frac{1}{8} \sum_{j=-4}^{4} \alpha^{i'j} \exp(\iota\boldsymbol{\xi}^k\cdot\mathbf{x}^j)\right\} \right\}$$

$$= \sum_{\boldsymbol{\xi}^4}{}^{+} 2^K \left\{ \frac{1}{8} \sum_{k=-4}^{4} (\alpha^{i'k})^2 \mathcal{F}T_{q_1 \dots q_n}(\boldsymbol{\xi}^k) \left\{\frac{1}{8} \sum_{j=-4}^{4} \alpha^{i'j} \exp(\iota\boldsymbol{\xi}\cdot(\mathbf{x}^j)^k)\right\} \right\}$$

$$= \sum_{\xi^4}^{+} 2^K \left\{ \frac{1}{8} \sum_{k=-4}^{4} \alpha^{i'k} \mathcal{F}T_{q_1 \dots q_n}(\boldsymbol{\xi}^k) \left\{ \frac{1}{8} \sum_{j'=-4}^{4} \alpha^{i'j'} \exp(\iota\boldsymbol{\xi} \cdot \mathbf{x}^{j'}) \right\} \right\}$$

$$= \sum_{\xi^4}^{+} 2^K \left\{ \frac{1}{8} \sum_{k=-4}^{4} \alpha^{i'k} \mathcal{F}T_{q_1 \dots q_n}(\boldsymbol{\xi}^k) \right\} \exp^{i'}(\iota\boldsymbol{\xi}^4 \cdot \mathbf{x}). \tag{14.3.9c}$$

Finally, using (14.3.7a) and (14.3.8), obtain

$$T^{i}_{q_1 \dots q_n}(\mathbf{x}) = \sum_{\xi^4}^{+} 2^K \mathcal{F}T^{i}_{q_1 \dots q_n}(\boldsymbol{\xi}^4) \exp^{i'}(\iota\boldsymbol{\xi}^4 \cdot \mathbf{x}). \tag{14.3.9d}$$

As is seen, the components of $\mathbf{T}^i$ can be expanded by the corresponding kernel $\exp^{i'}(\iota\boldsymbol{\xi} \cdot \mathbf{x})$, and the coefficient of the expansion is given by $2^K \mathcal{F}\mathbf{T}^i$, with $\mathcal{F}\mathbf{T}^i$ being the Fourier series expansion of $\mathbf{T}^i$ with respect to $\exp(\iota\boldsymbol{\xi} \cdot \mathbf{x})$.

Kernel function $\exp(\boldsymbol{\xi} \cdot \mathbf{x})$ satisfies the following orthonormality:

$$< \exp(\boldsymbol{\xi} \cdot \mathbf{x}) \exp(\boldsymbol{\zeta} \cdot \mathbf{x}) > = \begin{cases} 1 & \text{if } \boldsymbol{\xi} + \boldsymbol{\zeta} = \mathbf{0} \\ 0 & \text{otherwise.} \end{cases} \tag{14.3.11}$$

On the other hand, its ith MI sym/ant part, $\exp^{i}(\boldsymbol{\xi} \cdot \mathbf{x})$, satisfies a slightly more complicated orthogonality. Indeed, for $\boldsymbol{\xi}$ and $\boldsymbol{\zeta}$ with non-negative components,

$$< \exp^{i}(\boldsymbol{\xi} \cdot \mathbf{x}) \exp^{i}(\boldsymbol{\zeta} \cdot \mathbf{x}) > = 0 \quad \text{for } \boldsymbol{\xi} \neq \boldsymbol{\zeta} \quad \text{(i not summed)}, \tag{14.3.12a}$$

and

$$< (\exp^{i}(\boldsymbol{\xi} \cdot \mathbf{x}))^2 > = \begin{cases} 1/2^K & \text{if } \exp^{i}(\boldsymbol{\xi} \cdot \mathbf{x}) \neq 0 \\ 0 & \text{otherwise.} \end{cases} \tag{14.3.12b}$$

Therefore, from (14.3.12a,b), the Fourier series coefficient $\mathcal{F}T^{i}_{q_1 \dots q_n}(\boldsymbol{\xi}^4)$ is given by

$$\mathcal{F}T^{i}_{q_1 \dots q_n}(\boldsymbol{\xi}) = < T^{i}_{q_1 \dots q_n}(\mathbf{x}) \exp^{i'}(\iota\boldsymbol{\xi} \cdot \mathbf{x}) >. \tag{14.3.13}$$

As is seen, the factor 2^K for $\mathcal{F}\mathbf{T}^i$ in summation (14.3.9d) corresponds to the above properties of $\exp^{i}(\boldsymbol{\xi} \cdot \mathbf{x})$.

The real-valued tensor field $\mathbf{T}$ admits the following decomposition:

$$T_{q_1 \dots q_n}(\mathbf{x}) = \sum_{i=-4}^{4} T^{i}_{q_1 \dots q_n}(\mathbf{x})$$

$$= \sum_{i=-4}^{4} \left\{ \sum_{\xi^4}^{+} 2^K \mathcal{F}T^{i}_{q_1 \dots q_n}(\boldsymbol{\xi}^4) \exp^{i'}(\iota\boldsymbol{\xi}^4 \cdot \mathbf{x}) \right\}, \tag{14.3.14}$$

where the Fourier coefficients in the right-hand side are given by (14.3.13). As shown in Subsection 14.3.1, the MI sym/ant part of the kernel $\exp^{i}(\iota\boldsymbol{\xi} \cdot \mathbf{x})$ is either real or purely imaginary. Therefore, the Fourier coefficients $\mathcal{F}T^{i}_{q_1 \dots q_n}$ are real (imaginary) if $\exp^{i}(\iota\boldsymbol{\xi} \cdot \mathbf{x})$ is real (imaginary). Indeed, since $\mathcal{F}\mathbf{T}(-\boldsymbol{\xi})$ is the complex conjugate of $\mathcal{F}\mathbf{T}(\boldsymbol{\xi})$, and the reflection coefficients satisfy

$$\alpha^{i(-j)} = (\alpha^{i1}\alpha^{i2}\alpha^{i3})\,\alpha^{ij} = \alpha^{i(-4)}\,\alpha^{ij} \quad \text{(i not summed)}, \tag{14.3.15a}$$

for all j = 1, 2, 3, and 4, definition (14.3.7a) can be rewritten as

$$ {}_F T^{i}_{q_1 \dots q_n}(\boldsymbol{\xi}) = \frac{1}{8} \sum_{j=-4}^{4} \frac{\alpha^{i'j}}{2} \left\{ {}_F T_{q_1 \dots q_n}(\boldsymbol{\xi}^j) + \alpha^{i'-4} {}_F T_{q_1 \dots q_n}(-\boldsymbol{\xi}^j) \right\}. \tag{14.3.15b} $$

Therefore, the components of ${}_F\mathbf{T}^i$ are either real or purely imaginary,[6] depending on whether $\alpha^{i'-4}$ is $+1$ or -1.

14.4. BOUNDARY CONDITIONS FOR A UNIT CELL

In Subsections 14.1 and 14.2, mirror images are defined for position vectors and vectors, and MI sym/ant parts of a tensor field are introduced in terms of these mirror images. The MI sym/ant decomposition is applied to periodic field variables[7] $\mathbf{u}$, $\boldsymbol{\varepsilon}$, and $\boldsymbol{\sigma}$, defined in a unit cell.

14.4.1. Symmetry of Unit Cell

In Subsection 12.2, it is shown that the location of the unit cell relative to the coordinate axes can be chosen arbitrarily, while the shape of the unit cell (parallelepiped) is uniquely determined by the regularity vectors, $\mathbf{a}^{\alpha}$'s. *The location of the unit cell is chosen such that the eight parts of any field variable, decomposed through the* MI sym/ant operation *within the unit cell, are linearly independent.*

Based on the results prescribed in Subsection 14.2.4, the following governing field equations are immediately derived from (12.2.3a) and (12.2.3b), for the ith MI sym/ant parts of the field variables, $\mathbf{u}^i$, $\boldsymbol{\varepsilon}^i$, and $\boldsymbol{\sigma}^i$:

$$ \boldsymbol{\varepsilon}^i(\mathbf{x}) = \frac{1}{2}\{\nabla \otimes \mathbf{u}^i(\mathbf{x}) + (\nabla \otimes \mathbf{u}^i(\mathbf{x}))^T\}, \qquad \nabla \cdot \boldsymbol{\sigma}^i(\mathbf{x}) = \mathbf{0}. \tag{14.4.1a,b} $$

Let a general heterogeneous elasticity tensor, $\mathbf{C}' = \mathbf{C}'(\mathbf{x})$, be decomposed into eight MI sym/ant parts, as

$$ \mathbf{C}'(\mathbf{x}) = \sum_{j=-4}^{4} \mathbf{C}'^j(\mathbf{x}). \tag{14.4.2a} $$

Then, the second-order contraction between $\mathbf{C}'$ and $\boldsymbol{\varepsilon}^i$ is

$$ \mathbf{C}'(\mathbf{x}) : \boldsymbol{\varepsilon}^i(\mathbf{x}) = \sum_{j=-4}^{4} \mathbf{C}'^j(\mathbf{x}) : \boldsymbol{\varepsilon}^i(\mathbf{x}), \tag{14.4.2b} $$

where $\mathbf{C}'^j : \boldsymbol{\varepsilon}^i$ satisfies a distinct MI sym/ant condition depending on the value of j. In order that $\mathbf{C}' : \boldsymbol{\varepsilon}^i$ satisfy the ith MI sym/ant condition, $\mathbf{C}'$ must satisfy the fourth MI sym/ant condition, i.e., $\mathbf{C}'^i = \mathbf{0}$ for $i \neq 4$, and hence $\mathbf{C}' = \mathbf{C}'^4$. This

---

[6] It is also possible to show that ${}_F T^{i}_{q_1 \dots q_n}$ is either purely real or imaginary from its definition, without using the corresponding kernel function $\exp^i(\iota\boldsymbol{\xi}.\mathbf{x})$; the proof is straightforward if a fact that ${}_F T_{q_1 \dots q_n}(-\boldsymbol{\xi})$ is a complex conjugate of ${}_F T_{q_1 \dots q_n}(\boldsymbol{\xi})$ is used.

[7] For simplicity, superscript p for periodic fields is omitted in this subsection.

means that the unit cell must be both geometrically and elastically symmetric with respect to single, double, or triple reflections about all three coordinate planes. It then follows from (14.4.2b) that

$$\mathbf{C}'(\mathbf{x}) : \boldsymbol{\varepsilon}^{\mathrm{i}}(\mathbf{x}) = \boldsymbol{\sigma}^{\mathrm{i}}(\mathbf{x}), \tag{14.4.1c}$$

for i = ±1, ±2, ±3, and ±4.

As is clear from the three governing field equations, (14.4.1a~c), if the unit cell is such that $\mathbf{C}' \equiv \mathbf{C}'^4$, i.e., it is symmetric with respect to all three coordinate planes, then the ith MI sym/ant part of the stress field will depend only on the corresponding ith MI sym/ant part of the displacement and strain fields, i.e., no other MI sym/ant parts of these fields will be involved. In this case, the set of fields, $\{\mathbf{u}, \boldsymbol{\varepsilon}, \boldsymbol{\sigma}\}$, can be decomposed into eight mutually independent sets, $\{\mathbf{u}^{\mathrm{i}}, \boldsymbol{\varepsilon}^{\mathrm{i}}, \boldsymbol{\sigma}^{\mathrm{i}}\}$ for i = ±1, ±2, ±3, and ±4, with each set satisfying its own MI sym/ant condition.

It should be noted that decomposition to mutually independent sets of fields is still possible when $\mathbf{C}'$ satisfies a certain MI sym/ant condition other than $\mathbf{C}' = \mathbf{C}'^4$, but then the stress field corresponding to $\boldsymbol{\varepsilon}^{\mathrm{i}}$ (or $\mathbf{u}^{\mathrm{i}}$) will no longer be $\boldsymbol{\sigma}^{\mathrm{i}}$.

14.4.2. MI Sym/Ant Fields for a Symmetric Unit Cell

Now consider an alternative expression for a set of field variables which satisfy a particular MI sym/ant condition. To this end, for an arbitrary set of field variables, denoted by $G = G(\mathbf{x}) = \{\mathbf{u}(\mathbf{x}), \boldsymbol{\varepsilon}(\mathbf{x}), \boldsymbol{\sigma}(\mathbf{x})\}$, define the following four MI operators, $M^{(\mathrm{i})}$, which determine a set of *new* field variables from G:

$$M^{(\mathrm{i})}(G) \equiv M^{(\mathrm{i})}(\mathbf{x}; G) \equiv \{\mathbf{u}^{(\mathrm{i})}(\mathbf{x}), \boldsymbol{\varepsilon}^{(\mathrm{i})}(\mathbf{x}), \boldsymbol{\sigma}^{(\mathrm{i})}(\mathbf{x})\}, \tag{14.4.3a}$$

for i = 1, 2, 3, 4, where

$$\mathrm{u}_{\mathrm{p}}^{(\mathrm{i})}(\mathbf{x}) \equiv \alpha^{\mathrm{ip}}\, \mathrm{u}_{\mathrm{p}}(\mathbf{x}^{\mathrm{i}}),$$

$$\varepsilon_{\mathrm{pq}}^{(\mathrm{i})}(\mathbf{x}) \equiv \alpha^{\mathrm{ip}}\alpha^{\mathrm{iq}}\, \varepsilon_{\mathrm{pq}}(\mathbf{x}^{\mathrm{i}}),$$

$$\sigma_{\mathrm{pq}}^{(\mathrm{i})}(\mathbf{x}) \equiv \alpha^{\mathrm{ip}}\alpha^{\mathrm{iq}}\, \sigma_{\mathrm{pq}}(\mathbf{x}^{\mathrm{i}}), \qquad \text{(i, p, and q not summed)}. \tag{14.4.3b~d}$$

As is seen, $M^{(4)}(G)$ gives the original set of field variables, G. For i = 1, 2, 3, the set of field variables, $M^{(\mathrm{i})}(G)$, is the reflection of G with respect to the plane $\mathrm{x_i} = 0$. The set $M^{(\mathrm{i})}(G)$ is called the i*th MI reflection* of the set of the original fields G. By connecting (14.2.9) and the MI operation, it is proved that the ith MI sym/ant part of the original set of field variables is constructed by a suitable combination of its reflections.

Suppose that the three field variables $\{\mathbf{u}, \boldsymbol{\varepsilon}, \boldsymbol{\sigma}\}$ in U are such that they remain unchanged upon reflection about the plane $\mathrm{x_i} = 0$, i.e.,

$$\{\mathrm{u_p}(\mathbf{x}), \varepsilon_{\mathrm{pq}}(\mathbf{x}), \sigma_{\mathrm{pq}}(\mathbf{x})\}$$

$$= \{+\alpha^{ip}\,u_p(\mathbf{x}^i),\; +\alpha^{ip}\alpha^{iq}\,\varepsilon_{pq}(\mathbf{x}^i),\; +\alpha^{ip}\alpha^{iq}\,\sigma_{pq}(\mathbf{x}^i)\}, \tag{14.4.4a}$$

where i, p, and q are not summed; see Figure 14.4.1. Using the MI operation defined by (14.4.3), abbreviate this by $M^{(i)}(G) = +M^{(4)}(G)$.

Figure 14.4.1

Fields satisfying MI symmetry with respect to ($x_1 = 0$)-plane

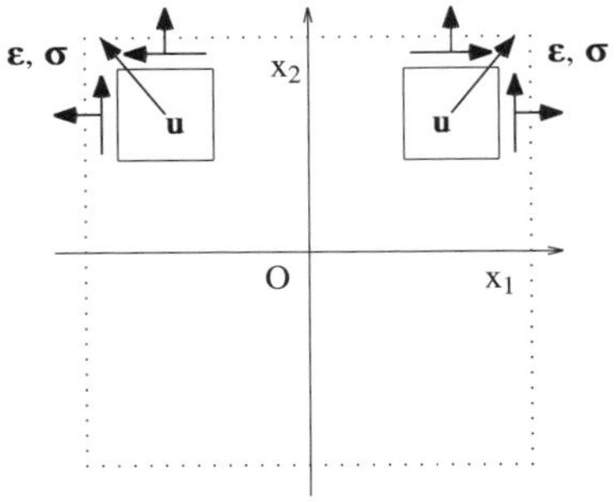

Similarly, if the three fields are such that their signs are reversed upon reflection about the plane $x_i = 0$, i.e., if

$$\{u_p(\mathbf{x}),\; \varepsilon_{pq}(\mathbf{x}),\; \sigma_{pq}(\mathbf{x})\}$$

$$= \{-\alpha^{ip}\,u_p(\mathbf{x}^i),\; -\alpha^{ip}\alpha^{iq}\,\varepsilon_{pq}(\mathbf{x}^i),\; -\alpha^{ip}\alpha^{iq}\,\sigma_{pq}(\mathbf{x}^i)\}, \tag{14.4.4b}$$

where i, p, and q are not summed, then abbreviate it by $M^{(i)}(G) = -M^{(4)}(G)$; see Figure 14.4.2. Therefore, *the notation* $M^{(4)}(G) = +\mathrm{M}^{(i)}(G)$, ($M^{(4)}(G) = -\mathrm{M}^{(i)}(G)$), means that the set of field variables, $G = \{\mathbf{u}, \boldsymbol{\varepsilon}, \boldsymbol{\sigma}\}$, is symmetric (antisymmetric), upon reflection with respect to the plane $x_i = 0$.

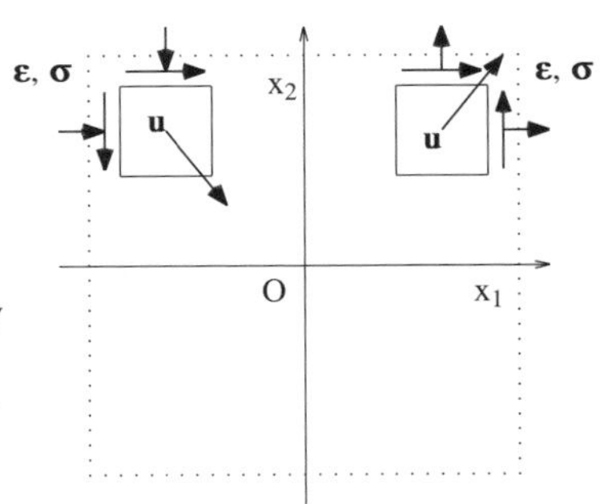

Figure 14.4.2

Fields satisfying MI anti-symmetry with respect to ($x_1 = 0$)-plane

Now, examine the relation between the ith reflection $M^{(i)}(G) = \{\mathbf{u}^{(i)}, \boldsymbol{\varepsilon}^{(i)}, \boldsymbol{\sigma}^{(i)}\}$ of the fields $G = \{\mathbf{u}, \boldsymbol{\varepsilon}, \boldsymbol{\sigma}\}$, and the ith MI sym/ant part of $G^i \equiv \{\mathbf{u}^i, \boldsymbol{\varepsilon}^i, \boldsymbol{\sigma}^i\}$ of these fields. Note that[8] the superscript i *without parentheses* refers to the ith MI sym/ant part of the corresponding fields, whereas superscript i *with parentheses* refers to their ith reflection, i.e., the reflection with respect to the plane $\mathbf{x}_i = 0$. From (14.2.5a) and (14.4.3b~d) it is easily seen that the ith MI sym/ant G^i of G is given by

$$G^i \equiv \frac{1}{8} \sum_{j=-4}^{4} \alpha^{ij} M^{(j)}(G), \tag{14.4.5a}$$

where the addition of a set of field variables is defined by a set of field variables which is obtained by adding the corresponding fields. Conversely, from $\sum_{j=-4}^{4} \alpha^{ij} \alpha^{kj} = 8\,\delta_{ik}$,

$$M^{(i)}(G) = \sum_{j=-4}^{4} \alpha_{ij}\, G^j. \tag{14.4.5b}$$

Therefore, it follows that the MI sym/ant part of a set of field variables, defined purely *mathematically*, coincides with the corresponding results of the MI operations. These results emerge from the *physical* operation of taking mirror images of actual field variables.

In particular, the following identity holds:

$$M^{(k)}(G^i) = \alpha^{ik}\, G^i \qquad \text{(i not summed)}. \tag{14.4.5c}$$

That is, the kth reflection of the ith MI sym/ant part G^i of the set of fields G, is given by $\alpha^{ik} G^i$ (i not summed). Furthermore, since $(\alpha^{ik})^2 = 1$, multiplying both sides of (14.4.5c) by α^{ik}, gives

$$G^i = \alpha^{ik} M^{(k)}(G^i), \tag{14.5.5d}$$

or

$$G^i = M^{(4)}(G^i) = \alpha^{i1} M^{(1)}(G^i) = \alpha^{i2} M^{(2)}(G^i) = \alpha^{i3} M^{(3)}(G^i) \quad \text{(i not summed)}. \tag{14.4.5e}$$

Hence, G^i is essentially unchanged through the MI operations.

14.4.3. Surface Data for MI Sym/Ant Set of Periodic Fields in a Symmetric Unit Cell

From now on, it is assumed that the unit cell is completely symmetric with respect to all three coordinate planes. Then, $\mathbf{C}'$ satisfies the fourth MI sym/ant condition, i.e., $\mathbf{C}' = \mathbf{C}'^4$. Now consider the surface tractions and displacements of the MI sym/ant sets of periodic fields in a *fully symmetric unit cell*. For simplicity, the implications of MI sym/ant conditions for periodic fields are examined with respect to, say, the plane $x_1 = 0$, i.e., the implications of $M^{(4)}(G) = +M^{(1)}(G)$ or $M^{(4)}(G) = -M^{(1)}(G)$ are examined. Let $\mathbf{u}$ and $\boldsymbol{\sigma}$ be the periodic

---

[8] To clarify these differences, consider a one-dimensional scalar-valued function $f = f(x)$. The MI reflection corresponds to defining new functions, say, $M^{(+)}(f) \equiv f(+x)$ and $M^{(-)}(f) \equiv f(-x)$, while the MI sym/ant part corresponds to the symmetric and antisymmetric parts, $\{f(+x)+f(-x)\}/2$ and $\{f(+x)-f(-x)\}/2$, which are denoted by f^{sym} and f^{ant}, respectively. Then, $f^{sym} = (M^{(+)}(f)+M^{(-)}(f))/2$ or $f^{ant} = (M^{(+)}(f)-M^{(-)}(f))/2$, and $M^{(+)}(f) = f^{sym}+f^{ant}$ and $M^{(-)}(f) = f^{sym}-f^{ant}$. For the set of three-dimensional tensor fields considered here, similar relations between the MI reflection and MI sym/ant parts are obtained by using the MI operation defined by (14.4.3).

displacement and stress fields, and $\mathbf{t} = \boldsymbol{\nu} \cdot \boldsymbol{\sigma}$ be the periodic tractions on the surface boundaries $x_1 = \pm a_1$, where $\boldsymbol{\nu} = \pm \mathbf{e}_1$. For $\mathbf{x}$ on the plane $x_1 = a_1$, and since $\mathbf{x}^4 - \mathbf{x}^1 = 2a_1 \mathbf{e}_1$, the periodicity of the boundary data requires $\mathbf{u}$ and $\boldsymbol{\sigma}$ to satisfy $\mathbf{u}(\mathbf{x}^1) = \mathbf{u}(\mathbf{x}^4)$ and $\boldsymbol{\sigma}(\mathbf{x}^1) = \boldsymbol{\sigma}(\mathbf{x}^4)$. Hence the tractions $\mathbf{t}$ satisfy $\mathbf{t}(\mathbf{x}^1) = -\mathbf{t}(\mathbf{x}^4)$. In component form, this can be written as

$$(u_1, u_2, u_3)(\mathbf{x}^1) = (+u_1, +u_2, +u_3)(\mathbf{x}^4),$$

$$(t_1, t_2, t_3)(\mathbf{x}^1) = (-t_1, -t_2, -t_3)(\mathbf{x}^4). \tag{14.4.6a,b}$$

From the definition of $M^{(4)}(G) = +M^{(1)}(G)$ and $M^{(4)}(G) = -M^{(1)}(G)$, the following relations for the displacements and tractions on the boundary surfaces are obtained: for the case when $M^{(4)}(G) = +M^{(1)}(G)$,

$$(u_1, u_2, u_3)(\mathbf{x}^1) = (-u_1, +u_2, +u_3)(\mathbf{x}^4),$$

$$(t_1, t_2, t_3)(\mathbf{x}^1) = (-t_1, +t_2, +t_3)(\mathbf{x}^4), \tag{14.4.7a,b}$$

and for the case when $M^{(4)}(G) = -M^{(1)}(G)$,

$$(u_1, u_2, u_3)(\mathbf{x}^1) = (+u_1, -u_2, -u_3)(\mathbf{x}^4),$$

$$(t_1, t_2, t_3)(\mathbf{x}^1) = (+t_1, -t_2, -t_3)(\mathbf{x}^4), \tag{14.4.7c,d}$$

for $\mathbf{x}$ on the plane $x_1 = a_1$.

From comparison of (14.4.6a,b) with (14.4.7a,b) and (14.4.7c,d), the following components of the periodic displacements and tractions must vanish on the boundary surfaces $x_1 = \pm a_1$:

$$u_1 = t_2 = t_3 = 0 \qquad \text{for } M^{(4)}(G) = +M^{(1)}(G),$$

$$t_1 = u_2 = u_3 = 0 \qquad \text{for } M^{(4)}(G) = -M^{(1)}(G). \tag{14.4.8a,b}$$

Similar boundary conditions can be obtained on the boundary surfaces $x_2 = \pm a_2$ and $x_3 = \pm a_3$ from the MI sym/ant conditions, by appropriately relating to $M^{(4)}(G) = +M^{(2)}(G)$ or $M^{(4)}(G) = -M^{(2)}(G)$ and $M^{(4)}(G) = +M^{(3)}(G)$ or $M^{(4)}(G) = -M^{(3)}(G)$, respectively. Indeed, denoting the normal and tangential components of the displacements and tractions by

$$\mathbf{u}^n \equiv (\boldsymbol{\nu} \cdot \mathbf{u})\,\boldsymbol{\nu}, \qquad \mathbf{u}^t \equiv \mathbf{u} - \mathbf{u}^n, \qquad \mathbf{t}^n \equiv (\boldsymbol{\nu} \cdot \mathbf{t})\,\boldsymbol{\nu}, \qquad \mathbf{t}^t \equiv \mathbf{t} - \mathbf{t}^n, \tag{14.4.9a~d}$$

the periodic boundary tractions and displacements must be such that

$$\mathbf{u}^n = \mathbf{0} \quad \text{and} \quad \mathbf{t}^t = \mathbf{0}, \tag{14.4.10a,b}$$

or

$$\mathbf{t}^n = \mathbf{0} \quad \text{and} \quad \mathbf{u}^t = \mathbf{0}, \tag{14.4.10c,d}$$

on the boundary surfaces of the unit cell. Note that (14.4.10a,b) correspond to the case of MI sym, and (14.4.10c,d) to the case of MI ant.

The surface data (14.4.10a,b) or (14.4.10c,d) produce unique displacement, strain, and stress fields in a unit cell, U, as well as in a finite body bounded by ∂U. Since the periodic field variables in an MI sym/ant set satisfy either

(14.4.10a,b) or (14.4.10c,d) on the boundary surfaces, these field variables are uniquely defined. It should be noted that even though each part of the decomposed set of the periodic fields has null surface data, the original periodic fields, obtained by the sum of the decomposed parts, may have more complex boundary conditions. The uniqueness of the solution of the original periodic fields is then guaranteed by the uniqueness of the decomposition of these fields into eight MI sym/ant sets, and by the uniqueness of the solution of each set.

14.4.4. Homogeneous Fields

As mentioned in Section 12, the homogeneous field variables are excluded from the periodic fields, since the corresponding (continuous) linear displacement field does not satisfy periodicity. However, the uniform strain or stress field, $\boldsymbol{\varepsilon}^{\mathrm{o}}$ or $\boldsymbol{\sigma}^{\mathrm{o}}$, can be regarded as input data which produce the periodic (disturbance) stress and strain fields in a periodically heterogeneous solid.[9] The uniqueness of these periodic fields, produced by the prescribed homogeneous stress or strain field, is also guaranteed. Since $\boldsymbol{\varepsilon}^{\mathrm{o}}$ and $\boldsymbol{\sigma}^{\mathrm{o}}$ produce linear displacements $\mathbf{u}^{\mathrm{o}} = \mathbf{x} \cdot \boldsymbol{\varepsilon}^{\mathrm{o}}$ and uniform tractions $\mathbf{t}^{\mathrm{o}} = \boldsymbol{\nu} \cdot \boldsymbol{\sigma}^{\mathrm{o}}$, respectively, the surface data (14.4.10a~d) must then be replaced by

$$\mathbf{u}^{\mathrm{n}} - \mathbf{u}^{\mathrm{on}} = \mathbf{0} \quad \text{and} \quad \mathbf{t}^{\mathrm{t}} - \mathbf{t}^{\mathrm{ot}} = \mathbf{0}, \tag{14.4.11a,b}$$

or

$$\mathbf{t}^{\mathrm{n}} - \mathbf{t}^{\mathrm{on}} = \mathbf{0} \quad \text{and} \quad \mathbf{u}^{\mathrm{t}} - \mathbf{u}^{\mathrm{ot}} = \mathbf{0}, \tag{14.4.11c,d}$$

where

$$\mathbf{u}^{\mathrm{on}} \equiv (\boldsymbol{\nu} \cdot \mathbf{u}^{\mathrm{o}})\, \boldsymbol{\nu}, \qquad \mathbf{u}^{\mathrm{ot}} \equiv \mathbf{u}^{\mathrm{o}} - \mathbf{u}^{\mathrm{on}},$$

$$\mathbf{t}^{\mathrm{on}} \equiv (\boldsymbol{\nu} \cdot \mathbf{t}^{\mathrm{o}})\, \boldsymbol{\nu}, \qquad \mathbf{t}^{\mathrm{ot}} \equiv \mathbf{t}^{\mathrm{o}} - \mathbf{t}^{\mathrm{on}}. \tag{14.4.12a~d}$$

These surface data may also be associated with the boundary of a finite body which is represented by the unit cell.

Now consider the MI sym/ant set of fields that corresponds to, say, a homogeneous strain field $\boldsymbol{\varepsilon}^{\mathrm{o}}$. From the definition of the MI sym/ant of a second-order tensor field, and since *the unit cell is assumed to be completely symmetric,* the components of the prescribed uniform strain field produce the following MI sym/ant sets:

$\varepsilon^{\mathrm{o}}_{11}\ \varepsilon^{\mathrm{o}}_{22}\ \varepsilon^{\mathrm{o}}_{33}$: G^4 — 4th MI sym/ant set
$M^{(4)}(G) = +M^{(1)}(G) = +M^{(2)}(G) = +M^{(3)}(G)$,

$\varepsilon^{\mathrm{o}}_{23}$: G^{-1} — $-$ 1st MI sym/ant set
$M^{(4)}(G) = +M^{(1)}(G) = -M^{(2)}(G) = -M^{(3)}(G)$,

---

[9] Recall that the homogeneous strain $\boldsymbol{\varepsilon}^{\mathrm{o}}$ and the homogeneous stress $\boldsymbol{\sigma}^{\mathrm{o}}$ are related through the overall elasticity or compliance tensor, $\bar{\mathbf{C}}$ or $\bar{\mathbf{D}}$. Hence, if $\boldsymbol{\varepsilon}^{\mathrm{o}}$ is prescribed, the traction boundary data are yet to be determined, since the corresponding homogeneous stress $\boldsymbol{\sigma}^{\mathrm{o}}$ is not known until $\bar{\mathbf{C}}$ is calculated.

$$\varepsilon^{o}_{31}: \qquad G^{-2} \qquad \begin{array}{l}\text{– 2nd MI sym/ant set}\\ M^{(4)}(G) = -M^{(1)}(G) = +M^{(2)}(G) = -M^{(3)}(G)\,,\end{array}$$

$$\varepsilon^{o}_{12}: \qquad G^{-3} \qquad \begin{array}{l}\text{– 3rd MI sym/ant set}\\ M^{(4)}(G) = -M^{(1)}(G) = -M^{(2)}(G) = +M^{(3)}(G)\,.\end{array}$$

The four other remaining sets are always zero for homogeneous strain fields. Furthermore, only the above four indicated sets can have nonvanishing volume averages. It must be emphasized here again, that each of the eight MI sym/ant sets of periodic strain and stress fields is mutually independent if the elasticity tensor field $\mathbf{C}'$ satisfies the fourth MI sym/ant condition, i.e., if there is complete symmetry with respect to all three coordinate planes; otherwise each of the eight MI sym/ant parts of, say, the stress field, may depend on some or all MI sym/ant parts of the strain field.

14.5. FOURIER SERIES EXPANSION OF MI SYM/ANT SET OF PERIODIC FIELDS

In the equivalent homogeneous solid, obtained by the introduction of suitable periodic eigenstrains or eigenstresses, all the periodic fields can be decomposed into eight MI sym/ant parts. It will be shown that for each MI sym/ant part, the governing equations for the corresponding Fourier series coefficients reduce to a simple real-valued form.

14.5.1. MI Sym/Ant Decomposition of Governing Field Equations

In Section 12, the equivalent homogeneous solid with a constant elasticity tensor $\mathbf{C}$ (= $\mathbf{D}^{-1}$) is introduced for a solid with a periodic elasticity tensor $\mathbf{C}'$ (= $\mathbf{D}'^{-1}$), through a periodic eigenstress field $\boldsymbol{\sigma}^*$ (or eigenstrain field $\boldsymbol{\varepsilon}^*$)[10]. The displacement field $\mathbf{u}$ produced by $\boldsymbol{\sigma}^*$, is defined by the following governing equation:

$$\boldsymbol{\nabla}\boldsymbol{\cdot}\{\mathbf{C}:(\boldsymbol{\nabla}\otimes\mathbf{u}(\mathbf{x}))\} + \boldsymbol{\nabla}\boldsymbol{\cdot}\boldsymbol{\sigma}^*(\mathbf{x}) = \mathbf{0}. \tag{14.5.1a}$$

The Fourier series coefficients of $\mathbf{u}$ are given in terms of those of $\boldsymbol{\sigma}^*$, through

$$-\boldsymbol{\xi}\boldsymbol{\cdot}\{\mathbf{C}:(\boldsymbol{\xi}\otimes F\mathbf{u}(\boldsymbol{\xi}))\} + \iota\,\boldsymbol{\xi}\boldsymbol{\cdot}F\boldsymbol{\sigma}^*(\boldsymbol{\xi}) = \mathbf{0}, \tag{14.5.1b}$$

for $\boldsymbol{\xi} \neq \mathbf{0}$; see (12.4.5a,b).

Now consider the MI sym/ant decomposition of the periodic fields in the equivalent homogeneous solid with periodic eigenstress. Choose the uniform elasticity tensor $\mathbf{C}$ of the equivalent homogeneous solid such that it satisfies the

[10] As mentioned, a uniform eigenstress or eigenstrain does not produce periodic fields in the equivalent homogeneous solid, although they are related to the homogeneous fields. In this section, attention is focused on the periodic fields, and the homogeneous fields are not considered.

fourth MI sym/ant condition; note that this choice is always possible whatever the nature of the symmetry of $\mathbf{C}'(\mathbf{x})$. However, a completely symmetric unit cell continues to be assumed. Therefore, the eight MI sym/ant sets of periodic fields, $\{\mathbf{u}^i, \boldsymbol{\varepsilon}^i, \boldsymbol{\sigma}^i\}$ are mutually independent.

Since $\mathbf{C}$ satisfies the fourth MI sym/ant condition, its components satisfy

$$C_{pqrs} = \frac{1}{8} \sum_{j=-4}^{4} \alpha^{4j} \alpha^{jp} \alpha^{jq} \alpha^{jr} \alpha^{js} C_{pqrs}$$

$$= \frac{1}{8} \sum_{j=-4}^{4} \alpha^{jp} \alpha^{jq} \alpha^{jr} \alpha^{js} C_{pqrs}, \tag{14.5.2a}$$

for p, q, r, s = 1, 2, 3. Since C_{pqrs} is constant and α^{jp} is either $+1$ or -1, (14.5.2a) is satisfied if

$$\alpha^{jp} \alpha^{jq} \alpha^{jr} \alpha^{js} C_{pqrs} = C_{pqrs} \qquad \text{(p, q, r, s not summed)}, \tag{14.5.2b}$$

which means that

$$C_{pqrs} \begin{cases} \neq 0 & \text{if } \alpha^{jp} \alpha^{jq} \alpha^{jr} \alpha^{js} = +1 \\ = 0 & \text{if } \alpha^{jp} \alpha^{jq} \alpha^{jr} \alpha^{js} = -1. \end{cases} \tag{14.5.2c}$$

As shown in Subsection 14.3, if (14.5.2c) is satisfied, the governing equations (14.5.1a,b) for the field variables $\mathbf{u}$ and $\boldsymbol{\sigma}^*$ can be decomposed into those for each MI sym/ant part of these fields, $\mathbf{u}^i$ and $\boldsymbol{\sigma}^{*i}$, and the corresponding Fourier series coefficients $F\mathbf{u}$ and $F\boldsymbol{\sigma}^*$ can be decomposed into $F\mathbf{u}^i$ and $F\boldsymbol{\sigma}^{*i}$.

Now this decomposition is demonstrated in greater detail. The governing field equation (14.5.1a) may be regarded as a first-order tensor operator which assigns zero to a vector. From definition (14.2.5a), the ith MI sym/ant of (14.5.1a) is

$$\frac{1}{8} \sum_{j=-4}^{4} \alpha^{ij} \Big\{ \partial_p \{C_{pqrs}(\partial_r u_s(\mathbf{x}^j)\} + \partial_p \{\sigma^*_{pq}(\mathbf{x}^j)\} \Big\} v^j_q = 0, \tag{14.5.3}$$

where $\mathbf{v}$ is an arbitrary vector and $\partial_p \equiv \partial/\partial x_p$. The corresponding MI sym/ant part of the Fourier series expansion of (14.5.3) is

$$\frac{1}{8} \sum_{\boldsymbol{\xi}}{}' \sum_{j=-4}^{4} \alpha^{ij} \{-C_{pqrs}\, \xi_p \xi_r\, Fu_s(\boldsymbol{\xi}) + \iota\, \xi_p\, F\sigma^*_{pq}(\boldsymbol{\xi})\} \exp(\iota\, \boldsymbol{\xi} \cdot \mathbf{x}^j)\, v^j_q$$

$$= \sum_{\boldsymbol{\xi}}{}' \{-C_{pqrs}\, \xi_p \xi_r\, Fu_s(\boldsymbol{\xi}) + \iota\, \xi_p\, F\sigma^*_{pq}(\boldsymbol{\xi})\} \{\frac{1}{8} \sum_{j=-4}^{4} \alpha^{ij} (\alpha^{qj} v^4_q) \exp(\iota\, \boldsymbol{\xi} \cdot \mathbf{x}^j)\}$$

$$= \sum_{\boldsymbol{\xi}}{}' \{-C_{pqrs}\, \xi_p \xi_r\, Fu_s(\boldsymbol{\xi}) + \iota\, \xi_p\, F\sigma^*_{pq}(\boldsymbol{\xi})\}\, v^4_q \exp^{i'}(\iota\, \boldsymbol{\xi} \cdot \mathbf{x}), \tag{14.5.4a}$$

where i' is given by

$$i' = i'(i;\, q); \quad \alpha^{i'j} = \alpha^{ij} \alpha^{qj} \qquad \text{(j not summed)}, \tag{14.5.5}$$

and $\exp^i(\iota\, \boldsymbol{\xi} \cdot \mathbf{x}^4)$ is the ith MI sym/ant part of $\exp(\iota\, \boldsymbol{\xi} \cdot \mathbf{x})$ defined by (14.3.3).

The uniqueness of the solution has been proved in Subsection 14.4. The value of the infinite sum (14.5.4a) does not change with a change in the order of summation with respect to $\boldsymbol{\xi}$. As explained in Subsection 14.3, for the right side of (14.5.4a), taking the summation first with respect to $\boldsymbol{\xi}$ with nonnegative components, together with a finite summation with respect to the distinct images, $\boldsymbol{\xi}^{\mathrm{k}}$'s, associated with $\boldsymbol{\xi}$, obtain

$$\sum_{\xi^4}^{+\prime} \sum_{k=-4}^{4}{}^{*} \{-C_{pqrs}\,\xi_p^k\,\xi_r^k\,Fu_s(\boldsymbol{\xi}^k) + \iota\,\xi_p^k\,F\sigma_{pq}^*(\boldsymbol{\xi}^k)\}\; v_q^4\,\exp^{i'}(\iota\,\boldsymbol{\xi}^k\boldsymbol{.}\mathbf{x})$$

$$= \sum_{\xi^4}^{+\prime} 2^K \Big\{ \frac{1}{8} \sum_{k=-4}^{4} \{-C_{pqrs}\,\xi_p^k\,\xi_r^k\,Fu_s(\boldsymbol{\xi}^k) + \iota\,\xi_p^k\,F\sigma_{pq}^*(\boldsymbol{\xi}^k)\}\; v_q^4\,\exp^{i'}(\iota\,\boldsymbol{\xi}^k\boldsymbol{.}\mathbf{x}) \Big\}$$

$$= \sum_{\xi^4}^{+\prime} 2^K \Big\{ \frac{1}{8} \sum_{k=-4}^{4} \{-C_{pqrs}\,\xi_p^k\,\xi_r^k\,Fu_s(\boldsymbol{\xi}^k) + \iota\,\xi_p^k\,F\sigma_{pq}^*(\boldsymbol{\xi}^k)\}\; v_q^4\,\alpha^{i'k}\exp^{i'}(\iota\,\boldsymbol{\xi}^4\boldsymbol{.}\mathbf{x}) \Big\}. \tag{14.5.4b}$$

From definition (14.5.5) of i′, summation with respect to k in (14.5.4b) produces,

$$\frac{1}{8} \sum_{k=-4}^{4} \sum_{p,r,s} \{-C_{pqrs}\,(\alpha^{kp}\,\xi_p^4)\,(\alpha^{kr}\,\xi_r^4)\,Fu_s(\boldsymbol{\xi}^k) + \iota(\alpha^{kp}\,\xi_p^4)F\sigma_{pq}^*(\boldsymbol{\xi}^k)\}\,\alpha^{ik}\,\alpha^{qk}$$

$$= \frac{1}{8} \sum_{k=-4}^{4} \sum_{p,r,s} \{-\xi_p^4\xi_r^4\,(C_{pqrs}\,\alpha^{kp}\,\alpha^{kq}\,\alpha^{kr}\,\alpha^{ks})\,(\alpha^{ik}\,\alpha^{ks}\,Fu_s(\boldsymbol{\xi}^k))$$
$$+\,\iota\,\xi_p^4\,(\alpha^{ik}\,\alpha^{kp}\,\alpha^{kq}\,F\sigma_{pq}^*(\boldsymbol{\xi}^k))\}. \tag{14.5.4c}$$

Since C_{pqrs} satisfies (14.5.2b), the above sum equals

$$-\,C_{pqrs}\,\xi_p^4\,\xi_r^4\,(Fu_s)^i(\boldsymbol{\xi}^4) + \iota\,\xi_p^4\,(F\sigma_{pq}^*)^i(\boldsymbol{\xi}^4),$$

where $\boldsymbol{\xi}^4$ has nonnegative components.

As shown in Subsection 14.3.2, $(F\mathbf{u})^i(\boldsymbol{\xi}^4) = F(\mathbf{u}^i)(\boldsymbol{\xi}^4) = F\mathbf{u}^i(\boldsymbol{\xi}^4)$ and $(F\boldsymbol{\sigma}^*)^i(\boldsymbol{\xi}^4) = F(\boldsymbol{\sigma}^{*i})(\boldsymbol{\xi}^4) = F\boldsymbol{\sigma}^{*i}(\boldsymbol{\xi}^4)$, for $\boldsymbol{\xi}^4$ with nonnegative components. Therefore, it follows from (14.5.4a~c) that the ith MI sym/ant part of the governing equation (14.5.1a) is[11]

$$\boldsymbol{\nabla}\boldsymbol{.}\{\mathbf{C} : (\boldsymbol{\nabla}\otimes\mathbf{u}^i(\mathbf{x}))\} + \boldsymbol{\nabla}\boldsymbol{.}\boldsymbol{\sigma}^{*i}(\mathbf{x}) = \mathbf{0}, \tag{14.5.6a}$$

and is expanded in the Fourier series, as follows:

$$-\,\boldsymbol{\xi}^4\boldsymbol{.}\{\mathbf{C} : (\boldsymbol{\xi}^4\otimes F\mathbf{u}^i(\boldsymbol{\xi}^4))\} + \iota\,\boldsymbol{\xi}^4\boldsymbol{.}F\boldsymbol{\sigma}^{*i}(\boldsymbol{\xi}^4) = \mathbf{0}, \tag{14.5.6b}$$

for $\boldsymbol{\xi}^4$ with nonnegative components. Comparison of (14.5.1b) and (14.5.6b) shows that the governing equation for $F\mathbf{u}^i$ with $F\boldsymbol{\sigma}^{*i}$ is identical with that for $F\mathbf{u}$ with $F\boldsymbol{\sigma}^*$.

The governing equation (14.5.6a) of the ith MI sym/ant part of the periodic fields has the two following significant advantages over the governing equation (14.5.1a) of the original periodic fields:

[11] Note that (14.5.6a) and (14.5.6b) are obtained from (14.5.1a) and (14.5.1b), with the aid of the results obtained in Subsection 14.2.4.

1) $F\mathbf{u}^{\mathrm{i}}$ and $F\boldsymbol{\sigma}^{*\mathrm{i}}$ are either real or purely imaginary, while $F\mathbf{u}$ and $F\boldsymbol{\sigma}^{*}$ are, in general, complex-valued. And (14.5.6b) is either real or purely imaginary, while (14.5.1b) is complex-valued;

2) (14.5.6b) holds for $\boldsymbol{\xi}^4$ with nonnegative components, while (14.5.1b) holds for all $\boldsymbol{\xi}$. Hence, if the summation is truncated to N, there are $(\mathrm{N}+1)^3-1$ terms for $\boldsymbol{\xi}^4$ in (14.5.6b), but $(2\mathrm{N}+1)^3-1$ terms in (14.5.1b).

Therefore, the MI sym/ant decomposition of the periodic fields considerably reduces the numerical computation, although the form of the governing equation remains the same.

14.5.2. Isotropic Equivalent Homogeneous Solid

The elasticity tensor $\mathbf{C}$ of the equivalent homogeneous solid is required to be uniform and to satisfy the fourth MI sym/ant condition. From (14.5.2c), if $\mathbf{C}$ is

$$C_{pqrs}=0 \text{ except for } C_{ppqq},\ C_{pqpq}, \text{ and } C_{pqqp} \quad (\mathrm{p},\ \mathrm{q} \text{ not summed}), \tag{14.5.7}$$

then the above requirements are satisfied. An isotropic elasticity tensor $\mathbf{C}$ apparently satisfies (14.5.7). For the isotropic case, the equations defining the Fourier series coefficients of the displacement field are given in Subsection 12.4. The corresponding ith MI sym/ant part is

$$-\{(\lambda+\mu)\,\boldsymbol{\xi}^4\otimes\boldsymbol{\xi}^4+\mu\,(\boldsymbol{\xi}^4\boldsymbol{.}\boldsymbol{\xi}^4)\,\mathbf{1}^{(2)}\}\boldsymbol{.}F\mathbf{u}^{\mathrm{i}}(\boldsymbol{\xi}^4)+\iota\,\boldsymbol{\xi}^4\boldsymbol{.}F\boldsymbol{\sigma}^{*\mathrm{i}}(\boldsymbol{\xi}^4)=\mathbf{0}, \tag{14.5.8a}$$

for $\boldsymbol{\xi}^4$ with nonnegative components. The coefficient of the periodic strain field $F\boldsymbol{\varepsilon}^{\mathrm{i}}$ is given by

$$F\boldsymbol{\varepsilon}^{\mathrm{i}}(\boldsymbol{\xi}^4)=-\frac{1}{\mu}\,\{sym\,(\bar{\boldsymbol{\xi}}^4\otimes\mathbf{1}^{(2)}\otimes\bar{\boldsymbol{\xi}}^4)-\frac{\lambda+\mu}{\lambda+2\mu}\,\bar{\boldsymbol{\xi}}^4\otimes\bar{\boldsymbol{\xi}}^4\otimes\bar{\boldsymbol{\xi}}^4\otimes\bar{\boldsymbol{\xi}}^4\}:F\boldsymbol{\sigma}^{*\mathrm{i}}(\boldsymbol{\xi}^4), \tag{14.5.8b}$$

where $\bar{\boldsymbol{\xi}}\equiv\boldsymbol{\xi}/|\boldsymbol{\xi}|$.

For illustration, consider two cases: (1) full symmetry with respect to all three coordinate planes, i.e., the fourth MI sym/ant condition is satisfied, and hence $M^{(4)}(G)=+M^{(1)}(G)=+M^{(2)}(G)=+M^{(3)}(G)$; and (2) symmetry with respect to the plane $\mathrm{x}_1=0$, and antisymmetry with respect to the planes $\mathrm{x}_2=0$, $\mathrm{x}_3=0$, i.e., the $-$1st MI sym/ant condition is satisfied, and hence $M^{(4)}(G)=+M^{(1)}(G)=-M^{(2)}(G)=-M^{(3)}(G)$; the reference elasticity tensor $\mathbf{C}$ is isotropic.

Case (1), $M^{(4)}(G)=+M^{(1)}(G)=+M^{(2)}(G)=+M^{(3)}(G)$: As shown in Subsection 14.2, the components u_p^4 and $\sigma_{\mathrm{pq}}^{*4}$ satisfy their own MI sym/ant conditions. u_p^4 and $\sigma_{\mathrm{pq}}^{*4}$ can be expanded in the following real-valued Fourier series:

$$\mathrm{u}_1^4(\mathbf{x})=\sum_{\boldsymbol{\xi}^4}{}^{+'}\ F^{\mathrm{R}}\mathrm{u}_1^4(\boldsymbol{\xi}^4)\,\mathrm{s}_1\mathrm{c}_2\mathrm{c}_3,$$

$$u_2^4(\mathbf{x}) = \sum_{\xi^4}{}^{+\prime}\; {}^{F\mathrm{R}}u_2^4(\boldsymbol{\xi}^4)\, c_1 s_2 c_3,$$

$$u_3^4(\mathbf{x}) = \sum_{\xi^4}{}^{+\prime}\; {}^{F\mathrm{R}}u_3^4(\boldsymbol{\xi}^4)\, c_1 c_2 s_3,$$

$$\sigma_{11}^{*4}(\mathbf{x}) = \sum_{\xi^4}{}^{+\prime}\; {}^{F\mathrm{R}}\sigma_{11}^{*4}(\boldsymbol{\xi}^4)\, c_1 c_2 c_3, \qquad \sigma_{23}^{*4}(\mathbf{x}) = \sum_{\xi^4}{}^{+\prime}\; -{}^{F\mathrm{R}}\sigma_{23}^{*4}(\boldsymbol{\xi}^4)\, c_1 s_2 s_3,$$

$$\sigma_{22}^{*4}(\mathbf{x}) = \sum_{\xi^4}{}^{+\prime}\; {}^{F\mathrm{R}}\sigma_{22}^{*4}(\boldsymbol{\xi}^4)\, c_1 c_2 c_3, \qquad \sigma_{31}^{*4}(\mathbf{x}) = \sum_{\xi^4}{}^{+\prime}\; -{}^{F\mathrm{R}}\sigma_{31}^{*4}(\boldsymbol{\xi}^4)\, s_1 c_2 s_3,$$

$$\sigma_{33}^{*4}(\mathbf{x}) = \sum_{\xi^4}{}^{+\prime}\; {}^{F\mathrm{R}}\sigma_{33}^{*4}(\boldsymbol{\xi}^4)\, c_1 c_2 c_3, \qquad \sigma_{12}^{*4}(\mathbf{x}) = \sum_{\xi^4}{}^{+\prime}\; -{}^{F\mathrm{R}}\sigma_{12}^{*4}(\boldsymbol{\xi}^4)\, s_1 s_2 c_3,$$

(14.5.9a~i)

where ${}^{F\mathrm{R}}(\ )$ stands for the real part of the corresponding Fourier series expansion of ( ).

Case (2), $M^{(4)}(G) = +M^{(1)}(G) = -M^{(2)}(G) = -M^{(3)}(G)$: In a similar manner, u_p^{-1} and σ_{pq}^{*-1} can be expanded in the following real-valued Fourier series:

$$u_1^{-1}(\mathbf{x}) = \sum_{\xi^4}{}^{+\prime}\; {}^{F\mathrm{R}}u_1^{-1}(\boldsymbol{\xi}^4)\, s_1 s_2 s_3,$$

$$u_2^{-1}(\mathbf{x}) = \sum_{\xi^4}{}^{+\prime}\; -{}^{F\mathrm{R}}u_2^{-1}(\boldsymbol{\xi}^4)\, c_1 c_2 s_3,$$

$$u_3^{-1}(\mathbf{x}) = \sum_{\xi^4}{}^{+\prime}\; -{}^{F\mathrm{R}}u_3^{-1}(\boldsymbol{\xi}^4)\, c_1 s_2 c_3,$$

$$\sigma_{11}^{*-1}(\mathbf{x}) = \sum_{\xi^4}{}^{+\prime}\; {}^{F\mathrm{R}}\sigma_{11}^{*-1}(\boldsymbol{\xi}^4)\, c_1 s_2 s_3, \qquad \sigma_{23}^{*-1}(\mathbf{x}) = \sum_{\xi^4}{}^{+\prime}\; -{}^{F\mathrm{R}}\sigma_{23}^{*-1}(\boldsymbol{\xi}^4)\, c_1 c_2 c_3,$$

$$\sigma_{22}^{*-1}(\mathbf{x}) = \sum_{\xi^4}{}^{+\prime}\; {}^{F\mathrm{R}}\sigma_{22}^{*-1}(\boldsymbol{\xi}^4)\, c_1 s_2 s_3, \qquad \sigma_{31}^{*-1}(\mathbf{x}) = \sum_{\xi^4}{}^{+\prime}\; {}^{F\mathrm{R}}\sigma_{31}^{*-1}(\boldsymbol{\xi}^4)\, s_1 s_2 c_3,$$

$$\sigma_{33}^{*-1}(\mathbf{x}) = \sum_{\xi^4}{}^{+\prime}\; {}^{F\mathrm{R}}\sigma_{33}^{*-1}(\boldsymbol{\xi}^4)\, c_1 s_2 s_3, \qquad \sigma_{12}^{*-1}(\mathbf{x}) = \sum_{\xi^4}{}^{+\prime}\; {}^{F\mathrm{R}}\sigma_{12}^{*-1}(\boldsymbol{\xi}^4)\, s_1 c_2 s_3.$$

(14.5.10a~i)

The governing equation for ${}^{F\mathrm{R}}\mathbf{u}^{i}$ (i = 4, − 1) then becomes

$$-\{(\lambda+\mu)\,\boldsymbol{\xi}^4\otimes\boldsymbol{\xi}^4 + \mu\,(\boldsymbol{\xi}^4\cdot\boldsymbol{\xi}^4)\,\mathbf{1}^{(2)}\}\cdot{}^{F\mathrm{R}}\mathbf{u}^{i}(\boldsymbol{\xi}^4) - \boldsymbol{\xi}^4\cdot{}^{F\mathrm{R}}\boldsymbol{\sigma}^{*i}(\boldsymbol{\xi}^4) = \mathbf{0}. \qquad (14.5.11a)$$

The kernel functions of the first, second, and third displacement components are $s_1c_2c_3$, $c_1s_2c_3$, and $c_1c_2s_3$ for i = 4; and $s_1s_2s_3$, $c_1c_2s_3$, and $c_1s_2c_3$ for i = − 1. These kernels correspond to $\exp^{i'}(\iota\,\boldsymbol{\xi}\cdot\mathbf{x})$ with i′ = i′(i; 1), i′ = i′(i; 2), and i′ = i′(i; 3) for i = 4 and i = − 1; see (14.3.3b~i) and definition (14.5.5) of i′(i; q).

Then, the real-valued coefficients of the expansion of the strain field, ${}^{F\mathrm{R}}\boldsymbol{\varepsilon}^{i}$, are

$$
{}^{F^R}\boldsymbol{\varepsilon}^{\mathrm{i}}(\boldsymbol{\xi}^4) = -\frac{1}{\mu}\{sym\,(\bar{\boldsymbol{\xi}}^4 \otimes \mathbf{1}^{(2)} \otimes \bar{\boldsymbol{\xi}}^4) - \frac{\lambda+\mu}{\lambda+2\mu}\bar{\boldsymbol{\xi}}^4 \otimes \bar{\boldsymbol{\xi}}^4 \otimes \bar{\boldsymbol{\xi}}^4 \otimes \bar{\boldsymbol{\xi}}^4\} : {}^{F^R}\boldsymbol{\sigma}^{*\mathrm{i}}(\boldsymbol{\xi}^4) \tag{14.5.11b}
$$

which is exactly the same as (14.5.8b). Hence, it can be concluded that there is one-to-one correspondence between $\{F\mathbf{u}^{\mathrm{i}}, F\boldsymbol{\varepsilon}^{\mathrm{i}}, F\boldsymbol{\sigma}^{\mathrm{i}}\}$ and $\{F^R\mathbf{u}^{\mathrm{i}}, F^R\boldsymbol{\varepsilon}^{\mathrm{i}}, F^R\boldsymbol{\sigma}^{\mathrm{i}}\}$, for[12] $\mathrm{i} = \pm 1, \pm 2, \pm 3$, and ± 4.

14.6. APPLICATION OF HASHIN-SHTRIKMAN VARIATIONAL PRINCIPLE

In Subsection 13, upper and lower bounds for the overall moduli of a periodic structure have been obtained, by applying the Hashin-Shtrikman variational principle to the eigenstress field in the equivalent homogenized solid, representing the periodic microstructure. In the periodic case, the bounds are *exact*. To calculate sharp bounds, a relatively large number of Fourier terms must be used. As mentioned before, the MI sym/ant decomposition of the periodic field variables is effective in reducing the numerical effort. Since a homogeneous strain $\boldsymbol{\varepsilon}^{\mathrm{o}}$ has only four sym/ant parts (see Subsection 14.4), only these need to be considered in computing bounds. Furthermore, the number of terms in the sums in the Fourier series expansions can be reduced by about a factor of eight. In this subsection, the MI sym/ant decomposition is applied to the eigenstress field, and the energy functional J and the associated quadratic forms defined in Subsection 13.4 are computed.

14.6.1. Inner Product of Stress and Strain

As shown in Subsection 13.2.2, the key to the applicability of the Hashin-Shtrikman variational principle is the vanishing of the average of the inner product of the stress and strain fields produced by the homogenizing eigenstress. Due to periodicity, however, the following identity results from (13.2.2), for any arbitrary statically admissible periodic stress field, $\boldsymbol{\sigma}^{\mathrm{p}}$, and any arbitrary kinematically admissible periodic strain field, $\boldsymbol{\varepsilon}^{\mathrm{p}}$,

$$
< \boldsymbol{\sigma}^{\mathrm{p}} : \boldsymbol{\varepsilon}^{\mathrm{p}} > = 0, \tag{14.6.1a}
$$

whether these fields are related through the constitutive relations or not. Hence, the ith and jth MI sym/ant parts of these periodic stress and strain fields, $\boldsymbol{\sigma}^{\mathrm{pi}}$ and $\boldsymbol{\varepsilon}^{\mathrm{pj}}$, satisfy

$$
< \boldsymbol{\sigma}^{\mathrm{pi}} : \boldsymbol{\varepsilon}^{\mathrm{pj}} > = 0 \tag{14.6.1b}
$$

for i, j = ±1, ±2, ±3, and ±4, due to their periodicity.

---

[12] From comparison of (14.5.9) or (14.5.10) with (14.3.9d), it is seen that $F^R(\ldots)^{\mathrm{i}}$ coincides with $\pm 2^{\mathrm{K}} F(\ldots)^{\mathrm{i}}$ or $\pm \iota 2^{\mathrm{K}} F(\ldots)^{\mathrm{i}}$.

From the requirements of periodicity and MI sym/ant, each MI sym/ant part of the periodic fields must satisfy certain null conditions on the boundary of the unit cell and admit a certain real-valued Fourier series expansion. With the aid of these properties, (14.6.1b) can be proved in an alternative manner. First, consider the case of i = j. It is shown in Subsection 13.4 that, on the boundary of the unit cell, the periodic field variables of each MI sym/ant set satisfy either: 1) zero normal tractions and zero tangential displacements, or 2) zero tangential tractions and zero normal displacements. Hence,

$$< \boldsymbol{\sigma}^{pi} : \boldsymbol{\varepsilon}^{pi} > = \frac{1}{U}\int_{\partial U} \mathbf{t}^{pi} \cdot \mathbf{u}^{pi}\, dS = 0 \qquad \text{(i not summed)} \tag{14.6.2}$$

which proves (14.6.1b) for i = j.

Next, consider the case of i ≠ j. For simplicity, compute the volume average of $F\sigma^{pi}_{pq}\exp^{i'}(\iota\boldsymbol{\xi}\cdot\mathbf{x})$ and $F\varepsilon^{pi}_{pq}\exp^{i'}(\iota\boldsymbol{\zeta}\cdot\mathbf{x})$, for each pair of p and q. As discussed in Subsection 13.2, i′ and j′ are given by i′(i; p, q) and j′(j; p, q). If i ≠ j, then, i′ and j′ are also different, and

$$< \exp^{i'}(\iota\boldsymbol{\xi}\cdot\mathbf{x}) \exp^{j'}(\iota\boldsymbol{\zeta}\cdot\mathbf{x}) > = 0, \tag{14.6.3a}$$

for any $\boldsymbol{\xi}$ and $\boldsymbol{\zeta}$, including $\boldsymbol{\xi} = \mathbf{0}$ or $\boldsymbol{\zeta} = \mathbf{0}$. This is because the volume average taken over the unit cell vanishes for any sine function. As given by (14.3.3), each kernel $\exp^{i}(\iota\boldsymbol{\xi}\cdot\mathbf{x})$ is a particular combination of cosine and sine functions. Since

$$\cos(\xi_p x_p)\sin(\zeta_p x_p) = \frac{1}{2}\{\sin((\xi_p+\zeta_p)x_p) - \sin((\xi_p-\zeta_p)x_p)\} \qquad \text{(p not summed)}, \tag{14.6.3b}$$

the product $\exp^{i}(\iota\boldsymbol{\xi}\cdot\mathbf{x})\exp^{j}(\iota\boldsymbol{\zeta}\cdot\mathbf{x})$ includes a sine function of at least one coordinate variable as a factor, if i ≠ j. Hence, (14.6.1b) is proved for the case of i ≠ j.

While (14.6.1b) is only for periodic fields, (14.6.3a) also applies to homogeneous fields. Indeed, for the ith and jth parts of the homogeneous stress and strain, $\boldsymbol{\sigma}^{oi}$ and $\boldsymbol{\varepsilon}^{oj}$,

$$< \boldsymbol{\sigma}^{oi} : \boldsymbol{\varepsilon}^{oj} > = \boldsymbol{\sigma}^{oi} : \boldsymbol{\varepsilon}^{oj} = 0, \tag{14.6.1c}$$

if i ≠ j. Furthermore, (14.6.3b) applies to the MI sym/ant part of other field variables. For example, the inner product of the ith MI sym/ant part of the eigenstress field, $\mathbf{s}^{*i}$, and the jth MI sym/ant part of the strain field, $\boldsymbol{\varepsilon}^{j}$, satisfies

$$< \mathbf{s}^{*i} : \boldsymbol{\varepsilon}^{j} > = 0 \qquad \text{for } i \neq j, \tag{14.6.4}$$

which holds, even if $\mathbf{s}^{*i}$ and $\boldsymbol{\varepsilon}^{j}$ include homogeneous parts.

14.6.2. Application of MI Sym/Ant Decomposition to Energy Functional

In the same manner as shown in Subsection 13.2.2, three conclusions on the properties of energy functional J for the ith MI sym/ant part of the eigenstress field can be drawn, from the fact that $< \,:\boldsymbol{\Gamma}^{\mathrm{P}}:\, >$ is self-adjoint. Then:[13]

1) The variation of J is given by

$$\delta \mathrm{J}(\mathbf{s}^{*\mathrm{i}};\, \boldsymbol{\varepsilon}^{\mathrm{oi}}) = < \delta\mathbf{s}^{*\mathrm{i}} : \{(\mathbf{C}' - \mathbf{C})^{-1} : \mathbf{s}^{*\mathrm{i}} + \boldsymbol{\Gamma}^{\mathrm{P}} : \mathbf{s}^{*\mathrm{i}} - \boldsymbol{\varepsilon}^{\mathrm{oi}}\} >. \tag{14.6.5a}$$

Hence, the Euler equations of J,

$$(\mathbf{C}' - \mathbf{C})^{-1} : \mathbf{s}^{*\mathrm{i}} + \boldsymbol{\Gamma}^{\mathrm{P}}(\mathbf{s}^{*\mathrm{i}}) - \boldsymbol{\varepsilon}^{\mathrm{oi}} = \mathbf{0}, \tag{14.6.5b}$$

coincide with the ith MI sym/ant part of consistency conditions (13.1.14b) for the ith MI sym/ant part of the *exact* eigenstress field $\boldsymbol{\sigma}^{*\mathrm{i}}$.

2) The stationary value of the functional $\mathrm{J}(\boldsymbol{\sigma}^{*\mathrm{i}};\, \boldsymbol{\varepsilon}^{\mathrm{oi}})$, is given by

$$\mathrm{J}(\boldsymbol{\sigma}^{*\mathrm{i}};\, \boldsymbol{\varepsilon}^{\mathrm{oi}}) = -\frac{1}{2} < \boldsymbol{\sigma}^{*\mathrm{i}} > : \boldsymbol{\varepsilon}^{\mathrm{oi}} = \frac{1}{2}\boldsymbol{\varepsilon}^{\mathrm{oi}} : (\mathbf{C} - \overline{\mathbf{C}}) : \boldsymbol{\varepsilon}^{\mathrm{oi}}, \tag{14.6.6}$$

where $\overline{\mathbf{D}}$ is the overall elasticity tensor of the unit cell.

3) If tensor field $\mathbf{C}' - \mathbf{C}$ (if $\mathbf{D}' - \mathbf{D}$) is positive-definite in the whole domain of the unit cell, functional J becomes positive-definite (negative-definite), i.e., for the ith MI sym/ant part of any arbitrary eigenstress $\mathbf{s}^{*\mathrm{i}}$,

$$\mathrm{J}(\boldsymbol{\sigma}^{*\mathrm{i}};\, \boldsymbol{\varepsilon}^{\mathrm{oi}}) \leq (\geq)\, \mathrm{J}(\mathbf{s}^{*\mathrm{i}};\, \boldsymbol{\varepsilon}^{\mathrm{oi}}), \tag{14.6.7}$$

where equality holds only when $\mathbf{s}^{*\mathrm{i}} = \boldsymbol{\sigma}^{*\mathrm{i}}$; see Section 9.

Similar conclusions are also obtained for energy functional I and the ith MI sym/ant part of an eigenstress field, $\mathbf{e}^{*\mathrm{i}}$. Furthermore, the equivalence relations obtained in Subsection 13.2.3 are still valid for $\mathrm{J}(\mathbf{s}^{*\mathrm{i}};\, \boldsymbol{\varepsilon}^{\mathrm{oi}})$ and $\mathrm{I}(\mathbf{e}^{*\mathrm{i}};\, \boldsymbol{\sigma}^{\mathrm{oi}})$, as follows:

$$\begin{aligned}\mathrm{J}(\mathbf{s}^{*\mathrm{i}};\, \boldsymbol{\varepsilon}^{\mathrm{oi}}) &= -\,\mathrm{I}(-\,\mathbf{D} : \mathbf{s}^{*\mathrm{i}};\, \mathbf{C} : \boldsymbol{\varepsilon}^{\mathrm{oi}} + < \mathrm{s}^{*\mathrm{i}} >) + \frac{1}{2} < \mathrm{s}^{*\mathrm{i}} > : \mathbf{D} : < \mathrm{s}^{*\mathrm{i}} >,\\ &= -\,\mathrm{I}(-\,\mathbf{D} : \mathbf{s}^{*\mathrm{i}};\, \mathbf{C} : \boldsymbol{\varepsilon}^{\mathrm{oi}}) - \frac{1}{2} < \mathrm{s}^{*\mathrm{i}} > : \mathbf{D} : < \mathrm{s}^{*\mathrm{i}} >,\end{aligned} \tag{14.6.8}$$

for $\mathrm{i} = \pm 1, \pm 2, \pm 3$, and ± 4.

From the results obtained in Subsection 14.6.1, it is proved that $\mathrm{J}(\mathbf{s}^{*};\, \boldsymbol{\varepsilon}^{\mathrm{o}})$ is given by the sum of $\mathrm{J}(\mathbf{s}^{*\mathrm{i}};\, \boldsymbol{\varepsilon}^{\mathrm{oi}})$ for $\mathrm{i} = \pm 1, \pm 2, \pm 3$, and, ± 4. Indeed, since any arbitrary eigenstress field, $\mathbf{s}^{*}$, can be decomposed as

$$\mathbf{s}^{*}(\mathbf{x}) = \sum_{\mathrm{i}=-4}^{4} \mathbf{s}^{*\mathrm{i}}(\mathbf{x}), \tag{14.6.9}$$

direct substitution of (14.6.9) into J yields

$$\mathrm{J}(\mathbf{s}^{*};\, \boldsymbol{\varepsilon}^{\mathrm{o}}) = \frac{1}{2} < \Big\{ \sum_{\mathrm{i}=-4}^{4} \mathbf{s}^{*\mathrm{i}} \Big\} : \Big\{ (\mathbf{C}' - \mathbf{C})^{-1} : \Big\{ \sum_{\mathrm{j}=-4}^{4} \mathbf{s}^{*\mathrm{j}} \Big\}$$

$$+\, \boldsymbol{\Gamma}^{\mathrm{P}}\Big(\Big\{ \sum_{\mathrm{j}=-4}^{4} \mathbf{s}^{*\mathrm{j}} \Big\}\Big) - 2\,\boldsymbol{\varepsilon}^{\mathrm{o}} \Big\} >$$

---

[13] For simplicity, repeated superscripts designating MI sym/ant parts *are not summed* in the remainder of this subsection, unless stated otherwise.

$$= \sum_{i=-4}^{4} \left\{ \frac{1}{2} < \mathbf{s}^{*i} : \{(\mathbf{C}' - \mathbf{C})^{-1} : \mathbf{s}^{*i} + \mathbf{\Gamma}^{P}(\mathbf{s}^{*i}) - 2\,\boldsymbol{\varepsilon}^{oi}\} > \right\}$$

$$= \sum_{i=-4}^{4} J(\mathbf{s}^{*i}; \boldsymbol{\varepsilon}^{oi}). \tag{14.6.10}$$

Hence, the contribution of each MI sym/ant part of the eigenstress to the energy functional can be separated, and is given by $J(\mathbf{s}^{*i}; \boldsymbol{\varepsilon}^{oi})$.

14.6.3. Application of MI Sym/Ant Decomposition to Quadratic Forms[14]

In Subsections 13.3 and 13.5, three quadratic forms, J, $\hat{J}$, and J′ have been defined from the energy functional J. These quadratic forms are obtained by substituting a particular class of eigenstress fields into J; J is for an infinite number of Fourier series coefficients of the eigenstress fields; $\hat{J}$ is for a finite number of Fourier series coefficients of the eigenstress fields; and J′ is for a piecewise constant eigenstress field. Therefore, the MI sym/ant decomposition can be applied to the arguments of the quadratic forms. Indeed, in the same manner as (14.6.10),

$$\left\{ \begin{array}{l} J(\{F\mathbf{s}^{*}\}; \boldsymbol{\varepsilon}^{o}) \\ \hat{J}(\{F\mathbf{s}^{*}\}; \boldsymbol{\varepsilon}^{o}) \\ J'(\{\mathbf{s}^{*\alpha}\}; \boldsymbol{\varepsilon}^{o}) \end{array} \right\} = \sum_{i=-4}^{4} \left\{ \begin{array}{l} J(\{F\mathbf{s}^{*i}\}; \boldsymbol{\varepsilon}^{oi}) \\ \hat{J}(\{F\mathbf{s}^{*i}\}; \boldsymbol{\varepsilon}^{oi}) \\ J'(\{\mathbf{s}^{*\alpha i}\}; \boldsymbol{\varepsilon}^{oi}) \end{array} \right\}, \tag{14.6.11}$$

where $F\mathbf{s}^{*i}$ and $\mathbf{s}^{*\alpha i}$ are the ith MI sym/ant parts of the Fourier series coefficient, $F\mathbf{s}^{*}$, and the constant eigenstress, $\mathbf{s}^{*\alpha}$, respectively.

Besides separating the contribution of each MI sym/ant part of the eigenstress field to the energy functional and the associated quadratic form, the MI sym/ant decomposition can reduce the numerical effort required to compute the Fourier series expansions. The number of summed terms in the Fourier series is reduced to almost one-eighth, if kernel $\exp(\iota\boldsymbol{\xi}.\mathbf{x})$ is replaced by $\exp^{i}(\iota\boldsymbol{\xi}.\mathbf{x})$. For example, $\boldsymbol{\sigma}^{*i}(\mathbf{x})$ and $\mathbf{\Gamma}^{P}(\mathbf{x}; \boldsymbol{\sigma}^{*i})$ are expanded in terms of the ith MI sym/ant part of kernel function, $\exp^{i}(\iota\boldsymbol{\xi}.\mathbf{x})$, as

$$\sigma^{*i}_{pq}(\mathbf{x}) = \sum_{\boldsymbol{\xi}}{}^{+}\, 2^{K(\boldsymbol{\xi})}\, F\sigma^{*i}_{pq}(\boldsymbol{\xi}) \exp^{i'}(\iota\boldsymbol{\xi}.\mathbf{x}),$$

$$\Gamma^{P}_{pq}(\mathbf{x}; \boldsymbol{\sigma}^{*i}) = \sum_{\boldsymbol{\xi}}{}^{+}\, 2^{K(\boldsymbol{\xi})}\, F\Gamma^{P}_{pqrs}(\boldsymbol{\xi})\, F\sigma^{*i}_{rs}(\boldsymbol{\xi}) \exp^{i'}(\iota\boldsymbol{\xi}.\mathbf{x}), \tag{14.6.12a,b}$$

where superscript + on the summation emphasizes that the sum is taken for $\boldsymbol{\xi}$'s with nonnegative components, and i′ is determined by the subscripts p and q of s^{*i}_{pq} or Γ^{P}_{pq}, i.e., $i' = i'(i; p, q)$, and $K(\boldsymbol{\xi})$ denotes the number of nonzero components[15] of $\boldsymbol{\xi}$. Although superscript i′ is not fixed, for simplicity, write

---

[14] As before, a quadratic form with linear terms is involved.

[15] For simplicity, superscript 4 for $\boldsymbol{\xi}$ or $\boldsymbol{\zeta}$ of nonnegative components is omitted in this section.

(14.6.12a,b) in tensor form,

$$\boldsymbol{\sigma}^{*i}(\mathbf{x}) = \sum_{\xi}{}^{+}\, 2^{K(\xi)}\, F\boldsymbol{\sigma}^{*i}(\boldsymbol{\xi}) \exp^{i'}(\iota\boldsymbol{\xi}\cdot\mathbf{x}),$$

$$\boldsymbol{\Gamma}^{P}(\mathbf{x};\, \boldsymbol{\sigma}^{*i}) = \sum_{\xi}{}^{+}\, 2^{K(\xi)}\, F\boldsymbol{\Gamma}^{P}(\boldsymbol{\xi}) : F\boldsymbol{\sigma}^{*i}(\boldsymbol{\xi}) \exp^{i'}(\iota\boldsymbol{\xi}\cdot\mathbf{x}), \qquad (14.6.12c,d)$$

if the dependence of i′ on the tensor components is clear from the context. The detailed derivation of (14.6.12) is given in Subsection 13.5.

When the ith MI sym/ant part of kernel function, $\exp^{i}(\iota\boldsymbol{\xi}\cdot\mathbf{x})$ is used, the associated g-integrals need to be computed. Note that unlike $\exp(\iota\boldsymbol{\xi}\cdot\mathbf{x})$ which satisfies $\exp(\iota\boldsymbol{\xi}\cdot\mathbf{x})\exp(\iota\boldsymbol{\zeta}\cdot\mathbf{x}) = \exp(\iota(\boldsymbol{\xi}+\boldsymbol{\zeta})\cdot\mathbf{x})$, $\exp^{i}(\iota\boldsymbol{\xi}\cdot\mathbf{x})$ does not satisfy $\exp^{i}(\iota\boldsymbol{\xi}\cdot\mathbf{x})\exp^{i}(\iota\boldsymbol{\xi}\cdot\mathbf{x}) \neq \exp^{i}(\iota(\boldsymbol{\xi}+\boldsymbol{\zeta})\cdot\mathbf{x})$. Define $g_{U}^{ij}(\boldsymbol{\xi};\, \boldsymbol{\zeta})$ for $\exp^{i}(\iota\boldsymbol{\xi}\cdot\mathbf{x})$ and $\exp^{j}(\iota\boldsymbol{\zeta}\cdot\mathbf{x})$, corresponding to the g_U-integral for $\exp(\iota\boldsymbol{\xi}\cdot\mathbf{x})$, by

$$g_{U}^{ij}(\boldsymbol{\xi};\, \boldsymbol{\zeta}) \equiv \langle \exp^{i}(\iota\boldsymbol{\xi}\cdot\mathbf{x}) \exp^{j}(\iota\boldsymbol{\zeta}\cdot\mathbf{x}) \rangle$$

$$= \begin{cases} 2^{K(\xi)-3} & \text{for } \boldsymbol{\xi} = \boldsymbol{\zeta},\ i = j, \text{ and } \exp^{i}(\iota\boldsymbol{\xi}\cdot\mathbf{x}) \neq 0 \\ 0 & \text{otherwise.} \end{cases} \qquad (14.6.13a)$$

In a similar manner, corresponding to the g-integral, define $g_{\alpha}^{ij}(\boldsymbol{\xi};\, \boldsymbol{\zeta})$ by

$$g_{\alpha}^{ij}(\boldsymbol{\xi};\, \boldsymbol{\zeta}) \equiv \langle \exp^{i}(\iota\boldsymbol{\xi}\cdot\mathbf{x}) \exp^{j}(\iota\boldsymbol{\zeta}\cdot\mathbf{x}) \rangle_{\alpha}$$

$$= \begin{cases} \langle \exp^{i}(\iota\boldsymbol{\xi}\cdot\mathbf{x}) \exp^{j}(\iota\boldsymbol{\zeta}\cdot\mathbf{x}) \rangle_{\alpha} & \text{for } i = j \\ 0 & \text{otherwise;} \end{cases} \qquad (14.6.13b)$$

see[16] Subsection 14.6.1.

Using (14.6.12) and (14.6.13), obtain new quadratic forms, J^{i}, $\hat{J}^{i}$, and J'^{i}, for the ith MI sym/ant part of the eigenstresses of a particular form, in the same manner as J, $\hat{J}$, and J', respectively, are defined. First, consider J and $\hat{J}$. Substituting the ith MI sym/ant part of the Fourier series expansion of an eigenstress field of the same class as that used for J and $\hat{J}$, and paying attention to the orthogonality of $\exp^{i}(\iota\boldsymbol{\xi}\cdot\mathbf{x})$'s, define

$$J^{i}(\{F\mathbf{s}^{*i}\};\, \boldsymbol{\varepsilon}^{oi}) \equiv J(\{\sum_{\xi}{}^{+}\, 2^{K(\xi)}\, F\mathbf{s}^{*i}(\boldsymbol{\xi}) \exp^{i'}(\iota\boldsymbol{\xi}\cdot\mathbf{x})\};\, \boldsymbol{\varepsilon}^{o})$$

$$= \frac{1}{2}\sum_{\xi}{}^{+}\, F\mathbf{s}^{*i}(\boldsymbol{\xi}) : \{\frac{\partial J^{i}}{\partial F\mathbf{s}^{*i}(-\boldsymbol{\xi})}\} - \frac{1}{2} F\mathbf{s}^{*i}(\mathbf{0}) : \boldsymbol{\varepsilon}^{oi},$$

$$\hat{J}^{i}(\{F\mathbf{s}^{*i}\};\, \boldsymbol{\varepsilon}^{oi}) \equiv J(\{\sum_{\xi}^{N}{}^{+}\, 2^{K(\xi)}\, F\mathbf{s}^{*i}(\boldsymbol{\xi}) \exp^{i'}(\iota\boldsymbol{\xi}\cdot\mathbf{x})\};\, \boldsymbol{\varepsilon}^{o})$$

$$= \frac{1}{2}\sum_{\xi}^{N}{}^{+}\, F\mathbf{s}^{*i}(\boldsymbol{\xi}) : \{\frac{\partial J^{i}}{\partial F\mathbf{s}^{*i}(\mathbf{x})}\} - \frac{1}{2} F\mathbf{s}^{*i}(\mathbf{0}) : \boldsymbol{\varepsilon}^{oi}, \qquad (14.6.14a,15a)$$

---

[16] The proof of $g_{\alpha}^{ij}(\boldsymbol{\xi};\, \boldsymbol{\zeta}) = 0$ is essentially similar to (14.6.3); the geometry (shape and location) of Ω_{α} must be fully symmetric with respect to all three coordinates, and hence, the volume integral of a sine function over Ω_{α} vanishes.

where the derivative of these quadratic forms are expressed in component form as

$$\frac{\partial J^{i}}{\partial Fs^{*i}_{pq}(\boldsymbol{\xi})} = \sum_{\boldsymbol{\zeta}}{}^{+} \left\{ \sum_{\alpha=0}^{n} \{f_{\alpha}\, g_{\alpha}^{i'i''}(\boldsymbol{\xi};\, \boldsymbol{\zeta})\, (\mathbf{C}^{\alpha} - \mathbf{C})^{-1}_{pqrs}\} \right\} Fs^{*i}_{rs}(\boldsymbol{\zeta})$$

$$+ F\Gamma^{P}_{pqrs}\, Fs^{*i}_{rs}(\boldsymbol{\xi}) - g_{U}^{i'i'}(\boldsymbol{\xi};\, \mathbf{0})\, \varepsilon^{oi}_{pq},$$

$$\frac{\partial \hat{J}^{i}}{\partial Fs^{*i}_{pq}(\boldsymbol{\xi})} = \sum_{\boldsymbol{\zeta}}^{N}{}^{+} \left\{ \sum_{\alpha=0}^{n} f_{\alpha}\, g_{\alpha}^{i'i''}(\boldsymbol{\xi};\, \boldsymbol{\zeta})\, (\mathbf{C}^{\alpha} - \mathbf{C})^{-1}_{pqrs}\} \right\} Fs^{*i}_{rs}(\boldsymbol{\zeta})$$

$$+ F\Gamma^{P}_{pqrs}\, Fs^{*i}_{rs}(\boldsymbol{\xi}) - g_{U}^{i'i'}(\boldsymbol{\xi};\, \mathbf{0})\, \varepsilon^{oi}_{pq}, \qquad (14.6.14b,15b)$$

where $i' = i'(i;\, p,\, q)$ and $i'' = i''(i;\, r,\, s)$. In the same manner as the derivatives of J and $\hat{J}$ correspond to the Fourier series expansion of the consistency conditions and the truncated Fourier series expansion of the consistency conditions, (14.6.14b) and (14.6.15b) correspond to the ith MI sym/ant part of the Fourier series expansion of the consistency conditions and the truncated Fourier series expansion of the consistency conditions, respectively.

In order to obtain the extremum (stationary) value of the above-defined quadratic form, $\hat{J}^{i}$, which is obtained by truncating the infinite Fourier series expansion in J^{i} up to N, $(N+1)^3$ terms have to be summed. On the other hand, to obtain the extremum value of the original $\hat{J}$, $(2N+1)^3$ terms must be summed. Furthermore, the g-integrals involved in $\hat{J}^{i}$ are either real or purely imaginary, while the g-integrals involved in J are complex-valued. Besides the decomposition of the quadratic form $\hat{J}$ into eight mutually independent $\hat{J}^{i}$'s, the MI sym/ant decomposition of the field variables reduces the numerical effort by almost an order of magnitude, without changing the mathematical structure of the original quadratic form.

Next, consider J′. From the definition[17] of the correlation tensor $\boldsymbol{\Gamma}^{P\alpha\beta}$, (13.4.17a), it follows that

$$< \boldsymbol{\Gamma}^{P}(H_{\beta}\mathbf{s}^{*\beta i}) >_{\alpha} = f_{\beta}\, \boldsymbol{\Gamma}^{P\alpha\beta} : \mathbf{s}^{*\beta i} \qquad (\beta \text{ not summed}). \qquad (14.6.16)$$

Hence, the quadratic form for $\mathbf{s}^{*\alpha i}$ coincides with that for $\mathbf{s}^{*\alpha}$, i.e.,

$$J'(\{\mathbf{s}^{*\alpha i}\};\, \boldsymbol{\varepsilon}^{oi}) \equiv J(\{\sum_{\alpha=0}^{n} H_{\alpha}\mathbf{s}^{*\alpha i}\};\, \boldsymbol{\varepsilon}^{o})$$

$$= \frac{1}{2} \sum_{\alpha=0}^{n} \mathbf{s}^{*\alpha i} : \left\{ (\mathbf{C}^{\alpha} - \mathbf{C})^{-1} : \mathbf{s}^{*\alpha i} + \right.$$

[17] Since $\mathbf{C}'$ satisfies the fourth MI sym/ant condition, the computation of $\boldsymbol{\Gamma}^{P\alpha\beta}$ is simplified by applying the MI sym/ant decomposition to the g-integrals; see Subsection 14.6.3.

$$+\sum_{\beta=0}^{n} f_\beta\, \mathbf{\Gamma}^{P\alpha\beta} : \mathbf{s}^{*\beta i} - 2\,\boldsymbol{\varepsilon}^{oi}\Big\}; \tag{14.6.17}$$

see Subsection 13.4.4. Therefore, from (14.6.17), it is seen that for piecewise constant eigenstresses, the same quadratic form J′ is obtained, even if the ith MI sym/ant part of the eigenstress field is used.

14.6.4. Two-Phase Periodic Structure

As a simple example, consider a two-phase periodic structure consisting of the matrix phase M and the inclusion phase Ω, to demonstrate the effectiveness of the MI sym/ant decomposition. For $\mathbf{C}'$ to satisfy the fourth MI sym/ant condition, the geometry (shape and location) of Ω must be fully symmetric with respect to all three coordinates. Hence, as shown in Subsection 14.6.3, the g-integral for Ω must satisfy

$$g^{ij}(\boldsymbol{\xi}; \boldsymbol{\zeta}) = 0 \qquad \text{if } i \neq j. \tag{14.6.18}$$

For simplicity, subscript Ω is omitted in $g^{ij}(\boldsymbol{\xi}; \boldsymbol{\zeta})$.

First, consider the case of the (truncated) Fourier series expansion of the eigenstress field. In expressing a set of tensorial equations in matrix form, as in Subsection 13.4.3, with the aid of (14.6.18), certain components of the matrix may be set to be zero. To illustrate this, for a fourth-order tensor, say, $(\mathbf{C}^\Omega - \mathbf{C})^{-1}$ which appears in (14.6.17), express the product with $g^{i'i''}$ in a six by six matrix. From the definition of i′, i′(i; 1, 1) = i′(i; 2, 2) = i′(i; 3, 3), i′(i; 2, 3), i′(i; 3, 1), and i′(i; 1, 2) are different for i = ±1, ±2, ±3, and ±4. Indeed, for the case of i = 4,

$$i'(4; 1, 1) = i'(4; 2, 2) = i'(4; 3, 3) = 4,$$

$$i'(4; 2, 3) = -1, \qquad i'(4; 3, 1) = -2, \qquad i'(4; 1, 2) = -3. \tag{14.6.19a~d}$$

Then, denoting $(\mathbf{C}^\alpha - \mathbf{C})^{-1}$ by $\mathbf{\Delta}$, express $g^{i'i''}\,\mathbf{\Delta}$ as[18]

$$[g^{i'i''}\Delta_{ab}] = \begin{bmatrix} g^{i^4 i^4}\Delta_{1111} & g^{i^4 i^4}\Delta_{1122} & g^{i^4 i^4}\Delta_{1133} & 0 & 0 & 0 \\ g^{i^4 i^4}\Delta_{2211} & g^{i^4 i^4}\Delta_{2222} & g^{i^4 i^4}\Delta_{2233} & 0 & 0 & 0 \\ g^{i^4 i^4}\Delta_{3311} & g^{i^4 i^4}\Delta_{3322} & g^{i^4 i^4}\Delta_{3333} & 0 & 0 & 0 \\ 0 & 0 & 0 & g^{i^{-1} i^{-1}}\Delta_{2323} & 0 & 0 \\ 0 & 0 & 0 & 0 & g^{i^{-2} i^{-2}}\Delta_{3131} & 0 \\ 0 & 0 & 0 & 0 & 0 & g^{i^{-3} i^{-3}}\Delta_{1212} \end{bmatrix}, \tag{14.6.20}$$

where

[18] This also applies to the case of multi-phase periodic structures, if each inclusion phase geometrically satisfies the fourth MI sym/ant condition.

$$i^4 \equiv i'(4;\, 1,\, 1) \equiv i'(4;\, 2,\, 2) \equiv i'(4;\, 3,\, 3),$$

$$i^{-1} \equiv i'(i;\, 2,\, 3), \qquad i^{-2} \equiv i'(i;\, 3,\, 1), \qquad i^{-3} \equiv i'(i;\, 1,\, 2); \tag{14.6.21a~d}$$

see (14.6.19). Therefore, in expressing the quadratic forms in matrix form, two three by three matrices need to be considered, instead of a six by six matrix, because the six by six matrix corresponding to tensorial equations may be written as

$$[(...)] = \begin{bmatrix} [(...)_{ab}]^{(1)} & [0_{ab}] \\ [0_{ab}] & [(...)_{ab}]^{(2)} \end{bmatrix}, \tag{14.6.22}$$

where $[(...)_{ab}]^{(1)}$ and $[(...)_{ab}]^{(2)}$ are three by three matrices, and $[(...)_{ab}]^{(2)}$ is diagonal.

Next, consider the case of the piecewise constant distribution of an eigenstress field. As shown in Subsection 14.6.3, the reduction of the Fourier series expansion due to the MI sym/ant decomposition does not appear in the average consistency conditions nor in the quadratic forms which are defined from the energy functional. However, the six by six matrix representation of the correlation tensor, say, $\mathbf{\Gamma}^{P\Omega\Omega}$, attains the form given by (14.6.22), if the geometry of Ω is fully symmetric with respect to all three coordinate planes. In this case, the g-integral satisfies

$$g(\boldsymbol{\xi}^i) = g(\boldsymbol{\xi}), \tag{14.6.23a}$$

for $i = \pm 1, \pm 2, \pm 3$, and ± 4, and components of $F\mathbf{\Gamma}^P$ satisfy

$$F\Gamma^P_{pqrs}(\boldsymbol{\xi}^i) = \begin{cases} +F\Gamma^P_{pqrs}(\boldsymbol{\xi}) & \text{for } i'(i;\, p,\, q) = +i'(i;\, r,\, s) \\ -F\Gamma^P_{pqrs}(\boldsymbol{\xi}) & \text{for } i'(i;\, p,\, q) = -i'(i;\, r,\, s). \end{cases} \tag{14.6.23b}$$

The proof is straightforward: since $\mathbf{C}$ satisfies the fourth MI sym/ant condition, from the definition of $F\Gamma^P_{pqrs}$ and $\boldsymbol{\xi}^i$,

$$F\Gamma^P_{pqrs}(\boldsymbol{\xi}^i) = sym\,\{\xi^i_p\,(\xi^i_t\,C_{tqru}\,\xi^i_u)^{-1}\,\xi^i_s\}$$

$$= \alpha^{ip}\,\alpha^{iq}\,\alpha^{ir}\,\alpha^{is}\,sym\,\{\xi^4_p\,(\xi^4_t\,C_{tqru}\xi^4_u)^{-1}\,\xi^4_s\}. \tag{14.6.23c}$$

Since α^{ij} takes on either $+1$ or -1, obtain (14.6.23b). As shown in (13.4.17b), $\mathbf{\Gamma}^{P\Omega\Omega}$ is the sum of $g_\Omega(-\boldsymbol{\xi})\,g_\Omega(\boldsymbol{\xi})\,F\mathbf{\Gamma}^P(\boldsymbol{\xi})$ for all $\boldsymbol{\xi}$'s, and hence, the matrix representation of $\mathbf{\Gamma}^{P\Omega\Omega}$ is

$$[\Gamma^{P\Omega\Omega}_{ab}] = \begin{bmatrix} [\Gamma^{P\Omega\Omega}_{ab}]^{(1)} & [0_{ab}] \\ [0_{ab}] & [\Gamma^{P\Omega\Omega}_{ab}]^{(2)} \end{bmatrix}, \tag{14.6.24}$$

where $[\Gamma^{P\Omega\Omega}_{ab}]^{(1)}$ and $[\Gamma^{P\Omega\Omega}_{ab}]^{(2)}$ are three by three matrices, and $[\Gamma^{P\Omega\Omega}_{ab}]^{(2)}$ is diagonal.[19] The other correlation tensors, $\mathbf{\Gamma}^{P\Omega M} = \mathbf{\Gamma}^{PM\Omega}$ and $\mathbf{\Gamma}^{PMM}$, can be expressed in the same matrix form as (14.6.24).

---

[19] The same comments apply to the approximate correlation tensors obtained by truncating the infinite sum of $g_\Omega(-\boldsymbol{\xi})\,g_\Omega(\boldsymbol{\xi})\,F\mathbf{\Gamma}^P(\boldsymbol{\xi})$. Furthermore, for a multi-phase periodic structure, either the exact or the approximate correlation tensors for any two Ω_α and Ω_β, can be expressed in the form (14.6.24).

In terms of these correlation tensors, the average consistency conditions, or the tensorial equations which optimize the energy functional , $\partial J'/\partial \mathbf{s}^{*M} = \mathbf{0}$ and $\partial J'/\partial \mathbf{s}^{*\Omega} = \mathbf{0}$, for the two-phase periodic structure, are expressed as

$$(\mathbf{C}^{M} - \mathbf{C})^{-1} : \boldsymbol{\sigma}^{*M} + \{\, (1-f)\, \boldsymbol{\Gamma}^{PMM} : \boldsymbol{\sigma}^{*M} + f\, \boldsymbol{\Gamma}^{PM\Omega} : \boldsymbol{\sigma}^{*\Omega} \,\} - \boldsymbol{\varepsilon}^{o} = \mathbf{0},$$

$$(\mathbf{C}^{\Omega} - \mathbf{C})^{-1} : \boldsymbol{\sigma}^{*\Omega} + \{\, (1-f)\, \boldsymbol{\Gamma}^{P\Omega M} : \boldsymbol{\sigma}^{*M} + f\, \boldsymbol{\Gamma}^{P\Omega\Omega} : \boldsymbol{\sigma}^{*\Omega} \,\} - \boldsymbol{\varepsilon}^{o} = \mathbf{0}; \tag{14.6.25a,b}$$

see Subsection 14.6.3. Therefore, if $\mathbf{C}$, $\mathbf{C}^{M}$, and $\mathbf{C}^{\Omega}$ are expressed in a matrix similar to (14.6.24)[20], then, the matrix form of (14.6.25a,b) becomes

$$([C_{ab}^{M(I)}] - [C_{ab}^{(I)}])^{-1}\, [\sigma_b^{*M(I)}] + \{(1-f)\, [\Gamma_{ab}^{PMM(I)}\, [\sigma_b^{*M(I)}]$$
$$+ f\, [\Gamma_{ab}^{PM\Omega(I)}][\sigma_b^{*\Omega(I)}]\} - [\varepsilon_a^{o(I)}] = [0_a],$$

$$([C_{ab}^{\Omega(I)}] - [C_{ab}^{(I)}])^{-1}\, [\sigma_b^{*\Omega(I)}] + \{(1-f)\, [\Gamma_{ab}^{P\Omega M(I)}\, [\sigma_b^{*M(I)}]$$
$$+ f\, [\Gamma_{ab}^{P\Omega\Omega(I)}][\sigma_b^{*\Omega(I)}]\} - [\varepsilon_a^{o(I)}] = [0_a], \tag{14.6.25c,d}$$

for $I = 1, 2$, where $[\varepsilon_a^{o(I)}]$ is a three by one vector, defined from a six by one vector $[\varepsilon_a]$ through $[\,[\varepsilon_a^{o(1)}]^T\, [\varepsilon^{a o(2)}]^T\,]^T \equiv [\varepsilon_a]$, and $[0_a]$ is a three by one zero vector. The solution to (14.6.25c,d) is

$$[\sigma_a^{*M(I)}] = \Big\{\, [\Omega_{ac}^{(I)}] - f(1-f)\, [\Gamma_{ad}^{M\Omega(I)}]\, [M_{de}^{(I)}]^{-1}\, [\Gamma_{ec}^{M\Omega(I)}] \,\Big\}^{-1}$$
$$\Big\{\, [1_{cb}] - f\, [\Gamma_{cf}^{M\Omega(I)}]\, [M_{fb}^{(I)}]^{-1} \,\Big\}\, [\varepsilon_b^{*(I)}],$$

$$[\sigma_a^{*\Omega(I)}] = \Big\{\, [M_{ac}^{(I)}] - f(1-f)\, [\Gamma_{ad}^{M\Omega(I)}]\, [\Omega_{de}^{(I)}]^{-1}\, [\Gamma_{ec}^{M\Omega(I)}] \,\Big\}^{-1}$$
$$\Big\{\, [1_{cb}] - (1-f)\, [\Gamma_{cf}^{M\Omega(I)}]\, [\Omega_{fb}^{(I)}]^{-1} \,\Big\}\, [\varepsilon_b^{*(I)}], \tag{14.6.26a,b}$$

where the three by three matrices, $[M_{ab}^{(I)}]$ and $[\Omega_{ab}^{(I)}]$, are defined by

$$[M_{ab}^{(I)}] \equiv [C_{ab}^{M(I)} - C_{ab}^{(I)}]^{-1} + (1-f)\, [\Gamma_{ab}^{MM(I)}],$$

$$[\Omega_{ab}^{(I)}] \equiv [C_{ab}^{\Omega(I)} - C_{ab}^{(I)}]^{-1} + f\, [\Gamma_{ab}^{\Omega\Omega(I)}], \tag{14.6.26c,d}$$

and $[1_{ab}]$ is a three by three identity matrix. It should be noted that since $[C_{ab}^{(2)}]$, $[C_{ab}^{M(2)}]$, and $[C_{ab}^{\Omega(2)}]$ are diagonal, (14.6.26) for $I = 2$ reduces to the corresponding scalar equation for the diagonal components.

---

[20] As shown in Section 3, if the matrix or the inclusion satisfies a certain material symmetry, the matrix form of its elasticity tensor, $\mathbf{C}^{M}$ or $\mathbf{C}^{\Omega}$, is given by (14.6.24). Note that the reference elasticity, $\mathbf{C}$, can always be chosen to satisfy a certain desired symmetry.

As shown in Subsection 14.6.3, (14.6.26) can be applied to any MI sym/ant part of the homogeneous strain and the corresponding eigenstress. Since the fourth MI sym/ant part of $\boldsymbol{\varepsilon}^{\rm o}$ is $\varepsilon^{\rm o}_{\rm pp}$, obtain the corresponding fourth MI sym/ant part of the eigenstresses, $\sigma^{*\rm M}_{\rm pp}$ and $\sigma^{*\Omega}_{\rm pp}$, from (14.6.26) for I = 1 (p not summed). For the other MI sym/ant parts of $\boldsymbol{\varepsilon}^{\rm o}$, $\varepsilon^{\rm o}_{23}$, $\varepsilon^{\rm o}_{31}$, or $\varepsilon^{\rm o}_{12}$, (corresponding to the – 1st, – 2nd, – 3rd MI sym/ant part), obtain[21] the corresponding eigenstresses from (14.6.26) for I = 2.

14.7. REFERENCES

Fotiu, P. and S. Nemat-Nasser (1995), Overall properties of elastic viscoplastic periodic composites, *Int'l J. Plasticity*, Vol. 12, No. 2, 163-190.

[21] Since (14.6.26) for I = 2 can be reduced to scalar equations for the diagonal components of the matrices, the three MI sym/ant parts of the eigenstresses are obtained independently.

APPENDIX A APPLICATION TO INELASTIC HETEROGENEOUS SOLIDS

The material presented in Part 1 provides a fundamental framework for quantitative evaluation of overall properties and failure modes of a broad class of solids with microheterogeneities and defects. While the concepts are presented and illustrated in the context of linear elasticity, essentially the entire formulation translates directly and applies to *inelastic* solids within the framework of small-deformation theories, provided expressions are properly interpreted in terms of appropriate strain and stress rates, or strain and stress increments. Indeed, with suitable interpretation of the rate quantities, the theory also applies to geometrically nonlinear problems involving finite deformations and rotations. This appendix provides a brief guide for the application of the basic results presented in this Part 1, to heterogeneous solids with inelastic constituents.

A.1. SOURCES OF INELASTICITY

Geometrical changes at the microscale often produce irreversible deformations and hence, serve as sources of inelasticity. Microcracking and microcrack growth are examples which are treated in some detail in Section 6. Cavitation and void growth, dislocation, twinning, and phase transformation are other examples of micro-events which lead to a macroscopic inelastic response of materials. These processes are generally highly history-, temperature-, and rate-dependent. Since a detailed examination of any of these basic issues will require a treatise in its own right, here attention is limited to a cursory examination of a few basic concepts which are presented to provide guidance for the reader in seeking to apply the general theories of the preceding sections to problems of this kind.

To be specific, two classes of inelastic solids are briefly examined: (1) solids whose constituents may be characterized through *phenomenological* inelastic models such as rate-independent plasticity and rate-dependent viscoplasticity; and (2) inelastic solids whose constituent properties require description at a yet-smaller length-scale. An example of this second class is slip-induced plastic deformation of single-crystal constituents in a polycrystal aggregate. Examples of the first class of problems include metal-matrix composites, such as aluminum matrix-alumina inclusions, aluminum matrix-silicon inclusions, intermetallics with various toughening or hardening precipitates and inclusions, such as titanium-aluminide, nickel-aluminide, niobium-silicide, and related alloys with toughening ductile or hardening brittle reinforcements, and finally, ceramic matrix-metal composites, such as boron carbide-aluminum cermets.

In the remainder of this appendix, illustrative constitutive relations for both classes of problems mentioned above are briefly presented, and their implementation in terms of the general theories of the preceding sections is pointed out. First, certain rate-independent phenomenological plasticity theories are outlined, with a brief examination of slip-induced crystal plasticity. Then their interpretation in terms of rate-dependent processes is mentioned. A general review of large-deformation phenomenological plasticity with an extensive list of references is given by Nemat-Nasser (1992), and a comprehensive account of single-crystal plasticity has been provided by Havner (1992). In view of these timely expositions, no attempt is made in this appendix to provide a comprehensive literature account.

A.2. RATE-INDEPENDENT PHENOMENOLOGICAL PLASTICITY

The classical rate-independent plasticity is based on the concept of the yield surface, either in stress space or in strain space. Traditionally, these theories have been developed for pressure-insensitive plastically incompressible solids, approximating in general the plastic response of metals. For application to porous metals and other materials such as granular masses where pressure sensitivity and inelastic volumetric strain are significant and often dominating features, the classical theories have been modified to include pressure and volumetric effects.

In the absence of microstructural changes, the response of elastoplastic materials is purely elastic; see, e.g., Havner (1982, 1992), Hill (1950, 1972, 1978), Naghdi (1960), and Nemat-Nasser (1983, 1992). The *rate* of change of the strain decomposes *exactly* into an elastic, $\dot{\boldsymbol{\varepsilon}}^{el}$, and a plastic, $\dot{\boldsymbol{\varepsilon}}^{pl}$, contribution as[1]

$$\dot{\boldsymbol{\varepsilon}} = \dot{\boldsymbol{\varepsilon}}^{el} + \dot{\boldsymbol{\varepsilon}}^{pl}. \tag{A.2.1a}$$

The stress rate relates to the elastic part of the strain rate by the elasticity relation

$$\dot{\boldsymbol{\sigma}} = \mathbf{C} : \dot{\boldsymbol{\varepsilon}}^{el}, \tag{A.2.1b}$$

where $\dot{\boldsymbol{\sigma}}$ is the stress rate,[2] and $\mathbf{C}$ is the current elasticity tensor, with inverse $\mathbf{D}$, the corresponding compliance.[3]

---

[1] This holds even at large strains and rotations, independently of the particular material strain measure or the reference state which may be employed, as long as the same reference state is used to measure the elastic and inelastic rates; Nemat-Nasser (1979, 1982).

[2] For finite deformations, an objective stress rate must be used.

[3] Note that both of these tensors depend on and vary with the strain measure and the reference state, and both are symmetric, e.g., $C_{ijkl} = C_{klij} = C_{jikl} = C_{ijlk}$, where a fixed rectangular Cartesian coordinate system is used.

The classical theory of elastoplasticity is based on the concept of the yield surface which, either in the stress space or in the strain space, defines a region within which the material response is elastic and on which the response may be elastoplastic. The shape and size of the yield surface vary with the history of the inelastic deformation. It may be smooth, with a continuously turning tangent plane, or it may possess corners or vertices, each consisting of a set of smooth intersecting surfaces. The existence of such vertices is an integral part of the physics of crystal plasticity. It plays a dominant role in phenomena such as instability by localized deformation, necking, and shear banding. In the sequel, first, cases associated with a smooth yield surface are considered, and then the response at a vertex is briefly discussed.

A.2.1. Constitutive Relations: Smooth Yield Surface

Rate-independent phenomenological plasticity essentially generalizes the results of uniaxial and torsional deformation of metals, where over a range of stresses (or strains) the sample behaves essentially elastically, but once a critical stress (strain) is reached, the response becomes elastoplastic in continued loading, and elastic upon unloading. The theory has been formulated in both stress and strain space; Hill (1967a, 1978).

In stress space the yield surface marks elastoplastic states. For stress points within the yield surface the material response is elastic. For points on the yield surface the response is elastoplastic. When the change in the stress state tends to lead into the yield surface, the material element undergoes elastic *unloading*. When the stress point moves on the current yield surface, the material element undergoes *neutral loading*. When the change in the stress state tends to lead out of the current yield surface, the material element undergoes *loading*. In this case the yield surface moves with the stress point and the stress point remains always on the yield surface. The yield surface may expand, or simply move, or do both. This is called *work-hardening*. When the yield surface expands self-similarly, the work-hardening is called *isotropic hardening*. On the other hand, when the yield surface does not change size but simply moves with the stress point, the material is said to be *kinematically hardening*. In general, a combined *isotropic-kinematic hardening* can be considered. These are idealizations, since in general, the yield surface changes shape in a complex manner, as is the case for polycrystalline metals; Hutchinson (1970), Kocks (1970), Iwakuma and Nemat-Nasser (1984), and Nemat-Nasser and Obata (1986).

Let $\boldsymbol{\mu}$ be normal to the yield surface in the stress space, and let $\mathbf{m}$ define the direction of the plastic strain. Then,

$$\dot{\boldsymbol{\varepsilon}} = \dot{\boldsymbol{\varepsilon}}^{el} + \dot{\boldsymbol{\varepsilon}}^{pl} = \boldsymbol{D}' : \dot{\boldsymbol{\sigma}}, \qquad \boldsymbol{D}' = \mathbf{D} + \mathbf{m} \otimes \boldsymbol{\mu}, \tag{A.2.2a,b}$$

where $\boldsymbol{D}'$ is the *instantaneous elastoplastic compliance tensor*.

In a similar manner, let $\boldsymbol{\lambda}$ be in the direction of the normal to the yield surface at a point in the strain space, and denote by $\boldsymbol{l}$ the direction of the inelastic stress "decrement". Then

$$\dot{\boldsymbol{\sigma}} = \boldsymbol{C}' : \dot{\boldsymbol{\varepsilon}}, \qquad \boldsymbol{C}' = \mathbf{C} - \boldsymbol{l} \otimes \boldsymbol{\lambda}, \tag{A.2.3a,b}$$

where $\boldsymbol{C}'$ is the *instantaneous elastoplastic modulus* tensor. It is required that $\boldsymbol{C}'$ and $\boldsymbol{D}'$ be each other's inverse. From this and the symmetry of the elastic modulus and compliance tensors, it follows that (Hill, 1967b)

$$\boldsymbol{\lambda} = \mathbf{C} : \boldsymbol{\mu}, \qquad \boldsymbol{\mu} = \mathbf{D} : \boldsymbol{\lambda}, \qquad \boldsymbol{l} = \frac{\mathbf{C} : \mathbf{m}}{1 + \boldsymbol{\lambda} : \mathbf{m}}, \qquad \mathbf{m} = \frac{\mathbf{D} : \boldsymbol{l}}{1 - \boldsymbol{\mu} : \boldsymbol{l}},$$

$$(1 + \boldsymbol{\lambda} : \mathbf{m})(1 - \boldsymbol{\mu} : \boldsymbol{l}) = 1. \tag{A.2.4a~e}$$

The conditions for loading or unloading are: $\boldsymbol{\lambda} : \dot{\boldsymbol{\varepsilon}} > 0$ for loading, $= 0$ for neutral loading, and < 0 for unloading. The corresponding conditions in the stress space are obtained by direct substitution. In addition, the conditions for *strain-hardening, perfect plasticity*, or *strain-softening* respectively, correspond to $\boldsymbol{\mu} : \boldsymbol{l} < 1$, $= 1$, and > 1.

A.2.2. Flow Potential and Associative Flow Rule

In the stress space, consider the following yield and flow potential surfaces:

$$\mathrm{f} = \mathrm{f}(\boldsymbol{\sigma}, \ldots) = 0, \qquad \mathrm{g} = \mathrm{g}(\boldsymbol{\sigma}, \ldots), \tag{A.2.5a,b}$$

where dots stand for temperature and for internal variables that characterize material hardening. The plastic strain rate is

$$\dot{\boldsymbol{\varepsilon}}^{pl} = \dot{\gamma} \frac{\partial \mathrm{g}}{\partial \boldsymbol{\sigma}}. \tag{A.2.5c}$$

Comparison with (A.2.2) shows that $\mathbf{m} = \partial \mathrm{g}/\partial \boldsymbol{\sigma}$ and $\dot{\gamma} = \boldsymbol{\mu} : \dot{\boldsymbol{\sigma}}$. For continued plastic flow, $\dot{\mathrm{f}} = (\partial \mathrm{f}/\partial \boldsymbol{\sigma}) : \dot{\boldsymbol{\sigma}} + \ldots = 0$, from which it follows that $\dot{\gamma} = \mathrm{H}^{-1}(\partial \mathrm{f}/\partial \boldsymbol{\sigma}) : \dot{\boldsymbol{\sigma}}$ and $\boldsymbol{\mu} = \mathrm{H}^{-1}(\partial \mathrm{f}/\partial \boldsymbol{\sigma})$, where H, the work-hardening parameter, depends on the manner by which the yield surface changes in the course of plastic flow. Hence, the instantaneous compliance tensor is defined by

$$\boldsymbol{D}' \equiv \mathbf{D} + \frac{1}{\mathrm{H}} \frac{\partial \mathrm{g}}{\partial \boldsymbol{\sigma}} \otimes \frac{\partial \mathrm{f}}{\partial \boldsymbol{\sigma}}. \tag{A.2.6a}$$

The inverse relation is easily obtained by calculating $\boldsymbol{\lambda}$ and $\boldsymbol{l}$ in terms of $\boldsymbol{\mu}$ and $\mathbf{m}$,

$$\boldsymbol{C}' = \mathbf{C} - \left\{\mathrm{H} + \frac{\partial \mathrm{f}}{\partial \boldsymbol{\sigma}} : \mathbf{C} : \frac{\partial \mathrm{g}}{\partial \boldsymbol{\sigma}}\right\}^{-1} \left\{\mathbf{C} : \frac{\partial \mathrm{g}}{\partial \boldsymbol{\sigma}}\right\} \otimes \left\{\mathbf{C} : \frac{\partial \mathrm{f}}{\partial \boldsymbol{\sigma}}\right\}. \tag{A.2.6b}$$

Equations (A.2.6a,b) are the required instantaneous compliance and modulus tensors of the material. As is seen, these are dependent on the stress state at each material point. They apply to plastically compressible or incompressible cases. Note that the dependence of the yield function and flow potential on the internal state variables can be quite general, including any desired work-hardening rules.

When the functions f and g are identical, tensors $\boldsymbol{\mu}$ and $\mathbf{m}$ become unidirectional in the stress space. In this case the plasticity is said to be governed by an *associative flow rule*, otherwise the flow rule is said to be *nonassociative*.

Some examples are given in the following subsections.

A.2.3. The J_2-Flow Theory with Isotropic Hardening

A widely used phenomenological plasticity is the so-called J_2-flow theory. The yield function and the flow potential are defined by

$$f \equiv g \equiv \tau - F(\gamma),$$

$$\tau = (\tfrac{1}{2}\boldsymbol{\sigma}' : \boldsymbol{\sigma}')^{1/2} \equiv \sqrt{J_2}, \qquad \dot{\gamma} = (2\dot{\boldsymbol{\varepsilon}}^{pl\,\prime} : \dot{\boldsymbol{\varepsilon}}^{pl\,\prime})^{1/2}, \tag{A.2.7a~c}$$

where prime denotes the deviatoric part and F is an arbitrary function.

Here the plastic flow is volume-preserving and pressure-independent. Hence, hydrostatic pressure or tension induces only elastic deformation. The deviatoric plastic deformation rate is given by

$$\dot{\boldsymbol{\varepsilon}}^{pl\,\prime} = \frac{(\boldsymbol{\mu} : \boldsymbol{\sigma}')}{2H}\,\boldsymbol{\mu}, \qquad \frac{\boldsymbol{\mu}}{\sqrt{2}} = \frac{\boldsymbol{\sigma}'}{2\tau}, \qquad H = \frac{\partial F}{\partial \gamma}. \tag{A.2.8a~c}$$

In the deviatoric stress space, $f = 0$ for F positive, is a sphere of squared radius $2F^2$. The quantity τ is the *effective stress*, and γ is the *effective plastic strain*. In pure shearing, τ is the shear stress and γ is the corresponding shear strain. Then $d\tau/d\gamma = F'$ defines the slope of the shear stress versus the plastic shear strain curve. Depending on whether F' is positive, zero, or negative, the radius of the yield surface increases (work-hardening), remains constant (perfect plasticity), or decreases (work-softening).

The final instantaneous compliance tensor is

$$\boldsymbol{D}' = \mathbf{D} + \frac{\boldsymbol{\sigma}' \otimes \boldsymbol{\sigma}'}{4\tau^2 H}, \tag{A.2.9a}$$

with the following inverse:

$$\boldsymbol{C}' = \mathbf{C} - \frac{(\mathbf{C} : \boldsymbol{\sigma}') \otimes (\mathbf{C} : \boldsymbol{\sigma}')}{4\tau^2 H + \boldsymbol{\sigma}' : \mathbf{C} : \boldsymbol{\sigma}'}. \tag{A.2.9b}$$

It is seen from (A.2.9a,b) that for elastoplastic materials, the instantaneous compliance and modulus tensors are dependent on the stress state and the history of deformation. Hence, the eigenstrain or eigenstress fields necessary for homogenization of a composite with elastoplastic constituents in general will have nonuniform spatial variation, and change as the deformation develops. In this case, the composite may be subdivided into elements, and piecewise constant eigenstrains or eigenstresses may be considered for homogenization. This procedure has been used by Accorsi and Nemat-Nasser (1986) and Nemat-Nasser *et al.* (1986) for periodic microstructures. Since in the periodic case, the unit cell is finite, at least in principle it is possible to homogenize the unit cell to any desired degree of accuracy by piecewise constant fields.

Alternative approximate procedures have been developed and applied to calculate the overall elastic-plastic response of composites consisting of an inelastic matrix and inelastic inclusions. A procedure which seeks to include both the inclusion-matrix and the inclusion-inclusion interactions, has emerged through the work of several authors, e.g., Taya and Chou (1981), Taya and Mura (1981), and Weng (1981, 1982, 1984, 1990). The method exploits a simplified version of Hill's (1965) self-consistent formulation of the overall response of elastic-plastic aggregates (Berveiller and Zaoui, 1979) in order to account in an approximate manner for the inclusion-matrix interaction, and the Mori-Tanaka (1973) mean field approach in order to incorporate the inclusion-inclusion interaction. The approximation involved in the modification of the self-consistent method limits the theory to proportional loading. The method has been applied to other problems, e.g., to study the inclusion-shape effect, again for proportional loading. Other related studies are Tandon and Weng (1988), Teply and Dvorak (1988), and Dvorak (1992).

A separate and rigorous approach to nonlinear composites has been initiated by Willis (1983), where the Hashin-Shtrikman variational principle is extended for application to nonlinear problems. This line of thought has been continued by Talbot and Willis (1985, 1986, 1987, 1992, 1997), Duva (1984, 1986), Duva and Hutchinson (1984), Ponte Castañeda and Willis (1988, 1995), Willis (1989, 1991, 1994), and Lee and Mear (1992a,b), who use the extended variational principles to produce bounds on the overall parameters of nonlinear composites; for an account and references to this line of approach, see Ponte Castañeda and Suquet (1998).

A.2.4. The J_2-Flow Theory with Kinematic Hardening

For a purely kinematic hardening model, the radius of the yield surface in the deviatoric stress space remains constant, while its center moves as plastic flow takes place. Let $\boldsymbol{\beta}$ be a symmetric *deviatoric* second-order tensor defining the location of the center of the yield surface in the deviatoric stress space. The yield surface is then expressed as

$$f \equiv \hat{\tau} - \tau_Y^0, \tag{A.2.10a}$$

where τ_Y^0 is a constant defining the yield stress in pure shear, and

$$\hat{\tau} = \{\tfrac{1}{2}(\boldsymbol{\sigma}' - \boldsymbol{\beta}) : (\boldsymbol{\sigma}' - \boldsymbol{\beta})\}^{1/2}. \tag{A.2.10b}$$

The compliance tensor is now given by

$$\boldsymbol{D}' = \mathbf{D} + \frac{1}{2\hat{H}} \hat{\boldsymbol{\mu}} \otimes \hat{\boldsymbol{\mu}}, \tag{A.2.11a}$$

where

$$\frac{\hat{\boldsymbol{\mu}}}{\sqrt{2}} = \frac{\boldsymbol{\sigma}' - \boldsymbol{\beta}}{2\hat{\tau}}, \tag{A.2.11b}$$

and $\hat{H}$ is obtained from the condition of continued plastic flow, i.e., from the so-called consistency condition $f = 0$ which yields,

$$\hat{H} = \frac{\hat{\boldsymbol{\mu}} : \dot{\boldsymbol{\beta}}}{\sqrt{2}\dot{\gamma}}. \tag{A.2.11c}$$

Additional assumptions are required to define the evolution of the parameter $\boldsymbol{\beta}$ and hence the work-hardening quantity $\hat{H}$. For this it is observed that the parameter $\boldsymbol{\beta}$ is a macroscopic *back stress* representing the effect of local residual stresses and strains on the overall plastic flow of the material. For the evolution of the back stress $\boldsymbol{\beta}$, it is often assumed

$$\dot{\boldsymbol{\beta}} = \dot{\gamma}\, \mathrm{A}\, \hat{\boldsymbol{\mu}}, \qquad \mathrm{A} = \mathrm{A}(\gamma). \tag{A.2.11d}$$

For combined isotropic-kinematic hardening, the yield function and the flow potential in the J_2-flow theory become

$$\mathrm{f} \equiv \mathrm{g} \equiv \hat{\tau} - \mathrm{F}(\gamma). \tag{A.2.12a}$$

The hardening parameter $\overline{\mathrm{H}}$ in this case is given by

$$\overline{\mathrm{H}} = \mathrm{H} + \hat{\mathrm{H}}. \tag{A.2.12b}$$

The compliance tensor is

$$\boldsymbol{D}' = \mathbf{D} + \frac{(\boldsymbol{\sigma}' - \boldsymbol{\beta}) \otimes (\boldsymbol{\sigma}' - \boldsymbol{\beta})}{4\hat{\tau}^2\, \overline{\mathrm{H}}}. \tag{A.2.12c}$$

The inverse relation is obtained from (A.2.9b): replace in the right-hand side, τ by $\hat{\tau}$, H by $\overline{\mathrm{H}}$, and $\boldsymbol{\sigma}'$ by $(\boldsymbol{\sigma}' - \boldsymbol{\beta})$.

A.2.5. The J_2-Flow Theory with Dilatancy and Pressure Sensitivity

For geomaterials, volumetric strains, frictional effects, and pressure sensitivity are of prime importance, especially in the study of localized deformation. The volumetric strain is also of importance in porous metals. The pressure sensitivity and plastic volumetric strains may stem from a variety of possibly interacting micromechanisms.

Classical plasticity has been modified for application to problems which involve inelastic volumetric strains as well as pressure effects. This has been done by including in the yield surface the dependence on the first stress invariant; see, e.g. Drucker and Prager (1952), and Berg (1970). A derivation of the constitutive relations based on the concepts of yield surface and flow potential has been given by Nemat-Nasser and Shokooh (1980) for large-deformation problems. In this formulation the effects of inelastic dilatation (or densification) and friction (pressure sensitivity) are delineated and discussed. The theoretical predictions have been compared with experimental results showing reasonably good success; see Dorris and Nemat-Nasser (1982), and Rowshandel and Nemat-Nasser (1987). Another feature of geomaterials, which stems from their frictional property, is the noncoaxiality of the plastic strain and the stress deviator; Mandel (1947), Spencer (1964, 1982), Rudnicki and Rice (1975), Mehrabadi and Cowin (1978), Nemat-Nasser *et al.* (1981), and Nemat-Nasser (1983).

As an illustration, consider isotropic and kinematic hardening with the following yield surface and flow potential:

$$\mathrm{f} \equiv \hat{\tau} - \mathrm{F}(\mathrm{I}, \Delta, \gamma), \qquad \mathrm{g} \equiv \hat{\tau} + \mathrm{G}(\mathrm{I}, \Delta, \gamma), \tag{A.2.13a,b}$$

where

$$\Delta = \int_0^t \dot{\varepsilon}_{kk}^{pl}\, dt, \qquad I = \sigma_{kk}, \tag{A.2.13c,d}$$

and where $\hat{\tau}$ is defined by (A.2.10b) and γ (as before) is the accumulated effective plastic strain. The quantity Δ is the total accumulated plastic volumetric strain.

Following the procedure of the previous subsections it is easy to show that the corresponding compliance tensor (including dilatancy or densification) is given by

$$\boldsymbol{D}' = \mathbf{D} + \frac{1}{\overline{H}}\left\{\frac{\hat{\boldsymbol{\mu}}}{\sqrt{2}} + \frac{\partial G}{\partial I}\,\mathbf{1}\right\} \otimes \left\{\frac{\hat{\boldsymbol{\mu}}}{\sqrt{2}} - \frac{\partial F}{\partial I}\,\mathbf{1}\right\}, \tag{A.2.13e}$$

where the hardening parameter $\overline{H}$ is

$$\overline{H} = \frac{\partial F}{\partial \Delta}\frac{\dot{\Delta}}{\dot{\gamma}} + \frac{\partial F}{\partial \gamma} + \frac{\hat{\boldsymbol{\mu}} : \dot{\boldsymbol{\beta}}}{\sqrt{2}\dot{\gamma}}. \tag{A.2.13f}$$

A.2.6. Constitutive Relations: Yield Vertex

As pointed out before, vertices or corners are an integral part of crystal plasticity. They play a central role in proper modeling of material instability by localization. Therefore, effective phenomenological models must also include vertices or, at least, account in some sense for the effect of the vertex structure in the elastoplastic response of materials. Since the vertices are part of single-crystal plasticity, (Hill, 1966; Hill and Rice, 1972; and Havner, 1992), many ideas and concepts may be borrowed from the theory of single-crystal plasticity, although a certain degree of arbitrariness is inherent in the phenomenological developments, since these are not as closely related to the physics and micromechanics of the phenomena, as are the concepts in slip-induced crystal plasticity.

A corner on the yield surface is formed by the intersection of several smooth surfaces, each characterizing a portion of the overall yield surface. In the strain space, consider a corner formed by the intersection of n surfaces, $\phi^\alpha = \phi^\alpha(\boldsymbol{\varepsilon}, \boldsymbol{\gamma})$ ($\alpha = 1, 2, ..., n$), where $\boldsymbol{\gamma}$ with components γ^α represents n scalar parameters characterizing the work-hardening associated with each segment of the yield surface. The stress rate for continued elastic-plastic flow is expressed as

$$\dot{\boldsymbol{\sigma}} = \mathbf{C} : \dot{\boldsymbol{\varepsilon}} - \dot{\gamma}^\alpha \boldsymbol{\lambda}^\alpha \qquad (\alpha \text{ summed; } \alpha = 1, 2, ..., n), \tag{A.2.14a}$$

where $\boldsymbol{\lambda}^\alpha$ is normal to the α'th yield surface, and $\dot{\gamma}^\alpha$ is calculated from the consistency conditions which state that, for continued plastic flow with the state remaining at the corner, $\dot{\phi}^\alpha = 0$. It thus follows that

$$-\boldsymbol{\lambda}^\alpha : \dot{\boldsymbol{\varepsilon}} + g^{\alpha\beta}\dot{\gamma}^\beta = 0 \qquad (\beta \text{ summed}), \tag{A.2.14b}$$

where

$$\boldsymbol{\lambda}^\alpha = -\partial\phi^\alpha/\partial\boldsymbol{\varepsilon}, \qquad g^{\alpha\beta} = \partial\phi^\alpha/\partial\gamma^\beta. \tag{A.2.14c}$$

When the $n \times n$ matrix with components $g^{\alpha\beta}$ admits an inverse with components

$g_{-1}^{\alpha\beta}$, then (A.2.14b) is solved to obtain

$$\dot{\gamma}^{\alpha} = g_{-1}^{\alpha\beta}\,\boldsymbol{\lambda}^{\beta} : \dot{\boldsymbol{\varepsilon}}. \tag{A.2.14d}$$

Hence, the modulus tensor becomes

$$\boldsymbol{C}' = \mathbf{C} - g_{-1}^{\alpha\beta}\,\boldsymbol{\lambda}^{\alpha} \otimes \boldsymbol{\lambda}^{\beta}. \tag{A.2.14e}$$

In a similar manner, consider a typical corner of the yield surface in the stress space, defined by the intersection of n surfaces $\psi^{\alpha} = \psi^{\alpha}(\boldsymbol{\sigma}, \boldsymbol{\gamma})$. The strain rate for a continued plastic flow, with the stress point remaining at the corner, becomes

$$\dot{\boldsymbol{\varepsilon}} = \mathbf{D} : \dot{\boldsymbol{\sigma}} + \dot{\gamma}^{\alpha}\,\boldsymbol{\mu}^{\alpha}, \tag{A.2.15a}$$

where $\boldsymbol{\mu}^{\alpha}$ is normal to the α'th surface at the corner, and

$$\boldsymbol{\mu}^{\alpha} : \dot{\boldsymbol{\sigma}} + h^{\alpha\beta}\,\dot{\gamma}^{\beta} = 0, \tag{A.2.15b}$$

where

$$\boldsymbol{\mu}^{\alpha} = \partial\psi^{\alpha}/\partial\boldsymbol{\sigma}, \qquad h^{\alpha\beta} = \partial\psi^{\alpha}/\partial\gamma^{\beta}. \tag{A.2.15c}$$

Again, when the matrix of $h^{\alpha\beta}$ admits an inverse with components $h_{-1}^{\alpha\beta}$, from (A.2.15c) obtain

$$\dot{\gamma}^{\alpha} = -\,h_{-1}^{\alpha\beta}\,\boldsymbol{\mu}^{\beta} : \dot{\boldsymbol{\sigma}}. \tag{A.2.15d}$$

The compliance tensor is

$$\boldsymbol{D}' = \mathbf{D} - h_{-1}^{\alpha\beta}\,\boldsymbol{\mu}^{\alpha} \otimes \boldsymbol{\mu}^{\beta}. \tag{A.2.15e}$$

For consistency, $\dot{\boldsymbol{\gamma}}$ obtained from (A.2.14d) and (A.2.15d) must be the same. Since (A.2.14a) and (A.2.15a) must be consistent, it follows that

$$\boldsymbol{\lambda}^{\alpha} = \mathbf{C} : \boldsymbol{\mu}^{\alpha}, \qquad \boldsymbol{\mu}^{\alpha} = \mathbf{D} : \boldsymbol{\lambda}^{\alpha}, \tag{A.2.16a}$$

and (A.2.14b) and (A.2.15b) then yield

$$\{(g^{\alpha\beta} + h^{\alpha\beta}) - (\boldsymbol{\mu}^{\alpha} : \mathbf{C} : \boldsymbol{\mu}^{\beta})\}\,\dot{\gamma}^{\beta} = 0. \tag{A.2.16b}$$

This is automatically satisfied if

$$g^{\alpha\beta} + h^{\alpha\beta} = \boldsymbol{\mu}^{\alpha} : \mathbf{C} : \boldsymbol{\mu}^{\beta}, \tag{A.2.16c}$$

in which case, while neither $g^{\alpha\beta}$ nor $h^{\alpha\beta}$ need to be symmetric, their sum must be.

In general, the matrix of $g^{\alpha\beta}$ may be singular and may not admit an inverse. The situation is similar to that of crystal plasticity. Here, however, there is little physically-based guidance for further development. Some simplifying assumptions can be made. For example, if each surface, ϕ^{α} or ψ^{α}, is regarded to depend only on one scalar parameter γ^{α}, then the matrices of both $g^{\alpha\beta}$ and $h^{\alpha\beta}$ are diagonal and can easily be inverted. Other assumptions may be made. These have been discussed in connection with crystal plasticity by Hill and Rice (1972), and in the phenomenological context by Sewell (1972, 1974), and summarized by Hill (1978).

A.2.7. Crystal Plasticity

As an illustration, the results of the preceding subsection are applied to examine a simple version of constitutive relations for plastic flow of single crystals which deform by slip on crystallographic planes.[4] Both rate-independent and rate-dependent plastic flow by slip are included; see also Subsection A.3.4. A basic feature of crystal plasticity is that material flows through the crystal lattice via dislocation motion, and the lattice itself can deform only elastically, in addition to possible rigid rotations. Thus, two different mechanisms are involved in the deformation: (1) plastic slip, and (2) elastic lattice deformation. For fcc (face-centered cubic) crystals considered here, there are a total of four crystallographic planes, each containing three slip orientations. When slip rates are viewed as nonnegative quantities, then there are a total of 24 slip systems in fcc crystals, in which no more (and generally less) than 12 can be active at each instant; see, e.g., Havner (1992) for a comprehensive account.

Let the orthogonal unit vectors $\mathbf{s}^\alpha$ and $\mathbf{n}^\alpha$, respectively define the αth slip direction and its plane, and set

$$\boldsymbol{\mu}^\alpha = \frac{1}{2}(\mathbf{s}^\alpha \otimes \mathbf{n}^\alpha + \mathbf{n}^\alpha \otimes \mathbf{s}^\alpha) \qquad (\alpha \text{ not summed; } \alpha = 1, 2, ..., \mathrm{n}), \qquad \text{(A.2.17a)}$$

where n is the number of slip systems; a slip system is defined by a pair of $\mathbf{s}^\alpha$ and $\mathbf{n}^\alpha$, for fixed α. The plastic strain rate is then defined by

$$\dot{\boldsymbol{\varepsilon}}^{pl} \equiv \dot{\gamma}^\alpha \boldsymbol{\mu}^\alpha, \qquad \text{(A.2.17b)}$$

where α is summed on all *active* slip systems; $\dot{\gamma}^\alpha$ is the rate of slip, and a slip system is regarded as active when its corresponding slip rate is positive. Note that with this convention, $\mathbf{s}^\alpha$ and $-\mathbf{s}^\alpha$ constitute two separate slip directions, since $\dot{\gamma}^\alpha$ is always nonnegative. Comparison with (A.2.15a) now shows a direct correspondence with the phenomenological theory.

The condition for slip is defined in terms of the resultant shear stress in the direction of slip, i.e., in terms of the *resolved shear stress*

$$\tau^\alpha = \boldsymbol{\sigma} : \boldsymbol{\mu}^\alpha. \qquad \text{(A.2.18a)}$$

In the rate-independent theory, the *Schmid law* is often used to define the flow condition. According to this law, yielding may begin on the slip system α, when τ^α reaches the current value of the slip system's *flow stress* τ_r^α which is determined by the current dislocation density and the corresponding substructure.

The set of systems for which $\tau^\alpha = \tau_r^\alpha$ is called *potentially active* or *critical*. For the system α to remain active, τ^α must increase to, and remain at the critical value τ_r^α. A hardening law similar to (A.2.15b) may be used to define changes of the critical values of the resolved shear stress. Thus, for active slip systems,

---

[4] For detailed discussions of slip-induced inelastic strain in crystalline solids, see Taylor (1934, 1938), Cottrell (1953), Hershey (1954), Lin (1954), Kocks (1958, 1970), Budiansky and Wu (1962), Mandel (1966), Hill (1967a,b), Rice (1970), Hutchinson (1970, 1976), Zarka (1972, 1973), Havner and Shalaby (1977), Nemat-Nasser *et al.* (1981), Nemat-Nasser (1983, 1986), and Havner (1992) who also provides a comprehensive historical account.

$$\dot{\tau}^\alpha = \dot{\tau}_r^\alpha = h^{\alpha\beta}\,\dot{\gamma}^\beta \qquad \text{for } \dot{\gamma}^\alpha > 0, \tag{A.2.18b}$$

where β is summed over all *active* slip systems. When a *critical system* is *inactive*,

$$\dot{\tau}^\alpha \le \dot{\tau}_r^\alpha = h^{\alpha\beta}\,\dot{\gamma}^\beta \qquad \text{for } \dot{\gamma}^\alpha = 0. \tag{A.2.18c}$$

Finally the noncritical systems are characterized by the inequality

$$\tau^\alpha < \tau_r^\alpha \qquad \text{for } \dot{\gamma}^\alpha = 0. \tag{A.2.18d}$$

The results of this subsection apply to both small and large deformations of single crystals. For large deformations, the crystal *spin* which leads to texturing of a polycrystal aggregate, must be included. This is defined by

$$\dot{\boldsymbol{\omega}} = \dot{\gamma}^\alpha\,\mathbf{r}^\alpha \qquad (\alpha \text{ summed}), \tag{A.2.19a}$$

where

$$\mathbf{r}^\alpha = \frac{1}{2}(\mathbf{s}^\alpha \otimes \mathbf{n}^\alpha - \mathbf{n}^\alpha \otimes \mathbf{s}^\alpha) \qquad (\alpha \text{ not summed; } \alpha = 1, 2, ..., n). \tag{A.2.19b}$$

The plastic part of the velocity gradient, $\boldsymbol{\nabla} \otimes \dot{\mathbf{u}}$, becomes

$$\boldsymbol{l}^{pl} \equiv (\boldsymbol{\nabla} \otimes \dot{\mathbf{u}})^{pl} = \sum_{\alpha=1}^{n} \dot{\gamma}^\alpha\,\mathbf{s}^\alpha \otimes \mathbf{n}^\alpha, \tag{A.2.19c}$$

and the total velocity gradient is

$$\boldsymbol{l} = \boldsymbol{l}^{el} + \boldsymbol{l}^{pl}. \tag{A.2.19d}$$

A.2.8. Aggregate Properties

It is reasonable to regard the stress and deformation fields to be uniform within each crystal in a suitably large aggregate representing an RVE of a polycrystal, since, by necessity, the grains must be very small relative to the size of the RVE. As discussed before, such an assumption is in accord with the usual continuum formulation of the flow and deformation of matter. It is also the basis of the general theory of single crystals.

The procedure for calculating the overall elastic-plastic moduli of an aggregate of single crystals, follows the one outlined in Subsection 7.5.4 for the corresponding overall elasticity tensor, except a rate formulation is now involved. For the self-consistent method, for example, (7.5.16) may be used. When an aggregate of essentially the same crystals of different orientations is considered, it is more transparent to use the averaging operator $< >$, to denote the average over all orientations, shapes, and sizes of the crystal constituents. Denote the *concentration tensors* of a typical crystal, Ω, by $\boldsymbol{A}^\Omega$ and $\boldsymbol{B}^\Omega$. These are defined as

$$\dot{\boldsymbol{\varepsilon}}^\Omega \equiv \boldsymbol{A}^\Omega : \dot{\boldsymbol{\varepsilon}}^o, \tag{A.2.20a}$$

when the macrostrain rate, $\dot{\boldsymbol{\varepsilon}}^o$, is prescribed, and

$$\dot{\boldsymbol{\sigma}}^{\Omega} \equiv \boldsymbol{B}^{\Omega} : \dot{\boldsymbol{\sigma}}^{\text{o}}, \tag{A.2.20b}$$

when the macrostress rate, $\dot{\boldsymbol{\sigma}}^{\text{o}}$, is prescribed. As discussed in Section 10, the functional form of these concentration tensors depends on the type of averaging method which is employed.

Now, suppose that the current instantaneous modulus and compliance tensors of Ω are $\boldsymbol{C}^{\Omega}$ and $\boldsymbol{D}^{\Omega}$, respectively. Then, it immediately follows that the overall instantaneous modulus and compliance tensors are respectively given by

$$\bar{\boldsymbol{C}} = <\boldsymbol{C}^{\Omega} : \boldsymbol{A}^{\Omega}> \quad \text{and} \quad \bar{\boldsymbol{D}} = <\boldsymbol{D}^{\Omega} : \boldsymbol{B}^{\Omega}>, \tag{A.2.20c,d}$$

where the summation is over all crystal orientations, shapes, and sizes in the aggregate. For self-consistency and upon averaging (A.2.20a,b), observe that the concentration tensors must satisfy

$$<\boldsymbol{A}^{\Omega}> = \mathbf{1}^{(4s)} \quad \text{and} \quad <\boldsymbol{B}^{\Omega}> = \mathbf{1}^{(4s)}. \tag{A.2.21a,b}$$

When an individual grain in an aggregate is regarded to be an ellipsoid embedded in a homogeneous matrix with moduli $\bar{\boldsymbol{C}}$, then the concentration tensors $\boldsymbol{A}^{\Omega}$ and $\boldsymbol{B}^{\Omega}$ can be calculated from the formulae given in the preceding sections. This tensor then depends on the orientation of the ellipsoid, as well as on its aspect ratios, but not on its size. Unless all the grains are aligned, the concentration tensors may not satisfy the consistency conditions (A.2.21a,b). This is an inherent problem also shared by linearly elastic composites; Hill (1965) and Walpole (1969). To remedy this, the concentration tensors must be normalized. This normalization, however, can be implemented in different ways; Nemat-Nasser and Obata (1986). A technique proposed by Walpole (1969) is to replace the concentration tensors by

$$\bar{\boldsymbol{A}}^{\Omega} \equiv \boldsymbol{A}^{\Omega} : <\boldsymbol{A}^{\Omega}>^{-1}, \qquad \bar{\boldsymbol{B}}^{\Omega} \equiv \boldsymbol{B}^{\Omega} : <\boldsymbol{B}^{\Omega}>^{-1}, \tag{A.2.21c,d}$$

which automatically satisfies (A.2.21a,b). Iwakuma and Nemat-Nasser (1984), and Nemat-Nasser and Obata (1986), in their self-consistent computation of both rate-independent and rate-dependent finite deformation of polycrystalline solids, observe that this normalization leads to a stable computational procedure.

A.3. RATE-DEPENDENT THEORIES

Inelastic deformation of solids and geomaterials in general is rate-dependent. Rate-independent plasticity theories represent idealizations with limited applicability. The rate-dependency is especially dominant at high strain-rate deformations.

To deal with rate effects within a phenomenological framework, two classes of constitutive models may be identified: (1) fully rate-dependent plasticity; and (2) viscoplasticity. In the first approach, the deformation rate consists of an elastic and an inelastic constituent, throughout the entire deformation history, i.e. there is no yield surface, whereas in the second approach, a yield surface is considered within which the response is elastic and on which the response may be elastic-viscoplastic, where the plastic flow is rate-dependent. In this latter case, loading and unloading, analogous to the rate-independent theories, are included.

A.3.1. Rate Dependent J_2-Plasticity

The simplest model is obtained when the plastic part of the deformation rate is expressed as

$$\dot{\boldsymbol{\varepsilon}}^{pl} = \frac{1}{\sqrt{2}}\dot{\gamma}\,\boldsymbol{\mu}, \tag{A.3.1}$$

where $\boldsymbol{\mu}$ is defined by (A.2.8b). The difference between the rate-dependent and the rate-independent J_2-plasticity then is in the manner by which $\dot{\gamma}$ is characterized. Whereas in the rate-independent case, $\dot{\gamma}$ is proportional to the rate of change of the deviatoric stress, for the rate-dependent case $\dot{\gamma}$ is expressed in terms of the effective stress τ and relevant parameters which characterize the rate-controlling processes and the current microstructure. Therefore, depending on the history and the deformation regime, different descriptions of $\dot{\gamma}$, in terms of τ and parameters which define the microstructure, may be used. Some commonly used models are examined in what follows.

A.3.2. Empirical Models

Most empirical models express the effective strain rate,[5] $\dot{\gamma}$, in terms of the corresponding effective stress, τ, strain, γ, and temperature, T, as follows:

$$\dot{\gamma} = \dot{\gamma}_0\, g(\tau, \gamma, T), \tag{A.3.2a}$$

where g is some suitable function, $\dot{\gamma}_0$ is a reference strain rate, and T is temperature, measured from some suitable reference value. An example is the power-law,

$$\dot{\gamma} = \dot{\gamma}_0 \left[\frac{\tau}{\tau_r}\right]^m, \tag{A.3.2b}$$

where m is a fixed parameter, usually very large for most metals, e.g., of the order of 100 at strain rates less than about $10^4\,s^{-1}$, and τ_r is the reference shear stress. In this representation, the reference stress τ_r includes the essential physics of the deformation. In general, τ_r is dependent on the accumulated dislocations and defects (often empirically represented by the plastic strain, γ), on the temperature, and on other parameters which define the current microstructure. An often used, simple representation is

$$\tau_r = \tau_0\,(1 + \gamma/\gamma_0)^N \exp\{-\lambda_0\,(T - T_0)\}, \tag{A.3.2c}$$

where τ_0 and γ_0 are normalizing stress and strain measures, N and λ_0 are material parameters, and T_0 is a reference temperature. Here, N is the strain-

[5] Usually, the effective stress, τ, is expressed in terms of $\dot{\gamma}$, γ, and T, but most existing relations can be inverted explicitly.

hardening parameter, whereas λ_0 is a parameter characterizing the thermal softening effects. From (A.3.2b), with m of the order of 100, it is clear that the plastic strain rate is negligibly small if the effective stress τ is less than the reference stress τ_r. Once τ reaches the value of τ_r, and especially when it tends to exceed this value, the plastic strain rate dominates, and therefore, the overall response for large values of m resembles that of the elastic-plastic rate-independent model.

Another often used model is proposed by Johnson and Cook (1985),

$$\frac{\dot{\gamma}}{\dot{\gamma}_0} = \exp\{a_0 \frac{\tau - \tau_r}{\tau_r}\}, \tag{A.3.3a}$$

where τ_r is defined as

$$\tau_r = \tau_0 \{1 + A\,(\gamma/\gamma_0)^n\} \left\{ 1 - \{(T - T_0)/(T_m - T_0)\}^m \right\}. \tag{A.3.3b}$$

With redefinition of the free parameters, the Johnson-Cook model can be written as

$$\tau = \tau_0 \{1 + A(\gamma/\gamma_0)^n\} \{1 + B\, ln\,(\dot{\gamma}/\dot{\gamma}_0)\} \left\{ 1 - \{(T - T_0)/(T_m - T_0)\}^m \right\}, \tag{A.3.3c}$$

where T_m is the melting temperature.

Finally, consider models proposed by Zerilli and Armstrong (1987, 1990, 1992), for fcc and bcc crystal structures. For fcc crystals, their model may be expressed as

$$\tau = \tau_0 + C_1\,(\gamma/\gamma_0)^{1/2}\,(\dot{\gamma}/\dot{\gamma}_0)^{C_2 T}\exp\{-C_3\,T\}, \tag{A.3.4a}$$

where $\tau_0 = A_0 + k\,d_G^{-1/2}$, with d_G being the average grain size in the polycrystal, and the constants C_i, i = 1,3, are the free parameters to be evaluated empirically. For the bcc crystalline materials, these authors' model is expressed as

$$\tau = \tau_0 + C_4\,(\dot{\gamma}/\dot{\gamma}_0)^{C_2 T}\exp\{-C_3\,T\} + C_5\,(\gamma/\gamma_0)^n. \tag{A.3.b}$$

Here, again, C's are the free parameters of the model.

A.3.3. Physically-based Models

Most metals deform plastically essentially through dislocation motion. Models have been developed based on the notion of thermally activated dislocation kinetics, for moderate strain rates (say, less than 10^4/s), and the notion of dislocation-drag mechanism for deformations at greater strain rates.[6]

---

[6] See Gilman (1969), Kocks *et al.* (1975), Clifton (1983), Follansbee (1986), Klepaczko and Chiem (1986), Regazzoni *et al.* (1987), Follansbee and Kocks (1988), Follansbee and Gray (1991), Nemat-Nasser and Isaacs (1997), and Nemat-Nasser and Li (1998).

The motion of dislocations is opposed by both short-range and long-range obstacles. The short-range barriers may be overcome by thermal activation, whereas the resistance due to long-range obstacles is essentially independent of the temperature, i.e., it is *athermal*. The short-range barriers may include the Peierls stress (which is the resistance due to the lattice structure itself), point defects such as vacancies and self-interstitials, other dislocations which intersect the slip plane, alloying elements, solute atoms (interstitials and substitutional), and any other sources which can pin down the dislocations. The long-range barriers may include grain boundaries, farfield forests of dislocations, and other structural elements which produce elastic stress fields opposing the dislocations in their motion.

As a simple model, consider

$$\tau(\gamma, \dot{\gamma}, T) = \tau^* + \tau_a, \tag{A.3.5}$$

where τ^* and τ_a are the thermal and athermal parts of the resistance to the dislocation motion, respectively. The stress τ^* is a decreasing function of temperature, T, and an increasing function of strain rate, $\dot{\gamma}$. The athermal part, τ_a, increases with increasing accumulated dislocations whose elastic field hinders the motion of mobile dislocations. In a suitable temperature range and for suitable strain rates, it is possible for impurity atoms to move to dislocations and pin them down, resulting in an additional hardening which is often referred to as *dynamic strain aging*. This phenomenon will not be examined in what follows.

Based on experimental results, and using the fundamental work of Kocks *et al.* (1975), Nemat-Nasser and Isaacs (1997)[7] propose the following model for bcc polycrystals (e.g., tantalum and molybdenum):

$$\tau(\gamma, \dot{\gamma}, T) = \begin{cases} \hat{\tau} \left\{ 1 - \left[-\dfrac{kT}{G_0} \ln \dfrac{\dot{\gamma}}{\dot{\gamma}_r} \right]^{1/q} \right\}^{1/p} + \tau_a & \text{for } T \le T_c \\ \tau_a & \text{for } T > T_c, \end{cases} \tag{A.3.6a}$$

where T_c is the critical temperature above which the dislocations can overcome the short-range barriers without the assistance of the applied loads. In (A.3.6a), p and q are parameters which define the profile of the energy barrier;[8] k is the Boltzman constant; G_0 is the magnitude of the energy barrier that the dislocation must overcome; and $\dot{\gamma}_0$ is a reference strain rate related to the density and the average velocity of the mobile dislocations and the barrier spacing by the following relation:

---

[7] Quasi-static and high-strain-rate compression experiments are performed at the first author's laboratories, at University of California, San Diego (UCSD), on several bcc and fcc metals, over a broad range of strains, strain rates, and temperatures. For the high-strain-rate tests, a recovery Hopkinson technique, developed by Nemat-Nasser *et al.* (1991, 1994), and Nemat-Nasser and Isaacs (1997), is used; this method has been called *the UCSD recovery Hopkinson technique*. Strains close to 100% are achieved in these tests, over a temperature range of 77 to 1,100K, and strain rates as high as 5×10^4/s. Perhaps the most important feature of this technique is that a sample can be subjected to a single predefined stress pulse and recovered without it having been subjected to any other loads, and, that this can be performed at high strain rates over a broad temperature range.

[8] see Ono (1968) who shows that p = 2/3 and q = 2 are suitable values for most cases.

$$\dot{\gamma}_r = b\, d\, \rho_m\, \omega_0, \tag{A.3.6b}$$

where b is the magnitude of the Burgers vector (which defines a unit dislocation displacement); d is the barrier spacing; ρ_m is the density of the mobile dislocations; and ω_0 is the attempt frequency which, among other factors, depends on the structure and composition of the crystal, as well as on the core structure of the dislocation. The critical temperature, T_c, is given by

$$T_c = -\frac{G_0}{k}\left\{ln\frac{\dot{\gamma}}{\dot{\gamma}_r}\right\}^{-1}. \tag{A.3.6c}$$

Also, the *threshold* stress,[9] $\hat{\tau}$, relates to the energy barrier, G_0, by

$$G_0 = \hat{\tau}\, b\, \lambda\, l, \tag{A.3.6d}$$

where λ represents the "size" of the barrier, and l, its spacing. For the Peierls barriers, both λ and l are of the order of the lattice dimensions; l is constant and λ may be interpreted as the average width of the barrier. For dislocations as the barriers, on the other hand, l = d directly depends on the density of dislocations and hence on the temperature and the plastic strain *histories*. The quantity $V^* \equiv b\lambda l$ is often called *activation volume*.

For $T < T_c$, (A.3.6a) can be solved to obtain $\dot{\gamma}$ explicitly as follows:

$$\dot{\gamma} = \dot{\gamma}_r \exp\left\{-\frac{G_0}{kT}\left[1 - \left(\frac{\tau - \tau_a}{\hat{\tau}}\right)^p\right]^q\right\}. \tag{A.3.6e}$$

The athermal part, τ_a, in this expression may be empirically related to the plastic strain, γ, the average grain size, d_G, and other microstructural parameters, by

$$\tau_a = \tau_a^0\, g_1(\gamma, d_G, \ldots). \tag{A.3.6f}$$

As shown by Nemat-Nasser and Isaacs (1997), this relation may often be approximated by a simple power law,

$$\tau_a \approx \tau_a^0 \gamma^{n_1}. \tag{A.3.6g}$$

For fcc polycrystals, the energy required to cross the Peierls barrier is rather small (less than 0.2eV), so that only at very low temperatures does the Peierls mechanism provide significant resistance to the dislocation motion. In this case, the dislocation forests which intersect the slip planes may be the essential barriers to the motion of the dislocations lying on the slip planes. Then the average spacing of the barriers along the dislocation line, l, and the average distance the dislocation travels before encountering the next barrier, d, are basically the same, relating to the *current* density of the dislocations, ρ, by

$$l = d \approx \rho^{-1/2}. \tag{A.3.7a}$$

If it is assumed that the barrier energy G_0 remains unchanged in the considered application, then the threshold stress may be expressed as

[9] $\hat{\tau}$ is the shear stress above which the barrier is crossed by a dislocation without any assistance from thermal activation.

$$\hat{\tau} = \tau^0 \frac{l_0}{l}, \quad \tau^0 \equiv \frac{G_0}{b\lambda l_0}, \tag{A.3.7b,c}$$

where l_0 and l are the initial (reference) and current average dislocation spacings,

$$\frac{l_0}{l} = (\rho/\rho_0)^{1/2}, \tag{A.3.7d}$$

with ρ_0 being the initial dislocation density. Hence, when the dislocation density increases by a factor of, say, 10^3 due to plastic deformation, the threshold stress for fcc metals may increase by a factor of about 30, as is observed by comparing the responses of the annealed and as-received samples of, say, OFHC (oxygen free, high conductivity) copper; see Nemat-Nasser and Li (1998).

Since, for fcc metals, both d and l are dependent on the density of dislocations, the model now yields

$$\dot{\gamma} = \frac{\dot{\gamma}_0}{(\rho/\rho_0)^{1/2}} \exp\left\{-\frac{G_0}{kT}\left\{1 - \left[\frac{\tau - \tau_a^0\, g_1(\gamma, d_G, \ldots)}{(\rho/\rho_0)^{1/2}\tau^0}\right]^p\right\}^q\right\}, \tag{A.3.7e}$$

where

$$\dot{\gamma}_0 \equiv b\lambda l_0 \omega_0. \tag{A.3.7e}$$

As illustrations, consider commercially pure tantalum (bcc) and OFHC copper (fcc). For the OFHC copper, Nemat-Nasser and Li (1998) show that a large body of the experimental results in monotonic loadings is modeled accurately by using the following representation:

$$(\rho/\rho_0)^{1/2} \approx 1 + a(T)\,\gamma^{1/2}, \quad a(T) \geq 0, \quad \frac{\partial a(T)}{\partial T} \leq 0. \tag{A.3.8a~c}$$

It turns out that a(T) may be chosen to have the following simple form:

$$a(T) = a_0\,\{1 - (T/T_m)^2\}, \tag{A.3.8d}$$

where T_m, the melting temperature for copper is about 1,350K, and a_0 depends on the initial average dislocation spacing. For the annealed samples, it is expected that a_0 should be an order of magnitude greater than that for the as-received samples, since the dislocation density increases with plastic straining at much greater rates for the annealed samples (with $\rho_0 \approx 10^7 \text{cm}^{-2}$). a_0 is considered as an adjustable parameter to be fixed empirically. In the OFHC Cu case, Nemat-Nasser and Li (1998) show that Table A.3.1 gives the value of the constitutive parameters reported by Nemat-Nasser and coworkers. Figures A.3.1 and A.3.2 illustrate some of the results presented by these authors for tantalum and OFHC copper. Note that, for the athermal part of the flow stress, the approximation (A.3.6g) is used.

A.3.4. Drag-controlled Plastic Flow

A simple model is often used to explain the flow stress of metals at high strain rates, where the viscous drag is the dominant resisting force to the motion of dislocation.[10] The model is used to motivate the derivation of a general constitutive expression which is then viewed in a phenomenological context.

Table A.3.1

Values of Various Constitutive Parameters for Annealed and As-received OFHC Copper, and for Commercially Pure Tantalum

Parameter	p	q	k/G_0	$\dot{\gamma}_0$	a_0	τ^0	τ_a^0	n_1
Annealed cu	2/3	2	4.9×10^{-5}/K	2×10^{10}/s	20	46MPa	220MPa	0.3
As-received cu	2/3	2	4.9×10^{-5}/K	2×10^{10}/s	1.8	400MPa	220MPa	0.3
Tantalum	2/3	2	8.62×10^{-5}/K	5.46×10^{8}/s	0	1,100MPa	473MPa	0.2

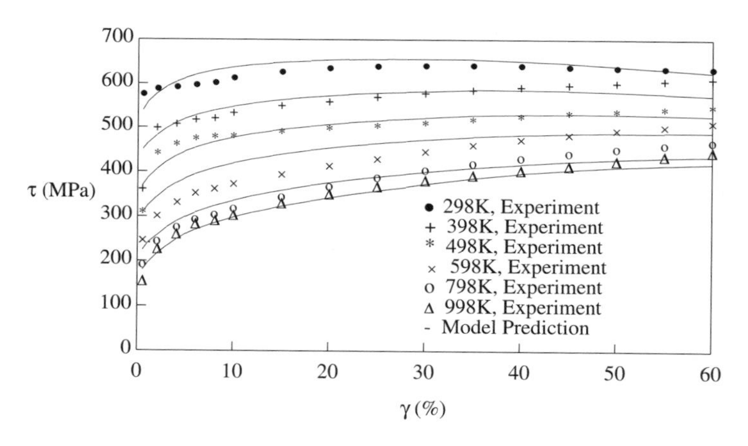

Figure A.3.1

Comparison of the adiabatic flow stress of commercially pure Ta at a 5,000/s strain rate, with model prediction (from Nemat-Nasser and Isaacs, 1997)

As a first step in developing a model, assume that the average time required for dislocations to move from one set of barriers to the next set, consists of two parts, a waiting period, t_w, and a running period, t_r. Then, the *average* dislocation velocity can be estimated by

$$\bar{v} = \frac{d+\lambda}{t_w+t_r} = \frac{(1+\lambda')\,d}{t_w+t_r}, \qquad \lambda' = \lambda/d, \tag{A.3.9a-c}$$

where, as before, d is the average travel distance between barriers and λ is the average barrier width, usually assumed to be much smaller than d, i.e., $\lambda << d$; this assumption is not used in what follows. As a second step, assume that the force resisting the dislocation motion, $\tau_D^* b$, measured per unit dislocation length,

[10] See, *e.g.*, Kocks *et al.* (1975), Clifton (1983), Regazzoni *et al.* (1987), and references cited

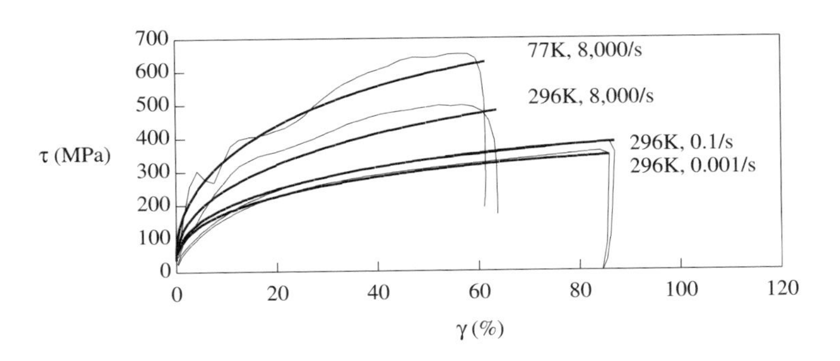

Figure A.3.2

Comparison of model predictions (thick curves) with experimental results (thin curves) for annealed OFHC copper at indicated strain rates and temperatures (from Nemat-Nasser and Li, 1998)

is linearly related to the *average running velocity* of the dislocation,[11] $\bar{v}_r$,

$$\tau_D^* b = D\,\bar{v}_r, \qquad \bar{v}_r = \frac{d}{t_r}, \tag{A.3.9d,e}$$

where D is the *drag* coefficient, and, in general, $\bar{v} \neq \bar{v}_r$. Between the barriers, the frictional stress, $\tau_D^* = D\,\bar{v}_r/b$, must be overcome by the net driving stress, $\tau^* = \tau - \tau_a$, so that $\tau_D^* = \tau^*$. In a more general setting, this net driving force may exceed the drag force, in which case the dislocation accelerates to the next barrier, and may even "overshoot" that barrier. A systematic analysis of this phenomenon requires consideration of the "mass" of a dislocation and other subtle concepts which are not considered in the present work; see Kocks *et al.* (1975).

Now, if the waiting period, t_w, is estimated by

$$t_w = \omega_0^{-1} \exp\{\Delta G/kT\}, \qquad \Delta G = G_0\left\{1 - \left(\frac{\tau^*}{\hat{\tau}}\right)^p\right\}^q, \tag{A.3.9f,g}$$

and the running time, t_r, is obtained from (A.3.9e), then, it follows that

$$\frac{\dot{\gamma}}{\dot{\gamma}_r} = \frac{(1+\lambda')\,\tau^*}{\tau_D^0}\left\{1 + \frac{\tau^*}{\tau_D^0}\exp\{\Delta G/kT\}\right\}^{-1}, \qquad \tau_D^0 = \omega_0\, D\, d/b, \tag{A.3.10a,b}$$

where $\dot{\gamma}_r$ is given by (A.3.6b), i.e., $\dot{\gamma}_r = b\,\rho_m\,\omega_0\, d$, and τ_D^0 may be viewed as a

therein.

[11] In crystals with strong Peierls resistance, this drag may be associated with the lattice friction. Then the drag coefficient D in (A.3.9d) may be a function of the average running velocity of the dislocations, $\bar{v}_r$. When the drag is due to the interaction of the moving dislocation with phonons and, at low temperatures, with electrons, D may be assumed to be a constant; see Kocks *et al.* (1975).

reference drag stress. Note that ΔG in these equations is a function of $\tau^* = \tau - \tau_a$, since the drag resistance acts on the dislocations while they travel between the barriers. When $\tau > \hat{\tau} = \tau^*(0)$, then the resistance to the flow is solely due to the athermal, τ_a, and frictional, τ_D^*, parts of the process, so that $\Delta G \approx 0$. In this case, only viscous drag and the stress field due to farfield inhomogeneities resist the motion of dislocations. Since, then, each dislocation vibrational "attempt" is successful, it follows that $t_w = \omega_0^{-1} << t_r$.

A.3.5. Viscoplastic J_2-Flow Theory

In the preceding subsection, the plastic part of the deformation rate is always nonzero, although it may be negligibly small, when τ/τ_r is less than 1. A viscoplasticity model is obtained if a rate-dependent yield stress is introduced, so that, whenever the effective stress τ falls below the current value of the yield stress, the plastic strain rate vanishes. There are a number of phenomenological models of this kind discussed in the literature; see, for example, Perzyna (1980, 1984).

As an illustration, consider the plastic deformation rate given by (A.3.1), with $\dot{\gamma}$ defined by (A.3.2a). The viscoplastic version of this simple model would be

$$\dot{\boldsymbol{\varepsilon}}^{pl} = \begin{cases} \dot{\gamma}\boldsymbol{\mu}/\sqrt{2} & \text{when } \tau > \tau_Y \\ 0 & \text{when } \tau < \tau_Y, \end{cases} \tag{A.3.11a}$$

where

$$\tau_Y = \tau_r\,(\dot{\gamma}/\dot{\gamma}_0)^{1/m}, \tag{A.3.11b}$$

and τ_r is given, e.g., by (A.3.2c).

A.3.6. Nonlinear Viscoplastic Model

For application of the general homogenization theory to nonlinear viscoplastic materials, consider an incremental formulation in terms of the *strain-rate increment*, i.e., $d\dot{\boldsymbol{\varepsilon}}$. From the models illustrated in this subsection, the strain rate for materials of this kind may be expressed as

$$\dot{\boldsymbol{\varepsilon}} = \mathbf{F}(\boldsymbol{\sigma}, \ldots), \tag{A.3.12a}$$

where $\mathbf{F}$ is some suitable function of the stress and other relevant state variables. For example, in the power-law model, $\dot{\boldsymbol{\varepsilon}}$ depends on both the stress through the effective stress τ, and on the stress rate. In the absence of elasticity, the dependence on the stress rate disappears, and an expression similar to (A.3.12a) results. The incremental relation then becomes

$$d\dot{\boldsymbol{\varepsilon}} = (\partial\mathbf{F}/\partial\boldsymbol{\sigma}) : d\boldsymbol{\sigma}, \tag{A.3.12b}$$

or

$$d\dot{\boldsymbol{\varepsilon}} = \boldsymbol{D}' : d\boldsymbol{\sigma}. \tag{A.3.12c}$$

The pseudo-compliance $\boldsymbol{D}'(\boldsymbol{\sigma})$ now is a function of stress, but the relation between the strain-rate increment and the stress increment is linear. This means that the results of the theories of the preceding sections also apply to this class of problems; see Nemat-Nasser *et al.* (1986) for worked out examples of composites with periodic microstructures.

A.3.7. Rate-Dependent Crystal Plasticity

For crystals with rate-dependent plastic slip, the slip rates are given as functions of the current stress state, temperature, and the current state of material hardness. To model this, let the slip rate $\dot{\gamma}^\alpha$ in (A.2.17b) be represented as

$$\dot{\gamma}^\alpha = \dot{\gamma}_0^\alpha \, g^\alpha(\tau^\alpha, \tau_r^\alpha, T) \qquad (\alpha \text{ not summed}), \tag{A.3.13}$$

where g^α is some suitable function of its arguments, $\dot{\gamma}_0^\alpha$ is a reference strain rate for the αth slip system, τ_r^α defines the hardening state of this slip system, and T is temperature. As an example, consider the power-law,

$$\dot{\gamma}^\alpha = \dot{\gamma}_0^\alpha \left[\frac{\tau^\alpha}{\tau_r^\alpha} \right]^m, \tag{A.3.14a}$$

where m, again here, is a fixed parameter, usually very large, similar to the phenomenological case, and the reference stress τ_r^α depends on the accumulated dislocations on all slip systems, on the temperature, and on other parameters which define the current state of the crystal. These effects can be included empirically through a hardening rule. For example, consider

$$\dot{\tau}_r^\alpha = h^{\alpha\beta} \dot{\gamma}^\beta + a \dot{T}, \tag{A.3.14b}$$

where $h^{\alpha\beta}$ is the strain-hardening matrix, and $\dot{T}$ is the rate of change of temperature. The parameter a represents strain softening due to temperature increase. At high strain rates, where an almost adiabatic regime may prevail, strain softening due to heating caused by plastic flow may play an important role in inducing localized deformations and hence, failure. For quasi-static deformation, however, an isothermal condition is often assumed, in which case $\dot{T} = 0$.

For rate-dependent slip, it may be more effective to directly ensure the coincidence of the signs of the slip rate and the corresponding driving shear stress. For the present illustration, this may be accomplished as follows:

$$\dot{\gamma}^\alpha = \dot{\gamma}_0^\alpha \, \mathrm{sign}\,(\tau^\alpha) \left[\frac{|\tau^\alpha|}{\tau_r^\alpha} \right]^m. \tag{A.3.14c}$$

In a similar manner, other models considered in Subsection A.3.2 and A.3.3 may be used to describe the slip rates. For application to bcc metals, Nemat-Nasser *et al.* (1998) use the model defined by (A.3.6e,g) and show that it nicely represents the experimental results for commercially pure tantalum.

A.4. REFERENCES

Accorsi, M. L. and Nemat-Nasser, S. (1986), Bounds on the overall elastic and instantaneous elastoplastic moduli of periodic composites, *Mech. Matr.*, Vol. 5, No. 3, 209-220.

Berveiller, M. and Zaoui, A. (1979), An extension of the self-consistent scheme to plastically-flowing polycrystals, *J. Mech. Phys. Solids*, Vol. 26, 325-344.

Berg, C. A. (1970), Plastic dilation and void interaction, in *Inelastic behavior of solids*, McGraw-Hill, New York, 171.

Budiansky, B. and Wu T. T. (1962), Theoretical prediction of plastic strains of polycrystals, *Proc. 4th U.S. National Congress of Applied Mechanics*, 1175-1185.

Clifton, R. J. (1983), Dynamic plasticity, *J. Appl. Mech.*, Vol. 50, 941-952.

Cottrell, A. H. (1953), *Dislocations and plastic flow in crystals*, Clarendon Press.

Dorris, J. F. and Nemat-Nasser, S. (1982), A plasticity model for flow of granular materials under triaxial stress states, *Int. J. Solids. Struct.* Vol. 18 , No. 6, 497-531.

Dvorak, G. J. (1992), Transformation field analysis of inelastic composite materials, *Proc. Royal Soc. Lond. A*, Vol. 437, 311-327.

Drucker, D. and Prager, W. (1952), Soil mechanics and plastic analysis for limit design, *Q. Appl. Math.,* Vol. 10, 157-165.

Duva, J. M. (1984), A self-consistent analysis of the stiffening effect of rigid inclusions on a power-law material, *J. Engng. Mater. Technol.*, Vol. 106, 317.

Duva, J. M. (1986), A constitutive description of nonlinear materials containing voids, *Mech. Matr.*, Vol. 5, 137.

Duva, J. M. and Hutchinson, J. W. (1984), Constitutive potentials for dilutely voided nonlinear materials, *Mech. Matr.*, Vol. 3, 41.

Follansbee, P. S. (1986), High-strain-rate deformation of fcc metals and alloys, in *Metallurgical Applications of Shock-Wave and High-Strain-Rate Phenomena.,* L. E. Murr, K. P. Staudhammer and M. A. Meyers (eds), Marcel Dekker, New York, 451-480.

Follansbee, P. S. and G. T. Gray (1991), The response of single crystal and polycrystal nickel to quasistatic and shock deformation, *Int. J. Plasticity*, Vol. 7, No. 7, 651-660.

Follansbee, P. S. and U. F. Kocks (1988), A constitutive description of the deformation of copper based on the use of the mechanical threshold stress as an internal state variable, *Acta Metal.*, Vol. 36, 81-93.

Gilman, J. J. (1969), A unified view of flow mechanisms in materials, in *Strength and Plasticity*, A. S. Argon (ed), MIT Press, 3-14.

Havner, K. S. (1982), The theory of finite plastic deformation of crystalline solids, in *Mechanics of solids (The Rodney Hill 60th anniversary volume)*, Pergamon Press, Oxford, 265-302.

Havner, K. S. (1992), *Finite plastic deformation of crystalline solids*, Cambridge University Press, Cambridge.

Havner, K. S. and Shalaby, A. H. (1977), A simple mathematical theory of finite distortional latent hardening in single crystals, *Proc. Roy. Soc. London*, Ser. A, Vol. 358, 47-70.

Hershey, A. V. (1954), The plasticity of an isotropic aggregate of anisotropic face-centered cubic crystals, *J. Appl. Mech.*, Vol. 21, 241-249.

Hill, R. (1950), *The mathematical theory of plasticity*, The Oxford Engineering Science Series, Clarendon Press, London.

Hill, R. (1965), Continuum micro-mechanics of elastoplastic polycrystals, *J. Mech. Phys. Solids*, Vol. 13, 89-101.

Hill, R. (1966), Generalized constitutive relations for incremental deformation of metal crystals by multislip, *J. Mech. Phys. Solids*, Vol. 14, 95-102.

Hill, R. (1967a), On the classical constitutive laws for elastic/plastic solids, in *Recent Progress in Applied Mechanics - The Folke Odqvist Volume* (Broberg, B., Hult, J., Niordson, F., eds.), Almqvist and Wiksell, Stockholm Wiley, New York 241-249.

Hill, R. (1967b), The essential structure of constitutive laws for metal composites and polycrystals, *J. Mech. Phys. Solids*, Vol. 15, 79-95.

Hill, R. (1972), On constitutive macro-variables for heterogeneous solids at finite strain, *Proc. Roy. Soc. Lond.*, Vol. A326, 131.

Hill, R. (1978), Aspects of invariance in solid mechanics, *Adv. Appl. Mech.* (Yih, C.-S., ed.), Academic Press, New York Vol. 18, 1-75.

Hill, R. and Rice, J. R. (1972), Constitutive analysis of elastic-plastic crystals at arbitrary strain, *J. Mech. Phys. Solids*, Vol. 20, 401-413.

Hutchinson, J. W. (1970), Elastic-plastic behavior of polycrystalline metals and composites, *Proc. Roy. Soc. London*, Ser. A, Vol. 319, 247-272.

Hutchinson, J. W. (1976), Bounds and self-consistent estimates for creep of polycrystalline materials, *Proc. Roy. Soc. London*, Ser. A, Vol. 325, 101-127.

Iwakuma, T. and Nemat-Nasser, S. (1984), Finite elastic-plastic deformation of polycrystalline metals, *Proc. Royal Soc. Lond. A*, Vol. 394, 87-119.

Johnson, G. R. and W. H. Cook (1985), Fracture characteristics of three metals subjected to various strains, strain rates, temperatures and pressures, *Engineering Fracture Mechanics*, Vol. 21, No. 1, 31-48.

Klepaczko, J. and C. Y. Chiem (1986), On rate sensitivity of fcc metals, instantaneous rate sensitivity and rate sensitivity of strain hardening, *J. Mech. Phys. Solids*, Vol. 34, 29-34.

Kocks, U. F. (1958), Polyslip in polycrystals *J Acta Met.*, Vol. 6, 85-94.

Kocks, U. F. (1970), The relation between polycrystal deformation and single-crystal deformation, *Metallurgical Transactions*, Vol. 1, 1121-1143.

Kocks, U. F., A. S. Argon, and M. F. Ashby (1975), Thermodynamics and kinetics of slip, *Progress in Materials Science*, Pergamon Press, New York, Vol. 19, 68-170.

Lee, B. J. and Mear, M. E. (1992a), Effective properties of power-law solids containing elliptical inhomogeneities Part I: rigid inclusions, *Mech. Matr.*, Vol. 13, 313-335.

Lee, B. J. and Mear, M. E. (1992b), Effective properties of power-law solids containing elliptical inhomogeneities Part II: voids, *Mech. Matr.*, 337-356.

Lin, T. H. (1954), A proposed theory of plasticity based on slips, *J Proc. U.S. Nat. Congr. Appl. Mech.*, Vol. 2, 461-468.

Mandel, J. (1947), Sur les lignes de glissement et le calcul des deplacements dans la deformation plastique, *Comptes rendus de l'Academie des Sciences*, Vol. 225, 1272-1273.

Mandel, J. (1966), Sur les equations d'ecoulement des sols ideaux en deformation plane et le concept du double glissement, *J. Mech. Phys. Solids*, Vol. 14, 303-308.

Mehrabadi, M. M., and Cowin, S. C. (1978), Initial planar deformation of dilatant granular materials, *J. Mech. Phys. Solids,* Vol. 26, 269-284.

Mori, T. and Tanaka, K. (1973), Average stress in matrix and average elastic energy of materials with misfitting inclusions, *Acta Met.*, Vol. 21, 571-574.

Naghdi, P. M. (1960), Stress-strain relations in plasticity and thermoplasticity, in *Plasticity, Proc. Symp. Nav. Struct. Mech. 2nd*, 121-167.

Nemat-Nasser, S. (1979), Decomposition of strain measures and their rates in finite deformation elastoplasticity, *Int. J. Solids Structures*, Vol. 15, 155-166.

Nemat-Nasser, S. (1982), On finite deformation elasto-plasticity, *Int. J. Solids. Struc.,* , Vol.18, 857-872.

Nemat-Nasser, S. (1983), On finite plastic flow of crystalline solids and geomaterials, *J. Appl. Mech. (50th Anniversary Issue)*, Vol. 50, No. 4b, 1114-1126.

Nemat-Nasser, S. (1986), Overall stresses and strains in solids with microstructure, in *Modeling small deformations of polycrystals*, (J. Gittus and J. Zarka, eds.), Elsevier Publishers, Netherlands, 41-64.

Nemat-Nasser, S. (1992), Phenomenological theories of elastoplasticity and strain localization at high strain rates, *Appl. Mech. Rev.*, Vol. 45, No. 3, Part 2, 19-45.

Nemat-Nasser, S. and Isaacs, J. (1997), Direct measurement of isothermal flow stress of metals at elevated temperatures and high strain rates with application to Ta and Ta-W alloys, *Acta Materialia*, Vol. 45, No.3, 907-919.

Nemat-Nasser, S., Isaacs, J. and Starrett, J. E. (1991), Hopkinson techniques for dynamic recovery experiments, *Proc. Royal Soc. Lond. A*, Vol. 435 (1991), 371-391.

Nemat-Nasser, S., Iwakuma, T., and Accorsi, M. (1986), Cavity growth and grain boundary sliding in polycrystalline solids, *Mech. Matr.*, Vol. 5, No. 4, 317-329.

Nemat-Nasser, S. and Li Y. (1998), Flow stress of fcc polycrystals with application to OFHC cu, *Acta Materialia*, Vol.46, No. 2, 565-577.

Nemat-Nasser, S., Li, Y. F., and Isaacs, J. B. (1994), Experimental/computational evaluation of flow stress at high strain rates with application to adiabatic shearbanding, *Mech. Mat.*, Vol. 17, 111-134.

Nemat-Nasser, S., Mehrabadi, M. M., and Iwakuma, T. (1981), On certain macroscopic and microscopic aspects of plastic flow of ductile materials, in *Three-Dimensional constitutive relations and ductile fracture* (Nemat-Nasser, S., ed.), North-Holland, Amsterdam, 157-172.

Nemat-Nasser, S. and Obata, M. (1986), Rate dependent, finite elasto-plastic deformation of polycrystals, *Proc. Roy. Soc. Lond. Ser. A*, Vol. 407, 343-375.

Nemat-Nasser, S., Okinaka, T., and Ni, L. (1998), A physically-based constitutive model for bcc crystals with application to polycrystalline tantalum, *J. Mech. Phys. Solids,* Vol. 46, No. 6, 1009-1038.

Nemat-Nasser, S. and Shokooh, A. (1980), On finite plastic flow of compressible materials with internal friction, *Int. J. Solids. Struc.,* Vol. 16, No. 6, 495-514.

Ono, K. (1968), Temperature dependence of dispersed barrier hardening, *J.*

Appl. Physics, Vol. 39, 1803-1806.
Perzyna, P. (1980), Modified theory of viscoplasticity. Application to advanced flow and instability phenomena, *Arch. Mech.*, Vol. 32, 403-420.
Perzyna, P. (1984), Constitutive modeling of dissipative solids for postcritical behaviour and fracture, *ASME J. Eng. Materials and Technology*, 410-419.
Ponte Castañeda, P. and Suquet, P. (1998), Nonlinear composites, in *Advances in applied mechanics,* Vol. 34, Academic Press, 171-302.
Ponte Castañeda, P. and Willis, J. R. (1988), On the overall properties of nonlinearly viscous composites, *Proc. R. Soc. Lond. A.*, Vol. 416, 217-244.
Ponte Castañeda, P. and Willis, J. R. (1995), The effect of spacial distribution on the effective behavior of composite materials and cracked media, *J. Mech. Phys. Solids*, Vol. 43, 1919-1951.
Regazzoni G., U. F. Kocks and P. S. Follansbee (1987), Dislocation kinetics at high strain rates, *Acta Metall.*, Vol. 35, 2865-2875.
Rice, J. R. (1970), On the structure of stress-strain relations for time-dependent plastic deformation in metals, *J. Appl. Mech.*, Vol. 37, 728-737.
Rowshandel, B. and Nemat-Nasser, S. (1987), Finite strain rock plasticity: Stress triaxiality, pressure and temperature effects, *Soil Dynamics and Earthquake Engineering*, Vol. 6, No. 4, 203-219.
Rudnicki, J. W. and Rice, J. R. (1975), Conditions for the localization of deformation in pressure-sensitive dilatant materials, *J. Mech. Phys. Solids*, Vol. 23, 371-394.
Sewell, M. J. (1972), A survey of plastic buckling, in *Stability*, University of Waterloo Press, Ontario, 85-197.
Sewell, M. J. (1974), On applications of saddle-shaped and convex generating functionals, in *Physical Structure in Systems Theory*, Academic Press, London, 219-245.
Spencer, A. J. M. (1964), A theory of kinematics of ideal soil under plane strain conditions, *J. Mech. Phys. Sol.*, Vol. 12, 337-351.
Spencer, A. J. M. (1982), Deformation of ideal granular materials, in *Mechanics of solids*, The Rodney Hill 60th Anniversary Volume, Pergamon Press, Oxford, 607-652.
Talbot, D. R. S. and Willis, J. R. (1985), Variational principles for inhomogeneous nonlinear media, *IMA J. Appl. Math.*, VOl. 35, 39-54.
Talbot, D. R. S. and Willis, J. R. (1986), A variational approach to the overall sink strength of a nonlinear lossy composite medium, *Proc. R. Soc. A,* Vol. 405, 159-180.
Talbot, D. R. S. and Willis, J. R. (1987), Bounds and self-consistent estimates for the overall properties of nonlinear composites, *IMA J. Appl. Math.*, Vol. 39, 215-240.
Talbot, D. R. S. and Willis, J. R. (1992), Some explicit bounds for the overall behavior of nonlinear composites, *Int. J. Solids Structures*, Vol. 29, 1981-1987.
Talbot, D. R. S. and Willis, J. R. (1997), Bounds of third order for the overall response of nonlinear composites, *J. Mech. Phys. Solids*, Vol. 45, 87-111.
Tandon, G. P. and Weng, G. J. (1988), A theory of particle-reinforced plasticity, *J. Appl. Mech.*, Vol. 110, 126-135.
Taya, M. and Chou, T. W. (1981), On two kinds of ellipsoidal inhomogeneities in an infinite elastic body: an application to a hybrid composite, *Int. J. Solids*

Structures, Vol. 17, 553-563.

Taya, M. and Mura, T. (1981), On stiffness and strength of an aligned short-fiber reinforced composite containing fiber-end cracks under uni-axial applied stress, *J. Appl. Mech.*, Vol. 48, 361-367.

Taylor, G. I. (1934), The mechanism of plastic deformation of crystals, *Proc. Roy. Soc.*, A145, 362-415.

Taylor, G. I. (1938), Analysis of plastic strain in a cubic crystal, in *S. Timoshenko 60th Anniversary Volume*, Macmillan, 218-224.

Teply, J. L. and Dvorak, G. J. (1988), Bounds on overall instantaneous properties of elastic-plastic composites, *J. Mech. Phys. Solids*, Vol. 36, 29-58.

Walpole, L. J. (1969), On the overall elastic moduli of composite materials, *J. Mech. Phys. Solids*, Vol. 17, 235-251.

Weng, G. J. (1981), Self-consistent determination of time-dependent behavior of metals, *J. Appl. Mech.*, Vol. 48, 41-46.

Weng, G. J. (1982), A unified, self-consistent theory for the plastic-creep deformation of metals, *J. Appl. Mech.*, Vol. 104,728-734.

Weng, G. J. (1984), Some elastic properties of reinforced solids, with special reference to isotropic ones containing spherical inclusions, *Int. J. Eng. Sci.*, Vol. 22, 845-856.

Weng, G. J. (1990), The overall elastoplastic stress-strain relations of dual metals, *J. Mech. Phys. Solids*, Vol. 38, 419-441.

Willis, J. R. (1983), The overall elastic response of composite materials, *J. Appl. Mech.*, Vol. 50, 1202-1209.

Willis, J. R. (1989), The structure of overall constitutive relations for a class of nonlinear composites, *J. Appl. Math.*, Vol. 43, 231-242.

Willis, J. R. (1991), On methods for bounding the overall properties nonlinear composites, *J. Mech. Phys. Solids*, Vol. 39, 73-86.

Willis, J. R. (1994), Upper and lower bounds for nonlinear composite behavior, *Mater. Sci. Engrg.*, A 175, 7-14.

Zarka, J. (1972), Généralization de la theorie du potential plastique multiple en viscoplasticité, *J. Mech. Phys. Solids*, Vol. 20, 179-195.

Zarka, J. (1973), Étude du comportement des monocristaux métalliques - Application àla traction du monocristal C.F.C., *J. Mécanique*, Vol. 12, 275-318.

Zerilli, F. J. and R. W. Armstrong (1987), Dislocation-mechanics-based constitutive relations for material dynamics calculations, *J. Appl. Physics*, Vol. 61, 1816-1825.

Zerilli, F. J. and R. W. Armstrong (1990), Description of tantalum deformation behavior by dislocation mechanics based constitutive relations, *J. Appl. Phys.*, Vol. 68, No. 4, 1580-1591.

Zerilli, F. J. and R. W. Armstrong (1992), The effect of dislocation drag on the stress-strain behavior of fcc metals, *Acta Metallurgica et Materialia*, Vol. 40, No.8, 1803-1808.

APPENDIX B HOMOGENIZATION THEORY

It is shown in Section 2 that the volume average of the microfields[1] provides the corresponding macrofields, and that the effective material properties can be defined as relations among the volume averages of the microfields. The averaging theorems of Section 2 guarantee that these relations are consistent with the experimentally measured material properties. Hence, this micromechanics theory, which may be called the *average field theory*, is based on the physics and experimental definition of the overall properties of microscopically heterogeneous solids.

Alternatively, it is possible to establish mathematical relations between the microfields and the corresponding macrofields, using a *multi-scale perturbation method*. The effective properties then naturally emerge as a consequence of these relations, without accounting for any specific physical measurements. This procedure is called the *homogenization theory*[2] in the present context.

The homogenization theory is usually formulated by using a periodic structure as a model of the microstructure. However, it is possible to formulate the homogenization theory in a more general setting, without assuming periodicity, and to show that it can produce effective properties identical to those obtained through the application of the average field theory. Furthermore, the general theory is capable of rigorously predicting the effective properties even when the strain gradients are very large and the average field theory may no longer apply.

B.1. SUMMARY OF AVERAGE FIELD THEORY

Consider a linearly elastic body, B, which consists of heterogeneous materials. As in Section 1, let D and d be the macro- and micro-length-scales. The length scale of B is orders[3] of magnitude greater than that of the macro-length-scale, D. Define a relative length-scale parameter,

---

[1] When the spatial variation of a field quantity is measured at the microscale, then the field is referred to as a "microfield".

[2] The homogenization theory has been extensively studied in the field of applied mathematics; see, for instance, Bensoussan *et al.* (1978), Sanchez-Palencia (1981), Backhvalov and Panasenko (1984), Nunan and Keller (1984), and, more recent works, Meguid and Kalamkarov (1994), Yi *et al.* (1998), and Castillero *et al.* (1998), where references to other contributions can be found. The application of the homogenization theory can be found in Walker *et al.* (1991), Oleinik *et al.* (1992), Hornung (1996), Kevorkina and Cole (1996), and Terada *et al.* (1996).

$$\varepsilon = \frac{d}{D} << 1. \tag{B.1.1}$$

The elasticity tensor field is $\mathbf{C}^\varepsilon(\mathbf{X})$, where $\mathbf{X}$ denotes a continuum point in B.

When B is subjected to, say, surface displacements, $\mathbf{u} = \mathbf{u}^o$ on ∂B, the resulting displacement, strain, and stress fields are denoted by $\mathbf{u}^\varepsilon = \mathbf{u}^\varepsilon(\mathbf{X})$, $\boldsymbol{\varepsilon}^\varepsilon = \boldsymbol{\varepsilon}^\varepsilon(\mathbf{X})$, and $\boldsymbol{\sigma}^\varepsilon = \boldsymbol{\sigma}^\varepsilon(\mathbf{X})$). These fields satisfy

$$\boldsymbol{\varepsilon}^\varepsilon(\mathbf{X}) = \frac{1}{2}\left\{ \boldsymbol{\nabla} \otimes \mathbf{u}^\varepsilon(\mathbf{X}) + (\boldsymbol{\nabla} \otimes \mathbf{u}^\varepsilon(\mathbf{X}))^T \right\},$$

$$\boldsymbol{\nabla} \boldsymbol{\cdot} \boldsymbol{\sigma}^\varepsilon(\mathbf{X}) = \mathbf{0},$$

$$\boldsymbol{\sigma}^\varepsilon(\mathbf{X}) = \mathbf{C}^\varepsilon(\mathbf{X}) : \boldsymbol{\varepsilon}^\varepsilon(\mathbf{X}). \tag{B.1.2a~c}$$

These expressions define a boundary-value problem for $\mathbf{u}^\varepsilon$. This boundary-value problem, however, cannot be solved in its present form, because $\mathbf{C}^\varepsilon$ changes within a micro-length-scale (d) while B is orders of magnitude greater than d, requiring an unreasonable numerical effort.

As shown in Subsection 2.6, the average field theory defines the macrofield variables through weighted averages,

$$\left\{ \begin{matrix} \mathbf{U}^\varepsilon \\ \mathbf{E}^\varepsilon \\ \boldsymbol{\Sigma}^\varepsilon \end{matrix} \right\}(\mathbf{X}) = \int_B \phi(\mathbf{X} - \mathbf{Y}) \left\{ \begin{matrix} \mathbf{u}^\varepsilon \\ \boldsymbol{\varepsilon}^\varepsilon \\ \boldsymbol{\sigma}^\varepsilon \end{matrix} \right\}(\mathbf{Y})\, dV_{\mathbf{Y}}, \tag{B.1.3}$$

where ϕ is a suitable weight function satisfying

$$\int_B \phi(\mathbf{X})\, dV = 1, \qquad \phi(\mathbf{X}) = 0 \quad \text{for } |\mathbf{X}| > D. \tag{B.1.4a,b}$$

As is seen, averaging using this weight function ϕ cancels the high oscillations of the field variables which occur at the micro-length-scale.

It follows from (B.1.2a,b) that

$$\mathbf{E}^\varepsilon(\mathbf{X}) = \frac{1}{2}\left\{ \boldsymbol{\nabla} \otimes \mathbf{U}^\varepsilon(\mathbf{X}) + (\boldsymbol{\nabla} \otimes \mathbf{U}^\varepsilon(\mathbf{X}))^T \right\},$$

$$\boldsymbol{\nabla} \boldsymbol{\cdot} \boldsymbol{\Sigma}^\varepsilon(\mathbf{X}) = \mathbf{0}. \tag{B.1.5a,b}$$

Therefore, if a relation between $\mathbf{E}^\varepsilon$ and $\boldsymbol{\Sigma}^\varepsilon$ is given, a governing equation for $\mathbf{U}^\varepsilon$ can be derived. For instance, if an overall elasticity tensor $\overline{\mathbf{C}}$ is found such that

$$\boldsymbol{\Sigma}^\varepsilon(\mathbf{X}) = \overline{\mathbf{C}} : \mathbf{E}^\varepsilon(\mathbf{X}), \tag{B.1.5c}$$

then, the three field equations, (B.1.5a~c), yield the following governing

[3] To give an insight into the length scale, consider a case when B is analyzed by the finite-element method. In the three-dimensional analysis, B is discretized by using $(10^1 \sim 10^3)^3$ elements. Each element corresponds to an RVE. Since D corresponds to the element size, d becomes $\varepsilon \times (10^{-1} \sim 10^{-3})$ of the size of B. As is seen, a huge computer capacity is required to compute the response of B and each microconstituent simultaneously.

equation for $\mathbf{U}^\varepsilon$:

$$\nabla \cdot (\overline{\mathbf{C}} : (\nabla \otimes \mathbf{U}^\varepsilon(\mathbf{X}))) = \mathbf{0} \qquad \text{in B.} \tag{B.1.6a}$$

Hence, a boundary-value problem for $\mathbf{U}^\varepsilon$ is obtained by assuming $\mathbf{u}^\varepsilon \approx \mathbf{U}^\varepsilon$ near ∂B, i.e.,

$$\mathbf{U}^\varepsilon(\mathbf{X}) = \mathbf{u}^o(\mathbf{X}) \qquad \text{on } \partial\text{B.} \tag{B.1.6b}$$

As is discussed in Section 2 and is illustrated throughout this book, the overall elasticity tensor, $\overline{\mathbf{C}}$, can be estimated from surface data and some knowledge of the microstructure.

B.2. SUMMARY OF HOMOGENIZATION THEORY

While the average field theory starts with three sets of field equations, the homogenization theory starts with a *single* set of the governing equations for $\mathbf{u}^\varepsilon$, deduced from (B.1.1a~c), i.e.,

$$\nabla \cdot (\mathbf{C}^\varepsilon(\mathbf{X}) : (\nabla \otimes \mathbf{u}^\varepsilon(\mathbf{X}))) = \mathbf{0} \qquad \text{in B.} \tag{B.2.1}$$

In order to express the changes in $\mathbf{C}^\varepsilon$ over the micro-length-scale, replace this elasticity tensor field by

$$\mathbf{C}^\varepsilon(\mathbf{X}) \sim \mathbf{C}(\mathbf{x}), \tag{B.2.2a}$$

where

$$\mathbf{x} = \frac{1}{\varepsilon}\mathbf{X}. \tag{B.2.2b}$$

Then, consider the following multi-scale or singular-perturbation representation[4] of $\mathbf{u}^\varepsilon$:

$$\mathbf{u}^\varepsilon(\mathbf{X}) \sim \sum_{n=0} \varepsilon^n \, \mathbf{u}^n(\mathbf{X}, \mathbf{x}). \tag{B.2.3}$$

Substituting (B.2.2a) and (B.2.3) into (B.2.1), and replacing ∇ by $\nabla_X + \varepsilon^{-1}\nabla_x$, obtain

$$\varepsilon^{-2}\left\{ \nabla_x \cdot (\mathbf{C}(\mathbf{x}) : (\nabla_x \otimes \mathbf{u}^0(\mathbf{X}, \mathbf{x}))) \right\}$$

[4] In (B.2.3), the perturbation parameter ε is defined as the ratio of the macro- and micro-length-scale. While this parameter is simple, it neglects the change in the magnitude of $\mathbf{C}$; it is intuitively expected that the perturbation parameter should decrease as, say, the ratio of the maximum and minimum values of a typical component of $\mathbf{C}$ decreases. It may be worth paying some attention to the following relation which shows how a change in the length scale can affect the variation of the elastic moduli: if $\mathbf{X}$ in (B.2.1) is replaced by $\mathbf{Y} = \mathrm{A}^{-1}\mathbf{X}$, (B.2.1) becomes $\nabla_Y \cdot (\mathrm{A}^2 \mathbf{C} : (\nabla_Y \otimes \mathbf{u}^\varepsilon)) = \mathbf{0}$. It appears that the change in the magnitude of the elastic moduli is then magnified by the factor A^2.

$$+\varepsilon^{-1}\Big\{\ \boldsymbol{\nabla}_X\boldsymbol{\cdot}(\mathbf{C}(\mathbf{x}):(\boldsymbol{\nabla}_x\otimes\mathbf{u}^0(\mathbf{X},\,\mathbf{x})))$$

$$+\,\boldsymbol{\nabla}_x\boldsymbol{\cdot}(\mathbf{C}(\mathbf{x}):(\boldsymbol{\nabla}_X\otimes\mathbf{u}^0(\mathbf{X},\,\mathbf{x})+\boldsymbol{\nabla}_x\otimes\mathbf{u}^1(\mathbf{X},\,\mathbf{x})))\Big\}$$

$$+\sum_{n=0}\varepsilon^n\Big\{\ \boldsymbol{\nabla}_X\boldsymbol{\cdot}(\mathbf{C}(\mathbf{x}):(\boldsymbol{\nabla}_X\otimes\mathbf{u}^n(\mathbf{X},\,\mathbf{x})+\boldsymbol{\nabla}_x\otimes\mathbf{u}^{n+1}(\mathbf{X},\,\mathbf{x})))$$

$$+\,\boldsymbol{\nabla}_x\boldsymbol{\cdot}(\mathbf{C}(\mathbf{x}):(\boldsymbol{\nabla}_X\otimes\mathbf{u}^{n+1}(\mathbf{X},\,\mathbf{x})+\boldsymbol{\nabla}_x\otimes\mathbf{u}^{n+2}(\mathbf{X},\,\mathbf{x})))\Big\}. \tag{B.2.4}$$

To solve (B.2.4), first assume

$$\boldsymbol{\nabla}_x\otimes\mathbf{u}^0(\mathbf{X},\,\mathbf{x})=\mathbf{0}, \tag{B.2.5}$$

or that $\mathbf{u}^0$ is a function of only continuum variable $\mathbf{X}$. Then, terms of $O(\varepsilon^{-2})$ vanish in (B.2.4).

Now assume that $\mathbf{u}^1$ is of the following form:

$$\mathbf{u}^1(\mathbf{X},\,\mathbf{x})=\boldsymbol{\chi}^1(\mathbf{x}):(\boldsymbol{\nabla}_X\otimes\mathbf{u}^0(\mathbf{X})), \tag{B.2.6a}$$

or in component form,

$$u_i^1(\mathbf{X},\,\mathbf{x})=\chi^1_{ikl}(\mathbf{x})\,D_k u_l^0(\mathbf{X}), \tag{B.2.6b}$$

where $D_i=\partial/\partial X_i$. Then, terms of $O(\varepsilon^{-1})$ become

$$\Big\{\ \boldsymbol{\nabla}_x\boldsymbol{\cdot}(\mathbf{C}(\mathbf{x}):(\boldsymbol{\nabla}_x\otimes\boldsymbol{\chi}^1(\mathbf{x})+\mathbf{1}^{(4s)}))\Big\}:(\boldsymbol{\nabla}_X\otimes\mathbf{u}^0(\mathbf{X}))=\mathbf{0}, \tag{B.2.7a}$$

or

$$\Big\{\ d_i\,(C_{ijkl}(\mathbf{x})\,(d_k\chi^1_{lmp}(\mathbf{x})+1^{(4s)}_{klmp}))\Big\}\,D_p u_m^0(\mathbf{X})=0, \tag{B.2.7b}$$

where $d_i=\partial/\partial x_i$. The physical interpretation of (B.2.7) is that $\boldsymbol{\chi}^1:(\boldsymbol{\nabla}_X\otimes\mathbf{u}^0)$ represents a microscopic displacement field produced by the presence of the stress field, $\mathbf{C}:(\boldsymbol{\nabla}_X\otimes\mathbf{u}^0)=\mathbf{C}(\mathbf{x}):(\boldsymbol{\nabla}_X\otimes\mathbf{u}^0(\mathbf{X}))$. Since this stress field does not satisfy equilibrium, oscillating microstrains and associated stresses are created. In other words, $\boldsymbol{\chi}^1(\mathbf{x})$ encompasses the response of the microstructure to the macroscopic strain field, $(\boldsymbol{\nabla}_X\otimes\mathbf{u}^0+(\boldsymbol{\nabla}_X\otimes\mathbf{u}^0)^T)/2$. Therefore, if a suitable domain with suitable boundary conditions is introduced, $\boldsymbol{\chi}^1$ can be determined.

Once $\boldsymbol{\chi}^1$ is determined, terms of $O(\varepsilon^0)$ become

$$\boldsymbol{\nabla}_X\boldsymbol{\cdot}\Big\{\ \mathbf{C}(\mathbf{x}):((\boldsymbol{\nabla}_x\otimes\boldsymbol{\chi}^1(\mathbf{x})+\mathbf{1}^{(4s)}):(\boldsymbol{\nabla}_X\otimes\mathbf{u}^0(\mathbf{X})))\Big\}$$

$$+\,\boldsymbol{\nabla}_x\boldsymbol{\cdot}\Big\{\ \mathbf{C}(\mathbf{x}):(\boldsymbol{\nabla}_X\otimes\mathbf{u}^1(\mathbf{X},\,\mathbf{x}))+\boldsymbol{\nabla}_x\otimes\mathbf{u}^2(\mathbf{X},\,\mathbf{x}))\Big\}=\mathbf{0}. \tag{B.2.8a}$$

The first term is the *macroscopic* equilibrium for $\mathbf{C}:(\boldsymbol{\nabla}_x\otimes\boldsymbol{\chi}^1+\mathbf{1}^{(4s)}):(\boldsymbol{\nabla}_X\otimes\mathbf{u}^0)$, i.e., the stress of $O(\varepsilon^0)$, while the second term is the *microscopic* equilibrium for $\mathbf{C}:(\boldsymbol{\nabla}_X\otimes\mathbf{u}^1+\boldsymbol{\nabla}_x\otimes\mathbf{u}^2)$, the stress of $O(\varepsilon^1)$. Now, introduce a constant tensor, $\overline{\mathbf{C}}^0$, such that

$$\nabla_X \boldsymbol{\cdot} (\overline{\mathbf{C}}^0 : (\nabla_X \otimes \mathbf{u}^0(\mathbf{X})))$$

$$+ \left\{ \nabla_x \boldsymbol{\cdot} \left\{ \mathbf{C}(\mathbf{x}) : (\nabla_X \otimes \mathbf{u}^1(\mathbf{X}, \mathbf{x})) + \nabla_x \otimes \mathbf{u}^2(\mathbf{X}, \mathbf{x})) \right\} \right.$$

$$\left. + \nabla_X \boldsymbol{\cdot} \left\{ (\mathbf{C}(\mathbf{x}) : (\nabla_x \otimes \boldsymbol{\chi}^1(\mathbf{x}) + \mathbf{1}^{(4s)}) - \overline{\mathbf{C}}^0) : (\nabla_X \otimes \mathbf{u}^0(\mathbf{X})) \right\} \right\} = \mathbf{0}. \tag{B.2.8b}$$

Then, consider the governing equation for $\mathbf{u}^0$, as

$$\nabla_X \boldsymbol{\cdot} (\overline{\mathbf{C}}^0 : (\nabla_X \otimes \mathbf{u}^0(\mathbf{X}))) = \mathbf{0} \quad \text{in B}, \tag{B.2.8c}$$

which requires that $\mathbf{u}^2$ satisfy

$$\nabla_x \boldsymbol{\cdot} (\mathbf{C}(\mathbf{x}) : (\nabla_x \otimes \mathbf{u}^2(\mathbf{X}, \mathbf{x}))) + \mathbf{f}^1(\mathbf{X}, \mathbf{x}) = \mathbf{0}, \tag{B.2.8d}$$

where

$$\mathbf{f}^1(\mathbf{X}, \mathbf{x}) = \nabla_x \boldsymbol{\cdot} \left\{ \mathbf{C}(\mathbf{x}) : (\nabla_X \otimes (\boldsymbol{\chi}^1(\mathbf{x}) : (\nabla_X \otimes \mathbf{u}^0(\mathbf{X})))) \right\}$$

$$+ \nabla_X \boldsymbol{\cdot} \left\{ (\mathbf{C}(\mathbf{x}) : (\nabla_x \otimes \boldsymbol{\chi}^1(\mathbf{x}) + \mathbf{1}^{(4s)}) - \overline{\mathbf{C}}^0) : (\nabla_X \otimes \mathbf{u}^0(\mathbf{X})) \right\}. \tag{B.2.8e}$$

The governing equation (B.2.3) can be satisfied up to $O(\varepsilon)$, for any arbitrary $\overline{\mathbf{C}}^0$, since $\mathbf{u}^2$ can be determined from (B.2.8d) for any $\overline{\mathbf{C}}^0$. However, the most suitable $\overline{\mathbf{C}}^0$ that minimizes the error of $O(\varepsilon^1)$ is probably[5] given by taking the volume average of $\mathbf{C} : (\nabla_x \otimes \boldsymbol{\chi}^1 + \mathbf{1}^{(4s)})$ over the domain where $\boldsymbol{\chi}^1$ is determined. Indeed, it is shown that when a periodic-structure model of the microstructure is used, $\overline{\mathbf{C}}^0$ is given by

$$\overline{\mathbf{C}}^0 = < \mathbf{C} : (\nabla_x \otimes \boldsymbol{\chi}^1 + \mathbf{1}^{(4s)}) >. \tag{B.2.9}$$

Once $\overline{\mathbf{C}}^0$ is determined, (B.2.4) is solved up to $O(\varepsilon^0)$. The solution is given by $\mathbf{u}^0$ which satisfies

$$\nabla_X \boldsymbol{\cdot} (\overline{\mathbf{C}}^0 : (\nabla_X \otimes \mathbf{u}^0(\mathbf{X}))) = \mathbf{0} \quad \text{in B}. \tag{B.2.10a}$$

It follows from $\mathbf{u}^\varepsilon \approx \mathbf{u}^0$ that the boundary conditions are

$$\mathbf{u}^0(\mathbf{X}) = \mathbf{u}^o(\mathbf{X}) \quad \text{on } \partial\text{B}. \tag{B.2.10b}$$

Suppose that the microstructure models (the arrangement of the microconstituents and the boundary conditions) considered in the average field theory and the homogenization theory are the same. Then, $\overline{\mathbf{C}}$ and $\overline{\mathbf{C}}^0$, which yield the overall stress of the microstructure when the overall strain is given, coincide. Hence, it follows from the comparison of (B.1.6a,b) and (B.2.10a,b) that

---

[5] According to the average field theory, the volume average of $\mathbf{C} : (\nabla_x \otimes \boldsymbol{\chi}^1 + \mathbf{1}^{(4s)})$ is the most suitable overall modulus tensor. From a mathematical point of view, however, the most suitable moduli can be defined by minimizing a suitable perturbation-error measure.

$$\mathbf{U}^{\varepsilon}(\mathbf{X}) = \mathbf{u}^{0}(\mathbf{X}). \tag{B.2.11}$$

Expressions (B.2.9)~(B.2.11) provide consistent relations between the average field theory and the homogenization theory.

B.3. EXTENSION OF HOMOGENIZATION THEORY

Compared with the average field theory, the homogenization theory has one advantage: it can provide higher-order terms in the singular-perturbation solution. In the linear case, these higher-order terms may not be significant, since, as shown later, their volume average, taken over the microstructure, vanishes. In non-linear cases, however, these terms need to be included, especially when the macroscopic strain gradients are large.

To show this, compute higher-order terms, $\mathbf{u}^n$, by solving (B.2.4). In this subsection, tensors of higher order appear, and index notation is used. Like $\mathbf{u}^1$, decompose $\mathbf{u}^n$ (n = 2, 3, ...) as

$$u_i^n(\mathbf{x}, \mathbf{X}) = \chi^n_{imp_1p_2...p_n}(\mathbf{x})\,(D_{p_1}D_{p_2}...D_{p_n}u^0_m(\mathbf{X})). \tag{B.3.1}$$

As is seen, $\chi^n_{kmp_1...p_n}$ is a tensor of n+2-nd order, and $D_{p_1}D_{p_2}...D_{p_n}u^0_m$ is the n-th order gradient of $\mathbf{u}^0$.

Substituting (B.3.1) into (B.2.4), write terms of $O(\varepsilon^n)$ as

$$\begin{aligned}
&D_i\,C_{ijkl}\left\{ D_l\,(\chi^n_{kmp_1...p_n}(D_{p_1}...D_{p_n}u^0_m)) + d_l\,(\chi^{n+1}_{kmp_1...p_{n+1}}(D_{p_1}D..._{p_{n+1}}u^0_m))\right\} \\
&\quad + d_i\,C_{ijkl}\left\{ D_l\,(\chi^{n+1}_{kmp_1...p_{n+1}}(D_{p_1}...D_{p_{n+1}}u^0_m)) + d_l\,(\chi^{n+2}_{kmp_1...p_{n+2}}(D_{p_1}...D_{p_{n+2}}u^0_m))\right\} \\
&= D_i\left\{ C_{ijkl}(\mathbf{x})\,(\delta_{lp_{n+1}}\chi^n_{kmp_1...p_n}(\mathbf{x}) + d_l\chi^{n+1}_{kmp_1...p_{n+1}}(\mathbf{x}))\,(D_{p_1}...D_{p_{n+1}}u^0_m(\mathbf{X}))\right\} \\
&\quad + \left\{ d_i\,(C_{ijkl}(\mathbf{x})\,(\delta_{lp_{n+2}}\chi^{n+1}_{kmp_1...p_{n+1}}(\mathbf{x}) + d_l\chi^{n+2}_{kmp_1...p_{n+2}}(\mathbf{x}))\right\}(D_{p_1}...D_{p_{n+2}}u^0_m(\mathbf{X})).
\end{aligned} \tag{B.3.2}$$

In the right-hand side, the first term is for the macroscopic equilibrium of the stress field of $O(\varepsilon^n)$, and the second term for the microscopic equilibrium of the stress field of $O(\varepsilon^{n+1})$.

Introducing a suitable constant tensor, $\overline{\mathbf{C}}^n_{ijkp_1p_2...p_n}$ (or $\overline{\mathbf{C}}^n$), assume[6] that the macroscopic equilibrium of the stress field of $O(\varepsilon^n)$ is attained when the the stress field of $O(\varepsilon^n)$,

$$C_{ijkl}(\mathbf{x})\,(\,\delta_{lp_{n+1}}\,\chi^{n}_{kmp_1\ldots p_n}(\mathbf{x})+d_l\chi^{n+1}_{kmp_1\ldots p_{n+1}}(\mathbf{x})\,)\,(D_{p_1}\ldots D_{p_{n+1}}u^0_m(\mathbf{X})),$$

is replaced by its volume average,

$$\bar{\mathbf{C}}^{n}_{ijklp_1\ldots p_n}\,(D_{p_1}\ldots D_{p_n}D_l u^0_k(\mathbf{X})).$$

Then require that the two equations of equilibrium in (B.3.2) hold separately. The first is for the macroscopic equilibrium, i.e.,

$$\bar{\mathbf{C}}^{n}_{ijklp_1\ldots p_n}\,(\,D_iD_lD_{p_1}\ldots D_{p_n}u^0_k(\mathbf{X})\,)=0 \qquad \text{in B}, \tag{B.3.3}$$

where

$$\bar{C}^{n}_{ijklp_1\ldots p_n}=<C_{ijmn}\,(\,\delta_{nl}\,\chi^{n}_{mkp_1\ldots p_n}+d_n\chi^{n+1}_{mkp_1\ldots p_np_l}\,)>. \tag{B.3.4}$$

The second is for the microscopic equilibrium, i.e.,

$$d_i\Big\{\,C_{ijkl}(\mathbf{x})\,(\,d_l\chi^{n+2}_{kmp_1\ldots p_{n+2}}(\mathbf{x})\,)\Big\}+f^{n+2}_{jmp_1\ldots p_{n+2}}(\mathbf{x})=0, \tag{B.3.5a}$$

where

$$\begin{aligned}
f^{n+2}_{jmp_1\ldots p_{n+2}}(\mathbf{x}) &= d_i\,(\,C_{ijkl}(\mathbf{x})\,(\,\delta_{lp_{n+2}}\,\chi^{n+1}_{kmp_1\ldots p_{n+1}}(\mathbf{x})\,)\,)\\
&\quad+\delta_{ip_{n+2}}\,C_{ijkl}(\mathbf{x})\,(\,d_l\chi^{n+1}_{kmp_1\ldots p_{n+1}}(\mathbf{x})+\delta_{lp_{n+1}}\,\chi^{n}_{kmp_1\ldots p_n}(\mathbf{x})\,)\\
&\quad-\delta_{ip_{n+2}}\,\delta_{lp_{n+1}}\,\bar{C}^{n}_{ijmlp_1\ldots p_n}.
\end{aligned} \tag{B.3.5b}$$

It should be noted that the right-hand side of (B.3.2) is decomposed as

$$\begin{aligned}
&\bar{C}^{n}_{ijklp_1\ldots p_n}\,(D_iD_lD_{p_1}\ldots D_{p_n}u^0_k(\mathbf{X}))\\
&\qquad+\Big\{\,d_i\,(\,C_{ijkl}(\mathbf{x})\,d_l\chi^{n+2}_{kmp_1\ldots p_{n+2}})(\mathbf{x})+f^{n+2}_{jmp_1\ldots p_{n+2}}(\mathbf{x})\Big\}\,(D_{p_1}\ldots D_{p_{n+2}}u^0_m(\mathbf{X})).
\end{aligned}$$

As is seen, (B.3.5) gives a recursive form for $\{\boldsymbol{\chi}^n\}$; $\boldsymbol{\chi}^{n+2}$ is determined from $\boldsymbol{\chi}^{n+1}$ and $\boldsymbol{\chi}^n$, since $\boldsymbol{\chi}^0$ and $\boldsymbol{\chi}^1$ are already determined ($\chi^0_{km}=\delta_{km}$). As in the case of $\boldsymbol{\chi}^1$, $\boldsymbol{\chi}^{n+2}$ can be determined by introducing a suitable model of the RVE.

The average response associated with $\boldsymbol{\chi}^n$ can be determined from the boundary conditions prescribed for the microstructural model. Suppose that zero surface displacements are prescribed for all $\boldsymbol{\chi}^n$'s (n = 1, 2, ...). The average strain associated with $\boldsymbol{\chi}^n$ ($<(\boldsymbol{\nabla}_x\otimes\boldsymbol{\chi}^n+(\boldsymbol{\nabla}_x\otimes\boldsymbol{\chi}^n)^T)/2>$) then vanishes. The associated average stress ($<\mathbf{C}:(\boldsymbol{\nabla}_x\otimes\boldsymbol{\chi}^n)>$), however, does not necessarily vanish; it is assumed in (B.3.4) that the macroscopic stress of $O(\varepsilon^n)$ is produced by the n-th gradient of $\mathbf{u}^0$. Therefore, the effects of higher-order strain gradients appear in these macrostresses. It should be recalled that even though the macrostress of

[6] The validity of this assumption is well understood in the context of the average field theory. By definition, $D_{p_1}\ldots D_{p_n}u^0_k$ is the n-th macroscopic strain gradient, which induces a microscopic stress, as it produces microscopically non-equilibrating stresses through C_{ijkl}. Due to the linearity, the overall behavior of the induced microscopic stress is related to the n-th macroscopic strain gradient. The assumed $\bar{\mathbf{C}}^n$ gives this relation, as the overall behavior is given by taking the volume average.

$O(\varepsilon^n)$ exists, the associated macro-strain-energy of $O(\varepsilon^n)$, i.e., $< (\nabla \otimes \boldsymbol{\chi}^n) : \mathbf{C} : (\nabla \otimes \boldsymbol{\chi}^n) /2 >$, vanishes, since the assumed boundary conditions are consistent such that the average strain energy is given by the product of the average stress and the average strain which vanishes.[7]

Through $\boldsymbol{\chi}^n$, therefore, the microscopic response can be predicted when the (n–1)-th-order macroscopic strain gradients are given. For instance, like $\boldsymbol{\chi}^1$ which yields a microscopic displacement field for a macroscopic strain, $\boldsymbol{\chi}^2$ gives a microscopic displacement field in the presence of the macroscopic strain gradient, $\nabla \otimes (\nabla \otimes \mathbf{u}^0)$. Furthermore, as shown in (B.3.4), these $\boldsymbol{\chi}^n$'s determine the overall higher-order responses through the overall moduli $\overline{\mathbf{C}}^n$'s.

In order to solve (B.3.3) for n = 1, 2, ..., consider another (regular) perturbation for $\mathbf{u}^0$,

$$\mathbf{u}^0(\mathbf{X}) \sim \sum_{n=0} \varepsilon^n \, \mathbf{U}^n(\mathbf{X}), \tag{B.3.6}$$

and rewrite (B.3.3) as

$$\overline{C}^0_{ijkl} D_i D_l U^n_k(\mathbf{X}) + \sum_{m=1}^{n} \overline{C}^m_{ijklp_1 \ldots p_m} D_i D_l D_{p_1} \ldots D_{p_m} U^{n-m}_k(\mathbf{X}) = 0 \qquad \text{in B.} \tag{B.3.7a}$$

The boundary conditions on ∂B are

$$U^n_i(\mathbf{X}) = \begin{cases} u^o_i(\mathbf{X}) & n = 0, \\ 0 & \text{otherwise,} \end{cases} \qquad \text{on } \partial B. \tag{B.3.7b}$$

Hence, the n-th order perturbation term can be determined once $\mathbf{U}^1$, $\mathbf{U}^2$, ..., $\mathbf{U}^{(n-1)}$ are given.

B.4. EFFECT OF STRAIN GRADIENT

It is of interest to examine the effects of the second-order terms, i.e., the effects of the macroscopic strain gradient. Hence, follow the steps previously outlined, and obtain the governing equations for the second-order terms. In order to show the procedure in a transparent manner, reformulate the homogenization theory with the help of the average field theory. To simplify the expressions, arguments of functions are omitted in the following discussion.

First, write the perturbation representation of the displacement field as

$$u^\varepsilon_i \sim \chi^0_{ih} \, u^0_h + \varepsilon \, \chi^1_{ihp} \, D_p u^0_h + \ldots, \tag{B.4.1}$$

where $\chi^0_{ih} = \delta_{ih}$. Each term of $O(\varepsilon^n)$ is decomposed into the n-th-order gradient of $\mathbf{u}^0$, a function of the slow variable $\mathbf{X}$, and the n+2-th-order tensor $\boldsymbol{\chi}^n$, a function of the fast variable $\mathbf{x} = \varepsilon^{-1} \mathbf{X}$.

[7] This means that the associated strain energy is of a higher order in ε.

The corresponding perturbation representation of the strain field then becomes

$$\varepsilon_{ij}^{\varepsilon} \sim \varepsilon_{ij}^{X\,0} + \varepsilon_{ij}^{x\,0} + \varepsilon\,(\,\varepsilon_{ij}^{X\,1} + \varepsilon_{ij}^{x\,1}\,) + \dots, \tag{B.4.2a}$$

where

$$\varepsilon_{ij}^{X\,0} = \frac{1}{2}\,(\,\delta_{im}\,\delta_{jn} + \delta_{in}\,\delta_{jm}\,)\,\chi_{mh}^{0}\,(D_n u_h^0),$$

$$\varepsilon_{ij}^{x\,0} = \frac{1}{2}\,(\,\delta_{im}\,\delta_{jn} + \delta_{in}\,\delta_{jm}\,)\,(d_n \chi_{mhp}^{1})\,(D_p u_h^0), \tag{B.4.2b,c}$$

and so on. As is seen, $\boldsymbol{\varepsilon}^{X\,n}$ and $\boldsymbol{\varepsilon}^{x\,n}$ include the n+1-th-order gradient of $\mathbf{u}^0$, i.e., $D_{p_1}...D_{p_n} u_h^0$.

Since $\mathbf{C}$ depends only on $\mathbf{x}$, write the corresponding perturbation representation of the stress field, as follows:

$$\sigma_{ij}^{\varepsilon} \sim \sigma_{ij}^{X\,0} + \sigma_{ij}^{x\,0} + \varepsilon\,(\,\sigma_{ij}^{X\,1} + \sigma_{ij}^{x\,1}\,) + \dots, \tag{B.4.3a}$$

where

$$\sigma_{ij}^{X\,0} = <C_{ijkl}\,(\,\delta_{lp}\,\chi_{kh}^{0} + d_l \chi_{khp}^{1}\,)>(D_p u_h^0),$$

$$\sigma_{ij}^{x\,0} = C_{ijkl}\,(\,\delta_{lp}\,\chi_{kh}^{0} + d_l \chi_{khp}^{1}\,)\,(D_p u_h^0) - \sigma_{ij}^{X\,0}, \tag{B.4.3b,c}$$

and so on. Again, $\boldsymbol{\sigma}^{X\,n}$ and $\boldsymbol{\sigma}^{x\,n}$ include the n+1-th-order gradient of $\mathbf{u}^0$.

According to the average field theory, $\boldsymbol{\sigma}^{X\,0}$ is regarded as the macrostress of $O(\varepsilon^0)$, since it is a function of $\mathbf{X}$ only. Hence, $\boldsymbol{\sigma}^{x\,0}$ is regarded as the microstress of $O(\varepsilon^0)$, and it vanishes when averaged over the microstructure. Similarly, $\boldsymbol{\sigma}^{X\,n}$ and $\boldsymbol{\sigma}^{x\,n}$ are regarded as the macrostress and microstress of $O(\varepsilon^n)$.

The same interpretation can be applied to the strain field, i.e., the macrostrain of $O(\varepsilon^0)$ is given by

$$<\varepsilon_{ij}^{X\,0} + \varepsilon_{ij}^{x\,0}> = \frac{1}{2}\,(\,\delta_{im}\,\delta_{jn} + \delta_{in}\,\delta_{jm}\,)<\delta_{np}\,\chi_{mh}^{0} + d_n \chi_{mhp}^{1}>D_p u_h^0. \tag{B.4.2d}$$

Note that the contribution of $\boldsymbol{\chi}^1$ vanishes if zero surface displacements are prescribed as its boundary conditions. The remainder of the strain field of $O(\varepsilon^n)$, $\boldsymbol{\varepsilon}^{X\,0} + \boldsymbol{\varepsilon}^{x\,0} - <\boldsymbol{\varepsilon}^{X\,0} + \boldsymbol{\varepsilon}^{x\,0}>$ $(= \boldsymbol{\varepsilon}^{x\,0})$, is regarded as the microstrain of $O(\varepsilon^0)$. The macrostrain and microstrain of $O(\varepsilon^n)$ are defined in the same manner, although $<\boldsymbol{\varepsilon}^{n\,x}>$ may vanish if $\boldsymbol{\chi}^n = \mathbf{0}$ is prescribed as the boundary condition for $\boldsymbol{\chi}^n$.

The perturbation representation of the stress field, (B.4.3a), leads to the following perturbation representation of the stress gradient:

$$\begin{aligned} \sigma_{ij,i}^{\varepsilon} &\sim \varepsilon^{-1}\,d_i \sigma_{ij}^{x\,0} \\ &+ \varepsilon^{0}\,(\,D_i \sigma_{ij}^{X\,0} + D_i \sigma_{ij}^{x\,0} + d_i \sigma_{ij}^{x\,1}\,) \\ &+ \varepsilon^{1}\,(\,D_i \sigma_{ij}^{X\,1} + D_i \sigma_{ij}^{x\,1} + d_i \sigma_{ij}^{x\,2}\,) + \dots . \end{aligned} \tag{B.4.4}$$

For the right-hand side of (B.4.4) to vanish, the macrostresses are required to satisfy

$$D_i\sigma_{ij}^{X0} + \varepsilon\, D_i\sigma_{ij}^{X1} + ... = 0. \tag{B.4.5a}$$

In view of (B.4.3b,c), rewrite this condition as

$$\bar{C}^0_{ijhp}\,(D_iD_pu^0_h) + \varepsilon\,\bar{C}^1_{ijhpq}\,(D_iD_pD_qu^0_h) + ... = 0, \tag{B.4.5b}$$

where

$$\bar{C}^0_{ijhp} = < C_{ijkl}\,(\,\delta_{lp}\,\chi^0_{kh} + d_l\chi^1_{khp}\,) >,$$

$$\bar{C}^1_{ijhpq} = < C_{ijkl}\,(\,\delta_{lq}\,\chi^1_{khp} + d_l\chi^2_{khpq}\,) >. \tag{B.4.5c,d}$$

Also, the microstresses are required to satisfy $d_i\sigma_{ij}^{x0} = 0$ and $D_i\sigma_{ij}^{X0} + d_i\sigma_{ij}^{x1} = 0$, i.e.,

$$d_i(\,C_{ijkl}\,(\,d_l\chi^1_{khp} + \delta_{lp}\,\chi^0_{kh}\,)\,) = 0,$$

$$d_i(\,C_{ijkl}\,(\,d_l\chi^2_{khpq} + \delta_{lq}\,\chi^1_{khp}\,)\,) + \delta_{iq}\,(\,C_{ijkl}\,(\,d_l\chi^1_{khp} + \delta_{lp}\,\chi^0_{kh}\,) - \bar{C}^0_{ijhp}\,) = 0. \tag{B.4.5e,f}$$

Functions $\boldsymbol{\chi}^1$ and $\boldsymbol{\chi}^2$ can be determined[8] by respectively solving (B.4.5e) and (B.4.5f) with suitable boundary conditions.

Since $\boldsymbol{\chi}^0$ is given by $\chi^0_{ih} = \delta_{ih}$, (B.4.5e) leads to the following symmetry properties of $\boldsymbol{\chi}^1$:

$$\chi^1_{ihp} = \chi^1_{iph}. \tag{B.4.6a}$$

Furthermore, the similar symmetry properties of $\boldsymbol{\chi}^2$ are derived from (B.4.5f) and the definition of $\mathbf{u}^2$ as $u^2_i = \chi^2_{ihpq}\,D_pD_qu^0_h$, i.e.,

$$\chi^2_{ihpq} = \chi^2_{iphq} = \chi^2_{ihqp}. \tag{B.4.6b}$$

Hence, for the three-dimensional case, $\boldsymbol{\chi}^1$ and $\boldsymbol{\chi}^2$ have 3×6 and 3×7 independent components. That is,

$$\boldsymbol{\chi}^1: \ \ \frac{1}{2}\,(\,D_pu^0_h + D_hu^0_p\,) \qquad \Rightarrow \ u^1_i,$$

$$\boldsymbol{\chi}^2: \ \ \frac{1}{3}\,(\,D_pD_qu^0_h + D_qD_hu^0_p + D_hD_pu^0_q\,) \ \Rightarrow \ u^2_i.$$

It should be noted that the symmetry properties of $\boldsymbol{\chi}^1$ and $\boldsymbol{\chi}^2$ imply the symmetry properties for $\bar{\mathbf{C}}^0$ and $\bar{\mathbf{C}}^1$, i.e.,

$$\bar{C}^0_{ijhp} = \bar{C}^0_{ijph}, \qquad \bar{C}^1_{ijhpq} = \bar{C}^1_{ijphq} = \bar{C}^1_{ijhqp}, \tag{B.4.7a,b}$$

[8] Theorems I and II can be used in the homogenization theory, to determine the microscopic response for an assumed microstructural model. It can be proved that if $\boldsymbol{\chi}^1$ and $\boldsymbol{\chi}^2$ are determined based on either zero surface displacements or zero surface tractions, the resulting $\bar{\mathbf{C}}^0$ and $\bar{\mathbf{C}}^1$ are bounds on all other overall moduli computed by using other $\boldsymbol{\chi}^1$'s and $\boldsymbol{\chi}^2$'s with different (consistent) boundary conditions. For instance, when a unit cell is used as a microstructural model, the overall moduli computed from the periodic boundary conditions lie between those computed for zero surface displacements and zero surface tractions.

with $\bar{C}^0_{ijhp} = \bar{C}^0_{jihp}$ and $\bar{C}^1_{ijhpq} = \bar{C}^1_{jihpq}$.

Once $\bar{\mathbf{C}}^0$ and $\bar{\mathbf{C}}^1$ are determined, it is possible to solve (B.4.5b) using another (regular) perturbation representation of $\mathbf{u}^0$,

$$u_i^0 \sim U_i^0 + \varepsilon\, U_i^1 + \dots . \tag{B.4.8}$$

This representation leads to a recursive formula of partial differential equations for $\mathbf{U}^n$, i.e.,

$$\bar{C}^0_{ijhp}\,(D_iD_pU_h^0) = 0, \qquad \bar{C}^0_{ijhp}\,(D_iD_pU_h^1) + \bar{C}^1_{ijhpq}\,(D_iD_pD_qU_h^0) = 0, \tag{B.4.9a,b}$$

and so on. Assuming that $\bar{\mathbf{C}}^0$ is invertible, rewrite the left side of (B.4.9b) as

$$D_i\Big\{\, \bar{C}^0_{ijhp}\,(D_pU_h^1) + \bar{C}^0_{ijhp}\,(D_pD_qU_h^0) \Big\}$$

$$= D_i\,\bar{C}^0_{ijhp}\Big\{\, D_pU_h^1 + ((\bar{\mathbf{C}}^0)^{-1} : \bar{\mathbf{C}}^1)_{hprst}\,(D_sD_tU_r^0) \Big\}$$

$$= D_i\,\bar{C}^0_{ijhp}\,D_t\Big\{\, \delta_{pt}\,U_h^1 + ((\bar{\mathbf{C}}^0)^{-1} : \bar{\mathbf{C}}^1)_{hprst}\,(D_sU_r^0) \Big\}.$$

Hence, there exists a suitable function $\mathbf{A}$ which satisfies

$$D_i\,\bar{C}^0_{ijhp}\,D_t\,A_{hpt} = 0, \tag{B.4.10a}$$

and yields $\mathbf{U}^1$ as

$$\delta_{pt}\,U_h^1 = A_{hpt} - ((\bar{\mathbf{C}}^0)^{-1} : \bar{\mathbf{C}}^1)_{hprst}\,D_sU_r^0, \tag{B.4.10b}$$

or, summing over p and t,

$$U_i^1 = \frac{1}{3}\Big\{\, A_{ipp} - ((\bar{\mathbf{C}}^0)^{-1} : \bar{\mathbf{C}}^1)_{ipklp}\,(D_lU_k^0) \Big\}. \tag{B.4.10c}$$

While $\mathbf{A}$ should be determined such that the boundary conditions prescribed for $\mathbf{U}^1$ are satisfied, it may be possible to neglect $\mathbf{A}$ and locally relate $\mathbf{U}^1$ to the gradient of $\mathbf{U}^0$, $\nabla_X \otimes \mathbf{U}^0$, by

$$U_i^1 \approx \Xi_{ikl}\,(D_lU_k^0), \tag{B.4.11a}$$

where Ξ is a third-order constant tensor given by

$$\Xi_{ikl} = -\frac{1}{3}\,((\bar{\mathbf{C}}^0)^{-1} : \bar{\mathbf{C}}^1)_{ipklp} = -\frac{1}{3}\,(\bar{\mathbf{C}}^0)^{-1}_{ipmn}\,\bar{C}^1_{mnklp}. \tag{B.4.11b}$$

Since $\bar{\mathbf{C}}^0$ and $\bar{\mathbf{C}}^1$ are determined from $\boldsymbol{\chi}^0$, $\boldsymbol{\chi}^1$, and $\boldsymbol{\chi}^2$ ((B.4.5c,d)), Ξ can be explicitly determined once the microstructural model is established. It should be noted that, by definition, this Ξ satisfies the following symmetry property:

$$\Xi_{ikl} = \Xi_{ilk}. \tag{B.4.11c}$$

Substitution of (B.4.8) and (B.4.11a) into (B.4.1) leads to

$$u_i^\varepsilon \sim U_i^0 + \varepsilon\,(\,\chi^1_{ikl} + \Xi_{ikl}\,)\,(D_kU_l^0) + \dots . \tag{B.4.12}$$

The inverse relation between $\mathbf{u}^\varepsilon$ and $\mathbf{U}^0$ is obtained, up to $O(\varepsilon)$, as follows:

$$U_i^0 \sim u_i^\varepsilon - \varepsilon\,(\chi_{ikl}^1 + \Xi_{ikl})\,D_l u_k^\varepsilon + \dots . \tag{B.4.13}$$

In this equation, $\boldsymbol{\chi}^1$ is the only term which changes within the micro-length-scale, or the only function which depends on $\mathbf{x}$. If, say, the volume average of $\boldsymbol{\chi}^1$ is taken in the domain where $\boldsymbol{\chi}^1$ is determined, then, the right-hand side of (B.4.13) is approximated by functions of only $\mathbf{X}$, i.e.,

$$U_i^0 \sim u_i^\varepsilon - \varepsilon\,(<\chi_{ikl}^1> + \Xi_{ikl})\,D_l u_k^\varepsilon + \dots . \tag{B.4.14}$$

Hence, the following equations are obtained from (B.4.9a):

$$D_i\,(\bar{C}_{ijkl}^0\,D_l u_k^\varepsilon + \bar{\bar{C}}_{ijklm}^0\,D_l D_m u_k^\varepsilon) = 0, \tag{B.4.15a}$$

where

$$\bar{\bar{C}}_{ijklm}^0 = -\bar{C}_{ijpl}^0\,(<\chi_{pkm}^1> + \Xi_{pkm}). \tag{B.4.15b}$$

As is seen, $\bar{\bar{\mathbf{C}}}^0$ produces stresses due to the following strain gradient:

$$\frac{1}{3}\,(D_l D_m u_k^\varepsilon + D_m D_k u_l^\varepsilon + D_k D_l u_m^\varepsilon),$$

since $\bar{\bar{\mathbf{C}}}^0$ satisfies the symmetry property,

$$\bar{\bar{C}}_{ijklm}^0 = \bar{\bar{C}}_{ijkml}^0 = \bar{\bar{C}}_{ijmlk}^0 = \bar{\bar{C}}_{ijlkm}^0,$$

with $\bar{\bar{C}}_{ijklm}^0 = \bar{\bar{C}}_{jiklm}^0$. It should be noted that both $\bar{\mathbf{C}}^0$ and $\bar{\bar{\mathbf{C}}}^0$ are determined from the assumed microstructural model. The effects of $\bar{\bar{\mathbf{C}}}^0$, however, can be neglected as they are multiplied by ε. In addition to this, $\bar{\bar{\mathbf{C}}}^0$ is small for linear microstructural models, because: 1) $<\boldsymbol{\chi}^1>$ is small since it gives the average displacement when the microstructural model is subjected to zero surface boundary data, and 2) Ξ is small, since it gives the overall displacement when a linearly varying strain field with zero volume average is prescribed. Therefore, it is concluded that the effects of $\bar{\bar{\mathbf{C}}}^0$ or the strain gradient can be omitted unless the microstructural model is non-linear and there are large strain gradients present.

B.5. REFERENCES

Bakhvalov, N. S. and Panasenko, G. (1984), *Homogenization: averaging processes in periodic media*, Kluwer, New York.

Bensoussan, A., Lions, J. L., and Papanicolaou, G. (1978), *Asymptotic Analysis for Periodic Structures*, North-Holland, New York.

Castillero, J. B., Otero, J. A., and Ramos, R. R. (1998), Asymptotic homogenization of laminated piezocomposite materials, *Int. J. Solids Structures*, Vol. 35, No.s 5-6, 527-541.

Hornung, U. (1996), *Homogenization and porous media*, Springer-Verlag, Berlin.

Kevorkina, J. and Cole, J. D. (1996), *Multiple scale and singular perturbation methods*, Springer-Varlag, Berlin.

Meguid, S. A. and Kalamkarov, A. L. (1994), Asymptotic homogenization of elastic composite materials with regular structure, *Int. J. Solids Structures*, Vol. 31, No.3, 303-316.

Nunan, K. C. and Keller, J. B. (1984), Effective elasticity tensor of a periodic composite, *J. Mech. Phys. Solids*, Vol. 32, 259-280.

Oleinik, O. A., Shamaev, A. S., and Yosifian, G. A. (1992), *Mathematical problems in elasticity and homogenization*, North-Holland, New York.

Sanchez-Palencia, E. (1981), *Non-homogeneous media and vibration theory*, Lecture Note in Physics, No. 127, Springer, Berlin.

Terada, K., Miura, T., and Kikuchi, N. (1996), Digital image-based modeling applied to the homogenization analysis of composite materials, *Computational Mechanics*, 188-202.

Walker, K. P, Freed, A. D., and Jordan, E. H. (1991), Microstress analysis of periodic composites, *Composite Engineering*, Vol. 1, 29-40.

Yi, Y-M., Park, S-H., and Youn, S-K. (1998), Asymptotic homogenization of viscoelastic composites with periodic microstructures, *Int. J. Solids Structures*, Vol. 35, No.17, 2039-2055.

APPENDIX C UNIFORM FIELD THEORY

The effect of the microstructure on the overall properties of heterogeneous solids, is the central theme of the present book. The overall behavior of an RVE is strongly influenced by the microstructure's morphology, such as the shape, orientation, and spatial arrangement of the microconstituents. The effect of the interaction among the microconstituents, which determines the level of strain and stress concentration, changes depending on the geometry, and hence the overall behavior depends on the microstructure. However, there exist particular sets of loads which do not cause strain and stress concentration. The resulting fields are uniform for any arbitrary geometry of the microstructure.[1] This is the essence of the *uniform field theory*.

Unlike the average field theory or the homogenization theory, the uniform field theory does not seek to determine the overall material properties. Instead, the uniform field theory provides conditions that must be satisfied by the overall moduli which may have been predicted by applying some other theoretical methods or may have been measured experimentally. Thus, it provides a mean of assessing whether or not the results of a given model or a set of experiments are reasonable. It is emphasized that the conditions which emerge by applying the uniform field theory to a composite do not depend on the microstructure of the RVE, because the uniform fields must be independent of the microstructure.

C.1. APPLICATION OF UNIFORM FIELD THEORY TO THERMOELASTICITY OF HETEROGENEOUS SOLIDS

Consider an RVE with a two-phase microstructure. The thermoelastic constitutive relations of the phases are given by

$$\boldsymbol{\varepsilon}(\mathbf{x}) = \mathbf{D}^{\alpha} : \boldsymbol{\sigma}(\mathbf{x}) + \mathbf{m}^{\alpha}\,\theta(\mathbf{x}), \tag{C.1.1a}$$

or

$$\varepsilon_{ij}(\mathbf{x}) = D^{\alpha}_{ijkl}\,\sigma_{kl}(\mathbf{x}) + m^{\alpha}_{ij}\,\theta(\mathbf{x}), \tag{C.1.1b}$$

where $\mathbf{D}^{\alpha}$ and $\mathbf{m}^{\alpha}$ are the compliance tensor and the thermal expansion coefficient tensor for the α-th phase ($\alpha = 1, 2$), and $\theta(\mathbf{x})$ is the change in the temperature from a certain reference value; in what follows, θ is simply called "temperature".

---

[1] The uniform field theory dates back to the work of Levin (1967) and Cibb (1968). More recently, the idea has been used extensively by Benveniste and Dvorak (1990), Dvorak (1990, 1992), Dvorak and Benveniste (1992), and Dunn (1993) to obtain relations among the thermoelastic or piezoelectric moduli of composites; for a review, see Benveniste and Dvorak (1997).

Suppose that the RVE is subjected to uniform stress and temperature boundary conditions, i.e., $(\mathbf{t}, \theta) = (\boldsymbol{\nu} \cdot \boldsymbol{\sigma}^o, \theta^o)$ on ∂V with constant $\boldsymbol{\sigma}^o$ and θ^o. With θ^o fixed, $\boldsymbol{\sigma}^o$ may be suitably adjusted to render the resulting field variables within V uniform. Indeed, such uniform fields are attained when $\boldsymbol{\sigma}^o$ is defined by

$$\boldsymbol{\sigma}^o = \boldsymbol{\sigma}^\theta(\theta^o), \tag{C.1.2a}$$

where

$$\boldsymbol{\sigma}^\theta(\theta^o) = -(\mathbf{D}^2 - \mathbf{D}^1)^{-1} : (\mathbf{m}^2 - \mathbf{m}^1)\, \theta^o. \tag{C.1.2b}$$

The proof is straightforward: if stress and temperature are uniform, $\boldsymbol{\sigma}(\mathbf{x}) = \boldsymbol{\sigma}^\theta$ and $\theta(\mathbf{x}) = \theta^o$, then the strain in each phase is uniform. Continuity of the displacement on the phase interface is always satisfied when the strains in the corresponding phases are the same, i.e.,

$$\mathbf{D}^1 : \boldsymbol{\sigma}^\theta(\theta^o) + \mathbf{m}^1\, \theta^o = \mathbf{D}^2 : \boldsymbol{\sigma}^\theta(\theta^o) + \mathbf{m}^2\, \theta^o. \tag{C.1.3}$$

This leads to (C.1.2b). Since the uniform fields are determined by only the material properties, they do *not* depend on the microstructure.

Now, let $\overline{\mathbf{D}}$ and $\overline{\mathbf{m}}$ be the overall moduli of this RVE, relating the average field quantities through

$$\bar{\boldsymbol{\varepsilon}} = \overline{\mathbf{D}} : \bar{\boldsymbol{\sigma}} + \overline{\mathbf{m}}\, \bar{\theta}. \tag{C.1.4}$$

For an RVE of any arbitrary microstructure, the uniform fields considered above are the solutions of particular suitable boundary-value problems[2] defined for the RVE. Hence, (C.1.4) must hold when the average field quantities are replaced by the uniform field quantities, i.e.,

$$\overline{\mathbf{D}} : \boldsymbol{\sigma}^\theta(\theta^o) + \overline{\mathbf{m}}\, \theta^o = \mathbf{D}^1 : \boldsymbol{\sigma}^\theta(\theta^o) + \mathbf{m}^1\, \theta^o, \tag{C.1.5a}$$

for any θ^o. It follows from (C.1.2b) that (C.1.5a) can be rewritten as

$$(\overline{\mathbf{D}} - \mathbf{D}^1) : (\mathbf{D}^2 - \mathbf{D}^1)^{-1} : (\mathbf{m}^2 - \mathbf{m}^1) - (\overline{\mathbf{m}} - \mathbf{m}^1) = \mathbf{0}. \tag{C.1.5b}$$

Expression (C.1.5b) is the required condition for the uniform field theory. That is, (C.1.5b) must necessarily be satisfied by any predicted or measured overall compliance tensor and thermal expansion coefficient tensor of any two-phase composite. Note that an expression dual to (C.1.5b) results if instead of uniform stress boundary data, linear displacements are considered. With $\mathbf{C}^\alpha$ and $\overline{\mathbf{C}}$ denoting the corresponding elasticity tensors, and with $\mathbf{l}^\alpha = -\mathbf{C}^\alpha : \mathbf{m}^\alpha$, this then leads to

$$(\overline{\mathbf{C}} - \mathbf{C}^1) : (\mathbf{C}^2 - \mathbf{C}^1)^{-1} : (\mathbf{l}^2 - \mathbf{l}^1) - (\overline{\mathbf{l}} - \mathbf{l}^1) = \mathbf{0}. \tag{C.1.5c}$$

---

[2] A boundary-value problem for the RVE consists of the governing equations for the displacement and temperature fields and the prescribed boundary conditions, as well as the continuity conditions on the phase interfaces, i.e., both tractions and displacements must be continuous across interfaces. Uniform fields always satisfy the continuity conditions, regardless of the geometry of the phase boundaries.

C.2. VERIFICATION OF AVERAGE FIELD THEORY

It is of interest to examine whether the conditions required by the *uniform* field theory are satisfied by the overall moduli which are predicted by applying the *average* field theory which has been the main topic of this book. The average field theory considers a general boundary-value problem for an RVE, and evaluates the average stress or strain within microconstituents using a microstructural *model* with specified geometry. Hence, the requirement of the uniform field theory should be automatically satisfied, since the general boundary-value problem is formally solved without violating any physical conditions, and the solution of a particular model of the microstructure is applied.

The average field theory guarantees that when the two-phase RVE is subjected to any arbitrary boundary conditions, the following relations hold for the resulting average field quantities:

$$\bar{\boldsymbol{\varepsilon}} = \mathbf{D}^1 : \bar{\boldsymbol{\sigma}} + \mathrm{f}\,(\mathbf{D}^2 - \mathbf{D}^1) : < \boldsymbol{\sigma} >_2 + \mathbf{m}^1\,\bar{\theta} + \mathrm{f}\,(\mathbf{m}^2 - \mathbf{m}^1) < \theta >_2, \tag{C.2.1}$$

where $\bar{\boldsymbol{\varepsilon}}$, $\bar{\boldsymbol{\sigma}}$, and $\bar{\theta}$ are the volume average of $\boldsymbol{\varepsilon}$, $\boldsymbol{\sigma}$, and θ, taken over the RVE, i.e., $\bar{\boldsymbol{\varepsilon}} = < \boldsymbol{\varepsilon} >$, $\bar{\boldsymbol{\sigma}} = < \boldsymbol{\sigma} >$, and $\bar{\theta} = < \theta >$, and $< >_\alpha$ stands for the volume average taken over the α-th phase. As shown in Section 10, (C.2.1) is obtained using the volume average of the strain field and the constitutive relations of the phases, i.e.,

$$\begin{aligned} < \boldsymbol{\varepsilon} > &= (1 - \mathrm{f}) < \boldsymbol{\varepsilon} >_1 + \mathrm{f} < \boldsymbol{\varepsilon} >_2 \\ &= (1 - \mathrm{f})\,(\mathbf{D}^1 : < \boldsymbol{\sigma} >_1 + \mathbf{m}^1 < \theta >_1) + \mathrm{f}\,(\mathbf{D}^2 : < \boldsymbol{\sigma} >_2 + \mathbf{m}^2 < \theta >_2) \\ &= \mathbf{D}^1 : < \boldsymbol{\sigma} > + \mathrm{f}\,(\mathbf{D}^2 - \mathbf{D}^1) : < \boldsymbol{\sigma} >_2 + \mathbf{m}^1 < \theta > + \mathrm{f}\,(\mathbf{m}^2 - \mathbf{m}^1) < \theta >_2, \end{aligned} \tag{C.2.2}$$

where $< > = (1 - \mathrm{f}) < >_1 + \mathrm{f} < >_2$ is used. Hence, (C.2.1) is *exact*.

When the temperature within the RVE is zero, (C.2.1) yields the overall compliance tensor,

$$\bar{\mathbf{D}} = \mathbf{D}^1 + \mathrm{f}\,(\mathbf{D}^2 - \mathbf{D}^1) : \mathbf{F}^2, \tag{C.2.3}$$

where the fourth-order tensor[3] $\mathbf{F}^2$ is the (stress) concentration tensor for the second phase, i.e., $\mathbf{F}^2$ yields $< \boldsymbol{\sigma} >_2 = \mathbf{F}^2 : \bar{\boldsymbol{\sigma}}$. Similarly, when the temperature within the RVE is uniform, θ^o, and the average stress is zero, (C.2.1) yields the overall thermal expansion coefficient tensor,

$$\bar{\mathbf{m}} = \mathbf{m}^1 + \mathrm{f}\,(\mathbf{m}^2 - \mathbf{m}^1) + \mathrm{f}\,(\mathbf{D}^2 - \mathbf{D}^1) : \mathbf{f}^2, \tag{C.2.4}$$

where the second-order tensor $\mathbf{f}^2$ is the stress concentration tensor for the second phase corresponding to the thermal effects, i.e., $< \boldsymbol{\sigma} >_2 = \mathbf{f}^2\,\theta^o$. For uniform boundary tractions, $\bar{\boldsymbol{\sigma}} = \boldsymbol{\sigma}^o$, (C.2.3) and (C.2.4) are exact for any two-phase composite.

In the average field theory, $\mathbf{F}^2$ and $\mathbf{f}^2$ are estimated using suitable models which *represent* the microstructure of the RVE. As explained in Section 11, an ellipsoidal inclusion Ω embedded in an unbounded body B is usually used as a *model* of the microstructure; the first and second phases may be identified as the matrix and the inclusion, respectively. Using this model, estimate $\mathbf{F}^2$ by considering a case when the unbounded body is subjected to farfield stress $\boldsymbol{\sigma}^\infty$ only, i.e., neglect the thermal effects. The disturbance fields are produced by the mismatch in the strain in the inclusion and the surrounding matrix,

$$\boldsymbol{\varepsilon}^{\text{stress}} = (\mathbf{D}^2 - \mathbf{D}^1) : \boldsymbol{\sigma}^\infty, \tag{C.2.5}$$

where the superscript "stress" indicates that the corresponding strain is due to an imposed farfield stress only, i.e., no thermal effects are being considered at this point. According to Eshelby's results, this uniform mismatch strain in an ellipsoidal Ω produces uniform disturbance strains and stresses in Ω, since the mismatch strain can be regarded as an eigenstrain uniformly distributed in the ellipsoidal region. The uniform disturbance stress in Ω is given by $\boldsymbol{\Lambda} : \boldsymbol{\varepsilon}^{\text{stress}}$, where $\boldsymbol{\Lambda}$ is expressed in terms of Eshelby's tensor, $\mathbf{S}$, for the ellipsoidal domain, as

$$\boldsymbol{\Lambda} = (\mathbf{B} : \mathbf{T}^{-1} - \mathbf{1}^{(4s)}) : (\mathbf{D}^2 - \mathbf{D}^1)^{-1}, \tag{C.2.6}$$

where[4] $\mathbf{B} = (\mathbf{D}^1 - \mathbf{D}^2)^{-1} : \mathbf{D}^1$ and $\mathbf{T}$ is the dual Eshelby tensor, given by $\mathbf{T} = \mathbf{1}^{(4s)} - (\mathbf{D}^1)^{-1} : \mathbf{S} : \mathbf{D}^1$; see Subsection 7.3.3, page 217. Since the total stress in Ω is $\boldsymbol{\sigma}^\infty + \boldsymbol{\Lambda} : \boldsymbol{\varepsilon}^{\text{stress}}$, the (stress) concentration tensor becomes

$$\mathbf{F}^2 = \mathbf{1}^{(4s)} + \boldsymbol{\Lambda} : (\mathbf{D}^2 - \mathbf{D}^1). \tag{C.2.7}$$

Next, estimate $\mathbf{f}^2$ considering the case when the uniform temperature θ^{o} is given in the body. Again, the disturbance fields are produced by the mismatch strain between the inclusion and the matrix,

$$\boldsymbol{\varepsilon}^{\text{temperature}} = (\mathbf{m}^2 - \mathbf{m}^1)\, \theta^{\text{o}}, \tag{C.2.8}$$

where the superscript "temperature" has a similar meaning as in (C.2.5). The uniform disturbance stress in Ω is now given by $\boldsymbol{\Lambda} : \boldsymbol{\varepsilon}^{\text{temperature}}$. Hence, the stress concentration tensor due to the temperature effects is

$$\mathbf{f}^2 = \boldsymbol{\Lambda} : (\mathbf{m}^2 - \mathbf{m}^1). \tag{C.2.9a}$$

This equation defines the concentration tensor $\mathbf{f}^2$ in terms of the concentration

---

[3] See Subsection 10.1 for a more detailed explanation of various stress concentration tensors; $\mathbf{F}^2$ used in (C.2.3) is the same as that defined by (10.1.2b). Note that while other forms of concentration tensors are used in Sections 4 and 7, they can be uniquely related to the concentration tensors presented in Section 10.

[4] In Subsection 7.3, the stress concentration tensor is expressed in a different manner; for instance, (7.3.21a) is rewritten as $(\mathbf{D}^2 - \mathbf{D}^1) : <\boldsymbol{\sigma}>_2 = \mathbf{H}^2 : \bar{\boldsymbol{\sigma}}$ with Ω being the second phase. The tensor $\mathbf{H}^2$ is given by (7.3.22a), and it relates to $\mathbf{F}^2$ or $\boldsymbol{\Lambda}$ through $\mathbf{H}^2 = (\mathbf{D}^2 - \mathbf{D}^1) : \mathbf{F}^2$ or $\mathbf{H}^2 = (\mathbf{D}^2 - \mathbf{D}^1) : (\mathbf{1}^{(4s)} + \boldsymbol{\Lambda} : (\mathbf{D}^2 - \mathbf{D}^1))$, respectively. Section 10 uses the same stress concentration tensor, $\mathbf{F}^2$, which relates $<\boldsymbol{\sigma}>_2 = \mathbf{F}^2 : \bar{\boldsymbol{\sigma}}$.

tensor $\mathbf{F}^2$. Indeed, from (C.2.7) and (C.2.9a), it follows that

$$\mathbf{f}^2 = (\mathbf{F}^2 - \mathbf{1}^{(4s)}) : (\mathbf{D}^2 - \mathbf{D}^1)^{-1} : (\mathbf{m}^2 - \mathbf{m}^1) \tag{C.2.9b}$$

which is the same as equation (2.12) of Benveniste and Dvorak (1990).[5]

As a specific case, consider the dilute distribution model; see Subsection 7.4, page 225. Substitution of (C.2.7) and (C.2.9a) into (C.2.3) and (C.2.4) then yields the overall moduli predicted according to the dilute distribution assumption,[6] i.e.,

$$\overline{\mathbf{D}} = \mathbf{D}^1 + f\,(\mathbf{D}^2 - \mathbf{D}^1) : (\mathbf{1}^{(4s)} + \mathbf{\Lambda} : (\mathbf{D}^2 - \mathbf{D}^1)),$$

$$\overline{\mathbf{m}} = \mathbf{m}^1 + f\,(\mathbf{m}^2 - \mathbf{m}^1) + f\,(\mathbf{D}^2 - \mathbf{D}^1) : \mathbf{\Lambda} : (\mathbf{m}^2 - \mathbf{m}^1). \tag{C.2.10a,b}$$

It is easily seen that $\overline{\mathbf{D}}$ and $\overline{\mathbf{m}}$ given by (C.2.10a,b) satisfy (C.1.5b) for any arbitrary $\mathbf{\Lambda}$. It is also seen that (C.1.5b) is satisfied by the overall moduli which evaluate the stress concentration tensors in a form similar to (C.2.7) and (C.2.9a).

As mentioned before, the *average* field scheme formally considers a general boundary-value problem for the RVE, which does include a particular boundary-value problem that leads to *uniform* fields in the RVE. The condition required by the uniform field theory, therefore, is always satisfied by the overall moduli predicted by the *average* field theory, as long as a *rational*[7] microstructural model is employed in estimating the (stress or strain) concentration tensors. In this sense, the results of the uniform field theory provide a check on whether or not the employed model is reasonable.

It is shown in Section 11 that *all conventional averaging schemes, such as the self-consistent and the Mori-Tanaka methods, or even the bounds obtained by the Hashin-Shtrikman variational principle, can be interpreted as special cases of the double-inclusion method, using particular geometries*[8] *for the inclusion and the matrix.* It can be shown that the estimate of the (stress) concentration tensors obtained by the double-inclusion method, has the same form as

---

[5] In a similar manner, the concentration tensor $\mathbf{e}^2$ in the relation $< \boldsymbol{\varepsilon} >_2 = \mathbf{e}^2\,\theta^{\mathrm{o}}$ can be expressed in terms of the concentration tensor $\mathbf{E}^2$ in the relation $< \boldsymbol{\varepsilon} >_2 = \mathbf{E}^2 : \bar{\boldsymbol{\varepsilon}}$ by $\mathbf{e}^2 = (\mathbf{E}^2 - \mathbf{1}^{(4s)}) : (\mathbf{C}^2 - \mathbf{C}^1)^{-1} : (\mathbf{l}^2 - \mathbf{l}^1)$, where $\mathbf{l}^\alpha = -\mathbf{C}^\alpha : \mathbf{m}^\alpha$; this is the same as equation (2.18) of Benveniste and Dvorak (1990).

[6] In Subsection 10, the dilute assumption uses a (stress) concentration tensor which is determined by the ellipsoidal inclusion model. The concentration tensor is denoted by $\mathbf{F}^{DD2} \approx \mathbf{F}^\infty(\mathbf{C}^2, \Omega; \mathbf{C}^1)$, which coincides with $\mathbf{1}^{(4s)} + \mathbf{\Lambda} : (\mathbf{D}^2 - \mathbf{D}^1)$ in (C.2.10a); see (10.1.7b), page 355.

[7] In the present example, the mechanism that causes stress concentration is the strain mismatch between the two phases. In the ellipsoidal inclusion model, the average stress in an isolated inclusion is computed exactly. This results in a consistent estimate of $\mathbf{F}^2$ and $\mathbf{f}^2$ in terms of $\mathbf{\Lambda}$, as given by (C.2.7) and (C.2.9a). It should be emphasized that all physical conditions in the inclusion *model* are fully satisfied by solving the corresponding "inclusion" (but not the actual) boundary-value problem without making any approximation.

(C.2.7) and (C.2.9), for any geometry consistent with the double-inclusion method. Therefore, these conventional methods predict the overall moduli that always satisfy (C.1.5b) or (C.1.5c).

C.3. APPLICATION OF UNIFORM FIELD THEORY TO COMPOSITES WITH ALIGNED FIBERS

As shown in Subsection C.1, uniform fields are solutions of boundary-value problems with special boundary conditions assigned to an RVE. Such solutions exist when the number of quantities which are used to prescribe the boundary conditions exceeds the number of independent fields which are to be uniform. For instance, the thermoelasticity problem of a two-phase composite has 7 quantities which are used to prescribe the boundary conditions (6 components of stress and 1 of temperature) and there are 6 uniform fields (6 components of strain). Hence, one free quantity exists which can be used to produce a uniform strain, in a manner similar to (C.1.2).

The condition of the existence of the uniform fields can be satisfied for a two-phase composite which consists of aligned microconstituents, such as long fibers or layers, without any reference to temperature. Let the x_3-axis be the axis of the alignment and assume that all interfaces between the phases are parallel to the x_3-axis. The constitutive relations of the phases,

$$\boldsymbol{\sigma}(\mathbf{x}) = \mathbf{C}^{\alpha} : \boldsymbol{\varepsilon}(\mathbf{x}), \tag{C.3.1}$$

are now rewritten in the following form:

$$\boldsymbol{\sigma}'(\mathbf{x}) = \mathbf{C}^{\alpha'} : \boldsymbol{\varepsilon}'(\mathbf{x}) + \mathbf{C}^{\alpha''} \varepsilon_{33}(\mathbf{x}),$$

$$\sigma_{33}(\mathbf{x}) = \mathbf{C}^{\alpha''} : \boldsymbol{\varepsilon}'(\mathbf{x}) + C^{\alpha}_{3333}\, \varepsilon_{33}(\mathbf{x}), \tag{C.3.2a,b}$$

or

$$\sigma'_{ij}(\mathbf{x}) = C^{\alpha'}_{ijkl}\, \varepsilon'_{kl}(\mathbf{x}) + C^{\alpha''}_{ij}\, \varepsilon_{33}(\mathbf{x}),$$

$$\sigma_{33}(\mathbf{x}) = C^{\alpha''}_{kl}\, \varepsilon'_{kl}(\mathbf{x}) + C^{\alpha}_{3333}\, \varepsilon_{33}(\mathbf{x}), \tag{C.3.2c,d}$$

where the components of $\boldsymbol{\sigma}'$ and $\boldsymbol{\varepsilon}'$ are $[\sigma_{11}, \sigma_{22}, \sigma_{12}, \sigma_{31}, \sigma_{32}]$ and $[\varepsilon_{11}, \varepsilon_{22}, \varepsilon_{12}, \varepsilon_{31}, \varepsilon_{32}]$, and $\mathbf{C}^{\alpha'}$ and $\mathbf{C}^{\alpha''}$ are the corresponding elasticity tensors, obtained from $\mathbf{C}^{\alpha}$ accordingly. The form of (C.3.2a) is the same as (C.1.1a) if $(\boldsymbol{\varepsilon}', \varepsilon_{33}, \boldsymbol{\sigma}')$ and $(\mathbf{C}^{\alpha'}, \mathbf{C}^{\alpha''})$ are replaced by $(\boldsymbol{\sigma}, \theta, \boldsymbol{\varepsilon})$ and $(\mathbf{D}^{\alpha}, \mathbf{m}^{\alpha})$, respectively.

---

[8] Note that, since the differential scheme is obtained by considering a sequence of dilute distribution cases, all comments about this latter method also apply to the results of the differential scheme. Hence, the moduli obtained by the differential scheme also satisfy (C.1.5b) or (C.1.5c) automatically.

The interface of the phases is arbitrary except that it is parallel to the x_3-axis. The continuity of tractions requires $\mathbf{n} \cdot [\boldsymbol{\sigma}] = \mathbf{0}$ for any unit vector $\mathbf{n}$ which is normal to the x_3-axis, i.e., $n_3 = 0$, where $[\boldsymbol{\sigma}]$ denotes the stress jump; see Subsection 2.4, page 35. This condition is satisfied if $\boldsymbol{\sigma}'$ is uniform. Therefore, there can exist a set of uniform fields, $(\boldsymbol{\varepsilon}', \varepsilon_{33}, \boldsymbol{\sigma}')$, which is the solution of a particular boundary-value problem for the RVE.

Now, suppose that the RVE is subjected to the linear displacement boundary conditions, $\mathbf{u}' = \mathbf{x} \cdot \boldsymbol{\varepsilon}'^{o}$ and $u_3 = x_3 \, \varepsilon_{33}^{o}$, where the components of $\mathbf{u}'$ are $[u_1, u_2]$. As in the case of the thermoelasticity problem, it can be seen that the uniform fields are attained when $\boldsymbol{\varepsilon}'^{o}$ is given as a function of ε_{33}^{o},

$$\boldsymbol{\varepsilon}'^{o} = \boldsymbol{\varepsilon}'^{N}(\varepsilon_{33}^{o}), \tag{C.3.3a}$$

where

$$\boldsymbol{\varepsilon}'^{N}(\varepsilon_{33}^{o}) = -(\mathbf{C}^{2'} - \mathbf{C}^{1'})^{-1} : (\mathbf{C}^{2''} - \mathbf{C}^{1''}) \, \varepsilon_{33}^{o}. \tag{C.3.3b}$$

The overall moduli, predicted or measured, are then required to be such that the uniform fields produced by the above particular boundary conditions are indeed a solution of this problem. The uniform strain and stress[9] fields are $\boldsymbol{\varepsilon}'^{N}(\bar{\varepsilon}_{33})$, $\bar{\varepsilon}_{33}$, and $\mathbf{C}^{1'} : \boldsymbol{\varepsilon}'^{N} + \mathbf{C}^{1''} \bar{\varepsilon}_{33}$. If $\bar{\mathbf{C}}'$ and $\bar{\mathbf{C}}''$ are defined based on the overall elasticity tensor, $\bar{\mathbf{C}}$, this requirement is expressed as

$$(\bar{\mathbf{C}}' - \mathbf{C}^{1'}) : (\mathbf{C}^{2'} - \mathbf{C}^{1'})^{-1} : (\mathbf{C}^{2''} - \mathbf{C}^{1''}) - (\bar{\mathbf{C}}'' - \mathbf{C}^{1''}) = \mathbf{0}, \tag{C.3.4}$$

which corresponds to (C.1.5c).

In essentially the same manner as in Subsection C.2, it is shown that the overall moduli predicted by the average field theory satisfy (C.3.4), provided that a rational model of the microstructure is used to compute the strain concentration tensors for the second phase. For instance, an inclusion model can be used by taking the limit as an ellipsoidal inclusion becomes infinitely long. Since the strain concentration is due to the mismatch in the stress in the inclusion and the matrix, the resulting disturbance can be computed using Eshelby's tensor, $\mathbf{S}$. For instance, the dilute distribution assumption regards $(\mathbf{C}^{2'} - \mathbf{C}^{1'}) : \boldsymbol{\varepsilon}^{\infty'}$ and $(\mathbf{C}^{2''} - \mathbf{C}^{1''}) \, \varepsilon_{33}^{\infty}$ as the mismatch in the stress when the farfield strain is prescribed by $\boldsymbol{\varepsilon}^{\infty'}$ and $\varepsilon_{33}^{\infty}$. The corresponding strain concentration tensors then are $\mathbf{E}^2 = \mathbf{1}^{(4s)} + \boldsymbol{\Gamma} : (\mathbf{C}^{2'} - \mathbf{C}^{1'})$ and $\mathbf{e}^2 = \boldsymbol{\Gamma} : (\mathbf{C}^{2''} - \mathbf{C}^{1''})$, where

$$\boldsymbol{\Gamma} = (\mathbf{A} : \mathbf{S}^{-1} - \mathbf{1}^{(4s)})^{-1} : (\mathbf{C}^{2'} - \mathbf{C}^{1'}), \tag{C.3.5}$$

$\mathbf{S}$ is Eshelby's tensor, and[10] $\mathbf{A} = (\mathbf{C}^1 - \mathbf{C}^2)^{-1} : \mathbf{C}^1$. $\mathbf{E}^2$ and $\mathbf{e}^2$ are related to each

[9] Note that σ_{33} does not have to be uniform. However, the uniform strain in the RVE produces uniform σ_{33} in each phase, and the average of σ_{33} can be computed as the volume average of the corresponding stress fields of the two phases, using (C.3.2b) with $\boldsymbol{\varepsilon}'^{N}(\bar{\varepsilon}_{33})$ and $\bar{\varepsilon}_{33}$.

[10] In Subsection 7.3, the tensor $\mathbf{J}$ is used to determine the (strain) concentration. Rewrite (7.3.21b) as $(\mathbf{D}^2 - \mathbf{D}^1) : \mathbf{C}^2 : <\boldsymbol{\varepsilon}>_2 = \mathbf{J}^2 : \boldsymbol{\varepsilon}^{\infty}$, where $\mathbf{J}^2$ is given by (7.3.22b), and can be expressed in terms of Eshelby's tensor, $\mathbf{S}$, as $\mathbf{J}^2 = (\mathbf{D}^2 - \mathbf{D}^1) : \mathbf{C}^2 : \mathbf{A} : (\mathbf{A} - \mathbf{S})^{-1}$. This then leads to the definition of $\boldsymbol{\Gamma}$.

other through $\mathbf{e}^2 = (\mathbf{E}^2 - \mathbf{1}^{(4s)}) : (\mathbf{C}^{2'} - \mathbf{C}^{1'})^{-1} : (\mathbf{C}^{2''} - \mathbf{C}^{1''})$. Therefore, the condition of the uniform field theory, (C.3.4), is satisfied by the overall moduli predicted by the dilute distribution assumption, as well as those predicted from other averaging schemes which have been examined in this book, e.g., in Section 10.

C.4. REFERENCES

Benveniste, Y. and Dvorak, G. J. (1990), On a correspondence between mechanical and thermal effects in two-phase composites, in *Micromechanics and inhomogeneity: The Toshio Mura anniversary volume*, (ed. by G.J. Weng, M. Taya and H. Abe), Springer-Verlag, New York, 65-81.

Benveniste, Y. and Dvorak, G. J. (1997), On micromechanics of inelastic and piezoelectric composites, Proceedings of the ICTAM Kyoto, August 25-31, 1996, Elsevier, 217-237.

Cibb, J. L. (1968), Shrinkage and thermal expansion of a two phase material, *Nature*, Vol. 220, 576-577.

Dunn, M. L. (1993), Exact relations between the thermoelectroelastic moduli of heterogeneous materials, *Proc. R. Soc. Lond.* A 441, 549-557.

Dvorak, G. J. (1990), On uniform fields in heterogeneous media, *Proc. R. Soc. Lond.* A 431, 89-110.

Dvorak, G. J. (1992), On some exact results in thermoplasticity of composite materials, *J. Thermal Stresses,* Vol. 15, 211-228.

Dvorak, G. J. and Benveniste, Y. (1992), On transformation strains and uniform fields in multiphase elastic media, *Proc. R. Soc. Lond.* A437, 291-310.

Levin, V. M. (1967), Thermal expansion coefficients of heterogeneous materials, *Izv. AN SSSR, Mekh. Tverd. Tela,* Vol. 2, 88-94.

APPENDIX D IMPROVABLE BOUNDS ON OVERALL PROPERTIES OF HETEROGENEOUS FINITE SOLIDS

The universal theorems of Subsection 2.5 state that, for a finite solid consisting of a general heterogeneous nonlinearly elastic material, among all consistent boundary data which yield the same overall average strain (stress), the strain (stress) field produced by uniform boundary tractions (linear boundary displacements), renders the elastic strain (complementary strain) energy an absolute minimum. Similar results are obtained when the material of the composite is viscoplastic with convex potentials, such that the stress, $\boldsymbol{\sigma}$, (the strain rate, $\boldsymbol{\varepsilon}$) is given by the gradient with respect to the strain rate, $\boldsymbol{\varepsilon}$, (the stress, $\boldsymbol{\sigma}$) of a convex potential, $\phi(\boldsymbol{\varepsilon})$, (a convex potential, $\psi(\boldsymbol{\sigma})$).[1] Based on these general results, computable bounds are developed in this appendix for the overall stress and strain (strain-rate) potentials of solids of any shape and inhomogeneity, subjected to any set of consistent boundary data. Statistical homogeneity and isotropy are not required. It is shown that the bounds can be improved by incorporating additional material and geometric data specific to the given finite heterogeneous solid. These bounds are not based on the equivalent homogenized reference solid (discussed in Section 9). They remain nonzero and finite even when cavities or rigid inclusions are present.[2] Illustrative examples with explicit results are given in Subsection D.2.

D.1. BOUNDS ON POTENTIALS FOR GENERAL BOUNDARY DATA

In this Subsection, general bounds are developed for the stress and strain (strain rate) potentials of a finite nonlinear solid, following Nemat-Nasser and Hori (1995) and Nemat-Nasser *et al.* (1995). In Subsection D.2, these are then specialized to produce explicit bounds for the corresponding elastic moduli and compliances of linearly elastic solids, following Balendran and Nemat-Nasser (1995); see also Huet (1990) for an alternative approach.

---

[1] To discuss both nolinear elasticity and viscoplasticity at the same time, the same notation, i.e., $\boldsymbol{\varepsilon}$, is used to represent strain and strain rate. The meaning should be clear from the context.

[2] Note that the bounds presented in Subsection 9 depend on the reference elasticity and compliance tensors. The resulting bounds on the moduli, therefore, are zero (the lower bound) or infinity (the upper bound) when the solid contains cavities or rigid inclusions.

D.1.1. Weak Kinematical or Statical Admissibility

In Subsection 2.4, expressions are developed for the average strain and stress tensors in a finite heterogeneous solid when the displacements or the tractions may be discontinuous across some interior surfaces.[3] Consider a finite heterogeneous solid which consists of various phases, and let S be the collection of all (isolated) surfaces across which certain components of the displacements (velocities) may suffer jump discontinuities, while everywhere else the displacement (velocity) field is continuous and continuously differentiable, including in the neighborhood of the discontinuity surfaces, where the field is then one-sided continuously differentiable. Denote this displacement (velocity) discontinuity by $\Delta\mathbf{u}$. The strain (strain rate) field obtained from this displacement (velocity) field is called *weakly compatible* if on each (isolated) discontinuity surface S_i,

$$\int_{S_i} \frac{1}{2}(\mathbf{n}\otimes\Delta\mathbf{u}+\Delta\mathbf{u}\otimes\mathbf{n})\, dS = \mathbf{0}. \tag{D.1.1}$$

If S_i is the boundary of an inclusion within the solid, then (D.1.1) requires that the average strain (strain rate) of the inclusion equals the corresponding *cavity strain (strain rate),* when the inclusion is removed; for the definition of cavity strain, see Subsection 5.4, page 118.

Similarly, a stress field is called *weakly self-equilibrating* if on each (isolated) discontinuity surface S_i,

$$\int_{S_i} \boldsymbol{\xi}\otimes\Delta\mathbf{t}\, dS = \mathbf{0}, \tag{D.1.2}$$

where $\Delta\mathbf{t}$ is the traction discontinuity, and $\boldsymbol{\xi}$ is a typical point on S_i. When S_i is the boundary of an inclusion within the solid, then (D.1.2) requires that over the inclusion boundary, the solid be subjected to tractions equivalent to the average stress of the inclusion.

Consider now the concept of *weak admissibility*, as follows:

Weakly Kinematically Admissible Strain (Strain Rate) Fields: A weakly compatible strain (strain rate) field *with a prescribed average value* is called *weakly kinematically admissible.*

Weakly Statically Admissible Stress Fields: A weakly self-equilibrating stress field *with a prescribed average value*, is called *weakly statically admissible.*

These fields need not satisfy any prescribed boundary data. Only the corresponding volume average is required to equal a given value. When prescribed displacement- (traction-) boundary conditions are satisfied and the displacement field and the tractions are everywhere continuous, then the *admissibility* is called *strong*.

---

[3] See Figure 2.4.1 and expressions (2.4.2), (2.4.4), and (2.4.5), pages 36 and 37.

D.1.2. Bounds on Potentials

To examine both rate-independent nonlinearly elastic and rate-dependent viscoplastic cases simultaneously, let $\boldsymbol{\varepsilon}$ stand for either the strain or the strain rate, depending on the considered case. Consider an *arbitrary* finite sample, M, of such nonlinear heterogeneous materials, and denote its boundary by ∂M; see Figure D.1.1. Let $\phi = \phi(\mathbf{x};\, \boldsymbol{\varepsilon})$ be the corresponding *stress potential* such that, at each point $\mathbf{x}$ in M, the stress, $\boldsymbol{\sigma}$, is given by $\boldsymbol{\sigma}(\mathbf{x}) = \partial\phi(\mathbf{x};\, \boldsymbol{\varepsilon})/\partial\boldsymbol{\varepsilon}$. The strain (strain rate) potential function, $\psi = \psi(\mathbf{x};\, \boldsymbol{\sigma})$, is such that $\phi(\mathbf{x};\, \boldsymbol{\varepsilon}) + \psi(\mathbf{x};\, \boldsymbol{\sigma}) = \boldsymbol{\sigma} : \boldsymbol{\varepsilon}$, and it follows that $\boldsymbol{\varepsilon}(\mathbf{x}) = \partial\psi(\mathbf{x};\, \boldsymbol{\sigma})/\partial\boldsymbol{\sigma}$. The stress potential ϕ and the strain (strain rate) potential ψ are convex, satisfying the convexity condition (2.5.31) and (2.5.36a) of pages 46 and 48, respectively.

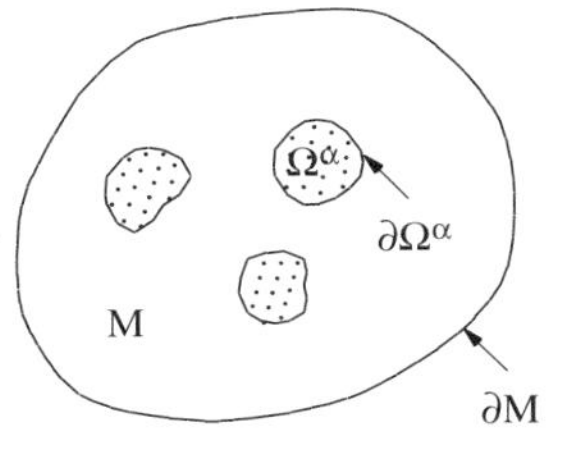

Figure D.1.1

A finite composite consisting of several heterogeneous inclusions and a matrix, all with pointwise convex stress and strain (strain rate) potentials

As in Subsection 2.5, define the overall potentials by $\Phi \equiv \,<\phi>_M$ and $\Psi \equiv \,<\psi>_M$. Then, applying expressions (2.5.34a) and (2.5.36a), obtain the following general results for nonlinearly elastic (viscoplastic) composites of any convex constituents:

I. Among all weakly kinematically admissible strain (strain-rate) fields, that which corresponds to uniform boundary tractions renders the overall stress potential, Φ, an absolute minimum, i.e.,

$$\Phi^G \geq \Phi^\Sigma \quad \text{when} \quad <\boldsymbol{\varepsilon}^G>_M \,=\, <\boldsymbol{\varepsilon}^\Sigma>_M; \tag{D.1.2a}$$

II. Among all weakly statically admissible stress fields, that which corresponds to the linear displacement (velocity) field, renders the overall strain (strain-rate) potential, Ψ, an absolute minimum, i.e.,

$$\Psi^G \geq \Psi^E \quad \text{when} \quad <\boldsymbol{\sigma}^G>_M \,=\, <\boldsymbol{\sigma}^E>_M. \tag{D.1.2b}$$

In (D.1.2a,b), superscripts G, Σ, and E, denote that the corresponding quantity is being associated respectively with general, uniform traction, and linear displacement (velocity) boundary data.

Consider now a special class of boundary data for which (2.5.38a) of page 49, is satisfied. Then

$$<\boldsymbol{\sigma} : \boldsymbol{\varepsilon}>_M \,=\, <\boldsymbol{\sigma}>_M : \,<\boldsymbol{\varepsilon}>_M. \tag{D.1.3}$$

Denote the stress and strain (strain rate) fields associated with these boundary data by $\boldsymbol{\sigma}^S$ and $\boldsymbol{\varepsilon}^S$, and the corresponding overall potentials by Φ^S and Ψ^S, respectively. It now follows that, *for the same overall strain (strain rate),*

$$\Phi^E \geq \Phi^S \geq \Phi^\Sigma, \tag{D.1.4a}$$

and, *for the same overall stress,*

$$\Psi^\Sigma \geq \Psi^S \geq \Psi^E. \tag{D.1.4b}$$

Note that, in general, Φ^E and Ψ^Σ are the overall stress and strain (strain rate) potentials, in the sense that their gradient with respect to $\mathbf{E}$ and $\mathbf{\Sigma}$, respectively, give the overall stress and strain (strain rate) tensors. In particular, for the special class of boundary data which satisfy (2.5.38a), it follows that

$$< \boldsymbol{\sigma} >_M = \partial\Phi^S/\partial\mathbf{E}, \quad < \boldsymbol{\varepsilon} >_M = \partial\Psi^S/\partial\mathbf{\Sigma}, \quad \Psi^S + \Phi^S = \mathbf{\Sigma} : \mathbf{E}. \tag{D.1.5a~c}$$

To apply these results to a finite composite, consider the method illustrated in Figure D.1.2. Here, the inclusions are removed and the resulting cavities and the isolated inclusions are subjected to uniform boundary data such that the resulting overall displacement (velocity) field is weakly kinematically admisssible when the overall stress field is self-equilibrating, or the stress field is weakly statically admisssible when the displacement (velocity) field is kinematically admisssible. This general approach can be used to construct analytical bounds for Φ^Σ and Ψ^E, as is shown in the sequel.

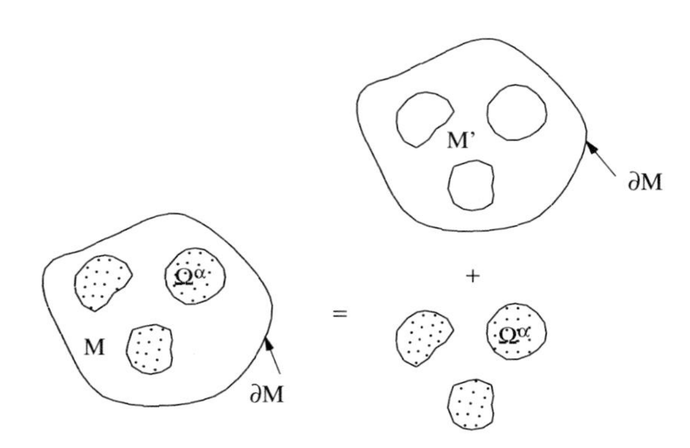

Figure D.1.2

To construct analytical bounds for the overall potentials, Φ^Σ and Ψ^E, remove the inclusions and subject the resulting cavities and the isolated inclusions to uniform boundary data such that the resulting overall displacement (velocity) field is weakly kinematically admisssible when the overall stress field is self-equilibrating, or the stress field is weakly statically admisssible when the overall displacement (velocity) field is kinematically admisssible

D.1.3. Calculation of Bounds on Overall Potentials

The calculation of the bounds on the overall stress potential for *an arbitrary heterogeneous solid* is examined first. The calculation of the strain (strain-rate) potential is then outlined.

The calculation of the lower bound for Φ^{Σ} is based on the following general observation which is valid for any finite M, of any heterogeneity, consisting of materials with pointwise convex potentials. Consider a finite composite, M, as shown in Figure D.1.3, and let M_0 with boundary ∂M_0 be a region containing Ω and being totally contained in M, and Ω_0 with boundary $\partial\Omega_0$ be a region containing Ω and being totally contained in M_0. Assume that uniform stress $\hat{\boldsymbol{\Sigma}}$ is prescribed in $M - M_0$, while Ω_0 is removed and instead, *uniform* tractions $-\mathbf{n}\cdot\hat{\boldsymbol{\sigma}}^0$ are applied on the boundary $\partial\Omega_0$, where $\mathbf{n}$ is the outward unit normal on $\partial\Omega_0$. Calculate $\hat{\boldsymbol{\sigma}}^0$ (constant) such that the *average* "cavity strain (strain rate)," $\bar{\boldsymbol{\varepsilon}}^C$, equals the average strain, $\bar{\boldsymbol{\varepsilon}}^{\Omega_0}$, of the Ω_0, when uniform tractions $\mathbf{n}\cdot\hat{\boldsymbol{\sigma}}^0$ are prescribed on $\partial\Omega_0$ of the isolated Ω_0, where

$$\bar{\boldsymbol{\varepsilon}}^C \equiv \frac{1}{\Omega_0}\int_{\partial\Omega_0}\frac{1}{2}(\mathbf{n}\otimes\mathbf{u}+\mathbf{u}\otimes\mathbf{n})\,dS. \tag{D.1.6}$$

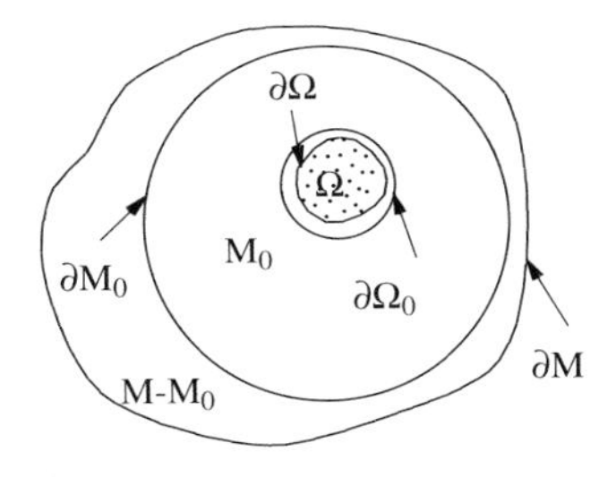

Figure D.1.3

A finite composite with an arbitrary boundary ∂M, contains an arbitrary heterogeneous inclusion Ω. Encase Ω in a sphere, $\partial\Omega_0$, which is contained in a sphere ∂M_0 which itself is contained in ∂M. To obtain computable bounds, remove $\partial\Omega_0$, and apply uniform tractions $-\mathbf{n}\cdot\hat{\boldsymbol{\sigma}}^0$ on the resulting cavity and $\mathbf{n}\cdot\hat{\boldsymbol{\sigma}}^0$ on the isolated $\partial\Omega_0$ such that the average cavity strain equals the average strain of Ω_0, while uniform stress $\hat{\boldsymbol{\Sigma}}$ is prescribed in $M - M_0$

The displacement (velocity) field $\mathbf{u}$ may not be continuous across ∂M_0. Denote the stress and strain (strain rate) fields in M by $\hat{\boldsymbol{\sigma}}(\mathbf{x})$ and $\hat{\boldsymbol{\varepsilon}}(\mathbf{x})$, respectively, and observe that

$$\hat{\boldsymbol{\sigma}} \equiv \begin{cases} \hat{\boldsymbol{\Sigma}} & \text{in } M - M_0 \\ \hat{\boldsymbol{\sigma}}(\mathbf{x};\, \hat{\boldsymbol{\Sigma}},\, \hat{\boldsymbol{\sigma}}^0) & \text{in } M_0 - \Omega_0 \\ \hat{\boldsymbol{\sigma}}(\mathbf{x};\, \hat{\boldsymbol{\sigma}}^0) & \text{in } \Omega_0, \end{cases} \tag{D.1.7a}$$

where $\hat{\boldsymbol{\Sigma}}$ and $\hat{\boldsymbol{\sigma}}^0$ are constant tensors. Also, whatever the composition of $M - M_0$, $M_0 - \Omega_0$, and Ω_0, and whatever the $\hat{\boldsymbol{\sigma}}^0$, the average stresses in M and Ω_0 are given by

$$<\hat{\boldsymbol{\sigma}}>_M = \hat{\boldsymbol{\Sigma}}, \quad <\hat{\boldsymbol{\sigma}}>_{\Omega_0} = \hat{\boldsymbol{\sigma}}^0, \tag{D.1.7b,c}$$

respectively. Similarly, since ∂M_0 is subjected to uniform tractions $\mathbf{n}\cdot\hat{\boldsymbol{\Sigma}}$, it also follows that

$$< \hat{\boldsymbol{\sigma}} >_{M_0} = \hat{\boldsymbol{\Sigma}}. \tag{D.1.7d}$$

The corresponding strain (strain-rate) field depends on the composition of the composite. It is defined by

$$\hat{\boldsymbol{\varepsilon}} \equiv \begin{cases} \hat{\boldsymbol{\varepsilon}}(\mathbf{x};\, \hat{\boldsymbol{\Sigma}}) & \text{in } \mathrm{M} - \mathrm{M}_0 \\ \hat{\boldsymbol{\varepsilon}}(\mathbf{x};\, \hat{\boldsymbol{\Sigma}},\, \hat{\boldsymbol{\sigma}}^0) & \text{in } \mathrm{M}_0 - \Omega_0 \\ \hat{\boldsymbol{\varepsilon}}(\mathbf{x};\, \hat{\boldsymbol{\sigma}}^0) & \text{in } \Omega_0, \end{cases} \tag{D.1.8a}$$

with the requirement that

$$< \hat{\boldsymbol{\varepsilon}} >_{\Omega_0} = \bar{\boldsymbol{\varepsilon}}^{\mathrm{C}} \tag{D.1.8b}$$

which then yields the required $\hat{\boldsymbol{\sigma}}^0$, where $\bar{\boldsymbol{\varepsilon}}^{\mathrm{C}}$ is defined by (D.1.6).

Now, let $\boldsymbol{\varepsilon}^{(1)}$ in (2.5.31), page 46, be the strain (strain-rate) field in M, which is produced by uniform tractions $\mathbf{n} \cdot \boldsymbol{\Sigma}$ applied on ∂M. Denote this strain (strain rate) field by $\boldsymbol{\varepsilon}^{\Sigma} \equiv \boldsymbol{\varepsilon}(\mathbf{x};\, \boldsymbol{\Sigma})$; this is a *compatible* field. Let $\boldsymbol{\varepsilon}^{(2)}$ and $\boldsymbol{\sigma}^{(2)}$ in (2.5.31) be the fields defined by (D.1.8a) and (D.1.7a), respectively. While the strain (strain-rate) field (D.1.8a) is not compatible, the stress field (D.1.7a) *is* self-equilibrating everywhere within M. Hence, the corresponding overall potentials satisfy

$$\Phi^{\Sigma} - \hat{\Phi}^{\hat{\Sigma}} \geq 0 \quad \text{when} \quad < \boldsymbol{\varepsilon}^{\Sigma} >_{\mathrm{M}} = < \hat{\boldsymbol{\varepsilon}} >_{\mathrm{M}}. \tag{D.1.9}$$

The overall stresses, $\boldsymbol{\Sigma}$ and $\hat{\boldsymbol{\Sigma}}$, are not, in general, the same. The overall stress $\hat{\boldsymbol{\Sigma}}$ must be adjusted such that the corresponding average strain (strain-rate) $< \hat{\boldsymbol{\varepsilon}} >_{\mathrm{M}}$ equals the average strain (strain-rate) $< \boldsymbol{\varepsilon}^{\Sigma} >_{\mathrm{M}}$ which is produced by uniform tractions $\mathbf{n} \cdot \boldsymbol{\Sigma}$ applied on ∂M of the original composite.

The calculation of the overall bounds on the strain (strain rate) potential for *an arbitrary heterogeneous solid* follows the same procedure, but, instead of the uniform stress $\hat{\boldsymbol{\Sigma}}$, the uniform *strain (strain rate)* $\hat{\mathbf{E}}$ is prescribed in $\mathrm{M} - \mathrm{M}_0$ of Figure D.1.3, and Ω_0 is removed and in its place, the *linear displacement (velocity)* $\mathbf{x} \cdot \hat{\boldsymbol{\varepsilon}}^0$ is applied on the boundary $\partial\Omega_0$. The constant strain (strain rate) $\hat{\boldsymbol{\varepsilon}}^0$ is then calculated such that the *average stress* in the inclusion Ω_0 is the same as the *average resistance* provided by the matrix. In general, the resulting tractions are not continuous across $\partial\Omega_0$. The strain (strain rate) field $\hat{\boldsymbol{\varepsilon}}(\mathbf{x})$ is

$$\hat{\boldsymbol{\varepsilon}} \equiv \begin{cases} \hat{\mathbf{E}} & \text{in } \mathrm{M} - \mathrm{M}_0 \\ \hat{\boldsymbol{\varepsilon}}(\mathbf{x};\, \hat{\mathbf{E}},\, \hat{\boldsymbol{\varepsilon}}^0) & \text{in } \mathrm{M}_0 - \Omega_0 \\ \hat{\boldsymbol{\varepsilon}}(\mathbf{x};\, \hat{\boldsymbol{\varepsilon}}^0) & \text{in } \Omega_0. \end{cases} \tag{D.1.10a}$$

Here, again, whatever the composition of $\mathrm{M} - \mathrm{M}_0$, $\mathrm{M}_0 - \Omega_0$, and Ω_0, and whatever the constant $\hat{\boldsymbol{\varepsilon}}^0$, the average strains (strain rates) in M and Ω_0 are given by

$$< \hat{\boldsymbol{\varepsilon}} >_{\mathrm{M}} = \hat{\mathbf{E}}, \qquad < \hat{\boldsymbol{\varepsilon}} >_{\Omega_0} = \hat{\boldsymbol{\varepsilon}}^0, \tag{D.1.10b,c}$$

respectively. Similarly, since $\partial\mathrm{M}_0$ is subjected to linear displacements (velocities) $\mathbf{x} \cdot \hat{\mathbf{E}}$, it also follows that

$$< \hat{\boldsymbol{\varepsilon}} >_{M_0} = \hat{\mathbf{E}}. \tag{D.1.10d}$$

It now readily follows that

$$\Psi^E - \hat{\Psi}^{\hat{E}} \geq 0 \;\text{ when }\; < \boldsymbol{\sigma}^E >_M = < \hat{\boldsymbol{\sigma}} >_M, \tag{D.1.11}$$

where $\hat{\Psi}^{\hat{E}}$ is the overall potential corresponding to the strain (strain rate) field defined by (D.1.10a). Note that the overall strains (strain rates) $\mathbf{E}$ and $\hat{\mathbf{E}}$, are not, in general, the same.

Bounds (D.1.9) and (D.1.11) remain valid when the matrix M contains several irregularly-shaped inclusions, as illustrated in Figure D.1.4. Each inclusion, Ω^α, may be surrounded by a regularly-shaped volume Ω_0^α, bounded by $\partial\Omega_0^\alpha$, and then subjected to uniform boundary data. As is shown in the following subsection, this procedure can be used to obtain explicit bounds for the overall energies of a composite with complex irregularly-shaped inclusions (or collections of inclusions), in terms of the solution of the boundary-value problem in which the matrix material is contained within two concentric spheres (two concentric circles, in two dimensions) and is then subjected to uniform boundary data.

Figure D.1.4

Replace a composite with irregularly-shaped boundary, ∂M, containing irregularly-shaped inclusions, $\partial\Omega^\alpha$, by the regularly-shaped ∂M_0, containing regularly-shaped inclusions, $\partial\Omega_0^\alpha$

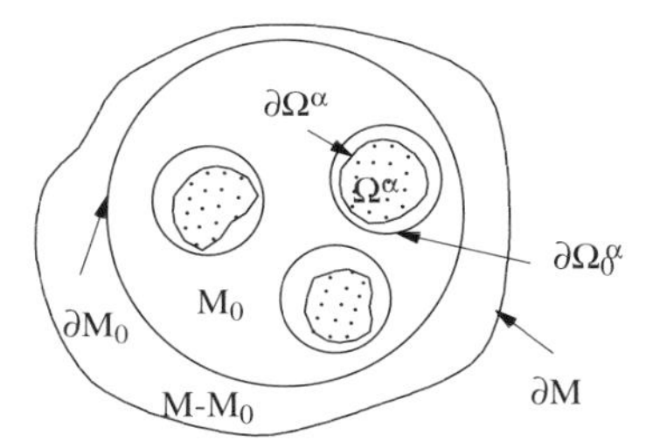

Each Ω_0^α in Figure D.1.4 is a composite consisting of the matrix material and the inclusion material which may have variable properties. Effective properties must be assigned to each Ω_0^α. A simple approach is to assume uniform stress or strain (strain rate), and estimate the overall properties of the inclusion Ω_0^α. For example, in the linear case, set

$$\bar{\mathbf{D}}^\alpha = \frac{1}{\Omega_0^\alpha}\int_{\partial\Omega_0^\alpha} \mathbf{D}^\alpha(\mathbf{x})\,dV, \quad \bar{\mathbf{C}}^\alpha = \frac{1}{\Omega_0^\alpha}\int_{\partial\Omega_0^\alpha} \mathbf{C}^\alpha(\mathbf{x})\,dV, \tag{D.1.12a,b}$$

to obtain estimates of the corresponding compliance and elasticity tensors. In particular, when the inclusion in Ω_0^α is uniform, (D.1.12a,b) reduce to

$$\bar{\mathbf{D}}^\alpha = (\Omega^\alpha/\Omega_0^\alpha)\,\mathbf{D}^\alpha + (1 - \Omega^\alpha/\Omega_0^\alpha)\,\mathbf{D},$$

$$\bar{\mathbf{C}}^\alpha = (\Omega^\alpha/\Omega_0^\alpha)\,\mathbf{C}^\alpha + (1 - \Omega^\alpha/\Omega_0^\alpha)\,\mathbf{C}. \tag{D.1.12c,d}$$

Since (D.1.12a,b) correspond to uniform stresses and uniform strains applied over the entire Ω_0^α, they are upper bounds for the corresponding tensors.

D.1.4. Bounds by Discretization

From the above results it now can be seen that the given composite can be subdivided into a "matrix" M′ and a set of non-intersecting subregions, M^α, $\alpha = 1, 2, ..., n$, all totally contained within M′, where $M' = M - \bigcup_{\alpha=1}^{n} M^\alpha$, as shown in Figure D.1.5. Each M^α may contain its own set of inhomogeneities. In the matrix M′, a uniform stress $\hat{\mathbf{\Sigma}}$ is prescribed such that the final overall average strain (strain-rate) $< \boldsymbol{\varepsilon}^\Sigma >_M$ is attained. Each M^α can now be treated in the manner discussed before, and the corresponding stress potential $\hat{\Phi}^{\hat{\Sigma}}$ can then be computed, leading to a lower bound for Φ^Σ, and hence for Φ^G; see (D.1.2a). The poorest bound is obtained when a uniform stress $\hat{\mathbf{\Sigma}}$ is assumed for the entire composite. This then leads to the lower bound associated with the Reuss model; Reuss (1929). Similarly, if a uniform strain (strain rate) $\hat{\mathbf{E}}$ is applied on M′, a lower bound $\hat{\Psi}^{\hat{E}}$, for Ψ^E and hence for Ψ^G is obtained. If a constant $\hat{\mathbf{E}}$ is used for the entire composite, then the poorest bound for Ψ^E is obtained. This corresponds to the Voigt model; Voigt (1889). All other bounds which may result from the procedure outlined in this work will be better than the Reuss and Voigt bounds. For the linear case, $\hat{\Phi}^{\hat{\Sigma}}$ provides a lower bound (upper bound) for the effective elasticity (compliance) tensor $\bar{\mathbf{C}}^\Sigma$ (tensor $\bar{\mathbf{D}}^\Sigma = (\bar{\mathbf{C}}^\Sigma)^{-1}$), while $\hat{\Psi}^{\hat{E}}$ yields a lower bound (upper bound) for the effective compliance (elasticity) tensor $\bar{\mathbf{D}}^E$ (tensor $\bar{\mathbf{C}}^E = (\bar{\mathbf{D}}^E)^{-1}$), respectively. In general, $\bar{\mathbf{C}}^\Sigma$ and $\bar{\mathbf{D}}^E$ are not each other's inverse.

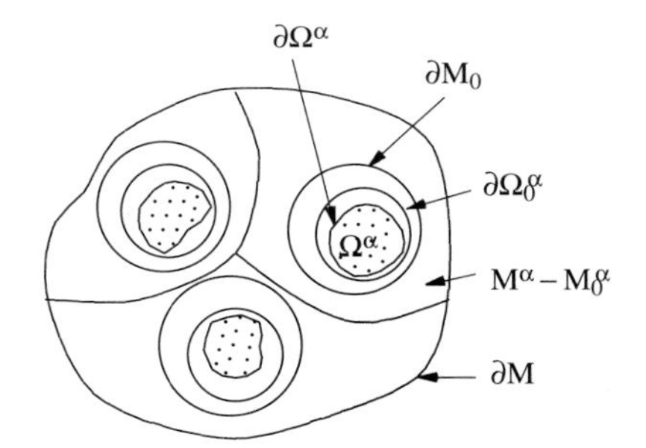

Figure D.1.5

A finite composite consisting of several heterogeneous inclusions and a matrix, is divided into several subregions, and each subregion (containing only one inclusion) is treated as in Figure D.1.3

D.2. LINEAR COMPOSITES

The method outlined in this appendix has been applied by Balendran and Nemat-Nasser (1995) to linear composites, obtaining explicit bounds. The advantage of these bounds is that they are *finite* for composites with cavities, rigid inclusions, or both cavities *and* rigid inclusions, provided that each cavity or inclusion is totally surrounded by the matrix material. The bounds obtained in Section 9 use comparison uniform solids. These bounds are zero (the lower bound) and infinity (the upper bound) when both cavities and rigid inclusions coexist.

Let the composite shown in Figure D.1.5 consist of linear materials. Divide the composite into a finite number of non-intersecting subregions, as illustrated in the figure. Consider a typical subregion M^α bounded by ∂M^α. As is suggested in the figure, consider in the subregion M^α, a regularly-shaped region (spherical in three dimensions and circular in two dimensions) M_0^α which is completely contained within M^α. Also, encase the inclusion Ω^α in a region Ω_0^α and remove Ω_0^α together with the inclusion. Apply uniform tractions $\mathbf{t} = -\mathbf{n} \cdot \hat{\boldsymbol{\sigma}}^\alpha$ on $\partial\Omega_0^\alpha$ while keeping the boundary ∂M_0^α traction free. The resulting average cavity strain then is

$$\hat{\boldsymbol{\varepsilon}}^\alpha = -\mathbf{H}_0^\alpha : \hat{\boldsymbol{\sigma}}^\alpha, \tag{D.2.1a}$$

where $\mathbf{H}_0^\alpha$ is an effective (fourth-order) influence tensor. The constant stress $\hat{\boldsymbol{\sigma}}^\alpha$ is calculated such that it produces in Ω_0^α an average strain equal to the cavity strain, $\hat{\boldsymbol{\varepsilon}}^\alpha$. Similarly, when ∂M_0^α is fixed and $\partial\Omega_0^\alpha$ is subjected to linear displacements $\mathbf{u} = \mathbf{x} \cdot \hat{\boldsymbol{\varepsilon}}^\alpha$, the resulting average stress is given by

$$\hat{\boldsymbol{\sigma}}^\alpha = -\mathbf{L}_0^\alpha : \hat{\boldsymbol{\varepsilon}}^\alpha, \tag{D.2.1b}$$

where $\mathbf{L}_0^\alpha$ is an effective (fourth-order) influence tensor. The constant strain $\hat{\boldsymbol{\varepsilon}}^\alpha$ in then calculated such that the average stress in Ω_0^α under boundary data $\mathbf{u} = \mathbf{x} \cdot \hat{\boldsymbol{\varepsilon}}^\alpha$ equals $\hat{\boldsymbol{\sigma}}^\alpha$.

Applying uniform tractions or linear displacements on the boundary ∂M_0^α of each subregion M_0^α, now obtain the following upper bounds for the effective modulus tensors of the corresponding subregion:

$$\hat{\mathbf{D}}^{\alpha\hat{\Sigma}} = (M_0^\alpha/M^\alpha)\,\hat{\mathbf{D}}_0^{\alpha\hat{\Sigma}} + (1 - M_0^\alpha/M^\alpha)\,\mathbf{D},$$

$$\hat{\mathbf{C}}^{\alpha\hat{E}} = (M_0^\alpha/M^\alpha)\,\hat{\mathbf{C}}_0^{\alpha\hat{E}} + (1 - M_0^\alpha/M^\alpha)\,\mathbf{C},$$

$$\hat{\mathbf{D}}_0^{\alpha\hat{\Sigma}} = \mathbf{D} + \rho_0^\alpha(\overline{\mathbf{D}}^\alpha - \mathbf{D}) - \rho_0^\alpha(\overline{\mathbf{D}}^\alpha - \mathbf{D}) : (\overline{\mathbf{D}}^\alpha - \mathbf{H}_0^\alpha)^{-1} : (\overline{\mathbf{D}}^\alpha - \mathbf{D}),$$

$$\hat{\mathbf{C}}_0^{\alpha\hat{\Sigma}} = \mathbf{C} + \rho_0^\alpha(\overline{\mathbf{C}}^\alpha - \mathbf{C}) - \rho_0^\alpha(\overline{\mathbf{C}}^\alpha - \mathbf{C}) : (\overline{\mathbf{C}}^\alpha - \mathbf{L}_0^\alpha)^{-1} : (\overline{\mathbf{C}}^\alpha - \mathbf{C}), \tag{D.2.2a~d}$$

where

$$\rho_0^\alpha = \Omega_0^\alpha/M_0^\alpha, \tag{D.2.2e}$$

and $\overline{\mathbf{C}}^\alpha$ and $\overline{\mathbf{D}}^\alpha$ are defined by (D.1.12c,d). Here, $\hat{\mathbf{D}}_0^{\alpha\hat{\Sigma}}$ and $\hat{\mathbf{C}}_0^{\alpha\hat{\Sigma}}$ are the bounds for the overall compliance and elasticity tensors of the subregion M_0^α.

To obtain the final bounds for the entire finite composite, calculate the overall strain $\hat{\mathbf{E}}$ (stress $\hat{\boldsymbol{\Sigma}}$) when uniform stress $\hat{\boldsymbol{\Sigma}}$ (strain $\hat{\mathbf{E}}$) is prescribed on $M' = M - \bigcup_{\alpha=1}^{n} M^\alpha$; see Figure D.1.5. This then leads to the following expressions:

$$\hat{\mathbf{D}}^{\hat{\Sigma}} = \sum_{\alpha=1}^{n} f_\alpha\,\hat{\mathbf{D}}^{\alpha\hat{\Sigma}}, \quad \hat{\mathbf{C}}^{\hat{E}} = \sum_{\alpha=1}^{n} f_\alpha\,\hat{\mathbf{C}}^{\alpha\hat{E}} \quad f_\alpha = M^\alpha/M. \tag{D.2.3a~c}$$

As can easily be deduced, $\hat{\mathbf{D}}^{\hat{\Sigma}}$ and $\hat{\mathbf{C}}^{\hat{E}}$ are upper bounds for $\overline{\mathbf{D}}^{\Sigma}$ and $\overline{\mathbf{C}}^{E}$, respectively. These bounds can be calculated using the solution for concentric spheres (circles in two dimensions), as is illustrated in the sequel.

In the case where the inclusions of the same size, shape, and material are distributed periodically in an infinitely extended matrix, each unit cell has the same upper bounds for the compliance and elasticity tensors. These bounds are given by (D.2.2a,b), i.e.,

$$\hat{\mathbf{D}}^{\hat{\Sigma}} = \hat{\mathbf{D}}^{\alpha\hat{\Sigma}}, \quad \hat{\mathbf{C}}^{\hat{E}} = \hat{\mathbf{C}}^{\alpha\hat{E}}. \tag{D.2.4a,b}$$

When $\partial\Omega_0^\alpha$ and ∂M_0^α are concentric spheres or circles, then the tensors $\mathbf{H}_0^\alpha$ and $\mathbf{L}_0^\alpha$ of (D.2.1a,b) can be computed. Using the closed-form solutions for concentric spheres in three dimensions and concentric circles in two dimensions, Hashin (1962) and Hashin and Rosen (1964) have obtained closed-form bounds for the overall bulk modulus of elastic composites. For spherical or cylindrical cavities (rigid inclusions), their resulting lower (upper) bound is the same as that obtained using the procedure outlined in this appendix.

D.2.1. Examples of Closed-form Bounds

To illustrate the application of the bounds (D.2.3a,b), consider fibrous composites with fibers extending in the x_3-direction. The fibers may be weaker or stronger than the matrix, with the limiting cases of cavities or rigid fibers. The cross-sectional shape may be arbitrary.

The solutions are obtained using the plane (strain or stress) elasticity problem of an annulus bounded by two concentric circles with radii R_1 (which defines $\partial\Omega_0$) and R_2 (which defines ∂M_0).[4] When this annulus is subjected to the boundary conditions

$$\mathbf{t} = \begin{cases} \mathbf{n} \cdot \boldsymbol{\sigma} = 0 & \text{on } \partial M_0 \\ -\mathbf{n} \cdot \boldsymbol{\sigma} = -\mathbf{n} \cdot \boldsymbol{\sigma}^0 & \text{on } \partial\Omega_0 \end{cases}, \tag{D.2.5}$$

the average strain in Ω_0 is given by

$$< \varepsilon_{kk} >_{\Omega_0} = -\frac{p}{2B}\,\sigma^0_{kk}, \quad < \varepsilon'_{ij} >_{\Omega_0} = -\frac{q}{2\mu}\,\sigma^{0'}_{ij},$$

$$p = \frac{2}{(\kappa-1)}\left[1 + \frac{(\kappa+1)\,\rho_0}{2\,(1-\rho_0)}\right],$$

$$q = (\kappa+1)\,\frac{1-\rho_0^3}{(1-\rho_0)^4} - 1, \quad \rho_0 = (R_1/R_2)^2, \tag{D.2.6a~e}$$

where

$$\varepsilon'_{ij} = \varepsilon_{ij} - \tfrac{1}{2}\,\delta_{ij}\,\varepsilon_{kk},$$

$$\sigma^{0'}_{ij} = \sigma^0_{ij} - \tfrac{1}{2}\,\delta_{ij}\,\sigma^0_{kk}, \quad i, j, k = 1, 2, \tag{D.2.7a,b}$$

and where μ is the in-plane shear modulus, $B = 2\mu/(\kappa-1)$ is the in-plane bulk modulus with $\kappa = (3-4\nu)$ for plane strain, and $\kappa = (3-\nu)/(1+\nu)$ for plane

[4] See Figure D.1.3 and note that the circles $\partial\Omega_0$ and ∂M_0 are now concentric.

stress. Expressions (D.2.6a~e) give the components of the influence tensor $\mathbf{H}_0^\alpha \equiv \mathbf{H}_0$ of (D.2.1a). In the present case, there is only one inclusion and $\mathbf{H}_0$ is isotropic.

Similarly, to obtain the influence tensor $\mathbf{L}_0^\alpha \equiv \mathbf{L}_0$ of (D.2.1b), subject the annulus to the boundary conditions

$$\mathbf{u} = \begin{cases} 0 & \text{on } \partial \mathrm{M}_0 \\ \mathbf{x} \cdot \boldsymbol{\varepsilon}^0 & \text{on } \partial \Omega_0 \end{cases}, \tag{D.2.8}$$

and note that the average stress in Ω_0 is given by

$$< \sigma_{kk} >_{\Omega_0} = -2\, r\, B\, \varepsilon_{kk}^0, \qquad < \sigma'_{ij} >_{\Omega_0} = -2\, s\, \mu\, \varepsilon_{ij}^{0\prime},$$

$$r = \frac{1}{2} \left[\kappa - 1 + (\kappa + 1) \frac{\rho_0}{1 - \rho_0} \right],$$

$$s = \frac{\kappa(\kappa+1)(1-\rho_0^3)}{(1-\rho_0)^2 \{\kappa^2 (1 + \rho_0 + \rho_0^2) - 3\rho_0\}} - 1. \tag{D.2.9a~d}$$

These results can now be used to construct bounds on moduli of composites with various microstructures, as shown in the sequel.

Circular Cylindrical Assemblages: Consider a unidirectionally reinforced *model* composite, consisting of cylindrical assemblages of *varying sizes* but the same fiber volume fraction, say, ρ_0, packed to completely fill the volume of an RVE. This model has been considered by Hashin (1962) and Hashin and Rosen (1964). The fibers may be heterogeneous. When the fibers are homogeneous, the response of an RVE of such a composite is transversely isotropic. For in-plane problems, the overall moduli of the RVE are defined through the following potentials:

$$\Phi(\bar{\boldsymbol{\varepsilon}}) = \frac{1}{2} \bar{B} (\bar{\varepsilon}_{kk})^2 + 2\, \bar{\mu}\, (\bar{\varepsilon}_{12})^2 + 2\, \bar{\mu}' (\bar{\varepsilon}'_{11})^2,$$

$$\Psi(\bar{\boldsymbol{\sigma}}) = \frac{1}{8\bar{B}} (\bar{\varepsilon}_{kk})^2 + \frac{1}{2\bar{\mu}} (\bar{\varepsilon}_{12})^2 + \frac{1}{2\bar{\mu}'} (\bar{\varepsilon}'_{11})^2, \tag{D.2.10a,b}$$

where quantities denoted by a superimposed bar are averages taken over a typical concentric cylindrical composite.

In view of (D.2.6) and (D.2.9), the influence tensors $\mathbf{H}_0$ and $\mathbf{L}_0$, for a typical concentric cylindrical composite, are transversely isotropic. Therefore, the bounds $\hat{\mathbf{D}}^{\hat{\Sigma}}$ and $\hat{\mathbf{C}}^{\hat{E}}$ resulting from (D.2.2) are also transversely isotropic. The bounds for the in-plane shear and bulk moduli are given by

$$\mu/\hat{\mu}_0^{\hat{\Sigma}} = 1 + \rho_0 (\mu/\bar{\mu}^D - 1)(1 + q)/(\mu/\bar{\mu}^D + q),$$

$$B/\hat{B}_0^{\hat{\Sigma}} = 1 + \rho_0 (B/\bar{B}^D - 1)(1 + p)/(B/\bar{B}^D + p),$$

$$\hat{\mu}_0^{\hat{E}}/\mu = 1 + \rho_0 (\bar{\mu}^C/\mu - 1)(1 + s)/(\bar{\mu}^C/\mu + s),$$

$$\hat{B}_0^{\hat{E}}/B = 1 + \rho_0 (\bar{B}^C/B - 1)(1 + r)/(\bar{B}^C/B + r), \tag{D.2.11a~d}$$

where

$$\bar{\mu}^C = \frac{1}{\Omega_0} \int_{\Omega_0} \mu^{\Omega_0} \, dV, \quad \bar{B}^C = \frac{1}{\Omega_0} \int_{\Omega_0} B^{\Omega_0} \, dV,$$

$$\frac{1}{\bar{\mu}^D} = \frac{1}{\Omega_0} \int_{\Omega_0} \frac{1}{\mu^{\Omega_0}} \, dV, \quad \frac{1}{\bar{B}^D} = \frac{1}{\Omega_0} \int_{\Omega_0} \frac{1}{B^{\Omega_0}} \, dV. \tag{D.2.12a~d}$$

In these expressions, μ^{Ω_0} and B^{Ω_0} are the shear and bulk moduli of the material in Ω_0 which may be heterogeneous. Note that, $\hat{\mu}_0^{\hat{\Sigma}}$ and $\hat{\mu}_0^{\hat{E}}$ bound both $\bar{\mu}$ and $\bar{\mu}'$.

In the case of homogeneous fibers, i.e., $\bar{B}^D = \bar{B}^C = B^{\Omega_0}$ and $\bar{\mu}^D = \bar{\mu}^C = \bar{\mu}^{\Omega_0}$, expressions (D.2.11b,d) reduce to

$$\hat{B}_0^{\hat{\Sigma}}/B = \hat{B}_0^{\hat{E}}/B = 1 + \rho_0 \left\{ \frac{2(1-\rho_0)}{\kappa+1} + \frac{1}{B^{\Omega_0}/B - 1} \right\}^{-1}. \tag{D.2.13}$$

This coincides with the result obtained by Hashin and Rosen (1964). The evaluation of the bounds for the shear modulus by the Hashin-Rosen method, requires solving of six linear equations for six unknowns. However, when the composite cylinders are hollow, the Hashin-Rosen bounds become

$$\bar{\mu}_0^E/\mu = 1 - \rho_0 \frac{(\kappa+1)(\kappa+\rho_0^3)}{(\kappa+\rho_0^3)(1+\kappa\rho_0) + \rho_0^2(\rho_0^2-3)(1-\rho_0)},$$

$$\mu/\bar{\mu}_0^\Sigma = 1 + \rho_0(\kappa+1)(1+\rho_0+\rho_0^2)/(1-\rho_0)^3. \tag{D.2.14a,b}$$

For this case, substitute $\bar{\mu}^D = \bar{\mu}^C = 0$ into (D.2.11a,c) to obtain

$$\hat{\mu}_0^{\hat{E}}/\mu = 1 - \rho_0 \frac{\kappa(\kappa+1)(1+\rho_0+\rho_0^2)}{\kappa(1+\kappa\rho_0)(1+\rho_0+\rho_0^2) + 3\rho_0(1-\rho_0)},$$

$$\mu/\hat{\mu}_0^{\hat{\Sigma}} = 1 + \rho_0(\kappa+1)(1+\rho_0+\rho_0^2)/(1-\rho_0)^3. \tag{D.2.14c,d}$$

For the composite cylinders with rigid fibers, the bounds by Hashin and Rosen (1964) are

$$\bar{\mu}_0^E/\mu = 1 + \rho_0 \frac{\kappa(\kappa+1)(1+\rho_0+\rho_0^2)}{(1-\rho_0)\{\kappa^2(1+\rho_0+\rho_0^2) - 3\rho_0\}},$$

$$\mu/\bar{\mu}_0^\Sigma = 1 - \rho_0 \frac{(\kappa+1)(1+\kappa\rho_0^3)}{(1+\kappa\rho_0^3)(\kappa+\rho_0) + 3\rho_0(1-\rho_0)^2}. \tag{D.2.15a,b}$$

For this case, the bounds from the present approach are obtained by substituting $1/\bar{\mu}^D = 1/\bar{\mu}^C = 0$ into (D.2.11a,c), arriving at

$$\hat{\mu}_0^{\hat{E}}/\mu = 1 + \rho_0 \frac{\kappa(\kappa+1)(1+\rho_0+\rho_0^2)}{(1-\rho_0)\{\kappa^2(1+\rho_0+\rho_0^2) - 3\rho_0\}},$$

$$\mu/\hat{\mu}_0^{\hat{\Sigma}} = 1 - \rho_0 \frac{(\kappa+1)(1+\rho_0+\rho_0^2)}{(\kappa+1)(1+\rho_0+\rho_0^2) - (1-\rho_0)^3}. \qquad \text{(D.2.15c,d)}$$

It is seen that when the inclusions of the composite cylinders are tubular cavities (rigid fibers), the lower (upper) bound from the present method is identical to that obtained by Hashin and Rosen (1964). However, the upper (lower) bound of the present method is greater (smaller) than that of Hashin and Rosen (1964); see Figure D.2.1.

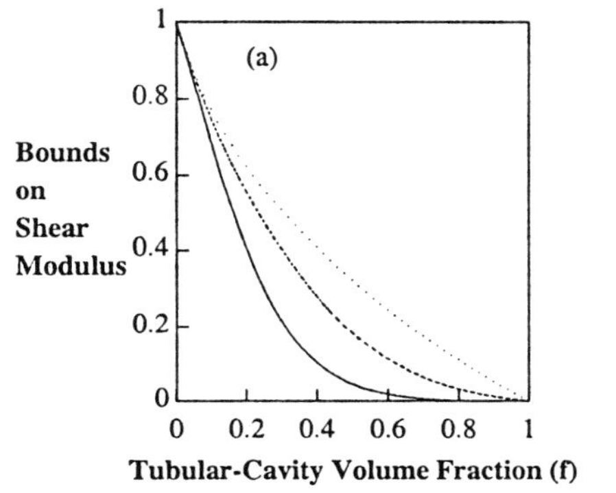

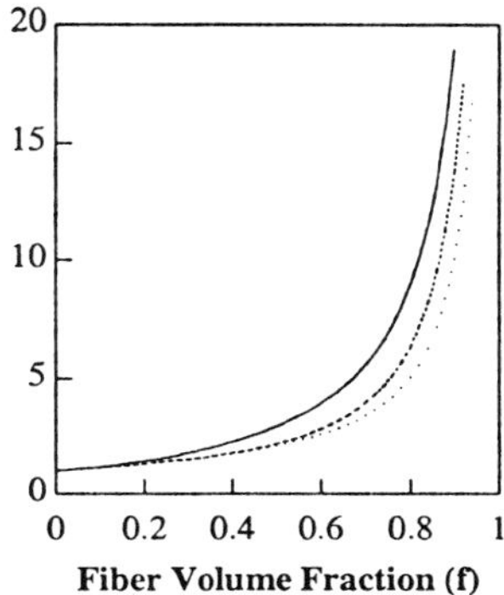

Figure D.2.1

Bounds on overall shear modulus of (a) hollow cylindrical assemblages, and (b) cylindrical assemblages with rigid fibers; curves denoted by (____, ...) are Hashin and Rosen (1964) bounds ($\bar{\mu}_0^E/\mu$ and $\bar{\mu}_0^\Sigma/\mu$), equations (D.2.14a,b) and (D.2.15a,b), and those denoted by (____, ...) are ($\hat{\mu}_0^{\hat{\Sigma}}/\mu$ and $\hat{\mu}_0^{\hat{E}}/\mu$), equations (D.2.14c,d) and (D.2.15c,d), Balendran and Nemat-Nasser (1995) bounds

Periodic Composites: As the second example, consider a fibrous composite where fibers are periodically distributed on the plane normal to their axes. The fibers may be weaker or stronger than the matrix, including the limiting cases where fibers are cavities or rigid inclusions. The fiber packing may be either triangular, square, or hexagonal; see Figure D.2.2a~c. The cross-sectional shape of the fibers may be arbitrary and the fibers may be heterogeneous. In this case, the boundaries ∂M_0^α, for the unit cells are the largest circles drawn within the unit cells while $\partial\Omega_0$ are the smallest circles drawn outside the fibers. Then the bounds for the in-plane shear and bulk moduli given by (D.2.4) are

$$1/\hat{\mu}^\Sigma = f_0/\hat{\mu}_0^{\hat{\Sigma}} + (1-f_0)/\mu,$$

$$1/\hat{B}^\Sigma = f_0/\hat{B}_0^{\hat{\Sigma}} + (1-f_0)/B,$$

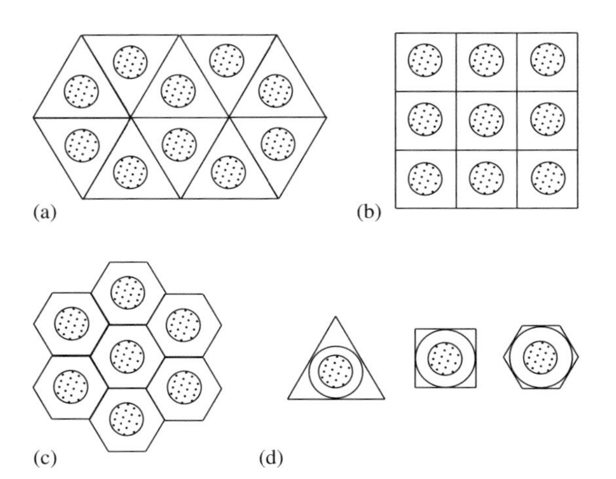

Figure D.2.2

Periodic distribution of cylindrical fibers with circular cross section: (a) triangular packing; (b) square packing; (c) hexagonal packing; and (d) typical unit cells

$$\hat{\mu}^{E} = f_0\,\hat{\mu}_0^{\hat{E}} + (1 - f_0)\,\mu,$$

$$\hat{B}^{E} = f_0\,\hat{B}_0^{\hat{E}} + (1 - f_0)\,B, \tag{D.2.16a~d}$$

where

$$f_0 \equiv M_0/M = \begin{cases} \pi/(3\sqrt{3}) & \text{for triangular} \\ \pi/4 & \text{for square} \\ \pi/(2\sqrt{3}) & \text{for hexagonal fiber packing.} \end{cases} \tag{D.2.17a}$$

Then the volume fraction f of the fibers is related to f_0 by

$$f \equiv \frac{\Omega}{M} = \frac{\Omega}{\Omega_0}\,\frac{\Omega_0}{M_0}\,\frac{M_0}{M} = \rho\,\rho_0\,f_0, \qquad \rho \equiv \frac{\Omega}{\Omega_0}. \tag{D.2.17b,c}$$

Substitute (D.2.11) into (D.2.16a~d), and in view of ((D.2.17b,c) obtain

$$\mu/\hat{\mu}^{\Sigma} = 1 + (f/\rho)\,(\mu/\bar{\mu}^{D} - 1)\,(1 + q)/(\mu/\bar{\mu}^{D} + q),$$

$$B/\hat{B}^{\Sigma} = 1 + (f/\rho)\,(B/\bar{B}^{D} - 1)\,(1 + p)/(B/\bar{B}^{D} + p),$$

$$\hat{\mu}^{E}/\mu = 1 + (f/\rho)\,(\bar{\mu}^{C}/\mu - 1)\,(1 + s)/(\bar{\mu}^{C}/\mu + s),$$

$$\hat{B}^E/B = 1 + (f/\rho)(\bar{B}^C/B-1)(1+r)/(\bar{B}^C/B + r). \tag{D.2.18a-d}$$

For homogeneous fibers,

$$\bar{\mu}^C = \rho\,\mu^{\Omega_0} + (1-\rho)\,\mu, \quad \bar{B}^C = \rho\,B^{\Omega_0} + (1-\rho)\,B,$$

$$1/\bar{\mu}^D = \rho\,1/\mu^{\Omega_0} + (1-\rho)\,1/\mu,$$

$$1/\bar{B}^D = \rho\,1/B^{\Omega_0} + (1-\rho)\,1/B. \tag{D.2.19a-d}$$

As an example consider fibers with a common circular cross section. In this case $\rho = 1$; see Balendran and Nemat-Nasser (1995) for other examples. The bounds on the bulk and shear moduli for this case are shown in Figures D.2.3 and D.2.4 for the modulus ratios ($\mu^{\Omega_0}/\mu = B^{\Omega_0}/B$) of 0.1 and 25. Note that, volume fractions of $\pi\sqrt{3}/9$, $\pi/4$, and $\pi\sqrt{3}/6$, in triangular, square, and hexagonal packing, denote the respective close pack of circular fibers. These Figures show that the bounds are reasonably close when the volume fraction of the fibers is less than about 80% of the close packing.

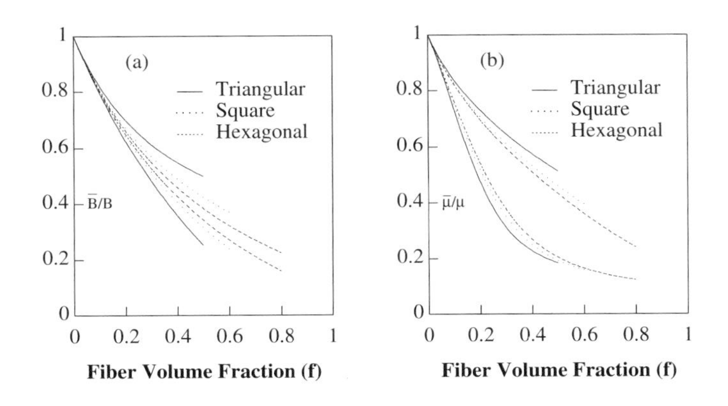

Figure D.2.3

Bounds on overall bulk (a), and shear (b) moduli with $\mu^{\Omega_0}/\mu = B^{\Omega_0}/B = 0.1$

The bounds for elastic solids with tubular cavities are shown in Figure D.2.5. It is seen in these figures that the lower bounds are nonzero. They become zero only for the close packing, which is the exact value.

The bounds for elastic solids with rigid fibers are shown in Figure D.2.6. It is seen that upper bounds are always finite. These become unbounded only for the close packing, which is exact.

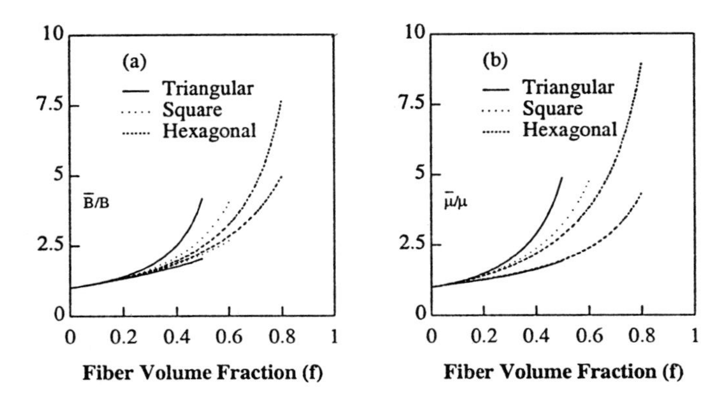

Figure D.2.4

Bounds on overall bulk (a), and shear (b) moduli with $\mu^{\Omega_0}/\mu = B^{\Omega_0}/B = 25$

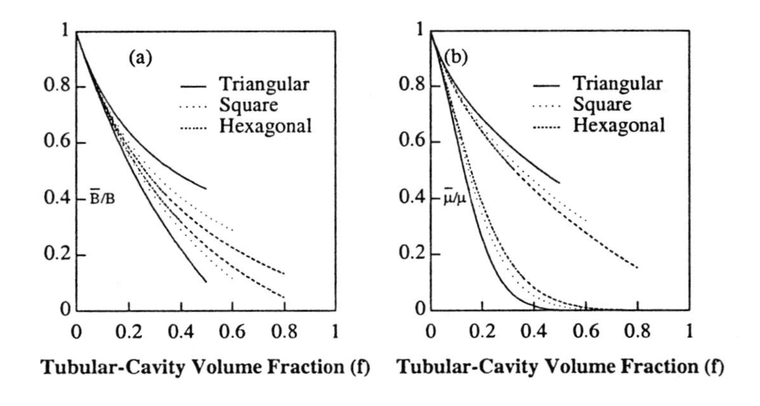

Figure D.2.5

Bounds on overall bulk (a), and shear (b) moduli of elastic solids with tubular cavities

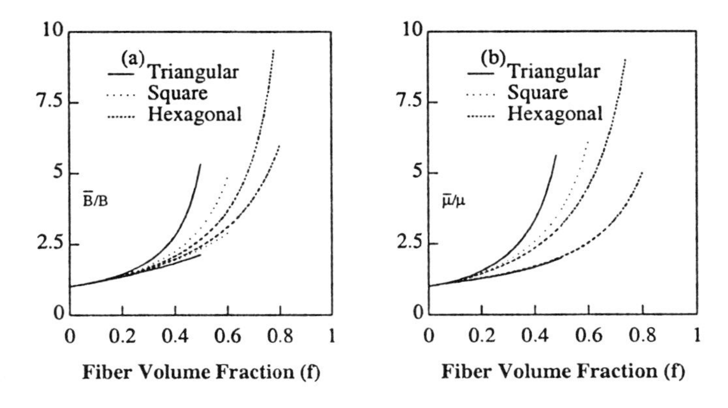

Figure D.2.6

Bounds on overall bulk (a), and shear (b) moduli of elastic solids with rigid fibers

D.3. REFERENCES

Balendran, B. and Nemat-Nasser, S. (1995), Bounds on elastic moduli of composites, *J. Mech. Phys. Solids*, Vol. 43, 1825-1853.

Hashin, Z. (1962), The elastic moduli of heterogeneous materials, *ASME J. Appl. Mech.*, Vol. 29, 143-150.

Hashin, Z. and Rosen, B. W. (1964), The elastic moduli of fiber-reinforced materials, *J. App. Mech.*, Vol. 31, 223-232.

Huet, C. (1990), Application of variational concepts to size effects in elastic heterogeneous bodies, *J. Mech. Phys. Solids*, Vol. 38, No. 6, 813-841.

Nemat-Nasser, S. and Hori, M. (1995), Universal bounds for overall properties of linear and nonlinear heterogeneous solids, *J. Engineering Mat. Tech.*, Vol. 117, 412-432.

Nemat-Nasser, S., Balendran, B., and Hori, M. (1995), Bounds for overall nonlinear elastic or viscoplastic properties of heterogeneous solids, *Microstructure property interactions in composite materials:* IUTAM Symposium in Aalborg, Denmark, 1994, R. Pyrz (ed.), Kluwer Academic Publishers, 215-221.

Reuss, A. (1929), Berechnung der Fliessgrenze von Mischkristallen auf Grund der Plastizitätsbedingung für Einkristalle, *Z. Angew. Math. Mech.*, Vol. 9, 49-58.

Voigt, W. (1889), Über die Beziehung zwischen den beiden Elastizitätskonstanten isotroper Körper, *Wied. Ann. Physik*, Vol. 38, 573-587.

PART 2

INTRODUCTION TO BASIC ELEMENTS OF ELASTICITY THEORY

PRÉCIS: PART 2

Basic elements of the *mathematical theory of elasticity*, are reviewed here in Part 2, rendering the book essentially self-contained. These elements form an essential background for the micromechanics covered in Part 1.

The theory of elasticity is concerned with the *response* of an ideal *model material*, called *elastic solid*, to applied loads or imposed surface displacements. Depending on the rate at which these loads or displacements are imposed and varied in time, the inertia of the elements of the elastic solid may or may not be relevant to the determination of its response. Accordingly, the considered problem may be in elastodynamic or elastostatic deformation states.

The *response* may refer to one or all of the following fields: the *stress field,* the *strain field,* or the *displacement field.* These fields are specified by sets of functions of position and time, describing the spacewise and timewise variations of *internal forces, local deformations*, and *particle displacements.*

The theory of elasticity is a *continuum theory*. The term *internal forces*, therefore, does not refer to forces at the atomic, molecular, or even crystalline levels, but rather to forces between adjacent *macroscopic elements* that, though small in comparison with typical dimensions of the considered body, are large in comparison with typical dimensions of crystals. Here, the terms *local deformation* and *particle displacement* also refer to macroscopic elements.

Even in this macroscopic setting, the mechanical properties of *real solids* are very complex. An attempt to include a wide range of properties leads to a theory too unwieldy for the practical analysis of stresses and strains in many applications. Instead of a comprehensive but unwieldy theory, simpler *models* have been developed to describe a limited range of mechanical behavior. Depending on the problem at hand, the use of only one model may be adequate, or a switch from one model to another may be necessary at some stage of the loading process. Each theory of this kind describes the behavior of an *ideal* (rather than an actual) *material.*

The theory of elasticity is concerned with such an ideal material, the *perfectly elastic solid.* Starting from the *natural state,* where there are no loads or internal forces (stresses), let such a solid be brought to a *final deformed state* by the quasi-static application of a set of loads. By following different loading paths, the same final state may be reached. *Each set of loads then does the same amount of work in the transition that it produces from the natural to the final state, if the solid is perfectly elastic. Moreover, this work is fully recovered on any return to the natural state by a slow removal of the loads.* This defines an ideal elastic solid.

There are no real solids that exhibit this perfectly elastic behavior under all conditions of loadings and deformation, but for many such solids an *elastic range* may be specified in which they *essentially* behave in this manner. As is implied by the word *essentially*, the specification of an elastic range is to some extent a matter of convention: what is a negligible departure from perfectly elastic behavior in one context may well become a significant departure in another context.

For some materials, for instance unvulcanized rubber, the elastic range extends to very large deformations; for others, for instance structural steel, it is restricted to very small deformations. Within this latter kind of elastic range, a typical deformation component is found to grow in proportion to a typical load component. The *classical theory of linear elasticity,* is exclusively concerned with mechanical behavior characterized by a *homogeneous linear relation between stresses and extremely small (infinitesimal) strains.* This is the subject of the present introductory account which is divided into two chapters; *i.e.,* Chapters V and VI. Chapter V covers the basic geometric, kinematic, dynamic, and constitutive fundamentals of the theory, while Chapter VI addresses the formulation of boundary-value problems, variational principles, and the solution methods.

More specifically, necessary aspects of three-dimensional vector and tensor analysis are examined in Section 15, where a tensor quantity is viewed as a geometrical or physical entity that exists independently of its coordinate representation. This permits an easy extension of the results to coordinate systems other than the rectangular Cartesian (Subsection 15.7). The vector space (Subsection 15.1), Cartesian coordinate transformation rules (Subsection 15.2), and the transformation rules for the tensor components (Subsection 15.3) are introduced. The del operator and several commonly used volume- and surface-integral identities are summarized in Subsection 15.4. Subsection 15.5 is devoted to an outline of various useful properties of second- and fourth-order tensors, and the matrix representation of their Cartesian components. The spectral analysis of symmetric second- and fourth-order tensors, respectively, in terms of the corresponding eigenvectors and symmetric second-order eigentensors, is given in Subsection 15.6.

The kinematic foundations are discussed in Section 16, giving an account of general strain measures (Subsection 16.1), infinitesimal strain and rotation measures (Subsection 16.2), a brief review of the principal and basic invariants of the linearized strain tensor, the compatibility conditions, and the plane strain conditions.

Section 17 addresses the basic instantaneous (dynamic) equilibrium equations for any solid, elastic or inelastic. The balance laws are introduced (Subsection 17.1), the concepts of *internal surface tractions* transmitted across any elementary material surface and the associated *stress tensor* are introduced and discussed (Section 17.2), followed by an examination of the principal and the basic invariants of the stress tensor, including its geometric representation (Subsection 17.3).

The constitutive equations for elastic materials are presented in Section 18, starting with the conservation of energy and relating the strain and stress in the continuum to the *internal energy* stored in the material (Subsections 18.1 and 18.2). This section is completed by a brief account of the elasticity and compliance tensors (Subsections 18.3) which are discussed in great detail in Section 3, Part 1, of this book.

Chapter VI focuses on general boundary-value problems of elastostatics and elastodynamics. The displacement and traction boundary-value problems, together with variational principles are formulated in Section 19. The field equations are presented in Subsection 19.1. The concepts of kinematically admissible displacement field and statically admissible stress field are introduced (Subsection 19.2), and potential (Subsection 19.3) and complementary potential (Subsection 19.4) energies are examined, together with the corresponding minimum principles. Then an account of general variational principles, including the effects of possible discontinuities in the field variables, is given (Subsection 19.5). The most general form of variational principles is considered, where any or all field variables are allowed to vary independently, leading to various weak statements of the field equations.

The integration of the elastokinematic field equations is considered in Section 20, where equations governing the propagation of dilatational and rotational waves are obtained based on the Helmholtz decomposition theorem, and their properties are discussed (Subsections 20.1 and 20.2). Elastostatics is reviewed in Subsection 20.3, giving an account of the representation of the displacement field in terms of the Papkovich-Neuber and the Galerkin-vector potentials. The Green functions of an unbounded and a half-space elastic solid are developed in Subsection 20.4.

Section 21 is devoted to the analysis of elasticity problems whose field variables are functions of only two spatial coordinates; this includes plane-strain and plane-stress problems. Emphasis is placed on anisotropic elasticity problems involving singularities, but starting with the isotropic case first. Airy's stress function (Subsection 21.1) and Muskhelishvili's complex potentials (Subsection 21.2) are introduced. The Green function and the fields produced by a center of dilatation and a dislocation are examined in this subsection. The Hilbert problem and its solution are reviewed in Subsection 21.3, illustrating the results by the solution of a straight crack in an unbounded isotropic solid. Subsection 21.4 deals with two-dimensional crack problems, their formulation in terms of both Cauchy's integral and Hadamard's *finite-part* integral. Plane anisotropic elasticity is considered in Subsection 21.5, expressing the stress, strain, and displacement fields in terms of Muskhelishvili's two complex potentials. The results are then illustrated using an unbridged and fully or partially bridged single straight crack in an unbounded anisotropic solid. A new subsection (Subsection 21.6) is added to this second edition, summarizing recent results on *duality principles* in anisotropic elasticity.

CHAPTER V

FOUNDATIONS

The fundamental concepts and relations used in the theory of elasticity are classified as geometric, kinematic, dynamic, and constitutive. Three-dimensional vector and tensor algebra and analysis form the geometric foundations. The term "kinematic" implies a study of the motion of a solid without reference to the forces that cause this motion. The term "dynamic" is used here to indicate a study of the external and internal forces without reference to the motion they produce. Finally, the kinematic and dynamic ingredients of the theory are related by constitutive equations which are used to characterize the constitution of the material comprising the considered solid. Below is an outline of the topics covered in this chapter.

The elements of three-dimensional vector and tensor analysis are examined in Section 15, including vector space (Subsection 15.1), Cartesian coordinate transformation rules (Subsection 15.2), and the transformation rules for the tensor components (Subsection 15.3). The del operator, and the volume- and surface-integral identities are outlined in Subsection 15.4. Various properties of second- and fourth-order tensors, and the matrix representation of their Cartesian components are given in Subsection 15.5, with their spectral representation covered in Subsection 15.6. Subsection 15.7 provides a summary of the basic field equations in the cylindrical and spherical coordinate systems.

Section 16 considers general strain measures (Subsection 16.1), infinitesimal strain and rotation measures (Subsection 16.2), the principal and basic invariants of the linearized strain tensor, the compatibility conditions, and the plane-strain conditions.

Section 17 addresses the instantaneous (dynamic) equilibrium equations for any solid, elastic or inelastic, the balance laws (Subsection 17.1), the concepts of internal surface tractions and the associated stress tensor (Section 17.2), and the principal and the basic invariants of the stress tensor, including its geometric representation (Subsection 17.3).

The generalized Hooke law is presented in Section 18, based on the conservation of energy (Subsections 18.1 and 18.2), leading to a brief account of the elasticity and compliance tensors (Subsections 18.3) which are more thoroughly covered in Section 3, Part 1, of this book.

SECTION 15 GEOMETRIC FOUNDATIONS

Certain aspects of three-dimensional vector and tensor analysis that are essential for the present study are reviewed in this section. Only the description of tensorial quantities in a *right-handed system of rectangular Cartesian coordinates* is considered. A tensor quantity is, however, viewed as a geometrical or physical entity that exists independently of any coordinate system to which it may be referred for the purpose of specification of its components. This permits an easy extension of the results to coordinate systems other than the rectangular Cartesian, as is briefly discussed in Subsection 15.7 at the end of this chapter.

Once a suitable system of physical units is selected, certain physical quantities can be placed into one-to-one correspondence with real numbers. Quantities of this kind are called *scalar.* The length l of a bar, the mass m of a piece of metal, and the distance d between two points in space are familiar examples of scalar quantities. There are other physical quantities, however, such as force, state of stress in a deformed solid, etc., whose mathematical description requires more than the field of real numbers. A subclass of such quantities comprises those physical or geometrical quantities which can be represented by vectors, i.e., directed line segments in space. The magnitude of the vector quantity is represented by the length of the line segment whose direction indicates the direction of the vector quantity. With this definition, two line segments of equal length that are parallel and have the same direction represent the same vector. In the following, a somewhat more precise description of these concepts is presented. This account, of necessity, is brief and the interested reader should consult standard texts[1].

15.1. VECTOR SPACE

The real number system is denoted by R. With an arbitrary origin and an arbitrary unit, elements in R are identified with points on the real number line which defines a one-dimensional Euclidean space. Elements in this space are called points. The set of all ordered pairs of real numbers is denoted by R^2. Each element in this set is associated with a point in a two-dimensional Euclidean space. An orthogonal pair of real lines intersecting at the origin form a Euclidean plane. These lines are called the coordinate axes. Again, with an arbitrary unit of length, points in this plane are identified by ordered pairs of real numbers. In a similar manner, one can define an n-dimensional Euclidean

[1] See, for example, Kellogg (1953), Halmos (1958), Coburn (1960), Borisenko and Tarapov (1968), and Williamson *et al.* (1972).

space, each point of which is associated with an element in R^n, i.e., a set of n ordered real numbers. Such a Euclidean space is denoted by E^n. For the most part, the two- and three-dimensional Euclidean spaces, E^2 and E^3, are considered. However, for the time being, a general n-dimensional space is considered.

A point in E^n is also associated with a vector **x** which connects the origin to that point and its *components* with respect to the n mutually orthogonal coordinate axes, when properly ordered, define the corresponding element in R^n. In this manner geometric points in E^n are associated with elements in R^n. The quantity **x** is called the *position vector* of the corresponding point in E^n. Denote the components of **x** by x_i ($i = 1, 2, ..., n$). Let **x** and **y** with components x_i and y_i be position vectors in E^n. Then, $\mathbf{v} = \mathbf{x} - \mathbf{y}$ defines a vector in the sense that it can be identified by an ordered set of n real numbers, $v_i = x_i - y_i$, namely an element in R^n. The collection of all such vectors is denoted by V and it is called an n-dimensional vector space. Note that the set of all such **v**'s is translation-invariant, that is, for any **z** in E^n, $\mathbf{v} = \mathbf{y} - \mathbf{x} = (\mathbf{y} + \mathbf{z}) - (\mathbf{x} + \mathbf{z})$, so that parallel translation of **v** leaves its components unchanged. This can be most clearly understood in terms of two- and three-dimensional vector spaces in which translation of a vector parallel to itself does not affect the values of its components.

A vector **v** is also called a *first-order tensor* when viewed as an *operator* in a *linear transformation* or a *linear mapping* of vectors to real numbers, i.e., mapping of elements in V to elements in R. This is done through the *inner* or *dot* product of **v** and any element in V, which yields a real number. If a general tensor is defined as an *operator* in a linear transformation of vector fields to real numbers, then vectors may be regarded as special tensors.

Let **B** be an operator with its domain in V. The corresponding transformation is called *linear* if and only if for every pair **v** and **w** in the domain of **B**, and every pair of real numbers a and b,

$$\mathbf{B}*(\mathrm{a}\,\mathbf{v} + \mathrm{b}\,\mathbf{w}) = \mathrm{a}\,\mathbf{B}*\mathbf{v} + \mathrm{b}\,\mathbf{B}*\mathbf{w}, \tag{15.1.1}$$

where $*$ defines the nature of the transformation which need not be specified at this point. *Henceforth only linear transformations are considered.*

15.2. ELEMENTARY CONCEPTS IN THREE-DIMENSIONAL SPACE

15.2.1. Rectangular Cartesian Coordinates

In a three-dimensional Euclidean space E^3, a vector is defined by three real numbers, i.e., its components with respect to a system of coordinates. Consider a right-handed system of rectangular Cartesian coordinates x_1, x_2, x_3. A point P is represented by its *position vector* **x**, the directed line segment OP. Let $\mathbf{e}_1$, $\mathbf{e}_2$, and $\mathbf{e}_3$, be three *unit base vectors,* that is, vectors of unit length in the positive coordinate directions. By the definition of vector summation and scalar multiplication (see Figure 15.2.1),

$$\mathbf{x} = x_1\,\mathbf{e}_1 + x_2\,\mathbf{e}_2 + x_3\,\mathbf{e}_3 = \sum_{i=1}^{3} x_i\,\mathbf{e}_i, \tag{15.2.1a}$$

where the components of **x** are

$$x_1 = \mathbf{x}\cdot\mathbf{e}_1, \qquad x_2 = \mathbf{x}\cdot\mathbf{e}_2, \qquad x_3 = \mathbf{x}\cdot\mathbf{e}_3. \tag{15.2.1b}$$

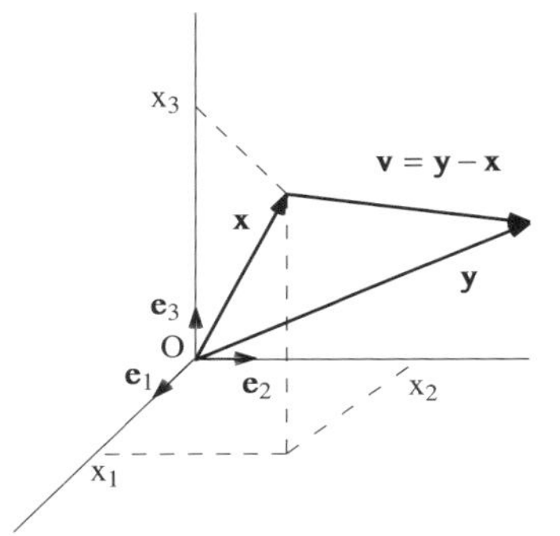

Figure 15.2.1

Three-dimensional Euclidean space and vector addition

Accordingly, (15.2.1a) may be written in the form

$$\mathbf{x} = \sum_{i=1}^{3} (\mathbf{x}\cdot\mathbf{e}_i)\,\mathbf{e}_i. \tag{15.2.1c}$$

Equations (15.2.1a,b) are more compactly written with the aid of the following summation convention:

> *Summation Convention.* A repeated subscript in a monomial represents the sum of the three terms which are obtained by setting this subscript equal to 1, 2, and 3.

For this rule to be meaningful, a subscript can occur at most twice in each monomial. In view of this rule, (15.2.1a,c) becomes,

$$\mathbf{x} = x_i\,\mathbf{e}_i = (\mathbf{x}\cdot\mathbf{e}_i)\,\mathbf{e}_i, \qquad i = 1, 2, 3. \tag{15.2.2}$$

The summation convention can be used for vectors in the vector space V in the same manner as for points in E^3. Let v_i and w_i denote the components of vectors $\mathbf{v} = v_i\,\mathbf{e}_i$ and $\mathbf{w} = w_j\,\mathbf{e}_j$ (i, j = 1, 2, 3). The dot product of **v** and **w** then is

$$\mathbf{v}\cdot\mathbf{w} = (v_i\,\mathbf{e}_i)\cdot(w_j\,\mathbf{e}_j) = (v_i\,w_j)\,\mathbf{e}_i\cdot\mathbf{e}_j, \qquad i, j = 1, 2, 3. \tag{15.2.3a}$$

From the definition of the unit base vectors it follows that

$$\mathbf{e}_i\cdot\mathbf{e}_j = \begin{cases} 0 & \text{if } i \neq j \\ 1 & \text{if } i = j. \end{cases} \tag{15.2.4}$$

It is convenient to introduce a new symbol δ_{ij}, called the *Kronecker delta,* defined by

$$\delta_{ij} = \begin{cases} 0 & \text{if } i \neq j \\ 1 & \text{if } i = j. \end{cases} \tag{15.2.5}$$

Equation (15.2.3a) then becomes

$$\mathbf{v} \bullet \mathbf{w} = (v_i\, w_j)\, \delta_{ij} = v_i\, w_i = v_1\, w_1 + v_2\, w_2 + v_3\, w_3. \tag{15.2.3b}$$

The magnitude v of the vector $\mathbf{v}$ is given by

$$v \equiv \sqrt{v_i\, v_i}. \tag{15.2.6}$$

The cross product of $\mathbf{v}$ and $\mathbf{w}$ is

$$\mathbf{v} \times \mathbf{w} = (v_i\, \mathbf{e}_i) \times (w_j\, \mathbf{e}_j) = (v_i\, w_j)\, \mathbf{e}_i \times \mathbf{e}_j. \tag{15.2.7a}$$

The cross products of the unit base vectors are

$$\begin{array}{lll} \mathbf{e}_1 \times \mathbf{e}_2 = \mathbf{e}_3, & \mathbf{e}_2 \times \mathbf{e}_3 = \mathbf{e}_1, & \mathbf{e}_3 \times \mathbf{e}_1 = \mathbf{e}_2, \\ \mathbf{e}_2 \times \mathbf{e}_1 = -\mathbf{e}_3, & \mathbf{e}_3 \times \mathbf{e}_2 = -\mathbf{e}_1, & \mathbf{e}_1 \times \mathbf{e}_3 = -\mathbf{e}_2, \\ \mathbf{e}_1 \times \mathbf{e}_1 = \mathbf{0}, & \mathbf{e}_2 \times \mathbf{e}_2 = \mathbf{0}, & \mathbf{e}_3 \times \mathbf{e}_3 = \mathbf{0}. \end{array} \tag{15.2.8a~i}$$

It is convenient to introduce a new symbol e_{ijk}, called the *permutation symbol*, defined as follows:

$$e_{ijk} = \begin{Bmatrix} +1 \\ -1 \\ 0 \end{Bmatrix} \text{ if i, j, k } \begin{Bmatrix} \text{form an even} \\ \text{form an odd} \\ \text{do not form} \end{Bmatrix} \text{ permutation of 1, 2, 3.} \tag{15.2.9}$$

Based on this definition, relations (15.2.8a~i) are concisely written as

$$\mathbf{e}_i \times \mathbf{e}_j = e_{ijk}\, \mathbf{e}_k. \tag{15.2.8j}$$

Accordingly, (15.2.7a) yields

$$\begin{aligned} \mathbf{v} \times \mathbf{w} &= (v_i\, w_j)\, e_{ijk}\, \mathbf{e}_k \\ &= (v_2 w_3 - v_3 w_2)\, \mathbf{e}_1 + (v_3 w_1 - v_1 w_3)\, \mathbf{e}_2 + (v_1 w_2 - v_2 w_1)\, \mathbf{e}_3. \end{aligned} \tag{15.2.7b}$$

If $\mathbf{u} = u_k\, \mathbf{e}_k = \mathbf{v} \times \mathbf{w}$, it then follows from (15.2.7b) that

$$u_k = v_i\, w_j\, e_{ijk} = e_{kij}\, v_i\, w_j. \tag{15.2.7c}$$

The following relations are direct consequences of the summation convention and definitions (15.2.5) and (15.2.9):

$$e_{ijk}\, e_{ijk} = 6, \qquad e_{ijk}\, e_{ijl} = 2\, \delta_{kl}, \qquad e_{ijk}\, e_{ilm} = \delta_{jl}\, \delta_{km} - \delta_{jm}\, \delta_{kl}. \tag{15.2.10a~c}$$

The repeated subscripts are called *dummy subscripts.* A dummy subscript may be replaced by any other subscript letter which is not otherwise used in the same relation. The subscripts that occur only once in each monomial of an equation are called *live subscripts.* Equation (15.2.10a), for example, contains only dummy subscripts, while in (15.2.10b), l and k are both live subscripts and i and j are both dummy subscripts.

15.2.2. Transformation of Coordinates

Consider now a coordinate transformation. Let x_1, x_2, x_3 and x_1', x_2', x_3', be the coordinates of a point P with position vector **x** in two systems of right-handed rectangular Cartesian coordinates that have the same origin O. Let $\mathbf{e}_j$ and $\mathbf{e}_j'$ (j = 1, 2, 3) denote the base vectors of the unprimed and primed coordinates, respectively. The vector **x** is expressed as

$$\mathbf{x} = x_j\,\mathbf{e}_j = x_j'\,\mathbf{e}_j', \qquad j = 1, 2, 3. \tag{15.2.11}$$

The dot product of this equation, first with $\mathbf{e}_i$ and then with $\mathbf{e}_i'$, yields

$$x_i = (\mathbf{e}_i \cdot \mathbf{e}_j')\,x_j', \qquad x_i' = (\mathbf{e}_i' \cdot \mathbf{e}_j)\,x_j, \qquad i, j = 1, 2, 3. \tag{15.2.12a,b}$$

Since $\mathbf{e}_i$ and $\mathbf{e}_j'$ are unit vectors, $\mathbf{e}_i \cdot \mathbf{e}_j'$ is the cosine of the angle formed by the x_i- and x_j'-axes. The cosine of this angle is denoted by Q_{ij}, i.e., $Q_{ij} = \mathbf{e}_i \cdot \mathbf{e}_j'$, and (15.2.12a,b) is rewritten as

$$x_i = Q_{ij}\,x_j', \qquad x_i' = Q_{ji}\,x_j, \qquad i, j = 1, 2, 3. \tag{15.2.12c,d}$$

In accord with the summation convention, for fixed i, the terms on the right-hand sides of (15.2.12c,d) are summed on j, for j = 1, 2, 3.

Equations (15.2.12c,d) define the transformation for components of **x** in E^3. These rules also govern the transformation of the base vectors,

$$\mathbf{e}_i = (\mathbf{e}_i \cdot \mathbf{e}_j')\,\mathbf{e}_j' = Q_{ij}\,\mathbf{e}_j', \qquad \mathbf{e}_i' = (\mathbf{e}_i' \cdot \mathbf{e}_j)\,\mathbf{e}_j = Q_{ji}\,\mathbf{e}_j, \qquad i, j = 1, 2, 3. \tag{15.2.13a,b}$$

Substitution of (15.2.12c) into (15.2.12d) yields the identity $x_i' = Q_{ji}\,Q_{jk}\,x_k'$, so that

$$Q_{ji}\,Q_{jk} = \delta_{ik}. \tag{15.2.14a}$$

Similarly, substitution of (15.2.12d) into (15.2.12c) results in

$$Q_{ij}\,Q_{kj} = \delta_{ik}. \tag{15.2.14b}$$

15.3. TENSORS IN THREE-DIMENSIONAL VECTOR SPACE

A vector space V is defined as a set of all vectors which can be expressed in terms of the difference of any two position vectors in E^3, i.e., $\mathbf{v} = \mathbf{y} - \mathbf{x}$. A right-handed rectangular Cartesian coordinate system is used to represent the components of any element **v** in V.

15.3.1. Vector as First-Order Tensor

Consider a vector **v** with components v_i and v_i' in the unprimed and primed coordinate systems, respectively, i.e.,

$$\mathbf{v} = v_i\,\mathbf{e}_i = v_i'\,\mathbf{e}_i'. \tag{15.3.1}$$

The dot product of both sides of (15.3.1), first with $\mathbf{e}_i$ and then with $\mathbf{e}_i'$, yields

$$v_i = Q_{ij}\,v_j', \qquad v_i' = Q_{ji}\,v_j. \tag{15.3.2a,b}$$

With reference to the unprimed coordinate system, the vector $\mathbf{v}$ is specified by its components v_i. These components change to v_i', as a new (primed) coordinate system is introduced. The rules (15.3.2a,b) links the components of $\mathbf{v}$ in the two coordinate systems. Therefore, *a vector may be defined by the triple components* v_i *which transform in accordance with rules* (15.3.2a,b) *under the coordinate transformations* (15.2.12c,d).

As pointed out before, it is customary to refer to vectors as *tensors of order 1*, in the sense that through dot or inner product, a vector $\mathbf{v}$ maps vectors in V to real numbers. In this sense, scalars are *tensors of order 0.*

15.3.2. Second-Order Tensor

The definitions of tensors of order 0 (scalars) and 1 (vectors) are easily generalized to tensors of higher order. A tensor of order 2 is a geometrical or physical quantity which may be specified by its 3^2 components with respect to a given coordinate system. Let $\mathbf{T}$ be a second-order tensor with components T_{ij} and T_{ij}' in the unprimed and primed coordinate systems, respectively. Express $\mathbf{T}$ in terms of its components and the corresponding coordinate base vectors, as

$$\mathbf{T} = T_{ij}\,\mathbf{e}_i \otimes \mathbf{e}_j = T_{ij}'\,\mathbf{e}_i' \otimes \mathbf{e}_j', \tag{15.3.3a}$$

where repeated indices are summed. In this manner, to each ordered pair of unit vectors, say, the dyad $\mathbf{e}_1 \otimes \mathbf{e}_2$, the corresponding component of $\mathbf{T}$, here T_{12} for $\mathbf{e}_1 \otimes \mathbf{e}_2$, is associated. The entire tensor $\mathbf{T}$ is then expressed by

$$\begin{aligned}\mathbf{T} &= T_{11}\,\mathbf{e}_1 \otimes \mathbf{e}_1 + T_{12}\,\mathbf{e}_1 \otimes \mathbf{e}_2 + \ldots + T_{33}\,\mathbf{e}_3 \otimes \mathbf{e}_3 \\ &= T_{11}'\,\mathbf{e}_1' \otimes \mathbf{e}_1' + T_{12}'\,\mathbf{e}_1' \otimes \mathbf{e}_2' + \ldots + T_{33}'\,\mathbf{e}_3' \otimes \mathbf{e}_3'\end{aligned} \tag{15.3.3b}$$

which is the *dyadic* representation of this tensor.

Let $\mathbf{v}$ and $\mathbf{w}$ be any two vectors in V. Their ordered dyadic product is defined by $\mathbf{v} \otimes \mathbf{w}$, which has the following coordinate representation:

$$\mathbf{v} \otimes \mathbf{w} = v_i\,w_j\,\mathbf{e}_i \otimes \mathbf{e}_j = v_i'\,w_j'\,\mathbf{e}_i' \otimes \mathbf{e}_j'. \tag{15.3.4}$$

The collection of all such dyadic pairs of vectors forms a linear space which is denoted by $V \times V$.

For dyads $\mathbf{e}_i \otimes \mathbf{e}_j$ and $\mathbf{e}_k \otimes \mathbf{e}_l$, the *double dot* operator, :, is defined by

$$\begin{aligned}(\mathbf{e}_i \otimes \mathbf{e}_j) : (\mathbf{e}_k \otimes \mathbf{e}_l) &\equiv (\mathbf{e}_i \cdot \mathbf{e}_k)\,(\mathbf{e}_j \cdot \mathbf{e}_l) = \delta_{ik}\,\delta_{jl}, \\ (\mathbf{e}_i' \otimes \mathbf{e}_j') : (\mathbf{e}_k' \otimes \mathbf{e}_l') &\equiv (\mathbf{e}_i' \cdot \mathbf{e}_k')\,(\mathbf{e}_j' \cdot \mathbf{e}_l') = \delta_{ik}\,\delta_{jl}.\end{aligned} \tag{15.3.5a,b}$$

Now a second-order tensor $\mathbf{T}$ is viewed as a *linear operator* which maps an element in $V \times V$ to a real number, i.e.,

$$\mathbf{T} : (\mathbf{v} \otimes \mathbf{w}) \in R \qquad \text{for } \mathbf{v} \otimes \mathbf{w} \text{ in } V \times V. \tag{15.3.6}$$

For components in the unprimed and primed coordinate systems, (15.3.6) is written as

$$(T_{ij}\, \mathbf{e}_i \otimes \mathbf{e}_j) : (v_k\, w_l\, \mathbf{e}_k \otimes \mathbf{e}_l) = (T_{ij}\, v_k\, w_l)\, \delta_{ik}\, \delta_{jl} = T_{ij}\, v_i\, w_j,$$

$$(T'_{ij}\, \mathbf{e}'_i \otimes \mathbf{e}'_j) : (v'_k\, w'_l\, \mathbf{e}'_k \otimes \mathbf{e}'_l) = (T'_{ij}\, v'_k\, w'_l)\, \delta_{ik}\, \delta_{jl} = T'_{ij}\, v'_i\, w'_j. \tag{15.3.7a,b}$$

In particular, if $\mathbf{v} \otimes \mathbf{w}$ is chosen to be $\mathbf{e}_i \otimes \mathbf{e}_j$ or $\mathbf{e}'_i \otimes \mathbf{e}'_j$, the components, T_{ij} and T'_{ij} in the corresponding coordinate system, are obtained,

$$T_{ij} \equiv \mathbf{T} : (\mathbf{e}_i \otimes \mathbf{e}_j), \qquad T'_{ij} \equiv \mathbf{T} : (\mathbf{e}'_i \otimes \mathbf{e}'_j). \tag{15.3.8a,b}$$

Furthermore, the relation between the unit vectors $\mathbf{e}_i$ and $\mathbf{e}'_i$ (i = 1, 2, 3) leads to the following component transformation rule for any second-order tensor, $\mathbf{T}$:

$$T_{ij} = Q_{ik}\, Q_{jl}\, T'_{kl}, \qquad T'_{ij} = Q_{ki}\, Q_{lj}\, T_{kl}. \tag{15.3.9a,b}$$

From (15.3.2a,b) and (15.3.9a,b) it follows that

$$\mathbf{T} : (\mathbf{v} \otimes \mathbf{w}) = T_{ij}\, v_i\, w_j = T'_{ij}\, v'_i\, w'_j. \tag{15.3.10}$$

Therefore, the linear transformation of $V \times V$ to real numbers through the second-order tensor $\mathbf{T}$ is independent of the coordinate system which may be used to represent the corresponding components.

Tensor $\mathbf{T}$ may be viewed as a linear operator which maps the vector space V to itself, i.e.,

$$\left\{ \begin{matrix} \mathbf{T} \cdot \mathbf{v} \\ \mathbf{v} \cdot \mathbf{T} \end{matrix} \right\} \in V \qquad \text{for } \mathbf{v} \text{ in } V, \tag{15.3.11}$$

where the dot product between the dyad $\mathbf{e}_i \otimes \mathbf{e}_j$ and the unit vector $\mathbf{e}_k$ is defined by

$$(\mathbf{e}_i \otimes \mathbf{e}_j) \cdot \mathbf{e}_k \equiv \mathbf{e}_i\, (\mathbf{e}_j \cdot \mathbf{e}_k) = \delta_{jk}\, \mathbf{e}_i, \qquad \mathbf{e}_i \cdot (\mathbf{e}_j \otimes \mathbf{e}_k) \equiv (\mathbf{e}_i \cdot \mathbf{e}_j)\, \mathbf{e}_k = \delta_{ij}\, \mathbf{e}_k. \tag{15.3.12a,b}$$

In general, $\mathbf{T} \cdot \mathbf{v}$ and $\mathbf{v} \cdot \mathbf{T}$ are two distinct vectors. In terms of the components in the unprimed and primed coordinate systems, (15.3.11) is written as

$$(\mathbf{T} \cdot \mathbf{v})_i = T_{ij}\, v_j, \qquad (\mathbf{v} \cdot \mathbf{T})_j = v_i\, T_{ij},$$

$$(\mathbf{T} \cdot \mathbf{v})'_i = T'_{ij}\, v'_j, \qquad (\mathbf{v} \cdot \mathbf{T})'_j = v'_i\, T'_{ij}. \tag{15.3.13a~d}$$

Again, with the aid of (15.3.2a,b) and (15.3.9a,b), it follows that

$$(\mathbf{T} \cdot \mathbf{v})_i = Q_{ij}\, (\mathbf{T} \cdot \mathbf{v})'_j, \qquad (\mathbf{T} \cdot \mathbf{v})'_i = Q_{ji}\, (\mathbf{T} \cdot \mathbf{v})_j,$$

$$(\mathbf{v} \cdot \mathbf{T})_i = Q_{ij}\, (\mathbf{v} \cdot \mathbf{T})'_j, \qquad (\mathbf{v} \cdot \mathbf{T})'_i = Q_{ji}\, (\mathbf{v} \cdot \mathbf{T})_j. \tag{15.3.13e~h}$$

This linear transformation of V to V through $\mathbf{T}$ is also independent of the coordinate system.

15.3.3. Higher-Order Tensors

Equations (15.3.9a,b) are used to define a tensor of order 2 as follows: with reference to the coordinate system x_i, the 3^2 components T_{ij} define a second-order tensor if these components transform according to (15.3.9a,b) under the coordinate transformation (15.2.12a,b). Similarly, a tensor of order 3 is specified by 3^3 components, T_{ijk} and T'_{ijk}, in the unprimed and primed coordinates, as

$$\mathbf{T} = T_{ijk}\,\mathbf{e}_i \otimes \mathbf{e}_j \otimes \mathbf{e}_k = T'_{ijk}\,\mathbf{e}'_i \otimes \mathbf{e}'_j \otimes \mathbf{e}'_k. \tag{15.3.14}$$

The components of **T** transform according to

$$T_{ijk} = Q_{il}\,Q_{jm}\,Q_{kn}\,T'_{lmn}. \tag{15.3.15}$$

The tensor **T** maps the triple product of V, $V^3 \equiv V \times V \times V$, to real numbers,

$$T_{ijk}\,v_i\,w_j\,u_k \in R \qquad \text{for every } \mathbf{v}, \mathbf{w}, \mathbf{u} \text{ in } V. \tag{15.3.16}$$

This is independent of the corresponding coordinate system.

In general, a tensor of order n is expressed as

$$\mathbf{T} = T_{i_1 i_2 \dots i_n}\mathbf{e}_{i_1} \otimes \mathbf{e}_{i_2} \otimes \dots \otimes \mathbf{e}_{i_n} = T'_{i_1 i_2 \dots i_n}\mathbf{e}'_{i_1} \otimes \mathbf{e}'_{i_2} \otimes \dots \mathbf{e}'_{i_n}$$

$$i_1, i_2, \dots, i_n = 1, 2, 3, \tag{15.3.17}$$

where the 3^n components $T_{i_1 i_2 \dots i_n}$'s in the unprimed coordinates are linked with $T'_{i_1 i_2 \dots i_n}$'s in the primed coordinates by the following transformation:

$$T_{i_1 i_2 \dots i_n} = Q_{i_1 j_1} Q_{i_2 j_2} \dots Q_{i_n j_n} T'_{j_1 j_2 \dots j_n},$$

$$i_1, i_2, \dots, i_n, j_1, j_2, \dots, j_n = 1, 2, 3. \tag{15.3.18}$$

The tensor **T** maps the n product of V, $V^n \equiv V \times V \times \dots \times V$, to real numbers,

$$T_{i_1 i_2 \dots i_n} v^1_{i_1} v^2_{i_2} \dots v^n_{i_n} \in R \qquad \text{for } \mathbf{v}^1, \mathbf{v}^2, \dots, \mathbf{v}^n \text{ in } V, \tag{15.3.19}$$

which is also independent of the corresponding coordinate system.

15.3.4. Remarks on Second-Order Tensors

As explained in Subsection 15.3.2, a second-order tensor **T** is regarded as a linear operator which transforms a vector to another vector. Transformation (15.3.11) is called *orthogonal* if **T** is an orthogonal tensor, that is, if

$$T_{ij}\,T_{kj} = \delta_{ik}. \tag{15.3.20}$$

The tensor $\mathbf{Q} = Q_{ij}\,\mathbf{e}_i \otimes \mathbf{e}_j$ with components $Q_{ij} = \mathbf{e}_i \cdot \mathbf{e}'_j$ satisfies (15.3.20). It is an orthogonal tensor. In the dyadic representation, **Q** (or any orthogonal tensor) is expressed as

$$\mathbf{Q} = \mathbf{e}'_k \otimes \mathbf{e}_k = \mathbf{e}'_1 \otimes \mathbf{e}_1 + \mathbf{e}'_2 \otimes \mathbf{e}_2 + \mathbf{e}'_3 \otimes \mathbf{e}_3, \tag{15.3.21}$$

whose components, Q_{ij} and Q'_{ij}, in the unprimed and primed coordinate systems are

$$Q_{ij} = \mathbf{Q} : (\mathbf{e}_i \otimes \mathbf{e}_j) = (\mathbf{e}_k' \cdot \mathbf{e}_i)(\mathbf{e}_k \cdot \mathbf{e}_j) = \mathbf{e}_i \cdot \mathbf{e}_j',$$

$$Q_{ij}' = \mathbf{Q} : (\mathbf{e}_i' \otimes \mathbf{e}_j') = (\mathbf{e}_k' \cdot \mathbf{e}_i')(\mathbf{e}_k \cdot \mathbf{e}_j') = \mathbf{e}_i \cdot \mathbf{e}_j'. \tag{15.3.22a,b}$$

Hence, $Q_{ij} = Q_{ij}'$. As is seen, $\mathbf{Q}$ transforms the unit base vector $\mathbf{e}_i$ of the unprimed coordinate system to the unit base vector $\mathbf{e}_i'$ of the primed one,

$$\mathbf{e}_i' = \mathbf{Q} \cdot \mathbf{e}_i, \qquad \mathbf{e}_i = \mathbf{e}_i' \cdot \mathbf{Q} = \mathbf{Q}^T \cdot \mathbf{e}_i', \qquad i = 1, 2, 3, \tag{15.3.23a,b}$$

where $\mathbf{Q}^T \equiv \mathbf{e}_k \otimes \mathbf{e}_k'$. In this context, (15.3.23a) may be interpreted as follows: through $\mathbf{Q} \cdot \mathbf{v}$, vector $\mathbf{v}$ with components v_i in the unprimed coordinate system is rotated into vector $\mathbf{w}$ with components $w_i' \equiv v_i$ in the primed coordinate system.

An orthogonal tensor that transforms a vector $\mathbf{v}$ into itself is called the *unit tensor*. From definition (15.3.21),

$$\mathbf{1}^{(2)} \equiv \mathbf{e}_i \otimes \mathbf{e}_i \equiv \delta_{ij}\, \mathbf{e}_i \otimes \mathbf{e}_j, \tag{15.3.24}$$

is the unit tensor. Or, from the definition of the Kronecker delta, $v_i = \delta_{ij}\, v_j$. *The superscript* (2) in (15.3.24) *denotes that this is a second-order unit tensor.* Definition (15.2.4) now yields

$$(\mathbf{e}_i' \otimes \mathbf{e}_i) \cdot (\mathbf{e}_j \otimes \mathbf{e}_j') = (\mathbf{e}_i \cdot \mathbf{e}_j)\, \mathbf{e}_i' \otimes \mathbf{e}_j' = \mathbf{e}_i' \otimes \mathbf{e}_i' = \mathbf{1}^{(2)},$$

$$(\mathbf{e}_i \otimes \mathbf{e}_i') \cdot (\mathbf{e}_j' \otimes \mathbf{e}_j) = (\mathbf{e}_i' \cdot \mathbf{e}_j')\, \mathbf{e}_i \otimes \mathbf{e}_j = \mathbf{e}_i \otimes \mathbf{e}_i = \mathbf{1}^{(2)}. \tag{15.3.25a,b}$$

A second-order tensor is called *symmetric* if it is equal to its transpose, $\mathbf{T} = \mathbf{T}^T$, where if T_{ij} are components of $\mathbf{T}$, then those of $\mathbf{T}^T$ are T_{ji}, i.e.,

$$\mathbf{T}^T \equiv (T_{ij} \mathbf{e}_i \otimes \mathbf{e}_j)^T \equiv T_{ij}\, \mathbf{e}_j \otimes \mathbf{e}_i = T_{ji}\, \mathbf{e}_i \otimes \mathbf{e}_j. \tag{15.3.26}$$

A tensor $\mathbf{T}$ is called *antisymmetric* (or *skew symmetric*) if $\mathbf{T} = -\mathbf{T}^T$. In general, any second-order tensor $\mathbf{T}$ can be written as a sum of two parts, a symmetric and an antisymmetric part, i.e.,

$$T_{ij} = \frac{1}{2}(T_{ij} + T_{ji}) + \frac{1}{2}(T_{ij} - T_{ji}) = T_{(ij)} + T_{[ij]}, \tag{15.3.27}$$

where the components of the symmetric and antisymmetric parts of $\mathbf{T}$ are denoted by $T_{(ij)}$ and $T_{[ij]}$, respectively. It is, therefore, important to keep in mind that, in the dyadic representation of a tensor, the order of various dyads must remain intact, unless the corresponding symmetry conditions are satisfied.

The *tensor* (or dyadic) product of a second-order tensor $\mathbf{T}$ and a vector $\mathbf{v}$ is a tensor of order 3, defined as

$$\mathbf{T} \otimes \mathbf{v} = (T_{ij}\, v_k)\, \mathbf{e}_i \otimes \mathbf{e}_j \otimes \mathbf{e}_k, \qquad \mathbf{v} \otimes \mathbf{T} = (v_i\, T_{jk})\, \mathbf{e}_i \otimes \mathbf{e}_j \otimes \mathbf{e}_k. \tag{15.3.28a,b}$$

For any two tensors $\mathbf{T}$ and $\mathbf{S}$ of any order, say, orders n and m, the $(n+m)$th-order tensors $\mathbf{T} \otimes \mathbf{S}$ and $\mathbf{S} \otimes \mathbf{T}$ are distinct quantities, i.e., in general $\mathbf{T}$ and $\mathbf{S}$ are *not* commutative, $\mathbf{T} \otimes \mathbf{S} \neq \mathbf{S} \otimes \mathbf{T}$.

Setting $j = k$ in the expression $T_{ij}\, v_k$, the *dot* product of $\mathbf{T}$ and $\mathbf{v}$, namely a vector, say, $\mathbf{u}$, is obtained. This is called *contraction.* In general, if two letter indices of the components of a tensor of order k are made identical, the components of a tensor of order $k-2$ are obtained. Contraction of the indices of a second-order tensor T_{ij} results in a scalar which is called the *trace* of $\mathbf{T}$ and is denoted by

$$\mathrm{tr}\mathbf{T} \equiv T_{ii}. \tag{15.3.29}$$

15.4. DEL OPERATOR AND THE GAUSS THEOREM

This section addresses certain elementary topics in tensor calculus. In particular, attention is focused on the Gauss theorem, since it plays a central role in computing the volume average of various physical field quantities, in terms of their boundary data.

Let $\mathbf{T} = \mathbf{T}(\mathbf{x})$ be an nth-order tensor-valued function of position $\mathbf{x}$ in a finite region D in E^3; in short, $\mathbf{T}(\mathbf{x})$ is called an nth-order tensor field in D. In terms of the base vectors $\mathbf{e}_i$, $\mathbf{T}(\mathbf{x})$ is expressed as

$$\mathbf{T}(\mathbf{x}) = T_{i_1 i_2 \dots i_n}(\mathbf{x})\, \mathbf{e}_{i_1} \otimes \mathbf{e}_{i_2} \otimes \dots \otimes \mathbf{e}_{i_n}, \tag{15.4.1}$$

where each component of $\mathbf{T}(\mathbf{x})$ is a scalar-valued function of $\mathbf{x}$ in D. When all components of $\mathbf{T}(\mathbf{x})$ are continuous functions of $\mathbf{x}$ in D, the tensor field $\mathbf{T}(\mathbf{x})$ is said to be *continuous* in D. Similarly, when all components of $\mathbf{T}(\mathbf{x})$ are differentiable in D, $\mathbf{T}(\mathbf{x})$ is said to be *differentiable* in D. In this manner, various mathematical properties of $\mathbf{T}(\mathbf{x})$ are defined in terms of the properties of its components. In general, tensor fields considered in micromechanics are assumed to be suitably smooth everywhere in their domain of definition, except on some planes such as interfaces between microconstituents or cracks. These surfaces are called *discontinuity surfaces.* Across a discontinuity surface, the tensor-valued function or some of its derivatives may not be continuous.

Differentiation of a suitably smooth tensor field produces another tensor field. The *del operator* is used to express such differentiation. The del operator is a vectorial differential operator, denoted by

$$\boldsymbol{\nabla} \equiv \partial_i\, \mathbf{e}_i, \tag{15.4.2}$$

where $\partial_i \equiv \partial/\partial x_i$ is the partial differential operator. The differential operator ∂_i transforms according to the component transformation rule, given by (15.3.2a,b). The proof is straightforward. With respect to the base vectors $\mathbf{e}_i'$, any point $\mathbf{x}$ is expressed by

$$\mathbf{x} = x_i\, \mathbf{e}_i = x_i'\, \mathbf{e}_i'. \tag{15.4.3a}$$

Then, since $Q_{ij} = \mathbf{e}_i \cdot \mathbf{e}_j'$, differentiation of $\mathbf{x}$ with respect to x_i gives

$$\frac{\partial \mathbf{x}}{\partial x_i}\ (= \mathbf{e}_i = Q_{ij}\, \mathbf{e}_j') = \frac{\partial x_j'}{\partial x_i}\, \mathbf{e}_j'. \tag{15.4.3b}$$

Hence, $\partial x_j'/\partial x_i = Q_{ij}$. According to the chain rule,

$$\partial_i \equiv \frac{\partial}{\partial x_i} \equiv \frac{\partial x_j'}{\partial x_i}\frac{\partial}{\partial x_j'} \equiv Q_{ij}\,\partial_j', \tag{15.4.3c}$$

where $\partial_i' \equiv \partial/\partial x_i'$. Relation (15.4.3c) is the same as the vector component transformation given by (15.3.2a).

The del operator $\mathbf{\nabla}$ is independent of the particular coordinates which may be used to express the differential operator ∂_i. Since $Q_{ij}\,Q_{ik} = \delta_{jk}$, the del operator satisfies

$$\mathbf{\nabla} = \partial_i\,\mathbf{e}_i = (Q_{ik}\,\partial_k')\,(Q_{il}\,\mathbf{e}_l') = \partial_i'\,\mathbf{e}_i'. \tag{15.4.4}$$

The del operator is a *vector operator* which may act on suitably smooth tensor fields. Its tensor (dot) product with a suitably smooth tensor field produces a tensor field one order higher (lower). The dyadic symbol "$\otimes$" and dot "$\bullet$" are used to denote the tensor and dot product operations. Assume that $\mathbf{T} = \mathbf{T}(\mathbf{x})$ is suitably smooth in its domain D. The two operations, "$\otimes$" and "$\bullet$", of $\mathbf{\nabla}$ with $\mathbf{T}$ are denoted by $\mathbf{\nabla}\otimes\mathbf{T}$ and $\mathbf{\nabla}\bullet\mathbf{T}$, i.e.,

$$\mathbf{\nabla}\otimes\mathbf{T} = (\mathbf{\nabla}\otimes\mathbf{T})(\mathbf{x}) = (\partial_{i_1}T_{i_2\ldots i_{n+1}})(\mathbf{x})\,\mathbf{e}_{i_1}\otimes\mathbf{e}_{i_2}\ldots\mathbf{e}_{i_{n+1}},$$

$$\mathbf{\nabla}\bullet\mathbf{T} = (\mathbf{\nabla}\bullet\mathbf{T})(\mathbf{x}) = (\partial_j T_{j\,i_1\ldots i_{n-1}})(\mathbf{x})\,\mathbf{e}_{i_1}\otimes\mathbf{e}_{i_2}\ldots\mathbf{e}_{i_{n-1}}. \tag{15.4.5a,b}$$

The first operation yields a tensor field one order higher than $\mathbf{T}$, and the second operation, a tensor field one order lower.

In vector analysis, gradient and divergence are defined, respectively, for suitably smooth scalar-valued and vector-valued functions. Let $f = f(\mathbf{x})$ and $\mathbf{F} = \mathbf{F}(\mathbf{x})$, respectively, be such scalar- and vector-valued functions. The gradient of f, and the divergence of $\mathbf{F}$ are defined by

$$grad\,f = \mathbf{\nabla}f = (\partial_i f)\,\mathbf{e}_i, \qquad div\,\mathbf{F} = \mathbf{\nabla}\bullet\mathbf{F} = \partial_i\,F_i, \tag{15.4.6a,b}$$

respectively. Comparison of (15.4.5a) with (15.4.6a), and (15.4.5b) with (15.4.6b), lead to the introduction of *gradient* and *divergence* of the tensor field $\mathbf{T} = \mathbf{T}(\mathbf{x})$, respectively, by

$$grad\,\mathbf{T} = \mathbf{\nabla}\otimes\mathbf{T}, \qquad div\,\mathbf{T} = \mathbf{\nabla}\bullet\mathbf{T}. \tag{15.4.7a,b}$$

For the nth-order tensor field $\mathbf{T}$, $grad\,\mathbf{T}$ and $div\,\mathbf{T}$ are the $(n+1)$th- and the $(n-1)$th-order tensor fields, respectively.

The permutation symbol, e_{ijk}, can be used to obtain the *curl* of a suitably smooth vector field $\mathbf{F}$,

$$curl\,\mathbf{F} = \mathbf{\nabla}\times\mathbf{F} = (e_{ijk}\,\partial_j\,F_k)\,\mathbf{e}_i. \tag{15.4.6c}$$

Similarly, the *curl* of the smooth tensor field $\mathbf{T}$ of order n is given by

$$curl\,\mathbf{T} = \mathbf{\nabla}\times\mathbf{T} = \mathbf{\nabla}\times\mathbf{T}(\mathbf{x}) = \{e_{i_1jk}\,(\partial_j\,T_{k\,i_2\ldots i_n})\}(\mathbf{x})\,\mathbf{e}_{i_1}\otimes\mathbf{e}_{i_2}\ldots\mathbf{e}_{i_n}, \tag{15.4.7c}$$

which is also an nth-order tensor field.

Similarly to differentiation, the integration of a tensor field $\mathbf{T}$ on a domain D can be defined in terms of the integration of its components. If the tensor field is integrable on its domain D in E^3, the volume integral of $\mathbf{T}$ over D becomes

$$\int_D \mathbf{T}\, dV = \{\int_D T_{i_1 i_2 \ldots i_n}\, dV\}\, \mathbf{e}_{i_1} \otimes \mathbf{e}_{i_2} \ldots \mathbf{e}_{i_n}. \tag{15.4.8a}$$

Here, again, advantage is taken of the fact that base vectors $\mathbf{e}_i$ are fixed. In the same manner, if the tensor field $\mathbf{T}$ is defined on some suitable surface A in E^3, the area integral of $\mathbf{T}$ over A is

$$\int_A \mathbf{T}\, dS = \{\int_A T_{i_1 i_2 \ldots i_n}\, dS\}\, \mathbf{e}_{i_1} \otimes \mathbf{e}_{i_2} \ldots \mathbf{e}_{i_n}. \tag{15.4.8b}$$

Note that integration of an nth-order *tensor field* produces an nth-order *tensor.*

The *Gauss theorem* or the *divergence theorem* relates the divergence of a vector-valued function $\mathbf{F}$ in a domain D, to its flux across the boundary ∂D of this domain, under suitable smoothness conditions. Let $\mathbf{F} = \mathbf{F}(\mathbf{x})$ be a smooth vector-valued function in domain D which is bounded by a suitably smooth ∂D. Then, the Gauss theorem states

$$\int_D \mathit{div}\, \mathbf{F}\, dV = \int_{\partial D} \boldsymbol{\nu} \cdot \mathbf{F}\, dS, \tag{15.4.9a}$$

or

$$\int_D \partial_i F_i\, dV = \int_{\partial D} \nu_i F_i\, dS, \tag{15.4.9b}$$

where $\boldsymbol{\nu}$ is the exterior unit normal on ∂D. A suitably smooth scalar-valued function f also satisfies a similar transformation,

$$\int_D \mathit{grad}\, f\, dV = \int_{\partial D} f\, \boldsymbol{\nu}\, dS, \tag{15.4.10a}$$

or

$$\int_D \partial_i f\, dV = \int_{\partial D} f\, \nu_i\, dS. \tag{15.4.10b}$$

Moreover, it can be shown that

$$\int_D \mathit{curl}\, \mathbf{F}\, dV = \int_{\partial D} \boldsymbol{\nu} \times \mathbf{F}\, dS, \tag{15.4.11a}$$

or

$$\int_D e_{ijk}\, \partial_j F_k\, dV = \int_{\partial D} e_{ijk}\, \nu_j F_k\, dS. \tag{15.4.11b}$$

These theorems may be applied to any smooth tensor field, say, $\mathbf{T} = \mathbf{T}(\mathbf{x})$; for fixed $i_1, i_2, \ldots, i_{n-1}$, the divergence of the components of $\mathbf{T}$ satisfies

$$\int_D \partial_j T_{j i_1 \ldots i_{n-1}}\, dV = \int_{\partial D} \nu_j T_{j i_1 \ldots i_{n-1}}\, dS; \tag{15.4.12a}$$

for fixed $i_1, i_2, \ldots, i_{n+1}$, the gradient of the components of $\mathbf{T}$ satisfies

$$\int_D \partial_{i_1} T_{i_2 \ldots i_{n+1}}\, dV = \int_{\partial D} \nu_{i_1} T_{i_2 \ldots i_{n+1}}\, dS; \tag{15.4.12b}$$

and for fixed i_1, i2, ..., i_n, the curl of the components of $\mathbf{T}$ satisfies

$$\int_D e_{i_1 jk}\, \partial_j T_{k i_2 \ldots i_n}\, dV = \int_{\partial D} e_{i_1 jk}\, \nu_j \mathbf{T}_{k i_2 \ldots i_n}\, dS. \tag{15.4.12c}$$

Therefore,

$$\int_D div\,\mathbf{T}\,dV = \int_D \boldsymbol{\nabla}\boldsymbol{.}\mathbf{T}\,dV = \int_{\partial D} \boldsymbol{\nu}\boldsymbol{.}\mathbf{T}\,dS,$$

$$\int_D grad\,\mathbf{T}\,dV = \int_D \boldsymbol{\nabla}\otimes\mathbf{T}\,dV = \int_{\partial D} \boldsymbol{\nu}\otimes\mathbf{T}\,dS,$$

$$\int_D curl\,\mathbf{T}\,dV = \int_D \boldsymbol{\nabla}\times\mathbf{T}\,dV = \int_{\partial D} \boldsymbol{\nu}\times\mathbf{T}\,dS. \tag{15.4.13a~c}$$

These are statements of the *Gauss theorem.* They can be summarized by

$$\int_D \boldsymbol{\nabla}*\mathbf{T}\,dV = \int_{\partial D} \boldsymbol{\nu}*\mathbf{T}\,dS, \tag{15.4.14}$$

where $*$ may stand for "$\boldsymbol{.}$", "$\otimes$", or "$\times$". Note that for the Gauss theorem to apply, certain smoothness conditions must be satisfied by the tensor field *and* the domain D.

15.5. SPECIAL TOPICS IN TENSOR ALGEBRA

This subsection focuses on certain special properties of second- and fourth-order tensors. *Second-order base tensors are introduced,* and certain symmetry properties of fourth-order tensors are examined, together with their corresponding matrix representation. Attention is again confined to quantities in three-dimensional vector space, and fixed rectangular Cartesian coordinate systems are used to represent tensor components.

15.5.1. Second-Order Base Tensors

In terms of the base vectors $\mathbf{e}_i$ (i = 1, 2, 3), it is convenient to define the *second-order base tensors* by

$$\mathbf{e}_{ij} \equiv \mathbf{e}_i \otimes \mathbf{e}_j, \tag{15.5.1}$$

and their *second-order contraction* by

$$\mathbf{e}_{ij} : \mathbf{e}_{kl} \equiv \delta_{ik}\delta_{jl}; \tag{15.5.2}$$

see (15.3.5). The second-order identity tensor $\mathbf{1}^{(2)}$, (15.3.24), now becomes

$$\mathbf{1}^{(2)} = \delta_{ij}\,\mathbf{e}_{ij} = \mathbf{e}_{ii}. \tag{15.5.3}$$

As pointed out before, this identity tensor maps a vector, i.e., a first-order tensor, to itself.

Let $\mathbf{S} = S_{ij}\,\mathbf{e}_i \otimes \mathbf{e}_j$ be a second-order tensor and $\mathbf{T} = T_{ijkl}\,\mathbf{e}_i \otimes \mathbf{e}_j \otimes \mathbf{e}_k \otimes \mathbf{e}_l$ be a fourth-order one. In terms of the second-order base tensors, they become

$$\mathbf{S} = S_{ij}\,\mathbf{e}_{ij}, \qquad \mathbf{T} = T_{ijkl}\,\mathbf{e}_{ij} \otimes \mathbf{e}_{kl}. \tag{15.5.4a,b}$$

Denote the collection of all such second-order tensors by $T^{(2)}$, and that of all fourth-order tensors by $T^{(4)}$, i.e.,

$$\mathcal{T}^{(2)} = \{\mathbf{S} \mid \mathbf{S} = \mathrm{S}_{ij}\,\mathbf{e}_{ij}\}, \qquad \mathcal{T}^{(4)} = \{\mathbf{T} \mid \mathbf{T} = \mathrm{T}_{ijkl}\,\mathbf{e}_{ij} \otimes \mathbf{e}_{kl}\}. \tag{15.5.5a,b}$$

$\mathcal{T}^{(2)}$ and $\mathcal{T}^{(4)}$ are referred to as second-order and fourth-order tensor spaces, respectively. Note that, for every $\mathbf{v}$ in V and $\mathbf{S}$ in $\mathcal{T}^{(2)}$, it follows that $\mathbf{S} \bullet \mathbf{v}$ is in V; see (15.3.11). Similarly, from the second-order contraction (15.5.2), it is observed that, for every $\mathbf{S}$ in $\mathcal{T}^{(2)}$, $\mathbf{T} : \mathbf{S}$ is in $\mathcal{T}^{(2)}$. Hence, an element of $\mathcal{T}^{(4)}$ transforms an element in $\mathcal{T}^{(2)}$ to an element in $\mathcal{T}^{(2)}$.

15.5.2. Matrix Operations for Second- and Fourth-Order Tensors

With the aid of the components in a rectangular Cartesian coordinate system, a vector, $\mathbf{v}$, and a second-order tensor, $\mathbf{S}$, may be represented respectively, by a three by one column matrix and a three by three matrix. Denote the column of the components of $\mathbf{v}$ by $[\mathrm{v_i}]$, and the matrix of the components of $\mathbf{S}$ by $[\mathrm{S_{ij}}]$. Dot operation between $\mathbf{v}$ and $\mathbf{S}$ is now reduced to matrix multiplication of $[\mathrm{v_i}]$ and $[\mathrm{S_{ij}}]$. The correspondences between some elementary tensorial and matrix operations are,

$$\mathbf{v} \Longleftrightarrow [\mathrm{v_i}], \qquad \mathbf{S} \Longleftrightarrow [\mathrm{S_{ij}}], \qquad \mathbf{S} \bullet \mathbf{v} \Longleftrightarrow [\mathrm{S_{ij}}][\mathrm{v_j}],$$

$$\mathbf{S}^{\mathrm{T}} \Longleftrightarrow [\mathrm{S_{ji}}], \qquad \mathbf{S}^{-1} \Longleftrightarrow [\mathrm{S_{ij}}]^{-1}, \qquad \mathbf{S}^{-\mathrm{T}} \Longleftrightarrow [\mathrm{S_{ij}}]^{-\mathrm{T}}. \tag{15.5.6a~f}$$

Here, superscript T denotes the transpose, and superscripts -1 and $-\mathrm{T}$, respectively, denote the inverse and the transposed inverse (when they exist).

The nine second-order base tensors are ordered as a one by nine row vector,

$$\{\mathbf{e}_{ij}\} = \{\mathbf{e}_{11}, \mathbf{e}_{12}, \mathbf{e}_{13}, \mathbf{e}_{21}, \mathbf{e}_{22}, \mathbf{e}_{33}, \mathbf{e}_{31}, \mathbf{e}_{32}, \mathbf{e}_{33}\}. \tag{15.5.7}$$

This arrangement is collectively denoted by the subscript $\{ij\}$. Then, tensorial operations between elements in $\mathcal{T}^{(2)}$ and $\mathcal{T}^{(4)}$ are reduced to nine-dimensional matrix operations. A second-order, $\mathbf{S}$, and a fourth-order, $\mathbf{T}$, tensor is now represented, respectively, by a nine by one column matrix and a nine by nine matrix, consisting of the corresponding components in a rectangular Cartesian coordinate system. The column of the components of $\mathbf{S}$ is denoted by $[\mathrm{S}_{\{ij\}}]$, and the matrix of the components of $\mathbf{T}$ by $[\mathrm{T}_{\{ij\}\{kl\}}]$. As before, superscripts T, -1, and $-\mathrm{T}$ denote the transpose, the inverse, and the transposed inverse, respectively. The following correspondence between tensorial and matrix representations is obtained:

$$\mathbf{S} \Longleftrightarrow [\mathrm{S}_{\{ij\}}], \qquad \mathbf{T} \Longleftrightarrow [\mathrm{T}_{\{ij\}\{kl\}}], \qquad \mathbf{T} : \mathbf{S} \Longleftrightarrow [\mathrm{T}_{\{ij\}\{kl\}}][\mathrm{S}_{\{kl\}}],$$

$$\mathbf{T}^{\mathrm{T}} \Longleftrightarrow [\mathrm{T}_{\{kl\}\{ij\}}], \qquad \mathbf{T}^{-1} \Longleftrightarrow [\mathrm{T}_{\{ij\}\{kl\}}]^{-1}, \qquad \mathbf{T}^{-\mathrm{T}} \Longleftrightarrow [\mathrm{T}_{\{ij\}\{kl\}}]^{-\mathrm{T}}. \tag{15.5.8a~f}$$

The *fourth-order unit tensor,* $\mathbf{1}^{(4)}$, with the components

$$1^{(4)}_{ijkl} \equiv \delta_{ik}\delta_{jl} \tag{15.5.9a}$$

is expressed by

$$\mathbf{1}^{(4)} \equiv \mathbf{e}_{ij} \otimes \mathbf{e}_{ij}. \tag{15.5.9b}$$

For any $\mathbf{S}$ in $T^{(2)}$, it follows that $\mathbf{1}^{(4)} : \mathbf{S} = \mathbf{S} : \mathbf{1}^{(4)} = \mathbf{S}$. In view of (15.5.8), the inverse of $[\mathrm{T}_{\{ij\}\{kl\}}]$, i.e., $[\mathrm{T}_{\{ij\}\{kl\}}]^{-1}$, satisfies

$$\mathbf{T}^{-1} : \mathbf{T} = \mathbf{T} : \mathbf{T}^{-1} = \mathbf{1}^{(4)}. \tag{15.5.10}$$

The magnitude (or norm) of a vector $\mathbf{v}$ is $|\mathbf{v}| \equiv \sqrt{\mathbf{v} \cdot \mathbf{v}}$. The magnitude (or norm) of a second-order tensor $\mathbf{S}$ is similarly defined by

$$|\mathbf{S}| = \sqrt{\mathbf{S} : \mathbf{S}} = \sqrt{\mathrm{S}_{ij}\, \mathrm{S}_{ij}}. \tag{15.5.11}$$

If a fourth-order tensor $\mathbf{T}$ satisfies[2] $\mathbf{T}^{\mathrm{T}} : \mathbf{T} = \mathbf{1}^{(4)}$, or $\mathrm{T}_{ijkl}\, \mathrm{T}_{ijmn} = \delta_{km}\, \delta_{ln}$, then transformation $\mathbf{T} : \mathbf{S}$ preserves the norm of $\mathbf{S}$, i.e.,

$$(\mathbf{T} : \mathbf{S}) : (\mathbf{T} : \mathbf{S}) = \mathbf{S} : (\mathbf{T}^{\mathrm{T}} : \mathbf{T}) : \mathbf{S} = \mathbf{S} : \mathbf{S}, \tag{15.5.12a}$$

or in component form,

$$(\mathrm{T}_{ijkl}\, \mathrm{S}_{kl})\, (\mathrm{T}_{ijmn}\, \mathrm{S}_{mn}) = \mathrm{S}_{kl}\, (\mathrm{T}_{ijkl}\, \mathrm{T}_{ijmn})\, \mathrm{S}_{mn} = \mathrm{S}_{kl}\, \mathrm{S}_{kl}. \tag{15.5.12b}$$

15.5.3. Second-Order Symmetric Base Tensors

In general, a tensor $\mathbf{S}$ in $T^{(2)}$ is called symmetric if $\mathbf{S}^{\mathrm{T}} = \mathbf{S}$. To represent symmetric tensors in a concise manner, the *second-order symmetric base tensors*, $\mathbf{e}^{s}_{ij}$, are introduced as follows:

$$\mathbf{e}^{s}_{ij} \equiv \frac{1}{2}(\mathbf{e}_{ij} + \mathbf{e}^{\mathrm{T}}_{ij}) = \frac{1}{2}(\mathbf{e}_i \otimes \mathbf{e}_j + \mathbf{e}_j \otimes \mathbf{e}_i). \tag{15.5.13}$$

While the second-order base tensors, $\{\mathbf{e}_{ij}\}$, are orthonormal, the symmetric base tensors, $\{\mathbf{e}^{s}_{ij}\}$, are only orthogonal, i.e., they are not unimodular,

$$\mathbf{e}^{s}_{ij} : \mathbf{e}^{s}_{kl} = \frac{1}{2}(\delta_{ik}\, \delta_{jl} + \delta_{il}\, \delta_{jk}); \tag{15.5.14}$$

for example, $\mathbf{e}^{s}_{11} : \mathbf{e}^{s}_{11}$ is 1 but $\mathbf{e}^{s}_{12} : \mathbf{e}^{s}_{12}$ is 1/2.

Now consider the set of all second-order symmetric tensors, $T^{(2s)}$, and the set of all fourth-order symmetric tensors, $T^{(4s)}$, which map second-order symmetric tensors to second-order symmetric tensors. In terms of $\mathbf{e}^{s}_{ij}$, these two sets are

$$T^{(2s)} \equiv \{\mathbf{S} \,|\, \mathbf{S} = \mathrm{S}_{ij}\, \mathbf{e}^{s}_{ij},\ \mathrm{S}_{ij} = \mathrm{S}_{ji}\},$$

$$T^{(4s)} \equiv \{\mathbf{T} \,|\, \mathbf{T} = \mathrm{T}_{ijkl}\, \mathbf{e}^{s}_{ij} \otimes \mathbf{e}^{s}_{kl},\ \mathrm{T}_{ijkl} = \mathrm{T}_{jikl} = \mathrm{T}_{ijlk}\}. \tag{15.5.15a,b}$$

---

[2] Denoting $\mathbf{e}'_i \otimes \mathbf{e}'_j$ by $\mathbf{e}'_{ij}$, one can define a fourth-order tensor $\mathbf{R} \equiv \mathbf{e}'_{ij} \otimes \mathbf{e}_{ij}$, which satisfies $\mathbf{R}^{\mathrm{T}} : \mathbf{R} = \mathbf{1}^{(4)}$. Tensor $\mathbf{R}$ transforms a second-order tensor $\mathbf{S}$ (with components S_{ij} in the $\mathbf{e}_i$-bases) to another second-order tensor whose components in the $\mathbf{e}'_i$-bases are also S_{ij}; compare fourth-order tensor $\mathbf{R}$ with second-order tensor $\mathbf{Q}$ of Subsection 15.3.

Note that, while $\mathbf{S} = \mathbf{S}^T$, in general $\mathbf{T}$ need not equal $\mathbf{T}^T$.

Associated with the second-order base tensors there are three (nonzero) *second-order antisymmetric base tensors* defined by

$$\mathbf{e}_{ij}^{a} \equiv \frac{1}{2}(\mathbf{e}_{ij} - \mathbf{e}_{ij}^{T}) = \frac{1}{2}(\mathbf{e}_i \otimes \mathbf{e}_j - \mathbf{e}_j \otimes \mathbf{e}_i). \tag{15.5.16}$$

The double contraction of $\mathbf{e}_{ij}^{s}$ and $\mathbf{e}_{kl}^{a}$ satisfies $\mathbf{e}_{ij}^{s} : \mathbf{e}_{kl}^{a} = \mathbf{e}_{ij}^{a} : \mathbf{e}_{kl}^{s} = 0$, for all i, j, k, l. Therefore, a fourth-order tensor $\mathbf{T}$ in $T^{(4s)}$ maps $\mathbf{e}_{ij}^{a}$ to $\mathbf{0}$, i.e.,

$$\mathbf{T} : \mathbf{e}_{ij}^{a} = \mathbf{0}, \qquad \text{i, j} = 1, 2, 3. \tag{15.5.17}$$

Hence, the mapping of $T^{(2)}$ to $T^{(2)}$ by an element $\mathbf{T}$ in $T^{(4s)}$ is not unique. The nine by nine matrix of $\mathbf{T}$ in $T^{(4s)}$, i.e., $[T_{\{ij\}\{kl\}}]$, is always singular.

15.5.4. Matrix Operations for Second- and Fourth-Order Symmetric Tensors

Since the second-order symmetric base tensors $\{\mathbf{e}_{ij}^{s}\}$ and their tensor product $\{\mathbf{e}_{ij}^{s} \otimes \mathbf{e}_{kl}^{s}\}$ span $T^{(2s)}$ and $T^{(4s)}$, respectively, tensorial operations between $T^{(2s)}$ and $T^{(4s)}$ can be related to *six-dimensional matrix operations.* For simplicity, denote the six second-order symmetric base tensors $\{\mathbf{e}_{ij}^{s}\}$ by $\{\mathbf{b}_a\}$ (a = 1, 2, ..., 6), as follows:

$$\mathbf{b}_1 = \mathbf{e}_{11}^{s}, \qquad \mathbf{b}_2 = \mathbf{e}_{22}^{s}, \qquad \mathbf{b}_3 = \mathbf{e}_{33}^{s},$$

$$\mathbf{b}_4 = \mathbf{e}_{23}^{s}, \qquad \mathbf{b}_5 = \mathbf{e}_{31}^{s}, \qquad \mathbf{b}_6 = \mathbf{e}_{12}^{s}. \tag{15.5.18a~f}$$

The double contraction of elements of $\{\mathbf{e}_{ij}^{s}\}$ is then reduced to the dot product of elements of $\{\mathbf{b}_a\}$. From (15.5.14), $\mathbf{b}_a$ is an orthogonal set. It is convenient to define a six by six matrix $[W] = [W_{ab}]$ by

$$[W] = \begin{bmatrix} 1\,0\,0\,0\,0\,0 \\ 0\,1\,0\,0\,0\,0 \\ 0\,0\,1\,0\,0\,0 \\ 0\,0\,0\,2\,0\,0 \\ 0\,0\,0\,0\,2\,0 \\ 0\,0\,0\,0\,0\,2 \end{bmatrix}, \tag{15.5.19}$$

and express the dot product, $\mathbf{b}_a \bullet \mathbf{b}_b$ (or the corresponding $\mathbf{e}_{ij}^{s} \bullet \mathbf{e}_{kl}^{s}$), in terms of the elements W_{ab} of [W], as

$$\mathbf{b}_a \bullet \mathbf{b}_b = \begin{cases} 1/W_{ab} & \text{if a = b} \\ 0 & \text{if a} \neq \text{b.} \end{cases} \tag{15.5.20a}$$

Hence,

$$W_{ab}\, \mathbf{b}_a \bullet \mathbf{b}_b = \delta_{ab} \qquad \text{(a, b not summed; a, b = 1, 2, ..., 6),} \tag{15.5.20b}$$

which shows the *weighted orthonormality* of base tensors $\mathbf{b}_a$ with the weighting factor W_{ab}. This then results in the corresponding weighted matrix multiplication with the weighting matrix $[W_{ab}]$, as shown below.

Let **S** and **T** be tensors in $T^{(2s)}$ and $T^{(4s)}$, respectively. Their components in a rectangular Cartesian coordinate system are used to express **S** and **T** in terms of $\{\mathbf{b}_a\}$, as

$$\mathbf{S} = S_{11}\,\mathbf{b}_1 + S_{22}\,\mathbf{b}_2 + S_{33}\,\mathbf{b}_3 + 2S_{23}\,\mathbf{b}_4 + 2S_{31}\,\mathbf{b}_5 + 2S_{12}\,\mathbf{b}_6,$$

$$\mathbf{T} = T_{1111}\,\mathbf{b}_1 \otimes \mathbf{b}_1 + T_{1122}\,\mathbf{b}_1 \otimes \mathbf{b}_2 + \ldots + 2T_{1131}\,\mathbf{b}_1 \otimes \mathbf{b}_5 + 2T_{1112}\,\mathbf{b}_1 \otimes \mathbf{b}_6$$

$$+ 2T_{1211}\,\mathbf{b}_6 \otimes \mathbf{b}_1 + 2T_{1222}\,\mathbf{b}_6 \otimes \mathbf{b}_2 + \ldots + 4T_{1231}\,\mathbf{b}_6 \otimes \mathbf{b}_5 + 4T_{1212}\,\mathbf{b}_6 \otimes \mathbf{b}_6. \tag{15.5.21a,b}$$

From these expressions, a six by one column matrix $[S_a]$ for **S** is defined as

$$[S_a] = \begin{bmatrix} S_{11} \\ S_{22} \\ S_{33} \\ S_{23} \\ S_{31} \\ S_{12} \end{bmatrix}, \tag{15.5.22a}$$

and a six by six matrix $[T_{ab}]$ for **T**, as

$$[T_{ab}] = \begin{bmatrix} T_{1111} & T_{1122} & T_{1133} & T_{1123} & T_{1131} & T_{1112} \\ T_{2211} & T_{2222} & T_{2233} & T_{2223} & T_{2231} & T_{2212} \\ T_{3311} & T_{3322} & T_{3333} & T_{3323} & T_{3331} & T_{3312} \\ T_{2311} & T_{2322} & T_{2333} & T_{2323} & T_{2331} & T_{2312} \\ T_{3111} & T_{3122} & T_{3133} & T_{3123} & T_{3131} & T_{3112} \\ T_{1211} & T_{1222} & T_{1233} & T_{1223} & T_{1231} & T_{1212} \end{bmatrix}. \tag{15.5.22b}$$

Then, with the aid of matrix $[W_{ab}]$, (15.5.19), the second-order contraction of **S** and **T** becomes

$$\mathbf{T} : \mathbf{S} \Longleftrightarrow [T_{ab}][W_{bc}][S_c]. \tag{15.5.23a}$$

Note the correspondence between the tensorial double contraction and the weighted matrix multiplication. This arises from the weighted orthonormality of the base tensors, $\mathbf{b}_a$; see (15.5.20b). Hence, if tensors $\mathbf{S}'$ and $\mathbf{T}'$ are in $T^{(2s)}$ and $T^{(4s)}$, respectively, then $\mathbf{S} : \mathbf{S}'$ and $\mathbf{T} : \mathbf{T}'$ become

$$\mathbf{S} : \mathbf{S}' \Longleftrightarrow [S_a][W_{ab}][S'_b], \qquad \mathbf{T} : \mathbf{T}' \Longleftrightarrow [T_{ab}][W_{bc}][T'_{cd}]. \tag{15.5.23b,c}$$

As is seen, the double contraction reduces to a weighted matrix multiplication, with the weighting factor $[W_{ab}]$.

From the fourth-order identity tensor, $\mathbf{1}^{(4)}$, the *fourth-order symmetric identity tensor,* $\mathbf{1}^{(4s)}$, is defined such that $\mathbf{1}^{(4s)}$ maps a second-order tensor in $T^{(2s)}$ to itself. This may be defined as

$$\mathbf{1}^{(4s)} \equiv \mathbf{e}^s_{ij} \otimes \mathbf{e}^s_{ij} \equiv \frac{1}{2}(\mathbf{e}_{ij} \otimes \mathbf{e}_{ij} + \mathbf{e}_{ij} \otimes \mathbf{e}_{ji}), \tag{15.5.24a}$$

with components, $1^{(4s)}_{ijkl}$, given by

$$1^{(4s)}_{ijkl} \equiv \frac{1}{2}(\delta_{ik}\,\delta_{jl} + \delta_{il}\,\delta_{jk}). \tag{15.5.24b}$$

Therefore, in terms of $\{\mathbf{b}_a\}$ tensor $\mathbf{1}^{(4s)}$ is expressed as

$$\mathbf{1}^{(4s)} = \mathbf{b}_1 \otimes \mathbf{b}_1 + \mathbf{b}_2 \otimes \mathbf{b}_2 + \mathbf{b}_3 \otimes \mathbf{b}_3 + 2\mathbf{b}_4 \otimes \mathbf{b}_4 + 2\mathbf{b}_5 \otimes \mathbf{b}_5 + 2\mathbf{b}_6 \otimes \mathbf{b}_6. \tag{15.5.25}$$

Hence, the corresponding six by six matrix, $[1^{(4s)}_{ab}]$, is the inverse of matrix $[W_{ab}]$,

$$[1^{(4s)}_{ab}] = [W_{ab}]^{-1} = \begin{bmatrix} 1 & 0 & 0 & 0 & 0 & 0 \\ 0 & 1 & 0 & 0 & 0 & 0 \\ 0 & 0 & 1 & 0 & 0 & 0 \\ 0 & 0 & 0 & 1/2 & 0 & 0 \\ 0 & 0 & 0 & 0 & 1/2 & 0 \\ 0 & 0 & 0 & 0 & 0 & 1/2 \end{bmatrix}, \tag{15.5.26}$$

which is *not* the six by six unit matrix; note that $\mathbf{1}^{(4s)}$ *is,* in fact, the *identity tensor,* in the sense that it maps any element in $T^{(4s)}$ to the same element.

As pointed out before, fourth-order tensors $\mathbf{T}$ in $T^{(4s)}$ are singular. However, for a $\mathbf{T}$ in $T^{(4s)}$, there may exist another element in $T^{(4s)}$ such that its second-order contraction with $\mathbf{T}$ yields the fourth-order symmetric identity tensor $\mathbf{1}^{(4s)}$. Then, this element is denoted by $\mathbf{T}^{-1}$, and is called the *inverse* of $\mathbf{T}$ in $T^{(4s)}$, i.e.,

$$\mathbf{T}^{-1} : \mathbf{T} = \mathbf{T} : \mathbf{T}^{-1} = \mathbf{1}^{(4s)}. \tag{15.5.27}$$

From (15.5.27), the six by six matrix corresponding to $\mathbf{T}^{-1}$, i.e., $[\mathrm{T}^{-1}_{ab}]$, must satisfy

$$[\mathrm{T}^{-1}_{ap}][W_{pq}][\mathrm{T}_{qb}] = [\mathrm{T}_{ap}][W_{pq}][\mathrm{T}^{-1}_{qb}] = [W_{ab}]^{-1}. \tag{15.5.28a}$$

Multiply (15.5.28a) by $[W_{ab}]$ either from the left or right, and using (15.5.26) obtain

$$([W_{ap}][\mathrm{T}^{-1}_{pq}][W_{qb}])\,[\mathrm{T}_{bc}] = [1_{ac}], \quad \text{or} \quad [\mathrm{T}_{ab}]\,([W_{bp}][\mathrm{T}^{-1}_{pq}][W_{qc}]) = [1_{ac}], \tag{15.5.28b,c}$$

where $[1_{ab}]$ is the *six by six identity* matrix. From (15.5.26),

$$[1_{ab}] = [W_{ap}][1^{(4s)}_{pb}]\,(= [1^{(4s)}_{ap}][W_{pb}]). \tag{15.5.29}$$

As is seen, matrix $[W_{ap}][\mathrm{T}^{-1}_{pq}][W_{qb}]$ is the inverse of $[\mathrm{T}_{ab}]$, and, similarly, matrix $[W_{ap}][\mathrm{T}_{pq}][W_{qb}]$ is the inverse of[3] $[\mathrm{T}^{-1}_{ab}]$.

The second-order strain and stress tensors, $\boldsymbol{\varepsilon}$ and $\boldsymbol{\sigma}$, are both symmetric and belong to $T^{(2s)}$. The corresponding six by one column matrices, $[\varepsilon_a]$ and $[\sigma_a]$, are given by (15.5.22a), if the letter S is replaced by ε and σ, respectively. The fourth-order elasticity and compliance tensors, $\mathbf{C}$ and $\mathbf{D}$, respectively, map $\boldsymbol{\varepsilon}$ to $\boldsymbol{\sigma}$ and $\boldsymbol{\sigma}$ to $\boldsymbol{\varepsilon}$, and hence belong to $T^{(4s)}$. The corresponding six by six matrices, $[C_{ab}]$ and $[D_{ab}]$, are given by (15.5.22b), if the character T is replaced

[3] Attention should be paid to the fact that in matrix form, $\mathbf{T}^{-1}$, the inverse tensor of $\mathbf{T}$, is expressed as $[\mathrm{T}^{-1}_{ab}] = [W_{ap}]^{-1}[\mathrm{T}_{pq}]^{-1}[W_{qb}]^{-1} = [1^{(4s)}_{ap}][\mathrm{T}_{pq}]^{-1}[1^{(4s)}_{bp}]$ with $[\mathrm{T}_{ab}]^{-1}$ being the inverse matrix of $[\mathrm{T}_{ab}]$.

by C and D, respectively. In Section 3, the engineering strain and stress are introduced by six by one column matrices, $[\gamma_a]$ and $[\tau_a]$, which are expressed in terms of $[\varepsilon_a]$ and $[\sigma_a]$, as

$$[\gamma_a] \equiv [W_{ab}][\varepsilon_b], \qquad [\tau_a] \equiv [\sigma_a]. \tag{15.5.30a,b}$$

Then, in terms of the elasticity and compliance tensors, **C** and **D**, the matrices which relate $[\gamma_a]$ and $[\tau_a]$ are defined by[4]

$$[C_{ab}] \equiv [\mathrm{C}_{ab}], \qquad [D_{ab}] \equiv [W_{ap}][\mathrm{D}_{pq}][W_{qb}], \tag{15.5.30c,d}$$

where $[\tau_a] = [C_{ab}][\gamma_b]$ and $[\gamma_a] = [D_{ab}][\tau_b]$.

From (15.5.23a~c) and (15.5.30a~d), the following relations between the tensor and matrix representations of $\boldsymbol{\varepsilon}$, $\boldsymbol{\sigma}$, **C**, and **D** are obtained:

$$\boldsymbol{\sigma} = \mathbf{C} : \boldsymbol{\varepsilon} \Longleftrightarrow [\sigma_a] = [\mathrm{C}_{ap}][W_{pb}][\varepsilon_b] \Longleftrightarrow [\tau_a] = [C_{ab}][\gamma_b],$$

$$\boldsymbol{\varepsilon} = \mathbf{D} : \boldsymbol{\sigma} \Longleftrightarrow [\varepsilon_a] = [\mathrm{D}_{ap}][W_{pb}][\sigma_b] \Longleftrightarrow [\gamma_a] = [D_{ab}][\tau_b]; \tag{15.5.31a,b}$$

and, the inverse relations between **C** and **D** are,

$$\mathbf{C} : \mathbf{D} = \mathbf{1}^{(4s)} \Longleftrightarrow [\mathrm{C}_{ap}][W_{pq}][\mathrm{D}_{qb}] = [1_{ab}^{(4s)}] \Longleftrightarrow [C_{ac}][D_{cb}] = [1_{ab}],$$

$$\mathbf{D} : \mathbf{C} = \mathbf{1}^{(4s)} \Longleftrightarrow [\mathrm{D}_{ap}][W_{pq}][\mathrm{C}_{qb}] = [1_{ab}^{(4s)}] \Longleftrightarrow [D_{ac}][C_{cb}] = [1_{ab}]. \tag{15.5.31c,d}$$

The matrix notation used in Part 1 follows the above definitions for $[\gamma_a]$, $[\tau_a]$, $[C_{ab}]$, and $[D_{ab}]$.

The tensor **J** is defined in Section 4, such that the cavity strain $\boldsymbol{\varepsilon}^c$ is determined, for a prescribed overall strain $\boldsymbol{\varepsilon}^o$, by $\boldsymbol{\varepsilon}^c = \mathbf{J} : \boldsymbol{\varepsilon}^o$. This tensorial equation has the following matrix representation:

$$[\varepsilon_a^c] = [\mathrm{J}_{ap}][W_{pb}][\varepsilon_b^o], \tag{15.5.32}$$

where the matrix $[\mathrm{J}_{ab}]$ is given by (15.5.22b), if the letter T is replaced by J. Therefore, in matrix form,

$$[\gamma_a^c] = [W_{ab}][\varepsilon_b^c] = [W_{ab}]\,([\mathrm{J}_{bc}][W_{cd}][\varepsilon_d^o])$$

$$= ([W_{ab}][\mathrm{J}_{bc}])\,([W_{cd}][\varepsilon_d^o]) = [J_{ab}][\gamma_b^o], \tag{15.5.33a}$$

where matrix $[J_{ab}]$ is

$$[J_{ab}] = [W_{ap}][\mathrm{J}_{pb}]. \tag{15.5.33b}$$

Hence, the matrix representation of tensor **J** may not result in a symmetric matrix, even if matrix $[\mathrm{J}_{ab}]$ (and hence tensor **J**) is symmetric. Note that the fourth-order symmetric identity tensor $\mathbf{1}^{(4s)}$ and the corresponding six by six identity matrix $[1_{ab}]$ are also related in the same manner as are **J** and $[J_{ab}]$; see (15.5.29).

---

[4] In this notation, the matrix formed from the components of the tensor **D** is denoted by $[\mathrm{D}_{ab}]$, and the inverse of $[C_{ab}]$ is denoted by $[D_{ab}]$.

15.6. SPECTRAL REPRESENTATION OF FOURTH-ORDER SYMMETRIC TENSORS

A second-order symmetric tensor, say, $\boldsymbol{\varepsilon}$, can be expressed in terms of its three principal values ε^α and the corresponding principal directions, $\mathbf{N}^\alpha$, as (see Subsection 16.2)

$$\boldsymbol{\varepsilon} = \sum_{\alpha=1}^{3} \varepsilon^\alpha \, \mathbf{N}^\alpha \otimes \mathbf{N}^\alpha. \tag{15.6.1a}$$

The principal directions are mutually orthogonal, forming an orthogonal triad,

$$\mathbf{N}^\alpha \boldsymbol{\cdot} \mathbf{N}^\beta = \delta_{\alpha\beta}. \tag{15.6.1b}$$

Hence, for a vector $\mathbf{V}$,

$$\boldsymbol{\varepsilon} \boldsymbol{\cdot} \mathbf{V} = \sum_{\alpha=1}^{3} \varepsilon^\alpha \, (\mathbf{V} \boldsymbol{\cdot} \mathbf{N}^\alpha) \, \mathbf{N}^\alpha. \tag{15.6.1c}$$

The principal values and directions of $\boldsymbol{\varepsilon}$ are the same as the eigenvalues and eigenvectors of the corresponding three by three matrix $[\varepsilon_{ab}]$.

In a manner similar to (15.6.1), a fourth-order symmetric tensor, say, $\mathbf{C} = \mathbf{C}^T$, can be expressed in terms of six principal values C^I, and the corresponding principal second-order symmetric tensors $\mathbf{E}^I$, as follows:

$$\mathbf{C} = \sum_{I=1}^{6} C^I \, \mathbf{E}^I \otimes \mathbf{E}^I, \tag{15.6.2a}$$

where the principal second-order symmetric tensors are orthonormal in the sense that

$$\mathbf{E}^I : \mathbf{E}^J = \delta_{IJ}. \tag{15.6.2b}$$

Hence, for an arbitrary second-order symmetric tensor $\boldsymbol{\varepsilon}$,

$$\mathbf{C} : \boldsymbol{\varepsilon} = \sum_{I=1}^{6} C^I \, (\boldsymbol{\varepsilon} : \mathbf{E}^I) \, \mathbf{E}^I. \tag{15.6.2c}$$

The principal values and principal second-order symmetric tensors of $\mathbf{C}$ are *related* to the eigenvalues and eigenvectors of the corresponding six by six matrix $[C_{ab}]$, but unlike in the case of a second-order symmetric tensor, they are not the same. The spectral decomposition of a fourth-order symmetric tensor (15.6.2a) is examined in this subsection.

In Section 3 and above, it is pointed out that a fourth-order tensor $\mathbf{C}$ satisfying $C_{ijkl} = C_{jikl} = C_{ijlk} = C_{klij}$, and a second-order symmetric tensor $\boldsymbol{\varepsilon}$ satisfying $\varepsilon_{ij} = \varepsilon_{ji}$, can be expressed in terms of a six by six square matrix $[C_{ab}]$, and a six by one column matrix $[\varepsilon_a]$, respectively. Their second-order contraction $\mathbf{C} : \boldsymbol{\varepsilon}$ then is given by a weighted matrix product involving the six by six square matrix $[W_{ab}]$ defined by (3.1.6c), i.e.,

$$\boldsymbol{\sigma} = \mathbf{C} : \boldsymbol{\varepsilon} \Longleftrightarrow [\sigma_a] = [C_{ab}][W_{bc}][\varepsilon_c], \tag{15.6.3}$$

where the matrix components, σ_a, ε_a, and C_{ab}, are the same as the corresponding tensor components, σ_{ij}, ε_{ij}, and C_{ijkl}; see (3.1.3a) and (3.1.3c). Therefore, there exists the following correspondence between tensor and matrix operations associated with the principal values and the principal second-order symmetric tensors of **C**:

$$\mathrm{C}^{\mathrm{I}}\,\mathbf{E}^{\mathrm{I}} = \mathbf{C} : \mathbf{E}^{\mathrm{I}} \Longleftrightarrow \mathrm{C}^{\mathrm{I}}\,[\mathrm{E}^{\mathrm{I}}_{\mathrm{a}}] = [\mathrm{C}_{\mathrm{ab}}][W_{\mathrm{bc}}]\,[\mathrm{E}^{\mathrm{I}}_{\mathrm{c}}] \qquad \text{(I not summed)}, \tag{15.6.4a}$$

and

$$\mathbf{E}^{\mathrm{I}} : \mathbf{E}^{\mathrm{J}} = \delta_{\mathrm{IJ}} \Longleftrightarrow [\mathrm{E}^{\mathrm{I}}_{\mathrm{a}}][W_{\mathrm{ab}}][\mathrm{E}^{\mathrm{J}}_{\mathrm{b}}] = \delta_{\mathrm{IJ}}. \tag{15.6.4b}$$

Equation (15.6.4b) shows the orthonormality of the principal (second-order symmetric) tensors of **C**.

Since $[W_{\mathrm{ab}}]$ is a positive diagonal matrix, its square root is given by a diagonal matrix consisting of the square roots of the elements of $[W_{\mathrm{ab}}]$, i.e., $[\sqrt{W_{\mathrm{ab}}}]$. Then, the matrix relation in (15.6.4a) may be expressed as

$$\mathrm{C}^{\mathrm{I}}\,([\sqrt{W_{\mathrm{ap}}}][\mathrm{E}^{\mathrm{I}}_{\mathrm{p}}]) = \{\,[\sqrt{W_{\mathrm{ap}}}][\mathrm{C}_{\mathrm{pq}}][\sqrt{W_{\mathrm{qb}}}]\}\,([\sqrt{W_{\mathrm{br}}}][\mathrm{E}^{\mathrm{I}}_{\mathrm{r}}]), \tag{15.6.5}$$

where I is not summed. Since the **C**-tensor is symmetric with respect to the first and last pairs of its indices, $\mathrm{C}_{\mathrm{ijkl}} = \mathrm{C}_{\mathrm{klij}}$, the corresponding $[\mathrm{C}_{\mathrm{ab}}]$-matrix is symmetric, $\mathrm{C}_{\mathrm{ab}} = \mathrm{C}_{\mathrm{ba}}$. Defining a six by six symmetric matrix $[\hat{\mathrm{C}}_{\mathrm{ab}}]$,

$$[\hat{\mathrm{C}}_{\mathrm{ab}}] \equiv [\sqrt{W_{\mathrm{ap}}}][\mathrm{C}_{\mathrm{pq}}][\sqrt{W_{\mathrm{qb}}}], \tag{15.6.6a}$$

compute the (real-valued) eigenvalues and eigenvectors from $[\hat{\mathrm{C}}_{\mathrm{ab}}][\hat{\mathrm{E}}_{\mathrm{b}}] = \mathrm{C}[\hat{\mathrm{E}}_{\mathrm{a}}]$. There are, in general, six pairs of $\{\mathrm{C}^{\mathrm{I}}, [\hat{\mathrm{E}}^{\mathrm{I}}_{\mathrm{a}}]\}$ (I = 1, 2, ..., 6), for this matrix, although the eigenvalues may not be distinct. The $[\hat{\mathrm{C}}_{\mathrm{ab}}]$-matrix and the orthonormality of $[\hat{\mathrm{E}}_{\mathrm{a}}]$'s may be expressed by

$$[\hat{\mathrm{C}}_{\mathrm{ab}}] = \sum_{\mathrm{I}=1}^{6} \mathrm{C}^{\mathrm{I}}\,[\hat{\mathrm{E}}^{\mathrm{I}}_{\mathrm{a}}][\hat{\mathrm{E}}^{\mathrm{I}}_{\mathrm{b}}]^{\mathrm{T}}, \qquad [\hat{\mathrm{E}}^{\mathrm{I}}_{\mathrm{a}}]\,[\hat{\mathrm{E}}^{\mathrm{J}}_{\mathrm{a}}] = \delta_{\mathrm{IJ}}. \tag{15.6.6b,c}$$

Note that if the eigenvalues are not distinct, the associated eigentensors in general, are not uniquely defined. Nevertheless, suitable orthonormal eigenvectors can always be chosen to complete this spectral representation. Therefore, if $[\mathrm{E}^{\mathrm{I}}_{\mathrm{a}}]$ is defined by

$$[\mathrm{E}^{\mathrm{I}}_{\mathrm{a}}] \equiv [\sqrt{W_{\mathrm{ab}}}][\hat{\mathrm{E}}_{\mathrm{b}}], \tag{15.6.7}$$

then the following matrix relation holds:

$$[\hat{\mathrm{C}}_{\mathrm{ab}}][\hat{\mathrm{E}}^{\mathrm{I}}_{\mathrm{b}}] = \mathrm{C}^{\mathrm{I}}\,[\hat{\mathrm{E}}^{\mathrm{I}}_{\mathrm{a}}] \Longleftrightarrow [\mathrm{C}_{\mathrm{ab}}][W_{\mathrm{bc}}][\mathrm{E}^{\mathrm{I}}_{\mathrm{c}}] = \mathrm{C}^{\mathrm{I}}\,[\mathrm{E}^{\mathrm{I}}_{\mathrm{a}}] \qquad \text{(I not summed)}. \tag{15.6.8}$$

From comparison of (15.6.4) and (15.6.8), it follows that the principal values of **C** are the eigenvalues of $[\hat{\mathrm{C}}_{\mathrm{ab}}]$, and the associated principal second-order symmetric tensors, $\mathbf{E}^{\mathrm{I}}$, are given by the associated eigentensors through (15.6.7). Hence, the following spectral representation of the **C**-tensor is obtained:

$$\mathbf{C} = \sum_{\mathrm{I}=1}^{6} \mathrm{C}^{\mathrm{I}}\,\mathbf{E}^{\mathrm{I}} \otimes \mathbf{E}^{\mathrm{I}}. \tag{15.6.9a}$$

Furthermore, since $\mathbf{E}^{\mathrm{I}}$'s are second-order symmetric tensors, they admit their own spectral decomposition similar to (15.6.1a). Denoting the three principal values and directions of $\mathbf{E}^{\mathrm{I}}$ by $E^{\mathrm{I}\alpha}$ and $\mathbf{N}^{\mathrm{I}\alpha}$ for $\alpha = 1, 2, 3$, observe that

$$\mathbf{C} = \sum_{\mathrm{I}=1}^{6} \sum_{\alpha=1}^{3} C^{\mathrm{I}} E^{\mathrm{I}\alpha} \mathbf{N}^{\mathrm{I}\alpha} \otimes \mathbf{N}^{\mathrm{I}\alpha} \otimes \mathbf{N}^{\mathrm{I}\alpha} \otimes \mathbf{N}^{\mathrm{I}\alpha}. \tag{15.6.9b}$$

Since the elastic strain energy density e is given by

$$e = \frac{1}{2} \boldsymbol{\varepsilon} : \mathbf{C} : \boldsymbol{\varepsilon} = \frac{1}{2} \sum_{\mathrm{I}=1}^{6} C^{\mathrm{I}} (\mathbf{E}^{\mathrm{I}} : \boldsymbol{\varepsilon})^2, \tag{15.6.10}$$

it is concluded that $\mathbf{C}$ *is positive-definite, if and only if* C^{I} *is positive for all* I, *leading to a non-negative elastic strain energy density.* Furthermore, from the normality of the eigentensors, it follows that the elastic energy associated with any eigentensor (mode) is decoupled from that associated with each of the remaining eigentensors. These issues have been examined by Mehrabadi and Cowin (1990) using a different approach;[5] see also Kelvin (1856).

As an example, consider the isotropic tensor $\mathbf{C} = \lambda \mathbf{1}^{(2)} \otimes \mathbf{1}^{(2)} + 2\mu \mathbf{1}^{(4s)}$. The associated $[\hat{C}_{ab}]$-matrix is

$$[\hat{C}_{ab}] = \begin{bmatrix} \lambda+2\mu & \lambda & \lambda & 0 & 0 & 0 \\ \lambda & \lambda+2\mu & \lambda & 0 & 0 & 0 \\ \lambda & \lambda & \lambda+2\mu & 0 & 0 & 0 \\ 0 & 0 & 0 & 2\mu & 0 & 0 \\ 0 & 0 & 0 & 0 & 2\mu & 0 \\ 0 & 0 & 0 & 0 & 0 & 2\mu \end{bmatrix}, \tag{15.6.11}$$

and the six pairs of eigenvalues and eigentensors are

$$(3\lambda+2\mu, \frac{1}{\sqrt{3}} \begin{bmatrix} 1 \\ 1 \\ 1 \\ 0 \\ 0 \\ 0 \end{bmatrix}), \quad (2\mu, \frac{1}{\sqrt{2}} \begin{bmatrix} 1 \\ -1 \\ 0 \\ 0 \\ 0 \\ 0 \end{bmatrix}), \quad (2\mu, \frac{1}{\sqrt{2}} \begin{bmatrix} 0 \\ 1 \\ -1 \\ 0 \\ 0 \\ 0 \end{bmatrix}),$$

$$\left(\text{or } (2\mu, \frac{1}{\sqrt{2}} \begin{bmatrix} -1 \\ 0 \\ 1 \\ 0 \\ 0 \\ 0 \end{bmatrix}) \right),$$

$$(2\mu, \begin{bmatrix} 0 \\ 0 \\ 0 \\ 1 \\ 0 \\ 0 \end{bmatrix}), \quad (2\mu, \begin{bmatrix} 0 \\ 0 \\ 0 \\ 0 \\ 1 \\ 0 \end{bmatrix}), \quad (2\mu, \begin{bmatrix} 0 \\ 0 \\ 0 \\ 0 \\ 0 \\ 1 \end{bmatrix}). \tag{15.6.12a~f}$$

Note that the $[1, 1, 1, 0, 0, 0]^{\mathrm{T}}$-matrix corresponds to dilatation, and the

[5] The material presented in Subsections 15.5 and 15.6 was completed in 1988 and used for class instruction; Nemat-Nasser and Hori (1989).

$[1, -1, 0, 0, 0, 0]^T$- and $[0, 0, 0, 1, 0, 0]^T$-matrices correspond to shearing. Hence, the isotropic tensor $\mathbf{C}$ may be expressed by

$$\mathbf{C} = \sqrt{3}K\,(\mathbf{e}_1\otimes\mathbf{e}_1 + \mathbf{e}_2\otimes\mathbf{e}_2 + \mathbf{e}_3\otimes\mathbf{e}_3)\otimes(\mathbf{e}_1\otimes\mathbf{e}_1 + \mathbf{e}_2\otimes\mathbf{e}_2 + \mathbf{e}_3\otimes\mathbf{e}_3)$$
$$+ \sqrt{2}\mu \Big[(\mathbf{e}_1\otimes\mathbf{e}_1 - \mathbf{e}_2\otimes\mathbf{e}_2)\otimes(\mathbf{e}_1\otimes\mathbf{e}_1 - \mathbf{e}_2\otimes\mathbf{e}_2)$$
$$+ (\mathbf{e}_2\otimes\mathbf{e}_2 - \mathbf{e}_3\otimes\mathbf{e}_3)\otimes(\mathbf{e}_2\otimes\mathbf{e}_2 - \mathbf{e}_3\otimes\mathbf{e}_3)$$
$$+ (\mathbf{e}_2\otimes\mathbf{e}_3 + \mathbf{e}_3\otimes\mathbf{e}_2)\otimes(\mathbf{e}_2\otimes\mathbf{e}_3 + \mathbf{e}_3\otimes\mathbf{e}_2)$$
$$+ (\mathbf{e}_3\otimes\mathbf{e}_1 + \mathbf{e}_1\otimes\mathbf{e}_3)\otimes(\mathbf{e}_3\otimes\mathbf{e}_1 + \mathbf{e}_1\otimes\mathbf{e}_3)$$
$$+ (\mathbf{e}_1\otimes\mathbf{e}_2 + \mathbf{e}_2\otimes\mathbf{e}_1)\otimes(\mathbf{e}_1\otimes\mathbf{e}_2 + \mathbf{e}_2\otimes\mathbf{e}_1) \Big], \tag{15.6.13}$$

where K is the bulk modulus given by $(3\lambda + 2\mu)/3$.

15.7. CYLINDRICAL AND SPHERICAL COORDINATES

The treatment of a special problem may often be simplified by the use of a suitable coordinate system that reflects the particular symmetry of the problem. For problems with cylindrical or spherical symmetry, it is appropriate to employ cylindrical or spherical coordinates.

First, consider the cylindrical coordinates; see Figure 15.7.1. At a generic point P, consider a right-handed triad defined by the unit base vectors $\mathbf{e}_r$, $\mathbf{e}_\theta$, and $\mathbf{e}_3$, which, respectively, are in the radial, circumferential, and axial directions.

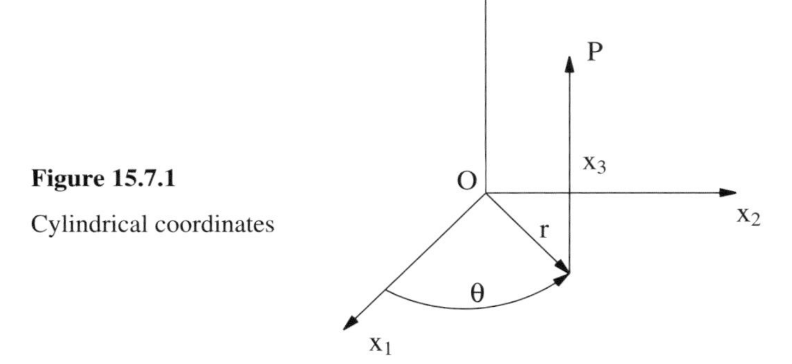

Figure 15.7.1

Cylindrical coordinates

From

$$\mathbf{e}_r = \mathbf{e}_1 \cos\theta + \mathbf{e}_2 \sin\theta, \qquad \mathbf{e}_\theta = -\,\mathbf{e}_1 \sin\theta + \mathbf{e}_2 \cos\theta, \tag{5.7.1a,b}$$

it follows that

$$\frac{\partial \mathbf{e}_r}{\partial r} = \frac{\partial \mathbf{e}_r}{\partial x_3} = \mathbf{0}, \qquad \frac{\partial \mathbf{e}_r}{\partial \theta} = \mathbf{e}_\theta, \qquad \frac{\partial \mathbf{e}_r}{\partial r} = \frac{\partial \mathbf{e}_\theta}{\partial x_3} = \mathbf{0}, \qquad \frac{\partial \mathbf{e}_\theta}{\partial \theta} = -\,\mathbf{e}_r. \tag{15.7.2a~d}$$

At a point P with coordinates r, θ, and x_3, the coordinate curves are: a straight line normal to the x_3-axis (θ = constant, x_3 = constant), a circle parallel to the x_1,x_2-plane (r = constant, x_3 = constant), and a straight line parallel to the x_3-axis (r = constant, θ = constant). A neighboring point P′ with coordinates $r + dp$, $\theta + d\theta$, and $x_3 + dx_3$, is at a distance ds from P, where

$$ds^2 = dr^2 + r^2 d\theta^2 + dx_3^2. \tag{15.7.3}$$

The vector operator $\boldsymbol{\nabla}$ is given by

$$\begin{aligned} \boldsymbol{\nabla} &\equiv \mathbf{e}_i \frac{\partial}{\partial x_i} \\ &\equiv (\mathbf{e}_i \frac{\partial r}{\partial x_i}) \frac{\partial}{\partial r} + (\mathbf{e}_i \frac{\partial \theta}{\partial x_i}) \frac{\partial}{\partial \theta} + \mathbf{e}_3 \frac{\partial}{\partial x_3} \\ &\equiv \mathbf{e}_r \frac{\partial}{\partial r} + \mathbf{e}_\theta \frac{\partial}{r\partial \theta} + \mathbf{e}_3 \frac{\partial}{\partial x_3}, \end{aligned} \tag{15.7.4}$$

where, in addition to (15.7.1), the relations $r = (x_1^2 + x_2^2)^{1/2}$ and $\theta = \tan^{-1}(x_2/x_1)$ are used.

Now, let $\mathbf{u}(r, \theta, x_3) = u_r \mathbf{e}_r + u_\theta \mathbf{e}_\theta + u_3 \mathbf{e}_3$ be the displacement field of the solid. Using (15.7.2) and (15.7.4), obtain the displacement gradient

$$\begin{aligned} \boldsymbol{\nabla} \otimes \mathbf{u} = {} & \mathbf{e}_r \otimes \mathbf{e}_r \frac{\partial u_r}{\partial r} + \mathbf{e}_r \otimes \mathbf{e}_\theta \frac{\partial u_\theta}{\partial r} + \mathbf{e}_r \otimes \mathbf{e}_3 \frac{\partial u_3}{\partial r} \\ & + \mathbf{e}_\theta \otimes \mathbf{e}_r \frac{1}{r}(\frac{\partial u_r}{\partial \theta} - u_\theta) + \mathbf{e}_\theta \otimes \mathbf{e}_\theta \frac{1}{r}(\frac{\partial u_\theta}{\partial \theta} + u_r) + \mathbf{e}_\theta \otimes \mathbf{e}_3 \frac{1}{r}\frac{\partial u_3}{\partial \theta} \\ & + \mathbf{e}_3 \otimes \mathbf{e}_r \frac{\partial u_r}{\partial x_e} + \mathbf{e}_3 \otimes \mathbf{e}_\theta \frac{\partial u_\theta}{\partial x_3} + \mathbf{e}_3 \otimes \mathbf{e}_3 \frac{\partial u_3}{\partial x_3} \qquad \text{(r, } \theta \text{ are not summed).} \end{aligned} \tag{15.7.5}$$

From this, *div* $\mathbf{u} = \boldsymbol{\nabla} \cdot \mathbf{u}$ and *curl* $\mathbf{u} = \boldsymbol{\nabla} \times \mathbf{u}$ are readily calculated. For example, *div* $\mathbf{u}$ is

$$\boldsymbol{\nabla} \cdot \mathbf{u} = \frac{\partial u_r}{\partial r} + \frac{1}{r}(\frac{\partial u_\theta}{\partial \theta} + u_r) + \frac{\partial u_3}{\partial x_3}. \tag{15.7.6}$$

Since the strain tensor is defined by the symmetric part of the displacement gradient, the components of this tensor may be written down from (15.7.5); they are

$$\varepsilon_{rr} = \frac{\partial u_r}{\partial r}, \qquad \varepsilon_{\theta r} = \varepsilon_{\theta r} = \frac{1}{2}\{\frac{1}{r}\frac{\partial u_r}{\partial \theta} + \frac{\partial u_\theta}{\partial r} - \frac{u_\theta}{r}\},$$
$$\varepsilon_{\theta\theta} = \frac{1}{r}(\frac{\partial u_\theta}{\partial \theta} + u_r), \qquad \varepsilon_{r3} = \varepsilon_{3r} = \frac{1}{2}\{\frac{\partial u_r}{\partial x_3} + \frac{\partial u_3}{\partial_r}\}, \tag{15.7.7a~f}$$
$$\varepsilon_{33} = \frac{\partial u_3}{\partial x_3}, \qquad \varepsilon_{\theta 3} = \varepsilon_{3\theta} = \frac{1}{2}\{\frac{\partial u_\theta}{\partial x_3} + \frac{1}{r}\frac{\partial u_3}{\partial \theta}\}.$$

Now, consider the conservation of linear momentum. In an invariant form, this equation reads

$$\nabla \cdot \boldsymbol{\sigma} + \rho_0 \mathbf{f} = \rho_0 \ddot{\mathbf{u}}, \tag{15.7.8}$$

where ρ_0 is the mass-density of the solid. The stress tensor $\boldsymbol{\sigma}$ in cylindrical coordinates is expressed as follows:

$$\boldsymbol{\sigma} = \sigma_{AB}\, \mathbf{e}_A \otimes \mathbf{e}_B \qquad (A, B = r, \theta, 3), \tag{15.7.9}$$

where the upper case Roman subscripts have the range, r, θ, and 3, and the summation convention applies to these subscripts. Using (15.7.2) and (15.7.4), obtain from (15.7.8)

$$\frac{\partial \sigma_{rr}}{\partial r} + \frac{1}{r}\frac{\partial \sigma_{\theta r}}{\partial \theta} + \frac{\partial \sigma_{3r}}{\partial x_3} + \frac{1}{r}(\sigma_{rr} - \sigma_{\theta\theta}) + \rho_0 f_r = \rho_0 \ddot{u}_r,$$

$$\frac{\partial \sigma_{r\theta}}{\partial r} + \frac{1}{r}\frac{\partial \sigma_{\theta\theta}}{\partial \theta} + \frac{\partial \sigma_{3\theta}}{\partial x_3} + \frac{2}{r}\sigma_{r\theta} + \rho_0 f_\theta = \rho_0 \ddot{u}_\theta,$$

$$\frac{\partial \sigma_{r3}}{\partial r} + \frac{1}{r}\frac{\partial \sigma_{\theta 3}}{\partial \theta} + \frac{\partial \sigma_{33}}{\partial x_3} + \frac{1}{r}\sigma_{r3} + \rho_0 f_3 = \rho_0 \ddot{u}_3. \tag{15.7.10a~c}$$

Note that, as a consequence of the conservation of angular momentum, the stress tensor is symmetric.

Finally, the generalized Hooke Law becomes

$$\sigma_{AB} = \lambda\varepsilon\delta_{AB} + 2\mu\varepsilon_{AB} \qquad (A, B = r, \theta, 3), \tag{15.7.11}$$

where $\delta_{AB} = 0$ if $A \neq B$, and $\delta_{AB} = 1$ if $A = B$.

Consider next the spherical coordinates, r, φ, θ; see Figure 15.7.2. The unit base vectors, in this case, are $\mathbf{e}_r$, $\mathbf{e}_\phi$, and $\mathbf{e}_\theta$, which respectively, define the radial, the meridional, and the circumferential directions. In terms of the rectangular Cartesian base vectors $\mathbf{e}_i$ (i = 1, 2, 3), the spherical base vectors are $\mathbf{e}_r = (\mathbf{e}_1 \cos\theta + \mathbf{e}_2 \sin\theta)\sin\phi + \mathbf{e}_3 \cos\phi$, $\mathbf{e}_\phi = (\mathbf{e}_1 \cos\theta + \mathbf{e}_2 \sin\theta)\cos\phi - \mathbf{e}_3 \sin\phi$, and $\mathbf{e}_\theta$ is the same as for the cylindrical coordinates. The element of length is

$$ds^2 = dr^2 + r^2 d\phi^2 + r^2 \sin^2\phi \, d\theta^2, \tag{15.7.12}$$

and the operator ∇ is given by

$$\nabla \equiv \mathbf{e}_i \frac{\partial}{\partial x_i}$$
$$\equiv (\mathbf{e}_i \frac{\partial r}{\partial x_i})\frac{\partial}{\partial r} + (\mathbf{e}_i \frac{\partial \phi}{\partial x_i})\frac{\partial}{\partial \phi} + (\mathbf{e}_i \frac{\partial \theta}{\partial x_i})\frac{\partial}{\partial \theta}$$

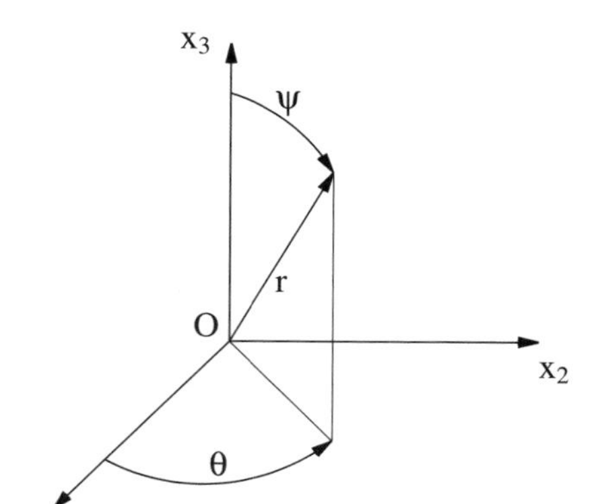

Figure 15.7.2

Spherical coordinates

$$\equiv \mathbf{e}_r \frac{\partial}{\partial r} + \mathbf{e}_\phi \frac{1}{r}\frac{\partial}{\partial \phi} + \mathbf{e}_\theta \frac{1}{r\sin\phi}\frac{\partial}{\partial \theta}. \tag{15.7.13}$$

Using the same line of reasoning as in the case of the cylindrical coordinates, obtain the following strain-components:

$$\varepsilon_{rr} = \frac{\partial u_r}{\partial r}, \qquad \varepsilon_{r\phi} = \frac{1}{2}\{\frac{1}{r}\frac{\partial u_r}{\partial \phi} + \frac{\partial u_\phi}{\partial r} - \frac{u_\phi}{r}\},$$

$$\varepsilon_{\phi\phi} = \frac{1}{r}\frac{\partial u_\phi}{\partial \phi} + \frac{u_r}{r}, \qquad \varepsilon_{r\theta} = \frac{1}{2}\{\frac{1}{r\sin\phi}\frac{\partial u_r}{\partial \theta} + \frac{\partial u_\theta}{\partial r} - \frac{u_\theta}{r}\},$$

$$\varepsilon_{\theta\theta} = \frac{1}{r\sin\phi}\frac{\partial u_\theta}{\partial \theta} + \frac{u_r}{r} + \frac{\cot\phi}{r}\,u_\phi,$$

$$\varepsilon_{\phi\theta} = \frac{1}{2}\{\frac{1}{r}\frac{\partial u_\theta}{\partial \phi} + \frac{1}{r\sin\phi}\frac{\partial u_\phi}{\partial \theta} - \frac{\cot\phi}{r}\,u_\theta\}. \tag{15.7.14a~f}$$

Similarly, the equations of motion become

$$\frac{\partial \sigma_{rr}}{\partial r} + \frac{1}{r}\frac{\partial \sigma_{\phi r}}{\partial \phi} + \frac{1}{r\sin\phi}\frac{\partial \sigma_{\theta\phi}}{\partial \theta}$$

$$+ \frac{1}{r}\,(2\sigma_{rr} - \sigma_{\phi\phi} - \sigma_{\theta\theta} + \sigma_{r\phi}\cot\phi) + \rho_0 f_r = \rho_0 \ddot{u}_r,$$

$$\frac{\partial \sigma_{r\phi}}{\partial r} + \frac{1}{r}\frac{\partial \sigma_{\phi\phi}}{\partial \phi} + \frac{1}{r\sin\phi}\frac{\partial \sigma_{\theta\phi}}{\partial \theta}$$

$$+ \frac{1}{r}\,\{3\sigma_{r\phi} + (\sigma_{\phi\phi} - \sigma_{\theta\theta})\cot\phi\} + \rho_0 f_\phi = \rho_0 \ddot{u}_\phi,$$

$$\frac{\partial \sigma_{r\theta}}{\partial r} + \frac{1}{r}\frac{\partial \sigma_{\phi\theta}}{\partial \phi} + \frac{1}{r\sin\phi}\frac{\partial \sigma_{\theta\theta}}{\partial \theta}$$

$$+\frac{1}{r}(3\sigma_{r\theta}+2\sigma_{\phi\theta}-\cot\phi)+\rho_0 f_\theta=\rho_0\ddot{u}_\theta. \qquad (15.7.15a\text{~}c)$$

15.8. REFERENCES

Borisenko, A. I. and Tarapov I. E. (1968), *Vector and tensor analysis with applications*, Translated from Russian by R. A. Silverman, Prentice-Hall, Inc., Englewood Cliffs.

Coburn, N. (1960), *Vector and Tensor Analysis*, Macmillan Company, New York.

Halmos, P. R. (1958), *Finite-Dimensional Vector Spaces*, D. Van Nostrand Co., Inc., New Jersey.

Kellogg, O. D. (1953), *Foundations of potential theory*, Dover, New York.

Kelvin, Lord (1856), *Phil. Trans. R. Soc.*, Vol. 166, 481.

Mehrabadi, M. M. and Cowin, S. C. (1990), Eigentensors of linear anisotropic elastic materials, *Q J. Mech. Appl. Math.*, Vol. 43, Pt. 1.

Nemat-Nasser, S. and Hori, M. (1989), *Micromechanics: Overall properties of heterogeneous elastic solids*, Lecture Notes Initiated at UCSD.

Williamson, R. E., Crowell, R. H., and Trotter, H. F. (1972), *Calculus of Vector Functions*, Prentice-Hall, Inc., Englewood Cliffs.

SECTION 16 KINEMATIC FOUNDATIONS

In this section the deformation of certain solid bodies is studied without reference to the forces that cause the deformation. The molecular structure of matter is disregarded and it is assumed that matter is continuously distributed throughout the space occupied by the solid.[1]

16.1. DEFORMATION AND STRAIN MEASURES

In the mechanics of solids, material points or *particles* of a solid are distinguished from the spatial points or positions that they happen to occupy at a given instant. In the undeformed configuration C_o of the solid, let a typical particle P occupy the position with position vector $\mathbf{X}$ (the position $\mathbf{X}$ for short). When the solid has moved to a deformed configuration C, the typical particle P occupies a new position $\mathbf{x}$. The deformation is described in terms of either the initial position $\mathbf{X}$ or the final position $\mathbf{x}$ of a typical particle P as the independent variable, obtaining the so-called *Lagrangian* or *Eulerian* formulation of the motion. *As will be seen later, in the theory of infinitesimal deformations, the two formulations lead to identical results.* For immediate use, however, adopt the convention of designating the coordinates of the particles in their undeformed state by capital letters with capital subscripts, and denote the positions of the particles in the deformed state by lower case letters carrying lower case subscripts. Since all quantities will be referred to a fixed system of right-handed rectangular Cartesian coordinates, no confusion will arise if the tensorial quantities are identified by their components; for example the brief expression "tensor E_{IJ}" is used instead of the more precise but longer expression "tensor $\mathbf{E}$ with components E_{IJ}".

The mapping of particles from C_o to C is expressed by

$$\mathbf{x} = \boldsymbol{\phi}(\mathbf{X}) \quad \text{or} \quad x_i = \phi_i(X_1, X_2, X_3) \tag{16.1.1a,b}$$

which must be one-to-one and hence invertible, so that

$$\mathbf{X} = \boldsymbol{\phi}^{-1}(\mathbf{x}) \quad \text{or} \quad X_I = \phi_I^{-1}(x_1, x_2, x_3). \tag{16.1.2a,b}$$

Assume that the single-valued functions (16.1.1) and (16.1.2) are as many times differentiable as is required in the context in which they are considered. The existence of a single-valued, continuously differentiable inverse of (16.1.1) is

[1] For introduction to fundamentals of continuum mechanics, see, e.g., Truesdell and Toupin (1960), Prager (1961), Fung (1965), Jaunzemis (1967), Green and Zerna (1968), Malvern (1969), Gurtin (1972), Eringen and Suhubi (1975), Chadwick (1976), and Spencer (1980).

actually guaranteed if the functions $\phi_i(X_1, X_2, X_3)$ are once continuously differentiable and the following determinant,

$$J \equiv \det\left|\frac{\partial \phi_i}{\partial X_I}\right| \equiv \det|\phi_{i,I}| = \frac{1}{6} e_{ijk}\, e_{IJK}\, \phi_{i,I}\, \phi_{j,J}\, \phi_{k,K},$$

$$\text{for } i, j, k, I, J, K = 1, 2, 3, \tag{16.1.3}$$

called *Jacobian,*[2] does not vanish for any $\mathbf{X}$ in C_o; without loss in generality, J will be taken positive.

The change of length of a material line element emanating from a generic particle P, and the change of the angle between two such elements are commonly used as measures of the deformation of the neighborhood of P. Let dX and dY be two material line elements, emanating from P and having spatial representations $d\mathbf{X}$ and $d\mathbf{Y}$ in C_o, and $d\mathbf{x}$ and $d\mathbf{y}$ in C, respectively. The change in the scalar product of $d\mathbf{X}$ and $d\mathbf{Y}$ is given by

$$\begin{aligned} d\mathbf{x}\cdot d\mathbf{y} - d\mathbf{X}\cdot d\mathbf{Y} &= \{d\mathbf{X}\cdot(\boldsymbol{\nabla}_X\otimes\boldsymbol{\phi})\}\cdot\{d\mathbf{Y}\cdot(\boldsymbol{\nabla}_X\otimes\boldsymbol{\phi})\} - d\mathbf{X}\cdot d\mathbf{Y} \\ &= (d\mathbf{X}\otimes d\mathbf{Y}) : \{(\boldsymbol{\nabla}_X\otimes\boldsymbol{\phi})\cdot(\boldsymbol{\nabla}_X\otimes\boldsymbol{\phi})^T - \mathbf{1}^{(2)}\} \\ &= d\mathbf{x}\cdot d\mathbf{y} - \{d\mathbf{x}\cdot(\boldsymbol{\nabla}_x\otimes\boldsymbol{\phi}^{-1})\cdot\{d\mathbf{y}\cdot(\boldsymbol{\nabla}_x\otimes\boldsymbol{\phi}^{-1})\} \\ &= (d\mathbf{x}\otimes d\mathbf{y}) : \{\mathbf{1}^{(2)} - (\boldsymbol{\nabla}_x\otimes\boldsymbol{\phi}^{-1})\cdot(\boldsymbol{\nabla}_x\otimes\boldsymbol{\phi}^{-1})^T\}, \end{aligned} \tag{16.1.4a}$$

or in component form,

$$\begin{aligned} dx_i\, dy_i - dX_I\, dY_I &= (\phi_{i,I}\,\phi_{i,J} - \delta_{IJ})\, dX_I\, dY_J \\ &= (\delta_{ij} - \phi^{-1}_{I,i}\,\phi^{-1}_{I,j})\, dx_i\, dy_j, \end{aligned} \tag{16.1.4b}$$

where $\boldsymbol{\nabla}_X$ and $\boldsymbol{\nabla}_x$ are the $\boldsymbol{\nabla}$ operators (differentiation) associated with $\mathbf{X}$ and $\mathbf{x}$, respectively; see (15.4.2) and (15.4.5). Since the left-hand side of (16.1.4) is scalar, and since this equation is valid for all choices of $d\mathbf{X}$ and $d\mathbf{Y}$, or $d\mathbf{x}$ and $d\mathbf{y}$, the deformation measures

$$\mathbf{E} \equiv \frac{1}{2}\{(\boldsymbol{\nabla}_X\otimes\boldsymbol{\phi})\cdot(\boldsymbol{\nabla}_X\otimes\boldsymbol{\phi})^T - \mathbf{1}^{(2)}\},$$

$$\mathbf{e} \equiv \frac{1}{2}\{\mathbf{1}^{(2)} - (\boldsymbol{\nabla}_x\otimes\boldsymbol{\phi}^{-1})\cdot(\boldsymbol{\nabla}_x\otimes\boldsymbol{\phi}^{-1})^T\}, \tag{16.1.5a,b}$$

or in component form,

$$E_{IJ} \equiv \frac{1}{2}(\phi_{i,I}\,\phi_{i,J} - \delta_{IJ}), \qquad e_{ij} \equiv \frac{1}{2}(\delta_{ij} - \phi^{-1}_{I,i}\,\phi^{-1}_{I,j}), \tag{16.1.5c,d}$$

---

[2] A comma followed by a subscript denotes partial differentiation with respect to the corresponding coordinate of the initial or final position of the particle, i.e., $\phi_{i,I} \equiv \partial\phi_i/\partial X_I$ and $\phi^{-1}_{I,i} \equiv \partial\phi^{-1}_I/\partial x_i$. In terms of the del operators, $\boldsymbol{\nabla}_X \equiv \partial_I\mathbf{e}_I$ and $\boldsymbol{\nabla}_x \equiv \partial_i\mathbf{e}_i$, associated with coordinates $\mathbf{X}$ and $\mathbf{x}$, note that $(\boldsymbol{\nabla}_X\otimes\boldsymbol{\phi})_{Ii} = \partial_I\phi_i$ and $(\boldsymbol{\nabla}_x\otimes\boldsymbol{\phi}^{-1})_{iI} = \partial_i\phi^{-1}_I$.

are second-order symmetric tensors. $\mathbf{E}$ is known as the *Lagrangian* and $\mathbf{e}$ as the *Eulerian*[3] strain tensor, respectively.

If dX and dY are one and the same material line element with initial length dS and final length ds, (16.1.4) reduces to

$$ds^2 - dS^2 = 2(d\mathbf{X} \otimes d\mathbf{X}) : \mathbf{E} = 2(d\mathbf{x} \otimes d\mathbf{x}) : \mathbf{e}$$

$$= 2E_{IJ}\, dX_I\, dX_J = 2e_{ij}\, dx_i\, dx_j, \qquad (16.1.6)$$

which defines the change in the square of the element's length. Let $d\mathbf{X}$ and $d\mathbf{Y}$ be two orthogonal elements, denote by γ the *decrease* of the angle between them, and obtain

$$ds\, ds' \sin\gamma = 2(d\mathbf{X} \otimes d\mathbf{Y}) : \mathbf{E} = 2(d\mathbf{x} \otimes d\mathbf{y}) : \mathbf{e}$$

$$= 2E_{IJ}\, dX_I\, dY_J = 2e_{ij}\, dx_i\, dy_j, \qquad (16.1.7)$$

where ds' is the final length of the element dY.

The strain measures (16.1.5a,b) are now expressed in terms of the displacement field of the solid. Let $\mathbf{x} - \mathbf{X}$ be the displacement of a typical particle P, and express it as $\mathbf{U}(\mathbf{X}) \equiv \boldsymbol{\phi}(\mathbf{X}) - \mathbf{X}$ and $\mathbf{u}(\mathbf{x}) \equiv \mathbf{x} - \boldsymbol{\phi}^{-1}(\mathbf{x})$ for $\mathbf{X}$ and $\mathbf{x}$, in the configurations C_o and C, respectively. Here, $\mathbf{U}$ is for the Lagrangian formulation while $\mathbf{u}$ is for the Eulerian formulation. From definitions of $\boldsymbol{\phi}$ and $\boldsymbol{\phi}^{-1}$, (16.1.1) and (16.1.2), $\mathbf{U}(\mathbf{X}) \equiv \mathbf{u}(\boldsymbol{\phi}(\mathbf{X}))$ and $\mathbf{u}(\mathbf{x}) \equiv \mathbf{U}(\boldsymbol{\phi}^{-1}(\mathbf{x}))$. Then, equations (16.1.1a,b) and (16.1.2a,b) may be written as

$$\boldsymbol{\phi}(\mathbf{X}) = \mathbf{X} + \mathbf{U}(\mathbf{X}) \quad \text{or} \quad \phi_i(\mathbf{X}) = \delta_{iI}\,\{X_I + U_I(\mathbf{X})\}, \qquad (16.1.8a,b)$$

and

$$\boldsymbol{\phi}^{-1}(\mathbf{x}) = \mathbf{x} - \mathbf{u}(\mathbf{x}) \quad \text{or} \quad \phi_I^{-1}(\mathbf{x}) = \delta_{Ii}\,\{x_i - u_i(\mathbf{x})\}, \qquad (16.1.9a,b)$$

from which it follows that

$$\boldsymbol{\nabla}_X \otimes \boldsymbol{\phi} = \mathbf{1}^{(2)} + \boldsymbol{\nabla}_X \otimes \mathbf{U}, \qquad \boldsymbol{\nabla}_x \otimes \boldsymbol{\phi}^{-1} = \mathbf{1}^{(2)} - \boldsymbol{\nabla}_x \otimes \mathbf{u}, \qquad (16.1.10a,b)$$

or in component form,

$$\phi_{i,I} = \delta_{iJ}\,(\delta_{JI} + U_{J,I}), \qquad \phi_I^{-1}{}_{,\,i} = \delta_{Ij}\,(\delta_{ji} - u_{j,i}). \qquad (16.1.10c,d)$$

Substitution into (16.1.5a,b) now yields

$$\mathbf{E} = \frac{1}{2}\{(\boldsymbol{\nabla}_X \otimes \mathbf{U}) + (\boldsymbol{\nabla}_X \otimes \mathbf{U})^T + (\boldsymbol{\nabla}_X \otimes \mathbf{U}) \cdot (\boldsymbol{\nabla}_X \otimes \mathbf{U})^T\},$$

$$\mathbf{e} = \frac{1}{2}\{(\boldsymbol{\nabla}_x \otimes \mathbf{u}) + (\boldsymbol{\nabla}_x \otimes \mathbf{u})^T - (\boldsymbol{\nabla}_x \otimes \mathbf{u}) \cdot (\boldsymbol{\nabla}_x \otimes \mathbf{u})^T\}, \qquad (16.1.11a,b)$$

or in component form,

$$E_{IJ} = \frac{1}{2}(U_{I,J} + U_{J,I} + U_{K,I}\, U_{K,J}), \qquad e_{ij} = \frac{1}{2}(u_{i,j} + u_{j,i} - u_{k,i}\, u_{k,j}). \qquad (16.1.11c,d)$$

---

[3] This is also called *Almansi's* strain tensor; see Truesdell and Toupin (1960).

16.2. INFINITESIMAL STRAIN MEASURE

The infinitesimal deformation theories are based on the assumption that the displacements as well as their gradients are infinitesimal quantities. With this assumption, the quadratic terms in (16.1.11a,b) can be neglected. In addition, the difference between the displacement gradients formed with respect to the Lagrangian and the Eulerian variables disappears. Therefore, both strain tensors E_{IJ} and e_{ij} may be expressed by the following symmetric strain tensor:

$$\boldsymbol{\varepsilon} = \frac{1}{2}\{(\boldsymbol{\nabla} \otimes \mathbf{u}) + (\boldsymbol{\nabla} \otimes \mathbf{u})^T\}, \tag{16.2.1a}$$

or in component form,

$$\varepsilon_{ij} = \frac{1}{2}(u_{j,i} + u_{i,j}) = u_{(i,j)}, \tag{16.2.1b}$$

where, *here and throughout the rest of this book no distinction is made between the Lagrangian and the Eulerian descriptions; all quantities are expressed in terms of* $\mathbf{x}$ *which is interpreted as the initial particle position.* The right-hand side of (16.2.1) is the symmetric part of the displacement gradient, $\boldsymbol{\nabla} \otimes \mathbf{u}$ or $u_{i,j} = \partial u_i/\partial x_j$. Thus, in the infinitesimal theory, the deformation of a material neighborhood of a particle is completely described by the spatial gradient of the displacement vector at that particle. To demonstrate the role that the displacement gradient $\boldsymbol{\nabla} \otimes \mathbf{u}$ plays in describing the motion of a material neighborhood, let $\mathbf{u} = \mathbf{u}(\mathbf{x})$ be the displacement of a particle P situated at $\mathbf{x}$, and consider a neighboring particle P′ at $\mathbf{x} + d\mathbf{x}$. For a sufficiently smooth displacement field, the displacement of P′ can be expressed as

$$\mathbf{u}(\mathbf{x} + d\mathbf{x}) = \mathbf{u}(\mathbf{x}) + d\mathbf{x} \cdot \boldsymbol{\nabla} \otimes \mathbf{u}(\mathbf{x}) + \dots . \tag{16.2.2}$$

Hence, within the context of the considered infinitesimal theory, the displacement of P′ relative to P is given by

$$\mathbf{u}(\mathbf{x} + d\mathbf{x}) - \mathbf{u}(\mathbf{x}) = d\mathbf{x} \cdot \{\boldsymbol{\nabla} \otimes \mathbf{u}(\mathbf{x})\} = d\mathbf{x} \cdot (\boldsymbol{\varepsilon} + \boldsymbol{\omega}), \tag{16.2.3a}$$

or in component form,

$$u_i(\mathbf{x} + d\mathbf{x}) - u_i(\mathbf{x}) = u_{i,j}(\mathbf{x})\, dx_j = (\varepsilon_{ij} + \omega_{ji})\, dx_j, \tag{16.2.3b}$$

where the symmetric part $\boldsymbol{\varepsilon}$ is defined by (16.2.1), and the antisymmetric part $\boldsymbol{\omega}$ is given by

$$\boldsymbol{\omega} = \frac{1}{2}\{(\boldsymbol{\nabla} \otimes \mathbf{u}) - (\boldsymbol{\nabla} \otimes \mathbf{u})^T\}, \tag{16.2.4a}$$

or in component form,

$$\omega_{ij} = \frac{1}{2}(u_{j,i} - u_{i,j}) = u_{[i,j]}. \tag{16.2.4b}$$

Since it follows from (16.1.6), (16.1.7), and (16.2.2), that all changes of length and angle in the neighborhood of P are specified by $\boldsymbol{\varepsilon}$, the tensor $\boldsymbol{\omega}$ represents a rigid-body rotation of this neighborhood, as is further discussed below in connection with (16.2.5) and (16.2.6).

16.2.1. Extension, Shear Strain, and Rotation

To bring out the physical significance of the strain tensor $\boldsymbol{\varepsilon}$, denote the unit vector along the material element dX by $\mathbf{m}$, and reduce (16.1.6) to

$$\boldsymbol{\varepsilon} : (\mathbf{m} \otimes \mathbf{m}) = \frac{1}{2} \frac{ds^2 - dS^2}{dS^2} = \frac{1}{2} \frac{ds - dS}{dS} \frac{ds + dS}{dS} \approx \frac{ds - dS}{dS} \qquad (16.2.5a)$$

which defines the *extension* of the material line element dX. The extension, in general, depends on the direction of the considered material line element. The notation $\lambda_{(\mathbf{m})}$ is used to denote this quantity,

$$\boldsymbol{\varepsilon} : (\mathbf{m} \otimes \mathbf{m}) = \varepsilon_{ij}\, m_i\, m_j = \lambda_{(\mathbf{m})}, \qquad (16.2.5b)$$

where the subscript $\mathbf{m}$ in the parentheses is to indicate that the extension of a line element with the direction of the unit vector $\mathbf{m}$ is being considered. In general, $\lambda_{(\mathbf{m})}$ is also a function of position $\mathbf{x}$ (and also time t for dynamic problems). The left-hand side of (16.2.5b) is called the normal component of the strain tensor $\boldsymbol{\varepsilon}$ in the direction of the considered material line element. It is therefore concluded that *the extension of a material element is equal to the normal component of the strain tensor* $\boldsymbol{\varepsilon}$ *in the direction of this element.* For an element along the x_1-axis, ε_{11} represents its extension. Similar remarks apply to ε_{22} and ε_{33}.

Next, consider two orthogonal material elements dX and dY with directions defined by the unit vectors $\mathbf{m}$ and $\mathbf{n}$, respectively. Equation (16.1.7) now becomes

$$2\boldsymbol{\varepsilon} : (\mathbf{m} \otimes \mathbf{n}) = \sin\gamma. \qquad (16.2.6a)$$

Since γ is infinitesimal,

$$\boldsymbol{\varepsilon} : (\mathbf{m} \otimes \mathbf{n}) \approx \frac{1}{2}\gamma, \qquad (16.2.6b)$$

where $\gamma/2$ is called the *shear strain*[4] for the orthogonal directions $\mathbf{m}$ and $\mathbf{n}$. The dependency of the shear strain on the considered directions is brought out by the notation $\gamma_{(\mathbf{m},\,\mathbf{n})}/2$ for this quantity. Hence,

$$\boldsymbol{\varepsilon} : (\mathbf{m} \otimes \mathbf{n}) = \varepsilon_{ij}\, m_i\, n_j = \frac{1}{2}\gamma_{(\mathbf{m},\,\mathbf{n})}. \qquad (16.2.6c)$$

The left-hand side of this equation is called the tangential component of the strain tensor $\boldsymbol{\varepsilon}$ in the orthogonal directions $\mathbf{m}$ and $\mathbf{n}$. It therefore follows that *the shear strain for two orthogonal material line elements is equal to the tangential component of the strain tensor* $\boldsymbol{\varepsilon}$ *with respect to the directions of these elements.* In particular, $\varepsilon_{12} = \varepsilon_{21}$, $\varepsilon_{23} = \varepsilon_{32}$, and $\varepsilon_{31} = \varepsilon_{13}$ are the shear strains corresponding to the three pairs of coordinate directions.

From (16.2.5) and (16.2.6) it follows that a material neighborhood will only undergo (infinitesimal) rigid-body translation and rotation if the strain

---

[4] In the engineering literature the total decrease in an initially right material angle is commonly called the shear strain. However, in mathematical treatments, half of this decrease, which is equal to the tangential component of the tensor $\boldsymbol{\varepsilon}$ in the considered directions, is commonly termed the shear strain. The first is convenient in matrix representation, and the second, in tensorial representation; see Section 15.

tensor $\boldsymbol{\varepsilon}$ vanishes there. The infinitesimal rotation is completely defined by the antisymmetric tensor $\boldsymbol{\omega}$. Denote by $d\mathbf{u}^r$ the displacement of P′ relative to P that corresponds to this rigid-body rotation,

$$d\mathbf{u}^r = d\mathbf{x} \cdot \boldsymbol{\omega}, \tag{16.2.7a}$$

or in component form,

$$du_i^r = \omega_{ji}\, dx_j. \tag{16.2.7b}$$

If $\boldsymbol{\Omega}$ is the *dual vector* of the antisymmetric tensor $\boldsymbol{\omega}$, namely,

$$\Omega_i \equiv \frac{1}{2} e_{ijk}\, \omega_{jk}, \tag{16.2.8}$$

then,

$$d\mathbf{u}^r = \boldsymbol{\Omega} \times d\mathbf{x} \qquad \text{or} \qquad du_i^r = e_{ijk}\, \Omega_j\, dx_k. \tag{16.2.9a,b}$$

This equation states that P′ rotates relative to P by an infinitesimal angle equal to the magnitude of $\boldsymbol{\Omega}$. The sense of this rotation is that of a right-handed screw which progresses along the positive direction of $\boldsymbol{\Omega}$.

16.2.2. Pure Deformation

If the rotation tensor $\boldsymbol{\omega}$ is zero, the neighborhood of P, in general, undergoes a rigid-body translation and a *pure deformation.* Corresponding to the pure deformation, the displacement $d\mathbf{u}^d$ of the particle P′ relative to P is given by

$$d\mathbf{u}^d = d\mathbf{x} \cdot \boldsymbol{\varepsilon} = ds\, (\mathbf{m} \cdot \boldsymbol{\varepsilon}), \tag{16.2.10a}$$

where $d\mathbf{x} = ds\, \mathbf{m}$. Thus a pure deformation, in general, changes the direction of the material line elements. There are, however, certain line elements whose directions are unaltered in a pure deformation, and whose extensions have extreme values. Indeed, if the pure deformation is to leave the direction of the element PP′ unchanged, then

$$d\mathbf{u}^d = ds\, (\boldsymbol{\varepsilon} \cdot \mathbf{m}) = \lambda\, ds\, \mathbf{m}, \tag{16.2.10b}$$

where λ is a scalar. Hence,

$$(\boldsymbol{\varepsilon} - \lambda \mathbf{1}^{(2)}) \cdot \mathbf{m} = \mathbf{0}, \tag{16.2.10c}$$

or in component form

$$(\varepsilon_{ij} - \lambda \delta_{ij})\, m_j = 0, \qquad i, j = 1, 2, 3. \tag{16.2.10d}$$

Consider now the extension in the direction defined by a unit vector $\mathbf{m}$; see (16.2.5). To obtain a direction for which the extension has an extreme value, introduce the constraint

$$\mathbf{m} \cdot \mathbf{m} - 1 = 0, \tag{16.2.11}$$

and let λ denote the Lagrangian multiplier. Now seek the stationary values of the quadratic form

$$\mathrm{I}(\mathbf{m}) \equiv \boldsymbol{\varepsilon} : (\mathbf{m} \otimes \mathbf{m}) - \lambda\,(\mathbf{m} \cdot \mathbf{m} - 1), \tag{16.2.12}$$

by setting its derivative with respect to **m** equal to **0**, obtaining (16.2.10c). It is therefore concluded that *pure deformation leaves unchanged the directions of those material line elements whose extensions have stationary values.*

System (16.2.10d) is a set of three linear, homogeneous equations in m_j, $j = 1, 2, 3$. Nontrivial solutions for these equations exist if and only if the determinant of the coefficients of the unknowns m_j is zero, that is, if

$$\det \mid \varepsilon_{ij} - \lambda \delta_{ij} \mid = 0 \tag{16.2.13a}$$

which defines a polynomial of third degree in λ. Expanding (16.2.13a), obtain

$$\lambda^3 - \mathrm{I}_\varepsilon \lambda^2 + \mathrm{II}_\varepsilon \lambda - \mathrm{III}_\varepsilon = 0, \tag{16.2.13b}$$

where the coefficients I_ε, II_ε, and III_ε are scalars; they are called *basic invariants* of the second-order symmetric tensor $\boldsymbol{\varepsilon}$, and are given by

$$\mathrm{I}_\varepsilon = \mathrm{tr}\,\boldsymbol{\varepsilon} = \varepsilon_{ii},$$

$$\mathrm{II}_\varepsilon = \frac{1}{2} e_{ijk}\, e_{ilm}\, \varepsilon_{jl}\, \varepsilon_{km} = \frac{1}{2}(\mathrm{I}_\varepsilon^2 - \boldsymbol{\varepsilon} : \boldsymbol{\varepsilon}),$$

$$\mathrm{III}_\varepsilon = \det \boldsymbol{\varepsilon} = \frac{1}{6}\,(2\varepsilon_{ij}\,\varepsilon_{jk}\,\varepsilon_{ki} - 3\mathrm{I}_\varepsilon\, \boldsymbol{\varepsilon} : \boldsymbol{\varepsilon} + \mathrm{I}_\varepsilon^3). \tag{16.2.14a~c}$$

Equation (16.2.13b) has three roots, λ_{I}, λ_{II}, and λ_{III}, which are called *principal values* (or proper numbers, or characteristic values) of the second-order tensor $\boldsymbol{\varepsilon}$. To each principal value λ_{J}, there corresponds a principal direction $\mathbf{m}^{\mathrm{J}}$, i.e., for J = I, II, and III ,

$$\boldsymbol{\varepsilon} \cdot \mathbf{m}^{\mathrm{J}} - \lambda_{\mathrm{J}}\, \mathbf{m}^{\mathrm{J}} = \mathbf{0} \qquad (\mathrm{J\ not\ summed}), \tag{16.2.13c}$$

which defines the first, second, and third principal directions, respectively, as J takes on values I, II, and III.

Since $\boldsymbol{\varepsilon}$ is a symmetric tensor, all its principal values are real. This can be shown as follows: The polynomial (16.2.13b), whose coefficients are real, has, at least, one real root; the other two roots are either both real or are complex conjugates of each other. Let us assume the latter and show that this leads to a contradiction. If λ_{I} is the real root, then[5] $\lambda_{\mathrm{II}} = a + \iota b$ and $\lambda_{\mathrm{III}} = \bar{\lambda}_{\mathrm{II}} = a - \iota b$, where $\iota = \sqrt{-1}$, and superposed bar denotes complex conjugate. The corresponding principal directions thus are m_j^{II} and $m_j^{\mathrm{III}} = \bar{m}_j^{\mathrm{II}}$, and from (16.2.10d),

$$\varepsilon_{jk}\, m_k^{\mathrm{II}} = \lambda_{\mathrm{II}}\, m_j^{\mathrm{II}}, \qquad \varepsilon_{jk}\, \bar{m}_k^{\mathrm{II}} = \bar{\lambda}_{\mathrm{II}}\, \bar{m}_j^{\mathrm{II}}. \tag{16.2.15a,b}$$

Multiplying the first equation by $\bar{m}_j^{\mathrm{II}}$, and the second by m_j^{II}, and then taking their difference, arrive at

$$2\iota b (m_j^{\mathrm{II}}\, \bar{m}_j^{\mathrm{II}}) = 0. \tag{16.2.15c}$$

---

[5] A superposed bar denotes complex conjugate.

This is possible if and only if b = 0.

Multiplying both sides of (16.2.13c) by $\mathbf{m}^J$, obtain

$$\boldsymbol{\varepsilon} : (\mathbf{m}^J \otimes \mathbf{m}^J) = \lambda_J \qquad (\text{J not summed}), \tag{16.2.16}$$

for J = I, II, and III. This result shows that the principal values, λ_I, λ_{II}, and λ_{III} are the extremum values of the extension; they are called *principal extensions.* It will now be shown that the principal directions corresponding to distinct principal extensions are orthogonal. To this end, write

$$\boldsymbol{\varepsilon} \cdot \mathbf{m}^I = \lambda_I \, \mathbf{m}^I, \qquad \boldsymbol{\varepsilon} \cdot \mathbf{m}^{II} = \lambda_{II} \, \mathbf{m}^{II}, \tag{16.2.17a,b}$$

which define the principal directions $\mathbf{m}^I$ and $\mathbf{m}^{II}$, respectively. Now, multiplying (16.2.17a) by $\mathbf{m}^{II}$ and (16.2.17b) by $\mathbf{m}^I$, and then taking their difference, obtain

$$(\lambda_I - \lambda_{II}) \, \mathbf{m}^I \cdot \mathbf{m}^{II} = 0. \tag{16.2.17c}$$

Therefore, if $\lambda_I \neq \lambda_{II}$, then $\mathbf{m}^I$ is orthogonal to $\mathbf{m}^{II}$. Similar results hold relative to the other principal direction $\mathbf{m}^{III}$. Thus, if $\lambda_I \neq \lambda_{II} \neq \lambda_{III}$, the principal directions $\mathbf{m}^J$, J = I, II, III, form an orthogonal triad, that is,

$$\mathbf{m}^J \cdot \mathbf{m}^K = \delta_{JK} = \begin{cases} 1 & \text{if J = K} \\ 0 & \text{if J} \neq \text{K.} \end{cases} \tag{16.2.18}$$

This and (16.2.10c) show that the *shear strains for the pairs of the principal directions are zero.* If two of the principal extensions are equal, then there is only one unique principal direction which corresponds to the distinct principal extension. Normal to this direction, any two principal directions may be taken as principal directions. If all three principal extensions are equal, any orthogonal triad constitutes a principal triad. In this case, the deformation is locally isotropic, i.e., $\boldsymbol{\varepsilon} = \lambda \mathbf{1}^{(2)}$, where λ denotes the extension common to all elements emanating from the considered point.

Since the principal values of $\boldsymbol{\varepsilon}$ are the roots of (16.2.13b), this equation may also be expressed as

$$\lambda^3 - I_\varepsilon \lambda^2 + II_\varepsilon \lambda - III_\varepsilon = (\lambda - \lambda_I)(\lambda - \lambda_{II})(\lambda - \lambda_{III}) = 0. \tag{16.2.19}$$

In terms of the principal extensions, the basic invariants of $\boldsymbol{\varepsilon}$ now become

$$I_\varepsilon = \lambda_I + \lambda_{II} + \lambda_{III}, \qquad II_\varepsilon = \lambda_I \lambda_{II} + \lambda_{II} \lambda_{III} + \lambda_{III} \lambda_I,$$

$$III_\varepsilon = \lambda_I \lambda_{II} \lambda_{III}. \tag{16.2.20a~c}$$

Finally, the strain tensor $\boldsymbol{\varepsilon}$ may be expressed as

$$\boldsymbol{\varepsilon} = \sum_{J=I}^{III} \lambda_J \, \mathbf{m}^J \otimes \mathbf{m}^J, \tag{16.2.21a}$$

or in component form,

$$\varepsilon_{ij} = \sum_{J=I}^{III} \lambda_J \, m_i^J \, m_j^J \tag{16.2.21b}$$

which shows that, in the principal triad, the matrix of the strain tensor is diagonal.

Next, examine those pairs of orthogonal directions for which the shear strains have extreme values called *principal shear strains;* the corresponding directions are called *principal shear directions.* Let **m** and **n** be two orthogonal unit vectors,

$$\mathbf{m}\cdot\mathbf{m}-1=0, \qquad \mathbf{n}\cdot\mathbf{n}-1=0, \qquad \mathbf{m}\cdot\mathbf{n}=0. \qquad (16.2.22\text{a}\sim\text{c})$$

The shear strain for these directions is $\boldsymbol{\varepsilon}:(\mathbf{m}\otimes\mathbf{n})$. This is to be maximized (minimized) subject to the constraints (16.2.22a~c). Let μ, η, and ν denote the Lagrangian multipliers, and consider the following expression:

$$\mathrm{J}(\mathbf{m},\mathbf{n}) \equiv \boldsymbol{\varepsilon}:(\mathbf{m}\otimes\mathbf{n}) - \frac{1}{2}\mu\,(\mathbf{m}\cdot\mathbf{m}-1) - \frac{1}{2}\eta\,(\mathbf{n}\cdot\mathbf{n}-1) - \nu\,\mathbf{m}\cdot\mathbf{n}. \qquad (16.2.23)$$

Setting $\partial\mathrm{J}/\partial\mathbf{m}$ and $\partial\mathrm{J}/\partial\mathbf{n}$ equal to zero, obtain

$$\boldsymbol{\varepsilon}\cdot\mathbf{n} = \mu\,\mathbf{m}+\nu\,\mathbf{n}, \qquad \boldsymbol{\varepsilon}\cdot\mathbf{m} = \eta\,\mathbf{n}+\nu\,\mathbf{m}, \qquad (16.2.24\text{a,b})$$

from which it immediately follows that

$$\mu = \eta \quad \text{and} \quad \boldsymbol{\varepsilon}:(\mathbf{m}\otimes\mathbf{n}) = \mu. \qquad (16.2.24\text{c,d})$$

Equations (16.2.24a,b) then yield

$$\boldsymbol{\varepsilon}\cdot(\mathbf{m}+\mathbf{n}) = (\nu+\mu)\,(\mathbf{m}+\mathbf{n}), \qquad \boldsymbol{\varepsilon}\cdot(\mathbf{m}-\mathbf{n}) = (\nu-\mu)\,(\mathbf{m}-\mathbf{n}), \qquad (16.2.25\text{a,b})$$

which state that $\nu+\mu$ and $\nu-\mu$ are the principal extensions; the corresponding principal directions are given by $\mathbf{m}+\mathbf{n}$ and $\mathbf{m}-\mathbf{n}$, respectively. Thus set

$$\nu+\mu = \lambda_{\mathrm{J}}, \qquad \mathbf{m}^{\mathrm{J}} = \frac{\mathbf{m}+\mathbf{n}}{|\mathbf{m}+\mathbf{n}|} = \frac{1}{\sqrt{2}}\,(\mathbf{m}+\mathbf{n}),$$

$$\nu-\mu = \lambda_{\mathrm{K}}, \qquad \mathbf{m}^{\mathrm{K}} = \frac{\mathbf{m}-\mathbf{n}}{|\mathbf{m}-\mathbf{n}|} = \frac{1}{\sqrt{2}}\,(\mathbf{m}-\mathbf{n})$$

$$\mathrm{J},\mathrm{K} = \mathrm{I},\mathrm{II},\mathrm{III}, \quad \mathrm{J}\neq\mathrm{K}, \qquad (16.2.26\text{a}\sim\text{d})$$

and conclude that the principal shear strains are given by

$$\mu = \frac{1}{2}(\lambda_{\mathrm{J}}-\lambda_{\mathrm{K}}), \qquad \mathrm{J}\neq\mathrm{K}, \qquad (16.2.27\text{a})$$

with the corresponding principal shear directions defined by

$$\mathbf{m} = \frac{1}{\sqrt{2}}\,(\mathbf{m}^{\mathrm{J}}+\mathbf{n}^{\mathrm{K}}), \quad \text{and} \quad \mathbf{n} = \frac{1}{\sqrt{2}}\,(\mathbf{m}^{\mathrm{J}}-\mathbf{n}^{\mathrm{K}}), \qquad \mathrm{J}\neq\mathrm{K}, \qquad (16.2.27\text{b,c})$$

or in component form,

$$\mathrm{m_i} = \frac{1}{\sqrt{2}}\,(\mathrm{m_i^J}+\mathrm{n_i^K}), \quad \text{and} \quad \mathrm{n_i} = \frac{1}{\sqrt{2}}\,(\mathrm{m_i^J}-\mathrm{n_i^K}), \qquad \mathrm{J}\neq\mathrm{K}. \qquad (16.2.27\text{d,e})$$

The principal shear directions **m** and **n**, therefore, lie in the plane of the principal directions $\mathbf{m}^{\mathrm{J}}$ and $\mathbf{n}^{\mathrm{K}}$, and bisect the angle formed by these latter vectors; half of the difference between the corresponding principal extensions is equal to the principal shear strain.

It is often useful to decompose the strain tensor $\boldsymbol{\varepsilon}$ into two parts, a *dilatational* part and a *distortional* part. The dilatation is defined by the first invariant I_ε of the strain tensor. For the considered infinitesimal deformation, I_ε

represents the expansion of a unit volume. *Instead of* I_ε, *the symbol* ε will be used to denote the dilatation, i.e.,

$$\varepsilon \equiv I_\varepsilon = \varepsilon_{ii} = \lambda_I + \lambda_{II} + \lambda_{III}. \tag{16.2.28}$$

To show that ε represents the expansion of a unit volume, consider a small rectangular parallelepiped whose edges are directed along the principal directions of the strain tensor, having the lengths l_1, l_2, and l_3, respectively. Since the shear strains for the principal directions vanish, the infinitesimal deformation maps this element into another rectangular parallelepiped whose edges are of lengths $l_1(1+\lambda_I)$, $l_2(1+\lambda_{II})$, and $l_3(1+\lambda_{III})$, respectively. Therefore, to the first order of approximation in λ_I, λ_{II}, λ_{III}, the change in the volume per unit initial volume is equal to $\varepsilon = \lambda_I + \lambda_{II} + \lambda_{III}$. Obviously, the tensor $\varepsilon \mathbf{1}^{(2)}/3$, which has equal principal values of magnitude $\varepsilon/3$, involves the same volume change; it is called the *dilatational part* of the strain tensor $\boldsymbol{\varepsilon}$.

The *distortional part* of the strain tensor is called the *strain deviator;* it is defined by

$$\boldsymbol{\varepsilon}' \equiv \boldsymbol{\varepsilon} - \frac{1}{3}\,\varepsilon \mathbf{1}^{(2)}, \tag{16.2.29a}$$

or in component form,

$$\varepsilon'_{ij} \equiv \varepsilon_{ij} - \frac{1}{3}\,\varepsilon\,\delta_{ij}. \tag{16.2.29b}$$

Since $\boldsymbol{\varepsilon}'$ does not involve a volume change, it is a measure of distortion, and, for this reason, it is often called the *distortion tensor.*

16.2.3. Compatibility Conditions

Having studied certain general properties of the strain tensor, now consider conditions for the *compatibility* of a given strain field, that is, *seek conditions under which an arbitrarily prescribed strain field in a simply connected region*[6] *corresponds to a continuous single-valued displacement field.* At the outset, note that such a displacement field can be unique only to within a rigid-body displacement, because rigid-body translations and rotations do not affect the strains. Thus, a "unique displacement field" shall mean that it is unique to within a rigid-body displacement.

For a prescribed displacement field, the strains may be calculated by simple differentiation; see (16.2.1). Given a strain field, on the other hand, six differential equations

$$\frac{1}{2}(u_{j,i} + u_{i,j}) = \varepsilon_{ij}, \tag{16.2.30}$$

yield continuous single-valued solutions for the three functions u_i, only when certain restrictions are met by the six components of the strain tensor ε_{ij}. These

[6] A region is called simply connected if, by a continuous deformation, every closed curve in this region can be reduced to a point without crossing the boundaries of the region.

restrictions are known as *integrability* or *compatibility conditions.*

Let the displacement of a point $\mathbf{x}^0$ be denoted by $\mathbf{u}^0$. If, to a given sufficiently smooth strain field, there corresponds a unique, continuous, single-valued displacement field $\mathbf{u} = \mathbf{u}(\mathbf{x})$, then the displacement $\mathbf{u}^1$ of a point $\mathbf{x}^1$ can be expressed as

$$\mathbf{u}^1 - \mathbf{u}^0 = \int_{\mathbf{x}^0}^{\mathbf{x}^1} d\mathbf{u} = \int_{\mathbf{x}^0}^{\mathbf{x}^1} (\boldsymbol{\nabla} \otimes \mathbf{u})^{\mathrm{T}} \cdot d\mathbf{x} = \int_{\mathbf{x}^0}^{\mathbf{x}^1} (\boldsymbol{\varepsilon} - \boldsymbol{\omega}) \cdot d\mathbf{x}, \tag{16.2.31a}$$

or in component form,

$$u_i^1 - u_i^0 = \int_{\mathbf{x}^0}^{\mathbf{x}^1} (\varepsilon_{ij} - \omega_{ij})\, dx_j, \tag{16.2.31b}$$

where the integration path, (assumed to be a rectifiable curve) may be selected arbitrarily in the considered simply connected region R. Note that the assumption of R being simply connected is rather essential here, since for a multiply connected region the displacement field may turn out to be multiple-valued.

Equation (16.2.31b) may be written as

$$u_i^1 = u_i^0 + (x_j^1 - x_j^0)\, \omega_{ji}^0 + \int_{\mathbf{x}^0}^{\mathbf{x}^1} \{\varepsilon_{ij} + (x_k^1 - x_k)\, \omega_{ki,j}\}\, dx_j, \tag{16.2.32a}$$

where integration by parts is used, and where $\boldsymbol{\omega}^0$ is the value of the rotation tensor at point $\mathbf{x}^0$. Noting that

$$\omega_{ki,j} = \frac{1}{2}(u_{i,k} - u_{k,i})_{,j} = \varepsilon_{ij,k} - \varepsilon_{kj,i}, \tag{16.2.32b}$$

reduce (16.2.32a) to

$$u_i^1(\mathbf{x}^1) = u_i^0 + (x_j^1 - x_j^0)\, \omega_{ji}^0 + \int_{\mathbf{x}^0}^{\mathbf{x}^1} \{\varepsilon_{ij} + (x_k^1 - x_k)\, (\varepsilon_{ij,k} - \varepsilon_{kj,i})\}\, dx_j. \tag{16.2.32c}$$

For a smooth, single-valued displacement field, $\mathbf{u}^1$ is uniquely defined by (16.2.32c), independently of a particular path of integration from $\mathbf{x}^0$ to $\mathbf{x}^1$. Therefore, the integrand

$$U_{ij} \equiv \varepsilon_{ij} + (x_k^1 - x_k)\, (\varepsilon_{ij,k} - \varepsilon_{kj,i}) \tag{16.2.33a}$$

in (16.2.32c) must be an exact differential, that is, it must follow that

$$U_{ij}\, dx_j = d\Phi_i, \qquad i = 1, 2, 3. \tag{16.2.33b}$$

Necessary and sufficient conditions for this are

$$\Phi_{i,jk} = \Phi_{i,kj} \Longleftrightarrow U_{ij,k} = U_{ik,j} \tag{16.2.33c}$$

from which it follows that

$$(x_k^1 - x_k)\, (\varepsilon_{ij,kl} + \varepsilon_{kl,ij} - \varepsilon_{ik,jl} - \varepsilon_{jl,ik}) = 0. \tag{16.2.33d}$$

This is satisfied for all x_k^1 in R, if

$$\varepsilon_{ij,kl} + \varepsilon_{kl,ij} - \varepsilon_{ik,jl} - \varepsilon_{jl,ik} = 0 \tag{16.2.34}$$

which are the desired integrability conditions. Note that if (16.2.24) is satisfied, the integrand in (16.2.32c) becomes an exact differential, rendering the displacement $\mathbf{u}^1$ independent of the considered path of integration. This implies that (16.2.34) are sufficient conditions for the integrability of (16.2.30). To show that they are also necessary, assume the existence of a single-valued continuous displacement field $\mathbf{u}(\mathbf{x})$ of class[7] C^3, and by successive differentiation obtain

$$\varepsilon_{ij,kl} = \frac{1}{2}(u_{i,jkl} + u_{j,ikl}). \tag{16.2.35a}$$

Now, interchanging the subscripts, obtain

$$\varepsilon_{kl,ij} = \frac{1}{2}(u_{k,lij} + u_{l,kij}). \tag{16.2.35b}$$

Interchanging i and k in (16.2.35a), and k and i in (16.2.35b), now combine the results to arrive at (16.2.34).

Therefore, in a simply connected region, conditions (16.2.34) are both necessary and sufficient to ensure the existence of a single-valued continuous displacement field. There are 81 equations expressed by (16.2.34), out of which only 6 are independent; the others are either identities or repetitions of these six equations. The six independent equations are

$$\varepsilon_{11,23} = (\varepsilon_{12,3} + \varepsilon_{31,2} - \varepsilon_{23,1})_{,1}, \quad \varepsilon_{12,12} = \frac{1}{2}(\varepsilon_{11,22} + \varepsilon_{22,11}),$$

$$\varepsilon_{22,31} = (\varepsilon_{23,1} + \varepsilon_{12,3} - \varepsilon_{31,2})_{,2}, \quad \varepsilon_{23,23} = \frac{1}{2}(\varepsilon_{22,33} + \varepsilon_{33,22}),$$

$$\varepsilon_{33,12} = (\varepsilon_{31,2} + \varepsilon_{23,1} - \varepsilon_{12,3})_{,3}, \quad \varepsilon_{31,31} = \frac{1}{2}(\varepsilon_{33,11} + \varepsilon_{11,33}). \tag{16.2.36a~f}$$

When R is a multiply connected region, conditions (16.2.36a~f), while still necessary, are no longer sufficient for the existence of a continuous single-valued displacement field. Since a multiply connected region can be reduced to a simply connected one by the introduction of suitable cuts, (16.2.32c) yields a unique displacement $\mathbf{u}^1$, if the integration path does not cross any one of these cuts and (16.2.36a~f) are also satisfied. Additional conditions are now obtained by the requirement that the displacement should be continuous across the cuts. In general, $m - 1$ cuts are needed to render an m-tuply connected region simply connected, and for each cut three conditions are required to ensure the continuity of three displacement components. Indeed, these three continuity conditions are satisfied if and only if the following integration taken along an arbitrary loop L_n surrounding the nth "hole" in R vanishes:

$$\int_{L_n} \{\varepsilon_{ij} + (x_k^1 - x_k)(\varepsilon_{ij,k} - \varepsilon_{kj,i})\}\, dx_n = 0, \quad j = 1, 2, 3, \tag{16.2.37}$$

where $\mathbf{x}^1$ is on L_n. Equation (16.2.37) states that the displacement has zero jump when its gradient is integrated around the nth hole; see Figure 16.2.1. A

[7] A function is of class C^n in a given region if it is continuous there, together with all of its derivatives up to and including the nth order.

multiply connected region can be reduced to a simply connected region by the introduction of suitable cuts. Therefore, there are, in general, $3(m-1)$ additional conditions which must be met, if the displacement field in an m-tuply connected region is to be single-valued and continuous.

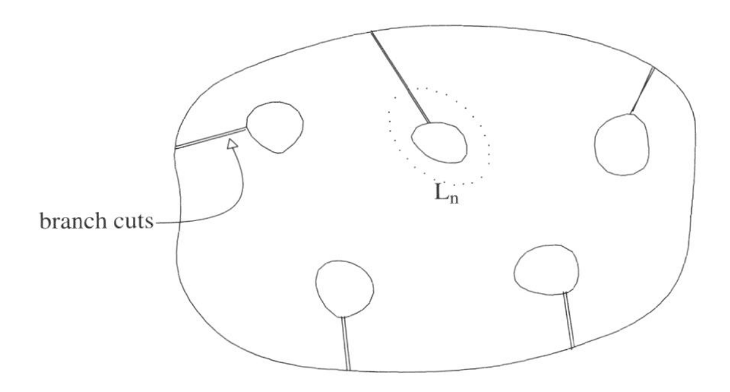

Figure 16.2.1

Branch cuts for m-tuply connected body, and integral path L_n

16.2.4. Two-Dimensional Case

Consider a special deformation called *plane.* A deformation is said to be plane if, with a suitable choice of the rectangular Cartesian coordinate system, the displacement field can be written in the form

$$u_1 = u_1(x_1, x_2), \qquad u_2 = u_2(x_1, x_2), \qquad u_3 = \text{constant}. \tag{16.2.38a~c}$$

The rotation vector $\mathbf{\Omega}$, with the components,

$$\Omega_1 = 0, \qquad \Omega_2 = 0, \qquad \Omega_3 = \frac{1}{2}(u_{2,1} - u_{1,2}), \tag{16.2.39a~c}$$

is parallel to the x_3-axis. A positive value of Ω_3 represents a counter-clockwise rotation of the considered material neighborhood about an axis that is parallel to the x_3-axis. The non-vanishing components of the strain tensor are

$$\varepsilon_{\alpha\beta} = \frac{1}{2}(u_{\beta,\,\alpha} + u_{\alpha,\,\beta}). \tag{16.2.40}$$

Here and throughout the rest of this subsection, Greek subscripts have the range 1, 2.

The extension in the direction of the unit vector $\mathbf{m}$ with components

$$m_1 = \cos\phi, \qquad m_2 = \sin\phi, \qquad m_3 = 0, \tag{16.2.41a~c}$$

is given by

$$\lambda_{(\mathbf{m})} = \varepsilon_{11}\cos^2\phi + \varepsilon_{22}\sin^2\phi + \varepsilon_{12}\sin^2\phi$$

$$= \frac{1}{2}(\varepsilon_{11}+\varepsilon_{22}) + \frac{1}{2}(\varepsilon_{11}-\varepsilon_{22})\cos2\phi + \varepsilon_{12}\sin2\phi, \tag{16.2.42}$$

where ϕ is the angle formed by the direction of **m** and the positive x_1-direction. If **n** is a unit vector in the x_1,x_2-plane, chosen in such a manner that **m**, **n**, and the x_3-direction form a right-handed orthogonal triad, then

$$n_1 = -\sin\phi, \qquad n_2 = \cos\phi, \qquad n_3 = 0. \tag{16.2.43a~c}$$

The shear strain for the direction **m** and **n** then is

$$\frac{1}{2}\gamma_{(\mathbf{m},\,\mathbf{n})} = -\frac{1}{2}(\varepsilon_{11}-\varepsilon_{22})\sin2\phi + \varepsilon_{12}\cos2\phi. \tag{16.2.44}$$

In the plane $x_3 = 0$, and along the principal directions, now choose a new system of rectangular coordinates x_1', x_2', and $x_3' = x_3$, and label them in such a manner that the principal extensions λ_{I} of the x_1'-direction and λ_{II} of the x_2'-direction, satisfy the condition $\lambda_{\mathrm{I}} \geq \lambda_{\mathrm{II}}$. Denoting by ϕ' the angle formed by the direction of **m** and the positive x_1'-direction, reduce (16.2.42) and (16.2.44) to

$$\lambda_{(\mathbf{m})} = \frac{1}{2}(\lambda_{\mathrm{I}}+\lambda_{\mathrm{II}}) + \frac{1}{2}(\lambda_{\mathrm{I}}-\lambda_{\mathrm{II}})\cos2\phi', \tag{16.2.45}$$

and

$$\frac{1}{2}\gamma_{(\mathbf{m},\,\mathbf{n})} = -\frac{1}{2}(\lambda_{\mathrm{I}}-\lambda_{\mathrm{II}})\sin2\phi'. \tag{16.2.46}$$

A line element parallel to the x_1',x_2'-plane undergoes an infinitesimal rotation Ω parallel to this plane and equal to the sum of (16.2.39c) and (16.2.46), i.e.,

$$\Omega = \Omega_3 - \frac{1}{2}(\lambda_{\mathrm{I}}-\lambda_{\mathrm{II}})\sin2\phi'. \tag{16.2.47}$$

In the λ, Ω-plane, consider a circle defined by

$$\lambda = \frac{1}{2}(\lambda_{\mathrm{I}}+\lambda_{\mathrm{II}}) + \frac{1}{2}(\lambda_{\mathrm{I}}-\lambda_{\mathrm{II}})\cos2\phi',$$

$$\Omega = \Omega_3 - \frac{1}{2}(\lambda_{\mathrm{I}}-\lambda_{\mathrm{II}})\sin2\phi', \qquad 0 \leq \phi' < \pi. \tag{16.2.48a,b}$$

Since ϕ' is the direction of a typical material element along the unit vector **m**, points on this circle define the strain and rotation of material-line elements of the corresponding orientations, measured from the x_1'-axis. If Ω_3 is excluded in (16.2.48b), then the construction known as *Mohr's strain circle* is obtained.

16.3. REFERENCES

Chadwick, P. (1976), *Continuum mechanics, concise theory and problems*, George Allen and Unwin.

Eringen, A. C. and Suhubi, E. S. (1975), *Elastodyamics*, Vol. II *linear theory*, Academic Press, New York.
Fung, Y. C. (1965), *Foundations of solid mechanics*, Englewood Cliffs, N. J.; Prentice-Hall.
Green, A. E. and Zerna, W. (1968), *Theoretical elasticity* (2nd ed.), Oxford University Press, Oxford.
Gurtin, M. E. (1972), The linear theory of elasticity, in *Handbuch der Physik* (Flügge, S., ed.), Springer-Verlag, New York, Vol. VIa/2.
Jaunzemis, W. (1967), *Continuum mechanics*, The Macmillan Company, New York.
Malvern, L. E. (1969), *Introduction to the mechanics of a continuous medium*, Prentice Hall.
Prager, W. (1961), An elementary discussion of definitions of stress rate, *J Q. Appl. Math.*, 403-407.
Spencer, A. J. M. (1980), *Continuum mechanics*, Longman, London.
Truesdell, C. and Toupin, R. A. (1960), The classical field theories, in *Handbuch der Physik* (Flügge, S., ed.), Springer-Verlag, New York, Vol. III/1.

SECTION 17 DYNAMIC FOUNDATIONS

In this section the basic equations that govern the instantaneous (dynamic) equilibrium of a solid which may or may not be elastic are considered. From the balance of linear and angular momenta (Euler's laws), applied to an arbitrary part of a solid, the local field equations (Cauchy's laws) are obtained. Although in this book attention is focused on small-deformation theories and, particularly, elasticity, these balance laws are valid for any continuum and for small, as well as large, deformations, provided they are interpreted in an appropriate manner.

17.1. EULER'S LAWS

In continuum mechanics one is concerned with the response of continua to applied forces (loads) and imposed surface displacements. Attention is confined to solid bodies, where the applied forces act over a part, or the entire surface of the solid in its contact with other bodies. These forces are usually specified per unit area of the surface element upon which they act, and are called *surface tractions* (or *traction vectors*). Hydrostatic pressure on submerged bodies is an example of surface tractions. The symbol $\mathbf{t}$ is used to denote the applied surface tractions. The physical dimension of $\mathbf{t}$ is force divided by squared length.

In a gravitational field, particles of a body are subjected to forces which are proportional to their mass. Since the mass of a continuum is regarded as continuously distributed throughout the space occupied by the body, the forces which relate to mass are likewise continuously distributed. These types of forces are defined per unit mass of the body, and are called *body forces.* The symbol $\mathbf{f}$ is used to designate body forces. The physical dimension of $\mathbf{f}$ is force divided by mass. Body forces may also stem from the interaction of pairs of particles forming the solid, as in the case of a solid under its own gravitation. Since forces of this kind, called *mutual loads,* may be prescribed *a priori,* it is assumed that they are accounted for in the specification of $\mathbf{f}$.

In addition to surface tractions and body forces, one may consider surface and body couples acting on a solid. For example, material points of a polarized continuum may be subjected to body couples, in addition to possible body forces, when this continuum is placed in an electromagnetic field.[1] However, the

[1] Note that surface and body couples should not be viewed as moments of surface and body forces. The existence of these couples may be postulated independently of, and in addition to the other types of forces and their corresponding moments.

study of the couple-stress theory is not pursued in this book.[2]

The forces defined above present the mechanical environment of the solid. They are prescribed *a priori,* and thus may be termed *applied loads.* Therefore, at each instant t, a considered solid that occupies the region R with surface ∂R may be subjected to the following loads: a resultant force **F** given by

$$\mathbf{F} = \int_{\partial R} \mathbf{t}\, dS + \int_R \rho\, \mathbf{f}\, dV, \tag{17.1.1}$$

and a resultant moment (torque) **L** (taken with respect to the origin O of the coordinate system) defined as

$$\mathbf{L} = \int_{\partial R} \mathbf{x} \times \mathbf{t}\, dA + \int_R \rho\, \mathbf{x} \times \mathbf{f}\, dV, \tag{17.1.2}$$

where ρ is the mass-density of the solid. Note that (17.1.1) and (17.1.2) are not restricted to elastic continua but are valid for continua of all types.

The motion of a solid is assumed to be governed by *Euler's laws* which are assertions regarding the manner by which external loads affect linear and angular momenta of the bodies. The first law is concerned with linear momentum. It states that the *instantaneous rate of change of the linear momentum* **P** *of a body R is equal to the resultant external force* **F** *that acts on the body at the considered instant.* Let **v**(**x**, t) be the velocity field of the solid at instant t. Euler's first law may be written as

$$\dot{\mathbf{P}} = \int_{\partial R} \mathbf{t}\, dS + \int_R \rho\, \mathbf{f}\, dV, \tag{17.1.3a}$$

or in component form,

$$\dot{P}_i = \int_{\partial R} t_i\, dS + \int_R \rho\, f_i\, dV, \tag{17.1.3b}$$

where

$$\dot{\mathbf{P}} \equiv \frac{d}{dt}\mathbf{P} = \frac{d}{dt}\{\int_R \rho\, \mathbf{v}\, dV\} \tag{17.1.3c}$$

or, in component form,

$$\dot{P}_i \equiv \frac{d}{dt}P_i = \frac{d}{dt}\{\int_R \rho\, v_i\, dV\}. \tag{17.1.3d}$$

Here a superposed dot (≡ d/dt) is the *material time-derivative* which for the infinitesimal theory may be interpreted as the partial time-derivative, since in that case there exists no difference between the *material* and *spatial* time-derivatives.

Euler's second law is concerned with angular momentum or moment of momentum. It states that the *instantaneous rate of change of angular momentum* **H** *of a body R is equal to the resultant external torque* **L** *that acts on the body at the considered instant.* The second law becomes

$$\dot{\mathbf{H}} = \int_{\partial R} \mathbf{x} \times \mathbf{t}\, dS + \int_R \rho\, \mathbf{x} \times \mathbf{f}\, dV, \tag{17.1.4a}$$

[2] Interested readers will find a simple account in a paper by Koiter (1964).

or in component form,

$$\dot{H}_i = \int_{\partial R} e_{ijk}\, x_j\, t_k\, dS + \int_R \rho\, e_{ijk}\, x_j\, f_k\, dV, \tag{17.1.4b}$$

where the angular momentum relative to the origin of coordinates is defined by

$$\mathbf{H} = \int_R \rho\, \mathbf{x} \times \mathbf{v}\, dV \qquad \text{or} \qquad H_i = \int_R \rho\, e_{ijk}\, x_j\, v_k\, dV. \tag{17.1.4c,d}$$

17.2. TRACTION VECTORS AND STRESS TENSOR

Two key continuum concepts are introduced in this section. They are the concept of *internal* surface tractions and the concept of *stress tensor* which are then used to express tractions transmitted across any elementary material surface. Then applying Euler's laws to an arbitrary region of a continuum, local field equations of equilibrium, called Cauchy's laws, are obtained.

17.2.1. Traction Vectors

A basic concept which characterizes the continuum theory, that is, the concept of *internal surface tractions* is now introduced. This concept may be stated as follows: the mechanical action of the material points which are situated on one side of an arbitrary material surface within a body, upon those on the other side, can be completely accounted for by prescribing a suitable set of traction vectors on this surface. With this concept (assumption), a part of a body can be removed and in its place suitable surface tractions prescribed on the newly formed boundaries, without affecting the motion and deformation of the remaining part of the body; such a removal of matter must not, of course, alter body forces which may include the mutual loads acting on the remaining part of the solid.

At a generic point $\mathbf{x}$, consider a surface element dS and let $\mathbf{n}$ denote a unit vector normal to this element. The mechanical action of the material points situated on the side of dS toward which $\mathbf{n}$ is pointing, upon those on the other side, is represented by surface tractions $\mathbf{t}^{(\mathbf{n})}$ acting on dS; see Figure 17.2.1. Clearly enough, one expects that these surface tractions should, in general, depend on the orientation of the element dS, specified by $\mathbf{n}$, as well as on its position $\mathbf{x}$, but not on its shape. Since infinitely many directions can be identified at a given point, it is clear that at each point, infinitely many traction vectors can also be identified, each acting on an element with a given unit normal. *These traction vectors are not, however, all independent, and according to Cauchy's theorem, they may all be expressed in terms of traction vectors on three distinct planes that pass through the considered point.*

At point $\mathbf{x}$, choose three orthogonal planes which are parallel to the coordinate planes and have $-\mathbf{e}_i$ (i = 1, 2, 3), as unit normals. Denote by $-\mathbf{t}^{(\mathbf{e}_i)}$

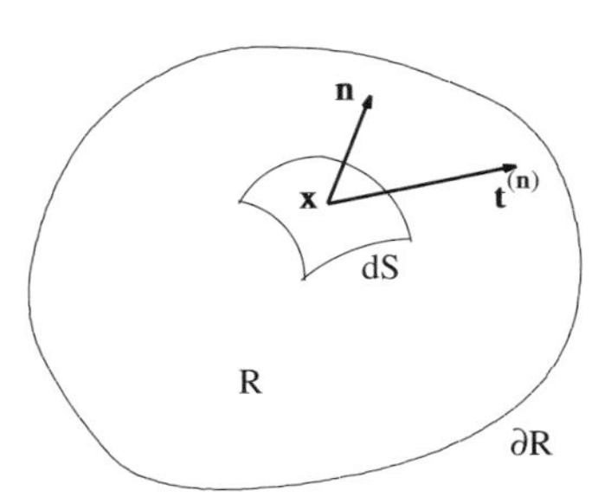

Figure 17.2.1

Surface traction $\mathbf{t}^{(\mathbf{n})}$ acting on elementary surface dS of body R

(i = 1, 2, 3), the traction vector on the plane whose unit normal is $-\mathbf{e}_i$. Then, consider a small tetrahedron with vertex at $\mathbf{x}$, and three faces that pass through $\mathbf{x}$ parallel to the coordinate planes and have $-\mathbf{e}_i$ (i = 1, 2, 3), as *exterior* unit normals. Let the fourth face of this tetrahedron be at a distance h from $\mathbf{x}$, and denote its area by dS, its exterior unit normal by $\mathbf{n}$, and the traction vectors acting on it by $\mathbf{t}^{(\mathbf{n})}$ which represents the action exerted by the material outside of the tetrahedron upon that inside. Now this tetrahedron can be isolated from the rest of the body and its motion can be studied. Apply Euler's first law (17.1.3), and, as height h shrinks to zero, obtain (see Figure 17.2.2)

$$-(\mathbf{t}^{(\mathbf{e}_1)}n_1 + \mathbf{t}^{(\mathbf{e}_2)}n_2 + \mathbf{t}^{(\mathbf{e}_3)}n_3) + \mathbf{t}^{(\mathbf{n})} = \mathbf{0}, \tag{17.2.1}$$

where inertia and body forces are neglected because they are proportional to the volume which, compared with the surface area of the element, constitutes an infinitesimal quantity of a higher order.

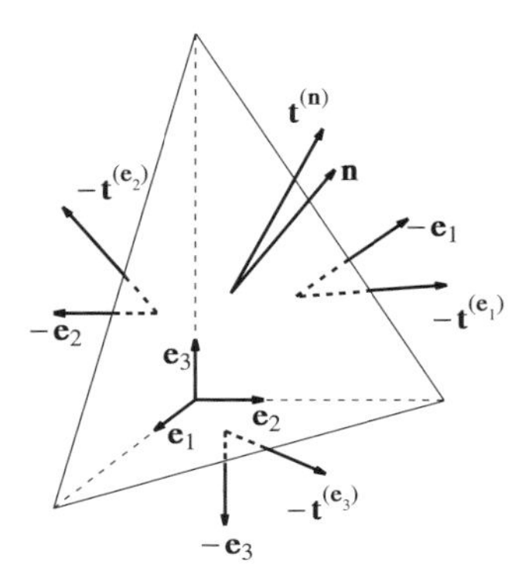

Figure 17.2.2

Tetrahedron with tractions acting on its surfaces, $\mathbf{t}^{(\mathbf{n})}$, $\mathbf{t}^{(\mathbf{e}_i)}$ (i = 1, 2, 3)

17.2.2. Stress Tensor

Equation (17.2.1) may be written as

$$\mathbf{t}^{(\mathbf{n})} = \mathbf{t}^{(\mathbf{e}_i)} n_i, \qquad (i = 1, 2, 3;\ i \text{ summed}), \tag{17.2.2}$$

which states that, at a typical point **x**, the traction vector on an arbitrary plane with exterior unit normal **n** is given as a linear and homogeneous vector-valued function of the direction cosines n_i; the coefficients in this linear relation are the traction vectors $\mathbf{t}^{(\mathbf{e}_i)}$ of the three orthogonal planes which pass through the considered point and are parallel to the coordinate planes. This is *Cauchy's theorem.*

If a plane with unit normal $\mathbf{m} = -\,\mathbf{n}$ is chosen, then

$$\mathbf{t}^{(\mathbf{m})} = \mathbf{t}^{(-\,\mathbf{n})} = -\,\mathbf{t}^{(\mathbf{e}_i)} n_i = -\,\mathbf{t}^{(\mathbf{n})} \tag{17.2.3}$$

which states that *traction vectors acting on opposite sides of the same surface element are equal in magnitude but opposite in direction.*

Let σ_{ij} denote the jth component of the traction vector on the surface with unit normal $\mathbf{e}_i$. Then

$$\sigma_{ij} = \mathbf{t}^{(\mathbf{e}_i)} \boldsymbol{\cdot} \mathbf{e}_j. \tag{17.2.4}$$

The traction vector $\mathbf{t}^{(\mathbf{n})}$ can also be expressed in terms of its components along the coordinate axes, i.e.,

$$\mathbf{t}^{(\mathbf{n})} = t_j^{(\mathbf{n})} \mathbf{e}_j, \tag{17.2.5a}$$

and thus (17.2.3) yields

$$t_j^{(\mathbf{n})} = n_i \sigma_{ij}. \tag{17.2.5b}$$

The quantity $\boldsymbol{\sigma} = \sigma_{ij} \mathbf{e}_i \otimes \mathbf{e}_j$ is called the *stress tensor.*[3] At a typical point, the traction vector is given by

$$\mathbf{t}^{(\mathbf{n})} = \mathbf{n} \boldsymbol{\cdot} \boldsymbol{\sigma} = (n_i \sigma_{ij}) \mathbf{e}_j \tag{17.2.6}$$

which is the dot product of **n** with $\boldsymbol{\sigma}$. The component σ_{ij} of the stress tensor represents the orthogonal projection along the x_j-axis of the traction vector that acts on the plane normal to the x_i-direction. For the positive face[4] of this plane, positive σ_{ij} denotes a component that is directed toward the positive x_j-direction. This commonly employed sign convention is illustrated in Figure 17.2.3, where it is assumed that the components of $\boldsymbol{\sigma}$ are all positive; they are shown for all positive faces, and also for the negative face that is normal to the x_2-direction. The components σ_{11}, σ_{22}, and σ_{33}, which are normal to the corresponding coordinate planes, are called *normal stresses.* The tangential components σ_{12}, σ_{23}, and σ_{31} are called *shearing stresses.* Note that the normal stress σ_{11}, for

[3] The tensor character of $\boldsymbol{\sigma}$ follows from the fact that the left-hand side of (17.2.5b) is a vector for all choices of the unit vector **n**.

[4] The positive face of the plane normal to the x_i-direction is the one whose unit normal points in the positive x_i-direction.

example, is the normal component of the stress tensor along the x_1-direction, while the shearing stress σ_{12} is the tangential component of this tensor along the x_1,x_2-directions.

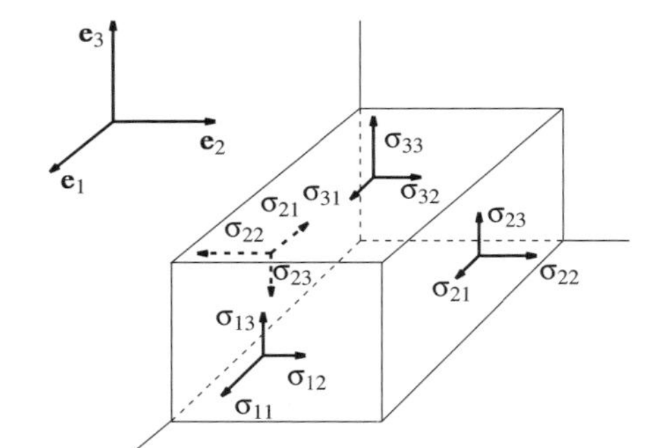

Figure 17.2.3

Components of Cauchy stress tensor

The normal stress on an element of area dS with unit normal $\mathbf{n}$ is the normal component of the corresponding traction vector $\mathbf{t}^{(\mathbf{n})}$ in the direction of $\mathbf{n}$. Denote this quantity by $N^{(\mathbf{n})}$, and obtain

$$N^{(\mathbf{n})} = \mathbf{t}^{(\mathbf{n})} \cdot \mathbf{n} = \boldsymbol{\sigma} : (\mathbf{n} \otimes \mathbf{n}) = \sigma_{ij}\, n_i\, n_j. \tag{17.2.7a}$$

The shearing stress on dS is the tangential component $S^{(\mathbf{n})}$ of $\mathbf{t}^{(\mathbf{n})}$; it is given by

$$(S^{(\mathbf{n})})^2 = \mathbf{t}^{(\mathbf{n})} \cdot \mathbf{t}^{(\mathbf{n})} - (N^{(\mathbf{n})})^2. \tag{17.2.7b}$$

Note that $S^{(\mathbf{n})}$ is defined as an unconditionally positive quantity.

17.2.3. Cauchy's Laws

Consider now the motion of a part B_1 of the solid. Denote by R_1 the instantaneous region occupied by B_1, and let ∂R_1 with the exterior unit normal $\mathbf{n}$ be its boundary surface. According to basic assumptions, the effect of the material points outside of this region upon those within the region is completely defined by the specification of surface tractions $\mathbf{t}^{(\mathbf{n})}$ on ∂R_1. Euler's laws of motion then yield

$$\int_{R_1} \rho\, (\mathbf{f} - \dot{\mathbf{v}})\, dV + \int_{\partial R_1} \mathbf{n} \cdot \boldsymbol{\sigma}\, dS = \mathbf{0},$$

$$\int_{R_1} \rho\, \mathbf{x} \times (\mathbf{f} - \dot{\mathbf{v}})\, dV + \int_{\partial R_1} \mathbf{x} \times (\mathbf{n} \cdot \boldsymbol{\sigma})\, dS = \mathbf{0}, \tag{17.2.8a,9a}$$

or in component form,

$$\int_{R_1} \rho\, (f_j - \dot{v}_j)\, dV + \int_{\partial R_1} n_i\, \sigma_{ij}\, dS = 0,$$

$$\int_{R_1} \rho\, e_{ijk}\, x_j\, (f_k - \dot{v}_k)\, dV + \int_{\partial R_1} e_{ijk}\, x_j\, (n_l \sigma_{lk})\, dS = 0, \qquad (17.2.8b,9b)$$

where (17.2.4) is used, and f_i is the component of the body force. Using the Gauss theorem,[5] these equations are reduced to

$$\int_{R_1} \{\sigma_{ij,i} + \rho\, (f_j - \dot{v}_j)\}\, dV = 0 \qquad (17.2.8c)$$

and

$$\int_{R_1} e_{ijk}\, \sigma_{jk}\, dV = 0, \qquad (17.2.9c)$$

where (17.2.8c) is used for deriving (17.2.9c). Since these equations are valid for an arbitrary part B_1 of the solid, the integrand must vanish, yielding the *Cauchy laws*

$$\nabla \boldsymbol{\cdot} \boldsymbol{\sigma} + \rho\, \mathbf{f} = \rho\, \dot{\mathbf{v}}, \qquad \boldsymbol{\sigma}^T = \boldsymbol{\sigma}, \qquad (17.2.10a,11a)$$

or in component form,

$$\sigma_{ij,\,i} + \rho\, f_j = \rho\, \dot{v}_j, \qquad \sigma_{ji} = \sigma_{ij}. \qquad (17.2.10b,11b)$$

Equation (17.2.10) expresses the conservation of linear momentum, while (17.2.11) is the statement of the conservation of angular momentum, when surface and body couples are absent; (17.2.11) shows that the stress tensor $\boldsymbol{\sigma}$ is symmetric.

For a surface element with the exterior unit normal $\mathbf{v}$, the traction vector $\mathbf{t}^{(\mathbf{v})} = \mathbf{v} \boldsymbol{\cdot} \boldsymbol{\sigma}$ must equal the externally applied surface traction $\mathbf{t}^o$. Thus

$$\mathbf{v} \boldsymbol{\cdot} \boldsymbol{\sigma} = \mathbf{t}^o \quad \text{or} \quad v_i\, \sigma_{ij} = t_j^o \quad \text{on } \partial R_\sigma, \qquad (17.2.12a,b)$$

where ∂R_σ denotes that part of the boundary of ∂R, upon which the surface tractions are applied. If ∂R_σ is traction free, then $\mathbf{v} \boldsymbol{\cdot} \boldsymbol{\sigma} = \mathbf{0}$ on ∂R_σ.

17.2.4. Principal Stresses

Since $\boldsymbol{\sigma}$ is a real, symmetric, second-order tensor, it has all the properties of such a tensor. Therefore, at each point $\mathbf{x}$ at time t, there exist three orthogonal planes upon which the traction vectors $\mathbf{t}^{(\mathbf{n})}$ are normal, the corresponding shearing stresses are zero, and the normal stresses are extrema. The directions of these planes are given by the principal directions of $\boldsymbol{\sigma}$. These orthogonal directions are called the *principal directions of stress.* The corresponding principal values N_J, J = I, II, III, are the principal stresses which are the roots of the equation

$$\det |\, \sigma_{ij} - N\, \delta_{ij} \,| = 0, \qquad (17.2.13a)$$

or, equivalently, the equation

[5] Note that the Gauss theorem can be used only if the stress tensor is sufficiently smooth; see Section 15. Hence, (17.2.8) is actually more than just a statement of the conservation of linear momentum.

$$\mathrm{N}^3 - \mathrm{I}_\sigma \mathrm{N}^2 + \mathrm{II}_\sigma \mathrm{N} - \mathrm{III}_\sigma = 0, \tag{17.2.13b}$$

where

$$\mathrm{I}_\sigma = \mathrm{tr}\,\boldsymbol{\sigma} = \sigma_{ii} = \mathrm{N_I} + \mathrm{N_{II}} + \mathrm{N_{III}},$$

$$\mathrm{II}_\sigma = \frac{1}{2} e_{ijk}\, e_{ilm}\, \sigma_{jl}\, \sigma_{km} = \frac{1}{2}(\mathrm{I}_\sigma^2 - \boldsymbol{\sigma} : \boldsymbol{\sigma}) = \mathrm{N_I N_{II}} + \mathrm{N_{II} N_{III}} + \mathrm{N_{III} N_I},$$

$$\mathrm{III}_\sigma = \det|\boldsymbol{\sigma}| = \frac{1}{6}(2\sigma_{ij}\,\sigma_{jk}\,\sigma_{ki} - 3\mathrm{I}_\sigma \boldsymbol{\sigma} : \boldsymbol{\sigma} + \mathrm{I}_\sigma^3) = \mathrm{N_I N_{II} N_{III}}, \tag{17.2.13c~e}$$

are the *basic invariants* of the stress tensor $\boldsymbol{\sigma}$. The principal stresses may be positive, negative, or zero; a positive value denotes tension, and a negative value, compression. The principal directions of the stress tensor are defined by

$$\mathbf{n}^J \boldsymbol{\cdot} \boldsymbol{\sigma} = \mathrm{N_J}\, \mathbf{n}^J, \qquad \mathrm{J = I, II, III} \qquad (\mathrm{J\ not\ summed}), \tag{17.2.14}$$

which shows that the corresponding shearing stresses are zero, and that the principal triad $\mathbf{n}^J$, J = I, II, III, is orthogonal, i.e.,

$$\mathbf{n}^J \boldsymbol{\cdot} \mathbf{n}^K = \begin{cases} 1 & \text{if } \mathrm{J = K} \\ 0 & \text{if } \mathrm{J \neq K}. \end{cases} \tag{17.2.15}$$

With respect to the principal triad, the matrix of the stress tensor is diagonal, and this tensor can be expressed as

$$\boldsymbol{\sigma} = \sum_{\mathrm{J=I}}^{\mathrm{III}} \mathrm{N_J}\, \mathbf{n}^J \otimes \mathbf{n}^J, \tag{17.2.16a}$$

or in component form,

$$\sigma_{ij} = \sum_{\mathrm{J=I}}^{\mathrm{III}} \mathrm{N_J}\, n_i^J\, n_j^J. \tag{17.2.16b}$$

Employing an analysis similar to that which led to (16.2.27), it is concluded that

$$\mathrm{S} = \frac{1}{2}\,|\mathrm{N_J} - \mathrm{N_K}| \qquad \mathrm{J \neq K}, \tag{17.2.17a}$$

defines the extremum values of the shearing stresses acting on the orthogonal planes with unit normals

$$\mathbf{m} = \frac{1}{\sqrt{2}}\,(n^J \pm n^K) \qquad \mathrm{J \neq K}. \tag{17.2.17b}$$

17.3. GEOMETRICAL REPRESENTATION OF STRESS TENSOR

Two geometrical representations of the second-order symmetric stress tensor $\boldsymbol{\sigma}$ are often considered. They are Mohr's circle and a quadratic form. These representations apply to any second-order symmetric tensor, e.g., the strain tensor.

17.3.1. Mohr's Circles

Tractions transmitted across a plane at a point can be represented by the coordinates of a point in a two-dimensional space, with the aid of the Mohr circle. Let the directions of the coordinate axes x_1, x_2, and x_3 coincide with the principal directions of the stress, and label them in such a manner that $N_I > N_{II} > N_{III}$. The normal stress $N^{(\mathbf{n})}$ transmitted across an element with the unit vector $\mathbf{n}$ is

$$N^{(\mathbf{n})} = N_I n_I^2 + N_{II} n_2^2 + N_{III} n_3^2, \tag{17.3.1a}$$

and the corresponding shearing stress is given by

$$(S^{(\mathbf{n})})^2 = N_I^2 n_I^2 + N_{II}^2 n_2^2 + N_{III}^2 n_3^2 - (N^{(\mathbf{n})})^2. \tag{17.3.1b}$$

To simplify the notation, set $N \equiv N^{(\mathbf{n})}$ and $S \equiv S^{(\mathbf{n})}$, and using the condition

$$n_I^2 + n_2^2 + n_3^2 = 1, \tag{17.3.2}$$

solve (17.3.1a) and (17.3.1b) for the direction cosines n_i. Thus,

$$n_I^2 = \frac{(N_{II} - N)(N_{III} - N) + S^2}{(N_{II} - N_I)(N_{III} - N_I)},$$

$$n_2^2 = \frac{(N_{III} - N)(N_I - N) + S^2}{(N_{III} - N_{II})(N_I - N_{II})},$$

$$n_3^2 = \frac{(N_I - N)(N_{II} - N) + S^2}{(N_I - N_{III})(N_{II} - N_{III})}. \tag{17.3.3a~c}$$

Since the denominators of (17.3.3a) and (17.3.3c) are positive and that of (17.3.3b) is negative, the admissible values of N and S must satisfy the following inequalities:

$$S^2 + \left[N - \frac{1}{2}(N_{II} + N_{III})\right]^2 - \frac{1}{4}(N_{II} - N_{III})^2 > 0,$$

$$S^2 + \left[N - \frac{1}{2}(N_{III} + N_I)\right]^2 - \frac{1}{4}(N_{III} - N_I)^2 < 0,$$

$$S^2 + \left[N - \frac{1}{2}(N_I + N_{II})\right]^2 - \frac{1}{4}(N_I - N_{II})^2 > 0. \tag{17.3.4a~c}$$

In the N,S-plane, a stress point with abscissa N and ordinate S corresponds to a real direction $\mathbf{n}$ if (17.3.4a~c) is satisfied. Setting the left-hand side of (17.3.4a~c) equal to zero, the equations of three circles which are labeled as I, II, and III, respectively are obtained. The centers of these circles lie on the N-axis at points $C_I = (N_{II} + N_{III})/2$, $C_{II} = (N_{III} + N_I)/2$, and $C_{III} = (N_I + N_{II})/2$, respectively, and their respective radii are $R_I = (N_{II} - N_{III})/2$, $R_{II} = (N_{III} - N_I)/2$, and $R_{III} = (N_I - N_{II})/2$. All admissible stress points are in the upper half-plane (since S is by definition nonnegative), inside or on circle II, and outside or on circles I and III. The stress points on circle I correspond to the directions that lie in the x_2,x_3-plane, since for these directions $n_1 = 0$. Similarly, stress points on circles II and III correspond to the directions for which $n_2 = 0$ and $n_3 = 0$, respectively. The stress point associated with a plane defined by any two directions can now be easily obtained; see Prager (1961) for details.

When two of the principal stresses vanish at a point, the state of stress is said to be locally *uniaxial*, with the axis of stress defined by the principal axis that is associated with the nonzero principal stress. The stress field is said to be uniaxial, if the state of stress is uniaxial everywhere in R.

The state of stress is said to be *plane* at a point, if one principal stress vanishes there. The plane of the stress is specified by the principal axes which correspond to the nonzero principal stresses. The stress field is said to be plane if the state of stress is plane everywhere in R.

Suppose that the state of plane stress at a point is specified by three stress components, σ_{11}, σ_{22}, and $\sigma_{12} = \sigma_{21}$, where σ_{11} and σ_{22} are not the principal stresses. To obtain the corresponding stress circle, locate in the N,S-plane two points P and P′ with coordinates $(\sigma_{11}, -\sigma_{12})$ and $(\sigma_{22}, \sigma_{12})$, respectively. The line PP′ intersects the N-axis at the center $C = (\sigma_{11} + \sigma_{22})/2$ of the stress circle whose radius is defined by $R^2 = \sigma_{12}^2 + (\sigma_{11} - \sigma_{22})^2/4$. The principal stresses then are

$$N_{I,II} = \frac{1}{2}(\sigma_{11} + \sigma_{22}) \pm \{\sigma_{12}^2 + \frac{1}{4}(\sigma_{11} - \sigma_{22})^2\}^{1/2}, \tag{17.3.5}$$

and the angle between the orientation of the plane of the major principal stress and the plane with traction $(\sigma_{11}, \sigma_{22})$, ψ, is defined by[6]

$$\tan 2\psi = -\frac{2\sigma_{12}}{\sigma_{11} - \sigma_{22}}. \tag{17.3.6}$$

17.3.2. Quadratic Form

There exists another geometric representation for a second-order symmetric tensor that may be used to study the stress or strain at a point in the solid. This is briefly discussed in connection with the stress tensor $\boldsymbol{\sigma}$. At a point $\mathbf{x}$ in the solid, a right-handed rectangular Cartesian coordinate system with the axes y_i, $i = 1, 2, 3$, is introduced parallel to the corresponding x_i-axes. The quadratic surface

$$\pm\phi = \boldsymbol{\sigma} : (\mathbf{y} \otimes \mathbf{y}) = \sigma_{ij}\, y_i\, y_j \tag{17.3.7}$$

is called *Cauchy's stress quadric,* where the sign of the left side is chosen so that (17.3.7) represents a real surface. The normal stress, transmitted at this

[6] Note that the sign convention for Mohr's circle is usually not the same as that defined in Figure 17.2.3.

point across an element with the unit vector **n**, is

$$N^{(\mathbf{n})} = \boldsymbol{\sigma} : (\mathbf{n} \otimes \mathbf{n}) = \frac{1}{r^2}\, \boldsymbol{\sigma} : (\mathbf{y} \otimes \mathbf{y}) = \pm \frac{1}{r^2}\, \phi, \tag{17.3.8a}$$

where $\mathbf{y} = r\,\mathbf{n}$, and r is the length of the radius vector of the quadratic surface measured along the unit normal **n**. The traction vector acting on this element is

$$\mathbf{t}^{(\mathbf{n})} = \mathbf{n} \cdot \boldsymbol{\sigma} = \frac{1}{r}\, \mathbf{y} \cdot \boldsymbol{\sigma} = \frac{1}{2r}\, \boldsymbol{\nabla}_y(\pm\,\phi). \tag{17.3.8b}$$

Since the right-hand side of (17.3.8b) is proportional to the gradient of ϕ, it is concluded that the direction of the traction vector on the considered element is perpendicular to a plane tangent to the quadratic surface at the terminus of the radius vector normal to that element. The magnitude of the traction vector is inversely proportional to the length of this radius vector.

A quadratic form has three principal axes which are orthogonal, and correspond to radius vectors with stationary lengths. These directions coincide with the principal directions of stress, and, as is clear from (17.3.8a), define the orientation of surface elements with stationary normal stresses.

The stress quadric may be an ellipsoid, a hyperboloid, or any one of their degenerate forms. The quadric is ellipsoidal if the principal stresses are distinct and of the same sign, and it is a hyperboloid if these stresses are distinct but one has a different sign from the other two. When two of the principal stresses are equal at a point, there is only one unique principal axis for the stress quadric which, in this case, reduces to a surface of revolution. The axis of revolution is defined by the direction of the distinct principal stress. Normal to this axis, any two orthogonal directions may be taken as the principal axes. If all three principal stresses are equal at a particle, then the stress quadric reduces to a sphere. In this case, the state of stress is isotropic, i.e., $\sigma_{ij} = p\,\delta_{ij}$, where p denotes the normal stress common to all elements of the area passing through the considered point. A state of stress of this type is called *spherical.* Note that, in general, the stress tensor $\boldsymbol{\sigma}$ may be decomposed into a spherical part and a *deviator* $\boldsymbol{\sigma}'$ whose trace is zero, i.e.,

$$\boldsymbol{\sigma} = \boldsymbol{\sigma}' + \frac{1}{3}\, I_\sigma\, \mathbf{1}^{(2)}, \tag{17.3.9}$$

where $I_\sigma = \sigma_{ii}$ is the first stress-invariant. The spherical part of the stress tensor is also called the *mean normal stress.*

17.4. REFERENCES

Koiter, W. T. (1964), Couple stresses in the theory of elasticity, Parts I and II, *Akademe van Wetenshappen Series B*, Vol. 67, 17-44.

Prager, W. (1961), An elementary discussion of definitions of stress rate, *J Q. Appl. Math.*, 403-407.

SECTION 18 CONSTITUTIVE RELATIONS

Within the context of the considered infinitesimal theory, certain aspects of the kinematics and dynamics of deformable continua are discussed in the preceding two sections. With the stated qualifications, these results apply to continuous media of various kinds that consist of materials of diverse constitutions. To characterize the constitution of a material that comprises a continuum, the *constitutive relations* which relate the kinematic and dynamic ingredients of the theory are introduced. In this section constitutive relations of certain solid bodies whose behavior can be characterized with sufficient accuracy by assuming that they consist of a *perfectly elastic* material are considered.

In order to define the constitutive equations for elastic materials, start from the conservation of energy and relate the strain and stress in the continuum to the *internal energy* stored in the material. The elastic strain energy and its conjugate, the complementary elastic energy, are then defined in terms of the strain and stress as a special case. For linear elasticity, the elastic strain (complementary) energy is quadratic in strain (stress), with the fourth-order elasticity (compliance) tensor defining the corresponding coefficients. The symmetry properties of the material are then represented by those of the fourth-order elasticity (compliance) tensor; see Section 3.

18.1. STRAIN ENERGY DENSITY

18.1.1. Conservation Laws

There are four conservation laws governing the motion and deformation of a continuum: 1) conservation of mass, 2) conservation of linear momentum, 3) conservation of angular momentum, and 4) conservation of energy. For the infinitesimal theory of elasticity, the first conservation law implies that the mass-density of the solid does not change in time, i.e.,

$$\rho = \rho(\mathbf{x};\, t) = \rho_0(\mathbf{x}), \tag{18.1.1}$$

where ρ_0 and ρ are the initial and instantaneous mass-densities, respectively. This is because the mass of an arbitrary part of the continuum, R_1, is given by

$$\int_{R_1} \rho(\mathbf{x};\, t)\, J(\mathbf{x};\, t)\, dV = \int_{R_1} \rho_0(\mathbf{x})\, dV, \tag{18.1.2}$$

where J is the Jacobian, and the left-hand side of (18.1.2) is the mass of R_1 at the initial state. Since J equals 1 for an infinitesimally small deformation, and since the above integral remains constant for arbitrary R_1 and t, the field equation

(18.1.1) must hold for every point in the continuum.

The second and third conservation laws are discussed in Section 17 in the form of Euler's laws. Both are stated for a finite part of the continuum. In the same manner that (18.1.1) is derived from (18.1.2), the two field equations, (17.2.10) and (17.2.11), which hold at every point of the continuum, are derived from Euler's laws. These field equations are Cauchy's laws, or the equations of motion.

Now consider the last conservation law, namely, the conservation of energy.[1] To this end, the instantaneous *specific internal energy density* (per unit mass), denoted by e, which is the measure of the stored energy in the continuum, is introduced. Let $\mathbf{v}$ be the instantaneous velocity field of an arbitrary region R_1 with a regular surface ∂R_1 within the continuum R. The internal and kinetic energies contained within R_1 then are

$$E = \int_{R_1} \rho\, e \, dV, \qquad K = \int_{R_1} \frac{1}{2} \rho\, \mathbf{v} \cdot \mathbf{v} \, dV. \tag{18.1.3a,b}$$

If the rate of heat supply is L, the conservation of energy asserts that

$$\dot{K} + \dot{E} = F + L, \tag{18.1.4}$$

where

$$F = \int_{\partial R_1} \mathbf{t}^{(\mathbf{n})} \cdot \mathbf{v} \, dS + \int_{R_1} \rho\, \mathbf{f} \cdot \mathbf{v} \, dV \tag{18.1.3c}$$

is the rate at which the prescribed surface tractions $\mathbf{t}^{(\mathbf{n})}$ and body forces $\mathbf{f}$ do work on the body R_1. Note that, (18.1.4) may be viewed as defining the internal energy E, since the heat supply L is supposed to be known.

Let $\mathbf{q}$ denote the *heat flux* through the surface ∂R_1, and h the heat created per unit mass in the body; h may be, for example, created by radiation-absorption. The heat supply L then is

$$L = -\int_{\partial R_1} \mathbf{q} \cdot \mathbf{n} \, dS + \int_{R_1} \rho\, h \, dV = \int_{R_1} (-\boldsymbol{\nabla} \cdot \mathbf{q} + \rho\, h) \, dV, \tag{18.1.3d}$$

where $\mathbf{n}$ denotes the exterior unit normal to ∂R_1. Substitution of (17.2.6) into $\mathbf{t}^{(\mathbf{n})}$ in (18.1.3c) and then (18.1.3a~d) into (18.1.4) now yields

$$\int_{R_1} \left\{ \{\rho\, \dot{e} - (\boldsymbol{\sigma} : \dot{\boldsymbol{\varepsilon}} - \boldsymbol{\nabla} \cdot \mathbf{q} + \rho\, h)\} - \{\boldsymbol{\sigma} : \dot{\boldsymbol{\omega}}\} - \mathbf{v} \cdot \{\boldsymbol{\nabla} \cdot \boldsymbol{\sigma} + \rho\, \mathbf{f} - \rho\, \dot{\mathbf{v}}\} \right\} dV = 0, \tag{18.1.5a}$$

where $\dot{\boldsymbol{\varepsilon}}$ and $\dot{\boldsymbol{\omega}}$ are the strain- and rotation-rate tensors, respectively, i.e., $\dot{\boldsymbol{\varepsilon}} = \{(\boldsymbol{\nabla} \otimes \mathbf{v}) + (\boldsymbol{\nabla} \otimes \mathbf{v})^{T}\}/2$ and $\dot{\boldsymbol{\omega}} = \{(\boldsymbol{\nabla} \otimes \mathbf{v}) - (\boldsymbol{\nabla} \otimes \mathbf{v})^{T}\}/2$. Since the body R_1 is arbitrary,

$$\left\{ \rho\, \dot{e} - (\boldsymbol{\sigma} : \dot{\boldsymbol{\varepsilon}} - \boldsymbol{\nabla} \cdot \mathbf{q} + \rho\, h) \right\} - \left\{ \boldsymbol{\sigma} : \dot{\boldsymbol{\omega}} \right\} - \mathbf{v} \cdot \left\{ \boldsymbol{\nabla} \cdot \boldsymbol{\sigma} + \rho\, \mathbf{f} - \rho\, \dot{\mathbf{v}} \right\} = 0. \tag{18.1.5b}$$

---

[1] In a purely thermomechanical system, the conservation of energy asserts the equivalence of heat energy and mechanical work.

The quantities in the last two sets of curly brackets vanish because of the conservation of linear and angular momenta, (17.2.10) and (17.2.11), leading to

$$\rho\,\dot{e} = \boldsymbol{\sigma} : \dot{\boldsymbol{\varepsilon}} - \boldsymbol{\nabla}\boldsymbol{\cdot}\mathbf{q} + \rho\,h\,, \tag{18.1.6a}$$

or in component form,

$$\rho\,\dot{e} = \sigma_{ij}\,\dot{\varepsilon}_{ij} - q_{i,i} + \rho\,h\,, \tag{18.1.6b}$$

which is the statement of local energy conservation. It states that the change of internal energy density is due to the rate of stress work, $\boldsymbol{\sigma} : \dot{\boldsymbol{\varepsilon}} = \sigma_{ij}\,\dot{\varepsilon}_{ij}$, and heating.

It is of interest to note that (18.1.4) includes, in addition to the energy conservation (18.1.6), statements of the balance of linear and angular momenta. (For finite deformation, it also contains the conservation of mass.) Indeed, it is argued by some authors that Cauchy's laws can be *deduced* from the energy balance (18.1.4), using the *invariance* of the quantities e, $\boldsymbol{\sigma}$, and $\mathbf{f}$ under superposed rigid-body motions. To this end, one writes (18.1.5) for a body that is in the same configuration at time t, but has a velocity field $\mathbf{v} + \mathbf{v}^{\mathrm{o}}$, where $\mathbf{v}^{\mathrm{o}}$ is a constant velocity field, and then subtracts the resulting equation from (18.1.5), to obtain

$$\mathbf{v}^{\mathrm{o}}\boldsymbol{\cdot}\left\{\,\boldsymbol{\nabla}\boldsymbol{\cdot}\boldsymbol{\sigma} + \rho\,\mathbf{f} - \rho\,\dot{\mathbf{v}}\,\right\} = 0. \tag{18.1.7}$$

This must hold for all $\mathbf{v}^{\mathrm{o}}$, yielding (17.2.10). One now repeats the same argument, using the velocity field $\mathbf{v} + \mathbf{x}\boldsymbol{\cdot}\boldsymbol{\omega}^{\mathrm{o}}$, where $\boldsymbol{\omega}^{\mathrm{o}}$ is a constant antisymmetric tensor, and obtains

$$\boldsymbol{\sigma} : \boldsymbol{\omega}^{\mathrm{o}} = 0. \tag{18.1.8}$$

This must hold for all $\boldsymbol{\omega}^{\mathrm{o}}$, yielding (17.2.11). Finally, (18.1.4), (18.1.7), and (18.1.8) yield the local energy equation (18.1.6).

18.1.2. Strain Energy Density Function w

When the heat supply L is zero, and the heat produced in the solid by the deformation is neglected (uncoupled thermoelasticity theory), (18.1.6) becomes

$$\rho\,\dot{e} = \boldsymbol{\sigma} : \dot{\boldsymbol{\varepsilon}}. \tag{18.1.9}$$

From (18.1.9), and for elastic materials, define the *strain energy density function* w such that the rate of internal energy density $\rho\dot{e}$ is the rate of change of w, i.e.,

$$\dot{\mathrm{w}} \equiv \rho\,\dot{e}. \tag{18.1.10}$$

Equation (18.1.4) then takes on the form

$$F = \frac{\mathrm{d}}{\mathrm{dt}}\left[\,\int_{\mathrm{R}_1} \mathrm{w}\;\mathrm{dV}\,\right] + \dot{K} \tag{18.1.11}$$

which states that the rate of work of the applied loads is equal to the sum of the rates of change of the strain and kinematic energies of the solid. This equation is a consequence of, and can be deduced from the conservation of linear momentum (17.2.10). To this end, multiply both sides of (17.2.10) by **v**, integrate the results over R_1, and using the Gauss theorem, obtain (18.1.11).

If the deformation is infinitesimally small, integrating (18.1.10) with respect to time t, obtain

$$w = \int_L \boldsymbol{\sigma} : d\boldsymbol{\varepsilon} = \int_L \sigma_{ij}\, d\varepsilon_{ij}, \tag{18.1.12}$$

where L stands for the integration path in the strain space.

18.2. LINEAR ELASTICITY

18.2.1. Elasticity

First, assume that the integral (18.1.12) is path-independent. The change in the strain energy density function from an initial strain $\boldsymbol{\varepsilon}^0$ to a final strain $\boldsymbol{\varepsilon}^1$ over any path in the strain-space then is given by

$$w(\boldsymbol{\varepsilon}^1) - w(\boldsymbol{\varepsilon}^0) = \int_{\boldsymbol{\varepsilon}^0}^{\boldsymbol{\varepsilon}^1} \boldsymbol{\sigma} : d\boldsymbol{\varepsilon}. \tag{18.2.1}$$

Since, for an elastic material the stress tensor $\boldsymbol{\sigma}$ depends only on the strain tensor $\boldsymbol{\varepsilon}$ (thermal effects are neglected), then

$$\boldsymbol{\sigma} = \boldsymbol{\sigma}(\boldsymbol{\varepsilon}), \tag{18.2.2a}$$

and hence

$$w(\boldsymbol{\varepsilon}) = \int_0^{\boldsymbol{\varepsilon}} \boldsymbol{\sigma}(\boldsymbol{\varepsilon}) : d\boldsymbol{\varepsilon}. \tag{18.2.2b}$$

Here, $\boldsymbol{\varepsilon} = \mathbf{0}$ at the initial state corresponds to a *preferred natural state,* to which the body returns whenever all external loads and imposed surface displacements that maintain the body in a deformed state are slowly released. The strain energy density at the preferred natural state is set at zero. From (18.2.2b) it follows that

$$\frac{\partial w}{\partial \boldsymbol{\varepsilon}}(\boldsymbol{\varepsilon}) = \boldsymbol{\sigma}(\boldsymbol{\varepsilon}), \tag{18.2.2c}$$

or in component form,

$$\frac{\partial w}{\partial \varepsilon_{ij}}(\boldsymbol{\varepsilon}) = \sigma_{ij}(\boldsymbol{\varepsilon}), \qquad i, j = 1, 2, 3. \tag{18.2.2d}$$

By means of a Legendre transformation, the relation (18.2.2) can be inverted and the strain expressed in terms of the stress. To this end, define w^c by

$$\mathrm{w^c} \equiv \mathrm{w^c}(\boldsymbol{\sigma}) \equiv \boldsymbol{\sigma} : \boldsymbol{\varepsilon} - \mathrm{w}(\boldsymbol{\varepsilon}), \tag{18.2.3a}$$

and by direct differentiation, obtain, with the aid of (18.2.2c,d),

$$\frac{\partial \mathrm{w^c}}{\partial \boldsymbol{\sigma}}(\boldsymbol{\sigma}) = \boldsymbol{\varepsilon}(\boldsymbol{\sigma}), \tag{18.2.3b,c}$$

or

$$\frac{\partial \mathrm{w^c}}{\partial \sigma_{ij}}(\boldsymbol{\sigma}) = \varepsilon_{ij}(\boldsymbol{\sigma}), \qquad i,j = 1, 2, 3.$$

The function $\mathrm{w^c}$ is called the *complementary strain energy density.* Since

$$\mathrm{d}(\boldsymbol{\sigma} : \boldsymbol{\varepsilon}) = \boldsymbol{\sigma} : \mathrm{d}\boldsymbol{\varepsilon} + \mathrm{d}\boldsymbol{\sigma} : \boldsymbol{\varepsilon} = \mathrm{dw}(\boldsymbol{\varepsilon}) + \mathrm{dw^c}(\boldsymbol{\sigma}), \tag{18.2.4}$$

the complementary strain energy density is expressed by the following integral:

$$\mathrm{w^c}(\boldsymbol{\sigma}) = \int_0^{\boldsymbol{\sigma}} \boldsymbol{\varepsilon}(\boldsymbol{\sigma}) : \mathrm{d}\boldsymbol{\sigma}. \tag{18.2.5}$$

18.2.2. Linear Elasticity

The simplest form of (18.2.2a) is *linear elasticity* which is characterized by the assumption that the stress tensor $\boldsymbol{\sigma}$ is connected to the strain tensor $\boldsymbol{\varepsilon}$ by a linear relation[2] called *the generalized Hooke's Law,*[3] i.e.,

$$\boldsymbol{\sigma} = \mathbf{C} : \boldsymbol{\varepsilon}, \tag{18.2.6a}$$

or in component form,

$$\sigma_{ij} = C_{ijkl}\, \varepsilon_{kl}. \tag{18.2.6b}$$

The fourth-order tensor $\mathbf{C}$ is called the *elasticity tensor,* and its components, the *elasticity coefficients.* For heterogeneous solids the elasticity tensor depends on the position $\mathbf{x}$. For a *homogeneous* material, $\mathbf{C}$ is constant; the elastic coefficients in this case are referred to as *elastic constants.*

The strain energy density function w now becomes

$$\mathrm{w}(\boldsymbol{\varepsilon}) = \frac{1}{2}\boldsymbol{\varepsilon} : \mathbf{C} : \boldsymbol{\varepsilon} = \frac{1}{2}\varepsilon_{ij}\, C_{ijkl}\, \varepsilon_{kl}. \tag{18.2.7}$$

Hence, without any loss in generality, regard $\mathbf{C}^T = \mathbf{C}$, i.e., C_{ijkl} is symmetric with respect to the exchange of ij and kl. Therefore, $\mathbf{C}$ belongs to $\mathcal{T}^{(4s)}$ (Section 15), and its corresponding six by six matrix is symmetric.

The compliance tensor $\mathbf{D}$ is defined as the inverse of $\mathbf{C}$ in $\mathcal{T}^{(4s)}$, i.e.,

$$\mathbf{D} : \mathbf{C} = \mathbf{C} : \mathbf{D} = \mathbf{1}^{(4s)}. \tag{18.2.8}$$

Hence, strain $\boldsymbol{\varepsilon}$ is given by the linear transformation of stress $\boldsymbol{\sigma}$,

[2] The relation must be homogeneous, since zero strain is to correspond to zero stress.

[3] Possible thermal stresses and strains are excluded from consideration, and an isothermal condition is assumed.

$$\boldsymbol{\varepsilon} = \mathbf{D} : \boldsymbol{\sigma}, \qquad \text{or} \qquad \varepsilon_{ij} = D_{ijkl}\,\sigma_{kl}. \tag{18.2.9a,b}$$

From (18.2.5), the complementary strain energy density function w^c becomes

$$w^c(\boldsymbol{\sigma}) = \frac{1}{2}\boldsymbol{\sigma} : \mathbf{D} : \boldsymbol{\sigma} = \frac{1}{2}\sigma_{ij}\,D_{ijkl}\,\sigma_{kl}. \tag{18.2.10}$$

18.3. ELASTICITY AND COMPLIANCE TENSORS

In this subsection the positive-definiteness and ellipticity of the fourth-order tensors **C** and **D**, which guarantee the uniqueness of the solution and material stability, are briefly examined.

18.3.1. Positive-Definiteness

For any arbitrary nonzero strain and stress, measured from the preferred natural state, the elastic energies w and w^c must be nonnegative, i.e.,

$$w(\boldsymbol{\varepsilon}) = \frac{1}{2}\boldsymbol{\varepsilon} : \mathbf{C} : \boldsymbol{\varepsilon} = \frac{1}{2}C_{ijkl}\,\varepsilon_{ij}\,\varepsilon_{kl} \geq 0,$$

$$w^c(\boldsymbol{\sigma}) = \frac{1}{2}\boldsymbol{\sigma} : \mathbf{D} : \boldsymbol{\sigma} = \frac{1}{2}D_{ijkl}\,\sigma_{ij}\,\sigma_{kl} \geq 0, \tag{18.3.1a,b}$$

where equalities hold only for $\boldsymbol{\varepsilon} = \mathbf{0}$ and $\boldsymbol{\sigma} = \mathbf{0}$, respectively.

In the six-dimensional matrix representation used in Section 15, (18.3.1a,b) are expressed as

$$w = \frac{1}{2}[\varepsilon_a]^T[W_{ap}][C_{pq}][W_{qb}][\varepsilon_b] = \frac{1}{2}([W_{pa}][\varepsilon_a])^T[C_{pq}]([W_{qb}][\varepsilon_b]),$$

$$w^c = \frac{1}{2}[\sigma_a]^T[W_{ap}][D_{pq}][W_{qb}][\sigma_b] = \frac{1}{2}([W_{pa}][\sigma_a])^T[D_{pq}]([W_{qb}][\sigma_b]). \tag{18.3.2a,b}$$

Therefore, the six by six symmetric matrices $[C_{ab}]$ and $[D_{ab}]$ are *positive-definite*. Their eigenvalues and eigenvectors are all real. Furthermore, for $[C_{ab}]$ and $[D_{ab}]$ to be positive-definite, it is necessary and sufficient that their six eigenvalues be positive. This is guaranteed when the six principal minors of these matrices are all positive, i.e., when:

$$C_{11}, \qquad \det\begin{bmatrix} C_{11} & C_{12} \\ C_{21} & C_{22} \end{bmatrix}, \qquad \det\begin{bmatrix} C_{11} & C_{12} & C_{13} \\ C_{21} & C_{22} & C_{23} \\ C_{31} & C_{32} & C_{33} \end{bmatrix}, \qquad \ldots, \qquad \det[C_{ab}],$$

$$D_{11}, \qquad \det\begin{bmatrix} D_{11} & D_{12} \\ D_{21} & D_{22} \end{bmatrix}, \qquad \det\begin{bmatrix} D_{11} & D_{12} & D_{13} \\ D_{21} & D_{22} & D_{23} \\ D_{31} & D_{32} & D_{33} \end{bmatrix}, \qquad ..., \qquad \det[D_{ab}], \tag{18.3.3a,b}$$

are all positive; see Fedorove (1968). Note that the positive-definiteness of $\mathbf{C}$ (or $[C_{ab}]$) is equivalent to that of $\mathbf{D}$ (or $[D_{ab}]$).

The positive-definiteness of $\mathbf{C}$ implies the uniqueness of the solution of the boundary-value problem for a finite body. The proof is straightforward. Let R be a linearly elastic solid with boundary ∂R, and $\mathbf{C}$ be its elasticity tensor. Using the averaging theorem, express the volume integral of the elastic strain energy density by

$$\int_R 2\,w\,dV = \int_R \boldsymbol{\sigma} : \boldsymbol{\varepsilon}\,dV = \int_{\partial R} \mathbf{t} \cdot \mathbf{u}\,dS, \tag{18.3.4}$$

where $\mathbf{t}$ is the surface traction on ∂R. If the integrand of the right-hand side of (18.3.4) satisfies

$$\mathbf{t} \cdot \mathbf{u} = 0 \qquad \text{on } \partial R, \tag{18.3.5a}$$

then, from the positive-definiteness of w,

$$\boldsymbol{\varepsilon} = \mathbf{0} \qquad \text{and} \qquad \boldsymbol{\sigma} = \mathbf{0}. \tag{18.3.5b,c}$$

Therefore, if there are two solutions for the same boundary data, the difference solution corresponds to zero strain and zero stress fields.

18.3.2. Strong Ellipticity

Excluding body forces, from the equations of motion (Cauchy's laws), (17.2.10,11), and Hooke's law (18.2.6), obtain

$$\boldsymbol{\nabla} \cdot (\mathbf{C} : \boldsymbol{\nabla} \otimes \mathbf{u}) = \rho \, \frac{\partial^2 \mathbf{u}}{\partial t^2}, \tag{18.3.6}$$

where the symmetry of $\mathbf{C}$ is also used. For an infinitely extended solid with constant $\mathbf{C}$, (18.3.6) admits plane-wave solutions of the form $\mathbf{u} = \mathbf{u}(\mathbf{w} \cdot \mathbf{x} - Vt)$, where $\mathbf{w}$ is the unit normal of the plane, and V is the wave speed. For the wave speed V to be nonzero and real, the elasticity tensor $\mathbf{C}$ must satisfy the following condition: for arbitrary nonzero vectors $\mathbf{v}$ and $\mathbf{w}$, it must follow that

$$(\mathbf{v} \otimes \mathbf{w}) : \mathbf{C} : (\mathbf{v} \otimes \mathbf{w}) = C_{ijkl}\, v_i\, v_k\, w_j\, w_l > 0. \tag{18.3.7a}$$

This is called the *strong ellipticity condition.*[4]

Since the strain tensor $\boldsymbol{\varepsilon}$ can be expressed in terms of its principal values and directions by $\boldsymbol{\varepsilon} = \sum_{J=I}^{III} \lambda_J\, \mathbf{m}^J \otimes \mathbf{m}^J$, (16.2.21a), the strong ellipticity (18.3.7a)

[4] If inequality > is replaced by ≥ in (18.3.7), *Hadamard's inequality* is obtained, which is the basic stability theorem for infinitesimally small deformations; see Truesdell and Noll (1965).

implies the positive-definiteness of **C**, and, conversely, the positive-definiteness of **C** implies the strong ellipticity. Because of the symmetry of **C**, (18.3.7a) can also be written as

$$\left[\frac{1}{2}(\mathbf{v}\otimes\mathbf{w}+\mathbf{w}\otimes\mathbf{v})\right]:\mathbf{C}:\left[\frac{1}{2}(\mathbf{v}\otimes\mathbf{w}+\mathbf{w}\otimes\mathbf{v})\right]>0. \tag{18.3.7b}$$

The tensor $(\mathbf{v}\otimes\mathbf{w}+\mathbf{w}\otimes\mathbf{v})/2$ belongs to $T^{(2s)}$. Hence if **C** is positive-definite, then the strong ellipticity is satisfied. Since the positive-definiteness of **C** is equivalent to that of **D**, the strong ellipticity of **C** is also equivalent to that of **D**. Furthermore, the strong ellipticity implies the uniqueness of the solution to elastostatic boundary-value problems.

18.4. REFERENCES

Fedorove, F. I. (1968), *Theory of elastic waves in crystals*, Plenum Press, New York.

Truesdell, C. and Noll, W. (1965), The non-linear field theories of mechanics in *Handbuch der Physik* (Flugge, S. ed.), Vol. III/3, Springer-Verlag, Berlin, 1-602.

CHAPTER VI

ELASTOSTATIC PROBLEMS OF LINEAR ELASTICITY

In this chapter, attention is focused on certain general properties of boundary-value problems in elastostatics. First, the displacement and traction boundary-value problems are formulated (Subsection 19.1), the concepts of kinematically admissible displacement field and statically admissible stress field are introduced (Subsection 19.2), and the field equations and the corresponding solutions are related to fundamental extremum principles of linear elasticity, i.e., the elastic energy potential (Subsection 19.3) and the complementary energy potential (Subsection 19.4). In addition, general variational principles which contain all basic kinematical, constitutive, and balance field equations as the associated Euler equations, are formulated and discussed (Subsection 19.5).

The elastokinematic field equations are considered in Section 20. The propagation of dilatational and rotational waves is formulated based on the Helmholtz decomposition theorem (Subsections 20.1 and 20.2). General solutions for three-dimensional problems are examined in terms of harmonic potentials, and the Green functions are developed for special problems (Subsections 20.3 and 20.4).

Then, attention is focused on two-dimensional problems in Section 21, where, with the aid of Airy's stress function (Subsection 21.1) and Muskhelishvili's complex potentials (Subsection 21.2), general solutions for a broad class of problems are derived and discussed, examining the Green functions associated with a point load, a center of dilatation, and a dislocation. The Hilbert problem is considered and illustrated in Subsection 21.3. The two-dimensional crack problems are discussed in Subsection 21.4 in terms of both Cauchy's and Hadamard's finite-part integrals. General solutions of plane anisotropic elasticity problems are presented in Subsection 21.5, using Muskhelishvili's two complex potentials. For illustration, unbridged and fully or partially bridged single straight cracks in an unbounded anisotropic solid are discussed. In Subsection 21.6, recent results on duality principles in anisotropic elasticity are reviewed.

SECTION 19 BOUNDARY-VALUE PROBLEMS AND EXTREMUM PRINCIPLES

First, a general boundary-value problem of an elastic solid R, is considered and formulated for either a displacement field or a stress field which satisfies the governing field equations in R and the prescribed boundary conditions on the surface ∂R. Next, an equivalent variational formulation of the boundary-value problem is given. To this end, two functionals, the total potential and complementary energies, are defined for the displacement and stress fields, respectively. The Euler equations for these functionals coincide with certain governing field equations in the boundary-value problem, and the natural boundary conditions for the Euler equations correspond to the prescribed boundary conditions on ∂R. Then the extremum principles for these variational problems are examined: the potential and complementary energies are not only stationary but are also extrema, if and only if the field variables satisfy all corresponding field equations.

19.1. BOUNDARY-VALUE PROBLEMS

For an elastic solid with a given geometry, the displacement, stress, and strain fields that correspond to prescribed sets of body forces $\mathbf{f}^o = \rho\,\mathbf{f}$, surface tractions $\mathbf{t}^o$, or surface displacements $\mathbf{u}^o$ (or a suitable combination of them) are sought. All three components of body forces $\mathbf{f}^o$ must be prescribed (they may be zero) everywhere within the region R that is occupied by the solid. On the surface ∂R which bounds the solid, on the other hand, some components of the surface tractions and the complementary components of the surface displacements can be defined, resulting in the so-called *mixed boundary-value problem,* e.g., either

$$u_i = u_i^o, \qquad \text{or} \qquad t_i = t_i^o, \qquad i = 1, 2, 3 \text{ on } \partial R, \tag{19.1.1}$$

is satisfied.[1] Envisaged also are problems in which surface tractions $\mathbf{t}^o$ are given on a part ∂R_T and surface displacements $\mathbf{u}^o$ on the remaining part $\partial R_U = \partial R - \partial R_T$ of the surface boundary ∂R. For *stress boundary-value problems,* all three components of $\mathbf{t}^o$ are prescribed on the entire[2] ∂R, while for *displacement boundary-value problems,* three displacement components u_i^o are specified on

[1] For example, t_1, t_3, and u_2 may be prescribed on a part of ∂R.

[2] These forces, together with the body forces, must constitute a self-equilibrating system of forces.

the entire ∂R.

The stress, strain, and displacement fields are to be determined from the following field equations:

$$\boldsymbol{\nabla}\boldsymbol{\cdot}\boldsymbol{\sigma} + \mathbf{f}^{o} = \mathbf{0}, \qquad \text{or} \qquad \sigma_{ij,i} + f_j^{o} = 0,$$

$$\boldsymbol{\sigma} = \mathbf{C} : \boldsymbol{\varepsilon}, \qquad \text{or} \qquad \sigma_{ij} = C_{ijkl}\, \varepsilon_{kl}, \qquad (19.1.2a\text{~}4b)$$

$$\boldsymbol{\varepsilon} = \frac{1}{2}\{(\boldsymbol{\nabla}\otimes\mathbf{u}) + (\boldsymbol{\nabla}\otimes\mathbf{u})^{T}\}, \qquad \text{or} \qquad \varepsilon_{ij} = \frac{1}{2}(u_{j,i} + u_{i,j}).$$

They must satisfy the prescribed boundary conditions on ∂R. For displacement boundary-value problems it is convenient to express (19.1.2) in terms of the displacement vector and its derivatives. Substitution from (19.1.4) and (19.1.3) into (19.1.2) yields

$$\boldsymbol{\nabla}\boldsymbol{\cdot}\{\mathbf{C} : (\boldsymbol{\nabla}\otimes\mathbf{u})\} + \rho\,\mathbf{f} = \mathbf{0}, \qquad (19.1.5a)$$

or in component form,

$$C_{ijkl}\, u_{l,ki} + \rho\, f_j = 0. \qquad (19.1.5b)$$

These are Navier's equations. Note that (19.1.5) requires that $\mathbf{u}$ and $\mathbf{f}^{o}$ be of class C^2 and C^0 respectively, while in (19.1.2), $\boldsymbol{\sigma}$ must be of class C^1 in R. If $\mathbf{f}^{o}$, for example, is not continuous in R, then derivatives of $\mathbf{u}$ may not be defined everywhere in R. If this occurs at a finite number of isolated points or surfaces in R, then (19.1.5) may still be satisfied "almost" everywhere in R. Navier's equations, (19.1.5), must be solved for $\mathbf{u}$, subject to the boundary conditions

$$\mathbf{u} = \mathbf{u}^{o} \quad \text{on } \partial R, \qquad (19.1.6a)$$

or

$$u_i = u_i^{o} \quad \text{on } \partial R. \qquad (19.1.6b)$$

The compatibility conditions (16.2.36) are redundant in the present case, since the displacement field $\mathbf{u}$ is being sought directly. Having obtained $\mathbf{u}$, the strain and stress fields are readily calculated from (19.1.4) and (19.1.3), respectively.

Consider now a traction boundary-value problem. In this case, (19.1.2) must be solved for the stress field $\boldsymbol{\sigma}$, subject to the boundary conditions

$$\boldsymbol{\nu}\boldsymbol{\cdot}\boldsymbol{\sigma} = \mathbf{t}^{o} \quad \text{on } \partial R, \qquad (19.1.7a)$$

or

$$\nu_i\, \sigma_{ij} = t_j^{o} \quad \text{or } \partial R. \qquad (19.1.7b)$$

The strain field is then given by

$$\boldsymbol{\varepsilon} = \mathbf{C}^{-1} : \boldsymbol{\sigma} = \mathbf{D} : \boldsymbol{\sigma} \qquad (19.1.8)$$

which must satisfy the following compatibility conditions,[3] in order to correspond to a unique displacement field:

---

[3] These conditions are obtained from (16.2.34) by setting k = l.

$$\varepsilon_{ij,kk} + \varepsilon_{kk,ij} = \varepsilon_{ik,jk} + \varepsilon_{jk,ik}. \tag{19.1.9a}$$

Note that since the stress and strain fields are unaffected by rigid-body displacements, uniqueness here is implied to within such rigid-body displacements. For the considered traction boundary-value problem, it is convenient to express (19.1.9) in terms of the stress components. In this case, the strain field given by $\mathbf{D} : \boldsymbol{\sigma}$ must be compatible, i.e., $\boldsymbol{\sigma}$ must satisfy

$$(D_{ijkl}\,\sigma_{kl})_{,mm} + (D_{mmkl}\,\sigma_{kl})_{,ij} = (D_{imkl}\,\sigma_{kl})_{,jm} + (D_{jmkl}\,\sigma_{kl})_{,im}. \tag{19.1.9b}$$

In particular, assuming isotropy, substitute (19.1.2) into (19.1.9b), and after simple manipulations, obtain

$$\sigma_{ij,kk} + \frac{2(\lambda+\mu)}{3\lambda+2\mu}\,\sigma_{kk,ij} + \frac{\lambda}{\lambda+2\mu}\,f^{o}_{k,k}\,\delta_{ij} + f^{o}_{i,j} + f^{o}_{j,i} = 0 \tag{19.1.10a}$$

which, because of the symmetry with respect to k and l, denotes six distinct equations that are known as the *Beltrami-Michell compatibility conditions.*[4] A stress field satisfying the equilibrium equations (19.1.2) and boundary condition (19.1.7) must also satisfy (19.1.9b), if it is to correspond to a continuous displacement field. Note that in terms of the Poisson ratio ν, (19.1.10a) becomes

$$\sigma_{ij,kk} + \frac{1}{1+\nu}\,\sigma_{ii,kl} + \frac{\nu}{1-\nu}\,f^{o}_{k,k}\,\delta_{ij} + f^{o}_{i,j} + f^{o}_{j,i} = 0. \tag{19.1.10b}$$

While it is of interest to establish the precise conditions under which the general elastostatic problems formulated above admit solutions, such an existence consideration is outside the scope of this book. Instead, it is assumed that the problems are suitably posed so that solutions exist. It should be recalled that, as shown in Section 18, *if the elasticity tensor* **C** (*or the compliance tensor* **D**) *is positive-definite, or strongly elliptic, then the solution to the boundary-value problem for a finite elastic solid is unique.*

19.2. KINEMATICALLY AND STATICALLY ADMISSIBLE FIELDS

19.2.1. Kinematically Admissible Displacement Field

A sufficiently smooth (actually of class C^1) displacement field that complies with all the geometrical (displacement) boundary conditions of a given elastostatic boundary-value problem is referred to as a *kinematically admissible displacement field.* Denote the set of these displacement fields by V^k, i.e.,

---

[4] Note that if (19.1.10a) is to be satisfied strictly *everywhere* in R, then $\mathbf{f}^o$ must be of class C^1 in this region. The stress field then is of class C^2, resulting in a displacement field which is of class C^3 everywhere in R. A function $f(\mathbf{x})$ is called of class C^n in R if the function and all its derivatives up to and including the nth derivative are continuous in R.

$$V^k \equiv \{\mathbf{u} \mid \mathbf{u} \in C^1 \text{ in R and } \mathbf{u} = \mathbf{u}^o \text{ on } \partial R_U\}, \tag{19.2.1}$$

where ∂R_U is the part of the surface ∂R on which $\mathbf{u}^o$ is prescribed.

The kinematically admissible displacement field, $\mathbf{u}$, is accompanied by the strain and stress fields, $\boldsymbol{\varepsilon}$ and $\boldsymbol{\sigma}$, through the strain-displacement and constitutive relations. That is

$$\boldsymbol{\varepsilon} = \boldsymbol{\varepsilon}(\mathbf{u}) = \frac{1}{2}\{(\boldsymbol{\nabla} \otimes \mathbf{u}) + (\boldsymbol{\nabla} \otimes \mathbf{u})^T\},$$

$$\boldsymbol{\sigma} = \boldsymbol{\sigma}(\mathbf{u}) = \mathbf{C} : \boldsymbol{\varepsilon} = \mathbf{C} : (\boldsymbol{\nabla} \otimes \mathbf{u}). \tag{19.2.2a,b}$$

Both the strain and stress tensors are symmetric,

$$\boldsymbol{\varepsilon}^T(\mathbf{u}) = \boldsymbol{\varepsilon}(\mathbf{u}), \qquad \boldsymbol{\sigma}^T(\mathbf{u}) = \boldsymbol{\sigma}(\mathbf{u}). \tag{19.2.2c,d}$$

Although $\boldsymbol{\varepsilon}(\mathbf{u})$ is compatible, $\boldsymbol{\sigma}(\mathbf{u})$ may not necessarily satisfy the equations of equilibrium (19.1.2). Since a properly posed elastostatic problem admits a unique solution, there exists, for a given problem, only one kinematically admissible displacement field that possesses the stress field which satisfies the equations of equilibrium and the traction boundary conditions.

The difference between two (arbitrary) kinematically admissible displacement fields is called a *virtual displacement field,* and is denoted by $\delta\mathbf{u}$. Hence, the virtual displacement field $\delta\mathbf{u}$ satisfies the following boundary condition: *at any point on ∂R_U where some components of the displacement vector $\mathbf{u}$ are prescribed by $\mathbf{u}^o$, the corresponding components of the virtual displacement vector $\delta\mathbf{u}$ vanish,* i.e.,

$$\delta\mathbf{u} = \mathbf{0} \qquad \text{on } \partial R_U. \tag{19.2.3}$$

From the virtual displacement field, obtain the virtual strain field,

$$\delta\boldsymbol{\varepsilon} \equiv \delta\boldsymbol{\varepsilon}(\delta\mathbf{u}) = \frac{1}{2}\{(\boldsymbol{\nabla} \otimes \delta\mathbf{u}) + (\boldsymbol{\nabla} \otimes \delta\mathbf{u})^T\}, \tag{19.2.4}$$

and the corresponding virtual stress field, $\delta\boldsymbol{\sigma} = \delta\boldsymbol{\sigma}(\mathbf{u}) = \mathbf{C} : \delta\boldsymbol{\varepsilon}$.

19.2.2. Statically Admissible Stress Field

A sufficiently smooth (actually of class C^1) stress field that satisfies the equations of equilibrium and all the stress boundary conditions of the considered problem is referred to as *statically admissible.* Denote the class of all such stress tensors by V^s, i.e.,

$$V^S \equiv \{\boldsymbol{\sigma} \mid \boldsymbol{\sigma} \in C^1,\ \boldsymbol{\nabla} \cdot \boldsymbol{\sigma} + \mathbf{f}^o = \mathbf{0} \text{ in R, and } \boldsymbol{\nu} \cdot \boldsymbol{\sigma} = \mathbf{t}^o \text{ on } \partial R_T\}, \tag{19.2.5}$$

where ∂R_T is the part of the surface ∂R on which $\mathbf{t}^o$ is prescribed.[5]

[5] Note that, in general, $\partial R_U + \partial R_T$ is not necessarily equal to ∂R for a mixed boundary-value problem.

From a statically admissible stress field, obtain the corresponding strain field through Hooke's law,

$$\boldsymbol{\varepsilon} \equiv \boldsymbol{\varepsilon}(\boldsymbol{\sigma}) = \mathbf{D} : \boldsymbol{\sigma}. \tag{19.2.6a}$$

From the symmetry of the compliance tensor **D**, this strain tensor is symmetric,

$$\boldsymbol{\varepsilon}^{\mathrm{T}}(\boldsymbol{\sigma}) = \boldsymbol{\varepsilon}(\boldsymbol{\sigma}). \tag{19.2.6b}$$

However, $\boldsymbol{\varepsilon}(\boldsymbol{\sigma})$ is not necessarily compatible. From the uniqueness of the solution of the elastostatic boundary-value problem, for a given problem, there exists only one statically admissible stress field that possesses the strain field which satisfies the compatibility conditions.

Like the virtual displacement field $\delta\mathbf{u}$, the virtual stress field $\delta\boldsymbol{\sigma}$ can be defined by the difference of two (arbitrary) statically admissible stress fields. The virtual stress field $\delta\boldsymbol{\sigma}$ satisfies the following boundary conditions: *at any point on $\partial\mathrm{R}_\mathrm{T}$ where some components of the traction vector $\mathbf{t}^\mathrm{o}$ are given, the corresponding components of the virtual traction $\boldsymbol{\nu} \cdot \delta\boldsymbol{\sigma}$ vanish,* i.e.,

$$\boldsymbol{\nu} \cdot \delta\boldsymbol{\sigma} = \mathbf{0} \qquad \text{on } \partial\mathrm{R}_\mathrm{T}. \tag{19.2.7}$$

In the region R, $\delta\boldsymbol{\sigma}$ is divergent-free,

$$\boldsymbol{\nabla} \cdot \delta\boldsymbol{\sigma} = \mathbf{0}, \tag{19.2.8}$$

and is accompanied by the virtual strain field $\delta\boldsymbol{\varepsilon}$,

$$\delta\boldsymbol{\varepsilon} \equiv \delta\boldsymbol{\varepsilon}(\delta\boldsymbol{\sigma}) = \mathbf{D} : \delta\boldsymbol{\sigma}, \tag{19.2.9}$$

which, from the symmetry of **D**, is symmetric.

19.3. POTENTIAL ENERGY

In this subsection the variational principle for kinematically admissible displacement fields is considered. First the virtual work principle is obtained. Then the total potential energy of the elastic solid is introduced, and the minimum potential energy theorem is discussed.

19.3.1. Virtual Work Principle

Consider a solid occupying a region R bounded by a regular surface $\partial\mathrm{R}$. Let it be in equilibrium under prescribed body forces $\mathbf{f}^\mathrm{o}$ and suitable mixed boundary conditions $\mathbf{t}^\mathrm{o}$ or $\mathbf{u}^\mathrm{o}$. The virtual work of the applied loads, $\mathbf{f}^\mathrm{o}$ and $\mathbf{t}^\mathrm{o}$, acting through the virtual displacement $\delta\mathbf{u}$, is defined by

$$\delta\mathrm{E} \equiv \int_\mathrm{R} \mathbf{f}^\mathrm{o} \cdot \delta\mathbf{u}\, \mathrm{dV} + \int_{\partial\mathrm{R}}' \mathbf{t}^\mathrm{o} \cdot \delta\mathbf{u}\, \mathrm{dS}, \tag{19.3.1a}$$

where the prime on $\int'_{\partial R}$ shows that the integration is taken over points on ∂R where some components of $\mathbf{t}^o$ are prescribed. From the Gauss theorem, the *actual stress field* $\boldsymbol{\sigma}$ satisfies

$$0 = \int_R \{\boldsymbol{\nabla} \cdot \boldsymbol{\sigma} + \mathbf{f}^o\} \cdot \delta\mathbf{u} \, dV$$

$$= \int_R \{\mathbf{f}^o \cdot \delta\mathbf{u} - \boldsymbol{\sigma} : \delta\boldsymbol{\varepsilon}\} \, dV + \int_{\partial R} (\boldsymbol{\nu} \cdot \boldsymbol{\sigma}) \cdot \delta\mathbf{u} \, dS. \tag{19.3.1b}$$

Since $\delta\mathbf{u} = \mathbf{0}$ at points where $\mathbf{u} = \mathbf{u}^o$, the surface integral in the right side of (19.3.1b) is replaced by $\int'_{\partial R} (\boldsymbol{\nu} \cdot \boldsymbol{\sigma}) \cdot \delta\mathbf{u} \, dS$. Hence, the virtual work δE becomes

$$\delta E = \int_R \boldsymbol{\sigma} : \delta\boldsymbol{\varepsilon} \, dV. \tag{19.3.2}$$

Note that *the virtual work theorem* (19.3.2) *applies to continua of all kinds, since its derivation does not rest on the particular constitutive relations that may be involved.*

19.3.2. Variational Principle for Kinematically Admissible Displacement Fields

The elastic strain energy is given by $w \equiv \boldsymbol{\varepsilon} : \mathbf{C} : \boldsymbol{\varepsilon}/2$. The right side of (19.3.2), therefore, is the change in the total stored elastic strain energy corresponding to the virtual strain field $\delta\boldsymbol{\varepsilon}$. Defining the total strain energy W as a functional of the strain field, $\boldsymbol{\varepsilon}$, by

$$W(\boldsymbol{\varepsilon}) \equiv \int_R w(\boldsymbol{\varepsilon}) \, dV = \int_R \frac{1}{2} \boldsymbol{\varepsilon} : \mathbf{C} : \boldsymbol{\varepsilon} \, dV, \tag{19.3.3a}$$

observe that

$$\delta W(\boldsymbol{\varepsilon}) = \int_R \delta \{\frac{1}{2} \boldsymbol{\varepsilon} : \mathbf{C} : \boldsymbol{\varepsilon}\} \, dV = \int_R (\mathbf{C} : \boldsymbol{\varepsilon}) : \delta\boldsymbol{\varepsilon} \, dV = \int_R \boldsymbol{\sigma} : \delta\boldsymbol{\varepsilon} \, dV. \tag{19.3.3b}$$

It therefore follows that *the potential energy* Π defined by

$$\Pi \equiv \Pi(\mathbf{u}; \mathbf{t}^o, \mathbf{f}^o) \equiv W(\boldsymbol{\varepsilon}) - \int_R \mathbf{f}^o \cdot \mathbf{u} \, dV - \int_{\partial R} \mathbf{t}^o \cdot \mathbf{u} \, dS, \tag{19.3.4}$$

is stationary, i.e.,

$$\delta\Pi(\mathbf{u}) = 0, \tag{19.3.5}$$

for the set of virtual variations $\delta\mathbf{u}$ *of the equilibrium displacement field* $\mathbf{u}$.

19.3.3. Minimum Potential Energy

The following minimum theorem is now proved.

Theorem of Minimum Potential Energy: Among all (infinitesimal) kinematically admissible displacement fields, that which is also statically admissible renders the potential energy Π *an absolute minimum.*

To prove this assertion, let $\mathbf{u}$ denote the actual displacement field, and designate by $\mathbf{u}'$ a displacement field which is kinematically admissible but is not identical to $\mathbf{u}$, i.e., $\mathbf{u}' \in V^k$ and $\mathbf{u}' \neq \mathbf{u}$. Then calculate the stored elastic strain energy for the difference displacement field, $\mathbf{u}' - \mathbf{u}$, which can be regarded as a virtual field $\delta\mathbf{u}$. Since $\mathbf{C}$ is positive-definite, $W(\delta\boldsymbol{\varepsilon})$ becomes

$$W(\delta\boldsymbol{\varepsilon}) = W(\boldsymbol{\varepsilon}') + W(\boldsymbol{\varepsilon}) - \int_R \boldsymbol{\varepsilon} : \mathbf{C} : \boldsymbol{\varepsilon}' \, dV > 0, \tag{19.3.6a}$$

where the strain fields corresponding to the displacement fields $\mathbf{u}$, $\mathbf{u}'$, and $\delta\mathbf{u}$ are denoted by $\boldsymbol{\varepsilon}$, $\boldsymbol{\varepsilon}'$, and $\delta\boldsymbol{\varepsilon}$, respectively. From definition (19.3.3a), (19.3.6a) leads to

$$W(\boldsymbol{\varepsilon}') - W(\boldsymbol{\varepsilon}) > \int_R (\mathbf{C} : \boldsymbol{\varepsilon}) : (\boldsymbol{\varepsilon}' - \boldsymbol{\varepsilon}) \, dV. \tag{19.3.6b}$$

Noting that $\mathbf{C} : \boldsymbol{\varepsilon}$ for the actual strain field belongs to V^s, rewrite the right side of (19.3.6b), as

$$\int_R (\mathbf{C} : \boldsymbol{\varepsilon}) : (\boldsymbol{\varepsilon}' - \boldsymbol{\varepsilon}) \, dV = \int_R \mathbf{f}^o \cdot (\mathbf{u}' - \mathbf{u}) \, dV + \int_{\partial R} \mathbf{t} \cdot (\mathbf{u}' - \mathbf{u}) \, dS, \tag{19.3.6c}$$

where the Cauchy laws and the Gauss theorem are employed, and $\mathbf{t} = \boldsymbol{\nu} \cdot \boldsymbol{\sigma}$. Inequality (19.3.6b) now becomes

$$W(\boldsymbol{\varepsilon}') - \int_R \mathbf{f}^o \cdot \mathbf{u}' \, dV - \int_{\partial R}' \mathbf{t}^o \cdot \mathbf{u}' \, dS > W(\boldsymbol{\varepsilon}) - \int_R \mathbf{f}^o \cdot \mathbf{u} \, dV - \int_{\partial R}' \mathbf{t}^o \cdot \mathbf{u} \, dS, \tag{19.3.7}$$

where traction vectors $\mathbf{t}$ and the surface integral are replaced by $\mathbf{t}^o$ and $\int_{\partial R}'$; this is because $\mathbf{t}$ $(= (\boldsymbol{\nu} \cdot (\mathbf{C} : \boldsymbol{\varepsilon}))$ $= \mathbf{t}^o$ at points where $\mathbf{t}^o$ is prescribed, and $\mathbf{u}' = \mathbf{u} = \mathbf{u}^o$ at points where $\mathbf{t}^o$ is not prescribed. The left side of (19.3.7) is the potential energy corresponding to the kinematically admissible displacement field $\mathbf{u}'$, and the right side corresponds to that of both the kinematically and statically admissible field $\mathbf{u}$. Hence the asserted theorem is proved.

The converse theorem also holds, that is, *a kinematically admissible displacement field which renders* Π *minimum is statically admissible.* To prove this, let $\mathbf{u}$ be a displacement field which makes Π minimum, and take the virtual displacement $\delta\mathbf{u}$ from $\mathbf{u}$. Then, defining the strain field associated with $\mathbf{u}$ by $\boldsymbol{\varepsilon} = ((\boldsymbol{\nabla} \otimes \mathbf{u}) + (\boldsymbol{\nabla} \otimes \mathbf{u})^T)/2$,

$$\begin{aligned} \Pi(\mathbf{u} + \delta\mathbf{u}) - \Pi(\mathbf{u}) &= W(\boldsymbol{\varepsilon} + \delta\boldsymbol{\varepsilon}) - W(\boldsymbol{\varepsilon}) - \int_R \mathbf{f}^o \cdot \delta\mathbf{u} \, dV - \int_{\partial R}' \mathbf{t}^o \cdot \delta\mathbf{u} \, dS \\ &= \delta\Pi(\mathbf{u}) + \delta^2\Pi(\mathbf{u}), \end{aligned} \tag{19.3.8a}$$

where

$$\delta\Pi(\mathbf{u}) = -\int_R \{\boldsymbol{\nabla}\cdot(\mathbf{C}:\boldsymbol{\varepsilon})+\mathbf{f}^o\}\cdot\delta\mathbf{u}\,dV + \int_{\partial R}' \{\boldsymbol{\nu}\cdot(\mathbf{C}:\boldsymbol{\varepsilon})-\mathbf{t}^o\}\cdot\delta\mathbf{u}\,dS,$$

$$\delta^2\Pi(\mathbf{u}) = W(\delta\boldsymbol{\varepsilon}). \tag{19.3.8b,c}$$

Note that the first variation of Π equals the virtual work associated with $\delta\mathbf{u}$ and the second variation of Π is positive-definite for nonzero $\delta\boldsymbol{\varepsilon}$. Therefore, in order that the strict inequality $\Pi(\mathbf{u}+\delta\mathbf{u}) - \Pi(\mathbf{u}) > 0$ holds for nontrivial $\delta\mathbf{u}$, it is necessary and sufficient that

$$\boldsymbol{\nabla}\cdot(\mathbf{C}:\boldsymbol{\varepsilon})+\mathbf{f}^o = \mathbf{0} \qquad \text{in R},$$

$$\boldsymbol{\nu}\cdot(\mathbf{C}:\boldsymbol{\varepsilon})-\mathbf{t}^o = \mathbf{0} \qquad \text{on } \partial R_T, \tag{19.3.9a,b}$$

which states that the corresponding stress field $\mathbf{C}:\boldsymbol{\varepsilon} = \boldsymbol{\sigma}$ is statically admissible. Note that (19.3.9) can be obtained as the Euler equations and the corresponding natural boundary conditions associated with a variational problem in which the potential energy functional Π is minimized over all kinematically admissible displacement fields, i.e., over V^k.

19.4. COMPLEMENTARY ENERGY

Whereas in the virtual work theorem, in the corresponding variational principle, and in the theorem of minimum potential energy considered in Subsection 19.3, the displacement is allowed to vary over all kinematically admissible displacement fields, it is sometimes convenient to consider a variation of the statically admissible stress field. In this subsection the virtual work theorem, the variational principle, and the theorem of minimum complementary energy for statically admissible stress fields are considered.

19.4.1. Virtual Work Principle for Virtual Stress

Consider the boundary-value problem stated in Subsection 19.3.1. The *virtual work* of the arbitrary tractions acting through the prescribed surface displacement $\mathbf{u}^o$ is

$$\delta E^c \equiv \int_{\partial R}'' \delta\mathbf{t}\cdot\mathbf{u}^o\,dS, \tag{19.4.1a}$$

where the double prime on $\int_{\partial R}$ stands for the integration over those points at which $\mathbf{u}^o$ (or some of its components) is prescribed. The virtual tractions $\delta\mathbf{t}$ are determined by the virtual stress $\delta\boldsymbol{\sigma}$ through $\boldsymbol{\nu}\cdot\delta\boldsymbol{\sigma}$. Since the virtual stress $\delta\boldsymbol{\sigma}$ is divergence-free,

$$\int_R \delta\boldsymbol{\sigma}:\boldsymbol{\varepsilon}\,dV = \int_{\partial R} (\boldsymbol{\nu}\cdot\delta\boldsymbol{\sigma})\cdot\mathbf{u}\,dS = \int_{\partial R}'' \delta\mathbf{t}\cdot\mathbf{u}^o\,dS, \tag{19.4.1b}$$

where the facts that $\mathbf{v} \cdot \delta\boldsymbol{\sigma} = \mathbf{0}$ at points where $\mathbf{t}^{\circ}$ is prescribed, and $\mathbf{u} = \mathbf{u}^{\circ}$ at points where $\mathbf{u}^{\circ}$ is prescribed have been used. Therefore, the virtual work δE^c becomes

$$\delta E^c = \int_R \delta\boldsymbol{\sigma} : \boldsymbol{\varepsilon} \, dV. \tag{19.4.2}$$

Note that, similarly to (19.3.2), *the virtual work theorem* (19.4.2) *applies to continua of all kinds.*

19.4.2. Variational Principle for Statically Admissible Stress Fields

For an elastic solid, the complementary strain energy is $w^c = \boldsymbol{\sigma} : \mathbf{D} : \boldsymbol{\sigma}/2$. Then the total complementary strain energy becomes

$$W^c \equiv W^c(\boldsymbol{\sigma}) = \int_R w^c(\boldsymbol{\sigma}) \, dV = \int_R \frac{1}{2} \boldsymbol{\sigma} : \mathbf{D} : \boldsymbol{\sigma} \, dV, \tag{19.4.3a}$$

and the right-hand side of (19.4.2) can be expressed as

$$\delta W^c(\boldsymbol{\sigma}) = \int_R \delta\{\frac{1}{2} \boldsymbol{\sigma} : \mathbf{D} : \boldsymbol{\sigma}\} \, dV = \int_R \boldsymbol{\sigma} : \mathbf{D} : \delta\boldsymbol{\sigma} \, dV. \tag{19.4.3b}$$

It then follows that *the complementary energy* Π^c, *defined by*

$$\Pi^c(\boldsymbol{\sigma};\, \mathbf{u}^{\circ}) \equiv W^c(\boldsymbol{\sigma}) - \int_{\partial R}'' (\mathbf{v} \cdot \boldsymbol{\sigma}) \cdot \mathbf{u}^{\circ} \, dS, \tag{19.4.4a}$$

is stationary, i.e.,

$$\delta \Pi^c(\boldsymbol{\sigma}) = 0, \tag{19.4.5}$$

for the set of statically admissible variations $\delta\boldsymbol{\sigma}$ *of the actual stress field.*

To emphasize that only statically admissible stress fields are considered, the complementary energy is redefined as

$$\Pi^c(\boldsymbol{\sigma};\, \mathbf{u}^0;\, \mathbf{f}^0) \equiv W^c(\boldsymbol{\sigma}) - \int_{\partial R}'' (\mathbf{v} \cdot \boldsymbol{\sigma}) \cdot \mathbf{u}^{\circ} dS + \int_R \boldsymbol{\lambda} \cdot \{\nabla \cdot \boldsymbol{\sigma} + \mathbf{f}^{\circ}\} \, dV, \tag{19.4.4b}$$

where $\boldsymbol{\lambda}$ is a Lagrange multiplier (a vector field). In (19.4.4b), the stress field need no longer satisfy (19.1.2). However, for statically admissible stress fields, functional Π^c defined by (19.4.4b) reduces to (19.4.4a). Therefore, (19.4.5) still holds, i.e., the complementary energy (19.4.4b) is stationary for the set of statically admissible variations of the actual stress field.

19.4.3. Minimum Complementary Energy

An absolute minimum principle known as *the theorem of minimum complementary energy* is now obtained. To this end, in addition to the actual stress field $\boldsymbol{\sigma}$, consider a statically admissible stress field $\boldsymbol{\sigma}'$, i.e., $\boldsymbol{\sigma}' \in V^s$, and $\boldsymbol{\sigma}' \neq \boldsymbol{\sigma}$. Employing an argument similar to that which led to (19.3.6b), obtain

$$W^c(\boldsymbol{\sigma}') - W^c(\boldsymbol{\sigma}) > \int_R \boldsymbol{\varepsilon} : \delta\boldsymbol{\sigma} \, dV, \tag{19.4.6}$$

where $\delta\boldsymbol{\sigma} = \boldsymbol{\sigma}' - \boldsymbol{\sigma}$ is regarded as a virtual stress field. This inequality may be written as

$$W^c(\boldsymbol{\sigma}') - \int''_{\partial R} (\boldsymbol{\nu} \cdot \boldsymbol{\sigma}') \cdot \mathbf{u}^o \, dS > W^c(\boldsymbol{\sigma}) - \int''_{\partial R} (\boldsymbol{\nu} \cdot \boldsymbol{\sigma}) \cdot \mathbf{u}^o \, dS, \tag{19.4.7a}$$

or, taking advantage of the statical admissibility of $\boldsymbol{\sigma}'$ and $\boldsymbol{\sigma}$,

$$W^c(\boldsymbol{\sigma}') - \int''_{\partial R} (\boldsymbol{\nu} \cdot \boldsymbol{\sigma}') \cdot \mathbf{u}^o \, dS + \int_R \boldsymbol{\lambda}' \cdot \{\boldsymbol{\nabla} \cdot \boldsymbol{\sigma}' + \mathbf{f}^o\} \, dV$$

$$> W^c(\boldsymbol{\sigma}) - \int''_{\partial R} (\boldsymbol{\nu} \cdot \boldsymbol{\sigma}) \cdot \mathbf{u}^o \, dS + \int_R \boldsymbol{\lambda} \cdot \{\boldsymbol{\nabla} \cdot \boldsymbol{\sigma} + \mathbf{f}^o\} \, dV. \tag{19.4.7b}$$

Thus follows the extremum theorem:

> *Theorem of Minimum Complementary Energy: Among all statically admissible stress fields, the actual stress field renders* Π^c *an absolute minimum.*

The converse theorem also holds, that is, *a statically admissible stress field which renders* Π^c *minimum satisfies the compatibility conditions.*

The proof of the minimum complementary energy and its converse is straightforward. The potential energy Π^c is defined by (19.4.4b). The variation of Π^c for a general (symmetric) stress field $\boldsymbol{\sigma}$ is considered first, and then it is assumed that $\boldsymbol{\sigma}$ actually belongs to[6] V^s. Let the general and associated virtual stress fields be $\boldsymbol{\sigma}$ (= $\boldsymbol{\sigma}^T$) and $\delta\boldsymbol{\sigma}$ (= $\delta\boldsymbol{\sigma}^T$), respectively, and denote the variation of the Lagrange multiplier for the equilibrium conditions by $\delta\boldsymbol{\lambda}$. Then, the first variation of Π^c becomes

$$\delta\Pi^c = \int_R \{\delta\boldsymbol{\sigma} : (\mathbf{D} : \boldsymbol{\sigma}) + \boldsymbol{\lambda} \cdot (\boldsymbol{\nabla} \cdot \delta\boldsymbol{\sigma}) + \delta\boldsymbol{\lambda} \cdot (\boldsymbol{\nabla} \cdot \boldsymbol{\sigma} + \mathbf{f}^o)\} \, dV$$

$$- \int''_{\partial R} (\boldsymbol{\nu} \cdot \delta\boldsymbol{\sigma}) \cdot \mathbf{u}^o \, dS$$

$$= \int_R \{\delta\boldsymbol{\sigma} : \{\mathbf{D} : \boldsymbol{\sigma} - \boldsymbol{\nabla} \otimes \boldsymbol{\lambda}\} + \delta\boldsymbol{\lambda} \cdot \{\boldsymbol{\nabla} \cdot \boldsymbol{\sigma} + \mathbf{f}^o\}\} \, dV$$

$$+ \int''_{\partial R} (\boldsymbol{\nu} \cdot \delta\boldsymbol{\sigma}) \cdot (\boldsymbol{\lambda} - \mathbf{u}^o) \, dS, \tag{19.4.8a}$$

and the second variation of Π^c becomes

$$\delta^2\Pi^c = W^c(\delta\boldsymbol{\sigma}) + \int_R \delta\boldsymbol{\lambda} \cdot (\boldsymbol{\nabla} \cdot \delta\boldsymbol{\sigma}) \, dV. \tag{19.4.8b}$$

Since the virtual stress field is symmetric,

---

[6] Functional Π^c, defined for a general stress field, is stationary and minimum for the actual stress field, $\boldsymbol{\sigma}$, since $\delta^2\Pi^c$ is positive-definite in this case.

$$\delta\boldsymbol{\sigma} : (\mathbf{D} : \boldsymbol{\sigma} - \boldsymbol{\nabla} \otimes \boldsymbol{\lambda}) = \delta\boldsymbol{\sigma} : \left\{\mathbf{D} : \boldsymbol{\sigma} - \frac{1}{2}\{(\boldsymbol{\nabla} \otimes \boldsymbol{\lambda}) + (\boldsymbol{\nabla} \otimes \boldsymbol{\lambda})^{\mathrm{T}}\}\right\}. \tag{19.4.9}$$

Therefore, in view of $(\mathbf{D} : \boldsymbol{\sigma})^{\mathrm{T}} = \mathbf{D} : \boldsymbol{\sigma}$, the Euler equation for Π^{c} for the statically admissible $\boldsymbol{\sigma}$ ensures the kinematic admissibility of $\mathbf{D} : \boldsymbol{\sigma}$, i.e., it follows that

$$\mathbf{D} : \boldsymbol{\sigma} = \frac{1}{2}\{(\boldsymbol{\nabla} \otimes \boldsymbol{\lambda}) + (\boldsymbol{\nabla} \otimes \boldsymbol{\lambda})^{\mathrm{T}}\} \qquad \text{in R}, \tag{19.4.10a}$$

with the corresponding natural boundary conditions,

$$\boldsymbol{\lambda} = \mathbf{u}^{\mathrm{o}} \qquad \text{on } \partial\mathrm{R}_{\mathrm{U}}. \tag{19.4.10b}$$

Equation (19.4.10a) is the compatibility condition for the stress field in R; it ensures that $\mathbf{D} : \boldsymbol{\sigma}$ is the symmetric part of the gradient of a vector field, and hence is compatible. Equation (19.4.10b) is the corresponding kinematic boundary condition on ∂R. Since the virtual stress $\delta\boldsymbol{\sigma}$ associated with the statically admissible stress $\boldsymbol{\sigma}$ in V^{s} is divergence-free, the second variation of Π^{c} becomes

$$\delta^2\Pi^{\mathrm{c}} = \mathrm{W}^{\mathrm{c}}(\delta\boldsymbol{\sigma}) > 0, \tag{19.4.11}$$

for nonzero $\delta\boldsymbol{\sigma}$. Therefore, the asserted theorem and its converse are proved.

19.5. GENERAL VARIATIONAL PRINCIPLES

In this subsection a new energy functional is defined from the sum of the total potential and complementary energies, and the sequences of the resulting general variational principles are formulated. Furthermore, including the effects of possible discontinuities in the field variables, the class of admissible functions is broadened to sectionally continuous fields. General variational principles of this kind have been discussed by Prange (1916), Reissner (1950), Hu (1955), and Washizu (1955), for continuous fields, and by Prager (1967) and Nemat-Nasser (1972), for fields admitting discontinuities; see Washizu (1968) and Nemat-Nasser (1974) for comprehensive accounts. Here, the most general form of such variational principles is presented.

19.5.1. General Potential Energy

In the preceding subsections, kinematically admissible displacement fields and statically admissible stress fields have been considered. Then, the strain field is expressed either in terms of the displacement field through the strain-displacement relations, or in terms of the stress field through Hooke's law. As shown in Subsection 19.4, for statically admissible stress fields, the strain field may be regarded as an independent field subject to arbitrary variations if the corresponding kinematical or constitutive relations are included in the corresponding variational statement by means of suitable side conditions, or constraints. The resulting variational principles, therefore, take on a slightly different form. This is illustrated for the general case of the sum of the potential and complementary energies, as follows.

To this end, the potential energy Π^* is regarded as a functional of the independent field variables $\mathbf{u}$, $\boldsymbol{\varepsilon}$, and $\boldsymbol{\sigma}$, and its stationary value is sought,[7] subject to the following side conditions: in the region R,

$$\boldsymbol{\nabla}\boldsymbol{\cdot}\boldsymbol{\sigma} + \mathbf{f}^0 = \mathbf{0}, \qquad \boldsymbol{\varepsilon} = \frac{1}{2}\{(\boldsymbol{\nabla}\otimes\mathbf{u}) + (\boldsymbol{\nabla}\otimes\mathbf{u})^{\mathrm{T}}\}; \tag{19.5.1a,b}$$

and on the boundary ∂R,

$$\begin{Bmatrix} \mathbf{u} = \mathbf{u}^{\mathrm{o}} \\ \boldsymbol{\nu}\boldsymbol{\cdot}\boldsymbol{\sigma} = \mathbf{t}^{\mathrm{o}} \end{Bmatrix} \text{ at points where } \begin{Bmatrix} \mathbf{u}^{\mathrm{o}} \\ \mathbf{t}^{\mathrm{o}} \end{Bmatrix} \text{ is prescribed.} \tag{19.5.2a,b}$$

To include these conditions in the variational statement, introduce the following four Lagrangian multipliers: a vector field $\boldsymbol{\lambda}^{\mathrm{u}}$, and a second-order symmetric tensor field $\boldsymbol{\Lambda}^{\sigma}$, both of class C^1 in R, for the side conditions (19.5.1a,b); and two vector fields, $\boldsymbol{\mu}^{\mathrm{u}}$ and $\boldsymbol{\mu}^{\mathrm{t}}$, of class C^0 on ∂R, for the boundary conditions (19.5.2a,b). Taking the sum of Π and Π^{c}, define functional Π^* by

$$\begin{aligned}
\Pi^* &\equiv \Pi^*(\mathbf{u}, \boldsymbol{\varepsilon}, \boldsymbol{\sigma}, \boldsymbol{\lambda}^{\mathrm{u}}, \boldsymbol{\Lambda}^{\sigma}, \boldsymbol{\mu}^{\mathrm{u}}, \boldsymbol{\mu}^{\mathrm{t}}; \mathbf{f}^{\mathrm{o}}, \mathbf{t}^{\mathrm{o}}, \mathbf{u}^{\mathrm{o}}) \\
&= \int_{\mathrm{R}} \{\mathrm{w}(\boldsymbol{\varepsilon}) + \mathrm{w}^{\mathrm{c}}(\boldsymbol{\sigma})\}\, \mathrm{dV} - \int_{\mathrm{R}} \mathbf{f}^{\mathrm{o}}\boldsymbol{\cdot}\mathbf{u}\, \mathrm{dV} - \int_{\partial\mathrm{R}}' \mathbf{t}^{\mathrm{o}}\boldsymbol{\cdot}\mathbf{u}\, \mathrm{dS} - \int_{\partial\mathrm{R}}'' (\boldsymbol{\nu}\boldsymbol{\cdot}\boldsymbol{\sigma})\boldsymbol{\cdot}\mathbf{u}^{\mathrm{o}}\, \mathrm{dS} \\
&\quad + \int_{\mathrm{R}} \left\{ \boldsymbol{\lambda}^{\mathrm{u}}\boldsymbol{\cdot}\{\boldsymbol{\nabla}\boldsymbol{\cdot}\boldsymbol{\sigma} + \mathbf{f}^{\mathrm{o}}\} - \boldsymbol{\Lambda}^{\sigma} : \left[\frac{1}{2}\{(\boldsymbol{\nabla}\otimes\mathbf{u}) + (\boldsymbol{\nabla}\otimes\mathbf{u})^{\mathrm{T}}\} - \boldsymbol{\varepsilon}\right] \right\} \mathrm{dV} \\
&\quad - \int_{\partial\mathrm{R}}' \boldsymbol{\mu}^{\mathrm{u}}\boldsymbol{\cdot}\{\boldsymbol{\nu}\boldsymbol{\cdot}\boldsymbol{\sigma} - \mathbf{t}^{\mathrm{o}}\}\, \mathrm{dS} - \int_{\partial\mathrm{R}}'' \boldsymbol{\mu}^{\mathrm{t}}\boldsymbol{\cdot}\{\mathbf{u} - \mathbf{u}^{\mathrm{o}}\}\, \mathrm{dS},
\end{aligned} \tag{19.5.3}$$

where the independent fields subject to variation, now are the field variables $\mathbf{u}$, $\boldsymbol{\varepsilon}$, and $\boldsymbol{\sigma}$, and the Lagrange multipliers, $\boldsymbol{\lambda}^{\mathrm{u}}$, $\boldsymbol{\Lambda}^{\sigma}$, $\boldsymbol{\mu}^{\mathrm{u}}$, and $\boldsymbol{\mu}^{\mathrm{t}}$. There are no side conditions imposed on these independent fields, except the symmetry of $\boldsymbol{\varepsilon}$, $\boldsymbol{\sigma}$, and $\boldsymbol{\Lambda}^{\sigma}$ (even this symmetry can be accounted for through Lagrange multipliers).

Setting $\delta\Pi^* = 0$, and integrating by parts the terms that involve $\boldsymbol{\nabla}\boldsymbol{\cdot}\delta\boldsymbol{\sigma}$ and $\boldsymbol{\nabla}\otimes\delta\mathbf{u}$, obtain[8]

$$\delta\Pi^* = \int_{\mathrm{R}} \delta\pi_1^*\, \mathrm{dV} + \int_{\partial\mathrm{R}}' \delta\pi_2^*\, \mathrm{dS} + \int_{\partial\mathrm{R}}'' \delta\pi_3^*\, \mathrm{dS}, \tag{19.5.4}$$

where the integrand $\delta\pi_1^*$ in the volume integral is given by

[7] Note that only a stationary value of Π^* is being sought, that is, a *minimum* principle is no longer considered.

[8] Variational theorems of this kind but not with the same generality have been considered by, for example, Hellinger (1914), Prange (1916), Reissner (1950), Hu (1955), Washizu (1955, 1968), and Prager (1967).

$$\begin{aligned}
\delta\pi_1^* = \quad & \{\boldsymbol{\nabla}\boldsymbol{.}\boldsymbol{\Lambda}^\sigma + \mathbf{f}^o\}\boldsymbol{.}(-\,\delta\mathbf{u}) && \text{(dual equilibrium)}\\
+\; & \{\mathbf{C}:\boldsymbol{\varepsilon} - \boldsymbol{\Lambda}^\sigma\} : \delta\boldsymbol{\varepsilon} && \text{(Hooke's law)}\\
+\; & \left\{\mathbf{D}:\boldsymbol{\sigma} - \frac{1}{2}\{(\boldsymbol{\nabla}\otimes\boldsymbol{\lambda}^u) + (\boldsymbol{\nabla}\otimes\boldsymbol{\lambda}^u)^T\}\right\} : \delta\boldsymbol{\sigma} && \text{(compatibility)}\\
+\; & \{\boldsymbol{\nabla}\boldsymbol{.}\boldsymbol{\sigma} + \mathbf{f}^o\}\boldsymbol{.}(\delta\boldsymbol{\lambda}^u) && \text{(equilibrium)}\\
+\; & \left\{\frac{1}{2}\{(\boldsymbol{\nabla}\otimes\mathbf{u}) + (\boldsymbol{\nabla}\otimes\mathbf{u})^T\} - \boldsymbol{\varepsilon}\right\} : \delta\boldsymbol{\Lambda}^\sigma && \text{(strain–displacement);}
\end{aligned} \tag{19.5.5a}$$

and the integrands $\delta\pi_2^*$ and $\delta\pi_3^*$ in the surface integrals are given by

$$\begin{aligned}
\delta\pi_2^* = \quad & \{\boldsymbol{\nu}\boldsymbol{.}\boldsymbol{\Lambda}^\sigma - \mathbf{t}^o\}\boldsymbol{.}\delta\mathbf{u} && \text{(dual tractions B.C.)}\\
+\; & \{\boldsymbol{\lambda}^u - \boldsymbol{\mu}^u\}\boldsymbol{.}(\boldsymbol{\nu}\boldsymbol{.}\delta\boldsymbol{\sigma}) && \text{(dual displacements B.C.)}\\
+\; & \{\boldsymbol{\nu}\boldsymbol{.}\boldsymbol{\sigma} - \mathbf{t}^o\}\boldsymbol{.}(-\,\delta\boldsymbol{\mu}^u) && \text{(tractions B.C.),}
\end{aligned} \tag{19.5.5b}$$

and

$$\begin{aligned}
\delta\pi_3^* = \quad & \{\boldsymbol{\nu}\boldsymbol{.}\boldsymbol{\Lambda}^\sigma - \boldsymbol{\mu}^t\}\boldsymbol{.}\delta\mathbf{u} && \text{(dual tractions B.C.)}\\
+\; & \{\boldsymbol{\lambda}^u - \mathbf{u}^o\}\boldsymbol{.}(\boldsymbol{\nu}\boldsymbol{.}\delta\boldsymbol{\sigma}) && \text{(dual displacements B.C.)}\\
+\; & \{\mathbf{u} - \mathbf{u}^o\}\boldsymbol{.}(-\,\delta\boldsymbol{\mu}^t) && \text{(displacements B.C.),}
\end{aligned} \tag{19.5.5c}$$

where B.C. stands for "boundary conditions". The consequence of the arbitrariness of the variation of each field is stated on the right of the corresponding term.

As is seen from (19.5.4) and (19.5.5a~c), $\delta\Pi^* = 0$ yields the proper kinematical and constitutive relations among the displacement, strain, and stress fields, and gives the appropriate field equations and boundary conditions. Furthermore, each Lagrange multiplier in Π^* is associated with a field variable, and indeed is *the corresponding dual field quantity*. This leads to the following correspondence between field variables and their dual fields:

$$\begin{aligned}
&\mathbf{u} \Longleftrightarrow \boldsymbol{\lambda}^u, \qquad \boldsymbol{\sigma} \Longleftrightarrow \boldsymbol{\Lambda}^\sigma \qquad \text{in R,}\\
&\mathbf{u} \Longleftrightarrow \boldsymbol{\mu}^u, \qquad \boldsymbol{\nu}\boldsymbol{.}\boldsymbol{\sigma} \Longleftrightarrow \boldsymbol{\mu}^t \qquad \text{on } \partial\text{R.}
\end{aligned} \tag{19.5.6a~d}$$

Therefore, the vanishing of the integrand of the right side of (19.5.5a) for arbitrary field variations $\delta\mathbf{u}$ (with $\delta\boldsymbol{\varepsilon}$) and $\delta\boldsymbol{\sigma}$, implies the field equations which govern the dual field variables $\boldsymbol{\Lambda}^\sigma$ and $\boldsymbol{\lambda}^u$. Conversely, arbitrary variations of the dual field variables $\delta\boldsymbol{\lambda}^u$ and $\delta\boldsymbol{\Lambda}^\sigma$ give the governing field equations for the field variables $\boldsymbol{\sigma}$ and $\mathbf{u}$. Similarly, the vanishing of the integrand of the right-hand side of (19.5.5b,c) provides the corresponding boundary conditions.

19.5.2. Jump Conditions at Discontinuity Surfaces

Thus far, certain rather strong continuity requirements have been imposed on the displacement, strain, or stress fields which enter the statement of the variational principles. In application,[9] however, it may prove useful to relax some of these continuity requirements and admit, for example, strain or stress fields which may be discontinuous across certain isolated surfaces. In such a case, the energy functional Π must then be suitably modified to account for such discontinuities.[10] This may be accomplished as follows.

Let the region R be divided into a finite number of subregions, within each of which the admissible displacement, strain, or stress fields are sufficiently continuous and differentiable, as is required by the basic elastostatic field equations. Refer to these subregions as *the domains of regularity* which are separated from each other by discontinuity surfaces across which some components of the traction and the complementary components of the displacement vector may suffer finite discontinuities or *jumps*. Let S be the collection of surfaces of this kind. Denote by $\mathbf{n}$ the unit normal which points outward from one subdomain of regularity, say, subdomain +, toward the adjacent subdomain, say, subdomain −. The jump $\Delta\mathbf{q}$ of a field quantity $\mathbf{q}$ at P on S may now be defined by

$$\Delta\mathbf{q} = \mathbf{q}^+ - \mathbf{q}^-, \tag{19.5.7a}$$

where $\mathbf{q}^+$ and $\mathbf{q}^-$ are the limiting values of $\mathbf{q}$ at P, as this point is approached along $\mathbf{n}$ from the interior of the domains + and −, respectively. For simplicity, denote

$$\bar{\mathbf{q}} = \frac{1}{2}(\mathbf{q}^+ + \mathbf{q}^-), \tag{19.5.7b}$$

which is the average value of $\mathbf{q}$ across S.

A jump in the displacement components across an interior surface may be viewed as a geometric boundary condition similar to (19.1.1), and, therefore,

$$\Delta\mathbf{u} = \Delta\mathbf{u}^{\mathrm{o}} \qquad \text{on S}, \tag{19.5.8a}$$

where $\Delta\mathbf{u}^{\mathrm{o}}$ denotes the jump in the displacements across S. Let the jump across S of the tractions be denoted by $\Delta\mathbf{t}^{\mathrm{o}}$, that is,

$$\Delta(\mathbf{n}\boldsymbol{\cdot}\boldsymbol{\sigma}) = \Delta\mathbf{t}^{\mathrm{o}} \qquad \text{on S}. \tag{19.5.8b}$$

It should be noted that jump conditions (19.5.8a,b) are sufficient conditions to determine uniquely the field variables in R. This is proved as follows: Prescribe displacement boundary conditions on both sides of S, $\mathbf{u} = \mathbf{u}^{\mathrm{o}+}$ on S^+, and $\mathbf{u} = \mathbf{u}^{\mathrm{o}-}$ on S^-, where $\mathbf{u}^{\mathrm{o}+}$ and $\mathbf{u}^{\mathrm{o}-}$ satisfy $\mathbf{u}^{\mathrm{o}+} - \mathbf{u}^{\mathrm{o}-} = \Delta\mathbf{u}^{\mathrm{o}}$. Since sufficient boundary conditions are prescribed for regions + and −, the corresponding boundary-value problem can be solved separately. Using the stress fields of the subregions + and −, then compute the tractions acting on both sides of S, i.e., on S^+, denoted by $\mathbf{n}\boldsymbol{\cdot}\boldsymbol{\sigma}^+$, and on S^-, denoted by $\mathbf{n}\boldsymbol{\cdot}\boldsymbol{\sigma}^-$. If particular $\mathbf{u}^{\mathrm{o}+}$ and $\mathbf{u}^{\mathrm{o}-}$

---

[9] For example, when approximate solutions of elastostatic boundary-value problems are being sought, see Prager (1967, 1968), or when eigenfrequencies of a composite elastic solid are being established; Nemat-Nasser and Lee (1973), Nemat-Nasser *et al.* (1975), Minagawa and Nemat-Nasser (1976), and Minagawa *et al.* (1992).

[10] A systematic treatment of this problem was first presented by Prager (1967, 1968). Our presentation here closely follows his, and Nemat-Nasser (1972, 1974).

satisfying $\mathbf{u}^{o+} - \mathbf{u}^{o-} = \Delta\mathbf{u}$ are chosen, the traction jump $\mathbf{n}\cdot(\boldsymbol{\sigma}^+ - \boldsymbol{\sigma}^-)$ equals $\Delta\mathbf{t}^o$. Hence, the resulting field variables in R satisfy both jump conditions (19.5.8a) and (19.5.8b).

On the discontinuity surface S, the jump in the product $\mathbf{t}\cdot\mathbf{u}$ is given by

$$\Delta(\mathbf{t}\cdot\mathbf{u}) = \mathbf{t}^+\cdot\mathbf{u}^+ - \mathbf{t}^-\cdot\mathbf{u}^- = \Delta\mathbf{t}\cdot\bar{\mathbf{u}} + \bar{\mathbf{t}}\cdot\Delta\mathbf{u}. \tag{19.5.9}$$

Taking advantage of (19.5.9), consider now the energy functional

$$\Pi^{**} \equiv \Pi^* - \int_S \{\Delta\mathbf{t}^o\cdot\bar{\mathbf{u}} + (\mathbf{n}\cdot\bar{\boldsymbol{\sigma}})\cdot\Delta\mathbf{u}^o\}\, dS$$

$$- \int_S \{\boldsymbol{\mu}^{u*}\cdot\{\mathbf{n}\cdot\Delta\boldsymbol{\sigma} - \Delta\mathbf{t}^o\} + \boldsymbol{\mu}^{t*}\cdot\{\Delta\mathbf{u} - \Delta\mathbf{u}^o\}\}\, dS, \tag{19.5.10}$$

where the two vector fields $\boldsymbol{\mu}^{t*}$ and $\boldsymbol{\mu}^{u*}$ on S are the Lagrange multipliers, and Π^* is given by (19.5.3). From (19.5.4),[11] now obtain

$$\delta\Pi^{**} = \delta\Pi^* + \int_S \Big\{ \{\mathbf{n}\cdot\Delta\boldsymbol{\Lambda}^\sigma - \Delta\mathbf{t}^o\}\cdot\delta\bar{\mathbf{u}} + \{\mathbf{n}\cdot\bar{\boldsymbol{\Lambda}}^\sigma - \boldsymbol{\mu}^{t*}\}\cdot\delta\Delta\mathbf{u}$$

$$+ \{\Delta\boldsymbol{\lambda}^u - \Delta\mathbf{u}^o\}\cdot(\mathbf{n}\cdot\delta\bar{\boldsymbol{\sigma}}) + \{\bar{\boldsymbol{\lambda}}^u - \boldsymbol{\mu}^{u*}\}\cdot(\mathbf{n}\cdot\delta\Delta\boldsymbol{\sigma})$$

$$+ \{\mathbf{n}\cdot\Delta\boldsymbol{\sigma} - \Delta\mathbf{t}^o\}\cdot(-\delta\boldsymbol{\mu}^{u*}) + \{\Delta\mathbf{u} - \Delta\mathbf{u}^o\}\cdot(-\delta\boldsymbol{\mu}^{t*}) \Big\}\, dS, \tag{19.5.11a}$$

where $\delta\Pi^*$ is given by (19.5.4) and (19.5.5).

To clarify the meaning of $\delta\Pi^{**}$, assume that either tractions or displacements are continuous across S, i.e., $\Delta\mathbf{t}^o\cdot\Delta\mathbf{u}^o = 0$ on S. Let $\int_S'$ denote the integration on parts of S where $\Delta\mathbf{t}^o$ is prescribed but $\mathbf{u}$ is continuous, and $\int_S''$ denote the integration on parts of S where $\Delta\mathbf{u}^o$ is prescribed but $\mathbf{n}\cdot\boldsymbol{\sigma}$ is continuous. Then, (19.5.11a) becomes

$$\delta\Pi^{**} = \delta\Pi^* + \int_S' \Big\{ \{\mathbf{n}\cdot\Delta\boldsymbol{\Lambda}^\sigma - \Delta\mathbf{t}^o\}\cdot\delta\bar{\mathbf{u}} + \{\Delta\boldsymbol{\lambda}^u\}\cdot(\mathbf{n}\cdot\delta\bar{\boldsymbol{\sigma}})$$

$$+ \{\bar{\boldsymbol{\lambda}}^u - \boldsymbol{\mu}^{u*}\}\cdot(\mathbf{n}\cdot\delta\Delta\boldsymbol{\sigma}) + \{\mathbf{n}\cdot\Delta\boldsymbol{\sigma} - \Delta\mathbf{t}^o\}\cdot(-\delta\boldsymbol{\mu}^{u*}) \Big\}\, dS$$

$$+ \int_S'' \Big\{ \{\mathbf{n}\cdot\Delta\boldsymbol{\Lambda}^\sigma\}\cdot\delta\bar{\mathbf{u}} + \{\mathbf{n}\cdot\bar{\boldsymbol{\lambda}} - \boldsymbol{\mu}^{t*}\}\cdot\delta\Delta\mathbf{u}$$

$$+ \{\Delta\boldsymbol{\lambda}^u - \Delta\mathbf{u}^o\}\cdot(\mathbf{n}\cdot\Delta\bar{\boldsymbol{\sigma}}) + \{\Delta\mathbf{u} - \Delta\mathbf{u}^o\}\cdot(-\delta\boldsymbol{\mu}^{t*}) \Big\}\, dS. \tag{19.5.11b}$$

[11] Care must be taken in applying the Gauss theorem to the volume integral of $\boldsymbol{\nabla}\cdot\delta\boldsymbol{\sigma}$ or $\boldsymbol{\nabla}\otimes\delta\mathbf{u}$.

The integrand of the surface integrals on S in the right-hand side of (19.5.11b) displays the duality between the Lagrange multipliers and the average field variables across S, i.e., $\boldsymbol{\mu}^{u*}$ and $\boldsymbol{\mu}^{t*}$ correspond to $\bar{\mathbf{u}}$ and $\mathbf{n}.\bar{\boldsymbol{\sigma}}$, respectively. Furthermore, it is seen that even if the components of $\Delta\mathbf{u}^o$ vanish on S, the variation $\delta\bar{\mathbf{u}}$ is still arbitrary there, and hence $\mathbf{n}.\Delta\boldsymbol{\Lambda}^{\sigma} - \Delta\mathbf{t}^o = \mathbf{0}$ provides the jump condition for the corresponding components of the dual tractions. Similar remarks apply to the case when the components of $\Delta\mathbf{t}^o$ vanish. On the other hand, due to the arbitrariness of $\delta\bar{\boldsymbol{\sigma}}$ or $\delta\bar{\mathbf{u}}$ on S, the jump of the dual displacements, $\Delta\boldsymbol{\lambda}^u$, or the dual tractions, $\mathbf{n}.\Delta\boldsymbol{\Lambda}^{\sigma}$, is zero, where $\Delta\mathbf{u}^o = \mathbf{0}$ or where $\Delta\mathbf{t}^o = \mathbf{0}$, respectively.

19.6. REFERENCES

Hellinger, E. (1914), Die allegemeinen Ansätse der Mechanik der Kontinua, *Enz. Math. Wiss.*, Vol. 4, 602-694.

Hu, H. (1955), On some variational principles in the theory of elasticity and plasticity, *Scintia Sinica*, Vol. 4, 33-54.

Minagawa, S. and Nemat-Nasser, S.(1976), Harmonic waves in three-dimensional elastic composites, *Int,l J. Solids Struct.*, Vol. 12, 769-777.

Minagawa, S., Yoshihara, K., and Nemat-Nasser, S. (1992), Analysis of harmonic waves in a composite material with piezoelectric effect, in *Int. J. Solids Structures*, 1901-1906.

Nemat-Nasser, S. (1972), General variational methods for elastic waves in composites, *J. Elasticity*, Vol. 2, 73-90.

Nemat-Nasser, S. (1974), General variational principles in nonlinear and linear elasticity with applications, in *Mechanics Today* (Nemat-Nasser, S. ed.), Pergamon Press, New York, 214-261.

Nemat-Nasser, S. and Lee, K. N. (1973), Application of general variational methods with discontinuous fields to bending, buckling, and vibration of beams, *Comp. Meth. Appl. Mech. Eng.* Vol. 2, 33-41.

Nemat-Nasser, S., Fu, F. C. L., and Minagawa, S. (1975), Harmonic waves in one, two, and three-dimensional composites, *Int. J. Solids Struct.*, Vol. 11, 617-642.

Prager, W. (1967), Variational principles of linear elastostatics for discontinuous displacements, strains, and stresses, in *Recent progress in applied mechanics* (Broberg, B., Hult, J., and Niordson, F., eds.), 463 -474.

Prager, W. (1968), Variational principles for elastic plates with relaxed continuity requirements, *Int. J. Solids, Struct.*, Vol. 4, 837-844.

Prange, G. (1916), unpublished *Habilitationansschrift*, Hannover.

Reissner, E. (1950), On a variational theorem in elasticity, *J. Math. Phys.*, Vol. 29, 90-95.

Washizu, K. (1955), On the variational principles of elasticity and plasticity, Tech. Rept. 25-18, Cont. N5ori-07833, MIT, March.

Washizu, K. (1968), *Variational methods in elasticity and plasticity*, Pergamon Press, New York.

SECTION 20 THREE-DIMENSIONAL PROBLEMS

In this section, integration of the elastokinematic field equations is considered for three-dimensional bodies, using the apparatus of the potential theory. Equations governing the propagation of dilatational and rotational waves are obtained and their solutions are discussed. For elastostatic problems, the Papkovich-Neuber and the Galerkin representations of the solutions are developed. Several examples are worked out in detail to illustrate the basic approach.

20.1. HELMHOLTZ'S DECOMPOSITION THEOREM

Let $\mathbf{u}$ be a sufficiently smooth vector-valued function defined in a convex region R that is bounded by a regular surface ∂R. The *Helmholtz decomposition theorem* states that $\mathbf{u}$ in R can be written as a sum of the gradient of a scalar potential U, and the curl of a vector potential $\mathbf{V}$ whose divergence vanishes, i.e.,

$$\mathbf{u}(\mathbf{x}) = \boldsymbol{\nabla}\mathrm{U}(\mathbf{x}) + \boldsymbol{\nabla} \times \mathbf{V}(\mathbf{x}), \qquad \boldsymbol{\nabla} \cdot \mathbf{V}(\mathbf{x}) = 0 \tag{20.1.1a,2a}$$

or in component form,

$$\mathrm{u}_i(\mathbf{x}) = \mathrm{U}_{,i}(\mathbf{x}) + e_{ijk}\,\mathrm{V}_{k,j}(\mathbf{x}), \qquad \mathrm{V}_{i,i}(\mathbf{x}) = 0. \tag{20.1.1b,2b}$$

To show this, consider the *Newtonian potential* given by

$$\mathbf{W}(\mathbf{x}) = -\frac{1}{4\pi}\int_{\mathrm{R}} \frac{\mathbf{u}(\boldsymbol{\xi})}{\mathrm{r}(\mathbf{x}, \boldsymbol{\xi})}\,\mathrm{dV}_{\xi}, \tag{20.1.3a}$$

or

$$\mathrm{W}_i(\mathbf{x}) = -\frac{1}{4\pi}\int_{\mathrm{R}} \frac{\mathrm{u}_i(\boldsymbol{\xi})}{\mathrm{r}(\mathbf{x}, \boldsymbol{\xi})}\,\mathrm{dV}_{\xi}, \tag{20.1.3b}$$

where $\mathrm{r} = \sqrt{(\mathbf{x}-\boldsymbol{\xi})\cdot(\mathbf{x}-\boldsymbol{\xi})}$ and the integration is with respect to $\boldsymbol{\xi} = \xi_i\,\mathbf{e}_i$ (i = 1, 2, 3), with $\mathbf{x}$ fixed. Operating on (20.1.3a,b) by the Laplacian, $\nabla^2 \equiv \partial^2/\partial x_1^2 + \partial^2/\partial x_2^2 + \partial^2/\partial x_3^2$, obtain[1]

$$\nabla^2\mathbf{W} = \mathbf{u}, \tag{20.1.4a}$$

or

$$\mathrm{W}_{i,jj} = \mathrm{u}_i. \tag{20.1.4b}$$

Now, with the aid of identity

[1] Note that in terms of $\boldsymbol{\nabla}$, $\nabla^2\mathbf{W}$ is given by $\nabla^2\mathbf{W} = \boldsymbol{\nabla}\cdot(\boldsymbol{\nabla}\otimes\mathbf{W})$.

$$\mathbf{\nabla}^2\mathbf{W} = \mathbf{\nabla}(\mathbf{\nabla}\cdot\mathbf{W}) - \mathbf{\nabla}\times\mathbf{\nabla}\times\mathbf{W} \tag{20.1.5a}$$

or

$$W_{i,jj} = W_{j,ji} - e_{ijk}\, e_{klm}\, W_{m,lj}, \tag{20.1.5b}$$

arrive at

$$\mathbf{u} = \mathbf{\nabla}\otimes(\mathbf{\nabla}\cdot\mathbf{W}) + \mathbf{\nabla}\times(-\mathbf{\nabla}\times\mathbf{W}) \tag{20.1.6}$$

which is equivalent to (20.1.1a), provided that

$$U = \mathbf{\nabla}\cdot\mathbf{W}, \qquad \mathbf{V} = -\mathbf{\nabla}\times\mathbf{W}. \tag{20.1.7a,b}$$

Note that region R need not be convex as long as it is "star-shaped"[2] with respect to an interior point. Moreover, unbounded regions are admitted, provided that $|\mathbf{x}|^2\,|\mathbf{u}(\mathbf{x})|$ remains bounded as $|\mathbf{x}|$ becomes large.

20.2. WAVE EQUATIONS

For simplicity, assume isotropic elasticity, $\mathbf{C} = \lambda\mathbf{1}^{(2)}\otimes\mathbf{1}^{(2)} + 2\mu\mathbf{1}^{(4s)}$. In the absence of body forces and thermal effects, but when inertia forces are included, Navier's equations become

$$\mu\,\nabla^2\mathbf{u} + (\lambda+\mu)\,\mathbf{\nabla}(\mathbf{\nabla}\cdot\mathbf{u}) = \rho\,\ddot{\mathbf{u}}. \tag{20.2.1}$$

Assume $\mathbf{u}(\mathbf{x}, t)$ is sufficiently smooth for $\mathbf{x}$ in R and $-\infty < t < +\infty$. Operating on both sides of (20.2.1) first by $\mathbf{\nabla}\cdot$ and then by $\mathbf{\nabla}\times$, obtain

$$\nabla^2(\mathbf{\nabla}\cdot\mathbf{u}) = \frac{1}{C_1^2}\,\frac{\partial^2}{\partial t^2}\,(\mathbf{\nabla}\cdot\mathbf{u}) \tag{20.2.2}$$

and

$$\nabla^2(\mathbf{\nabla}\times\mathbf{u}) = \frac{1}{C_2^2}\,\frac{\partial^2}{\partial t^2}\,(\mathbf{\nabla}\times\mathbf{u}), \tag{20.2.3}$$

where

$$C_1^2 = \frac{\lambda+2\mu}{\rho}, \qquad C_2^2 = \frac{\mu}{\rho} \tag{20.2.4a,b}$$

are the speed of the *dilatation and shear waves*, respectively. Set $\mathbf{\nabla}\cdot\mathbf{u} = \varepsilon$ and $\mathbf{\nabla}\times\mathbf{u} = 2\,\mathbf{\Omega}$, and write (20.2.2) and (20.2.3) as

$$\Box_1^2\,\varepsilon = 0, \qquad \Box_2^2\,\mathbf{\Omega} = \mathbf{0}, \tag{20.2.5a,6a}$$

where

---

[2] R is said to be "star-shaped" if there exists a point in R from which all other points in R can be reached by means of straight lines.

$$\Box_1^2 \equiv \nabla^2 - \frac{1}{C_1^2}\frac{\partial^2}{\partial t^2}, \qquad \Box_2^2 \equiv \nabla^2 - \frac{1}{C_2^2}\frac{\partial^2}{\partial t^2}. \tag{20.2.5b,6b}$$

Equations (20.2.5) and (20.2.6) are a pair of wave equations governing the propagation of the dilatational, $\varepsilon = \boldsymbol{\nabla} \boldsymbol{\cdot} \mathbf{u}$, and rotational, $\boldsymbol{\Omega} = \boldsymbol{\nabla} \times \mathbf{u}/2$, waves in an isotropic elastic body. These equations are obtained by Stokes (1851).

Consider a decomposition of $\mathbf{u}(\mathbf{x}, t)$ in accordance with Helmholtz's theorem. Set $\mathbf{u} = \boldsymbol{\nabla} U + \boldsymbol{\nabla} \times \mathbf{V}$, where $\boldsymbol{\nabla} \boldsymbol{\cdot} \mathbf{V} = 0$, and substituting into (20.2.1), obtain

$$(\lambda + 2\mu)\, \boldsymbol{\nabla}(\Box_1^2 U) + \mu\, \boldsymbol{\nabla} \times (\Box_2^2 \mathbf{V}) = 0, \tag{20.2.7}$$

where it is assumed that U and $\mathbf{V}$ are sufficiently smooth, so that the order of differentiation with respect to the coordinate variables and the time can be interchanged. From (20.2.7), deduce that if

$$\Box_1^2 U = 0, \qquad \Box_2^2 \mathbf{V} = \mathbf{0}, \tag{20.2.8a,b}$$

then the Navier equations (20.2.1) are identically satisfied.

On the other hand, it is not immediately obvious that every solution of the Navier equations admits a representation in terms of certain scalar, say, ϕ, and vector, say, $\boldsymbol{\psi}$, potentials which satisfy, respectively, the wave equations (20.2.8a,b); it must also be required that $\boldsymbol{\nabla} \boldsymbol{\cdot} \boldsymbol{\psi} = 0$. The proof of this converse assertion, which is known as the completeness of the considered representation, has been discussed by a number of authors; for a historical account and detailed discussion, see Sternberg (1959). To establish this proof, operate on (20.2.7) by $\boldsymbol{\nabla} \boldsymbol{\cdot}$ and $\boldsymbol{\nabla} \times$, to arrive at

$$\nabla^2 (\Box_1^2 U) = 0, \qquad \nabla^2 (\Box_2^2 \mathbf{V}) = \mathbf{0}, \tag{20.2.9a,b}$$

respectively. Integrating these equations, write

$$\Box_1^2 U = a(\mathbf{x}, t), \qquad \Box_2^2 \mathbf{V} = \mathbf{b}(\mathbf{x}, t), \tag{20.2.10a,b}$$

where it must be required that

$$\nabla^2 a = 0, \qquad \nabla^2 \mathbf{b} = \mathbf{0}, \qquad \boldsymbol{\nabla} \boldsymbol{\cdot} \mathbf{b} = 0. \tag{20.2.10c~e}$$

Define $A(\mathbf{x}, t)$ and $\mathbf{B}(\mathbf{x}, t)$ such that

$$A(\mathbf{x}, t) = C_1^2 \int_{t_0}^{t} d\tau \left\{ \int_{t_0}^{\tau} a(\mathbf{x}, \lambda)\, d\lambda \right\},$$

$$\mathbf{B}(\mathbf{x}, t) = C_2^2 \int_{t_0}^{t} d\tau \left\{ \int_{t_0}^{\tau} \mathbf{b}(\mathbf{x}, \lambda)\, d\lambda \right\}, \tag{20.2.11a,b}$$

and obtain

$$a = \frac{1}{C_1^2}\frac{\partial^2 A}{\partial t^2}, \qquad \mathbf{b} = \frac{1}{C_2^2}\frac{\partial^2 \mathbf{B}}{\partial t^2}. \tag{20.2.12a,b}$$

Then, in view of (20.2.10c~e),

$$\nabla^2 A = 0, \qquad \nabla^2 \mathbf{B} = \mathbf{0}, \qquad \boldsymbol{\nabla} \boldsymbol{\cdot} \mathbf{B} = 0. \tag{20.2.12c~e}$$

Hence, set

$$\phi_1 = U + A, \qquad \boldsymbol{\psi}_1 = \mathbf{V} + \mathbf{B}, \tag{20.2.13a,b}$$

and obtain

$$\Box_1^2 \phi_1 = 0, \qquad \Box_2^2 \boldsymbol{\psi}_1 = \mathbf{0}. \tag{20.2.14a,b}$$

Now, substitution from (20.2.13a,b) into $\mathbf{u} = \boldsymbol{\nabla} U + \boldsymbol{\nabla} \times \mathbf{V}$, yields

$$\mathbf{u} = \boldsymbol{\nabla}\phi_1 + \boldsymbol{\nabla} \times \boldsymbol{\psi}_1 + \mathbf{u}^*, \tag{20.2.15a}$$

where

$$\mathbf{u}^* = -\boldsymbol{\nabla} A - \boldsymbol{\nabla} \times \mathbf{B}. \tag{20.2.15b}$$

From (20.2.15a,b) and identity (20.1.4),

$$\boldsymbol{\nabla} \cdot \mathbf{u}^* = 0, \qquad \boldsymbol{\nabla} \times \mathbf{u}^* = \mathbf{0}, \tag{20.2.16a,b}$$

from which it immediately follows that $\mathbf{u}^*$ is the gradient of a scalar field whose Laplacian vanishes in R, i.e.,[3]

$$\mathbf{u}^* = \boldsymbol{\nabla}\phi_2(\mathbf{x}, t), \qquad \nabla^2 \phi_2 = 0 \qquad \text{in R}. \tag{20.2.16c,d}$$

Substitution from (20.2.16c,d) into (20.2.15a), and then from (20.2.15a) into the Navier equations, (20.2.1), now yields the following differential equation for $\phi_2(\mathbf{x}, t)$:

$$\boldsymbol{\nabla} \frac{\partial^2 \phi_2}{\partial t^2} = \mathbf{0} \tag{20.2.17a}$$

which upon integration gives

$$\phi_2 = \alpha(t) + t\,\beta(\mathbf{x}) + \gamma(\mathbf{x}), \tag{20.2.17b}$$

where

$$\nabla^2 \beta = 0, \qquad \nabla^2 \gamma = 0. \tag{20.2.17c,d}$$

Hence, if

$$\phi = \phi_1(\mathbf{x}, t) + \phi_2(\mathbf{x}, t) - \alpha(t), \qquad \boldsymbol{\psi} = \boldsymbol{\psi}_1(\mathbf{x}, t), \tag{20.2.18a,b}$$

then,

$$\mathbf{u} = \boldsymbol{\nabla} \otimes \phi + \boldsymbol{\nabla} \times \boldsymbol{\psi}, \qquad \boldsymbol{\nabla} \cdot \boldsymbol{\psi} = 0, \tag{20.2.19a,b}$$

where

$$\Box_1^2 \phi = 0, \qquad \Box_2^2 \boldsymbol{\psi} = \mathbf{0}. \tag{20.2.19c,d}$$

This completes the proof. The above theorem is named after Clebesh, but its proof has been given by Duhem; see Sternberg (1959) for references.

---

[3] Note that the Laplacian of ϕ_2 is given by $\nabla^2 \phi_2 = \boldsymbol{\nabla} \cdot (\boldsymbol{\nabla}\phi_2)$ in terms of $\boldsymbol{\nabla}$.

20.3. PAPKOVICH-NEUBER REPRESENTATION

Consider now the solution of Navier's equation for elastostatic problems.

20.3.1. Papkovich-Neuber Representation

Since thermal effects can be included through equivalent body forces and surface tractions, they are not considered explicitly. Hence, examine the solution of

$$\nabla^2\mathbf{u} + \frac{\lambda+\mu}{\mu}\,\boldsymbol{\nabla}(\boldsymbol{\nabla}\boldsymbol{\cdot}\mathbf{u}) + \frac{1}{\mu}\,\mathbf{f}^{\mathrm{o}} = \mathbf{0}, \qquad \mathbf{f}^{\mathrm{o}} = \rho\,\mathbf{f}, \tag{20.3.1a,b}$$

which satisfies certain appropriate boundary conditions. To this end, decompose $\mathbf{u}$ in accordance with Helmholtz's theorem,

$$\mathbf{u} = \boldsymbol{\nabla}\Phi^* + \boldsymbol{\nabla}\times\boldsymbol{\Psi}^*, \qquad \boldsymbol{\nabla}\boldsymbol{\cdot}\boldsymbol{\Psi}^* = 0, \tag{20.3.2a,b}$$

and substituting into (20.3.1a), obtain

$$\nabla^2\{\alpha\,\boldsymbol{\nabla}\Phi^* + \boldsymbol{\nabla}\times\boldsymbol{\Psi}^*\} + \frac{1}{\mu}\,\mathbf{f}^{\mathrm{o}} = \mathbf{0}, \tag{20.3.3a}$$

where

$$\alpha = \frac{\lambda+2\mu}{\mu} = \frac{2(1-\nu)}{1-2\nu}, \tag{20.3.3b}$$

with ν being the Poisson ratio.

To obtain the *Papkovich-Neuber representation,* set

$$\boldsymbol{\Psi} = \alpha\,\boldsymbol{\nabla}\Phi^* + \boldsymbol{\nabla}\times\boldsymbol{\Psi}^*, \tag{20.3.4}$$

and substituting into (20.3.3a), arrive at

$$\nabla^2\boldsymbol{\Psi} = -\frac{1}{\mu}\,\mathbf{f}^{\mathrm{o}}, \tag{20.3.5a}$$

where

$$\boldsymbol{\nabla}\boldsymbol{\cdot}\boldsymbol{\Psi} = \alpha\,\nabla^2\Phi^*. \tag{20.3.5b}$$

Equation (20.3.5b) admits a solution in the form

$$\Phi^* = \frac{1}{2\alpha}\,(\mathbf{x}\boldsymbol{\cdot}\boldsymbol{\Psi} + \Phi), \tag{20.3.6a}$$

where the scalar potential Φ is given by

$$\nabla^2\Phi = \frac{1}{\mu}\,\mathbf{x}\boldsymbol{\cdot}\mathbf{f}^{\mathrm{o}}. \tag{20.3.6b}$$

To verify this assertion, substitute from (20.3.6a) into (20.3.5b), to obtain

$$\alpha\,\nabla^2\Phi^* - \boldsymbol{\nabla}\boldsymbol{\cdot}\boldsymbol{\Psi} = \frac{1}{2}\nabla^2(\mathbf{x}\boldsymbol{\cdot}\boldsymbol{\Psi} + \Phi) - \boldsymbol{\nabla}\boldsymbol{\cdot}\boldsymbol{\Psi}$$

$$= \frac{1}{2}\mathbf{x}\cdot\nabla^2\mathbf{\Psi} + \frac{1}{2\mu}\,\mathbf{x}\cdot\mathbf{f}^o$$

$$= \frac{1}{2}\mathbf{x}\cdot\{\nabla^2\mathbf{\Psi} + \frac{1}{\mu}\,\mathbf{f}^o\}, \tag{20.3.7}$$

where the identity $\nabla^2(\mathbf{x}\cdot\mathbf{\Psi}) = 2\mathbf{\nabla}\cdot\mathbf{\Psi} + \mathbf{x}\cdot\nabla^2\mathbf{\Psi}$ is also used. Note that in view of (20.3.5a), (20.3.7) is identically satisfied.

The displacement $\mathbf{u}$ may now be expressed as

$$\mathbf{u} = \mathbf{\Psi} - \frac{1}{4(1-\nu)}\,\mathbf{\nabla}(\mathbf{x}\cdot\mathbf{\Psi} + \Phi), \tag{20.3.8a}$$

where

$$\nabla^2\Phi = \frac{1}{\mu}\,\mathbf{x}\cdot\mathbf{f}^o, \qquad \nabla^2\mathbf{\Psi} = -\frac{1}{\mu}\,\mathbf{f}^o \qquad \text{in R.} \tag{20.3.8b,c}$$

Hence, the solution of Navier's equation (20.3.1) is reduced to the solution of Poisson's equation (20.3.8b,c). When the body forces are zero, these equations reduce to Laplace's equation,

$$\nabla^2\Phi = 0, \qquad \nabla^2\mathbf{\Psi} = 0 \qquad \text{in R.} \tag{20.3.8d,e}$$

The relations between the potentials ϕ and $\boldsymbol{\psi}$ considered in Subsection 20.2, and Φ and $\mathbf{\Psi}$ considered above, have been obtained by Sternberg (1959).

A basic question regarding the solution (20.3.8a~c) of (20.3.1) is that, inasmuch as there are only three displacement components involved in (20.3.1), the four unknown potentials, namely three components of $\mathbf{\Psi}$ and the scalar Φ, cannot be entirely independent. Hence, it is expected that one of these potentials can be defined arbitrarily, without impairing the validity or the completeness of the representation (20.3.8a). This question has been the subject of a number of papers, and has been satisfactorily resolved by Eubanks and Sternberg (1956). These authors proved that when R is "star-shaped" with respect to an interior point which is to be taken as the origin of the coordinates, and when 4ν is not an integer, then Φ in (20.3.8a) can be set equal to zero without a loss in the completeness of the representation. Moreover, if R is "star-shaped" with respect to an axis, say, the x_3-axis, and if the distance of any point of R from this axis is bounded, then Ψ_3 can be taken as zero everywhere in R, without impairing the completeness of the Papkovich-Neuber representation. For problems which possess an axis of symmetry, and for which the displacement field in polar coordinates can be expressed as

$$u_r = u_r(r,\, x_3), \qquad u_\theta = 0, \qquad u_3 = u_3(r,\, x_3), \tag{20.3.9a~c}$$

Ψ_1 and Ψ_2 may be taken to be identically zero, without a loss in completeness. Problems of this latter kind are known as "torsionless axisymmetric problems."

For the solution of boundary-value problems in which tractions are prescribed on a portion of the boundary of the solid, express the stress tensor in terms of the potential functions Φ and $\mathbf{\Psi}$. To this end, use (20.3.8a), to obtain the strain tensor, as

$$\boldsymbol{\varepsilon} = \frac{1}{2}\{(\boldsymbol{\nabla}\otimes\boldsymbol{\Psi}) + (\boldsymbol{\nabla}\otimes\boldsymbol{\Psi})^{\mathrm{T}}\} - \frac{1}{4(1-\nu)}\,\boldsymbol{\nabla}\otimes\{\boldsymbol{\nabla}(\mathbf{x}\,.\,\boldsymbol{\Psi}+\Phi)\}, \tag{20.3.10a}$$

or in component form,

$$\varepsilon_{ij} = \frac{1}{2}(\Psi_{j,i}+\Psi_{i,j}) - \frac{1}{4(1-\nu)}\,(x_k\Psi_k+\Phi)_{,ij}. \tag{20.3.10b}$$

The dilatation field is thus given by

$$\varepsilon = \frac{1-2\nu}{2(1-\nu)}\,\boldsymbol{\nabla}\,.\,\boldsymbol{\Psi} = \frac{1-2\nu}{2(1-\nu)}\,\Psi_{k,k}, \tag{20.3.11}$$

where (20.3.8b,c) is also used. Hence, from Hooke's law,

$$\boldsymbol{\sigma} = \lambda\,\varepsilon\,\mathbf{1}^{(2)} + 2\mu\,\boldsymbol{\varepsilon}, \tag{20.3.12a}$$

obtain

$$\boldsymbol{\sigma} = \mu\left\{\frac{\nu}{1-\nu}(\boldsymbol{\nabla}\,.\,\boldsymbol{\Psi})\,\mathbf{1}^{(2)} + \{\boldsymbol{\nabla}\otimes\boldsymbol{\Psi} + (\boldsymbol{\nabla}\otimes\boldsymbol{\Psi})^{\mathrm{T}}\}\right.$$

$$\left. - \frac{1}{2(1-\nu)}\boldsymbol{\nabla}\otimes\{\boldsymbol{\nabla}(\mathbf{x}\,.\,\boldsymbol{\Psi}+\Phi)\}\right\}, \tag{20.3.12b}$$

or in component form,

$$\sigma_{ij} = \mu\left\{\frac{\nu}{1-\nu}\Psi_{k,k}\,\delta_{ij} + (\Psi_{j,i}+\Psi_{i,j}) - \frac{1}{2(1-\nu)}(x_k\Psi_k+\Phi)_{,ij}\right\}. \tag{20.3.12c}$$

20.3.2. Galerkin Vector

Before closing this section, consider another representation of the solution of Navier's equations in terms of a vector-valued function known as the *Galerkin vector*. To this end, noting (20.1.7) and (20.3.4), set

$$\Phi^* = \frac{1}{\alpha}\,\boldsymbol{\nabla}.\mathbf{G}, \qquad \boldsymbol{\Psi}^* = -\,\boldsymbol{\nabla}\times\mathbf{G}, \tag{20.3.13a,b}$$

and from (20.3.3) obtain

$$\nabla^4\,\mathbf{G} = -\,\frac{1}{\mu}\,\mathbf{f}^{\mathrm{o}}, \tag{20.3.14}$$

where ∇^4 is the biharmonic operator, $\nabla^4 \equiv \nabla^2\nabla^2$. In terms of $\mathbf{G}$, the displacement field is given by

$$\mathbf{u} = \nabla^2\,\mathbf{G} - \frac{1}{2(1-\nu)}\,\boldsymbol{\nabla}(\boldsymbol{\nabla}.\mathbf{G}). \tag{20.3.15}$$

Relations between the Galerkin vector and the Papkovich-Neuber potential have been discussed by Mindlin (1936).

20.4. CONCENTRATED FORCE IN INFINITE AND SEMI-INFINITE SOLIDS

Consider solutions for the problem of a semi-infinite elastic body subjected to a concentrated force in the interior or on its plane boundary. To this end, use the systematic method developed by Mindlin (1953). In this method, potentials $\mathbf{\Phi}$ and $\mathbf{\Psi}$ are obtained which satisfy (20.3.8b,c) and the corresponding boundary conditions on the plane boundary of the solid. When a concentrated force acts in the interior of an infinitely extended elastic body (unbounded solid), there are no boundary conditions to consider except for the regularity requirements at infinity. The solution of this problem was obtained by Kelvin in 1848. The solution of the problem of a semi-infinite body subjected to a normal concentrated force on its plane boundary was obtained by Boussinesq in 1885, and the same problem, except for a tangential force, was solved by Cerruti in 1882; see Love (1944), for a detailed discussion; see also Westergaard (1952). Mindlin (1936) obtains the solution for the problem of a single force in the interior of a semi-infinite solid bounded by a plane.

20.4.1. Green's Second Identity

Consider *Green's second identity,* and choose the function w such that: 1) $\nabla^2 w = 0$ in R; 2) $w = -1/r$ on ∂R; and 3) w is regular everywhere in $R + \partial R$, where $r = \sqrt{(\mathbf{x} - \boldsymbol{\xi}) \cdot (\mathbf{x} - \boldsymbol{\xi})}$, with $\mathbf{x}$ being the position vector of a fixed point P in R, and $\boldsymbol{\xi}$ being that of the variable point Q. Green's second identity now becomes

$$p\,u = -\int_R G(\mathbf{x}, \boldsymbol{\xi})\, \nabla^2 u(\boldsymbol{\xi})\, dV_\xi - \int_{\partial R} u(\boldsymbol{\xi}) \frac{\partial G}{\partial \nu}(\mathbf{x}, \boldsymbol{\xi})\, dS_\xi, \tag{20.4.1}$$

where $G = (1/r) + w$ is Green's function, and where p is equal to 4π, 2π, or zero according to whether P is taken in R, on ∂R, or outside of the region R. In (20.4.1), $\partial/\partial\nu \equiv \nu_i\, \partial/\partial\xi_i \equiv \boldsymbol{\nu} \cdot \boldsymbol{\nabla}$, where $\boldsymbol{\nu}$ is the exterior unit normal to ∂R. From the assumed regularity of u at infinity, for unbounded regions, instead of (20.4.1), it follows that

$$4\pi\, u = -\int_R \frac{\nabla^2 u}{r}\, dV \tag{20.4.2}$$

which is a consequence of Green's third identity.

20.4.2. Infinitely Extended Solid

As an example, consider an infinitely extended body with a concentrated force $\mathbf{p}$ applied at a point, say, the origin O, in a given direction. Consider a small region R_0 with boundary ∂R_0 about the origin O, and distribute $\mathbf{f}^o$ in R_0 such that

$$\lim_{R_0 \to 0} \int_{R_0} \mathbf{f}^o \, dV = \mathbf{p}, \qquad \lim_{R_0 \to 0} \int_{R_0} \boldsymbol{\xi} \times \mathbf{f}^o \, dV = \mathbf{0}, \tag{20.4.3a,b}$$

where $\mathbf{f}^o$ is the intensity of the distributed load per unit volume. From (20.3.8b,c), it follows that (see Figure 20.4.1)

$$\nabla^2 \Phi = \begin{cases} 1/\mu \, \boldsymbol{\xi} \cdot \mathbf{f}^o & \text{for Q in } R_0 \\ 0 & \text{for Q in } R - R_0, \end{cases} \tag{20.4.4}$$

and

$$\nabla^2 \boldsymbol{\Psi} = \begin{cases} -1/\mu \, \mathbf{f}^o & \text{for Q in } R_0 \\ 0 & \text{for Q in } R - R_0. \end{cases} \tag{20.4.5}$$

Substitution from (20.4.4) and (20.4.5) into (20.4.2) now results in

$$\Phi = 0, \qquad \boldsymbol{\Psi} = \frac{1}{4\pi\mu} \frac{\mathbf{p}}{|\mathbf{x}|}, \tag{20.4.6a,b}$$

where, in addition to (20.4.3), the mean-value theorem of the potential theory is also used; see Kellogg (1953). The displacement field now becomes

$$\mathbf{u}(\mathbf{x}) = \frac{1}{4\pi\mu} \left\{ \frac{1}{r} \mathbf{p} - \frac{1}{4(1-\nu)} \nabla \{ \frac{1}{r} \mathbf{x} \cdot \mathbf{p} \} \right\}, \tag{20.4.7}$$

where $r = |\mathbf{x}|$. Suppose now that $\mathbf{p}$ acts along the x_3-axis, $\mathbf{p} = p\,\mathbf{e}_3$. In polar coordinates

$$\mathbf{u} = \frac{p}{4\pi\mu} \left\{ \frac{1}{4(1-\nu)} \frac{\rho\, x_3}{r^3} \mathbf{e}_\rho + \{ \frac{1}{r} - \frac{1}{4(1-\nu)} (\frac{1}{r} - \frac{x_3^2}{r^3}) \} \mathbf{e}_3 \right\}, \tag{20.4.8a}$$

where $\rho^2 = x_1^2 + x_2^2$. The corresponding displacement components in the rectangular coordinates are

$$u_i = \frac{p}{16\pi\mu(1-\nu)} \left\{ \frac{3-4\nu}{r} \delta_{i3} + \frac{x_3 x_i}{r^3} \right\}, \tag{20.4.8b}$$

for i = 1, 2, 3.

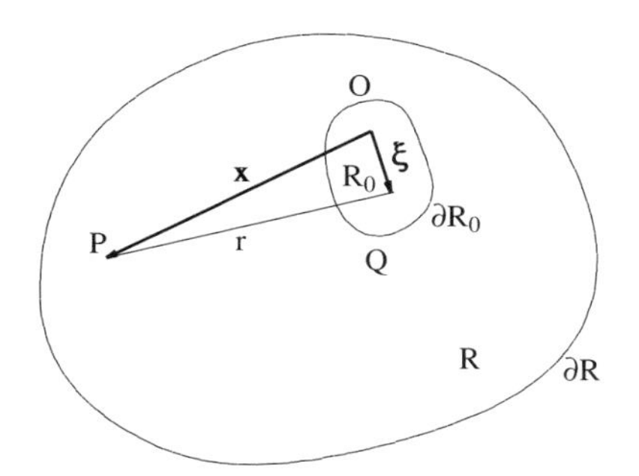

Figure 20.4.1

Small region R_0 around origin where $\mathbf{f}^o$ is distributed

20.4.3. Semi-Infinite Body with Normal Concentrated Forces

Now consider the problem of a semi-infinite elastic body which is loaded on its plane boundary or in its interior by arbitrary distributed forces. For this class of problems, it suffices to obtain a solution for concentrated loads applied on the boundary or in the interior of the body, and then employ a superposition procedure to arrive at the solution for other distributed loading situations.

Consider then the semi-infinite (half-space) body shown in Figure 20.4.2, where the x_3-axis is taken perpendicular to the plane boundary of the region, and the x_2-axis is normal to the plane of the figure. Let P with the position vector $\mathbf{x}$ be a fixed point, and denote the position vector of the variable point Q by $\boldsymbol{\xi}$. In addition to these points, consider the *mirror image* Q′ of Q with respect to the plane boundary Ox_1x_2, and denote its position vector by $\boldsymbol{\xi}'$. Let Q_0 be the intersection of QQ′ with the Ox_1x_2-plane, and $\boldsymbol{\xi}_0$ be its position vector.

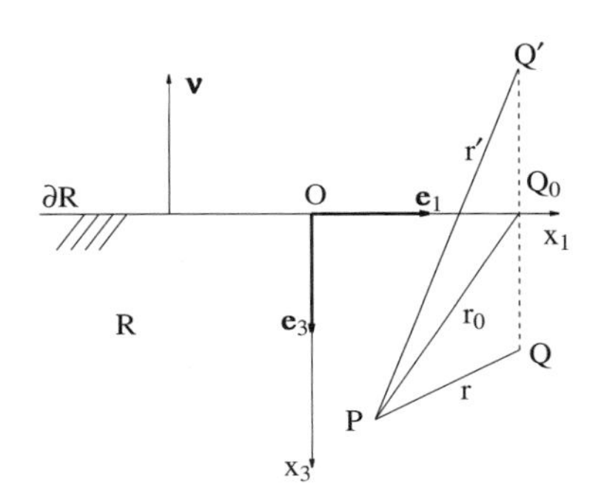

Figure 20.4.2

Semi-infinite body R and points P, Q, and Q′

Referring to Figure 20.4.2, it follows that

$$r = |\mathbf{x} - \boldsymbol{\xi}|, \qquad r' = |\mathbf{x} - \boldsymbol{\xi}'|, \qquad r_0 = |\mathbf{x} - \boldsymbol{\xi}_0|. \tag{20.4.9a~c}$$

Now examine the consequence of choosing the Green function G in (20.4.1), as

$$G(\mathbf{x}, \boldsymbol{\xi}) = \frac{1}{r} - \frac{1}{r'}. \tag{20.4.10}$$

Since the exterior unit normal to ∂R is $\boldsymbol{\nu} = -\mathbf{e}_3$,

$$\frac{\partial G}{\partial \nu} = -\frac{\partial G}{\partial \xi_3} = -\frac{\partial}{\partial \xi_3}\left\{\frac{1}{r} - \frac{1}{r'}\right\} = -\frac{x_3 - \xi_3}{r^3} - \frac{x_3 + \xi_3}{r'^3} \tag{20.4.11a}$$

which on ∂R ($\xi_3 = 0$) reduces to

$$\frac{\partial G}{\partial \nu} = -\frac{2x_3}{r_0^3} = 2\frac{\partial}{\partial x_3}\left\{\frac{1}{r_0}\right\}. \tag{20.4.11b}$$

Substitution from this equation into (20.4.1), for P inside R, now yields

$$4\pi u = -\int_R \left\{\frac{1}{r} - \frac{1}{r'}\right\} \nabla^2 u \, dV_\xi - 2\frac{\partial}{\partial x_3}\left\{\int_{\partial R} \frac{u}{r_0}\, dS_\xi\right\}, \tag{20.4.12}$$

where the integrations are carried out with respect to $\boldsymbol{\xi}$, while $\mathbf{x}$ is held fixed.

In the sequel, (20.4.12) is used to obtain the appropriate potential functions $\boldsymbol{\Phi}$ and $\boldsymbol{\Psi}$ for two problems: (1) a half-space subjected to a concentrated interior force which acts perpendicular to the plane boundary, and (2) a half-space subjected to a concentrated interior force which acts parallel to the plane boundary. The solution for a half-space subjected on its plane boundary to a concentrated force can then be obtained as a special case.

Consider first a concentrated interior force normal to and at a distance c from the Ox_1x_2-plane which forms the boundary of the half-space $x_3 \geq 0$; see Figure 20.4.3. Let $\mathbf{p} = p\,\mathbf{e}_3$ be applied at a point on the x_3-axis, a distance c from the origin O. Consider a small region R_0 with the boundary ∂R_0 about the point of the application of $\mathbf{p}$, and distribute $\mathbf{f}^o$ within this region in accordance with (20.4.3).

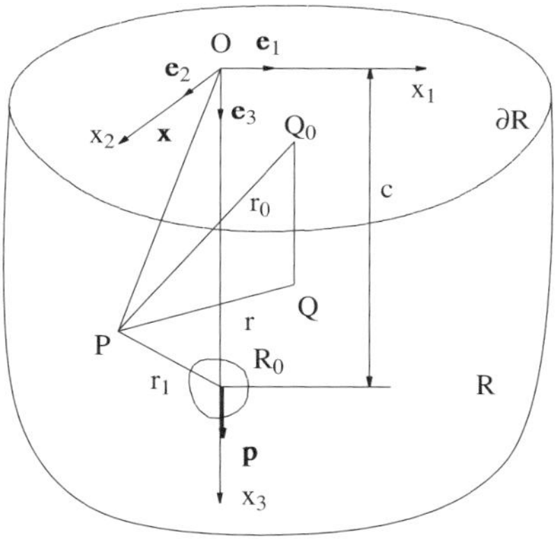

Figure 20.4.3

Concentrated force normal to boundary of half-space

Since this problem is torsionless and axisymmetric, set $\Psi_1 = \Psi_2 = 0$, and from (20.3.8a) obtain

$$\mathbf{u} = \Psi_3\,\mathbf{e}_3 - \frac{1}{4(1-\nu)}\boldsymbol{\nabla}\otimes(x_3\,\Psi_3 + \Phi). \tag{20.4.13}$$

Equation (20.3.8b,c) now reduces to[4]

$$\nabla^2\Phi = \begin{cases} 1/\mu\,\xi_3\,f_3^o & \text{for Q in } R_0 \\ 0 & \text{for Q in } R - R_0, \end{cases} \tag{20.4.14}$$

and

$$\nabla^2\Psi_3 = \begin{cases} -1/\mu\,f_3^o & \text{for Q in } R_0 \\ 0 & \text{for Q in } R - R_0. \end{cases} \tag{20.4.15}$$

From (20.3.12), moreover, the stress components are

---

[4] Note that $p = \lim_{R_0 \to 0} \int_{R_0} f_3^o\, dV$.

$$\sigma_{3i} = \frac{\mu}{\alpha}\,\{\Psi_{3,i} - \frac{1}{1-2\nu}\,(x_3\Psi_{3,3i} + \Phi_{,3i})\} \qquad (i = 1, 2),$$

$$\sigma_{33} = \frac{\mu}{\alpha}\,\{\alpha\,\Psi_{3,3} - \frac{1}{1-2\nu}\,(x_3\Psi_{3,33} + \Phi_{,33})\}, \tag{20.4.16a,b}$$

which, because ∂R is traction-free, reduce to

$$\sigma_{3i} = \frac{\mu}{\alpha}\,\{\Psi_{3,i} - \frac{1}{1-2\nu}\,\Phi_{,3i}\} = 0 \qquad (i = 1, 2),$$

$$\sigma_{33} = \frac{\mu}{\alpha}\,\{\alpha\,\Psi_{3,3} - \frac{1}{1-2\nu}\,\Phi_{,33}\} = 0, \tag{20.4.17a,b}$$

where $\alpha = 2(1-\nu)/(1-2\nu)$.

Inspection of (20.4.17) now reveals that the expressions $2(1-\nu)\Psi_{3,3} - \Phi_{,33}$ and $(1-2\nu)\Psi_3 - \Phi_{,3}$ have known Laplacians in R and vanish on ∂R; the vanishing of $(1-2\nu)\Psi_3 - \Phi_{,3}$ on ∂R follows from the fact that $\int_{-\infty}^{x} \sigma_{3i}\, dx_i = 0$ on ∂R. It therefore appears reasonable to choose for u in (20.4.12) quantities defined by $2(1-\nu)\Psi_{3,3} - \Phi_{,33}$ and $(1-2\nu)\Psi_3 - \Phi_{,3}$. Substitute from $2(1-\nu)\Psi_{3,3} - \Phi_{,33}$ into (20.4.12), and noting (20.4.14) and (20.4.15), obtain

$$2(1-\nu)\,\Psi_{3,3} - \Phi_{,33} = \frac{1}{4\pi\mu}\int_{R_0} \{\frac{1}{r} - \frac{1}{r'}\}$$

$$\times\,\{2(1-\nu)\,\frac{\partial f_3^o}{\partial \xi_3} + \frac{\partial^2(\xi_3\, f_3^o)}{\partial \xi_3^2}\}\, dV. \tag{20.4.18}$$

Now integrate the first integral corresponding to the first term inside the brackets in the right-hand side of (20.4.18) once by parts, the second integral corresponding to the second term inside of the brackets twice by parts, and taking the limit as R_0 goes to zero, obtain

$$\lim_{R_0 \to 0} \int_{R_0} \{\frac{1}{r} - \frac{1}{r'}\}\,\frac{\partial f_3^o}{\partial \xi_3}\, dV = p\,\frac{\partial}{\partial x_3}\,\{\frac{1}{r'} + \frac{1}{r_2}\} \tag{20.4.19}$$

and

$$\lim_{R_0 \to 0} \int_{R_0} \{\frac{1}{r} - \frac{1}{r'}\}\,\frac{\partial^2(\xi_3\, f_3^o)}{\partial \xi_3^2}\, dV = c\,p\,\frac{\partial^2}{\partial x_3^2}\,\{\frac{1}{r_1} - \frac{1}{r_2}\}, \tag{20.4.20}$$

where

$$r_1^2 = x_1^2 + x_1^2 + (x_3 - c)^2, \qquad r_2^2 = x_1^2 + x_1^2 + (x_3 + c)^2, \tag{20.4.21,22}$$

and where the following results, as R_0 approaches zero, are used:

$$\frac{\partial G}{\partial \xi_i} = -\frac{\partial}{\partial x_i}\,\{\frac{1}{r_1} - \frac{1}{r_2}\} \qquad (i = 1, 2), \qquad \frac{\partial G}{\partial \xi_3} = -\frac{\partial}{\partial x_3}\,\{\frac{1}{r_1} + \frac{1}{r_2}\}$$

$$\frac{\partial^2 G}{\partial \xi_3^2} = \frac{\partial^2}{\partial x_3^2}\,\{\frac{1}{r_1} - \frac{1}{r_2}\}. \tag{20.4.23a~c}$$

From (20.4.19), (20.4.20), and (20.4.21), it follows that

$$2(1-\nu)\,\Psi_3 - \Phi_{,3} = \frac{p}{4\pi\mu}\left\{2(1-\nu)\,\{\frac{1}{r_1} + \frac{1}{r_2}\} + c\,\{\frac{1}{r_1} - \frac{1}{r_2}\}_{,33}\right\}. \quad (20.4.24a)$$

Next, set $u = (1-2\nu)\Psi_3 - \Phi_{,3}$ in (20.4.12), and arrive at

$$2(1-2\nu)\,\Psi_3 - \Phi_{,3} = \frac{p}{4\pi\mu}\left\{(1-2\nu)\,\{\frac{1}{r_1} - \frac{1}{r_2}\} + c\,\{\frac{1}{r_1} - \frac{1}{r_2}\}_{,3}\right\}. \quad (20.4.24b)$$

Combining (20.4.24a) and (20.4.24b), finally obtain

$$\Psi_3 = \frac{p}{4\pi\mu}\left\{\frac{1}{r_1} + \frac{3-4\nu}{r_2} + \frac{2c\,(x_3+c)}{r_2^3}\right\},$$

$$\Phi = \frac{p}{4\pi\mu}\left\{4(1-\nu)\,(1-2\nu)\ln(r_2+x_3+c) - \frac{c}{r_1} - \frac{(3-4\nu)\,c}{r_2}\right\}, \quad (20.4.25a,b)$$

which completes the solution.

The displacement field can be calculated from (20.3.8) and the stress field from (20.3.12). The solution for the case of a concentrated load acting normal to the plane boundary of a half-space (Boussinesq's problem) corresponds to $c = 0$ in (20.4.25a,b). This gives

$$\Psi_3 = \frac{(1-\nu)\,p}{\pi\mu}\,\frac{1}{r} \quad (20.4.26a)$$

and

$$\Phi = \frac{(1-\nu)\,(1-2\nu)}{\pi\mu}\,p\ln(r+x_3), \quad (20.4.26b)$$

where, as before, $r = |\mathbf{x}|$. Using the cylindrical coordinates, obtain the following displacement components:

$$u_r = \frac{p\,\rho}{4\pi\mu\,r}\left\{\frac{x_3}{r^2} - \frac{1-2\nu}{r+x_3}\right\}, \qquad u_\theta = 0,$$

$$u_3 = \frac{p}{4\pi\mu\,r}\,\{2(1-\nu) + \frac{x_3^2}{r^2}\}, \quad (20.4.27a\sim c)$$

where $\rho^2 = x_1^2 + x_2^2$. The corresponding nonzero stress components are

$$\sigma_{rr} = \frac{p}{2\pi\,r}\,\{\frac{1-2\nu}{r+x_3} - \frac{3\rho^2 x_3}{r^4}\}, \qquad \sigma_{\theta\theta} = \frac{p\,(1-2\nu)}{2\pi\,r}\,\{\frac{x_3}{r^2} - \frac{1}{r+x_3}\},$$

$$\sigma_{33} = -\frac{\pi}{2\pi}\,\frac{3x_3^3}{r^5}, \qquad \sigma_{r3} = \sigma_{3r} = -\frac{p}{2\pi}\,\frac{3\rho x_3^2}{r^5}. \quad (20.4.28a\sim d)$$

20.4.4. Semi-Infinite Body with Tangential Concentrated Forces

Now consider a half-plane subjected in its interior to a concentrated force applied parallel to it, and at a distance c from its plane boundary ∂R. Choose the x_1-axis parallel to the load $\mathbf{q} = q\,\mathbf{e}_1$, as in Figure 20.4.4. Distribute $\mathbf{f}^o$ in a small region R_0 in the manner defined by (20.4.3), and note that $f_2^o = f_3^o = 0$. Choosing $\Psi_2 = 0$, write the boundary conditions on ∂R, as

$$\sigma_{31} = \frac{\mu}{2(1-\nu)} \{(1-2\nu)(\Psi_{3,1} + \Psi_{1,3}) - x_1 \Psi_{1,31} - \Phi_{,31}\} = 0,$$

$$\sigma_{32} = \frac{\mu}{2(1-\nu)} \{(1-2\nu)\Psi_{3,2} - x_1 \Psi_{1,32} - \Phi_{,32}\} = 0,$$

$$\sigma_{33} = \frac{\mu}{2(1-\nu)} \{2(1-\nu)\Psi_{3,3} + 2\nu\Psi_{1,1} - x_1 \Psi_{1,33} - \Phi_{,33}\} = 0 \qquad \text{on } \partial R. \tag{20.4.29a~c}$$

Differentiate (20.4.29a) with respect to x_2, and (20.4.29b) with respect to x_1, and forming the difference of the resulting equations, arrive at

$$\Psi_{1,23} = 0 \qquad \text{on } \partial R. \tag{20.4.30a}$$

Integrating this equation with respect to x_2, obtain $\Psi_{1,3} = f(x_1)$ on ∂R, where, without a loss in generality, it is assumed that $f(x_1) = 0$. It now appears reasonable to choose u in (20.4.12) equal to $\Psi_{1,3}$ which vanishes on ∂R and has a known Laplacian in R. Equation (20.4.12) thus yields

$$\Psi_{1,3} = \frac{1}{4\pi\mu}\int_{R_0} \{\frac{1}{r} - \frac{1}{r'}\} \frac{\partial f_1^o}{\partial \xi_3} dV = \frac{q}{4\pi\mu} \{\frac{1}{r_1} + \frac{1}{r_2}\}_{,3} \qquad (\text{as } R_0 \to 0). \tag{20.4.30b}$$

Upon integration with respect to x_3, this equation becomes

$$\Psi_1 = \frac{q}{4\pi\mu} \{\frac{1}{r_1} + \frac{1}{r_2}\}. \tag{20.4.31}$$

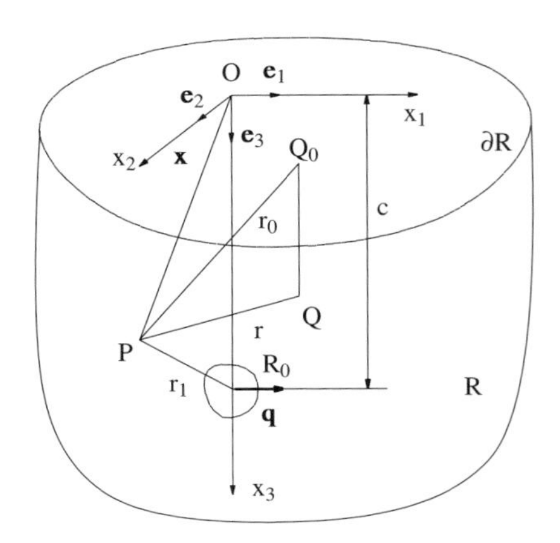

Figure 20.4.4

Concentrated force parallel to boundary of half-space

Now consider (20.4.29b). Noting that, $\int_{-\infty}^{x_2} \sigma_{32}\, dx_2 = 0$ since ∂R is traction-free, obtain

$$(1-2\nu)\,\Psi_3 - \Phi_{,3} = x_1\,\Psi_{1,3} = 0, \tag{20.4.32a}$$

where the fact that $\Psi_{1,3} = 0$ on ∂R is also used. Therefore choose u in (20.4.12) equal to $(1-2\nu)\Psi_3 - \Phi_{,3}$, and employing (20.4.32a), arrive at

$$(1-2\nu)\,\Psi_3 - \Phi_{,3} = \frac{1}{4\pi\mu}\int_{R_0} \{\frac{1}{r} - \frac{1}{r'}\}\,\xi_1 \frac{\partial f_1^o}{\partial \xi_3}\, dV. \tag{20.4.32b}$$

At the limit when R_0 goes to zero, (20.4.32b) yields

$$(1-2\nu)\,\Psi_3 - \Phi_{,3} = 0. \tag{20.4.32c}$$

Finally, write (20.4.29c) as

$$2(1-\nu)\,\Psi_{3,3} - \Phi_{,33} + 2\nu\,\Psi_{1,1} - x_1\,\Psi_{1,33} = 0, \tag{20.4.33a}$$

and note from (20.4.31) that on ∂R,

$$2\nu\,\Psi_{1,1} - x_1\,\Psi_{1,33} = \frac{(1-2\nu)\,x_1\,q}{2\pi\mu\, r_0^3} - \frac{3q\,c^2\,x_1}{2\pi\mu\, r_0^5}, \tag{20.4.33b}$$

where $r_0^2 = x_1^2 + x_2^2 + c^2$. Moreover, $-(1-2\nu)\,\Psi_{1,1} = (1-2\nu)\,x_1\,q/\pi\mu\, r_0^3$ on ∂R, $q\,c/2\pi\,\mu\,(1/r_2)_{,13} = 3q\,c^2 x_1/2\pi\mu\, r_0^5$, and hence (20.4.33a) becomes

$$\chi = 2(1-\nu)\,\Psi_{3,3} - \Phi_{,33} - (1-2\nu)\,\Psi_{1,1} - \frac{q\,c}{2\pi\,\mu}\{\frac{1}{r_2}\}_{,13} = 0, \tag{20.4.34a}$$

on ∂R. Now, in R_0,

$$\nabla^2\chi = -\frac{1}{\mu}\,x_1\,f_{1,33}^o + \frac{1-2\nu}{\mu}\,f_{1,1}^o, \tag{20.4.34b}$$

and therefore (20.4.12) yields

$$\chi = \frac{1}{4\pi\mu}\int_{R_0} \{\frac{1}{r} - \frac{1}{r'}\}\,\{\xi_1\,\frac{\partial^2 f_1^o}{\partial \xi_3^2} - (1-2\nu)\,\frac{\partial f_1^o}{\partial \xi_1}\}\, dV \tag{20.4.34c}$$

which vanishes on ∂R, and which, as R_0 goes to zero, reduces to

$$\chi = -\frac{(1-2\nu)\,q}{4\pi\mu}\,\{\frac{1}{r_1} - \frac{1}{r_2}\}_{,1}. \tag{20.4.34d}$$

Combining (20.4.34a) and (20.4.34c,d), obtain

$$2(1-\nu)\,\Psi_{3,3} - \Phi_{,33} = -\frac{(1-2\nu)\,q\,x_1}{2\pi\,\mu\, r_2^3} + \frac{3q\,c\,x_1\,(x_3+c)}{2\pi\,\mu\, r_2^5} \tag{20.4.35a}$$

which, upon integration with respect to x_3, yields

$$2(1-\nu)\,\Psi_3 - \Phi_{,3} = -\frac{(1-2\nu)\,q\,x_1}{2\pi\,\mu\, r_2(r_2+x_3+c)} - \frac{q\,c\,x_1}{2\pi\,\mu\, r_2^3}. \tag{20.4.35b}$$

From this and (20.4.32c),

$$\Psi_3 = \frac{(1-2\nu)\,q\,x_1}{2\pi\,\mu\,r_2(r_2+x_3+c)} - \frac{q\,c\,x_1}{2\pi\,\mu\,r_2^3},$$

$$\Phi = -\frac{(1-2\nu)^2\,q\,x_1}{2\pi\,\mu\,(r_2+x_3+c)} - \frac{(1-2\nu)\,q\,x_1}{2\pi\,\mu\,r_2\,(r_2+x_3+c)}, \tag{20.4.30c,d}$$

which, together with (20.4.31), completes the solution.

It is interesting to note that when the Poisson ratio ν equals 1/2 (incompressible elastic materials), Φ vanishes and Ψ_3 reduces to $-qc\,x_1/2\pi\mu\,r_2^3$; a similar observation can be made regarding the solution (20.4.26a,b). This fact has been exploited by Westergaard (1952) to simplify the solution of this class of problems. Observe that when $c = 0$ (Cerruti's problem),

$$u_1 = \frac{q}{4\pi\,\mu\,r}\left\{1 + \frac{x_1^2}{r^2} + (1-2\nu)\,\{\frac{r}{r+x_3} - \frac{x_1^2}{(r+x_3)^2}\}\right\},$$

$$u_2 = \frac{q\,x_1\,x_2}{4\pi\,\mu\,r}\left\{\frac{1}{r^2} - \frac{1-2\nu}{(r+x_3)^2}\right\},$$

$$u_3 = \frac{q\,x_1}{4\pi\,\mu\,r}\left\{\frac{x_3}{r^2} - \frac{1-2\nu}{r+x_3}\right\}, \tag{20.4.36a~c}$$

and the corresponding stress components are given by

$$\sigma_{11} = \frac{q\,x_1}{2\pi\,r^3}\left\{-\frac{3x_1^2}{r^2} + \frac{1-2\nu}{(r+x_3)^2}\,\{r^2 - x_2^2 - \frac{2r\,x_2^2}{r+x_3}\}\right\},$$

$$\sigma_{22} = \frac{q\,x_1}{2\pi\,r^3}\left\{-\frac{3x_1^2}{r^2} + \frac{1-2\nu}{(r+x_3)^2}\,\{3r^2 - x_2^2 - \frac{2r\,x_2^2}{r+x_3}\}\right\},$$

$$\sigma_{12} = \frac{q\,x_2}{2\pi\,r^3}\left\{-\frac{3x_1^2}{r^2} + \frac{1-2\nu}{(r+x_3)^2}\,\{-r^2 - x_1^2 - \frac{2r\,x_2^2}{r+x_3}\}\right\},$$

$$\sigma_{31} = \sigma_{13} = -\frac{3q\,x_1^2\,x_3}{2\pi\,r^5}, \qquad \sigma_{32} = \sigma_{23} = -\frac{3q\,x_1\,x_2\,x_3}{2\pi\,r^5},$$

$$\sigma_{33} = -\frac{3q\,x_1\,x_3^2}{2\pi\,r^5}. \tag{20.4.37a~f}$$

20.5. REFERENCES

Eubanks, R. A. and Sternberg, E. (1956), On the completeness of the Boussinesq-Papkovich stress function, *J. Rat. Mech. Anal.*, Vol. 5, 735-746.

Kellogg, O. D. (1953), *Foundations of potential theory*, Dover, New York.

Love, A. E. H. (1944), *A Treatise on the mathematical theory of elasticity*, Dover Publications, New York.

Mindlin, R. D. (1936), Force at a point in the interior of a semi-infinite solid, in *Proc. 1st Midwest Conference on Solid Mech.*, 56-59.

Mindlin, R. D. (1953), Force at a point in the interior of a semi-infinite solid, in *Proceedings of First Midwestern Conference on Solid Mechanics*, April 1953, 56-59.

Sternberg, E. (1959), On the integration of the equation of motion in the classical theory of elasticity, *Arch. Rat. Mech. Anal.*, Vol. 6, 34-50.

Stokes, G. G. (1851) On the dynamical theory of diffraction, *Trans. Comb. Phil. Soc.*, Vol. 9, 1.

Westergaard, H. M. (1952), *Theory of elasticity and plasticity*, Harvard University Press, Cambridge.

SECTION 21 SOLUTIONS OF SINGULAR PROBLEMS

In this section certain basic mathematical foundations of two-dimensional elasticity problems are summarized. Airy's stress function and Muskhelishvili's (1956) complex potentials are introduced. Emphasis is placed on problems involving singularities. Included also are certain basic topics in two-dimensional anisotropic elasticity.

21.1. AIRY'S STRESS FUNCTION

Airy's stress function is a scalar potential which is introduced to satisfy the two-dimensional equations of equilibrium in the absence of body forces. The compatibility requirement then leads to a differential equation for Airy's stress function, which, together with the boundary conditions, defines this function uniquely. For isotropic elasticity, a bi-harmonic equation is obtained for Airy's stress function.

The introduction of complex-valued functions considerably simplifies the solution to this bi-harmonic equation and the expressions of the corresponding stress, strain, and displacement fields. Airy's stress function can be expressed in terms of Muskhelishvili's complex potentials, with immediate application to anisotropic two-dimensional problems. This and related issues are briefly discussed in this and subsequent subsections.

21.1.1. Solution to Equilibrium Equations

In the absence of body forces,[1] the equations of equilibrium in two dimensions are given by

$$\sigma_{ij,i} = 0 \qquad \text{for i, j} = 1, 2 \tag{21.1.1a}$$

or, more explicitly,

$$\sigma_{11,1} + \sigma_{21,2} = 0, \qquad \sigma_{12,1} + \sigma_{22,2} = 0, \tag{21.1.1b,c}$$

where $\sigma_{12} = \sigma_{21}$, and comma followed by indices denotes partial differentiation

[1] If body forces are conservative or given by the gradient of a certain potential, then the field variables can be formulated in a manner similar to the one presented here.

with respect to the corresponding coordinates; see Subsection 17.2. Consider a stress function, $U = U(x_1, x_2)$, such that

$$\sigma_{11} = U_{,22}, \qquad \sigma_{22} = U_{,11}, \qquad \sigma_{12} = \sigma_{21} = -U_{,12}. \tag{21.1.2a~c}$$

These stress components *always* satisfy the equations of equilibrium. Function U is called *Airy's stress function.*

Conversely, if a stress field satisfies (21.1.1), then the stress components are given by (21.1.2) in terms of a certain function, $U = U(x_1, x_2)$. The proof is straightforward. From (21.1.1b), there exists a certain function, F_1, such that

$$\sigma_{11} = F_{1,2} \quad \text{and} \quad \sigma_{12} = -F_{1,1}. \tag{21.1.3a,b}$$

Similarly, from (21.1.1c), there exists another function, F_2, such that

$$\sigma_{22} = F_{2,1} \quad \text{and} \quad \sigma_{21} = -F_{2,2}. \tag{21.1.3c,d}$$

Since $\sigma_{12} = \sigma_{21}$ leads to $F_{1,1} = F_{2,2}$, functions F_1 and F_2 are expressed as

$$F_1 = F_{,2} \quad \text{and} \quad F_2 = F_{,1}, \tag{21.1.4a,b}$$

in terms of a certain function, F, which is identified with the Airy stress function, U.

21.1.2. Governing Equation for Airy's Stress Function

First, for simplicity, consider an isotropic material. The constitutive relations for linear elasticity in two dimensions then are

$$\begin{bmatrix} \varepsilon_{11} \\ \varepsilon_{22} \\ 2\varepsilon_{12} \end{bmatrix} = \frac{1}{\mu} \begin{bmatrix} (\kappa+1)/8 & (\kappa-3)/8 & 0 \\ (\kappa-3)/8 & (\kappa+1)/8 & 0 \\ 0 & 0 & 1 \end{bmatrix} \begin{bmatrix} \sigma_{11} \\ \sigma_{22} \\ \sigma_{12} \end{bmatrix}, \tag{21.1.5}$$

where μ is the shear modulus, $\kappa = 3 - 4\nu$ for plane strain and $(3-\nu)/(1+\nu)$ for plane stress, with ν being the Poisson ratio; see Subsection 3.1 or 5.1.

The compatibility condition for two-dimensional plane problems is given by

$$\varepsilon_{11,22} - 2\,\varepsilon_{12,12} + \varepsilon_{22,11} = 0; \tag{21.1.6}$$

see Subsection 16.2. From (21.1.2), (21.1.5), and (21.1.6), Airy's stress function must satisfy

$$U_{,1111} + 2\,U_{,1212} + U_{,2222} = 0 \tag{21.1.7a}$$

or

$$\nabla^4 U = 0, \tag{21.1.7b}$$

where $\nabla^2 \equiv \partial^2/\partial x_1^2 + \partial^2/\partial x_2^2$. Hence, U is *bi-harmonic.*

Note that (21.1.7) is restricted to isotropic materials, although the equation of equilibrium, (21.1.1), and the compatibility condition, (21.1.6), hold for any material, when there are no body forces and only small deformations are considered.

21.1.3. Analytic Functions

A complex number, z, is defined in terms of a pair of real numbers, x_1 and x_2, as

$$z \equiv x_1 + \iota\, x_2, \tag{21.1.8}$$

where $\iota \equiv \sqrt{-1}$ is the imaginary number. A point in a two-dimensional plane is given by its position vector measured from the origin, **x**, or by the corresponding Cartesian coordinates, x_1 and x_2. This provides a unique correspondence between complex number z and point **x** in the two-dimensional plane. The two-dimensional plane is then identified with the complex plane.

The real part of a complex number z is defined as $Re\,z \equiv x_1$, and the imaginary part as $Im\,z \equiv x_2$. The complex number conjugate of z is defined by $\bar{z} \equiv x_1 - \iota\, x_2$. Hence, from $Re\,z = Re\,\bar{z}$ and $Im\,z = -Im\,\bar{z}$, it follows that

$$\begin{aligned} z &= x_1 + \iota\, x_2 & & x_1 = \frac{1}{2}(z + \bar{z}) \\ & & \Longleftrightarrow & \\ \bar{z} &= x_1 - \iota\, x_2 & & x_2 = -\frac{\iota}{2}(z - \bar{z}). \end{aligned} \tag{21.1.9}$$

A two-dimensional vector-valued function is written as

$$\mathbf{f} = \mathbf{f}(\mathbf{x}), \tag{21.1.10a}$$

with components,

$$f_1 = f_1(x_1, x_2), \qquad f_2 = f_2(x_1, x_2). \tag{21.1.10b,c}$$

As mentioned, the complex number z corresponds to vector **x**, if the two-dimensional plane is regarded as a complex plane. In a similar manner, a complex number can be associated with the value of the vector-valued function **f** if the domain of **f** is regarded as a region in another complex plane. Hence, defining the complex-valued function as $f \equiv f_1 + \iota\, f_2$, (2.1.9) leads to

$$\mathbf{f} = \mathbf{f}(\mathbf{x}) \quad \Longleftrightarrow \quad f = f(z, \bar{z}). \tag{21.1.11}$$

The complex-valued function f is simply called a *complex function*.[2]

Taking advantage of (21.1.9), consider the following relation between differentiation with respect to z and $\bar{z}$ and that with respect to x_1 and x_2:

$$\begin{aligned} \frac{\partial}{\partial z} &\equiv \frac{\partial x_1}{\partial z}\frac{\partial}{\partial x_1} + \frac{\partial x_2}{\partial z}\frac{\partial}{\partial x_2} \equiv \frac{1}{2}\frac{\partial}{\partial x_1} - \frac{\iota}{2}\frac{\partial}{\partial x_2}, \\ \frac{\partial}{\partial \bar{z}} &\equiv \frac{\partial x_1}{\partial \bar{z}}\frac{\partial}{\partial x_1} + \frac{\partial x_2}{\partial \bar{z}}\frac{\partial}{\partial x_2} \equiv \frac{1}{2}\frac{\partial}{\partial x_1} + \frac{\iota}{2}\frac{\partial}{\partial x_2}. \end{aligned} \tag{21.12a,b}$$

Hence, it follows that

---

[2] Usually, a complex function is regarded as a function of a complex number, z, and is denoted by f(z). This expression means that f is a function of two independent variables, $Re\,z = x_1$ and $Im\,z = x_2$.

$$\frac{\partial^2}{\partial z \partial \bar{z}} \equiv \frac{\partial^2}{\partial \bar{z} \partial z} \equiv \frac{1}{4} \{ \frac{\partial^2}{\partial x_1^2} + \frac{\partial^2}{\partial x_2^2} \} \equiv \frac{1}{4} \nabla^2. \tag{21.1.12c}$$

From (21.1.12c), it is seen that if complex function $f = f(z, \bar{z})$ satisfies $\partial^2 f/\partial z \partial \bar{z} = 0$, then its real and imaginary parts, $Re f = f_1$ and $Im f = f_2$, are harmonic. Since $\partial^2 f/\partial z \partial \bar{z} = 0$ means that f is a function of either z or $\bar{z}$ but not both, assume $f = f(z)$ and obtain

$$\frac{\partial f}{\partial \bar{z}} = \frac{1}{2} \{ \frac{\partial}{\partial x_1} + \iota \frac{\partial}{\partial x_2} \} (f_1 + \iota f_2)$$

$$= \frac{1}{2} \{ \frac{\partial f_1}{\partial x_1} - \frac{\partial f_2}{\partial x_2} \} + \iota \frac{1}{2} \{ \frac{\partial f_2}{\partial x_1} + \frac{\partial f_1}{\partial x_2} \} = 0. \tag{21.1.13a}$$

Hence,

$$\frac{\partial Re f}{\partial x_1} - \frac{\partial Im f}{\partial x_2} = 0, \qquad \frac{\partial Im f}{\partial x_1} + \frac{\partial Re f}{\partial x_2} = 0. \tag{21.1.13b,c}$$

If complex function f satisfies (21.1.13), and the indicated partial derivatives are continuous, then, this function is called *analytic*,[3] and its real and imaginary parts are harmonic; the *Cauchy-Riemann* theorem. Note that a harmonic real-valued function may be regarded as the real (or, equivalently, the imaginary) part of a complex analytic function.

Differentiation of an analytic function, say, f, with respect to z is denoted by $f' = df/dz$. It can be proved that an analytic function can be differentiated as many times as desired, and the resulting derivatives are also analytic. Furthermore, it can also be proved that integration of an analytic function with respect to z also yields another analytic function. An analytic function may not be single-valued. For example, $\ln z$ is analytic but *not* single-valued in an annulus with $z = 0$ located within the inner region. In order to express physical field variables in terms of analytic functions, special attention must be paid to address the question of the single-valuedness[4] of the analytic function.

21.1.4. Bi-Harmonic Functions

Based on the above summary, consider expressing Airy's stress function in terms of complex analytic functions. From (21.1.12),

---

[3] See, e.g., Hille (1959). In the present book, a general complex function is expressed by $f(z, \bar{z})$, and a general analytic function by $f(z)$, to emphasize the difference between a complex function and an analytic function.

[4] A single-valued analytic function is called *holomorphic*; see Subsection 21.3.1.

$$\{\frac{\partial^2}{\partial z \partial \overline{z}}\}^2 \equiv \{\frac{\partial^2}{\partial \overline{z} \partial z}\}^2 \equiv \frac{1}{16}\{\frac{\partial^2}{\partial x_1^2} + \frac{\partial^2}{\partial x_2^2}\}^2 \equiv \frac{1}{16}\nabla^4. \tag{21.1.14}$$

Hence, a bi-harmonic function is given by the real or the imaginary part of a complex function $f = f(z, \overline{z})$ which satisfies $(\partial^2/\partial z \partial \overline{z})^2 f = 0$.

A bi-harmonic function can be expressed in terms of two complex analytic functions, in the same manner as a harmonic function is given by the real or imaginary part of a complex analytic function. From $(\partial^2/\partial z \partial \overline{z})^2 f = 0$, it follows that $f = f(z, \overline{z})$ can be expressed as

$$f(z, \overline{z}) = g(z) + \overline{z}\, h(z) + \hat{g}(\overline{z}) + z\, \hat{h}(\overline{z}), \tag{21.1.15}$$

where $g(z)$ and $h(z)$, and $\hat{g}(\overline{z})$ and $\hat{h}(\overline{z})$ are complex analytic functions of z and $\overline{z}$.

In view of (21.1.15), Airy's stress function, $U = U(x_1, x_2)$, is expressed in terms of two analytic complex functions, $\phi(z)$ and $\chi(z)$, as

$$U(x_1, x_2) = Re\{\overline{z}\,\phi(z) + \chi(z)\} \tag{21.1.16a}$$

which is equivalent to (21.1.15) when the last two terms in the right side of (21.1.15) are taken to be the complex conjugate of the first two corresponding terms. Complex function χ is often replaced by $\int \psi\, dz$, such that ϕ and ψ have the same physical dimensions, since $\overline{z}$ has a dimension of length. Note that ψ is analytic, since differentiation or integration of an analytic function results in an analytic function. Hence, (21.1.16a) is replaced by

$$U(x_1, x_2) = Re\left\{\overline{z}\,\phi(z) + \int^{z} \psi(z')\, dz'\right\}. \tag{21.1.16b}$$

The two analytic functions, ϕ and ψ, are *Muskhelishvili's complex potentials*.

Once Airy's stress function is expressed in terms of Muskhelishvili's complex potentials, the resulting stress, strain, and displacement fields are obtained by simple manipulation. These fields satisfy the three governing field equations, namely, the equations of equilibrium, the constitutive relations for isotropic materials, and the compatibility conditions (or the strain-displacement relations). Hence, a boundary-value problem of two-dimensional elasticity can be solved by choosing a suitable pair of Muskhelishvili's complex potentials, such that the resulting fields satisfy the prescribed boundary conditions.

In Table 21.1.1, the stress and displacement components, σ_{ij} and u_i, are given in terms of Muskhelishvili's complex potentials. This table also gives the resultant force and moment, F_i and M, transmitted across any simple arc which connects points A and B,

$$F_i = \int_A^B \sigma_{ji}\, \nu_j\, dl,$$

$$M = \int_A^B \{x_1(\sigma_{12}\nu_1 + \sigma_{22}\nu_2) - x_2(\sigma_{11}\nu_1 + \sigma_{12}\nu_2)\}\, dl, \tag{21.1.17a,b}$$

where $\boldsymbol{\nu}$ is the unit normal of the arc pointing from the – to the + side of AB, as shown in Figure 21.1.1.

Table 21.1.1

Expressions for stress, strain, resultant force, and the resultant moment in terms of Muskhelishvili's complex potentials

field variable	expression
$\sigma_{11}+\sigma_{22}$	$4Re\phi'$
$\sigma_{22}-\sigma_{11}+2\iota\,\sigma_{12}$	$2\,(\bar{z}\,\phi''+\psi')$
$2\mu\,(u_1+\iota\,u_2)$	$\kappa\phi-z\,\overline{\phi'}-\overline{\psi}$
$F_1+\iota\,F_2$	$-\iota\,(\phi+z\,\overline{\phi'}+\overline{\psi})$
M	$Re\{\int\psi\,dz-\bar{z}\,\psi-z\,\bar{z}\,\overline{\phi'}\}$

Figure 21.1.1

Arc AB with unit normal **ν**; resultant force and moment pointing from – to + side

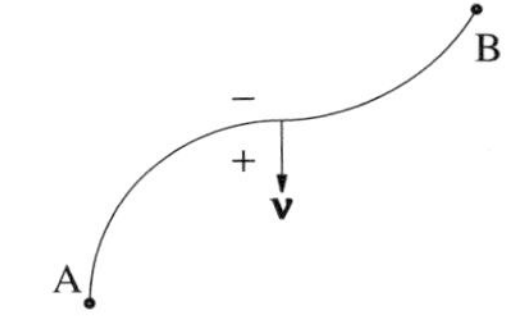

21.2. GREEN'S FUNCTION AND DISLOCATION

The two-dimensional Green function for an infinitely extended, homogeneous, isotropic, linearly elastic solid will be derived in terms of Muskhelishvili's complex potentials. Associated with this, and preliminary to formulating the crack problem, the complex representation of a dislocation is also obtained.

21.2.1. Green's Function

To solve a boundary-value problem in two-dimensional elasticity, a suitable Airy stress function, or equivalently, a suitable pair of Muskhelishvili's complex potentials must be constructed such that the prescribed boundary conditions are satisfied, since all other governing field equations are already satisfied by these potentials.

Suppose that only a concentrated force, $\mathbf{P} = (P_1, P_2)$, acts at the origin, O; see Figure 21.2.1. The force condition at the origin and the boundary conditions at infinity then are: 1) in the neighborhood of the origin, the stress field must be in equilibrium with P_1 and P_2; and 2) far from the origin, the stress field must vanish.

The resultant force transmitted over a closed loop which surrounds the origin is used as the prescribed boundary condition, i.e., for any small loop surrounding the origin, the resultant force is required to equal **P**. Hence,

$$[F_1 + \iota F_2] = P_1 + \iota P_2, \tag{21.2.1}$$

where [ ] denotes the corresponding resultant quantity around an arbitrary loop surrounding the origin. The farfield boundary conditions are

$$\lim_{|z| \to \infty} \sigma_{ij}(z) = 0 \quad \text{for } i, j = 1, 2, \tag{21.2.2}$$

where $|z| = \sqrt{x_1^2 + x_2^2}$.

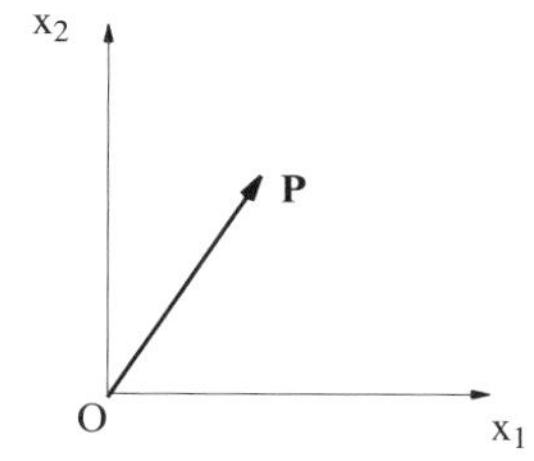

Figure 21.2.1

Point force **P** at origin of infinitely extended solid

In view of (21.2.1), set $\phi = A \ln z$ and $\psi = B \ln z$, with unknown complex constants, $A = A_1 + \iota A_2$ and $B = B_1 + \iota B_2$. These complex potentials satisfy (21.2.2). Since (21.2.1) is equivalent to two independent real equations, two other real equations are needed in order to determine the four real unknowns, A_1, A_2, B_1, and B_2. Consider the continuity of the displacement around a loop which surrounds the origin, i.e.,

$$2\mu\,[u_1 + \iota u_2] = [\kappa\,\phi(z) - z\,\overline{\phi'(z)} - \overline{\psi(z)}] = 0. \tag{21.2.3}$$

This provides the additional two equations.

From $[\ln z] = 2\pi\iota$ and $[\overline{\ln z}] = -\,2\pi\iota$ for any loop surrounding the origin, it follows that

$$[F_1 + \iota F_2] = 2\pi\,(A - \bar{B})$$

$$= 2\pi\,(A_1 - B_1) + 2\pi\iota\,(A_2 + B_2), \tag{21.2.4a}$$

and

$$2\mu\,[u_1 + \iota u_2] = 2\pi\,\iota\,(\kappa A + \bar{B})$$

$$= 2\pi\,(-\,\kappa A_2 + B_2) + 2\pi\iota\,(\kappa A_1 + B_1). \tag{21.2.4b}$$

Substituting (21.2.4a,b) into (21.2.1) and (21.1.3), obtain A_1, A_2, B_1, and B_2, as

$$A_1 = \frac{P_1}{2\pi(1+\kappa)}, \qquad A_2 = \frac{P_2}{2\pi(1+\kappa)},$$
$$B_1 = -\frac{\kappa P_1}{2\pi(1+\kappa)}, \qquad B_2 = \frac{\kappa P_2}{2\pi(1+\kappa)}. \tag{21.2.5a~d}$$

Hence, complex potentials, ϕ and ψ, for force **P** concentrated at the origin are given by

$$\phi(z) = \frac{P}{2\pi(1+\kappa)} \ln z, \qquad \psi(z) = -\frac{\kappa \overline{P}}{2\pi(1+\kappa)} \ln z, \tag{21.2.5e,f}$$

where $P = P_1 + \iota P_2$.

The Green function for a concentrated force acting at a typical point, $\mathbf{x}^o = (x_1^o, x_2^o)$ is obtained through a translation. To this end, let $U(x_1, x_2)$ be Airy's stress function defined by ϕ and ψ of (21.2.5), and denote by $\hat{U}(\hat{x}_1, \hat{x}_2)$ Airy's stress function with the origin of the coordinate system at $-\mathbf{x}^o$, where $\hat{\mathbf{x}} = \mathbf{x} + \mathbf{x}^o$, and $\hat{\mathbf{x}} = (\hat{x}_1, \hat{x}_2)$ or $\hat{z} = \hat{x}_1 + \iota \hat{x}_2$. Then, $\hat{U}$ is given by

$$\hat{U}(\hat{x}_1, \hat{x}_2) = U(x_1, x_2) = U(\hat{x}_1 - x_1^o, \hat{x}_2 - x_2^o). \tag{21.2.6a}$$

Denote the complex potentials for $U(x_1, x_2)$ by $\phi(z)$ and $\chi(z)$, and those for $\hat{U}(\hat{x}_1, \hat{x}_2)$ by $\hat{\phi}(\hat{z})$ and $\hat{\chi}(\hat{z})$, where $\chi'(z) = \psi(z)$ and $\hat{\chi}'(\hat{z}) = \hat{\psi}(\hat{z})$. Rewrite (21.2.6a) as

$$Re\{\overline{\hat{z}}\,\hat{\phi}(\hat{z}) + \hat{\chi}(\hat{z})\} = Re\{\overline{z}\,\phi(z) + \chi(z)\}$$
$$= Re\{\overline{\hat{z}}\,\phi(\hat{z} - z^o) + \chi(\hat{z} - z^o) - \overline{z^o}\,\phi(\hat{z} - z^o)\}. \tag{21.2.6b}$$

Hence, $\hat{\phi}(\hat{z})$ and $\hat{\chi}(\hat{z})$ are

$$\hat{\phi}(\hat{z}) = \phi(\hat{z} - z^o), \qquad \hat{\chi}(\hat{z}) = \chi(\hat{z} - z^o) - \overline{z^o}\,\phi(\hat{z} - z^o). \tag{21.2.7a,b}$$

From (21.2.7b) and $d/d\hat{z} \equiv d/dz$, $\hat{\psi}(\hat{z}) = \hat{\chi}'(\hat{z})$ is given by

$$\hat{\psi}(\hat{z}) = \psi(\hat{z} - z^o) - \overline{z^o}\,\phi'(\hat{z} - z^o). \tag{21.2.7c}$$

This is *the coordinate transformation rule* for Muskhelishvili's complex potentials, which is different from that for Airy's stress function, (21.2.6a).

From (21.2.7), the complex potentials for a concentrated force **P** applied at $\mathbf{x}^o$ are obtained, as follows:

$$\phi(z; P, z^o) = \frac{P}{2\pi(1+\kappa)} \ln(z - z^o),$$
$$\psi(z; P, z^o) = -\frac{\kappa \overline{P}}{2\pi(1+\kappa)} \ln(z - z^o) + \frac{P}{2\pi(1+\kappa)} \frac{\overline{z^o}}{z - z^o}, \tag{21.2.8a,b}$$

where $P = P_1 + \iota P_2$ and $z^o = x_1^o + \iota x_2^o$.

21.2.2. Dislocation

The Green function is obtained under the assumption that the resultant force transmitted across a small loop is prescribed, while there are no displacement gaps across this loop, i.e., $[F] = P$ and $[u] = 0$, where $F = F_1 + \iota F_2$ is the resultant force and $[u] = [u_1] + \iota [u_2]$ is the displacement *gap*. It is of interest to consider the physical phenomenon corresponding to a case where the resultant force across a small loop is zero while a nonzero displacement gap is prescribed, i.e., when $[F] = 0$ and $[u] = b$ with $b = b_1 + \iota b_2$. This defines a *dislocation*.

The displacement jump caused by a dislocation is interpreted as follows: since the displacement field must be continuous, a dislocation produces a dislocated line in the plane, across which the displacement field is *discontinuous* while the tractions are *continuous*; see Figure 21.2.2.

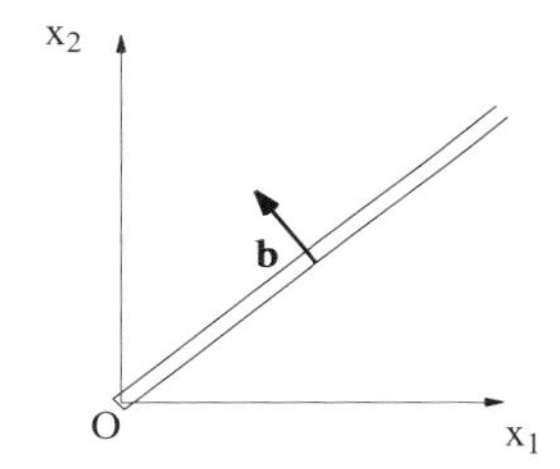

Figure 21.2.2

Dislocation with Burgers vector **b** at origin of infinitely extended solid

A dislocation is characterized by the Burgers vector which determines the direction and the magnitude of the slip induced by the dislocation. In two-dimensional elasticity, the Burgers vector is expressed by $\mathbf{b} = (b_1, b_2)$. In view of (21.2.4a,b), the complex potentials ϕ and ψ for a dislocation with Burgers' vector **b** at the origin become

$$\phi(z) = -\frac{\iota \mu b}{\pi(\kappa+1)} \ln z, \qquad \psi(z) = \frac{\iota \mu \bar{b}}{\pi(\kappa+1)} \ln z. \tag{21.2.9a,b}$$

Furthermore, from (21.2.7), the complex potentials for a dislocation with Burgers' vector **b** at z^o are

$$\phi(z;\, b,\, z^o) = -\frac{\iota \mu b}{\pi(\kappa+1)} \ln (z - z^o),$$

$$\psi(z;\, b,\, z^o) = \frac{\iota \mu \bar{b}}{\pi(\kappa+1)} \ln (z - z^o) + \frac{\iota \mu b}{\pi(\kappa+1)} \frac{\bar{z}^o}{z - z^o}. \tag{21.2.10a,b}$$

Note that in (21.2.10) or equivalently, in (21.2.9), the line of the displacement discontinuity is not specified, although its magnitude and orientation are given by Burgers vector **b**. Since $\ln (z - z^o)$ is a multi-valued function, a suitable *cut* is required to render it single-valued. This cut starts form z^o and extends to infinity without intersecting itself. It corresponds to the line of discontinuity. Hence, (21.2.10) corresponds to various dislocations with different lines of discontinuity.

21.2.3. Center of Dilatation and Disclination

Muskhelishvili's complex potentials with elementary functions (or, equivalently, the corresponding Airy's stress function) may be used to examine fields associated with interesting physical phenomena. As an example of such elementary functions, consider $\ln z$ which produces fields with a singularity at the origin. As shown in Subsections 21.2.2 and 21.2.3, a point force and a dislocation can be expressed in terms of $\ln z$. The resulting stress, strain, and displacement fields are not bounded at the origin in this case.

A center of dilatation and a disclination are other phenomena that produce fields with singularities. Here, a center of dilatation is a point with finite expansion, or the limit of a circular region undergoing finite expansion, as its radius vanishes, with the product of the expansion and the area of the region remaining constant; see Figure 21.2.3. A disclination is a semi-infinite cut, with the displacement jump across the cut increasing linearly with the distance from the origin of the cut. A center of dilatation corresponds to the case where a region with finite expansion is viewed from far away, and a disclination corresponds to the case where a sharp wedge is driven into a plate; see Figure 21.2.4.

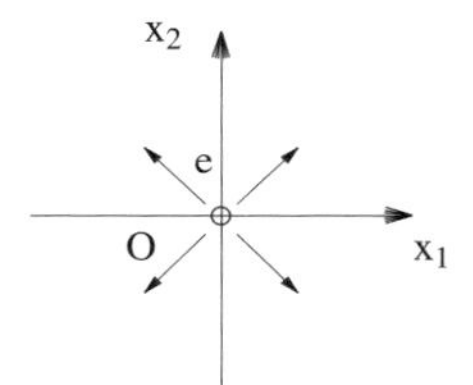

Figure 21.2.3

Center of dilatation with expansion e at origin of infinitely extended solid

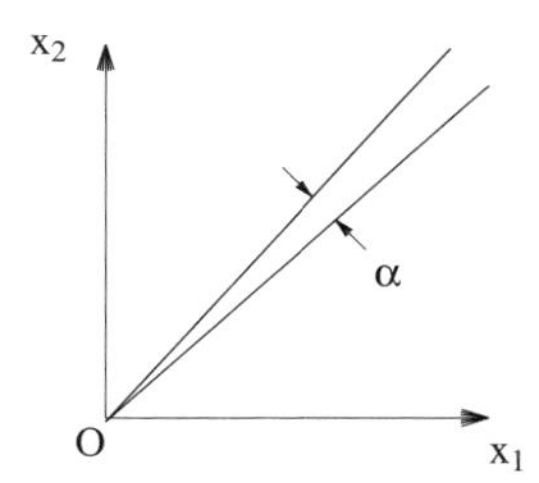

Figure 21.2.4

Disclination with magnitude α at origin of infinitely extended solid

In the polar coordinate system, (r, θ), Airy's stress function for a center of dilatation located at the origin is given by

$$U(r, \theta) = e \ln r, \tag{21.2.11}$$

where e is the amount of volume expansion; see Figure 21.2.3. Similarly, Airy's stress function for a disclination starting from the origin is given by

$$U(r, \theta) = \gamma r^2 \ln r, \tag{21.2.12}$$

where $\gamma = \alpha\mu/\{\pi(\kappa+1)\}$ and α is the disclination angle; see Figure 21.2.4. These Airy stress functions do not depend on θ.

In terms of Muskhelishvili's complex potentials, (21.2.11) is expressed as

$$U(r,\,\theta) = Re\{e\,\ln z\} \quad \Longleftrightarrow \quad \phi = 0, \quad \psi = \frac{e}{z}, \tag{21.2.13}$$

and (21.2.12) is expressed as

$$U(r,\,\theta) = Re\{\bar{z}(\gamma z \ln z)\} \quad \Longleftrightarrow \quad \phi = \gamma z \ln z, \quad \psi = 0. \tag{21.2.14}$$

The complex potentials for the dislocation can be obtained by taking the limit of a doublet of disclinations as they approach each other, leaving an infinitesimally small distance. This can be shown directly from the coordinate transformation rule (21.2.7). Suppose that a pair of disclinations with the disclination angles γ and $-\gamma$ are located at the origin and at z^o, respectively. Superposition of these two disclinations yields

$$\phi = \gamma z \ln z - \gamma(z - z^o)\ln(z - z^o),$$

$$\psi = -\overline{z^o}\gamma\{\ln(z - z^o) - 1\}. \tag{21.2.15a,b}$$

In the limit as z^o approaches 0 with $\gamma z^o = C$ fixed, these potentials become

$$\lim_{z^o\to 0}\phi = \lim_{z^o\to 0}\frac{C}{z^o}\{z\ln z - (z - z^o)\ln(z - z^o)\} = C(\ln z - 1),$$

$$\lim_{z^o\to 0}\psi = \lim_{z^o\to 0}\frac{\bar{C}}{\overline{z^o}}\{\overline{z^o}(\ln(z - z^o) - 1)\} = \bar{C}(\ln z - 1). \tag{21.2.15c,d}$$

Since a constant term in the complex potentials does not produce any fields, (21.2.15c,d) coincide with (21.2.9) if C is replaced by $-\iota\mu b/\pi(\kappa + 1)$. Hence, it follows that a dislocation is obtained by taking a suitable limit of a doublet of disclinations.

In a similar manner, a center of dilatation can be obtained by taking the limit of a doublet of dislocations. If two dislocations with Burgers' vectors b and $-b$ are located at the origin and at z^o, respectively, the corresponding Muskhelishvili potentials are given by

$$\phi = C\ln z - C\ln(z - z^o),$$

$$\psi = \bar{C}\ln z + \{-\bar{C}\ln(z - z^o) + C\frac{\overline{z^o}}{z - z^o}\}, \tag{21.2.16a,b}$$

where $C = -\iota\mu b/\pi(\kappa + 1)$. As z^o approaches 0 with $C\overline{z^o} = e$ (real) fixed, these potentials become

$$\lim_{z^o\to 0}\phi = \lim_{z^o\to 0}\frac{z^o}{\overline{z^o}}\frac{e}{z},$$

$$\lim_{z^o\to 0}\psi = \lim_{z^o\to 0}\frac{z^o}{\overline{z^o}}\frac{e}{z} + \frac{e}{z}. \tag{21.2.16c,d}$$

If the orientation of z^o is fixed as Θ, then, the limit of $z^o/\overline{z^o}$ is defined by $e^{2\iota\Theta}$. Hence, superposition of potentials for another doublet of dislocations with the same magnitude C but opposite orientation, $\Theta \pm \pi/2$, yields

$$\phi = 0, \qquad \psi = 4\,\frac{e}{z}. \tag{21.2.16e,f}$$

As is seen, these potentials are the same as those for a center of dilatation, (21.2.13), except for a nonessential factor of 4.

21.3. THE HILBERT PROBLEM

In two-dimensional elasticity, a crack is viewed as an arc in a plane. Since boundary conditions for the crack are prescribed on both upper and lower surfaces of the crack, this problem can be stated as a Hilbert problem. Here, a Hilbert problem is a boundary-value problem with boundary data prescribed on both sides of an arc. In this subsection, holomorphic functions which are used in the Hilbert problem are briefly reviewed as a mathematical preliminary, and then the Hilbert problem is stated. Finally, a traction-free crack in an infinite domain subjected to uniform tension is examined as an example of the Hilbert problem.

21.3.1. Holomorphic Functions

A complex-valued function, $f(x_1, x_2)$, in a region, D, is *analytic*, if it satisfies the *Cauchy-Riemann equations*, $\partial \mathit{Re} f / \partial x_1 = \partial \mathit{Im} f / \partial x_2$ and $\partial \mathit{Re} f / \partial x_2 = -\partial \mathit{Im} f / \partial x_1$, and the partial derivatives are continuous. Complex function f is called *holomorphic* in D if it is analytic and single-valued everywhere in D. The single-valuedness is important, as it is necessary that the physical fields defined by these functions be single-valued.

The basic properties of a holomorphic function in D are as follows:

1) derivatives of a holomorphic function are holomorphic in D;
2) the integral of a holomorphic function, however, is not[5] necessarily holomorphic in D;
3) a holomorphic function admits a unique convergent power series, called *Laurent's series*, over the corresponding annulus;
4) both real and imaginary parts of a holomorphic function are harmonic and single-valued; and
5) if two functions holomorphic in D_1 and D_2 take the same value on the intersection of D_1 and D_2, then a new holomorphic function can be defined in $D_1 + D_2$, which coincides with each function in its own domain (*continuation theorem*).

---

[5] The integral of a holomorphic function is analytic, but not necessarily single-valued; for example, $1/z$ is holomorphic in the region excluding the origin, but its integral, $\ln z$, is not single-valued, though it is analytic there.

Holomorphic functions are used[6] to determine Muskhelishvili's complex potentials, since these potentials must produce analytic and single-valued fields, in order to correspond to the physical quantities.

21.3.2. The Cauchy Integral

Let C be a finite (open) simply-connected part of an *unbounded* domain, D, and denote its boundary by ∂C; see Figure 21.3.1. If $\mathrm{f} = \mathrm{f}(\mathrm{z})$ is holomorphic *inside* of C and continuous on ∂C, then the values of f on ∂C completely define this function within C by the following integral:

$$\frac{1}{2\pi\iota}\int_{\partial C}\frac{\mathrm{f(t)}}{\mathrm{t-z}}\,\mathrm{dt} = \begin{cases} \mathrm{f(z)} & \text{if } \mathrm{z} \in C \\ 0 & \text{if } \mathrm{z} \in D - C, \end{cases} \tag{21.3.1a}$$

where $D - C$ is the remaining part[7] of D, excluding C. If, on the other hand, f is holomorphic *outside* of C and continuous on ∂C, it then follows that

$$\frac{1}{2\pi\iota}\int_{\partial C}\frac{\mathrm{f(t)}}{\mathrm{t-z}}\,\mathrm{dt} = \begin{cases} \mathrm{f}(\infty) & \text{if } \mathrm{z} \in C \\ \mathrm{f}(\infty) - \mathrm{f(z)} & \text{if } \mathrm{z} \in D - C, \end{cases} \tag{21.3.1b}$$

where $\mathrm{f}(\infty)$ is the farfield value of f. Equations (21.3.1a) and (21.3.1b) are the *Cauchy integral theorems*.

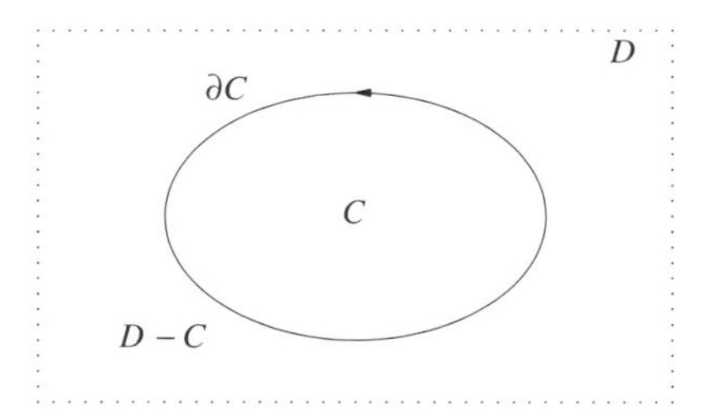

Figure 21.3.1

Simply connected region C for Cauchy integral theorem

Next, consider a sufficiently smooth complex-valued function g defined on a nonintersecting arc, L, in D; see Figure 21.3.2. The Cauchy integral of g over L,

$$\mathrm{G(z)} \equiv \frac{1}{2\pi\iota}\int_{L}\frac{\mathrm{g(t)}}{\mathrm{t-z}}\,\mathrm{dt}, \tag{21.3.2a}$$

---

[6] Precisely speaking, Muskhelishvili's complex potentials need not be given in terms of holomorphic functions; see Subsection 21.2 where a logarithmic function is used to determine Muskhelishvili's complex potentials.

[7] The boundary of C, ∂C, is not included in $D - C$ nor in C.

is holomorphic in D, except possibly on arc L. Function G(z) is continuous on both sides of arc L, although it may be discontinuous across L. Let a and b denote the end points of arc L; see Figure 21.3.2. Then,

$$\lim_{z \to a} (z-a)\, G(z) = 0, \qquad \lim_{z \to b} (z-b)\, G(z) = 0. \tag{21.3.2b,c}$$

Function G = G(z) is the *Cauchy integral of* g(s) *on arc* L. Note that g(s) is defined for s on L only.

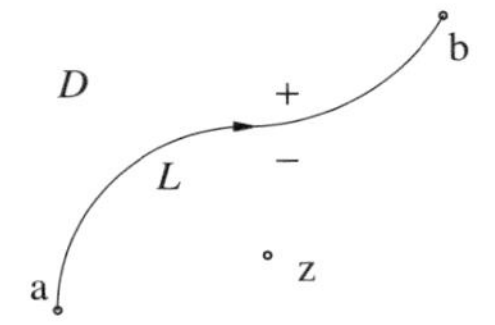

Figure 21.3.2

Cauchy integral on arc L

On arc L in D, a plus (+) and a minus (−) sign can be defined, as shown in Figure 21.3.2. The limiting values of G(z) as z approaches a point s on L (excluding the end points of L) from the + and the − sides, are denoted by $G^+(s)$ and $G^-(s)$, respectively, i.e.,

$$G^+(s) \equiv \lim_{z^+ \to s} G(z^+), \qquad G^-(s) \equiv \lim_{z^- \to s} G(z^-), \tag{21.3.3a,b}$$

where z^+ and z^- are in the + and − sides of L. For G^+ and G^-, the following equalities hold:

$$G^+(s) - G^-(s) = g(s)$$

$$G^+(s) + G^-(s) = \frac{1}{\pi\iota} \int_L \frac{g(t)}{t-s}\, dt. \tag{21.3.4a,b}$$

Equations (21.3.4a,b) are called the *Plemelj formulae*.

21.3.3. The Hilbert Problem

For a given function, g(s), on an arc, L, consider the following boundary-value problem with a given constant, α, for a complex-valued potential, G(z):

$$G^+(s) - \alpha G^-(s) = g(s) \qquad s \in L, \tag{21.3.5}$$

where G^+ and G^- are defined by (21.3.3). Consider a solution which is holomorphic in $D - L$ and behaves as $O(z^n)$ at infinity, with n being a positive integer, i.e., as $|z| \to \infty$, the function becomes unbounded like z^n.

The general solution of (21.3.5) is then given by

$$G(z) = \frac{X(z)}{2\pi\iota} \int_L \frac{g(t)}{X^+(t)\,(t-z)}\, dt + X(z)\, P(z). \tag{21.3.6}$$

Here, P(z) is an arbitrary polynomial of degree n + 1, and X(z) is defined by

$$X(z) \equiv (z-a)^{-\xi}(z-b)^{\xi-1}, \tag{21.3.7a}$$

where a and b are the end points of L, and ξ is given by

$$\xi \equiv \frac{1}{2\pi\iota} \log \alpha. \tag{21.3.7b}$$

A detailed proof of (21.3.6) can be found in Muskhelishvili (1956). An outline is as follows: from definitions (21.3.7a,b), it follows that

$$X^+(s) = \alpha X^-(s) \qquad \text{for } s \in L. \tag{21.3.8a}$$

Hence, (21.3.5) can be written as

$$\left[\frac{G(s)}{X(s)}\right]^+ - \left[\frac{G(s)}{X(s)}\right]^- = \frac{g(s)}{X^+(s)} \qquad \text{for } s \in L. \tag{21.3.8b}$$

Thus, $(G(s)/X(s))$ and $g(s)/X^+(s)$ in (21.3.8b) are viewed as G(s) and g(s) in Plemelj formula (21.3.4a). When the domain D is unbounded and simple poles of order less than $n+1$ are admitted at infinity, then (21.3.2a) yields (21.3.6).

Kernel X(z) satisfies

$$\lim_{|z| \to \infty} z\, X(z) = 1. \tag{21.3.9a}$$

Since $X^+(s) = \alpha X^-(s)$ for s on L, the line integral in the right side of (21.3.6) can be replaced by the following contour integral:

$$\int_L \frac{g(s)}{X^+(s)(s-z)}\, ds = \frac{1}{1-\alpha} \int_{L'} \frac{g(t)}{X(t)(t-z)}\, dt, \tag{21.3.9b}$$

where L' is a contour which surrounds arc L but does not contain point z; see Figure 21.3.3. If g on L is smoothly connected to an analytic function in L', the contour integral in the right side of (21.3.9b) can be evaluated by using (21.3.1a,b). Indeed, if $g = 1$ on L and $b = -a$, then,

$$\frac{1}{1-\alpha} \int_{L'} \frac{g(t)}{X(t)(t-z)}\, dt = \frac{2\pi\iota}{1-\alpha} \left\{\frac{1}{X(z)} - (z - (2\xi - 1)\, a)\right\}, \tag{21.3.10}$$

where ξ is given by (21.3.7b). In this manner, the general Hilbert problem defined by (21.3.5) can be solved for a broad class of functions g(s) defined on arbitrary smooth arcs. The application to two-dimensional crack problems is examined below; see Erdogan (1977).

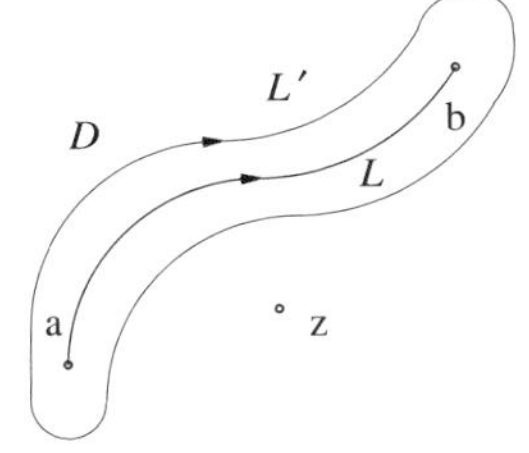

Figure 21.3.3

Contour L' around arc L

21.3.4. Examples

As an illustration of the application of the Hilbert problem, consider a slit crack in an unbounded isotropic linearly elastic solid, subjected to uniform tension σ^o at infinity; see Figure 21.3.4. From Table 21.1.1, analytic functions $\phi(z)$ and $\psi(z)$ must be obtained such that the following boundary conditions are satisfied:

$$\sigma_{11},\, \sigma_{12} \rightarrow 0 \quad \text{and} \quad \sigma_{22} \rightarrow \sigma^o \quad \text{as } |z| \rightarrow \infty,$$

$$\sigma_{12} = \sigma_{22} = 0 \quad \text{on both sides of } L, \tag{21.3.11a,b}$$

where L is a straight line from $(-a, 0)$ to $(+a, 0)$.

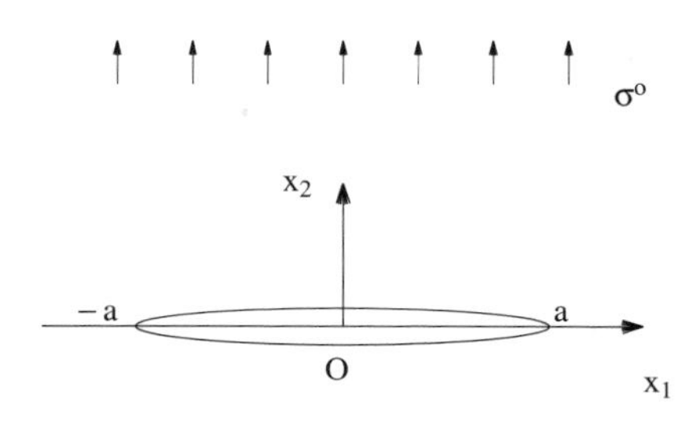

Figure 21.3.4

A single crack in an infinitely extended solid subjected to uniform tension at infinity

From real-valued equations (21.3.11b), complex-valued boundary conditions are constructed, as

$$(\sigma_{12} - \iota\,\sigma_{22})^+ = 0 \quad \text{and} \quad (\sigma_{12} - \iota\,\sigma_{22})^- = 0 \quad \text{on } L. \tag{21.3.11c,d}$$

The first two equations in Table 21.1.1 give the left side of (21.3.11c,d), as

$$\frac{1}{2}\,(\sigma_{22} - \iota\,\sigma_{12}) = \phi'(z) + \overline{\phi'(z)} + z\,\overline{\phi''(z)} + \overline{\psi'(z)}. \tag{21.3.12a}$$

In view of (21.3.11c,d) and (21.3.12a), the following complex potential is defined:

$$\Omega(z) \equiv \overline{\phi'(\overline{z})} + z\,\overline{\phi''(\overline{z})} + \overline{\psi'(\overline{z})}. \tag{21.3.13}$$

Here, say, $\overline{\psi'(\overline{z})}$ is analytic, though $\overline{\psi'(z)}$ is not. Hence, Ω defined by (21.3.13) is analytic. Substitution of Ω for ψ in (21.3.12a) yields

$$\frac{1}{2}\,(\sigma_{22} - \iota\,\sigma_{12}) = \phi'(z) + (z - \overline{z})\,\overline{\phi''(z)} + \Omega(\overline{z}). \tag{21.3.12b}$$

Since $z - \overline{z}$ vanishes on L, $\sigma_{22} - \iota\,\sigma_{12}$ can be expressed in terms of the two holomorphic complex functions, ϕ' and Ω.

For simplicity define $\phi'(z) \equiv \Phi(z)$, and substitute (21.3.12b) into boundary conditions (21.3.11c,d) to arrive at

$$\Phi^+(s) + \Omega^-(s) = 0 \qquad z \to s \text{ with } x_2 > 0$$
$$\Phi^-(s) + \Omega^+(s) = 0 \qquad z \to s \text{ with } x_2 < 0, \tag{21.3.14a,b}$$

where s is a point on the crack. It thus follows that

$$\{\Phi(s) + \Omega(s)\}^+ + \{\Phi(s) + \Omega(s)\}^- = 0,$$
$$\{\Phi(s) - \Omega(s)\}^+ - \{\Phi(s) - \Omega(s)\}^- = 0. \tag{21.3.14c,d}$$

The crack problem is now reduced to two Hilbert problems for two holomorphic functions, $\Phi(z) + \Omega(z)$ and $\Phi(z) - \Omega(z)$, with an additional restriction that $\Phi(z)$ and $\Omega(z)$ must be bounded as $|z| \to \infty$, since the stress field is bounded at infinity.

Equation (21.3.14d) requires that holomorphic function $\Phi(z) - \Omega(z)$ be continuous across the crack. Since this function is bounded at infinity, it must be a complex constant,

$$\Phi(z) - \Omega(z) = c_1. \tag{21.3.15a}$$

Equation (21.3.14c) defines a Hilbert problem for $\Phi + \Omega$, (21.3.5), with $g(s) \equiv 0$ and $\alpha = -1$, and (21.3.6) leads to

$$\Phi(z) + \Omega(z) = \frac{c_2 + c_3 z}{\sqrt{z^2 - a^2}}, \tag{21.3.15b}$$

where polynomial P(z) in (21.3.6) is set to be $c_2 + c_3 z$, to render Φ and Ω bounded at infinity. From (21.3.15a,b), it now follows that

$$\phi'(z) \equiv \Phi(z) = \frac{1}{2}\left\{ c_1 + \frac{c_2 + c_3 z}{\sqrt{z^2 - a^2}} \right\},$$

$$\Omega(z) = \frac{1}{2}\left\{ -c_1 + \frac{c_2 + c_3 z}{\sqrt{z^2 - a^2}} \right\}. \tag{21.3.15c,d}$$

From the traction-free conditions on L, and the symmetry of the stress field about $x_1 = 0$, conclude that $Re\, c_1 = -\sigma^o/2$, $c_2 = 0$, and $c_3 = \sigma^o$. Since only $Re\, c_1$ enters the stress field, this completes the solution. Note that $c_2 = 0$ also follows from the requirement of the single-valuedness of the displacement field.

21.4. TWO-DIMENSIONAL CRACK PROBLEMS

In this subsection, an alternative method to solve a crack problem in two-dimensional elasticity is presented. The results obtained in Subsection 21.2.2 are used to formulate integral equations which define the distributed dislocations necessary to satisfy the traction-free (or other) boundary conditions on a crack.

21.4.1. Crack and Dislocations

In two-dimensional elasticity, a crack may be regarded as an (straight or curved) arc across which a displacement field is suitably discontinuous, and the continuity of tractions is maintained in the absence or presence of internal forces or pressures in the crack. If concentrated or distributed forces act on the surfaces of the crack, the corresponding traction boundary conditions must be satisfied on each of the crack surfaces.

It has been emphasized that boundary-value problems in elasticity can be solved by applying the Green function for a concentrated force on the boundary: (1) for traction boundary-value problems, the field variables are obtained by integration of the Green function weighted with the prescribed tractions, and (2) for displacement or mixed boundary-value problems, the specification of unknown boundary tractions or displacements in terms of the known data leads to integral equations which ensure that the field variables satisfy the required boundary conditions.

The application of the Green function to the crack problem, however, is not quite straightforward, since the displacement field has discontinuities. The dislocation formulation, however, provides advantages when the tractions are continuous. An effective approach is to obtain the required dislocations which produce suitable discontinuous displacement fields. A crack is then expressed in terms of these dislocations distributed on the arc where the crack lies.

21.4.2. Integral Equation for Dislocation Density

Consider a straight crack of length 2a, lying on the x_1-axis from $(-a, 0)$ to $(+a, 0)$. Suppose that a single dislocation with Burgers vector $\mathbf{b} = (b_1, b_2)$ exists at $x_1 = x^o$ $(|x^o| < a)$. Referring to Subsection 21.2.2, let the cut for $\ln(z - x^o)$ be the semi-infinite line starting from $(x^o, 0)$ to $(\infty, 0)$. The displacement discontinuity across the x_1-axis then is

$$[u](x) = [u_1](x) + \iota\,[u_2](x) = \begin{cases} -b & \text{if } x > x^o \\ 0 & \text{if } x < x^o, \end{cases} \tag{21.4.1}$$

where $b = b_1 + \iota\, b_2$. The tractions on the x_1-axis are

$$t(x) = \sigma_{12}(x) + \iota\,\sigma_{22}(x) = \frac{2\mu b}{\pi(\kappa+1)}\,\frac{1}{x - x^o}. \tag{21.4.2}$$

Note that the same tractions act on the upper and lower edges of the cut, since the stress field due to the dislocation is continuous.

According to the principle of superposition, the desired displacement gap across L or the desired tractions acting on L, can be obtained by distributing suitable dislocations there. Thus, with b as a function defined on L, i.e., $b(x) = b_1(x) + \iota\, b_2(x)$ for x on L, obtain the following expression for the displacement gap and traction:

$$[u](x) = -\int_{-a}^{x} b(\zeta)\,d\zeta \qquad \text{for x on } L, \tag{21.4.3}$$

and

$$t(x) = \int_{-a}^{a} \frac{2\mu b(\zeta)}{\pi(\kappa+1)} \frac{1}{x-\zeta} \, d\zeta \qquad \text{for x on } x_1\text{–axis.} \tag{21.4.4}$$

Function $b = b(x)$ is called the *dislocation density*, and the displacement gap given by (21.4.3) is the *crack-opening-displacement* (COD). It should be recalled that for a dislocation at ζ, the cut for $\ln(z-\zeta)$ is a semi-infinite line on the x_1-axis, starting from $(\zeta, 0)$ to $(\infty, 0)$.

If the left side of (21.4.3) or (21.4.4) is prescribed as the boundary conditions, each equation can be regarded as an integral equation for $b(\zeta)$. For example, if the displacement gap across the crack surface is given by $[u^o]$, (21.4.3) yields

$$\int_{-a}^{x} b(\zeta) \, d\zeta = -[u^o](x), \tag{21.4.5}$$

and if the tractions on the crack surfaces are t^o, (21.4.3) yields

$$\int_{-a}^{a} \frac{2\mu b(\zeta)}{\pi(\kappa+1)} \frac{1}{x-\zeta} \, d\zeta = -t^o(x). \tag{21.4.6}$$

Integral equation (21.4.6) is called the *Cauchy integral equation of the first kind.* Although the form of (21.4.6) is similar to the Cauchy integral on arc L, (21.3.2), $b(\zeta)$ is the distribution of the dislocation density defined only on L, and it may not be possible to extend it as a holomorphic function in the neighborhood of L.

Solutions of crack problems with prescribed traction conditions can be obtained from solving the integral equation (21.4.6).[8] To this end, the following additional condition is required:[9] the COD must be zero at the ends of the crack, such that the displacement discontinuity is limited to $|x_1| < a$. From (21.4.3), this condition is written as

$$\int_{-a}^{a} b(\zeta) \, d\zeta = 0. \tag{21.4.7}$$

This is the *consistency condition* for the dislocation density function $b(\zeta)$.

21.4.3. Example

Consider the example studied in Subsection 21.3.4. By superposing a uniform compression, $-\sigma^o$, the traction-free crack in a plane under uniform tension represents a crack subjected to a uniform compression on its upper and lower

---

[8] Equation (21.4.6) is also applicable to the case when prescribed tractions are not the same on the upper and lower surfaces of the crack. In this case, suppose that suitable forces are distributed on say, the upper crack surface only. Then, the traction boundary conditions on both crack surfaces can be satisfied by the dislocations (which produce the same tractions on both surfaces) and the distributed forces (which produce the different tractions). Note that the tractions due to the distributed forces can be expressed in terms of Green's function.

[9] In general, the solution of the Cauchy integral equation is not unique.

surfaces. Since tangential tractions acting on the crack surfaces are zero, the tangential Burgers vector vanishes. Integral equation (21.4.6) is reduced to the following real-valued expression:

$$\int_{-a}^{a} \frac{2\mu b_2(\zeta)}{\pi(\kappa+1)} \frac{1}{x-\zeta}\, d\zeta = \sigma^o; \tag{21.4.8a}$$

and consistency condition (21.4.7) becomes

$$\int_{-a}^{a} b_2(\zeta)\, d\zeta = 0. \tag{21.4.8b}$$

The solution of (21.4.8a) with (21.4.8b) is

$$b_2(x) = \frac{-\sigma^o(\kappa+1)}{2\mu} \frac{x}{\sqrt{a^2-x^2}}. \tag{21.4.9}$$

As is seen, the dislocation density function is singular at the ends of the crack, so that the corresponding stress field similarly becomes singular there. For example, σ_{22} on $x_2 = 0$ is

$$\sigma_{22}(x) = \begin{cases} -\sigma^o & |x| < a \\ -\sigma^o(|x|/\sqrt{x^2-a^2}-1) & |x| > a, \end{cases} \tag{21.4.10a}$$

and from (21.4.3) and (21.4.9), it follows that

$$[u](x) = \frac{-\sigma^o(\kappa+1)}{2\mu}\sqrt{a^2-x^2} \qquad |x| < a. \tag{21.4.10b}$$

21.4.4. Alternative Integral Equation for Crack Problem

It is shown in Section 6 that the COD plays a direct role in estimating the overall strains due to the presence of microcracks in an elastic RVE. It is, therefore, effective to formulate the crack problem in linear elasticity, directly in terms of COD, $[\mathbf{u}](\zeta)$. This approach is considered briefly in this subsection in terms of Hadamard's "finite part integral"; Hadamard (1952), and Kaya and Erdogan (1987). The method also applies to the case when a crack is partially or fully bridged due to the presence of whiskers, fibers, or second-phase inclusions, as well as unbroken material ligaments; see Nemat-Nasser and Hori (1987). Mechanisms of this kind are considered as sources of toughening in the design of ceramic composites; see, for example, Evans (1990), and Li (1992).

In view of (21.4.3), the dislocation density at x is given by the derivative of COD at x, i.e.,

$$b(x) = -\frac{d[u]}{dx}(x). \tag{21.4.11}$$

The consistency condition (21.4.7) ensures that the COD is zero at the ends of the crack. Thus, from $[u](x) = 0$ at $x = \pm a$ and integration by parts, rewrite the right side of (21.4.4) in terms of $[u]$, as

$$\int_{-a}^{a} \frac{2\mu b(\zeta)}{\pi(\kappa+1)} \frac{1}{x-\zeta} d\zeta = \fint_{-a}^{a} \frac{2\mu\,[u](\zeta)}{\pi(\kappa+1)} \frac{1}{(x-\zeta)^2} d\zeta. \tag{21.4.12}$$

As is seen, the kernel of the integral in the right side of (21.4.12) has a squared singularity. Therefore, this integral must be interpreted in a special manner which has been considered by Hadamard (1952) in great detail; see also Kaya and Erdogan (1987). As it stands, the integral is only meaningful when viewed in terms of Hadamard's "finite part integral". A brief outline is given in Subsection 21.4.5. The notation $\fint$ in this integral is used to emphasize that the finite part integral is involved.

With the aid of (21.4.12), the integral equation for the dislocation density, (21.4.6), is rewritten as an integral equation for the COD,

$$\fint_{-a}^{a} \frac{2\mu\,[u](\zeta)}{\pi(\kappa+1)} \frac{1}{(x-\zeta)^2} d\zeta = -\,t^o(x), \tag{21.4.13}$$

where t^o is the given tractions. An additional advantage of formulation (21.4.12) compared with (21.4.6) is that the consistency condition (21.4.7) is reduced to simple boundary conditions, $[u] = 0$ at the ends of L.

Although the singularity of the kernel is higher than that in (21.4.6), the strongly singular integral equation (21.4.13) can be solved directly. Indeed, expansion of the COD in terms of the Chebychev polynomials of the second kind reduces (21.4.13) to a set of linear equations for the unknown coefficients in this expansion. The nth order Chebychev polynomial of the second kind, U_n, is defined by (Luke, 1969)

$$U_n(x) \equiv \csc\theta \sin\{(n+1)\theta\} \qquad (\theta = \arccos x;\ |x| \le 1), \tag{21.4.14}$$

and has the following orthogonality:

$$\frac{2}{\pi}\int_{-1}^{1} W(x)\, U_n(x)\, U_m(x)\, dx = \begin{cases} 1 & \text{if } n = m \\ 0 & \text{otherwise,} \end{cases} \tag{21.4.15a}$$

where $W(x) = \sqrt{1-x^2}$. Furthermore, U_n satisfies

$$\fint_{-1}^{1} \frac{1}{(y-x)^2} W(y)\, U_n(y)\, dy = -\,\pi\,(n+1)\, U_n(x). \tag{21.4.15b}$$

In view of (21.4.15a), expand $[u](\zeta)$ as

$$[u](\zeta) = W(\zeta/a)\left\{ \sum_{n=0}^{\infty} u_n\, U_n(\zeta/a) \right\}, \tag{21.4.16a}$$

where

$$u_n = \frac{2}{\pi\, a}\int_{-a}^{a} [u](\zeta)\, U_n(\zeta/a)\, d\zeta. \tag{21.4.16b}$$

Now, multiply both sides of (21.4.13) by $2W(x/a)\, U_n(x/a)/\pi a$, and integrate it from $-a$ to a, to obtain

$$\frac{2\mu(n+1)}{a(\kappa+1)}\, u_n = \frac{2}{\pi a}\int_{-a}^{a} t^o(x)\, W(x/a)\, U_n(x/a)\, dx. \tag{21.4.17}$$

In this manner, $[u](\zeta)$ can be computed directly. Note that term $W(x/a)$ in the expansion of $[u](\zeta)$ corresponds to the singularity of the dislocation density function.

21.4.5. Finite-Part Integral

The integrand in the integral

$$I = \int_a^b \frac{f(t)}{t - t_o}\, dt \qquad \text{for } a < t_o < b, \tag{21.4.18a}$$

is unbounded at $t = t_o$ when $f(t_o) \neq 0$. Nevertheless, this integral exists in the sense of the Cauchy principal value when $f(t)$ is suitably smooth; see Muskhelishvili (1956). Indeed, from

$$I = \int_a^b \frac{f(t) - f(t_o)}{t - t_o}\, dt + f(t_o) \int_a^b \frac{dt}{t - t_o}, \tag{21.4.18b}$$

it follows that when

$$\lim_{t \to t_o} | f(t) - f(t_o) | \leq A \lim_{t \to t_o} | t - t_o |^{\alpha}, \tag{21.4.19}$$

for $0 < \alpha < 1$ and a constant A, then the first integral in the right side of (21.4.18b) can be evaluated directly, and the last integral in the sense of the Cauchy principal value becomes

$$f(t_o) \int_a^b \frac{dt}{t - t_o} = f(t_o) \lim_{\varepsilon \to 0} \left\{ \int_a^{t_o - \varepsilon} \frac{dt}{t - t_o} + \int_{t_o + \varepsilon}^{b} \frac{dt}{t - t_o} + \int_{t_o - \varepsilon}^{t_o + \varepsilon} \frac{dt}{t - t_o} \right\}$$

$$= f(t_o) \left\{ \ln(-1) + \ln\left[\frac{b - t_o}{a - t_o} \right] \right\}, \tag{21.4.20}$$

where the value $\ln(-1) = \iota\pi$ may be used; see Figure 21.4.1.

Figure 21.4.1

Integration path for Cauchy principal value

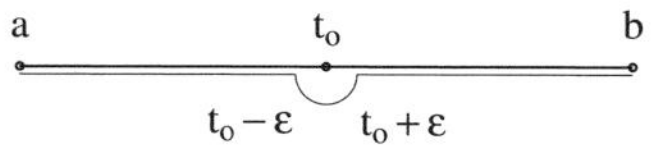

In a similar manner, the integral

$$J = ⨍_a^b \frac{f(t)}{(t - t_o)^2}\, dt \tag{21.4.21a}$$

which is clearly unbounded, may be interpreted to equal its *finite part*, or to be identified as a *finite-part integral*; see Hadamard (1952). For illustration, assume that $f(t)$ and $f'(t)$ both exist and are suitably well-behaved within the interval (a, b). Then, integral J may be rewritten as

$$J = \int_a^b \frac{f(t) - f(t_o) - (t - t_o)\, f'(t_o)}{(t - t_o)^2}\, dt$$

$$+ f(t_o) \left[- \frac{1}{b - t_o} + \frac{1}{a - t_o} \right] + f'(t_o) \left\{ \ln(-1) + \ln\left[\frac{b - t_o}{a - t_o} \right] \right\}. \qquad (21.4.21b)$$

Since the first integral in the right side of (21.4.21b) is bounded for suitably smooth f(t), the finite-part integral has a finite value and is well-defined.

Recently, Kaya and Erdogan (1987) have examined various properties of the finite-part integral, particularly in relation to problems in linear fracture mechanics. These authors list a number of interesting properties of finite-part integrals, when the function f(t) is identified with various special functions. As illustrations, consider the Legendre polynomials of the first and second kind, $P_n(t)$ and $Q_n(t)$, and the Chebychev polynomials of the first and second kind, $T_n(t)$ and $U_n(t)$, with weighting functions $(1 - t^2)^{-1/2}$ and $(1 - t^2)^{1/2}$, respectively. Table 21.4.1 gives the corresponding finite-part integral from -1 to 1.

Table 21.4.1

Finite-part integral of some special functions

Function f(t)	Finite-Part Integral $⨍_{-1}^{1} \frac{f(t)}{(t - t_o)^2}\, dt$
$P_n(t)$	$\frac{2(n+1)}{(1 - t_o^2)} \{t_o Q_n(t_o) - Q_{n+1}(t_o)\}$
$T_n(t)/\sqrt{1 - t^2}$	$\frac{\pi}{2(1 - t_o^2)} \{-(n-1)\, U_n(t_o) + (n+1)\, U_{n-2}(t_o)\}$
$U_n(t)\sqrt{1 - t^2}$	$-\pi(n+1)\, U_n(t_o)$

As pointed out before, the finite-part integral is effective when dealing with calculation of the COD under complicated conditions, e.g., bridged cracks, as is briefly examined by Nemat-Nasser and Hori (1987).

21.5. ANISOTROPIC CASE

The preceding subsections deal with isotropic materials, for which constitutive relation (21.1.5) is assumed. For general anisotropic materials, Airy's stress function and the associated Muskhelishvili's complex potentials require some modification. The formulation based on these potentials still enables one to solve two-dimensional anisotropic elasticity problems, as effectively as the isotropic ones. In this subsection, a basic formulation of cracks in anisotropic materials is presented in terms of these potentials, and solutions to illustrative problems are given.

21.5.1. Airy's Stress Function and Muskhelishvili's Complex Potentials for Anisotropic Materials

Consider a general two-dimensional anisotropic material with the following constitutive relations:

$$\begin{bmatrix} \varepsilon_{11} \\ \varepsilon_{22} \\ 2\varepsilon_{12} \end{bmatrix} = \begin{bmatrix} D_{11} & D_{12} & D_{16} \\ D_{12} & D_{22} & D_{26} \\ D_{16} & D_{26} & D_{66} \end{bmatrix} \begin{bmatrix} \sigma_{11} \\ \sigma_{22} \\ \sigma_{12} \end{bmatrix}; \tag{21.5.1}$$

see Subsection 3.1.4. Equations of equilibrium, (21.1.1), are satisfied by the Airy stress function, U. Compatibility condition (21.1.6) with constitutive relations (21.5.1) now leads to

$$D_{22}\,U_{,1111} - 2D_{26}\,U_{,1112} + (2D_{12} + D_{66})\,U_{,1122} - 2\,D_{16}\,U_{,1222} + D_{11}\,U_{,2222} = 0. \tag{21.5.2a}$$

This is the governing equation for Airy's stress function in general (two-dimensional) anisotropic materials.

A general solution of (21.5.2a) can be obtained in the form of f(X) with $X \equiv x_1 + s\,x_2$. Then, $\partial/\partial x_1$ and $\partial/\partial x_2$ in (21.5.2) are replaced by d/dX and s d/dX, respectively, and (21.5.2) is rewritten as

$$\left\{ D_{22} - 2\,D_{26}\,s + (2D_{12} + D_{66})\,s^2 - 2\,D_{16}\,s^3 + D_{11}\,s^4 \right\} \frac{d^4U}{dX^4} = 0 \tag{21.5.3}$$

which leads to the *characteristic equation*,

$$D_{11}\,s^4 - 2\,D_{16}\,s^3 + (2D_{12} + D_{66})\,s^2 - 2\,D_{26}\,s + D_{22} = 0. \tag{21.5.4}$$

Denoting the four roots of (21.5.4) by s_1, s_2, s_3, and s_4, consider a general solution of (21.5.2a) in the form

$$U(x_1, x_2) = \sum_{i=1}^{4} f_i(x_1 + s_i\,x_2), \tag{21.5.2b}$$

where f_1, f_2, f_3, and f_4 are suitably smooth but arbitrary functions.

Due to the positive-definiteness of the elastic complementary strain energy, there are no real roots for (21.5.4). Hence, two of the four roots, say, s_1 and s_2, are complex conjugates of the other two, s_3 and s_4, since D_{ij}'s are real; e.g., Eshelby *et al.* (1953), Lekhnitskii (1963), or Dundurs (1968). As in the case of isotropic materials, general solution (21.5.2b) can be expressed in terms of the real or imaginary part of analytic functions of suitable complex variables. Define two complex numbers, z_1 and z_2, as

$$z_I = x_1 + s_I x_2 \qquad \qquad x_1 = \frac{s_I \bar{z}_I - \bar{s}_I z_I}{s_I - \bar{s}_I}$$

$$\Longleftrightarrow \qquad \text{(I not summed)} \qquad (21.5.5a,b)$$

$$\bar{z}_I = x_1 + \bar{s}_I x_2 \qquad \qquad x_2 = \frac{z_I - \bar{z}_I}{s_I - \bar{s}_I}$$

for I = 1 and 2. Define differentiation with respect to z_I and $\bar{z}_I$ by

$$\frac{\partial}{\partial z_I} = \frac{1}{s_I - \bar{s}_I} \left[-\bar{s}_I \frac{\partial}{\partial x_1} + \frac{\partial}{\partial x_2} \right],$$

$$\frac{\partial}{\partial \bar{z}_I} = \frac{1}{s_I - \bar{s}_I} \left[s_I \frac{\partial}{\partial x_1} - \frac{\partial}{\partial x_2} \right] \qquad \text{(I not summed).} \qquad (21.5.5c,d)$$

Complex number z_I corresponds to $z = x_1 + \iota x_2$ when the material is isotropic; see Subsection 21.1.

Since the left side of (21.5.2a) can be expressed in terms of s_I's and $\bar{s}_I$'s, as

$$D_{11} \left\{ \prod_{I=1}^{2} \left[s_I \frac{\partial}{\partial x_1} - \frac{\partial}{\partial x_2} \right] \left[\bar{s}_I \frac{\partial}{\partial x_1} - \frac{\partial}{\partial x_2} \right] \right\} U,$$

in view of (21.5.5c) and (21.5.5d), (21.5.2a) is rewritten as

$$\frac{\partial^2}{\partial z_1 \partial \bar{z}_1} \left[\frac{\partial^2}{\partial z_2 \partial \bar{z}_2} U \right] = 0. \qquad (21.5.6)$$

A complex function, $f(z, \bar{z})$, which satisfies $\partial^2 f/\partial z \partial \bar{z} = 0$, is analytic. Thus, a general solution of (21.5.6) can now be expressed in terms of two complex analytic functions. Indeed, if s_1 and s_2 are distinct,

$$U(x_1, x_2) = Re\{\chi_1(z_1) + \chi_2(z_2)\} \qquad (21.5.7a)$$

or

$$U(x_1, x_2) = Re\{\int^{z_1} \psi_1(z_1')\, dz_1' + \int^{z_2} \psi_2(z_2')\, dz_2'\}, \qquad (21.5.7b)$$

where $\psi_1(z_1)$ and $\psi_2(z_2)$, or $\chi_1(z_1)$ and $\chi_2(z_2)$, are analytic with respect to their arguments. If s_1 and s_2 are equal, (21.5.6) becomes

$$\left[\frac{\partial^2}{\partial z_1 \partial \bar{z}_1} \right]^2 U = 0, \qquad (21.5.8)$$

and a solution is given by

$$U(x_1, x_2) = Re\{\bar{z}_1 \phi(z_1) + \chi(z_1)\}$$

$$= Re\{\bar{z}_1 \phi(z_1) + \int^{z_1} \psi(z_1')\, dz_1'\}, \qquad (21.5.9)$$

where ψ, χ, and ϕ are analytic functions of z_1. An isotropic medium corresponds to this special case, since the roots of the characteristic equation for isotropic D_{ij} are $\pm\iota$.

For simplicity, it is assumed that $s_1 \neq s_2$, and two complex potentials $\psi_1(z_1)$ and $\psi_2(z_2)$ are used from now on. Table 21.5.1 presents the stress, displacement, and resultant force components, produced by ψ_1 and ψ_2.

Table 21.5.1

Expression of stress, displacement, resultant force, and resultant moment in terms of Muskhelishvili's complex potentials for anisotropic materials

field variable	expression
σ_{11}	$Re\{s_1^2\,\psi_1' + s_2^2\,\psi_2'\}$
σ_{22}	$Re\{\psi_1' + \psi_2'\}$
σ_{12}	$-Re\{s_1\,\psi_1' + s_2\,\psi_2'\}$
u_1	$Re\{(D_{11}\,s_1^2 - D_{16}\,s_1 + D_{12})\,\psi_1 + (D_{11}\,s_2^2 - D_{16}\,s_2 + D_{12})\,\psi_2\}$
u_2	$Re\{(D_{12}\,s_1^2 - D_{26}\,s_1 + D_{22})\,\psi_1/s_1 + (D_{12}\,s_2^2 - D_{26}\,s_2 + D_{22})\,\psi_2/s_2\}$
F_1	$Re\{s_1\,\psi_1 + s_2\,\psi_2\}$
F_2	$-Re\{\psi_1 + \psi_2\}$
M	$-x_1\,Re\psi_1 - x_2\,Re\{s_1\,\psi_1\} + Re\{\int \psi_1\,dz_1\}$ $-x_1\,Re\psi_2 - x_2\,Re\{s_2\,\psi_2\} + Re\{\int \psi_2\,dz_2\}$

21.5.2. Dislocation in Anisotropic Medium

As shown in Subsection 21.2, a dislocation at the origin of the coordinate system in a two-dimensional isotropic medium is obtained by enforcing: 1) zero resultant forces around a loop surrounding the origin; and 2) a displacement jump around the loop equal to Burgers' vector **b**. Essentially the same formulation applies when a dislocation in a two-dimensional anisotropic material is considered. Indeed, $\psi_1 = A_1 \ln z_1$ and $\psi_2 = A_2 \ln z_2$ produce the following components of the resultant force and displacement jump:

$$[F_1] = -2\pi\,Im\{s_1\,A_1 + s_2\,A_2\},$$

$$[F_2] = 2\pi\,Im\{A_1 + A_2\}, \tag{21.5.10a,b}$$

and

$$[u_1] = -2\pi\, Im\{p_1 A_1 + p_2 A_2\},$$

$$[u_2] = -2\pi\, Im\{q_1 A_1 + q_2 A_2\}, \tag{21.5.10c,d}$$

where p_I and q_I are complex constants defined by

$$p_I \equiv D_{11} s_I^2 - D_{16} s_I + D_{12}, \qquad q_I \equiv \frac{D_{12} s_I^2 - D_{26} s_I + D_{22}}{s_I}, \tag{21.5.11a,b}$$

for I = 1, 2.[10]

Unknown constants A_1 and A_2 are determined such that boundary conditions $[F_i] = 0$ and $[u_i] = b_i$ are satisfied, i.e.,

$$\begin{bmatrix} Im s_1 & Re s_1 & Im s_2 & Re s_2 \\ 0 & 1 & 0 & 1 \\ Im p_1 & Re p_1 & Im p_2 & Re p_2 \\ Im q_1 & Re q_1 & Im q_2 & Re q_2 \end{bmatrix} \begin{bmatrix} Re A_1 \\ Im A_1 \\ Re A_2 \\ Im A_2 \end{bmatrix} = \begin{bmatrix} 0 \\ 0 \\ -b_1/2\pi \\ -b_2/2\pi \end{bmatrix} \tag{21.5.12}$$

which leads to

$$Re A_1 = \frac{1}{2\pi D}\left\{ \{Im q_2 - \frac{Re q_1 - Re q_2}{Re s_1 - Re s_2} Im s_2\}\, b_1 - \{Im p_2 - \frac{Re p_1 - Re p_2}{Re s_1 - Re s_2} Im s_2\}\, b_2 \right\},$$

$$Re A_2 = \frac{1}{2\pi D}\left\{ -\{Im q_1 - \frac{Re q_1 - Re q_2}{Re s_1 - Re s_2} Im s_1\}\, b_1 + \{Im p_1 - \frac{Re p_1 - Re p_2}{Re s_1 - Re s_2} Im s_1\}\, b_2 \right\},$$

$$Im A_1 = -\frac{1}{Re s_1 - Re s_2}\{Im s_1\, Re A_1 + Im s_2\, Re A_2\},$$

$$Im A_2 = \frac{1}{Re s_1 - Re s_2}\{Im s_1\, Re A_1 + Im s_2\, Re A_2\}, \tag{21.5.13a~d}$$

where

$$D = \{Im p_1 - \frac{Re p_1 - Re p_2}{Re s_1 - Re s_2} Im s_1\}\{Im q_2 - \frac{Re q_1 - Re q_2}{Re s_1 - Re s_2} Im s_2\} - \{Im p_2 - \frac{Re p_1 - Re p_2}{Re s_1 - Re s_2} Im s_2\}\{Im q_1 - \frac{Re q_1 - Re q_2}{Re s_1 - Re s_2} Im s_1\}. \tag{21.5.13e}$$

Denoting the above solution by $A_I(b) = Re A_I(b) + \iota Im A_I(b)$ with $b = b_1 + \iota b_2$, the complex potentials for the dislocation with Burgers vector **b** at the origin are expressed as

[10] At this point, assume $Re s_1 - Re s_2$ is nonzero.

$$\psi_I(z_I) = A_I(b) \ln z_I \tag{21.5.14a}$$

for I = 1 and 2.

In anisotropic materials, Airy's stress function is given by the real part of Muskhelishvili's potentials; see (21.5.7). Hence, the coordinate transformation rule for complex potentials in an anisotropic medium is the same as that for Airy's stress function of the isotropic medium. Thus, for a dislocation with Burgers' vector **b** at $\mathbf{x}^o = (x_1^o, x_2^o)$, the complex potentials are given by

$$\psi_I(z_I; b, z_I^o) = A_I(b) \ln (z_I - z_I^o), \tag{21.5.14b}$$

where $z_I^o = x_1^o + s_I x_2^o$, for I = 1 and 2. In terms of the roots of the characteristic equation, the derivatives of these complex potentials are expressed in the following simple form which has been obtained by Obata *et al.* (1989):

$$\psi_1'(z) = \frac{1}{\pi\iota}\frac{1}{D_{11}}\frac{s_1 b_1 - b_2}{(s_1 - s_2)(s_1 - \bar{s}_1)(s_1 - \bar{s}_2)}\frac{1}{z_1 - z_1^o},$$

$$\psi_2'(z) = \frac{1}{\pi\iota}\frac{1}{D_{11}}\frac{s_2 b_1 - b_2}{(s_2 - s_1)(s_2 - \bar{s}_1)(s_2 - \bar{s}_2)}\frac{1}{z_2 - z_2^o}. \tag{21.5.14c,d}$$

As an illustration, consider the special case of a transversely isotropic material, where $D_{16} = D_{26} = 0$ in (21.5.1); see Section 3.1. The characteristic equation now becomes

$$D_{11}\, s^4 + 2\,(\tfrac{1}{2} D_{66} + D_{12})\, s^2 + D_{22} = 0, \tag{21.5.15a}$$

and assuming that $(\tfrac{1}{2} D_{66} + D_{12})^2 - D_{11} D_{22} > 0$ and $\tfrac{1}{2} D_{66} + D_{12} > 0$, note that the roots of the characteristic equation are

$$s = \pm\iota\,\beta_1, \quad \pm\iota\,\beta_2 \quad (\beta_1, \beta_2 > 0). \tag{21.5.15b}$$

Now, compute the Airy stress function for a dislocation with the Burgers vector $b = b_1 + \iota b_2$. Since s_1 and s_2 are purely imaginary, the matrix in the left side of (21.5.12) is replaced by

$$\begin{bmatrix} \beta_1 & 0 & \beta_2 & 0 \\ 0 & 1 & 0 & 1 \\ 0 & -D_{11}\beta_1^2 + D_{12} & 0 & -D_{11}\beta_2^2 + D_{12} \\ (D_{12}\beta_1^2 - D_{22})/b\eta_1 & 0 & (D_{12}\beta_2^2 - D_{22})/b\eta_2 & 0 \end{bmatrix}.$$

Hence, the Airy stress function is given by

$$U(x_1, x_2) = \frac{1}{2\pi(\beta_1^2 - \beta_2^2)}\, Re\Big\{ \Big[-\beta_1\beta_2^2 \frac{b_2}{D_{22}} + \iota\frac{b_1}{D_{11}} \Big]\, z_1\,(\ln z_1 - 1)$$

$$+ \Big[\beta_1^2\beta_2 \frac{b_2}{D_{22}} - \iota\frac{b_1}{D_{11}} \Big]\, z_2\,(\ln z_2 - 1) \Big\}. \tag{21.5.16}$$

The corresponding stress field then is

$$\sigma_{ij}(z) = \frac{1}{r}\, f_{ijk}(\theta)\, b_k \qquad \text{for i, j, k} = 1, 2, \tag{21.5.17a}$$

where polar coordinate $z = r\,e^{\iota\theta}$ is used, and the coefficients f_{ijk} are given by

$$f_{111} = k_1 (-\beta_1^3 \gamma_1 + \beta_2^3 \gamma_2) \sin\theta \qquad f_{112} = k_2 \beta_1^2 \beta_2^2 (\beta_1 \gamma_1 - \beta_2 \gamma_2) \cos\theta,$$

$$f_{221} = k_1 (\beta_1 \gamma_1 - \beta_2 \gamma_2) \sin\theta \qquad f_{222} = k_2 \beta_1 \beta_2 (-\beta_2 \gamma_1 + \beta_1 \gamma_2) \cos\theta,$$

$$f_{121} = f_{211} = k_1 (\beta_1 \gamma_1 - \beta_2 \gamma_2) \cos\theta,$$

$$f_{122} = f_{212} = k_2 \beta_1^2 \beta_2^2 (\beta_1 \gamma_1 - \beta_2 \gamma_2) \sin\theta, \tag{21.5.17b~g}$$

with $k_1 = 1/2\pi(\beta_1^2 - \beta_2^2) D_{11}$, $k_2 = 1/2\pi(\beta_1^2 - \beta_2^2) D_{22}$, and

$$\gamma_I = \frac{1}{\cos^2\theta + \beta_I^2 \sin^2\theta} \quad \text{for } I = 1, 2. \tag{21.5.17h}$$

21.5.3. Crack in Anisotropic Medium

In anisotropic materials, a Hilbert problem can be formulated to solve the crack problem. As explained in Subsection 21.3, this formulation is effective when conditions on the crack surfaces are simple such that the required Cauchy integral can be computed. An alternative method[11] is a formulation in terms of a suitable distribution of dislocations. This formulation leads to a singular integral equation for the dislocation density function. Since Muskhelishvili's complex potentials for a dislocation in anisotropic materials are given by a logarithmic function, as in isotropic materials, the corresponding integral equations also have a form similar to that in the isotropic case.

Consider the isolated crack, L, examined in Subsection 21.4, i.e., a straight crack lying on the x_1-axis from $(-a, 0)$ to $(+a, 0)$. If dislocation density function, $b(\zeta) = b_1(\zeta) + \iota b_2(\zeta)$, is prescribed on L, the COD, $[u](x) = [u_1](x) + \iota [u_2](x)$, becomes

$$[u](x) = -\int_{-a}^{x} b(\zeta)\, d\zeta, \tag{21.5.18}$$

and tractions on the crack surfaces, $t(x) = t_1(x) + \iota t_2(x)$, are given by

$$t(x) = \int_{-a}^{a} \frac{1}{x-\zeta} \Big[Re\{A_1(b(\zeta)) + A_2(b(\zeta))\} - \iota\, Re\{s_1 A_1(b(\zeta)) + s_2 A_2(b(\zeta))\} \Big]\, d\zeta, \tag{21.5.19}$$

where A_1 and A_2 are given by (21.5.13).

If the COD is prescribed as the boundary conditions, $b(\zeta)$ is directly obtained from (21.5.18). If tractions acting on the upper and lower surfaces of the crack are the same and given by $t^o(x) = t_1^o(x) + \iota t_2^o(x)$, then, (21.5.19) yields the following integral equation for $b(\zeta)$:

[11] This method applies to general boundary conditions. It has been used to examine crack kinking and other phenomena in anisotropic solids by Obata *et al.* (1989), and Azhdari and Nemat-Nasser (1996a,b, 1998); see also Nemat-Nasser and Azhdari (1997), and Azhdari *et al.* (1998).

$$\int_{-a}^{a} \frac{1}{x-\zeta} \Big[Re\{ A_1(b(\zeta)) + A_2(b(\zeta)) \}$$

$$- \imath\, Re\{ s_1 A_1(b(\zeta)) + s_2 A_2(b(\zeta)) \} \Big] \, d\zeta = - t^o(x). \qquad (21.5.20)$$

Similarly to (21.4.6), this equation is a Cauchy integral equation of the first kind, and the consistency condition is required to obtain $b(\zeta)$ uniquely, i.e.,

$$\int_{-a}^{a} b(\zeta) \, d\zeta = 0, \qquad (21.5.21)$$

which ensures that the crack is closed at its ends.

As pointed out in Subsection 21.4.4, an alternative integral equation is formulated from (21.5.20), if dislocation density function, $b(\zeta)$, is replaced by COD, $[u](x)$. From the linear dependence of $A_I(b)$ on b, and from $b(x) = - d[u]/dx(x)$, it follows that

$$A_I(b(x)) = A_I(-\frac{d[u]}{dx}(x)) = -\frac{d}{dx} A^I([u](x)). \qquad (21.5.22)$$

Therefore, with the aid of consistency condition (21.5.19), or equivalently, $[u](\pm a) = 0$, an integral equation for $[u](x)$ is derived from (21.5.18), as follows:

$$⨎_{-a}^{a} \frac{1}{(x-\zeta)^2} \Big[Re\{ A_1([u](\zeta)) + A_2([u](\zeta)) \}$$

$$- \imath\, Re\{ s_1 A_1([u](\zeta)) + s_2 A_2([u](\zeta)) \} \Big] \, d\zeta = - t^o(x). \qquad (21.5.23)$$

As shown in Subsection 21.4.4, integral equation (21.5.23) with a squared singularity can be solved directly for $[u](x)$ by expanding $[u](x)$ in terms of the Chebychev polynomials of the second kind.

As an illustration, consider a crack in a transversely isotropic medium, normal to the axis of isotropy, lying on the x_1-axis. The corresponding stress fields associated with a single dislocation are given by (21.5.19). Since the stress field of the dislocation on the x_1-axis is written as

$$\sigma_{ij}(x) = f_{ijk} \int_{-a}^{a} \frac{b_k(\zeta)}{x-\zeta} \, d\zeta \qquad \text{for } i, j = 1, 2, \qquad (21.5.24a)$$

it follows that

$$f_{2ij} \int_{-a}^{a} \frac{b_j(\zeta)}{x-\zeta} \, d\zeta = - t_i^o(x) \qquad \text{for } i = 1, 2, \qquad (21.5.24b)$$

defines the required dislocation density function $b_i(x)$ such that the crack remains traction-free. In terms of the finite-part integral, (21.5.24b) becomes

$$f_{2ij} ⨎_{-a}^{a} \frac{[u_j](\zeta)}{(x-\zeta)^2} \, d\zeta = - t_i^o(x) \qquad \text{for } i = 1, 2. \qquad (21.5.24c)$$

21.5.4. Full or Partial Crack Bridging

The formulation of crack problems in anisotropic solids, in terms of a finite-part integral, directly involves the COD, $[u_i](x)$. This then provides a convenient technique to solve a problem of partially or fully bridged cracks in reinforced composites. Figure 21.5.1 shows a crack in a unidirectionally reinforced composite. The crack is bridged by unbroken fibers. It is expected that with a suitable design of the interface between the fiber and the matrix, improved toughness can be attained; see, e.g., Budiansky *et al.* (1988), Rose (1987), and Evans (1990).

As a simple example, assume that the bridging force is linearly dependent on the COD, i.e.,

$$p_i(x) = \begin{cases} -K_o\,[u_i](x) & \text{for } b < |x| < a \\ 0 & \text{for } |x| < b. \end{cases} \tag{21.5.25}$$

In view of notation (21.5.25), the integral equation (21.5.24c) becomes

$$f_{2ij} \fint_{-a}^{a} \frac{[u_j](\zeta)}{(x-\zeta)^2}\, d\zeta - K_o\, H(b - |x|)\, [u_i](x) = -t_i^o(x), \tag{21.5.26a}$$

for $|x| < a$, where $H(x) = 0$ for $x < 0$ and 1 for $x > 0$. For simplicity, set

$$X \equiv x/a, \qquad B \equiv b/a, \qquad F_{ij} \equiv f_{2ij}/E_m, \qquad a_o \equiv E_m/K_o, \qquad l \equiv a/a_o,$$

$$U_i(X) \equiv [u_i](x)/a, \qquad T_i(X) \equiv t_i^o(x)/E_m, \tag{21.5.27a~g}$$

where E_m is a certain reference Young modulus, which might be identified as Young's modulus of the matrix of the fiber-reinforced composite. Then, (21.5.26a) is rewritten as

$$-F_{ij} \fint_{-1}^{1} \frac{U_j(Z)}{(X-Z)^2}\, dZ + l\, H(B - |X|)\, U_i(X) = T_i(X), \tag{21.5.26b}$$

for $|X| < 1$. The solution of this equation in terms of the Chebychev

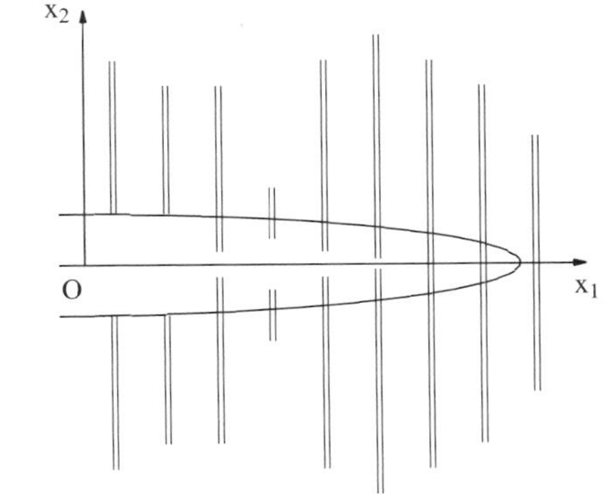

Figure 21.5.1

Crack which is partially bridged by unidirectional reinforcing fibers

polynomials of the second kind, has been discussed by Nemat-Nasser and Hori (1987) who also consider other examples of bridging forces, and provide numerical illustrations, for fully and partially bridged cases.

Rose (1987) formulates a partially bridged crack in terms of the distributed dislocations, for both a linear and a nonlinear relation between the bridging forces and the COD. Solutions are then given through an asymptotic approach for "small" cracks, i.e., $l \equiv a/a_o < 1$. Budiansky (1986) and Horii *et al.* (1987) have examined the effect of small-scale bridging in a narrow zone close to the crack tip for isotropic materials; Horii *et al.* also have studied the finite-crack problem.

When the crack is "very small", i.e., $l \equiv a/a_o << 1$, the standard perturbation may be used to construct the solution for both linear and nonlinear cases; see Rose (1987) and Hori and Nemat-Nasser (1990). On the other hand, when $l >> 1$, the crack is considered to be "very big" and the usual perturbation methods do not apply. Set $\varepsilon \equiv 1/l$, and rewrite (21.5.26b) as

$$-\varepsilon F_{ij} \,⨎_{-1}^{1} \frac{U_j(Z)}{(X-Z)^2}\, dZ + H(B - |X|)\, U_i(X) = \varepsilon T_i(X). \tag{21.5.26c}$$

A singular perturbation solution technique is given by Willis and Nemat-Nasser (1990) for this class of strongly singular integral equations. The method also applies to the case when the bridging force is nonlinear.

21.6. DUALITY PRINCIPLES IN ANISOTROPIC ELASTICITY

In this subsection, anisotropic elasticity problems are considered, where the displacement field, $\mathbf{u}$, and hence the strain, $\boldsymbol{\varepsilon}$, and stress, $\boldsymbol{\sigma}$, fields are functions of only two space variables, say, the rectangular Cartesian coordinates, x_1, and x_2. In such a case, equilibrium equations, in the absence of body forces, reduce to[12]

$$\sigma_{j1,1} + \sigma_{j2,2} = 0, \qquad j = 1, 2, 3. \tag{21.6.1a}$$

These equations are identically satisfied if the stress components are expressed as suitable gradients of a vector potential, $\boldsymbol{\phi}$, as follows:

$$\sigma_{j1} = \frac{\partial \phi_j}{\partial x_2}, \qquad \sigma_{j2} = -\frac{\partial \phi_j}{\partial x_1}, \tag{21.6.1b,c}$$

where ϕ_j ($j = 1, 2, 3$) are the components of $\boldsymbol{\phi}$. The basic field equations can now be expressed in terms of a six-dimensional vector field $\boldsymbol{\eta} = [\mathbf{u}, \boldsymbol{\phi}]^T$, with components ($u_1$, u_2, u_3, ϕ_1, ϕ_2, ϕ_3), as follows:

---

[12] To facilitate the interpretation of the results, it will be assumed that all quantities, e.g., the coordinate variables, the displacement vector, the stress and elasticity tensors, and the stress-potential vector, are rendered physically dimensionless, using suitable length, stress, and force scales.

$$\frac{\partial \boldsymbol{\eta}}{\partial x_2} = \boldsymbol{N} \frac{\partial \boldsymbol{\eta}}{\partial x_1}. \tag{21.6.2}$$

The six by six matrix[13] $\boldsymbol{N}$ is called the *fundamental elasticity matrix*. This matrix is completely defined in terms of the elasticity tensor of the material, $\mathbf{C}$, as is discussed later on in this subsection. First, however, note the following simple observation.

Dual Solutions: Consider an arbitrary six by six nonsingular constant matrix $\boldsymbol{K}$, and note from (21.6.2) that

$$\frac{\partial \boldsymbol{K\eta}}{\partial x_2} = \boldsymbol{KNK}^{-1} \frac{\partial \boldsymbol{K\eta}}{\partial x_1}. \tag{21.6.3a}$$

Thus, provided that

$$\boldsymbol{N}^K = \boldsymbol{KNK}^{-1} \tag{21.6.3b}$$

corresponds to a real positive-definite elasticity tensor, say, $\mathbf{C}^K$, a general solution of the original system of equations (21.6.2) may be used to generate other general solutions for elastic solids with the *same geomerty but different elastitcity tensors*. Once the boundary conditions of the original problem are prescribed, then the transformation $\boldsymbol{K\eta}$ defines the corresponding boundary conditions of the second (called dual) problem. Note that the matrix $\boldsymbol{K}$ cannot be completely arbitrary if $\mathbf{C}^K$ is to correspond to a realistic elasticity tensor, e.g., to be positive-definite and possess all the usual symmetries discussed in Section 3. Indeed, from (21.6.3b), it follows that, for any nonsingular matrix $\boldsymbol{K}$, the matrices $\boldsymbol{N}$ and $\boldsymbol{N}^K$ have the same eigenvalues (see also (21.6.18b), leaving only the eigenvectors as potential variables, subject to symmetry requirements.

It has been shown by Nemat-Nasser and Ni (1995) and Ni and Nemat-Nasser (1996) that when the matrix $\boldsymbol{K}$ is chosen to be

$$\boldsymbol{K} = \begin{bmatrix} \mathbf{0}^{(3)} & \mathbf{1} \\ \mathbf{1} & \mathbf{0}^{(3)} \end{bmatrix}, \tag{21.6.4a}$$

where, as before, $\mathbf{1}$ is the three by three identity matrix, and $\mathbf{0}^{(3)}$ is the three by three zero matrix, then a dual solution *for the same material with the same elasticity tensor* can be generated through simple parameter interchange in the solution of the original probelm. The boundary conditions of the two problems are then dual of one another; see Subsection 21.6.3.

In order to distinguish between other dual solutions (with different elastic properties) defined by possibly arbitrary $\boldsymbol{K}$ through transformation (21.6.3), and the duality principle associated with dual solutions *with the same elasticity tensor*, obtained when the matrix $\boldsymbol{K}$ is restricted to that given by (21.6.4a), use the following notation for (21.6.4a):

---

[13] The matrix $\boldsymbol{N}$ was introduced by Ingebrigtsen and Tonning (1969). Its eigenvalue problem for anisotropic elasticity is examined by Malen (1971), and Barnett and Lothe (1973); see also Ni and Nemat-Nasser (1996), and Ting (1996) who gives a comprehensive account and review. The six-dimensional elastic field equations in terms of $\boldsymbol{N}$ are studied by Chadwick and Smith (1977), and applied to angularly inhomogeneous media by Kirchner (1989), and Ting (1989).

$$\boldsymbol{I} = \begin{bmatrix} \mathbf{0}^{(3)} & \mathbf{1} \\ \mathbf{1} & \mathbf{0}^{(3)} \end{bmatrix}, \tag{21.6.4b}$$

and note that

$$\boldsymbol{I} = \boldsymbol{I}^{\mathrm{T}} = \boldsymbol{I}^{-1}. \tag{21.6.4c}$$

In this subsection, first the properties of the fundamental elasticity matrix $\boldsymbol{N}$ are briefly reviewed. Then the duality principle is discussed, an alternative proof to that originally provided by Ni and Nemat-Nasser (1996), is given, and the results are illustrated.

Fundamental Elasticity Matrix: From the constitutive relations and the equations of equilibrium,

$$\sigma_{ij} = C_{ijkl}\,\varepsilon_{kl} = C_{ijkl}\,u_{k,l},$$

$$\sigma_{j\beta,\beta} = C_{i\beta k\alpha}\,u_{k,\alpha\beta} = 0, \tag{21.6.5a,b}$$

where the Latin subscripts $i, j, k, l = 1, 2, 3,$ and the Greek subscripts $\alpha, \beta = 1, 2$, it follows that

$$\frac{\partial \phi_j}{\partial x_2} = C_{1jk1}\frac{\partial u_k}{\partial x_1} + C_{1jk2}\frac{\partial u_k}{\partial x_2},$$

$$-\frac{\partial \phi_j}{\partial x_1} = C_{2jk1}\frac{\partial u_k}{\partial x_1} + C_{2jk2}\frac{\partial u_k}{\partial x_2}. \tag{21.6.6a,b}$$

Therefore, setting[14]

$$\mathbf{Q} \equiv [C_{j1k1}], \qquad \mathbf{R} \equiv [C_{j1k2}], \qquad \mathbf{T} \equiv [C_{j2k2}], \tag{21.6.7a~c}$$

it follows that

$$\boldsymbol{N} = \begin{bmatrix} \mathbf{N}_{11} & \mathbf{N}_{12} \\ \mathbf{N}_{21} & \mathbf{N}_{11}^{\mathrm{T}} \end{bmatrix}, \tag{21.6.8a}$$

where the following notation is used:

$$\mathbf{N}_{11} = -\mathbf{T}^{-1}\mathbf{R}^{\mathrm{T}}, \quad \mathbf{N}_{21} = \mathbf{Q} - \mathbf{R}\,\mathbf{T}^{-1}\mathbf{R}^{\mathrm{T}}, \quad \mathbf{N}_{12} = -\mathbf{T}^{-1}. \tag{21.6.8b~d}$$

Note that $\mathbf{N}_{12}$ and $\mathbf{N}_{21}$ are symmetric, $\mathbf{N}_{12} = \mathbf{N}_{12}^{\mathrm{T}}$ and $\mathbf{N}_{21} = \mathbf{N}_{21}^{\mathrm{T}}$.

It can be shown that the fundamental elasticity matrix $\boldsymbol{N}$ has no real-valued eigenvalues; see Chadwick and Smith (1977). Hence, since it is real-valued, it has three pairs of complex conjugate eigenvalues. Consider the case where there are distinct eigenvalues.[15] Denote by p_k, $k = 1, 2, 3$, the first three eigenvalues with positive imaginary parts. Then, the other three eigenvalues are $p_{k+3} = \bar{p}_k$, $k = 1, 2, 3$.

---

[14] The three by three matrices are denoted by bold roman letters.

[15] The cases where there are repeated eigenvalues are examined in Subsection 21.6.4; see also Ni and Nemat-Nasser (1996) and Ting (1996) for comments and references.

Since $\boldsymbol{N}$ is not symmetric, it has two sets of eigenvectors, the right- and the left-eigenvectors. Let the right-eigenvector ζ_k correspond to the eigenvalue p_k, $k = 1, 2, ..., 6$. These eigenvalues (assumed to be distinct) and eigenvectors are defined by the following system of homogeneous linear equations:

$$\boldsymbol{N}\,\zeta_k = p_k\,\zeta_k, \quad \text{(k not summed)}. \tag{21.6.9a}$$

Now, denote the left-eigenvectors by ξ_k, and observe that these are defined by

$$\boldsymbol{N}^T\xi_k = p_k\,\xi_k, \quad \text{(k not summed)}. \tag{21.6.9b}$$

The first three components of the right-eigenvector ζ_k correspond to the displacement components, (u_1, u_2, u_3), while the second three are associated with the components of the stress-potential vector, (ϕ_1, ϕ_2, ϕ_3). Set

$$\zeta_k = \begin{bmatrix} \mathbf{a}_k \\ \boldsymbol{l}_k \end{bmatrix}. \tag{21.6.9c}$$

Then, as shown below, obtain

$$\xi_k = \begin{bmatrix} \boldsymbol{l}_k \\ \mathbf{a}_k \end{bmatrix}. \tag{21.6.9d}$$

Hence, the first three components of the left-eigenvector ξ_k correspond to the stress-potential vector, (ϕ_1, ϕ_2, ϕ_3), while the second three are associated with the components of the displcement field, (u_1, u_2, u_3), i.e., ζ_k and ξ_k are each other's dual. This is established as follows.

Since the matrix $\boldsymbol{IN}$ is symmetric,

$$\boldsymbol{IN} = \begin{bmatrix} \mathbf{N}_{21} & \mathbf{N}_{11}^T \\ \mathbf{N}_{11} & \mathbf{N}_{12} \end{bmatrix} = (\boldsymbol{IN})^T = \boldsymbol{N}^T\boldsymbol{I}, \tag{21.6.10a}$$

multiply (21.6.9a) from left by $\boldsymbol{I}$ to obtain

$$\boldsymbol{N}^T(\boldsymbol{I}\,\zeta_k) = p_k\,(\boldsymbol{I}\,\zeta_k), \quad \text{(k not summed)}. \tag{21.6.10b}$$

Thus, the left-eigenvectors are related to the right-eigenvectors by

$$\xi_k = \boldsymbol{I}\,\zeta_k = \begin{bmatrix} \boldsymbol{l}_k \\ \mathbf{a}_k \end{bmatrix}, \quad k = 1, 2, 3.$$

Therefore, *the right- and left-eigenvectors of* $\boldsymbol{N}$ are each other's dual. As is shown below, this feature leads to a general duality principle for the solution of a class of elasticity problems when *the geometry and the elasticity of the domain is fixed, and the auxiliary or boundary conditions of the two problems are each other's dual;* this duality of the boundary conditions is discussed in Subsection 21.6.3.

Define the six by six matrix $\boldsymbol{Z}$ by

$$\boldsymbol{Z} \equiv [\zeta_1, \zeta_2, \zeta_3, \zeta_4, \zeta_5, \zeta_6] = \begin{bmatrix} \mathbf{A} & \bar{\mathbf{A}} \\ \mathbf{L} & \bar{\mathbf{L}} \end{bmatrix}, \tag{21.6.11a}$$

where $\mathbf{A}$ and $\mathbf{L}$ are given by

$$\mathbf{A} = [\mathbf{a}_1, \mathbf{a}_2, \mathbf{a}_3], \quad \mathbf{L} = [\boldsymbol{l}_1, \boldsymbol{l}_2, \boldsymbol{l}_3]. \tag{21.6.11b,c}$$

In view of (21.6.9d), introduce the dual of $\boldsymbol{Z}$ by

$$\boldsymbol{Y} \equiv [\boldsymbol{\xi}_1, \boldsymbol{\xi}_2, \boldsymbol{\xi}_3, \boldsymbol{\xi}_4, \boldsymbol{\xi}_5, \boldsymbol{\xi}_6] = \begin{bmatrix} \mathbf{L} & \overline{\mathbf{L}} \\ \mathbf{A} & \overline{\mathbf{A}} \end{bmatrix}. \tag{21.6.12a}$$

Hence, $\boldsymbol{Z}$ and $\boldsymbol{Y}$ are related by

$$\boldsymbol{Z} = \boldsymbol{I}\boldsymbol{Y}, \quad \boldsymbol{Y} = \boldsymbol{I}\boldsymbol{Z}. \tag{21.6.12b,c}$$

The dual eigenvectors corresponding to different eigenvalues are orthogonal. This follows from (21.6.9a,b) by multiplying (21.6.9a) from the left by $\boldsymbol{\xi}_j^{\mathrm{T}}$ and multiply the transpose of (21.6.9b) from the right by $\boldsymbol{\zeta}_j$, and subtract the results from each other to arrive at

$$(p_k - p_j)\, \boldsymbol{\xi}_j^{\mathrm{T}} \boldsymbol{\zeta}_k = 0. \tag{21.6.13a}$$

Hence, if the eigenvectors are properly normalized, it follows that

$$\boldsymbol{\xi}_j^{\mathrm{T}} \boldsymbol{\zeta}_k = \delta_{jk}, \quad \text{when } p_k \neq p_j, \tag{21.6.13b}$$

or, in vectorial notation,

$$\boldsymbol{\xi}_j \cdot \boldsymbol{\zeta}_k = \delta_{jk} \quad \text{when } p_k \neq p_j. \tag{21.6.13c}$$

Indeed, (21.6.13c) defines the normalization procedure, and it can be shown that

$$\boldsymbol{\xi}_k \cdot \boldsymbol{\zeta}_k = 1. \tag{21.6.13d}$$

This orthonormality leads to the conclusion that

$$\boldsymbol{Y}^{\mathrm{T}}\boldsymbol{Z} = \begin{bmatrix} \mathbf{L}^{\mathrm{T}} & \mathbf{A}^{\mathrm{T}} \\ \overline{\mathbf{L}}^{\mathrm{T}} & \overline{\mathbf{A}}^{\mathrm{T}} \end{bmatrix} \begin{bmatrix} \mathbf{A} & \overline{\mathbf{A}} \\ \mathbf{L} & \overline{\mathbf{L}} \end{bmatrix} = \begin{bmatrix} \mathbf{1} & \mathbf{0}^{(3)} \\ \mathbf{0}^{(3)} & \mathbf{1} \end{bmatrix} = \mathbf{1}^{(6)}, \tag{21.6.14a}$$

where $\mathbf{1}^{(6)}$ is the six by six identity matrix. Since $\boldsymbol{Y}^{\mathrm{T}}$ and $\boldsymbol{Z}$ are each other's inverse, it also follows that

$$\boldsymbol{Z}\boldsymbol{Y}^{\mathrm{T}} = \begin{bmatrix} \mathbf{A} & \overline{\mathbf{A}} \\ \mathbf{L} & \overline{\mathbf{L}} \end{bmatrix} \begin{bmatrix} \mathbf{L}^{\mathrm{T}} & \mathbf{A}^{\mathrm{T}} \\ \overline{\mathbf{L}}^{\mathrm{T}} & \overline{\mathbf{A}}^{\mathrm{T}} \end{bmatrix} = \mathbf{1}^{(6)}. \tag{21.6.14b}$$

This is known as the closure relation, see, e.g., Ting (1996).

The fundamental elasticity matrix $\boldsymbol{N}$ can now be expressed as

$$\boldsymbol{N} = \boldsymbol{Z}\boldsymbol{P}\boldsymbol{Y}^{\mathrm{T}}, \tag{21.6.15a}$$

where $\boldsymbol{P}$ is a six by six diagonal matrix consisting of the eigenvalues of $\boldsymbol{N}$, i.e.,

$$\boldsymbol{P} = \operatorname{diag}[p_1, p_2, p_3, \bar{p}_1, \bar{p}_2, \bar{p}_3]. \tag{21.6.15b}$$

The basic field equation (21.6.2) now becomes

$$\frac{\partial \boldsymbol{\eta}}{\partial x_2}(\mathbf{x}; \boldsymbol{P}, \mathbf{A}, \mathbf{L}) = \boldsymbol{Z}\boldsymbol{P}\boldsymbol{Y}^{\mathrm{T}} \frac{\partial \boldsymbol{\eta}}{\partial x_1}(\mathbf{x}; \boldsymbol{P}, \mathbf{A}, \mathbf{L}), \tag{21.6.16}$$

where the fact that the solution $\boldsymbol{\eta}$ depends on the elastic properties of the material through the matrices $\boldsymbol{P}$, $\mathbf{A}$, and $\mathbf{L}$, is displayed explicitly. From this general solution specific solutions can be obtained once suitable boundary conditions are prescribed.

In view of (21.6.14a), it is useful to rewrite (21.6.16) as

$$\frac{\partial \tilde{\boldsymbol{\eta}}}{\partial x_2} = \boldsymbol{P}\,\frac{\partial \tilde{\boldsymbol{\eta}}}{\partial x_1}, \quad \tilde{\boldsymbol{\eta}} = \boldsymbol{Y}^{\mathrm{T}}\boldsymbol{\eta}. \tag{21.6.17a,b}$$

These define six uncoupled first-order partial differential equations for the six components of the vector field $\tilde{\boldsymbol{\eta}}$ as functions of the eigenvalues, p_k. The general solution therefore is

$$\tilde{\boldsymbol{\eta}} = [f_1(z_1),\, f_2(z_2),\, f_3(z_3),\, f_4(\overline{z}_1),\, f_5(\overline{z}_2),\, f_6(\overline{z}_3)]^{\mathrm{T}}$$

$$\equiv \boldsymbol{f}(\mathbf{x};\, \boldsymbol{P}), \quad z_k = x_1 + p_k x_2, \tag{21.6.17c,d}$$

where the functions f_k, $k = 1, 2, ..., 6$, are arbitrary, subject to the requirement that they must lead to real-valued final solutions. As is seen, this solution does not involve the matrices **A** and **L**. It thus remains unchanged if these matrices are interchanged. This fact is used in the following subsection to prove the duality principle.

The general solution for $\boldsymbol{\eta}$ is given by

$$\boldsymbol{\eta} = \hat{\boldsymbol{\eta}}(\mathbf{x};\, \boldsymbol{P},\, \mathbf{A},\, \mathbf{L}) = \mathbf{Z}\tilde{\boldsymbol{\eta}}(\mathbf{x};\, \boldsymbol{P}). \tag{21.6.17e}$$

The matrices **A** and **L** enter this solution through $\boldsymbol{Z}$ only. Thus, noting (21.6.12), observe that

$$\boldsymbol{\eta}' = \hat{\boldsymbol{\eta}}(\mathbf{x};\, \boldsymbol{P},\, \mathbf{L},\, \mathbf{A}) = \boldsymbol{Y}\tilde{\boldsymbol{\eta}}(\mathbf{x};\, \boldsymbol{P}) \tag{21.6.18a}$$

is the general solution of

$$\frac{\partial(\mathbf{Z}^{\mathrm{T}}\boldsymbol{\eta}')}{\partial x_2} = \boldsymbol{P}\,\frac{(\partial \mathbf{Z}^{\mathrm{T}}\boldsymbol{\eta}')}{\partial x_1}. \tag{21.6.18b}$$

21.6.1. A General Duality Principle

Examine the *general solution* of the system of partial differential equations (21.6.16), given by (21.6.17d) for any (given) piecewise constant elasticity tensor **C** but without regard to any specific boundary data that must be considered in order to complete the solution in specific cases.

Consider now another general solution of (21.6.16), obtained from (21.6.18a), as follows

$$\boldsymbol{\eta}^{\mathrm{D}} = \boldsymbol{I}\hat{\boldsymbol{\eta}}(\mathbf{x};\, \boldsymbol{P},\, \mathbf{L},\, \mathbf{A}) = \mathbf{Z}\tilde{\boldsymbol{\eta}}(\mathbf{x};\, \boldsymbol{P}). \tag{21.6.18c}$$

As is seen, this new solution is obtained by simply multiplying (21.6.18a) from the left by $\boldsymbol{I}$. This operation interchanges the expression for the displacement field **u** with the corresponding expression for the stress-potential vector $\boldsymbol{\phi}$.

Ni and Nemat-Nasser (1996) prove that (21.6.18c) satisfies (21.6.16) *for the same fundamental elasticity matrix,* $\boldsymbol{N} = \boldsymbol{Z}\boldsymbol{P}\boldsymbol{Y}^{\mathrm{T}}$, and hence the same elasticity tensor, **C**. For an alternative proof, rewrite (21.6.18b) as

$$\frac{\partial \hat{\boldsymbol{\eta}}}{\partial x_2}(\mathbf{x};\, \boldsymbol{P},\, \mathbf{L},\, \mathbf{A}) = \boldsymbol{YPZ}^{\mathrm{T}} \frac{\partial \hat{\boldsymbol{\eta}}}{\partial x_1}(\mathbf{x};\, \boldsymbol{P},\, \mathbf{L},\, \mathbf{A}). \tag{21.6.18d}$$

Now, multiply both sides of this equation by $\boldsymbol{I}$, and since $\boldsymbol{II} = \mathbf{1}^{(6)}$, the six by six identity matrix, obtain

$$\frac{\partial \boldsymbol{I}\hat{\boldsymbol{\eta}}}{\partial x_2}(\mathbf{x};\, \boldsymbol{P},\, \mathbf{L},\, \mathbf{A}) = \boldsymbol{IYPZ}^{\mathrm{T}}\boldsymbol{I} \frac{\partial \boldsymbol{I}\hat{\boldsymbol{\eta}}}{\partial x_1}(\mathbf{x};\, \boldsymbol{P},\, \mathbf{L},\, \mathbf{A}),$$

$$= \boldsymbol{ZPY}^{\mathrm{T}} \frac{\partial \boldsymbol{I}\hat{\boldsymbol{\eta}}}{\partial x_1}(\mathbf{x};\, \boldsymbol{P},\, \mathbf{L},\, \mathbf{A}), \tag{21.6.18e}$$

or, in terms of the original fundamental elasticity matrix,

$$\frac{\partial \boldsymbol{\eta}^{\mathrm{D}}}{\partial x_2}(\mathbf{x};\, \boldsymbol{P},\, \mathbf{L},\, \mathbf{A}) = \boldsymbol{N} \frac{\partial \boldsymbol{\eta}^{\mathrm{D}}}{\partial x_1}(\mathbf{x};\, \boldsymbol{P},\, \mathbf{L},\, \mathbf{A}). \tag{21.6.19}$$

Thus, if $\boldsymbol{\eta} = \hat{\boldsymbol{\eta}}(\mathbf{x};\, \boldsymbol{P},\, \mathbf{A},\, \mathbf{L})$ is a general solution of (21.6.16), then *the dual field* $\boldsymbol{\eta}^{\mathrm{D}} = \boldsymbol{I}\hat{\boldsymbol{\eta}}(\mathbf{x};\, \boldsymbol{P},\, \mathbf{L},\, \mathbf{A})$ *also is a general solution of the same basic field equations with the same elasticity tensor,* **C**. This duality of the solutions of the two elasticity problems with the same elasticity and geometry, should be distinguished from the duality given by (21.6.3a) where the elasticity of the solid is changed by the operation $\boldsymbol{KNK}^{-1}$, even though the eigenvalues are preserved by this class of transformations, as pointed out before.

Since the expressions for the displacement **u**, and the stress-potential vector $\boldsymbol{\phi}$, in the original solution $\boldsymbol{\eta}$, are interchanged in order to obtain the corresponding expressions for the dual solution $\boldsymbol{\eta}^{\mathrm{D}}$, the boundary or other auxiliary conditions which may be defined for these fields, must accordingly be interchangeable. The two sets of boundary (or auxiliary) conditions of this kind (one set for the original, and the other for the dual problem) must therefore be dual, as is further discussed in Subsection 21.6.3. The following simple example illustrates this and related issues.

21.6.2. An Example

A Line Dislocation: As an illustration, consider the problem of a line dislocation in an infinite anisotropic homogeneous elastic medium. Assume that a straight dislocation with Burgers vector **b** is situated at the origin of the coordinate system, having an infinitely long dislocation line in the x_3-direction; see Figure 21.6.1. This dislocation represents a discontinuity in the displacement field, described by the Burgers vector **b**. For a pure screw dislocation, the only nonzero component of **b** is b_3, while for a pure edge dislocation, $b_3 = 0$, and the Burgers vector **b** is orthogonal to the x_3-axis. The stress field is continuous everywhere (except at the dislocation core, where it is not defined). Thus the following two auxiliary conditions must be satisfied:

$$\mathbf{u}(x_1,\, 0^+) - \mathbf{u}(x_1,\, 0^-) = -H(x_1)\,\mathbf{b},$$

$$\sigma_{i2}(x_1,\, 0^+) = \sigma_{i2}(x_1,\, 0^-), \quad i = 1, 2, 3, \tag{21.6.20a,b}$$

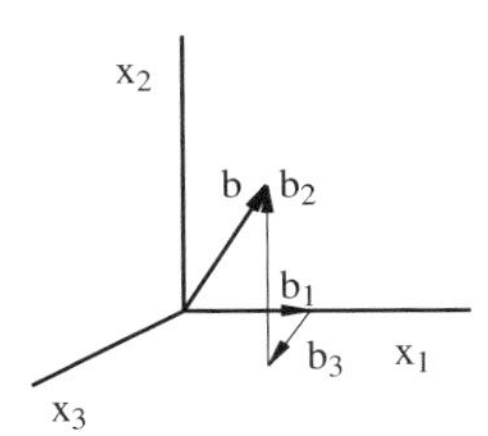

Figure 21.6.1

Dislocation with Burgers vector **b** at origin of coordinates in an infinitely extended solid

where the superscripts + and - denote the values of the corresponding quantity evaluated at the upper and lower faces of the plane $x_2 = 0$, respectively; the upper face at $x_2 = 0$ is defined by $x_2 > 0$, $x_2 \to 0$, and the lower face by $x_2 < 0$, $x_2 \to 0$; and H is the Heaviside step function. In terms of the vector potential $\boldsymbol{\phi}$ defined in (21.6.1), (21.6.20b) becomes

$$\boldsymbol{\phi}(x_1, 0^+) = \boldsymbol{\phi}(x_1, 0^-). \tag{21.6.20c}$$

In addition, it is required that the strain and rotation vanish at infinity,

$$\frac{\partial \mathbf{u}}{\partial x_j} \to \mathbf{0}, \quad \text{as } |x_k| \to \infty \text{ for } j, k = 1, 2. \tag{21.6.20d}$$

With the aid of a Fourier transform, the basic six-dimensional equation (21.6.2) with the auxiliary conditions (21.6.20) can be solved explicitly, leading to (Stroh, 1958; Barnett and Lothe 1973)

$$\mathbf{u}(x_1, x_2) = \frac{1}{\pi} \, Im\{\mathbf{A}[\sum_{k=1}^{3} \log(x_1 + p_k x_2)\, \mathbf{J}_k]\, \mathbf{L}^T\}\, \mathbf{b},$$

$$\boldsymbol{\phi}(x_1, x_2) = \frac{1}{\pi} \, Im\{\mathbf{L}[\sum_{k=1}^{3} \log(x_1 + p_k x_2)\, \mathbf{J}_k]\, \mathbf{L}^T\}\, \mathbf{b}, \tag{21.6.21a,b}$$

where the principal value of the logarithmic function $\log z$ is defined by $\log z = \log|z| + i\arg(z)$, with $0 \le \arg(z) < 2\pi$, and $\mathbf{J}_k = [\delta_{ik}\, \delta_{jk}]$ (k not summed) is a three by three matrix with zero elements except for the k-th diagonal element which is one.

A Line Force: Now, examine the problem of a line force of constant intensity in an infinitely extended elastic solid with the *same elasticity tensor.* Let this line force act along the x_3-axis; replace **b** by **f** in Figure 21.6.1. Denote its constant intensity (measured per unit length of the x_3-axis) by $\mathbf{f} = (f_1, f_2, f_3)$. Everywhere in the x_1, x_2-plane, the displacement field is continuous, whereas the tractions suffer a jump discontinuity at the origin. These conditions are expressed as

$$\mathbf{u}(x_1,\ 0^+) = \mathbf{u}(x_1,\ 0^-),$$

$$\sigma_{j2}(x_1,\ 0^+) - \sigma_{j2}(x_1,\ 0^-) = \delta(x_1)\, f_j. \tag{21.6.22a,b}$$

In terms of the vector potential $\boldsymbol{\phi}$, the last condition becomes

$$\boldsymbol{\phi}(x_1,\ 0^+) - \boldsymbol{\phi}(x_1,\ 0^-) = -H(x_1)\, \mathbf{f}. \tag{21.6.22c}$$

At infinity, it is required that

$$\sigma_{j2}(x_1,\ x_2),\ \sigma_{j1}(x_1,\ x_2) \to 0 \tag{21.6.22d}$$

or

$$\frac{\partial \boldsymbol{\phi}}{\partial x_j} \to \mathbf{0},\ \text{as } |x_k| \to \infty \text{ and } x_2 \neq 0. \tag{21.6.22e}$$

Again, with the aid of a Fourier transform, the solution of this problem is obtained to be (Stroh, 1958; Barnett and Lothe 1973)

$$\mathbf{u}(x_1,\ x_2) = \frac{1}{\pi}\, Im\{\mathbf{A}[\sum_{k=1}^{3} \log(x_1 + p_k x_2)\, \mathbf{J}_k]\, \mathbf{A}^T\}\, \mathbf{f},$$

$$\boldsymbol{\phi}(x_1,\ x_2) = \frac{1}{\pi}\, Im\{\mathbf{L}[\sum_{k=1}^{3} \log(x_1 + p_k x_2)\, \mathbf{J}_k]\, \mathbf{A}^T\}\, \mathbf{f}. \tag{21.6.23a,b}$$

Duality of Solutions: It is seen that a duality exists between solutions (21.6.21) and (21.6.23), as well as between the corresponding auxiliary conditions (21.6.20) and (21.6.22). This is summarized in Table 21.6.1.

The coefficients of the Burgers vector **b** in the expressions for the solution of the line dislocation problem, can be viewed as the corresponding Green function. Similarly, the coefficients of the line force **f** in Table 21.6.1, define the associated Green function. It is seen that these Green functions are each other's dual. A transformation defined by

$$\mathcal{L}\,(\boldsymbol{\eta}) \equiv \boldsymbol{I}\hat{\boldsymbol{\eta}}(\mathbf{x};\ \boldsymbol{P},\ \mathbf{L},\ \mathbf{A}) = \boldsymbol{\eta}^D(\mathbf{x};\ \boldsymbol{P},\ \mathbf{L},\ \mathbf{A}) \tag{21.6.24a}$$

produces the dual solution from the original solution, for the same geometry and material properties. Conversely, the original solution can be obtained from $\boldsymbol{\eta}^D$ by the same transformation,

$$\mathcal{L}\,(\boldsymbol{\eta}^D) \equiv \boldsymbol{I}(\boldsymbol{I}\hat{\boldsymbol{\eta}}(\mathbf{x};\ \boldsymbol{P},\ \mathbf{A},\ \mathbf{L})) = \boldsymbol{\eta}(\mathbf{x};\ \boldsymbol{P},\ \mathbf{A},\ \mathbf{L}), \tag{21.6.24b}$$

as can readily be seen from Table 21.6.1.

It is important to note that, once the *nature* of the boundary data for one of the problems in fixed, then that for the dual problem is also fixed. In the example of Table 21.6.1, a discontinuity is defined for the displacement field, **u**, in the first problem. This requires a similar condition (discontinuity) to be assigned to the stress-potential vector field, $\boldsymbol{\phi}$, for the dual problem. The Green functions obtained by letting both **b** and **f** be unit vectors, are each other's dual, and can be used to generate other dual solutions. Certain simple dual boundary conditions are examined below.

Table 21.6.1:

Duality between a line dislocation and a line force

Line Dislocation	Line Force
Auxiliary Conditions	
$\mathbf{u}(x_1, 0^+) - \mathbf{u}(x_1, 0^-) = -H(x_1)\,\mathbf{b}$	$\boldsymbol{\phi}(x_1, 0^+) - \boldsymbol{\phi}(x_1, 0^-) = -H(x_1)\,\mathbf{f}$
$\boldsymbol{\phi}(x_1, 0^+) = \boldsymbol{\phi}(x_1, 0^-)$	$\mathbf{u}(x_1, 0^+) = \mathbf{u}(x_1, 0^-)$
$\dfrac{\partial \mathbf{u}}{\partial x_j} \to \mathbf{0}$ as $\lvert x_k \rvert \to \infty$ for j, k = 1, 2, $x_2 \neq 0$	$\dfrac{\partial \boldsymbol{\phi}}{\partial x_j} \to \mathbf{0}$ as $\lvert x_k \rvert \to \infty$ for j, k = 1, 2, $x_2 \neq 0$
Solutions	
$\mathbf{u}(x_1, x_2) = \dfrac{1}{\pi}\, Im\{\mathbf{A}[\sum_{k=1}^{3} \log(x_1 + p_k x_2)\,\mathbf{J}_k]\,\mathbf{L}^T\}\,\mathbf{b}$	$\boldsymbol{\phi}(x_1, x_2) = \dfrac{1}{\pi}\, Im\{\mathbf{L}[\sum_{k=1}^{3} \log(x_1 + p_k x_2)\,\mathbf{J}_k]\,\mathbf{A}^T\}\,\mathbf{f}$
$\boldsymbol{\phi}(x_1, x_2) = \dfrac{1}{\pi}\, Im\{\mathbf{L}[\sum_{k=1}^{3} \log(x_1 + p_k x_2)\,\mathbf{J}_k]\,\mathbf{L}^T\}\,\mathbf{b}$	$\mathbf{u}(x_1, x_2) = \dfrac{1}{\pi}\, Im\{\mathbf{A}[\sum_{k=1}^{3} \log(x_1 + p_k x_2)\,\mathbf{J}_k]\,\mathbf{A}^T\}\,\mathbf{f}$

21.6.3. Dual Boundary Conditions

For the class of considered problems, the elastic solid is cylindrical in geometry, having an axis parallel to the x_3-direction. It thus suffices to consider a typical cross section perpendicular to the x_3-direction. Let this cross section be denoted by S, being bounded by ∂S. Define by $\boldsymbol{\nu} = \nu_1\mathbf{e}_1 + \nu_2\mathbf{e}_2$ the exterior unit normal on ∂S.

Duality of Traction- and Displacement-boundary Data: When tractions $\mathbf{t}^o(x_1, x_2)$ are prescribed on ∂S, the boundary conditions become

$$\boldsymbol{\nu} \cdot \boldsymbol{\sigma} = \mathbf{t}^o \quad \text{or} \quad \nu_i\,\sigma_{ij} = t_j^o \qquad \text{on } \partial S. \tag{21.6.25a,b}$$

Let l measure length along ∂S, as shown in Figure 21.6.2, and denote the unit tangent vector of ∂S by $\mathbf{t} = t_1\,\mathbf{e}_1 + t_2\,\mathbf{e}_2$. Since $\nu_1 = t_2$ and $\nu_2 = -t_1$, using (21.6.1b,c), it follows that

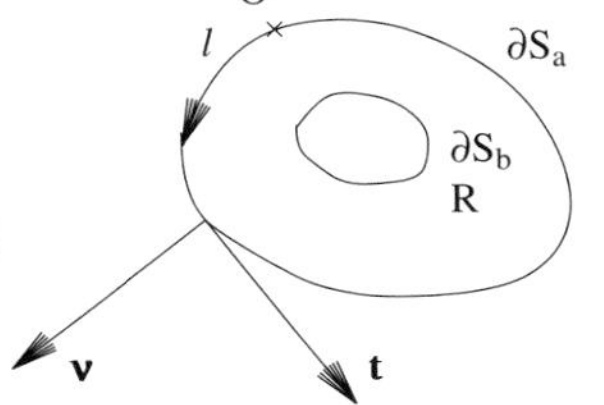

Figure 21.6.2

A doubly-connected region R subjected to mixed data on its boundary, $\partial S = \partial S_a + \partial S_b$

$$\mathbf{t}^o = \nu_1 \frac{\partial \boldsymbol{\phi}}{\partial x_2} - \nu_2 \frac{\partial \boldsymbol{\phi}}{\partial x_1} = (\frac{\partial \boldsymbol{\phi}}{\partial x_1} t_1 + \frac{\partial \boldsymbol{\phi}}{\partial x_2} t_2) = \frac{\partial \boldsymbol{\phi}}{\partial l}. \tag{21.6.25c}$$

Thus, on a traction-free boundary, $\partial\boldsymbol{\phi}/\partial l = \mathbf{0}$. Since on a fixed boundary $\partial\mathbf{u}/\partial l = \mathbf{0}$, it follows that when a part, say, ∂S_a, of the boundary ∂S of an original problem is fixed, then ∂S_a is traction-free in the dual problem, and vice versa, i.e.,

$$\frac{\partial \mathbf{u}}{\partial l} = \mathbf{0} \quad \Longleftrightarrow \quad \frac{\partial \boldsymbol{\phi}}{\partial l} = \mathbf{0} \quad \text{on } \partial S_a. \tag{21.6.26}$$

When the displacement vector $\mathbf{u}$ is given on ∂S_a, then $\partial \mathbf{u}/\partial l$ can be calculated on ∂S_a. Similarly, when $\boldsymbol{\phi}$ is prescribed on ∂S_a, then $\partial \boldsymbol{\phi}/\partial l$ can be calculated there. In general, the following duality holds between the displacement and the stress-potential boundary data:

$$\mathbf{u} = \mathbf{f}(x_1, x_2) \quad \Longleftrightarrow \quad \boldsymbol{\phi} = \mathbf{g}(x_1, x_2) \quad \text{on } \partial S_a. \tag{21.6.27a}$$

The functions $\mathbf{f}$ and $\mathbf{g}$, with components f_1, f_2, f_3 and g_1, g_2, g_3, are arbitrary, except for some minimal smoothness requirements, and can be prescribed as desired. This means that, to obtain the solution of the dual problem for given $\mathbf{g}$, switch $\mathbf{A}$ and $\mathbf{L}$ in the solution of the original problem, and replace $\mathbf{f}$ by $\mathbf{g}$.

It is important to note that the boundary tractions prescribed on the boundary ∂S_a, define only $\partial \boldsymbol{\phi}/\partial l$ and not $\boldsymbol{\phi}$, as is seen from (21.6.25c). Integration of the prescribed tractions with respecct to length, l, along the boundary ∂S_a, then yields $\boldsymbol{\phi}$ to within a constant. In addition to this, the question of potential rigid-body motion may have to be addressed in special cases. An example is the duality of the solution of a crack and a rigid line inclusion in an unbounded region; see Subsection 21.6.5.

More general mixed boundary conditions can be included. Let $\boldsymbol{H}$ be a one by six matrix consisting of three zeros and three ones, in an order which produces a consistent set of boundary conditions; see Section 19. Then, duality exists between any two sets of boundary data which satisfy

$$\boldsymbol{H}\boldsymbol{\eta} = \mathbf{f}(x_1, x_2) \quad \Longleftrightarrow \quad \boldsymbol{H}\boldsymbol{\eta}^{D} = \mathbf{g}(x_1, x_2) \quad \text{on } \partial S_a, \tag{21.6.27b}$$

where $\boldsymbol{\eta}$ and $\boldsymbol{\eta}^{D}$ are related by the transformation (21.6.24). For example, $\boldsymbol{H} = [1, 1, 1, 0, 0, 0]$ corresponds to the dual boundary data (21.6.27a).

General Mixed Boundary Data: Consider two elasticity problems, say, problems 1 and 2, for the *same elastic domain with the same elasticity tensor*, $\mathbf{C}$. *Suppose the solution of the first problem is defined by*

$$\boldsymbol{\eta}^{(1)}(\mathbf{x}) = [\mathbf{u}^{(1)}(\mathbf{x};\, \mathbf{A}, \mathbf{L};\, \mathbf{f}_a, \mathbf{f}_b),\ \boldsymbol{\phi}^{(1)}(\mathbf{x};\, \mathbf{A}, \mathbf{L};\, \mathbf{f}_a, \mathbf{f}_b)]^T \tag{21.6.28a}$$

which satisfies the following (self-consistent) boundary conditions:

$$\boldsymbol{\pi}(\mathbf{u}^{(1)}, \boldsymbol{\phi}^{(1)};\, \mathbf{f}_a, \mathbf{f}_b) = \mathbf{0}, \tag{21.6.28b}$$

where the functions[16] $\mathbf{f}_a$ *and* $\mathbf{f}_b$ *are prescribed and define the spatial variation of the appropriate field quantity (e.g.,* $\mathbf{f}_a$ *defines* $\mathbf{u}^{(1)}$ *on* ∂S_a*, and* $\mathbf{f}_b$ *defines* $\boldsymbol{\phi}^{(1)}$ *on* ∂S_b*) on the boundary of the domain. The notation in the right-hand side of (21.6.28a) shows that the displacement,* $\mathbf{u}^{(1)}$*, and the stress potential,* $\boldsymbol{\phi}^{(1)}$*, depend parametrically on the matrices* $\mathbf{A}$ *and* $\mathbf{L}$*, and are functionals* of the boundary data, defined by the functions: $\mathbf{f}_a(\mathbf{x})$ for $\mathbf{x}$ on ∂S_a, and $\mathbf{f}_b(\mathbf{x})$ for $\mathbf{x}$ on ∂S_b; see (21.6.35a,b) for an example.

---

[16] It is possible to establish the duality for a case when the boundary has several subdomains, on each of which a different mixed boundary condition is prescribed.

Suppose now that the solution of a second elasticity problem with the following boundary conditions, is required:

$$\boldsymbol{\pi}(\boldsymbol{\phi}^{(2)}, \mathbf{u}^{(2)}; \mathbf{g}_a, \mathbf{g}_b) = \mathbf{0}. \tag{21.6.29a}$$

Here, the operator $\boldsymbol{\pi}$ is the same as that in (21.6.28b), but $\mathbf{u}$ and $\boldsymbol{\phi}$ are switched, and the functions $\mathbf{g}_a$ and $\mathbf{g}_b$ are arbitrary. As an example, if in (21.6.28b) $\mathbf{f}_a$ defines $\mathbf{u}^{(1)}$ on ∂S_a, and $\mathbf{f}_b$ defines $\boldsymbol{\phi}^{(1)}$ on ∂S_b, then, in (21.6.29a) $\mathbf{g}_a$ defines $\boldsymbol{\phi}^{(2)}$ on ∂S_a, and $\mathbf{g}_b$ defines $\mathbf{u}^{(2)}$ on ∂S_b.

To obtain the solution of this new problem for the *same geometry and material properties*, from the solution of the first problem, simply set

$$\mathbf{u}^{(2)} \equiv \boldsymbol{\phi}^{(1)}(\mathbf{x}; \mathbf{L}, \mathbf{A}; \mathbf{g}_a, \mathbf{g}_b),$$

$$\boldsymbol{\phi}^{(2)} \equiv \mathbf{u}^{(1)}(\mathbf{x}; \mathbf{L}, \mathbf{A}; \mathbf{g}_a, \mathbf{g}_b). \tag{21.6.29b,c}$$

This yields the solution of the dual problem, i.e.,

$$\boldsymbol{\eta}^{(2)}(\mathbf{x}) = [\mathbf{u}^{(2)}, \boldsymbol{\phi}^{(2)})]^T. \tag{21.6.29d}$$

This solution is a functional of the arbitrary functions $\mathbf{g}_a$ and $\mathbf{g}_b$ which define the spatial variation of the approperiate field quantities. These functions need not be the same as the functions $\mathbf{f}_a$ and $\mathbf{f}_b$ of problem 1. To satisfy duality, the boundary data for the second solution must, however, have the general *form* defined by (21.6.29a).

The duality established above can more appropriately be expressed in terms of the *structure of the operators which act on functions that define the boundary conditions*. Let $\boldsymbol{\Pi}_u$ and $\boldsymbol{\Pi}_\phi$ be the operators which produce the final displacement and stress potential, $\mathbf{u}$ and $\boldsymbol{\phi}$, from the boundary data; e.g., $\boldsymbol{\Pi}_u$ and $\boldsymbol{\Pi}_\phi$ produce $\mathbf{u}^{(1)}$ and $\boldsymbol{\phi}^{(1)}$ in (21.6.28a) by operating on prescribed boundary functions $\mathbf{f}_a$ and $\mathbf{f}_b$, as

$$[\mathbf{u}^{(1)}, \boldsymbol{\phi}^{(1)}] = [\boldsymbol{\Pi}_u(\mathbf{A}, \mathbf{L}; \mathbf{f}_a, \mathbf{f}_b), \boldsymbol{\Pi}_\phi(\mathbf{A}, \mathbf{L}; \mathbf{f}_a, \mathbf{f}_b)]. \tag{21.6.28c}$$

Then, *exactly the same operators*, $\boldsymbol{\Pi}_u$ and $\boldsymbol{\Pi}_\phi$, produce the dual solution, $\mathbf{u}^{(2)}$ and $\boldsymbol{\phi}^{(2)}$, for the boundary conditions (21.6.29a), as follows:

$$[\mathbf{u}^{(2)}, \boldsymbol{\phi}^{(2)}] = [\boldsymbol{\Pi}_\phi(\mathbf{L}, \mathbf{A}; \mathbf{g}_a, \mathbf{g}_b), \boldsymbol{\Pi}_u(\mathbf{L}, \mathbf{A}; \mathbf{g}_a, \mathbf{g}_b)]; \tag{21.6.29e}$$

see for illustration (21.6.35a,b) and compare with (21.6.37a,b).

21.6.4. Fundamental Elasticity Matrix with Repeated Eigenvalues

When the fundamental elasticity matrix, $\boldsymbol{N}$, admits multiple eigenvalues, the problem is degenerate. If the multiplicity of the eigenvalue p_k is m_k, with $1 \leq m_k \leq 3$, then the corresponding generalized right-eigenvector $\boldsymbol{\zeta}_k$ of $\boldsymbol{N}$, is defined by

$$(\boldsymbol{N} - p_k \mathbf{1}^{(6)})^{m_k} \boldsymbol{\zeta}_k = \mathbf{0}, \quad \text{with} \quad (\boldsymbol{N} - p_k \mathbf{1}_{(6)})^{m_k - 1} \boldsymbol{\zeta}_k \neq \mathbf{0}. \tag{21.6.30a,b}$$

If there are six distinct eigenvalues, then $m_k = 1$.

For the repeated eigenvalues, two cases can occur: (i) Semi-simple degenerate case, for which there still exist six linearly independent, say, right-eigenvectors $\boldsymbol{\zeta}_k$, $k = 1, 2, ..., 6$; and (ii) Non-semi-simple degenerate case, which does not involve six such linearly independent eigenvectors. A typical example for this case is the isotropic material, where the first three eigenvalues are all equal to ι and there are only four linearly independent eigenvectors, i.e., $\boldsymbol{\zeta}_1$, $\boldsymbol{\zeta}_2$, and their complex conjugates. The first three generalized eigenvectors in this case satisfy (21.6.30a,b) for $m_1 = 1$, $m_2 = 1$, and $m_3 = 2$, i.e., $\boldsymbol{N}\,\boldsymbol{\zeta}_1 = \iota\,\boldsymbol{\zeta}_1$, $\boldsymbol{N}\,\boldsymbol{\zeta}_2 = \iota\,\boldsymbol{\zeta}_2$, and $\boldsymbol{N}\,\boldsymbol{\zeta}_3 = \boldsymbol{\zeta}_2 + \iota\,\boldsymbol{\zeta}_3$.

For the semi-simple degenerate case, there are six linearly independent right- and six linearly independent left-eigenvectors. These are given by the normalized solutions of (21.6.9a,b), and the proof of the duality follows as before.

When the eigenvalues are repeated and the eigenvectors are not all linearly independent, the corresponding non-semi-simple degenerate case is classified according to the multiplicities m_k, defined in (21.6.30a,b), of the first three generalized eigenvectors, as: (A) $m_1 = 1$, $m_2 = 1$, $m_3 = 2$ (the isotropic case is in this subcase); and (B) $m_1 = 1$, $m_2 = 2$, $m_3 = 3$. For subcase (A),

$$(\boldsymbol{N} - p_1\,\mathbf{1}^{(6)})\,\boldsymbol{\zeta}_1 = \mathbf{0}, \quad (\boldsymbol{N} - p_2\,\mathbf{1}^{(6)})\,\boldsymbol{\zeta}_2 = \mathbf{0}, \quad (\boldsymbol{N} - p_2\,\mathbf{1}^{(6)})\,\boldsymbol{\zeta}_3 = \boldsymbol{\zeta}_2, \tag{21.6.31a~c}$$

and $\boldsymbol{\zeta}_{k+3} = \bar{\boldsymbol{\zeta}}_k$, $k = 1, 2, 3$. From (21.6.31a~c), it follows that

$$\boldsymbol{N}\,\boldsymbol{Z} = \boldsymbol{Z}\,\boldsymbol{W}_A, \tag{21.6.31d}$$

where the six by six matrix $\boldsymbol{W}_A$ is defined by

$$\boldsymbol{W}_A = \begin{bmatrix} p_1 & 0 & 0 & 0 & 0 & 0 \\ 0 & p_2 & 1 & 0 & 0 & 0 \\ 0 & 0 & p_2 & 0 & 0 & 0 \\ 0 & 0 & 0 & \bar{p}_1 & 0 & 0 \\ 0 & 0 & 0 & 0 & \bar{p}_2 & 1 \\ 0 & 0 & 0 & 0 & 0 & \bar{p}_2 \end{bmatrix}. \tag{21.6.31e}$$

The corresponding left-eigenvectors are now calculated such that the normality conditions are satisfied. These eigenvectors are the solutions of the following set of linear equations:

$$(\boldsymbol{N}^T - p_1\,\mathbf{1}^{(6)})\,\boldsymbol{\xi}_1 = \mathbf{0}, \quad (\boldsymbol{N}^T - p_2\,\mathbf{1}^{(6)})\,\boldsymbol{\xi}_3 = \mathbf{0}, \quad (\boldsymbol{N}^T - p_2\,\mathbf{1}^{(6)})\,\boldsymbol{\xi}_2 = \boldsymbol{\xi}_3. \tag{21.6.32a~c}$$

Consider the subcase (B) now. In this case,

$$(\boldsymbol{N} - p\,\mathbf{1}^{(6)})\,\boldsymbol{\zeta}_1 = \mathbf{0}, \quad (\boldsymbol{N} - p\,\mathbf{1}^{(6)})\,\boldsymbol{\zeta}_2 = \boldsymbol{\zeta}_1, \quad (\boldsymbol{N} - p\,\mathbf{1}^{(6)})\,\boldsymbol{\zeta}_3 = \boldsymbol{\zeta}_2, \tag{21.6.33a~c}$$

and $\boldsymbol{\zeta}_{k+3} = \bar{\boldsymbol{\zeta}}_k$, $k = 1, 2, 3$. From (21.6.33a~c), it follows that

$$\boldsymbol{N}\boldsymbol{Z} = \boldsymbol{Z}\boldsymbol{W}_{\mathrm{B}}, \tag{21.6.33d}$$

where the six by six matrix $\boldsymbol{W}_{\mathrm{B}}$ is defined by

$$\boldsymbol{W}_{\mathrm{B}} = \begin{bmatrix} p & 1 & 0 & 0 & 0 & 0 \\ 0 & p & 1 & 0 & 0 & 0 \\ 0 & 0 & p & 0 & 0 & 0 \\ 0 & 0 & 0 & \bar{p} & 1 & 0 \\ 0 & 0 & 0 & 0 & \bar{p} & 1 \\ 0 & 0 & 0 & 0 & 0 & \bar{p} \end{bmatrix}. \tag{21.6.33e}$$

The corresponding left-eigenvectors are then obtained from

$$(\boldsymbol{N}^{\mathrm{T}} - p\,\mathbf{1}^{(6)})\,\boldsymbol{\xi}_1 = \mathbf{0}, \quad (\boldsymbol{N}^{\mathrm{T}} - p\,\mathbf{1}^{(6)})\,\boldsymbol{\xi}_1 = \boldsymbol{\xi}_2, \quad (\boldsymbol{N}^{\mathrm{T}} - p\,\mathbf{1}^{(6)})\,\boldsymbol{\xi}_2 = \boldsymbol{\xi}_3, \tag{21.6.34a~c}$$

while enforcing the orthonormality.

For the above degenerate cases, the decomposition (21.6.15) now applies, and the duality of the solutions follows.

21.6.5. Examples of Duality

Ni and Nemat-Nasser (1996) present a number of examples, illustrating the duality principle. These examples include line dislocations and line forces acting on the interface between two dissimilar bonded half spaces, inclusions and cavities, the Green functions for half spaces, image forces in half spaces, and cracks and deformable line inclusions (anti-cracks).

As an illustration consider a crack located at $|x_1| < a$, $x_2 = 0$, $-\infty \leq x_3 \leq \infty$, and let its faces be subjected to tractions of equal magnitude and orientation but opposite sense, defined by $\sigma_{j2}(x_1, 0^+) = t_j(x_1)$ for $|x_1| < a$, where $\mathbf{t}(x_1)$ is a prescribed suitably smooth vector-valued function. The final solution[17] is given by

$$\mathbf{u}(\mathbf{x}) = -\frac{1}{\pi^2} Re\{\mathbf{A}\,[\sum_{k=1}^{3} \log(x_1 + p_k x_2)\,\mathbf{J}_k]\,\mathbf{L}^{-1}\} * \frac{1}{\sqrt{a^2 - x_1^2}} \int_{-a}^{a} \frac{\sqrt{a^2 - \xi^2}\,\mathbf{t}(\xi)}{x_1 - \xi}\,d\xi,$$

$$\boldsymbol{\phi}(\mathbf{x}) = -\frac{1}{\pi^2} Re\{\mathbf{L}\,[\sum_{k=1}^{3} \log(x_1 + p_k x_2)\,\mathbf{J}_k]\,\mathbf{L}^{-1}\} * \frac{1}{\sqrt{a^2 - x_1^2}} \int_{-a}^{a} \frac{\sqrt{a^2 - \xi^2}\,\mathbf{t}(\xi)}{x_1 - \xi}\,d\xi, \tag{21.6.35a,b}$$

where the asterisk, *, denotes the convolution integral. This solution is valid for any suitably smooth function $\mathbf{t}(\mathbf{x})$.

---

[17] The right side of (21.6.35a) and (21.6.35b) are examples of the operators, $\mathbf{\Pi}_u$ and fPi$_\phi$, which were discussed in (21.6.28c).

As a dual to the crack problem, consider a line inclusion, i.e., a flat surface $|x_1| < a$, $x_2 = 0$, $-\infty < x_3 < \infty$, along which the displacement gradient is prescribed to be

$$\frac{\partial \mathbf{u}}{\partial x_1}(x_1, 0) + \mathbf{d}(x_1) = \mathbf{0}, \qquad \text{for } |x_1| < a, \tag{21.6.36}$$

where $\mathbf{d}(x_1)$ for $|x_1| < a$, is given. Note here that, unlike the strain, $\partial u_1/\partial x_1$ and $\partial u_2/\partial x_1$ are not invariant under a rigid-body rotation. Hence, the possible contribution of a rigid-body displacement must be included. This leads to an additional term $\mathbf{r} = (0, \omega_{21}, \omega_{31})$, to be included in the solution of this problem,[18] where the rotation components, ω_{21} and ω_{31}, are to be determined. This solution is given by (Ni and Nemat-Nasser, 1996)

$$\mathbf{u}(\mathbf{x}) = \frac{1}{\pi^2} Re\{\mathbf{A}\,[\sum_{k=1}^{3} \log(x_1 + p_k x_2)\,\mathbf{J}_k]\,\mathbf{A}^{-1}\} * \frac{1}{\sqrt{a^2 - x_1{}^2}} \int_{-a}^{a} \frac{\sqrt{a^2 - \xi^2}\,\hat{\mathbf{d}}(\xi)}{x_1 - \xi}\, d\xi,$$

$$\boldsymbol{\phi}(\mathbf{x}) = \frac{1}{\pi^2} Re\{\mathbf{L}\,[\sum_{k=1}^{3} \log(x_1 + p_k x_2)\,\mathbf{J}_k]\,\mathbf{A}^{-1}\} * \frac{1}{\sqrt{a^2 - x_1{}^2}} \int_{-a}^{a} \frac{\sqrt{a^2 - \xi^2}\,\hat{\mathbf{d}}(\xi)}{x_1 - \xi}\, d\xi,$$

(21.6.37a,b)

where $\hat{\mathbf{d}}(x_1) = \mathbf{d}(x_1) - \mathbf{r}$, with $\mathbf{r}$ to be obtained from the side conditions (Dundurs and Markenscoff, 1989)

$$(\iota\,\mathbf{A}\,\mathbf{A}^T)^{-1} \int_{-a}^{a} \sqrt{a^2 - \xi^2}\,[\mathbf{d}(\xi) - \mathbf{r}]\, d\xi = (c, 0, 0)^T, \tag{21.6.37c}$$

where c is an immaterial constant.

The functions $\mathbf{t}(\mathbf{x})$ in (21.6.35) and $\hat{\mathbf{d}}(\mathbf{x})$ in (21.6.37) can be prescribed essentially arbitrarily. This does not affect the duality of the solutions, since these solutions hold for any suitable prescribed tractions on the crack and distortion of the line inclusion. Indeed, from (21.6.35) and (21.6.37), the corresponding Green functions can easily be extracted, and these Green functions are each other's dual, independently of the specific data prescribed on $|x_1| < a$, $x_2 = 0$, $-\infty \le x_3 \le \infty$. The operators $\mathbf{\Pi}_\phi$ and $\mathbf{\Pi}_u$ in this example, are convolution integrals.

21.7. REFERENCES

Azhdari, A. and Nemat-Nasser, S. (1996a), Hoop stress intensity factor and crack-kinking in anisotropic brittle solids, *Int. J. Solids Structures*, Vol. 33, No. 14, 2023-2037.

Azhdari, A. and Nemat-Nasser, S. (1996b), Energy-release rate and crack kinking in anisotropic brittle solids, *J. Mech. Phys. Solids*, Vol. 44, No. 6, 929-

[18] See Dundurs and Markenscoff (1989) and, also Hasebe *et al.* (1984a,b); for a discussion of cavities and rigid inclusions, see Dundurs (1989).

951.

Azhdari, A., and Nemat-Nasser, S. (1998), Experimental and computational study of fracturing in an anisotropic brittle solid, *Mech. Mat.*, Vol. 28, 247-262.

Azhdari, A., Nemat-Nasser, S., and Rome, J. (1998), Experimental observation and computational modeling of fracturing in an anisotropic brittle crystal (Sapphire), submitted 3/30/98.

Barnett, D.M. and Lothe, J. (1973), Synthesis of the sextic and the integral formalism for dislocations, Green's functions, and surface waves in anisotropic elastic solids, *Physica Norvegica* 7, 13-19.

Budiansky, B. (1986), Micromechanics II, in *Proceedings of the Tenth U.S. National Congress of Applied Mechanics,* Austin, Texas.

Budiansky, B., Amazigo, J. C., and Evans, A. G. (1988), Small-scale crack bridging and the fracture toughness of particle reinforced ceramics, *J. Mech. Phys. Solids*, Vol. 36, 167-187.

Chadwick, P. and Smith, G.D. (1977), Foundations of the theory of surface waves in anisotropic elastic materials, in: C.-S. Yih, ed., *Advances in Applied Mechanics*, Vol. 17, 303-376, Academic Press, New York.

Dundurs, J. (1968), Elastic interaction of dislocations with inhomogeneities, in *Mathematical theory of dislocations*, ASME, New York, 70-115.

Dundurs, J. (1989), Cavities vis-à-vis rigid inclusions and some related general results in plane elasticity, *J. Appl. Mech.*, Vol. 56, 780-790.

Dundurs, J. and Markenscoff X. (1989), A Green's function formulation of anticracks and their interaction with load-induced singularities, ASME *J. appl. Mech.*, Vol. 56, 550-555.

Erdogan, F. (1977), Mixed boundary-value problems in Mechanics, in *Mechanics Today* (Nemat-Nasser, S., ed.), Vol. 4, 1-86.

Eshelby, J. D., Read, W. T., and Shockley, W. (1953), Anisotropy elasticity with applications to dislocation theory, *Acta Metall.*, Vol. 1, 251-259.

Evans, A. G. (1990), Perspective on the development of high-toughness ceramics, *J. Am. Ceram. Soc.*, Vol. 73, 187-206.

Hadamard, J. (1952), *Lectures on Cauchy's problem in linear partial differential equations*, Dover, New York.

Hasebe, N., Keer, L. M., and Nemat-Nasser, S. (1984a), Stress analysis of a kinked crack initiating from a rigid line inclusion. Part I: formulation," *Mech. Mat.*, Vol. 3, No. 2, 131-145.

Hasebe, N., Nemat-Nasser, S., and Keer, L. M. (1984b), Stress analysis of a kinked crack initiating from a rigid line inclusion. Part II: direction of propagation, *Mech. Mat.*, Vol. 3, No. 2, 147-156.

Hille, E., (1959), *Analytic function theory*, Vol. I, Ginn and Company, Boston.

Hori, M. and Nemat-Nasser, S. (1990), Asymptotic solution of a class of strongly singular integral equations, *SIAMM, J. Appl. Math.*, Vol. 50, No. 3, 716-725.

Horii, H., Hasegawa, A., and Nishino, F. (1987), Process zone model and influencing factors in fracture of concrete, in *SEM/RILEM International Conference on Fracture of Cements and Rocks* (Shah, S. and Swartz, S. E., eds.), 299.

Ingebrigtsen, K.A. and Tonning, A. (1969), Elastic surface waves in crystals, *Phys. Rev.* 184, 942-951.

Kaya, A. C. and Erdogan, F. (1987), On the solution of integral equations with strong singularities, in *Numerical solution of singular integral equations* (Gerasoulis, A. and Vichnevestsky, R., eds.), IMACS, New Brunswick, 54.

Kirchner, H.O.K. (1989), Elastically anisotropic inhomogeneous media, I. a new formalism, *Phil. Mag.* A 60, 423-432.

Lekhnitskii, A. G. (1963), *Theory of elasticity of an anisotropic elastic body* (Trans. from Russian by Fern, P.), Holder Day, San Francisco.

Li, V. C., guest ed., (1992), *Micromechanical modeling of quasi-brittle materials behavior, Appl. Mech. Rev.*, Vol. 45, No. 8.

Luke, Y. L. (1969), *The special functions and their approximations, Vol. I*, Academic Press, New York.

Malen, K. (1971), A unified six-dimensional treatment of elastic Green's functions and dislocations, *Phys. Stat. Sol.* B 44, 661-672.

Muskhelishvili, N. I. (1956), *Some basic problems of the mathematical theory of elasticity,* Translated from the 3rd. Russian edition by J. R. M. Radok, Noordhoff, Groningen.

Nemat-Nasser, S. and Azhdari, A. (1997), Fracturing in anisotropic brittle solids: theory and some preliminary experimental results, *Proceedings 14th US Army Symposium on Solid Mechanics*, Batelle Press, 17-28.

Nemat-Nasser, S. and Hori, M. (1987), Toughening by partial or full bridging of cracks in ceramics and fiber reinforced composites, *Mech. Matr.*, Vol. 6, 245-269.

Nemat-Nasser, S. and Ni, L. (1995), A duality principle and correspondence relations in elasticity, *Int. J. Solids Structures* Vol. 32, 467-472.

Ni, L. and Nemat-Nasser, S. (1996), A general duality principle in elasticity, *Mech. Mat.*, Vol. 24, 87-123.

Obata, M., Nemat-Nasser, S., and Goto, Y. (1989), Branched cracks in anisotropic elastic solids, *J. Appl. Mech.*, Vol. 56, No. 4, 858-864.

Roach, G. F. (1982) *Green's functions* (second edition), Cambridge University Press.

Rose, L. R. F. (1987), Crack reinforcement by distributed springs, *J. Mech. Phys. Solids*, Vol. 35, 383.

Stroh, A. N. (1958), Dislocations and cracks in anisotropic elasticity, *Phil. Mag.*, Vol. 3, 625-646.

Ting, T.C.T. (1989), Line forces and dislocations in angularly inhomogeneous anisotropic elastic wedges and spaces, *Q. Appl. Math.* 47, 123-128.

Ting, T.C.T. (1996), *Anisotropic elasticity*, Oxford University Press, Oxford.

Willis, J. R. and Nemat-Nasser, S. (1990), Singular perturbation solution of a class of singular integral equations, *Quarterly of Appl. Math.*, Vol. 43, 741-753.

AUTHOR INDEX

L

M

N

Z

SUBJECT INDEX

C

D

E

F

G

H

I

J

K

L

M

N

O

P

Q

R

S

T